T0400136

Quaternary Alloys Based on IV–VI and IV–VI$_2$ Semiconductors

IV–VI and IV–VI$_2$ semiconductors have attracted considerable attention due to their applications in the fabrication of electronic and optoelectronic devices as light-emitting diodes and solar cells. The electrical properties of these semiconductors can also be tuned by adding impurity atoms. Because of their wide application in various devices, the search for new semiconductor materials and the improvement of existing materials is an important field of study. Doping with impurities is a common method of modifying and diversifying the properties of physical and chemical semiconductors. This book covers all known information about phase relations in quaternary systems based on IV–VI and IV–VI$_2$ semiconductors, providing the first systematic account of phase equilibria in quaternary systems based on IV–VI and IV–VI$_2$ semiconductors and making research originally published in Ukrainian and Russian accessible to the wider scientific community. This book will be of interest to undergraduate and graduate students studying materials science, solid-state chemistry, and engineering. It will also be relevant for researchers at industrial and national laboratories, in addition to phase diagram researchers, inorganic chemists, and solid-state physicists.

Key Features:

- Provides up-to-date experimental and theoretical information.
- A source of information for synthesizing semiconducting materials with predetermined properties.
- Delivers a critical evaluation of many industrially important systems presented in the form of two-dimensional sections for the condensed phases.

Vasyl Tomashyk is a leading researcher at the V.Ye. Lashkaryov Institute for Semiconductor Physics of the National Academy of Sciences of Ukraine. After graduating from Chernivtsi State University in 1972 (master of chemistry), he became a doctor of chemical sciences (1992), professor (1999), and author of about 690 publications in scientific journals and conference proceedings, including 12 books (seven of which are published by CRC Press) devoted to physical-chemical analysis, chemistry of semiconductors, and chemical treatment of semiconductor surfaces.

Prof. Tomashyk is a specialist at a high international level in the field of solid-state and semiconductor chemistry, including physical-chemical analysis and the technology of semiconductor materials. He was head of research topics within the international program "Copernicus". He is a member of the Materials Science International Team (Stuttgart, Germany, since 1999), which prepares a series of prestigious reference books under the titles of *Ternary Alloys* and *Binary Alloys*, and has had 35 articles published in the Landolt–Börnstein New Series. Prof. Tomashyk actively works with young researchers and graduate students and 22 Ph.D. theses have been prepared under his supervision. For many years, he has been a professor at Ivan Franko Zhytomyr State University.

Quaternary Alloys Based on IV–VI and IV–VI$_2$ Semiconductors

Vasyl Tomashyk

CRC Press
Taylor & Francis Group
Boca Raton New York London

CRC Press is an imprint of the
Taylor & Francis Group, an **informa** business

First edition published 2024
by CRC Press
6000 Broken Sound Parkway NW, Suite 300, Boca Raton, FL 33487-2742

and by CRC Press
4 Park Square, Milton Park, Abingdon, Oxon, OX14 4RN

CRC Press is an imprint of Taylor & Francis Group, LLC

© 2024 Vasyl Tomashyk

ISBN: 978-0-367-63925-9 (hbk)
ISBN: 978-0-367-64221-1 (pbk)
ISBN: 978-1-003-12348-4 (ebk)

DOI: 10.1201/9781003123484

Typeset in Times
by SPi Technologies India Pvt Ltd (Straive)

This book is dedicated to the memory of my teacher, who gave me the first knowledge of physical-chemical analysis and phase equilibria in the binary and ternary systems, Professor Dmytro Belotskyi (1918–2020).

Contents

Preface

A significant volume of semiconductor devices and circuits employs IV–VI and IV–VI$_2$ semiconductors, the most commonly used crystal material for integrated circuits. The IV–VI semiconductors are among the most interesting materials in solid-state physics. Many of them crystallize in the rock-salt structure, and structural transitions are common for them. The most widely studied compounds in this group are PbTe, PbSe, PbS, SnTe, and GeTe. These materials have small gaps which are usually less than 0.5 eV, hence they are good candidates for devices like infrared lasers and detectors. Despite their simple crystal structure, some of these compounds exhibit ferroelectric, paraelectric, and superconducting behavior. In addition, the temperature dependence of the energy gaps, the energy positions of impurity levels, the high doping levels found, the static dielectric constants, and the electronic structure of some of the alloys of these compounds appear to be anomalous compared to the "conventional behavior" of the diamond and zinc-blende semiconductors. These semiconductors either have already found use or are promising materials for infrared sensors and sources, thermoelectric elements, solar cells, memory elements, etc. The basic characteristics of these compounds, namely, narrow band gap, high permittivity, relatively high radiation resistance, high mobility of charge carriers, and high bond ionicity, are unique among semiconductor substances.

Though ternary phase diagrams based on Si, Ge, Sn, and Pb chalcogenides have been published in many issues and collected in the handbook by V. Tomashyk, *Ternary Alloys Based on IV–VI and IV–VI$_2$ Semiconductors* (London: Taylor & Francis, 2022), data pertaining to diagrams of quaternary systems based on these semiconductors are preferentially dispersed in scientific literature. This reference book is intended to describe and illustrate the up-to-date experimental and theoretical information about phase relations in based on IV–VI and IV–VI$_2$ semiconductors systems with four components. The book critically evaluates many industrially significant systems presented in the form of two-dimensional sections for the condensed phases.

In most cases the properties of IV–VI and IV–VI$_2$ semiconductors can be modified by doping with isovalent or heterovalent foreign impurities (E_1, E_2), or by interaction with other binary compounds, which can lead to the formation of a solid solution or the formation of quaternary compounds. Such quaternary materials have expanded and improved the properties of semiconductors that can be used to produce work items to create new electronic devices.

The present book aims to collect and systematize all available data on the IV–E_1–E_2–VI quaternary systems. It includes more than 1500 quaternary systems based on IV–VI and IV–VI$_2$ semiconductors, including the literature data from 1726 papers; the data are illustrated by almost 500 figures. The information is divided into 12 chapters according to the number of possible combinations of Si, Ge, Sn or Pb with S, Se or Te. The chapters are structured in the order at first of IV Group element number in the Periodic system increasing, i.e., from Si to Pb compounds, and then in order of the chalcogen number increasing, i.e., from sulfides to tellurides. The same principle is used for further description of the systems in every chapter, i.e., in order of the third and fourth component number in the Periodic system increasing.

The homogeneity range is of great importance for the crystal defect structure governing. Therefore, the reference book collects all such data accessible by now. Besides, this book presents the data on the baric and temperature dependencies of the impurities' solubility, both in the semiconductors' lattice and liquid phase, as well as the pressure-composition relationship. As semiconductors and metal mutual solubility are usually small values, the illustrating figure presents a restricted concentrations range (in mol%).

Most of the figures are presented in their original form, although some are a little corrected. If the published data varied essentially, several versions were presented in comparison. The content of system components is presented in mol% (this is not indicated on the figures). If the original phase diagram is given with mass%, this is indicated on the figure.

The book is meant for researchers at industrial and national laboratories and for university and graduate students majoring in materials science, solid-state chemistry, semiconductor chemistry, and engineering. It is also suitable for phase relation researchers, inorganic chemists, and solid-state physicists.

About the Author

Vasyl Tomashyk is a leading researcher at the V.E. Lashkaryov Institute of Semiconductor Physics of the National Academy of Sciences of Ukraine. He graduated from Chernivtsi State University in 1972 (master of chemistry). He is a doctor of chemical sciences (1992), a professor (1999), and author of about 690 publications in scientific journals and conference proceedings, and 12 books (seven of them were published by CRC Press), which are devoted to physical-chemical analysis, the chemistry of semiconductors, and the chemical treatment of semiconductor surfaces.

Tomashyk is a high-level international specialist in the field of solid-state and semiconductor chemistry, including physical-chemical analysis and technology of semiconductor materials. He was head of research topics within the international program "Copernicus". He is a member of Materials Science International Team (Stuttgart, Germany, since 1999), which prepares a series of prestigious reference books under the titles *Ternary Alloys* and *Binary Alloys*, and has had 12 chapters published in this series and 35 chapters published in the Landolt–Börnstein New Series. Tomashyk actively works with young researchers and graduate students and under his supervision 22 Ph.D. theses have been prepared. For many years, he has been a professor at Ivan Franko Zhytomyr State University in Ukraine.

List of Symbols and Acronyms

DMF	Dimethylformamide		**ppm**	Parts per million
DSC	Differential scanning calorimetry		**rpm**	Revolutions per minute
DTA	Differential thermal analysis		**SEM**	Scanning electron microscopy
EMF	Electromotive force		**HRTEM**	High-resolution transmission electron microscopy
EPMA	Electron probe microanalysis		**(X)**	Solid solution based on X
M	mol		**XRD**	X-ray diffraction
mM	mmol			

1

Systems Based on Silicon Sulfides

1.1 Silicon–Hydrogen–Nitrogen–Sulfur

When SiS_2 reacts with liquid NH_3 at $-33°C$, the **SiNH(SNH)$_2$**, **SiS(NH$_2$)$_2$** and **SiSNH** quaternary compound can be detected (Behrens and Ostermeier 1962).

1.2 Silicon–Lithium–Magnesium–Sulfur

The **Li$_2$Mg$_2$Si$_2$S$_6$** quaternary compound, which crystallizes in the tetragonal structure with the lattice parameters $a = 616.79 \pm 0.02$, $c = 662.37 \pm 0.02$ pm, a calculated density of 2.367 g·cm^{-3}, and an energy gap 3.24 eV, is formed in the Si–Li–Mg–S system (Barton et al. 2022). To prepare this compound, stoichiometric amounts of the reagents were weighed and ground together with an agate mortar and pestle in the glove box. The reaction mixture was added to a graphite crucible that was later introduced to a fused silica tube and sealed at a pressure of $\sim 10^{-2}$ Pa. The sealed reaction vessel was seated in a furnace with a temperature controller. The furnace was heated to 300°C for 3 h, held at 300°C for 10 h, heated to 650°C for 5 h, held at that temperature for 10 h, heated to 900°C for 10 h, and kept at 900°C for 4 days, followed by slow cooling to 500°C at a rate of 7°C·h^{-1} and finally rapid cooling to room temperature. Due to the high oxophilicity of Li and Mg metals, an inner graphite crucible was used to prevent the reaction of the starting materials with the fused silica outer vessel. After the reaction, the graphite crucible was broken in order to retrieve the reaction products, some of which stuck to the crucible. Therefore, the losses of some amount of the product generally preclude a regrind and reheat strategy to obtain higher phase purity of the target compound.

1.3 Silicon–Lithium–Strontium–Sulfur

The **Li$_2$SrSiS$_4$** quaternary compound, which crystallizes in the tetragonal structure with the lattice parameters $a = 646.9 \pm 0.3$, $c = 768.9 \pm 0.7$ pm, a calculated density of 2.661 g·cm^{-3}, and an energy gap of 3.94 eV, is formed in the Si–Li–Sr–S system (Yang et al. 2020b). Its whole preparation process is completed in an Ar-filled glove box. The spontaneous crystallization of the mixture of SrS, Li$_2$S and SiS$_2$ (molar ratio 1:3:1) was completed in the vacuum-sealed silica tube. Firstly, the tube was heated to 600°C in 30 h and kept at this temperature for about 100 h, then quickly down to room temperature. The product was ground and then loaded into the vacuum-sealed silica tube.

It was further sintered again at 1000°C for 10 days and slowly cooled to room temperature at 5°C·h^{-1}. Finally, many colorless crystals were found in the tube after washing with DMF solvent, and it was stable in the air within several weeks.

1.4 Silicon–Lithium–Zinc–Sulfur

The **Li$_2$ZnSiS$_4$** quaternary compound, which crystallizes in the orthorhombic structure with the lattice parameters $a = 1289.2 \pm 0.2$, $b = 777.39 \pm 0.12$, $c = 614.51 \pm 0.10$ pm, and an energy gap of 3.90 eV, is formed in the Si–Li–Zn–S system (Li et al. 2018a). It was synthesized using the mixture of Li, Zn, Si, and S with the stoichiometric ratio at 900°C, and then cooling down to room temperature in one week. Air-stable single crystals were obtained by spontaneous crystallization.

1.5 Silicon–Lithium–Cadmium–Sulfur

The **Li$_2$CdSiS$_4$** quaternary compound, which crystallizes in the orthorhombic structure with the lattice parameters $a = 761.1 \pm 0.3$, $b = 679.3 \pm 0.2$, $c = 630.4 \pm 0.2$ pm, a calculated density of 2.880 g·cm^{-3}, and an energy gap of 3.76 eV [3.09 eV (Zhang et al. 2020b)], is formed in the Si–Li–Cd–S system (Li et al. 2020a). This compound was prepared via high-temperature solid-state synthesis (Zhang et al. 2020b). Stoichiometric quantities of Li$_2$S, Cd, Si, and S were weighed and combined in an Ar-filled glove box. The mixture was loaded into a graphite tube that was loosely capped and subsequently inserted into a fused-silica tube, which was then flame-sealed under vacuum (~ 0.1 Pa). Next, the tube was placed into a furnace. The reaction was heated to 850°C in 12 h, held at that temperature for 7 days, cooled at a rate of 2°C·h^{-1} to 650°C, and finally cooled to room temperature in 24 h. Colorless crystals with a faint tint of yellow were obtained.

Single crystals of this compound were obtained as follows: the mixture of Li, CdS, Si, and S about 0.5 g was weighted in a molar ratio of 2:0.55:1:3.5 and loaded into a graphite crucible in an Ar-filled glove box (Li et al. 2020a). Then, the crucible was sealed in a silicon tube under a vacuum (10^{-4} Pa). The tube was heated to 850°C, kept there for 30 h, slowly cooled at 3°C·h^{-1} to 300°C, and cooled to room temperature naturally in a furnace. The resultant sample was washed with distilled water and dried in air. The colorless transparent crystals of the title compound were obtained. They are not moisture sensitive.

DOI: 10.1201/9781003123484-1

1.6 Silicon–Lithium–Mercury–Sulfur

The **Li₂HgSiS₄** quaternary compound, which crystallizes in the orthorhombic structure with the lattice parameters $a = 759.2 \pm 0.2$, $b = 676.25 \pm 0.19$, $c = 632.95 \pm 0.18$ pm, a calculated density of 3.790 g·cm⁻³, and an energy gap of 2.68 eV (Wu and Pan 2017) [$a = 752.26*/756.24**$, $b = 671.25*/674.67**$ and $c = 630.42*/633.82**$ pm according to the calculations with (*) and without (**) including the spin-orbit coupling and an energy gap of 3.195 eV (Alnujaim et al. 2020)], is formed in the Si–Li–Hg–S system. The target compound was prepared with a mixture Li, HgS, Si, and S (molar ratio 2:1:1:3) (Wu and Pan 2017). In the preparation process, a graphite crucible was added to the vacuum-sealed silica tube to avoid the reaction between Li and the silica tube at a high temperature. The temperature process was set as follows: first, it was heated to 700°C for 2 days, and kept at this temperature for about four days, then slowly cooled down to 300°C within four days, and finally quickly cooled to room temperature by turning off the furnace. The obtained product was washed by DMF to remove the other byproducts. Yellow crystals of Li₂HgSiS₄ were obtained, and they remained stable in the air for over half a year.

1.7 Silicon–Lithium–Aluminum–Sulfur

Li₄SiS₄–Li₅AlS₄. The **Li₄₊ₓSi₁₋ₓAlₓS₄** solid solution in the whole composition range is formed in this system (Murayama et al. 2002). The lattice parameters change continuously with composition. According to the data of Huang et al. (2019c), the Li₄.₁Al₀.₁Si₀.₉S₄ solid solution has two polymorphic modifications. High-temperature modification crystallizes in the cubic structure with the lattice parameter $a = 1006.328 \pm 0.010$ pm. To prepare these solid solutions, the Li₂S, Al₂S₃, and Si₂S powders were weighed in stoichiometric ratios in an Ar-filled glove box and then mechanically mixed in a sealed stainless-steel pot for 30 min at a speed of 370 rpm using a vibration milling apparatus. The obtained mixtures were pelletized, placed in a carbon tube, and sealed at 15 Pa in a carbon-coated quartz tube, followed by annealing at 1000°C for 5 h or at 700°C for 8 h (Murayama et al. 2002; Huang et al. 2019c). The annealing alloys were quenched in ice water or slowly cooled to room temperature.

1.8 Silicon–Lithium–Gallium–Sulfur

The **Li₂GaSiS₄** quaternary compound, which crystallizes in the monoclinic structure with the lattice parameters $a = 680.01 \pm 0.05$, $b = 6006.7 \pm 0.4$, $c = 679.48 \pm 0.07$ pm, $\beta = 119.580 \pm 0.002°$, a calculated density of 2.565 g·cm⁻³, and an energy gap of 4.11 eV, is formed in the Si–Li–In–S system (Chen et al. 2022b). The single crystal sample synthesis of LiGaSiS₄ consisted of two steps. The first step was the synthesis of binary materials Li₂S and Ga₂S₃. The light yellow Li₂S was prepared by dissolving Li (2 mM) and S (1 mM) in a liquid ammonia solution, and Ga₂S₃ was prepared by a high-temperature solid phase reaction of Ga (3 mM) and S (2 mM). Then a mixture

of Li₂S (1 mM), Si (2 mM), Ga₂S₃ (1 mM), and S (2 mM) was thoroughly ground and mixed in a glove box filled with argon gas protection. To avoid reacting with air, it was packed into quartz tubes, pumped to vacuum and sealed, and finally placed in a muffle furnace. Since the sulfides containing lithium generally have relatively high melting points, the high temperature gives rise to the sulfur vapor pressure in the quartz tube rising rapidly, which often leads to the explosion of the tube and the failure of crystal growth. Therefore, the heating process was divided into two steps: the temperature was increased to 700°C within 20 h at the beginning, and cooled to 200°C within 10 h. The sample was taken out at room temperature. The resealed sample was reheated to 900°C at a heating rate of 35°C·h⁻¹ after the second grinding, kept at this temperature for five days to ensure sufficient reaction, and slowly cooled down to 600°C and then naturally cooled to room temperature. In the end, a large quantity of off-white LiGaSiS₄ was obtained. The crystals existed stably in water and air.

1.9 Silicon–Lithium–Indium–Sulfur

The **Li₂In₂SiS₆** quaternary compound, which melts congruently at 858°C and crystallizes in the monoclinic structure with the lattice parameters $a = 1207.41 \pm 0.11$, $b = 702.21 \pm 0.05$, $c = 1208.02 \pm 0.09$ pm, $\beta = 110.060 \pm 0.005°$, a calculated density of 3.203 g·cm⁻³, and an energy gap of 3.61 eV, is formed in the Si–Li–In–S system (Yin et al. 2012a). The structural, electronic, and optical properties of this compound have been investigated by *ab initio* calculations based on the density functional theory by Wong et al. (2018) using the full potential linearized augmented plane wave method. The next values of the lattice parameters were determined: $a = 1207.27$, $b = 702.13$, $c = 1207.88$ pm (the value of β is not calculated).

To prepare this compound, the mixtures of LiInS₂ (1 mM) and SiS₂ (0.5 mM) were ground, loaded into a fused silica tube under an Ar atmosphere in a glove box, then moved to a high-vacuum line, and flame-sealed under a high vacuum of 10⁻³ Pa (Yin et al. 2012a). The tube was then placed in a furnace and heated to 900°C for 15 h, left for 48 h, cooled to 320°C at a rate of 3°C·h⁻¹, and finally cooled to room temperature by switching off the furnace. Many light-yellow block-shaped crystals were found in the ampoule. The crystals are stable in the air. Polycrystalline samples of Li₂In₂SiS₆ were synthesized by solid-state reaction techniques. The mixtures of Li₂S, SiS₂, and In₂S₃ (molar ratio 1:1:1) were heated to 700°C for 15 h, kept at that temperature for 72 h, and then the furnace was turned off.

1.10 Silicon–Lithium–Lanthanum–Sulfur

The **LiLa₃SiS₇** quaternary compound, which crystallizes in the hexagonal structure with the lattice parameters $a = 1033.91 \pm 0.02$, $c = 575.053 \pm 0.009$ pm, a calculated density of 4.218 g·cm⁻³, and an energy gap of 2.75 eV, is formed in the Si–Li–La–S system (Craig et al. 2022). It was prepared via a direct combination of the elements or binary sulfides. This compound shows high thermal stability (>1000°C).

1.11 Silicon–Lithium–Cerium–Sulfur

The **LiCe$_3$SiS$_7$** quaternary compound, which crystallizes in the hexagonal structure with the lattice parameters $a = 1024.61 \pm 0.01$, $c = 572.46 \pm 0.01$ pm, and a calculated density of 4.338 g·cm^{-3}, is formed in the Si–Li–Ce–S system (Craig et al. 2022). It was prepared via a direct combination of the elements or binary sulfides. This compound is a semiconductor and shows high thermal stability (>1000°C).

1.12 Silicon–Lithium–Praseodymium–Sulfur

The **LiPr$_3$SiS$_7$** quaternary compound, which crystallizes in the hexagonal structure with the lattice parameters $a = 1017.05 \pm 0.01$, $c = 570.927 \pm 0.01$ pm, and a calculated density of 4.430 g·cm^{-3}, is formed in the Si–Li–Pr–S system (Craig et al. 2022). It was prepared via a direct combination of the elements or binary sulfides. This compound is a semiconductor and shows high thermal stability (>1000°C).

1.13 Silicon–Lithium–Neodymium–Sulfur

The **LiNd$_3$SiS$_7$** quaternary compound, which crystallizes in the hexagonal structure, is formed in the Si–Li–Nd–S system (Craig et al. 2022). It was prepared via a direct combination of the elements or binary sulfides. This compound is a semiconductor and shows high thermal stability (>1000°C).

1.14 Silicon–Lithium–Samarium–Sulfur

The **LiSm$_3$SiS$_7$** quaternary compound, which crystallizes in the hexagonal structure with the lattice parameters $a = 1000.7 \pm 0.2$, $c = 566.8 \pm 0.3$ pm, a calculated density of 4.800 g·cm^{-3}, and an energy gap of 2.83 eV, is formed in the Si–Li–Sm–S system (Zhen et al. 2016a). To synthesize this compound, Li, Sm, Si, and S were weighted at the stoichiometric ratio. All the raw materials were firstly loaded into a graphite crucible, which was then put into a silica tube, and the tube was flame-sealed under 10^{-3} Pa. The detailed temperature-setting curve for the muffle furnace was heated to 850°C in 50 h and kept at this temperature for about 100 h, then slowly cooled to 300°C by 80 h, finally quickly cooled to the room temperature. The product was washed with DMF to remove the byproducts. Pale-yellow crystals were found, which are stable in the air. Because Li metal is easily oxidized in the air, all the preparations were completed in an Ar-filled glove box.

1.15 Silicon–Lithium–Gadolinium–Sulfur

The **LiGd$_3$SiS$_7$** quaternary compound, which crystallizes in the hexagonal structure, is formed in the Si–Li–Gd–S system (Craig et al. 2022). It was prepared via direct combination of the elements or binary sulfides. This compound is a semiconductor and shows high thermal stability (>1000°C).

1.16 Silicon–Lithium–Dysprosium–Sulfur

The **LiDy$_3$SiS$_7$** quaternary compound, which crystallizes in the hexagonal structure, is formed in the Si–Li–Dy–S system (Craig et al. 2022). It was prepared via direct combination of the elements or binary sulfides. This compound is a semiconductor and shows high thermal stability (>1000°C).

1.17 Silicon–Lithium–Lead–Sulfur

The **Li$_2$PbSiS$_4$** quaternary compound, which crystallizes in the tetragonal structure with the lattice parameters $a = 646.18 \pm 0.01$, $c = 773.33 \pm 0.02$ pm, and an energy gap of 2.51 eV, is formed in the Si–Li–Pb–S system (Stoyko et al. 2021). This compound was prepared using stoichiometric amounts of Li$_2$S, Pb, Si, and S that were weighed and combined in an Ar-filled glove-box. The reagents were loaded into a graphite crucible that was subsequently placed into a fused silica tube. The tube was vacuum sealed at a pressure of ~0.1 Pa and inserted into a furnace. The tube was heated from room temperature to 800°C in 24 h, held at that temperature for 48 h, and cooled at a rate of 4°C·h^{-1} to 400°C, followed by cooling to room temperature over the course of 12 h. After this, the tube was cut open. The product mainly consisted of small yellow plate-like crystals. The compound was found to be relatively stable for a short period of time, on the order of several days; however, prolonged exposure to ambient conditions results in its degradation.

1.18 Silicon–Lithium–Nitrogen–Sulfur

SiS$_2$–Li$_3$N–Li$_2$S. The glasses were prepared by a melt-quenching procedure for the compositions $(60-3x/2)$Li$_2$S+40SiS$_2$+xLi$_3$N for $x = 0$, 3, 5 (Sakamoto et al. 1999). Li$_2$S, SiS$_2$, and Li$_3$N were mixed and put into a silica ampoule, the inside of which was carbon coated. The ampoule was then flame-sealed under vacuum, and the mixture was melted at 1000–1200°C for 1 h and quenched in ice water to obtain bulk glasses. The raw materials and the glasses obtained were handled in a dry N$_2$ filled glove box. Both temperatures of glass transition and crystallization increased with an increase in the Li$_3$N content.

1.19 Silicon–Lithium–Phosphorus–Sulfur

Li$_4$SiS$_4$–Li$_3$PS$_4$. The **Li$_{4-x}$Si$_{1-x}$P$_x$S$_4$** solid solution in the whole composition range is formed in this system (Murayama et al. 2002). According to the data of Hori et al. (2014), the range of the solid solution was determined to be $0.525 \leq x \leq 0.60$ ($0.20 < \delta < 0.43$ in Li$_{10+\delta}$Si$_{1+\delta}$P$_{2-\delta}$S$_{12}$). These solid solutions crystallize in the orthorhombic structure which was transformed into the tetragonal structure via high-pressure treatment with pressure in the range of 3–5 GPa (Kuhn et al. 2014). One of the compositions of the solid solution, **Li$_{11}$Si$_2$PS$_{12}$**, has the lattice parameters $a = 869.05 \pm 0.14$ and $c = 1257.03 \pm 0.20$ pm (Kuhn et al. 2014) and another, **Li$_{10.35}$Si$_{1.35}$P$_{1.65}$S$_{12}$**, is characterized by the lattice parameters $a = 866.9578 \pm 0.0018$ and $c = 1253.6058$

± 0.0048 pm (Hori et al. 2014) [a = 864.390 ± 0.003 and c = 1250.773 ± 0.008 pm at 17 K, a = 868.385 ± 0.005 and c = 1255.589 ± 0.007 pm at 300 K; and a = 876.517 ± 0.005 and c = 1267.003 ± 0.012 pm at 800 K (Hori et al. 2015b)].

All samples of the solid solutions were synthesized using solid-state reactions (Murayama et al. 2002; Kuhn et al. 2014; Hori et al. 2014, 2015b). The Li₂S, P₂S₅, and SiS₂ powders were weighed in stoichiometric ratios in an Ar-filled glove box and then mechanically mixed by planetary ball-milling for 40-96 h. The specimens were then pressed into pellets, sealed in a carbon-coated quartz ampoule at 10 Pa, and heated at a reaction temperature of 650°C for 72–96 h, applying a 2.2°C·min⁻¹ heating rate and a 0.60°C·min⁻¹ cooling rate.

1.20 Silicon–Lithium–Oxygen–Sulfur

SiS₂–Li₂S–Li₂SO₄ and **SiS₂–Li₂S–Li₄SiO₄**. The glasses (100-x)(0.6Li₂S·0.4SiS₂)·xLi₂SO₄(xLi₄SiO₄) were prepared by means of twin-roller rapid quenching (Hirai et al. 1995). SiS₂, Li₂S, and Li₂SO₄ or Li₄SiO₄ were mixed in a vitreous carbon crucible, and then the mixture was melted at 1000°C for 2 h in a furnace. The molten sample was dropped into a rotating twin-roller to obtain flake-like samples, the thickness of which was about 20 pm. These processes were carried out in a dry N₂-filled glove box. The glasses were obtained up to x = 10 for Li₂SO₄ and up to x = 20 for Li₄SiO₄. It was shown that 5 mol% Li₄SiO₄ improves the glass stability. Doping of Li₂SO₄ to the SiS₂–Li₂S system did not improve the glass stability.

1.21 Silicon–Lithium–Bromine–Sulfur

SiS₂–Li₂S–LiBr. The glass-forming region in this system, as determined by XRD, is given in Figure 1.1 (Kennedy and Yang 1987). All preparations of the samples were carried

out in a He-filled glove box because of the extremely hygroscopic nature of the starting materials as well as the glass products. The glasses were synthesized by thoroughly mixing 0.4SiS₂+0.6Li₂S glass with LiBr in the molar ratios desired, and heating at 950°C followed by quenching directly into liquid nitrogen. With the direct liquid nitrogen rapid quenching method, the boundaries of the glass-forming region were extended to higher Li-content compared to conventional quenching, and 30 mol% LiBr could be obtained in the glasses even with the Li₂S-rich base glass composition of 0.4SiS₂-0.6Li₂S.

1.22 Silicon–Lithium–Iodine–Sulfur

SiS₂–Li₂S–LiI. The glass-forming region in this system, as determined by XRD, is given in Figure 1.2 (Kennedy and Yang 1987). All preparations were the same as in the case of the SiS₂–Li₂S–LiBr system. With the direct liquid nitrogen rapid quenching method, the boundaries of the glass-forming region were extended to higher Li-content compared to conventional quenching, and 40 mol% LiI could be obtained in the glasses even with the Li₂S-rich base glass composition of 0.4SiS₂-0.6Li₂S.

1.23 Silicon–Sodium–Strontium–Sulfur

The **Na₂SrSiS₄** quaternary compound, which crystallizes in the trigonal structure with the lattice parameters a = 2348.6 ± 0.3, c = 687.64 ± 0.19 pm, a calculated density of 2.638 g·cm⁻³, and an energy gap of 3.87 eV, is formed in the Si–Na–Sr–S system (Yang et al. 2020b). It was prepared in the same way as Li₂SrSiS₄ was synthesized using Na₂S instead of Li₂S. This compound is stable in the air within several weeks.

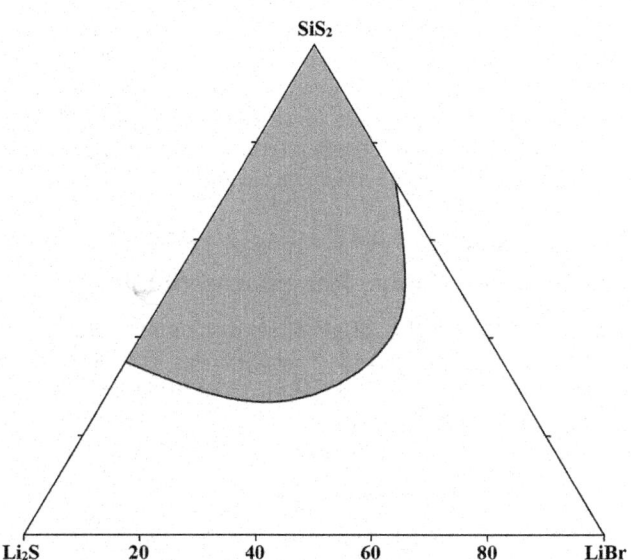

FIGURE 1.1 Glass-forming region in the SiS₂–Li₂S–LiBr quasiternary system. (From Kennedy, J.H., and Yang, Y., *J. Solid State Chem.*, **69**(2), 252, 1987.)

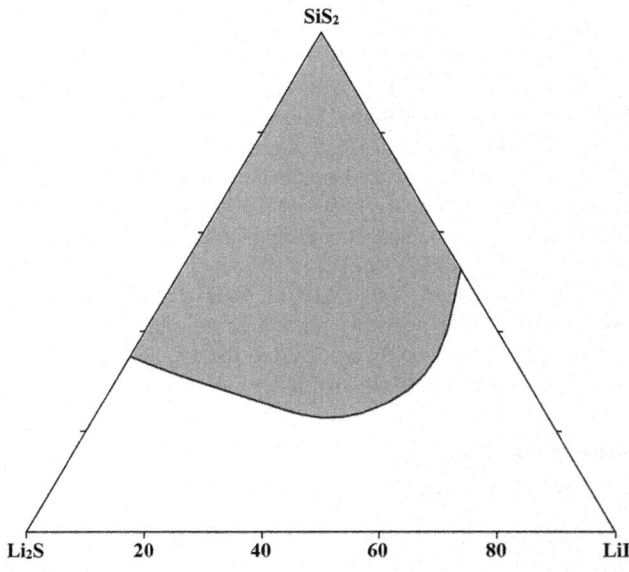

FIGURE 1.2 Glass-forming region in the SiS₂–Li₂S–LiI quasiternary system. (From Kennedy, J.H., and Yang, Y., *J. Solid State Chem.*, **69**(2), 252, 1987.)

1.24 Silicon–Sodium–Mercury–Sulfur

The **Na$_2$Hg$_3$Si$_2$S$_8$** quaternary compound, which crystallizes in the tetragonal structure with the lattice parameters $a = 872.5 \pm 0.5$, $c = 897.2 \pm 1.1$ pm, a calculated density of 4.670 g·cm^{-3}, and an energy gap of 2.86 eV, is formed in the Si–Na–Hg–S system (Wu et al. 2016c). To obtain this compound, an Ar-filled glove box was used. A mixture of Na (2 mM), HgS (3 mM), Si (2 mM), and excess sulfur (8 mM) were loaded into a graphite crucible, and then put into a tidy silica tube that was flame-sealed under 10^{-3} Pa. The tube was heated to 700°C for 30 h and kept at this temperature for about 60 h, then slowly cooled down to 300°C for 4 days, and finally quickly cooled to ambient temperature. The product was carefully washed with DMF and then dried at 100°C. Pale-yellow crystals of Na$_2$Hg$_3$Si$_2$S$_8$ were obtained. This compound is stable in the air after several months.

1.25 Silicon–Sodium–Indium–Sulfur

The **Na$_2$In$_2$SiS$_6$** quaternary compound, which melts congruently at ~763°C and crystallizes in the monoclinic structure with the lattice parameters $a = 1203.9 \pm 0.3$, $b = 788.77 \pm 0.14$, $c = 1216.5 \pm 0.3$ pm, $\beta = 108.341 \pm 0.005°$, a calculated density of 3.036 g·cm^{-3}, and an energy gap of 2.470 eV (Li et al. 2016b) [$a = 1198.3 \pm 0.4$, $b = 778.6 \pm 0.3$, $c = 1212.6 \pm 0.4$ pm, $\beta = 108.15 \pm 0.01°$, a calculated density of 3.065 g·cm^{-3}, and an energy gap of 3.27 eV (Yohannan and Vidyasagar 2016a)], is formed in the Si–Na–In–S system.

The title compound was synthesized by solid-state reaction of Na$_2$S, In, Si, and S mixed in a molar ratio of 3:1:1:5 (Li et al. 2016b). The mixture was finely ground, introduced into a quartz tube, and then flame-sealed under vacuum. This tube was put in a furnace, held at 450°C for 5 h, raised to 820°C for 10 h, retained there for 36 h, and then cooled to 400°C for 100 h and powered off. Yellow crystals of Na$_2$In$_2$SiS$_6$ were obtained. They are stable in moisture conditions and air for several months. The pure powder phase was prepared through a stoichiometry mixture of Na$_2$S, In, Si, and S molar ratio of 1:2:1:5, which were held at 620°C for 72 h and then decreased to room temperature for 20 h. The product was reground and reprocessed by the identical procedure to improve the purity.

Na$_2$In$_2$SiS$_6$ was also prepared in the form of pale yellow block-shaped single crystals, along with NaInS$_2$ as a secondary phase, when a stoichiometric reactant mixture of NaInS$_2$ (2 mM), Si (1 mM) and S (2 mM) was heated in an evacuated sealed tube, at 850°C for 2 days and then cooled to 650°C at 2°C·h^{-1} (Yohannan and Vidyasagar 2016a).

1.26 Silicon–Sodium–Yttrium–Sulfur

The **NaY$_3$SiS$_7$** quaternary compound, which crystallizes in the hexagonal structure with the lattice parameters $a = 976.57 \pm 0.07$, $c = 570.61 \pm 0.05$ pm, and a calculated density of 3.821 g·cm^{-3}, is formed in the Si–Na–Y–S system (Hartenbach and Schleid 2003a). The reaction of NaBr, Y, S, and SiS$_2$ in the molar ratio 1:4:6:2 with an excess of NaBr as a flux in evacuated silica ampoules at 850°C for 7 days leads to the formation of deep red, hexagonal pillar-shaped single crystals of the title compound.

1.27 Silicon–Sodium–Samarium–Sulfur

The **NaSm$_3$SiS$_7$** quaternary compound, which crystallizes in the hexagonal structure with the lattice parameters $a = 998.31 \pm 0.07$, $c = 566.55 \pm 0.04$ pm and a calculated density of 4.935 g·cm^{-3}, is formed in the Si–Na–Sm–S system (Hartenbach et al. 2007). It was prepared from a mixture of Sm, S, and SiS$_2$ (molar ratio 4:6:1) using an excess of NaCl as a flux and Na source. The annealing of the mixture at 850°C for 8 days in evacuated silica glass ampoule lead to dark red single crystals in the form of hexagonal columns. They are air and water resistant.

1.28 Silicon–Sodium–Europium–Sulfur

The **Na$_8$Eu$_2$(Si$_2$S$_6$)$_2$** quaternary compound, which crystallizes in the monoclinic structure with the lattice parameters $a = 677.8 \pm 0.2$, $b = 1168.0 \pm 0.2$, $c = 766.2 \pm 0.2$ pm, $\beta = 107.271 \pm 0.003°$, and a calculated density of 2.823 g·cm^{-3}, is formed in the Si–Na–Eu–S system (Choudhury et al. 2015). It has been prepared by a reactive flux method using a Na$_2$S$_2$ flux (0.4 mM) along with elemental Eu (0.2 mM), Si (0.5 mM), and S (0.6 mM) (approximate molar ratio of Na/Eu/Si/S was 8:2:5:14). The mixture was loaded into fused silica ampoule under a N$_2$ atmosphere which was then evacuated, flame-sealed, and placed in a furnace. The furnace temperature was increased to 750°C at a rate of 35°C·h^{-1}, held constant at 750°C for 150 h, and finally decreased to room temperature at a rate of 5°C·h^{-1}. After the product cooled, the ampoule was opened in the air and the solid yellow product was soaked in DMF, and sonicated for 30 s in order to both dissolve and remove any remaining flux and to loosen the product crystals. The product contained only yellow crystals; however, in the cold zone of the ampoule there was a small amount of a light brown deposit that presumably was SiS$_2$. The DMF-treated yellow product slowly changed color, darkening after several hours after being exposed to the atmosphere, indicating that the product was unstable in the air.

1.29 Silicon–Sodium–Oxygen–Sulfur

The **Na$_6$Si$_3$S$_8$O** quaternary compound, which crystallizes in the monoclinic structure with the lattice parameters $a = 680.06 \pm 0.08$, $b = 1434.14 \pm 0.18$, $c = 1620.7 \pm 0.2$ pm, $\beta = 90.561 \pm 0.008°$, a calculated density of 2.079 g·cm^{-3}, and an energy gap of 3.57 eV, is formed in the Si–Na–O–S system (Wu et al. 2017b). It was synthesized in a vacuum-sealed silica tube under the raw mixture of Na, Si, and S with Na$_2$CO$_3$ as a flux. Obtained products were repeatedly washed with the DMF to remove the flux and other byproducts. This compound is stable in the air for several months.

1.30 Silicon–Sodium–Bromine–Sulfur

The **Na$_3$SiS$_3$Br** quaternary compound is formed in the Si–Na–Br–S system (Feltz and Pfaff 1983). To prepare it, approximately 0.05 to 0.08 M solution of Na$_6$Si$_2$S$_6$, the equivalent amount of a methanolic bromine solution, was added dropwise. Na$_3$SiS$_3$Br crystallizes during slow concentration in vacuum as an almost colorless precipitate.

1.31 Silicon–Potassium–Yttrium–Sulfur

The **KYSiS$_4$** quaternary compound, which crystallizes in the monoclinic structure with the lattice parameters $a = 637 \pm 2$, $b = 656 \pm 2$, $c = 861 \pm 3$ pm, and $\beta = 108.1 \pm 0.4$, is formed in the Si–K–Y–S system (Wu and Ibers 1993a). It was prepared by the reaction of Si with K$_2$S$_5$ and Y$_2$S$_3$. The starting materials were placed in a quartz tube that was subsequently evacuated to 10^{-3} Pa and sealed. The molar ratio of K/Y/Si/S was 1:1:1:4. The quartz tube was heated gradually to 500°C, where it was kept for 24 h before being successively brought to 700°C for 24 h and to 1000°C for 150 h. The tube was then cooled (4°C·h^{-1}) to 300°C and the furnace was shut off. Colorless crystals of the title compound were obtained. KYSiS$_4$ appears to be modestly stable in the air but decomposes gradually in the presence of water. Bulk samples of this compound were prepared by reaction of stoichiometric amounts of starting materials at 1000°C for 10 days with intermittent grinding.

1.32 Silicon–Potassium–Lanthanum–Sulfur

The **KLaSiS$_4$** quaternary compound, which crystallizes in the monoclinic structure with the lattice parameters $a = 653.34 \pm 0.06$, $b = 657.23 \pm 0.06$, $c = 867.02 \pm 0.08$ pm, $\beta = 107.496 \pm 0.009°$, and a calculated density of 3.127 g·cm^{-3} (Hartenbach and Schleid 2005b) [$a = 657 \pm 2$, $b = 660 \pm 2$, $c = 869 \pm 3$ pm, and $\beta = 107.2 \pm 0.4$ (Wu and Ibers 1993a)], is formed in the Si–K–La–S system. It was prepared in the same way as **KYSiS$_4$** was synthesized using La$_2$S$_3$ instead of Y$_2$S$_3$ (Wu and Ibers 1993a). Colorless crystals of the title compound were obtained. KLaSiS$_4$ appear to be modestly stable in the air but decomposes gradually in the presence of water. Bulk samples of this compound were prepared by reaction of stoichiometric amounts of starting materials at 1000°C for 10 days with intermittent grinding.

Pale yellow, platelet-shaped, air- and water-resistant single crystals of this compound were derived from the reaction of La and S with SiS$_2$ in a molar ratio of 2:3:1 with an excess of KCl as a flux and a source of potassium ions, in evacuated silica ampoules at 850°C within 7 days (Hartenbach and Schleid 2005b).

1.33 Silicon–Potassium–Cerium–Sulfur

The **KCeSiS$_4$** quaternary compound, which crystallizes in the monoclinic structure with the lattice parameters $a = 649.15 \pm 0.06$, $b = 656.18 \pm 0.06$, $c = 863.96 \pm 0.08$ pm, $\beta = 107.531$ $\pm 0.009°$, and a calculated density of 3.176 g·cm^{-3} (Hartenbach and Schleid 2002b) [$a = 652.01 \pm 0.03$, $b = 659.82 \pm 0.03$, $c = 869.20 \pm 0.05$ pm, and $\beta = 107.701 \pm 0.003°$ (Gauthier et al. 1999)], is formed in the Si–K–Ce–S system. It was prepared by introducing Ce$_2$S$_3$ (1 mM), Si (2.3 mM), K$_2$S (1 mM), and S (4.6 mM) in a silica tube that was then sealed under vacuum (1 Pa) (Gauthier et al. 1999). The powder mixture was heated at 800°C for 7 days and slowly cooled to room temperature. The reaction product was then finely ground in a dry box and annealed at 700°C for 7 days. This last step not only favors a higher crystallization state of KCeSiS$_4$ but also makes possible the elimination of the SiS$_2$ excess by sublimation at the cold end of the tube. The resulting phase was a homogeneous bright yellow powder.

Brownish-yellow, plate-shaped single crystals of this compound were grown by the reaction of KCl, Ce$_2$S$_3$, and SiS$_2$ (molar ratio 1:1:1) using a six-fold molar amount of KCl as a flux in evacuated silica tubes by heating at 850°C for 7 days (Hartenbach and Schleid 2002b). They are resistant both to air and water.

1.34 Silicon–Potassium–Samarium–Sulfur

The **KSmSiS$_4$** quaternary compound, which crystallizes in the monoclinic structure with the lattice parameters $a = 642.6 \pm 0.6$, $b = 658.2 \pm 0.5$, $c = 860.2 \pm 0.8$ pm, $\beta = 107.90 \pm 0.10°$, a calculated density of 3.317 g·cm^{-3}, and an energy gap of 2.40 eV, is formed in the Si–K–Sm–S system (Guo et al. 2008). Single crystals of this compound were obtained by the solid-state reaction with KI as a flux. The starting materials were Sm$_2$O$_3$ (0.476 mM), S (3.805 mM), Si (0.961 mM), and B (2.868 mM) in a molar ratio of Sm/S/B/Si = 1:6:3:1 together with 400 mg KI. The mixture of starting materials was ground into fine powder in an agate mortar and pressed into pellets, followed by being loaded into a quartz tube, which was evacuated to 10^{-2} Pa and flame-sealed. The sample was placed into a furnace, heated from room temperature to 300°C for 5 h and kept at 300°C for 10 h, heated to 650°C for 5 h and kept at this temperature for 10 h, then heated to 950°C in 5 h and held this temperature for 10 days, and finally cooled down to 300°C in 5 days and powered off. Yellow plate-like crystals of KSmSiS$_4$ were observed after the product was washed with distilled water and dried with acetone. This compound is stable in the air and water with no signs of appreciable change in crystal quality over several months.

1.35 Silicon–Potassium–Europium–Sulfur

The **KEuSiS$_4$** quaternary compound, which crystallizes in the monoclinic structure with the lattice parameters $a = 642.6 \pm 0.4$, $b = 658.2 \pm 0.5$, $c = 856.6 \pm 0.7$ pm, $\beta = 107.83 \pm 0.06°$, a calculated density of 3.345 g·cm^{-3} at 170 K, and an energy gap of 1.72 eV, is formed in the Si–K–Eu–S system (Evenson IV and Dorhout 2001). To prepare this compound, a mixture of S (1.97 mM), K$_2$S$_2$ (0.563 mM), Si (0.288 mM), and Eu (0.288 mM) were loaded into a fused silica ampoule inside an inert atmosphere glove box. The ampoule was flame-sealed under vacuum and placed in a tube furnace. The temperature was ramped to 725°C where it remained for 150 h, and the furnace was allowed to cool back to room temperature at 4°C·h^{-1}. Washing the product with DMF gave brown plates of the title compound.

1.36 Silicon–Potassium–Ytterbium–Sulfur

The **KYbSiS$_4$** quaternary compound, which crystallizes in the monoclinic structure with the lattice parameters a = 632.44 ± 0.10, b = 655.52 ± 0.10, c = 857.01 ± 0.15 pm, β = 108.001 ± 0.013°, a calculated density of 3.621 g·cm^{-3}, and an energy gap of 2.34 eV, is formed in the Si–K–Yb–S system (Guo et al. 2008) [a = 631.6 ± 0.5, b = 655.7 ± 0.5, c = 857.1 ± 0.7 pm, β = 108.09 ± 0.01° (Gray et al. 2005)]. Single crystals of this compound were obtained by the solid-state reaction with KI as a flux (Guo et al. 2008). The starting materials were Yb$_2$O$_3$ (0.566 mM), S (4.523 mM), K$_2$S (0.571 mM), Si (1.139 mM), and B (3.423 mM) in a molar ratio of Yb/S/K/B/Si = 1:4.5:1:3:1 together with 300 mg KI. The mixture of starting materials was ground into fine powder in an agate mortar and pressed into pellets, followed by being loaded into a quartz tube which was evacuated to 10^{-2} Pa and flame-sealed. The sample was placed into a furnace, heated from room temperature to 300°C for 5 h and kept at 300°C for 10 h, heated to 650°C for 5 h and kept at this temperature for 10 h, then heated to 950°C in 5 h and held this temperature for 10 days, and finally cooled down to 300°C in 5 days and powered off. Orange block crystals of KYbSiS$_4$ were observed after the product was washed with distilled water and dried with acetone. This compound is stable in the air and water with no signs of appreciable change in crystal quality over several months.

Crystals of KYbSiS$_4$ were also formed from a molten chalcogenide flux reaction of 83.8 mg Yb, 12.8 mg Si, 30.8 mg S, and 68.9 mg K$_2$S$_2$ (Gray et al. 2005). These reactants were placed in a fused silica ampoule in an inert atmosphere glove box. The ampoule was sealed under vacuum, and heated to 750°C at a rate of 35°C·h^{-1}. After 150 h of annealing, the ampoule was cooled at 4°C·h^{-1} to room temperature. Orange plates of the title compound were obtained after the product was washed with DMF to dissolve the remaining flux.

1.37 Silicon–Potassium–Bismuth–Sulfur

The **KBiSiS$_4$** quaternary compound, which crystallizes in the monoclinic structure with the lattice parameters a = 647.69 ± 0.13, b = 673.71 ± 0.13, c = 1716.8 ± 0.4 pm, β = 108.14 ± 0.03°, a calculated density of 3.773 g·cm^{-3}, and an energy gap of 2.25 eV, is formed in the Si–K–Bi–S system (Mei et al. 2010). To synthesize this compound, a reaction mixture of K$_2$S$_3$ (1 mM), Si (1.5 mM), Bi (0.5 mM), and S (4.8 mM) was loaded into a fused silica tube under an Ar atmosphere in a glove box. This tube was sealed under vacuum (10^{-3} Pa) and then placed in a furnace. The sample was heated to 200°C over 20 h, kept at 200°C for 10 h, then heated to 600°C within 24 h, kept at 600°C for 72 h, slowly cooled at 4°C·h^{-1} to 100°C, and

then cooled to room temperature. The reaction mixture was washed free of flux with DMF and then dried with acetone. The product consists of red plates of KBiSiS$_4$. It is moderately stable in the air.

1.38 Silicon–Rubidium–Lanthanum–Sulfur

The **RbLaSiS$_4$** quaternary compound, which crystallizes in the orthorhombic structure with the lattice parameters a = 1728.4 ± 0.2, b = 667.23 ± 0.06, c = 652.89 ± 0.06 pm, and a calculated density of 3.358 g·cm^{-3}, is formed in the Si–Rb–La–S system (Hartenbach and Schleid 2005b). Yellow single crystals of this compound were derived from the reaction of La and S with SiS$_2$ in a molar ratio of 2:3:1 with an excess of RbCl as a flux and a source of rubidium ions, in evacuated silica ampoules at 850°C within 7 days.

1.39 Silicon–Rubidium–Europium–Sulfur

The **RbEuSiS$_4$** quaternary compound, which crystallizes in the orthorhombic structure with the lattice parameters a = 639.2 ± 0.1, b = 663.4 ± 0.2, c = 1700.1 ± 0.3 pm, and a calculated density of 3.628 g·cm^{-3}, is formed in the Si–Rb–Eu–S system (Choudhury et al. 2006). It was synthesized by combining Eu (0.250 mM), Si (0.278 mM), S (0.505 mM), and Rb$_2$S$_5$ (0.506 mM) with an approximately compositional ratio Rb/Eu/Si/S of 4:1:1:12. The mixture was directly loaded into fused silica ampoule inside a N$_2$-filled atmosphere glove box. The ampoule was flame-sealed under vacuum and placed in a furnace. The furnace was ramped to 725°C at a rate of 35°C·h^{-1}, and the temperature was held constant at 725°C for 150 h. Then the furnace was slowly cooled to ambient temperature at a rate of 3°C·h^{-1}. After the reaction product cooled, the ampoule was opened in an inert atmosphere glove box, and the solid product was soaked in DMF for 6 h in order to dissolve and wash away any remaining flux and loosen the product crystals. The product, after treatment with DMF, revealed red plates of RbEuSiS$_4$ along with few uncharacterized yellow crystals.

1.40 Silicon–Rubidium–Bismuth–Sulfur

The **RbBiSiS$_4$** quaternary compound, which crystallizes in the monoclinic structure with the lattice parameters a = 647.14 ± 0.04, b = 679.99 ± 0.04, c = 1790.58 ± 0.11 pm, β = 108.856 ± 0.001°, and a calculated density of 4.015 g·cm^{-3} at 153 K, is formed in the Si–Rb–Bi–S system (Yao et al. 2002). For the synthesis of this compound, the reaction mixture of Rb$_2$S$_3$ (0.5 mM), Si (1.0 mM), Bi (0.5 mM), and S (2.5 mM) was loaded into a fused silica tube under an Ar atmosphere in a glove box. This tube was sealed under a vacuum (10^{-2} Pa) and then placed in a furnace. The sample was heated to 620°C over 15 h, kept at this temperature for 84 h, and slowly cooled at 6°C·h^{-1} to 100°C, and then the furnace was turned off. The reaction mixture was washed free of flux with DMF and then dried with acetone. The product consisted of orange-red plates of RbSiBiS$_4$.

1.41 Silicon–Cesium–Cerium–Sulfur

The **CsCeSiS$_4$** quaternary compound, which crystallizes in the orthorhombic structure with the lattice parameters $a = 1787.41 \pm 0.09$, $b = 674.13 \pm 0.03$, $c = 647.32 \pm 0.03$ pm, and a calculated density of 3.656 g·cm^{-3}, is formed in the Si–Cs–Ce–S system (Hartenbach and Schleid 2006). Pale yellow, platelet-shaped single crystals of this compound were obtained after the reaction of Ce, S, and SiS$_2$ with an excess of CsCl both as a flux and a Cs$^+$ cation source (molar ratio 4:3:6:3) in evacuated silica ampoule at 850°C for 7 days. The crystals turn out to be air and water-resistant because of passivation with a thin oxosilicate layer caused by surface hydrolysis. So it was feasible to wash the bulk material with water in order to isolate appropriate crystals.

1.42 Silicon–Cesium–Samarium–Sulfur

The **CsSmSiS$_4$** quaternary compound, which crystallizes in the orthorhombic structure with the lattice parameters $a = 640.31 \pm 0.05$, $b = 672.69 \pm 0.05$, $c = 1777.5 \pm 0.2$ pm, and a calculated density of 3.814 g·cm^{-3}, is formed in the Si–Cs–Sm–S system (Hartenbach and Schleid 2003b). The reaction of Sm with S and SiS$_2$ with an excess of CsCl serving both as a flux medium and as a reactant (Cs source) in evacuated silica ampoule at 850°C for 7 days leads to air and water-resistant platelet-shaped single crystals of the title compound as a result of the next reaction: CsCl + 2Sm + 3S + SiS$_2$ = CsSmSiS$_4$ + CsSCl.

1.43 Silicon–Cesium–Europium–Sulfur

The **CsEuSiS$_4$** quaternary compound, which crystallizes in the orthorhombic structure with the lattice parameters $a = 638.75 \pm 0.06$, $b = 670.96 \pm 0.06$, $c = 1773.2 \pm 0.2$ pm, and a calculated density of 3.856 g·cm^{-3}, is formed in the Si–Cs–Eu–S system (Hartenbach and Schleid 2003b). It was synthesized in the same way as CsSmSiS$_4$ was prepared using Eu instead of Sm.

1.44 Silicon–Cesium–Gadolinium–Sulfur

The **CsGdSiS$_4$** quaternary compound, which crystallizes in the orthorhombic structure with the lattice parameters $a = 637.46 \pm 0.05$, $b = 669.77 \pm 0.06$, $c = 1770.9 \pm 0.2$ pm, and a calculated density of 3.922 g·cm^{-3}, is formed in the Si–Cs–Gd–S system (Hartenbach and Schleid 2003b). It was synthesized in the same way as CsSmSiS$_4$ was prepared using Gd instead of Sm.

1.45 Silicon–Cesium–Terbium–Sulfur

The **CsTbSiS$_4$** quaternary compound, which crystallizes in the orthorhombic structure with the lattice parameters $a = 635.06 \pm 0.05$, $b = 667.98 \pm 0.06$, $c = 1769.2 \pm 0.2$ pm, and a calculated density of 3.966 g·cm^{-3}, is formed in the Si–Cs–Tb–S system (Hartenbach and Schleid 2003b). It was synthesized in the same way as CsSmSiS$_4$ was prepared using Tb instead of Sm.

1.46 Silicon–Cesium–Dysprosium–Sulfur

The **CsDySiS$_4$** quaternary compound, which crystallizes in the orthorhombic structure with the lattice parameters $a = 634.32 \pm 0.05$, $b = 666.73 \pm 0.06$, $c = 1768.4 \pm 0.2$ pm, and a calculated density of 4.012 g·cm^{-3}, is formed in the Si–Cs–Dy–S system (Hartenbach and Schleid 2003b). It was synthesized in the same way as CsSmSiS$_4$ was prepared using Dy instead of Sm.

1.47 Silicon–Cesium–Holmium–Sulfur

The **CsHoSiS$_4$** quaternary compound, which crystallizes in the orthorhombic structure with the lattice parameters $a = 632.69 \pm 0.05$, $b = 665.71 \pm 0.05$, $c = 1767.1 \pm 0.2$ pm, and a calculated density of 4.053 g·cm^{-3}, is formed in the Si–Cs–Ho–S system (Hartenbach and Schleid 2003b). It was synthesized in the same way as CsSmSiS$_4$ was prepared using Ho instead of Sm.

1.48 Silicon–Cesium–Erbium–Sulfur

The **CsErSiS$_4$** quaternary compound, which crystallizes in the orthorhombic structure with the lattice parameters $a = 631.72 \pm 0.05$, $b = 665.34 \pm 0.06$, $c = 1765.9 \pm 0.2$ pm, and a calculated density of 4.085 g·cm^{-3}, is formed in the Si–Cs–Er–S system (Hartenbach and Schleid 2003b). It was synthesized in the same way as CsSmSiS$_4$ was prepared using Er instead of Sm.

1.49 Silicon–Cesium–Thulium–Sulfur

The **CsTmSiS$_4$** quaternary compound, which crystallizes in the orthorhombic structure with the lattice parameters $a = 630.28 \pm 0.05$, $b = 664.73 \pm 0.06$, $c = 1762.9 \pm 0.2$ pm, and a calculated density of 4.120 g·cm^{-3}, is formed in the Si–Cs–Tm–S system (Hartenbach and Schleid 2003b). It was synthesized in the same way as CsSmSiS$_4$ was prepared using Tm instead of Sm.

1.50 Silicon–Cesium–Bismuth–Sulfur

The **CsBiSiS$_4$** quaternary compound, which crystallizes in the monoclinic structure with the lattice parameters $a = 933.51 \pm 0.07$, $b = 693.13 \pm 0.05$, $c = 1281.15 \pm 0.10$ pm, $\beta = 109.096 \pm 0.001°$, and a calculated density of 4.225 g·cm^{-3} at 153 K, is formed in the Si–Cs–Bi–S system (Yao et al. 2002). It was prepared by the reaction of Cs$_2$S$_3$ (0.5 mM), Si (1.0 mM), Bi$_2$S$_3$ (0.5 mM), and S (2.0 mM). The mixture was loaded into a fused silica tube under an Ar atmosphere in a glove box. This tube was sealed under a vacuum (10^{-2} Pa) and then placed in a furnace. The sample was heated to 800°C over 20 h, kept at this temperature for 84 h, and slowly cooled at 6°C·h^{-1} to 100°C, and then the furnace was turned off. The reaction mixture was washed free of flux with DMF and then dried with acetone. The product consisted of yellow plates of $CsSiBiS_4$.

1.51 Silicon–Copper–Magnesium–Sulfur

The Cu_2MgSiS_4 quaternary compound, which crystallizes in the orthorhombic structure with the lattice parameters $a = 756.3 \pm 0.4$, $b = 644.8 \pm 0.3$, $c = 617.9 \pm 0.3$ pm, a calculated density of 3.392 g·cm^{-3}, and an energy gap of 3.20 eV, is formed in the Si–Cu–Mg–S system (Liu et al. 2013) [$a = 753.96$, $b = 635.44$, $c = 614.17$ pm, and a calculated density of 3.4732 g·cm^{-3} (calculated values) (Bedjaoui et al. 2017)]. To prepare this compound, the stoichiometric mixture with an overall weight of about 300 mg of Cu, Mg, Si, and S (molar ratio 2:1:1:4) was ground into fine powder in an agate mortar and pressed into pellets, then loaded into quartz tubes and finally flame-sealed under vacuum about 10^{-2} Pa. The sample was placed into a furnace, held at 300°C for 10 h, then heated to 650°C over 5 h and kept for 10 h, subsequently heated to 950°C at 10 h, kept for 5 days, and finally cooled down to 300°C over 8 days before switching off the furnace power. Good-quality crystals of the title compound were obtained. A polycrystalline sample could be synthesized by direct combination of the elements and they were air stable.

1.52 Silicon–Copper–Strontium–Sulfur

The Cu_2SrSiS_4 quaternary compound, which crystallizes in the trigonal structure with the lattice parameters $a = 607.10 \pm 0.03$, $c = 1513.9 \pm 0.2$ pm, a calculated density of 3.825 g·cm^{-3}, and an energy gap of 3.04 eV (Yang et al. 2020b) [$a = 606.874 \pm 0.006$, $c = 1513.73 \pm 0.01$ pm, and an energy gap of 3.4 eV (Sun et al. 2020a)], is formed in the Si–Cu–Sr–S system. This compound was prepared by combining reagents in stoichiometric amounts with a total mass of ~250–400 mg (Sun et al. 2020a). The mixture was weighed, ground with a mortar and pestle, and cold-pressed, all in a N_2-filled glove box. Then it was flame-sealed in a fused silica ampoule. The ampoule was annealed at 800°C for 96 h and allowed to cool naturally. The compound was ground, pressed, resealed, and annealed several times to achieve desired phase purity.

The whole process of its preparation can also be carried out in an argon-filled glove box, as follows (Yang et al. 2020b). The spontaneous crystallization of the mixture of SrS, Cu_2S, and SiS_2 (molar ratio 1:3:1) was completed in the vacuum-sealed silica tube. Firstly, the tube was heated to 600°C for 30 h and kept at this temperature for about 100 h, then quickly down to room temperature. The product was ground and then loaded into the vacuum-sealed silica tube. It was further sintered again at 1000°C for 10 days and slowly cooled to room temperature at 5°C·h^{-1}. Finally, many colorless crystals were found in the tube after washing with DMF solvent, and it was stable in the air within several weeks.

1.53 Silicon–Copper–Zinc–Sulfur

The Cu_2ZnSiS_4 quaternary compound is formed in the Si–Cu–Zn–S system. It melts incongruently at 1123°C ± 5°C (Schäfer and Nitsche 1977) [according to the data of Yao et al. (1987), it decomposes at 620°C ± 10°C] and has two polymorphic modifications (Rosmus et al. 2012). α-Cu_2ZnSiS_4 crystallizes in the orthorhombic structure with the lattice parameters $a = 743.74 \pm 0.01$, $b = 640.01 \pm 0.01$, $c = 613.94 \pm 0.01$ pm, and a calculated density of 3.964 g·cm^{-3} (Rosmus and Aitken 2011) [$a = 744$, $b = 639$, and $c = 613$ pm, the calculated and experimental densities of 3.97 and 3.94 g·cm^{-3}, respectively, and an energy gap of 3.25 eV (Schleich and Wold 1977); $a = 740$, $b = 640$, and $c = 608$ pm (Nitsche et al. 1967); $a = 743.5$, $b = 639.6$, and $c = 613.5$ pm (Schäfer and Nitsche 1974, 1977); $a = 743.6 \pm 0.1$, $b = 639.8 \pm 0.1$, $c = 613.7 \pm 0.2$ pm, and an energy gap of 3.04 eV (Yao et al. 1987); according to the data of Rosmus et al. (2012), an energy gap of α-Cu_2ZnSiS_4 is 1.3 eV].

β-Cu_2ZnSiS_4 also crystallizes in the orthorhombic structure with the lattice parameters $a = 613.4092 \pm 0.0006$, $b = 639.2752 \pm 0.0005$, $c = 742.1228 \pm 0.0008$ pm, a calculated density of 3.981 g·cm^{-3} at 180 K, and an energy gap of 1.7 eV (Rosmus et al. 2012).

High-temperature solid-state synthesis was used to prepare α- and β-Cu_2ZnSiS_4 (Rosmus and Aitken 2011; Rosmus et al. 2012). Stoichiometric ratios of the elements were weighed and then ground for 30 min in an Ar-filled glove box using an agate mortar and pestle. The sample was transferred into a graphite crucible, which was then inserted in a fused silica tube. The tube was flame-sealed under a vacuum of 0.1 Pa and transported to a furnace. The sample was heated to 1000°C in 12 h, held at this temperature for 96–168 h, and then cooled at 7.5°C·h^{-1} to room temperature. Its single crystals have grown by chemical vapor transport using I_2 as the transport agent (Nitsche et al. 1967; Schleich and Wold 1977; Yao et al. 1987).

1.54 Silicon–Copper–Cadmium–Sulfur

Cu_2SiS_3–CdS. The phase diagram is shown in Figure 1.3 (Piskach et al. 1997, 1999). The eutectic composition and temperature are 6 mol% CdS and 895°C, respectively. The Cu_2CdSiS_4 quaternary compound is formed in this system. It melts incongruently at 978°C [at 1016°C ± 5°C (Schäfer and Nitsche 1977); according to the data of Yao et al. (1987), it decomposes at 510°C ± 10°C] and has two polymorphous modifications (Piskach et al. 1997, 1999). The temperature of this transformation is 937°C from the CdS-rich side and 845°C from the Cu_2SiSe_3 side. This indicates the existence of a homogeneity region for the Cu_2CdSiS_4 compound (45–50 mol% CdS at the eutectic temperature). Low-temperature modification crystallizes in the orthorhombic structure with the lattice parameters 761.29 ± 0.02, $b = 648.13 \pm 0.02$, and $c = 625.02 \pm 0.02$ pm, and a calculated density of 4.2621 ± 0.0004 g·cm^{-3} (Sachanyuk et al. 2006c) [$a = 758$, $b = 644$, and $c = 617$ pm (Nitsche et al. 1967); $a = 760$, $b = 648$, and $c = 625$ pm (Chapuis and Niggli 1968); $a = 759.8 \pm 0.8$, $b = 648.6 \pm 0.06$, $c = 625.8 \pm 1.1$ pm,

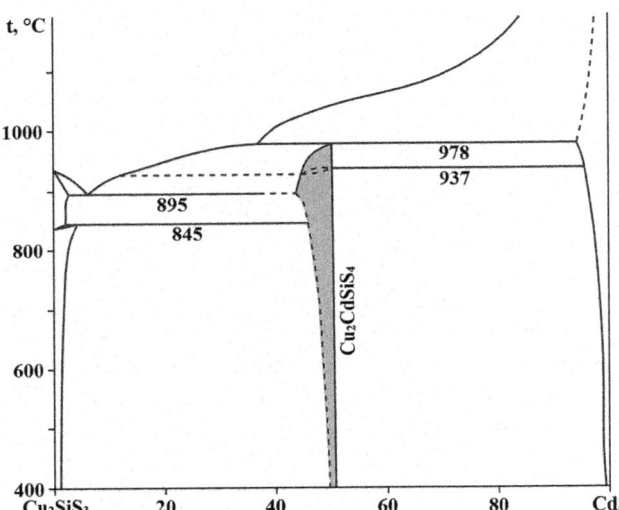

FIGURE 1.3 Phase diagram of the Cu$_2$SiS$_3$–CdS system. (From Piskach, L.V., et al., *Zhurn. neorgan. khimii*, **44**(5), 823, 1999.)

and a calculated density of 4.27 g·cm^{-3} (Chapuis and Niggli 1972); $a = 761.4$, $b = 648.9$, and $c = 625.4$ pm (Schäfer and Nitsche 1974, 1977); $a = 760.9 \pm 0.1$, $b = 648.5 \pm 0.1$, $c = 625.1 \pm 0.1$ pm, and an energy gap of 2.45 eV (Yao et al. 1987); $a = 759.7 \pm 0.3$, $b = 648.5 \pm 0.3$, and $c = 623.1$ pm (Piskach et al. 1999)].

Solubility of Cu$_2$SiS$_3$ in CdS at the peritectic temperature reaches 6 mol% and decreases to 2 mol% at 400°C (Piskach et al. 1999).

This system was investigated through DTA, metallography, and XRD. The ingots were annealed at 400°C for 250 h (Piskach et al. 1997, 1999). Cu$_2$CdSiS$_4$ was obtained by heating the mixture of Cu, Cd, and S in the stoichiometric ratio up to 1100°C (40°C·h^{-1}) with the next cooling to 400°C (10°C·h^{-1}) and annealing at this temperature for 250 h (Sachanyuk et al. 2006c). Single crystals of this compound were grown by the chemical transport reactions using I$_2$ as the transport agent (Nitsche et al. 1967; Chapuis and Niggli 1968, 1972; Schäfer and Nitsche 1974; Yao et al. 1987).

1.55 Silicon–Copper–Mercury–Sulfur

The **Cu$_2$HgSiS$_4$** and **Cu$_6$Hg$_{0.973}$SiS$_{5.973}$** quaternary compounds are formed in the Si–Cu–Hg–S system. First of them melts incongruently at 859°C ± 5°C and crystallizes in the orthorhombic structure with the lattice parameters $a = 757.81 \pm 0.02$, $b = 647.80 \pm 0.01$, $c = 626.36 \pm 0.02$ pm, and a calculated density of 5.2271 ± 0.0004 g·cm^{-3} (Sachanyuk et al. 2006c) [$a = 759.2$, $b = 648.4$, and $c = 626.9$ pm (Schäfer and Nitsche 1974, 1977)]. The second compound crystallizes in the cubic structure with the lattice parameter $a = 989.38 \pm 0.01$ pm and a calculated density of 5.416 ± 0.002 g·cm^{-3} (Gulay et al. 2002c).

Cu$_2$HgSiS$_4$ was obtained by heating the mixture of Cu, HgS, and S in the stoichiometric ratio up to 1100°C (40°C·h^{-1}) with the next cooling to 400°C (10°C·h^{-1}) and annealing at this

temperature for 250 h (Sachanyuk et al. 2006c). The synthesis was completed by quenching the sample in cold water. This compound can be also prepared by the chemical transport reactions (Schäfer and Nitsche 1974).

The synthesis of Cu$_6$Hg$_{0.973}$SiS$_{5.973}$ was carried out in evacuated quartz ampoules in two stages (Gulay et al. 2002c). At the first stage, the ampoules with charges were exposed to local heating in the flame of an O$_2$-gas burner till complete consumption of S. The second stage of the synthesis was realized by a single-temperature method in a furnace at 1200°C for 2 h (heating rate was 50°C·h^{-1}), and then the temperature was slowly decreased (10°C·h^{-1}) to 400°C. At this temperature, the sample was homogenized over 500 h, and then it was quenched in the air. The obtained sample was a tight alloy of black color and with a metallic luster.

1.56 Silicon–Copper–Aluminum–Sulfur

The **CuAlSiS$_4$** quaternary compound, which crystallizes in the tetragonal structure with the lattice parameters $a = 534.7$, $c = 1031$ pm, and the calculated and experimental densities of 2.78 and 2.85 g·cm^{-3}, respectively, are formed in the Si–Cu–Al–S system (Hahn and Strick 1967b). This compound was obtained by a long annealing of the elements' mixture or the mixture of binary components between 500°C and 1000°C. It has a light gray color.

1.57 Silicon–Copper–Gallium–Sulfur

The **CuGaSiS$_4$** quaternary compound, which crystallizes in the tetragonal structure with the lattice parameters $a = 534.7$, $c = 515.8$ pm and the calculated and experimental densities of 3.26 and 3.26 g·cm^{-3}, respectively, is formed in the Si–Cu–Ga–S system (Hahn and Strick 1967b). This compound was obtained by a long annealing of the elements' mixture or the mixture of binary components between 500°C and 1000°C. It has a yellow color.

1.58 Silicon–Copper–Indium–Sulfur

SiS$_2$–CuInS$_2$. The part of the phase diagram of this system is presented in Figure 1.4 (Sachanyuk et al. 2007). The eutectic contains 67 mol% SiS$_2$ and crystallizes at 905°C, the metatectic point corresponds to ~7 mol% SiS$_2$ and 1038°C, and the peritectic point is at 11 mol% SiS$_2$ and 986°C. The homogeneity region of α-CuInS$_2$ extends to 9 mol% SiS$_2$ at 400°C. The **Cu$_2$In$_2$SiS$_6$** quaternary compound, which melts incongruently at 960°C and decomposes at 834°C, is formed in this system. It crystallizes in the monoclinic structure with the lattice parameters $a = 1187.25 \pm 0.03$, $b = 678.80 \pm 0.02$, $c = 1197.18 \pm 0.02$ pm, $\beta = 110.224 \pm 0.002°$, and a calculated density of 4.2342 ± 0.004 g·cm^{-3}. The system was investigated through DTA and XRD, and the alloys were annealed at 400°C for 250 h. Some alloys for the refinement of the phase diagram were quenched from 950°C.

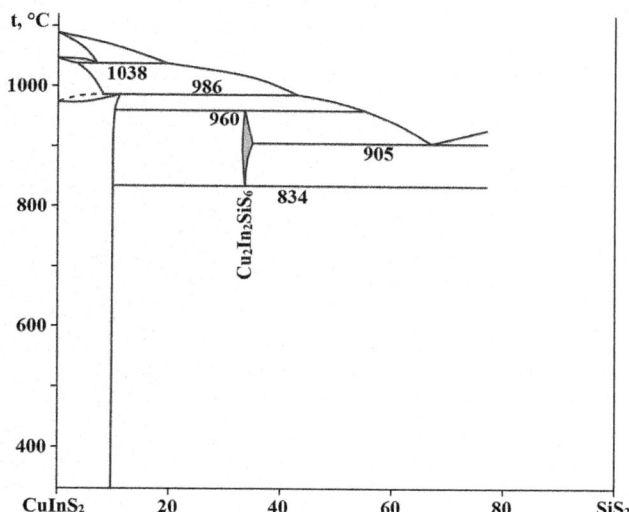

FIGURE 1.4 Part of the phase diagram of the SiS_2–$CuInS_2$ system. (From Sachanyuk, V.P., et al., *J. Alloys Compd.*, **443**, 61, 2007.)

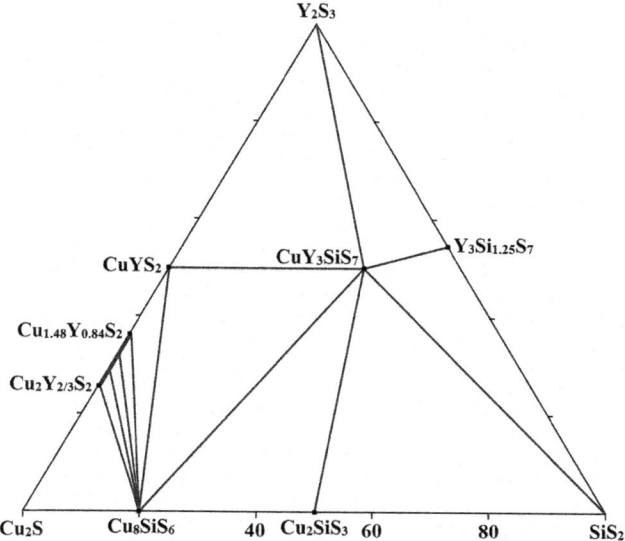

FIGURE 1.5 Isothermal section of the SiS_2–Cu_2S–Y_2S_3 quasiternary system at 600°C. (From Lychmanyuk, O.S., et al., *Nauk. Visnyk Volyns'k. Derzh. Univ. im. Lesi Ukrainky. Ser. Khim. nauky*, (4), 118, 2006.)

1.59 Silicon–Copper–Yttrium–Sulfur

SiS_2–Cu_2S–Y_2S_3. The isothermal section of this system at 600°C was constructed by Lychmanyuk et al. (2006a) using the alloys annealed at this temperature for 240 h and quenched in the air and is presented in Figure 1.5. The CuY_3SiS_7 quaternary compound is formed in this system. It crystallizes in the hexagonal structure with the lattice parameters $a = 985.4 \pm 0.1$, $c = 565.6 \pm 0.1$ pm, and a calculated density of 4.069 g·cm^{-3} (Gulay et al. 2005b) [$a = 984 \pm 1$ and $c = 568.0 \pm 0.5$ pm (Guittard et al. 1968; Guittard and Julien-Pouzol 1970)].

To synthesize this compound, the calculated amounts of the elements were sealed in an evacuated silica ampoule (Gulay et al. 2005b). The synthesis was realized in a tube furnace. The ampoule was heated with a heating rate of 30°C·h^{-1} to the maximal temperature, 1150°C. The sample was kept at this temperature for 4 h. After that, it was cooled slowly (10°C·h^{-1}) to 600°C and annealed at this temperature for 240 h. After annealing, the ampoule and the sample were quenched in cold water. The prepared product was a brown-colored compact alloy. CuY_3SiS_7 can also be prepared from a mixture of Y_2S_3, Cu_2S, Si, and S, which was pressed and introduced into the silica ampoule and gradually heats up to 800°C in about a week (Guittard and Julien-Pouzol 1970). At the end of the preparation, it is necessary to heat the sample to 1200°C to obtain the pure hexagonal phase.

1.60 Silicon–Copper–Lanthanum–Sulfur

The $CuLa_3SiS_7$ quaternary compound, which crystallizes in the hexagonal structure with the lattice parameters $a = 1031.0 \pm 0.5$ and $c = 579.4 \pm 0.2$ pm (Collin and Laruelle 1971) [$a = 1028 \pm 1$ and $c = 575.5 \pm 0.5$ pm (Guittard et al. 1968; Guittard and Julien-Pouzol 1970; Collin and Flahaut 1972)], is formed in the Si–Cu–La–S system.

$CuLa_3SiS_7$ was obtained from a mixture of La_2S_3, Cu_2S, Si, and S, which was pressed and introduced into the silica ampoule and gradually heated up to 800°C in about a week (Guittard and Julien-Pouzol 1970). At the end of the preparation, it is necessary to heat the sample to 1200°C to obtain the pure hexagonal phase. Single crystals of this compound were obtained by prolonged heating at 1250°C of a stoichiometric mixture of La_2S_3, Cu, Si, and S (Collin and Laruelle 1971).

1.61 Silicon–Copper–Cerium–Sulfur

The $CuCe_3SiS_7$ quaternary compound, which crystallizes in the hexagonal structure with the lattice parameters $a = 1023.6 \pm 0.1$, $c = 576.7 \pm 0.1$ pm, and a calculated density of 4.673 g·cm^{-3} (Gulay et al. 2007b) [$a = 1019 \pm 1$ and $c = 573.5 \pm 0.5$ pm (Guittard et al. 1968; Guittard and Julien-Pouzol 1970); $a = 1022.19 \pm 0.07$, $c = 576.28 \pm 0.04$ pm, and a calculated density of 4.690 g·cm^{-3} (Hartenbach et al. 2007)], is formed in the Si–Cu–Ce–S system.

The synthesis of this compound was realized using the mixture of the elements (Gulay et al. 2007b). The calculated amounts of the components were sealed in an evacuated silica ampoule. The ampoule was heated with a heating rate of 30°C·h^{-1} to the maximal temperature, 1150°C and kept at this temperature for 3 h. Then it was cooled slowly (10°C·h^{-1}) to 600°C and annealed at this temperature for 240 h. After annealing the ampoule with the samples were quenched in cold water.

$CuCe_3SiS_7$ was also obtained from a mixture of Ce_2S_3, Cu_2S, Si, and S, which was pressed and introduced into the silica ampoule and gradually heated up to 800°C in about a week (Guittard and Julien-Pouzol 1970). At the end of the preparation, it is necessary to heat the sample to 1200°C to obtain the pure hexagonal phase.

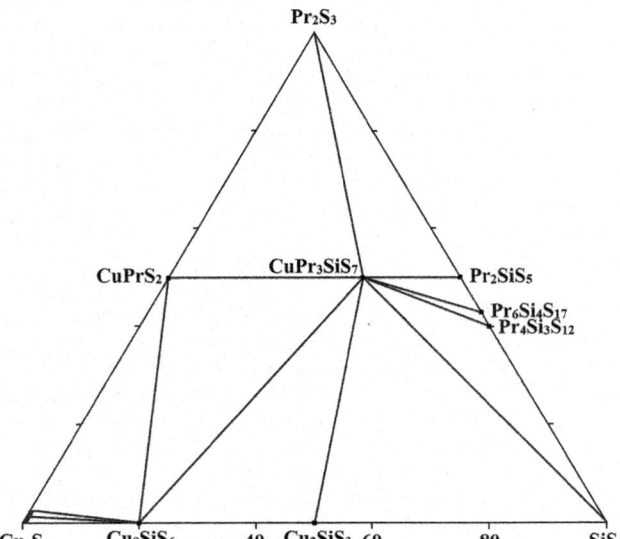

FIGURE 1.6 Isothermal section of the SiS$_2$–Cu$_2$S–Pr$_2$S$_3$ quasiternary system at 600°C. (From Lychmanyuk, O.S., et al., *Nauk. Visnyk Volyns'k. Derzh. Univ. im. Lesi Ukrainky. Ser. Khim. nauky*, (15), 10, 2007.)

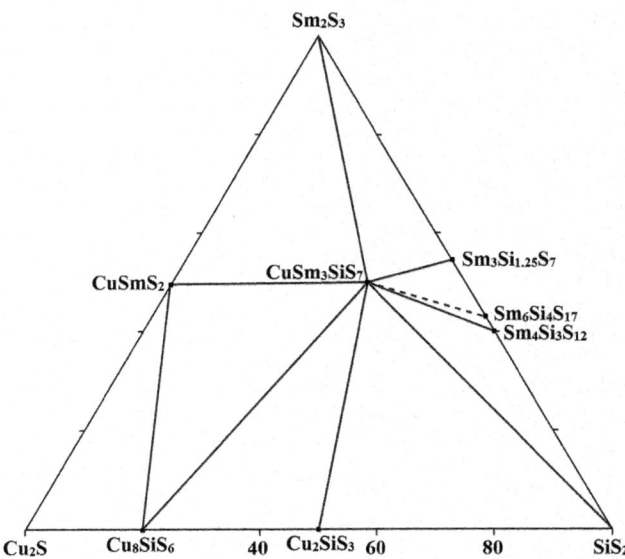

FIGURE 1.7 Isothermal section of the SiS$_2$–Cu$_2$S–Sm$_2$S$_3$ quasiternary system at 600°C. (From Gulay, L.D., and Lychmanyuk, O.S., *Nauk. vsnyk Volyns'k. Nats. Univ. im. Lesi Ukrainky. Ser. Khim. nauky*, (29), 29, 2009.)

It can also be prepared from a mixture of Cu, S, SiS$_2$, and Ce$_2$S$_3$ with an excess of CeCl$_3$ as a flux (Hartenbach et al. 2007). The annealing of the mixture at 850°C for 8 days in an evacuated silica glass ampoule led to dark red columnar single crystals of CuCe$_3$SiS$_7$.

1.62 Silicon–Copper–Praseodymium–Sulfur

SiS$_2$–Cu$_2$S–Pr$_2$S$_3$. The isothermal section of this system at 600°C was constructed by Lychmanyuk et al. (2007a) using the alloys annealed at this temperature for 240 h and quenched in the air and is presented in Figure 1.6. The **CuPr$_3$SiS$_7$** quaternary compound is formed in this system. It crystallizes in the hexagonal structure with the lattice parameters a = 1016.15 ± 0.02, c = 574.74 ± 0.02 pm, and a calculated density of 4.773 g·cm^{-3} (Gulay et al. 2007b) [a = 1015 ± 1 and c = 571.0 ± 0.5 pm (Guittard et al. 1968; Guittard and Julien-Pouzol 1970)]. This compound was prepared in the same way as CuCe$_3$SiS$_7$ was synthesized using Pr$_2$S$_3$ instead of Ce$_2$S$_3$ (Guittard and Julien-Pouzol 1970; Gulay et al. 2007b).

1.63 Silicon–Copper–Neodymium–Sulfur

The **CuNd$_3$SiS$_7$** quaternary compound, which crystallizes in the hexagonal structure with the lattice parameters a = 1011.6 ± 0.1, c = 572.5 ± 0.1 pm, and a calculated density of 4.901 g·cm^{-3} (Gulay et al. 2007b) [a = 1008 ± 1 and c = 570.0 ± 0.5 pm (Guittard et al. 1968; Guittard and Julien-Pouzol 1970)], is formed in the Si–Cu–Nd–S system. It was prepared in the same way as CuCe$_3$SiS$_7$ was synthesized using Nd$_2$S$_3$ instead of Ce$_2$S$_3$ (Guittard and Julien-Pouzol 1970; Gulay et al. 2007b).

1.64 Silicon–Copper–Samarium–Sulfur

SiS$_2$–Cu$_2$S–Sm$_2$S$_3$. The isothermal section of this system at 600°C was constructed by Gulay and Lychmanyuk (2009) using the alloys annealed at this temperature for 240 h and quenched in the air and is presented in Figure 1.7. The **CuSm$_3$SiS$_7$** quaternary compound is formed in this system. It crystallizes in the hexagonal structure with the lattice parameters a = 1001.93 ± 0.03, c = 569.68 ± 0.03 pm, and a calculated density of 5.143 g·cm^{-3} (Gulay et al. 2007b) [a = 1000 ± 1 and c = 568.0 ± 0.5 pm (Guittard et al. 1968; Guittard and Julien-Pouzol 1970)]. This compound was prepared in the same way as CuCe$_3$SiS$_7$ was synthesized using Sm$_2$S$_3$ instead of Ce$_2$S$_3$ (Guittard and Julien-Pouzol 1970; Gulay et al. 2007b).

1.65 Silicon–Copper–Europium–Sulfur

The **CuEu$_2$SiS$_4$** quaternary compound, which crystallizes in the trigonal structure with the lattice parameters a = 605.76 ± 0.10, c = 1510.07 ± 0.05 pm, a calculated density of 4.520 g·cm^{-3}, and an energy gap of 2.30 eV, is formed in the Si–Cu–Eu–S system (Sun et al. 2019). Single crystals of this compound were obtained by high-temperature solid-state reactions with KI as a flux. The reagents were weighted by stoichiometry, and the total reagents mass was 500 mg with additional 400 mg KI. Namely, 135.15 mg Cu, 187.12 mg Eu$_2$O$_3$, 136.37 mg S, 29.87 mg Si, and 11.50 mg B were used to obtain CuEu$_2$SiS$_4$. The mixture was ground into a fine powder using an agate mortar and pressed into pellets. Then pellets were loaded into quartz tube evacuated to 10^{-2} Pa and sealed using O$_2$/H$_2$ flame. The tube was put into a furnace and heated according to a profile as follow. The tube was heated from room temperature to 300°C by 5 h and kept for 5 h, then heated to 600°C by 5 h and kept for 5 h, then heated to 800°C by 5 h and kept for 90 h, and finally cooled down slowly to 300°C after 5 days and shut down the furnace. Yellow block crystals of the title compound, which were air and water stable were obtained. They were washed using hot distilled water and ethanol for several times under ultrasonic wave irradiation to eliminate KI and some impurities.

1.66 Silicon–Copper–Gadolinium–Sulfur

The $CuGd_3SiS_7$ quaternary compound, which crystallizes in the hexagonal structure with the lattice parameters $a = 994.59 \pm 0.07$ and $c = 567.03 \pm 0.04$ pm (Hartenbach et al. 2006) [$a = 993 \pm 1$ and $c = 567.0 \pm 0.5$ pm (Guittard et al. 1968; Guittard and Julien-Pouzol 1970)], is formed in this system. To prepare this compound, a mixture of Cu, S, Gd, and SiS_2 in stoichiometric ratio with $GdCl_3$ as a flux was put in an evacuated silica glass ampoule and heated at 800°C (Hartenbach et al. 2006). It was prepared in the same way as $CuCe_3SiS_7$ was synthesized using Gd_2S_3 instead of Ce_2S_3 (Guittard and Julien-Pouzol 1970).

1.67 Silicon–Copper–Terbium–Sulfur

The $CuTb_3SiS_7$ quaternary compound, which crystallizes in the hexagonal structure with the lattice parameters $a = 988.97 \pm 0.02$, $c = 565.82 \pm 0.02$ pm, and a calculated density of 5.493 $g \cdot cm^{-3}$ (Gulay et al. 2007b) [$a = 990 \pm 1$ and $c = 567.0 \pm 0.5$ pm (Guittard et al. 1968; Guittard and Julien-Pouzol 1970)], is formed in the Si–Cu–Tb–S system. It was prepared in the same way as $CuCe_3SiS_7$ was synthesized using Tb_2S_3 instead of Ce_2S_3 (Guittard and Julien-Pouzol 1970; Gulay et al. 2007b).

1.68 Silicon–Copper–Dysprosium–Sulfur

The $CuDy_3SiS_7$ quaternary compound, which crystallizes in the hexagonal structure with the lattice parameters $a = 986.89 \pm 0.02$, $c = 565.82 \pm 0.02$ pm, and a calculated density of 5.591 $g \cdot cm^{-3}$ (Gulay et al. 2007b) [$a = 986 \pm 1$ and $c = 566.5 \pm 0.5$ pm (Guittard et al. 1968; Guittard and Julien-Pouzol 1970)], is formed in the Si–Cu–Dy–S system. It was prepared in the same way as $CuCe_3SiS_7$ was synthesized using Dy_2S_3 instead of Ce_2S_3 (Guittard and Julien-Pouzol 1970; Gulay et al. 2007b).

1.69 Silicon–Copper–Holmium–Sulfur

$SiS_2–Cu_2S–Ho_2S_3$. The isothermal section of this system at 600°C was constructed by Lychmanyuk et al. (2007b) using the alloys annealed at this temperature for 240 h and quenched in cold water and is presented in Figure 1.8. The $CuHo_3SiS_7$ quaternary compound is formed in this system. It crystallizes in the hexagonal structure with the lattice parameters $a = 981.44 \pm 0.02$ and $c = 564.22 \pm 0.02$ pm (Lychmanyuk et al. 2007b) [$a = 981 \pm 1$ and $c = 566.0 \pm 0.5$ pm (Guittard et al. 1968; Guittard and Julien-Pouzol 1970)]. This compound was prepared in the

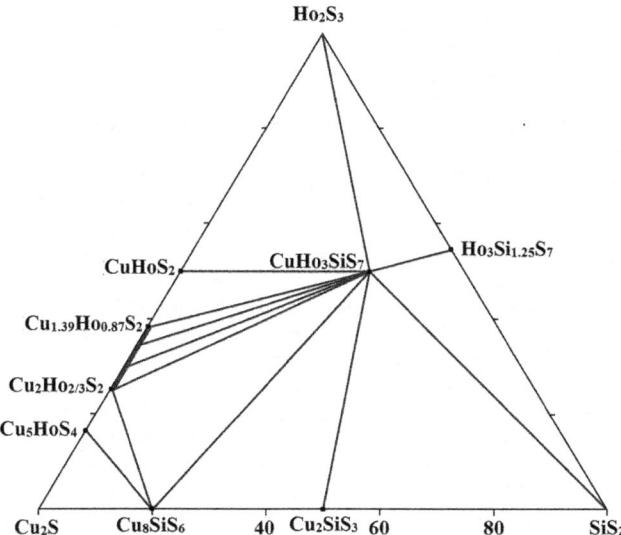

FIGURE 1.8 Isothermal section of the $SiS_2–Cu_2S–Ho_2S_3$ quasiternary system at 600°C. (From Lychmanyuk, O.S., et al., *Pol. J. Chem.*, **81**(3), 353, 2007.)

same way as $CuCe_3SiS_7$ was synthesized using Ho_2S_3 instead of Ce_2S_3 (Guittard and Julien-Pouzol 1970).

1.70 Silicon–Copper–Erbium–Sulfur

The $CuEr_3SiS_7$ quaternary compound, which crystallizes in the hexagonal structure with the lattice parameters $a = 979.29 \pm 0.02$, $c = 565.18 \pm 0.02$ pm, and a calculated density of 5.785 $g \cdot cm^{-3}$ (Gulay et al. 2007b) [$a = 978 \pm 1$ and $c = 566.0 \pm 0.5$ pm (Guittard et al. 1968; Guittard and Julien-Pouzol 1970)], is formed in the Si–Cu–Er–S system. It was prepared in the same way as $CuCe_3SiS_7$ was synthesized using Er_2S_3 instead of Ce_2S_3 (Guittard and Julien-Pouzol 1970; Gulay et al. 2007b).

1.71 Silicon–Copper–Lead–Sulfur

$Cu_2SiS_3–PbS$. The phase diagram of the system through DTA, XRD and measuring of the microhardness, is presented in Figure 1.9 (Olekseyuk et al. 2005c). The eutectic contains ~69 mol% PbS and crystallizes at 590°C. The phase transformation of Cu_2SiS_3 at 845°C takes place in this system. Two quaternary compounds, Cu_2PbSiS_4 and $\sim Cu_4PbSi_2S_7$, was discovered in the system. Both compounds form incongruently. The first compound melts at 689°C and is dimorphous. This compound has a narrow homogeneity region. $\sim Cu_4PbSi_2S_7$ also melts incongruently at 751°C and undergoes a polymorphous transformation at 660°C as well. The crystal structure of this compound was not determined. The alloys of the $Cu_2SiS_3–PbS$ section were synthesized in two sets. The first set includes alloys with 0–30 and 70–100 mol% PbS content. The maximum synthesis temperature for this set of alloys was based on the melting points of the system components and equaled 1150°C. The rest of the alloys were synthesized at 1000°C.

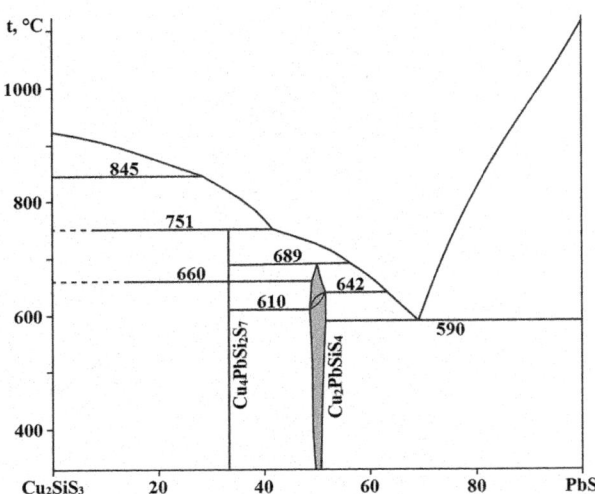

FIGURE 1.9 Phase diagram of the Cu₂SiS₃–PbS system. (From Olekseyuk, I.D., et al., *J. Alloys Compd.*, **399**(1–2), 149, 2005.)

α-Cu₂PbSiS₄ crystallizes in the trigonal structure with the lattice parameters $a = 605.08 \pm 0.06$, $c = 1517.4 \pm 0.7$ pm, and an energy gap of 1.69 eV (Nhalil et al. 2018) [$a = 605.65 \pm 0.02$, $c = 1518.41 \pm 0.05$ pm, and a calculated density of 5.0664 ± 0.0005 g·cm⁻³ (Olekseyuk et al. 2006c)]. To synthesize this compound, a stoichiometric amount of elements was weighed and mixed properly using a mortar and pestle before pelletizing and loading into a quartz ampoule in a N₂-filled glove box. The quartz tube was flame-sealed under a dynamic vacuum (<10⁻³ Pa). The reaction mixture was heated at a rate of 40°C·h⁻¹ to 1000°C, held for 24 h, and then cooled to room temperature at a rate of 100°C·h⁻¹ or cooled at a rate of 10°C·h⁻¹ to 600°C, annealed for 250 h and then quenched into cold water (Olekseyuk et al. 2006c; Nhalil et al. 2018). To improve the phase purity and crystallinity, the sample was reground, pelletized, and annealed under identical conditions as necessary. Cu₂PbSiS₄ is stable in ambient air for a period of 6 weeks.

1.72 Silicon–Copper–Chromium–Sulfur

The **Cu₄Si₂Cr₆S₁₅** quaternary compound, which decomposes at temperatures more than 1100°C, is formed in the Si–Cu–Cr–S system (Babitsyna and Novotortsev 1986). It was synthesized from the elements at 960–980°C for 100 h, with the next annealing at 900°C for 300 h.

1.73 Silicon–Copper–Chlorine–Sulfur

The **Cu₈₋ₓSiS₆₋ₓClₓ** quaternary phase, which has a polymorphic transformation at 60°C and crystallizes in the orthorhombic structure with the lattice parameters $a = 699.4 \pm 0.3$, $b = 690.4 \pm 0.3$, and $c = 977.3 \pm 0.4$ pm for $x = 0.1$ (Kuhs et al. 1979), is formed in the Si–Cu–Cl–S system. This phase was prepared by reacting stoichiometric amounts of elements in an

evacuated, sealed quartz ampoule for about 6 days at temperatures of 600°C to 700°C.

1.74 Silicon–Copper–Bromine–Sulfur

The **Cu₈₋ₓSiS₆₋ₓBrₓ** quaternary phase, which crystallizes in the cubic structure with the lattice parameter $a = 983.56 \pm 0.09$ and a calculated density of 5.093 g·cm⁻³ for $x = 0.18$ (Nilges and Pfitzner 2005) and $a = 987.6 \pm 0.3$ for $x = 0.6$ (Kuhs et al. 1979), is formed in the Si–Cu–Br–S system. The phase with $x = 0.18$ was obtained by the reaction of the stoichiometric mixture of elements in an evacuated, sealed silica ampoule at 1000°C (Nilges and Pfitzner 2005). After grinding, the phase was annealed at 920°C for 19 days. The phase with $x = 0.6$ was prepared by reacting stoichiometric amounts of elements in an evacuated, sealed quartz ampoule for about 6 days at temperatures of 600°C to 700°C (Kuhs et al. 1979).

1.75 Silicon–Copper–Iodine–Sulfur

The **Cu₈₋ₓSiS₆₋ₓIₓ** quaternary phase, which crystallizes in the cubic structure with the lattice parameter $a = 994.61 \pm 0.02$ and a calculated density of 5.130 g·cm⁻³ for $x = 1$ and $a = 989.67 \pm 0.02$ and a calculated density of 5.102 g·cm⁻³ for $x = 0.51$ (Nilges and Pfitzner 2005), and $a = 993.0 \pm 0.2$ for $x = 0.9$ (Kuhs et al. 1979), is formed in the Si–Cu–I–S system. The phases with $x = 1$ and $x = 0.51$ were obtained by the reaction of the stoichiometric mixtures of elements in an evacuated, sealed silica ampoule at 1100°C (Nilges and Pfitzner 2005). After grinding the phases were annealed at 900°C ($x = 1$) or 920°C ($x = 0.51$) for 19 days. The phase with $x = 0.9$ was prepared by reacting stoichiometric amounts of elements in an evacuated, sealed quartz ampoule for about 6 days at temperatures of 600°C to 700°C (Kuhs et al. 1979). Cu₇SiS₅I was also synthesized by chemical transport from Si, Cu, S, and CuI (Studenyak et al. 2007b).

1.76 Silicon–Copper–Manganese–Sulfur

The **Cu₂MnSiS₄** quaternary compound, which melts incongruently at 1020°C [at 1140°C (Schäfer and Nitsche 1977)] and crystallizes in the orthorhombic structure with the lattice parameters $a = 753.62 \pm 0.09$, $b = 644.16 \pm 0.04$, $c = 618.66 \pm 0.09$ pm, and a calculated density of 3.74 g·cm⁻³ (Quintero et al. 2014a, 2014b, 2017) [$a = 753.3$, $b = 643.5$ and $c = 617.9$ pm (Schäfer and Nitsche 1974, 1977); $a = 752.3$, $b = 643.3$, and $c = 617.8$ pm, and the calculated and experimental densities of 3.75 and 3.80 g·cm⁻³ (Guen et al. 1979; Guen and Glaunsinger 1980); $a = 753.4$, $b = 633.0$, $c = 620.5$ pm (Lamarche et al. 1991); $a = 754.3 \pm 0.2$, $b = 644.6 \pm 0.1$, $c = 619.3 \pm 0.1$ pm, and a calculated density of 3.732 g·cm⁻³ (Bernert and Pfitzner 2005); $a = 753.68 \pm 0.03$, $b = 643.82 \pm 0.02$, $c = 618.63 \pm 0.03$ pm, and a calculated density of 3.7431 ± 0.0005 g·cm⁻³ (Sachanyuk et al. 2006c)], is formed in the Si–Cu–Mn–S system.

This compound was produced by the melt and anneal technique (Lamarche et al. 1991; Bernert and Pfitzner 2005; Sachanyuk et al. 2006c; Quintero et al. 2014a, 2014b, 2017).

The mixture of the elements of the 1 g sample was sealed under vacuum ($\approx 10^{-3}$ Pa) in a small quartz ampoule, and then was heated up to 200°C and kept for about 1–2 h, then the temperature was raised to 500°C using a rate of 40°C·h^{-1}, and held at this temperature for 14 h. After that, the sample was heated from 500°C to 800°C at a rate of 30°C·h^{-1} and kept at this temperature for another 14 h. Then the temperature was raised to 1150°C at 60°C·h^{-1}, and the components were melted together at this temperature. The furnace temperature was brought slowly (4°C·h^{-1}) down to 600°C, and the sample was annealed at this temperature for 1 month. Then, the sample was slowly cooled to room temperature using a rate of about 2°C·h^{-1}. Single crystals of the title compound were grown through chemical transport reactions using iodine as a transport agent (Schäfer and Nitsche 1974; Guen et al. 1979; Guen and Glaunsinger 1980).

1.77 Silicon–Copper–Iron–Sulfur

The **Cu₂FeSiS₄** quaternary compound, which melts congruently at 1002°C [at 1023°C (Schäfer and Nitsche 1977)] and crystallizes in the orthorhombic structure with the lattice parameters $a = 742.10$, $b = 641.70$, $c = 614.09$ pm and a calculated density of 3.85 g·cm^{-3} (Quintero et al. 2014b, 2017) [$a = 743$, $b = 643$, and $c = 616$ pm (Nitsche et al. 1967); $a = 740.4$, $b = 641.1$, and $c = 614.0$ pm (Schäfer and Nitsche 1974, 1977), is formed in the Si–Cu–Fe–S system. This compound was synthesized in the same way as Cu₂MnSiS₄ was prepared (Nitsche et al. 1967; Schäfer and Nitsche 1974; Quintero et al. 2014b, 2017).

1.78 Silicon–Copper–Cobalt–Sulfur

The **Cu₂CoSiS₄** quaternary compound, which melts incongruently at 985°C [at 1071°C (Schäfer and Nitsche 1977)] and crystallizes in the tetragonal structure with the lattice parameters $a = 526.93 \pm 0.01$, $c = 1033.63 \pm 0.02$ pm and a calculated density of 3.96 g·cm^{-3} (Quintero et al. 2014a,b) [$a = 527.0$ and $c = 1032.7$ pm (Schäfer and Nitsche 1974, 1977); $a = 526.44 \pm 0.01$, $c = 1031.60 \pm 0.05$ pm, and a calculated density of 3.9764 g·cm^{-3} (Gulay et al. 2004a) or in the orthorhombic structure with the lattice parameters $a = 737.5$, $b = 638.2$, $c = 611.7$ pm (Schäfer and Nitsche 1977)], is formed in the Si–Cu–Co–S system. This compound was synthesized in the same way as Cu₂MnSiS₄ was prepared (Schäfer and Nitsche 1974; Gulay et al. 2004a; Quintero et al. 2014b, 2017).

1.79 Silicon–Copper–Nickel–Sulfur

Cu₂SiS₃–NiS. The phase diagram of this system is presented in Figure 1.10 (Nazarchuk et al. 2008). The eutectics contain 24 and 78 mol% NiS and crystallize at respectively 892°C and 875°C. The polymorphic transition of the solid solution based on NiS takes place at 355°C. The **Cu₄NiSi₂S₇** quaternary compound, which melts congruently at 954°C [incongruently at 896°C (Schäfer and Nitsche 1977)] and crystallizes in the orthorhombic structure with the lattice parameters $a = 1154.363 \pm 0.002$, $b = 531.86 \pm 0.02$, $c = 816.38 \pm 0.02$ pm

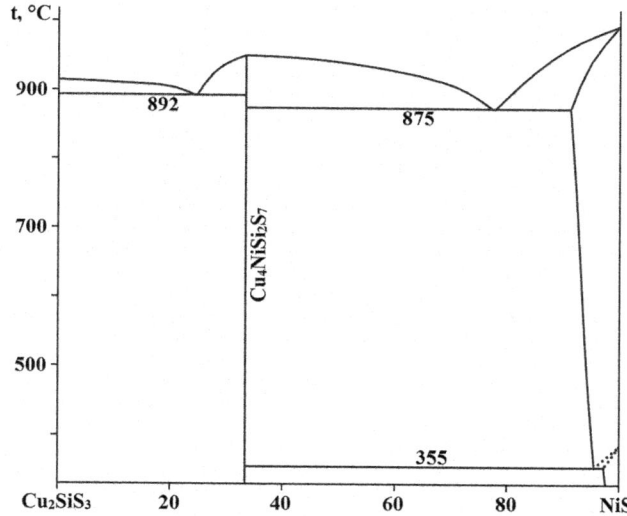

FIGURE 1.10 Phase diagram of the Cu₂SiS₃–NiS system. (From Nazarchuk, O.P., et al., *Nauk. Visnyk Volyns'k. Nats. Univ. im. Lesi Ukrainky. Ser. Khim. nauky*, (16), 21, 2008.)

and a calculated density of 5.5650 g·cm^{-3} [in the monoclinic structure with the lattice parameters $a = 1155.1$, $b = 531.3$, $c = 816.5$ pm, and $\beta = 98.72°$ (Schäfer and Nitsche 1977; Schäfer et al. 1980)], is formed in this system (Nazarchuk et al. 2008).

According to the data of Schäfer and Nitsche (1974), one more compound, **Cu₂NiSiS₄**, which crystallizes the monoclinic structure with the lattice parameters $a = 514.3$, $b = 531.1$, $c = 517.9$ pm, and $\beta = 89.60°$, was found in this system. However, the existence of this compound has not been confirmed by later studies (Nazarchuk et al. 2008).

Single crystals of these two compounds were grown through chemical transport reactions using iodine as a transport agent (Schäfer and Nitsche 1974; Schäfer et al. 1980).

SiS₂–Cu₂S–NiS. The isothermal section of this system at 400°C was constructed by Nazarchuk et al. (2008) using the alloys annealed at this temperature for 240 h and quenched in cold water and is given in Figure 1.11. Six three-phase fields exist in the system.

1.80 Silicon–Silver–Strontium–Sulfur

Two quaternary compounds, **Ag₂SrSiS₄** and **Ag₂Sr₃Si₂S₈**, are formed in the Si–Ag–Sr–S system (McKeown Wessler et al. 2021). The first of them crystallizes in the tetragonal structure with the lattice parameters $a = 676.69 \pm 0.02$, $c = 761.44 \pm 0.04$ pm, a calculated density of 4.379 g·cm^{-3}, and an energy gap of 2.08 eV. The bulk powder of the title compound was synthesized by mixing Ag, SrS, Si, and S powders. This was done with SrS held at 95% of the stoichiometric molar content to avoid the formation of impurity phases. The powder was ground and cold-pressed under a nitrogen atmosphere into a pellet. This pellet was then sealed in a prebaked quartz ampoule under a dynamic vacuum of roughly 5×10^{-4} Pa. The sealed ampoule was ramped up from room temperature to 700°C at a rate of 30°C·h^{-1} and annealed for 4 days before turning the furnace off to cool. After the first heat treatment, roughly 3–5% of the

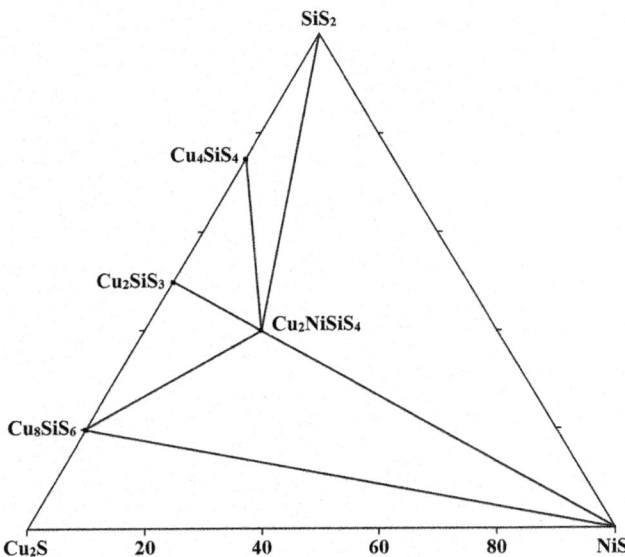

FIGURE 1.11 Isothermal section of the SiS₂–Cu₂S–NiS quasiternary system at 600°C. (From Nazarchuk, O.P., et al., *Nauk. Visnyk Volyns'k. Nats. Univ. im. Lesi Ukrainky. Ser. Khim. nauky*, (16), 21, 2008.)

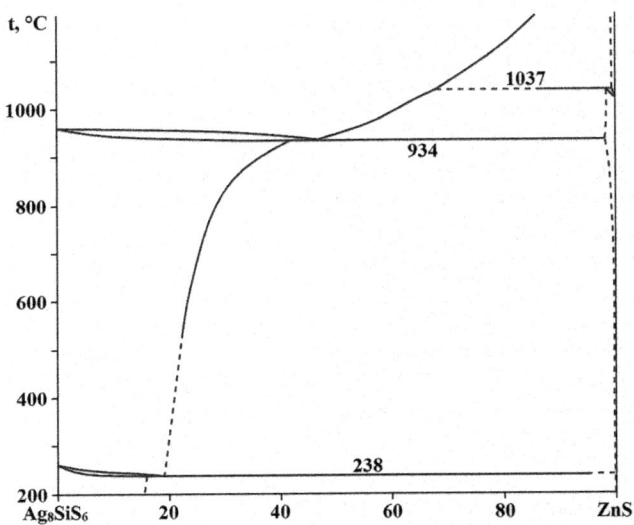

FIGURE 1.12 Phase diagram of the Ag₈SiS₆–ZnS system. (From Piskach, L.V. et al., *J. Alloys Compd.*, **421**(1–2), 98, 2006.)

overall pellet mass was lost, and a black material was left on the walls of the quartz ampoules. The grinding, pressing, and annealing process was repeated to improve phase homogeneity. The further annealing processes did not cause any significant mass or material loss, and the pellet remained intact during the further anneals, in contrast to the first anneal. High-quality, yellow single crystals of Ag₂SrSiS₄ separated naturally during the grinding of the pellet from the powder synthesis.

Ag₂Sr₃Si₂S₈ crystallizes in the cubic structure with the lattice parameter $a = 1402.67 \pm 0.01$ and a calculated density of 3.762 g·cm⁻³ (McKeown Wessler et al. 2021). Single crystals of this compound were grown in an attempt to obtain the single crystals of Ag₂SrSiS₄ using the recrystallization from its synthesized powder in a eutectic 0.4KI–0.6CsI flux. Synthesized powder of Ag₂SrSiS₄ was mixed with KI and CsI (molar ratio 1:12.5). The combined material was ground, pressed, and sealed as above. The sealed pellet was ramped to 700°C at a rate of 30°C·h⁻¹ and held at this temperature for 4 days before being cooled to 400°C at 4°C·h⁻¹. The ampoule was then allowed to cool naturally to room temperature. The product was washed with DMF, and sub-millimeter single crystals could be separated from any remaining halide flux material.

1.81 Silicon–Silver–Barium–Sulfur

The **Ag₂BaSiS₄** quaternary compound, which crystallizes in the tetragonal structure with the lattice parameters $a = 675.053 \pm 0.002$, $c = 799.643 \pm 0.003$ pm, and an energy gap of 2.2 eV, is formed in the Si-Ag-Ba-S system (Sun et al. 2020a). It was prepared by combining reagents in stoichiometric amounts with a total mass of ~250-400 mg. The mixture was weighed, ground with a mortar and pestle, and cold-pressed, all in a N₂-filled glove box. Then it was sealed under N₂ to prevent Ag₈SiS₆ from evaporating from the pellet. The ampoule was annealed at 800°C for 96 h and allowed to

cool naturally. The compound was ground, pressed, resealed, and annealed once again to achieve desired phase purity.

1.82 Silicon–Silver–Zinc–Sulfur

Ag₈SiS₆–ZnS. The phase diagram of this system is a eutectic type (Figure 1.12) (Piskach et al. 2006). The eutectic composition and temperature are 47 mol% ZnS and 934°C, respectively. The solubility of ZnS in Ag₈SiS₆ at the eutectic temperature is 42 mol%. The crystallization of the solid solutions occurs in a narrow temperature range that does not exceed 20°C. These solid solutions are characterized by the polymorph transformation at 238°C, and solid solutions based on ZnS have the transformation sphalerite/wurtzite at 1037°C. The system was investigated by DTA, XRD, and metallography.

The **Ag₂ZnSiS₄** quaternary compound, which crystallizes in the monoclinic structure with the lattice parameters $a = 640.52 \pm 0.01$, $b = 654.84 \pm 0.01$, $c = 793.40 \pm 0.01$ pm, $\beta = 90.455° \pm 0.001°$, and a calculated density of 4.366 g·cm⁻³, is formed in the Si–Ag–Zn–S system (Brunetta et al. 2012a). This compound is a direct band gap semiconductor with $E_g = 3.28$ eV. Single crystals of Ag₂ZnSiS₄ were produced by heating the powder mixture of Zn, Ag, Si, and S in a stoichiometric ratio in a fused silica tube under a vacuum up to 800°C for 12 h and held at that temperature for 96 h. After that, the sample was slowly cooled to 500°C at 5°C·h⁻¹ (60 h) and then allowed to cool radiatively to ambient temperature.

1.83 Silicon–Silver–Cadmium–Sulfur

Ag₂SiS₃–CdS. The phase diagram of this system is a eutectic type (Figure 1.13) (Parasyuk and Piskach 1999). The eutectic crystallizes at 668°C and contains 8 mol% CdS. The

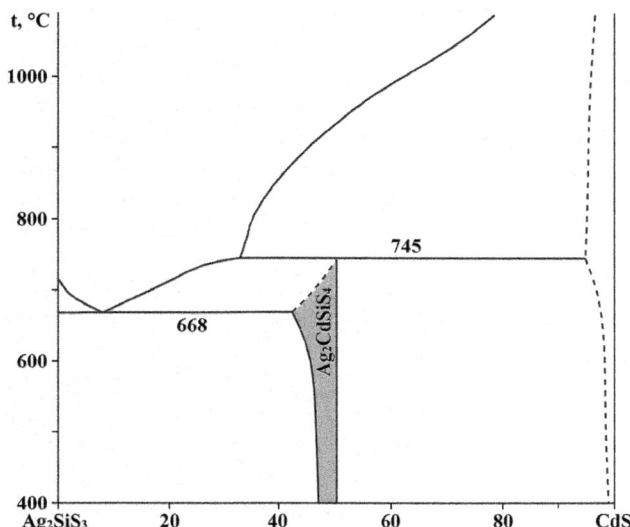

FIGURE 1.13 Phase diagram of the Ag_2SiS_3–CdS system. (From Parasyuk, O.V., and Piskach, L.V., *Zhurn. neorgan. khimii*, **44**(6), 1032, 1999.)

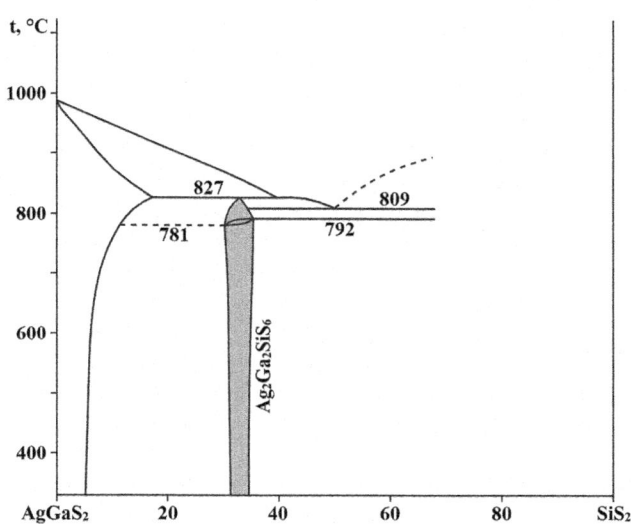

FIGURE 1.15 Part of the phase diagram of the SiS_2–$AgGaS_2$ system. (From Piasecki, M., et al., *J. Solid State Chem.*, **246**, 363, 2017.)

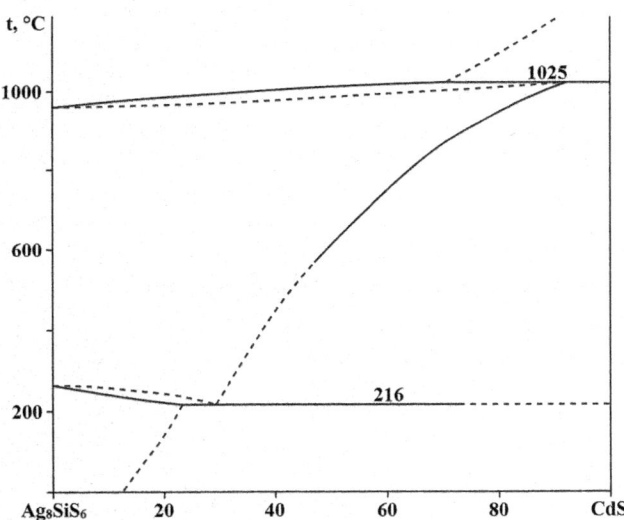

FIGURE 1.14 Phase diagram of the Ag_8SiS_6–CdS system. (From Piskach, L.V. et al., *J. Alloys Compd.*, **421**(1–2), 98, 2006.)

Ag_2CdSiS_4 quaternary compound is formed in this system. It melts incongruently at 745°C and crystallizes in the orthorhombic structure with the lattice parameters $a = 793.9 \pm 0.5$, $b = 679.1 \pm 0.4$, and $c = 652.7 \pm 0.5$ pm. The solubility of CdS in Ag_2SiS_3 is negligible, and the solid solution based on CdS does not exceed 2 mol% Ag_2SiS_3. The homogeneity region of Ag_2CdSiS_4 is situated only from the Ag_2SiS_3 side and is less than 5 mol% Ag_2SiS_3 at 400°C. This system was investigated through DTA, XRD, and metallography, and the alloys were annealed at 400°C for 250 h.

Ag_8SiS_6–CdS. The phase diagram of this system is a peritectic type (Figure 1.14) (Piskach et al. 2006). The peritectic composition and temperature are 70 mol% CdS and 1025°C, respectively. The solubility of Ag_8SiS_6 in CdS at the peritectic

temperature is 93 mol%. The crystallization of the solid solutions occurs in a narrow temperature range that does not exceed 20°C. These solid solutions are characterized by the polymorphic transformation at 216°C. The system was investigated through DTA, XRD, and metallography.

1.84 Silicon–Silver–Mercury–Sulfur

The $Ag_6Hg_{0.897}SiS_{5.897}$ quaternary compound, which crystallizes in the cubic structure with the lattice parameter $a = 1050.55 \pm 0.02$ pm, is formed in the Si–Ag–Hg–S system (Gulay et al. 2002c). The synthesis of this compound was carried out in evacuated quartz ampoules in two stages. At the first stage, the ampoules with charges were exposed to local heating in the flame of an O_2-gas burner till complete consumption of S. The second stage of the synthesis was realized by a single-temperature method in a furnace at 1000°C for 2 h (heating rate was 50°C·h⁻¹), and then the temperature was slowly decreased (10°C·h⁻¹) to 400°C. At this temperature, the sample was homogenized over 500 h, and then it was quenched in the air. The obtained sample was a tight alloy of black color and with metallic luster.

1.85 Silicon–Silver–Gallium–Sulfur

SiS_2–$AgGaS_2$. The phase diagram of this system was constructed through DTA and XRD in the 0-65 mol% SiS_2 range and is a presented in Figure 1.15 (Piasecki et al. 2017). The alloys for the investigations were annealed at 400°C for 500 h. The eutectic contains 50 mol% SiS_2 and crystallizes at 809°C. The composition of the peritectic point is 40 mol% SiS_2. Solid solutions based on $AgGaS_2$ are extended to ~17 mol% SiS_2 and are significantly narrower with the temperature decreasing (~5 mol% SiS_2 at 400°C). The **$Ag_2Ga_2SiS_6$** quaternary compound,

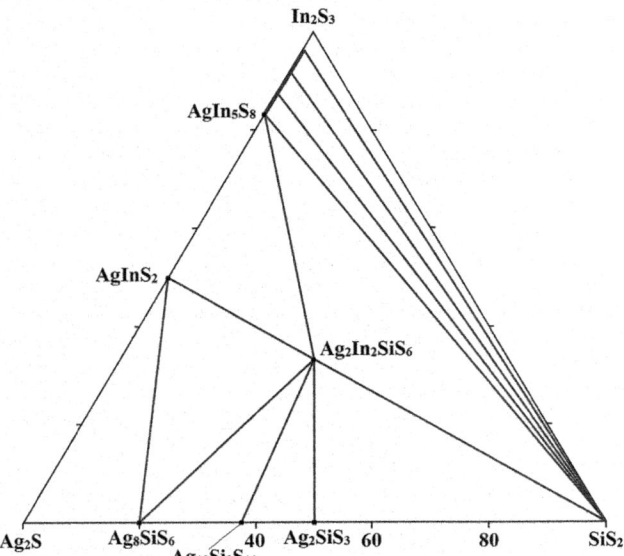

FIGURE 1.16 Isothermal section of the SiS_2–Ag_2S–In_2S_3 quasiternary system at room temperature. (From Sachanyuk, V.P., et al., *J. Alloys Compd.*, **452**(2), 348, 2008.)

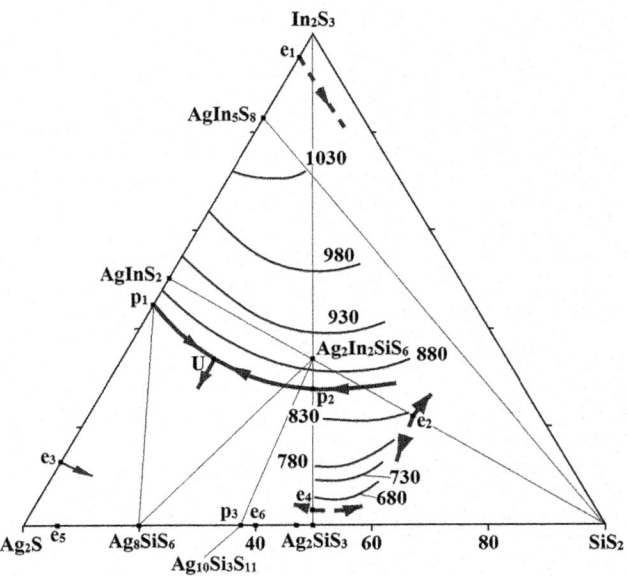

FIGURE 1.17 Part of the liquidus surface of the SiS_2–Ag_2S–In_2S_3 quasiternary system. (From Sachanyuk, V.P., et al., *J. Alloys Compd.*, **452**(2), 348, 2008.)

which melts incongruently at 827°C and reveals a polymorphic transformation within the temperature range of 781–792°C, is formed in this system. α-$Ag_2Ga_2SiS_6$ crystallizes in the tetragonal structure with the lattice $a = 571.64 \pm 0.04$, $c = 980.23 \pm 0.07$ pm, a calculated density of 3.9783 ± 0.0008 g·cm⁻³, and an energy gap of 2.35 eV.

1.86 Silicon–Silver–Indium–Sulfur

SiS_2–$AgInS_2$. This system is a nonquasibinary section of the SiS_2–Ag_2S–In_2S_3 quasiternary system since $AgInS_2$ melts incongruently (Sachanyuk et al. 2008). The eutectic coordinates are 57 mol% SiS_2 and 824°C. The homogeneity region of the solid solution based on $AgInS_2$ is < 5 mol% SiS_2. The $AgInS_2$ polymorphism is the reason for the peritectoid process at 738°C. At lower temperatures, the section is quasibinary, where the intermediate quaternary phase is in equilibrium with either $AgInS_2$ or SiS_2. The **$Ag_2In_2SiS_6$** quaternary compound, which melts incongruently at 860°C and crystallizes in the monoclinic structure with the lattice parameters $a = 1213.79 \pm 0.03$, $b = 716.81 \pm 0.02$, $c = 1211.71 \pm 0.04$ pm, $\beta = 109.252 \pm 0.002°$, and a calculated density of 4.4430 ± 0.0004 g·cm⁻³ (Sachanyuk et al. 2008) [$a = 1214.81 \pm 0.06$, $b = 717.26 \pm 0.05$, $c = 1212.63 \pm 0.07$ pm, $\beta = 109.255 \pm 0.003°$, and an energy gap of 2.0 eV (Reshak et al. 2013a)], is formed in this system. The results indicate the comparatively low hygroscopicity of $Ag_2In_2SiS_6$.

SiS_2–Ag_2S–In_2S_3. The isothermal section of this quasiternary system at room temperature is shown in Figure 1.16 (Sachanyuk et al. 2008). Five ternary compounds exist in the system, and $Ag_2In_2SiS_6$ forms equilibria with all available ternary sulfides and SiS_2. A significant homogeneity region is observed only for $AgIn_5S_8$ where it is localized along the Ag_2S–In_2S_3 section. Some vertical sections of this system were also constructed.

Partial liquidus surface projection of this quasiternary system near $Ag_2In_2SiS_6$ is given in Figure 1.17 (Sachanyuk et al. 2008). The field of primary crystallization of quaternary compound is shifted from the stoichiometric composition to lower In_2S_3 content.

1.87 Silicon–Silver–Lanthanum–Sulfur

The **$Ag_{1-\delta}La_3SiS_7$** quaternary compound, which crystallizes in the hexagonal structure with the lattice parameters $a = 1041.68 \pm 0.08$, $c = 578.25 \pm 0.06$ pm, and a calculated density of 4.684 g·cm⁻³ for $\delta = 0.1$ (Daszkiewicz et al. 2008a) [$a = 1039 \pm 1$, $c = 576.0 \pm 0.5$ pm for $\delta = 0$ (Guittard et al. 1968; Guittard and Julien-Pouzol 1970); $a = 1042.1 \pm 0.2$, $c = 578.5 \pm 0.3$ pm for $\delta = 0$ (Wu and Huang 2005)], is formed in the Si–Ag–La–S system.

This compound was obtained by fusion of a mixture of elements in evacuated silica ampoules in a furnace (Daszkiewicz et al. 2008a). The ampoule was heated with a heating rate of 30°C·h⁻¹ to 1150°C and was kept at this temperature for 3 h. Afterwards, it was cooled slowly (10°C·h⁻¹) to 600°C and annealed at this temperature for 240 h. After annealing the ampoule was quenched in cold water.

$Ag_{1-\delta}La_3SiS_7$ can be also prepared from a mixture of La_2S_3, Ag_2S, Si, and S, which was pressed and introduced into the silica ampoule and gradually heats up to 800°C in about a week (Guittard and Julien-Pouzol 1970). At the end of the preparation, it is necessary to heat the sample to 1200°C to obtain the pure hexagonal phase.

The title compound was also synthesized by the reaction of stoichiometric amounts of elements in a KBr flux (Wu and Huang 2005). The mixture was loaded under Ar and sealed under a vacuum (10^{-2} Pa) in a silica tube. The tube was placed and heated in a furnace to 900°C, kept at 850°C for 72 h, cooled at a rate of 0.05°C·min⁻¹ to 450°C, and finally cooled to room temperature. The reaction mixture was washed with distilled water and dried with acetone. The reaction product comprised hexagonal pale greenish-yellow needles, which were modestly stable in the air.

1.88 Silicon–Silver–Cerium–Sulfur

The $Ag_{1-\delta}Ce_3SiS_7$ quaternary compound, which crystallizes in the hexagonal structure with the lattice parameters $a = 1031.2 \pm 0.1$, $c = 573.95 \pm 0.07$ pm, and a calculated density of 4.786 g·cm⁻³ for $\delta = 0.18$ (Daszkiewicz et al. 2008a) [$a = 1030$, $c = 576$ pm for $\delta = 0$ (Guittard and Julien-Pouzol 1970)], is formed in the Si–Ag–Ce–S system. This compound was synthesized in the same way as $Ag_{1-\delta}La_3SiS_7$ was obtained using Ce_2S_3 instead of La_2S_3.

1.89 Silicon–Silver–Praseodymium–Sulfur

The $Ag_{1-\delta}La_3SiS_7$ quaternary compound, which crystallizes in the hexagonal structure with the lattice parameters $a = 1024.8 \pm 0.1$, $c = 572.23 \pm 0.05$ pm, and a calculated density of 4.971 g·cm⁻³ for $\delta = 0.15$ (Daszkiewicz et al. 2008a) [$a = 1020$, $c = 575$ pm for $\delta = 0$ (Guittard and Julien-Pouzol 1970)], is formed in the Si–Ag–Pr–S system. This compound was synthesized in the same way as $Ag_{1-\delta}La_3SiS_7$ was obtained using Pr_2S_3 instead of La_2S_3.

1.90 Silicon–Silver–Neodymium–Sulfur

The $Ag_{1-\delta}Nd_3SiS_7$ quaternary compound, which crystallizes in the hexagonal structure with the lattice parameters $a = 1019.2 \pm 0.1$, $c = 570.20 \pm 0.06$ pm, and a calculated density of 5.100 g·cm⁻³ for $\delta = 0.19$ (Daszkiewicz et al. 2008a) [$a = 1014$, $c = 575$ pm for $\delta = 0$ (Guittard and Julien-Pouzol 1970)], is formed in the Si–Ag–Nd–S system. This compound was synthesized in the same way as $Ag_{1-\delta}La_3SiS_7$ was obtained using Nd_2S_3 instead of La_2S_3.

1.91 Silicon–Silver–Samarium–Sulfur

The $Ag_{1-\delta}Sm_3SiS_7$ quaternary compound, which crystallizes in the hexagonal structure with the lattice parameters $a = 1010.0 \pm 0.1$, $c = 566.43 \pm 0.06$ pm, and a calculated density of 5.343 g·cm⁻³ for $\delta = 0.23$ (Daszkiewicz et al. 2008a) [$a = 1002$, $c = 574$ pm for $\delta = 0$ (Guittard and Julien-Pouzol 1970)], is formed in the Si–Ag–Sm–S system. This compound was synthesized in the same way as $La_3Ag_{1-\delta}SiS_7$ was obtained using Sm_2S_3 instead of La_2S_3.

1.92 Silicon–Silver–Lead–Sulfur

Two quaternary compounds, Ag_2PbSiS_4 and $Ag_2Pb_3Si_2S_8$, are formed in the Si–Ag–Pb–S system. First of them crystallizes in the orthorhombic structure with the lattice parameters $a =$

986.98 ± 0.07, $b = 986.97 \pm 0.05$, $c = 697.91 \pm 0.02$ pm, and an energy gap of 1.87 eV (Sun et al. 2020a, 2020b). It was prepared by combining reagents in stoichiometric amounts with a total mass of ~250–400 mg. The mixture was weighed, ground with a mortar and pestle, and cold-pressed, all in a N_2-filled glove box. Then it was sealed under N_2 to prevent Ag_8SiS_6 from evaporating from the pellet. The ampoule was annealed at 525°C for 96 h and allowed to cool naturally. The compound was ground, pressed, resealed, and annealed several times to achieve desired phase purity.

$Ag_2Pb_3Si_2S_8$ crystallizes in the cubic structure with the lattice parameters $a = 1391.67 \pm 0.01$, and a calculated density of 5.619 g·cm⁻³ at 80 K and $a = 1396.88 \pm 0.02$, a calculated density of 5.574 g·cm⁻³, and an energy gap of 1.95 eV at room temperature (McKeown Wessler et al. 2022).

1.93 Silicon–Silver–Oxygen–Sulfur

The $Ag_8SiO_4S_2$ quaternary compound, which crystallizes in the tetragonal structure with the lattice parameters $a = 700.5 \pm 0.5$, $c = 1775 \pm 3$ pm, and the calculated and experimental densities of 7.77 ± 0.03 and 7.3 g·cm⁻³, respectively, is formed in the Si–Ag–O–S system (Schultze-Rhonhof 1974).

1.94 Silicon–Silver–Tellurium–Sulfur

The $Ag_8SiS_4Te_2$ quaternary compound, which crystallizes in the hexagonal structure with the lattice parameters $a = 1319.7 \pm 0.2$ and $c = 1266.6 \pm 0.1$ pm, is formed in the Si–Ag–Te–S system (Frank and Pfitzner 2012).

1.95 Silicon–Silver–Chlorine–Sulfur

The $Ag_{8-x}SiS_{6-x}Cl_x$ quaternary phase (x value is not specified), which has a polymorphic transformation at 218°C and crystallizes in the orthorhombic structure with the lattice parameters $a = 1504.7 \pm 0.5$, $b = 743.4 \pm 0.3$, and $c = 1053.8 \pm 0.4$ pm (Kuhs et al. 1979), is formed in the Si–Ag–Cl–S system. This phase was prepared by reacting stoichiometric amounts of elements in an evacuated, sealed quartz ampoule for about 6 days at temperatures of 600°C to 700°C.

1.96 Silicon–Silver–Bromine–Sulfur

The $Ag_{8-x}SiS_{6-x}Br$ quaternary phase, which melts at 891°C and crystallizes in the cubic structure with the lattice parameter $a = 1058.36 \pm 0.01$ pm for $x = 1$ (Laqibi et al. 1987) [$a = 1058.8 \pm 0.7$ pm for $x = 1$ (Nagel and Range 1978); $a = 1047.9 \pm 0.4$ pm (x value is not specified) (Kuhs et al. 1979)], is formed in the Si–Ag–Br–S system. This phase was obtained by solid-state reactions as follows: $3Ag_2S + SiS_2 + AgBr = Ag_7SiS_5Br$ (Laqibi et al. 1987). Suitable quantities of the components were finely ground and mixed thoroughly and then treated in a quartz ampoule sealed under a secondary vacuum (10⁻² Pa). The mixture was heated to 700°C over a period of 3 days, kept at this temperature for 48 h, and then maintained at 580°C for 1 week. It can also be prepared by reacting stoichiometric amounts of elements in an evacuated, sealed quartz ampoule for about 6 days at temperatures of 600°C to 700°C (Kuhs et al. 1979).

1.97 Silicon–Silver–Iodine–Sulfur

The $Ag_{8-x}SiS_{6-x}I$ quaternary compound, which melts at 903°C and crystallizes in the cubic structure with the lattice parameter $a = 1064.43 \pm 0.01$ pm for $x = 1$ (Laqibi et al. 1987) [$a = 1072.0 \pm 0.9$ pm for $x = 1$ (Nagel and Range 1978); $a = 1059.6 \pm 0.4$ pm (x value is not specified) (Kuhs et al. 1979)], is formed in the Si–Ag–I–S system. It was synthesized in the same way as $Ag_{8-x}SiSe_{6-x}Br$ was prepared using AgI instead of AgBr (Laqibi et al. 1987) or by reacting stoichiometric amounts of elements in an evacuated, sealed quartz ampoule for about 6 days at temperatures of 600°C to 700°C (Kuhs et al. 1979).

1.98 Silicon–Silver–Iron–Sulfur

The Ag_2FeSiS_4 quaternary compound, which melts incongruently at 860°C and crystallizes in the monoclinic structure with the lattice parameters $a = 642.20 \pm 0.01$, $b = 661.85 \pm 0.01$, $c = 786.50 \pm 0.01$ pm, $\beta = 90.614 \pm 0.001°$, and a calculated density of 4.242 g·cm⁻³, is formed in the Si–Ag–Fe–S system (Brunetta et al. 2013). Single crystals of this compound were produced by weighing stoichiometric amounts of elements or binary starting material in an Ar-filled glove box. Reagents were combined in an agate mortar, ground for 20 min with a pestle, and then transferred to a graphite crucible, which was then placed in a fused silica tube. The tube was then flame-sealed under a vacuum of approximately 0.1 Pa and placed in a furnace, which was heated to 700°C over 12 h, held at that temperature for 96 h, and then cooled to 420°C over 50 h, after which the sample was allowed to cool radiatively.

1.99 Silicon–Magnesium–Lanthanum–Sulfur

The $MgLa_6Si_2S_{14}$ quaternary compound, which crystallizes in the hexagonal structure with the lattice parameters $a = 1036.3 \pm 0.3$ and $c = 574.2 \pm 0.1$ pm (Gitzendanner et al. 1997) [$a = 1035 \pm 2$ and $c = 571 \pm 2$ pm (Michelet and Flahaut 1969)], is formed in the Si–Mg–La–S system. This compound was obtained in a sealed silica ampoule at around 1000°C by the interaction of MgS and La_2S_3 with Si and the desired amount of S (Michelet and Flahaut 1969). It appears as a polycrystalline mass. Sufficiently prolonged heating makes it possible to obtain small single crystals.

This compound was also obtained from the reaction of stoichiometric amounts of La, Mg, Si, and S in a NaCl/MgCl₂ flux in a vitreous carbon crucible sealed in an evacuated fused silica tube (Gitzendanner et al. 1997). The tube was heated slowly to 600°C for 48 h and held there for 12 h to allow for complete prereaction of the sulfur. This reduces the sulfur vapor pressure that otherwise could lead to explosions at higher temperatures. The temperature was then increased to 1000°C over 12 h and

held there for 72 h. The tube was then slowly cooled (2°C·h⁻¹) to 500°C, at which point the furnace was shut off and allowed to cool to room temperature. The flux was washed away by soaking the product in distilled water for several hours. The major phase was $MgLa_6Si_2S_{14}$ in the form of light-orange transparent crystals; however, minor secondary products, including La_2S_3, were also obtained.

1.100 Silicon–Magnesium–Yttrium–Sulfur

The $MgY_6Si_2S_{14}$ quaternary compound, which crystallizes in the hexagonal structure with the lattice parameters $a = 984.0 \pm 0.5$, $c = 561.9 \pm 0.3$ pm, a calculated density of 3.745 g·cm⁻³, and an energy gap of 2.26 eV (Chi and Guo 2017) [$a = 986 \pm 2$ and $c = 562 \pm 2$ pm (Michelet and Flahaut 1969)], is formed in the Si–Mg–Y–S system. Single crystals of the title compound were obtained by solid-state reaction with KI as a flux (Chi and Guo 2017). The starting materials were Y_2O_3, S, Mg, Si, and boron powder. Each sample had a total mass of 500 mg and 400 mg KI additional, and the molar ratio of Y/S/Mg/Si was 6:14:1:2. The mixture of starting materials was ground into fine powder in an agate mortar and pressed into pellets, followed by being loaded into a quartz tube. The tube was evacuated to 10⁻² Pa and flame-sealed. The sample was placed into a muffle furnace, heated from room temperature to 300°C for 5 h, and equilibrated for 10 h, followed by heating to 700°C for 5 h and equilibrated for 10 h, then heated to 950°C over 5 h and maintained for 7 days, finally cooled down to 300°C in 5 days and powered off. The single crystals, which are stable in moisture and air, were obtained.

1.101 Silicon–Magnesium–Cerium–Sulfur

The $MgCe_6Si_2S_{14}$ quaternary compound, which crystallizes in the hexagonal structure with the lattice parameters $a = 1028.7 \pm 0.3$, $c = 569.8 \pm 0.4$ pm, and a calculated density of 4.356 g·cm⁻³ (Chi and Guo 2017) [$a = 1027.2 \pm 0.1$, $c = 570.65 \pm 0.07$ pm, and a calculated density of 4.363 g·cm⁻³ (Strok et al. 2013a)], is formed in the Si–Mg–Ce–S system.

This compound was obtained from a mixture of the chemical elements by direct single-temperature synthesis in graphitized and evacuated (0.013 Pa) quartz ampoule (Strok et al. 2013a). This mixture was heated to 400°C and annealed at this temperature for 5 days. The maximal temperature of the synthesis was 1150°C. Homogenization was made at 600°C, and after that, the sample was quenched in water. Single crystals of the title compound were obtained in the same way as in the case of $MgY_6Si_2S_{14}$ using CeO₂ instead of Y_2O_3 (Chi and Guo 2017).

1.102 Silicon–Magnesium–Praseodymium–Sulfur

The $MgPr_6Si_2S_{14}$ quaternary compound, which crystallizes in the hexagonal structure with the lattice parameters $a = 1017.5 \pm 0.2$ and $c = 567.8 \pm 0.2$ pm, is formed in the Si–Mg–Pr–S system (Chi and Guo 2017). Powder sample of this compound were obtained in the same way as single crystals of $MgY_6Si_2S_{14}$ were prepared using Pr_2O_3 instead of Y_2O_3.

1.103 Silicon–Magnesium–Neodymium–Sulfur

The $MgNd_6Si_2S_{14}$ quaternary compound, which crystallizes in the hexagonal structure with the lattice parameters $a = 1012.0 \pm 0.1$, $c = 567.5 \pm 0.1$ pm, and a calculated density of 4.602 $g \cdot cm^{-3}$, is formed in the Si–Mg–Nd–S system (Chi and Guo 2017). Single crystals of the title compound were obtained in the same way as in the case of $MgY_6Si_2S_{14}$ using Nd_2O_3 instead of Y_2O_3.

1.104 Silicon–Magnesium–Samarium–Sulfur

The $MgSm_6Si_2S_{14}$ quaternary compound, which crystallizes in the hexagonal structure with the lattice parameters $a = 1003.5 \pm 0.1$, $c = 564.7 \pm 0.1$ pm, and a calculated density of 4.826 $g \cdot cm^{-3}$, is formed in the Si–Mg–Sm–S system (Chi and Guo 2017). Single crystals of the title compound were obtained in the same way as in the case of $MgY_6Si_2S_{14}$ using Sm_2O_3 instead of Y_2O_3.

1.105 Silicon–Magnesium–Manganese–Sulfur

The $Mg_{0.6}Mn_{1.4}SiS_4$ quaternary phase, which crystallizes in the orthorhombic structure with the lattice parameters $a = 1267.6 \pm 0.1$, $b = 743.0 \pm 0.2$, and $c = 592.7 \pm 0.2$ pm (Fuhrmann and Pickardt 1990b), is formed in the Si–Mg–Mn–S system. Single crystals of this phase were obtained by chemical transport reactions.

1.106 Silicon–Magnesium–Iron–Sulfur

The $(Mg_{1-x}Fe_x)_2SiS_4$ quaternary phase, which crystallizes in the orthorhombic structure with the lattice parameters $a = 1267.7 \pm 0.4$, $b = 740.5 \pm 0.2$, $c = 591.3 \pm 0.2$ pm, and a calculated density of 2.505 $g \cdot cm^{-3}$ for $x = 0.07$ and $a = 1263.3 \pm 0.5$, $b = 734.8 \pm 0.3$, $c = 590.1 \pm 0.2$ pm, and a calculated density of 2.715 $g \cdot cm^{-3}$ for $x = 0.3$ (Fuhrmann and Pickardt 1990a) and $a = 1258.6 \pm 0.7$, $b = 732.9 \pm 0.3$, and $c = 587.0 \pm 0.3$ pm for $x = 0.43$ (Fuhrmann and Pickardt 1990b)], is formed in the Si–Mg–Fe–S system. Single crystals of this phase were obtained by chemical transport reactions carried out in sealed quartz ampoules at 900°C/800°C for two weeks using Mg_2Si, Fe, Si and S_8 as starting materials. The red transparent crystals for $x = 0.3$ and yellow transparent crystals $x = 0.07$ were obtained.

1.107 Silicon–Calcium–Oxygen–Sulfur

Two quaternary compounds, $Ca_5(SiO_4)_2(SO_4)$ and $Ca_{11}(SiO_4)_4O_2S$, are formed in the Si–Ca–O–S system. First of them crystallizes in the orthorhombic structure with the lattice parameters $a = 686.3 \pm 0.1$, $b = 1538.7 \pm 0.2$, $c = 1018.1 \pm 0.1$ pm, and the calculated and experimental densities of 2.97 and 2.94 ± 0.02 $g \cdot cm^{-3}$, respectively, for mineral ternesite (Irran et al. 1997; Jambor et al. 1998b) [$a = 1018.3 \pm 0.1$, $b = 1540.8 \pm 0.5$, $c = 682.5 \pm 0.1$ pm, and an experimental density of 2.95 ± 0.01 $g \cdot cm^{-3}$ (Pryce 1972); $a = 1018.2 \pm 0.1$, $b = 1539.8 \pm 0.2$, $c = 685.00 \pm 0.09$ pm, and a calculated density of 2.96 $g \cdot cm^{-3}$ (Brotherton et al. 1974)]. $Ca_5(SiO_4)_2(SO_4)$ was prepared from $CaCO_3$, crushed quartz, and $CaSO_4 \cdot 2H_2O$ by ignition at 1150°C for 150 h in Pt boats (Gutt and Smith 1966). For successful preparation, the mixture has to be finely ground between periods of heating. This compound is stable up to 1150°C.

$Ca_{11}(SiO_4)_4O_2S$ (mineral jasmundite) crystallizes in the tetragonal structure with the lattice parameters $a = 1046.1 \pm 0.1$, $c = 881.3 \pm 0.1$ pm, and the calculated and experimental densities of 3.23 and 3.03 $g \cdot cm^{-3}$, respectively (Glasser and Lee 1981; Hentschel et al. 1983; Dunn et al. 1984b).

1.108 Silicon–Barium–Gallium–Sulfur

Two quaternary compounds, $BaGa_2SiS_6$ and $Ba_2Ga_8SiS_{16}$, are formed in the Si–Ba–Ga–S system. The first of them melts congruently at 968°C and crystallizes in the trigonal structure with the lattice parameters $a = 955.44 \pm 0.02$, $c = 864.98 \pm 0.04$ pm, a calculated density of 3.622 $g \cdot cm^{-3}$, and an energy gap of 3.75 eV (Yin et al. 2012b).

$Ba_2Ga_8SiS_{16}$ crystallizes in the hexagonal structure with the lattice parameters $a = 1086.6 \pm 0.5$, $c = 1191.9 \pm 0.8$ pm, a calculated density of 3.743 $g \cdot cm^{-3}$, and an energy gap of ≈ 3.4 eV (Liu et al. 2015b).

Crystals of $BaGa_2SiS_6$ were grown by spontaneous nucleation of the pure polycrystalline sample, which was synthesized by stoichiometric reactions of BaS, Ga_2S_3, and SiS_2 (molar ratio 1:1:1) at 900°C for 48 h (Yin et al. 2012b). Then, 0.5 g of the as-prepared powder of $BaGa_2SiS_6$ was loaded into a fused silica tube under an Ar atmosphere in a glove box. The tube was sealed under a 10^{-3} Pa atmosphere and then placed in a furnace. The reaction mixture was heated to 1000°C for 15 h, kept at 1000°C for 48 h, followed by slow cooling to 320°C at a rate of 3°C·h⁻¹, and finally cooled to room temperature by switching off the furnace. The product consisted of light yellow crystals of the title compound.

For the synthesis of $Ba_2Ga_8SiS_{16}$, a stoichiometric mixture with an overall weight of about 300 mg of the starting materials Ba, Ga_2S_3, Si, and S (molar ratio 2:4:1:4) was loaded into a quartz tube and then flame-sealed under vacuum of ~ 10^{-2} Pa (Liu et al. 2015b). The tube was then placed in a furnace, held at 300°C for 5 h, then heated to 650°C for over 5 h and kept there for 10 h, subsequently heated to 980°C for over 10 h, dwelled for 5 days, and finally slowly cooled to 400°C before the furnace power was switched off. Transparent yellow single crystals of the title compound were obtained in high purity, which were stable in the air for several months. All starting reactants were handled inside an Ar-filled glove box with controlled oxygen and moisture levels below 0.1 ppm.

1.109 Silicon–Barium–Germanium–Sulfur

Ba_2SiS_4–Ba_2GeS_4. A continuous solid solution exists in this system (Dumail et al. 1971). The variation of crystalline lattice parameters obeys Vegard's law. The different compositions of the solid solution were obtained by mixing in suitable proportions either of BaS, SiS_2, and GeS_2, or of BaS, Si, Ge, and S. It should be noted that the solid solution has never been obtained by mixing Ba_2SiS_4 and Ba_2GeS_4.

1.110 Silicon–Zinc–Cadmium–Sulfur

SiS_2–ZnS–CdS. Solid solutions are formed in this system along the section Cd_4SiS_6–$(ZnS)_{0.8}(SiS_2)_{0.2}$ within the interval from 75 to 100 mol% Cd_4SiS_6 (Dubrovin et al. 1989). $(Zn_xCd_{1-x})_4SiS_6$ single crystals at $x = 0.05$, 0.13, 0.18, and 0.25 were grown using chemical transport reactions with iodine as a transport agent. Lattice parameters of obtained solid solutions change versus composition according to Vegard's law.

1.111 Silicon–Zinc–Yttrium–Sulfur

The $ZnY_6Si_2S_{14}$ quaternary compound, which crystallizes in the hexagonal structure with the lattice parameters $a = 980.7 \pm 0.1$, $c = 564.0 \pm 0.1$ pm, and a calculated density of 3.902 $g \cdot cm^{-3}$ (Guo et al. 2009b) [$a = 980 \pm 2$ and $c = 565 \pm 2$ pm system (Michelet and Flahaut 1969)], is formed in the Si–Zn–Y–S system. Single crystals of this compound were obtained by solid-state reactions with KI as a flux (Guo et al. 2009b). The reaction mixture contained ZnS (0.718 mM), Si (1.424 mM), S (5.708 mM), Y_2O_3 (0.713 mM), B (4.255 mM), and KI (2.410 mM). It was ground into fine powder in an agate mortar and pressed into pellets, followed by being loaded into a quartz tube. The tube was evacuated at 10^{-2} Pa and flame-sealed. The sample was placed into a furnace, heated from room temperature to 300°C over 5 h, kept at 300°C for 10 h, heated to 650°C over 5 h, kept at 650°C for 10 h, heated to 950°C over 5 h, kept at 950°C for 10 days, then cooled down to 300°C over 5 days, and powered off. Yellow block crystals of the title compound were hand-picked under a microscope. They are stable in the presence of air and water and show no signs of appreciable change in crystal quality over several months.

$ZnY_6Si_2S_{14}$ can also be obtained in a sealed silica ampoule at around 1000°C by interaction of ZnS and Y_2S_3 with Si and the desired amount of S (Michelet and Flahaut 1969). It appears as a polycrystalline mass. Sufficiently prolonged heating makes it possible to obtain small single crystals.

1.112 Silicon–Zinc–Lanthanum–Sulfur

The $ZnLa_6Si_2S_{14}$ quaternary compound, which crystallizes in the hexagonal structure with the lattice parameters $a = 1031.2 \pm 0.1$ and $c = 574.0 \pm 0.1$ pm (Zhou et al. 2019) [$a = 1034 \pm 2$ and $c = 575 \pm 2$ pm (Michelet and Flahaut 1969)], is formed in the Si–Zn–La–S system. It was synthesized from a stoichiometric mixture of the elements with a total mass of 0.30 g that was pressed into pellets and loaded into fused silica tube, which was evacuated and sealed (Zhou et al. 2019). The tube was heated at 600°C over 12 h and then heated at 1050°C for 4 days. The polycrystalline sample obtained is stable in the air for several months.

$ZnLa_6Si_2S_{14}$ can be also obtained in a sealed silica ampoule at around 1000°C by interaction of ZnS and La_2S_3 with Si and the desired amount of S (Michelet and Flahaut 1969). It appears as a polycrystalline mass. Sufficiently prolonged heating makes it possible to obtain small single crystals.

1.113 Silicon–Zinc–Cerium–Sulfur

The $ZnCe_6Si_2S_{14}$ quaternary compound, which crystallizes in the hexagonal structure with the lattice parameters $a = 1018.0 \pm 0.1$, $c = 570.1 \pm 0.1$ pm, and an energy gap of 1.8 eV (Zhou et al. 2019), is formed in the Si–Zn–Ce–S system. It was synthesized in the same way as $ZnLa_6Si_2S_{14}$ was prepared.

1.114 Silicon–Zinc–Praseodymium–Sulfur

The $ZnPr_6Si_2S_{14}$ quaternary compound, which crystallizes in the hexagonal structure with the lattice parameters $a = 1009.1 \pm 0.2$ and $c = 563.7 \pm 0.1$ pm (Zhou et al. 2019), is formed in the Si–Zn–Pr–S system. It was synthesized in the same way as $ZnLa_6Si_2S_{14}$ was prepared.

1.115 Silicon–Zinc–Neodymium–Sulfur

The $ZnNd_6Si_2S_{14}$ quaternary compound, which crystallizes in the hexagonal structure with the lattice parameters $a = 1001.1 \pm 0.1$ and $c = 563.7 \pm 0.1$ pm (Zhou et al. 2019), is formed in the Si–Zn–Nd–S system. It was synthesized in the same way as $ZnLa_6Si_2S_{14}$ was prepared.

1.116 Silicon–Zinc–Samarium–Sulfur

The $ZnSm_6Si_2S_{14}$ quaternary compound, which crystallizes in the hexagonal structure with the lattice parameters $a = 989.9 \pm 0.2$ and $c = 560.4 \pm 0.1$ pm (Zhou et al. 2019), is formed in the Si–Zn–Sm–S system. It was synthesized in the same way as $ZnLa_6Si_2S_{14}$ was prepared.

1.117 Silicon–Zinc–Gadolinium–Sulfur

The $ZnGd_6Si_2S_{14}$ quaternary compound, which crystallizes in the hexagonal structure with the lattice parameters $a = 984.9 \pm 0.3$ and $c = 561.1 \pm 0.2$ pm (Zhou et al. 2019), is formed in the Si–Zn–Gd–S system. It was synthesized in the same way as $ZnLa_6Si_2S_{14}$ was prepared.

1.118 Silicon–Zinc–Terbium–Sulfur

The $ZnTb_6Si_2S_{14}$ quaternary compound, which crystallizes in the hexagonal structure with the lattice parameters $a = 976.8 \pm 0.3$ and $c = 560.0 \pm 0.2$ pm (Zhou et al. 2019), is formed in the Si–Zn–Tb–S system. It was synthesized in the same way as $ZnLa_6Si_2S_{14}$ was prepared.

1.119 Silicon–Zinc–Dysprosium–Sulfur

The **ZnDy₆Si₂S₁₄** quaternary compound, which crystallizes in the hexagonal structure with the lattice parameters $a = 973.8 \pm 0.2$ and $c = 560.9 \pm 0.2$ pm (Zhou et al. 2019), is formed in the Si–Zn–Dy–S system. It was synthesized in the same way as ZnLa₆Si₂S₁₄ was prepared.

1.120 Silicon–Zinc–Holmium–Sulfur

The **ZnHo₆Si₂S₁₄** quaternary compound, which crystallizes in the hexagonal structure with the lattice parameters $a = 973.2 \pm 0.1$ and $c = 560.6 \pm 0.1$ pm (Zhou et al. 2019), is formed in the Si–Zn–Ho–S system. It was synthesized in the same way as ZnLa₆Si₂S₁₄ was prepared.

1.121 Silicon–Zinc–Erbium–Sulfur

The **ZnEr₆Si₂S₁₄** quaternary compound, which crystallizes in the hexagonal structure with the lattice parameters $a = 971.0 \pm 0.2$ and $c = 559.0 \pm 0.1$ pm (Zhou et al. 2019), is formed in the Si–Zn–Er–S system. It was synthesized in the same way as ZnLa₆Si₂S₁₄ was prepared.

1.122 Silicon–Cadmium–Lanthanum–Sulfur

SiS₂–CdS–La₂S₃. The isothermal section of this quasiternary system at 1050°C is shown in Figure 1.18 (Perez et al. 1970). The **CdLa₆Si₂S₁₄** quaternary compound, which crystallizes in the hexagonal structure with the lattice parameters $a = 1035.8 \pm 0.1$, $c = 573.7 \pm 0.1$ pm, and an energy gap of 2.7 eV (Zhou et al. 2019) [$a = 1038.0 \pm 0.4$ and $c = 575.0 \pm 0.2$ pm (Perez et al.

1970)], is formed in this system. It was synthesized in the same way as ZnLa₆Si₂S₁₄ was prepared (Zhou et al. 2019).

1.123 Silicon–Cadmium–Cerium–Sulfur

SiS₂–CdS–Ce₂S₃. The isothermal section of this quasiternary system at 1050°C is shown in Figure 1.19 (Perez et al. 1970). The **CdCe₆Si₂S₁₄** quaternary compound, which crystallizes in the hexagonal structure with the lattice parameters $a = 1022.5 \pm 0.1$ and $c = 568.8 \pm 0.1$ pm (Zhou et al. 2019) [$a = 1022.5 \pm 0.4$ and $c = 570.5 \pm 0.2$ pm (Perez et al. 1970)], is formed in this system. It was synthesized in the same way as ZnLa₆Si₂S₁₄ was prepared (Zhou et al. 2019).

1.124 Silicon–Cadmium–Praseodymium–Sulfur

SiS₂–CdS–Pr₂S₃. The isothermal section of this quasiternary system at 1050°C is analogous as one of the SiS₂–CdS–Ce₂S₃ quasiternary system (Perez et al. 1970). The **CdPr₆Si₂S₁₄** quaternary compound, which crystallizes in the hexagonal structure with the lattice parameters $a = 1015.8 \pm 0.2$ and $c = 567.0 \pm 0.1$ pm (Zhou et al. 2019) [$a = 1020.0 \pm 0.4$ and $c = 570.0 \pm 0.2$ pm (Perez et al. 1970)], is formed in this system. It was synthesized in the same way as ZnLa₆Si₂S₁₄ was prepared (Zhou et al. 2019).

1.125 Silicon–Cadmium–Neodymium–Sulfur

SiS₂–CdS–Nd₂S₃. The isothermal section of this quasiternary system at 1050°C is analogous as one of the SiS₂–CdS–Ce₂S₃ quasiternary system (Perez et al. 1970). The **CdNd₆Si₂S₁₄** quaternary compound, which crystallizes in the hexagonal

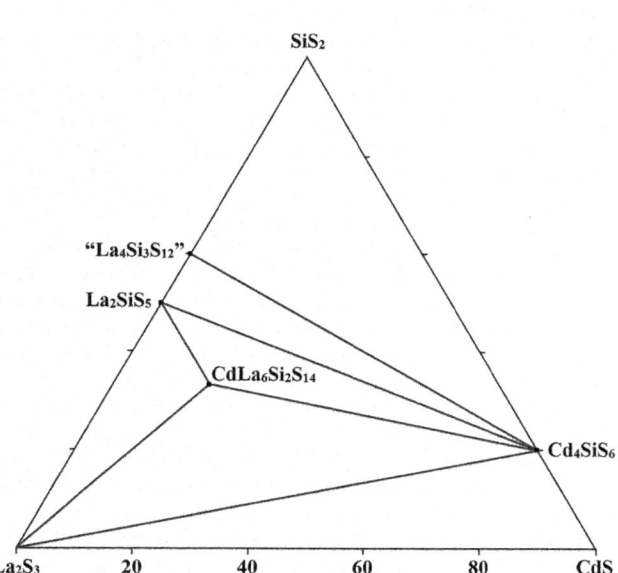

FIGURE 1.18 Isothermal section of the SiS₂–CdS–La₂S₃ quasiternary system at 1050°C. (From **Pe**rez, G. et al., *J. Solid State Chem.*, **2**(1), 42, 1970.)

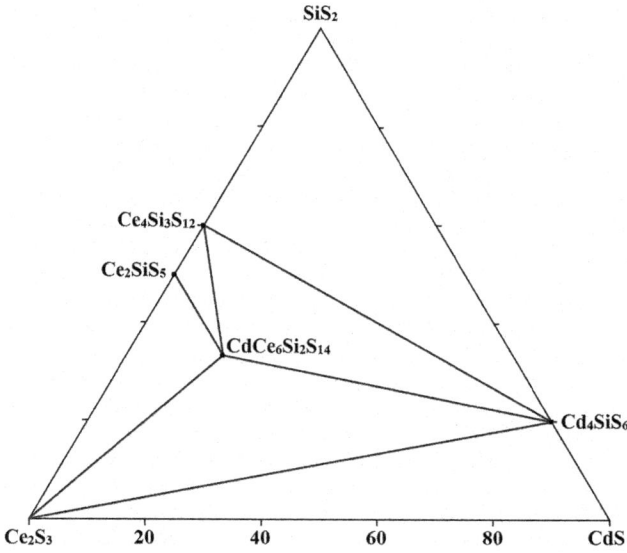

FIGURE 1.19 Isothermal section of the SiS₂–CdS–Ce₂S₃ quasiternary system at 1050°C. (From **Pe**rez, G. et al., *J. Solid State Chem.*, **2**(1), 42, 1970.)

structure with the lattice parameters $a = 1010.2 \pm 0.2$ and $c = 564.6 \pm 0.1$ pm (Zhou et al. 2019) [$a = 1016.0 \pm 0.4$ and $c = 568.5 \pm 0.2$ pm (Perez et al. 1970)], is formed in this system. It was synthesized in the same way as $ZnLa_6Si_2S_{14}$ was prepared (Zhou et al. 2019).

1.126 Silicon–Cadmium–Samarium–Sulfur

SiS_2–CdS–Sm_2S_3. The isothermal section of this quasiternary system at 1050°C is analogous as one of the SiS_2–CdS–Ce_2S_3 quasiternary system (Perez et al. 1970). The $CdSm_6Si_2S_{14}$ quaternary compound, which crystallizes in the hexagonal structure with the lattice parameters $a = 999.2 \pm 0.2$ and $c = 561.4 \pm 0.1$ pm (Zhou et al. 2019) [$a = 1006.0 \pm 0.4$ and $c = 566.0 \pm 0.2$ pm (Perez et al. 1970)], is formed in this system. It was synthesized in the same way as $ZnLa_6Si_2S_{14}$ was prepared (Zhou et al. 2019).

1.127 Silicon–Cadmium–Gadolinium–Sulfur

SiS_2–CdS–Gd_2S_3. The isothermal section of this quasiternary system at 1050°C is analogous as one of the SiS_2–CdS–Ce_2S_3 quasiternary system (Perez et al. 1970). The $CdGd_6Si_2S_{14}$ quaternary compound, which crystallizes in the hexagonal structure with the lattice parameters $a = 997.37 \pm 0.08$, $c = 562.87 \pm 0.04$ pm and a calculated density of 5.345 g·cm^{-3} (Zhou et al. 2019) [$a = 998.5 \pm 0.4$ and $c = 563.5 \pm 0.2$ pm (Perez et al. 1970)], is formed in this system. It was synthesized in the same way as $ZnLa_6Si_2S_{14}$ was prepared (Zhou et al. 2019).

1.128 Silicon–Cadmium–Terbium–Sulfur

The $CdTb_6Si_2S_{14}$ quaternary compound, which crystallizes in the hexagonal structure with the lattice parameters $a = 987.4 \pm 0.1$ and $c = 560.3 \pm 0.1$ pm (Zhou et al. 2019), is formed in the Si–Cd–Tb–S system. It was synthesized in the same way as $ZnLa_6Si_2S_{14}$ was prepared.

1.129 Silicon–Cadmium–Dysprosium–Sulfur

The $CdDy_6Si_2S_{14}$ quaternary compound, which crystallizes in the hexagonal structure with the lattice parameters $a = 983.1 \pm 0.2$ and $c = 560.5 \pm 0.1$ pm (Zhou et al. 2019), is formed in the Si–Cd–Dy–S system. It was synthesized in the same way as $ZnLa_6Si_2S_{14}$ was prepared.

1.130 Silicon–Cadmium–Holmium–Sulfur

The $CdHo_6Si_2S_{14}$ quaternary compound, which crystallizes in the hexagonal structure with the lattice parameters $a = 979.8 \pm 0.2$ and $c = 560.8 \pm 0.1$ pm (Zhou et al. 2019), is formed in the Si–Cd–Ho–S system. It was synthesized in the same way as $ZnLa_6Si_2S_{14}$ was prepared.

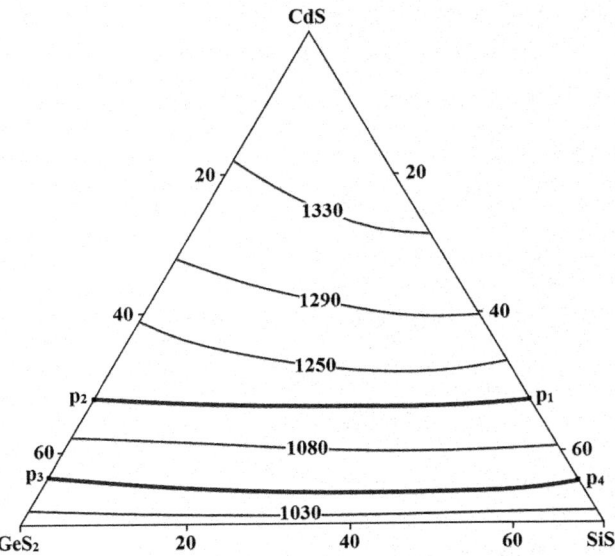

FIGURE 1.20 Part of the liquidus surface of the GeS$_2$–CdS–SiS$_2$ quasiternary system at the 30-100 mol% CdS. (From Odin, I.N., and Chukichev, M.V., *Zhurn. neorgan. khimii*, **45**(2), 255, 2000.)

1.131 Silicon–Cadmium–Germanium–Sulfur

SiS_2–CdS–GeS_2. The liquidus surface of this quasiternary system at the 70–100 mol% CdS is shown in Figure 1.20 (Odin and Chukichev 2000). It consists of three fields of the primary crystallization of the solid solutions based on CdS and α- and β-$Cd_4Si_xGe_{1-x}S_6$ solid solutions. The ingots were annealed at 20°C lower than the temperature of the nonvariant equilibrium with liquid participation for 1000 h. The system was investigated through DTA, XRD, and metallography.

1.132 Silicon–Cadmium–Selenium–Sulfur

SiS_2–$CdSe$. This section is a nonquasibinary section of the Si–Cd–Se–S quaternary system (Odin 1996). Solid solutions $CdSe_{1-x}S_x$ and $SiS_{2(1-x)}Se_{2x}$ crystallize primarily in this system. Phases based on β- and α-$Cd_4SiSe_{6-x}S_x$ also crystallize from the CdSe-rich side. The next peritectic reaction: L + $CdSe_{1-x}S_x$ ↔ β-$Cd_4SiSe_{6-x}S_x$ takes place in this system at 80 mol% CdSe. The system was investigated through DTA, metallography, and XRD. The ingots were annealed at 20°C below the temperatures of nonvariant equilibria with liquid participation for 1000 h.

CdS–$SiSe_2$. This section is also a nonquasibinary section of the Si–Cd–Se–S quaternary system (Odin 1996). It was investigated through DTA, metallography, and XRD. The ingots were annealed at 20°C below the temperatures of nonvariant equilibria with liquid participation for 1000 h.

$SiS_2 + 2CdSe$ ↔ $SiSe_2 + 2CdS$. The fields of primary crystallizations of $CdSe_{1-x}S_x$, $SiS_{2-x}Se_x$, β-$Cd_4SiSe_{6-x}S_x$, and α-$Cd_4SiSe_{6-x}S_x$ solid solutions exist on the liquidus surface of this ternary mutual system (Figure 1.21) (Odin 1996). The p_1p_3-, p_2p_4-, and e_1e_2-lines correspond to the peritectic

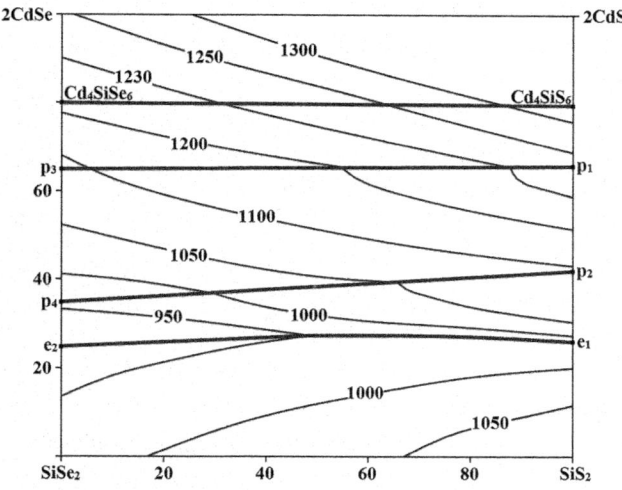

FIGURE 1.21 Liquidus surface of the SiS_2 +2CdSe $\leftrightarrow$ $SiSe_2$ + 2CdS ternary mutual system. (From Odin, I.N., *Zhurn. neorgan. khimii*, **41**(6), 941, 1996.)

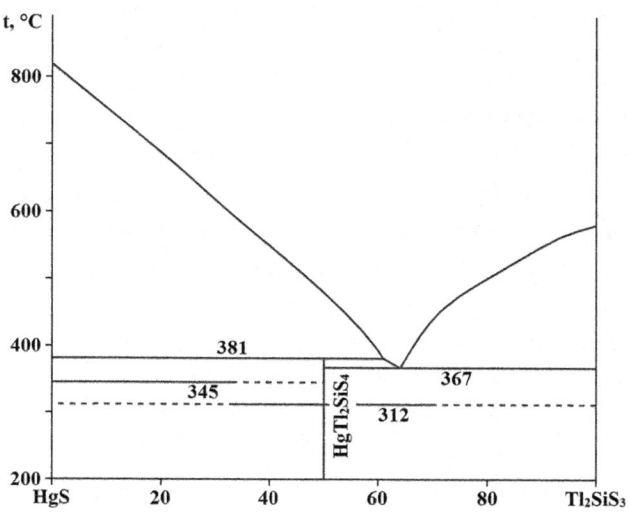

FIGURE 1.22 Phase diagram of the Tl_2SiS_3–HgS system. (From Olekseyuk, I.D. et al., *Nauk. Visnyk Volyns'k. Nats. Univ. im. Lesi Ukrainky. Khim. nauky*, [(17)242], 62, 2012.)

formation of the β-$Cd_4SiSe_{6-x}S_x$ solid solutions, to the polymorphous transformation α-$Cd_4SiSe_{6-x}S_x$ $\rightarrow$ β-$Cd_4SiSe_{6-x}S_x$ with liquid participation, and to the crystallization of $SiS_{2-x}Se_x$ + α-$Cd_4SiSe_{6-x}S_x$ mixtures from the liquid, respectively. These lines are monovariant and no monovariant points exist in this system. The system was investigated through DTA, metallography, and XRD. The ingots were annealed at 20°C below the temperatures of nonvariant equilibria with liquid participation for 1000 h.

1.133 Silicon–Mercury–Thallium–Sulfur

Tl_2SiS_3–HgS. The phase diagram of this system is a eutectic type (Figure 1.22) (Olekseyuk et al. 2012a). The eutectic composition and temperature are 36 mol% HgS and 367°C, respectively. Thermal effects at 343°C correspond to the polymorphous transformation of HgS. Solubility of HgS in Tl_2SiS_3 is not higher than 5 mol% at the eutectic temperature. The **$HgTl_2SiS_4$** quaternary compound is formed in this system. It melts incongruently at 381°C and has polymorphous transformation at 312°C. This system was investigated by DTA, XRD, and measuring of the microhardness. The ingots were annealed at 250°C for 250 h.

SiS_2–HgS–Tl_2S. Isothermal section of this system at 250°C is shown in Figure 1.23 (Olekseyuk et al. 2012a). Two quaternary compounds, $HgTl_2SiS_4$ and **$HgTl_2Si_3S_8$**, exist in this quasiternary system at this temperature. The alloys were annealed at 250°C for 250 h with the next quenching in water.

1.134 Silicon–Mercury–Fluorine–Sulfur

The **$Hg_3SiS_2F_6$** quaternary compound is formed in the Si–Hg–F–S system (Puff et al. 1969). This compound exists in two polymorphic modifications. α-$Hg_3SiS_2F_6$ crystallizes in the

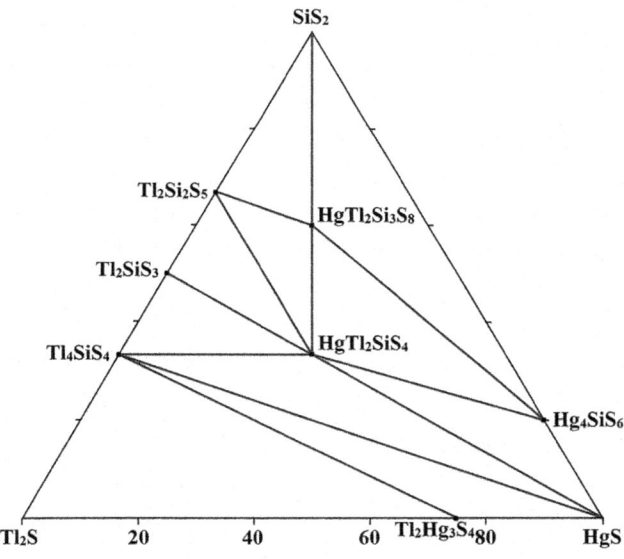

FIGURE 1.23 Isothermal section of the SiS_2–HgS–Tl_2S quasiternary system at 250°C. (From Olekseyuk, I.D. et al. *Nauk. Visnyk Volyns'k. Nats. Univ. im. Lesi Ukrainky. Khim. nauky*, [(17)242], 62, 2012.)

orthorhombic structure with the lattice parameters $a = 1221.9$, $b = 1490.2$, $c = 2233.4$ pm, and the calculated and experimental densities of 6.607 and 6.57 g·cm^{-3}, respectively. β-$Hg_3SiS_2F_6$ crystallizes in the hexagonal structure with the lattice parameters $a = 695.7$, $c = 458.4$ pm, and the calculated and experimental densities of 6.82 and 6.73 g·cm^{-3}, respectively. α-$Hg_3S_2SiF_6$ was obtained at temperatures below 120°C by the interaction of HgS solution in $Hg(CH_3COO)_2$/CH_3COOH and H_2SiF_6, or by the interaction of HgO solution in H_2SiF_6 and H_2S (or thioacetamide, or HgS), or by the interaction of the solution of $Hg_2S(CH_3COO)_2$ and H_2SiF_6, or by the interaction of HgO solution in H_2SiF_6 and CS_2. β-$Hg_3SiS_2F_6$ was prepared by the heating of α-Hg₃SiS₂F₆ at 140°C or by the interaction of HgS with the HgO solution in H₂SiF₆ at 120°C.

1.135 Silicon–Aluminum–Yttrium–Sulfur

The $Al_{0.50}Y_3(Si_{0.50}Al_{0.50})S_7$ quaternary compound, which crystallizes in the hexagonal structure with the lattice parameters $a = 962.1 \pm 0.1$, $c = 593.9 \pm 0.1$ pm, and a calculated density of 3.712 g·cm⁻³, is formed in the Si–Al–Y–S system (Guo et al. 2009a). Single crystals of this compound were obtained by solid-state reactions with KI as a flux. Total mass of the mixture was 500 mg and 400 mg KI additionally. The starting materials were Al, Si, S, Y₂O₃, and B (molar ratio 0.33:2:8:1:6). The mixture was ground into fine powders in an agate mortar, pressed into pellets, and then loaded into quartz tube. The tube was evacuated to 10⁻² Pa and flame-sealed. The sample was placed into a furnace, heated from the room temperature to 300°C in 5 h, kept at 300°C for 10 h, then heated to 650°C in 5 h, kept at 650°C for 10 h, then heated to 950°C in 5 h, kept at 950°C for 10 days, then cooled down to 300°C in 5 days, and powered off. Block crystals were hand-picked under a microscope. They are stable in the presence of the air and water.

1.136 Silicon–Aluminum–Lanthanum–Sulfur

The $Al_{0.44}La_3Si_{0.93}S_7$ quaternary compound, which crystallizes in the hexagonal structure with the lattice parameters $a = 1027.7 \pm 0.2$, $c = 579.3 \pm 0.1$ pm, and a calculated density of 4.256 g·cm⁻³ at 153 K, is formed in the Si–Al–La–S system (Yang and Ibers 2000). The reaction was carried out at 850–900°C for 7 days in unprotected fused silica tubes with the use of a BaBr₂/KBr eutectic flux (molar ratio 1.1:1). A few brown needles were obtained in the reaction of BaS, La, Al, and S (molar ratio 1:1:1:3). $Al_{0.44}La_3Si_{0.93}S_7$ can also be obtained through direct synthesis from La, Al, Si, and S at the same temperature.

1.137 Silicon–Aluminum–Samarium–Sulfur

The $Al_{0.33}Sm_3SiS_7$ quaternary compound, which crystallizes in the hexagonal structure with the lattice parameters $a = 997.9 \pm 0.1$, $c = 567.9 \pm 0.1$ pm, and a calculated density of 4.831 g·cm⁻³, is formed in the Si–Al–Sm–S system (Guo et al. 2009b). Single crystals of this compound were obtained by solid-state reactions with KI as a flux. It was crystallized from the reaction containing Al (0.222 mM), Si (1.353 mM), S (5.427 mM), Sm₂O₃ (0.680 mM), B (4.070 mM), and KI (2.410 mM). The mixture was ground into fine powder in an agate mortar and pressed into pellets, followed by being loaded into a quartz tube. The tube was evacuated at 10⁻² Pa and flame-sealed. The sample was placed into a furnace, heated from room temperature to 300°C in 5 h, kept at 300°C for 10 h, heated to 650°C over 5 h, kept at 650°C for 10 h, heated to 950°C over 5 h, kept at 950°C for 10 days, then cooled down to 300°C in 5 days, and powered off. Yellow block crystals of the title compound were hand-picked under a microscope. They are stable in the presence of air and water and show no signs of appreciable change in crystal quality over several months.

1.138 Silicon–Aluminum–Gadolinium–Sulfur

Two quaternary phases, $Al_{0.57}Gd_3(Si_{0.23}Al_{0.73})S_7$ and $Al_{0.44}Gd_3(Si_{0.70}Al_{0.30})S_7$, are formed in the Si–Al–Gd–S system (Guo et al. 2009a). The first of them crystallizes in the hexagonal structure with the lattice parameters $a = 982.8 \pm 0.2$, $c = 579.3 \pm 0.1$ pm, and a calculated density of 5.064 g·cm⁻³, and the second has an energy gap of 2.34 eV. Single crystals of these phases were prepared in the same way as $Al_{0.50}Y_3(Si_{0.50}Al_{0.50})S_7$ was obtained using Gd₂O₃ instead of Y₂O₃. They are stable in the presence of air and water.

1.139 Silicon–Aluminum–Dysprosium–Sulfur

Three quaternary phases, $Al_{0.38}Dy_3(Si_{0.85}Al_{0.15})S_7$, $Al_{0.5}Dy_3(Si_{0.50}Al_{0.50})S_7$, and $Al_{0.55}Dy_3(Si_{0.34}Al_{0.66})S_7$, which crystallize in the hexagonal structure with the lattice parameters $a = 977.6 \pm 0.1$, 963.6 ± 0.1, and 978.4 ± 0.3, $c = 568.2 \pm 0.1$, 594.4 ± 0.1, and 584.6 ± 0.4 pm, and a calculated density of 5.298, 5.232 and 5.168 g·cm⁻³, respectively, are formed in the Si–Al–Dy–S system (Guo et al. 2009a, 2009b). The first two phases were crystallized from separate reactions containing Al (0.222 mM), Si (1.317 mM), S (5.271 mM), Dy₂O₃ (0.660 mM), B (3.978 mM), and KI (2.410 mM). The preparation process was the same as for $Al_{0.33}Sm_3SiS_7$ quaternary compound. Single crystals of third phase were prepared in the same way as $Al_{0.50}Y_3(Si_{0.50}Al_{0.50})S_7$ was obtained using Dy₂O₃ instead of Y₂O₃. They are stable in the presence of air and water.

1.140 Silicon–Gallium–Thallium–Sulfur

The $GaTlSiS_4$ quaternary compound, which melts at 815°C and crystallizes in the orthorhombic structure with the lattice parameters $a = 1158 \pm 1$, $b = 1653.5 \pm 0.6$, $c = 703.6 \pm 0.3$ pm, and the calculated and experimental densities of 4.25 and 4.24 g·cm⁻³, respectively, is formed in the Si–Ga–Tl–S system (Nakamura et al. 1984b). Synthesis of this compound was carried out by the conventional evacuated silica tube method. The stoichiometric mixture of the elements (Tl was used as TlS) was sealed in an evacuated silica glass tube and heated in a furnace at 940°C for 3 days. The melt was then shaken vigorously a number of times to ensure thorough mixing, then it was annealed at 400°C for 3 days, and slowly cooled to room temperature by turning off the furnace power. The title compound with a cream-white color was obtained.

1.141 Silicon–Gallium–Lead–Sulfur

SiS_2–$PbGa_2S_4$. The phase diagram of this system, constructed through DTA, XRD, and metallography, is presented in Figure 1.24 (Piskach et al. 2019). The eutectic contains 60 mol% SiS₂ and crystallizes at 685°C. At 400°C PbGa₂S₄ dissolves 10 mol% SiS₂ and SiS₂ dissolves not higher than 5 mol% PbGa₂S₄. The Ga_2PbSiS_6 quaternary compound, which melts incongruently at 702°C and has a polymorphic transformation at 557°C, is formed

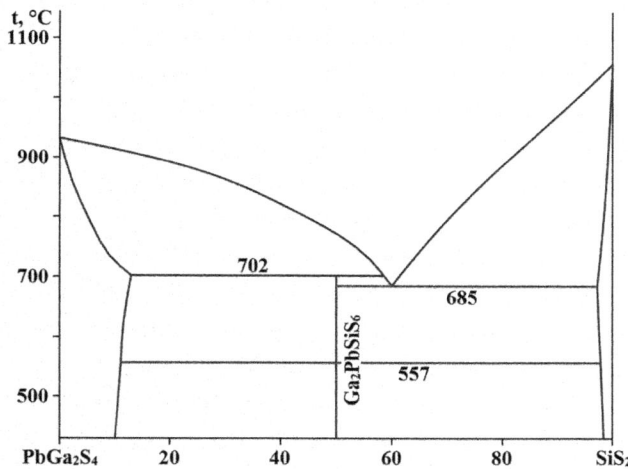

FIGURE 1.24 Phase diagram of the SiS_2–$PbGa_2S_4$ system. (From Piskach, L., et al., *Persp. tekhnol. ta prylady*, (14), 109, 2019.)

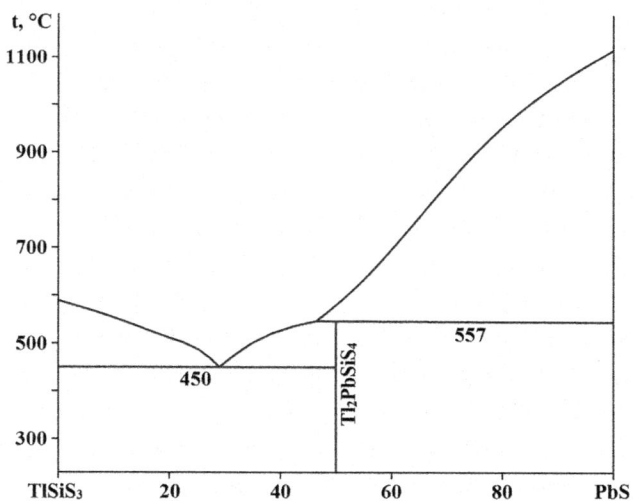

FIGURE 1.25 Phase diagram of the Tl_2SiS_3–PbS system. (From Mozolyuk, M.Yu., et al., *Mater. Chem. Phys.*, **195**, 132, 2017.)

in this system. The alloys for the investigations were annealed at 400°C for 500 h, with the next quenching in cold water.

1.142 Silicon–Indium–Thallium–Sulfur

The **InTlSiS₄** quaternary compound, which decomposes at 300°C and crystallizes in the orthorhombic structure with the lattice parameters $a = 1174 \pm 1$, $b = 1703.8 \pm 0.7$, $c = 728.5 \pm 0.3$ pm, and the calculated and experimental densities of 4.34 and 4.37 g·cm⁻³, respectively, is formed in the Si–In–Tl–S system (Nakamura et al. 1984a, 1984b). Synthesis of this compound was carried out by the conventional evacuated silica tube method. The stoichiometric mixture of the elements (Tl and In were used as TlS and In_2S_3, respectively) was sealed in an evacuated silica glass tube and heated in a furnace at 750°C for one day. The melt was then shaken vigorously a number of times to ensure thorough mixing, then it was annealed at 460°C for 5 days, and slowly cooled to room temperature by turning off the furnace power. The title compound with reddish brown color was obtained.

1.143 Silicon–Indium–Lanthanum–Sulfur

The **(In$_{2/3}$□$_{4/3}$)La₆Si₂S₁₄** quaternary compound, which crystallize in the hexagonal structure with the lattice parameters $a = 1034.4 \pm 0.5$ and 576.2 ± 0.3, is formed in the Si–In–La–S system (Collin et al. 1973). Yellow single crystals of this compound were grown by heating a mixture of La_2S_3, In_2S_3, Si, and S in an evacuated quartz ampoule at 1250°C.

1.144 Silicon–Indium–Samarium–Sulfur

The **In$_{0.33}$Sm₃SiS₇** quaternary compound, which crystallizes in the hexagonal structure, is formed in the Si–In–Sm–S system (Guo et al. 2009a). Single crystals of this compound were obtained by solid-state reactions with KI as a flux. The total

mass of the mixture was 500 mg and 400 mg KI additionally. The starting materials were In_2O_3, Si, S, Sm_2O_3, and B (molar ratio 1:2:8:1:10). The mixture was ground into fine powders in an agate mortar, pressed into pellets, and then loaded into a quartz tube. The tube was evacuated to 10^{-2} Pa and flame-sealed. The sample was placed into a furnace, heated from room temperature to 300°C in 5 h, kept at 300°C for 10 h, then heated to 650°C in 5 h, kept at 650°C for 10 h, then heated to 950°C in 5 h, kept at 950°C for 10 days, then cooled down to 300°C in 5 days, and powered off. Block crystals were hand-picked under a microscope. They are stable in the presence of air and water.

1.145 Silicon–Thallium–Lead–Sulfur

Tl₂SiS₃–PbS. The phase diagram of this system, constructed through DTA and XRD, is presented in Figure 1.25 (Olekseyuk et al. 2012a; Mozolyuk et al. 2017). The eutectic contains 29 mol% PbS and crystallizes 450°C. The **Tl₂PbSiS₄** quaternary compound, which melts incongruently at 545°C (peritectic point is at 46 mol% PbS) and crystallizes in the monoclinic structure with the lattice parameters $a = 881.41 \pm 0.04$, $b = 901.50 \pm 0.05$, $c = 1043.83 \pm 0.05$ pm, $\beta = 94.490 \pm 0.004°$, and a calculated density of 6.2030 ± 0.0009 g·cm⁻³, is formed in this system.

SiS₂–Tl₂S–PbS. The isothermal section of this quasiternary system at 250°C is shown in Figure 1.26 (Olekseyuk et al. 2012a; Mozolyuk et al. 2017). One more quaternary compound, **Tl₂PbSi₃S₈**, exists in this system. The alloys for investigations were annealed at 250°C for 250 h.

1.146 Silicon–Yttrium–Lead–Sulfur

SiS₂–Y₂S₃–PbS. The isothermal section of this quasiternary system at 500°C is shown in Figure 1.27 (Marchuk et al. 2008c). The **Y₂Pb(SiS₄)₂** quaternary compound, which crystallizes in the tetragonal structure with the lattice parameters $a =$

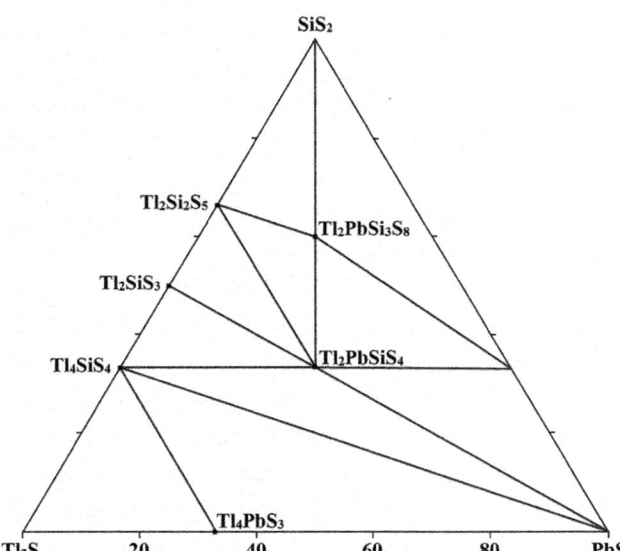

FIGURE 1.26 Isothermal section of the SiS_2–Tl_2S–PbS quasiternary system at 250°C. (From Mozolyuk, M.Yu., et al., *Mater. Chem. Phys.*, **195**, 132, 2017.)

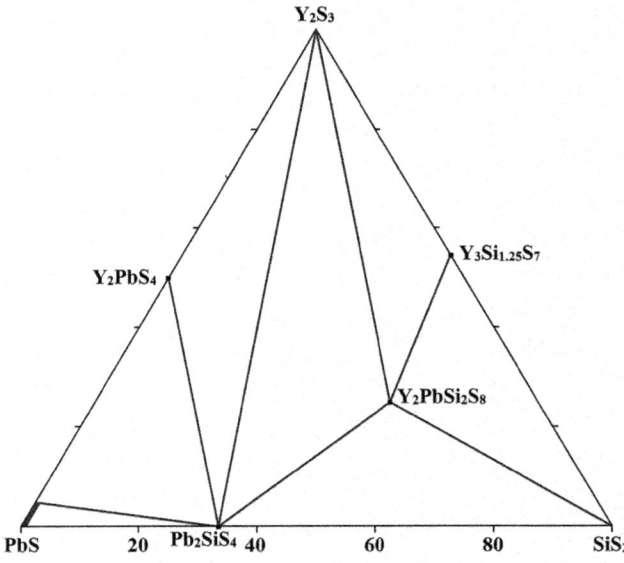

FIGURE 1.27 Isothermal section of the SiS_2–Y_2S_3–PbS quasiternary system at 500°C. (From Marchuk, O.V., et al., *Nauk. Visnyk Volyns'k. Nats. Univ. im. Lesi Ukrainky. Ser. Khim. nauky*, (13), 24, 2008.)

884.33 ± 0.02, $c = 2597.45 \pm 0.09$ pm, and a calculated density of 3.944 g·cm⁻³ (Daszkiewicz et al. 2012; Marchuk 2019) [$a = 880.58 \pm 0.02$ and $c = 2586.8 \pm 0.1$ pm (Marchuk et al. 2008c)], is formed in this system. The alloys for investigations were annealed at 500°C for 500 h with the next quenching in water (Marchuk et al. 2008c).

Powder sample of $Y_2Pb(SiS_4)_2$ was prepared by sintering the elemental constituents in an evacuated quartz tube (Daszkiewicz et al. 2012). The synthesis was carried out in a furnace. The ampoule was first heated with a rate of 30°C·h⁻¹ up to 1150°C, and then kept at this temperature for 3 h. Afterwards, the sample was cooled slowly (10°C·h⁻¹) down to 500°C and annealed at this temperature for 720 h. Subsequently, the ampoule was quenched in the air.

1.147 Silicon–Yttrium–Oxygen–Sulfur

Two quaternary compounds, **Y_2SiO_4S** and **$Y_4S_3(Si_2O_7)$**, are formed in the Si–Y–O–S system. The first crystallizes in the orthorhombic structure with the lattice parameters $a = 604.62 \pm 0.08$, $b = 689.76 \pm 0.09$, $c = 1065.58 \pm 0.13$ pm, and a calculated density of 4.513 g·cm⁻³ (Kiryakov et al. 2018). $Y_4S_3(Si_2O_7)$ crystallizes in the tetragonal structure with the lattice parameters $a = 1167.06 \pm 0.16$, $c = 1358.73 \pm 0.19$ pm, a calculated density of 4.451 g·cm⁻³ at 100 K, and an energy gap of ≈ 3.6 eV (Koscielski and Ibers 2011; Tarasenko et al. 2018).

To prepare Y_2SiO_4S, Y_2O_2S (0.8 mM) and Si (0.8 mM) were ground 2–3 times with acetone in a mortar and loaded in a quartz tube with 500 mg of CsCl flux (Kiryakov et al. 2018). The tube was evacuated to a residual pressure of 1.3 Pa and sealed. The synthesis was carried out in a muffle furnace. The tube was heated to 1100°C and kept for 12 h. The product obtained was washed with water from CsCl, then with alcohol, and dried in the air. Colorless prismatic crystals were obtained.

$Y_4S_3(Si_2O_7)$ was synthesized when amorphous SiO_2 (7.5 mM), Si (2.5 mM), S (5 mM), Y_2O_2S (10 mM), and CsCl (3 g) as a flux were ground in a mortar and were placed in a glassy carbon crucible to prevent variation of stoichiometry due to possible reactivity of silica ampoule (Tarasenko et al. 2018). The crucible was sealed under vacuum in silica ampoule and was heated to 1100°C in 2 h, kept there for 12 h, and then cooled to 25°C in 5 h. The reaction product was washed with distilled water to remove CsCl. The crystals were colorless well-shaped strangulated octahedrons.

1.148 Silicon–Yttrium–Chromium–Sulfur

The **$Y_6CrSi_2S_{14}$** quaternary compound, which crystallizes in the hexagonal structure with the lattice parameters $a = 974 \pm 2$ and $c = 568 \pm 2$ pm, is formed in the Si–Y–Cr–S system (Michelet and Flahaut 1969). It was obtained in a sealed silica ampoule at around 1000°C by the interaction of CrS and Y_2S_3 with Si and the desired amount of S.

1.149 Silicon–Yttrium–Manganese–Sulfur

The **$Y_6MnSi_2S_{14}$** quaternary compound, which crystallizes in the hexagonal structure with the lattice parameters $a = 985 \pm 2$ and $c = 563 \pm 2$ pm, is formed in the Si–Y–Mn–S system (Michelet and Flahaut 1969). It was synthesized in a sealed silica ampoule at around 1000°C by the interaction of MnS and Y_2S_3 with Si and the desired amount of S.

1.150 Silicon–Yttrium–Iron–Sulfur

The **$Y_6FeSi_2S_{14}$** quaternary compound, which crystallizes in the hexagonal structure with the lattice parameters $a = 980 \pm 2$ and $c = 565 \pm 2$ pm, is formed in the Si–Y–Fe–S system

(Michelet and Flahaut 1969). It was synthesized in a sealed silica ampoule at around 1000°C by the interaction of FeS and Y_2S_3 with Si and the desired amount of S.

1.151 Silicon–Yttrium–Palladium–Sulfur

The $Y_6PdSi_2S_{14}$ quaternary compound, which crystallizes in the hexagonal structure with the lattice parameters $a = 978.91 \pm 0.03$, $c = 568.40 \pm 0.04$ pm, and a calculated density of 4.030 g·cm^{-3}, is formed in the Si–Y–Pd–S system (Iyer et al. 2017b). It was synthesized from a stoichiometric mixture of the elements and a total mass of 0.2 g with the use of KI flux. The mixture was pressed into pellets and loaded into a fused silica tube, which was evacuated and sealed. The tube was heated at 1050°C for 4 days and cooled to 600°C over 4 days, and then the furnace was turned off.

1.152 Silicon–Lanthanum–Lutetium–Sulfur

The $La_6(Lu_{2/3}\square_{4/3})Si_2S_{14}$ quaternary compound is formed in the Si–La–Lu–S system (Collin et al. 1973). Single crystals of this compound were grown by heating a mixture of La_2S_3, Lu_2S_3, Si, and S in an evacuated quartz ampoule at 1250°C.

1.153 Silicon–Lanthanum–Tin–Sulfur

The $La_6(Sn_{1/2}\square_{3/2})Si_2S_{14}$ quaternary compound, which crystallizes in the hexagonal structure with the lattice parameters $a = 1032.9 \pm 0.5$ and $c = 578.4 \pm 0.2$ pm, is formed in the Si–La–Sn–S system (Collin et al. 1973). Single crystals of this compound were grown by heating a mixture of La_2S_3, SnS, Si, and S in an evacuated quartz ampoule at 1250°C.

1.154 Silicon–Lanthanum–Lead–Sulfur

SiS_2–La_2S_3–PbS. The isothermal section of this quasiternary system at 500°C is shown in Figure 1.28 (Marchuk and Gulay 2012). Seven single-phase, 13 two-phase, and 6 three-phase regions exist in the system. The solubility on the base of the starting components is insignificant (≈1–2 mol%). The highest solubility is observed on the basis of La_2PbS_4: a solid solution of composition $La_{2+2/3x}Pb_{1-x}S_4$ ($x = 0$–0.86) is localized along the PbS–La_2S_3 system. The $La_2Pb(SiS_4)_2$ quaternary compound, which crystallizes in the trigonal structure with the lattice parameters $a = 905.22 \pm 0.13$ and $c = 2696.4 \pm 0.5$ pm (Gulay et al. 2010; Marchuk 2019)], is formed in this system. The alloys for investigations were annealed at 500°C for 500 h, with the next quenching in water (Marchuk and Gulay 2012).

$La_2Pb(Si_2S_4)_2$ was prepared by sintering the elemental constituents in a stoichiometric ratio in an evacuated quartz ampoule in a tube furnace (Gulay et al. 2010). The ampoule was heated at a rate of 12–30°C·h^{-1} to a maximum temperature of 1100–1150°C and kept at this temperature for 4 h. It was then cooled slowly (12–30°C·h^{-1}) to 500°C and annealed at this

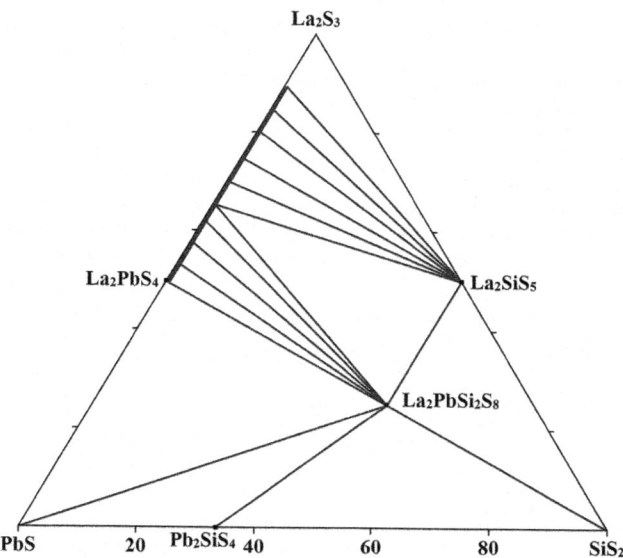

FIGURE 1.28 Isothermal section of the SiS_2–La_2S_3–PbS quasiternary system at 500°C. (From Marchuk, O.V., and Gulay, L.D., *Nauk. Visnyk Volyns'k. Nats. Univ. im. Lesi Ukrainky. Ser. Khim. nauky*, (17), 93, 2012.)

temperature for 500 h. After annealing, the ampoule with the sample was quenched in cold water.

1.155 Silicon–Lanthanum–Zirconium–Sulfur

The $La_6(Zr_{1/2}\square_{3/2})Si_2S_{14}$ quaternary compound is formed in this system (Collin et al. 1973). Single crystals of this compound were grown by heating a mixture of La_2S_3, ZrS_2, Si, and S in an evacuated quartz ampoule at 1250°C.

1.156 Silicon–Lanthanum–Antimony–Sulfur

The $La_3Sb_{0.33}SiS_7$ quaternary compound, which crystallizes in the hexagonal structure with the lattice parameters $a = 1037.0 \pm 0.3$, $c = 575.5 \pm 0.2$ pm, and a calculated density of 4.396 g·cm^{-3}, is formed in the Si–La–Sb–S system (Zhao 2015). To synthesize this compound, a stoichiometric mixture of the elements was loaded into a silica tube, which was evacuated and sealed. The sample was heated to 950°C for 24 h and kept at that temperature for 4 days. After that, it was cooled (2.5°C·h^{-1}) to 200°C, then the furnace was turned off. The red block crystals were obtained. The elements were stored in a N_2-filled glove box (moisture and oxygen level was less than 0.1 ppm), and all manipulations were performed inside the glove box.

1.157 Silicon–Lanthanum–Oxygen–Sulfur

The $La_4S_3(Si_2O_7)$ quaternary compound, which crystallizes in the tetragonal structure with the lattice parameters $a = 1209.8 \pm 0.3$, $c = 1437.9 \pm 0.5$ pm, and a calculated density of 5.18 g·cm^{-3}, is formed in the Si–La–O–S system (Zeng et al. 1999).

For preparing this compound, the starting materials La_2S_3 (0.385 mM), Na_2S (0.384 mM), Ge (0.386 mM), and S (0.780 mM) were ground thoroughly under a blanket of N_2 in a dry box and then pressed into a pellet, which was subsequently sealed in an evacuated quartz tube. It was heated at a rate of $15°C·h^{-1}$ to $500°C$ and kept for 27 h, then brought to $800°C$ for 120 h; thereafter, the furnace was shut off. A mixture of the above precursor and 0.75 g NaBr flux was ground together and pressed into a pellet, which was then sealed in an evacuated carbon-coated quartz tube. The tube was heated at a rate of $20°C·h^{-1}$ to $660°C$, soaked for 24 h, raised at a rate of $25°C·h^{-1}$ to $980°C$, held for 144 h, then cooled at a rate of $2°C·h^{-1}$ to $300°C$, and the furnace was shut off. The resultant mixture was washed with distilled water, and transparent yellow crystals were selected mechanically from the insoluble residual.

1.158 Silicon–Lanthanum–Chromium–Sulfur

The **$La_6CrSi_2S_{14}$** quaternary compound, which crystallizes in the hexagonal structure with the lattice parameters $a = 1034 ± 2$ and $c = 576 ± 2$ pm, is formed in the Si–La–Cr–S system (Michelet and Flahaut 1969). It was obtained in a sealed silica ampoule at around $1000°C$ by interaction of CrS and La_2S_3 with Si and the desired amount of S.

1.159 Silicon–Lanthanum–Chlorine–Sulfur

The **$La_3Cl(SiS_4)_2$** quaternary compound, which crystallizes in the monoclinic structure with the lattice parameters $a = 1567.2 ± 0.3$, $b = 777.8 ± 0.2$, $c = 1101.5 ± 0.2$ pm, $β = 96.88 ± 0.02°$, and a calculated density of 3.811 g·cm⁻³, is formed in the Si–La–Cl–S system (Hatscher and Urland 2002d). Single crystals of this compound were obtained by reaction of the elements. La chips, S and Si powders (molar ratio 1:3.3:1) were loaded into fused silica tube in a glove box under Ar atmosphere. About 0.3 mM (molar ratio approximately 0.3) of Cl_2 were condensed in the evacuated tube which were sealed afterwards. This tube was placed in a furnace and exposed to a temperature gradient of $1000–850°C$ for 10 days. Then the ampoule was quenched in water. In the middle of the tube air stable crystals of $La_3Cl(SiS_4)_2$ were obtained.

1.160 Silicon–Lanthanum–Bromine–Sulfur

The **$La_3Br(SiS_4)_2$** quaternary compound, which crystallizes in the monoclinic structure with the lattice parameters $a = 1583.3 ± 0.4$, $b = 783.0 ± 0.1$, $c = 1098.2 ± 0.3$ pm, $β = 97.33 ± 0.03°$, and a calculated density of 3.981 g·cm⁻³, is formed in the Si–La–Br–S system (Hatscher and Urland 2002c). Single crystals of this compound can be obtained from the elements. La chips, S and Si powders were weight in a quartz vial under a protective gas in a molar ratio of 1:3.25:1.5. This was evacuated under liquid N_2 cooling, and about 50 mg of Br_2 were condensed in. The ampoule was removed in vacuum and then subjected to a temperature gradient of $1000°C$ to $800°C$ for 10 days in a

two-zone furnace. Air-stable brown transparent crystals of $La_3Br(SiS_4)_2$ were obtained in the middle of the ampoule.

1.161 Silicon–Lanthanum–Iodine–Sulfur

The **$La_3I(SiS_4)_2$** quaternary compound, which crystallizes in the monoclinic structure with the lattice parameters $a = 1611.0 ± 0.3$, $b = 791.75 ± 0.12$, $c = 1093.1 ± 0.2$ pm, $β = 97.94 ± 0.02°$, and a calculated density of 4.119 g·cm⁻³ (Hatscher and Urland 2002a) [$a = 1609.20 ± 0.03$, $b = 791.17 ± 0.01$, $c = 1092.96 ± 0.02$ pm and $β = 97.959 ± 0.001°$ (Riccardi et al. 1999)], is formed in the Si–La–I–S system. Single crystals of this compound can be obtained from the elements (Hatscher and Urland 2002a). La chips, S and Si powders, and iodine were loaded into a quartz glass tube (molar ratio 1:3.3:1.03:0.4). The tube was evacuated, sealed, and heated for 10 days in a temperature gradient of $1000°C$ to $800°C$. After cooling, the molten reaction mixture was finely ground and once more sealed in an evacuated quartz glass tube with some iodine (molar ratio approximately 0.15). The ampoule was placed in a furnace, following the temperature program described above. This procedure had to be repeated a second time to obtain the air stable white crystals of $La_3I(SiS_4)_2$.

1.162 Silicon–Lanthanum–Manganese–Sulfur

The **$La_6MnSi_2S_{14}$** quaternary compound, which crystallizes in the hexagonal structure with the lattice parameters $a = 1035 ± 2$ and $c = 573 ± 2$ pm, is formed in the Si–La–Mn–S system (Michelet and Flahaut 1969; Collin and Laruelle 1970). It was synthesized in a sealed silica ampoule at $1000–1350°C$ by the interaction of MnS and La_2S_3 with SiS_2 or Si and the desired amount of S.

1.163 Silicon–Lanthanum–Iron–Sulfur

The **$La_6FeSi_2S_{14}$** quaternary compound, which crystallizes in the hexagonal structure with the lattice parameters $a = 1032 ± 2$ and $c = 576 ± 2$ pm, is formed in the Si–La–Fe–S system (Michelet and Flahaut 1969). It was synthesized in a sealed silica ampoule at around $1000°C$ by the interaction of FeS and La_2S_3 with Si and the desired amount of S.

1.164 Silicon–Lanthanum–Cobalt–Sulfur

The **$La_6CoSi_2S_{14}$** quaternary compound, which crystallizes in the hexagonal structure with the lattice parameters $a = 1029.0 ± 0.2$ and $c = 576.3 ± 0.2$ pm, is formed in the Si–La–Co–S system (Jin et al. 1985). Crystal structure of this compound is hexagonal within the pressure range from 0.1 MPa to 5.36 GPa (a and c change from $1027.9 ± 0.2$ and $574.9 ± 0.1$ pm to $1003.6 ± 0.4$ and $562.7 ± 0.4$ pm, respectively) (Meng et al. 1986). At 8.44 GPa this compound crystallizes in the orthorhombic structure with the lattice parameters $a = 496.2 ± 0.3$, $b = 557.1 ± 0.2$,

$c = 859.4 \pm 0.9$ pm. The bulk modulus of $La_6CoSi_2S_{14}$ is 79.21 ± 0.04 GPa. Single crystals of the title compound were obtained by a flux growth method using $LaCl_3$ as a flux (Jin et al. 1985).

1.165 Silicon–Lanthanum–Nickel–Sulfur

The $La_6NiSi_2S_{14}$ quaternary compound, which crystallizes in the hexagonal structure with the lattice parameters $a = 1029.3 \pm 0.1$ and $c = 577.4 \pm 0.1$ pm, is formed in the Si–La–Ni–S system (Jin et al. 1985). The crystal structure of this compound is hexagonal within the pressure range from 0.1 MPa to 4.71 GPa (a and c change from 1030.7 ± 0.2 and 576.2 ± 0.1 pm to 1011.0 ± 0.5 and 564.9 ± 0.2 pm, respectively) (Meng et al. 1986). The bulk modulus of $La_6CoSi_2S_{14}$ is 75.50 ± 0.05 GPa. Single crystals of the title compound were obtained by flux growth method using $LaCl_3$ as a flux (Jin et al. 1985).

1.166 Silicon–Lanthanum–Palladium–Sulfur

The $La_6PdSi_2S_{14}$ quaternary compound, which crystallizes in the monoclinic structure with the lattice parameters $a = 1000.53 \pm 0.09$, $b = 575.76 \pm 0.06$, $c = 1033.51 \pm 0.10$ pm, $\beta = 117.932 \pm 0.003°$, a calculated density of 4.548 g·cm^{-3} at 100 K, and an energy gap of 1.41 eV, is formed in the Si–La–Pd–S system (Akopov et al. 2022b). To obtain this compound, first La–Pd–Si precursor was prepared via arc-melting using pieces of La, Pd, and Si. Sample with a total mass of 0.6-1 g were weighed out in a ratio of La/Pd/Si = 6:1:2.05. Single crystals were grown using arc-melted precursor, S powder, and KI flux. The ratio of (precursor + S)/KI was kept at 1:30 by mass. The reactants and KI were added to a silica ampoule, which was sealed under vacuum. The reaction products were washed with deionized water to remove KI flux, the sample was then vacuum filtered and allowed to dry. Bulk powder of the material was synthesized using the appropriate arc-melted precursor and sulfur powder. Sample was prepared by loading the silicide binary and sulfur (1:14 ratio) in a fused silica ampoule, which was then flame sealed under vacuum. For both crystals and bulk powder, the samples were heated over 12 h to 1000–1050°C, dwelled for 72–120 h and then cooled to room temperature over 8 h.

1.167 Silicon–Cerium–Lead–Sulfur

The $Ce_2Pb(SiS_4)_2$ quaternary compound, which crystallizes in the tetragonal structure with the lattice parameters $a = 900.30 \pm 0.07$, $c = 2676.5 \pm 0.3$ pm, and a calculated density of 4.243 g·cm^{-3}, is formed in the Si–Ce–Pb–S system (Daszkiewicz et al. 2012; Marchuk 2019). Powder sample of the title compound was prepared by sintering the elemental constituents in an evacuated quartz tube. The synthesis was carried out in a furnace. The ampoule was first heated with a rate of 30°C·h^{-1} up to 1150°C, and then kept at this temperature for 3 h. Afterwards, the sample was cooled slowly (10°C·h^{-1}) down to 500°C and annealed at this temperature for 720 h. Subsequently, the ampoule was quenched in the air.

1.168 Silicon–Cerium–Oxygen–Sulfur

The $Ce_4S_3(Si_2O_7)$ quaternary compound, which crystallizes in the tetragonal structure with the lattice parameters $a = 1205.43 \pm 0.08$ and $c = 1423.51 \pm 0.09$ pm, is formed in the Si–Ce–O–S system (Hartenbach and Schleid 2002a). Colorless transparent single crystals of this compound were obtained as almost regular octahedra by the reaction of CeO_2 and Ce with S and SiO_2 (molar ratio 3:1:6:4) in an excess of molten $CeCl_3$ as a fluxing agent at 850°C for seven days in an evacuated silica ampoule. The crystals of the title compound emerged as main product (along with traces of CeOCl and Ce_2O_2S as by-products) and remain stable to air and water.

1.169 Silicon–Cerium–Chlorine–Sulfur

The $Ce_3Cl(SiS_4)_2$ quaternary compound, which crystallizes in the monoclinic structure with the lattice parameters $a = 1559.4 \pm 0.3$, $b = 770.2 \pm 0.2$, $c = 1096.9 \pm 0.2$ pm, $\beta = 97.07 \pm 0.02°$, and a calculated density of 3.904 g·cm^{-3} (Hatscher and Urland 2002d) [$a = 1561.87 \pm 0.06$, $b = 771.03 \pm 0.03$, $c = 1098.41 \pm 0.03$ pm, $\beta = 96.970 \pm 0.002°$, and an energy gap of 2.77 eV (Riccardi et al. 1999)], is formed in the Si–Ce–Cl–S system. Single crystals of this compound were obtained by reaction of the elements (Hatscher and Urland 2002d). Ce chips, S and Si powders (molar ratio 1:3.3:1) were loaded into fused silica tube in a glove box under Ar atmosphere. About 0.3 mM (molar ratio approximately 0.3) of Cl_2 were condensed in the evacuated tube, which was sealed afterward. This tube was placed in a furnace and exposed to a temperature gradient of 1000°C-850°C for 10 days. Then the ampoule was quenched in water. In the middle of the tube, air stable crystals of $Ce_3Cl(SiS_4)_2$ were obtained. $CeCl_3$ can also be used as a chlorine source (Riccardi et al. 1999)].

1.170 Silicon–Cerium–Bromine–Sulfur

The $Ce_3Br(SiS_4)_2$ quaternary compound, which crystallizes in the monoclinic structure with the lattice parameters $a = 1570.4 \pm 0.3$, $b = 776.5 \pm 0.2$, $c = 1092.2 \pm 0.2$ pm, $\beta = 97.28 \pm 0.02°$, and a calculated density of 4.087 g·cm^{-3} (Hatscher and Urland 2002c) [$a = 1571.50 \pm 0.04$, $b = 776.43 \pm 0.02$, $c = 1093.44 \pm 0.03$ pm, $\beta = 97.295 \pm 0.001°$, and an energy gap of 2.82 eV (Riccardi et al. 1999)], is formed in the Si–Ce–Br–S system. Single crystals of this compound were obtained by reaction of the elements (Hatscher and Urland 2002c). Ce chips, S and Si powders were weight in a quartz vial under a protective gas in a molar ratio of 1:3.25:1.5. This was evacuated under liquid N_2 cooling, and about 50 mg of Br_2 were condensed. The ampoule was removed in vacuum and then subjected a temperature gradient of 1000°C to 800°C for 10 days in a two-zone furnace. Air-stable greenish-green transparent crystals of $Ce_3Br(SiS_4)_2$ were obtained in the middle of the ampoule. $CeBr_3$ can also be used as a bromine source (Riccardi et al. 1999)].

1.171 Silicon–Cerium–Iodine–Sulfur

The **Ce$_3$I(SiS$_4$)$_2$** quaternary compound, which crystallizes in the monoclinic structure with the lattice parameters $a = 1596.34 \pm 0.05$, $b = 785.02 \pm 0.02$, $c = 1086.64 \pm 0.03$ pm, $\beta = 97.931 \pm 0.002°$, and a calculated density of 4.201 g·cm^{-3} (Gauthier et al. 1998a,b) [$a = 1597.35 \pm 0.05$, $b = 785.59 \pm 0.02$, $c = 1087.34 \pm 0.03$ pm, $\beta = 97.935 \pm 0.002°$, and an energy gap of 2.92 eV (Riccardi et al. 1999)], is formed in the Si–Ce–I–S system.

The title compound was prepared from a mixture of Si (3.15 mM), S (5.775 mM), I$_2$ (0.525 mM), and Ce$_2$S$_3$ (1.575 mM), handled under inert atmosphere (Gauthier et al. 1998a, b). The starting materials were placed in an evacuated quartz tube, then sealed and heated at 700°C for 4 days. The sample was then cooled slowly to room temperature, finely ground, and heated to 900°C. This two-step reaction led to air-stable yellowish crystals of Ce$_3$I(SiS$_4$)$_2$. The SiS$_2$ excess was eliminated by sublimation at the coldest end of the reaction tube.

1.172 Silicon–Cerium–Manganese–Sulfur

The **Ce$_6$MnSi$_2$S$_{14}$** quaternary compound, which crystallizes in the hexagonal structure with the lattice parameters $a = 1025 \pm 2$ and $c = 572 \pm 2$ pm, is formed in the Si–Ce–Mn–S system (Michelet and Flahaut 1969). It was synthesized in a sealed silica ampoule at around 1000°C by interaction of MnS and Ce$_2$S$_3$ with Si and the desired amount of S.

1.173 Silicon–Praseodymium–Lead–Sulfur

The **Pr$_2$Pb(SiS$_4$)$_2$** quaternary compound, which crystallizes in the tetragonal structure with the lattice parameters $a = 897.44 \pm 0.07$, $c = 2664.0 \pm 0.2$ pm, and a calculated density of 4.299 g·cm^{-3}, is formed in the Si–Pr–Pb–S system (Daszkiewicz et al. 2012; Marchuk 2019). A powder sample of this compound was prepared by sintering the elemental constituents in an evacuated quartz tube. The synthesis was carried out in a furnace. The ampoule was first heated with a rate of 30°C·h^{-1} up to 1150°C, and then kept at this temperature for 3 h. Afterwards, the sample was cooled slowly (10°C·h^{-1}) down to 500°C and annealed at this temperature for 720 h. Subsequently, the ampoule was quenched in the air.

1.174 Silicon–Praseodymium–Antimony–Sulfur

The **Pr$_3$Sb$_{0.33}$SiS$_7$** quaternary compound, which crystallizes in the hexagonal structure with the lattice parameters $a = 1022.4 \pm 0.4$, $c = 571.1 \pm 0.3$ pm, and a calculated density of 4.596 g·cm^{-3}, is formed in the Si–Pr–Sb–S system (Zhao 2015). To synthesize this compound, a stoichiometric mixture of the elements was loaded into a silica tube, which was evacuated and sealed. The sample was heated to 950°C for 24 h and kept at that temperature for 4 days. After that it was cooled (2.5°C·h^{-1}) to 200°C, then the furnace was turned off. The red block crystals were obtained. The elements were stored in a N$_2$-filled

glove box (moisture and oxygen level was less than 0.1 ppm), and all manipulations were performed inside the glove box.

1.175 Silicon–Praseodymium–Oxygen–Sulfur

The **Pr$_4$S$_3$(Si$_2$O$_7$)** quaternary compound, which crystallizes in the tetragonal structure with the lattice parameters $a = 1201.6 \pm 0.1$, $c = 1412.0 \pm 0.2$ pm, and a calculated density of 5.395 g·cm^{-3}, is formed in the Si–Pr–O–S system (Sieke and Schleid 2000). For synthesizing this compound, Pr, Pr$_6$O$_{11}$, and SiO$_2$ were brought to reaction with S in a suitable molar ratio (850°C, 7 days) in an evacuated silica tube. By using NaCl as a flux, Pr$_4$S$_3$(Si$_2$O$_7$) crystallizes as pale green, transparent single crystals with the appearance of slightly compressed octahedra.

1.176 Silicon–Praseodymium–Chlorine–Sulfur

The **Pr$_3$Cl(SiS$_4$)$_2$** quaternary compound, which crystallizes in the monoclinic structure with the lattice parameters $a = 1555.9 \pm 0.3$, $b = 764.2 \pm 0.1$, $c = 1093.2 \pm 0.2$ pm, $\beta = 97.40 \pm 0.02°$, and a calculated density of 3.972 g·cm^{-3}, is formed in the Si–Pr–Cl–S system (Hatscher and Urland 2002d). Single crystals of this compound were obtained by reaction of the elements. Pr chips, S and Si powders (molar ratio 1:3.3:1) were loaded into fused silica tube in a glove box under an Ar atmosphere. About 0.3 mM (molar ratio approximately 0.3) of Cl$_2$ were condensed in the evacuated tube, which was sealed afterwards. This tube was placed in a furnace and exposed to a temperature gradient of 1000–850°C for 10 days. Then the ampoule was quenched in water. In the middle of the tube air-stable crystals of Pr$_3$Cl(SiS$_4$)$_2$ were obtained.

1.177 Silicon–Praseodymium–Bromine–Sulfur

The **Pr$_3$Br(SiS$_4$)$_2$** quaternary compound, which crystallizes in the monoclinic structure with the lattice parameters $a = 1562.6 \pm 0.3$, $b = 770.1 \pm 0.2$, $c = 1088.9 \pm 0.2$ pm, $\beta = 97.50 \pm 0.02°$, and a calculated density of 4.168 g·cm^{-3} (Hatscher and Urland 2002c), is formed in the Si–Pr–Br–S system. Single crystals of this compound were obtained by reaction of the elements. Pr chips, S and Si powders were weight in a quartz vial under a protective gas in a molar ratio of 1:3.25:1.5. This was evacuated under liquid N$_2$ cooling, and about 50 mg of Br$_2$ was condensed. The ampoule was removed in vacuum and then subjected to a temperature gradient of 1000°C to 800°C for 10 days in a two-zone furnace. Air-stable green transparent crystals of Pr$_3$Br(SiS$_4$)$_2$ were obtained in the middle of the ampoule.

1.178 Silicon–Praseodymium–Iodine–Sulfur

The **Pr$_3$I(SiS$_4$)$_2$** quaternary compound, which crystallizes in the monoclinic structure with the lattice parameters $a = 1586.6 \pm 0.8$, $b = 779.3 \pm 0.4$, $c = 1081.2 \pm 0.8$ pm, $\beta = 97.99 \pm 0.05°$, and a calculated density of 4.326 g·cm^{-3}, is formed in the

Si–Pr–I–S system (Hatscher and Urland 2001). Single crystals of this compound were prepared by a two-step reaction of Pr, S, Si, and iodine (molar ratio 1:3.25:1:0.33) in an evacuated quartz tube. First, the mixture was subjected to a temperature gradient of 1000°C to 850°C for 10 days. The ampoule was then quenched; the obtained material was fine-grained, and again subjected to these conditions with additional iodine in an evacuated quartz ampoule. After 7 days, air-stable green transparent crystals of $Pr_3I(SiS_4)_2$ were obtained.

1.179 Silicon–Praseodymium–Manganese–Sulfur

The $Pr_6MnSi_2S_{14}$ quaternary compound, which crystallizes in the hexagonal structure with the lattice parameters $a = 1018 \pm 2$ and $c = 569 \pm 2$ pm, is formed in the Si–Pr–Mn–S system (Michelet and Flahaut 1969). It was synthesized in a sealed silica ampoule at around 1000°C by the interaction of MnS and Pr_2S_3 with Si and the desired amount of S.

1.180 Silicon–Neodymium–Lead–Sulfur

The $Nd_2Pb(SiS_4)_2$ quaternary compound, which crystallizes in the tetragonal structure with the lattice parameters $a = 894.2 \pm 0.1$, $c = 2649.2 \pm 0.4$ pm, and a calculated density of 4.389 $g \cdot cm^{-3}$, is formed in the Si–Nd–Pb–S system (Daszkiewicz et al. 2012; Marchuk 2019). A powder sample of this compound was prepared by sintering the elemental constituents in an evacuated quartz tube. The synthesis was carried out in a furnace. The ampoule was first heated with a rate of 30°C·h^{-1} up to 1150°C, and then kept at this temperature for 3 h. Afterwards, the sample was cooled slowly (10°C·h^{-1}) down to 500°C and annealed at this temperature for 720 h. Subsequently, the ampoule was quenched in the air.

1.181 Silicon–Neodymium–Oxygen–Sulfur

Two quaternary compounds, $Nd_4S_3(Si_2O_7)$ and $Nd_{4.667}S(SiO_4)_3$, are formed in the Si–Nd–O–S system. First of them crystallizes in the tetragonal structure with the lattice parameters $a = 1198.6 \pm 0.1$ and $c = 1406.4 \pm 0.2$ pm (Grupe et al. 1992). $Nd_{4.667}S(SiO_4)_3$ crystallizes in the hexagonal structure with the lattice parameters $a = 983.37 \pm 0.09$, $c = 684.04 \pm 0.06$ pm, and a calculated density of 5.690 $g \cdot cm^{-3}$ (Hartenbach and Schleid 2005a).

Oxidation of Nd with S (molar ratio S/Nd = 1:2) in an evacuated quartz ampoule at 850°C (heating rate 2°C·h^{-1}) and annealing at this temperature for 7 days, sometimes below addition of NaCl as a flux, produces black cake made of sesquisulfide and ternary chloride (Grupe et al. 1992). From these, very hard, transparent red-violet single crystals of $Nd_4S_3(Si_2O_7)$ can be isolated in low yield.

For synthesizing $Nd_{4.667}S(SiO_4)_3$, Nd_2O_3, Nd, SiO_2, S (molar ratio 6:2:3:9), and CsCl as a flux were used (Hartenbach and Schleid 2005a). After heating for 7 days at 850°C, in an evacuated silica tube, hard, lath-shaped crystals of this compound were obtained.

1.182 Silicon–Neodymium–Chlorine–Sulfur

The $Nd_3Cl(SiS_4)_2$ quaternary compound, which crystallizes in the orthorhombic structure with the lattice parameters $a = 1240.3 \pm 0.2$, $b = 1035.8 \pm 0.2$, $c = 1616.4 \pm 0.3$ pm, and a calculated density of 4.405 $g \cdot cm^{-3}$, is formed in the Si–Nd–Cl–S system (Hatscher and Urland 2003). Single crystals of this compound were obtained by reaction of the elements. Nd chips, S and Si powders (molar ratio 1.0:2.0:0.33) were loaded into fused silica tube in a glove box under Ar atmosphere. About 10 kPa (molar ratio approximately 0.3) of Cl_2 were condensed in the evacuated tube. After sealing, this tube was placed in a furnace and exposed to a temperature gradient of 1000–800°C for 10 days. Then the ampoule was quenched in water. In the middle of the tube air stable violet crystals of $Nd_3Cl(SiS_4)_2$ were obtained.

1.183 Silicon–Neodymium–Bromine–Sulfur

The $Nd_3Br(SiS_4)_2$ quaternary compound, which crystallizes in the monoclinic structure with the lattice parameters $a = 1561.4 \pm 0.4$, $b = 766.0 \pm 0.1$, $c = 1085.3 \pm 0.2$ pm, $\beta = 97.66 \pm 0.02°$, and a calculated density of 4.261 $g \cdot cm^{-3}$ (Hatscher and Urland 2002c), is formed in the Si–Nd–Br–S system. Single crystals of this compound were obtained by reaction of the elements. Nd chips, S and Si powders were weight in a quartz vial under a protective gas in a molar ratio of 1:3.25:1.5. This was evacuated under liquid N_2 cooling and about 50 mg of Br_2 was condensed. The ampoule was removed in vacuum and then subjected to a temperature gradient of 1000°C to 800°C for 10 days in a two-zone furnace. Air-stable violet transparent crystals of $Nd_3Br(SiS_4)_2$ were obtained in the middle of the ampoule.

1.184 Silicon–Neodymium–Iodine–Sulfur

The $Nd_3I(SiS_4)_2$ quaternary compound, which crystallizes in the monoclinic structure with the lattice parameters $a = 1583.7 \pm 0.3$, $b = 777.46 \pm 0.14$, $c = 1079.05 \pm 0.19$ pm, $\beta = 98.10 \pm 0.02°$, and a calculated density of 4.405 $g \cdot cm^{-3}$, is formed in the Si–Nd–I–S system (Hatscher and Urland 2001). Single crystals of the title compound were prepared by a two-step reaction of Nd, S, Si, and iodine (molar ratio 1:3.25:1:0.33) in an evacuated quartz tube. First, the mixture was subjected to a temperature gradient of 1000°C to 850°C for 10 days. The ampoule was then quenched; the obtained material was fine-grained, and again subjected to these conditions with additional iodine in an evacuated quartz ampoule. After 7 days, air-stable violet transparent crystals of $Nd_3I(SiS_4)_2$ were obtained.

1.185 Silicon–Neodymium–Manganese–Sulfur

The $Nd_6MnSi_2S_{14}$ quaternary compound, which crystallizes in the hexagonal structure with the lattice parameters $a = 1012 \pm 2$ and $c = 569 \pm 2$ pm, is formed in the Si–Nd–Mn–S system

(Michelet and Flahaut 1969). It was synthesized in a sealed silica ampoule at around 1000°C by the interaction of MnS and Nd$_2$S$_3$ with Si and the desired amount of S.

1.186 Silicon–Samarium–Lead–Sulfur

The **Sm$_2$Pb(SiS$_4$)$_2$** quaternary compound, which crystallizes in the tetragonal structure with the lattice parameters $a = 888.54 \pm 0.04$, $c = 2628.3 \pm 0.1$ pm, and a calculated density of 4.545 g·cm^{-3}, is formed in the Si–Sm–Pb–S system (Daszkiewicz et al. 2012; Marchuk 2019). A powder sample of this compound was prepared by sintering the elemental constituents in an evacuated quartz tube. The synthesis was carried out in a furnace. The ampoule was first heated with a rate of 30°C·h^{-1} up to 1150°C, and then kept at this temperature for 3 h. Afterwards, the sample was cooled slowly (10°C·h^{-1}) down to 500°C and annealed at this temperature for 720 h. Subsequently, the ampoule was quenched in the air.

1.187 Silicon–Samarium–Oxygen–Sulfur

Two quaternary compounds, **Sm$_4$S$_3$(Si$_2$O$_7$)** and **Sm$_{4.667}$S(SiO$_4$)$_3$**, are formed in the Si–Sm–O–S system. First of them crystallizes in the tetragonal structure with the lattice parameters $a = 1186.4 \pm 0.1$, $c = 1387.0 \pm 0.2$ pm, and a calculated density of 5.891 g·cm^{-3} (Sieke and Schleid 1999) [$a = 1183.9 \pm 0.5$, $c = 1392.8 \pm 0.5$ pm, and a calculated density of 5.89 g·cm^{-3} (Siegrist et al. 1982)]. Sm$_{4.667}$S(SiS$_4$)$_3$ crystallizes in the hexagonal structure with the lattice parameters $a = 975.02 \pm 0.09$, $c = 676.13 \pm 0.06$ pm, and a calculated density of 6.025 g·cm^{-3} (Hartenbach and Schleid 2005a).

Sm$_4$S$_3$(Si$_2$O$_7$) was synthesized as light yellow transparent crystals by reaction of Sm, Sm$_2$O$_3$, S, and SiO$_2$ with fluxing SmCl$_3$ in suitable molar ratios in fused silica tube at 850°C for 7 days (Siegrist et al. 1982; Grupe et al. 1992). This compound was also obtained in an attempt to grow SmS$_2$ single crystals by an iodine-transport reaction in a closed silica tube at 800°C to 1000°C. Its thin plates had a yellowish color, whereas bulky crystals looked greenish. Obviously, Sm$_4$S$_3$(Si$_2$O$_7$) decomposes peritectically.

For synthesizing Sm$_{4.667}$S(SiO$_4$)$_3$, Sm$_2$O$_3$, Sm, SiO$_2$, S (molar ratio 6:2:3:9), and CsCl as a flux were used (Hartenbach and Schleid 2005a). After heating for 7 days at 850°C in an evacuated silica tube, hard, lath-shaped crystals of this compound were obtained.

1.188 Silicon–Samarium–Chlorine–Sulfur

The **Sm$_3$S$_2$Cl(SiS$_4$)** quaternary compound, which apparently has two polymorphic modifications, is formed in the Si–Sm–Cl–S system. First of them crystallizes in the orthorhombic structure with the lattice parameters $a = 1230.0 \pm 0.2$, $b = 1029.0 \pm 0.2$, $c = 1606.1 \pm 0.3$ pm, and a calculated density of 4.620 g·cm^{-3} (Hatscher and Urland 2002b). Second modifications of this compound crystallize in the hexagonal structure with the lattice parameters $a = 993.63 \pm 0.07$,

$c = 571.76 \pm 0.04$ pm, and a calculated density of 4.796 g·cm^{-3} (Hartenbach et al. 2007).

Single crystals of the orthorhombic Sm$_3$S$_2$Cl(SiS$_4$) were prepared from the elements (Hatscher and Urland 2002b). Sm chips, S and Si silicon powders, and chlorine were added to a quartz glass tube in a molar ratio of 1:2:0.3:~0.3. The ampoule was evacuated, sealed, and heated for 10 days in a two-zone furnace with its ends held at temperatures of 1000°C and 800°C. After cooling, a few air-stable orange crystals were obtained.

A hexagonal modification was prepared from a mixture of Sm, S, and SiS$_2$ (molar ratio 4:6:1) using an excess of SmCl$_3$ as a flux. The annealing of the mixture at 850°C for 8 days in an evacuated silica glass ampoule led to dark red single crystals in the form of hexagonal columns. They are air and water-resistant.

1.189 Silicon–Samarium–Bromine–Sulfur

The **Sm$_3$Br(SiS$_4$)$_2$** quaternary compound, which crystallizes in the monoclinic structure with the lattice parameters $a = 1555.4 \pm 0.3$, $b = 758.5 \pm 0.2$, $c = 1079.9 \pm 0.2$ pm, $\beta = 98.28 \pm 0.02°$, and a calculated density of 4.445 g·cm^{-3} (Hatscher and Urland 2002c), is formed in the Si–Sm–Br–S system. Single crystals of this compound were obtained by reaction of the elements. Sm chips, S and Si powders were weight in a quartz vial under a protective gas in a molar ratio of 1:3.25:1.5. This was evacuated under liquid N$_2$ cooling, and about 50 mg of Br$_2$ was condensed. The ampoule was removed in a vacuum and then subjected to a temperature gradient of 1000°C to 800°C for 10 days in a two-zone furnace. Air-stable orange transparent crystals of Sm$_3$Br(SiS$_4$)$_2$ were obtained in the middle of the ampoule.

1.190 Silicon–Samarium–Iodine–Sulfur

The **Sm$_3$I(SiS$_4$)$_2$** quaternary compound, which crystallizes in the monoclinic structure with the lattice parameters $a = 1573.2 \pm 0.3$, $b = 770.97 \pm 0.19$, $c = 1073.3 \pm 0.2$ pm, $\beta = 98.19 \pm 0.03°$, and a calculated density of 4.591 g·cm^{-3}, is formed in the Si–Sm–I–S system (Hatscher and Urland 2001). Single crystals of the title compound were prepared by a two-step reaction of Sm, S, Si, and iodine (molar ratio 1:3.25:1:0.33) in an evacuated quartz tube. First, the mixture was subjected to a temperature gradient of 1000°C to 850°C for 10 days. The ampoule was then quenched; the obtained material was fine-grained, and again subjected to these conditions with additional iodine in an evacuated quartz ampoule. After 7 days, air-stable orange transparent crystals of Sm$_3$I(SiS$_4$)$_2$ were obtained.

1.191 Silicon–Samarium–Manganese–Sulfur

The **Sm$_6$MnSi$_2$S$_{14}$** quaternary compound, which crystallizes in the hexagonal structure with the lattice parameters $a = 1003 \pm 2$ and $c = 566 \pm 2$ pm, is formed in the Si–Sm–Mn–S system

(Michelet and Flahaut 1969). It was synthesized in a sealed silica ampoule at around 1000°C by interaction of MnS and Sm_2S_3 with Si and the desired amount of S.

1.192 Silicon–Samarium–Iron–Sulfur

The $Sm_6FeSi_2S_{14}$ quaternary compound, which crystallizes in the hexagonal structure with the lattice parameters $a = 998.2 \pm 0.1$ and $c = 566.0 \pm 0.1$ pm, is formed in the Si–Sm–Fe–S system (Sun et al. 2016). Single crystals of the title compound were obtained by solid-state reactions with KI as a flux. The starting materials were SmO, Fe, Si, S, and additional boron powder. The sample had a total mass of 500 mg and 400 mg KI additional (molar ratio of Sm/Fe/Si/S/B was 6:1:2:14:6). The mixture of the starting materials was ground into fine powder in an agate mortar and pressed into one pellet, followed by being loaded into one quartz tube. The tube was evacuated and flame-sealed. The sample was placed into a furnace, heated from room temperature to 300°C in 5 h, kept at 300°C for 10 h, then heated to 650°C in 5 h, kept at 650°C for 10 h, then heated to 950°C in 5 h, kept at 950°C for 7 days, then cooled down to 300°C in 5 days, and powered off. The block-black crystals of $Sm_6FeSi_2S_{14}$ stable in the air and water were obtained.

1.193 Silicon–Samarium–Cobalt–Sulfur

The $Sm_6CoSi_2S_{14}$ quaternary compound, which crystallizes in the hexagonal structure with the lattice parameters $a = 997.47 \pm 0.02$, $c = 565.68 \pm 0.01$ pm, and a calculated density of 4.9941 ± 0.0003 g·cm^{-3}, is formed in the Si–Sm–Co–S system (Melnychuk et al. 2017). It was synthesized from a stoichiometric mixture of the elements by heating in a quartz container up to 1150°C with the next annealing at 500°C for 500 h.

1.194 Silicon–Europium–Titanium–Sulfur

The $Eu_3Ti_{0.85}SiS_7$ quaternary compound, which crystallizes in the hexagonal structure with the lattice parameters $a = 1021.62 \pm 0.04$, $c = 598.58 \pm 0.06$ pm, and a calculated density of 4.598 g·cm^{-3}, is formed in the Si–Eu–Ti–S system (Yang et al. 2022a). This compound was prepared through a high-temperature solid-state reaction. Eu, Ti, Si, and S powder with KI as a flux were used as starting materials. The molar ratios of Eu/Ti/Si/S were 3:1:1:9, which were weighted with a total mass of 500 mg and an additional 400 mg KI. The mixture was firstly ground into fine powder in an agate mortar and pressed into a pellet, which was subsequently loaded into a quartz tube. The tube was evacuated to be 10^{-2} Pa and sealed, then the ampoule was put into a muffle furnace and heated from room temperature to 850°C, kept at 850°C for 2 days, then cooled down to 300°C slowly with a rate of 5°C·h^{-1}. After washing with deionized water and ethanol under ultrasound, black rod crystals of $Eu_3Ti_{0.85}SiS_7$ were obtained.

1.195 Silicon–Gadolinium–Lead–Sulfur

The $Gd_2Pb(SiS_4)_2$ quaternary compound, which crystallizes in the tetragonal structure with the lattice parameters $a = 886.33 \pm 0.06$, $c = 2618.5 \pm 0.3$ pm, and a calculated density of 4.663 g·cm^{-3}, is formed in the Si–Gd–Pb–S system (Daszkiewicz et al. 2012; Marchuk 2019). A powder sample of this compound was prepared by sintering the elemental constituents in an evacuated quartz tube. The synthesis was carried out in a furnace. The ampoule was first heated with a rate of 30°C·h^{-1} up to 1150°C, and then kept at this temperature for 3 h. Afterwards, the sample was cooled slowly (10°C·h^{-1}) down to 500°C and annealed at this temperature for 720 h. Subsequently, the ampoule was quenched in the air.

1.196 Silicon–Gadolinium–Oxygen–Sulfur

The $Gd_4S_3(Si_2O_7)$ quaternary compound, which crystallizes in the tetragonal structure with the lattice parameters $a = 1177.53 \pm 0.08$, $c = 1378.31 \pm 0.09$ pm, and a calculated density of 6.210 $\pm$ 0.004 g·cm^{-3}, is formed in the Si–Gd–O–S system (Sieke et al. 2002). It has been synthesized by reaction of Gd_2O_3, Gd, S, and SiO_2 (molar ratio 1:2:3:2) using GdCl$_3$ flux in an evacuated silica tube at 850°C for 7 days.

1.197 Silicon–Gadolinium–Bromine–Sulfur

The $Gd_3Br(SiS_4)_2$ quaternary compound, which crystallizes in the monoclinic structure with the lattice parameters $a = 1556.5 \pm 0.3$, $b = 750.8 \pm 0.1$, $c = 1074.5 \pm 0.2$ pm, $\beta = 99.26 \pm 0.02°$, and a calculated density of 4.632 g·cm^{-3} (Hatscher and Urland 2002c), is formed in the Si–Gd–Br–S system. Single crystals of this compound were obtained by reaction of the elements. Gd chips, S and Si powders were weight in a quartz vial under a protective gas in a molar ratio of 1:3.25:1.5. This was evacuated under liquid N_2 cooling and about 50 mg of Br_2 was condensed. The ampoule was removed in a vacuum and then subjected to a temperature gradient of 1000°C to 800°C for 10 days in a two-zone furnace. Air-stable colorless transparent crystals of $Gd_3Br(SiS_4)_2$ were obtained in the middle of the ampoule.

1.198 Silicon–Gadolinium–Manganese–Sulfur

The $Gd_6MnSi_2S_{14}$ quaternary compound, which crystallizes in the hexagonal structure with the lattice parameters $a = 995 \pm 2$ and $c = 564 \pm 2$ pm, is formed in the Si–Gd–Mn–S system (Michelet and Flahaut 1969). It was synthesized in a sealed silica ampoule at around 1000°C by interaction of MnS and Gd_2S_3 with Si and the desired amount of S.

1.199 Silicon–Terbium–Lead–Sulfur

The $Tb_2Pb(SiS_4)_2$ quaternary compound, which crystallizes in the tetragonal structure with the lattice parameters $a = 886.04 \pm 0.01$, $c = 2611.84 \pm 0.07$ pm, and a calculated density of 4.696 g·cm^{-3}, is formed in the Si–Tb–Pb–S system (Daszkiewicz et al. 2012; Marchuk 2019). A powder sample of this compound was prepared by sintering the elemental constituents in an evacuated quartz tube. The synthesis was carried out in a furnace. The ampoule was first heated with a rate of 30°C·h^{-1} up to 1150°C, and then kept at this temperature for 3 h. Afterwards, the sample was cooled slowly (10°C·h^{-1}) down to 500°C and annealed at this temperature for 720 h. Subsequently, the ampoule was quenched in the air.

1.200 Silicon–Terbium–Oxygen–Sulfur

The **Tb$_4$S$_3$(Si$_2$O$_7$)** quaternary compound, which crystallizes in the tetragonal structure with the lattice parameters $a = 1173.78 \pm 0.07$, $c = 1369.57 \pm 0.09$ pm, and a calculated density of 6.336 ± 0.004 g·cm^{-3}, is formed in the Si–Tb–O–S system (Sieke et al. 2002). It has been synthesized by reaction of Tb$_2$O$_3$, Tb, S, and SiO$_2$ (molar ratio 1:2:3:2) using TbCl$_3$ flux in an evacuated silica tube at 850°C for 7 days. The mixture of Tb$_4$O$_7$ and Tb (molar ratio 3:2) can be used instead of Tb$_2$O$_3$.

1.201 Silicon–Terbium–Iodine–Sulfur

The **Tb$_3$I(SiS$_4$)$_2$** quaternary compound, which crystallizes in the monoclinic structure with the lattice parameters $a = 1564.5 \pm 0.3$, $b = 761.36 \pm 0.16$, $c = 1064.66 \pm 0.18$ pm, $\beta = 98.94 \pm 0.02°$, and a calculated density of 4.858 g·cm^{-3}, is formed in the Si–Tb–I–S system (Hatscher and Urland 2001). Single crystals of the title compound were prepared by a two-step reaction of Tb, S, Si, and iodine (molar ratio 1:3.25:1:0.33) in an evacuated quartz tube. First, the mixture was subjected to a temperature gradient of 1000°C to 850°C for 10 days. The ampoule was then quenched; the obtained material was fine-grained, and again subjected to these conditions with additional iodine in an evacuated quartz ampoule. After 7 days, air-stable light-yellow transparent crystals of Tb$_3$I(SiS$_4$)$_2$ were obtained.

1.202 Silicon–Terbium–Cobalt–Sulfur

The **Tb$_6$CoSi$_2$S$_{14}$** quaternary compound, which crystallizes in the hexagonal structure with the lattice parameters $a = 983.37 \pm 0.02$, $c = 564.24 \pm 0.01$ pm, and a calculated density of 5.3322 ± 0.0003 g·cm^{-3}, is formed in the Si–Tb–Co–S system (Melnychuk et al. 2017). It was synthesized from a stoichiometric mixture of the elements by heating in a quartz container up to 1150°C with the next annealing at 500°C for 500 h.

1.203 Silicon–Dysprosium–Lead–Sulfur

The **Dy$_2$Pb(SiS$_4$)$_2$** quaternary compound, which crystallizes in the tetragonal structure with the lattice parameters $a = 884.22 \pm 0.01$, $c = 2600.33 \pm 0.05$ pm, and a calculated density of 4.777 g·cm^{-3}, is formed in the Si–Dy–Pb–S system (Daszkiewicz

et al. 2012; Marchuk 2019). Powder sample of this compound was prepared by sintering the elemental constituents in an evacuated quartz tube. The synthesis was carried out in a furnace. The ampoule was first heated with a rate of 30°C·h^{-1} up to 1150°C, and then kept at this temperature for 3 h. Afterwards, the sample was cooled slowly (10°C·h^{-1}) down to 500°C and annealed at this temperature for 720 h. Subsequently, the ampoule was quenched in the air.

1.204 Silicon–Dysprosium–Oxygen–Sulfur

The **Dy$_4$S$_3$(Si$_2$O$_7$)** quaternary compound, which crystallizes in the tetragonal structure with the lattice parameters $a = 1169.78 \pm 0.07$, $c = 1361.60 \pm 0.09$ pm, and a calculated density of 6.519 ± 0.004 g·cm^{-3} (Sieke et al. 2002) [$a = 1167.8$ and $c = 1359.7$ pm (Meetsma et al. 1991)], is formed in the Si–Dy–O–S system. It has been synthesized by reaction of Dy$_2$O$_3$, Dy, S, and SiO$_2$ (molar ratio 1:2:3:2) using DyCl$_3$ flux in an evacuated silica tube at 850°C for 7 days (Sieke et al. 2002). Colorless single crystals of unknown composition, which were formed as a byproduct during chemical transport of (DyS)$_{1.2}$TaS$_2$ with Cl$_2$ as a transport agent in evacuated quartz ampoules at 750–1600°C (the main products were Dy$_2$S$_3$ and TaS$_2$) most likely there were the crystals of Dy$_4$S$_3$(Si$_2$O$_7$) (Grupe et al. 1992).

1.205 Silicon–Dysprosium–Manganese–Sulfur

The **Dy$_6$MnSi$_2$S$_{14}$** quaternary compound, which crystallizes in the hexagonal structure with the lattice parameters $a = 984 \pm 2$ and $c = 563 \pm 2$ pm, is formed in the Si–Dy–Mn–S system (Michelet and Flahaut 1969). It was synthesized in a sealed silica ampoule at around 1000°C by the interaction of MnS and Dy$_2$S$_3$ with Si and the desired amount of S.

1.206 Silicon–Dysprosium–Cobalt–Sulfur

The **Dy$_6$CoSi$_2$S$_{14}$** quaternary compound, which crystallizes in the hexagonal structure with the lattice parameters $a = 978.74 \pm 0.01$, $c = 565.67 \pm 0.01$ pm, and a calculated density of 5.4451 ± 0.0003 g·cm^{-3}, is formed in the Si–Dy–Co–S system (Melnychuk et al. 2016a). It was synthesized from a stoichiometric mixture of the elements by heating in a quartz container up to 1150°C with the next annealing at 500°C for 500 h.

1.207 Silicon–Dysprosium–Nickel–Sulfur

The **Dy$_6$NiSi$_2$S$_{14}$** quaternary compound, which crystallizes in the hexagonal structure with the lattice parameters $a = 976.96 \pm 0.01$, $c = 569.50 \pm 0.01$ pm, and a calculated density of 5.4274 ± 0.0003 g·cm^{-3}, is formed in the Si–Dy–Ni–S system (Melnychuk et al. 2016a). It was synthesized from a stoichiometric mixture of the elements by heating in a quartz container up to 1150°C with the next annealing at 500°C for 500 h.

1.208 Silicon–Holmium–Lead–Sulfur

The **Ho$_2$Pb(SiS$_4$)$_2$** quaternary compound, which crystallizes in the tetragonal structure with the lattice parameters $a = 884.28 \pm 0.02$, $c = 2596.3 \pm 0.1$ pm, and a calculated density of 4.812 g·cm^{-3}, is formed in the Si–Ho–Pb–S system (Daszkiewicz et al. 2012; Marchuk 2019). A powder sample of this compound was prepared by sintering the elemental constituents in an evacuated quartz tube. The synthesis was carried out in a furnace. The ampoule was first heated with a rate of 30°C·h^{-1} up to 1150°C, and then kept at this temperature for 3 h. Afterwards, the sample was cooled slowly (10°C·h^{-1}) down to 500°C and annealed at this temperature for 720 h. Subsequently, the ampoule was quenched in the air.

1.209 Silicon–Holmium–Oxygen–Sulfur

Two quaternary compounds, **Ho$_2$S(SiO$_4$)** and **Ho$_4$S$_3$(Si$_2$O$_7$)**, are formed in the Si–Ho–O–S system. First of them crystallizes in the orthorhombic structure with the lattice parameters $a = 605.87 \pm 0.05$, $b = 690.41 \pm 0.06$, $c = 1064.95 \pm 0.09$ pm, and a calculated density of 6.770 g·cm^{-3} (Hartenbach et al. 2002). Ho$_4$S$_3$(Si$_2$O$_7$) crystallizes in the tetragonal structure with the lattice parameters $a = 1166.02 \pm 0.07$, $c = 1355.78 \pm 0.09$ pm, and a calculated density of 6.659 ± 0.004 g·cm^{-3} (Sieke et al. 2002).

The yellow platelet-shaped, air- and water-resistant crystals of Ho$_2$S(SiO$_4$) emerged as a single-crystalline byproduct, obtained during the synthesis of Ho$_2$OS$_2$ by reaction of a mixture of Ho$_2$O$_3$, Ho, and S with the wall of the evacuated silica tubes used as a container with an excess of CsCl as a flux at 800°C (Hartenbach et al. 2002).

Ho$_4$S$_3$(Si$_2$O$_7$) has been synthesized by reaction of Ho$_2$O$_3$, Ho, S, and SiO$_2$ (molar ratio 1:2:3:2) using HoCl$_3$ flux in evacuated silica tube at 850°C for 7 days (Sieke et al. 2002).

1.210 Silicon–Holmium–Nickel–Sulfur

The **Ho$_6$NiSi$_2$S$_{14}$** quaternary compound, which crystallizes in the hexagonal structure with the lattice parameters $a = 972.93 \pm 0.01$, $c = 567.40 \pm 0.01$ pm, and a calculated density of 5.5448 ± 0.0002 g·cm^{-3}, is formed in the Si–Ho–Ni–S system (Melnychuk et al. 2016b). It was synthesized from a stoichiometric mixture of the elements by the heating in a quartz container up to 1150°C with the next annealing at 500°C for 500 h.

1.211 Silicon–Erbium–Lead–Sulfur

The **Er$_2$Pb(SiS$_4$)$_2$** quaternary compound, which crystallizes in the tetragonal structure with the lattice parameters $a = 883.0$ and $c = 258.40$ pm, is formed in the Si–Er–Pb–S system (Marchuk 2019). Powder sample of this compound was prepared by sintering the elemental constituents in an evacuated quartz tube. The synthesis was carried out in a furnace. The ampoule was first heated with a rate of 12°C·h^{-1} up to 1150°C,

and then kept at this temperature for 4 h. Afterwards, the sample was cooled slowly (12°C·h^{-1}) down to 500°C and annealed at this temperature for 500 h with the next quenching in the air.

1.212 Silicon–Erbium–Oxygen–Sulfur

Two quaternary compounds, **Er$_2$S(SiO$_4$)** and **Er$_4$S$_3$(Si$_2$O$_7$)**, are formed in the Si–Er–O–S system. First of them crystallizes in the orthorhombic structure with the lattice parameters $a = 1070.02 \pm 0.08$, $b = 1235.48 \pm 0.09$, $c = 683.64 \pm 0.06$ pm, and a calculated density of 6.742 g·cm^{-3} (Hartenbach et al. 2004). Er$_4$S$_3$(Si$_2$O$_7$) crystallizes in the tetragonal structure with the lattice parameters $a = 1161.69 \pm 0.07$, $c = 1348.62 \pm 0.09$ pm, and a calculated density of 6.813 ± 0.004 g·cm^{-3} (Sieke et al. 2002) [$a = 1164.6 \pm 0.1$ and $c = 1347.3 \pm 0.2$ pm (Grupe et al. 1992)].

During the reaction of CdS with Er and S in evacuated silica ampoules at 900°C for 10 days, pink lath-shaped air- and water-resistant crystals of Er$_2$S(SiO$_4$) occur as a byproduct (Hartenbach et al. 2004).

The pale yellowish pink crystals of Er$_4$S$_3$(Si$_2$O$_7$) were obtained by reaction of Er$_2$O$_3$, Er, S, and SiO$_2$ (molar ratio 1:2:3:2) using ErCl$_3$ flux in evacuated silica tube at 850°C for 7 days (Grupe et al. 1992; Sieke et al. 2002).

1.213 Silicon–Erbium–Manganese–Sulfur

The **Er$_6$MnSi$_2$S$_{14}$** quaternary compound, which crystallizes in the hexagonal structure with the lattice parameters $a = 974 \pm 2$ and $c = 564 \pm 2$ pm, is formed in the Si–Er–Mn–S system (Michelet and Flahaut 1969). It was synthesized in a sealed silica ampoule at around 1000°C by interaction of MnS and Er$_2$S$_3$ with Si and the desired amount of S.

1.214 Silicon–Erbium–Cobalt–Sulfur

The **Er$_6$CoSi$_2$S$_{14}$** quaternary compound, which crystallizes in the hexagonal structure with the lattice parameters $a = 970.56 \pm 0.02$, $c = 565.99 \pm 0.02$ pm, and a calculated density of 5.6369 ± 0.0004 g·cm^{-3}, is formed in the Si–Ho–Ni–S system (Marchuk et al. 2013). This compound is in thermodynamic equilibrium with the binary compounds of the SiS$_2$–Er$_2$S$_3$–CoS quasiternary system, that is, the isothermal section of this quasiternary system at 500°C consists of 4 single-phase, 6 two-phase, and 3 three-phase fields. Solubility based on binary components and ternary compounds does not exceed 1–2 mol%. Er$_6$CoSi$_2$S$_{14}$ was synthesized from a stoichiometric mixture of the elements by heating in a quartz container up to 1150°C with the next annealing at 500°C for 500 h.

1.215 Silicon–Thulium–Oxygen–Sulfur

The **Tm$_4$S$_3$(Si$_2$O$_7$)** quaternary compound, which crystallizes in the tetragonal structure with the lattice parameters $a = 1158.14 \pm 0.07$, $c = 1341.25 \pm 0.08$ pm, and a calculated density of 6.942

± 0.004 g·cm⁻³, is formed in the Si–Tm–O–S system (Sieke et al. 2002). It has been synthesized by reaction of Tm_2O_3, Tm, S, and SiO_2 (molar ratio 1:2:3:2) using $TmCl_3$ flux in an evacuated silica tube at 850°C for 7 days.

1.216 Silicon–Thulium–Manganese–Sulfur

The $Tm_6MnSi_2S_{14}$ quaternary compound, which crystallizes in the hexagonal structure with the lattice parameters $a = 971 \pm 2$ and $c = 565 \pm 2$ pm, is formed in the Si–Tm–Mn–S system (Michelet and Flahaut 1969). It was synthesized in a sealed silica ampoule at around 1000°C by the interaction of MnS and Tm_2S_3 with Si and the desired amount of S.

1.217 Silicon–Ytterbium–Oxygen–Sulfur

Two quaternary compounds, $Yb_4S_3(Si_2O_7)$ and $Yb_5S(SiO_4)_3$, are formed in the Si–Yb–O–S system. First of them crystallizes in the tetragonal structure with the lattice parameters $a = 1154.3 \pm 0.1$ and $c = 1332.2 \pm 0.1$ pm (Range et al. 1996). $Yb_5S(SiO_4)_3$ crystallizes in the hexagonal structure with the lattice parameters $a = 972.36 \pm 0.09$, $c = 648.49 \pm 0.06$ pm, and a calculated density of 7.340 g·cm⁻³ (Wickleder et al. 2002).

Single crystals of $Yb_4S_3(Si_2O_7)$ could be isolated as a side product after reaction of a stoichiometric mixture of Yb_2S_3-I (normal pressure phase) and YbI_3 in a high-pressure apparatus (heating up to 1700°C at 2 GPa, releasing the pressure to 1.6 GPa and cooling with 10°C·min⁻¹ to 1500°C; after 4 minutes quenching to ambient conditions) (Range et al. 1996). The main products of this reaction were Yb_2S_3-II (high-pressure phase) and Yb_2OS_2.

Dark-red, pillar-shaped single crystals of $Yb_5S(SiO_4)_3$ were obtained by the reaction of Yb_2O_3, Yb and S with SiO_2 (molar ratio 2:1:1:3) in the presence CsBr as a flux in evacuated silica ampoules at 850°C and an annealing time of seven days (Wickleder et al. 2002). The subsequent cooling to room temperature took place at a cooling rate of approximately 18°C·h⁻¹.

1.218 Silicon–Tin–Lead–Sulfur

The $SnPbSiS_4$ quaternary phase, which crystallizes in the monoclinic structure with the lattice parameters $a = 643.34 \pm 0.08$, $b = 669.12 \pm 0.08$, $c = 850.37 \pm 0.10$ pm, $\beta = 108.576 \pm 0.004°$, and an energy gap of 1.82 eV, is formed in the Si–Sn–Pb–S system (Zhou et al. 2022). Millimeter-scale single crystals of this compound were obtained in a spontaneous crystallization way.

1.219 Silicon–Lead–Oxygen–Sulfur

The $Pb_8[O_2(SO_4)(Si_4O_{13})]$ or $8PbO·SO_3·4SiO_2$ quaternary compound, which melts congruently at 785°C, shows no phase transition, and crystallizes in the monoclinic structure with the

lattice parameters $a = 914.0 \pm 0.3$, $b = 1955.4 \pm 0.6$, $c = 1131.3 \pm 0.4$ pm, and $\beta = 89.68 \pm 0.03°$, is formed in the Si–Pb–O–S system (Fröhlich 1984). Single crystals of this compound can be prepared by slow cooling the stoichiometric mixture of PbO, $PbSO_4$, and $PbSiO_3$ from 800°C to 600°C and subsequent annealing at 600°C.

1.220 Silicon–Oxygen–Iron–Sulfur

SiO_2–FeO–Fe_3O_4–FeS. The liquid phase relations in this system were studied by MacLean (1969), who found that iron silicate melts coexisted with iron sulfide melts over a wide concentration range. The solubility of FeS in the iron silicate melt was at a maximum when in equilibrium with iron, and it decreased with increasing oxygen potential. An immiscible liquid formed from the excess FeS. The obtained results were combined with those of the earlier studies to construct the phase diagram of the SiO_2–FeO–FeS system at iron saturation.

SiO_2–Fe–O–S. This system was investigated by the interaction of iron in silica crucibles with an atmosphere of controlled oxygen and sulfur partial pressures at 1200°C (Li and Rankin 1994). The equilibrium compositions of the melts were determined over the range 10^{-7} to 10^{-4} Pa oxygen and $10^{2.25}$ to 10^4 Pa sulfur, and it was found that the system can exist as either a slag or oxysulfide. The oxysulfide contained appreciable quantities of dissolved oxygen and silica, although the levels decreased as the sulfur content was increased. Sulfur also had the effect of reducing the solubility of silica in the slag. When Cu was added to the system, the solubility of oxygen and silica in the oxysulfide phase decreased dramatically. The results were examined in terms of the thermodynamics of the relevant reactions, and the predominance area diagram for the Cu-free system was established by combining the present results with those of earlier investigations.

Equilibrium distributions of Cu, S, and O between silica-saturated SiO_2–Fe–O slags and Cu_2S–FeS (25–79 mass% Cu) have been examined experimentally under controlled partial pressures of SO_2 (Tavera and Davenport 1979). The temperature range of the experiments was 1150°C to 1300°C and $p(SO_2)$ was varied between 10 and 100 kPa. Concentrations of Cu in the experimental slags were found to be low (<1 mass% Cu) under these conditions, as long as the grade of the coexistent matte was below 60 mass% Cu. Copper in slag concentration rose dramatically, however, when the matte grade was increased above this level. It was also shown that whereas FeS has a high solubility for oxygen and is itself soluble in slag under oxidizing conditions, Cu_2S and slag are almost completely immiscible. Cu–Fe–S mattes behave in an intermediate manner.

1.221 Silicon–Manganese–Iron–Sulfur

The $Mn_{1.4}Fe_{0.6}SiS_4$ quaternary phase, which crystallizes in the orthorhombic structure with the lattice parameters $a = 1265.1 \pm 0.4$, $b = 736.1 \pm 0.2$, and $c = 589.9 \pm 0.3$ pm (Fuhrmann and Pickardt 1990b), is formed in the Si–Mn–Fe–S system. Single crystals of this phase were obtained by chemical transport reactions.

2

Systems Based on Silicon Selenides

2.1 Silicon–Lithium–Indium–Selenium

The $Li_2In_2SiSe_6$ quaternary compound, which melts congruently at 801°C and crystallizes in the monoclinic structure with the lattice parameters $a = 1262.10 \pm 0.04$, $b = 737.88 \pm 0.03$, $c = 1258.00 \pm 0.04$ pm, $\beta = 109.704 \pm 0.002°$, a calculated density of 4.489 g·cm^{-3}, and an energy gap of 2.54 eV, is formed in the Si–Li–In–Se system (Yin et al. 2012a). The structural, electronic, and optical properties of this compound have been investigated by *ab initio* calculations based on the density functional theory by Wong et al. (2018) using the full potential linearized augmented plane wave method. The next values of the lattice parameters were determined: $a = 1261.95$, $b = 737.79$, $c = 1257.86$ pm (the value of β is not calculated).

To prepare this compound, the mixture of LiInSe$_2$ (1 mM) and SiSe$_2$ (0.5 mM) was ground, loaded into a fused silica tube under an Ar atmosphere in a glove box, then moved to a high-vacuum line, and flame-sealed under a high vacuum of 10^{-3} Pa (Yin et al. 2012a). The tube was then placed in a furnace and heated to 850°C in 12 h, left for 48 h, cooled to 200°C at a rate of 4°C·h^{-1}, and finally cooled to room temperature by switching off the furnace. Many orange block-shaped crystals, which are stable in the air, were found in the ampoule. Polycrystalline samples of Li$_2$In$_2$SiSe$_6$ were synthesized by solid-state reaction techniques. The mixtures of Li$_2$Se, SiSe$_2$, and In$_2$Se$_3$ (molar ratio 1:1:1) were heated to 700°C for 15 h, kept at that temperature for 72 h, and then the furnace was turned off.

2.2 Silicon–Sodium–Manganese–Selenium

The $Na_4MgSi_2Se_6$ quaternary compound, which crystallizes in the monoclinic structure with the lattice parameters $a = 763.7 \pm 0.2$, $b = 1186.8 \pm 0.3$, $c = 705.2 \pm 0.3$ pm, $\beta = 106.729 \pm 0.014°$, a calculated density of 3.506 g·cm^{-3}, and an energy gap of 2.85 eV, is formed in the Si–Na–Mg–Se system (Wu et al. 2015b). It was synthesized using a mixture of Na, Mg, Si, and Se at stoichiometric proportions. Firstly, the mixture was filled into a graphite crucible in a vacuum-sealed quartz tube, then put the tube into a furnace with the following procedure: heated to 700°C within 40 h and left at this temperature for about 50 h, then slowly cooled to 300°C over 5 days, finally rapidly down to room temperature. Block single crystals with yellow color were found in the tube. All the preparation processes were completed in an Ar-filled glove box.

2.3 Silicon–Sodium–Samarium–Selenium

The $Na_9Sm(Si_2Se_6)_2$ quaternary compound, which crystallizes in the monoclinic structure with the lattice parameters $a = 700.52 \pm 0.11$, $b = 1208.52 \pm 0.18$, $c = 791.30 \pm 0.12$ pm, $\beta = 107.200 \pm 0.003°$, and a calculated density of 3.677 g·cm^{-3} (Martin et al. 2006) [$a = 701.1 \pm 0.1$, $b = 1209.8 \pm 0.2$, $c = 791.6 \pm 0.2$ pm, $\beta = 107.178 \pm 0.003°$ at 173 K (Knaust et al. 2005)], is formed in the Si–Na–Sm–Se system. Crystals of the title compound were formed from a molten chalcogenide flux reaction of Sm (29.9 mg), Si (33.6 mg), Se (62.6 mg), and Na$_2$Se$_2$ (201.0 mg) (Knaust et al. 2005; Martin et al. 2006). The reactants were placed into a fused silica ampoule in an inert atmosphere glove box. The ampoule was sealed under vacuum, and heated to 725°C at a rate of 35°C·h^{-1}. After 150 h of annealing, the ampoule was cooled at 3°C·h^{-1} to room temperature. Yellow thick plate crystals were obtained after the product was washed with DMF to dissolve the remaining flux.

2.4 Silicon–Sodium–Europium–Selenium

Two quaternary compounds, $Na_2EuSiSe_4$ and $Na_8Eu_2(Si_2Se_6)_2$, are formed in the Si–Na–Eu–Se system. First of them crystallizes in the trigonal structure with the lattice parameters $a = 2439.8 \pm 0.2$, $c = 714.7 \pm 0.4$ pm, and a calculated density of 4.396 g·cm^{-3} (Choudhury and Dorhout 2007). Na$_8$Eu$_2$(Si$_2$Se$_6$)$_2$ crystallizes in the monoclinic structure with the lattice parameters $a = 708.98 \pm 0.10$, $b = 1222.80 \pm 0.16$, $c = 794.99 \pm 0.11$ pm, $\beta = 107.427 \pm 0.002°$, and a calculated density of 3.908 g·cm^{-3} (Martin et al. 2006).

Na$_2$EuSiSe$_4$ was synthesized by combining Eu (0.4 mM), Si (1.0 mM), Se (1.1 mM), and Na$_2$Se$_2$ (0.8 mM). Reactants were loaded into a fused silica ampoule inside a N$_2$-filled glove box. The ampoule was flame-sealed under a vacuum and placed in a furnace. The furnace was ramped to 750°C at a rate of 35°C·h^{-1}, and the temperature was held constant at 750°C for 150 h. The furnace was then slowly cooled to ambient temperature at a rate of 5°C·h^{-1}. The ampoule was opened in the air, and the solid product contained pale orange needle-shaped crystals. The solid product was soaked in DMF and sonicated for 30 sec to loosen the crystals and dissolve the excess alkali selenide flux. The crystals are stable in dry DMF in a stoppered vial; however, they decompose when the vial is kept open in the air.

Na$_8$Eu$_2$(Si$_2$Se$_6$)$_2$ was obtained by combining Eu (0.301 mM), Na$_2$Se$_2$ (0.605 mM), Se (0.828 mM), and Si (0.773 mM). The mixture was heated up to 750°C at 35°C·h^{-1}, where it remained

DOI: 10.1201/9781003123484-2

for 150 h. The sample was cooled to room temperature at 3°C·h⁻¹, yielding orange-brown plate-like crystals of the title compound.

2.5 Silicon–Sodium–Lead–Selenium

The $Na_8Pb_2(Si_2Se_6)_2$ quaternary compound, which crystallizes in the monoclinic structure with the lattice parameters $a = 707.0 \pm 0.3$, $b = 1221.4 \pm 0.3$, $c = 796.9 \pm 0.3$ pm, $\beta = 107.29 \pm 0.03°$, and a calculated density of 4.190 g·cm⁻³, is formed in the Si–Na–Pb–Se system (Marking and Kanatzidis 1997). Red-orange chunky crystals of this compound were formed through basic non-oxidizing flux reactions. It was prepared by heating a mixture of Na_2Se, Pb, Si, and Se (molar ratio 2:2:5:9.5) at 850°C for 3 days and slowly cooling to 450°C. The title compound appeared to be stable in water and other solvents but showed partial surface decomposition when left under ambient atmospheric conditions overnight. All manipulations were carried out under a nitrogen atmosphere.

2.6 Silicon–Sodium–Bromine–Selenium

The Na_3SiSe_3Br quaternary compound is formed in the Si–Na–Br–Se system (Feltz and Pfaff 1983). To prepare this compound, to approximately 0.05 to 0.08 M solution of $Na_6Si_2Se_6$, the equivalent amount of a methanolic bromine solution was added dropwise. This compound crystallizes during slow concentration in a vacuum as an almost colorless precipitate.

2.7 Silicon–Potassium–Lanthanum–Selenium

The $KLaSiSe_4$ quaternary compound, which crystallizes in the monoclinic structure with the lattice parameters $a = 681 \pm 2$, $b = 694 \pm 2$, $c = 906 \pm 2$ pm, and $\beta = 108.4 \pm 0.3$, is formed in the Si–K–La–Se system (Wu and Ibers 1993a). It was prepared by the reaction of Si with K_2Se_5 and La_2Se_3. The starting materials were placed in a quartz tube that was subsequently evacuated to 10⁻³ Pa and sealed. The molar ratio of K/La/Si/Se was 1:1:1:4. The quartz tube was heated gradually to 500°C, where it was kept for 24 h before being successively brought to 700°C for 24 h and to 1000°C for 150 h. The tube was then cooled (4°C·h⁻¹) to 300°C, and the furnace was shut off. Yellow-brown crystals of the title compound were obtained. $KLaSiSe_4$ appears to be modestly stable in the air but decomposes gradually in the presence of water. Bulk samples of this compound were prepared by reaction of stoichiometric amounts of starting materials at 1000°C for 10 days with intermittent grinding.

2.8 Silicon–Potassium–Praseodymium–Selenium

The $KPrSiSe_4$ quaternary compound, which crystallizes in the monoclinic structure with the lattice parameters $a = 673.6 \pm 0.2$, $b = 694.3 \pm 0.2$, $c = 899.0 \pm 0.1$ pm, $\beta = 108.262 \pm 0.002°$, and a calculated density of 4.357 g·cm⁻³, is formed in the Si–K–Pr–Se system (Choudhury et al. 2006). It was synthesized by the reaction of Pr (0.35 mM), Si (0.18 mM), and Se (0.88 mM),

with KCl (5.37 mM) acting as a reactive flux in the molar ratio K/Pr/Si/Se/Cl of 30:2:1:5:30. The mixture was directly loaded into fused silica ampoule inside a N₂-filled atmosphere glove box. The ampoule was flame-sealed under a vacuum and placed in a furnace. The furnace was ramped to 850°C at a rate of 20°C·h⁻¹, and the temperature was held constant for 168 h and cooled to 500°C at a rate of 7°C·h⁻¹ and then switched off. After the reaction products cooled, the ampoule was opened in an inert atmosphere glove box, and the solid product was soaked in DMF for 6 h in order to dissolve and wash away any remaining flux and loosen the product crystals. Paraffin was poured over the product to prevent the air and water-sensitive crystals from hydrolysis. Green platelet-shaped single crystals of the title compound were obtained.

2.9 Silicon–Potassium–Europium–Selenium

The $K_2EuSiSe_5$ quaternary compound, which crystallizes in the monoclinic structure with the lattice parameters $a = 1166.9 \pm 0.3$, $b = 984.4 \pm 0.2$, $c = 891.7 \pm 0.2$ pm, $\beta = 91.583 \pm 0.005°$, a calculated density of 4.236 g·cm⁻³ at 171 K, and an energy gap of 2.00 eV, is formed in the Si–K–Eu–Se system (Evenson IV and Dorhout 2001). It was synthesized by combining Se (0.561 mM), K_2Se_4 (0.374 mM), Si (0.188 mM), and Eu (0.188 mM). Reactants were loaded into a fused silica ampoule inside an inert atmosphere glove box. The ampoule was flame-sealed under vacuum and placed in a tube furnace. The temperature was ramped to 525°C where it remained for 150 h, and the furnace was allowed to cool back to room temperature at 4°C·h⁻¹. Washing the product with DMF gave orange plates of the title compound and EuSe.

2.10 Silicon–Cesium–Cerium–Selenium

The $CsCeSiSe_4$ quaternary compound, which crystallizes in the orthorhombic structure with the lattice parameters $a = 667.65 \pm 0.03$, $b = 705.29 \pm 0.04$, $c = 1841.03 \pm 0.09$ pm, and a calculated density of 4.727 g·cm⁻³, is formed in the Si–Cs–Ce–Se system (Hartenbach and Schleid 2006). The reaction of Ce, Se, and Si with an excess of CsCl both as a flux and a Cs⁺ cation source (molar ratio 4:3:12:3) in evacuated silica ampoule at 850°C for 7 days yields yellow platelet-shaped single crystals of the title compound. It undergoes complete hydrolysis, so the bulk material was protected with paraffin.

2.11 Silicon–Copper–Magnesium–Selenium

The $Cu_2MgSiSe_4$ quaternary compound, which crystallizes in the orthorhombic structure with the lattice parameters $a = 795.3 \pm 0.5$, $b = 679.7 \pm 0.4$, $c = 650.7 \pm 0.7$ pm, and a calculated density of 4.676 g·cm⁻³, is formed in the Si–Cu–Mg–Se system (Liu et al. 2013). To prepare this compound, the stoichiometric mixture with an overall weight of about 300 mg of Cu, Mg, Si, and Se (molar ratio 2:1:1:4) was ground into fine powder in an agate mortar and pressed into pellets, then loaded into quartz tube and finally flame-sealed under vacuum about 10^{-2} Pa. The sample was placed into a furnace, held at 300°C for 10 h, then heated to 650°C in 5 h and kept for 10 h, subsequently heated to 950°C in 10 h, kept for 5 days, and finally cooled down to 300°C over 8 days before switching off the furnace power. Good-quality crystals of the $Cu_2MgSiSe_4$ were obtained. Attempts to prepare a single-phase polycrystalline sample of this compound failed as the powdered batches became air and moisture-sensitive.

2.12 Silicon–Copper–Barium–Selenium

The $Cu_2BaSiSe_4$ quaternary compound, which crystallizes in the trigonal structure with the lattice parameters $a = 641.88 \pm 0.19$, $c = 1608.3 \pm 1.0$ pm, a calculated density of 5.281 g·cm^{-3}, and an energy gap of 2.62 eV, is formed in the Si–Cu–Ba–Se system (Nian et al. 2017). A conventional high-temperature solid-state method was used to synthesize the title compound. A stoichiometric mixture of Cu, Ba, Si, and Se (molar ratio 1:2:1:4) was loaded into a silica tube and sealed under a high vacuum of 10^{-3} Pa. The furnace was programmed by the following steps: heated from room temperature to 600°C in 30 h and kept at this temperature for 40 h; then, heated to 1050°C in 20 h and left at this temperature for 100 h; finally, cooled to 400°C at a rate of 5°C·h^{-1} and then the furnace was shut down to room temperature. After the reaction, the product was washed with DMF to remove the unreacted sulfur and other byproducts. Many red crystals of $Cu_2BaSiSe_4$ were gained that can be stable in the air for several months. Since the Ba metal is easily oxidized in the air, an Ar-filled glove box was used to avoid the effects of oxygen and moisture in the preparation processes.

2.13 Silicon–Copper–Zinc–Selenium

The $Cu_2ZnSiSe_4$ quaternary compound is formed in the Si–Cu–Zn–Se system. It melts incongruently at 973°C (Schäfer and Nitsche 1977) [according to the data of Yao et al. (1987), it decomposes at 470°C] and crystallizes in the orthorhombic structure with the lattice parameters $a = 782.6 \pm 0.3$, $b = 672.7 \pm 0.2$, and $c = 644.5 \pm 0.1$ pm (Yao et al. 1987) [$a = 783$, $b = 673$, and $c = 644$ pm, the calculated and experimental densities of 5.25 and 5.22 g·cm^{-3}, respectively, and an energy gap of 2.33 eV [2.20 eV (Yao et al. 1987)] (Schleich and Wold 1977); $a = 782.3$, $b = 672.0$, and $c = 644.0$ pm (Schäfer and Nitsche 1974, 1977); $a = 780$, $b = 670$, and $c = 646$ pm (Nitsche et al. 1967)]. Single crystals of $Cu_2ZnSiSe_4$ were obtained by the chemical transport reactions using I_2 as the transport agent (Nitsche et al. 1967; Schäfer and Nitsche 1974; Schleich and Wold 1977; Yao et al. 1987).

2.14 Silicon–Copper–Cadmium–Selenium

Cu_2SiSe_3–CdSe. This section is nonquasibinary one of the Si–Cu–Cd–Se quaternary system since Cu_2SiSe_3 melts incongruently (Piskach et al. 1997; Olekseyuk et al. 1998b). The $Cu_2CdSiSe_4$ quaternary compound is formed in this system, which melts incongruently at 912°C [at 921°C (Schäfer and

Nitsche 1977)], has a polymorphous transformation, and its low-temperature modification crystallizes in the orthorhombic structure with the lattice parameters $a = 797.9 \pm 0.4$, $b = 684.1 \pm 0.3$, and $c = 653.9 \pm 0.5$ pm [$a = 799.0$, $b = 682.4$, and $c = 656.4$ pm (Schäfer and Nitsche 1974, 1977)] and an experimental density 4.96 g·cm^{-3} (Olekseyuk et al. 1998b).

Cu_8SiSe_6, β-$Cu_2CdSiSe_4$ and solid solutions based on CdSe primarily crystallize in this system. The narrow immiscibility region exists at 1058°C. The solubility of Cu_2SiSe_3 in CdSe at peritectic temperature is not higher than 3 mol% and the solubility of CdSe in Cu_2SiSe_3 is a little greater. The α-Cu_2SiSe_3 → β-Cu_2SiSe_3 peritectoid transformation takes place at 708°C. Thermal effects at 790°C correspond to the polymorphous transformation of $Cu_2CdSiSe_4$ and at 885°C and 892°C to the peritectic reaction L + Cu_8SiSe_6 ⇔ β-$Cu_2CdSiSe_4$ + α-$Cu2SiSe_3$ and to the L + Cu_8SiSe_6 + β-$Cu_2CdSiSe_4$ secondary crystallization, respectively (Piskach et al. 1997; Olekseyuk et al. 1998b). This system was investigated through DTA, metallography, XRD, and measuring of the microhardness and density (Piskach et al. 1997; Olekseyuk et al. 1998b). The ingots were annealed at 580°C for 1000 h. The crystals of $Cu_2CdSiSe_4$ were grown by the horizontal gradient freezing or iodine transport method (Schäfer and Nitsche 1974; Matsushita et al. 2005).

2.15 Silicon–Copper–Mercury–Selenium

Cu_2SiSe_3–HgSe. This section is a nonquasibinary one of the Si–Cu–Hg–Se quaternary system since Cu_8SiSe_6 primarily crystallizes from the Cu_2SiSe_3 side (Parasyuk et al. 1997, 1998). The eutectic composition and temperature are 93 mol% HgSe and 762°C, respectively. The $Cu_2HgSiSe_4$ quaternary compound is formed in this system. It melts congruently at 928°C [incongruently at 813°C (Schäfer and Nitsche 1977)], has polymorphous transformation at 685°C (Parasyuk et al. 1997, 1998), and crystallizes in the orthorhombic structure with the lattice parameters $a = 796.2$, $b = 682.3$, and $c = 656.9$ pm (Schäfer and Nitsche 1974, 1977). Cu_2SiSe_3 has a polymorphous transformation at 615°C, and the thermal effects at 658°C correspond to the polymorphous transformation of the solid solutions based on this compound.

The solubility of HgSe in the high-temperature modification of Cu_2SiSe_3 is equal to 7 mol% at 873°C, and the solubility of Cu_2SiSe_3 in HgSe at 400°C is not higher than 1 mol% (Parasyuk et al. 1997, 1998). This system was investigated through DTA, XRD, and metallography, and the ingots were annealed at 400°C for 250 h.

One more quaternary compound, $Cu_{5.976}Hg_{0.972}SiSe_6$, which crystallizes in the cubic structure with the lattice parameter $a = 1030.13 \pm 0.02$ pm and a calculated density of 6.5408 $\pm$ 0.0004 g·cm^{-3}, is formed in the Si–Cu–Hg–Se quaternary system (Gulay et al. 2004b).

2.16 Silicon–Copper–Aluminum–Selenium

The $CuAlSiSe_4$ quaternary compound, which crystallizes in the tetragonal structure with the lattice parameters $a = 560.4$, $c = 1086$ pm, and the calculated and experimental densities

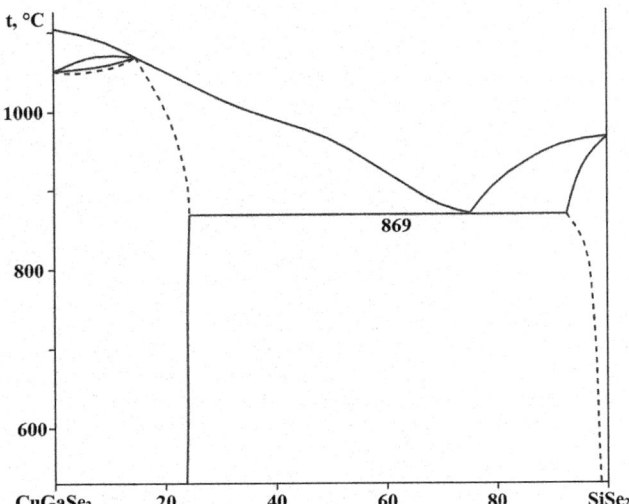

FIGURE 2.1 Phase equilibria in the SiSe₂–CuGaSe₂ system. (From Sachanyuk, V.P., et al., *J. Alloys Compd.*, **420**, 54, 2006.)

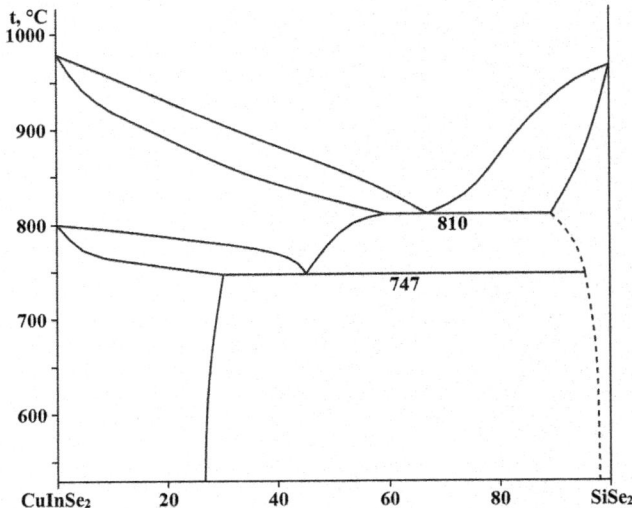

FIGURE 2.2 Phase diagram of the SiSe₂–CuInSe₂ system. (From Sachanyuk, V.P., et al., *J. Alloys Compd.*, **420**, 54, 2006.)

of 4.23 and 4.23 $g \cdot cm^{-3}$, respectively, is formed in the Si–Cu–Al–Se system (Hahn and Strick 1967b). This compound was obtained by long annealing of the elements' mixture or the mixture of binary components between 500°C and 1000°C. It has a brown color.

2.17 Silicon–Copper–Gallium–Selenium

SiSe₂–CuGaSe₂. The phase equilibria in this system, constructed through DTA and XRD, is presented in Figure 2.1 (Sachanyuk et al. 2006a). Since CuCaSe₂ forms according to the peritectic reaction L + (Ga₂Se₃) ↔ CuCaSe₂, the section is nonquasibinary in the 0 to ~13 mol% SiSe₂ range. The eutectic contains 25 mol% CuGaSe₂ and crystallizes at 869°C. The solubility of SiSe₂ in CuGaSe₂ reaches up to 24 mol% at 400°C. The boundary solid solution Cu₀.₇₆Ga₀.₇₆Si₀.₂₄Se₂ crystallizes in the tetragonal structure with the lattice parameters $a = 559.91 \pm 0.02$, $c = 1093.30 \pm 0.06$ pm, and a calculated density of 5.1636 ± 0.0006 $g \cdot cm^{-3}$. The solid solubility range of SiSe₂ is minor though it increases significantly with temperature.

The **CuGaSiSe₄** quaternary compound, which melts at 954°C and crystallizes in the tetragonal structure with the lattice parameters $a = 557.9$ and $c = 1072.1$ pm (Matsushita and Katsui 2005) [$a = 559.0$, $c = 538.0$ pm, and the calculated and experimental densities of 4.71 and 4.74 $g \cdot cm^{-3}$, respectively (Hahn and Strick 1967b)], is formed in the Si–Cu–Ga–Se system. This compound was obtained by long annealing of the elements' mixture or the mixture of binary components between 500°C and 1000°C (Hahn and Strick 1967b). It has a brown color.

2.18 Silicon–Copper–Indium–Selenium

SiSe₂–CuInSe₂. The phase diagram of this system, constructed through DTA and XRD, is given in Figure 2.2 (Sachanyuk et al. 2006a). It belongs to the eutectic type, and the eutectic

contains 33 mol% CuInSe₂ and crystallizes at 810°C. The solubility of SiSe₂ in CuInSe₂ reaches up to 59 mol% at the eutectic temperature and decreases to 25 mol% at 400°C. Solid solution based on CuInSe₂ has a polymorphic transformation at 747°C, and the eutectoid point contains ~ 45 mol% SiSe₂. The boundary solid solution Cu₀.₇₅Ga₀.₇₅Si₀.₂₅Se₂ crystallizes in the tetragonal structure with the lattice parameters $a = 577.49 \pm 0.04$, $c = 1151.9 \pm 0.2$ pm, and a calculated density of 5.136 ± 0.002 $g \cdot cm^{-3}$. The solid solubility range of SiSe₂ is minor though it increases significantly with temperature.

According to the data of Matsushita and Katsui (2005), the **CuInSiSe₄** quaternary compound, which melts at 836°C and crystallizes in the tetragonal structure with the lattice parameters $a = 557.8$, $c = 1151.0$ pm, is formed in the Si–Cu–In–Se system.

2.19 Silicon–Copper–Yttrium–Selenium

SiSe₂–Cu₂Se–Y₂Se₃. The isothermal section of this system at 600°C was constructed by Lychmanyuk et al. (2006a) using the alloys annealed at this temperature for 240 h and quenched in the air and is presented in Figure 2.3. The **CuY₃SiSe₇** quaternary compound is formed in this system. It crystallizes in the hexagonal structure with the lattice parameters $a = 1026.9 \pm 0.1$, $c = 595.4 \pm 0.1$ pm, and a calculated density of 5.565 $g \cdot cm^{-3}$ (Gulay et al. 2005b).

To synthesize this compound, the calculated amounts of the elements were sealed in an evacuated silica ampoule (Gulay et al. 2005b). The synthesis was realized in a tube furnace. The ampoule was heated with a heating rate of $30°C \cdot h^{-1}$ to the maximal temperature, 1150°C. The sample was kept at this temperature for 4 h. After that, it was cooled slowly ($10°C \cdot h^{-1}$) to 600°C and annealed at this temperature for 240 h. After annealing, the ampoule with the sample was quenched in cold water. The prepared product was a brown-colored compact alloy.

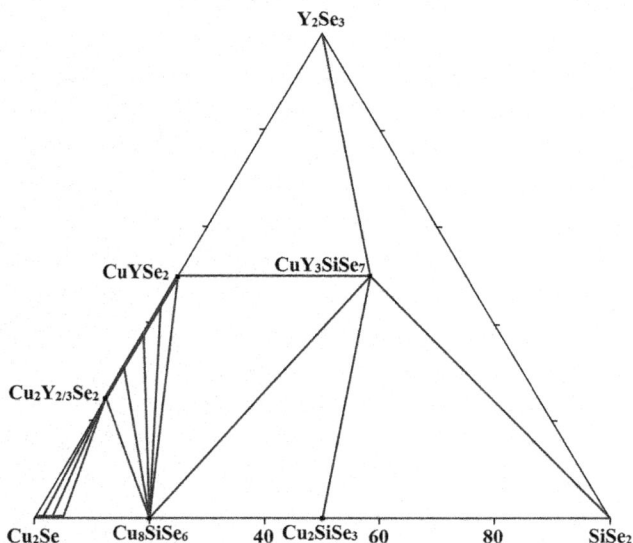

FIGURE 2.3 Isothermal section of the $SiSe_2$–Cu_2Se–Y_2Se_3 quasiternary system at 600°C. (From Lychmanyuk, O.S., et al., *Nauk. Visnyk Volyns'k. Derzh. Univ. im. Lesi Ukrainky. Ser. Khim. nauky*, (4), 118, 2006.)

FIGURE 2.4 Isothermal section of the $SiSe_2$–Cu_2Se–Pr_2Se_3 quasiternary system at 600°C. (From Lychmanyuk, O.S., et. al., *Nauk. Visnyk Volyns'k. Derzh. Univ. im. Lesi Ukrainky. Ser. Khim. nauky*, (15), 10, 2007.)

2.20 Silicon–Copper–Lanthanum–Selenium

The **$CuLa_3SiSe_7$** quaternary compound, which crystallizes in the hexagonal structure with the lattice parameters $a = 1071.16 \pm 0.01$, $c = 607.54 \pm 0.01$ pm, and a calculated density of 5.836 $g \cdot cm^{-3}$ (Gulay et al. 2007b) [$a = 1061 \pm 1$ and $c = 608.0 \pm 0.5$ pm (Guittard et al. 1968; Guittard and Julien-Pouzol 1970)], is formed in the Si–Cu–La–Se system.

The synthesis of this compound was realized using a mixture of the elements (Gulay et al. 2007b). The calculated amounts of the components were sealed in an evacuated silica ampoule. The ampoule was heated with a heating rate of $30°C \cdot h^{-1}$ to the maximal temperature, 1150°C, and kept at this temperature for 3 h. Then it was cooled slowly ($10°C \cdot h^{-1}$) to 600°C and annealed at this temperature for 240 h. After annealing, the ampoule with the sample was quenched in cold water.

$CuLa_3SiSe_7$ was also obtained from a mixture of La_2Se_3, Cu_2Se, Si, and S, which was pressed and introduced into the silica ampoule and gradually heated up to 800°C in about a week (Guittard and Julien-Pouzol 1970). At the end of the preparation, it is necessary to heat the sample to 1200°C to obtain the pure hexagonal phase.

2.21 Silicon–Copper–Cerium–Selenium

The **$CuCe_3SiSe_7$** quaternary compound, which crystallizes in the hexagonal structure with the lattice parameters $a = 1062.15 \pm 0.02$, $c = 603.86 \pm 0.02$ pm, and a calculated density of 5.993 $g \cdot cm^{-3}$ (Gulay et al. 2007b) [$a = 1054 \pm 1$ and $c = 606.0 \pm 0.5$ pm (Guittard et al. 1968; Guittard and Julien-Pouzol 1970)], is formed in the Si–Cu–Ce–Se system. This compound was synthesized in the same way as $CuLa_3SiSe_7$ was prepared using Ce_2Se_3 instead of La_2Se_3 (Guittard and Julien-Pouzol 1970; Gulay et al. 2007b).

2.22 Silicon–Copper–Praseodymium–Selenium

$SiSe_2$–Cu_2Se–Pr_2Se_3. The isothermal section of this system at 600°C was constructed by Lychmanyuk et al. (2007a) using the alloys annealed at this temperature for 240 h and quenched in the air is presented in Figure 2.4. The **$CuPr_3SiSe_7$** quaternary compound, which crystallizes in the hexagonal structure with the lattice parameters $a = 1058.54 \pm 0.02$, $c = 603.67 \pm 0.02$ pm, and a calculated density of 6.049 $g \cdot cm^{-3}$ (Gulay et al. 2007b) [$a = 1051 \pm 1$ and $c = 603.0 \pm 0.5$ pm (Guittard et al. 1968; Guittard and Julien-Pouzol 1970)], is formed in this system. This compound was synthesized in the same way as $CuLa_3SiSe_7$ was prepared using Pr_2Se_3 instead of La_2Se_3 (Guittard and Julien-Pouzol 1970; Gulay et al. 2007b).

2.23 Silicon–Copper–Neodymium–Selenium

The **$CuNd_3SiSe_7$** quaternary compound, which crystallizes in the hexagonal structure with the lattice parameters $a = 1056.00 \pm 0.02$, $c = 603.13 \pm 0.02$ pm, and a calculated density of 6.140 $g \cdot cm^{-3}$ (Gulay et al. 2007b) [$a = 1046 \pm 1$ and $c = 602.0 \pm 0.5$ pm (Guittard et al. 1968; Guittard and Julien-Pouzol 1970)], is formed in the Si–Cu–Nd–Se system. This compound was synthesized in the same way as $CuLa_3SiSe_7$ was prepared using Nd_2Se_3 instead of La_2Se_3 (Guittard and Julien-Pouzol 1970; Gulay et al. 2007b).

2.24 Silicon–Copper–Samarium–Selenium

The **$CuSm_3SiSe_7$** quaternary compound, which crystallizes in the hexagonal structure with the lattice parameters $a = 1050.42 \pm 0.02$, $c = 602.44 \pm 0.02$ pm, and a calculated density of 6.319 $g \cdot cm^{-3}$ (Gulay et al. 2007b) [$a = 1040 \pm 1$ and $c = 600.0$

± 0.5 pm (Guittard et al. 1968; Guittard and Julien-Pouzol 1970)], is formed in the Si–Cu–Sm–Se system. This compound was synthesized in the same way as CuLa$_3$SiSe$_7$ was prepared using Sm$_2$Se$_3$ instead of La$_2$Se$_3$ (Guittard and Julien-Pouzol 1970; Gulay et al. 2007b).

2.25 Silicon–Copper–Gadolinium–Selenium

The **CuGd$_3$SiSe$_7$** quaternary compound, which crystallizes in the hexagonal structure with the lattice parameters a = 1046.43 ± 0.03, c = 602.40 ± 0.03 pm, and a calculated density of 6.488 g·cm^{-3} (Gulay et al. 2007b) [a = 1034 ± 1 and c = 599.0 ± 0.5 pm (Guittard et al. 1968; Guittard and Julien-Pouzol 1970)], is formed in the Si–Cu–Gd–Se system. This compound was synthesized in the same way as CuLa$_3$SiSe$_7$ was prepared using Gd$_2$Se$_3$ instead of La$_2$Se$_3$ (Guittard and Julien-Pouzol 1970; Gulay et al. 2007b).

2.26 Silicon–Copper–Terbium–Selenium

The **CuTb$_3$SiSe$_7$** quaternary compound, which crystallizes in the hexagonal structure with the lattice parameters a = 1041.15 ± 0.03, c = 601.32 ± 0.03 pm, and a calculated density of 6.595 g·cm^{-3} (Gulay et al. 2007b) [a = 1031 ± 1 and c = 598.0 ± 0.5 pm (Guittard et al. 1968; Guittard and Julien-Pouzol 1970)], is formed in the Si–Cu–Tb–Se system. This compound was synthesized in the same way as CuLa$_3$SiSe$_7$ was prepared using Tb$_2$Se$_3$ instead of La$_2$Se$_3$ (Guittard and Julien-Pouzol 1970; Gulay et al. 2007b).

2.27 Silicon–Copper–Dysprosium–Selenium

The **CuTb$_3$SiSe$_7$** quaternary compound, which crystallizes in the hexagonal structure with the lattice parameters a = 1038.23 ± 0.03, c = 601.92 ± 0.03 pm, and a calculated density of

6.689 g·cm^{-3} (Gulay et al. 2007b) [a = 1028 ± 1 and c = 597.0 ± 0.5 pm (Guittard et al. 1968; Guittard and Julien-Pouzol 1970)], is formed in the Si–Cu–Dy–Se system. This compound was synthesized in the same way as CuLa$_3$SiSe$_7$ was prepared using Dy$_2$Se$_3$ instead of La$_2$Se$_3$ (Guittard and Julien-Pouzol 1970; Gulay et al. 2007b).

2.28 Silicon–Copper–Holmium–Selenium

SiSe$_2$–Cu$_2$Se–Ho$_2$Se$_3$. The isothermal section of this system at 600°C was constructed by Lychmanyuk et al. (2007b) using the alloys annealed at this temperature for 240 h and quenched in cold water and is presented in Figure 2.5. The **CuHo$_3$SiSe$_7$** quaternary compound is formed in this system. It crystallizes in the hexagonal structure with the lattice parameters a = 1022.94 ± 0.02 and c = 595.54 ± 0.02 pm.

2.29 Silicon–Copper–Lead–Selenium

SiSe$_2$–Cu$_2$Se–PbSe. The isothermal section of this system at 400°C was constructed by Shpak et al. (2014) using the alloys annealed at this temperature for 540 h and is given in Figure 2.6. Six three-phase fields and 12 two-phase equilibria were identified. The **CuPb$_{1.5}$SiSe$_4$** quaternary compound, which crystallizes in the cubic structure with the lattice parameters a = 1433.78 ± 0.06 pm and a calculated density of 6.4738 g·cm^{-3} is formed in this system.

According to the data of Gulay et al. (2005c), one more quaternary compound, **Cu$_{0.5}$Pb$_{1.75}$SiSe$_4$**, that crystallizes in the cubic structure with the lattice parameter a = 1434.5 ± 0.1 pm, is formed in the Si–Cu–Pb–Se system. It was prepared by fusion of the elements in an evacuated silica ampoule. The synthesis was carried out in a tube furnace with a heating rate of 20°C·h^{-1}. The ampoule was heated to 1150°C; the sample was

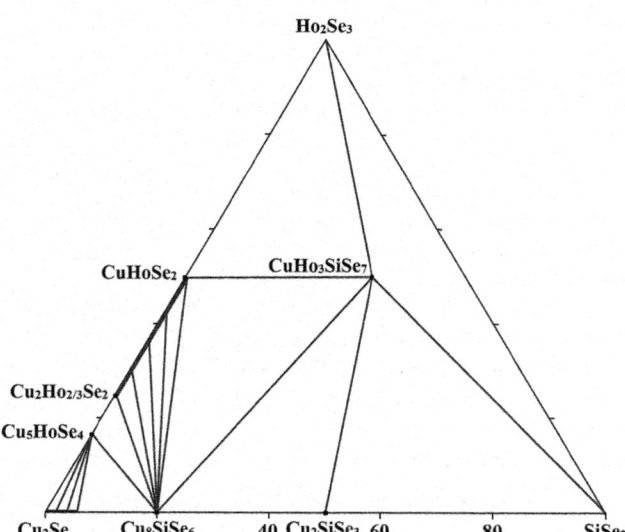

FIGURE 2.5 Isothermal section of the SiSe$_2$–Cu$_2$Se–Ho$_2$Se$_3$ quasiternary system at 600°C. (From Lychmanyuk, O.S., et al., *Pol. J. Chem.*, **81**(3), 353, 2007.)

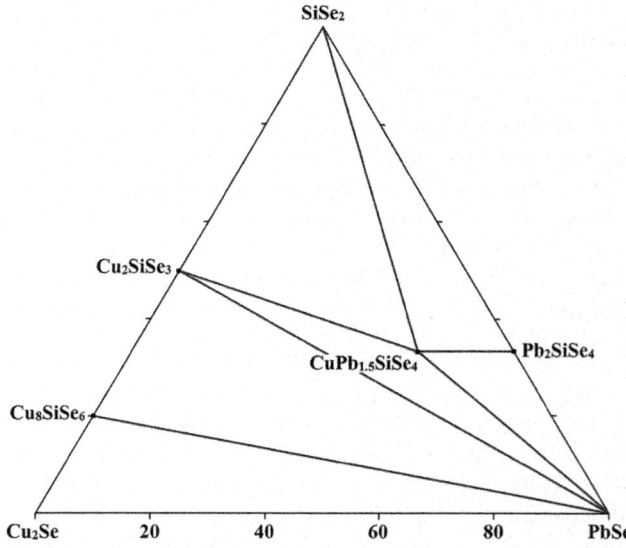

FIGURE 2.6 Isothermal section of the SiSe$_2$–Cu$_2$Se–PbSe quasiternary system at 400°C. (From Shpak, O., et al., *Nauk. Visnyk Skhidnoyevrop. Nats. Univ. im. Lesi Ukrainky. Ser. Khim. nauky*, [21(298)], 39, 2014.)

kept at that maximal temperature for 4 h and was then cooled slowly to 600°C with a rate of 10°C·h⁻¹ and annealed at this temperature for another 240 h. Finally, the furnace was turned off, and the ampoule was cooled to room temperature.

2.30 Silicon–Copper–Chromium–Selenium

The $Cu_4Si_2Cr_6Se_{15}$ quaternary compound, which decomposes at 1015°C, is formed in the Si–Cu–Cr–Se system (Babitsyna and Novotortsev 1986). It was synthesized from the elements at 960–980°C for 100 h, with the next annealing at 900°C for 300 h.

2.31 Silicon–Copper–Iodine–Selenium

The $Cu_{8-x}SiSe_{6-x}I_x$ quaternary phase, which crystallizes in the cubic structure with the lattice parameter $a = 1028.52 \pm 0.04$ and a calculated density of 6.112 g·cm⁻³ for $x = 0.56$ (Nilges and Pfitzner 2005), and $a = 1034.6 \pm 0.2$ for $x = 0.83$ (Kuhs et al. 1979), is formed in the Si–Cu–I–Se system. The phase with $x = 0.56$ was obtained by the reaction of the stoichiometric mixture of elements in an evacuated, sealed silica ampoule at 1000°C (Nilges and Pfitzner 2005). After grinding, the phase was annealed at 920°C for 19 days. The phase with $x = 0.83$ was prepared by reacting stoichiometric amounts of elements in an evacuated, sealed quartz ampoule for about 6 days at temperatures of 600°C to 700°C (Kuhs et al. 1979).

2.32 Silicon–Copper–Manganese–Selenium

The $Cu_2MnSiSe_4$ quaternary compound, which melts incongruently at 888°C and crystallizes in the orthorhombic structure with the lattice parameters $a = 791.4$, $b = 678.4$ and $c = 650.2$ pm (Schäfer and Nitsche 1974, 1977; Lamarche et al. 1991), is formed in the Si–Cu–Mn–Se system. This compound was produced by the melt and annealed technique (Lamarche et al. 1991). To obtain it, the components of 1.0-g sample were sealed under vacuum in a quartz capsule that had previously been coated with carbon in order to prevent a reaction of the charge with the quartz. The capsule was then raised to a temperature of 1150°C for approximately an hour to allow some reaction to occur. The sample was held at 1150°C for a time of 0.5 to 1 h and then cooled to room temperature. It was then annealed at 625°C for a week and then quenched to room temperature. Single crystals of the title compound were grown through chemical transport reactions using iodine as a transport agent (Schäfer and Nitsche 1974).

2.33 Silicon–Copper–Iron–Selenium

The $Cu_2FeSiSe_4$ quaternary compound, which melts incongruently at 781°C and has two polymorphic modifications, is formed in the Si–Cu–Fe–Se system (Schäfer and Nitsche 1974, 1977). The first modification crystallizes in the orthorhombic structure with the lattice parameters $a = 777.5$, $b = 673.1$, $c = 645.1$ pm, and the second modification crystallizes in the tetragonal structure with the lattice parameters $a = 554.9$ and $c = 1095.1$ pm (Schäfer and Nitsche 1974, 1977; Quintero et al. 1999).

The title compound was prepared by the melt and anneals technique (Quintero et al. 1999). The mixture was made from the appropriate amounts of the elements and was sealed under vacuum in a small quartz ampoule, which had previously been carbonized to prevent the interaction of the components with the quartz. The components were melted together at 1150°C for about an hour, annealed to equilibrium at 500°C, and then cooled to room temperature by leaving the ampoule in the switched-off furnace. Single crystals of the title compound were grown through chemical transport reactions using iodine as a transport agent (Schäfer and Nitsche 1974).

2.34 Silicon–Copper–Cobalt–Selenium

The $Cu_2CoSiSe_4$ quaternary compound, which melts incongruently at 839°C and crystallizes in the orthorhombic structure with the lattice parameters $a = 556.8 \pm 0.3$, $b = 550.1 \pm 0.2$, $c = 539.8 \pm 0.3$ pm, and a calculated density of 5.32 g·cm⁻³, is formed in the Si–Cu–Co–Se system (Quintero et al. 2014a, 2014b). To prepare this compound, the components of 1 g sample were sealed under vacuum (~ 10⁻³ Pa) in a small quartz ampoule, which had previously been carbonized to prevent interaction of the components with the quartz. The synthesis was realized inside a vertical furnace. The ampoule with the components was heated up to 200°C and kept for about 1–2 h, then the temperature was raised to 500°C using a rate of 40°C·h⁻¹, and held at this temperature for 14 h. After that, the sample was heated from 500°C to 800°C at a rate of 30°C·h⁻¹ and kept at this temperature for another 14 h. Then the temperature was raised to 1150°C at 60°C·h⁻¹, and the mixture was melted at this temperature for about 2–3 h. The furnace was brought slowly (4°C·h⁻¹) to 600°C, and the sample was annealed at this temperature for one month. Then the sample was slowly cooled to room temperature at a rate of about 2°C·h⁻¹.

2.35 Silicon–Copper–Nickel–Selenium

Cu_2SiSe_3–NiSe. The phase diagram of this system, constructed through DTA and XRD, is a eutectic type (Figure 2.7) (Nazarchuk et al. 2008). The eutectic contains 78 mol% NiSe and crystallizes at 802°C. The polymorphic transition of the solid solution based on NiSe takes place at 177°C.

$SiSe_2$–Cu_2Se–NiSe. The isothermal section of this system at 400°C was constructed by Nazarchuk et al. (2008) using the alloys annealed at this temperature for 240 h and quenched in cold water. Quaternary compounds are not formed in the system. The system is divided by the Cu_8SiSe_6–NiSe and Cu_2SiSe_3–NiSe quasibinary sections into three subsystems.

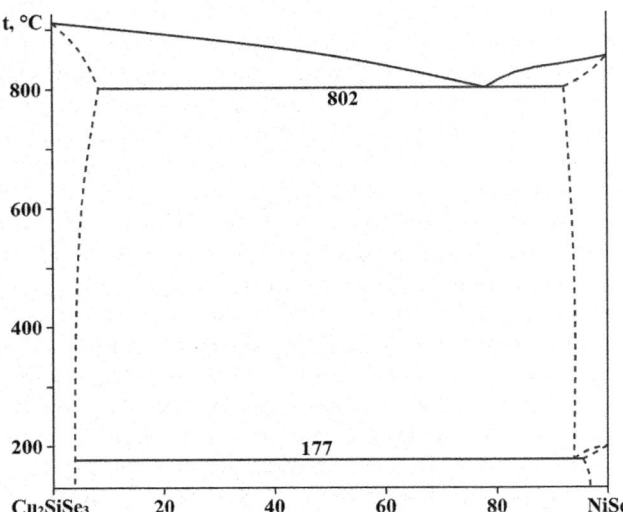

FIGURE 2.7 Phase diagram of the Cu₂SiSe₃–NiSe system. (From Nazarchuk, O.P., et al., *Nauk. Visnyk Volyns'k. Nats. Univ. im. Lesi Ukrainky. Ser. Khim. nauky*, (16), 21, 2008.)

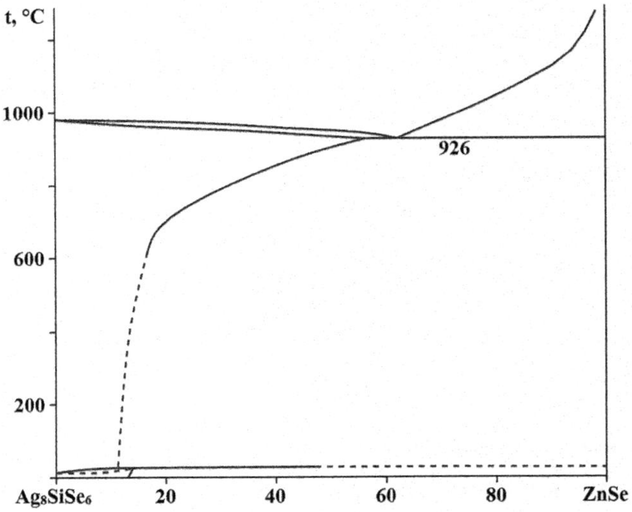

FIGURE 2.8 Phase diagram of the Ag₈SiSe₆–ZnSe system. (From Piskach, L.V. et al., *J. Alloys Compd.*, **421**(1–2), 98, 2006.)

2.36 Silicon–Silver–Barium–Selenium

The **Ag₂BaSiSe₄** quaternary compound, which crystallizes in the tetragonal structure with the lattice parameters $a = 706.6 \pm 0.3$, $c = 823.3 \pm 0.7$ pm, a calculated density of 5.631 g·cm⁻³, and an energy gap of 1.83 eV, is formed in the Si–Ag–Ba–Se system (Nian et al. 2018). For the syntheses of the target compound, a stoichiometric mixture of the starting materials Ag, Ba, Si, and Se (molar ratio 2:1:1:4) was loaded into a silica tube, which was flame-sealed under a high vacuum of 10⁻³ Pa. After that, the tube was moved into a furnace. The furnace was programmed by the following steps: heated to 600°C in 30 h and kept at this temperature for 40 h; then, heated to 1000°C within 20 h and dwelled at this temperature for 100 h; finally, followed by slow cooling to 400°C at a rate of 5°C·h⁻¹ and then cooled to room temperature with 10 h. After the reaction, Ag₂BaSiSe₄ with red color was successfully obtained, and it is stable in the air for several months.

2.37 Silicon–Silver–Zinc–Selenium

Ag₈SiSe₆–ZnSe. The phase diagram of this system, constructed through DTA, XRD, and metallography, is a eutectic type (Figure 2.8) (Piskach et al. 2006). The eutectic composition and temperature are 62 mol% ZnSe and 926°C, respectively. The solubility of ZnSe in Ag₈SiSe₆ at the eutectic temperature is 57 mol% and less than 2 mol% at room temperature (Parasyuk et al. 2003c; Piskach et al. 2006).

SiSe₂–Ag₂Se–ZnSe. The isothermal section of this quasiternary system at room temperature was constructed by Parasyuk et al. (2003c) using the ingots, annealed at 400°C for 250 h with further cooling by turning off the furnace. No quaternary phases were found. The system is divided by the Ag₈SiSe₆–ZnSe quasibinary system for two three-phase regions.

2.38 Silicon–Silver–Cadmium–Selenium

Ag₈SiSe₆–CdSe. The phase diagram of this system, constructed through DTA, XRD, and metallography, is a peritectic type (Piskach et al. 2006). The peritectic composition and temperature are 86 mol% CdSe and 1002°C, respectively. The solubility of CdSe in Ag₈SiSe₆ at the peritectic temperature is 96 mol%.

SiSe₂–Ag₂Se–CdSe. The isothermal section of this quasiternary system at room temperature was constructed by Parasyuk et al. (2003c) using the ingots, annealed at 400°C for 250 h with further cooling by turning off the furnace. No quaternary phases were found. The system is divided by the Ag₈SiSe₆–CdSe and Ag₈SiSe₆–Cd₄SiSe₆ quasibinary systems for three three-phase regions.

2.39 Silicon–Silver–Mercury–Selenium

SiSe₂–Ag₂Se–HgSe. The isothermal section of this quasiternary system at room temperature is shown in Figure 2.9 (Parasyuk et al. 2003b). The existence of several intermediate phases located in the Ag₈SiSe₆–Hg₄SiSe₆ section was established. The **Ag₆HgSiSe₆** quaternary compound is formed in this system with the homogeneity range from 23 to 43 mol% Hg₄SiSe₆ (Ag₆.₁₆₋₄.₅₆Hg₀.₉₂₋₁.₇₂SiSe₆, γ-phase). It melts congruently at 939°C (Li et al. 2019a) and crystallizes in the orthorhombic structure with the lattice parameters $a = 767.52 \pm 0.09$, $b = 767.72 \pm 0.09$, $c = 1085.4 \pm 0.1$ pm, a calculated density of 6.929 ± 0.002 g·cm⁻³ (Gulay et al. 2002b), and an energy gap of 1.00 eV (Li et al. 2019a).

The formation of three new intermediate phases Ag₋₇.₀₄₋₋₆.₃₂Hg₋₀.₄₈₋₋₀.₈₄SiSe₆ (β-phase), Ag₋₃.₄₄₋₋₂.₉₆Hg₋₂.₂₈₋₋₂.₅₂SiSe₆ (δ-phase), and Ag₂Hg₃SiSe₆ (ε-phase) was also found (Parasyuk et al. 2003b). The β-phase (of Ag₆.₄Hg₀.₈SiSe₆ composition) and δ-phase (of Ag₃.₂Hg₂.₄SiSe₆ composition) crystallize in

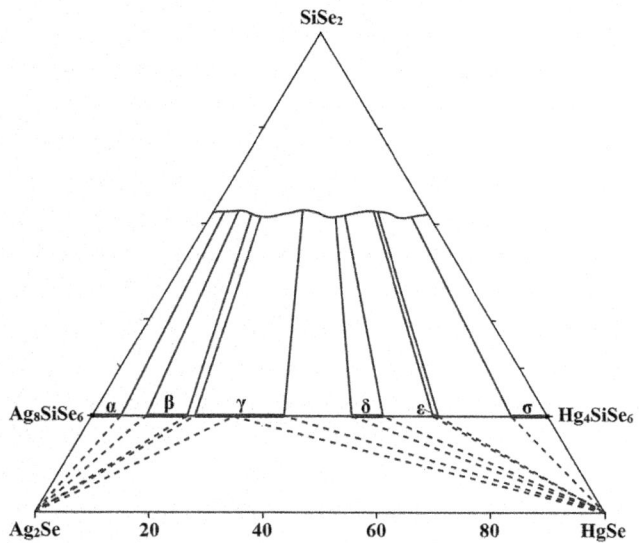

FIGURE 2.9 Isothermal section of the SiSe$_2$–Ag$_2$Se–HgSe quasiternary system at room temperature. (From Parasyuk, O.V., et al., *J. Alloys Compd.*, **348**(1–2), 157, 2003b.)

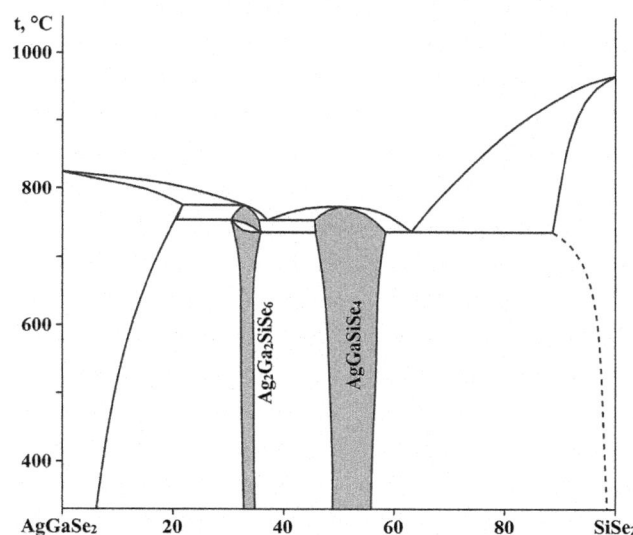

FIGURE 2.10 Phase diagram of the SiSe$_2$–AgGaSe$_2$ system. (From Parasyuk, O.V., et al., *RSC Adv.*, **6**(93), 90958, 2016.)

the cubic structure with the lattice parameters $a = 1088.06 \pm 0.04$ and 1071.08 ± 0.02 pm and the calculated and experimental densities of 6.9212 ± 0.0008 and 7.1479 ± 0.0004 g·cm^{-3}, respectively.

Polycrystalline sample of Ag$_6$HgSiSe$_6$ was obtained by traditional solid-state reactions in a stoichiometric mixture of Ag$_2$Se (1.5 mM), HgSe (0.5 mM), and SiSe$_2$ (0.5 mM) (Li et al. 2019a). The mixture was loaded into a fused silica tube, which was then evacuated to 10^{-3} Pa and sealed. Afterward, the sample was heated in a furnace to 750°C over 15 h and kept at that temperature for 120 h. Then the furnace was shut off. All manipulations were performed in an Ar-filled glove box with H$_2$O and O$_2$ contents of less than 0.1 ppm.

The alloys were annealed at 400°C for 500 h (Parasyuk et al. 2003b). After that the ampoules were cooled to 100°C at the rate of 10°C·h^{-1}, and at this temperature the furnace was turned off and the ampoules with the samples were cooled to room temperature.

2.40 Silicon–Silver–Gallium–Selenium

SiSe$_2$–AgGaSe$_2$. The phase equilibria in this system, constructed through DTA and XRD, is presented in Figure 2.10 (Parasyuk et al. 2016). The system is characterized by the formation of two quaternary compounds: **AgGaSiSe$_4$** and **Ag$_2$Ga$_2$SiSe$_6$**. The eutectics contain 37 and 63 mol% SiSe$_2$ and crystallize at 754°C and 737°C, respectively. At 400°C, AgGaSe$_2$ dissolves 9 mol% SiSe$_2$. The alloys for the investigations were annealed at 400°C for 500 h.

AgGaSiSe$_4$ melts congruently at 774°C (Parasyuk et al. 2016) [at 759°C (Zhang et al. 2016)]. Its homogeneity region is quite wide at the eutectic temperature (~46–58 mol% SiSe$_2$) and narrowed considerably at lower temperatures. This compound crystallizes in the orthorhombic structure (in the

tetragonal structure; see Goodchild et al. 1981) with the lattice parameters $a = 6290.5$, $b = 711.5$, $c = 1237.6$ pm, a calculated density of 5.003 g·cm^{-3} at 180 K, and an energy gap of 2.63 eV (Zhang et al. 2016). To prepare AgGaSiSe$_4$, the appropriate amounts of AgGaSe$_2$ and SiSe$_2$ (molar ratio 1:1) were mixed, and the reaction mixture was loaded into a silica tube, which was sealed under vacuum (10^{-3} Pa) and placed in a furnace. The tube was heated to 700°C in 15 h, kept at that temperature for 24 h, and then cooled to room temperature. The procedure was repeated three times with intermittent grindings. Single crystals were grown from as-prepared AgGaSiSe$_4$ powder, which was reloaded into a fused silica tube under Ar atmosphere in a glove box. The tube was sealed under a vacuum (10^{-3} Pa) and placed in a furnace. The sample was heated to 900°C by 20 h, kept at that temperature for 24 h, cooled at 3°C·h^{-1} to 600°C, and then cooled rapidly to room temperature. The product consisted of yellow crystals, which are stable in the air.

Ag$_2$Ga$_2$SiSe$_6$ melts incongruently at 769°C and exists in two polymorphic modifications (Parasyuk et al. 2016). The polymorphic transition takes place in the subsolidus region in the temperature range of 736–748°C, which indicates a certain homogeneity region for this compound, which does not exceed 2 mol% at 400°C. α-Ag$_2$Ga$_2$SiSe$_6$ crystallizes in the tetragonal structure with the lattice parameters $a = 590.21 \pm 0.01$, $c = 1041.12 \pm 0.10$ pm, and a calculated density of 5.344 g·cm^{-3}. An energy gap of this compound decreases with increasing temperature (100–300 K) from 2.13 to 1.97 eV.

According to the data of Mei et al. (2014), one more quaternary compound, **Ag$_2$Ga$_3$SiSe$_8$**, was found in the SiSe$_2$–AgGaSe$_2$ system. It melts congruently at 774°C and crystallizes in the tetragonal structure with the lattice parameters $a = 590.41 \pm 0.08$, $c = 1049.9 \pm 0.2$ pm, a calculated density of 5.411 g·cm^{-3}, and an energy gap of 2.30 eV. To synthesize this compound, appropriate amounts of AgGaSe$_2$ and SiSe$_2$ (molar ratio 3:1) were mixed, and the reaction mixture was loaded into

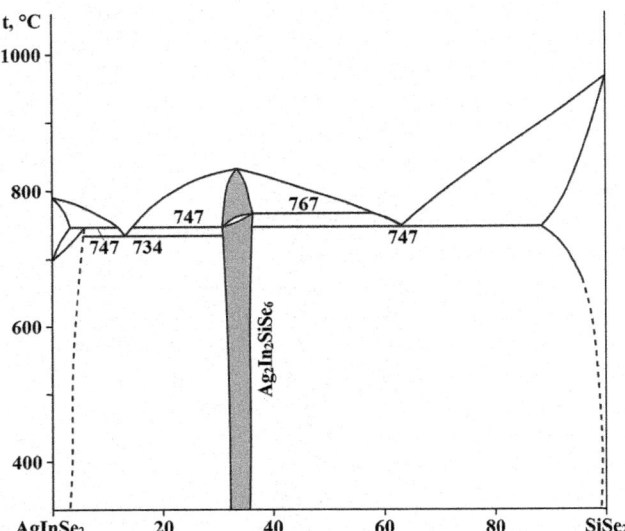

FIGURE 2.11 Phase diagram of the SiSe₂–AgInSe₂ system. (From Olekseyuk, I.D., et al., *J. Alloys Compd.*, **414**(1–2), 73, 2006.)

the silica tube. The tube was sealed under a vacuum (10^{-3} Pa) and then placed in a furnace. Then it was heated to 700°C in 15 h, kept at that temperature for 24 h, and cooled to room temperature. The reaction was repeated three times with intermittent grinding.

2.41 Silicon–Silver–Indium–Selenium

SiSe₂–AgInSe₂. The phase equilibria in this system, constructed through DTA and XRD, is given in Figure 2.11 (Olekseyuk et al. 2006c). The eutectics contain 13 and 62 mol% SiSe₂ and crystallize at 734°C and 747°C, respectively. On the AgInSe₂-side, the solid solutions with the tetragonal structure extend to 3 mol% SiSe₂ at 400°C. The polymorphous transition of AgInSe₂ takes place at 747°C. The extent of the solid solution based on SiSe₂ is insignificant, ≈3 mol% at 400°C, increasing somewhat with temperature. The existence of the **Ag₂In₂SiSe₆** quaternary compound was established in the system with a homogeneity region at 400°C ranging from 32 to 35 mol% SiSe₂. It crystallizes in the monoclinic structure with the lattice parameters a = 1266.83 ± 0.04, b = 745.65 ± 0.03, c = 1261.33 ± 0.04 pm, β = 109.286 ± 0.002°, and a calculated density of 5.5941 ± 0.0007 g·cm⁻³. The alloys for the investigations were annealed at 400°C for 250 h, with the next quenching into cold water.

2.42 Silicon–Silver–Lanthanum–Selenium

The **AgLa₃SiSe₇** quaternary compound, which crystallizes in the hexagonal structure with the lattice parameters a = 1081.7 ± 0.1 and c = 606.40 ± 0.04 pm (Daszkiewicz et al. 2009b) [a = 1070 ± 1 and c = 613.0 ± 0.5 pm (Guittard et al. 1968; Guittard and Julien-Pouzol 1970); a = 1057.6 ± 0.3, c = 598.7 ± 0.5 pm, and a calculated density of 6.33 g·cm⁻³ (Lin et al. 1997)], is formed in the Si–Ag–La–Se system. This compound was prepared by fusion of the elements (molar ratio La/Ag/Si/Se = 3:1:1:7) into

an evacuated silica ampoule in a tube furnace (Daszkiewicz et al. 2009b). The ampoule was heated with a heating rate of 30°C·h⁻¹ to the maximal temperature, 1150°C, and was kept at this temperature for 3 h. Afterwards, it was cooled slowly (10°C·h⁻¹) to 600°C and annealed at this temperature for 240 h. After annealing the ampoule was quenched in cold water.

AgLa₃SiSe₇ can also be synthesized another way (Lin et al. 1997). A mixture of LaOCl (300 mg) and boron powder (20 mg) was ground and pressed into a pellet, which was then sealed into an evacuated quartz tube. This tube was kept at 950°C for 24 h. The resultant mixture was then ground with Se (100 mg), Si (20 mg), and Ag (100 mg) and pressed into a pellet. This pellet and I₂ (20 mg) were sealed into a quartz tube under a vacuum. The mixture was allowed to react at 350°C for 24 h and then at 950°C for 96 h. After cooling, well-formed black prism-shaped crystals were obtained.

2.43 Silicon–Silver–Cerium–Selenium

The **AgCe₃SiSe₇** quaternary compound, which crystallizes in the hexagonal structure with the lattice parameters a = 1073.1 ± 0.1 and c = 603.1 ± 0.1 pm (Daszkiewicz et al. 2009b) [a = 1060 and c = 611 pm (Guittard and Julien-Pouzol 1970)], is formed in the Si–Ag–Ce–Se system. This compound was prepared in the same way as AgLa₃SiSe₇ was synthesized using Ce instead of La (Daszkiewicz et al. 2009b).

2.44 Silicon–Silver–Praseodymium–Selenium

The **AgPr₃SiSe₇** quaternary compound, which crystallizes in the hexagonal structure with the lattice parameters a = 1065.5 ± 0.1 and c = 600.5 ± 0.1 pm (Daszkiewicz et al. 2009b) [a = 1051 and c = 610 pm (Guittard and Julien-Pouzol 1970)], is formed in the Si–Ag–Pr–Se system. This compound was prepared in the same way as AgLa₃SiSe₇ was synthesized using Pr instead of La (Daszkiewicz et al. 2009b).

2.45 Silicon–Silver–Neodymium–Selenium

The **AgNd₃SiSe₇** quaternary compound, which crystallizes in the hexagonal structure with the lattice parameters a = 1060.2 ± 0.1 and c = 598.8 ± 0.1 pm (Daszkiewicz et al. 2009b) [a = 1045 and c = 610 pm (Guittard and Julien-Pouzol 1970)], is formed in the Si–Ag–Nd–Se system. This compound was prepared in the same way as AgLa₃SiSe₇ was synthesized using Nd instead of La (Daszkiewicz et al. 2009b).

2.46 Silicon–Silver–Samarium–Selenium

The **Ag₀.₈₅Sm₃SiSe₇** quaternary compound, which crystallizes in the hexagonal structure with the lattice parameters a = 1049.7 ± 0.1 and c = 594.66 ± 0.06 pm (Daszkiewicz et al. 2009b), is formed in the Si–Ag–Sm–Se system. This compound was prepared in the same way as AgLa₃SiSe₇ was synthesized using Sm instead of La.

2.47 Silicon–Silver–Gadolinium–Selenium

The $Ag_{0.81}Gd_3SiSe_7$ quaternary compound, which crystallizes in the hexagonal structure with the lattice parameters a = 1043.3 ± 0.1 and c = 593.67 ± 0.01 pm (Daszkiewicz et al. 2009b), is formed in the Si–Ag–Gd–Se system. This compound was prepared in the same way as $AgLa_3SiSe_7$ was synthesized using Gd instead of La.

2.48 Silicon–Silver–Terbium–Selenium

The $Ag_{0.70}Tb_3SiSe_7$ quaternary compound, which crystallizes in the hexagonal structure with the lattice parameters a = 1036.2 ± 0.1 and c = 592.96 ± 0.05 pm (Daszkiewicz et al. 2009b), is formed in the Si–Ag–Tb–Se system. This compound was prepared in the same way as $AgLa_3SiSe_7$ was synthesized using Tb instead of La.

2.49 Silicon–Silver–Dysprosium–Selenium

The $Ag_{0.72}Dy_3SiSe_7$ quaternary compound, which crystallizes in the hexagonal structure with the lattice parameters a = 1033.2 ± 0.1 and c = 594.94 ± 0.05 pm (Daszkiewicz et al. 2009b), is formed in the Si–Ag–Dy–Se system. This compound was prepared in the same way as $AgLa_3SiSe_7$ was synthesized using Dy instead of La.

2.50 Silicon–Silver–Lead–Selenium

The $Ag_{0.5}Pb_{1.75}SiSe_4$ quaternary compound, which crystallizes in the cubic structure with the lattice parameter a = 1449.95 ± 0.02 pm with a calculated density of 6.6274 g·cm^{-3}, is formed in the Si–Ag–Pb–Se system (Gulay et al. 2005c). It was prepared by fusion of the elements in an evacuated silica ampoule. The synthesis was carried out in a tube furnace with a heating rate of 20°C·h^{-1}. The ampoule was heated to 1150°C; the sample was kept at that maximal temperature for 4 h and was then cooled slowly to 600°C with a rate of 10°C·h^{-1} and annealed at this temperature for another 240 h. Finally, the furnace was turned off and the ampoule was cooled to room temperature.

2.51 Silicon–Silver–Bromine–Selenium

The $Ag_{8-x}SiSe_{6-x}Br_x$ quaternary phase (x value is not specified) is formed in the Si–Ag–Br–Se system (Kuhs et al. 1979). It was prepared by reacting stoichiometric amounts of the elements in an evacuated, sealed quartz ampoule for about 6 days at temperatures of 600°C to 700°C.

2.52 Silicon–Silver–Iodine–Selenium

The $Ag_{8-x}SiSe_{6-x}I_x$ quaternary phase (x value is not specified) is formed in the Si–Ag–I–Se system (Kuhs et al. 1979). It was prepared by reacting stoichiometric amounts of the elements in an evacuated, sealed quartz ampoule for about 6 days at temperatures of 600–700°C.

2.53 Silicon–Silver–Manganese–Selenium

The $Ag_2MnSiSe_4$ quaternary compound, which crystallizes in the orthorhombic structure with the lattice parameters a = 776.4, b = 664.4, and c = 626.1 pm (Lamarche et al. 1991), is formed in the Si–Ag–Mn–Se system. This compound was produced by the melt and annealed technique. To obtain it, the components of 1.0-g sample were sealed under vacuum in a quartz capsule that had previously been coated with carbon in order to prevent the reaction of the charge with the quartz. The capsule was then raised to a temperature of 1150°C for approximately an hour to allow some reaction to occur and then cooled to room temperature. The sample was annealed at 625°C for a week and then quenched to room temperature.

2.54 Silicon–Silver–Iron–Selenium

The $Ag_2FeSiSe_4$ quaternary compound, which melts at 569°C and crystallizes in the orthorhombic structure with the lattice parameters a = 765.3, b = 652.9, and c = 663.8 pm, is formed in the Si–Ag–Fe–Se system (Quintero et al. 1999). This compound was prepared by the melt and anneals technique. The mixture was made from the appropriate amounts of the elements and was sealed under vacuum in a small quartz ampoule, which had previously been carbonized to prevent the interaction of the components with the quartz. The components were melted together at 1150°C for about an hour, annealed to equilibrium at 500°C, and then cooled to room temperature by leaving the ampoule in the switched-off furnace.

2.55 Silicon–Magnesium–Cerium–Selenium

The $Mg_{0.5}Ce_3SiSe_7$ quaternary compound, which crystallizes in the hexagonal structure with the lattice parameters a = 1066.9 ± 0.1, c = 606.11 ± 0.08 pm, and a calculated density of 5.697 g·cm^{-3}, is formed in the Si–Mg–Ce–Se system (Strok et al. 2015). This compound was obtained by the fusion of the elements in an evacuated silica ampoule. In order to exclude the reaction of magnesium with quartz, the walls of the ampoule were lined with graphite. The ampoule was gradually heated in a furnace at a rate of 30°C·h^{-1} up to a maximal temperature of 1150°C, and kept at this temperature for 3 h. After that, it was slowly cooled (10°C·h^{-1}) to 600°C, and annealed at this temperature for 720 h. Subsequently, the ampoule was quenched in the air.

2.56 Silicon–Magnesium–Praseodymium–Selenium

The $Mg_{0.5}Pr_3SiSe_7$ quaternary compound, which crystallizes in the hexagonal structure with the lattice parameters a = 1053.1 ± 0.1, c = 604.20 ± 0.08 pm, and a calculated density of 5.812 g·cm^{-3}, is formed in the Si–Mg–Pr–Se system (Strok et al. 2015). This compound was prepared in the same way as Ce$_3$Mg$_{0.5}$SiSe$_7$ was synthesized.

2.57 Silicon–Barium–Zinc–Selenium

The **BaZnSiSe$_4$** quaternary compound, which crystallizes in the orthorhombic structure with the lattice parameters $a = 1130.55 \pm 0.08$, $b = 1123.44 \pm 0.07$, $c = 619.94 \pm 0.04$ pm, a calculated density of 4.611 g·cm^{-3}, and an energy gap of 2.71 eV, is formed in the Si–Ba–Zn–Se system (Yin et al. 2017). To prepare this compound, a mixture of BaSe (0.40 mM), ZnSe (0.40 mM), Si (0.40 mM), and Se (0.80 mM) were combined (molar ratio 1:1:1:2), finely ground, and loaded into fused silica tube which was evacuated and sealed. The tube was heated to 900°C over 15 h, kept at that temperature for 60 h, slowly cooled to 400°C over 144 h, and then cooled to room temperature by shutting off the furnace. The products contained light yellow crystals of BaZnSiSe$_4$, which are air stable. The powder sample was prepared by reaction of BaSe, ZnSe, Si, and Se (molar ratio 1:1:1:2) as before, except that two heat treatments were applied. The mixture was heated to 750°C over 24 h, kept at that temperature for 24 h, and then cooled by shutting off the furnace. The obtained sample was reground, and the heat treatment was repeated.

2.58 Silicon–Barium–Gallium–Selenium

The **BaGa$_2$SiSe$_6$** quaternary compound, which melts congruently at 923°C and crystallizes in the trigonal structure with the lattice parameters $a = 996.7 \pm 0.1$, $c = 904.7 \pm 0.2$ pm, a calculated density of 4.984 g·cm^{-3}, and an energy gap of 2.88 eV formed in the Si–Ba–Ga–Se system (Yin et al. 2012b). Crystals of this compound were grown by spontaneous nucleation of the pure polycrystalline sample, which was synthesized by stoichiometric reactions of BaSe, Ga$_2$Se$_3$, Si, and Se (molar ratio 1:1:1:2) at 700°C for 48 h. Then, 0.5 g of the as-prepared powder of BaGa$_2$SiSe$_6$ was loaded into a fused silica tube under an Ar atmosphere in a glove box. The tube was sealed under a high vacuum (10^{-3} Pa) and then placed in a furnace. The reaction mixture was heated to 1000°C for 15 h, kept at 1000°C for 48 h, followed by slow cooling to 320°C at a rate of 3°C·h^{-1}, and finally cooled to room temperature by switching off the furnace. The product consisted of light-yellow crystals of the title compound.

2.59 Silicon–Barium–Antimony–Selenium

The **Ba$_4$SiSb$_2$Se$_{11}$** quaternary compound, which decomposes at 522°C and crystallizes in the orthorhombic structure with the lattice parameters $a = 939.81 \pm 0.03$, $b = 2571.92 \pm 0.07$, $c = 877.48 \pm 0.03$ pm, a calculated density of 5.291 g·cm^{-3} at 170 K, and an energy gap of 1.43 eV, is formed in the Si–Ba–Sb–Se system (Choi and Kanatzidis 2001). Single crystals of this compound were initially discovered from a mixture of Ba (0.2 mM), Th (0.2 mM), Sb (0.2 mM), and Se (2 mM), which

was sealed under vacuum in a silica tube and heated to 720°C for 5 days followed by cooling to 150°C at 2°C·h^{-1}. The excess Ba$_x$Sb$_y$Se$_z$ flux was removed with degassed water to reveal silver needlelike crystals of Ba$_4$SiSb$_2$Se$_{11}$. The crystals are air and water stable. Bulk material was prepared rationally from a mixture of Ba (0.8 mM), Si (0.2 mM), Sb (0.4 mM), and Se (2.4 mM). The mixture was sealed under vacuum in a silica tube and heated to 600°C for 2 days, followed by cooling to 150°C at 2°C·h^{-1}.

2.60 Silicon–Barium–Chlorine–Selenium

The **Ba$_4$Si$_3$Se$_9$Cl$_2$** quaternary compound, which crystallizes in the hexagonal structure with the lattice parameters $a = 1000.24 \pm 0.16$, $c = 1240.6 \pm 0.3$ pm, and an energy gap of 1.76 eV, is formed in the Si–Ba–Cl–Se system (Liu et al. 2017). This compound was obtained from a mixture of BaCl$_2$, Ba, Si, and Se in a stoichiometric ratio of 1.2:3:3:9 with a slight excess of BaCl$_2$ as a reactive flux. The reactants were loaded into a fused silica tube under vacuum and then heated to 950°C within 50 h, kept at this temperature for 140 h, and subsequently cooled to 300°C at 3°C·h^{-1} before switching off the furnace. The product was washed with distilled water to eliminate excess BaCl$_2$ and then dried with ethanol. After this treatment, the compound was obtained as a pure phase with dark-red color. All manipulations were conducted in a glove box (moisture and oxygen levels less than 0.1 ppm) or under a vacuum.

2.61 Silicon–Barium–Bromine–Selenium

The **Ba$_4$Si$_3$Se$_9$Br$_2$** quaternary compound, which melts congruently at 778°C and crystallizes in the hexagonal structure with the lattice parameters $a = 1016.72 \pm 0.05$, $c = 1243.62 \pm 0.06$ pm, a calculated density of 4.487 g·cm^{-3}, and an energy gap of 2.96 eV, is formed in the Si–Ba–Br–Se system (Xing et al. 2020b). Single crystals of this compound were initially obtained from a reaction of BaSe, SiSe$_2$, and BaBr$_2$ (molar ratio 3:3:1). Mixture of finely ground reactants was loaded into a fused silica tube, which was evacuated and sealed. The tube was placed in a furnace where it was heated to 900°C over 20 h, kept at that temperature for 72 h, cooled to 200°C at a rate of 4°C·h^{-1}, and then cooled to room temperature by shutting off the furnace. Polycrystalline sample was prepared in the same manner except that the tube was heated at 750°C over 120 h. Ba$_4$Ge$_3$Se$_9$Br$_2$ is slightly air sensitive and decomposes in less than 1 week. All materials were stored or manipulated in an argon-filled glove box (with moisture and oxygen content of <0.1 ppm) or under a vacuum.

2.62 Silicon–Mercury–Thallium–Selenium

Tl$_2$SiSe$_3$–HgSe. The phase diagram of this system, constructed through DTA, XRD, and measuring of the microhardness, is a eutectic type (Figure 2.12) (Olekseyuk et al. 2012a). The eutectic composition and temperature are 35 mol% HgSe and

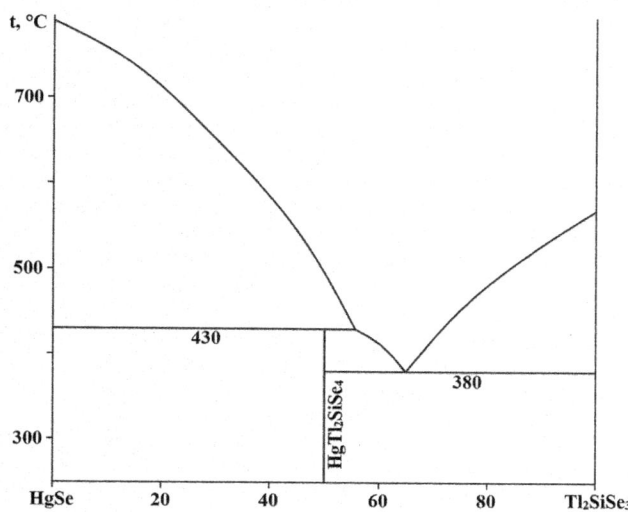

FIGURE 2.12 Phase diagram of the Tl₂SiSe₃–HgSe system. (From Olekseyuk, I.D. et al., *Nauk. Visnyk Volyns'k. Nats. Univ. im. Lesi Ukrainky. Khim. nauky*, [(17)242], 62, 2012.)

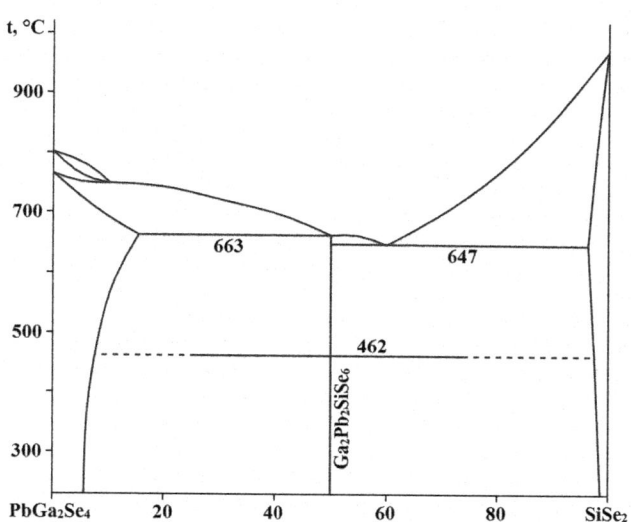

FIGURE 2.14 Phase diagram of the SiSe₂–PbGa₂Se₄ system. (From Piskach, L., et al., *Persp. tekhnol. ta prylady*, (14), 109, 2019.)

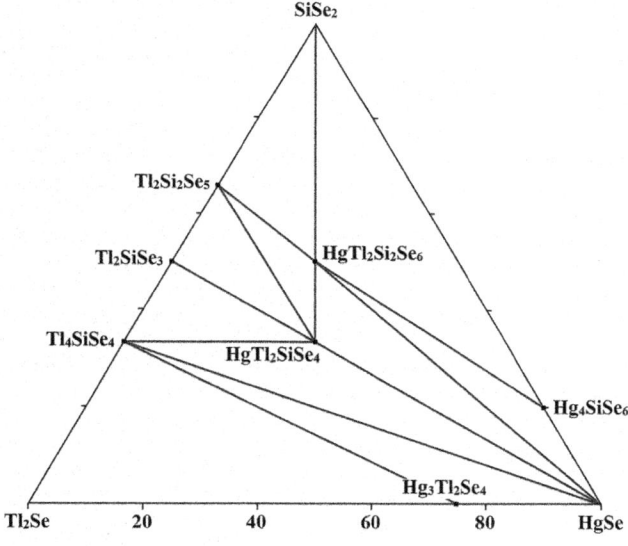

FIGURE 2.13 Isothermal section of the SiSe₂–HgSe–Tl₂Se quasiternary system at 250°C. (From Mozolyuk, M.Yu. et al., *Mater. Res. Bull.*, **47**(11), 3830, 2012.)

380°C, respectively. The **HgTl₂SiSe₄** quaternary compound is formed in this system. It melts incongruently at 430°C and crystallizes in the tetragonal structure with the lattice parameters $a = 800.32 \pm 0.03$, $c = 668.79 \pm 0.04$ pm, and calculated density 7.390 ± 0.001 g·cm⁻³ (Mozolyuk et al. 2012c; Olekseyuk et al. 2012a). The ingots for investigations were annealed at 250°C for 250 h.

SiSe₂–HgSe–Tl₂Se. The isothermal section of this system at 250°C is shown in Figure 2.13 (Mozolyuk et al. 2012c; Olekseyuk et al. 2012a). HgTl₂SiSe₄ and **HgTl₂Si₂Se₆** quaternary compounds exist in this quasiternary system at this temperature.

2.63 Silicon–Mercury–Fluorine–Selenium

The **Hg₃SiSe₂F₆** quaternary compound is formed in this system (Puff et al. 1969). It could be obtained by the interaction of HgSe with the solution of HgO in H₂SiF₆ at 120°C.

2.64 Silicon–Gallium–Lead–Selenium

SiSe₂–PbGa₂Se₄. The phase equilibria in this system, constructed through DTA, XRD, and metallography, are shown in Figure 2.14 (Piskach et al. 2019). This system is a nonquasibinary section of the Si–Ga–Pb–Se quaternary system as PbGa₂Se₄ melts incongruently. The eutectic contains 60 mol% SiSe₂ and crystallizes at 647°C. At 400°C PbGa₂Se₄ dissolves ~5 mol% SiSe₂ and SiSe₂ dissolves less than 3 mol% PbGa₂Se₄. The **Ga₂PbSiSe₆** quaternary compound, which melts incongruently at 663°C and has a polymorphic transformation at 462°C, is formed in this system. The alloys for the investigations were annealed at 400°C for 500 h, with the next quenching in cold water.

According to the data of Luo et al. (2015), Ga₂PbSiSe₆ is thermally stable up to 350°C in N₂ atmosphere and up to 330°C in the air. It crystallizes in the monoclinic structure with the lattice parameters $a = 718.8 \pm 0.5$, $b = 2317.1 \pm 1.9$, $c = 704.4 \pm 0.5$ pm, $\beta = 116.25 \pm 0.03°$, a calculated density of 5.356 g·cm⁻³, and an energy gap of 1.96 eV. It was synthesized from a mixture of PbSe (0.593 mM), Ga₂Se₃ (0.593 mM), Si (0.593 mM), and Se (1.186 mM). The reagents were mixed and ground carefully, homogenized thoroughly with ethanol in an agate mortar, and then pressed into sheets on the tablet machine. The prepared sheets were put in a silica tube and flame-sealed under a 10⁻² Pa. Subsequently, the silica tube was placed in a furnace and heated to 750°C for 48 h, dwelled at that temperature for 50 h, and then was cooled at a rate of 3°C·h⁻¹ to 300°C followed by rapid cooling to room temperature. A few red crystals of the

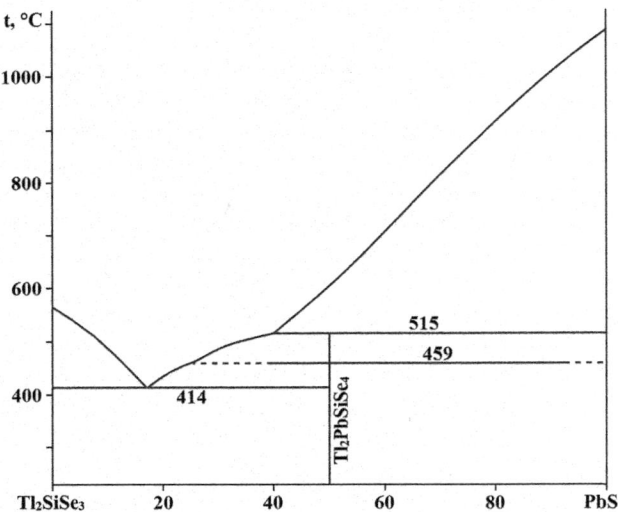

FIGURE 2.15 Phase diagram of the Tl₂SiSe₃–PbSe system. (From Olekseyuk, I.D., *Nauk. Visnyk Volyns'k. Nats. Univ. im. Lesi Ukrainky. Ser. Khim. nauky*, (17), 62, 2012.)

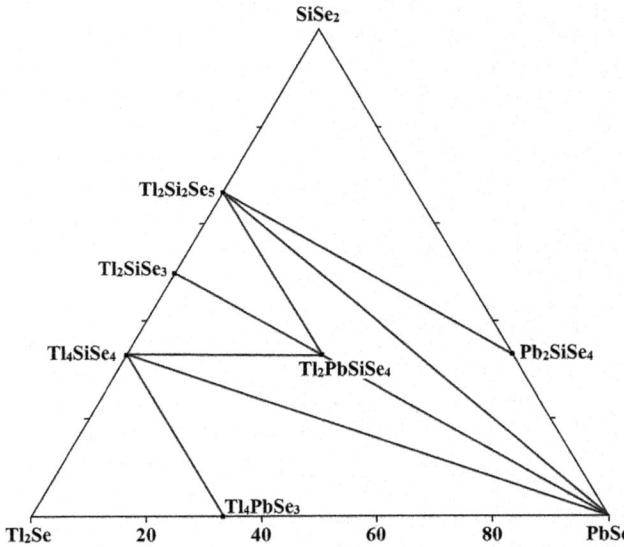

FIGURE 2.16 Isothermal section of the SiSe₂–Tl₂Se–PbSe quasiternary system at 250°C. (From Olekseyuk, I.D., *Nauk. Vsnyk Volyns'k. Nats. Univ. im. Lesi Ukrainky. Ser. Khim. nauky*, (17), 62, 2012.)

title compound were obtained. They are stable for at least 5 months in the air and stable under ethanol and water.

A pure phase of Ga₂PbSiSe₆ was produced by the stoichiometric mixture of the Pb, Ga, Si, and Se (molar ratio 1:2:1:6) (Luo et al. 2015). The mixture was heated to 600°C for 24 h, dwelled at that temperature for 50 h, and then cooled down to room temperature for 24 h. In order to improve the homogeneity and purity, the resulting product was reground by using a mortar and pestle and heated again at the same temperature.

2.65 Silicon–Thallium–Lead–Selenium

Tl₂SiSe₃–PbSe. The phase diagram of this system is presented in Figure 2.15 (Olekseyuk et al. 2012a). The eutectic contains 17 mol% PbSe and crystallizes at 414°C. The **Tl₂PbSiSe₄** quaternary compound, which melts incongruently at 515°C and has a polymorphic transformation at 459°C, is formed in this system. The peritectic contains 40 mol% PbSe.

SiSe₂–Tl₂Se–PbSe. The isothermal section of this system at 250°C is given in Figure 2.16 (Olekseyuk et al. 2012a). The alloys for the investigations were annealed at 250°C for 250 h, with the next quenching in cold water.

2.66 Silicon–Yttrium–Lanthanum–Selenium

The **Y₁.₅La₁.₅Si₁.₇₅Se₇** quaternary compound, which crystallizes in the hexagonal structure with the lattice parameters a = 1059.68 ± 0.02, c = 599.95 ± 0.02 pm, and a calculated density of 5.3707 ± 0.0004 g·cm⁻³, is formed in the Si–Y–La–Se system (Smitiukh et al. 2017b, 2018b). This compound was prepared by co-melting the elements in an evacuated quartz container (residual pressure 0.1 Pa). The total mass of the starting batch was 0.8 g. The synthesis was performed by heating

the mixture to 1150°C at a rate of 12°C·h⁻¹ with the next exposure at this temperature for 4 h, cooling to 500°C at a rate of 12°C·h⁻¹, annealing at this temperature for 240 h and quenching to room-temperature water.

2.67 Silicon–Yttrium–Praseodymium–Selenium

The **Y₁.₅Pr₁.₅Si₁.₇₅Se₇** quaternary compound, which crystallizes in the hexagonal structure with the lattice parameters a = 1048.25 ± 0.03, c = 596.83 ± 0.02 pm, and a calculated density of 5.4957 ± 0.0004 g·cm⁻³, is formed in the Si–Y–Pr–Se system (Smitiukh et al. 2016). This compound was synthesized in the same way as Y₁.₅La₁.₅Si₁.₇₅Se₇ was prepared.

2.68 Silicon–Yttrium–Lead–Selenium

SiSe₂–Y₂Se₃–PbSe. The isothermal section of this quasiternary system at 500°C was constructed by Marchuk et al. (2008c) using the alloys annealed at this temperature for 500 h with the next quenching in water. Quaternary compounds were not found in this system. The system is divided by the Pb₂SiSe₄–Y₆Pb₂Se₁₁ and Pb₂SiSe₄–Y₂Se₃ quasibinary section into three three-phase regions.

2.69 Silicon–Yttrium–Oxygen–Selenium

The **Y₂SiO₄Se** quaternary compound, which crystallizes in the orthorhombic structure with the lattice parameters a = 599.35 ± 0.07, b = 692.16 ± 0.08, c = 1076.88 ± 0.12 pm, and a calculated density of 5.187 g·cm⁻³ is formed in the Si–Y–O–Se system (Kiryakov et al. 2018). To prepare this compound, Y₂O₃ (0.66 mM), Se (0.66 mM), Si (0.33 mM), and SiO₂ (0.33 mM) were ground 2–3 times with acetone in a mortar and loaded into a quartz ampoule with 450 mg of CsCl flux. The ampoule was evacuated up to 1 Pa and sealed, heated to 1100°C and kept for 12 h. The product obtained was washed with water from the flux and dried with alcohol in the air. Transparent colorless needle-shaped plates were obtained.

2.70 Silicon–Lanthanum–Dysprosium–Selenium

The $La_{1.5}Dy_{1.5}Si_{1.66}Se_7$ quaternary compound, which crystallizes in the hexagonal structure with the lattice parameters $a = 1059.51 \pm 0.04$, $c = 599.82 \pm 0.03$ pm, and a calculated density of 5.992 g·cm^{-3}, is formed in the Si–La–Dy–Se system (Smitiukh et al. 2016). This compound was synthesized in the same way as $Y_{1.5}La_{1.5}Si_{1.75}Se_7$ was prepared.

2.71 Silicon–Lanthanum–Erbium–Selenium

The $La_{1.5}Er_{1.5}Si_{1.67}Se_7$ quaternary compound, which crystallizes in the hexagonal structure with the lattice parameters $a = 1058.90 \pm 0.04$, $c = 599.52 \pm 0.03$ pm, and a calculated density of 6.0402 ± 0.0007 g·cm^{-3}, is formed in the Si–La–Er–Se system (Smitiukh et al. 2017a). This compound was synthesized in the same way as $Y_{1.5}La_{1.5}Si_{1.75}Se_7$ was prepared.

2.72 Silicon–Lanthanum–Lead–Selenium

$SiSe_2$–La_2Se_3–$PbSe$. The isothermal section of this quasiternary system at 500°C is shown in Figure 2.17 (Marchuk and Gulay 2012). Six single-phase, 10 two-phase, and 5 three-phase regions exist in the system. The solubility on the base

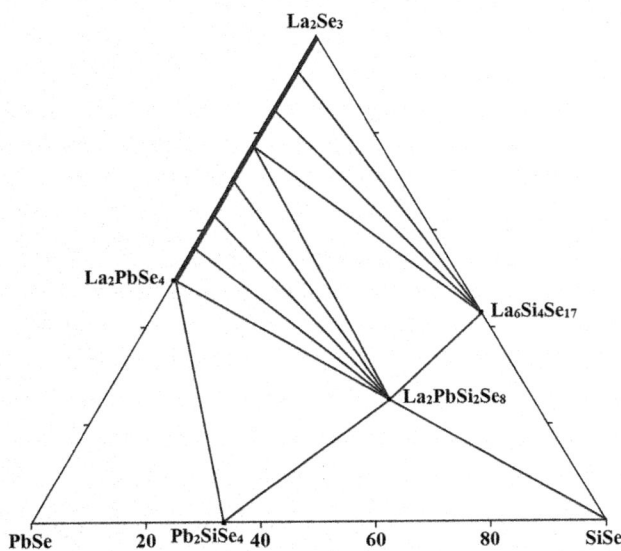

FIGURE 2.17 Isothermal section of the $SiSe_2$–La_2Se_3–$PbSe$ quasiternary system at 500°C. (From Marchuk, O.V., and Gulay, L.D., *Nauk. Visnyk Volyns'k. Nats. Univ. im. Lesi Ukrainky. Ser. Khim. nauky*, (17), 93, 2012.)

of the starting components is insignificant (≈1–2 mol%). The alloys for investigations were annealed at 500°C for 500 h, with the next quenching in water.

The $La_2PbSi_2Se_8$ quaternary compound, which crystallizes in the trigonal structure with the lattice parameters $a = 939.84 \pm 0.06$, $c = 2808.9 \pm 0.3$ pm, and a calculated density of 5.438 g·cm^{-3}, is formed in the Si–La–Pb–Se system (Daszkiewicz et al. 2012; Marchuk 2019). A powder sample of this compound was prepared by sintering the elemental constituents in an evacuated quartz tube. The synthesis was carried out in a furnace. The ampoule was first heated with a rate of 30°C·h^{-1} up to 1150°C, and then kept at this temperature for 3 h. Afterwards, the sample was cooled slowly (10°C·h^{-1}) down to 500°C and annealed at this temperature for 720 h. Subsequently, the ampoule was quenched in the air.

2.73 Silicon–Lanthanum–Oxygen–Selenium

Two quaternary compounds, La_2SiO_4Se and $La_4Si_2O_7Se_3$, are formed in the Si–La–O–Se system. First of them crystallizes in the orthorhombic structure with the lattice parameters $a = 627.9 \pm 0.4$, $b = 730.6 \pm 0.5$, $c = 1117.7 \pm 0.7$ pm, and a calculated density of 5.813 g·cm^{-3} at 111 K (Brennan and Ibers 1991). $La_4Si_2O_7Se_3$ crystallizes in the tetragonal structure with the lattice parameters $a = 1228.46 \pm 0.18$, $c = 1469.92 \pm 0.10$ pm, and a calculated density of 5.753 g·cm^{-3} (Deudon et al. 1993).

Single crystals of La_2SiO_4Se were isolated from reactions that produced single crystals of $La_2Ta_3O_8Se_2$ (Brennan and Ibers 1991). A combination of La_2Se_3, Ta_2O_5 and tantalum powders (molar ratio 1:1:1) was reacted for 3 days at 1200°C in evacuated sealed quartz tube to produce crystals of $La_2Ta_3O_8Se_2$ and La_2SiO_4Se, the source of silicon being the quartz tube. The title compound grew as clear colorless prisms. La_2SiO_4Se powder was prepared by the reaction of La_2O_2Se with H_2SiO_3 (molar ratio 1:1) at 1200°C for 3 days.

$La_4Si_2O_7Se_3$ was obtained as by-product in the preparation of the misfit selenide $(LaSe)_{1.14}(NbSe_2)_2$ by using La, Nb, and Se (molar ratio 1:2:5) as starting elements (Deudon et al. 1993). These elements were sealed in an evacuated (≈1 Pa) quartz tube protected by a thin carbon film. It was heated in a furnace at 1050°C for 8 days. The furnace was then cooled to room temperature within a day. In addition to a misfit ternary phase, some small, octahedral transparent crystals of the title compound were found. They could be easily separated from the major black product.

2.74 Silicon–Cerium–Lead–Selenium

The $Ce_2Pb(SiSe_4)_2$ quaternary compound, which crystallizes in the trigonal structure with the lattice parameters $a = 935.1 \pm 0.1$, $c = 2790.8 \pm 0.4$ pm, and a calculated density of 5.540 g·cm^{-3}, is formed in the Si–Ce–Pb–Se system (Daszkiewicz et al. 2012; Marchuk 2019). It was prepared in the same way as $La_2Pb(SiSe_4)_2$ was obtained.

2.75 Silicon–Cerium–Oxygen–Selenium

Two quaternary compounds, **Ce_2SiO_4Se** and **$Ce_4Si_2O_7Se_3$**, are formed in the Si–Ce–O–Se system. First of them crystallizes in the orthorhombic structure with the lattice parameters $a = 622.50 \pm 0.06$, $b = 723.54 \pm 0.07$, $c = 1107.39 \pm 0.10$ pm, and a calculated density of 6.010 g·cm⁻³ at 153 K (Deng et al. 2004). $Ce_4Si_2O_7Se_3$ crystallizes in the tetragonal structure with the lattice parameters $a = 1221.7 \pm 0.1$ and $c = 1456.6 \pm 0.2$ pm (Grupe and Urland 1989).

Clear light-red needles of Ce_2SiO_4Se were obtained accidentally in the reaction of Ce (71 mg), Se (40 mg), CsCl (150 mg), and Cs_2Se_3 (150 mg) (Deng et al. 2004). The materials were mixed and sealed in an unprotected fused silica tube that was then evacuated to 10⁻² Pa. The tube was heated to 900°C, kept at 900°C for 72 h, cooled at 4°C·h⁻¹ to 200°C, and then the furnace was turned off. The reaction mixture was washed with deionized water and finally dried with acetone.

To prepare $Ce_4Si_2O_7Se_3$, a mixture of Ce, Se, and I_2 (molar ratio 1.2:1:2.5) was filled into a quartz ampoule and melted under a vacuum (Grupe and Urland 1989). The powdery products were mixed with 300 mg of powdered quartz and brought into an evacuated quartz ampoule. After annealing at 950°C for 7 days, prism-shaped colorless crystals of the title compound were obtained.

2.76 Silicon–Cerium–Manganese–Selenium

The **$Ce_3Mn_{0.5}SiSe_7$** quaternary compound, which crystallizes in the hexagonal structure with the lattice parameters $a = 1066.84 \pm 0.02$, $c = 601.27 \pm 0.02$ pm, a calculated density of 4.2321 g·cm⁻³, and an energy gap of 1.81 eV, is formed in the Si–Ce–Mn–Se system (He et al. 2015). The crystals of the title compound were synthesized by the traditional molten salt method with KI acting as a flux. The elements and KI were mixed with a molar ratio of Ce/Mn/Ge/Se/KI = 3:0.5:1:7:30. The mixture was ground and sealed in carbon-coated fused silica tube under vacuum (0.1 Pa). Then, the tube was slowly heated to 1000°C in a furnace, followed by keeping at this temperature for 3 days. Afterwards, the tube was slowly cooled to 500°C at the rate of 2°C·h⁻¹, and then quenched in the air to room temperature. The final product was obtained by carefully breaking the tube, washed by distilled water several times, and dried by acetone to obtain red crystals of $Ce_3Mn_{0.5}SiSe_7$. The direct combination reactions of the starting materials, without the flux (KI), were used to synthesize the powder samples. All operations were carried out in an Ar-protected glove box.

2.77 Silicon–Cerium–Iron–Selenium

The **$Ce_3Fe_{0.5}SiSe_7$** quaternary compound, which crystallizes in the hexagonal structure with the lattice parameters $a = 1063.80 \pm 0.03$, $c = 600.79 \pm 0.03$ pm, a calculated density of 4.2726 g·cm⁻³, and an energy gap of 1.37 eV is formed in the Si–Ce–Fe–Se system (He et al. 2015). Black crystals of the title compound were obtained in the same way as $Ce_3Mn_{0.5}SiSe_7$ was synthesized using Fe instead of Mn.

2.78 Silicon–Praseodymium–Erbium–Selenium

The **$Pr_{1.5}Er_{1.5}Si_{1.67}Se_7$** quaternary compound, which crystallizes in the hexagonal structure with the lattice parameters $a = 1047.90 \pm 0.03$, $c = 597.35 \pm 0.02$ pm, and a calculated density of 6.2075 ± 0.0005 g·cm⁻³, is formed in the Si–Pr–Er–Se system (Smitiukh et al. 2017a). This compound was synthesized in the same way as $Y_{1.5}La_{1.5}Si_{1.75}Se_7$ was prepared.

2.79 Silicon–Praseodymium–Lead–Selenium

The **$Pr_2Pb(SiSe_4)_2$** quaternary compound, which crystallizes in the trigonal structure with the lattice parameters $a = 932.64 \pm 0.09$, $c = 2777.9 \pm 0.3$ pm, and a calculated density of 5.603 g·cm⁻³, is formed in the Si–Pr–Pb–Se system (Daszkiewicz et al. 2012; Marchuk 2019). It was prepared in the same way as $La_2Pb(SiSe_4)_2$ was obtained.

2.80 Silicon–Praseodymium–Nitrogen–Selenium

The **$Pr_3Se_2SiN_3$** or **$Pr_9Se_6[Si_3N_9]$** quaternary compound, which crystallizes in the trigonal structure with the lattice parameters $a = 1892.41 \pm 0.09$, $c = 1037.25 \pm 0.06$ pm, and a calculated density of 6.047 g·cm⁻³, is formed in the Si–Pr–N–Se system (Lissner and Schleid 2004). The oxidation of Pr with Se and NaN_3, and additives of $PrCl_3$ and NaCl as a flux, in an evacuated silica tube at 900°C for 7 days, assisted by reaction with container material (SiO_2) yielded dark yellow rods of the title compound, as a by-product, while actually attempting to synthesize praseodymium nitride selenide.

2.81 Silicon–Praseodymium–Oxygen–Selenium

The **$Pr_4S_3(Si_2O_7)$** quaternary compound, which crystallizes in the tetragonal structure with the lattice $a = 1215.4 \pm 0.1$ and $c = 1445.5 \pm 0.3$ pm, is formed in the Si–Pr–O–Se system (Grupe et al. 1992). The yellowish green crystals of this compound has been obtained by reaction of Pr, Se and iodine (molar ratio 1:1:2.5), which were filled under argon in a glove box into quartz ampoule and melted under vacuum. Heating for 7 days at 900°C provides praseodymium selenide and iodide, which were mixed with equimolar amount of finely ground quartz glass. This new mixture was heated again under vacuum in quartz ampoule at 900°C for 7 days. The end product was the single crystals of $Pr_4S_3(Si_2O_7)$.

2.82 Silicon–Neodymium–Lead–Selenium

The **$Nd_2Pb(SiSe_4)_2$** quaternary compound, which crystallizes in the trigonal structure with the lattice parameters $a = 929.98 \pm 0.09$, $c = 2767.0 \pm 0.2$ pm, and a calculated density of 5.690 g·cm⁻³, is formed in the Si–Nd–Pb–Se system (Daszkiewicz et al. 2012; Marchuk 2019). It was prepared in the same way as $La_2Pb(SiSe_4)_2$ was obtained.

2.83 Silicon–Neodymium–Oxygen–Selenium

Two quaternary compounds, **Nd$_2$SiO$_4$Se** and **Nd$_4$Si$_2$O$_7$Se$_3$**, are formed in the Si–Nd–O–Se system. First of them crystallizes in the orthorhombic structure with the lattice parameters $a = 618.2 \pm 0.1$, $b = 717.4 \pm 0.2$, $c = 1102.4 \pm 0.2$ pm, and a calculated density of 6.246 g·cm^{-3} at 111 K (Grupe and Urland 1990). Nd$_4$Si$_2$O$_7$Se$_3$ crystallizes in the tetragonal structure with the lattice parameters $a = 1209.9 \pm 0.1$ and $c = 1438.7 \pm 0.3$ pm (Grupe and Urland 1989).

Powdered samples of Nd$_2$SiO$_4$Se were obtained by sealing Nd$_2$Se$_3$ (500 mg) together with silica dust (170 mg) into an evacuated quartz ampoule with the next heating to 1000°C for 2 days (Grupe and Urland 1990). For the production of the single crystals, Nd metal shavings (1 g), Se powder, and iodine (molar ratio 1:1:2.5) were used. These three containers were then transferred to another quartz ampoule so that came to lie in the order of I$_2$, Se, Nd. The quartz ampoule was evacuated under cooling of liquid N$_2$, sealed, and placed in a two-zone furnace. First, I$_2$ and Nd were heated at 120°C and 300°C, respectively, for 2 days, and then exposed for a further 2 days of Se and Nd to temperatures of 330°C and 550°C, respectively. Finally, the ampoule was opened and the contents were sealed together with quartz glass shards into another quartz ampoule under vacuum (0.1 Pa). The vial was placed vertically in a tube furnace and heated to 850°C for 2 days. A gray-black melt cake was obtained, which included pale violet, rod-shaped crystals of Nd$_2$SiO$_4$Se.

To prepare Nd$_4$Si$_2$O$_7$Se$_3$, a mixture of Nd, Se, and I$_2$ (molar ratio 1.2:1:2.5) was filled into a quartz ampoule and melted under vacuum (Grupe and Urland 1989). The powdery products were mixed with 300 mg of powdered quartz and brought into an evacuated quartz ampoule. After annealing at 950°C for 7 days, prism-shaped violet crystals of the title compound were obtained.

2.84 Silicon–Samarium–Lead–Selenium

The **Sm$_2$Pb(SiSe$_4$)$_2$** quaternary compound, which crystallizes in the trigonal structure with the lattice parameters $a = 926.20 \pm 0.04$, $c = 2748.7 \pm 0.1$ pm, and a calculated density of 5.830 g·cm^{-3}, is formed in the Si–Sm–Pb–Se system (Daszkiewicz et al. 2012; Marchuk 2019). It was prepared in the same way as La$_2$Pb(SiSe$_4$)$_2$ was obtained.

2.85 Silicon–Samarium–Oxygen–Selenium

Two quaternary compounds, **Sm$_2$SiO$_4$Se** and **Sm$_4$Si$_2$O$_7$Se$_3$**, are formed in the Si–Sm–O–Se system. First of them crystallizes in the orthorhombic structure with the lattice parameters $a = 612.6 \pm 0.1$, $b = 709.0 \pm 0.1$, $c = 1094.0 \pm 0.2$ pm, and a calculated density of 6.595 g·cm^{-3} (Person et al. 2000). Sm$_4$Si$_2$O$_7$Se$_3$ crystallizes in the tetragonal structure with the lattice parameters $a = 1201.3 \pm 0.1$ and $c = 1421.1 \pm 0.2$ pm (Gruppe et al. 1992).

Powder-shaped samples of Sm$_2$SiO$_4$Se were obtained by using samarium polyselenide, obtained from Sm and Se (atomic ratio 1:2), together with SiO$_2$ dust (approximately 17 %) in a molar ratio 1:2 (Person et al. 2000). The mixture was melted into the evacuated quartz ampoule and then heated to 1000°C for 3 days. The obtained compound was contaminated by unreacted polyselenide. Single crystals of this compound were prepared through chemical transport reactions using iodine as a transport agent.

The orange crystals of Sm$_4$Si$_2$O$_7$Se$_3$ were prepared in the same way as the crystals of Pr$_4$Si$_2$O$_7$Se$_3$ were obtained using Sm instead of Pr (Gruppe et al. 1992).

2.86 Silicon–Samarium–Iron–Selenium

The **Sm$_3$Fe$_{0.5}$SiSe$_7$** quaternary compound, which crystallizes in the hexagonal structure with the lattice parameters $a = 1040.75 \pm 0.04$, $c = 595.66 \pm 0.04$ pm, a calculated density of 6.4646 g·cm^{-3}, and an energy gap of 1.13 eV is formed in the Si–Sm–Fe–Se system (He et al. 2015). Black crystals of the title compound were obtained in the same way as Ce$_3$Mn$_{0.5}$SiSe$_7$ was synthesized using Sm and Fe instead of Ce and Mn, respectively.

2.87 Silicon–Gadolinium–Lead–Selenium

The **Gd$_2$Pb(SiSe$_4$)$_2$** quaternary compound, which crystallizes in the trigonal structure with the lattice parameters $a = 923.20 \pm 0.07$, $c = 2732.9 \pm 0.3$ pm, and a calculated density of 5.971 g·cm^{-3}, is formed in the Si–Gd–Pb–Se system (Daszkiewicz et al. 2012; Marchuk 2019). It was prepared in the same way as La$_2$Pb(SiSe$_4$)$_2$ was obtained.

2.88 Silicon–Gadolinium–Oxygen–Selenium

The **Gd$_4$Si$_2$O$_7$Se$_3$** quaternary compound, which crystallizes in the tetragonal structure with the lattice parameters $a = 1194.0 \pm 0.1$ and $c = 1410.5 \pm 0.2$ pm, is formed in the Si–Gd–O–Se system (Gruppe et al. 1992). The orange crystals of Gd$_4$Si$_2$O$_7$Se$_3$ were prepared in the same way as the crystals of Pr$_4$Si$_2$O$_7$Se$_3$ were obtained using Gd instead of Pr.

2.89 Silicon–Terbium–Oxygen–Selenium

The **Tb$_2$SiO$_4$Se** quaternary compound, which crystallizes in the orthorhombic structure with the lattice parameters $a = 603.87 \pm 0.04$, $b = 698.55 \pm 0.05$, and $c = 1081.31 \pm 0.07$ pm at 153 K (Ijjaali and Ibers 2002). Clear needles of this compound were obtained accidentally in the reaction of Tb (0.547 mM), Cd (0.543 mM), Se (1.09 mM), CsCl (0.947 mM), and Cs$_2$Se$_3$ (150 mg). The materials were mixed and sealed in an unprotected and evacuated fused silica tube. The tube was heated to 1000°C, kept at 1000°C for 50 h, cooled at 4°C·h^{-1} to 200°C, and then the furnace was turned off. The reaction mixture was washed with water, then DMF, and finally dried with acetone. The compound resulted from the reaction of the starting components with the fused silica tube.

2.90 Silicon–Dysprosium–Oxygen–Selenium

The **Dy$_2$SiO$_4$Se** quaternary compound, which crystallizes in the orthorhombic structure with the lattice parameters $a = 603.6 \pm 0.1$, $b = 696.4 \pm 0.1$, $c = 1081.2 \pm 0.2$ pm, and a calculated density of 7.250 g·cm^{-3}, is formed in the Si–Dy–O–Se system (Person et al. 2000). Powder-shaped samples and single crystals of this compound were prepared in the same way as in the case of Sm$_2$SiO$_4$Se obtaining using dysprosium polyselenide instead of samarium polyselenide.

2.91 Silicon–Holmium–Oxygen–Selenium

The **Ho$_2$SiO$_4$Se** quaternary compound, which crystallizes in the orthorhombic structure with the lattice parameters $a = 601.0 \pm 0.1$, $b = 693.6 \pm 0.1$, $c = 1078.6 \pm 0.2$ pm, and a calculated density of 7.399 g·cm^{-3}, is formed in the Si–Dy–O–Se system (Person et al. 2000). Powder-shaped samples and single crystals of this compound were prepared in the same way as in the case of Sm$_2$SiO$_4$Se obtaining using holmium polyselenide instead of samarium polyselenide.

2.92 Silicon–Erbium–Oxygen–Selenium

The **Er$_2$SiO$_4$Se** quaternary compound, which crystallizes in the orthorhombic structure with the lattice parameters $a = 600.2 \pm 0.2$, $b = 688.0 \pm 0.2$, $c = 1075.2 \pm 0.2$ pm, and a calculated density of 7.562 g·cm^{-3}, is formed in the Si–Er–O–Se system (Stöwe 1994). Well-shaped brown crystals of this compound were obtained as byproducts of the synthesis of erbium selenides from the elements in evacuated and sealed silica ampoules with graphite inlets at 980°C.

3

Systems Based on Silicon Telluride

3.1 Silicon–Sodium–Aluminum–Tellurium

The **Na₃AlSiTe₄** quaternary compound, which crystallizes in the monoclinic structure with the lattice parameters $a = 747 \pm 1$, $b = 434 \pm 1$, $c = 852 \pm 1$ pm, $\beta = 107.4 \pm 0.1°$, and the calculated and experimental densities of 3.99 and 3.93 g·cm⁻³, respectively, is formed in the Si–Na–Al–Te system (Weiss and Schäfer 1979). To prepare this compound, a stoichiometric mixture of the elements was carefully heated in a quartz ampoule to 980°C under dry argon under 10 Pa, left at this temperature for about 1 h, and then cooled down to room temperature within 12 h. The result was dark reguli, the fragments of which were pale yellow and transparent. Amber-colored, translucent platelets with a hexagonal outline could be broken from cavities. These crystals were relatively soft and had a strong smell of H₂Te in the air. They immediately covered themselves with a dark, matt tellurium layer.

3.2 Silicon–Sodium–Europium–Tellurium

The **Na₈Eu₂(Si₂Te₆)₂** quaternary compound, which crystallizes in the monoclinic structure with the lattice parameters $a = 764.66 \pm 0.10$, $b = 1321.95 \pm 0.17$, $c = 843.17 \pm 0.11$ pm, $\beta = 107.662 \pm 0.002°$, and a calculated density of 4.355 g·cm⁻³ (Martin et al. 2006) [$a = 764.8 \pm 0.1$, $b = 1322.3 \pm 0.2$, $c = 843.9 \pm 0.1$ pm, $\beta = 107.667 \pm 0.003°$ (Knaust et al. 2005)], is formed in the Si–Na–Eu–Te system. Crystals of the title compound were formed from a molten chalcogenide flux reaction of Eu (37.5 mg), Si (17.6 mg), Te (149.8 mg), and Na₂Te (86.0 mg) (Knaust et al. 2005; Martin et al. 2006). The reactants were placed into a fused silica ampoule in an inert atmosphere glove box. The ampoule was sealed under a vacuum, and heated to 750°C at a rate of 35°C·h⁻¹. After 150 h of annealing, the ampoule was cooled at 3°C·h⁻¹ to room temperature. Orange prismatic crystals were obtained after the product was washed with DMF to dissolve the remaining flux.

3.3 Silicon–Copper–Barium–Tellurium

The **Cu₃.₇₅Ba₄Si₂Te₉** quaternary compound, which crystallizes in the orthorhombic structure with the lattice parameters $a = 1342.93 \pm 0.04$, $b = 2009.80 \pm 0.07$, $c = 856.99 \pm 0.03$ pm, a calculated density of 5.727 g·cm⁻³ and an energy gap of 0.89 eV, is formed in the Si–Cu–Ba–Te system (Cui et al. 2010). To obtain this compound, the starting materials, Ba, Cu, Si, and

Te (molar ratio 4:4:2:9), were loaded into a fused silica tube, which was then sealed under a high vacuum of approximately 1 Pa. The reaction mixture was heated to 750°C within 48 h in a furnace, kept at that temperature for 2 h, and then cooled to 200°C at a rate of 3°C·h⁻¹, followed by switching off the furnace. All chemical elements were stored in an argon-filled glove box with water and oxygen levels below 1 ppm.

3.4 Silicon–Copper–Zinc–Tellurium

The **Cu₂ZnSiTe₄** quaternary compound, which melts at 700°C and crystallizes in the tetragonal structure with the lattice parameters $a = 598$ and $c = 1178$ pm an and energy gap of 1.47 eV (Matsushita and Katsui 2005; Matsushita et al. 2005) [$a = 597.2 \pm 0.1$ and $c = 1179.7 \pm 0.4$ pm (Haeuseler et al. 1991)], is formed in the Si–Cu–Zn–Te system. Single crystals of this compound were obtained by the chemical transport reactions using I₂ as the transport agent or through the horizontal gradient freezing method (Matsushita et al. 2005).

3.5 Silicon–Copper–Cadmium–Tellurium

Cu₂SiTe₃–CdTe. This section is a nonquasibinary section of the Si–Cu–Cd–Te quaternary system since Si₂Te₃ primarily crystallizes from the Cu₂SiTe₃-rich side (Piskach et al. 1997; Olekseyuk et al. 1998b). The eutectic composition and temperature are 23 mol% CdTe and 539°C, respectively. The immiscibility region within the interval of 46–56 mol% CdTe exists in this system at a monotectic temperature of 762°C. The **Cu₂CdSiTe₄** quaternary compound is formed in this system (the peritectic composition is 40 mol% CdTe). It melts incongruently at 657°C (Piskach et al. 1997; Olekseyuk et al. 1998b) – or, according to Matsushita and Katsui 2005 and Matsushita et al. 2005, at 650°C – and crystallizes in the tetragonal structure with the lattice parameters $a = 610.11 \pm 0.02$, $c = 1180.15 \pm 0.05$ pm, and a calculated density of 5.8811 ± 0.0005 g·cm⁻³ (Parasyuk et al. 2006b) [$a = 611.0 \pm 0.1$ and $c = 1181.1 \pm 0.3$ pm (Haeuseler et al. 1991); $a = 607.3 \pm 0.2$ and $c = 1177.1 \pm 0.6$ pm, with the calculated and experimental densities of 5.64 and 5.57 g·cm⁻³, respectively (Piskach et al. 1997; Olekseyuk et al. 1998b); $a = 612$ and $c = 1179$ pm (Matsushita and Katsui 2005; Matsushita et al. 2005].

This system was investigated through DTA, XRD, metallography, and measuring of the microhardness and density (Piskach et al. 1997; Olekseyuk et al. 1998b). The ingots were annealed at 500°C for 1000 h.

DOI: 10.1201/9781003123484-3

$Cu_2CdSiTe_4$ was synthesized from the mixture of elements in an evacuated and sealed ampoule (Parasyuk et al. 2006b). The mixture was heated to 930°C at a rate of 30°C·h⁻¹, stayed at this temperature for 6 h, cooled at a rate of 10°C·h⁻¹ to 400°C, and annealed at this temperature for 250 h. Single crystals of this compound were obtained by the horizontal gradient freezing method (Matsushita et al. 2005).

3.6 Silicon–Copper–Mercury–Tellurium

Cu_2SiTe_3–$HgTe$. The $Cu_2HgSiTe_4$ quaternary compound, which melts congruently at 625°C (Parasyuk 1998) and crystallizes in the tetragonal structure with the lattice parameters $a = 609.522 \pm 0.009$, $c = 1183.52 \pm 0.03$ pm, and a calculated density of 6.5416 ± 0.0003 g·cm⁻³ (Parasyuk et al. 2006a) [$a = 609.2 \pm 0.1$ and $c = 1183.1 \pm 0.4$ pm (Haeuseler et al. 1991)], is formed in this system. The solid solution based on Cu_2SiTe_3 contains 7 mol% HgTe and the solid solubility in HgTe at 400°C is 2.5 mol% Cu_2SiTe_3 (Parasyuk 1998).

$Cu_2HgSiTe_4$ was synthesized from the mixture of Cu, Si, Te, and HgTe in an evacuated and sealed quartz ampoule (Parasyuk et al. 2006a). The mixture was heated to 750°C, remained at this temperature for 6 h, cooled at a rate of 10°C·h⁻¹ to 400°C, and then annealed at this temperature for 250 h.

3.7 Silicon–Copper–Aluminum–Tellurium

The $CuAlSiTe_4$ quaternary compound, which crystallizes in the tetragonal structure with the lattice parameters $a = 601.4$, $c = 593.7$ pm, and the calculated and experimental densities of 4.86 and 5.14 g·cm⁻³, respectively, is formed in the Si–Cu–Al–Te system (Hahn and Strick 1967b). This compound was obtained by long annealing of the elements' mixture or the mixture of binary components between 500°C and 1000°C. It has a brown-black color.

3.8 Silicon–Copper–Gallium–Tellurium

The $CuGaSiTe_4$ quaternary compound, which crystallizes in the tetragonal structure with the lattice parameters $a = 599.5$, $c = 590.9$ pm, and the calculated and experimental densities of 5.25 and 5.32 g·cm⁻³, respectively, is formed in the Si–Cu–Ga–Te system (Hahn and Strick 1967b). This compound was obtained by long annealing of the elements' mixture or the mixture of binary components between 500°C and 1000°C. It has a brown-black color.

3.9 Silicon–Copper–Manganese–Tellurium

The $Cu_2MnSiTe_4$ quaternary compound, which has two polymorphic modifications, is formed in the Si–Cu–Mn–Te system (Lamarche et al. 1991). The first modification crystallizes in the tetragonal structure with the lattice parameters $a = 598.3$ and $c = 1172.4$ pm, and the second one crystallizes in the cubic structure with the lattice parameter $a = 605$ pm. This compound was produced by the melt and anneals technique. To obtain it, the components of 1.0-g sample were sealed under a vacuum in a quartz capsule, which had previously been coated with carbon in order to prevent the reaction of the charge with the quartz. The capsule was then raised to a temperature of 1150°C for approximately an hour to allow some reaction to occur and then cooled to room temperature.

3.10 Silicon–Copper–Iron–Tellurium

The $Cu_2FeSiTe_4$ quaternary compound, which melts at 457°C and crystallizes in the orthorhombic structure with the lattice parameters $a = 739.3$, $b = 679.7$, and $c = 651.0$ pm, is formed in the Si–Cu–Fe–Te system (Quintero et al. 1999). This compound was prepared by the melt and anneals technique. The mixture was made from the appropriate amounts of the elements and was sealed under vacuum in small quartz ampoules, which had previously been carbonized to prevent interaction of the components with the quartz. The components were melted together at 1150°C for about an hour, annealed to equilibrium at 500°C, and then cooled to room temperature by leaving the ampoule in the switched-off furnace.

3.11 Silicon–Silver–Barium–Tellurium

The $Ag_{3.97}Ba_4Si_2Te_9$ quaternary compound, which crystallizes in the orthorhombic structure with the lattice parameters $a = 836.89 \pm 0.03$, $b = 1358.83 \pm 0.04$, $c = 1022.52 \pm 0.03$ pm, a calculated density of 6.037 g·cm⁻³ and an energy gap of 0.89 eV, is formed in the Si–Ag–Ba–Se system (Cui et al. 2010). The starting materials, Ba, Ag, Si, and Te (molar ratio 4:4:2:9), were loaded into a fused silica tube, which was then sealed under a high vacuum of approximately 1 Pa. The reaction mixture was heated to 750°C within 48 h in a furnace, kept at that temperature for 2 h, and then cooled to 200°C at a rate of 3°C·h⁻¹, followed by switching off the furnace. All chemical elements were stored in an argon-filled glove box with water and oxygen levels below 1 ppm.

3.12 Silicon–Silver–Germanium–Tellurium

Ag_8SiTe_6–Ag_8GeTe_6. This system, studied through DTA and XRD using the alloys annealed at 530°C for 500 h, is a non-quasibinary section of the Si–Ag–Ge–Te system as Ag_8GeTe_6 melts incongruently (Ashirov 2022). Due to the incongruent melting of Ag_8GeTe_6, the phase based on Ag_2Te primarily crystallizes on the Ag_8GeTe_6 side. It was established that below the liquidus, a continuous series of solid solutions with a cubic structure is formed in the system. The lattice parameter increases linearly with increasing Ge content (i.e., the concentration dependence of the lattice parameters follows Vegard's law).

3.13 Silicon–Silver–Manganese–Tellurium

The **Ag₂MnSiTe₄** quaternary compound, which has two polymorphic modifications, is formed in the Si–Ag–Mn–Te system (Lamarche et al. 1991). The first modification crystallizes in the orthorhombic structure with the lattice parameters $a = 812.0$, $b = 687.6$, and $c = 668.9$ pm, and the second one crystallizes in the cubic structure with the lattice parameter $a = 575$ pm. This compound was prepared in the same way as Cu₂MnSiTe₄ was synthesized. The obtained samples were annealed and quenched from various temperatures in the range 400°C–750°C. All showed mainly orthorhombic structure plus a little cubic and MnTe₂, the amounts of the last two varying somewhat with the different samples.

3.14 Silicon–Silver–Iron–Tellurium

The **Ag₂FeSiTe₄** quaternary compound, which melts at 569°C and crystallizes in the orthorhombic structure with the lattice parameters $a = 772.1$, $b = 665.7$, and $c = 658.8$ pm, is formed in the Si–Ag–Fe–Te system (Quintero et al. 1999). This compound was prepared in the same way as Cu₂FeSiTe₄ was obtained.

3.15 Silicon–Cadmium–Thallium–Tellurium

The **CdTl₂SiTe₄** quaternary compound, which crystallizes in the tetragonal structure with the lattice parameters $a = 841.21 \pm 0.06$, $c = 702.89 \pm 0.09$ pm, and a calculated density of 7.075 ± 0.002 g·cm⁻³, is formed in the Si–Cd–Tl–Te system (Selezen et al. 2020). This compound was synthesized in a furnace from Tl, Cd, Si, and Te in quartz containers that were evacuated to 0.1 Pa and sealed. The ampoule was heated to 680°C at the rate of 20°C·h⁻¹, kept at this temperature for 5 h, heated to 1180°C at 10°C·h⁻¹, and held for 5 h. Then it was cooled to 300°C at the rate of 10°C·h⁻¹ and annealed for 350 h. Finally, the sample was quenched into a 20% NaCl solution.

3.16 Silicon–Mercury–Thallium–Tellurium

The **HgTl₂SiTe₄** quaternary compound, which crystallizes in the tetragonal structure with the lattice parameters $a = 839.29 \pm 0.04$, $c = 703.96 \pm 0.05$ pm, and a calculated density of 7.687 ± 0.002 g·cm⁻³, is formed in the Si–Hg–Tl–Te system (Selezen et al. 2020). This compound was prepared in the same way as CdTl₂SiTe₄ was synthesized.

3.17 Silicon–Mercury–Fluorine–Tellurium

The **Hg₃SiTe₂F₆** quaternary compound is formed in the Si–Hg–F–Te system (Puff et al. 1969). This compound could be obtained by the interaction of HgTe with the solution of HgO in H₂SiF₆ at 120°C.

3.18 Silicon–Lanthanum–Oxygen–Tellurium

The **La₂(Si₆O₁₃)(TeO₃)₂** quaternary compound, which crystallizes in the monoclinic structure with the lattice parameters $a = 1476.6 \pm 0.9$, $b = 731.8 \pm 0.4$, $c = 809.9 \pm 0.5$ pm, $\beta = 103.05 \pm 0.01°$, and a calculated density of 3.917 g·cm⁻³, is formed in the Si–La–O–Te system (Kong et al. 2008). Colorless needle-shaped single crystals of the title compound were initially obtained by solid-state reaction of La₂O₃ (0.4 mM), SiO₂ (1.6 mM), and TeO₂ (0.8 mM) at 960°C for 6 days in attempts to replace all germanium in La₄(Si₅.₂Ge₂.₈O₁₈)(TeO₃)₄ by silicon. The highest purity was obtained by the reaction of a mixture of La₂O₃, SiO₂, and TeO₂ (molar ratio 1:1:6) at 960°C for 6 days.

3.19 Silicon–Cerium–Oxygen–Tellurium

The **Ce₂SiO₄Te** quaternary compound, which crystallizes in the orthorhombic structure with the lattice parameters $a = 636.47 \pm 0.06$, $b = 728.07 \pm 0.07$, $c = 1127.43 \pm 0.10$ pm, and a calculated density of 6.356 g·cm⁻³ at 153 K is formed in the Si–Ce–O–Te system. (Deng et al. 2004). The light-yellow needles of this compound were obtained accidentally in the reaction of Ce (70 mg), Ti (48 mg), TeO₂ (80 mg), and KCl (150 mg). The mixture was sealed in an unprotected fused silica tube that was then evacuated to 10⁻² Pa. The tube was heated to 800°C, kept at 800°C for 72 h, cooled at 4°C·h⁻¹ to 100°C, and then the furnace was turned off. The reaction mixture was washed with deionized water and dried with acetone.

3.20 Silicon–Praseodymium–Oxygen–Tellurium

The **Pr₂SiO₄Te** quaternary compound, which has two polymorphic modifications, is formed in the Si–Pr–O–Te system (Weber and Schleid 1999). α-Pr₂SiO₄Te crystallizes in the orthorhombic structure with the lattice parameters $a = 633.70 \pm 0.03$, $b = 724.42 \pm 0.04$, $c = 1125.13 \pm 0.08$ pm, and a calculated density of 6.449 g·cm⁻³, and β-Pr₂SiO₄Te crystallizes in the monoclinic structure with the lattice parameters $a = 989.90 \pm 0.07$, $b = 648.03 \pm 0.04$, $c = 870.68 \pm 0.06$ pm, $\beta = 94.307 \pm 0.008°$, and a calculated density of 5.981 g·cm⁻³. Light green transparent single crystals of both modifications were obtained at the interaction of Pr with TeO₂ and SiO₂ (molar ratio 2:1:1) in the presence of CsCl as a flux in an evacuated quartz ampoule at 950°C for 10 days. After washing out the flux, the two hydrolysis-resistant products were found to be free of foreign phases. The single crystals of α-Pr₂SiO₄Te, which usually make up 80–90 % of the total product, were only found in the regulus. They crystallize in the form of regular cuboids, while those of β-Pr₂SiO₄Te appeared as needles. The latter can be found almost exclusively outside the melting cake or, at most, on its surface.

3.21 Silicon–Neodymium–Oxygen–Tellurium

The **Nd$_2$SiO$_4$Te** quaternary compound, which has two poly-morphic modifications, is formed in the Si–Nd–O–Te system (Yang and Ibers 2000). One of the modifications crystallizes in the monoclinic structure with the lattice parameters $a = 982.3 \pm 0.2$, $b = 642.1 \pm 0.1$, $c = 867.6 \pm 0.2$ pm, $\beta = 94.60 \pm 0.03°$, and a calculated density of 6.188 g·cm^{-3} at 153 K. The second one crystallizes in the orthorhombic structure with the lattice parameters $a = 627.9 \pm 0.1$, $b = 718.9 \pm 0.1$, $c = 1116.8 \pm 0.2$ pm, and a calculated density of 6.696 g·cm^{-3} at 153 K. To prepare this compound, the reactions were carried out at 850–900°C for 7 days in unprotected fused silica tubes with the use of a BaBr$_2$/KBr eutectic flux (molar ratio 1.1:1). Two types of crystals, light-purple prismatic needles (orthorhombic) and flat yellow needles (monoclinic), were obtained in a yield of about 20% in the reaction of BaTe, Nd, Zn, and Te in a molar ratio of 1:1:1:3.

3.22 Silicon–Samarium–Oxygen–Tellurium

The **Sm$_2$SiO$_4$Te** quaternary compound, which has two poly-morphic modifications, is formed in the Si–Sm–O–Te system. The first modifications crystallize in the monoclinic structure with the lattice parameters $a = 976.0 \pm 0.2$, $b = 635.7 \pm 0.1$, $c = 860.1 \pm 0.2$ pm, $\beta = 94.87 \pm 0.03°$, and a calculated density of 6.500 g·cm^{-3} at 153 K (Yang and Ibers 2000). The second one crystallizes in the orthorhombic structure with the lattice parameters $a = 620.1 \pm 0.1$, $b = 709.1 \pm 0.1$, $c = 1107.7 \pm 0.2$ pm, and a calculated density of 7.096 g·cm^{-3} at 115 K (Yang and Ibers 2000) [$a = 623.82 \pm 0.08$, $b = 713.06 \pm 0.07$, $c = 1112.26 \pm 0.11$ pm, and a calculated density of 6.986 g·cm^{-3} (Person et al. 2000)].

To prepare this compound, the reactions were carried out at 850°C–900°C for 7 days in unprotected fused silica tubes with the use of a BaBr$_2$/KBr eutectic flux in a molar ratio of 1.1:1 (Yang and Ibers 2000). Two types of crystals, dark red prisms (orthorhombic) and vivid red flat needles (monoclinic), were obtained in a yield of about 20% in the reaction of BaTe, Sm, Zn, and Te in a molar ratio of 1:1:1:3. Single crystals of Sm$_2$SiO$_4$Te were prepared through chemical transport reactions using iodine as a transport agent (Person et al. 2000).

3.23 Silicon–Gadolinium–Oxygen–Tellurium

Two quaternary compounds, **Gd$_2$SiO$_4$Te** and **Gd$_4$(SiO$_4$)$_2$OTe**, are formed in the Si–Gd–O–Te system. Both crystallize in the orthorhombic structure with the lattice parameters $a = 616.96 \pm 0.04$, $b = 705.75 \pm 0.05$, and $c = 1105.79 \pm 0.07$ pm at 153 K for the first compound (Ijjaali et al. 2001) and $a = 1249.53 \pm 0.10$, $b = 1086.83 \pm 0.08$, $c = 680.75 \pm 0.05$ pm, and a calculated density of 6.874 g·cm^{-3} for the second one (Daszkiewicz and Gulay 2015).

White needles of Gd$_2$SiO$_4$Te were obtained accidentally in the reaction of Gd (0.095 g), Mn (0.020 g), and Te (0.185 g) in a fused silica tube with KBr (200 mg) added to promote

crystal growth (Ijjaali et al. 2001). The materials were mixed and sealed in the unprotected tube that was then evacuated to $6.7 \cdot 10^{-3}$ Pa. The tube was heated to 880°C at 0.3°C·min^{-1}, kept at 880°C for 4 days, cooled at 0.04°C·min^{-1} to 600°C, and then the furnace was turned off. The reaction mixture was washed free of bromide salts with water and then dried with acetone.

To synthesize Gd$_4$(SiO$_4$)$_2$OTe, the starting sample was pre-pared as a mixture of Gd (2.1 mM), Cu (2.1 mM), and Te (4.2 mM) for obtaining GdCuTe$_2$. The sample was sintered in an evacuated quartz ampoule, and the synthesis was carried out in a tube resistance furnace. The ampoule was heated at a rate of 30°C·h^{-1} to 1150°C and then kept at this temperature for 3 h. Afterwards, the sample was cooled slowly (30°C·h^{-1}) to 600°C and annealed at this temperature for 720 h. The furnace was turned off, and the ampoule was cooled to room tempera-ture. Eight colorless single crystals of the title compound were found in the prepared sample.

3.24 Silicon–Zirconium–Germanium–Tellurium

The **ZrSi$_{1-x}$Ge$_x$Te** solid solutions were prepared by directly mixing elements in stoichiometric proportions (Wang and Hughbanks 1995). The reaction temperature was raised to 550°C over 5 h and maintained at that temperature for 2 days. The temperature was then raised to 850°C over a 24 h interval, maintained at 850°C for another 4 days, and then quenched to room temperature. Since some of the products are sensitive to both moisture and oxygen, all operations were performed under a N$_2$ atmosphere.

3.25 Silicon–Zirconium–Arsenic–Tellurium

Procedures and conditions used in preparing **ZrSi$_{1-x}$As$_x$Te** solid solutions were the same as those for ZrSi$_{1-x}$Ge$_x$Te obtain-ing (Wang and Hughbanks 1995). One of the solid solution compositions at $x = 0.34$ crystallizes in the tetragonal structure with the lattice parameters $a = 371.10 \pm 0.03$, $c = 972.3 \pm 0.2$ pm, and a calculated density of 6.484 g·cm^{-3}.

Some other materials in the Si–Zr–As–Te system were also obtained (Wang and Hughbanks 1995). They were prepared in silica ampoules by mixing 0.3 g of a Zr, As, Si, and Te mixture in a molar ratio of 4:3:1:4 with 3 mg of TeCl$_4$. The reaction temperature was brought up to 550°C in 5 h and kept there for 2 days. It was then raised to 1000°C over 12 h, and a tempera-ture gradient was applied in the range of 1000°C–950°C for 3 weeks. Several plate crystals with the approximate compo-sition ZrSi$_{0.47\pm0.01}$As$_{0.70\pm0.01}$Te$_{0.79\pm0.03}$ were found in the product. Subsequently, this composition was used to obtain a single-phase product with the composition **ZrSi$_{0.5}$As$_{0.7}$Te$_{0.8}$** (tetrago-nal structure, $a = 381.16 \pm 0.01$, $c = 839.8 \pm 0.3$ pm and a calculated density of 7.151 g·cm^{-3}).

In an attempt to grow single crystals of this phase by use of an Sn flux, 0.6 g of a mixture of elemental Zr, Si, As, Te, and Sn in a molar ratio of 1:0.5:0.7:0.8:1 was mixed with 3 mg of TeCl$_4$. The reaction temperature was raised to 550°C over a 6 h interval and maintained there for 3 days. It was then raised to 900°C over 12 h, and a 900–950°C temperature gradient was applied for another 21 days before allowing the furnace to cool to room temperature. Microprobe analysis of selected plate crystals gave the approximate composition $ZrSi_{0.8\pm0.1}As_{0.47\pm0.02}Te_{0.74\pm0.02}$. A quantitative synthesis of **$ZrSi_{0.9}As_{0.5}Te_{0.7}$** (tetragonal structure, a = 374.64 ± 0.07, c = 854.9 ± 0.4 pm for **$ZrSi_{0.9}As_{0.4}Te_{0.7}$**) could be achieved by mixing elements in stoichiometric proportions.

In an attempt to synthesize **$ZrSi_{0.5}AsTe_{0.5}$**, a reaction created by mixing Zr, Si, As, and Te in a molar ratio of 1:0.5:1:0.5 with 3 mg of $TeCl_4$ was conducted in a silica ampoule. The reaction temperature was raised to 550°C over 1 h and kept there for 2 days. It was then raised to 900°C over 12 h, and a 900–950°C temperature gradient was applied for another 21 days before allowing the furnace to cool to room temperature. Microprobe analysis of selected plate crystals gave the approximate composition **$ZrSi_{0.7}AsTe_{0.3}$** (tetragonal structure, a = 370.20 ± 0.04, c = 834.6 ± 0.2 pm). Since some of the products are sensitive to both moisture and oxygen, all operations were performed under a N_2 atmosphere.

4

Systems Based on Germanium Sulfides

4.1 Germanium–Hydrogen–Nitrogen–Sulfur

While GeS does not react with liquid NH_3, GeS_2 dissolves with the formation of $NH_4(GeS_2NH_2)$ (Behrens and Ostermeier 1962). At the interaction of GeS_2 with the solution of $(NH_4)_2S$ in the ammonia, the $(NH_4)_2(GeS_3)$ quaternary compound is formed. This compound, like $NH_4(GeS_2NH_2)$, converts into GeS_2 at the heating up to 150°C with the release of NH_3 and H_2S.

4.2 Germanium–Lithium–Magnesium–Sulfur

Two quaternary compounds, $Li_2Mg_2Ge_2S_6$ and $Li_4MgGe_2S_7$, are formed in the Ge–Li–Mg–S system. First of them crystallizes in the tetragonal structure with the lattice parameters $a = 625.94 \pm 0.01$, $c = 662.98 \pm 0.02$ pm, a calculated density of 2.953 g·cm^{-3}, and an energy gap 3.18 eV (Barton et al. 2022). To prepare this compound, stoichiometric amounts of the reagents were weighed and ground together with an agate mortar and pestle in the glove box. The reaction mixture was added to a graphite crucible that was later introduced to a fused silica tube and sealed at a pressure of ~10^{-2} Pa. The sealed reaction vessel was seated in a furnace with a temperature controller. The furnace was heated to 300°C in 3 h, held at 300°C for 10 h, heated to 650°C over 5 h, held at that temperature for 10 h, heated to 900°C in 10 h, and kept at 900°C for 5 days, followed by slow cooling to 500°C at a rate of 5°C·h^{-1} and finally rapid cooling to room temperature. Due to the high oxophilicity of Li and Mg metal, an inner graphite crucible was used to prevent the reaction of the starting materials with the fused silica outer vessel. After the reaction, the graphite crucible was broken in order to retrieve the reaction products, some of which stuck to the crucible. Therefore, the losses of some amount of the product generally preclude a regrind and reheat strategy to obtain higher phase purity of the target compound.

$Li_4MgGe_2S_7$ crystallizes in the monoclinic structure with the lattice parameters $a = 1687.2 \pm 0.6$, $b = 671.1 \pm 0.2$, $c = 1015.6 \pm 0.4$ pm, $\beta = 95.169°$, a calculated density of 2.446 g·cm^{-3}, and an energy gap of 4.12 eV (Abudurusuli et al. 2021). The target compound was prepared through the conventional solid-state reaction method in the following way. The starting materials Li_2S, Mg, Ge, and S (molar ratio 2.5:1:2:5) were weighed and loaded into a graphite crucible, then put into a fused silica tube, which was sealed under a high vacuum of 10^{-3} Pa. The sealed tube was moved into a furnace and heated to 950°C for 60 h, left for 100 h to ensure the mixture completely melted, after that the temperature was cooled to 300°C at a rate of 3°C·h^{-1}, and the furnace was switched off. $Li_4MgGe_2S_7$ is highly corrosive to fused silica tubes, and it often causes tube explosions. So the carbon crucible was used to avoid corrosion of the silica tube during the reaction.

4.3 Germanium–Lithium–Strontium–Sulfur

The Li_2SrGeS_4 quaternary compound, which crystallizes in the tetragonal structure with the lattice parameters $a = 654.20 \pm 0.15$, $c = 775.1 \pm 0.3$ pm, a calculated density of 3.027 g·cm^{-3}, and an energy gap of 3.75 eV, is formed in the Ge–Li–Sr–S system (Wu et al. 2019). To prepare this compound, a mixture of Li (2 mM), SrS (1 mM), Ge (1 mM), and S (3 mM) was first loaded into a graphite crucible that avoids the reaction between an alkali metal and silica tube, then put it into the silica tube and flame-sealed under 10^{-3} Pa. The tube was heated to 850°C in 30 h and kept at this temperature for about 100 h, then cooled slowly down to ambient temperature for 5 days. DMF solvent was chosen to wash the products. Finally, many colorless crystals of the title compound were found, which are stable in the air after several months. The whole preparation process is completed in an Ar-filled glove box.

4.4 Germanium–Lithium–Barium–Sulfur

The Li_2BaGeS_4 quaternary compound, which crystallizes in the tetragonal structure with the lattice parameters $a = 663.8 \pm 0.4$, $c = 803.3 \pm 1.0$ pm, a calculated density of 3.303 g·cm^{-3}, and an energy gap of 3.66 eV, is formed in the Ge–Li–Ba–S system (Wu et al. 2017c). To synthesize the title compound, a mixture of Li (2 mM), Ba (1 mM), Ge (1 mM), and S (4 mM) was first loaded into a graphite crucible that avoids the reaction between Li and silica tube, then put it into the silica tube and flame-sealed under 10^{-3} Pa. The tube was heated to 800°C for 30 h and kept at this temperature for about 90 h, then cooled slowly down to ambient temperature for 5 days. The product was washed with DMF. Finally, many small colorless transparent crystals of Li_2BaGeS_4 were found, which are stable in the air after several months.

4.5 Germanium–Lithium–Zinc–Sulfur

Two quaternary phases, Li_2ZnGeS_4 and $Li_{4-2x}Zn_xGeS_4$ ($0 \leq x \leq 0.20$), are formed in the Ge–Li–Zn–S system. First of them has two polymorphic modifications. α-Li_2ZnGeS_4 crystallizes in the orthorhombic structure with the lattice parameters $a = 1306.76 \pm 0.02$, $b = 783.870 \pm 0.010$, $c = 621.640 \pm 0.010$ pm,

DOI: 10.1201/9781003123484-4

a calculated density of 2.922 g·cm^{-3}, and an energy gap of 4.07 eV (Zhang et al. 2020a) [a = 783, b = 653 and c = 621 pm (Kanno et al. 2000)]. β-Li$_2$ZnGeS$_4$ crystallizes in the monoclinic structure with the lattice parameters a = 621.3 ± 0.5, b = 652.5 ± 0.5, c = 783.0 ± 0.6 pm, $β$ = 90.70°, and an energy gap of 3.49 eV (Huang et al. 2019d).

Li$_{4-2x}$Zn$_x$GeS$_4$ is a solid solution of Li$_2$ZnGeS$_4$ in Li$_4$GeS$_4$ and crystallizes in the orthorhombic structure with the lattice parameters a = 1405, b = 775 and c = 616 pm (the value of x is not specified) (Kanno et al. 2000).

The colorless, prismatic crystals of α-Li$_2$ZnGeS$_4$ were originally obtained from the reaction of Li$_2$S, Zn, Ge, and S (molar ratio1:1:1:3) with a total weight of about 650 mg plus an extra 100 mg of Li$_2$S as a flux (Zhang et al. 2020a). The reaction mixture was thoroughly ground with an agate mortar and pestle in the glove box, transferred to a graphite crucible, and sealed in an evacuated fused silica tube (~0.1 Pa). The tube was heated to 700°C within 9 h, dwelled at that temperature for 96 h, and cooled to 400°C at 3°C·h^{-1} before switching off the furnace. The product was rinsed with (NH$_4$)$_2$S and distilled water to remove the excess flux and then dried with acetone. The final product consisted of long, block-like colorless crystals of α-Li$_2$ZnGeS$_4$. A phase-pure powder sample was produced from a stoichiometric mixture of Li$_2$S, Zn, Ge, and S with a total weight of 1.12 g. The reaction mixture was ground thoroughly in the glove box and loaded into a graphite crucible that was subsequently sealed in an evacuated fused silica tube. The tube was heated to 300°C for 10 h, held at that temperature for 24 h, heated to 680°C over 100 h, followed by a dwell period of 7 days, and subsequently cooled to 80°C at 2°C·h before switching off the furnace. All reagents were stored in an Ar-filled glove box (moisture and oxygen levels less than 0.1 ppm), and all manipulations were carried out in the glove box or under vacuum.

To prepare β-Li$_2$ZnGeS$_4$, a raw mixture of Li, ZnS, Ge, and S (molar ratio 2:1:1:3) was firstly loaded into the graphite crucible and then put into a silica tube (Huang et al. 2019d). The tube was under vacuum (10^{-3} Pa). A muffle furnace was used to complete the crystallization reaction, and the setting temperature process was shown as follows: firstly heated to 300°C in 10 h and left at this temperature for 20 h to make the partial S participate in the reaction, further heated to 900°C to ensure the mixture melt completely while kept at this temperature within 4 days, then slowly cooled down to 300°C in 100 h, finally quickly down to the room temperature in one day. Many air-stable colorless crystals of β-Li$_2$ZnGeS$_4$ were found under the optical microscope.

To obtain Li$_{4-2x}$Zn$_x$GeS$_4$, the starting materials (Li$_2$S, GeS$_2$, and ZnS) were weighed, mixed in appropriate molar ratios in an Ar-filled glove box, put into a carbon-coated quartz tube and heated at 600–800°C for 8 h (Kanno et al. 2000). After the reaction, the tube was slowly cooled to room temperature.

4.6 Germanium–Lithium–Cadmium–Sulfur

The **Li$_2$CdGeS$_4$** quaternary compound, which melts at 892.2°C (Brant et al. 2014a) and crystallizes in the orthorhombic structure with the lattice parameters a = 773.74 ± 0.01, b = 684.98 ±

0.01, c = 636.88 ± 0.01 pm, and an energy gap of 3.10 eV [2.78 eV (Li et al. 2011); 3.15 eV (Brant et al. 2014a)] (Lekse et al. 2009), is formed in the Ge–Li–Cd–S system. According to the first-principles calculation within the local density approximation (LDA) and generalized gradient approximation (GGA), the lattice parameters of this compound are a = 780.1, b = 682.7, and c = 641.6 pm and bulk modulus B_0 = 57.7 GPa and pressure derivative bulk modulus $B_0{}'$ = 4.31 (LDA), and a = 750.3, b = 651.2, and c = 619.7 pm and bulk modulus B_0 = 57.4 GPa and pressure derivative bulk modulus $B_0{}'$ = 4.53 (GGA) (Li et al. 2013) [a = 760.1, b = 665.4, and c = 625.3 pm (Li et al. 2011)]. Li$_2$CdGeS$_4$ is not mechanically stable above 8.6 GPa (Li et al. 2013). Polycrystalline powder of this compound was synthesized by heating stoichiometric amounts of Li$_2$S, Cd, Ge, and S at 525°C for 163.5 h and then quenching in ice water (Lekse et al. 2009). Its single crystals were grown from a lithium polysulfide flux at 650°C.

A quantitative yield of Li$_2$CdGeS$_4$ was obtained by grinding stoichiometric amounts of Cd, Ge, and S plus a slight excess of Li$_2$S that can act as a molten flux at elevated temperatures using an agate mortar and pestle in an Ar-filled glove box, and then placing the mixture into a graphite crucible inside a fused silica tube (Brant et al. 2014a). The tube was sealed under a pressure of approximately 0.1 Pa. The sample was heated at 750°C in 144 h, slowly cooled to 550°C for 200 h, and then allowed to cool to room temperature naturally. The tube was opened under ambient conditions, and the product was rinsed with methanol to remove the excess Li$_2$S$_x$. The pale-yellow crystals of the title compound were obtained.

4.7 Germanium–Lithium–Mercury–Sulfur

Two quaternary compounds, **Li$_2$HgGeS$_4$** and **Li$_4$HgGe$_2$S$_7$**, are formed in the Ge–Li–Hg–S system. Li$_2$HgGeS$_4$ crystallizes in the orthorhombic structure with the lattice parameters a = 770.9 ± 0.9, b = 681.2 ± 0.8, c = 638.4 ± 0.7 pm, a calculated density of 4.114 g·cm^{-3} and an energy gap of 2.46 eV (Wu and Pan 2017) [a = 766.07*/770.45**, b = 675.12*/678.13**, and c = 637.49*/640.34** pm according to the calculations with (*) and without (**) including the spin-orbit coupling and an energy gap of 2.953 eV (Alnujaim et al. 2020)]. Li$_4$HgGe$_2$S$_7$ crystallizes in the monoclinic structure with the lattice parameters a = 1687.6 ± 0.2, b = 677.64 ± 0.08, c = 1016.13 ± 0.13 pm, $β$ = 93.360 ± 0.007°, a calculated density of 3.424 g·cm^{-3}, and an energy gap of 2.75 eV (Wu et al. 2017a).

Initially, Li$_2$HgGeS$_4$ was prepared with a mixture Li, HgS, Si, and S (molar ratio 2:1:1:3) (Wu and Pan 2017). In the preparation process, a graphite crucible was added into the vacuum-sealed silica tube to avoid the reaction between Li and the silica tube at a high temperature. The temperature process was set as follows: first, it was heated to 700°C in 2 days, and kept at this temperature for about 4 days, then slowly cooled down to 300°C within 4 days, and finally quickly cooled to room temperature by turning off the furnace. Obtained product was washed by DMF to remove the other byproducts. The main product was Li$_4$HgGe$_2$S$_7$ and only small amount of Li$_2$HgGeS$_4$ was found. Pure phase of Li$_2$HgGeS$_4$ could be obtained while a molar ratio of Li/HgS was greater than 2:1. This compound is stable in the air.

Yellow air stable crystals of $Li_4HgGe_2S_7$ were synthesized by the solid-state reactions in the vacuum-sealed silica tubes (Wu et al. 2017a).

4.8 Germanium–Lithium–Gallium–Sulfur

GeS_2–Li_2S–Ga_2S_3. Two quaternary compounds, $LiGaGe_2S_6$ and $Li_2Ga_2GeS_6$, and the $Li_{4+x+δ}(Ge_{1-x-δ}Ga_x)S_4$ $(0.2 ≤ x ≤ 0.6)$ solid solution are formed in this system. $LiGaGe_2S_6$ melts at 663°C and crystallizes in the orthorhombic structure with the lattice parameters $a = 1192.5 ± 0.2$, $b = 2264.7 ± 0.5$, $c = 683.08 ± 0.14$ pm, a calculated density of 2.9827 g·cm⁻³, and an energy gap of 3.52 eV (Mei et al. 2017). $Li_2Ga_2GeS_6$ also crystallizes in the orthorhombic structure with the lattice parameters $a = 1207.96 ± 0.02$, $b = 2273.00 ± 0.04$, $c = 684.048 ± 0.011$ pm, and an energy gap of 3.72 and 3.51 eV at 80 K and 300 K, respectively (Isaenko et al. 2015) [$a = 1194.3 ± 0.5$, $b = 2259.0 ± 0.8$, $c = 680.5 ± 0.2$ pm, a calculated density of 2.976 g·cm⁻³, and an energy gap of $≈ 3.36$ eV (Kim et al. 2008)]. The $Li_{4+x+δ}(Ge_{1-x-δ}Ga_x)S_4$ solid solution also crystallizes in the orthorhombic structure with the lattice parameters $a = 689$, $b = 620$, and $c = 796$ pm at $x = 0.25$, $δ = 0.025$ and $δ' = 0.14$ (Kanno et al. 2000).

To prepare $LiGaGe_2S_6$, the mixture of Li_2S, Ga_2S_3, and GeS_2 (molar ratio 1:1:4) was ground and loaded into a carbon-coated silica tube under an Ar atmosphere in a glove box (Mei et al. 2017). The tube was placed into a furnace and heated to 700–800°C, kept there for 20 h, and the furnace quickly cooled to room temperature. Single crystals of this compound were grown as described above, but the mixture was heated to 750–850°C, kept there for 20 h, and then slowly cooled (3–5°C·h⁻¹) to room temperature.

$Li_2Ga_2GeS_6$ was obtained by pyrosynthesis from Li, Ga, and Ge, as well as sulfur in a stoichiometric ratio (Isaenko et al. 2015). The mixture was placed into a glass carbon crucible located inside a silica container. This container was sealed after evacuation and mounted in the two-zone furnace so that the lower part of the container was located in the hot zone with a temperature of ~1000°C, whereas its upper part was in the cold zone (400°C). Synthesis time was 5–6 h, and afterwards the furnace was switched off. The obtained charge was then displaced into a glass carbon crucible with a conic bottom and a press cap. This crucible was placed into a silica ampoule, evacuated, and sealed. Ampoule with the crucible was moved in the furnace at a rate of 1 mm·day⁻¹ from the upper hot zone with T = 900°C to the lower cold one (700°C). The temperature gradient in the crystallization area was 10°C·cm⁻¹. Thus, a single crystal was grown.

Single crystals of $Li_2Ga_2GeS_6$ were also obtained from 1-g batches of the reactants, Li_2S, GeS_2, Ga_2S_3 (molar ratio 1:4:1) (Kim et al. 2008). The mixture was placed inside a pre-dried carbon-coated quartz tube inside an oxygen- and water-free glove box (<1 ppm O_2 and H_2O), and then sealed under vacuum. The sealed tube and contents were heated according to the next temperature profile: room temperature → (10 h) 950°C for 5 h → (7 h) 750°C → (100 h) 500°C for 20 h → (air

quenching) room temperature. A mass of the crystalline phase was observed on the bottom of the quartz tube.

To obtain $Li_{4+x+δ}(Ge_{1-x-δ}Ga_x)S_4$, the starting materials (Li_2S, GeS_2, and Ga_2S_3) were weighed, mixed in appropriate molar ratios in an Ar-filled glove box, put into a carbon-coated quartz tube and heated at 600–800°C for 8 h (Kanno et al. 2000). After the reaction, the tube was slowly cooled to room temperature.

The glass region of the GeS_2–Li_2S–Ga_2S_3 quasiternary system could extend to the 0–20 mol% range of Li content with around 40 mol% 0.5 Ga_2S_3 (Liu and Zhang 2014) [the glass-forming region was observed at cation ratios of 0 to 40–70 mol% Li and less than 50 mol% Ga (Yamashita and Yamanaka 2003)]. The glasses were not obtained with more than 28 mol% Li-cation ratios (Liu and Zhang 2014). Glass transition temperature increased slightly with the addition of a moderate amount of Ga_2S_3 (Yamashita and Yamanaka 2003). The glasses were prepared from the mixtures of Li, Ge, Ga, and S by the conventional melt-quenching method (Liu and Zhang 2014). An explosion from Li reacting with liquid S was avoided by carefully controlling the heating rate. Firstly, the temperature was increased very slowly and started to rock the sample at 120°C, and kept this temperature for 1 h. Continually, the temperature was increased at the rate of 1°C·min⁻¹ from 120°C to 750°C and at a rate of 0.5°C·min⁻¹ from 750°C to 950°C, and finally was kept 10 h to make the raw materials react fully in the oven at 950°C. Bulk glasses were also melted in Si-coated silica tubes (Yamashita and Yamanaka 2003).

4.9 Germanium–Lithium–Indium–Sulfur

The $Li_2In_2GeS_6$ quaternary compound, which melts congruently at 780°C and crystallizes in the monoclinic structure with the lattice parameters $a = 1216.5 ± 0.2$, $b = 708.40 ± 0.14$, $c = 1213.1 ± 0.2$ pm, $β = 110.26 ± 0.03°$, a calculated density of 3.444 g·cm⁻³, and an energy gap of 3.45 eV, is formed in the Ge–Li–In–S system (Yin et al. 2012a). The structural, electronic, and optical properties of this compound have been investigated by *ab initio* calculations based on the density functional theory by Wong et al. (2018) using the full potential linearized augmented plane wave method. The next values of the lattice parameters were determined: $a = 1216.09$, $b = 708.16$, $c = 1212.96$ pm (the value of $β$ is not calculated).

To prepare this compound, the mixtures of $LiInS_2$ (1 mM) and GeS_2 (0.5 mM) were ground, loaded into fused silica tube under an Ar atmosphere in a glove box, then moved to a high-vacuum line, and flame-sealed under a high vacuum of 10⁻³ Pa (Yin et al. 2012a). The tube was then placed in a furnace and heated to 900°C in 15 h, left for 48 h, cooled to 320°C at a rate of 3°C·h⁻¹, and finally cooled to room temperature by switching off the furnace. Many block-shaped crystals with the color of light yellow were found in the ampoule. The crystals are stable in the air. Polycrystalline samples of $Li_2In_2GeS_6$ were synthesized by solid-state reaction techniques. The mixtures of Li_2S, GeS_2, and In_2S_3 (molar ratio 1:1:1) were heated to 700°C in 15 h, kept at that temperature for 72 h, and then the furnace was turned off.

4.10 Germanium–Lithium–Lanthanum–Sulfur

The **LiLa$_3$GeS$_7$** quaternary compound, which crystallizes in the hexagonal structure, is formed in the Ge–Li–La–S system (Craig et al. 2022). It was prepared via a direct combination of the elements or binary sulfides. This compound is a semiconductor and shows high thermal stability (>1000°C).

4.11 Germanium–Lithium–Cerium–Sulfur

The **LiCe$_3$GeS$_7$** quaternary compound, which crystallizes in the hexagonal structure, is formed in the Ge–Li–Ce–S system (Craig et al. 2022). It was prepared via a direct combination of the elements or binary sulfides. This compound is a semiconductor and shows high thermal stability (>1000°C).

4.12 Germanium–Lithium–Praseodymium–Sulfur

The **LiPr$_3$GeS$_7$** quaternary compound, which crystallizes in the hexagonal structure, is formed in the Ge–Li–Pr–S system (Craig et al. 2022). It was prepared via a direct combination of the elements or binary sulfides. This compound is a semiconductor and shows high thermal stability (>1000°C).

4.13 Germanium–Lithium–Neodymium–Sulfur

The **LiNd$_3$GeS$_7$** quaternary compound, which crystallizes in the hexagonal structure, is formed in the Ge–Li–Nd–S system (Craig et al. 2022). It was prepared via a direct combination of the elements or binary sulfides. This compound is a semiconductor and shows high thermal stability (>1000°C).

4.14 Germanium–Lithium–Samarium–Sulfur

The **LiSm$_3$GeS$_7$** quaternary compound, which crystallizes in the hexagonal structure with the lattice parameters $a = 999.1 \pm 0.3$, $c = 575.2 \pm 0.3$ pm, and a calculated density of 5.042 g·cm^{-3}, is formed in the Ge–Li–Sm–S system (Zhen et al. 2016a). To synthesize this compound, Li, Sm, Ge, and S were weighted at the stoichiometric ratio. All the raw materials were firstly loaded into a graphite crucible, which was then put into a silica tube, and the tube was flame-sealed under 10^{-3} Pa. The detailed temperature-setting curve for the muffle furnace was heated to 850°C for 50 h and kept at this temperature for about 100 h, then slowly cooled to 300°C by 80 h, and finally quickly cooled to room temperature. The product was washed with DMF to remove the byproducts. Yellow crystals were found, which would deliquesce when exposed to the air for a long time. Because Li metal is easily oxidized in the air, all the preparations were completed in an Ar-filled glove box.

4.15 Germanium–Lithium–Europium–Sulfur

The **Li$_2$EuGeS$_4$** quaternary compound, which crystallizes in the tetragonal structure with the lattice parameters $a = 654.47 \pm 0.04$, $c = 769.60 \pm 0.06$ pm, a calculated density of 3.694 g·cm^{-3} at 173 K, and an energy gap of 2.54 eV, is formed in the Ge–Li–Eu–S system (Aitken et al. 2001). To synthesize this compound, in a N$_2$-filled glove box, Eu (0.3 mM), Ge (0.3 mM), Li$_2$S (0.3 mM), and S (2.4 mM) were loaded into a graphite tube. The graphite tube was inserted into a silica tube and flame-sealed under vacuum (approximately 2·10^{-2} Pa). This tube was heated from 50°C to 650°C in 24 h. The reaction was kept at this temperature for 96 h and then cooled at a rate of 2°C·h^{-1} to 250°C followed by rapid cooling to 50°C in 3 h. DMF was used to remove the excess flux. As a result, the yellow polyhedra of the title compound were isolated. Li$_2$EuGeS$_4$ decomposes in water and is sensitive to moist air.

4.16 Germanium–Lithium–Gadolinium–Sulfur

The **LiGd$_3$GeS$_7$** quaternary compound, which crystallizes in the hexagonal structure with the lattice parameters $a = 990.0 \pm 0.7$, $c = 575.3 \pm 0.5$ pm, and a calculated density of 5.274 g·cm^{-3}, is formed in the Ge–Li–Gd–S system (Zhen et al. 2016a). This compound was synthesized in the same way as LiSm$_3$GeS$_7$ was prepared using Gd instead of Sm.

4.17 Germanium–Lithium–Dysprosium–Sulfur

The **LiDy$_3$GeS$_7$** quaternary compound, which crystallizes in the hexagonal structure, is formed in the Ge–Li–Dy–S system (Craig et al. 2022). It was prepared via a direct combination of the elements or binary sulfides. This compound is a semiconductor and shows high thermal stability (>1000°C).

4.18 Germanium–Lithium–Lead–Sulfur

Two quaternary compounds, **Li$_{0.5}$Pb$_{1.75}$GeS$_4$** and **Li$_2$PbGeS$_4$**, are formed in the Ge–Li–Pb–S system. First of them crystallizes in the cubic structure with the lattice parameter $a = 1401.63 \pm 0.06$, a calculated density of 5.470 g·cm^{-3} at 173 K, and an energy gap of 1.95 eV (Aitken et al. 2000). Li$_2$PbGeS$_4$ crystallizes in the tetragonal structure with the lattice parameters $a = 652.24 \pm 0.05$, $c = 776.03 \pm 0.08$ pm, a calculated density of 4.244 g·cm^{-3} at 173 K, and an energy gap of 2.41 eV (Aitken et al. 2001).

To prepare Li$_{0.5}$Pb$_{1.75}$GeS$_4$, in a N$_2$-filled glove box, Pb (0.45 mM), Ge (0.3 mM), Li$_2$S (0.15 mM), and S (2.4 mM) were loaded into a graphite tube (Aitken et al. 2000). This tube was inserted into a Pyrex tube and flame-sealed under vacuum (approximately 2·10^{-2} Pa). The Pyrex tube was then placed into a furnace and heated from 50°C to 500°C in 24 h. The reaction was kept at this temperature for 96 h and then cooled at 2.5°C·h^{-1} to 250°C followed by rapid cooling to 50°C in 2 h.

DMF was used to remove the excess flux. Washing with ether revealed pure orange/red crystalline chunks of $Li_{0.5}Pb_{1.75}GeS_4$. In order to obtain crystals suitable for single-crystal XRD, the same starting mixture was loaded into a graphite tube, which was inserted into a quartz tube and flame-sealed under vacuum (approximately $2 \cdot 10^{-2}$ Pa). This tube was heated from 50°C to 650°C in 12 h. The reaction was kept at this temperature for 72 h and then cooled at 2.75°C·h⁻¹ to 250°C followed by rapid cooling to 50°C in 2 h. The product was isolated in the same manner as previously to reveal beautiful, red plate-like single crystals as a pure phase.

To synthesize Li_2PbGeS_4, in a N_2-filled glove box, Pb (0.3 mM), Ge (0.3 mM), Li_2S (0.3 mM), and S (2.4 mM) were loaded into a graphite tube (Aitken et al. 2001). The graphite tube was inserted into a silica tube and flame-sealed under a vacuum (approximately $2 \cdot 10^{-2}$ Pa). This tube was heated from 50°C to 600°C in 24 h. The reaction was kept at this temperature for 72 h and then cooled at a rate of 2.5°C·h⁻¹ to 250°C followed by rapid cooling to 50°C in 2 h. DMF was used to remove the excess flux. Washing with ether revealed yellow polyhedra of the title compound. Li_2PbGeS_4 decomposes in water immediately and within a day upon exposure to moist air.

4.19 Germanium–Lithium–Phosphorus–Sulfur

Li_4GeS_4–Li_3PS_4. The phase diagram of this system, constructed through DTA, XRD, and metallography, is shown in Figure 4.1 (Hori et al. 2015a). The eutectic contains 2 mol% Li_4GeS_4 and crystallizes at ca. 550°C. The ranges of the solid solutions at room temperature were determined to be 0–20, 50–67, and 90–98 mol% Li_3PS_4. These solid-solution ranges were also found to expand at high temperatures. The $Li_{10}GeP_2S_{12}$ quaternary compound, which melts incongruently at ca. 650°C and has a wide homogeneity region, is formed in this system. It crystallizes in the tetragonal structure

with the lattice parameters $a = 871.87$, $c = 1263.85$ pm, and a calculated density of 2.035 g·cm⁻³ at room temperature and $a = 865.21$, $c = 1258.16$ pm, and a calculated density of 2.076 g·cm⁻³ at 100 K (Kuhn et al. 2013b) [$a = 871.771 \pm 0.005$ and $c = 1263.452 \pm 0.010$ pm (Kamaya et al. 2011); $a = 871.8 \pm 0.1$ and $c = 1266.0 \pm 0.1$ pm for $Li_{10}GeP_2S_{12}$ composition and $a = 871.4 \pm 0.2$ and $c = 1260.7 \pm 0.7$ pm for Li_7GePS_8 composition (Kuhn et al. 2013a); $a = 871.3023 \pm 0.0005$ and $c = 1263.9695 \pm 0.0011$ pm for $Li_{10.35}Ge_{1.35}P_{1.65}S_{12}$ composition (Kwon et al. 2015)].

Both Li_7GePS_8 and $Li_{10}GeP_2S_{12}$ were synthesized from stoichiometric amounts of Li_2S, Ge, and P (Kuhn et al. 2013a,b). A slight excess of sulfur yielding a vapor pressure of approximately 0.1 MPa under the reaction conditions was used in order to ensure complete oxidation of Ge and P. The starting materials were mechanically treated in a high-energy ball mill for 1 day. The obtained precursor powders were heated in evacuated quartz tubes according to the following temperature programs: Li_7GePS_8: 30°C·h⁻¹ → 450°C (4 h) → 550°C (1 h); $Li_{10}GeP_2S_{12}$: 30°C·h⁻¹ → 420°C (1 day). Due to the lower synthesis temperature, $Li_{10}GeP_2S_{12}$ shows a lower crystallinity. However, higher synthesis temperatures for the metastable tetragonal modification unavoidably led to the partial formation of the undesired orthorhombic modification.

$Li_{10}GeP_2S_{12}$ could also be prepared from a mixture of GeS_2, Li_2S, and P_2S_5, which were weight in an appropriate molar ratio (Kamaya et al. 2011; Kato et al. 2012; Kwon et al. 2015). The mixture was ground in a vibration mill for 30 min, pelletized under 220 MPa pressure, sealed in a silica tube at 30 Pa, and heated at 400°C for 8 h with 2.2°C·min⁻¹ heating rate and 0.60°C·min⁻¹ cooling rate. The resulting product was again subjected to grinding and mixing, after which the samples were once again pelletized under 220 MPa, followed by heat treatment at 550°C for 8 h with 3.0°C·min⁻¹ heating rate and 0.76°C·min⁻¹ cooling rate.

According to the data of Kanno and Murayama (2001), the solid solution in the whole composition range is formed in the Li_4GeS_4–Li_3PS_4 system. Careful analysis of the diffraction patterns, however, indicated that the system is divided into three composition regions depending on the appearance of superlattice reflections. The patterns in region I ($Li_{4-x}Ge_{1-x}P_xS_4$, $0 < x \leq 0.6$) were indexed assuming a monoclinic superlattice cell $a \times 3b \times 2c$, which is related to the parent lattice of the orthorhombic Li_4GeS_4. However, the superlattice reflections in region II ($0.6 < x < 0.8$) were indexed by a different monoclinic cell with $a \times 3b \times 3c$, and the sample at $x = 0.8$ in region III ($0.8 \leq x < 1.0$) showed the monoclinic $a \times 3b \times 2c$ cell. The three regions of the monoclinic superstructures correspond to different types of cation ordering.

Inoue et al. (2017) noted that one more quaternary compound, $Li_7Ge_3PS_{12}$, is formed in the Ge–Li–P–S system. It is stable up to 400°C and crystallizes in the cubic structure with the lattice parameter $a = 980.192 \pm 0.003$ pm. To prepare this compound, GeS_2, Li_2S, and P_2S_5 were mixed in the appropriate molar ratios in an Ar-filled glove box. All mixtures were pelletized, placed into carbon-coated quartz glass tubes, and sealed under vacuum (~10 Pa). The samples were then heated at 870°C for 8 h in a furnace and cooled slowly to 28°C.

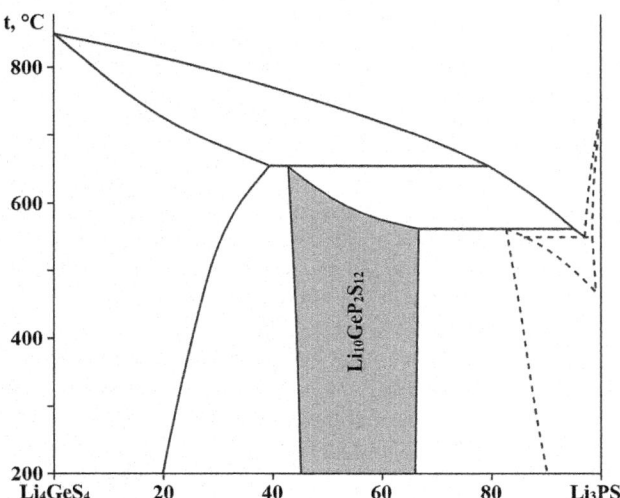

FIGURE 4.1 Phase diagram of the Li_4GeS_4–Li_2PS_4 system. (From Sachanyuk, V.P., et al., *J. Alloys Compd.*, **420**, 54, 2006.)

4.20 Germanium–Lithium–Manganese–Sulfur

Two quaternary compounds, **Li$_2$MnGeS$_4$** and **Li$_4$MnGe$_2$S$_7$**, are formed in the Ge–Li–Mn–S system. First of them crystallizes in the orthorhombic structure with the lattice parameters $a = 1335.46 \pm 0.02$, $b = 788.71 \pm 0.01$, $c = 628.06 \pm 0.01$ pm, a calculated density of 2.708 g·cm^{-3}, and an energy gap of 3.069 eV (Brant et al. 2015a). This compound was obtained by grinding in an agate mortar and pestle stoichiometric amounts of Mn chips, Ge pieces, and S powder plus a 20% excess of Li$_2$S that can act as a molten Li$_2$S$_x$ flux at elevated temperatures. The mixture was placed into a graphite crucible inside a fused silica tube that was sealed under vacuum (10^{-2} Pa). The reaction vessel was heated at 700°C within 144 h, slowly cooled to 650°C in 50 h, and then allowed to cool to room temperature naturally. The reaction vessel was opened under ambient conditions, and the product was rinsed with methanol to remove the excess Li$_2$S$_x$ flux. A pale-orange polycrystalline powder of Li$_2$MnGeS$_4$ was obtained.

Li$_4$MnGe$_2$S$_7$ crystallizes in the monoclinic structure with the lattice parameters $a = 1683.3 \pm 0.3$, $b = 670.92 \pm 0.13$, $c = 1012.1 \pm 0.2$ pm, and $\beta = 94.76 \pm 0.03°$ and a calculated density of 2.637 g·cm^{-3} at 100 K (Kaib et al. 2013). To prepare light pink block crystals of the title compound, a stoichiometric mixture of Li$_2$S (15.36 mM), Ge (15.36 mM), S (61.41 mM), and Mn (3.84 mM) was melted together in a furnace with a definite temperature program: heating up to 650°C with a heating rate of 18°C·h^{-1}, keeping for 24 h at 650°C, and cooling down to room temperature with a rate of 6°C·h^{-1}. To obtain the phase-pure product, the prepared mixture was washed with water to remove the soluble byproducts, and the surplus of sulfur was removed via sublimation at 320°C and reduced pressure of ≈10^{-6} Pa. All reaction steps were performed with strong exclusion of air and external moisture under Ar atmosphere at a high-vacuum, double-manifold Schlenk line, or N$_2$ atmosphere in a glove box.

4.21 Germanium–Lithium–Iron–Sulfur

The **Li$_2$FeGeS$_4$** quaternary compound, which crystallizes in the monoclinic structure with the lattice parameters $a = 622.94 \pm 0.01$, $b = 661.07 \pm 0.01$, $c = 780.81 \pm 0.02$ pm, and $\beta = 90.311 \pm 0.002°$ at room temperature and $a = 622.38 \pm 0.01$, $b = 660.36 \pm 0.01$, $c = 778.59 \pm 0.02$ pm, and $\beta = 90.328 \pm 0.002°$ at 20 K (Brant et al. 2014b) [$a = 622.86 \pm 0.02$, $b = 660.29 \pm 0.02$, $c = 779.38 \pm 0.02$ pm, $\beta = 90.047 \pm 0.002°$, and a calculated density of 2.803 g·cm^{-3} at 173 K (Brunetta et al. 2013)], is formed in the Ge–Li–Fe–S system.

This compound was prepared by grinding a mixture of Li$_2$S (1.2 mM), Fe (1 mM), Ge (1 mM), and S (3 mM) using an agate mortar and pestle in an Ar-filled glove box (Brunetta et al. 2013; Brant et al. 2014b). The reactants were placed into a graphite crucible inside a fused silica tube. The tube was sealed under a pressure of approximately 0.1 Pa. The sample was heated at 50°C·h^{-1} to 600°C, held at 600°C for 96 h, slowly cooled to 350°C over 50 h, and then allowed to cool to room temperature radiatively. The tube was opened under ambient conditions. The excess Li$_2$S$_x$ flux was rinsed with DMF and hexane. Dark-red polycrystalline Li$_2$FeGeS$_4$ was obtained.

4.22 Germanium–Lithium–Cobalt–Sulfur

The **Li$_2$CoGeS$_4$** quaternary compound, which crystallizes in the orthorhombic structure with the lattice parameters $a = 619.33 \pm 0.02$, $b = 652.11 \pm 0.02$, $c = 783.60 \pm 0.02$ pm, and a calculated density of 2.872 g·cm^{-3}, is formed in the Ge–Li–Co–S system (Brant et al. 2015b). This compound was prepared by grinding Li$_2$S (1.2 mM), Co (1 mM), Ge (1 mM), and S (3 mM) using an agate mortar and pestle in an Ar-filled glove box, and then placing the mixture into a graphite crucible inside a fused silica tube. The tube was sealed under a pressure of approximately 0.1 Pa. The sample was heated at 650°C for 144 h, slowly cooled to 550°C in 100 h and then allowed to cool to room temperature naturally. The tube was opened under ambient conditions. Excess Li$_2$S$_x$ was removed by washing the product with DMF, methanol, and hexane. A bright-green polycrystalline Li$_2$CoGeS$_4$ was obtained.

4.23 Germanium–Sodium–Silver–Sulfur

The **Na$_5$AgGe$_2$S$_7$** quaternary compound, which crystallizes in the monoclinic structure with the lattice parameters $a = 914.56 \pm 0.18$, $b = 1026.11 \pm 0.19$, $c = 1486.3 \pm 0.3$ pm, $\beta = 104.315°$, and a calculated density of 2.912 g·cm^{-3}, is formed in the Ge–Na–Ag–S system (Zhang et al. 2013). It was prepared from a mixture of GeO$_2$ (0.02 mM), AgNO$_3$ (0.03 mM), S (0.44 mM), CH$_3$COONa (0.12 mM), and about 0.4 mL of 1,2-diaminopropane. The mixture was placed in a Pyrex glass tube, which was sealed with ca. 10% filling under an air atmosphere, placed in a stainless-steel autoclave, and heated at 170°C for 5 days. Cooled naturally to ambient temperature, final products shaped as colorless blocks were obtained after washing several times with ethanol and dried at room temperature. This compound is water sensitive.

4.24 Germanium–Sodium–Strontium–Sulfur

The **Na$_2$SrGeS$_4$** quaternary compound, which crystallizes in the trigonal structure with the lattice parameters $a = 2354.27 \pm 0.14$, $c = 694.58 \pm 0.08$ pm, a calculated density of 2.998 g·cm^{-3}, and an energy gap of 3.80 eV, is formed in the Ge–Na–Sr–S system (Wu et al. 2019). To synthesize the title compound, a mixture of Na (2 mM), SrS (1 mM), Ge (1 mM), and S (3 mM) was firstly loaded into a graphite crucible that avoids the reaction between alkali metal and silica tube, then put it into the silica tube and flame-sealed under 10^{-3} Pa. The tube was heated to 850°C in 30 h and kept at this temperature for about 100 h, then cooled slowly down to ambient temperature over 5 days. DMF solvent was chosen to wash the products. Finally, many colorless crystals of Na$_2$SrGeS$_4$ were found, which are stable in the air after several months. The whole preparation process is completed in an Ar-filled glove box.

4.25 Germanium–Sodium–Barium–Sulfur

The **Na$_2$BaGeS$_4$** quaternary compound, which crystallizes in the trigonal structure with the lattice parameters a = 2520.0 ± 0.5, c = 752.5 ± 0.4 pm, a calculated density of 4.462 g·cm^{-3}, and an energy gap of 3.7 eV, is formed in the Ge–Na–Ba–S system (Wu et al. 2016a, b). The crystals of this compound were obtained using the mixture of Na, Ba, Ge, and S (molar ratio 1:6:2:8), which was firstly loaded into a graphite crucible that avoids the reaction between Na and silica tube, then put into the silica tube and flame-sealed under vacuum (10^{-3} Pa). The tube was heated to 600–700°C in 30 h and kept there for about 90 h, then slowly cooled down to ambient temperature within 5 days. Nearly colorless transparent crystals of Na$_2$BaGeS$_4$ were found after washing with DMF. They are stable in the air. Owing to Na and Ba metals being easily oxidized in the air, an Ar-filled glove box was used to complete the preparation processes.

4.26 Germanium–Sodium–Zinc–Sulfur

The **Na$_2$ZnGe$_2$S$_6$** quaternary compound, which melts incongruently at 650°C and crystallizes in the monoclinic structure with the lattice parameters a = 726.30, b = 1234.04, c = 1130.57 pm, β = 99.959°, and a calculated density of 2.988 g·cm^{-3} (Yohannan and Vidyasagar 2016a) [a = 728.4 ± 0.9, b = 1234.6 ± 1.5, c = 1144.7 ± 1.4 pm, β = 100.627 ± 0.013°, a calculated density of 2.947 g·cm^{-3}, and an energy gap of 3.25 eV (Li et al. 2016a)], is formed in the Ge–Na–Zn–S system.

This compound was obtained as single gray crystals by the Na$_2$S$_2$O$_3$ flux method (Yohannan and Vidyasagar 2016a). A mixture of Na$_2$CO$_3$ (1.7 mM), Zn (1.1 mM), Ge (2.2 mM), and S (7.8 mM) was heated at 550°C for 2 days and then cooled to 450°C at 1.4°C·h^{-1}. The crystals of Na$_2$ZnGe$_2$S$_6$ were synthesized by solid-state reaction as follows (Li et al. 2016a). The mixture of Na, Zn, Ge, and S with stoichiometric ratio was loaded into a graphite crucible, which was removed into a fused silica tube under an Ar atmosphere in a glove box. The tube was flame-sealed under a high vacuum (10^{-3} Pa) and placed in a furnace, heated to 700°C in 50 h, dwelled at 700°C for 100 h, cooled to 200°C at a rate of 5°C·h^{-1}, and finally cooled to room temperature by switching off the furnace. After that, colorless single crystals were obtained. This compound is stable in moisture conditions and air for several months.

4.27 Germanium–Sodium–Cadmium–Sulfur

The **Na$_2$CdGe$_2$S$_6$** quaternary compound, which crystallizes in the monoclinic structure with the lattice parameters a = 731.2 ± 0.3, b = 1267.2 ± 0.5, c = 1152.8 ± 0.4 pm, β = 98.869 ± 0.005°, and a calculated density of 3.121 g·cm^{-3} (Li et al. 2017a) [a = 730.38, b = 1265.73, c = 1152.97 pm, β = 98.8702°, and a calculated density of 3.128 g·cm^{-3} (Yohannan and Vidyasagar 2016a)], is formed in the Ge–Na–Cd–S system.

The crystals of this compound were synthesized by solid-state reaction. First, Na, CdS, Ge, and S (molar ratio 2:1:2:5) were placed into the graphite crucible, which was flame-sealed into a fused silica tube under a high vacuum (10^{-3} Pa). The tube was moved in a furnace, heated to 850°C in 60 h, dwelled there for 30 h, cooled to 450°C at a rate of 5°C·h^{-1}, then rapidly cooled to room temperature by switching off the furnace. After washing with DMF and deionized water, pale-yellow crystals of Na$_2$CdGe$_2$S$_6$ were obtained.

A polycrystalline sample of the title compound was synthesized by heating a stoichiometric mixture of Na$_2$CO$_3$ (1 mM), CdS (1 mM), GeO$_2$ (2 mM), in a continuous stream of CS$_2$ vapor, at 500°C for 24 h with one intermittent grinding (Yohannan and Vidyasagar 2016a). Gray block-shaped single crystals were obtained along with a very small amount of GeS$_2$, from the Na$_2$S$_2$O$_3$ flux method, by heating a reactant mixture of Na$_2$CO$_3$ (1.5 mM), Ge (2 mM), Cd (1 mM), and S (7 mM) at 550°C for 2 days and then cooling to 450°C at 1.4°C·h^{-1}.

4.28 Germanium–Sodium–Mercury–Sulfur

The **Na$_2$Hg$_3$Ge$_2$S$_8$** quaternary compound, which crystallizes in the tetragonal structure with the lattice parameters a = 884.8 ± 0.2, c = 903.2 ± 0.4 pm, a calculated density of 4.929 g·cm^{-3}, and an energy gap of 2.68 eV, is formed in the Ge–Na–Hg–S system (Wu et al. 2016c). To prepare this compound, an Ar-filled glove box was used. A mixture of Na (2 mM), HgS (3 mM), Ge (2 mM), and excess sulfur (8 mM) were loaded into a graphite crucible, and then put into a tidy silica tube that was flame-sealed under 10^{-3} Pa. The tube was heated to about 600°C within 30 h and kept at this temperature for about 60 h, then slowly cooled down to 300°C in 4 days, and finally quickly cooled to ambient temperature. The product was carefully washed with DMF and then dried at 100°C. Yellow crystals of Na$_2$Hg$_3$Ge$_2$S$_8$ were obtained. This compound is stable in the air.

4.29 Germanium–Sodium–Aluminum–Sulfur

Two quaternary compounds, **Na(AlS$_2$)(GeS$_2$)** and **Na(AlS$_2$)(GeS$_2$)$_4$**, are formed in the Ge–Na–Al–S system. First of them crystallizes in the tetragonal structure with the lattice parameters a = 742.74 ± 0.11, c = 585.60 ± 0.12 pm, a calculated density of 2.578 g·cm^{-3} at 200 K, and an energy gap of 3.6 eV (Al-Bloushi et al. 2015). Na(AlS$_2$)(GeS$_2$)$_4$ is thermally stable up to 650°C and crystallizes in the monoclinic structure with the lattice parameters a = 680.3 ± 0.3, b = 3820.7 ± 0.2, c = 694.7 ± 0.4 pm, β = 119.17 ± 0.03°, a calculated density of 2.784 g·cm^{-3} at 200 K, and an energy gap of 3.1 eV (Alahmari et al. 2018).

Na(AlS$_2$)(GeS$_2$) was prepared from a stoichiometric mixture of Al (3.7 mM), Ge (3.7 mM), S (13.1 mM), and Na$_2$S (1.85 mM) by direct combination reaction (Al-Bloushi et al. 2015). The starting materials were weighed and mixed inside the glove box and sealed in a fused silica ampoule. The sample was heated at 850°C for 4 days and cooled down to room temperature within 4 days. The colorless needle crystals of the title compound were obtained after washing the reaction products with DMF and diethyl ether.

Na(AlS$_2$)(GeS$_2$)$_4$ was prepared by a direct combination reaction of Na$_2$S (0.37 mM), Al (0.75 mM), Ge (3.02 mM), and S (7.18 mM) (Alahmari et al. 2018). The mixture was filled and hand-pressed into a glassy carbon crucible, which was placed into a silica tube and flame-sealed under vacuum. A phase pure yellowish-orange crystalline material was obtained after heating the sample at 850°C for 72 h followed by slow cooling to 400°C with a rate of 2°C·h^{-1}, before it was cooled to room temperature. This compound partially decomposes after exposure to air for one week.

All preparations and handling for both compounds were carried out in an Ar- or N$_2$-filled glove box due to the air and moisture sensitivity of the starting materials.

4.30 Germanium–Sodium–Gallium–Sulfur

The **Na$_2$Ga$_2$GeS$_6$** quaternary compound, which melts congruently at 811°C and crystallizes in the orthorhombic structure with the lattice parameters $a = 1240.5 \pm 0.1$, $b = 2273.0 \pm 0.1$, $c = 727.37 \pm 0.05$ pm, a calculated density of 2.918 g·cm^{-3}, and an energy gap of 3.29 eV (Yohannan and Vidyasagar 2016a) [$a = 1243.2 \pm 0.4$, $b = 2258.4 \pm 0.7$, $c = 723.9 \pm 0.2$ pm, a calculated density of 2.944 g·cm^{-3}, and an energy gap of 3.10 eV (Li et al. 2018c)], is formed in the Ge–Na–Ga–S system.

To synthesize this compound, a mixture of Na$_2$S (0.666 mM), Ga (1.332 mM), Ge (0.666 mM), and S (3.331 mM) was sealed in the evacuated glass ampoule, heated at 800°C for 4 days, and then cooled to 350°C at 4°C·h^{-1} (Li et al. 2018c). After washing with degassed DMF, light-yellow crystals of Na$_2$Ga$_2$GeS$_6$ were obtained. They are stable in the air.

This compound was also obtained in the form of gray block-shaped single crystals, when a mixture of Na$_2$CO$_3$ (1.7 mM), Ga$_2$S$_3$ (1.1 mM), Ge (1.1 mM), and S (4.4 mM) was heated in a sealed quartz tube at 825°C for 3 days and then cooled to 525°C at 4°C·h^{-1} and the furnace was finally turned off (Yohannan and Vidyasagar 2016a). The product was isolated by washing off the Na$_2$S$_3$O$_3$ flux formed.

4.31 Germanium–Sodium–Indium–Sulfur

The **Na$_2$In$_2$GeS$_6$** quaternary compound, which melts congruently at ~705°C [incongruently at 755°C (Yohannan and Vidyasagar 2016a)] and crystallizes in the monoclinic structure with the lattice parameters $a = 1216.9 \pm 0.5$, $b = 781.9 \pm 0.3$, $c = 1223.5 \pm 0.6$ pm, $\beta = 108.645 \pm 0.006°$, a calculated density of 3.255 g·cm^{-3}, and an energy gap of 2.417 eV (Li et al. 2016b) [$a = 1216.51$, $b = 782.12$, $c = 1223.55$ pm, $\beta = 108.655°$, and a calculated density of 3.256 g·cm^{-3} (Yohannan and Vidyasagar 2016a)], is formed in the Ge–Na–In–S system. The title compound was synthesized by solid-state reaction of Na$_2$S, In, Ge, and S mixed in a molar ratio of 3:1:1:5. The mixture was finely ground, introduced into a quartz tube, and then flame-sealed under vacuum. This tube was put in a furnace, held at 450°C for 5 h, raised to 820°C in 10 h, retained there for 36 h, and then cooled to 400°C in 100 h, and powered off. Orange crystals of Na$_2$In$_2$GeS$_6$ were obtained. They are stable in moisture conditions and air for

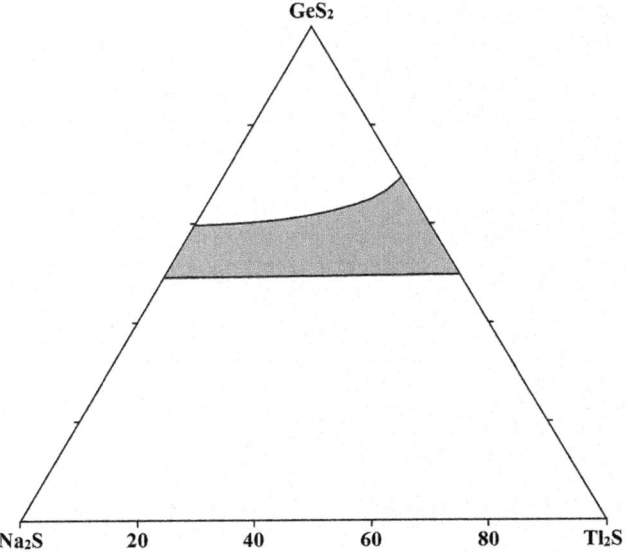

FIGURE 4.2 Glass-forming region in the GeS$_2$–Na$_2$S–Tl$_2$S quasiternary system. (From Imaoka, M. and Jamadzaki, T. [in Japanese], *Monthly J. Inst. Industr. Sci. Univ. Tokyo*, **19**(9), 261, 1967.) Open access

several months. The pure powder phase was prepared through a stoichiometry mixture of Na$_2$S, In, Ge, and S, which were held at 620°C for 72 h and then cooled to room temperature at 20 h. The product was reground and reprocessed by the identical procedure to improve the purity.

Na$_2$In$_2$GeS$_6$ was also obtained in the form of pale-yellow block-shaped single crystals, when a stoichiometric mixture of NaInS$_2$ (1.85 mM), Ge (0.92 mM), and S (1.85 mM) was heated in an evacuated sealed tube at 700°C for 2 days and then cooled to 500°C at 3°C·h^{-1} (Yohannan and Vidyasagar 2016a).

4.32 Germanium–Sodium–Thallium–Sulfur

The glass-forming region in the **GeS$_2$–Na$_2$S–Tl$_2$S** quasiternary system is presented in Figure 4.2 (Imaoka and Jamadzaki 1967).

4.33 Germanium–Sodium–Cerium–Sulfur

The **NaCe$_3$GeS$_7$** quaternary compound, which crystallizes in the hexagonal structure with the lattice parameters $a = 1022.48 \pm 0.03$, $c = 580.35 \pm 0.04$ pm, and a calculated density of 4.679 g·cm^{-3}, is formed in the Ge–Na–Ce–S system (Choudhury and Dorhout 2015). Single crystals of this compound were obtained from the reaction of the anhydrous CeCl$_3$ and Na$_2$GeS$_3$. Typically, CeCl$_3$ (0.36 mM) and Na$_2$GeS$_3$ (0.28 mM) were mixed and ground in an agate mortar inside a N$_2$-filled glove box. The mixture was transferred to a graphite crucible, which was then placed inside a fused silica tube, evacuated, flame-sealed, placed vertically in a furnace, and heated at 825°C for 96 h. Only a few high-quality crystals of NaCe$_3$GeS$_7$ were obtained. The major phase for this reaction was Ce$_4$(GeS$_4$)$_3$.

4.34 Germanium–Sodium–Neodymium–Sulfur

The **NaNd₃GeS₇** quaternary compound, which crystallizes in the hexagonal structure with the lattice parameters $a = 1008.20 \pm 0.04$, $c = 578.97 \pm 0.05$ pm, and a calculated density of 4.905 g·cm⁻³, is formed in the Ge–Na–Nd–S system (Choudhury and Dorhout 2015). Single crystals of this compound were obtained in the same way as in the case of NaCe₃GeS₇ preparation using anhydrous NdCl₃ instead of anhydrous CeCl₃. Only a few high-quality crystals of NaNd₃GeS₇ were obtained. The major phase for this reaction was Nd₄(GeS₄)₃.

4.35 Germanium–Sodium–Samarium–Sulfur

The **NaSm₃GeS₇** quaternary compound, which crystallizes in the hexagonal structure with the lattice parameters $a = 993.66 \pm 0.07$, $c = 580.18 \pm 0.08$ pm, a calculated density of 5.162 g·cm⁻³ at 100 K, and an energy gap of 1.33 eV, is formed in the Ge–Na–Sm–S system (Choudhury and Dorhout 2015). Single crystals of this compound were obtained in the same way as in the case of NaCe₃GeS₇ preparation using anhydrous SmCl₃ instead of anhydrous CeCl₃. This reaction yielded only the title compound as the major product.

4.36 Germanium–Sodium–Europium–Sulfur

Two quaternary compounds, **Na₁.₅₁₅EuGeS₄** and **Na₈Eu₂(Ge₂S₆)₂**, are formed in the Ge–Na–Eu–S system. First of them crystallizes in the tetragonal structure with the lattice parameters $a = 2332.2 \pm 0.3$, $c = 683.85 \pm 0.16$ pm, and a calculated density of 3.597 g·cm⁻³ (Choudhury et al. 2012). Na₈Eu₂(Ge₂S₆)₂ crystallizes in the monoclinic structure with the lattice parameters $a = 684.8 \pm 0.2$, $b = 1181.2 \pm 0.2$, $c = 773.5 \pm 0.2$ pm, $\beta = 107.164 \pm 0.003°$, a calculated density of 3.230 g·cm⁻³, and an energy gap of 2.15 eV (Choudhury et al. 2015).

Na₁.₅₁₅EuGeS₄ has been prepared as the result of an attempt to synthesize **Na₂EuGeS₄** (Choudhury et al. 2012). In a N₂-filled glove box, a stoichiometric mixture of Ge, Eu, S, and Na₂S was placed in a fused silica ampoule. The ampoule was flame-sealed under vacuum and placed in a furnace. The furnace temperature was increased to 750°C at a rate of 35°C·h⁻¹, and then held constant at 750°C for 150 h. The furnace temperature was then slowly reduced to ambient temperature at a rate of 5°C·h⁻¹. The cooled ampoule was opened in the air, and the reaction product contained bundles of dark-red long hexagonal rods of the title compound.

Na₈Eu₂(Ge₂S₆)₂ was synthesized by a reactive flux method using a Na₂S₂ flux (0.8 mM) along with elemental Eu (0.4 mM), Ge (1.0 mM), and S (1.2 mM) (approximate molar ratio of Na/Eu/Ge/S = 8:2:5:14) (Choudhury et al. 2015). The mixture was loaded into a fused silica ampoule under a N₂ atmosphere which was then evacuated, flame-sealed, and placed in a furnace. The furnace temperature was increased to 750°C at a rate of 35°C·h⁻¹, held constant at 750°C for 150 h, and finally decreased to room temperature at a rate of 5°C·h⁻¹. The heat-treated ampoule was opened in the air, and the solid orange-red product was soaked in DMF and sonicated for several minutes in order to both dissolve and remove any remaining flux and loosen the product crystals. The product contained only orange-red crystals of Na₈Eu₂(Ge₂S₆)₂. However, in the cold zone of the ampoule there was a small amount of GeS₂. This compound is very stable in the air.

4.37 Germanium–Sodium–Gadolinium–Sulfur

The **NaGd₃GeS₇** quaternary compound, which crystallizes in the hexagonal structure with the lattice parameters $a = 986.46 \pm 0.04$, $c = 580.07 \pm 0.05$ pm, a calculated density of 5.379 g·cm⁻³, and an energy gap of 1.30 eV, is formed in the Ge–Na–Gd–S system (Choudhury and Dorhout 2015). Single crystals of this compound were obtained in the same way as in the case of NaCe₃GeS₇ preparation using anhydrous GdCl₃ instead of anhydrous CeCl₃. This reaction yielded only the title compound as the major product.

4.38 Germanium–Sodium–Ytterbium–Sulfur

The **NaYb₃GeS₇** quaternary compound, which crystallizes in the hexagonal structure with the lattice parameters $a = 954.21 \pm 0.06$, $c = 588.33 \pm 0.07$ pm, a calculated density of 6.007 g·cm⁻³, and an energy gap of 2.18 eV, is formed in the Ge–Na–Yb–S system (Choudhury and Dorhout 2015). Single crystals of this compound were obtained in the same way as in the case of NaCe₃GeS₇ preparation using anhydrous YbCl₃ instead of anhydrous CeCl₃. This reaction yielded only the title compound as the major product.

4.39 Germanium–Sodium–Tin–Sulfur

The **Na₈Sn₂(Ge₂S₆)₂** quaternary compound, which crystallizes in the monoclinic structure with the lattice parameters $a = 684.9 \pm 0.1$, $b = 1177.7 \pm 0.1$, $c = 770.5 \pm 0.1$ pm, $\beta = 106.44 \pm 0.01°$, a calculated density of 3.054 g·cm⁻³, and an energy gap of 2.06 eV, is formed in the Ge–Na–Sn–S system (Marking and Kanatzidis 1997). Deep ruby-red chunky crystals of this compound were formed through basic non-oxidizing flux reactions. It was prepared by heating a mixture of Na₂S, Sn, Ge, and S (molar ratio 2:2:5:5) at 750°C for 3 days and slowly cooling to 350°C. The title compound is relatively stable in water and other common solvents, although some decomposition was detected in samples left under ambient atmospheric conditions for periods exceeding several weeks. All manipulations were carried out under a nitrogen atmosphere.

4.40 Germanium–Sodium–Lead–Sulfur

Two quaternary compounds, **Na₀.₅Pb₁.₇₅GeS₄** and **Na₈Pb₂(Ge₂S₆)₂**, are formed in the Ge–Na–Pb–S system. First of them crystallizes in the cubic structure with the lattice parameter $a = 1411.5 \pm 0.1$, a calculated density of 5.430 g·cm⁻³, and an energy gap of 2.08 eV (Aitken et al. 2000). $Na_8Pb_2(Ge_2S_6)_2$ quaternary compound crystallizes in the monoclinic structure with the lattice parameters $a = 683.9 \pm 0.2$, $b = 1178.3 \pm 0.2$, $c = 774.4 \pm 0.2$ pm, $\beta = 107.00 \pm 0.02°$, a calculated density of 3.543 g·cm⁻³, and an energy gap of 2.43 eV (Marking and Kanatzidis 1997).

To prepare $Na_{0.5}Pb_{1.75}GeS_4$, in a N_2-filled glove box Pb (0.25 mM), Ge (0.75 mM), Na_2S (0.50 mM), and S (2 mM) were loaded into a Pyrex tube (Aitken et al. 2000). This tube was flame-sealed under vacuum (approximately 0.02 Pa) and inserted into a furnace. The temperature was raised from 50°C to 530°C in 20 h. It was kept at 530°C for 34 h, and then cooled at 5°C·h⁻¹ to 50°C. DMF was used to remove the excess flux. Washing with ether revealed small orange/red crystals of this compound as a pure phase.

Orange-yellow plates of $Na_8Pb_2(Ge_2S_6)_2$ were formed through basic non-oxidizing flux reactions (Marking and Kanatzidis 1997). They were prepared by heating a mixture of Na_2S, Pb, Ge, and S (molar ratio 4:2:5:9.5) at 850°C for 3 days and slowly cooled to 450°C over 80 h. The title compound is relatively stable in water and other common solvents although some decomposition was detected in samples left under ambient atmospheric conditions for periods exceeding several weeks. All manipulations were carried out under a nitrogen atmosphere.

4.41 Germanium–Sodium–Arsenic–Sulfur

The glass-forming region in the $GeS_2–Na_2S–As_2S_3$ quasiternary system is presented in Figure 4.3 (Imaoka and Jamadzaki 1967).

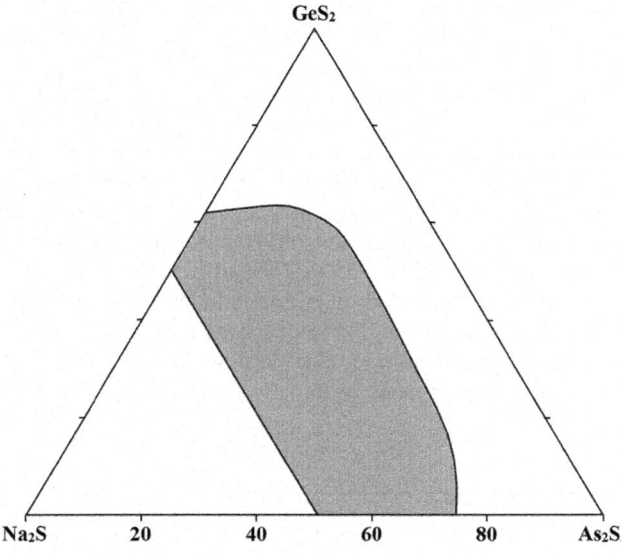

FIGURE 4.3 Glass-forming region in the $GeS_2–Na_2S–As_2S_3$ quasiternary system. (From Imaoka, M. and Jamadzaki, T. [in Japanese], *Monthly J. Inst. Industr. Sci. Univ. Tokyo*, **19**(9), 261, 1967.) Open access

4.42 Germanium–Sodium–Antimony–Sulfur

The $Na_9Sb(Ge_2S_6)_2$ quaternary compound, which crystallizes in the monoclinic structure with the lattice parameters $a = 758.57 \pm 0.07$, $b = 1157.43 \pm 0.11$, $c = 681.75 \pm 0.11$ pm, $\beta = 106.587 \pm 0.003°$, a calculated density of 2.905 g·cm⁻³, and an energy gap of ~2.58 eV, is formed in the Ge–Na–Sb–S system (Wu et al. 2016d). To synthesize this compound, a mixture of Na, Sb, Ge, and S (molar ratio 9:1:4:12) was loaded into a sealed silica tube evacuated to 10⁻³ Pa. The mixture was first heated to 600°C within 30 h, left at this temperature for 72 h, then cooled to 300°C at a rate of 3°C·h⁻¹, and finally cooled rapidly to room temperature by switching off the furnace. After repeatedly washing with DMF, many yellow crystals were found, which were stable in the air for several months. As for the easily oxidized Na element, the preparation process was completed in a glove box with an Ar atmosphere.

4.43 Germanium–Sodium–Bromine–Sulfur

The Na_3GeS_3Br quaternary compound is formed in the Ge–Na–Br–S system (Feltz and Pfaff 1980). To prepare this compound, 24.7 mL of bromine solution in CH_3OH (43.7 mg Br_2/mL) was added to a solution of $Na_6Ge_2S_6·4CH_3OH$ (4.09 g) in CH_3OH (100 mL) drop-wise while cooling at room temperature. Conversion occurs immediately, and a colorless, finely crystalline precipitate separates from the colorless solution after extensive concentration in vacuum. It is sensitive to hydrolysis and very readily soluble in methanol. Precipitation by drop-wise addition of CCl_4 increases the yield.

4.44 Germanium–Potassium–Copper–Sulfur

Two quaternary compounds, $K_2Cu_2GeS_4$ and $K_4Cu_8Ge_3S_{12}$, are formed in the Ge–K–Cu–S system. First of them crystallizes in the monoclinic structure with the lattice parameters $a = 706.3 \pm 0.3$, $b = 543.5 \pm 0.3$, $c = 1103.7 \pm 0.6$ pm, $\beta = 112.83 \pm 0.03°$, and an energy gap of 2.3 eV (Sun et al. 2017) [$a = 711.8$, $b = 538.8$, $c = 1121.2$ pm, β value not specified, a calculated density of 1.7 g·cm⁻³ and an energy gap of 2.167 eV (calculated values) (Ali et al. 2019)]. $K_4Cu_8Ge_3S_{12}$ is stable up to 400°C and crystallizes in the cubic structure with the lattice parameter $a = 1767.53$ pm, a calculated density of 3.049 g·cm⁻³, and an energy gap of 2.2 eV (Zhang et al. 2010).

Single crystals of $K_2Cu_2GeS_4$ were synthesized by using molten thiourea as a reactive flux (Sun et al. 2017). The mixture of Ge powder (5 mM), Cu powder (5 mM), thiourea (0.1 M), and KOH (10 g) was ground in an agate mortar. It was then transferred into a 50 mL Teflon-lined stainless-steel autoclave. The reaction was carried out at 220°C for 3 days. After the autoclave cooled down to room temperature, the as-reacted mixture was taken out and washed with deionized water several times. After drying in acetone, the final products were collected.

To prepare $K_4Cu_8Ge_3S_{12}$, Cu powder (0.11 mM), GeO_2 (0.04 mM), K_2CO_3 (0.09 mM), S powder (1.0 mM), and about 250 mg of diethylenetriamine and 170 mg of CH_3OH/H_2O (volume ratio 1:1) as solvent were mixed and stirred substantially (Zhang et al. 2010). The mixture was sealed in a Pyrex glass tube at air atmosphere, placed into a stainless-steel autoclave, and then heated at 160°C for 7 days. After being cooled naturally to ambient temperature, the tube was broken with caution. The products were washed with ethylenediamine and ethanol, respectively, and yellow block crystals of the title compound were obtained as pure phase.

4.45 Germanium–Potassium–Silver–Sulfur

The **K₂Ag₂GeS₄** quaternary compound, which crystallizes in the monoclinic structure with the lattice parameters $a = 1329.35 \pm 0.01$, $b = 635.90 \pm 0.09$, $c = 1240.10 \pm 0.02$ pm, and $\beta = 112.118 \pm 0.010°$, is formed in the Ge–K–Ag–S system (An et al. 2004). The synthesis of this compound was as follows: Ge (10.9 mg), AgNO₃ (50.7 mg), K₂CO₃ (103.5 mg), and S (18.3 mg) were put into a glass tube, to which 0.4 mL of ethanol/glycerol (volume ratio 2:1) was added. The glass tube was sealed (reagents filled about 10% of the tube), placed into a Teflon-lined stainless-steel autoclave, and heated at 120°C for 5 days. The products were washed with ethanol and water, respectively. Colorless crystals of the title compound were obtained.

4.46 Germanium–Potassium–Gold–Sulfur

The **K₂Au₂GeS₄** quaternary compound, which melts incongruently at 490°C and crystallizes in the monoclinic structure with the lattice parameters $a = 764.35 \pm 0.05$, $b = 1477.50 \pm 0.11$, $c = 899.21 \pm 0.06$ pm, $\beta = 114.327 \pm 0.005°$, a calculated density of 4.831 g·cm⁻³ at 200 K, and an energy gap of 2.9 eV, is formed in the Ge–K–Au–S system (Davaasuren et al. 2017). This compound was synthesized from a stoichiometric mixture of Au (2 mM), GeS (1 mM), and K₂S (1 mM) by direct combination reaction. The starting materials were weighed and mixed inside the glove box and sealed in a fused silica ampoule. The sample was heated at 650°C for 3 days and cooled down to room temperature within 5 days. Phase-pure K₂Au₂GeS₄ was obtained after washing the reaction products with DMF and diethyl ether. The sample was handled in a N₂-filled glove box due to the air and moisture sensitivity of the starting materials.

4.47 Germanium–Potassium–Zinc–Sulfur

Two quaternary compounds, **K₂ZnGe₃S₈** and **K₁₀Zn₄Ge₄S₁₇**, are formed in the Ge–K–Zn–S system. First of them melts congruently at around 750°C and crystallizes in the monoclinic structure with the lattice parameters $a = 710.27 \pm 0.14$, $b = 1193.6 \pm 0.2$, $c = 1674.2 \pm 0.3$ pm, $\beta = 96.20 \pm 0.03°$, a calculated density of 2.908 g·cm⁻³, and an energy gap of 3.36 eV (Luo et al. 2018).

K₁₀Zn₄Ge₄S₁₇ melts congruently at 656°C and crystallizes in the cubic structure with the lattice parameter $a = 981.7 \pm 0.4$ pm, a calculated density of 2.612 g·cm⁻³ at 153 K, and an energy gap of 3.34 eV (Palchik et al. 2004).

K₂ZnGe₃S₈ was synthesized from a mixture of K₂S₃ (0.5 mM), ZnS (0.5 mM), and GeS₂ (0.15 mM) (Luo et al. 2018). The reagents were well mixed and put in a quartz tube in a glove box filled with Ar. The tube was vacuum-sealed and then placed in a furnace. The mixture was heated from 20°C to 950°C at a heating rate of 1°C·min⁻¹, kept 48 h at 950°C, and then cooled to room temperature at a cooling rate of 4°C·h⁻¹. Following the steps above, many transparent lamellar crystals of the title compound were found in the tube.

K₁₀Zn₄Ge₄S₁₇ was synthesized from a mixture of Ge (0.5 mM), Zn (0.5 mM), K₂S (2.5 mM), and S (6 mM) (Palchik et al. 2004). The reagents were mixed, sealed in an evacuated silica tube, and heated at 500°C for 4 days. This was followed by cooling to room temperature at 5°C·h⁻¹. The excess flux was removed with methanol to reveal yellow-brown transparent rectangular crystals that are air sensitive. All preparation was done under a N₂ atmosphere.

The glass-forming region in the **GeS₂–K₂S–ZnS** quasiternary system of the Ge–K–Zn–S system is shown in Figure 4.4 (Imaoka and Jamadzaki 1967).

4.48 Germanium–Potassium–Cadmium–Sulfur

The glass-forming region in the **GeS₂–K₂S–CdS** quasiternary system is presented in Figure 4.5 (Imaoka and Jamadzaki 1967).

4.49 Germanium–Potassium–Mercury–Sulfur

The **K₂Hg₃Ge₂S₈** quaternary compound is formed in the Ge–K–Cd–S system. It melts incongruently at ~576°C [580°C (Kanatzidis et al. 1997a, 1997b)], and has two polymorphic modifications (Liao et al. 2003). One of them crystallizes in the

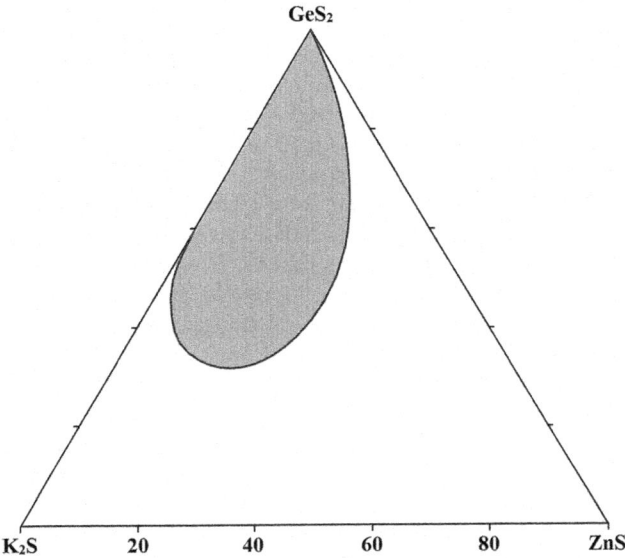

FIGURE 4.4 Glass-forming region in the GeS₂–K₂S–ZnS quasiternary system. (From Imaoka, M. and Jamadzaki, T. [in Japanese], *Monthly J. Inst. Industr. Sci. Univ. Tokyo*, **19**(9), 261, 1967.) Open access

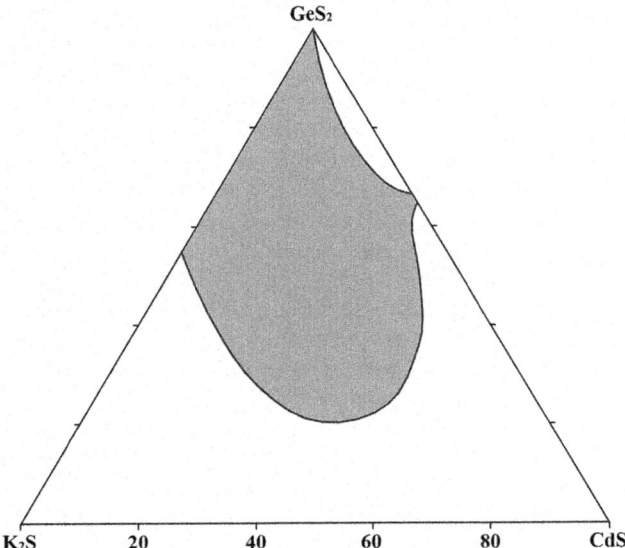

GeS$_2$

K$_2$S   20   40   60   80   CdS

FIGURE 4.5   Glass-forming region in the GeS$_2$–K$_2$S–CdS quasiternary system. (From Imaoka, M. and Jamadzaki, T. [in Japanese], *Monthly J. Inst. Industr. Sci. Univ. Tokyo*, **19**(9), 261, 1967.) Open access

orthorhombic structure with the lattice parameters $a = 1908.2 \pm 0.2$, $b = 955.1 \pm 0.1$, $c = 828.71 \pm 0.08$ pm, a calculated density of 9.514 g·cm^{-3}, and an energy gap of 2.64 eV (Liao et al. 2003) [$a = 1918.8 \pm 0.3$, $b = 961.8 \pm 0.2$, $c = 832.8 \pm 0.5$ pm, and a calculated density of 4.674 g·cm^{-3} (Kanatzidis et al. 1997a, 1997b)]. Other modification crystallizes in the monoclinic structure with the lattice parameters $a = 959.48 \pm 0.07$, $b = 836.08 \pm 0.06$, $c = 966.38 \pm 0.07$ pm, and $\beta = 94.637 \pm 0.001°$, a calculated density of 9.322 g·cm^{-3}, and an energy gap of 2.70 eV (Liao et al. 2003).

α-K$_2$Hg$_3$Ge$_2$S$_8$ was prepared by the reaction of the mixture of Ge (0.25 mM), HgS (0.38 mM), K$_2$S (0.50 mM), and S (4.00 mM) in evacuated glass ampoule at 400°C for 96 h upon a cooling rate of 4°C·h^{-1} (Kanatzidis et al. 1997a, 1997b; Liao et al. 2003) The obtained yellow crystals were washed with degassed DMF. This modification is stable in the air and insoluble in water.

β-K$_2$Hg$_3$Ge$_2$S$_8$ was synthesized by the reaction of Ge (1.01 mM), HgS (1.50 mM), K$_2$S (0.50 mM), and excess S (4.02 mM) in evacuated glass ampoule at 520°C over 6 h with the next holding at that temperature for 1 h and cooling to 480°C at a rate of 1°C·h^{-1} (Liao et al. 2003). The product was washed with degassed methanol, deionized water, and CS$_2$ to remove excess sulfur and then dried with acetone and ether. A pure, yellow microcrystalline phase was obtained, which is stable in the air and insoluble in water.

This modification could be also obtained by heating the mixture up to 400–500°C over a period of 8–12 h with the next holding at that temperature for 48–96 h and cooling to about 200°C at a rate of 4°C·h^{-1} (Liao et al. 2003). Excess flux was dissolved using methanol under N$_2$ atmosphere, and the crystals were washed with water to dissolve any ternary phase present and then dried with acetone and ether. Excess S was dissolved using CS$_2$. All of the time periods in the temperature profile can be shortened without drastically reducing the yield, but the product becomes increasingly microcrystalline. Relatively large (larger

than a millimeter on edge) optical quality pieces of K$_2$Hg$_3$Ge$_2$S$_8$ crystals can be synthesized by recrystallization using slow cooling and/or temperature cycling of K$_2$Hg$_3$Ge$_2$S$_8$/K$_2$S$_8$ flux reaction (Kanatzidis et al. 1997a, 1997b; Liao et al. 2003).

4.50 Germanium–Potassium–Aluminum–Sulfur

The **K(AlS$_2$)(GeS$_2$)** quaternary compound, which crystallizes in the tetragonal structure with the lattice parameters $a = 788.26 \pm 0.02$, $c = 586.42 \pm 0.04$ pm, a calculated density of 2.433 g·cm^{-3} at 200 K, and an energy gap of 3.5 eV, is formed in the Ge–K–Al–S system (Al-Bloushi et al. 2015). It was prepared from a stoichiometric mixture of Al (3.7 mM), Ge (3.7 mM), S (13.1 mM), and K$_2$S (1.85 mM) by direct combination reaction. The starting materials were weighed and mixed inside the glove box and sealed in a fused silica ampoule. The sample was heated at 850°C for 4 days and cooled down to room temperature within 4 days. The colorless needle crystals of the title compound were obtained after washing the reaction products with DMF and diethyl ether. The samples were handled in an Ar-filled glove box due to the air and moisture sensitivity of the starting materials.

4.51 Germanium–Potassium–Gallium–Sulfur

Two quaternary compounds, **KGaGeS$_4$** and **K$_3$Ga$_3$Ge$_7$S$_{20}$**, are formed in the Ge–K–Ga–S system. First of them crystallizes in the monoclinic structure with the lattice parameters $a = 685.3 \pm 0.2$, $b = 1595.6 \pm 0.5$, $c = 710.8 \pm 0.2$ pm, and $\beta = 111.95 \pm 0.01°$ at 115 K (Wu et al. 1992). K$_3$Ga$_3$Ge$_7$S$_{20}$ melts congruently at 787°C and also crystallizes in the monoclinic structure with the lattice parameters $a = 676.65 \pm 0.04$, $b = 3752.7 \pm 0.2$, $c = 667.96 \pm 0.04$ pm, $\beta = 90.802 \pm 0.005°$, a calculated density of 2.890 g·cm^{-3} at 153 K, and an energy gap of 3.25 eV (Li et al. 2016c).

KGaGeS$_4$ was prepared by the reaction of K$_2$S$_5$ with Ga, Ge, and S (molar ratio 1:2:2:3) in a quartz tube (Wu et al. 1992). The quartz tube was evacuated (~10^{-3} Pa), sealed, and then heated gradually to 500°C, where it was kept for 24 h before being successively brought to 700°C for 24 h and to 900°C for 100 h. Next, the tube was cooled at a rate of 4°C·h^{-1} to 300°C, and then the furnace was shut off. Colorless plate-like crystals formed in the tube. They are stable in the air and do not decompose significantly in water.

A polycrystalline sample of K$_3$Ga$_3$Ge$_7$S$_{20}$ was synthesized using a solid-state reaction technique (Li et al. 2016c). A mixture of K$_2$S, Ga$_2$S$_3$, and GeS$_2$ according to the stoichiometric ratio was ground and loaded into fused silica tubes under an Ar atmosphere in a glove box, which was sealed under vacuum (10^{-3} Pa) and then placed in a furnace. The sample was heated to 800°C over 20 h, kept at that temperature for 72 h, and then the furnace was turned off. Single crystals of this compound were obtained from a mixture of K$_2$S$_3$ (0.15 M), Ga$_2$S$_3$ (0.15 M), and GeS$_2$ (0.60 M), which was ground and loaded into a fused silica tube under an Ar atmosphere in a glove box. The tube was sealed under vacuum (10^{-3} Pa) and then placed in a furnace. The sample was heated to 1000°C over 20 h and kept at that temperature for 48 h, then cooled at a slow rate (2.5°C·h⁻¹) to 300°C, and finally cooled to room temperature. The resultant orange and light-yellow crystals were manually selected.

4.52 Germanium–Potassium–Indium–Sulfur

The **KInGeS₄** quaternary compound, which has two polymorphic modifications, is formed in the Ge–K–In–S system. First modification crystallizes in the cubic structure with the lattice parameter $a = 1303.66 \pm 0.05$, a calculated density of 3.191 g·cm⁻³, and an energy gap of 3.1 eV [3.34 eV (Friedrich et al. 2021)] (Yohannan and Vidyasagar 2016b). It was synthesized by heating in a continuous stream of CS_2 vapor appropriate stoichiometric mixture of K_2CO_3, CH_3COOK, In_2O_3, and GeO_2 with one intermittent grinding. An alumina boat was loaded with the reactant mixture and placed in a ceramic tube in a horizontal tubular furnace. A continuous stream of N_2 was bubbled through the CS_2 liquid and then passed through the ceramic tube to provide CS_2 vapor for the synthetic reaction. The block-shaped yellow single crystals of this modification were obtained by heating a stoichiometric reactant mixture of $KInS_2$ (1.4 mM), Ge (1.4 mM), and S (2.8 mM) at 750°C for 2 days and then cooled to 550°C at 3°C·h⁻¹.

Second modification of $KInGeS_4$ crystallizes in the triclinic structure with the lattice parameters $a = 720.0 \pm 0.5$, $b = 773.9 \pm 0.5$, $c = 1503.7 \pm 1.0$ pm, $\alpha = 91.83 \pm 0.03°$, $\beta = 99.26 \pm 0.04°$, and $\gamma = 112.01 \pm 0.04°$ (Wu et al. 1992). It was obtained in the same way as $KGaGeS_4$ was prepared using In instead of Ga. This modification is stable in the air and does not decompose significantly in water.

4.53 Germanium–Potassium–Yttrium–Sulfur

The **KYGeS₄** quaternary compound, which crystallizes in the monoclinic structure with the lattice parameters $a = 642 \pm 2$, $b = 660 \pm 1$, $c = 849 \pm 2$ pm, and $\beta = 107.0 \pm 0.1°$ at 115 K, is formed in the Ge–K–Y–S system (Wu and Ibers 1993a). It was prepared by the reaction of Ge with K_2S_5 and Y_2S_3. The starting materials were placed in a quartz tube that was subsequently evacuated to 10⁻³ Pa and sealed. The molar ratio of K/Y/Ge/S was 1:1:1:4. The quartz tube was heated gradually to 500°C, where it was kept for 24 h before being successively brought to 700°C for 24 h and to 1000°C for 150 h. The tube was then cooled (4°C·h⁻¹) to 300°C, and the furnace was shut off. Colorless crystals of the title compound were obtained. KYGeS₄ appears to be modestly stable in the air but decomposes gradually in the presence of water. Bulk samples of this compound were prepared by reaction of stoichiometric amounts of starting materials at 1000°C for 10 days with intermittent grinding.

4.54 Germanium–Potassium–Lanthanum–Sulfur

The **KLaGeS₄** quaternary compound, which crystallizes in the monoclinic structure with the lattice parameters $a = 665.3 \pm 0.1$, $b = 667.9 \pm 0.2$, $c = 864.3 \pm 0.2$ pm, $\beta = 107.57 \pm 0.01°$ at 115 K, and an energy gap of 3.4 eV, is formed in the Ge–K–La–S system (Wu and Ibers 1993a). The crystals and bulk sample of this compound were synthesized in the same way as KYGeS₄ was prepared using La_2S_3 instead of Y_2S_3. Colorless crystals of the title compound were obtained. KLaGeS₄ appear to be modestly stable in the air but decomposes gradually in the presence of water.

4.55 Germanium–Potassium–Neodymium–Sulfur

The **KNdGeS₄** quaternary compound, which crystallizes in the monoclinic structure with the lattice parameters $a = 657 \pm 2$, $b = 669 \pm 2$, $c = 867 \pm 2$ pm, and $\beta = 107.8 \pm 0.3°$, is formed in the Ge–K–Nd–S system (Wu and Ibers 1993a). The crystals and bulk sample of this compound were synthesized in the same way as KYGeS₄ was prepared using Nd_2S_3 instead of Y_2S_3. The crystals of KNdGeS₄ appear green under normal fluorescent light, but the color nearly vanishes with illumination by intense incandescent light or sunlight. This compound is modestly stable in the air but decomposes gradually in the presence of water.

4.56 Germanium–Potassium–Europium–Sulfur

The **KEuGeS₄** quaternary compound, which crystallizes in the monoclinic structure with the lattice parameters $a = 651.0 \pm 0.2$, $b = 664.9 \pm 0.2$ $c = 860.3 \pm 0.3$ pm, $\beta = 107.80 \pm 0.02°$, a calculated density of 3.671 g·cm⁻³ at 167 K, and an energy gap of 1.71 eV, is formed in the Ge–K–Eu–S system (Evenson IV and Dorhout 2001). This compound was synthesized by combining S (1.34 mM), K_2S_2 (0.384 mM), Ge (0.191 mM), and Eu (0.191 mM) that were loaded into a fused silica ampoule inside an inert atmosphere glove box. The ampoule was flame-sealed under vacuum and placed in a tube furnace. The temperature was ramped to 725°C where it remained for 150 h, and the furnace was allowed to cool back to room temperature at 4°C·h⁻¹. The product was washed with DMF, yielding deep-red plates of the title compound.

4.57 Germanium–Potassium–Gadolinium–Sulfur

The **KGdGeS₄** quaternary compound, which crystallizes in the monoclinic structure with the lattice parameters $a = 650 \pm 2$, $b = 667 \pm 2$, $c = 864 \pm 3$ pm, and $\beta = 108.0 \pm 0.3°$, is formed in the Ge–K–Gd–S system (Wu and Ibers 1993a). The crystals and bulk sample of this compound were synthesized in the same way as KYGeS₄ was prepared using Gd_2S_3 instead of Y_2S_3. Colorless crystals of the title compound were obtained. KGdGeS₄ appear to be modestly stable in the air but decomposes gradually in the presence of water.

4.58 Germanium–Potassium–Terbium–Sulfur

The **KTbGeS₄** quaternary compound, which crystallizes in the monoclinic structure with the lattice parameters $a = 647.1 \pm 0.2$, $b = 664.5 \pm 0.2$ $c = 862.3 \pm 0.3$ pm, and $\beta = 107.992$

± 0.005°, is formed in the Ge–K–Tb–S system (Chan and Dorhout 2005). Crystals of this compound were formed from a molten chalcogenide flux reaction of Tb (51.5 mg), Ge (21.8 mg), S (94.0 mg), and K_2S_2 (40.6 mg). The reactants were combined in a fused silica ampoule in an inert atmosphere glove box, sealed under vacuum, and heated to 725°C at a rate of 35°C·h⁻¹. After 150 h of heating, the ampoule was cooled at 4°C·h⁻¹ to room temperature. DMF was added to dissolve the remaining potassium sulfide flux, resulting in well-formed plates of $KTbGeS_4$.

4.59 Germanium–Potassium–Tin–Sulfur

Two quaternary compounds, $K_4Sn_3Ge_3S_{14}$ and $K_8Sn_2(Ge_2S_6)_2$, are formed in the Ge–K–Sn–S system (Marking and Kanatzidis 1997; Palchik et al. 2005). First of them crystallizes in the monoclinic structure with the lattice parameters $a = 694.42 \pm 0.16$, $b = 1511.8 \pm 0.4$, $c = 1207.1 \pm 0.3$ pm, $\beta = 100.51 \pm 0.00°$, and an energy gap of 2.46 eV (Palchik et al. 2005). $K_4Sn_3Ge_3S_{14}$ was synthesized from a mixture of Sn (0.8 mM), Ge (0.8 mM), K_2S (0.8 mM), and S (9.6 mM), which was sealed in a silica tube under vacuum (0.1 Pa) and heated at 500°C for 96 h. The tube was then cooled to room temperature at 5°C·h⁻¹. The excess flux (K_2S_x) was removed with MeOH to reveal air-stable yellowish-white crystals.

4.60 Germanium–Potassium–Lead–Sulfur

Two quaternary compounds, $K_2PbGe_2S_6$ and $K_8Pb_2(Ge_2S_6)_2$, are formed in the Ge–K–Pb–S system (Marking and Kanatzidis 1997; Palchik et al. 2005). The first of them crystallizes in the monoclinic structure with the lattice parameters $a = 926.8 \pm 0.2$, $b = 1433.8 \pm 0.3$, $c = 891.9 \pm 0.2$ pm, $\beta = 90.56 \pm 0.00°$, and an energy gap of 2.83 eV (Palchik et al. 2005). To prepare $K_2PbGe_2S_6$, a mixture of Ge (1 mM), Pb (0.75 mM), K_2S (0.75 mM), and S (9.0 mM) was sealed in a silica tube under vacuum (0.1 Pa) and heated at 500°C for 96 h. The tube was then cooled to room temperature at 5°C·h⁻¹. The excess flux (K_2S_x) was removed with MeOH to reveal air-stable yellowish-white crystals.

The glass-forming region in the $GeS_2–K_2S–PbS$ quasiternary system is presented in Figure 4.6 (Imaoka and Jamadzaki 1967).

4.61 Germanium–Potassium–Phosphorus–Sulfur

The $K_4GeP_4S_{12}$ quaternary compound, which crystallizes in the monoclinic structure with the lattice parameters $a = 1216.95 \pm 0.08$, $b = 783.77 \pm 0.05$, $c = 2318.46 \pm 0.17$ pm, $\beta = 100.189 \pm 0.006°$, a calculated density of 2.251 g·cm⁻³, and an energy gap of 3.0 eV, is formed in the Ge–K–P–S system (Morris et al. 2012). It was synthesized by combining K_2S (0.5 mM), Ge (0.25 mM), P (1.00 mM), and S (2.50 mM) in a fused silica tube inside a N_2-filled glove box. The tube was flame-sealed under vacuum (10⁻² Pa), heated to 500°C within 6 h, held there

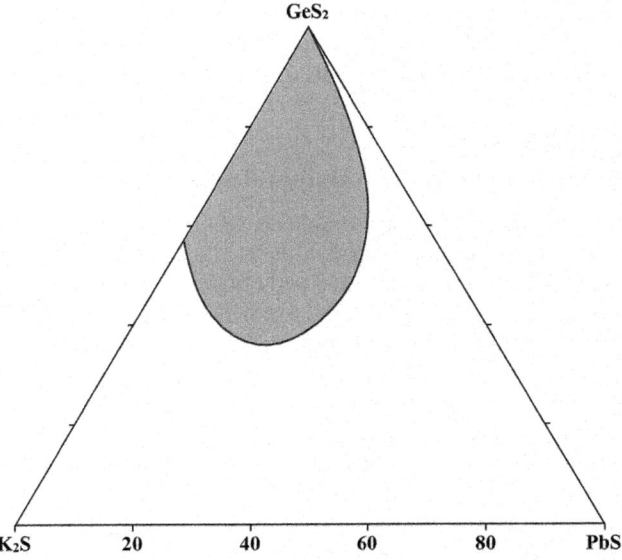

FIGURE 4.6 Glass-forming region in the $GeS_2–K_2S–PbS$ quasiternary system. (From Imaoka, M. and Jamadzaki, T. [in Japanese], *Monthly J. Inst. Industr. Sci. Univ. Tokyo*, **19**(9), 261, 1967.) Open access

for 24 h, cooled to 200°C in 24 h, and then to room temperature in 3 h. The resulting product contained pale-yellow, rod-like crystals of the target compound.

4.62 Germanium–Potassium–Bismuth–Sulfur

The $KBiGeS_4$ quaternary compound, which crystallizes in the monoclinic structure with the lattice parameters $a = 659.81 \pm 0.13$, $b = 681.49 \pm 0.14$, $c = 1728.4 \pm 0.4$ pm, $\beta = 108.46 \pm 0.04°$, and a calculated density of 4.045 g·cm⁻³, is formed in the Ge–K–Bi–S system (Mei et al. 2010). To synthesize this compound, a reaction mixture of K_2S_3 (1 mM), Ge (1.5 mM), Bi (0.5 mM), and S (4.8 mM) was loaded into a fused silica tube under an Ar atmosphere in a glove box. This tube was sealed under vacuum (10⁻³ Pa) and then placed in a furnace. The sample was heated to 200°C in 20 h, kept at 200°C for 10 h, then heated to 600°C in 24 h, kept at 600°C for 72 h, slowly cooled at 4°C·h⁻¹ to 100°C, and then cooled to room temperature. The reaction mixture was washed free of flux with DMF and then dried with acetone. The product consists of red plates of $KBiSiS_4$. It is moderately stable in the air.

4.63 Germanium–Potassium–Tantalum–Sulfur

The $KTaGeS_5$ quaternary compound, which crystallizes in the triclinic structure with the lattice parameters $a = 693.7 \pm 0.1$, $b = 695.0 \pm 0.2$, $c = 884.4 \pm 0.3$ pm, $\alpha = 71.07 \pm 0.02°$, $\beta = 78.56 \pm 0.02°$, and $\gamma = 75.75 \pm 0.02°$, is formed in the Ge–K–Ta–S system (Dong et al. 2009). The title compound was prepared by the reactions of Ta, Ge, and S with the use of KCl flux, which was previously dried at 120°C for 4 h. Stoichiometric combination of the elements were mixed in silica tube with the addition of KCl. Mass ratio of the reactants and KCl was

1:2. The tube was evacuated (~0.1 Pa), sealed, and heated in a furnace from the room temperature to 750°C over 14 h, where it was kept for 72 h. Then it was cooled to 100°C at 4°C·h⁻¹ and quenched to room temperature. After the reaction, excess flux was removed with distilled water. Dark-red platy crystals, which are stable in the air and water, were isolated.

4.64 Germanium–Potassium–Manganese–Sulfur

The **K₂MnGe₃S₈** quaternary compound, which melts congruently at ≈682°C and crystallizes in the monoclinic structure with the lattice parameters $a = 719.78 ± 0.14$, $b = 1195.3 ± 0.2$, $c = 3363.6 ± 0.7$ pm, $β = 96.13°$, a calculated density of 2.804 g·cm⁻³ at 153 K, and an energy gap of 2.95 eV is formed in the Ge–K–Mn–S system (Li et al. 2018d). Its single crystals were prepared by a spontaneous crystallization method. K₂S, MnS, and GeS₂ were mixed in the molar ratio of 1:1:3, ground, and loaded into a fused silica tube under an Ar atmosphere in a glove box. After that, it was flame-sealed under a high vacuum of 10⁻³ Pa and placed in a furnace. The mixture was heated to 800°C in 20 h and kept at that temperature for 72 h, then cooled at a rate of 3°C·h⁻¹ to 300°C, and ultimately cooled to room temperature. The resultant light-yellow crystals were manually picked. They are stable under air and moisture conditions for several months.

4.65 Germanium–Potassium–Iron–Sulfur

The **K₂FeGe₃S₈** quaternary compound, which crystallizes in the monoclinic structure with the lattice parameters $a = 710.89 ± 0.05$, $b = 1188.23 ± 0.08$, $c = 1675.88 ± 0.11$ pm, $β = 96.604 ± 0.002°$, a calculated density of 2.873 g·cm⁻³, and an energy gap of 2.6 eV is formed in the Ge–K–Fe–S system (Ji et al. 2021a). A nearly single-phase sample of this compound was obtained via a high-temperature solid-state reaction. K, Fe, Ge, and S were loaded (molar ratio 2:1:3:8) and sealed in a carbonized silica ampoule. This ampoule was slowly heated to 125°C in 10 h, then annealed at 125°C for 10 h to fully react K with S. After low-temperature annealing, the ampoule was heated to 800°C in 10 h and annealed at this temperature for 120 h, then cooled down to room temperature over 24 h. The sample was ground in a mortar in the glove box, resealed in carbonized silica ampoules, and annealed for a second time using the same temperature profile. After the second annealing process, the green crystals were found as major products.

4.66 Germanium–Potassium–Cobalt–Sulfur

The **K₂CoGe₃S₈** quaternary compound, which crystallizes in the triclinic structure with the lattice parameters $a = 701.55 ± 0.07$, $b = 776.97 ± 0.07$, $c = 1434.15 ± 0.13$ pm, $α = 96.796 ± 0.003°$, $β = 92.647 ± 0.003°$, $γ = 114.042 ± 0.003°$, a calculated density of 2.86 g·cm⁻³, and an energy gap of 2.1 eV is formed in the Ge–K–Co–S system (Ji et al. 2021a). This compound was prepared in the same way as in the case of K₂FeGe₃S₈ using

Co instead of Fe. After the second annealing process, the red crystals were found as major products.

4.67 Germanium–Rubidium–Copper–Sulfur

The **Rb₄Cu₈Ge₃S₁₂** quaternary compound, which is stable up to 360°C and crystallizes in the cubic structure with the lattice parameter $a = 1751.12$ pm, a calculated density of 3.594 g·cm⁻³, and an energy gap of 2.3 eV, is formed in the Ge–Rb–Cu–S system (Zhang et al. 2010). To prepare this compound, Cu powder (0.095 mM), GeO₂ (0.04 mM), Rb₂CO₃ (0.09 mM), S powder (0.53 mM), and about 250 mg of ethylenediamine and 170 mg CH₃OH/H₂O (volume ratio 1:1) as solvent were mixed and stirred substantially. The mixture was sealed in a Pyrex glass tube at air atmosphere, placed into a stainless-steel autoclave, and then heated at 160°C for 7 days. After being cooled naturally to ambient temperature, the tube was broken with caution. The products were washed with ethylenediamine and ethanol, respectively, and yellow block crystals of the title compound were obtained.

4.68 Germanium–Rubidium–Silver–Sulfur

The **Rb₂Ag₂GeS₄** quaternary compound, which crystallizes in the monoclinic structure with the lattice parameters $a = 747.74 ± 0.03$, $b = 2211.20 ± 0.07$, $c = 629.63 ± 0.02$ pm, $β = 115.664°$, and a calculated density of 4.159 g·cm⁻³ (Zhang et al. 2013) [$a = 747.42 ± 0.02$, $b = 2211.60 ± 0.05$, $c = 629.02 ± 0.01$ pm, and $β = 115.518 ± 0.002°$ (An et al. 2004)] is formed in the Ge–Rb–Ag–S system. It was synthesized using a mixture of Ge (0.01 mM), AgNO₃ (0.01 mM), S (0.19 mM), RbCl (0.04 mM), and about 0.4 mL of triethylenetetramine (Zhang et al. 2013). The mixture was placed in a Pyrex glass tube, which was sealed with ca. 10% filling under an air atmosphere, placed in a stainless-steel autoclave, and heated at 170°C for 5 days. Cooled naturally to ambient temperature, final products shaped as pale-yellow blocks were obtained after washing several times with ethanol and dried at room temperature. Rb₂Ag₂GeS₄ could be also prepared as follows: Ge (10.9 mg), AgNO₃ (50.7 mg), Rb₂CO₃ (173 mg), and S (18.3 mg) were put into a glass tube, to which 0.4 mL of ethanol/glycerol (volume ratio 2:1) was added. The glass tube was sealed (reagents filled about 10% of the tube), placed into a Teflon-lined stainless-steel autoclave, and heated at 120°C for 5 days. The products were washed with ethanol and water, respectively. The pure title compound was obtained.

4.69 Germanium–Rubidium–Mercury–Sulfur

The **Rb₂Hg₃Ge₂S₈** quaternary compound, which melts incongruently (Liao et al. 2003) and crystallizes in the monoclinic structure with the lattice parameters $a = 993.8 ± 0.3$, $b = 635.2 ± 0.2$, $c = 1311.7 ± 0.3$ pm, $β = 97.33 ± 0.03°$ (Marking et al. 1998), and an energy gap of 2.80 eV (Liao et al. 2003) [2.65 eV (Kanatzidis et al. 1997a, 1997b)], is formed in the Ge–Rb–Hg–S system. Small yellow crystals of the title compound were prepared with a slight excess of HgS powder in the mixture of Ge, HgS, Rb₂S, and S by heating at 500°C for 96 h and cooling to 200°C at 4°C/h (Liao et al. 2003) or by heating a mixture of 4Rb₂S + 4.5HgS + 2Sn + 32S at 350°C for 4 days and slowly cooling to 152°C over 99 h (Marking et al. 1998). This compound is stable in the air.

4.70 Germanium–Rubidium–Aluminum–Sulfur

The **Rb₃(AlS₂)₃(GeS₂)₇** quaternary compound, which melts congruently at 873°C and crystallizes in the monoclinic structure with the lattice parameters $a = 675.37 \pm 0.03$, $b = 3778.25 \pm 0.19$, $c = 675.15 \pm 0.03$ pm, $\beta = 90.655 \pm 0.004°$, a calculated density of 2.866 g·cm⁻³ at 190 K, and an energy gap of 3.1 eV, is formed in the Ge–Rb–Al–S system (Rothenberger et al. 2010). The synthesis of this compound was achieved by heating stoichiometric amounts of Rb₂S₅, Al, Ge, and S in a sealed and evacuated fused silica tube at 850°C for 3 days and cooled to 400°C in 4 days. Rb₃(AlS₂)₃(GeS₂)₇ is partially hydrolyzed when suspended in water at room temperature.

4.71 Germanium–Rubidium–Gallium–Sulfur

Two quaternary compounds, **RbGaGeS₄** and **Rb₃Ga₃Ge₇S₂₀**, are formed in the Ge–Rb–Ga–S system. First of them crystallizes in the orthorhombic structure with the lattice parameters $a = 1685.39 \pm 0.06$, $b = 713.30 \pm 0.03$, $c = 1214.10 \pm 0.05$ pm, a calculated density of 3.240 g·cm⁻³, and an energy gap of 3.26 eV (Friedrich et al. 2020). Rb₃Ga₃Ge₇S₂₀ crystallizes in the monoclinic structure with the lattice parameters $a = 677.04 \pm 0.05$, $b = 3767.0 \pm 0.3$, $c = 671.17 \pm 0.5$ pm, $\beta = 90.628 \pm 0.006°$, and an energy gap of 3.19 eV (Li and Yan 2021).

RbGaGeS₄ was prepared by reacting Rb₂S, Ga₂S₃, GeS₂ or Ge, and S in a sealed evacuated quartz tube (Friedrich et al. 2020). The reaction mixture was heated to 900°C with a heating rate of 50°C·h⁻¹, annealed at 900°C for 2 days and cooled to room temperature with a cooling rate of 20°C·min⁻¹. The resulting crude product was finely ground in an agate mortar under a N₂ atmosphere and annealed at 900°C for 2 additional days. White crystals were obtained. This compound is stable in moist air.

To prepare Rb₃Ga₃Ge₇S₂₀, RbBr, Ge, Ga, and S (molar ratio 3:7:3:20) were mixed and ground, therein RbBr flux is 5% excess (Li and Yan 2021). The mixture was transferred into a quartz ampoule, which was flame-sealed under a high vacuum (10⁻³ Pa) and put into a furnace. The sealed ampoule was sintered at 950°C for 4 days and then cooled to room temperature for 3 days. After washing with deionized water, the colorless and transparent crystals of the title compound were obtained.

4.72 Germanium–Rubidium–Indium–Sulfur

The **RbInGeS₄** quaternary compound, which apparently has two polymorphic modifications, is formed in the Ge–Rb–In–S system. α-CsInSnS₄ crystallizes in the orthorhombic structure with the lattice parameters $a = 1733.15 \pm 0.04$, $b = 738.84 \pm$

0.02, $c = 1224.63 \pm 0.03$ pm, a calculated density of 3.386 g·cm⁻³, and an energy gap of 3.33 eV (Friedrich et al. 2021). It was obtained by reacting Rb₂S (1.25 mM), In (2.49 mM), Ge (2.49 mM), and S (8.73 mM). The reaction mixture was heated to 1000°C with a heating rate of 3°C·min⁻¹. After annealing at 1000°C for 24 h the tube was cooled to room temperature with a cooling rate of 1°C·min⁻¹. The white polycrystalline product is stable in moist air and forms plate-like crystallites upon crushing in an agate mortar.

β-RbInGeS₄ crystallizes in the cubic structure with the lattice parameter $a = 1318.65 \pm 0.08$, a calculated density of 3.486 g·cm⁻³, and an energy gap of 3.1 eV (Yohannan and Vidyasagar 2016b). This modification was synthesized by heating in a continuous stream of CS₂ vapor appropriate stoichiometric mixture of Rb₂CO₃, In₂O₃, and GeO₂ with one intermittent grinding. An alumina boat was loaded with the reactant mixture and placed in a ceramic tube in a horizontal tubular furnace. A continuous stream of N₂ was bubbled through the CS₂ liquid and then passed through the ceramic tube to provide CS₂ vapor for the synthetic reaction. The block-shaped yellow single crystals of this compound were obtained from a reactant mixture of RbInS₂ (1.1 mM) and GeS₂ (1.1 mM). The mixture was heated, along with RbCl (3 mM) and RbI (4.2 mM), at 650°C for 2 days and then cooled to 550°C at 1.5°C·h⁻¹, before the furnace was finally switched off. The product was isolated by washing off the flux with water and then dried in the air.

4.73 Germanium–Rubidium–Europium–Sulfur

The **RbEuGeS₄** quaternary compound, which crystallizes in the monoclinic structure with the lattice parameters $a = 649.8 \pm 0.2$, $b = 668.9 \pm 0.3$, $c = 896.4 \pm 0.3$ pm, $\beta = 108.647 \pm 0.006°$, and a calculated density of 3.943 g·cm⁻³, is formed in the Ge–Rb–Eu–S system (Choudhury et al. 2006). It was synthesized by combining Eu (0.250 mM), Ge (0.260 mM), S (0.521 mM), and Rb₂S₅ (0.511 mM) with an approximately compositional ratio Rb/Eu/Ge/S of 4:1:1:12. The mixture was directly loaded into a fused silica ampoule inside a N₂-filled atmosphere glove box. The ampoule was flame-sealed under vacuum and placed in a furnace. The furnace was ramped to 725°C at a rate of 35°C·h⁻¹, and the temperature was held constant at 725 °C for 150 h. Then, the furnace was slowly cooled to ambient temperature at a rate of 3°C·h⁻¹. After the reaction product cooled, the ampoule was opened in an inert atmosphere glove box, and the solid product was soaked in DMF for 6 h in order to dissolve and wash away any remaining flux and loosen the product crystals. The product, after treatment with DMF, revealed red plates of RbEuGeS₄ with no other visible impurity.

4.74 Germanium–Rubidium–Phosphorus–Sulfur

The **Rb₄GeP₄S₁₂** quaternary compound, which crystallizes in the monoclinic structure with the lattice parameters $a = 1232.17 \pm 0.06$, $b = 804.62 \pm 0.03$, $c = 2357.43 \pm 0.11$ pm, $\beta = 100.622 \pm 0.004°$, a calculated density of 2.669 g·cm⁻³, and an energy gap of 3.0 eV, is formed in the Ge–Rb–P–S system (Morris et al. 2012). It was synthesized by combining Rb₂S

(0.5 mM), Ge (0.25 mM), P (1.00 mM), and S (2.50 mM) in a fused silica tube inside a N_2-filled glove box. The tube was flame-sealed under vacuum (10^{-2} Pa), heated to 500°C for 6 h, held there for 24 h, cooled to 200°C for 24 h, and then to room temperature in 3 h. The resulting product contained pale-yellow, rod-like crystals of the target compound.

4.75 Germanium–Rubidium–Bismuth–Sulfur

The **RbBiGeS₄** quaternary compound, which crystallizes in the monoclinic structure with the lattice parameters $a = 658.64 \pm 0.04$, $b = 685.59 \pm 0.04$, $c = 1798.10 \pm 0.12$ pm, $\beta = 109.075 \pm 0.001°$, a calculated density of 4.287 g·cm⁻³ at 153 K, and an energy gap of 2.23 eV, is formed in the Ge–Rb–Bi–S system (Yao et al. 2002). For the synthesis of this compound, the reaction mixture of Rb_2S_3 (0.5 mM), Ge (1.0 mM), Bi (0.5 mM), and S (2.5 mM) was loaded into a fused silica tube under an Ar atmosphere in a glove box. This tube was sealed under a vacuum (10^{-2} Pa) and then placed in a furnace. The sample was heated to 620°C for 15 h, kept at this temperature for 84 h, slowly cooled at 6°C·h⁻¹ to 100°C, and then the furnace was turned off. The reaction mixture was washed free of flux with DMF and then dried with acetone. The product consisted of orange-red plates of $RbGeBiS_4$.

4.76 Germanium–Rubidium–Tantalum–Sulfur

The **RbTaGeS₅** quaternary compound, which crystallizes in the triclinic structure with the lattice parameters $a = 699.6 \pm 0.3$, $b = 703.3 \pm 0.3$, $c = 898.5 \pm 0.4$ pm, $\alpha = 70.33 \pm 0.03°$, $\beta = 78.12 \pm 0.04°$, $\gamma = 75.63 \pm 0.04°$, a calculated density of 4.149 g·cm⁻³, and an energy gap of 1.92 eV, is formed in the Ge–Rb–Ta–S system (Dong et al. 2009). This compound was obtained by the reactions of Ta, Ge, and S with the use of RbCl flux, which was previously dried at 120°C for 4 h. A stoichiometric combination of the elements was mixed in a silica tube with the addition of RbCl. The mass ratio of the reactants and RbCl was 1:2. The tube was evacuated (~0.1 Pa), sealed, and heated in a furnace from room temperature to 750°C over 14 h, where it was kept for 72 h. Then it was cooled to 100°C at 4°C·h⁻¹ and quenched to room temperature. After the reaction, excess flux was removed with distilled water. Dark-red platy crystals, which are stable in the air and water, were isolated.

4.77 Germanium–Rubidium–Manganese–Sulfur

The **Rb₂MnGe₃S₈** quaternary compound, which crystallizes in the orthorhombic structure with the lattice parameters $a = 727.56 \pm 0.04$, $b = 1216.68 \pm 0.08$, $c = 1683.51 \pm 0.08$ pm, a calculated density of 3.121 g·cm⁻³, and an energy gap of 3.01 eV, is formed in the Ge–Rb–Mn–S system (Hu et al. 2018). To synthesize this compound, a mixture of RbCl, Mn, Ge, and S (molar ratio 4:2:3:8) was loaded into a silica tube inside a glove box. Subsequently, the assembly was flame-sealed under high vacuum (10^{-6} Pa), and then heated to 400°C within 20 h, dwelled for 10 h, then heated to 800°C in 20 h, maintained

for 100 h, and finally cooled to 350°C at a rate of 3°C·h⁻¹. The product was washed with distilled water to remove the chloride. Yellow plate-like crystals of $Rb_2MnGe_3S_8$ were obtained.

4.78 Germanium–Rubidium–Iron–Sulfur

The **Rb₂FeGe₃S₈** quaternary compound, which crystallizes in the monoclinic structure with the lattice parameters $a = 723.55 \pm 0.05$, $b = 1687.97 \pm 0.12$, $c = 1220.57 \pm 0.09$ pm, $\beta = 96.339° \pm 0.003°$, and an energy gap of 1.64 eV, is formed in the Ge–Rb–Fe–S system (Wang et al. 2022). To prepare this compound, it is necessary to weigh 1 g mixture of Rb_2S_3, Fe, Ge, and S with a molar ratio of 1:1:3:5, 1:2:3:5, 1:2:4:5 in the Ar-filled glove box without oxygen and moisture. After pouring into a mortar and grinding it evenly, put it into a quartz tube and draw a vacuum to make the pressure in the tube less than 10^{-3} Pa. After sealing the quartz tube, put it in the muffle furnace, and slowly raise the temperature to 800°C. After being kept for 20 h, the temperature is lowered to 400°C within 150 h, and then naturally lowered to room temperature.

4.79 Germanium–Cesium–Zinc–Sulfur

The **Cs₂ZnGe₃S₈** quaternary compound, which crystallizes in the monoclinic structure with the lattice parameters $a = 726.24 \pm 0.04$, $b = 1698.01 \pm 0.09$, $c = 1262.31 \pm 0.08$ pm, $\beta = 97.706° \pm 0.005°$, a calculated density of 3.468 g·cm⁻³, and an energy gap of 3.32 eV, is formed in the Ge–Cs–Zn–S system (Morris et al. 2013). To obtain this compound, a mixture of Cs_2S (0.25 mM), Zn (0.25 mM), Ge (0.75 mM), and S (1.75 mM) was added to a fused silica tube inside a N_2-filled glove box. The tube was then evacuated, flame-sealed, and heated gently at first in a weak methane/oxygen flame to start the reaction. It was then heated more strongly until a glowing liquid was obtained for a total of ~2 min. The tube was allowed to cool in the air at a rate of ~200°C·min⁻¹. Yellow plate-like crystals of $Cs_2ZnGe_3S_8$ were obtained from such a reaction.

4.80 Germanium–Cesium–Cadmium–Sulfur

The **Cs₂CdGe₃S₈** quaternary compound, crystallizes in the orthorhombic structure with the lattice parameters $a = 741.92 \pm 0.02$, $b = 1250.99 \pm 0.04$, and $c = 1708.43 \pm 0.06$ pm, a calculated density of 3.571 g·cm⁻³, and an energy gap of 3.38 eV, is formed in the Ge–Cs–Cd–S system (Morris et al. 2013). This compound was obtained by the interaction of Cs_2S, Cd, Ge, and S in stoichiometric ratio, which was placed in a fused silica tube and heated in a furnace to 750°C within 8 h, held there for 12 h, and cooled to room temperature in 24 h.

4.81 Germanium–Cesium–Mercury–Sulfur

The **Cs₂Hg₃Ge₂S₈** quaternary compound is formed in the Ge–Cs–Hg–S system. It crystallizes in the triclinic structure with the lattice parameters $a = 780.8 \pm 0.2$, $b = 916.4 \pm 0.2$, $c =$

661.2 ± 0.2 pm, α = 92.02 ± 0.2°, β = 108.65 ± 0.2°, and γ = 108.10 ± 0.2° (Marking et al. 1998). Yellow-orange crystals of $Cs_2Hg_3Ge_2S_8$ were obtained by heating a mixture of $4Cs_2S$ + $5HgS$ + $2Ge$ + $32S$ at 520°C for 4 days and slowly cooling to 180°C over 80 h.

4.82 Germanium–Cesium–Gallium–Sulfur

The **CsGaGeS$_4$** quaternary compound, which has three polymorphic modifications, is formed in the Ge–Cs–Ga–S system (Friedrich et al. 2020). The first modification crystallizes in the orthorhombic structure with the lattice parameters a = 1701.25 ± 0.08, b = 718.48 ± 0.03, c = 1250.38 ± 0.06 pm, a calculated density of 3.507 g·cm^{-3}, and an energy gap of 3.18 eV [3.01 eV (Friedrich et al. 2021)]. This modification was prepared by reacting Cs_2S, Ga_2S_3, GeS_2 or Ge, and S in a sealed evacuated quartz tube. The reaction mixture was heated to 900°C with a heating rate of 50°C·h^{-1}, annealed at 900°C for 2 days and cooled to room temperature with a cooling rate of 20°C·min^{-1}. The resulting crude product was finely ground in an agate mortar under N$_2$ atmosphere and annealed at 900°C for 2 additional days. White crystals were obtained. One single batch to prepare orthorhombic CsGaGeS$_4$ yielded a phase-pure sample of triclinic modification (a = 716.11 ± 0.01, b = 759.44 ± 0.02, c = 1463.45 ± 0.03 pm, α = 91.003 ± 0.002°, β = 92.314 ± 0.002°, γ = 106.680 ± 0.002°, a calculated density of 3.520 g·cm^{-3}, and an energy gap of 2.98 eV). This batch was annealed at 750°C for 21 days, and from it small quantities of monoclinic modification (a = 769.95 ± 0.05, b = 1637.21 ± 0.09, c = 689.30 ± 0.04 pm, β = 111.894 ± 0.004°, and a calculated density of 3.324 g·cm^{-3}) could be obtained by next annealing of the compound above 600°C and subsequent quenching. All of these modifications are stable in moist air.

4.83 Germanium–Cesium–Indium–Sulfur

The **CsInGeS$_4$** quaternary compound, which has two polymorphic modifications, is formed in the Ge–Cs–In–S system (Yohannan and Vidyasagar 2016b). The first modification crystallizes in the orthorhombic structure with the lattice parameters a = 1749.7 ± 0.2, b = 744.3 ± 0.1, c = 1250.8 ± 0.2 pm, a calculated density of 3.659 g·cm^{-3}, and an energy gap of 2.4 eV [3.23 eV (Friedrich et al. 2021)]. This modification was obtained, when a mixture of CsInS$_2$ (1.1 mM), Ge (1.1 mM), S (2.2 mM), and CsCl (5.9 mM) was heated, in an evacuated sealed silica tube, at 725°C for 2 days and then cooled to 525°C at 3°C·h^{-1} before the furnace was finally turned off. The solid product contents were washed with distilled water to dissolve away the CsCl flux, and the homogenous α-CsInGeS$_4$ was filtered and dried at room temperature in the open air.

The second modification (β-CsInGeS$_4$) crystallizes in the cubic structure with the lattice parameter a = 1338.32 ± 0.04 pm, a calculated density of 3.729 g·cm^{-3}, and an energy gap of 2.1 eV (Yohannan and Vidyasagar 2016b). It was obtained as a homogeneous phase, in the form of single crystals, from a reactant mixture of Cs$_2$CO$_3$ (0.84 mM), In (1.1 mM), Ge (1.1 mM), and S (5 mM). A quartz tube containing the mixture was initially heated at 300°C under dynamic vacuum for 10 min to expel the CO$_2$ generated and then sealed under vacuum. This sealed quartz tube was heated at 650°C for 3 days, then cooled to 550°C at 1.5°C·h^{-1} and the furnace was finally turned off. The solid product contents were washed with distilled water to dissolve away Cs$_2$S$_2$O$_3$ formed, and single crystals of β-CsInGeS$_4$ were isolated. Single crystals of α-CsInGeS$_4$ were obtained, when β-CsInGeS$_4$ (1.11 mM) was heated in CsCl (5.6 mM) flux, at 700°C for 2 days, and then cooled to 500°C at 3°C·h^{-1} before the furnace was finally turned off. The product was isolated by washing off the CsCl flux with water and then dried in the open air. The block-shaped yellow single crystals of this compound were obtained.

4.84 Germanium–Cesium–Samarium–Sulfur

The **CsSmGeS$_4$** quaternary compound, which melts congruently at ~1086°C and crystallizes in the orthorhombic structure with the lattice parameters a = 676.8 ± 0.2, b = 1775.2 ± 0.2, c = 648.8 ± 0.1 pm and, a calculated density of 4.12 g·cm^{-3}, is formed in the Ge–Cs–Sm–S system (Bucher and Hwu 1994). Pale-yellow transparent crystals of the title compound were produced via a two-step synthetic route. The CsCl salt was added as a flux to the pre-prepared AgSm$_3$GeS$_7$ mixture (mass ratio 10:1). The reaction was carried out in an evacuated, carbon-coated silica ampoule and heated for 7 days at ca. 750°C followed by slow cooling. Crystals of the title compound were isolated from the flux by washing the reaction product with deionized water, using the suction filtration method. CsSmGeS$_4$ seems to be stable in the air, showing no signs of appreciable degradation in crystal quality over a period of months.

The polycrystalline CsSmGeS$_4$ was synthesized from the stoichiometric reaction of corresponding elements in a fused silica tube. The reaction was slowly heated up from room temperature to 550°C in 3 days, followed by 750°C, and kept isotherm for 6 days with one grinding.

4.85 Germanium–Cesium–Lead–Sulfur

The **Cs$_4$Pb$_4$Ge$_5$S$_{16}$** quaternary compound, which crystallizes in the orthorhombic structure with the lattice parameters a = 3830.20 ± 0.19, b = 907.70 ± 0.05, c = 1944.2 ± 0.1 pm, and an energy gap of 2.61 eV, is formed in the Ge–Cs–Pb–S system (Palchik et al. 2005). To prepare this compound, a mixture of Ge (0.66 mM), Pb (0.5 mM), Cs$_2$S (0.5 mM), and S (6.0 mM) was sealed under vacuum (<0.1 Pa) in a silica tube and heated at 400°C for 96 h. This was followed by cooling to room temperature at 5°C·h^{-1}. The excess flux (Cs$_2$S$_x$) was removed with MeOH to reveal yellowish-white air-stable crystals of the title compound.

4.86 Germanium–Cesium–Phosphorus–Sulfur

The **Cs$_4$GeP$_4$S$_{12}$** quaternary compound, which crystallizes in the monoclinic structure with the lattice parameters a = 1264.75 ± 0.06, b = 830.20 ± 0.03, c = 2424.28 ± 0.13 pm,

$\beta = 102.012 \pm 0.004°$, a calculated density of 2.969 $g \cdot cm^{-3}$, and an energy gap of 3.1 eV, is formed in the Ge–Cs–P–S system (Morris et al. 2012). It was synthesized by combining Cs_2S (0.5 mM), Ge (0.25 mM), P (1.00 mM), and S (2.50 mM) in a fused silica tube inside a N_2-filled glove box. The tube was flame-sealed under a vacuum (10^{-2} Pa), heated to 500°C within 6 h, held there for 24 h, cooled to 200°C for 24 h, and then to room temperature in 3 h. The resulting product contained pale-yellow, rod-like crystals of the target compound.

4.87 Germanium–Cesium–Bismuth–Sulfur

The **CsBiGeS$_4$** quaternary compound, which crystallizes in the monoclinic structure with the lattice parameters $a = 654.74 \pm 0.04$, $b = 692.82 \pm 0.04$, $c = 1888.75 \pm 0.11$ pm, $\beta = 110.173 \pm 0.001°$, a calculated density of 4.482 $g \cdot cm^{-3}$ at 153 K, and an energy gap of 2.28 eV, is formed in the Ge–Cs–Bi–S system (Yao et al. 2002). For the synthesis of this compound, the reaction mixture of Cs_2S_3 (0.5 mM), Ge (1.0 mM), Bi (0.5 mM), and S (2.5 mM) was loaded into a fused silica tube under an Ar atmosphere in a glove box. This tube was sealed under a vacuum (10^{-2} Pa) and then placed in a furnace. The sample was heated to 600°C over 15 h, kept at this temperature for 84 h, slowly cooled at 6°C·h^{-1} to 100°C, and then the furnace was turned off. The reaction mixture was washed free of flux with DMF and then dried with acetone. The product consisted of orange-red plates of CsGeBiS$_4$.

4.88 Germanium–Cesium–Tantalum–Sulfur

The **CsTaGeS$_5$** quaternary compound, which crystallizes in the triclinic structure with the lattice parameters $a = 701.2 \pm 0.4$, $b = 720.2 \pm 0.3$, $c = 926.7 \pm 0.5$ pm, $\alpha = 68.55 \pm 0.03°$, $\beta = 77.27 \pm 0.04°$, $\gamma = 74.75 \pm 0.04°$, a calculated density of 4.363 $g \cdot cm^{-3}$, and an energy gap of 1.99 eV, is formed in the Ge–Cs–Ta–S system (Dong et al. 2009). This compound was obtained by the reactions of Ta, Ge, and S with the use of CsCl/LiCl flux, which was previously dried at 120°C for 4 h. Stoichiometric combination of the elements was mixed in a silica tube with the addition of flux. The mass ratio of the reactants and CsCl/LiCl was 1:3. The tube was evacuated (~0.1 Pa), sealed, heated at a rate of 50°C·h^{-1} to 850°C, and kept there for 72 h. Then it was cooled to room temperature at 3°C·h^{-1}. After the reaction, excess flux was removed with distilled water. Dark-red polyhedral crystals, which are stable in the air and water, were isolated from the reaction mixture.

4.89 Germanium–Cesium–Manganese–Sulfur

The **Cs$_2$MnGe$_3$S$_8$** quaternary compound, which crystallizes in the monoclinic structure with the lattice parameters $a = 737.21 \pm 0.02$, $b = 1711.42 \pm 0.05$, $c = 1264.81 \pm 0.04$ pm, $\beta = 97.4362 \pm 0.0011$, a calculated density of 3.337 $g \cdot cm^{-3}$, and an energy gap of 2.93 eV, is formed in the Ge–Cs–Mn–S system (Hu et al. 2018). To synthesize this compound, a mixture of CsCl, Mn, Ge, and S (molar ratio 4:2:3:8) was loaded into a silica tube inside a glove box. Subsequently, the assembly was flame-sealed under a high vacuum (10^{-6} Pa), and then heated to 400°C over 20 h, dwelled for 10 h, then heated to 800°C in 20 h, maintained for 100 h, and finally cooled to 350°C at a rate of 3°C·h^{-1}. The product was washed with distilled water to remove the chloride. The yellow plate-like crystals of $Rb_2MnGe_3S_8$ were obtained.

4.90 Germanium–Cesium–Iron–Sulfur

The **Cs$_2$FeGe$_3$S$_8$** quaternary compound, which crystallizes in the monoclinic structure and has an energy gap of 1.80 eV, is formed in the Ge–Cs–Fe–S system (Wang et al. 2022). To prepare this compound, it is necessary to weigh 1 g mixture of Cs_2S_3, Fe, Ge, and S with a molar ratio of 1:1:3:5, 1:2:3:5, 1:2:4:5 in the Ar-filled glove box without oxygen and moisture. Pour it into a mortar and grind it evenly, then put it into a quartz tube and draw a vacuum to make the pressure in the tube less than 10^{-3} Pa. After sealing the quartz tube, put it in the muffle furnace and slowly raise the temperature to 800°C. After being kept for 20 h, the temperature is lowered to 400°C within 150 h, and then naturally lowered to room temperature.

4.91 Germanium–Copper–Silver–Sulfur

Cu$_2$GeS$_3$–Ag$_2$GeS$_3$. The homogeneity regions of solid solutions based on Cu_2GeS_3 (30 mol%) and Ag_2GeS_3 (25 mol%) in this system were determined by measuring the electromotive force (EMF) of the concentration circuits (Alverdiev et al. 2018). The relative partial thermodynamic functions of Cu in the alloys and the standard thermodynamic functions of formation and standard entropies of the **Cu$_{2-x}$Ag$_x$GeS$_3$** solid solutions ($x = 0.2, 0.4, 0.6, 1.6,$ and 1.8) were calculated from the EMF data. Using the obtained thermodynamic functions and the corresponding functions of Ag_2GeS_3, the thermodynamic functions of mixing during the formation of these solid solutions from ternary compounds were calculated. The alloys were homogenized at 530°C and 300°C for 500 h with the next slow cooling to room temperature.

Cu$_8$GeS$_6$–Ag$_8$GeS$_6$. The experimental data obtained through the DTA, XRD, and EMF method confirmed that a continuous series of solid solutions are formed between high-temperature modifications of starting compounds (Abbasova et al. 2017a; Alverdiev et al. 2017b). Below 230°C, there are two two-phase regions in this system and wide regions of solid solutions based on the initial compounds. Within the homogeneity regions, with a decrease in the copper content, its partial entropy monotonically increases, and the partial enthalpy and Gibbs free energy decrease. At the interfaces between the regions of solid solutions with the two-phase regions, abrupt changes in the partial entropy and enthalpy were observed.

The **(Cu$_{4.7}$Ag$_{3.3}$)GeS$_6$** quaternary compound (mineral putzite), which crystallizes in the cubic structure with the lattice parameter $a = 1012.50 \pm 0.12$ pm and a calculated density of 5.788 $g \cdot cm^{-3}$, is formed in the Ge–Cu–Ag–S system (Paar et al. 2004; Piilonen et al. 2005a).

4.92 Germanium–Copper–Magnesium–Sulfur

The **Cu₂MgGeS₄** quaternary compound, which crystallizes in the orthorhombic structure with the lattice parameters $a = 763.8 \pm 0.4$, $b = 651.5 \pm 0.4$, $c = 622.5 \pm 0.3$ pm, a calculated density of 3.776 g·cm⁻³, and an energy gap of 2.36 eV, is formed in the Ge–Cu–Mg–S system (Liu et al. 2013) [$a = 761.42$, $b = 645.80$, $c = 620.44$ pm and a calculated density of 3.8343 g·cm⁻³ (calculated values) (Bedjaoui et al. 2017)]. To prepare this compound, the stoichiometric mixture with an overall weight of about 300 mg of Cu, Mg, Ge, and S (molar ratio 2:1:1:4) was ground into fine powder in an agate mortar and pressed into pellets, then loaded into quartz tube and finally flame-sealed under vacuum about 10⁻² Pa. The sample was placed into a furnace, held at 300°C for 10 h, then heated to 650°C within 5 h and kept for 10 h, subsequently heated to 950°C in 10 h, kept for 5 days, and finally cooled down to 300°C over 8 days before switching off the furnace. Good-quality crystals of the title compound were obtained. A polycrystalline sample could be synthesized by a direct combination of the elements, and they are air stable.

4.93 Germanium–Copper–Strontium–Sulfur

The **Cu₂SrGeS₄** quaternary compound, which crystallizes in the trigonal structure with the lattice parameters $a = 614.3 \pm 0.2$, $c = 1528.2 \pm 0.6$ pm, the calculated, and experimental densities of 4.12 and 4.00 g·cm⁻³, respectively, and an energy gap of 2.8 eV [2.49–2.50 eV (Zhu et al. 2017)], is formed in the Ge–Cu–Sr–S system (Teske 1979b; Llanos et al. 2003). Polycrystalline Cu₂SrGeS₄ was prepared from a thoroughly ground mixture of SrS, Cu, Ge, and S in stoichiometric proportions (Llanos et al. 2003). The mixture was placed in an alumina boat and heated in a CS₂/Ar stream at 450°C for about 36 h, with two intermittent grindings. The title compound was obtained, which was citreous in color and stable in the air.

4.94 Germanium–Copper–Barium–Sulfur

Two quaternary compounds, **Cu₂BaGeS₄** and **Cu₆BaGe₂S₈**, are formed in the Ge–Cu–Ba–S system. First of them crystallizes in the trigonal structure with the lattice parameters $a = 620.92 \pm 0.08$, $c = 1552.0 \pm 0.4$ pm, a calculated density of 4.473 g·cm⁻³, and an energy gap of 2.47 eV [2.46–2.47 eV (Zhu et al. 2017] (Nian et al. 2017) [$a = 621.5$, $c = 1553.4$ pm, and the calculated and experimental densities of 4.43 and 4.32 g·cm⁻³, respectively (Teske 1979b)]. Cu₆BaGe₂S₈ crystallizes in the orthorhombic structure with the lattice parameters $a = 612.2 \pm 0.1$, $b = 1208.4 \pm 0.3$, and $c = 1761.4 \pm 0.5$ pm (Tampier and Johrendt 1998).

The crystals of Cu₂BaGeS₄ were synthesized from a mixture of BaS, Cu, Ge, and S (molar ratio 1:2:1:3.5) (Nian et al. 2017). First, the reaction mixture was loaded into a silica tube and flame-sealed under a high vacuum of (10⁻³ Pa). Second, the furnace was programmed by the following steps: heated from room temperature to 600°C in 30 h and kept at this temperature for 40 h; then heated to 1050°C in 20 h and left at this temperature for 100 h; finally, cooled to 400°C at a rate of 5°C·h⁻¹ and then the furnace was shut down to room temperature. Third, the obtained products were washed with DMF to remove the unreacted sulfur and other byproducts. Finally, yellow crystals of Cu₂BaGeS₄ were obtained after drying in the air. They are stable in the air for several months. Since Ba is easily oxidized in the air, an Ar-filled glove box was used to avoid the effects of oxygen and moisture in the preparation processes.

Cu₆BaGe₂S₈ was synthesized by direct reaction of the elements at 750°C in a corundum crucible under an Ar atmosphere (Tampier and Johrendt 1998).

4.95 Germanium–Copper–Zinc–Sulfur

Cu₂GeS₃–ZnS. The phase diagram of this system, constructed through DTA, XRD, and scanning electron microscope (SEM) using the alloys annealed at 400°C for 500 h, is shown in Figure 4.7 (Parasyuk et al. 2005e). The peritectic point is at 43 mol% ZnS, and the thermal effects at 810°C that could correspond to the phase transition of **Cu₂ZnGeS₄** from tetragonal to the orthorhombic structure were not observed. The eutectic point coordinates are 969°C and 6 mol% ZnS. The homogeneity range of the Cu₂ZnGeS₄ is narrow, and the solid solubility in the components is less than 2 mol%.

There is one intermediate quaternary phase Cu₂ZnGeS₄ in this system that melts incongruently at 1086°C [1120°C (Ichikawa et al. 2000; Matsushita and Katsui 2005; Matsushita et al. 2005); 1107°C ± 5°C (Schäfer and Nitsche 1977), decomposes at 620°C ± 10°C (Yao et al. 1987)], and crystallizes in the tetragonal structure with the lattice parameters $a = 534.127 \pm 0.009$ and $c = 1050.90 \pm 0.02$ pm [$a = 534.5 \pm 0.3$ and $c = 1052.1 \pm 0.6$ pm (Ottenburgs and Goethals 1972); $a = 534.1$ and $c = 1051.2$ pm (Schäfer and Nitsche 1977); $a = 542.9$ and $c = 1084.7$ pm (Das et al. 2013)], and a calculated density of 4.3561 ± 0.0002 g·cm⁻³ (Parasyuk et al. 2005e).

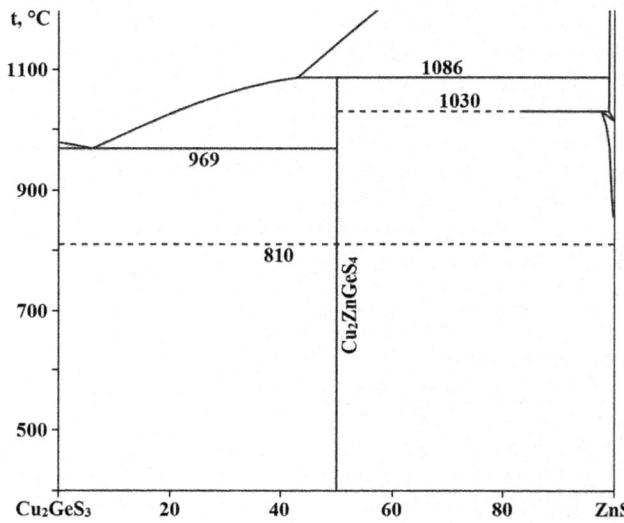

FIGURE 4.7 Phase diagram of the Cu₂GeS₃–ZnS system. (From Parasyuk, O.V. et al., *J. Alloys Compd.*, **397**(1–2), 85, 2005.)

This compound can also crystallizes in the orthorhombic structure with the lattice parameters a = 750.6 ± 0.3, b = 647.6 ± 0.4, c = 618.9 ± 0.2 pm, and an energy gap of 2.04 eV (Yao et al. 1987) [a = 747, b = 645, and c = 612 pm (Nitsche et al. 1967); a = 651 ± 1, b = 752 ± 1, c = 616 ± 1 pm, and an experimental density of 4.54 g·cm^{-3} (Ottenburgs and Goethals 1972); a = 750, b = 648, and c = 618 pm, the calculated and experimental densities of 4.35 and 4.37 g·cm^{-3}, respectively, and an energy gap of 2.1 eV (Schleich and Wold 1977); a = 750.4, b = 647.4, c = 618.5 pm, and the calculated and experimental densities of 4.35 and 4.34 g·cm^{-3}, respectively (Schäfer and Nitsche 1974, 1977, Guen et al. 1979, Guen and Glaunsinger 1980); a = 750.9 ± 0.1, b = 647.1 ± 0.1, c = 618.6 ± 0.1 pm, and an energy gap of 2.0 eV (Honig et al. 1988); a = 747–757, b = 645–647, c = 612–613 pm (Ichikawa et al. 2000); a = 757, b = 647, c = 613 pm (Matsushita and Katsui 2005); an energy gap of 2.2 eV (Tsuji et al. 2010)].

According to the data of Doverspike et al. (1990), Cu$_2$ZnGeS$_4$ crystallizes in the tetragonal structure below 790°C and in the orthorhombic structure above 790°C.

The structures of two polymorphs of Cu$_2$ZnGeS$_4$ have been determined by Moodie and Whitefield (1986). The first of them is a tetragonal polymorph with the lattice parameters a = 527 and c = 1054 pm. The second polymorph has a pseudorhombohedral structure and can be described in terms of a triply primitive cell with orthogonal axes a = 3660, b = 655, and c = 752 pm.

Because the energy between kesterite- and stannite-type structures is low, both structures can coexist in synthesized samples of Cu$_2$ZnGeS$_4$. According to the first-principles calculations, kesterite-type structure of this compound is characterized by the tetragonal structure with the lattice parameters a = 526.4 and c = 1084.3 pm [a = 535.8 and c = 1064.1 pm (Chen et al. 2010)], and an energy gap of 2.43 eV (Chen and Ravindra 2013), and stannite-type structure is also characterized by the tetragonal structure with the lattice parameters a = 532.8 and c = 1074.1 pm [a = 533.3 and c = 1074.1 pm (Chen et al. 2010)], and an energy gap of 2.14 eV (Chen and Ravindra 2013). According to the calculation of Chen et al. (2010), wurtzite-derived polytypes of kesterite (orthorhombic structure with the lattice parameters a = 754.4, b = 651.9, and c = 622.6 pm) and stannite (orthorhombic structure with the lattice parameters a = 750.3, b = 654.7, and c = 622.6 pm) could exist for Cu$_2$ZnGeS$_4$ compound.

Cu$_2$ZnGeS$_4$ was obtained by the chemical vapor reactions using I$_2$ as the transport agent (Nitsche et al. 1967; Schäfer and Nitsche 1974; Schleich and Wold 1977; Guen et al. 1979; Guen and Glaunsinger 1980; Yao et al. 1987; Matsushita et al. 2005) or via vertical or horizontal gradient freezing method (Ichikawa et al. 2000; Matsushita et al. 2005; Das et al. 2013). The powder of this compound was synthesized by solid-state reaction: the starting materials ZnS, Cu$_2$S, and GeS$_2$ were mixed with a 15% excess amount of ZnS and GeS$_2$ (Tsuji et al. 2010). The mixture was sealed in a quartz ampoule in a vacuum and heat-treated at 550–650°C for 10 h. Cu$_2$ZnGeS$_4$ synthesized at 650°C was the low-temperature phase of the stannite type. A wurtzite-type high-temperature phase was not confirmed.

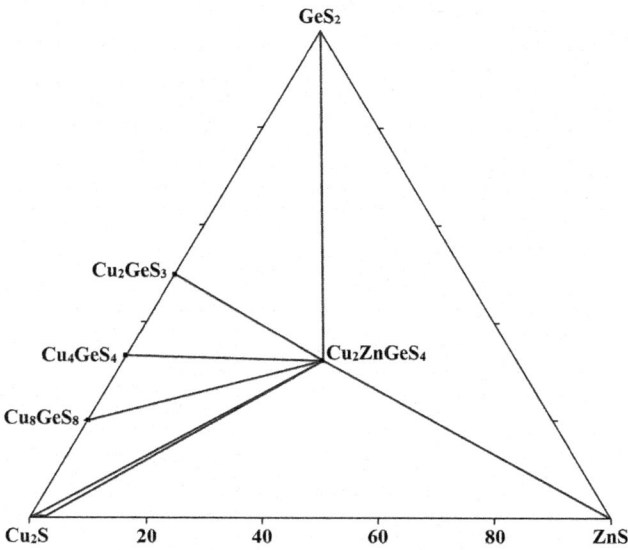

FIGURE 4.8 Isothermal section of the GeS$_2$–Cu$_2$S–ZnS quasiternary system at 400°C. (From Parasyuk, O.V. et al., *J. Alloys Compd.*, **397**(1–2), 85, 2005.)

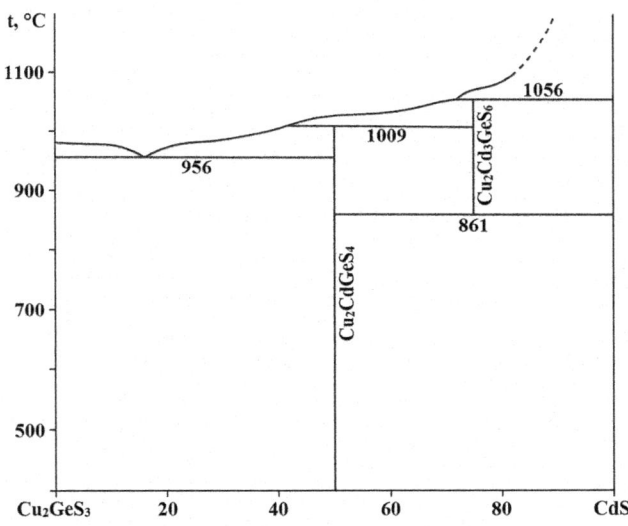

FIGURE 4.9 Phase diagram of the Cu$_2$GeS$_3$–CdS system. (From Piskach, L.V. et al., *J. Alloys Compd.*, **299**(1–2), 227, 2000.)

Single crystals of the orthorhombic Cu$_2$ZnGeS$_4$ could be synthesized at 810°C in a vacuum-sealed silica tube, either from synthetic tetragonal Cu$_2$ZnGeS$_4$, or either from pure elements (Ottenburgs and Goethals 1972).

GeS$_2$–Cu$_2$S–ZnS. The isothermal section of this system at 400°C is shown in Figure 4.8 (Parasyuk et al. 2005e).

4.96 Germanium–Copper–Cadmium–Sulfur

Cu$_2$GeS$_3$–CdS. This system is a quasibinary one containing two quaternary compounds: **Cu$_2$CdGeS$_4$** and **Cu$_2$Cd$_3$GeS$_6$** (Figure 4.9) (Piskach et al. 2000). The first one forms according to a peritectic reaction at 1009°C [at 1020°C (Ichikawa et al. 2000; Matsushita and Katsui 2005; Matsushita et al. 2005); at 957°C (Piskach et al. 1997); melts congruently at 1021°C (Schäfer and Nitsche 1977)] and possesses a narrow homogeneity region. A eutectic exists between Cu_2GeS_3 and Cu_2CdGeS_4 at 16 mol% CdS and 956°C (Piskach et al. 2000) [14 mol% CdS and 767°C (Piskach et al. 1997)].

Cu_2CdGeS_4 crystallizes in the orthorhombic structure with the lattice parameters $a = 770.24 \pm 0.03$, $b = 654.86 \pm 0.02$, $c = 629.28 \pm 0.03$ pm, and a calculated density of 4.6067 ± 0.0006 g·cm⁻³ (Parasyuk et al. 2005g) [$a = 772$, $b = 657$, and $c = 629$ pm (Nitsche et al. 1967); $a = 769.2 \pm 0.2$, $b = 655.5 \pm 0.2$, and $c = 629.9 \pm 0.2$ pm (Parthé et al. 1969, Schäfer and Nitsche 1974, 1977); $a = 770.5 \pm 0.8$, $b = 655.8 \pm 0.6$, and $c = 630.8 \pm 0.3$ pm (Filonenko et al. 1991); $a = 770$–772, $b = 655$–657, and $c = 628$–629 pm (Ichikawa et al. 2000); $a = 770.3 \pm 0.1$, $b = 655.49 \pm 0.09$, and $c = 631.2 \pm 0.1$ pm (Piskach et al. 2000); $a = 770$, $b = 655$, and $c = 628$ pm (Matsushita and Katsui 2005; Matsushita et al. 2005)]. A calculated density of this compound is 4.6 g·cm⁻³ (Parthé et al. 1969) and an energy gap is 2.05 eV (Davydyuk et al. 2002, 2005) [1.9–2.0 eV (Filonenko et al. 1991)].

According to the preliminary data of Piskach et al. (1997), the CdS–Cu_2GeS_3 system was considered as nonquasibinary due to the incongruent melting of Cu_2GeS_3 [later it was shown (Piskach et al. 2000) that this ternary compound melts congruently].

The composition of the second quaternary compound was estimated from the concentration-dependent size of the endothermal effects and corresponds to 75 mol% CdS ($Cu_2Cd_3GeS_6$) (Piskach et al. 2000). This phase is formed at 1056°C according to the peritectic reaction and is unstable and decomposes at 861°C. The solid solubility ranges of the components of this section do not exceed 2 mol%.

The investigation of this system was carried out using DTA, XRD, and metallography. All alloys were annealed at 400°C for 250 h, and then they were quenched in cold water (Piskach et al. 1997, 2000). Single crystals of the Cu_2CdGeS_4 quaternary compound were grown through chemical vapor transport reactions using I_2 as the transport agent and gradient freezing method (Nitsche et al. 1967; Parthé et al. 1969; Schäfer and Nitsche 1974; Yao et al. 1987; Filonenko et al. 1991; Ichikawa et al. 2000; Davydyuk et al. 2002, 2005; Matsushita et al. 2005). Single crystals could be also grown by the vertical Bridgman method with the next annealing for 100 h at 600°C and slowly cooling to room temperature (Parasyuk et al. 2005g).

GeS_2–Cu_2S–CdS. The isothermal section of this system at 400°C is shown in Figure 4.10 (Parasyuk et al. 2005e). Two quaternary compounds Cu_2CdGeS_4 and $\mathbf{Cu_6Cd_{0.83}Ge_{1.17}S_{6.57}}$ are formed in this system. The alloys for investigations were annealed at 400°C during 500 h.

4.97 Germanium–Copper–Mercury–Sulfur

$CuGeS_3$–HgS. The phase diagram of this system is shown in Figure 4.11 (Parasyuk et al. 2002d). Two quaternary compounds exist in this system. $\mathbf{Cu_2HgGeS_4}$ melts congruently at 936°C [at 907°C $\pm$ 5°C (Schäfer and Nitsche 1977)] and gives rise to a polymorphous transformation between 650°C

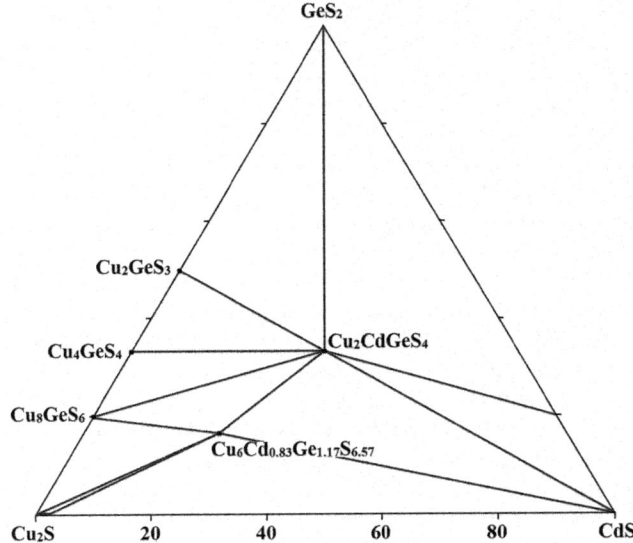

FIGURE 4.10 Isothermal section of the GeS_2–Cu_2S–CdS quasiternary system at 400°C. (From Parasyuk, O.V. et al., *J. Alloys Compd.*, **397**(1–2), 85, 2005.)

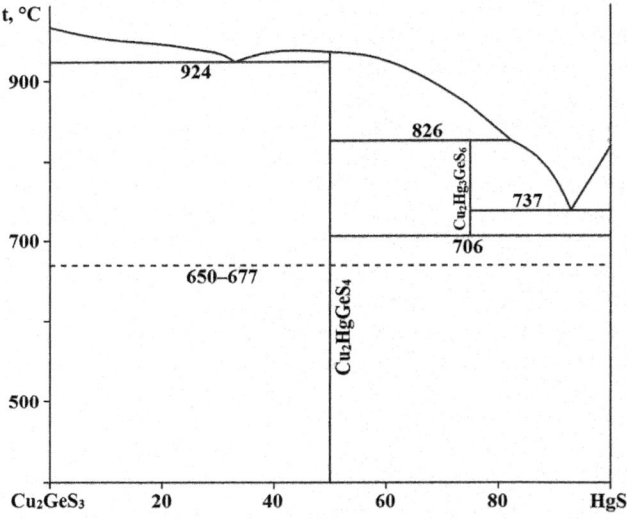

FIGURE 4.11 Phase diagram of Cu_2GeS_3–HgS system. (From Parasyuk, O.V. et al., *J. Alloys Compd.*, **334**(1–2), 143, 2002.)

and 677°C. The eutectic with a melting temperature of 924°C and a composition of 33 mol% HgS forms between Cu_2GeS_3 and Cu_2HgGeS_4 (Parasyuk et al. 2002d). The second eutectic is located between $\mathbf{Cu_2Hg_3GeS_6}$ and HgS. It melts at 737°C and corresponds to the composition with 93 mol% HgS. The solid solubility ranges of the initial compounds are negligible.

The low-temperature modification of Cu_2HgGeS_4 crystallizes in the tetragonal structure with the lattice parameters $a = 548.73 \pm 0.04$ and $c = 1054.23 \pm 0.02$ pm [$a = 549.0$ and $c = 1055.0$ pm (Schäfer and Nitsche 1974, 1977)] and a calculated density of 5.5339 ± 0.0006 g·cm⁻³ (Parasyuk et al. 2002d).

The high-temperature modification of Cu_2HgGeS_4 crystallizes in the orthorhombic structure with the lattice parameters

$a = 768.222 \pm 0.004$, $b = 655.608 \pm 0.003$, $c = 631.435 \pm 0.004$ pm, and an energy gap of 1.6 eV (Vu et al. 2022) [$a = 767.9$, $b = 652.2$, and $c = 632.5$ pm (Schäfer and Nitsche 1977); $a = 768.11 \pm 0.04$, $b = 655.46 \pm 0.02$, and $c = 631.44 \pm 0.04$ pm (Parasyuk et al. 2002d); $a = 768.22 \pm 0.03$, $b = 655.61 \pm 0.02$, $c = 631.44 \pm 0.03$ pm, and a calculated density 5.5185 g·cm^{-3} (Olekseyuk et al. 2005a)]. Orthorhombic Cu$_2$HgGeS$_4$ was derived by alloying a stoichiometric composition of Cu, Ge, S, and preliminary synthesized HgS (Vu et al. 2022). The alloying was performed in an evacuated quartz ampoule (residual pressure was less than 0.01 Pa). The synthesis was carried out in a furnace with a digital control system of the technological processes. In the first stage, the mixture was heated to 940°C at a rate of 24°C·h^{-1} and, then, on the second stage, it was annealed at 940°C over 2 h, followed by cooling to 750°C at a rate of 12°C·h^{-1}. Finally, the mixture was annealed at 750°C (over 10 h), followed by quenching the ampoule to room temperature water.

Cu$_2$Hg$_3$GeS$_6$ melts incongruently at 826°C and decomposes according to a eutectoid reaction at 706°C (Parasyuk et al. 2002d).

This system was investigated through DTA, XRD, and metallography, and the annealing of the ingots was carried out at 400°C for 250 h (Parasyuk et al. 2002d). After annealing, the ampoules with the samples were quenched in the air. The growing of the Cu$_2$HgGeS$_4$ single crystals was carried out by chemical vapor transport reactions and I$_2$ as a transport agent (Schäfer and Nitsche 1974; Parasyuk et al. 2002d).

GeS$_2$–Cu$_2$S–HgS. The isothermal section of this quasiternary system at 400°C is shown in Figure 4.12 (Marchuk et al. 2002). Equilibria exist between the quaternary compound Cu$_2$HgGeS$_4$ and the binary components HgS and GeS$_2$ and the ternary Cu$_2$GeS$_3$, Cu$_8$GeS$_6$, Hg$_4$GeS$_6$, and Cu$_4$GeS$_4$ compounds and the quaternary phase Cu$_6$Hg$_{0.92}$GeS$_{5.92}$, which crystallizes in the cubic structure with the lattice parameter $a = 999.88 \pm$

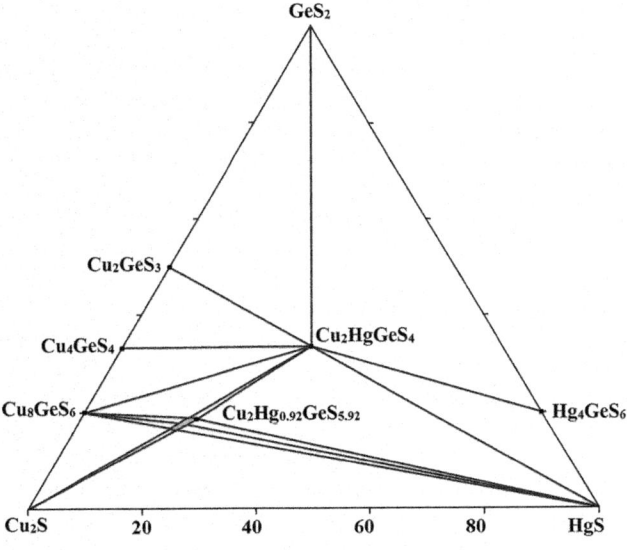

FIGURE 4.12 Isothermal section of the GeS$_2$–Cu$_2$S–HgS quasiternary system at 400°C. (From Marchuk, O.V. et al., *J. Alloys Compd.*, **333**(1–2), 143, 2002.)

0.01 pm and a calculated density of 5.494 ± 0.001 g·cm^{-3}. This compound has a homogeneity region within the interval of 63–66 mol% Cu$_2$S with changing of the lattice parameter from 1054.1 to 1048.5 pm. The ingots for the investigations were annealed at 400°C for 250 h.

4.98 Germanium–Copper–Boron–Sulfur

The **CuBGeS$_4$** quaternary compound, which crystallizes in the tetragonal structure with the lattice parameters $a = 534.3$, $c = 521.7$ pm, and the calculated and experimental densities of 3.07 and 2.90 g·cm^{-3}, respectively, is formed in the Ge–Cu–B–S system (Hahn and Strick 1967b). This compound was obtained by long annealing of the elements' mixture or the mixture of binary components between 500°C and 1000°C. It has a gray-black color.

4.99 Germanium–Copper–Aluminum–Sulfur

Two quaternary compounds, **CuAlGeS$_4$** and **Cu(AlS$_2$)(GeS$_2$)$_4$**, are formed in the Ge–Cu–Al–S system. First of them crystallizes in the tetragonal structure with the lattice parameters $a = 532.0$, $c = 1005$ pm, and the calculated and experimental densities of 3.40 and 3.35 g·cm^{-3}, respectively (Hahn and Strick 1967b). Cu(AlS$_2$)(GeS$_2$)$_4$ is thermally stable up to 650°C and crystallizes in the monoclinic structure with the lattice parameters $a = 679.6 \pm 0.1$, $b = 3762.8 \pm 0.8$, $c = 687.9 \pm 0.1$ pm, $\beta = 119.52 \pm 0.03°$, a calculated density of 3.044 g·cm^{-3} at 150 K, and an energy gap of 2.1 eV (Alahmari et al. 2018).

CuAlGeS$_4$ was obtained by long annealing of the elements mixture or the mixture of binary components between 500°C and 1000°C. It has a brown color (Hahn and Strick 1967b).

Cu(AlS$_2$)(GeS$_2$)$_4$ was prepared by the ion-exchange reaction of Na(AlS$_2$)(GeS$_2$)$_4$ with CH$_3$COOCu (molar ratio 1:1), which were carried out at room temperature in 20 mL glass vials (Alahmari et al. 2018). For this, Na(AlS$_2$)(GeS$_2$)$_4$ (50 mg) was added to a solution of 9.3 mg of CH$_3$COOCu in extra dry DMF (3 mL). The mixture was slowly stirred for 3 days to ensure the completeness of the ion-exchange reaction. Homogeneous ion-exchange product was isolated in the form of light brown block-shaped crystals after the reaction products were washed with DMF and dried with diethyl ether. This compound can also be obtained from acetonitrile or ethanol solutions using CH$_3$COOCu or CuBr. It is air-stable. All sample preparations and handling were carried out in a N$_2$-filled glove box.

4.100 Germanium–Copper–Gallium–Sulfur

The **CuGaGeS$_4$** quaternary compound, which melts at ~1000°C (Matsushita and Katsui 2005) and crystallizes in the tetragonal structure with the lattice parameters $a = 533.4$, $c = 1005$ pm, and the calculated and experimental densities of 3.88 and 3.83 g·cm^{-3}, respectively, are formed in the Ge–Cu–Ga–S system (Hahn and Strick 1967b). This compound was obtained

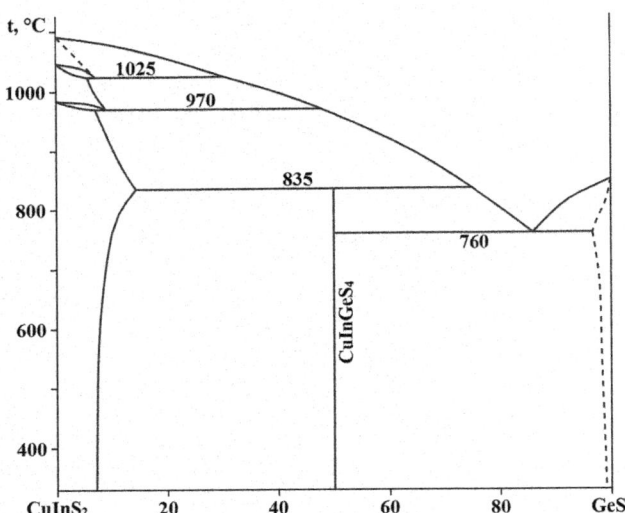

FIGURE 4.13 Phase diagram of the GeS_2–$CuInS_2$ system. (From Gorgut, G.P., et al., *J. Cryst. Growth*, **324**(1), 212, 2011.)

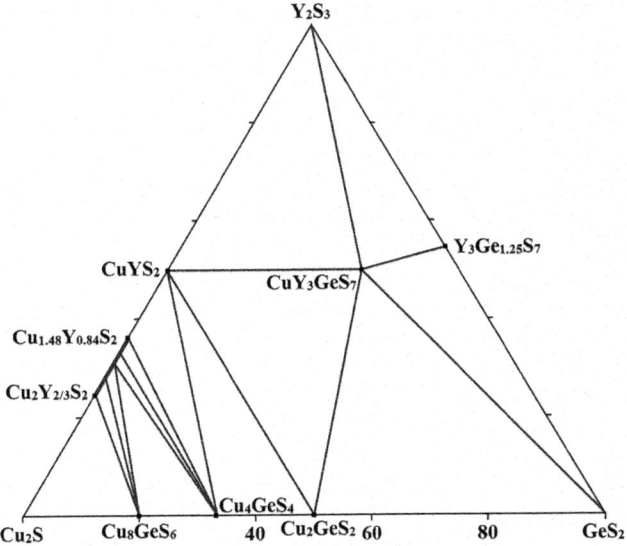

FIGURE 4.14 Isothermal section of the GeS_2–Cu_2S–Y_2S_3 quasiternary system at 600°C. (From Gulay, L.D., et al., *J. Alloys Compd.*, **414**(1–2), 113, 2006.)

by long annealing of the elements' mixture or the mixture of binary components between 500°C and 1000°C. It has an orange-black color.

4.101 Germanium–Copper–Indium–Sulfur

GeS_2–$CuInS_2$. The phase diagram of this system, constructed through DTA and XRD using the alloys annealed at 400°C for 250 h with the next quenching in cold water, is given in Figure 4.13 (Gorgut et al. 2011). The existence of polymorphous transformations of $CuInS_2$ complicates the interaction with invariant catatectic processes. The catatectic point coordinates are 7 mol% GeS_2 at 1025°C and 9 mol% GeS_2 at 970°C, respectively. The homogeneity region of α-$CuInS_2$ reaches up to 4 mol% GeS_2 at 400°C. The eutectic contains 14 mol% $CuInS_2$ and crystallizes at 760°C. According to the data of Pamplin et al. (1977), the solubility of GeS_2 in $CuInS_2$ is less than 25 mol%.

The **CuInGeS₄** quaternary compound, which melts incongruently at 835°C and crystallizes in the tetragonal structure with the lattice parameters $a = 554.92 \pm 0.02$, $c = 1002.82 \pm 0.06$ pm, and a calculated density of 4.0777 ± 0.0006 g·cm⁻³ (Gorgut et al. 2011) [$a = 553.9$, $c = 1004$ pm, and the calculated and experimental densities of 4.09 and 4.02 g·cm⁻³, respectively (Hahn and Strick 1967b)], is formed in this system. According to the data of Matsushita and Katsui (2005), this compound melts at 914°C and crystallizes in the cubic structure with the lattice parameter $a = 1055$ pm.

The synthesis of $CuInGeS_4$ was performed in a furnace by the single-temperature method realized by the gradual heating of the sample (heating rate of 10°C·h⁻¹) to 1100°C with an intermediate exposure to 550°C for 36 h (Gorgut et al. 2011). The annealing was performed at 400°C for 250 h. The synthesis process was finished after quenching the sample in cold water.

This compound was also obtained by long annealing of the elements' mixture or the mixture of binary components between 500°C and 1000°C (Hahn and Strick 1967b). It has a brown-green color.

4.102 Germanium–Copper–Yttrium–Sulfur

GeS_2–Cu_2S–Y_2S_3. The isothermal section of this system at 600°C is presented in Figure 4.14 (Gulay et al. 2006c). The formation of the **CuY₃GeS₇** quaternary compound, which crystallizes in the hexagonal structure with the lattice parameters $a = 983.5 \pm 0.1$, $c = 576.5 \pm 0.1$ pm, and a calculated density of 4.314 g·cm⁻³ [an energy gap is ~2.12 eV (Akopov et al. 2022a)], was established in this system. The samples for the investigations were prepared by melting a mixture of the elements in evacuated silica ampoules. The synthesis was realized in a tube furnace. The ampoules were heated with a heating rate of 30°C·h⁻¹ to the maximal temperature, 1150°C. The samples were kept at this temperature for 4 h. After that, they were cooled slowly (10°C·h⁻¹) to 600°C and annealed for 240 h. After annealing, the ampoule with the sample was quenched in cold water.

CuY_3GeS_7 was synthesized as follows (Akopov et al. 2022a). First, a precursor with an overall composition of Y_3CuGe was prepared using pieces of Y, Cu, and Ge. A sample with a total mass of 0.6–1.0 g was weighed out in a ratio of Y/Cu/Ge = 3:1:1.05 and arc-melted. The bulk powder of the title compound was synthesized using the arc-melted precursor and sulfur powder. The sample was prepared by loading the finely ground precursor and sulfur (molar ratio 1:7) in a fused silica ampoule, which was then flame-sealed under vacuum. The sample was heated over 12 h to 1050°C, dwelled for 72–120 h, and then cooled to room temperature over 8 h.

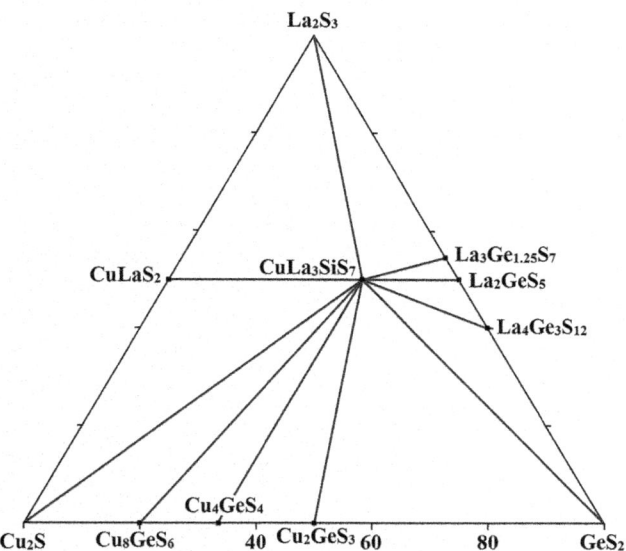

FIGURE 4.15 Isothermal section of the GeS_2–Cu_2S–La_2S_3 quasiternary system at 600°C. (From Olekseyuk, I.D., et al., *Nauk. Visnyk Volyns'k. Derzh. Univ. im. Lesi Ukrainky. Ser. Khim. nauky*, (29), 13, 2009.)

FIGURE 4.16 Isothermal section of the GeS_2–Cu_2S–Pr_2S_3 quasiternary system at 600°C. (From Lychmanyuk, O.S., et al., *Nauk. Visnyk Volyns'k. Derzh. Univ. im. Lesi Ukrainky. Ser. Khim. nauky*, (15), 10, 2007.)

4.103 Germanium–Copper–Lanthanum–Sulfur

GeS_2–Cu_2S–La_2S_3. The isothermal section of this system at 600°C is presented in Figure 4.15 (Olekseyuk et al. 2009e). The existence of seven ternary and one quaternary compound, **$CuLa_3GeS_7$**, in this system has been confirmed. This section is characterized by a small solubility based on all compounds, and $CuLa_3GeS_7$ forms equilibria with all binary and ternary compounds. Ten fields of three-phase equilibria exist in the system.

$CuLa_3GeS_7$ crystallizes in the hexagonal structure with the lattice parameters $a = 1029.4 \pm 0.2$, $c = 586.2 \pm 0.1$ pm, and an energy gap of ~2 eV (Poduska et al. 2002b) [$a = 1030 \pm 1$ and $c = 583.0 \pm 0.5$ pm (Guittard et al. 1968; Guittard and Julien-Pouzol 1970; Collin and Flahaut 1972); an energy gap of ~2.57 eV (Akopov et al. 2022a)].

A diffraction-quality crystal of this compound was synthesized unintentionally in a reaction between Bi_2Te_3, Pb_2GeS_4, and CuI powders (Poduska et al. 2002b). During manipulations of the reactants in an Ar-filled glove box in which lanthanide elements are frequently used, a small amount of La was also incorporated into the sample. A cold-pressed pellet of this mixture was sealed in an evacuated silica tube and heated at 800°C for 48 h with the next cooling over 12 h. The resulting ingot was silver-colored, with a few orange and red pieces. While most of the pieces had unit cells that matched that of monoclinic $Pb_5S_2I_6$, two crystals (which appeared dark brown in transmitted light) yielded a hexagonal unit cell, and these proved to have the $CuLa_3GeS_7$ structure. This compound is a small minority phase.

$CuLa_3GeS_7$ can also be prepared from a mixture of La_2S_3, Cu_2S, Ge, and S, which was pressed and introduced into the silica ampoule and gradually heats up to 800–850°C in about a week (Guittard and Julien-Pouzol 1970; Collin and Flahaut

1972). At the end of the preparation, it is necessary to heat the sample to 1200°C to obtain the pure hexagonal phase.

This compound can also be obtained in the same way as CuY_3GeS_7 was prepared (Akopov et al. 2022a).

4.104 Germanium–Copper–Cerium–Sulfur

The **$CuCe_3GeS_7$** quaternary compound, which crystallizes in the hexagonal structure with the lattice parameters $a = 1022.5 \pm 0.1$, $c = 583.50 \pm 0.07$ pm, and a calculated density of 4.909 $g \cdot cm^{-3}$ (Gulay et al. 2006d) [$a = 1021$ and $c = 580$ pm (Guittard and Julien-Pouzol 1970)], is formed in the Ge–Cu–Ce–S system. This compound was prepared by fusion of the elements in an evacuated silica ampoule in a tube furnace. The ampoule was heated at a rate of $30°C \cdot h^{-1}$ to the maximal temperature, 1150°C, kept at this temperature for 4 h, cooled slowly ($10°C \cdot h^{-1}$) down to 600°C, and annealed at this temperature for 240 h. After annealing, the ampoule was quenched in cold water.

4.105 Germanium–Copper–Praseodymium–Sulfur

GeS_2–Cu_2S–Pr_2S_3. The isothermal section of this system at 600°C was constructed by Lychmanyuk et al. (2007a) using the alloys annealed at this temperature for 240 h and quenched in the air and is presented in Figure 4.16. The **$CuPr_3GeS_7$** quaternary compound, which crystallizes in the hexagonal structure with the lattice parameters $a = 1016.8 \pm 0.1$, $c = 582.23 \pm 0.07$ pm, and a calculated density of 4.990 $g \cdot cm^{-3}$ (Gulay et al. 2006d) [$a = 1016$ and $c = 578$ pm (Guittard and Julien-Pouzol 1970)], is formed in this system. This compound was prepared in the same way as $CuCe_3GeS_7$ was synthesized.

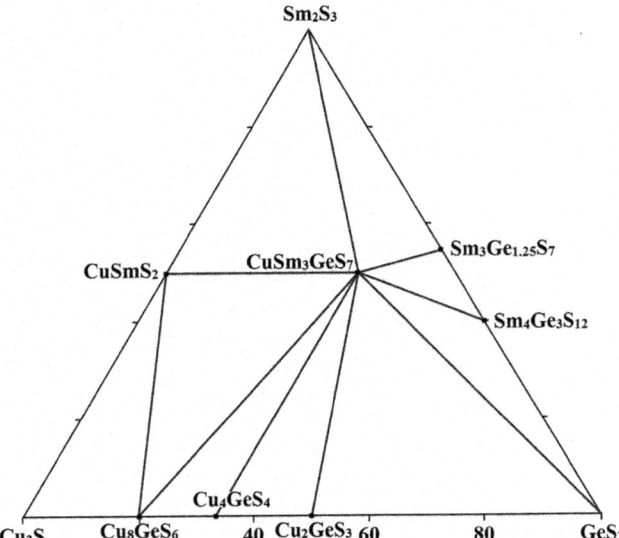

FIGURE 4.17 Isothermal section of the GeS_2–Cu_2S–Sm_2S_3 quasiternary system at 600°C. (From *Olekseyuk, I.D., et al., Nauk. Visnyk Volyns'k. Nats. Univ. im. Lesi Ukrainky. Ser. Khim. nauky*, (16), 54, 2010.)

4.106 Germanium–Copper–Neodymium–Sulfur

The **CuNd$_3$GeS$_7$** quaternary compound, which crystallizes in the hexagonal structure with the lattice parameters $a = 1012.3 \pm 0.1$, $c = 579.42 \pm 0.07$ pm, and a calculated density of 5.123 g·cm^{-3} (Gulay et al. 2006d) [$a = 1010$ and $c = 577$ pm (Guittard and Julien-Pouzol 1970)], is formed in the Ge–Cu–Nd–S system. This compound was prepared in the same way as CuCe$_3$GeS$_7$ was synthesized.

4.107 Germanium–Copper–Samarium–Sulfur

GeS$_2$–Cu$_2$S–Sm$_2$S$_3$. The isothermal section of this system at 600°C is given in Figure 4.17 (Olekseyuk et al. 2010c). The existence of six ternary and one quaternary, **CuSm$_3$GeS$_7$**, compounds has been confirmed in this system. CuSm$_3$GeS$_7$ crystallizes in the hexagonal structure with the lattice parameters $a = 1001.06 \pm 0.07$, $c = 577.02 \pm 0.06$ pm, and a calculated density of 5.382 g·cm^{-3} (Strok et al. 2010) [$a = 1001$ $c = 557$ pm (Guittard and Julien-Pouzol 1970); $a = 1001.44 \pm 0.01$, $c = 577.14 \pm 0.01$ pm, and a calculated density of 5.3768 g·cm^{-3} (Gulay et al. 2006d); an energy gap of ~2.29 eV (Akopov et al. 2022a)]. The existence of the Cu$_2$Sm$_{2/3}$S$_2$ was not confirmed. CuSm$_3$GeS$_7$ was prepared in the same way as CuCe$_3$GeS$_7$ was synthesized (Strok et al. 2010; Gulay et al. 2006d; Akopov et al. 2022a).

4.108 Germanium–Copper–Europium–Sulfur

The **Cu$_2$EuGeS$_4$** quaternary compound, which crystallizes in the trigonal structure with the lattice parameters $a = 611.89 \pm 0.03$, $c = 1525.82 \pm 0.09$ pm, a calculated density of 4.832 g·cm^{-3}, and an energy gap of 2.32 eV (Sun et al. 2019) [$a = $

610.2 ± 0.3, $c = 1522 \pm 8$ pm, and an energy gap of 1.6 eV (Llanos et al. 2003)], is formed in the Ge–Cu–Eu–S system.

Single crystals of Cu$_2$EuGeS$_4$ were obtained by high-temperature solid-state reactions with KI as a flux (Sun et al. 2019). The reagents were weighted by stoichiometry, and the total reagent mass was 500 mg with additional 400 mg KI. Namely, 123.46 mg Cu, 170.94 mg Eu$_2$O$_3$, 124.58 mg S, 70.52 mg Ge, and 10.50 mg B were used to obtain the title compound. The mixture was ground into a fine powder using an agate mortar and pressed into pellets. Then pellets were loaded into a quartz tube evacuated to 10^{-2} Pa and sealed using O$_2$/H$_2$ flame. The tube was put into a furnace and heated according to a profile as follows. It was heated from room temperature to 300°C by 5 h and kept for 5 h, then heated to 600°C by 5 h and kept for 5 h, then heated to 800°C by 5 h and kept for 90 h, and finally cooled down slowly to 300°C after 5 days and shut down the furnace. Yellow block crystals of Cu$_2$EuGeS$_4$, which are air and water stable, were obtained. They were washed using hot distilled water and ethanol several times under ultrasonic wave irradiation to eliminate KI and some impurities.

Polycrystalline Cu$_2$EuGeS$_4$ was prepared from a thoroughly ground mixture of EuS, Cu, Ge, and S in stoichiometric proportions (Llanos et al. 2003). The mixture was placed in an alumina boat and heated in a CS$_2$/Ar stream at 450°C for about 36 h, with two intermittent grindings. The title compound was obtained, green in color and stable in the air.

4.109 Germanium–Copper–Gadolinium–Sulfur

The **CuGd$_3$GeS$_7$** quaternary compound, which crystallizes in the hexagonal structure with the lattice parameters $a = 994.28 \pm 0.01$, $c = 575.92 \pm 0.01$ pm, and a calculated density of 5.6055 g·cm^{-3} (Gulay et al. 2006d) [$a = 995$ and $c = 577$ pm (Guittard and Julien-Pouzol 1970); an energy gap of ~2.67 eV (Akopov et al. 2022a)], is formed in the Ge–Cu–Gd–S system. This compound was prepared in the same way as CuCe$_3$GeS$_7$ was synthesized (Gulay et al. 2006d; Akopov et al. 2022a).

4.110 Germanium–Copper–Terbium–Sulfur

The **CuTb$_3$GeS$_7$** quaternary compound, which crystallizes in the hexagonal structure with the lattice parameters $a = 988.63 \pm 0.01$, $c = 575.35 \pm 0.01$ pm, and a calculated density of 5.7095 g·cm^{-3} (Gulay et al. 2006d) [$a = 992$ and $c = 576$ pm (Guittard and Julien-Pouzol 1970)], is formed in the Ge–Cu–Tb–S system. This compound was prepared in the same way as CuCe$_3$GeS$_7$ was synthesized.

4.111 Germanium–Copper–Dysprosium–Sulfur

The **CuDy$_3$GeS$_7$** quaternary compound, which crystallizes in the hexagonal structure with the lattice parameters $a = 983.71 \pm 0.03$, $c = 575.52 \pm 0.03$ pm, and a calculated density of 5.8390 g·cm^{-3} (Gulay et al. 2006d) [$a = 987$ and $c = 576$ pm (Guittard and Julien-Pouzol 1970)], is formed in the

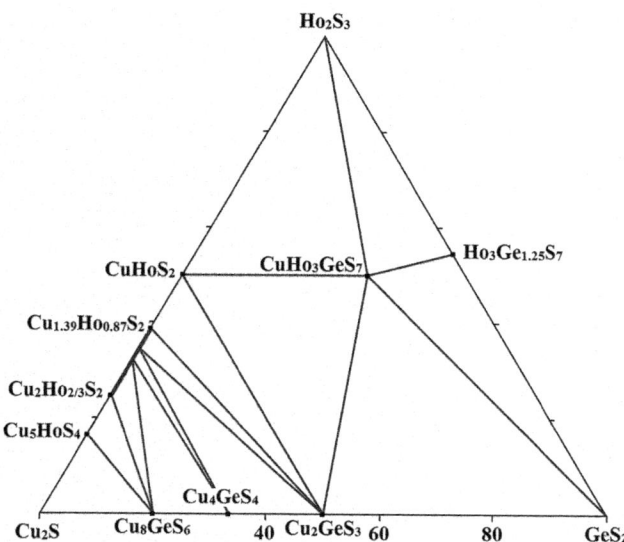

FIGURE 4.18 Isothermal section of the GeS₂–Cu₂S–Ho₂S₃ quasiternary system at 600°C. (From Lychmanyuk, O.S., et al., *Pol. J. Chem.*, **81**(3), 353, 2007.)

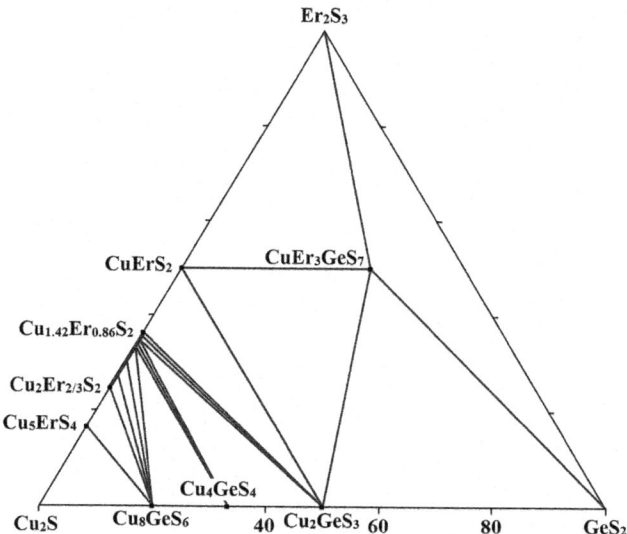

FIGURE 4.19 Isothermal section of the GeS₂–Cu₂S–Er₂S₃ quasiternary system at 600°C. (From Gulay, L.D., and Lychmanyuk, O.S., *Nauk. Visnyk Volyns'k. Nats. Univ. im. Lesi Ukrainky. Ser. Khim. nauky*, (29), 29, 2009.)

Ge–Cu–Dy–S system. This compound was prepared in the same way as $CuCe_3GeS_7$ was synthesized.

4.112 Germanium–Copper–Holmium–Sulfur

GeS_2–Cu_2S–Ho_2S_3. The isothermal section of this system at 600°C was constructed by Lychmanyuk et al. (2007b) using the alloys annealed at this temperature for 240 h and quenched in cold water and is presented in Figure 4.18. The **$CuHo_3GeS_7$** quaternary compound is formed in this system. It crystallizes in the hexagonal structure with the lattice parameters $a = 978.65 \pm 0.02$ and $c = 576.06 \pm 0.02$ pm.

4.113 Germanium–Copper–Erbium–Sulfur

GeS_2–Cu_2S–Er_2S_3. The isothermal section of this system at 600°C was constructed by Gulay and Lychmanyuk (2009) using the alloys annealed at this temperature for 240 h and quenched in the air, and is shown in Figure 4.19. The **$CuEr_3GeS_7$** quaternary compound is formed in this system. It crystallizes in the hexagonal structure with the lattice parameters $a = 974.15 \pm 0.01$, $c = 578.35 \pm 0.01$ pm, and a calculated density of 6.0248 g·cm⁻³ (Gulay et al. 2006d) [$a = 979$ and $c = 574$ pm (Guittard and Julien-Pouzol 1970)]. This compound was prepared in the same way as $CuCe_3GeS_7$ was synthesized.

4.114 Germanium–Copper–Tin–Sulfur

Cu_2GeS_3–Cu_2SnS_3. The phase diagram of this system, constructed through DTA, XRD, and metallography using the samples annealing at 400°C for 500 h, belongs to the type III according to the Roseboom's classification (Marushko and

Piskach 2006). Continuous solid solutions are formed in the system. There is a linear increase of the lattice parameters with increasing of the Cu_2SnS_3 content. $Cu_2Ge_{0.17}Sn_{0.83}S_3$ solid solution has been produced through powder technology using mixtures of elemental powders by combining mechanical alloying and spark plasma sintering (Neves et al. 2016). It has been found that solid solution is synthesized directly during the shot mechanical alloying step. It is quite stable upon spark plasma sintering at 600°C under a pressure of 50 MPa. The band gap energy was estimated to be in the range of 1.24 for Cu_2SnS_3 to 1.34 for $Cu_2Ge_{0.17}Sn_{0.83}S_3$.

Cu_4GeS_4–Cu_4SnS_4. $Cu_4Ge_xSn_{1-x}S_4$ solid solution up to $x = 0.30$ is formed in this system (Anzai and Fukazawa 1986). Its orthorhombic lattice parameters a, b, and c decrease with an increase in x, while the phase transition temperature increases. Later this system was studied in all concentration regions and it was shown that at room temperature, the orthorhombic structure changes to a monoclinic one at a transition concentration of $x = 0.6$ as x increases (Kamimura et al. 1990), and the transition temperature linearly increases with x.

To prepare $Cu_4Ge_xSn_{1-x}S_4$, the appropriate amounts of Cu, Sn, Ge, and S were sealed in an evacuated quartz tube, and the mixture was heated at 700°C for a week or more (Anzai and Fukazawa 1986; Kamimura et al. 1990).

4.115 Germanium–Copper–Lead–Sulfur

Two quaternary compounds, **$Cu_{0.5}Pb_{1.75}GeS_4$** and **Cu_2PbGeS_4**, are formed in the Ge–Cu–Pb–S system. The first of them crystallizes in the cubic structure with the lattice parameter $a = 1381.45 \pm 0.09$ pm and an energy gap of 1.48 eV (Iyer et al. 2004). It was prepared by a stoichiometric combination of the elements. The reactants were loaded in a fused silica tube and flame-sealed under vacuum (< 0.1 Pa). The tube was then put in a furnace, and the temperature was ramped to 650°C and

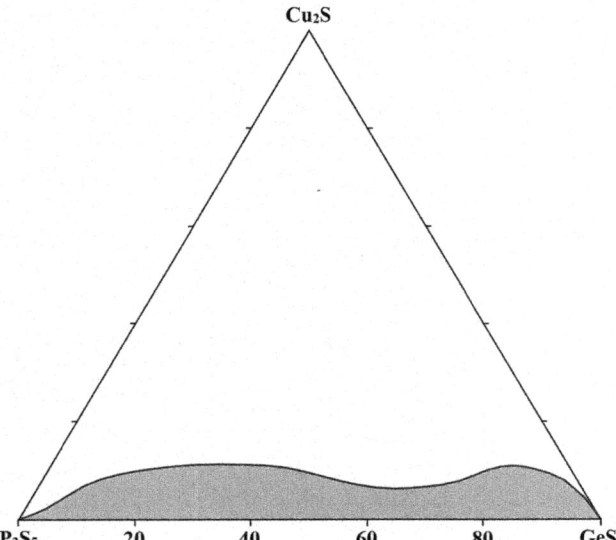

FIGURE 4.20 Glass-forming region in the GeS$_2$–Cu$_2$S–P$_2$S$_5$ quasiternary system. (From Bereznyuk, O.P., and Petrus', I.I., *Nauk. Visn. Uzhgorod. Univ., Ser. Khim.*, [2(44)], 41, 2020.)

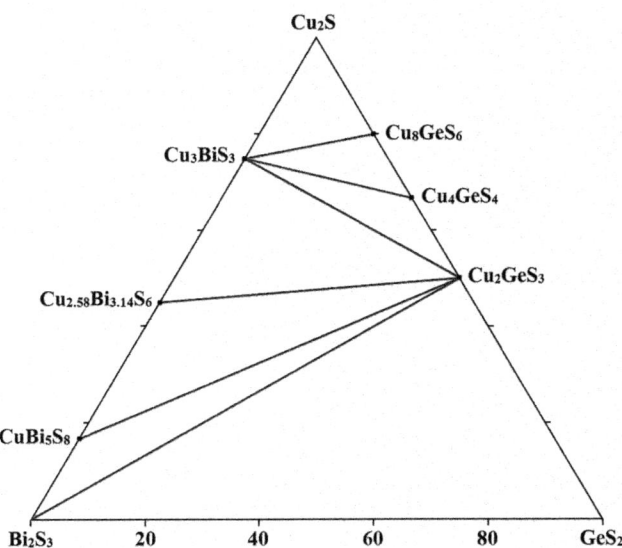

FIGURE 4.21 Isothermal section of the GeS$_2$–Cu$_2$S–Bi$_2$S$_3$ quasiternary system at 400°C. (From Olekseyuk, I., and Zhbankov O., *Visnyk L'viv. Univ. Ser. khim.*, (46), 53, 2005.)

kept at that temperature for 48 h and then cooled to room temperature at a rate of 30°C·h^{-1}. This compound was obtained as dark-red crystals.

Cu$_2$PbGeS$_4$ crystallizes in the trigonal structure with the lattice parameters $a = 612.52 \pm 0.01$, $c = 1533.32 \pm 0.02$ pm, and an energy gap of 1.55 eV (Sun et al. 2020a). It was prepared by combining reagents in stoichiometric amounts with a total mass of ~250–400 mg. The mixture was weighed, ground with a mortar and pestle, and cold pressed, all in a N$_2$-filled glove box. Then it was flame-sealed under vacuum (10^{-4} Pa). The ampoule was annealed at 575°C for 96 h and allowed to cool naturally. The compound was ground, pressed, resealed, and annealed several times to achieve desired phase purity.

4.116 Germanium–Copper–Phosphorus–Sulfur

Cu$_2$GeS$_3$–CuGe$_2$P$_3$. A complete solid solution for this system was found with the lattice parameters changing from 536.78 pm for CuGe$_2$P$_3$ to 528.95 pm for the alloys containing 90 mol% Cu$_2$GeS$_3$ and obeying Vegard's law (Omar 1990).

GeS$_2$–Cu$_2$S–P$_2$S$_5$. The glass-forming region of this system is presented in Figure 4.20 (Bereznyuk and Petrus 2020). When samples are quenched from 900°C, the maximum Cu$_2$S content in glasses is 10 mol%.

4.117 Germanium–Copper–Arsenic–Sulfur

The **CuGeAsS$_3$** quaternary compound, which crystallizes in the tetragonal structure with the lattice parameter $a = 376.66 \pm 0.01$ and $a = 521.2 \pm 0.1$ pm, is formed in the Ge–Cu–As–S system (Mel'nikova et al. 1995).

4.118 Germanium–Copper–Bismuth–Sulfur

GeS–Cu$_2$S–Bi$_2$S$_3$. The isothermal section of this quasiternary system at 400°C was constructed by Olekseyuk and Zhbankov (2005). It is characterized by the equilibria between GeS and CuBi$_5$S$_6$, Cu$_{2.58}$Bi$_{3.14}$S$_6$, and Cu$_3$BiS$_3$ ternary compounds. No quaternary significant solid solutions were found.

GeS$_2$–Cu$_2$S–Bi$_2$S$_3$. The isothermal section of this quasiternary system at 400°C is presented in Figure 4.21 (Olekseyuk and Zhbankov 2005). Seven three-phase regions divided by six quasibinary equilibria exist in this system. No quaternary compounds and significant solid solutions were found. The alloys for the investigations of both isothermal sections were annealed at 400°C for 300 h, with the next quenching in cold water.

4.119 Germanium–Copper–Vanadium–Sulfur

The **Cu$_{26}$V$_2$Ge$_6$S$_{32}$** quaternary compound, which crystallizes in the cubic structure with the lattice parameter $a = 1062.9 \pm 0.8$ pm, is formed in the Ge–Cu–V–S system (Suekuni et al. 2014). Samples of this compound were synthesized by direct reaction of Cu, V, Ge, and S. The stoichiometric quantities of the elements were sealed in an evacuated quartz tube, which was heated to 250°C over 2 h and kept at this temperature for 2 h, and then heated to 1100°C in 8 h. After being kept for 50 h, it was cooled to room temperature in the furnace. This compound is stable (no phase decomposition) up to at least 730°C (Lemoine et al. 2020).

4.120 Germanium–Copper–Niobium–Sulfur

The **Cu$_{26}$Nb$_2$Ge$_{6-x}$S$_{32}$** quaternary phase, which crystallizes in the cubic structure, is formed in the Ge–Cu–Nb–S system (Bouyrie et al. 2017). To prepare it, the mixtures of elements with appropriate ratios were loaded onto fused quartz tubes. The tubes were evacuated at a pressure of ~5·10^{-3} Pa and then flame-sealed. The mixtures were heated to 1050°C at a rate of ~0.3°C·min^{-1}, held for 12 h, and then cooled to room temperature at a rate of ~0.6°C·min^{-1}. The samples were sintered under uniaxial pressure to obtain dense compacts. The prepared ingots were then hand-ground into powders before being placed into graphite dies, which were inserted into a hot-press furnace. Sintering was performed at 750°C for 1 h under a uniaxial pressure of 70 MPa under Ar gas flow (300 mL·min^{-1}). The heating and cooling rates were 10°C·min^{-1} and ~0.3°C·min^{-1}, respectively. The lattice parameter *a* for as-prepared and sintered compact was 1063.8 and 1068.7 for *x* = 0 and 1063.4 and 1070.5 for *x* = 0.5, respectively.

4.121 Germanium–Copper–Tantalum–Sulfur

The $Cu_{26}Ta_2Ge_{6-x}S_{32}$ quaternary phase, which crystallizes in the cubic structure, is formed in the Ge–Cu–Ta–S system (Bouyrie et al. 2017). It was synthesized in the same way as $Cu_{26}Nb_2Ge_{6-x}S_{32}$ was prepared. The lattice parameter *a* for as-prepared and sintered compact was 1065.0 and 1068.8 pm for *x* = 0 and 1064.3 and 1071.2 pm for *x* = 0.5, respectively.

4.122 Germanium–Copper–Selenium–Sulfur

Cu_2GeS_3–Cu_2GeSe_3. The phase diagram of this system, constructed through DTA, XRD, SEM, and EMF of concentration chains (Figure 4.22), is characterized by continuous solubility both in the solid and liquid state (Bagheri et al. 2015). No maximum or minimum were observed on the liquidus and solidus curves.

$GeS_2 + 2Cu_2Se \leftrightarrow GeSe_2 + 2Cu_2S$. Some vertical and isothermal sections, as well as the liquidus surface of this system were constructed through DTA, XRD, SEM, and EMF of concentration chains (Bagheri et al. 2015; Alverdiev et al. 2017c). It was shown that continuous or wide areas of solid solutions are

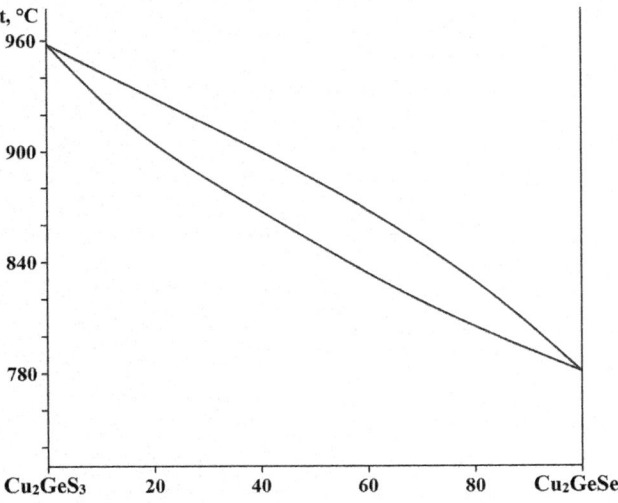

FIGURE 4.22 Phase diagram of the Cu₂GeS₃–Cu₂GeSe₃ system. (From Bagheri, S.M., et al., *J. Alloys Compd.*, **625**, 131, 2015.)

formed along the Cu_2S–Cu_2Se, Cu_2GeS_3–Cu_2GeSe_3, Cu_8GeS_6–Cu_8GeSe_6, and GeS_2–$GeSe_2$ quasibinary systems. Five three-phase regions exist in the system at 300°C (Figure 4.23a), and continuous solid solutions with three two-phase regions were found at 530°C (Figure 4.23b) (Alverdiev et al. 2017c). The isothermal sections of this system at 830°C and 930°C are presented in Figure 4.23c,d. The liquidus surface of this system (Figure 4.24) includes four fields of the primary crystallization of the $Cu_2S_xSe_{1-x}$, $Cu_2GeS_{3x}Se_{3(1-x)}$, $Cu_8GeS_{6x}Se_{6(1-x)}$ and $GeS_{2x}Se_{2(1-x)}$ solid solutions.

The partial molar thermodynamic functions of Cu for the $Cu_2GeS_{3x}Se_{3(1-x)}$ and $Cu_8GeS_{6x}Se_{6(1-x)}$ solid solutions were calculated on the EMF measurements by Bagheri et al. (2015) and Alverdiev et al. (2017b).

4.123 Germanium–Copper–Chromium–Sulfur

Two quaternary compounds, $Cu_4Ge_2Cr_6S_{15}$ and $Cu_{26}Cr_2Ge_6S_{32}$, are formed in the Ge–Cu–Cr–S system. The first decomposes at temperatures of more than 1100°C (Babitsyna and Novotortsev 1986). It was synthesized from the elements at 960–980°C for 100 h, with the next annealing at 900°C for 300 h.

$Cu_{26}Cr_2Ge_6S_{32}$ crystallizes in the cubic structure with the lattice parameter *a* = 1055.24 ± 0.1 pm (Lemoine et al. 2020). An almost linear thermal evolution of the unit cell parameter up to 560–570°C could be observed, followed by a relatively constant unit cell parameter and an increase of its standard deviation at higher temperatures. This compound is stable at least up to 430°C in a no oxidative atmosphere. It starts to decompose above 560°C into Cu_8GeS_6 and $CuCrS_2$. Above 660°C, $Cu_{26}Cr_2Ge_6S_{32}$ fully decomposed.

4.124 Germanium–Copper–Tungsten–Sulfur

The Cu_6GeWS_8 (mineral catamarcaite) which crystallizes in the hexagonal structure with the lattice parameters *a* = 752.38 ± 0.08, *c* = 1239.0 ± 0.3 pm, and a calculated density of 4.892 g·cm^{-3}, is formed in the Ge–Cu–W–S system (Putz et al. 2006).

4.125 Germanium–Copper–Chlorine–Sulfur

The $Cu_{8-x}GeS_{6-x}Cl_x$ quaternary phase, which has a polymorphic transformation at 36°C and crystallizes in the orthorhombic structure with the lattice parameters *a* = 703.9 ± 0.3, *b* = 696.8 ± 0.4, and *c* = 986.8 ± 0.4 pm for *x* = 0.07 (Kuhs et al. 1979), is formed in the Ge–Cu–Cl–S system. This phase was prepared by reacting stoichiometric amounts of the elements in an evacuated, sealed quartz ampoule for about 6 days at temperatures of 600°C to 700°C.

4.126 Germanium–Copper–Bromine–Sulfur

The $Cu_{8-x}GeS_{6-x}Br_x$ quaternary phase, which crystallizes in the cubic structure with the lattice parameter *a* = 993.98 ± 0.01 and a calculated density of 5.202 g·cm^{-3} for *x* = 0.25 (Nilges

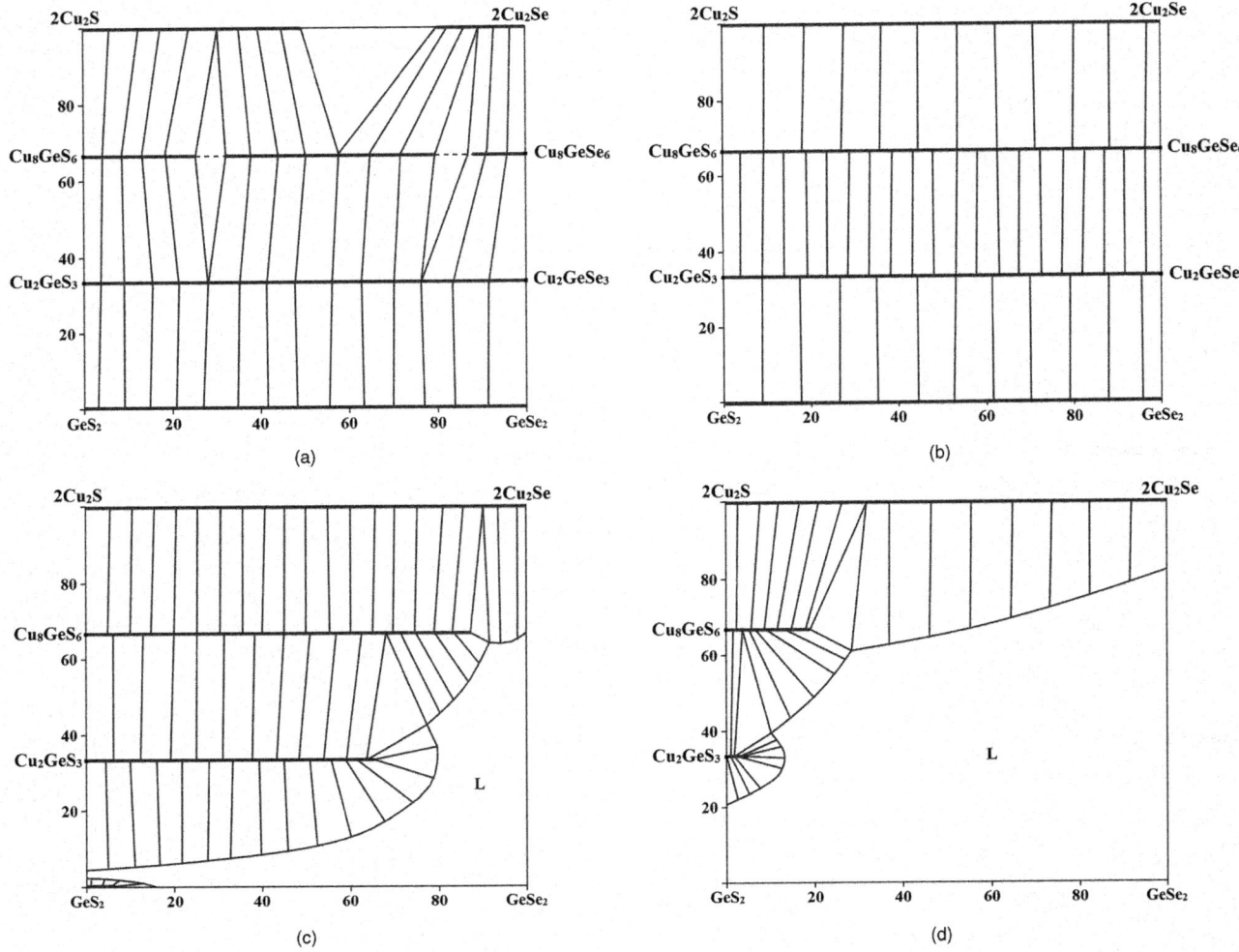

FIGURE 4.23 Isothermal section of the GeS₂ + 2Cu₂Se ↔ GeSe₂ + 2Cu₂S system at (a) room temperature, (b) at 530°C, (c) at 830°C and (d) at 930°C. (From Alverdiev, I.J., et al., *J. Alloys Compd.*, **691**, 255, 2017.)

and Pfitzner 2005) and $a = 995.3 \pm 0.3$ for $x = 0.7$ (Kuhs et al. 1979), is formed in the Ge–Cu–Br–S system. The phase with $x = 0.25$ was obtained by the reaction of the stoichiometric mixture of the respective elements in an evacuated, sealed silica ampoule at 1000°C (Nilges and Pfitzner 2005). After grinding, the phase was annealed at 880°C for 23 days. The phase with $x = 0.7$ was prepared by reacting stoichiometric amounts of the elements in an evacuated, sealed quartz ampoule for about 6 days at temperatures of 600°C to 700°C (Kuhs et al. 1979).

4.127 Germanium–Copper–Iodine–Sulfur

The $\mathbf{Cu_{8-x}GeS_{6-x}I_x}$ quaternary phase, which melts incongruently at 955°C (Studenyak et al. 2007a) and crystallizes in the cubic structure with the lattice parameter $a = 1001.81 \pm 0.02$ and a calculated density of 5.314 g·cm⁻³ for $x = 1$ (Nilges and Pfitzner 2005), $a = 996.0 \pm 0.1$ for $x = 0.5$ (Tomm et al. 2008) and $a = 1001.2 \pm 0.4$ for $x = 0.78$ (Kuhs et al. 1979), is formed in the Ge–Cu–I–S system. The phase with $x = 1$ was obtained by the reaction of the stoichiometric mixture of elements in an evacuated, sealed silica ampoule at 1100°C (Nilges

and Pfitzner 2005). After grinding, the phase was annealed at 850°C for 25 days. Single crystals of the phase with $x = 0.5$ were grown using chemical vapor transport with I₂ as a transport agent (Tomm et al. 2008). The composition of the obtained cubic crystals varied from $x = 0.07$ to $x = 0.84$ in dependence on the content of the transport agent. The phase with $x = 0.78$ was prepared by reacting stoichiometric amounts of the elements in an evacuated, sealed quartz ampoule for about 6 days at temperatures of 600°C to 700°C (Kuhs et al. 1979). Cu₇GeS₅I was also synthesized by a chemical transport from Ge, Cu, S, and CuI, and it melts incongruently at 955°C (Studenyak et al. 2007a, 2007b).

According to the data of Tomm et al. (2008), the composition of high-temperature single crystals with hexagonal structure can be written as $Cu_{7.69}GeS_{5.69}I_{0.31}$, e.g. $x = 0.31$.

4.128 Germanium–Copper–Manganese–Sulfur

Two quaternary compounds, $\mathbf{Cu_2MnGeS_4}$ and $\mathbf{Cu_4MnGe_2S_7}$, are formed in the Ge–Cu–Mn–S system. The first melts incongruently at 1011–1018°C [at 999°C (Quintero et al. 2014b)]

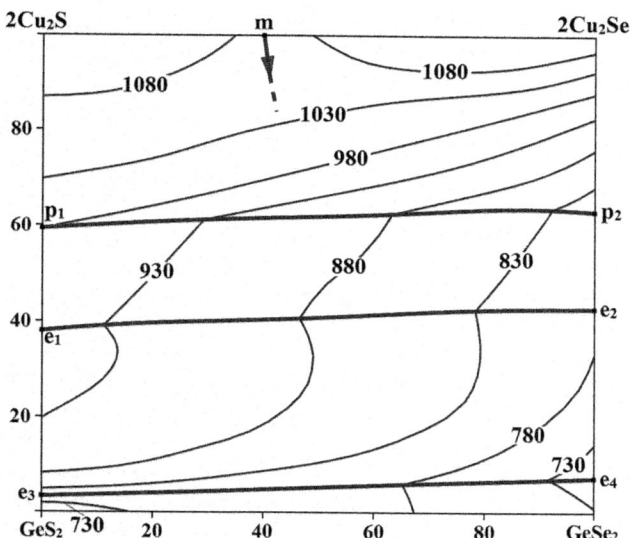

FIGURE 4.24 Liquidus surface of the $GeS_2 + 2Cu_2Se \leftrightarrow GeSe_2 + 2Cu_2S$ system. (From Alverdiev, I.J., et al., *J. Alloys Compd.*, **691**, 255, 2017.)

[melts congruently at 994°C (Schäfer and Nitsche 1977)] and crystallizes in the orthorhombic structure with the lattice parameters given in Table 4.1. This compound undergoes a phase transition and is characterized by an energy gap of 2.21 eV (Glenn et al. 2021) [1.70 eV (Beraich et al. 2020); 1.90 eV (Honig et al. 1988)].

$Cu_4MnGe_2S_7$ melts incongruently at 961°C, undergo a phase transition, and crystallizes in the monoclinic structure with the lattice parameters $a = 1673.32 \pm 0.03$, $b = 647.60 \pm 0.01$, $c = 980.22 \pm 0.02$ pm, $\beta = 93.1517° \pm 0.0009°$, a calculated density of 4.250 g·cm⁻³, and an energy gap of 1.98 eV (Glenn et al. 2021).

Cu_2MnGeS_4 was obtained by the chemical vapor reactions using I_2 as the transport agent (Nitsche et al. 1967; Allemand and Winterberger 1970; Schäfer and Nitsche 1974; Guen et al. 1979; Guen and Glaunsinger 1980; Honig et al. 1988) or by the melt and anneal technique using stoichiometric mixtures of the elements (Lamarche et al. 1991; Bernert and Pfitzner 2005, 2006; Sachanyuk et al. 2006c; Quintero et al.

2014a, b, 2017). Typically, the mixture of the elements of 1-g sample was sealed under vacuum ($\approx 10^{-3}$ Pa) in a small quartz ampoule, and then was heated up to 200°C and kept for about 1–2 h, then the temperature was raised to 500°C using a rate of 40°C·h⁻¹, and held at this temperature for 14 h. After that, the sample was heated from 500°C to 800°C at a rate of 30°C·h⁻¹ and kept at this temperature for another 14 h. Then the temperature was raised to 1150°C at 60°C·h⁻¹, and the components were melted together at this temperature. The temperature was brought slowly (4°C·h⁻¹) down to 600°C, and the sample was annealed at this temperature for 1 month. Then, the sample was slowly cooled to room temperature using a rate of about 2°C·h⁻¹.

The Cu–Mn–Ge precursor of Cu_2MnGeS_4 thin films have been deposited by spray ultrasonic, from an aqueous solution, prepared by dissolution of $CuCl_2·2H_2O$ (0.04 M), $MnCl_2·4H_2O$ (0.02 M), and GeO_2 (0.02 M) in deionized water (Beraich et al. 2020). After deposition, the film was allowed to cool to the ambient temperature, and then was annealed at 500°C in the presence of excess S powder for one hour inside a quartz tube of a tubular furnace under Ar flow, which allows the interdiffusion of sulfur gas and then the formation of Cu_2MnGeS_4. Finally, the system was naturally cooled down to room temperature by stopping the furnace heating and cutting off the Ar flow.

Single crystals of $Cu_4MnGe_2S_7$ were prepared from a stoichiometric mixture of the elements, enough material to prepare ~0.7 mM of product. The reagents were combined, ground, and pressed into a pellet, which was subsequently inserted into a fused silica tube, evacuated to ~10^{-2} Pa, and flame-sealed. The reaction was heated from room temperature to 900°C over 12 h, held at 900°C for 168 h, cooled from 900°C to 300°C at a rate of 5°C·h⁻¹, and cooled to room temperature radiatively. After breaking the sintered pellet apart, small, deep-red, polyhedral-shaped single crystals were found. Microcrystalline powder of this compound was prepared from a stoichiometric mixture of the elements, enough material to prepare 2 mM of product. The reagents were combined, shaken (not ground), and pressed into a pellet, which was subsequently inserted into a fused silica tube, evacuated to ~10^{-2} Pa, and flame-sealed. The reaction was heated from room temperature to 1075°C over 216 h, held at

Table 4.1

Crystallographic data for Cu_2MnGeS_4

a, pm	b, pm	c, pm	$d_{calc.}$ g·cm⁻³	$d_{meas.}$	References
761	650	618	–	–	Nitsche et al. (1967); Allemand and Winterberger (1970)
760.8	651.1	623.6	–	–	Schäfer and Nitsche (1974, 1977)
760.7	650.6	623.4	4.23	4.19	Guen et al. (1979); Guen and Glaunsinger (1980)
761.0 ± 0.1	651.3 ± 0.1	623.6 ± 0.1	–	–	Honig et al. (1988)
762.9	651.4	626.5	–	–	Lamarche et al. (1991)
763.5 ± 0.1	652.67 ± 0.07	624.38 ± 0.07	4.087	–	Bernert and Pfitzner (2005)
762.01 ± 0.03	651.82 ± 0.02	624.33 ± 0.03	4.0999 ± 0.0005	–	Sachanyuk et al. (2006c)
761.62 ± 0.04	651.67 ± 0.03	623.82 ± 0.03	4.11	–	Quintero et al. (2014a,b, 2017)
762	652	624	–	–	Beraich et al. (2020)

1075°C for 168 h, cooled from 1075°C to 875°C at a rate of 2°C·h⁻¹, and cooled to room temperature radiatively. The product consisted of a reddish-black ingot of $Cu_4MnGe_2S_7$.

4.129 Germanium–Copper–Iron–Sulfur

Two quaternary compounds **Cu_2FeGeS_4** and **$Cu_4FeGe_2S_7$**, are formed in the Ge–Cu–Fe–S system. Cu_2FeGeS_4 melts congruently at 996°C (Quintero et al. 2014b) [at 990°C (Moh 1975); at 983°C (Schäfer and Nitsche 1977)] and crystallizes in the tetragonal structure with the lattice parameters $a = 533.49 \pm 0.01$, $c = 1052.34 \pm 0.03$ pm, and a calculated density of 4.26 g·cm⁻³ (Quintero et al. 2014a,b 2017) [$a = 532.5$, $c = 1051$ pm (Nitsche et al. 1967; Wintenberger 1979); $a = 532.5 \pm 0.3$ and $c = 1052.7 \pm 0.6$ pm (Ottenburgs and Goethals 1972); $a = 533.0$ and $c = 1052.8$ pm (Schäfer and Nitsche 1974, 1977); $a = 532.7$, $c = 1052.2$ pm, and the calculated and experimental densities of 4.26 and 4.25 g·cm⁻³, respectively (Guen et al. 1979; Guen and Glaunsinger 1980)]. According to the data of Moh (1975), this compound has two polymorphs with a phase transition temperature of 635°C. Both polymorphs crystallize in the tetragonal structure with the lattice parameters $a = 532.8 \pm 0.4$, $c = 1053.3 \pm 1.0$ pm and 536.4 ± 0.4, 1066 ± 1 pm at 800°C for high- and low-temperature modifications of Cu_2FeGeS_4, respectively

$Cu_4FeGe_2S_7$ shows relatively high thermal stability, melts congruently, and crystallizes in the monoclinic structure with the lattice parameters $a = 1174.05 \pm 0.06$, $b = 535.89 \pm 0.02$, $c = 834.20 \pm 0.04$ pm, $\beta = 98.661° \pm 0.004°$, and a calculated density of 4.350 g·cm⁻³ (Craig et al. 2020).

Two more quaternary compounds, **$Cu_{13}Ge_2Fe_2S_{16}$** and **$Cu_{22}Fe_8Ge_4S_{32}$**, are also formed in this system. The first of them, mineral germanite, crystallizes in the cubic structure with the lattice parameter $a = 1058.62 \pm 0.05$ pm and a calculated density of 4.47 g·cm⁻³ (de Jong 1930; Tettenhorst and Corbato 1984) [$a = 1058.5$ pm and a calculated density of 4.46 g·cm⁻³ (Murdoch 1953)]. $Cu_{22}Fe_8Ge_4S_{32}$ also crystallizes in the cubic structure with the lattice parameter from $a = 1058.84$ to 1059.86 pm, depending on the preparation methods (Paradis-Fortin et al. 2020) [$a = 1058.9$ pm (Kumar et al. 2017)].

Cu_2FeGeS_4 was produced by the melt and anneal technique (Quintero et al. 2014a, b, 2017). The mixture of the elements from the 1-g sample was sealed under vacuum ($\approx 10^{-3}$ Pa) in a small quartz ampoule, and was heated up to 200°C and kept for about 1–2 h, then the temperature was raised to 500°C using a rate of 40°C·h⁻¹, and held at this temperature for 14 h. After that, the sample was heated from 500°C to 800°C at a rate of 30°C·h⁻¹ and kept at this temperature for another 14 h. Then the temperature was raised to 1150°C at 60°C·h⁻¹, and the components were melted together at this temperature. The furnace temperature was brought slowly (4°C·h⁻¹) down to 600°C, and the sample was annealed at this temperature for one month. Then, the sample was slowly cooled to room temperature using a rate of about 2°C·h⁻¹. Single crystals of the title compound were grown through chemical transport reactions using iodine as a transport agent (Nitsche et al. 1967; Wintenberger 1979; Guen et al. 1979; Guen and Glaunsinger 1980).

$Cu_4FeGe_2S_7$ was synthesized by combining stoichiometric amounts of Cu, Fe, Ge, and S powders (Craig et al. 2020). The powders were mixed and placed into a fused silica tube that was subsequently evacuated and flame-sealed. The reaction vessel was placed upright into a ceramic container inside a furnace, where it was heated to 1000°C in 24 h, held there for 48 h, cooled to 900°C over the course of 50 h, and held there for 96 h, before being allowed to cool to room temperature over a 24 h period. Subsequently, the reaction vessel was cut open, and the content consisted of loose silvery gray microcrystalline powder of the title compound.

Four samples of $Cu_{22}Fe_8Ge_4S_{32}$ were synthesized by combining two different powder syntheses (mechanical alloying or sealed tube samples) and two sintering processes (spark plasma sintering or hot pressing) (Paradis-Fortin et al. 2020). For both synthesis methods, Cu, Fe, S, and Ge powders were stored and manipulated in a glove box under an argon atmosphere. The starting materials for both mechanically alloyed and sealed tube samples were weighted in a stoichiometric ratio and ground in an agate mortar. For the synthesis of the mechanically alloyed samples, two batches of 4 g each were prepared and put into two 45 mL tungsten carbide jars along with a total of 14 balls with a diameter of 10 mm, for a 13:1 ball-to-powder weight ratio. The milling lasted for 360 min at 600 rpm decomposed in 24 cycles of 15 min each with 1 min pause and a reverse of the milling direction. For the synthesis of the sealed tube samples, 4 g batches of powder were pressed into eight pellets of ~0.5 g with a 5 mm die. The pellets were placed in sealed silica tubes evacuated down to a pressure of ~1 Pa from an Ar atmosphere. The reaction was performed in a vertical tubular furnace with a heating rate of 2°C·min⁻¹ and a plateau at 700°C for 24 h. The sample was cooled down to 500°C at a natural cooling rate and then air-quenched. Powders (ca. 3 g) from the sealed tube samples or mechanical alloying synthesis were weighted and put into a graphite die and densified by spark plasma sintering at 600°C under a uniaxial pressure of 64 MPa with a heating rate of 30°C·min⁻¹ and a holding time of 30 min in a spark plasma sintering furnace under static secondary vacuum. The hot-pressed samples (ca. 3 g) were sintered in a graphite die at 600°C under a uniaxial pressure of 64 MPa with a heating rate of 15°C·min⁻¹ and a holding time of 60 min under dynamic primary vacuum (about 0.3 mbar).

$CuFe_xGe_{1-x}S_2$ solid solution $(0.5 < x < 1.0)$ is also formed in the Ge–Cu–Fe–S quaternary system (Ackermann et al. 1976). It crystallizes in the tetragonal structure and the lattice parameters a and c, the calculated and experimental densities change from 533.2 ± 0.1 and 1053.1 ± 0.1 pm, 4.25 and 4.25 ± 0.1 g·cm⁻³ for x 0.53 to 528.0 ± 0.1 and 1040.9 ± 0.1 pm, 4.18 and 4.18 ± 0.1 g·cm⁻³ for $x = 1$. Single crystals of this solid solution series have been prepared by the chemical vapor transport technique using I_2 as a transport agent.

Polycrystalline samples of $Cu_{26-x}Fe_{4+x}Ge_4S_{32}$ $(0 \leq x \leq 4)$ were synthesized by mechanical alloying followed by spark plasma sintering (Kumar et al. 2017). Stoichiometric amounts of Cu, Fe, Ge, and S were loaded in a 25 mL tungsten carbide jar under an argon atmosphere. High-energy ball-milling was performed at room temperature at a disk rotation speed of 600 rpm. Powders were milled in an argon atmosphere for 6 h. The resulting powders were ground and sieved down to

150 µm. Powders were then placed in graphite dies and densified by spark plasma sintering at 600°C for 30 min under a pressure of 64 MPa (heating and cooling rate of 50°C·min⁻¹ and 20°C·min⁻¹, respectively). All sample preparations and handling of powders were performed in an argon-filled glove box with oxygen content <1 ppm.

4.130 Germanium–Copper–Cobalt–Sulfur

Two quaternary compounds, Cu_2CoGeS_4 and $Cu_4CoGe_2S_7$, are formed in the Ge–Cu–Co–S system. First of them melts congruently at 1050°C (Quintero et al. 2014b) [at 1031°C (Schäfer and Nitsche 1977)] and crystallizes in the tetragonal structure with the lattice parameters $a = 529.57 \pm 0.01$, $c = 1047.09 \pm 0.05$ pm, and a calculated density of 4.3747 g·cm⁻³ (Gulay et al. 2004a) [$a = 530$, $c = 1048$ pm, and the calculated and experimental densities of 4.36 and 4.37–4.38 g·cm⁻³, respectively (Allemand and Winterberger 1970; $a = 530.3$ and $c = 1049.2$ pm (Schäfer and Nitsche 1977); Guen et al. 1979; Guen and Glaunsinger 1980); $a = 530.7 \pm 0.2$, $c = 1049.3 \pm 0.5$ pm, and a calculated density of 4.348 g·cm⁻³ (Bernert and Pfitzner 2006); $a = 529.57$, $c = 1047.40$ pm, and a calculated density of 4.37 g·cm⁻³ (Quintero et al. 2014b); $a = 530.6 \pm 0.1$, $c = 1049.0 \pm 0.1$ pm, and an energy gap of 0.81 eV (Bourgès et al. 2020)].

$Cu_4CoGe_2S_7$ shows relatively high thermal stability, melts congruently, and crystallizes in the monoclinic structure with the lattice parameters $a = 1172.80 \pm 0.02$, $b = 533.987 \pm 0.010$, $c = 833.133 \pm 0.014$ pm, $\beta = 98.6680° \pm 0.0012°$, and a calculated density of 4.396 g·cm⁻³ (Craig et al. 2020).

Cu_2CoGeS_4 was obtained by the chemical vapor reactions using I_2 as the transport agent (Allemand and Winterberger 1970; Guen et al. 1979; Guen and Glaunsinger 1980) or by the melt and anneal technique using stoichiometric mixtures of the elements (Gulay et al. 2004a; Bernert and Pfitzner 2006; Quintero et al. 2014b). Typically, calculated amounts of the elements were sealed in an evacuated quartz container. The synthesis was realized in a shaft furnace. The ampoule with the components was heated at a rate of 30°C·h⁻¹ to 1100°C. The sample was kept at this temperature for 4 h. After that, it was cooled slowly to 400°C at a rate of 10°C·h⁻¹. Further annealing at 400°C during 240 h was applied. After annealing, the ampoule with the sample was quenched in cold water.

$Cu_4CoGe_2S_7$ was prepared in the same manner as $Cu_4FeGe_2S_7$ using Co instead of Fe (Craig et al. 2020).

The $Cu_{2+x}Co_{1-x}GeS_4$ solid solution with the tetragonal structure is formed in the Ge–Cu–Co–S system (Bourgès et al. 2020). Polycrystalline samples ($0 \leq x \leq 0.225$) were synthesized in a two-step process. A stoichiometric mixture of high-purity powders of Cu, Co, Ge, and S was ground in a mortar, sealed in evacuated silica tubes, heated up to 850°C for 24 h (100°C·h⁻¹), and then cooled down to room temperature at the same rate. The resulting powders were densified by spark plasma sintering process at 850°C for 30 min (heating and cooling rate of 100°C·min⁻¹) under a pressure of 50 MPa using graphite dies of 10 mm diameter and vacuum enclosure. It was determined that up to $x = 0.15$ the samples were composed only of the single phase of the solid solution. This solid solution is stable up to 450°C.

4.131 Germanium–Copper–Nickel–Sulfur

Two quaternary compounds, Cu_2NiGeS_4 and $Cu_4NiGe_2S_7$, are formed in the Ge–Cu–Ni–S system. First of them crystallizes in the orthorhombic structure with the lattice parameters $a = 533.2$, $b = 526.3$, and $c = 522.7$ pm (Schäfer and Nitsche 1974) [$a = 534$, $b = 527$, and $c = 1047$ pm (Nitsche et al. 1967)].

$Cu_4NiGe_2S_7$ melts incongruently at 886°C and crystallizes in the monoclinic structure with the lattice parameters $a = 1170.3$, $b = 533.3$, $c = 831.1$ pm, $\beta = 98.37°$ pm, and a calculated density of 4.44 g·cm⁻³ (Schäfer and Nitsche 1977; Schäfer et al. 1980).

Single crystals of both compounds were grown by the chemical vapor reactions using I_2 as the transport agent (Nitsche et al. 1967; Schäfer and Nitsche 1977; Schäfer et al. 1980).

4.132 Germanium–Silver–Strontium–Sulfur

Two quaternary compounds, Ag_2SrGeS_4 and $Ag_2Sr_3Ge_2S_8$, are formed in the Ge–Ag–Sr–S system.

Ag_2SrGeS_4 apparently has two polymorphic modifications: the first of them crystallizes in the tetragonal structure with the lattice parameters $a = 683.40 \pm 0.01$, $c = 764.70 \pm 0.02$ pm, a calculated density of 4.688 g·cm⁻³, and an energy gap of 1.73 eV (McKeown Wessler et al. 2021). The second modification crystallizes in the orthorhombic structure with the lattice parameters $a = 682.0$, $b = 697.3$, $c = 769.0$ pm, and an energy gap of 1.33 eV (Zhu et al. 2017).

The bulk powder of the tetragonal modification was synthesized by mixing Ag, SrS, Ge, and S powders. This was done with SrS held at 90% of the stoichiometric molar content to avoid the formation of impurity phases. The powder was ground and cold-pressed under a nitrogen atmosphere into a pellet. This pellet was then sealed in a prebaked quartz ampoule under a dynamic vacuum of roughly 5 ×10⁻⁴ Pa. The sealed ampoule was ramped up from room temperature to 600°C at a rate of 30°C·h⁻¹ and annealed for 4 days before turning the furnace off to cool. After the first heat treatment, roughly 3–5% of the overall pellet mass was lost, and a black material was left on the walls of the quartz ampoules. The grinding, pressing, and annealing process was repeated to improve phase homogeneity. The further annealing processes did not cause any significant mass or material loss, and the pellet remained intact during the further anneals, in contrast to the first anneal. Crystals of Ag_2SrGeS_4, suitably large for single-crystal XRD, could not be identified in this manner. Single crystals of the tetragonal modification were recrystallized from the synthesized powder in a eutectic 0.4KI–0.6CsI flux. Synthesized powder of Ag_2SrGeS_4 was mixed with KI and CsI in a 1:12.5 molar ratio. The combined material was ground, pressed, and sealed as above. The sealed pellet was ramped to 600°C at a rate of 30°C·h⁻¹ and held at this temperature for 4 days before being cooled to 400°C at 4°C·h⁻¹. The annealing temperature is above the melting point of the KI/CsI eutectic (~530°C), creating a liquid medium for the Ag_2SrGeS_4 crystal growth. The ampoule was then allowed to cool naturally to room temperature. The product was washed with DMF, and sub-millimeter-sized yellow-orange crystals could be separated from any remaining halide flux material.

$Ag_2Sr_3Ge_2S_8$ crystallizes in the cubic structure with the lattice parameter $a = 1409.25 \pm 0.15$ pm, a calculated density of 4.178 g·cm⁻³, and an energy gap of 2.62 eV (Yang et al. 2020a) [$a = 1413.1 \pm 0.3$ pm (McKeown Wessler et al. 2021)]. The whole preparation process of this compound was completed in an Ar-filled glove box (Yang et al. 2020a). The targeted product was prepared under the stoichiometric ratio of SrS, Ag_2S, and GeS_2 in the vacuum-sealed silica tube. Firstly, the tube was heated to 600°C for 30 h and kept at this temperature for about 100 h, then cooled quickly down to room temperature. The obtained product was ground and then loaded into the vacuum-sealed silica tube. It was further sintered again at 900°C for 10 days and cooled slowly down to room temperature at 5°C·h⁻¹. Finally, many millimeter-level high-quality crystals with yellow color were found in the tube after washing with DMF. They are stable in the air within several weeks.

4.133 Germanium–Silver–Barium–Sulfur

The Ag_2BaGeS_4 quaternary compound, which is thermally stable up to 901°C (Chen et al. 2018b) and crystallizes in the tetragonal structure with the lattice parameters $a = 683.4 \pm 0.3$, $c = 804.1 \pm 0.3$ pm, a calculated density of 4.898 g·cm⁻³, and an energy gap of 2.12 eV (Liu et al. 2020b) [$a = 682.8$, $c = 801.7$ pm and the calculated and experimental densities of 4.52 and 4.49 g·cm⁻³, respectively (Teske 1979a); $a = 682.0 \pm 0.9$, $c = 802.1 \pm 0.2$ pm, a calculated density of 4.93 g·cm⁻³, and an energy gap of 2.02 eV (Chen et al. 2018b; $a = 682.0$, $c = 802.1$ pm, and an energy gap of 1.691 eV (calculated values) (Naseri et al. 2020); an energy gap of 1.38 eV (Zhu et al. 2017)], is formed in the Ge–Ag–Ba–S system.

This compound can be obtained as an orange, microcrystalline powder from the intimate mixture of its binary starting sulfides at 550°C in a fused silica ampoule with a sintered corundum crucible and protective gas under reduced pressure (Teske 1979a). Recrystallization in bromide melts produces red-orange crystals of Ag_2BaGeS_4. The title compound is stable in the air and hydrolyzes very slowly in water.

Dark-red irregular bulk crystals of Ag_2BaGeS_4 can also be prepared from a mixture of Ba (1.0 mM), Ag (2.0 mM), Ge (1.0 mM), S (4.0 mM), and $BaCl_2$ (2.7 mM) (Chen et al. 2018b). The chemicals were loaded in a fused silica tube and sealed under vacuum (<10⁻³ Pa). The tube was heated at 300°C for 20 h, raised to 900°C and stayed for 100 h, then cooled to 200°C at 3°C·h⁻¹ before turning off the furnace. The product was washed with deionized water to remove the $BaCl_2$ flux and soluble by-product, and dried by ethanol in the air.

The next method can also be used to prepare Ag_2BaGeS_4 (Liu et al. 2020b). The mixture of $AgNO_3$ (19 mg), Ge (5 mg), S (27 mg), $Ba(OH)_2 \cdot 8H_2O$ (14 mg), and ethylenediamine (about 270 mg) was sealed in a Pyrex glass tube (about 10% filling volume of the tube) at air atmosphere, and then the glass tube was placed in a stainless-steel autoclave (about 80% filling volume of water to balance the pressure), and finally heated in the furnace at 170°C for 7 days. After being cooled to ambient

temperature naturally, the product was washed with water and ethanol, respectively, and orange block crystals were obtained.

The structural, electronic, optical, and thermoelectric properties of Ag_2BaGeS_4 were calculated using the first-principles calculations based on the full linearized augmented plane wave method and semi-classical transport Boltzmann theory (Naseri et al. 2020).

4.134 Germanium–Silver–Zinc–Sulfur

Ag_2GeS_3–ZnS. The phase diagram of this system, constructed through DTA and XRD, is presented in Figure 4.25 (Moroz et al. 2019). The eutectic contains 20 mol% ZnS and crystallizes at 562°C. Two quaternary compounds, Ag_2ZnGeS_4 and $Ag_4ZnGe_2S_7$, are formed in this system.

Ag_2ZnGeS_4 melts incongruently at 704°C, has a phase transformation at 596°C ($\Delta H_{tr} = 5.12 \pm 0.05$ kJ·M⁻¹), and crystallizes in the tetragonal structure with the lattice parameters $a = 574.996 \pm 0.009$, $c = 1034.34 \pm 0.03$ pm, and a calculated density of 4.6799 ± 0.0003 g·cm⁻³ (Parasyuk et al. 2010) [$a = 574.59 \pm 0.07$ and $c = 1033.2 \pm 0.1$ pm (Moroz et al. 2019)]. The energy gap of this compound is equal to 2.5 eV (Tsuji et al. 2010) and its thermodynamic functions are the next: $\Delta G^0_{f,298} = -464.1 \pm 2.5$ kJ·M⁻¹, $\Delta H^0_{f,298} = -449.2 \pm 2.6$ kJ·M⁻¹ and $S^0_{298} = 336.3 \pm 1.3$ J·(M·K)⁻¹ (Moroz et al. 2019).

Ag_8GeS_6–ZnS. The phase diagram of this system, constructed through DTA, XRD, and metallography, is a eutectic type (Figure 4.26) (Piskach et al. 2006). The eutectic composition and temperature are 42 mol% ZnS and 928°C, respectively. The solubility of ZnS in Ag_8GeS_6 at the eutectic temperature is 38 mol%. The crystallization of the solid solutions occurs in a narrow temperature range that does not exceed 20°C. These solid solutions are characterized by the polymorph transformation at 226°C and solid solutions based on ZnS have the transformation sphalerite-wurtzite at 1041°C.

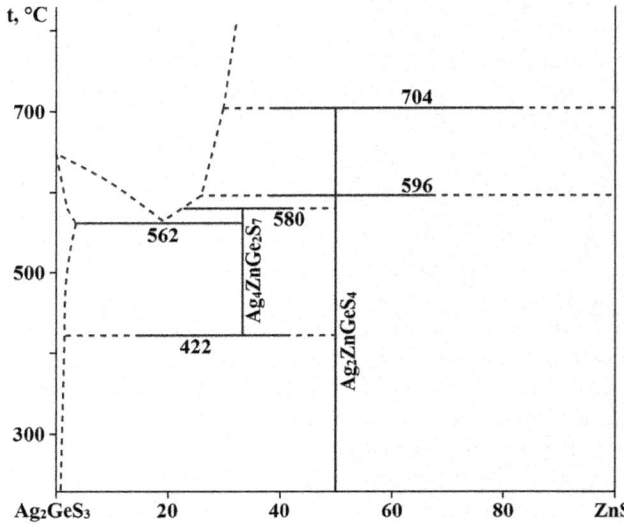

FIGURE 4.25 Phase diagram of the Ag_2GeSe_3–ZnS system. (From Moroz, M., et al., *Mater. Proc. Fund., Miner. Met. Mater. Ser.*, 215, 2019.)

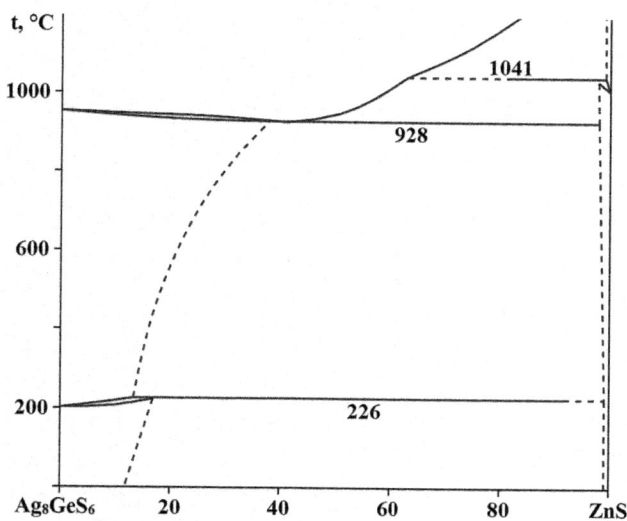

FIGURE 4.26 Phase diagram of the Ag$_8$GeS$_6$–ZnS system. (From Piskach, L.V., et al., *J. Alloys Compd.*, **421**(1–2), 98, 2006.)

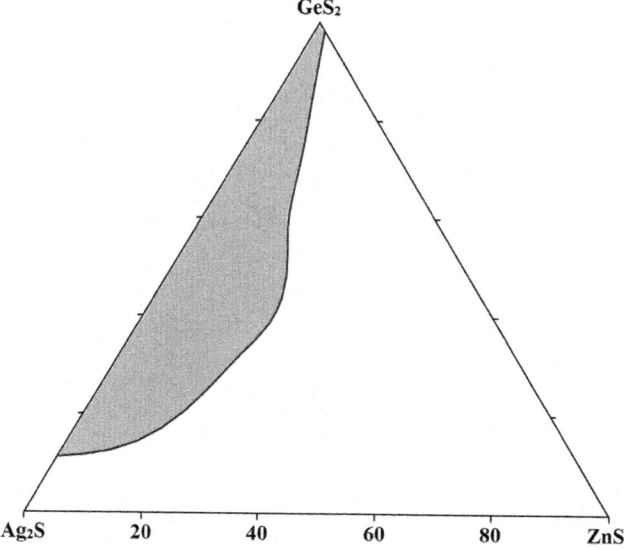

FIGURE 4.28 Glass-forming region in the GeS$_2$–Ag$_2$S–ZnS quasiternary system. (From Parasyuk, O.V. et al., *J. Alloys Compd.*, **500**(1), 26, 2010.)

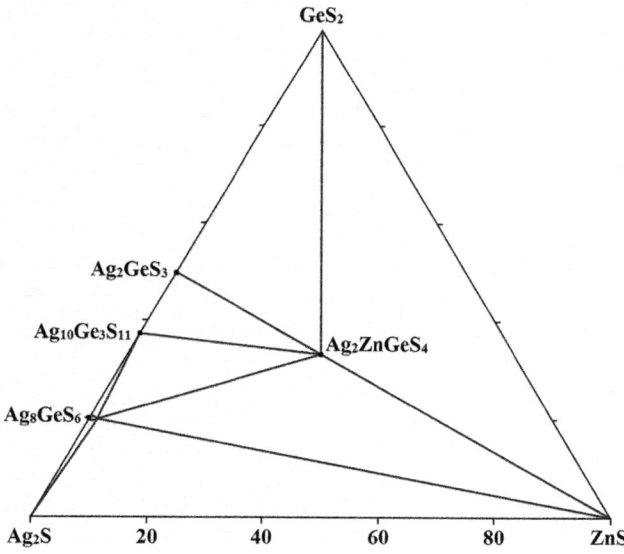

FIGURE 4.27 Isothermal section of GeS$_2$–Ag$_2$S–ZnS quasiternary system at room temperature. (From Parasyuk, O.V. et al., *J. Alloys Compd.*, **500**(1), 26, 2010.)

GeS$_2$–Ag$_2$S–ZnS. The isothermal section of this system at room temperature is shown in Figure 4.27 (Parasyuk et al. 2010).

The glass-formation region in the GeS$_2$–Ag$_2$S–ZnS quasiternary system is localized along the GeS$_2$–Ag$_2$S quasibinary system, and the maximum amount of ZnS in the glass phase is 12 mol% (Figure 4.28) (Parasyuk et al. 2010). It narrows with higher content of the glass-forming element. The glass transition temperature varies only narrowly for obtained glasses, likely because the amounts of ZnS that can be sustained by the glassy state are quite moderate. The glass transition

temperature values increase slightly from ~280°C for glasses of the GeS$_2$–Ag$_2$S system to ~300°C for the glasses containing the highest concentration of ZnS (10–12 mol%). The variations in the crystallization temperature are likewise minor.

The powder of Ag$_2$ZnGeS$_4$ was synthesized by solid-state reaction: the starting materials ZnS, Ag$_2$S, and GeS$_2$ were mixed with a 15% excess of ZnS and GeS$_2$ (Tsuji et al. 2010). The mixture was sealed in a quartz ampoule tube in a vacuum, and heat-treated at 550–650°C for 10 h.

Both quaternary compounds were also obtained through solid-state synthesis from finely dispersed mixtures of Ag$_2$GeS$_3$ and SnS at 480°C for 300 h (Moroz et al. 2019).

This system was investigated by DTA, and the alloys were annealed at 400°C for 800 h (Parasyuk et al. 2010).

The presence of four subsystems has been established in the ZnS–Ag$_2$GeS$_3$–Ge–GeS$_2$ part of the Ag–Zn–Ge–S system: Ag$_2$GeS$_3$–Ge–GeS–Ag$_2$ZnGeS$_4$, Ag$_2$GeS$_3$–GeS$_2$–GeS–Ag$_2$ZnGeS$_4$, Ag$_2$ZnGeS$_4$–GeS–GeS$_2$–ZnS, and Ag$_2$ZnGeS$_4$–GeS–Ge–ZnS (Moroz et al. 2019).

4.135 Germanium–Silver–Cadmium–Sulfur

Ag$_2$GeS$_3$–CdS. The phase diagram is a eutectic type (Figure 4.29) (Piskach and Parasyuk 1998). For the Ag$_2$GeS$_3$ compound, the congruent melting at 638°C has been confirmed. The eutectic crystallizes at 598°C and contains 4 mol% CdS. The **Ag$_2$CdGeS$_4$** and **Ag$_4$CdGe$_2$S$_7$** quaternary compounds are formed in this system. Ag$_2$CdGeS$_4$ melts incongruently at 768°C and exhibits polymorphic transformation with a temperature decrease from 632°C to 612°C at the increase in CdS content. It crystallizes in the orthorhombic structure with the lattice parameters for stoichiometric composition $a = 803.38 \pm 0.03$, $b = 686.80 \pm 0.02$, and $c = 658.66 \pm 0.03$ pm [$a = 801.6 \pm 0.5$, $b = 683.1 \pm 0.4$, and $c = 654.8 \pm 0.5$ (Piskach and

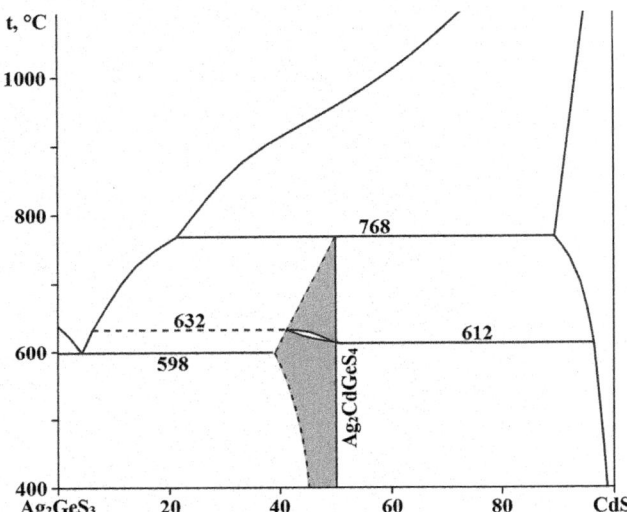

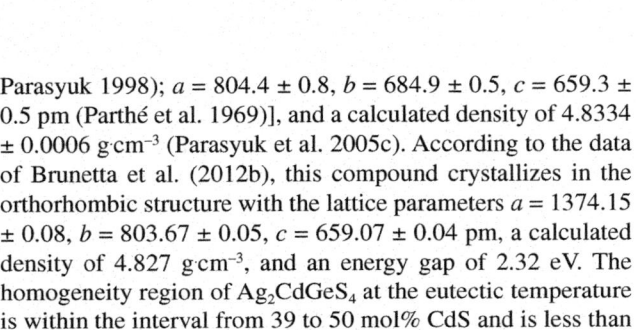

FIGURE 4.29 Phase diagram of the Ag₂GeS₃–CdS system. (From Piskach, L.V. et al., *Pol. J. Chem.*, **72**(6), 1112, 1998.)

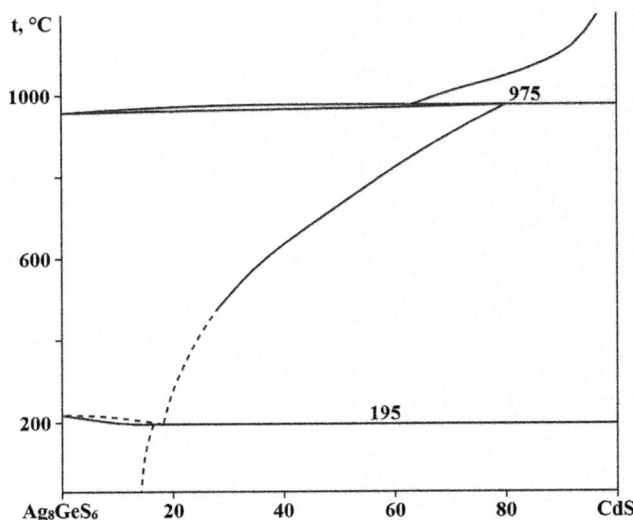

FIGURE 4.30 Phase diagram of the Ag₈GeS₆–CdS system. (From Parasyuk, O.V. et al., *J. Alloys Compd.*, **397**(1–2), 95, 2005b; Piskach L.V. et al., *J. Alloys Compd.*, **421**(1–2), 98, 2006.)

Parasyuk 1998); $a = 804.4 \pm 0.8$, $b = 684.9 \pm 0.5$, $c = 659.3 \pm 0.5$ pm (Parthé et al. 1969)], and a calculated density of 4.8334 ± 0.0006 g·cm⁻³ (Parasyuk et al. 2005c). According to the data of Brunetta et al. (2012b), this compound crystallizes in the orthorhombic structure with the lattice parameters $a = 1374.15 \pm 0.08$, $b = 803.67 \pm 0.05$, $c = 659.07 \pm 0.04$ pm, a calculated density of 4.827 g·cm⁻³, and an energy gap of 2.32 eV. The homogeneity region of Ag₂CdGeS₄ at the eutectic temperature is within the interval from 39 to 50 mol% CdS and is less than 5 mol% CdS at 400°C (Piskach and Parasyuk 1998).

Ag₄CdGe₂S₇ crystallizes in the monoclinic structure with the lattice parameters $a = 1743.64 \pm 0.08$, $b = 683.34 \pm 0.03$, $c = 1053.50 \pm 0.04$ pm, and $\beta = 93.589° \pm 0.003°$ (Gulay et al. 2002a).

The solubility of CdS in Ag₂GeS₃ is an insignificant and solid solution based on CdS contains 11 mol% Ag₂GeS₃ at the peritectic temperature and less than 2 mol% at 400°C (Piskach and Parasyuk 1998). This system was investigated through DTA, XRD, and metallography and the alloys were annealed at 400°C for 250 h.

Ag₂CdGeS₄ was synthesized via high-temperature solid-state synthesis (Brunetta et al. 2012b). Its single crystals were produced using stoichiometric quantities of Ag, Cd, and Ge with some excess of S. The samples were heated to 800°C for 12 h and held at this temperature for 96 h. After a slow cooling step of 5°C·h⁻¹ (60 h) to 500°C, the samples were allowed to cool to ambient temperature.

Ag₈GeS₆–CdS. The phase diagram, constructed through DTA, XRD, and metallography using the alloys annealed at 400°C for 250 h, is a peritectic type (Figure 4.30) (Parasyuk et al. 2005d; Piskach et al. 2006). The peritectic composition and temperature are ~63 mol% CdS and 975°C, respectively. The solid solution based on β-Ag₈GeS₆ contains up to 81 mol% CdS at the peritectic temperature. The region of its existence substantially narrows with temperature decreasing, and at 195°C, it decomposes according to a eutectoid reaction.

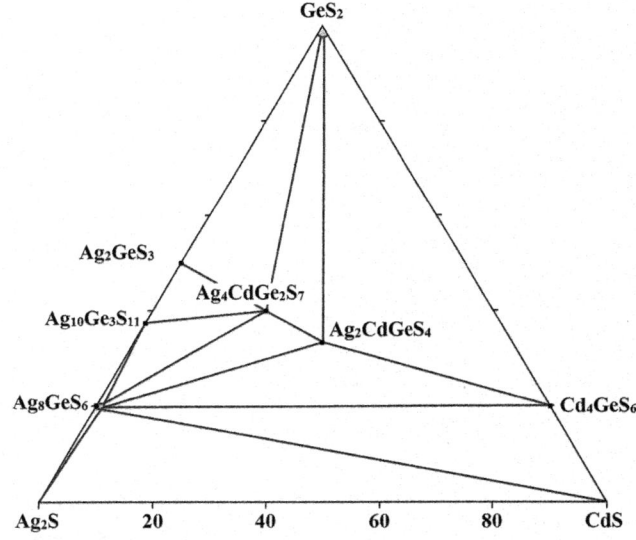

FIGURE 4.31 Isothermal section of the GeS₂–Ag₂S–CdS quasiternary system at 400°C. (From Parasyuk, O.V. et al., *J. Alloys Compd.*, **397**(1–2), 95, 2005.)

GeS₂–Ag₂S–CdS. The isothermal section of this quasiternary system at 400°C is presented in Figure 4.31 (Parasyuk et al. 2005d).

4.136 Germanium–Silver–Mercury–Sulfur

Ag₂GeS₃–HgS. Three quaternary compounds, **Ag₂HgGeS₄**, **Ag₂Hg₃GeS₆**, and **Ag₄HgGe₂S₇**, are formed in this system. Ag₂HgGeS₄ possesses a narrow homogeneity range and crystallizes in the orthorhombic structure with the lattice parameters $a = 802.47 \pm 0.04$, $b = 686.84 \pm 0.03$, and $c = 659.55 \pm 0.04$ pm, and a calculated density of 5.629 ± 0.002 g·cm⁻³

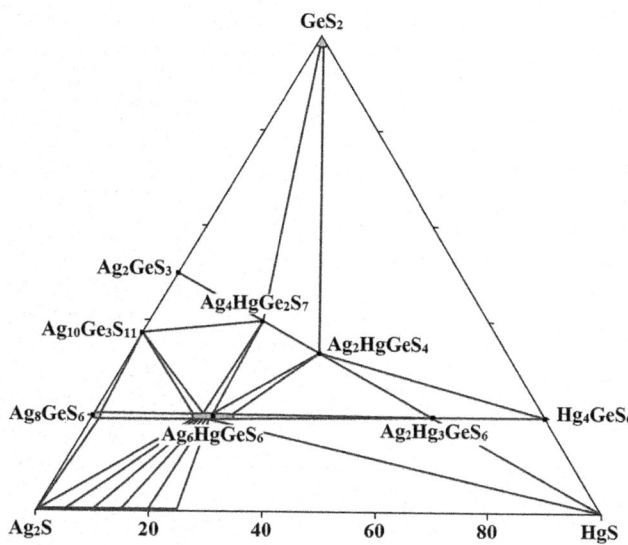

FIGURE 4.32 Isothermal section of the GeS$_2$–Ag$_2$S–HgS quasiternary system at 400°C. (From Parasyuk, O.V. et al., *J. Alloys Compd.*, **336**(1–2), 213, 2002.)

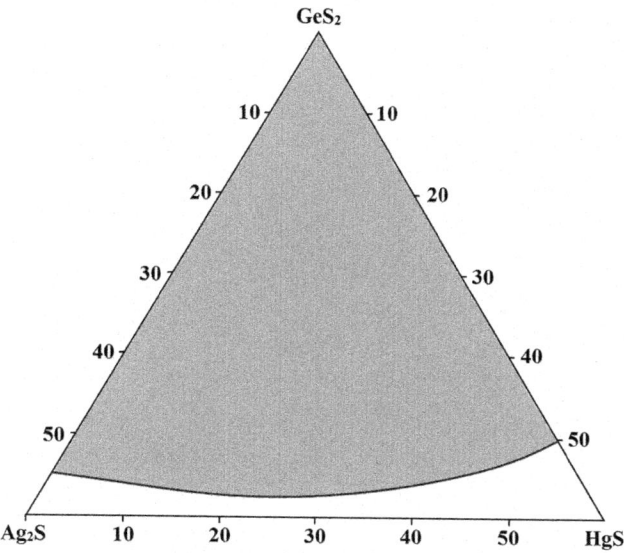

FIGURE 4.33 Glass-forming region in the GeS$_2$–Ag$_2$S–HgS quasiternary system. (From Olekseyuk, I.D. et al., *Chem. Met. Alloys*, **2**(1–2), 49, 2009.)

(Parasyuk et al. 2002a) [a = 800.8, b = 687.1, and c = 659.3 pm (Haeuseler and Himmrich 1989)]. Ag$_2$Hg$_3$GeS$_6$ crystallizes in the tetragonal structure with the lattice parameters a = 1461.9 ± 0.3 and c = 2079.6 ± 0.5 pm (Parasyuk et al. 2002a), and Ag$_4$HgGe$_2$S$_7$ crystallizes in the monoclinic structure with the lattice parameters a = 1745.46 ± 0.08, b = 680.93 ± 0.02, c = 1053.42 ± 0.03 pm, β = 93.398° ± 0.03°, and a calculated density of 5.321 ± 0.003 g·cm^{-3} (Gulay et al. 2002a).

Ag$_2$HgGeS$_4$ was obtained by the interaction of chemical elements at 700°C (Haeuseler and Himmrich 1989). This system was investigated by DTA, XRD, and metallography (Parasyuk et al. 2002a).

GeS$_2$–Ag$_2$S–HgS. The isothermal section of this quasiternary system at 400°C is shown in Figure 4.32 (Parasyuk et al. 2002a). Besides the quaternary compounds, forming in the Ag$_2$GeS$_3$–HgS section, one more quaternary compound **Ag$_6$Hg$_{0.82}$GeS$_{5.82}$** is obtained in the GeS$_2$–Ag$_2$S–HgS quasiternary system. It crystallizes in the cubic structure with the lattice parameter a = 1055.47 ± 0.02 pm and a calculated density of 5.9930 ± 0.0007 g·cm^{-3} (Gulay and Parasyuk 2001). This quaternary compound possesses a homogeneity region along the Ag$_8$GeS$_6$–Hg$_4$GeS$_6$ section in the concentration interval 22–31 mol% Hg$_4$GeS$_6$, and the lattice constant decreases from a = 1055.84 ± 0.07 to 1054.04 ± 0.07 pm in the homogeneity range (Parasyuk et al. 2002a).

The glass-forming region (Figure 4.33) covers the entire concentration triangle for GeS$_2$ concentrations over 50 mol% (Olekseyuk et al. 2009c). The glass transition temperature, changing from 232°C to 340°C, gradually falls with decreasing GeS$_2$ content. For a steady GeS$_2$ concentration, it decreases with increasing HgS content. The optical band gap energy for all glass samples varies from 1.7 to 2.3 eV. For the glassy sample with 50 mol% GeS$_2$, it increases from 1.72 to 2.07 eV as HgS content increases from 16.7 to 33.3 mol%.

This system was investigated by DTA, XRD, and metallography, and the alloys were annealed at 400°C during 250

h (Gulay and Parasyuk 2001, Parasyuk et al. 2002a). After annealing, the ampoules with the samples were quenched in the air. Ag$_6$Hg$_{0.82}$GeS$_{5.82}$ compound was prepared by fusion of Ag, Ge, S, and HgS (Gulay and Parasyuk 2001). Glass alloys were synthesized by the melt-quenching method from high-purity Ag, Ge, S, and HgS (Olekseyuk et al. 2009c).

4.137 Germanium–Silver–Aluminum–Sulfur

The Ag(AlS$_2$)(GeS$_2$)$_4$ quaternary compound, which is thermally stable up to 650°C and crystallizes in the monoclinic structure with the lattice parameters a = 679.8 ± 0.1, b = 3841.6 ± 0.8, c = 681.2 ± 0.1 pm, β = 119.65 ± 0.03°, a calculated density of 3.204 g·cm^{-3} at 200 K, and an energy gap of 1.0 eV, is formed in the Ge–Ag–Al–S system (Alahmari et al. 2018). This compound was prepared by the ion-exchange reaction of Na(AlS$_2$)(GeS$_2$)$_4$ with AgNO$_3$ (molar ratio 1:1), which was carried out at room temperature in 20 mL glass vials. For this, Na(AlS$_2$)(GeS$_2$)$_4$ (50 mg) was added to a solution of 12.8 mg of AgNO$_3$ in extra dry DMF (3 mL). The mixture was slowly stirred for 3 days to ensure the completeness of the ion-exchange reaction. Homogeneous ion-exchange product was isolated in the form of dark gray block-shaped crystals after the reaction products were washed with DMF and dried with diethyl ether. This compound can also be obtained from acetonitrile solution using AgNO$_3$ or AgCl. It is air stable. All sample preparations and handling were carried out in a N$_2$-filled glove box.

4.138 Germanium–Silver–Gallium–Sulfur

GeS$_2$–AgGaS$_2$. The more reliable phase diagram of this system constructed through DTA, XRD, and metallography using the alloys annealed at 450°C for 1000°C with the next quenching in cold water is presented in Figure 4.34 (Olekseyuk et

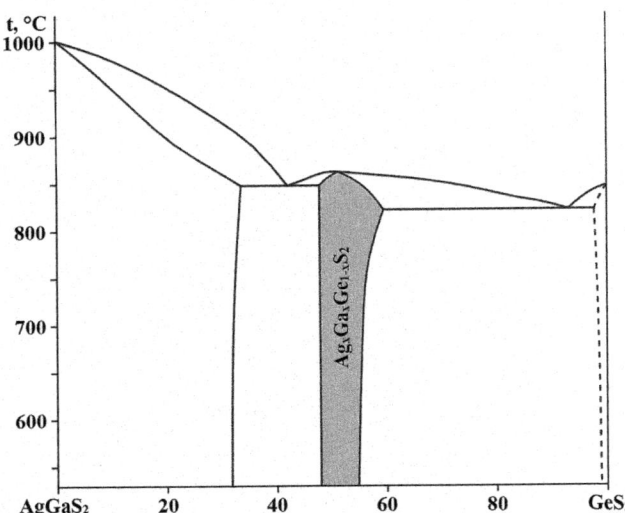

FIGURE 4.34 Phase diagram of the GeS₂–AgGaS₂ system. (From Olekseyuk, I.D., et al., *Pol. J. Chem.*, **76**(7), 915, 2002.)

al. 2002a). The eutectics contain 42 and 93 mol% GeS₂ and crystallize at 848°C and 822°C, respectively. The solid solution based on AgGaS₂ extends to ~30 mol% GeS₂, and the extent of the solid solution based on GeS₂ does not exceed 2 mol% AgGaS₂. In the concentration range of 48–55 mol% GeS₂ at 450°C, the Ag$_x$Ga$_x$Ge$_{1-x}$S₂ phase of variable compositions exists. Within the homogeneity region, the lattice parameters change from $a = 1202.8 \pm 0.3$, $b = 2295.0 \pm 0.1$, $c = 687.9 \pm 0.1$ pm for $x = 0.52$ to $a = 1200.23 \pm 0.03$, $b = 2288.17 \pm 0.07$, $c = 687.30 \pm 0.02$ pm for $x = 0.45$.

The phase diagram of the GeS₂–AgGaS₂ system was also constructed by Badikov et al. (1991) and Chbani et al. (1995).

The **AgGaGeS₄** quaternary compound at the equilibrium vapor pressure of 56.81 kPa melts congruently at 848°C (Vasil'eva and Nikolaev 2006) [melts incongruently at 845°C (Badikov et al. 1991); at 840°C (Chbani et al. 1995)] and crystallizes in the orthorhombic structure with the lattice parameters given in Table 4.2. Vapor pressure measurements demonstrate that the GeS₂–AgGaS₂ join is quasibinary and AgGaGeS₄ is a phase of variable composition with a homogeneity range shifted to AgGaS₂-rich compositions (Vasil'eva and Nikolaev 2006). In the temperature range of 500–870°C, crystalline and molten AgGaGeS₄ decomposes through vaporization of GeS₂, which dissociates in the vapor phase into GeS and S$_n$.

Polycrystalline samples of AgGaGeS₄ can be prepared by the melt and anneal technique from a mixture of the elements or AgGaS₂ and GeS₂ (Yurchenko et al. 2005; Ni et al. 2009; Huang et al. 2017, 2019a). In a typical experiment, the stoichiometric amounts of Ag, Ga, Ge, and S were weight and introduced into a double-wall quartz ampoule which was evacuated to 10⁻⁴ Pa (Huang et al. 2017). The synthesis was carried out in a two-zone furnace. The special process was employed to avoid quartz ampoule explosion, which could be due to the high vapor pressure of sulfur. In the first step, the temperature of the lower zone, where the chemical reaction takes place, is increased to 950°C, and the temperature of the upper zone is kept below 450°C. Gaseous sulfur and GeS₂ would condense and liquefy in the upper zone which reduced the pressure into the quartz reactor. Then, the liquid S and GeS₂ flowed back to the lower zone and involved into the reaction. To keep the upper zone temperature lower than 450°C, the quartz ampoule was pulled back to the furnace mouth. After 24 h at about 950°C in the lower zone, the temperature in the lower room is raised up to 1040°C, and the temperature in the upper zone is raised up to 1080°C in order to consume sulfides condensed on ampoule walls, and the quartz ampoule was placed back to the initial position. Then, after 12 h, the furnace was vertically positioned. At the same time, mechanical and temperature oscillations were adopted, and the ampoule was cooled down to room temperature. Yellow polycrystalline ingot of the title compound was obtained.

Single crystals of AgGaGeS₄ have been grown by the modified Bridgman–Stockbarger technique (Yurchenko et al. 2005; Khyzhun et al. 2014; Huang et al. 2019a) as well as by vapor transport in sealed quartz ampoule from the elements (Schunemann et al. 2006).

The phase identification of AgGaGeS₄·nGeS₂ ($n = 0$–4) crystals grown by vertical Bridgman–Stockbarger technique was carried out to find the boundary value n between a homogeneous solid solution and its mixture with GeS₂ (Nikolaev and Vasilyeva 2013). To obtain reliable results, the conventional methods of X-ray diffraction (XRD) and energy dispersive X-ray analysis (EDX) were completed by less common vapor pressure measurements, which are very sensitive to the detection of small amounts of crystalline and glassy GeS₂ and heterogeneous state of the crystals. The boundary value $n = 1.5$ at 772°C and the coexistence of the solid solution AgGaGeS₄·1.5GeS₂ with β-GeS₂ for $n > 1.5$ was found. Glassy

Table 4.2

Crystallographic data for AgGaGeS₄

a, pm	b, pm	c, pm	$d_{calc.}$	$d_{meas.}$	References
			g·cm⁻³		
1202.8 ± 0.3	2291.8 ± 0.8	687.4 ± 0.1	–	–	Pobedimskaya et al. (1981); Alimova et al. (1981)
1201.5	2290.4	687.4	–	–	Yurchenko et al. (2005)
1202.8 ± 1.0	2290.9 ± 2.0	687.8 ± 2.0	–	–	Ni et al. (2009)
1201.9 ± 0.2	2292.4 ± 0.3	688.37 ± 0.07	–	–	Khyzhun et al. (2014)
1204.1	2291.9	687.1	–	–	Huang et al. (2017)
1203.94 ± 0.02	2292.95 ± 0.05	687.67 ± 0.01	3.9726 ± 0.0007	–	Huang et al. (2019a)

GeS$_2$ (~2 mol%) was revealed by vapor pressure measurement and XRD studies in all the crystals.

Polycrystalline samples of AgGaGenS$_{2(n+1)}$ (n = 2, 3, 4, 5) were synthesized by vapor transport with mechanical oscillation method using different cooling processes, and there single crystals were grown by the modified Bridgman method (Huang et al. 2019b). All these phases crystallize in the orthorhombic structure with the lattice parameters a = 1193.39 ± 0.03, b = 2280.13 ± 0.05, c = 687.35 ± 0.01 pm, and a calculated density of 3.6594 ± 0.0006 g·cm^{-3} for n = 2; a = 1188.76 ± 0.03, b = 2272.25 ± 0.06, c = 687.58 ± 0.01 pm, and a calculated density of 3.4971 ± 0.0007 g·cm^{-3} for n = 3; a = 1185.45 ± 0.03, b = 2266.60 ± 0.05, c = 687.65 ± 0.01 pm, and a calculated density of 3.4020 ± 0.0007 g·cm^{-3} for n = 4; and a = 1183.88 ± 0.03, b = 2263.14 ± 0.05, c = 687.66 ± 0.01 pm, and a calculated density of 3.3373 ± 0.0006 g·cm^{-3} for n = 5. The energy gap of these phases is 2.85*/2.89** eV for n = 2, 2.89*/2.90** eV for n = 3, 2.90*/2.92** eV for n = 4, and 2.92*/2.96** eV for n = 5 for the samples before (*) and after (**) annealing. It is possible that all these phases form one continuous solid solution.

GeS$_2$–Ag$_2$S–Ga$_2$S$_3$. The liquidus surfaces, as well as some vertical sections of this quasiternary system were constructed by Chbani et al. (1995). Four ternary invariants were discovered: two eutectics and two transition points. The field of AgGaS$_2$ primary crystallization is very extensive and that of AgGaGeS$_4$ is very particular.

The glass-forming region of this system is presented in Figure 4.35 (Julien et al. 1994; Chbani et al. 1995). The glass transition temperature changes from 277°C to 414°C. The glasses were obtained by quenching from the liquid state. Homogeneous samples were prepared from the binary sulfides mixed according to convenient proportions. The mixtures were melted in evacuated silica ampoules at 1100°C for 24 h and then quenched in cold water. The glasses were yellow when lightly weighted with Ag$_2$S and became orange or even red when silver enriched.

4.139 Germanium–Silver–Indium–Sulfur

GeS$_2$–AgInS$_2$. According to the data of Pamplin et al. (1977), the phase diagram of this system is a eutectic type and the eutectic crystallizes at 718°C. The solubility of GeS$_2$ in AgInS$_2$ is negligible.

The **Ag$_2$In$_2$GeS$_6$** quaternary compound, which crystallizes in the monoclinic structure with the lattice parameters a = 1220.89 ± 0.04, b = 721.15 ± 0.03, c = 1219.78 ± 0.05 pm, β = 109.508 ± 0.002°, and a calculated density of 4.6604 ± 0.0006 g·cm^{-3} (Sachanyuk et al. 2008) [a = 1223.16 ± 0.05, b = 721.90 ± 0.04, c = 1221.24 ± 0.05 pm, β = 109.532 ± 0.003°, and an energy gap of 1.96 eV (Reshak et al. 2013a)], is formed in the Ge–Ag–In–S system. The results indicate the comparatively low hygroscopicity of Ag$_2$In$_2$GeS$_6$.

GeS$_2$–Ag$_2$S–In$_2$S$_3$. The isothermal section of this quasiternary system at room temperature is shown in Figure 4.36 (Sachanyuk et al. 2008). Five ternary compounds exist in the system, and Ag$_2$In$_2$GeS$_6$ forms equilibria with all available ternary sulfides and GeS$_2$. A significant homogeneity region is observed only for AgIn$_5$S$_8$, where it is localized along the Ag$_2$S–In$_2$S$_3$ section. Some vertical sections of this system were also constructed.

4.140 Germanium–Silver–Yttrium–Sulfur

The **Ag$_{0.50}$Y$_3$GeS$_7$** quaternary compound, which crystallizes in the hexagonal structure with the lattice parameters a = 980.90 ± 0.14, c = 580.59 ± 0.12 pm, and a calculated density of 4.240 g·cm^{-3} (Daszkiewicz et al. 2009a), is formed in the Ge–Ag–Y–S system. It was prepared by sintering the elemental constituents with the atomic ratio Y/Ag/Ge/S = 3:1:1:7 in an evacuated silica ampoule in a tube furnace. The ampoule was heated at a rate of 30°C·h^{-1} to a maximum temperature of

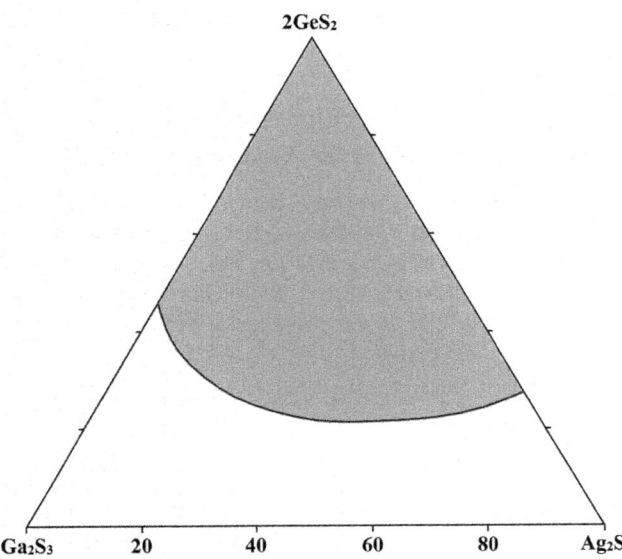

FIGURE 4.35 Glass-forming region in the GeS$_2$–Ag$_2$S–Ga$_2$S$_3$ quasiternary system. (From Julien, C., et al., *Mater. Sci. Eng.*, **B22**(2–3), 191, 1994.)

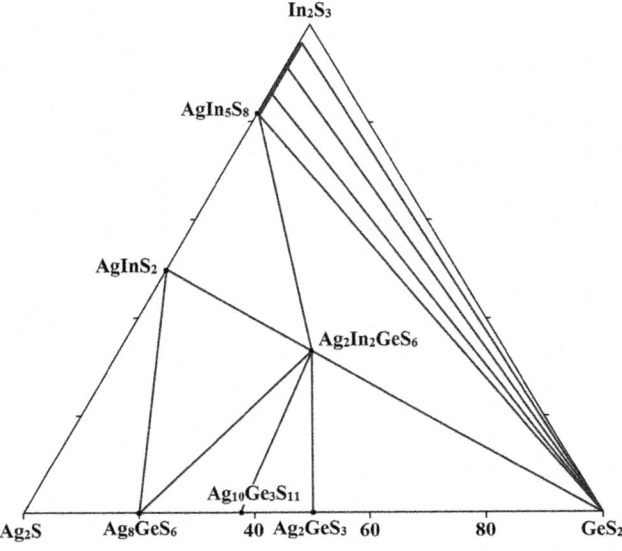

FIGURE 4.36 Isothermal section of GeS$_2$–Ag$_2$S–In$_2$S$_3$ quasiternary system at room temperature. (From Sachanyuk, V.P., et al., *J. Alloys Compd.*, **452**(2), 348, 2008.)

1150°C and kept at this temperature for 3 h. Afterwards, the ampoule was cooled slowly (10°C·h^{-1}) to 600°C, annealed at this temperature for 240 h, and then quenched in cold water.

4.141 Germanium–Silver–Lanthanum–Sulfur

The **AgLa$_3$GeS$_7$** quaternary compound, which melts congruently at ca. 1196°C (Hwu et al. 1995) and crystallizes in the hexagonal structure with the lattice parameters $a = 1040.56 \pm 0.15$, $c = 582.80 \pm 0.12$ pm, and a calculated density of 4.875 g·cm^{-3} for Ag$_{0.82}$La$_3$GeS$_7$ (Daszkiewicz et al. 2009a) [$a = 1036 \pm 1$ and $c = 585.0 \pm 0.5$ pm (Guittard et al. 1968; Guittard and Julien-Pouzol 1970); $a = 1042.3 \pm 0.1$, $c = 584.0 \pm 0.2$ pm, a calculated density of 4.966 g·cm^{-3}, and an energy gap of 2.50 eV (Hwu et al. 1995)], is formed in the Ge–Ag–La–S system.

AgLa$_3$GeS$_7$ can be prepared from a mixture of La$_2$S$_3$, Cu$_2$S, Ge, and S, which was pressed and introduced into the silica ampoule and gradually heats up to 800–850°C in about a week (Guittard and Julien-Pouzol 1970). At the end of the preparation, it is necessary to heat the sample to 1200°C to obtain the pure hexagonal phase. Single crystals of this compound have been grown using a KBr flux (Hwu et al. 1995). A stoichiometric mixture of the elements and Ag$_2$S with respect to the nominal composition "AgLaGeS$_4$" were sealed in an evacuated quartz ampoule and heated to 750°C for 6 days. Single crystals were grown from a KBr flux, with 10:1 charge to flux ratio by mass, heated to 1000°C and followed by slow cooling at 2°C·h^{-1} to 600°C. Crystals were isolated by washing with distilled water under suction filtration. The subsequent stoichiometric synthesis was carried out by a direct reaction of stoichiometric mixtures of Ag$_2$S, La, Ge, and S in a fused silica ampoule at 750°C for 5 days.

Ag$_{0.82}$La$_3$GeS$_7$ was prepared by sintering the elemental constituents with the atomic ratio La/Ag/Ge/S = 3:1:1:7 in an evacuated silica ampoule in a tube furnace (Daszkiewicz et al. 2009a). The ampoule was heated at a rate of 30°C·h^{-1} to a maximum temperature of 1150°C and kept at this temperature for 3 h. Afterwards the ampoule was cooled slowly (10°C·h^{-1}) to 600°C, annealed at this temperature for 240 h and then quenched in cold water.

4.142 Germanium–Silver–Cerium–Sulfur

The **AgCe$_3$GeS$_7$** quaternary compound, which crystallizes in the hexagonal structure with the lattice parameters $a = 1039.02 \pm 0.15$, $c = 584.25 \pm 0.12$ pm, and a calculated density of 4.939 g·cm^{-3} for Ag$_{0.88}$Ce$_3$GeS$_7$ (Daszkiewicz et al. 2009a) [$a = 1013$ and $c = 585$ pm (Guittard and Julien-Pouzol 1970)], is formed in the Ge–Ag–Ce–S system. AgCe$_3$GeS$_7$ and Ag$_{0.88}$Ce$_3$GeS$_7$ were synthesized in the same way as AgLa$_3$GeS$_7$ and Ag$_{0.82}$La$_3$GeS$_7$ were obtained.

4.143 Germanium–Silver–Praseodymium–Sulfur

The **AgPr$_3$GeS$_7$** quaternary compound, which crystallizes in the hexagonal structure with the lattice parameters $a = 1022.90 \pm 0.14$, $c = 577.60 \pm 0.11$ pm, and a calculated density of 5.176 g·cm^{-3} for Ag$_{0.90}$Pr$_3$GeS$_7$ (Daszkiewicz et al. 2009a) [$a = 1004$

and $c = 584$ pm (Guittard and Julien-Pouzol 1970)], is formed in the Ge–Ag–Pr–S system. AgPr$_3$GeS$_7$ and Ag$_{0.90}$Pr$_3$GeS$_7$ were synthesized in the same way as AgLa$_3$GeS$_7$ and Ag$_{0.82}$La$_3$GeS$_7$ were obtained.

4.144 Germanium–Silver–Neodymium–Sulfur

The **Ag$_{0.84}$Nd$_3$GeS$_7$** quaternary compound, which crystallizes in the hexagonal structure with the lattice parameters $a = 1019.30 \pm 0.14$, $c = 576.93 \pm 0.12$ pm, and a calculated density of 5.245 g·cm^{-3} (Daszkiewicz et al. 2009a), is formed in the Ge–Ag–Nd–S system. It was synthesized in the same way as Ag$_{0.82}$La$_3$GeS$_7$ was prepared.

4.145 Germanium–Silver–Samarium–Sulfur

The **Ag$_{0.74}$Sm$_3$GeS$_7$** quaternary compound, which crystallizes in the hexagonal structure with the lattice parameters $a = 1008.09 \pm 0.14$, $c = 576.04 \pm 0.12$ pm, and a calculated density of 5.420 g·cm^{-3} (Daszkiewicz et al. 2009a), is formed in the Ge–Ag–Sm–S system. It was synthesized in the same way as Ag$_{0.82}$La$_3$GeS$_7$ was prepared.

4.146 Germanium–Silver–Europium–Sulfur

The **Ag$_{0.5}$Eu$_{1.75}$GeS$_4$** quaternary compound, which crystallizes in the cubic structure with the lattice parameter $a = 1394.91 \pm 0.14$ pm, calculated density of 5.097 g·cm^{-3} at 173 K, and an energy gap of 2.14 eV, is formed in the Ge–Ag–Eu–S system (Iyer et al. 2004). It was synthesized from a reaction of Ag, Eu, Ge, and S (molar ratio 0.5:1.75:1:4). The reactants were loaded in a fused silica tube and flame-sealed under vacuum (< 0.1 Pa). The tube was then put in a furnace, and the temperature was ramped to 850°C and kept at that temperature for 48 h, and then cooled to room temperature. Black crystals of the target compound were obtained.

4.147 Germanium–Silver–Gadolinium–Sulfur

The **Ag$_{0.63}$Gd$_3$GeS$_7$** quaternary compound, which crystallizes in the hexagonal structure with the lattice parameters $a = 996.37 \pm 0.14$, $c = 576.60 \pm 0.14$ pm and a calculated density of 5.602 g·cm^{-3} (Daszkiewicz et al. 2009a), is formed in the Ge–Ag–Gd–S system. It was synthesized in the same way as Ag$_{0.82}$La$_3$GeS$_7$ was prepared.

4.148 Germanium–Silver–Terbium–Sulfur

The **Ag$_{0.59}$Tb$_3$GeS$_7$** quaternary compound, which crystallizes in the hexagonal structure with the lattice parameters $a = 990.03 \pm 0.14$, $c = 576.54 \pm 0.12$ pm, and a calculated density of 5.679 g·cm^{-3} (Daszkiewicz et al. 2009a), is formed in the Ge–Ag–Tb–S system. It was synthesized in the same way as Ag$_{0.82}$La$_3$GeS$_7$ was prepared.

4.149 Germanium–Silver–Dysprosium–Sulfur

The $Ag_{0.51}Dy_3GeS_7$ quaternary compound, which crystallizes in the hexagonal structure with the lattice parameters $a = 980.03 \pm 0.14$, $c = 578.79 \pm 0.12$ pm, and a calculated density of 5.795 g·cm^{-3} (Daszkiewicz et al. 2009a), is formed in the Ge–Ag–Dy–S system. It was synthesized in the same way as $Ag_{0.82}La_3GeS_7$ was prepared.

4.150 Germanium–Silver–Holmium–Sulfur

The $Ag_{0.50}Ho_3GeS_7$ quaternary compound, which crystallizes in the hexagonal structure with the lattice parameters $a = 974.01 \pm 0.14$, $c = 579.94 \pm 0.12$ pm, and a calculated density of 5.895 g·cm^{-3} (Daszkiewicz et al. 2009a), is formed in the Ge–Ag–Ho–S system. It was synthesized in the same way as $Ag_{0.82}La_3GeS_7$ was prepared.

4.151 Germanium–Silver–Erbium–Sulfur

The $Ag_{0.50}Er_3GeS_7$ quaternary compound, which crystallizes in the hexagonal structure with the lattice parameters $a = 969.21 \pm 0.14$, $c = 583.08 \pm 0.12$ pm, and a calculated density of 5.970 g·cm^{-3} (Daszkiewicz et al. 2009a), is formed in the Ge–Ag–Er–S system. It was synthesized in the same way as $Ag_{0.82}La_3GeS_7$ was prepared.

4.152 Germanium–Silver–Tin–Sulfur

$Ag_8GeS_6–Ag_8SnS_6$ The continuous solid solution series between low- and high-temperature modifications of both initial compounds were found in this system (Figure 4.37)

(Aliyeva et al. 2014). The linear increase of the lattice parameters was observed with increasing the Ag_8SnS_6 content.

$Ag_8GeS_6–Ag_8SnS_6–Ag_2S$. The isothermal section at room temperature, the liquidus surface, and two vertical sections of this system were constructed by Aliyeva et al. (2014). No quaternary compounds were detected. The liquidus surface (Figure 4.38) consists of two fields corresponding to the primary crystallization of the solid solution based on Ag_2S and high-temperature $Ag_8Ge_{1-x}Sn_xS_6$ solid solution. The partial molar thermodynamic functions of Ag were calculated for the low-temperature $Ag_8Ge_{1-x}Sn_xS_6$ solid solution based on the results of the EMF measurements.

The phase equilibria in the $Ag_8GeS_6–Ag_8SnS_6–Ag_2S$ system were investigated experimentally through DTA, XRD, SEM, and EMF measurements of the concentration chains.

4.153 Germanium–Silver–Lead–Sulfur

$PbS–Ag_8GeS_6$. The phase diagram of this system, constructed through DTA, XRD, and metallography, is a eutectic type (Figure 4.39) (Bilousov et al. 2006). The eutectic contains 72 mol% PbS and crystallizes at 678°C. The eutectoid transformation $\gamma\text{-}Ag_8GeS_6 \leftrightarrow \beta\text{-}Ag_8GeS_6 + PbS$ takes place at 205°C. The solubility of PbS in Ag_8GeS_6 at the eutectic temperature is 13 mol% and <5 mol% at room temperature.

$GeS_2–Ag_2S–PbS$. The isothermal section of this quasiternary system is shown in Figure 4.40 (Kogut et al. 2011). Two quaternary compounds, $Ag_{0.5}Pb_{1.75}GeS_4$ and Ag_2PbGeS_4, exist in this system. Ten three-phase fields were determined in the system. Additionally, 19 two-phase equilibria of binary, ternary, and quaternary compounds were identified. No significant solid solution ranges were found.

$Ag_{0.5}Pb_{1.75}GeS_4$ melts at 673°C and crystallizes in the cubic structure with the lattice parameter $a = 1402.91 \pm 0.05$ pm, calculated density of 5.940 g·cm^{-3} at 173 K, and an energy gap of

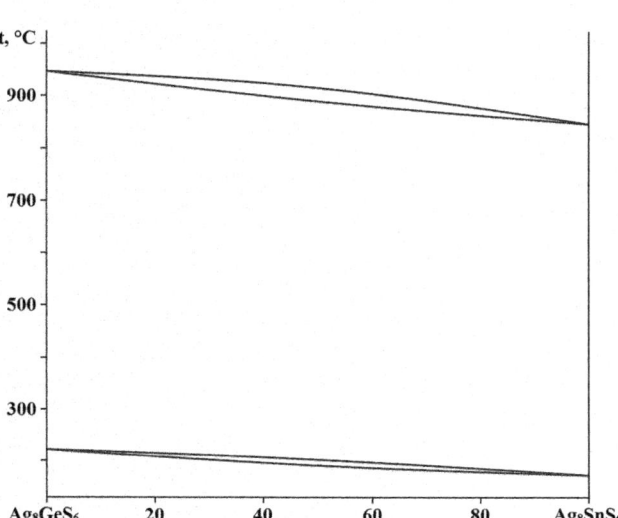

FIGURE 4.37 Phase diagram of the $Ag_8GeS_6–Ag_8SnS_6$ system. (From Aliyeva, Z.M., et al., *J. Alloys Compd.*, **611**, 395, 2014.)

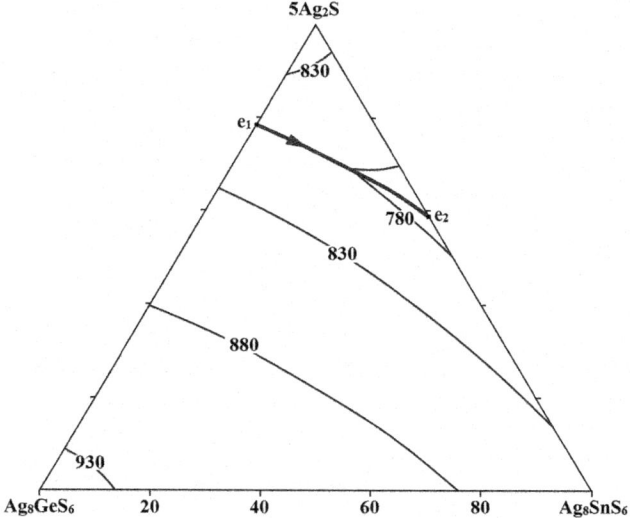

FIGURE 4.38 Liquidus surface of the $Ag_8GeS_6–Ag_8SnS_6–Ag_2S$ quasiternary system. (From Aliyeva, Z.M., et al., *J. Alloys Compd.*, **611**, 395, 2014.)

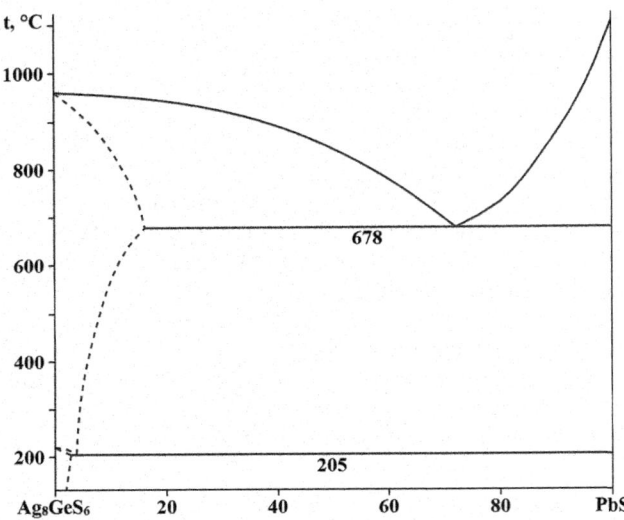

FIGURE 4.39 Phase diagram of the PbS–Ag$_8$GeS$_6$ system. (From Bilousov, O.V., et al., *Nauk. Visnyk Volyns'k. Derzh. Univ. im. Lesi Ukrainky. Ser. Khim. nauky*, (4), 128, 2006.)

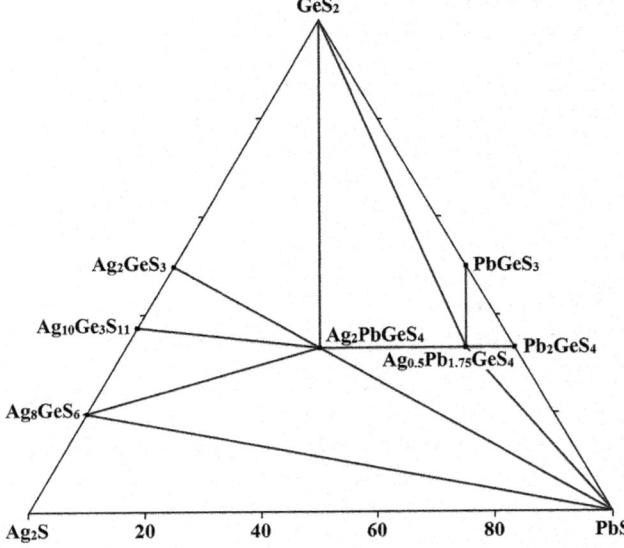

FIGURE 4.40 Isothermal section of GeS$_2$–Ag$_2$S–PbS quasiternary system at room temperature. (From Kogut, Yu. et al., *J. Alloys Compd.*, **509**(11), 4264, 2011.)

1.83 eV (Iyer et al. 2004). Ag$_2$PbGeS$_4$ crystallizes in the orthorhombic structure with the lattice parameters $a = 1023.13 \pm 0.08$, $b = 1025.58 \pm 0.05$, $c = 676.84 \pm 0.07$ pm, and an energy gap of 1.68 eV (Nhalil et al. 2018) [$a = 1023.90 \pm 0.04$, $b = 1025.87 \pm 0.05$, $c = 677.01 \pm 0.03$ pm, and a calculated density of 3.8257 ± 0.0008 g·cm^{-3} (Kogut et al. 2011)].

Ag$_{0.5}$Pb$_{1.75}$GeS$_4$ was prepared by a stoichiometric combination of the elements (Iyer et al. 2004). The reactants were loaded in a fused silica tube and flame-sealed under vacuum (< 0.1 Pa). The tube was then put in a furnace, and the temperature was ramped up to 250°C where it was maintained for 6 h. The tube was then heated to 650°C, kept at that temperature for 48 h, and then cooled to 250°C at a rate of 5°C·h^{-1} followed by rapid cooling to room temperature.

To synthesize Ag$_2$PbGeS$_4$, stoichiometric amount of elements were weighed and mixed properly using a mortar and pestle before pelletizing and loading into a quartz ampoule in a N$_2$-filled glove box (Nhalil et al. 2018). The quartz tube was flame-sealed under a dynamic vacuum (< 10^{-4} Pa). The reaction mixture was heated at 1000°C for 24 h, and then cooled to room temperature at a rate of 100°C·h^{-1}. To improve the phase purity and crystallinity, the sample was reground, pelletized, and annealed under identical conditions as necessary. This compound is stable in ambient air for a period of 6 weeks.

According to the data of Meng et al. (2022, 2023), one more quaternary compound (Ag$_2$□)Pb$_3$Ge$_2$S$_8$ (mineral ruizhongite), which crystallizes in the cubic structure with the lattice parameter $a = 1405.59 \pm 0.02$ pm, is formed in the GeS$_2$–Ag$_2$S–PbS quasiternary system. It is necessary to note, that this compound as well as Ag$_{0.5}$Pb$_{1.75}$GeS$_4$ are situated on the Pb$_2$GeS$_4$–Ag$_2$PbGeS$_4$ section.

As a result of the triangulation of the GeS$_2$–Ag$_8$GeS$_6$–GePbS part of the Ge–Ag–Pb–S system using DTA, XRD, and EMF methods, the analytical equations of the temperature

dependence of Gibbs energy of formation of the saturated solid solutions based on Ag$_{0.5}$Pb$_{1.75}$GeS$_4$, Ag$_2$PbGeS$_4$, and Ag$_{6.72}$Pb$_{0.16}$Ge$_{0.84}$S$_{5.20}$ in the range of 240–290°C were obtained (Moroz et al. 2017a). The values of the standard thermodynamic functions of the saturated solid solutions were calculated: $\Delta G^0_{f,298} = -388.2 \pm 2.3$ kJ·M^{-1}, $\Delta H^0_{f,298} = -350.0 \pm 10.2$ kJ·M^{-1} and $\Delta S^0_{f,298} = 128.2 \pm 3.4$ J·(M·K)$^{-1}$for (Ag$_{0.5}$Pb$_{1.75}$GeS$_4$); $\Delta G^0_{f,298} = -325.7 \pm 0.7$ kJ·M^{-1}, $\Delta H^0_{f,298} = -303.6 \pm 4.3$ kJ·M^{-1} and $\Delta S^0_{f,298} = 74.1 \pm 0.6$ J·(M·K)$^{-1}$ for (Ag$_2$PbGeS$_4$); $\Delta G^0_{f,298} = -401.8 \pm 1.5$ kJ·M^{-1}, $\Delta H^0_{f,298} = -273.5 \pm 1.9$ kJ·M^{-1} and $\Delta S^0_{f,298} = 430.6 \pm 1.8$ J·(M·K)$^{-1}$ for (Ag$_{6.72}$Pb$_{0.16}$Ge$_{0.84}$S$_{5.20}$).

4.154 Germanium–Silver–Phosphorus–Sulfur

GeS$_2$–Ag$_2$S–P$_2$S$_5$. The glass-forming region of this system is presented in Figure 4.41 (Bereznyuk and Petrus' 2020). The maximum content of Ag$_2$S, which is part of the glasses, is 70 mol%.

4.155 Germanium–Silver–Selenium–Sulfur

GeSe$_2$–Ag$_4$SSe. The phase diagram of this system, constructed through DTA, XRD, and measuring of the microhardness and density, is shown in Figure 4.42 (Vassiliev et al. 2012). The eutectics contain ~25 and 95 mol% Ag$_4$SSe and crystallize at 600°C and 797°C, respectively, and the eutectoid exists at 95 mol% Ag$_4$Sse and 65°C. The **Ag$_8$GeS$_2$Se$_4$** quaternary compound is formed in this system. It melts congruently at 910°C, has a polymorphic transformation at 715°C, and a wide homogeneity region at low temperatures, which is shifted toward the GeSe$_2$-rich side. This compound crystallizes in the orthorhombic structure with the lattice parameters $a = 623.05$, $b = 325.31$,

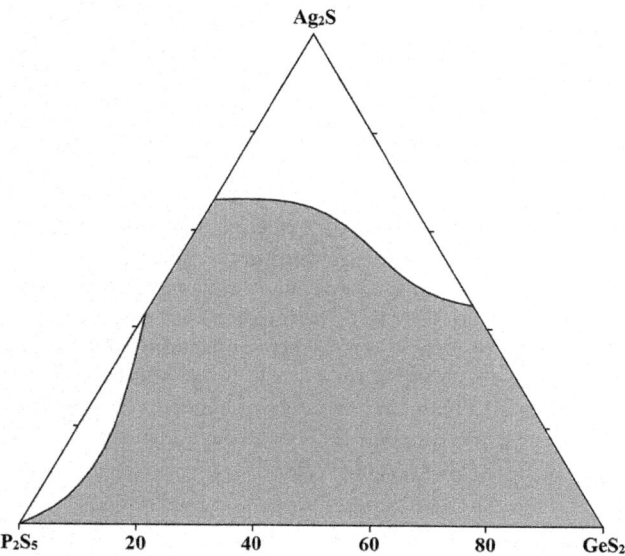

FIGURE 4.41 Glass-forming region in the GeS₂–Ag₂S–P₂S₅ quasiternary system. (From Bereznyuk, O.P., and Petrus', I.I., *Nauk. visn. Uzhgorod. Univ., Ser. Khim.*, [2(44)], 41, 2020.)

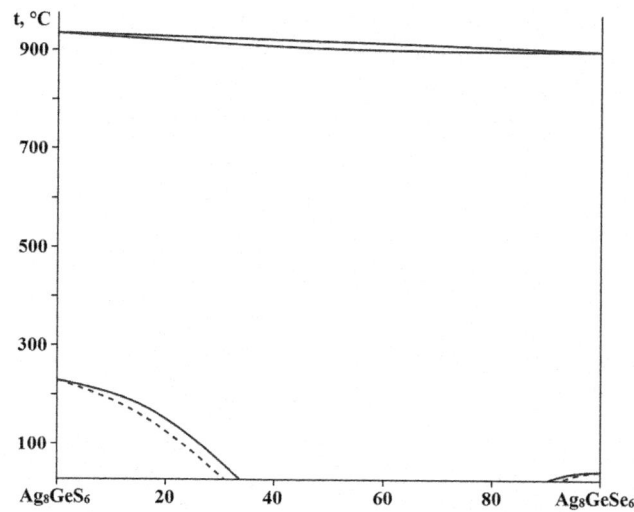

FIGURE 4.43 Phase diagram of the Ag₈GeS₆–Ag₈GeSe₆ system. (From Bagheri S.M., *Azerb. khim. zhurn.*, (3), 15, 2014.)

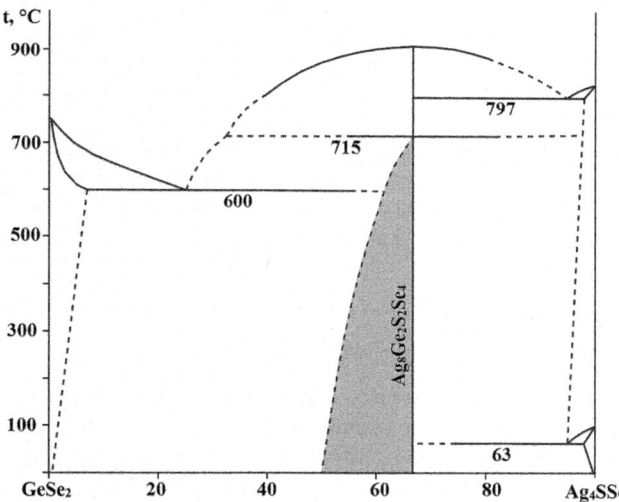

FIGURE 4.42 Phase diagram of the GeS₂–Ag₄SSe system. (From Vasiliev, V., et al., *J. Phase Equilib. Diffus.*, **33**(2), 106, 2012.)

and $c = 247.89$ pm. There is some doubt about the existence of the Ag₄SSe ternary compound.

Ag₈GeS₆–Ag₈GeSe₆. The phase diagram of this system, constructed through DTA, XRD, and EMF measurements with the solid electrolyte, is given in Figure 4.43 (Bagheri et al. 2014b). It was established that the system is characterized by continuous solubility in the liquid state and also between high-temperature cubic modifications of starting compounds. The formation of high-temperature solid solutions decreases the temperature of polymorphic transitions of both components. According to the data of Mikolaychuk and Moroz (1985a), both high- and low-temperature modifications of Ag₈GeSe₆ and Ag₈GeSe₆ form continuous solid solutions.

4.156 Germanium–Silver–Tellurium–Sulfur

The **Ag₈GeTe₂S₄** quaternary compound (mineral alburnite), which crystallizes in the cubic structure with the lattice parameter $a = 1040 \pm 10$ pm and a calculated density of 7.828 g·cm⁻³, is formed in the Ge–Ag–Te–S system (Tămaş et al. 2013, 2014).

4.157 Germanium–Silver–Chlorine–Sulfur

The **Ag₈₋ₓGeS₆₋ₓClₓ** quaternary phase is formed in the Ge–Ag–Cl–S system. It crystallizes in the orthorhombic structure with the lattice parameters $a = 643.0 \pm 0.1$, $b = 761.0 \pm 0.2$, $c = 2266.0 \pm 0.5$ pm, and calculated density of 5.505 g·cm⁻³ at $x = 2$ (Wagener et al. 2006). Ag₈₋ₓGeS₆₋ₓClₓ (x value is not specified) has a polymorphic transformation at 150°C and also crystallizes in the orthorhombic structure with the lattice parameters $a = 1510.9 \pm 0.9$, $b = 747.9 \pm 0.4$, and $c = 1059.9 \pm 0.7$ pm (Kuhs et al. 1979).

The phase with $x = 2$ was prepared by the reaction of intimately mixed samples of Ag, Ge, and S (molar ratio 6:1:4 and a typical mass sum of 1 g) together with 0.5 mL concentrated HCl (Wagener et al. 2006). Alternatively, the preparation can be carried out by starting from a stoichiometric mixture of Ag₂S, AgCl, and GeS₂. The hydrothermal reaction is carried out in a silica glass ampoule which is placed in an autoclave filled with an appropriate quantity of H₂O in order to generate a balancing counter pressure preventing the silica ampoule from exploding. The autoclave was heated for 2 weeks up to 230°C and subsequently cooled down to room temperature with a cooling rate of 2°C·h⁻¹. An inhomogeneous reaction product that contains black, red, and colorless crystalline parts with the Ag₆GeS₄Cl₂ as the main products were formed. The excess of

AgCl impurity was washed out with a diluted Na$_2$S$_2$O$_3$ solution (~ 0.5 M·L^{-1}). Subsequently, S was dissolved by washing with CS$_2$ or careful heating to 150°C, but the remaining GeS$_2$ cannot be removed by this procedure. The transparent red needles of the title compound can be separated under the light microscope. It decomposes irreversibly between 150°C and 500°C.

Ag$_{8-x}$GeS$_{6-x}$Cl$_x$ (x value is not specified) was synthesized by reacting stoichiometric amounts of the elements in an evacuated, sealed quartz ampoule for about 6 days at temperatures of 600°C to 700°C (Kuhs et al. 1979).

4.158 Germanium–Silver–Bromine–Sulfur

Three quaternary phases, **Ag$_3$GeS$_3$Br**, **Ag$_3$Ge$_2$S$_5$Br**, and **Ag$_{8-x}$GeS$_{6-x}$Br$_x$**, are formed in the Ge–Ag–Br–S system. The first is a glass-forming phase of variable composition (Moroz et al. 2013a). The crystallographic parameters of the saturated solid solution of GeS$_2$ in Ag$_3$GeS$_3$Br are as follows: cubic structure, $a = 1016.572 \pm 0.003$ pm. It melts incongruently at 440°C (Moroz et al. 2013b). Ag$_3$Ge$_2$S$_5$Br also melts incongruently at 454°C and also crystallizes in the cubic structure with the lattice parameter $a = 1016.702 \pm 0.007$ pm and calculated density of 4.481 g·cm^{-3} (Moroz et al. 2014a). This compound was also obtained in the glassy state. Ag$_{8-x}$GeS$_{6-x}$Br$_x$ ($x = 2$) crystallizes in the orthorhombic structure with the lattice parameters $a = 653.892 \pm 0.005$, $b = 772.656 \pm 0.005$ and $c = 2290.338 \pm 0.017$ pm (Moroz et al. 2010) [$a = 654.0 \pm 0.1$, $b = 772.7 \pm 0.2$, $c = 2289.9 \pm 0.5$ pm, and a calculated density of 5.785 g·cm^{-3} (Wagener et al. 2006)]. Ag$_{8-x}$GeS$_{6-x}$Br$_x$ (x value is not specified) crystallizes in the cubic structure with the lattice parameters $a = 1065.0 \pm 0.3$ pm (Kuhs et al. 1979) [melts at 896°C and crystallizes in the cubic structure with the lattice parameters $a = 1065.09 \pm 0.01$ pm at $x = 1$ (Laqibi et al. 1987)].

Ag$_{8-x}$GeS$_{6-x}$Br$_x$ with $x = 2$ and with x not specified were prepared in the same way as chlorine analogue compound were obtained using concentrated HBr and AgBr instead of concentrated HCl and AgCl (Kuhs et al. 1979; Wagener et al. 2006). The phase with $x = 2$ decomposes irreversibly between 150°C and 500°C. Ag$_{8-x}$GeS$_{6-x}$Br$_x$ with $x = 2$ can also be synthesized in the solid state at 177–403°C from appropriate quantities of Ag$_2$S, GrS$_2$, and AgBr or Ag$_3$SBr, Ag$_2$GeS$_3$, and AgBr (Moroz et al. 2010).

GeS$_2$–Ag$_2$S–AgBr. The homogeneous glassy alloys in this quasiternary system have been prepared along the Ag$_2$GeS$_3$–AgBr section in the concentration range of 0–53 mol% AgBr (Moroz et al. 2013b), along the Ag$_3$GeS$_3$Br–GeS$_2$ section in the region of the solid solution based on Ag$_3$GeS$_3$Br (Moroz et al. 2013a), and along the Ag$_{0.225}$Ge$_{0.260}$S$_{0.515}$–AgBr section in the composition range 1.0–9.6 mol% AgBr (Moroz et al. 2014b). Some of these glasses are made of superionic materials.

4.159 Germanium–Silver–Iodine–Sulfur

The **Ag$_{8-x}$GeS$_{6-x}$I** quaternary phase, which melts at 907°C and crystallizes in the cubic structure with the lattice parameter $a = 1071.16 \pm 0.01$ pm for $x = 1$ (Laqibi et al. 1987) [$a = 1071.3 \pm$ 0.5 pm (x value is not specified) (Kuhs et al. 1979); $a = 1072.2 \pm 0.7$ pm (Nagel and Range 1978, 1979); $a = 1071.16 \pm 0.01$ and the calculated and experimental densities of 6.02 and 5.96 g·cm^{-3}, respectively at room temperature and $a = 1067.4 \pm 0.2$ at –173 K (Cros et al. 1986)], is formed in the Ge–Ag–I–S system.

This phase was obtained by solid-state reactions as follows: 3Ag$_2$S + GeS$_2$ + AgI = Ag$_7$GeS$_5$I (Laqibi et al. 1987). Suitable quantities of the components were finely ground and mixed thoroughly and then treated in a quartz ampoule sealed under a secondary vacuum (10^{-2} Pa). The mixture was heated to 700°C over a period of 3 days, kept at this temperature for 48 h, and then maintained at 580°C for 1 week. It can also be prepared by reacting stoichiometric amounts of elements in an evacuated, sealed quartz ampoule for about 6 days at temperatures of 600°C to 700°C (Kuhs et al. 1979).

4.160 Germanium–Silver–Manganese–Sulfur

The **Ag$_2$MnGeS$_4$** quaternary compound, which crystallizes in the orthorhombic structure with the lattice parameters $a = 794.7$, $b = 689.2$, and $c = 652.7$, is formed in the Ge–Ag–Mn–S system (Lamarche et al. 1991). This compound was prepared by the melt and anneals technique using stoichiometric mixtures of the elements. The components were sealed under a vacuum in a quartz capsule that had previously been coated with carbon in order to prevent the reaction of the charge with the quartz. The capsule was then raised to a temperature of 1150°C, held at 1150°C for 0.5–1 h, and then cooled to room temperature. The sample was then annealed at 500°C for a week and quenched to room temperature.

4.161 Germanium–Gold–Cobalt–Sulfur

The **Au$_2$CoGeS$_4$** quaternary compound is formed in the Ge–Au–Co–S system (Davaasuren et al. 2017). To prepare this compound, K$_2$Au$_2$GeS$_4$ (0.146 mM) was dissolved in 7 mL of deionized and degassed water, inside the glove box. One equimolar aqueous solution of Co(CH$_3$COO) was immediately added drop-wise under continuous stirring, which resulted in the formation of a dark brownish-clear solution. The black chalcogels were formed after 7 days by heating the solutions at 40°C. The water-soluble products were removed together with the water by washing the gel with absolute ethanol more than 10 times. The gel was dried using the CO$_2$ critical point drier to remove the ethanol and retain the porous nature of the materials.

4.162 Germanium–Gold–Nickel–Sulfur

The **Au$_2$NiGeS$_4$** quaternary compound is formed in the Ge–Au–Ni–S system (Davaasuren et al. 2017). It was obtained in the same way as Au$_2$CoGeS$_4$ was prepared using an aqueous solution of NiCl$_2$ instead of Co(CH$_3$COO).

4.163 Germanium–Magnesium–Yttrium–Sulfur

The $MgY_6Ge_2S_{14}$ quaternary compound, which crystallizes in the hexagonal structure with the lattice parameters $a = 978.8 \pm 0.1$, $c = 577.45 \pm 0.05$ pm, and a calculated density of 3.992 $g\cdot cm^{-3}$ (Huch et al. 2006) [$a = 982 \pm 2$ and $c = 577 \pm 2$ pm (Michelet and Flahaut 1969)], is formed in the Ge–Mg–Y–S system.

This compound was prepared by fusion of the high-purity elements in an evacuated silica ampoule (Huch et al. 2006). In order to exclude the reaction of magnesium with quartz the ampoule was covered by graphite. The synthesis was realized in a tube furnace. The ampoule was heated with a heating rate of $30°C\cdot h^{-1}$ to a maximal temperature of 1150°C. The sample was kept at maximal temperature for 4 h. After that, they were cooled slowly ($10°C\cdot h^{-1}$) to 600°C and annealed at this temperature for 240 h. After annealing, the ampoule with the sample was quenched in cold water. The prepared product was a brown-colored compact sample.

$MgY_6Ge_2S_{14}$ was also obtained in a sealed silica ampoule at around 1000°C by the interaction of MgS and Y_2S_3 with Ge and the desired amount of S (Michelet and Flahaut 1969). It appears as a polycrystalline mass. Sufficiently prolonged heating makes it possible to obtain small single crystals.

4.164 Germanium–Magnesium–Lanthanum–Sulfur

The $MgLa_6Ge_2S_{14}$ quaternary compound, which crystallizes in the hexagonal structure with the lattice parameters $a = 1036.7 \pm 0.1$ and $c = 581.4 \pm 0.1$ pm (Gitzendanner et al. 1997) [$a = 1035 \pm 2$ and $c = 581 \pm 2$ pm (Michelet and Flahaut 1969)], is formed in the Ge–Mg–La–S system.

Crystals of this compound were isolated from the reaction of stoichiometric amounts of La, Mg, Ge, and S in a $CaCl_2/NaCl$ eutectic flux (Gitzendanner et al. 1997). This reaction was carried out in a vitreous carbon crucible sealed in an evacuated fused silica tube. The tube was heated slowly to 600°C over 48 h and held there for 12 h to allow for complete prereaction of the sulfur. This reduces the sulfur vapor pressure that otherwise could lead to explosions at higher temperatures. The temperature was then increased to 1000°C over 12 h and held there for 72 h. The tube was then slowly cooled ($2°C\cdot h^{-1}$) to 500°C at which point the furnace was shut off and allowed to cool to room temperature. The flux was washed away by soaking the product in distilled water for several hours. Transparent, orange-brown hexagonal needlelike crystals were obtained after filtration. The crystals were washed with more distilled water and dried in a desiccator. La_2S_3 was present as a minor phase.

$MgLa_6Ge_2S_{14}$ was also prepared in a sealed silica ampoule at around 1000°C by the interaction of MgS and La_2S_3 with Ge and the desired amount of S (Michelet and Flahaut 1969). It appears as a polycrystalline mass. Sufficiently prolonged heating makes it possible to obtain small single crystals.

4.165 Germanium–Magnesium–Cerium–Sulfur

The $MgCe_6Ge_2S_{14}$ quaternary compound, which crystallizes in the hexagonal structure with the lattice parameters $a = 1026.26 \pm 0.15$, $c = 576.79 \pm 0.12$ pm, and a calculated density of 4.605 $g\cdot cm^{-3}$ (Gulay et al. 2007a) [$a = 1026.2 \pm 0.2$, $c = 578.49 \pm 0.07$ pm, and a calculated density of 4.592 $g\cdot cm^{-3}$ (Huch et al. 2006)], is formed in the Ge–Mg–Ce–S system.

This compound was prepared in the same way as $MgY_6Ge_2S_{14}$ was synthesized (Huch et al. 2006). Single crystals of $MgCe_6Ge_2S_{14}$ were grown by fusion of the elemental constituents in the stoichiometric ratio in evacuated silica ampoule (Gulay et al. 2007a). In order to avoid the reaction of magnesium with SiO_2, the inner walls of the ampoule were covered with graphite. The ampoule was heated in a tube furnace with a heating rate of $30°C\cdot h^{-1}$ to 1150°C and was kept at this temperature for 3 h. It was then cooled down slowly ($10°C\cdot h^{-1}$) to 600°C, annealed at this temperature for a further 240 h, and finally quenched in cold water. The obtained red crystals had a prismatic habit.

4.166 Germanium–Magnesium–Praseodymium–Sulfur

The $MgPr_6Ge_2S_{14}$ quaternary compound, which crystallizes in the hexagonal structure with the lattice parameters $a = 1012.49 \pm 0.03$, $c = 580.40 \pm 0.02$ pm, and a calculated density of 4.716 $g\cdot cm^{-3}$, is formed in the Ge–Mg–Pr–S system (Huch et al. 2006). This compound was prepared in the same way as $MgY_6Ge_2S_{14}$ was synthesized.

4.167 Germanium–Magnesium–Neodymium–Sulfur

The $MgNd_6Ge_2S_{14}$ quaternary compound, which crystallizes in the hexagonal structure with the lattice parameters $a = 1007.01 \pm 0.03$, $c = 580.47 \pm 0.03$ pm, and a calculated density of 4.832 $g\cdot cm^{-3}$, is formed in the Ge–Mg–Nd–S system (Huch et al. 2006). This compound was prepared in the same way as $MgY_6Ge_2S_{14}$ was synthesized.

4.168 Germanium–Magnesium–Samarium–Sulfur

The $MgSm_6Ge_2S_{14}$ quaternary compound, which crystallizes in the hexagonal structure with the lattice parameters $a = 995.00 \pm 0.03$, $c = 578.72 \pm 0.03$ pm, and a calculated density of 5.087 $g\cdot cm^{-3}$, is formed in the Ge–Mg–Sm–S system (Huch et al. 2006). This compound was prepared in the same way as $MgY_6Ge_2S_{14}$ was synthesized.

4.169 Germanium–Magnesium–Gadolinium–Sulfur

The **MgGd₆Ge₂S₁₄** quaternary compound, which crystallizes in the hexagonal structure with the lattice parameters $a = 993.19 \pm 0.03$, $c = 570.83 \pm 0.03$ pm, and a calculated density of 5.317 g·cm⁻³, is formed in the Ge–Mg–Gd–S system (Huch et al. 2006). This compound was prepared in the same way as MgY₆Ge₂S₁₄ was synthesized.

4.170 Germanium–Magnesium–Terbium–Sulfur

The **MgTb₆Ge₂S₁₄** quaternary compound, which crystallizes in the hexagonal structure with the lattice parameters $a = 980.07 \pm 0.05$, $c = 578.43 \pm 0.04$ pm, and a calculated density of 5.424 g·cm⁻³, is formed in the Ge–Mg–Tb–S system (Huch et al. 2006). This compound was prepared in the same way as MgY₆Ge₂S₁₄ was synthesized.

4.171 Germanium–Magnesium–Dysprosium–Sulfur

The **MgDy₆Ge₂S₁₄** quaternary compound, which crystallizes in the hexagonal structure with the lattice parameters $a = 976.98 \pm 0.04$, $c = 576.75 \pm 0.05$ pm, and a calculated density of 5.529 g·cm⁻³, is formed in the Ge–Mg–Dy–S system (Huch et al. 2006). This compound was prepared in the same way as MgY₆Ge₂S₁₄ was synthesized.

4.172 Germanium–Magnesium–Holmium–Sulfur

The **MgHo₆Ge₂S₁₄** quaternary compound, which crystallizes in the hexagonal structure with the lattice parameters $a = 973.5 \pm 0.2$, $c = 579.41 \pm 0.07$ pm, and a calculated density of 5.614 g·cm⁻³, is formed in the Ge–Mg–Ho–S system (Huch et al. 2006). This compound was prepared in the same way as MgY₆Ge₂S₁₄ was synthesized.

4.173 Germanium–Magnesium–Erbium–Sulfur

The **MgEr₆Ge₂S₁₄** quaternary compound, which crystallizes in the hexagonal structure with the lattice parameters $a = 969.4 \pm 0.2$, $c = 579.94 \pm 0.07$ pm, and a calculated density of 5.706 g·cm⁻³, is formed in the Ge–Mg–Er–S system (Huch et al. 2006). This compound was prepared in the same way as MgY₆Ge₂S₁₄ was synthesized.

4.174 Germanium–Strontium–Cadmium–Sulfur

The **SrCdGeS₄** quaternary compound, which melts congruently at 958°C and crystallizes in the orthorhombic structure with the lattice parameters $a = 1033.52 \pm 0.14$, $b = 1023.35 \pm 0.14$, $c = 644.08 \pm 0.09$ pm, a calculated density of 3.908

g·cm⁻³, and an energy gap of 2.6 eV, is formed in the Ge–Sr–Cd–S system (Dou et al. 2019). Single crystals of this compound were grown from reactions involving slow cooling from the melt. A mixture of Sr (0.70 mM), Cd (0.70 mM), Ge (0.70 mM), and S (2.80 mM) was loaded into a fused silica tube, which was evacuated and sealed. The tube was heated to 950°C over 30 h, kept at that temperature for 96 h, and then slowly cooled to room temperature over 144 h. Light-yellow (almost colorless) air-stable crystals were obtained. Polycrystalline samples were prepared from reactions of the same mixture of starting materials as before. The tube was heated to 750°C in 24 h, kept at that temperature for 96 h, and then cooled by shutting off the furnace. The sample was reground and subjected to repeated heat treatment.

4.175 Germanium–Barium–Cadmium–Sulfur

The **BaCdGeS₄** quaternary compound, which melts congruently at 921°C and crystallizes in the orthorhombic structure with the lattice parameters $a = 2128.5 \pm 0.4$, $b = 2169.1 \pm 0.5$, $c = 1278.6 \pm 0.4$ pm, a calculated density of 4.056 g·cm⁻³, and an energy gap of 2.58 eV (Lin et al. 2019b) [$a = 2169$, $b = 2125$, and $c = 1278$ pm, and the calculated and experimental densities of 4.08 and 3.96 g·cm⁻³, respectively (Teske 1980d)], is formed in the Ge–Ba–Cd–S system. Light green crystals of this compound were crystallized from a reaction mixture containing Ba (0.66 mM), Cd (0.66 mM), Ge (0.66 mM), and S (2.65 mM), which was loaded into a quartz tube and flame-sealed under vacuum (~10⁻² Pa) (Lin et al. 2019b). The tube was placed into a furnace, heated up to 250°C and held at that temperature for 2 h, then heated up to 875°C in 24 h and kept for 96 h, and then cooled down to 300°C at 4°C·h⁻¹ before switching off the furnace. BaCdGeS₄ is stable in the air and moisture condition. Its single crystals can be relatively easily grown through the Bridgman–Stockbarger technique.

4.176 Germanium–Barium–Gallium–Sulfur

Two quaternary compounds, **BaGa₂GeS₆** and **Ba₂Ga₈GeS₁₆**, are formed in the Ge–Ba–Ga–S system. The first of them melts congruently at 941°C [at 983°C (Badikov et al. 2016); at 977°C for the Ba₂.₇Ga₅.₄Ge₃.₆S₁₈ composition (Haeuseler and Schmidt 1994)] and crystallizes in the trigonal structure with the lattice parameters $a = 960.20 \pm 0.01$, $c = 868.89 \pm 0.02$ pm, a calculated density of 3.890 g·cm⁻³, and an energy gap of 3.23 eV (Yin et al. 2012b) [$a = 959.67 \pm 0.11$, $c = 867.1 \pm 0.2$ pm, a calculated density of 3.902 g·cm⁻³, and an energy gap of 3.26 eV (Lin et al. 2012); $a = 960.2$, $c = 869.0$ pm, a calculated density of 3.890 g·cm⁻³, and an energy gap of 3.37 or 3.29 eV (Badikov et al. 2016); $a = 960.0 \pm 0.1$ and $c = 868.4 \pm 0.1$ pm for the Ba₂.₇Ga₅.₄Ge₃.₆S₁₈ composition (Haeuseler and Schmidt 1994)].

Crystals of BaGa₂GeS₆ were obtained by spontaneous nucleation of the pure polycrystalline sample, which was synthesized by stoichiometric reaction of BaS, Ga₂S₃, and GeS₂ (molar ratio 1:1:1) at 900°C for 48 h (Yin et al. 2012b). Then, 0.5 g of the as-prepared powder of BaGa₂GeS₆ was loaded into a fused silica tube under an Ar atmosphere in a glove box. The tube was sealed under a 10^{-3} Pa atmosphere and then placed in a furnace. The reaction mixture was heated to 1000°C in 15 h, kept at 1000°C for 48 h, followed by slow cooling to 320°C at a rate of $3°C\cdot h^{-1}$, and finally cooled to room temperature by switching off the furnace. The product consisted of light-yellow crystals of the title compound.

The synthesis of this compound can also be achieved using the solid-state reaction of BaS, Ga, Ge, and S (Lin et al. 2012). The stoichiometric mixture in evacuated under a high vacuum and the sealed ampoule was slowly heated to 700°C, held for about 70 h, and then was sintered at 1000°C for 3 days in a furnace before being slowly cooled to 500°C at a rate of $5°C\cdot h^{-1}$. Single crystals of $BaGa_2GeS_6$ were grown by the Bridgman–Stockbarger techniques (Lin et al. 2012; Badikov et al. 2016).

Single crystals of $Ba_{2.7}Ga_{5.4}Ge_{3.6}S_{18}$ were prepared by slowly cooling a melt of the compound with the following temperature profile: from 1030°C to 780°C at $1°C\cdot h^{-1}$ interrupted by isothermal tempering periods at 990°C for 28 h, at 1005°C for 36 h and at 960°C for 24 h; at 780°C the furnace was switched off and the sample allowed to cool in the furnace (Haeuseler and Schmidt 1994).

$Ba_2Ga_8GeS_{16}$ melts congruently at 1008°C and crystallizes in the hexagonal structure with the lattice parameters $a = 1088.26$ and $c = 1190.17$ pm (Badikov et al. 2022a, b) [$a = 1088.6 \pm 0.8$, $c = 1191.5 \pm 0.3$ pm, a calculated density of 3.851 $g\cdot cm^{-3}$, and an energy gap of ≈ 3.0 eV (Liu et al. 2015b)].

For the synthesis of $Ba_2Ga_8GeS_{16}$, a stoichiometric mixture with an overall weight of about 300 mg of the starting materials Ba, Ga_2S_3, Ge, and S (molar ratio 2:4:1:4) was loaded into a quartz tube, and then flame-sealed under vacuum of $\sim 10^{-2}$ Pa (Liu et al. 2015b). The tube was then placed in a furnace, held at 300°C for 5 h, heated to 650°C for over 5 h, and kept there for 10 h, subsequently heated to 980°C for over 10 h, dwelled for 5 days, and finally slowly cooled to 400°C before the furnace power was switched off. Transparent yellow single crystals of the title compound were obtained. They are stable in the air for several months. All starting reactants were handled inside an Ar-filled glove box with controlled oxygen and moisture levels below 0.1 ppm.

Colorless crystals of $Ba_2Ga_8GeS_{16}$ were synthesized in a graphitized quartz ampoule, filled with Ba and S (BaS composition), Ga_2S_3, and GeS_2 in a stoichiometric ratio. The ampoule was evacuated to 10^{-4} Pa, sealed, placed in a horizontal furnace, and heated to 1050°C for 12 h. The melt obtained was kept at this temperature for 24 h and stirred for complete homogenization. Growth was performed in a vertical furnace with a temperature gradient of $10–15°C\cdot cm^{-1}$ in the crystallization zone, with melt heated to a temperature exceeding the melting point by 30–40°C. The crystal growth rate was 6 mm per day. The grown crystals were cooled to room temperature with the switched-off furnace. High optical quality single crystals of this compound were grown by the vertical Bridgman–Stockbarger method (Badikov et al. 2022a).

GeS_2–BaS–Ga_2S_3. The glass-forming regions in this quasiternary system are presented in Figure 4.44 (Haeuseler and Schmidt 1994). The solid and dashed lines show the range of glass-formation for samples quenched from 1030°C and 830°C, respectively.

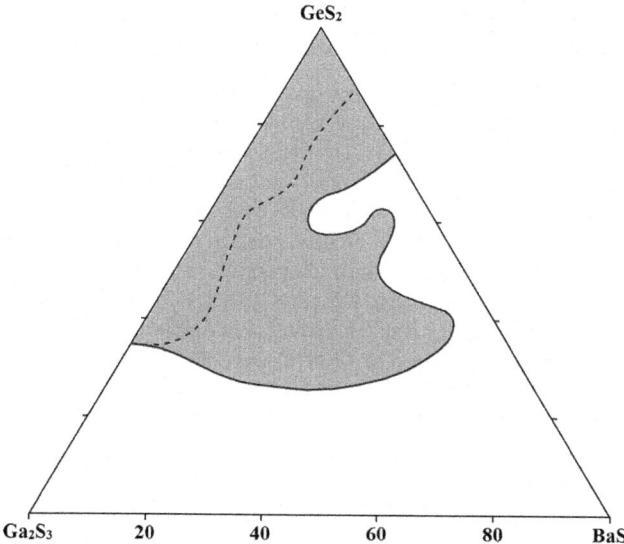

FIGURE 4.44 Glass-forming region in the GeS_2–BaS–Ga_2S_3 quasiternary system. (From Haeuseler, H., et al., *J. Alloys Compd.*, **204**(1–2), 209, 1994.)

4.177 Germanium–Barium–Antimony–Sulfur

The **$Ba_2Sb_4GeS_{10}$** quaternary compound, which crystallizes in the tetragonal structure with the lattice parameters $a = 1131.19 \pm 0.04$, $c = 1363.84 \pm 0.09$ pm, and a calculated density of 4.395 $g\cdot cm^{-3}$, is formed in the Ge–Ba–Sb–S system (Geng 2012). The title compound was synthesized through a high-temperature solid-state reaction in an evacuated and sealed silica tube. A mixture of BaS (0.5 mM), Sb (1 mM), Ge (0.25 mM), and S (2 mM) was loaded in a silica ampoule, sealed under vacuum (10^{-2} Pa), heated gradually to 900°C (holding for 10 h) in 60 h, and then cooled to room temperature in 300 h. Rod-shaped crystals of $Ba_2Sb_4GeS_{10}$ with a dark-red color were obtained. The crystals are stable under air and moisture conditions.

4.178 Germanium–Barium–Chlorine–Sulfur

The **$Ba_4Ge_3S_9Cl_2$** quaternary compound, which crystallizes in the hexagonal structure with the lattice parameters $a = 982.00 \pm 0.04$, $c = 1205.44 \pm 0.12$ pm, and an energy gap of 2.91 eV, is formed in the Ge–Ba–Cl–S system (Liu et al. 2017). This compound was obtained from a mixture of $BaCl_2$, Ba, Si, and S in a stoichiometric ratio of 1.2:3:3:9 with a slight excess of $BaCl_2$ as a reactive flux. The reactants were loaded into a fused silica tube under vacuum and then heated to 950°C within 50 h, kept at this temperature for 140 h, and subsequently cooled to 300°C at $3°C\cdot h^{-1}$ before switching off the furnace. The product was washed with distilled water to eliminate excess $BaCl_2$ and then dried with ethanol. After this treatment, the compound was obtained as a pure phase with light-yellow color. $Ba_4Ge_3S_9Cl_2$ is thermally stable up to at least 700°C under the N_2 atmosphere. All manipulations were conducted in a glove box (moisture and oxygen levels less than 0.1 ppm) or under a vacuum.

4.179 Germanium–Barium–Rhodium–Sulfur

The **BaRh₂Ge₄S₆** quaternary compound, which crystallizes in the orthorhombic structure with the lattice parameters $a = 595.12 \pm 0.01$, $b = 589.41 \pm 0.01$, $c = 2920.11 \pm 0.05$ pm, a calculated density of 5.36 g·cm⁻³, and an energy gap of 1.389 eV, is formed in the Ge–Ba–Rh–S system (Lei et al. 2014). A polycrystalline sample of this compound was synthesized using high-pressure and high-temperature methods. BaS was mixed with stoichiometric amounts of Rh, Ge, and S, ground well, and then pelletized. The pellet was loaded into an *h*-BN capsule and then heated to 1200°C and 5 GPa for 2 h using a belt-type high-pressure apparatus. All starting materials were prepared in a glove box filled with purified Ar. Single crystals were grown by prolonging the annealing time (12 h) under high pressure.

4.180 Germanium–Barium–Iridium–Sulfur

The **BaIr₂Ge₄S₆** quaternary compound, which crystallizes in the orthorhombic structure with the lattice parameters $a = 594.80 \pm 0.04$, $b = 591.48 \pm 0.04$, $c = 2915.2 \pm 0.2$ pm, and an energy gap of 1.539 eV, is formed in the Ge–Ba–Ir–S system (Lei et al. 2014). Polycrystalline sample and single crystals of this compound were obtained in the same way as BaRh₂Ge₄S₆ was prepared using Ir instead of Rh.

4.181 Germanium–Zinc–Cadmium–Sulfur

According to the data of XRD, the **(Zn_xCd₁₋ₓ)₄GeS₆** solid solution up to $x = 0.25$ is formed along the Cd₄GeS₆–(ZnS)₀.₈(GeS₂)₀.₂ section in the Ge–Zn–Cd–S system (Dubrovin et al. 1991). The lattice parameters of this solid solution change linearly with composition. The ingots for investigations were annealed at 600°C for 500 h. Single crystals of the $(Zn_xCd_{1-x})_4GeS_6$ solid solution were obtained by the chemical transport reactions.

4.182 Germanium–Zinc–Yttrium–Sulfur

The **ZnY₆Ge₂S₁₄** quaternary compound, which crystallizes in the hexagonal structure with the lattice parameters $a = 977 \pm 2$ and $c = 580 \pm 2$ pm, is formed in the Ge–Zn–Y–S system (Michelet and Flahaut 1969). It was obtained in a sealed silica ampoule at around 1000°C by an interaction of ZnS and Y₂S₃ with Ge and the desired amount of S. This compound appears as a polycrystalline mass. Sufficiently prolonged heating makes it possible to obtain small single crystals.

4.183 Germanium–Zinc–Lanthanum–Sulfur

The **ZnLa₆Ge₂S₁₄** quaternary compound, which melts incongruently above 1300°C and crystallizes in the hexagonal structure with the lattice parameters $a = 1034.5 \pm 0.1$ and $c = 582.6 \pm 0.1$ pm (Zhou et al. 2019) [$a = 1034 \pm 2$ and $c = 583 \pm 2$ pm

(Michelet and Flahaut 1969)], is formed in the Ge–Zn–La–S system. It was synthesized from a stoichiometric mixture of the elements with a total mass of 0.30 g that was pressed into pellets and loaded into a fused silica tube, which was evacuated and sealed (Zhou et al. 2019). The tube was heated at 600°C for 12 h and then heated at 1050°C for 4 days. The polycrystalline sample obtained is stable in the air for several months.

ZnLa₆Ge₂S₁₄ can also be obtained in a sealed silica ampoule at around 1000°C by the interaction of ZnS and La₂S₃ with Ge and the desired amount of S (Michelet and Flahaut 1969). It appears as a polycrystalline mass. Sufficiently prolonged heating makes it possible to obtain small single crystals.

4.184 Germanium–Zinc–Cerium–Sulfur

The **ZnCe₆Ge₂S₁₄** quaternary compound, which crystallizes in the hexagonal structure with the lattice parameters $a = 1016.4 \pm 0.2$, $c = 576.1 \pm 0.1$ pm, and an energy gap of 1.8 eV, is formed in the Ge–Zn–Ce–S system (Zhou et al. 2019). It was synthesized in the same way as ZnLa₆Ge₂S₁₄ was prepared.

4.185 Germanium–Zinc–Praseodymium–Sulfur

The **ZnPr₆Ge₂S₁₄** quaternary compound, which crystallizes in the hexagonal structure with the lattice parameters $a = 1009.0 \pm 0.2$ and $c = 573.5 \pm 0.1$ pm, is formed in the Ge–Zn–Pr–S system (Zhou et al. 2019). It was synthesized in the same way as ZnLa₆Ge₂S₁₄ was prepared.

4.186 Germanium–Zinc–Neodymium–Sulfur

The **ZnNd₆Ge₂S₁₄** quaternary compound, which crystallizes in the hexagonal structure with the lattice parameters $a = 999.6 \pm 0.2$ and $c = 569.7 \pm 0.1$ pm, is formed in the Ge–Zn–Nd–S system (Zhou et al. 2019). It was synthesized in the same way as ZnLa₆Ge₂S₁₄ was prepared.

4.187 Germanium–Zinc–Samarium–Sulfur

The **ZnSm₆Ge₂S₁₄** quaternary compound, which crystallizes in the hexagonal structure with the lattice parameters $a = 990.5 \pm 0.2$ and $c = 569.3 \pm 0.1$ pm, is formed in the Ge–Zn–Sm–S system (Zhou et al. 2019). It was synthesized in the same way as ZnLa₆Ge₂S₁₄ was prepared.

4.188 Germanium–Zinc–Europium–Sulfur

The **ZnEuGeS₄** quaternary compound, which crystallizes in the orthorhombic structure with the lattice parameters $a = 1223.4 \pm 0.4$, $b = 2039.8 \pm 0.6$, $c = 2068.2 \pm 0.6$ pm, a calculated density of 4.305 g·cm⁻³, and an energy gap of 2.26 eV (Chi et al. 2017). Single crystals of the title compound were obtained by solid-state reaction with KI as a flux. The starting materials were a stoichiometric mixture of Eu_2O_3, Zn, Ge, and S. A certain amount of boron powder and KI was added to the sample as a reducing reagent and a flux, respectively. The sample has a total mass of 500 mg and 400 mg KI additional. The mixture was ground into fine powder in an agate mortar and pressed into pellets, followed by being loaded into a quartz tube. The tube was evacuated to be 10^{-2} Pa and flame-sealed. The sample was placed into a furnace, heated from room temperature to 300°C in 5 h and equilibrated for 10 h, followed by heating to 650°C over 5 h and equilibrated for another 10 h, then heated to 950°C in 5 h and homogenized for 10 days, finally cooled down to 300°C in 5 days and powered off. The crystals of $ZnEuGeS_4$ stable in moisture and air were obtained and hand-picked under a microscope since the yield was not high, then washed using ethanol and water under ultrasonic wave.

4.189 Germanium–Zinc–Gadolinium–Sulfur

The $ZnGd_6Ge_2S_{14}$ quaternary compound, which crystallizes in the hexagonal structure with the lattice parameters $a = 997.39 \pm 0.03$, $c = 572.96 \pm 0.04$ pm, and a calculated density of 5.550 g·cm^{-3}, is formed in the Ge–Zn–Gd–S system (Zhou et al. 2019). It was synthesized in the same way as $ZnLa_6Ge_2S_{14}$ was prepared.

4.190 Germanium–Zinc–Terbium–Sulfur

The $ZnTb_6Ge_2S_{14}$ quaternary compound, which crystallizes in the hexagonal structure with the lattice parameters $a = 970.1 \pm 0.2$ and $c = 571.9 \pm 0.1$ pm, is formed in the Ge–Zn–Tb–S system (Zhou et al. 2019). It was synthesized in the same way as $ZnLa_6Ge_2S_{14}$ was prepared.

4.191 Germanium–Zinc–Dysprosium–Sulfur

The $ZnDy_6Ge_2S_{14}$ quaternary compound, which crystallizes in the hexagonal structure with the lattice parameters $a = 970.1 \pm 0.2$ and $c = 571.9 \pm 0.1$ pm, is formed in the Ge–Zn–Dy–S system (Zhou et al. 2019). It was synthesized in the same way as $ZnLa_6Ge_2S_{14}$ was prepared.

4.192 Germanium–Zinc–Holmium–Sulfur

The $ZnHo_6Ge_2S_{14}$ quaternary compound, which crystallizes in the hexagonal structure with the lattice parameters $a = 967.3 \pm 0.1$ and $c = 576.4 \pm 0.1$ pm, is formed in the Ge–Zn–Ho–S system (Zhou et al. 2019). It was synthesized in the same way as $ZnLa_6Ge_2S_{14}$ was prepared.

4.193 Germanium–Zinc–Erbium–Sulfur

The $ZnEr_6Ge_2S_{14}$ quaternary compound, which crystallizes in the hexagonal structure with the lattice parameters $a = 960.6 \pm 0.1$ and $c = 580.8 \pm 0.1$ pm, is formed in the Ge–Zn–Er–S

system (Zhou et al. 2019). It was synthesized in the same way as $ZnLa_6Ge_2S_{14}$ was prepared.

4.194 Germanium–Cadmium–Mercury–Sulfur

GeS_2–CdS–HgS. Using coprecipitation of CdS with HgS and GeS_2 by the hydrogen sulfide and then heating up to 300°C, any quaternary compounds were obtained (Kislinskaya 1974).

4.195 Germanium–Cadmium–Gallium–Sulfur

GeS_2–CdS–Ga_2S_3. The liquidus surface of this quasiternary system was depicted for the first time by Barnier et al. (1990). According to these data, the ternary system is triangulated by the quasibinary section GeS_2–$Cd_5Ga_2S_8$, though this section is quasibinary only in part of the subliquidus region because the $Cd_5Ga_2S_8$ phase is metastable.

Olekseyuk et al. (2006b) noted that the liquidus surface of the GeS_2–CdS–Ga_2S_3 quasiternary system (Figure 4.45) consists of 8 fields of the primary crystallization, which are separated by 14 monovariant lines and by 6 nonvariant points, of which 4 are the transition points and 2 are the ternary eutectics: U_1 (1012°C) – L + β-Cd_4GeS_6 ↔ CdS + α-Cd_4GeS_6; U_2 (900°C) – L + CdS ↔ α-Cd_4GeS_6 + $Cd_5Ga_2S_8$; U_3 (860°C) – L + $Cd_5Ga_2S_8$ ↔ α-Cd_4GeS_6 + $CdGa_2S_4$; U_4 (744°C) – L + β-Ga_2S_3 ↔ α-Ga_2S_3 + $CdGa_2S_4$; E_1 (735°C) – L ↔ GeS_2 + $CdGa_2S_4$ + α-Cd_4GeS_6; E_2 (726°C) – L ↔ α-Ga_2S_3 + $CdGa_2S_4$ + $CdGa_2S_4$.

The isothermal section of the GeS_2–CdS–Ga_2S_3 system at 400°C (Figure 4.46) has been constructed by Olekseyuk et al. (2006b). Minor solid solutions exist for every system component. In the middle of the concentration triangle, the largest solid solubility is based on $CdGa_2S_4$ that stretches along the GeS_2–$CdGa_2S_4$ section to 8 mol% GeS_2. The solid solution

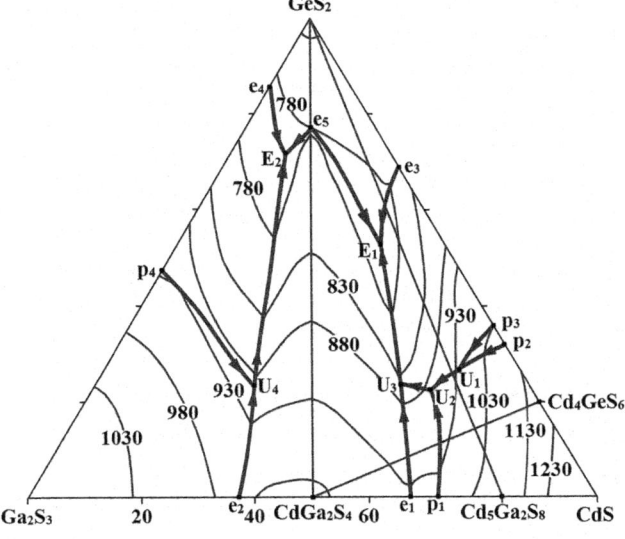

FIGURE 4.45 Liquidus surface of the GeS_2–CdS–Ga_2S_3 quasiternary system. (From Olekseyuk, I.D. et al., *J. Alloys Compd.*, **421**(1–2), 91, 2006.)

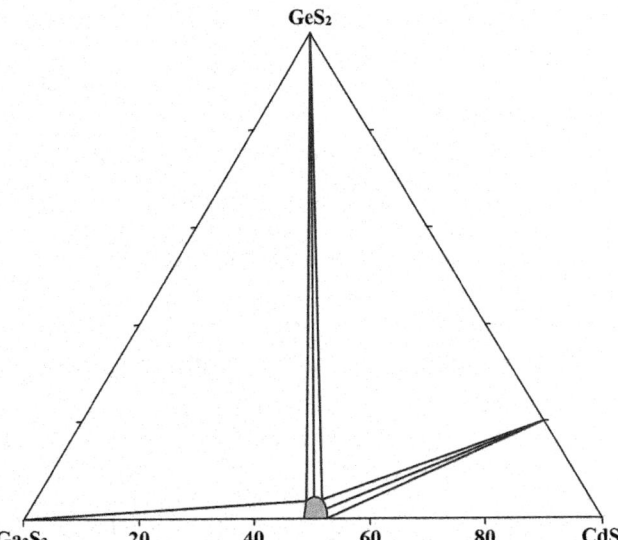

FIGURE 4.46 Isothermal section of the GeS₂–CdS–Ga₂S₃ quasiternary system at 400°C. (From Olekseyuk, I.D. et al., *J. Alloys Compd.*, **421**(1–2), 91, 2006.)

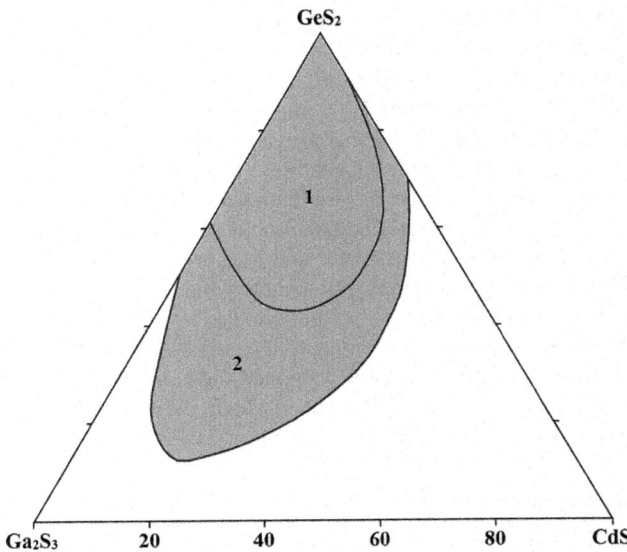

FIGURE 4.47 Glass-forming region in the GeS₂–CdS–Ga₂S₃ quasiternary system for the alloys quenched from 1000°C. (1 – From Wang, X. et al., *Mater. Chem. Phys.*, **83**(2–3), 284, 2004; 2 – From Barnier, S. et al., *Mater. Sci. Eng.*, **B7**(3), 209, 1990.)

range of GeS₂ incorporates up to 3 mol% CdGa₂S₄. On the Cd₄GeS₆–CdGa₂S₄ section, the region of the solid solution based on CdGa₂S₄ is 3 mol% and the solid solution range of Cd₄GeS₆ does not exceed 2 mol%.

The glasses exist in a wide region in the GeS₂–CdS–Ga₂S₃ system (Figure 4.47) (Barnier et al. 1990; Wang et al. 2004). The glassy region for the alloys quenched from 1000°C is widely extended toward Ga₂S₃ (Barnier et al. 1990). The formed glasses are yellow amber colored, and this color becomes redder with the simultaneous increase in the concentration of Ga₂S₃ and CdS. Minimum and maximum vitreous temperatures appear at 270°C and 400°C, respectively. According to the data of Wang et al. (2004), the glass-forming region for the samples heating at 970°C is mainly situated in the GeS₂-rich domain, and the amount of dissolved CdS is up to over 30 mol%. The obtained glasses have relatively high glass transition temperatures (T_g = 375–436°C) and good thermal stability. The experimental results indicate (Gu et al. 2005) that the GeS₂ acts in the GeS₂–CdS–Ga₂S₃ glasses as the network former, the Ga₂S₃ as the net intermediate, and the CdS as the net modifier.

This system was investigated through DTA, XRD, and metallography (Barnier et al. 1990; Olekseyuk et al. 2006b). The GeS₂–CdS–Ga₂S₃ glasses were prepared by conventional melt-quenching techniques (Wang et al. 2004; Gu et al. 2005).

4.196 Germanium–Cadmium–Lanthanum–Sulfur

GeS₂–CdS–La₂S₃. The isothermal section of this quasiternary system at 1050°C is shown in Figure 4.48 (Perez et al. 1970a). The region of the La₆Cd$_y$Ge$_{2.5-y/2}$□$_{1.5-y/2}$S₁₄ solid solutions with rhombohedral structure is strongly limited, and the formula of these solutions is closed to La₆Cd□Ge₂S₁₄ with the lattice

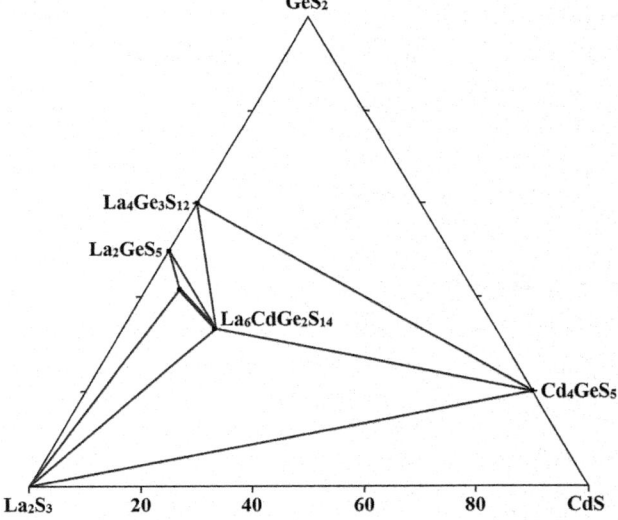

FIGURE 4.48 Isothermal section of the GeS₂–CdS–La₂S₃ quasiternary system at 1050°C. (From Perez, G. et al., *J. Solid State Chem.*, 2(1), 42, 1970.)

parameters a = 1038.9 ± 0.1, c = 581.4 ± 0.1 pm, and an energy gap of 2.3 eV (Zhou et al. 2019) [a = 1037.8 ± 0.4 and c = 580.8 ± 0.2 pm (Perez et al. 1970a)]. **La₆CdGe₂S₁₄** was synthesized from a stoichiometric mixture of the elements with a total mass of 0.30 g that was pressed into pellets and loaded into fused silica tube, which was evacuated and sealed (Zhou et al. 2019). The tube was heated at 600°C for 12 h and then heated at 1050°C for 4 days. The polycrystalline sample obtained is stable in the air for several months.

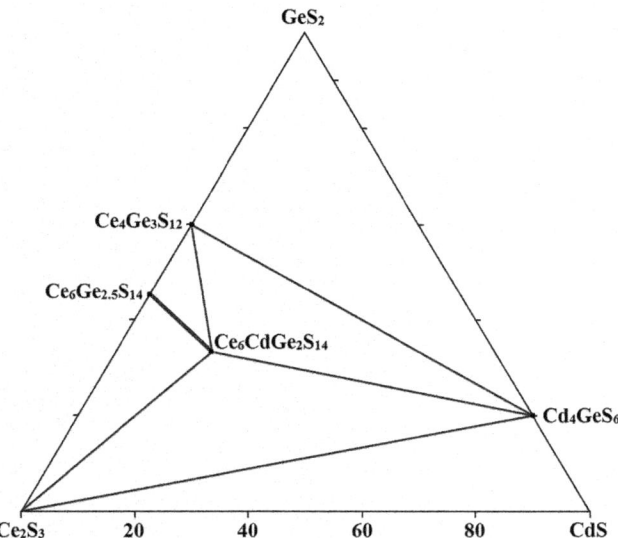

FIGURE 4.49 Isothermal section of the GeS_2–CdS–Ce_2S_3 quasiternary system at 1050°C. (From Perez, G. et al., *J. Solid State Chem.*, 2(1), 42, 1970.)

4.197 Germanium–Cadmium–Cerium–Sulfur

GeS_2–CdS–Ce_2S_3. The isothermal section of this quasiternary system at 1050°C is presented in Figure 4.49 (Perez et al. 1970). The region of the $Ce_6Cd_yGe_{2.5-y/2}\square_{1.5-y/2}S_{14}$ solid solutions with rhombohedral structure exists between the $Ce_6Ge_{2.5}S_{14}$ and $Ce_6CdGe_2S_{14}$; the lattice parameters of the quaternary compound are $a = 1022.0 \pm 0.1$ and $c = 575.5 \pm 0.1$ pm (Zhou et al. 2019) [$a = 1022.5 \pm 0.4$ and $c = 577.0 \pm 0.2$ pm (Perez et al. 1970)]. $Ce_6CdGe_2S_{14}$ was synthesized in the same way as $La_6CdGe_2S_{14}$ was prepared (Zhou et al. 2019).

4.198 Germanium–Cadmium–Praseodymium–Sulfur

GeS_2–CdS–Pr_2S_3. The isothermal section of this quasiternary system at 1050°C is analogous to one of the GeS_2–CdS–Ce_2S_3 quasiternary system (Perez et al. 1970a). The region of the $Pr_6Cd_yGe_{2.5-y/2}\square_{1.5-y/2}S_{14}$ solid solutions with rhombohedral structure exists between the $Pr_6Ge_{2.5}S_{14}$ and $Pr_6CdGe_2S_{14}$; the lattice parameters of the quaternary compound are $a = 1012.2 \pm 0.2$ and $c = 571.5 \pm 0.1$ pm (Zhou et al. 2019) [$a = 1021.0 \pm 0.4$ and $c = 577.0 \pm 0.2$ pm (Perez et al. 1970)]. $Pr_6CdGe_2S_{14}$ was synthesized in the same way as $La_6CdGe_2S_{14}$ was prepared (Zhou et al. 2019).

4.199 Germanium–Cadmium–Neodymium–Sulfur

GeS_2–CdS–Nd_2S_3. The isothermal section of this quasiternary system at 1050°C is analogous to one of the GeS_2–CdS–Ce_2S_3 quasiternary system (Perez et al. 1970). The region of the $Nd_6Cd_yGe_{2.5-y/2}\square_{1.5-y/2}S_{14}$ solid solutions with rhombohedral

structure exists between the $Nd_6Ge_{2.5}S_{14}$ and $Nd_6CdGe_2S_{14}$; the lattice parameters of the quaternary compound are $a = 1005.8 \pm 0.2$ and $c = 569.6 \pm 0.1$ pm (Zhou et al. 2019) [$a = 1015.4 \pm 0.4$ and $c = 575.7 \pm 0.2$ pm (Perez et al. 1970)]. $Nd_6CdGe_2S_{14}$ was synthesized in the same way as $La_6CdGe_2S_{14}$ was prepared (Zhou et al. 2019).

4.200 Germanium–Cadmium–Samarium–Sulfur

GeS_2–CdS–Sm_2S_3. The isothermal section of this quasiternary system at 1050°C is analogous to one of the GeS_2–CdS–Ce_2S_3 quasiternary system (Perez et al. 1970). The region of the $Sm_6Cd_yGe_{2.5-y/2}\square_{1.5-y/2}S_{14}$ solid solutions with rhombohedral structure exists between the $Sm_6Ge_{2.5}S_{14}$ and $Sm_6CdSGe_2S_{14}$; the lattice parameters of the quaternary compound are $a = 997.9 \pm 0.3$ and $c = 564.7 \pm 0.3$ pm (Zhou et al. 2019) [$a = 1005.2 \pm 0.4$ and $c = 573.3 \pm 0.2$ pm (Perez et al. 1970)]. $Sm_6CdGe_2S_{14}$ was synthesized in the same way as $La_6CdGe_2S_{14}$ was prepared (Zhou et al. 2019).

4.201 Germanium–Cadmium–Europium–Sulfur

The $EuCdGeS_4$ quaternary compound, which melts congruently at 997°C and crystallizes in the orthorhombic structure with the lattice parameters $a = 1027.31 \pm 0.04$, $b = 1018.27 \pm 0.04$, $c = 643.54 \pm 0.03$ pm, a calculated density of 4.590 g·cm^{-3}, and an energy gap of 2.5 eV, is formed in the Ge–Cd–Eu–S system (Xing et al. 2019). A polycrystalline sample of this compound was synthesized by a solid-state reaction using equimolar amounts of EuS, CdS, and GeS_2, which were mixed and ground in an Ar-filled glove box. Then the mixture was loaded into a quartz tube, sealed under vacuum, and heated to 800°C for 15 h, maintained at this temperature for 72 h, and cooled down to room temperature by turning off the furnace. In order to enhance their crystallinity and purity, regrinding and repeated heat treatment were necessary for the sample. Single crystals were grown using the same mixture of starting materials as before. The tube loading with starting materials was evacuated and heated in a furnace to 1000°C over 20 h and maintained at this temperature for 72 h. When the next procedure of cooling down to 400°C at a rate of 3°C·h^{-1} was completed, the furnace was shut off. Orange crystals of $EuCdGeS_4$ were obtained.

4.202 Germanium–Cadmium–Gadolinium–Sulfur

GeS_2–CdS–Gd_2S_3. The isothermal section of this quasiternary system at 1050°C is analogous to one of the GeS_2–CdS–Ce_2S_3 quasiternary system (Perez et al. 1970). The region of the $Gd_6Cd_yGe_{2.5-y/2}\square_{1.5-y/2}S_{14}$ solid solutions with rhombohedral structure exists between the $Gd_6Ge_{2.5}S_{14}$ and $Gd_6CdGe_2S_{14}$; the lattice parameters of the quaternary compound are $a = 992.1 \pm 0.2$ and $c = 569.3 \pm 0.2$ pm (Zhou et al. 2019) [$a = 997.0 \pm 0.4$ and $c = 572.5 \pm 0.2$ pm (Perez et al. 1970)]. $Gd_6CdGe_2S_{14}$ was synthesized in the same way as $La_6CdGe_2S_{14}$ was prepared (Zhou et al. 2019).

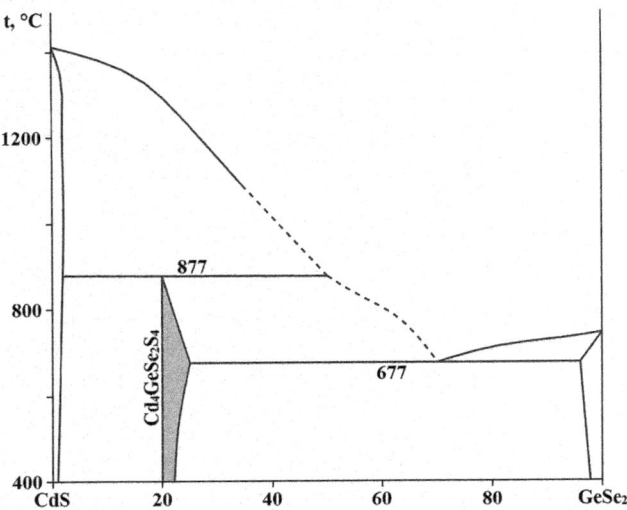

FIGURE 4.50 Phase diagram of the GeSe₂–CdS system. (From Mirzoyev, A.J. et al., *Zhurn. khim. problem*, (2), 322, 2006.)

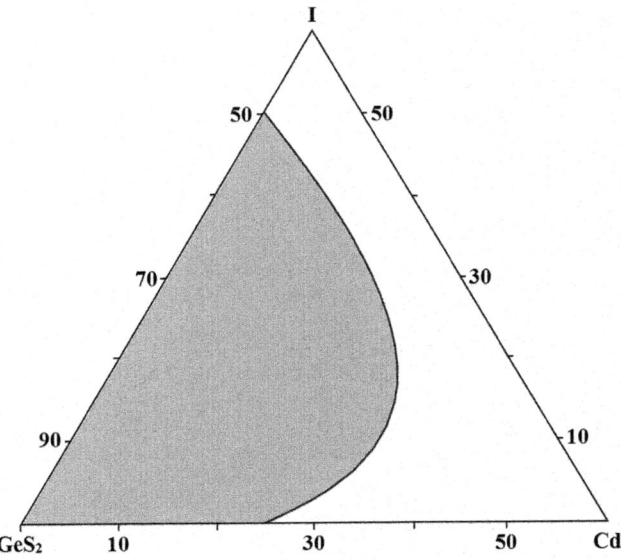

FIGURE 4.51 Glass-forming region in the GeS₂–Cd–I system. (From Vassilev, V.S., and Ivanova, Z.G., *J. Phys. Chem. Solids*, **58**(4), 573, 1997.)

4.203 Germanium–Cadmium–Terbium–Sulfur

The **Tb₆CdGe₂S₁₄** quaternary compound, which crystallizes in the hexagonal structure with the lattice parameters $a = 987.0 \pm 0.3$ and $c = 568.4 \pm 0.2$ pm is formed in the Ge–Cd–Tb–S system (Zhou et al. 2019). Tb₆CdGe₂S₁₄ was synthesized in the same way as La₆CdGe₂S₁₄ was prepared.

4.204 Germanium–Cadmium–Selenium–Sulfur

GeSe₂–CdS. The phase diagram of this system, constructed through DTA, XRD, metallography, and measuring of microhardness and EMF of concentration chains is a eutectic type (Figure 4.50) (Mirzoyev et al. 2006). The eutectic composition and temperature are 30 mol% CdS and 667°C, respectively. **Cd₄GeSe₂S₄** quaternary compound is formed in this system. It melts incongruently at 877°C (peritectic point contains 50 mol% CdS) and crystallizes in the monoclinic structure with the lattice parameters $a = 1261.8$, $b = 724.2$, $c = 1255.3$ pm, and $\beta = 110°80'$, and the calculated and experimental densities of 5.01 and 5.12 g·cm⁻³, respectively. The limited solid solutions based on CdS and GeSe₂ are formed in this system (Asadov 2006, Mirzoyev et al. 2006).

4.205 Germanium–Cadmium–Iodine–Sulfur

GeS₂–Cd–I. The glass-forming region in this system is situated between the GeS₂–Cd and GeS₂–I side of the Gibbs triangle and limited by the $(GeS_2)_{0.75}Cd_{0.25}$ and $(GeS_2)_{0.45}I_{0.55}$ compositions (Figure 4.51) (Vassilev and Ivanova 1997). The glass transition temperature for obtained glasses is within the interval from 86°C to 95°C. The bulk glasses were prepared by heating the mixture of constituents to approximately 900°C.

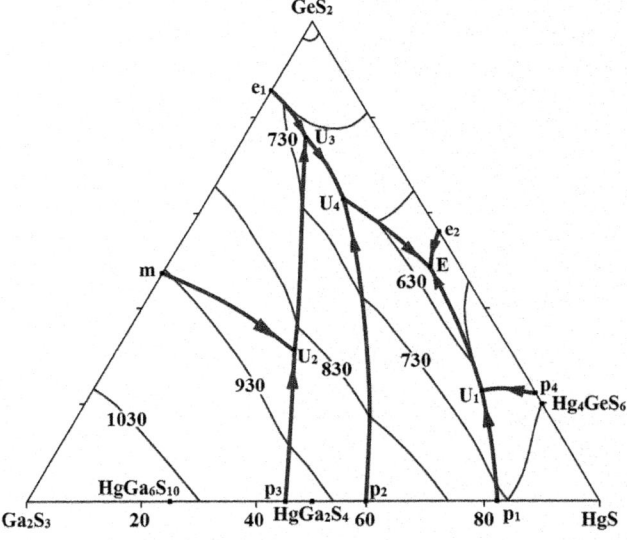

FIGURE 4.52 Liquidus surface of the GeS₂–HgS–Ga₂S₃ quasiternary system. (From Olekseyuk, I.D. et al., *J. Alloys Compd.*, **417**(1–2), 131, 2006.)

The synthesis was carried out in an evacuated silica ampoule in a rotating furnace and quenched in ice water. The melts were heated first to 200°C and held for 2 h, then to 400°C and 650°C for 7 h and after that to 900°C for 2 h.

4.206 Germanium–Mercury–Gallium–Sulfur

GeS₂–HgS–Ga₂S₃. Seven fields of primary crystallization form the liquidus surface of this quasiternary system (Figure 4.52) (Olekseyuk et al. 2006a). Four of them correspond to the components of the system: HgS, γ-Ga₂S₃, δ-Ga₂S₃, and GeS₂. The other three fields belong to HgGa₂S₄, HgGa₆S₁₀, and Hg₄GeS₆.

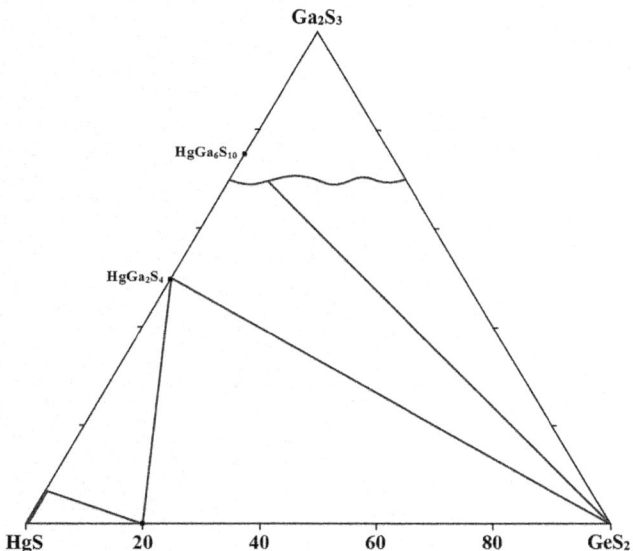

FIGURE 4.53 Isothermal section of the GeS$_2$–HgS–Ga$_2$S$_3$ quasiternary system at 400°C. (From Olekseyuk, I.D. et al., *Nauk. Visnyk Volyns'k. Derzh. Univ. im. Lesi Ukrainky. Khim. nauky*, (6), 38, 2001.)

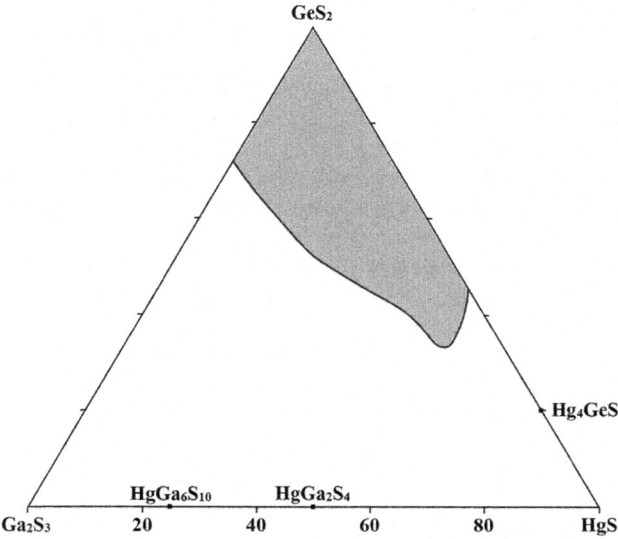

FIGURE 4.54 Glass-forming region in the GeS$_2$–HgS–Ga$_2$S$_3$ quasiternary system. (From Kevshyn, A.H., et al., *Glass Phys. Chem.*, **36**(1), 27, 2010.)

Eleven monovariant lines and five ternary invariant points separate the fields of primary crystallization. The next ternary invariant points exist in the system: E (556°C) – L ↔ HgGa$_2$S$_4$ + Hg$_4$GeS$_6$ + GeS$_2$; U_1 (643°C) – L + α-HgS ↔ HgGa$_2$S$_4$ + Hg$_4$GeS$_6$; U_2 (915°C) – L + γ-Ga$_2$S$_3$ ↔ HgGa$_6$S$_{10}$ + δ-Ga$_2$S$_3$; U_3 (680°C) – L + δ-Ga$_2$S$_3$ ↔ HgGa$_6$S$_{10}$ + GeS$_2$; U_4 (667°C) – L + HgGa$_6$S$_{10}$ ↔ HgGa$_2$S$_4$ + GeS$_2$. No quaternary phases were found in this system.

Isothermal section of this quasiternary system at 400°C within the interval 0–70 mol% Ga$_2$S$_3$ is shown in Figure 4.53 (Olekseyuk et al. 2001e). The existence of the HgGe$_2$S$_5$ ternary compound was not confirmed. The region of the solid solution based on HgS is not higher than 5 mol%, and the solubility of HgGa$_2$S$_4$ in Hg$_4$GeS$_6$ is equal to 2 mol% at this temperature.

This system was investigated by the XRD and metallography, and all alloys were annealed at 400°C for 250 h with the next quenching in the air or in cold water (Olekseyuk et al. 2001e, 2006a).

The glass-forming region in this system is presented in Figure 4.54 (Kevshyn et al. 2010). The glassy alloys in the GeS$_2$–HgS–Ga$_2$S$_3$ system belong to materials with a wide energy gap, which lies in the range ~2.68–2.94 eV, depending on the material composition. The glassy alloys were synthesized from a mixture of Ga, Ge, S, and HgS, which was heated at a rate of 30–50°C·h^{-1} to a temperature 200°C above the liquidus temperature. After holding for 6 h with periodic stirring, the melt was quenched to room temperature in a 25% NaCl solution.

4.207 Germanium–Mercury–Thallium–Sulfur

The **HgTl$_2$GeS$_4$** quaternary compound is formed in the Ge–Hg–Tl–S system (Mozolyuk et al. 2012a).

4.208 Germanium–Mercury–Europium–Sulfur

The **HgEuGeS$_4$** quaternary compound, which crystallizes in the orthorhombic structure with the lattice parameters a = 1029.67 ± 0.10, b = 1021.75 ± 0.08, c = 640.53 ± 0.05 pm, a calculated density of 5.454 g·cm^{-3}, and an energy gap of 2.04 eV, is formed in the Ge–Hg–Eu–S system (Yan et al. 2020d). The single crystals of EuHgGeS$_4$ were synthesized using a high-temperature solid-state reaction. The total mass of the mixture of Eu$_2$O$_3$, HgS, Ge, S, and B (molar ratio 1:2:2:6:2) was 500 mg, and another 400 mg KI was added as a flux. After ground, the fine powder of the mixture was pressed into a pellet, and then loaded into a quartz tube evacuated to be 10^{-2} Pa and sealed using oxyhydrogen flame. The tube was put into a muffle furnace and then heat-treated. The reaction was heated to 900°C slowly and maintained for 7 days, and then cooled down to room temperature with a rate of 2°C·h^{-1}. Finally, orange rod-shaped crystals of the title compound were obtained, which are stable in air and water.

4.209 Germanium–Mercury–Selenium–Sulfur

The **Hg$_2$GeSe$_2$S$_2$** and **Hg$_4$GeSe$_2$S$_4$** quaternary compounds, which melt incongruently at 707°C and 862°C, respectively, are formed in the Ge–Hg–Se–S system (Asadov 2006). The solubility of HgS in GeSe$_2$ and GeSe$_2$ in HgS at 600°C reaches 18 and 5 mol%, respectively. This system was investigated by DTA, XRD, metallography, and measuring of the microhardness and EMF of concentration chains.

4.210 Germanium–Aluminum–Thallium–Sulfur

The **AlTlGeS₄** quaternary compound, which melts at 775°C and crystallizes in the orthorhombic structure with the lattice parameters $a = 1159 \pm 1$, $b = 1650 \pm 1$, $c = 708.5 \pm 0.5$ pm, and the calculated and experimental densities of 4.24 and 4.24 g·cm⁻³, respectively, is formed in the Ge–Al–Tl–S system (Nakamura et al. 1984b). Synthesis of this compound was carried out by the conventional evacuated silica tube method. The stoichiometric mixture of the elements (Tl was used as TlS) was sealed in an evacuated silica glass tube and heated in a furnace at 960°C for 1 day. The melt was then shaken vigorously a number of times to ensure thorough mixing, then it was annealed at 600°C for 5 days, and slowly cooled to room temperature by turning off the furnace power. The title compound with brown color was obtained.

4.211 Germanium–Aluminum–Samarium–Sulfur

The **Al₀.₃₆Sm₃Ge₀.₉₈S₇** quaternary compound, which crystallizes in the hexagonal structure with the lattice parameters $a = 996.4 \pm 0.1$, $c = 576.8 \pm 0.2$ pm, a calculated density of 5.065 g·cm⁻³, and an energy gap of 2.18 eV, is formed in the Ge–Al–Sm–S system (Guo et al. 2009c). Crystals of this compound were obtained by a solid-state reaction with KI as a flux. It had a total mass of 500 mg and 400 mg KI additional. The mixture of Al, Ge, S, Sm₂O₃, and B (molar ratio 1:2:8:1:6) was ground into fine powder in an agate mortar and pressed into one pellet, which was loaded into a quartz tube. The tube was evacuated to 10⁻² Pa and flame-sealed. The sample was placed into a furnace, heated from room temperature to 300°C over 5 h and kept at this temperature for 10 h, then heated to 650°C in 5 h and kept constant for 10 h, then heated to 950°C over 5 h and maintained for 10 days, and finally cooled down to 300°C in 5 days and powdered off. Yellow-green block crystals of the title compound were obtained.

4.212 Germanium–Gallium–Thallium–Sulfur

GeS₂–TlGaS₂. The phase diagram of this system, constructed through DTA and XRD using the alloys annealed at 250°C for 240 h with the next quenching in the air, is shown in Figure 4.55 (Tsisar et al. 2017b). The eutectics contain 35 and 63 mol% GeS₂ and crystallize at 727°C and 767°C, respectively. The homogeneity region of the solid solution based on TlGaS₂ decreases with temperature decreasing and is < 10 mol% GeS₂ at 250°C. Two quaternary compounds, **TlGaGeS₄** and **TlGaGe₃S₈**, are formed in this system. TlGaGeS₄ melts at 802°C (Tsisar et al. 2017b) [at 807°C (Nakamura et al. 1984b)] and crystallizes in the orthorhombic structure with the lattice parameters $a = 1679.42 \pm 0.05$, $b = 710.27 \pm 0.02$, $c = 1151.93 \pm 0.04$ pm, a calculated density of 4.564 g·cm⁻³, and an energy gap of 2.18 eV (Friedrich et al. 2020) [$a = 1153 \pm 1$, $b = 1696 \pm 1$, $c = 719.5 \pm 0.7$ pm, and the calculated and experimental densities of 4.48 and 4.43 g·cm⁻³, respectively (Nakamura et al. 1984b)].

TlGaGeS₄ was prepared by reacting Tl, Ga₂S₃, GeS₂ or Ge, and S in a sealed evacuated quartz tube (Friedrich et al. 2020).

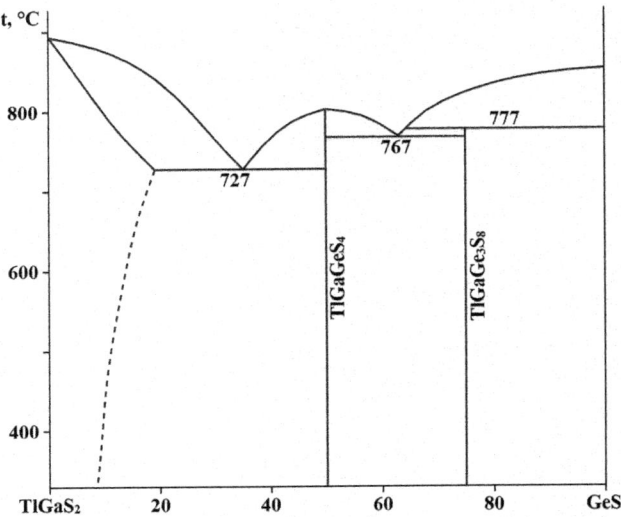

FIGURE 4.55 Phase diagram of the GeS₂–TlGaS₂ system. (From Tsisar, O.V., et al., *Nauk. Visn. Uzhgorod. Univ. Ser. Khimiya*, [2(38)], 26, 2017.)

The reaction mixture was heated to 900°C with a heating rate of 50°C·h⁻¹, annealed at 900°C for 2 days and cooled to room temperature with a cooling rate of 20°C·min⁻¹. The resulting crude product was finely ground in an agate mortar under N₂ atmosphere and annealed at 900°C for 2 additional days. Pale-yellow crystals were obtained. This compound is stable in moist air.

TlGaGe₃S₈ melts incongruently at 777°C (Tsisar et al. 2017b) and crystallizes in the monoclinic structure with the lattice parameters $a = 666.82 \pm 0.04$, $b = 372.95 \pm 0.02$, $c = 673.06 \pm 0.04$ pm, $\beta = 90.274 \pm 0.005°$, and a calculated density of 3.7383 ± 0.0006 g·cm⁻³ (Tsisar et al. 2017a). This compound was prepared by the reaction of stoichiometric mixture of Tl, Ga, Ge, and S in an evacuated quartz ampoule. The ampoule was heated to 400°C (30°C·h⁻¹), held at this temperature for 24 h, then heated to 950°C, held for 5 h, and cooled at a rate of 10–20°C·h⁻¹. The homogenization was made at 250°C for 240 h.

Tl₂GeS₃–TlGaS₂. The phase diagram of this system is a eutectic type (Figure 4.56) (Tsisar et al. 2017b). The eutectic contains 25 mol% TlGaS₂ and crystallizes at 359°C. At 250°C, the solubility of TlGaS₂ in Tl₂GeS₃ and Tl₂GeS₃ in TlGaS₂ reach 13 and 6 mol%, respectively.

GeS₂–Ga₂S₃–Tl₂S. The isothermal of this system at 250°C (Figure 4.57) includes 11 one-phase, 21 two-phase, and 11 three-phase regions (Tsisar et al. 2017b).

The glass-forming region in this system was investigated by quenching molten alloys from 1030°C and is presented in Figure 4.58 (Tsisar et al. 2017d). The minimum concentration of the glass-forming agent GeS₂ is 30 mol%. The glass transition temperature is within the interval from 170°C to 321°C.

4.213 Germanium–Gallium–Lanthanum–Sulfur

Two quaternary compounds, **GaLa₃Ge₀.₅S₇** and **Ga₂La₂GeS₈**, are formed in the Ge–Ga–La–S system. GaLa₃Ge₀.₅S₇ crystallizes in the hexagonal structure with the lattice parameters $a =$

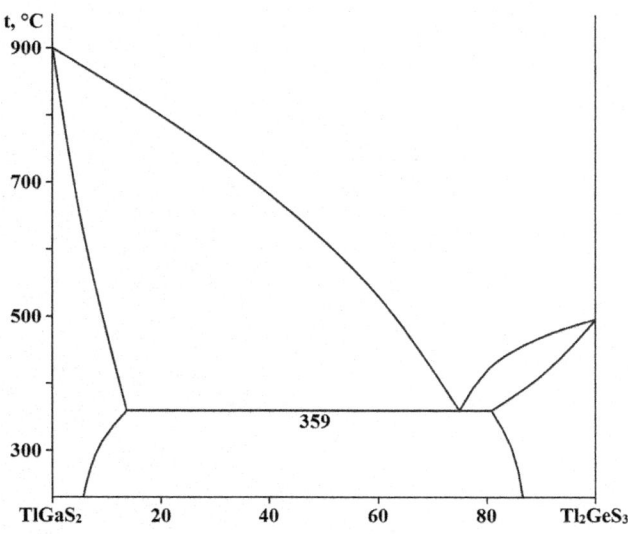

FIGURE 4.56 Phase diagram of the Tl_2GeS_3–$TlGaS_2$ system. (From Tsisar, O.V., et al., *Nauk. Visn. Uzhgorod. Univ. Ser. Khimiya*, [2(38)], 26, 2017.)

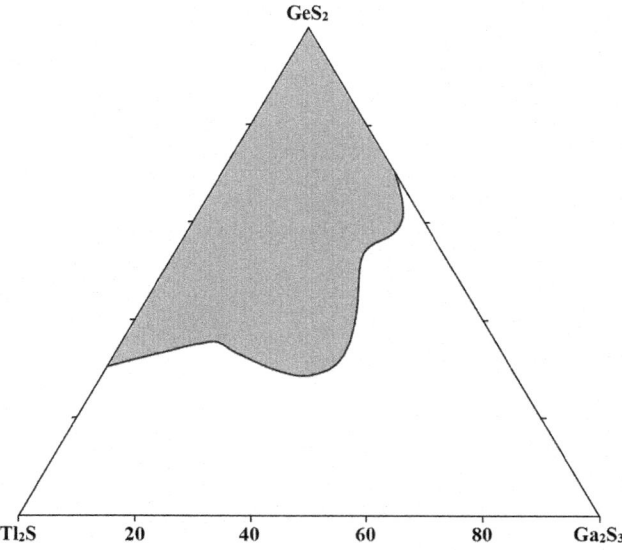

FIGURE 4.58 Glass-forming region in the GeS_2–Ga_2S_3–Tl_2S quasiternary system. (From Tsisar, O.V., et al., *J. Mater. Sci: Mater Electron.*, **28**(24), 19003, 2017.)

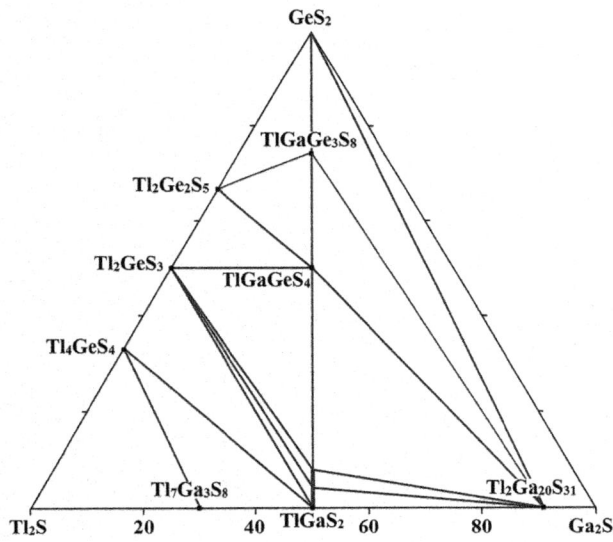

FIGURE 4.57 Isothermal section of the GeS_2–Ga_2S_3–Tl_2S quasiternary system at 250°C. (From Tsisar, O.V., et al., *Nauk. Visn. Uzhgorod. Univ. Ser. Khimiya*, [2(38)], 26, 2017.)

1025.60 ± 0.06, $c = 591.48 \pm 0.05$ pm, a calculated density of 4.605 g·cm⁻³, and an energy gap of 2.54 eV (Shi et al. 2015). $Ga_2La_2GeS_8$ crystallizes in the orthorhombic structure with the lattice parameters $a = 1769.6 \pm 0.7$, $b = 580.6 \pm 0.2$, $c = 1141.9 \pm 0.4$ pm, a calculated density of 4.225 g·cm⁻³, and an energy gap of 2.78 eV (Chen et al. 2011).

$GaLa_3Ge_{0.5}S_7$ was synthesized in the following way (Shi et al. 2015). The mixture of the elements with corresponding ratios was loaded into a silica crucible, which was then put inside a longer silica jacket. This assembly was then flame-sealed under a high vacuum of 10⁻³ Pa. The mixture was heated to 950°C in 35 h and annealed at that temperature for 120 h, then subsequently cooled at a rate of 5°C·h⁻¹ to 300°C, before switching off the furnace. Yellow single crystals were obtained. After the establishment of the formula, this compound could be prepared as a pure phase from the stoichiometric elements' mixtures at 950°C. It is stable in the air for more than 6 months. All acquired elements were stored inside an Ar-filled glove box (moisture and oxygen levels less than 0.1 ppm), and all manipulations were carried out in the glove box or under a vacuum.

To prepare $Ga_2La_2GeS_8$, the mixture of Ga, La, Ge, and S with a molar ratio 2:2:1:8 was loaded in a silica crucible jacketed by a silica tube, and then the tube was flame-sealed under a high vacuum of 10⁻³ Pa (Chen et al. 2011). It was heated at 40°C·h⁻¹ to 450°C and kept there for 10 h; then heated at 30°C·h⁻¹ to 1100°C and held for 120 h. The cooling process started with a rapid cool to 820°C, followed by a duration of 5 days at this temperature, and then cooled to 200°C at 8°C·h⁻¹. This reaction produced a small amount of light-yellow crystals that were refined as $Ga_2La_2GeS_8$. All elements were stored inside an Ar-filled glove box (moisture and oxygen levels less than 0.1 ppm), and loading manipulations were carried out in the glove box.

4.214 Germanium–Gallium–Samarium–Sulfur

The **$GaSm_3Ge_{0.5}S_7$** quaternary compound, which crystallizes in the hexagonal structure with the lattice parameters $a = 988.7 \pm 0.3$, $c = 588.1 \pm 0.3$ pm, a calculated density of 5.213 g·cm⁻³, and an energy gap of ~2.5 eV, is formed in the Ge–Ga–Sm–S system (Shi et al. 2015). It was prepared in the same way as $GaLa_3Ge_{0.5}S_7$ was synthesized heating the mixture up to 980°C

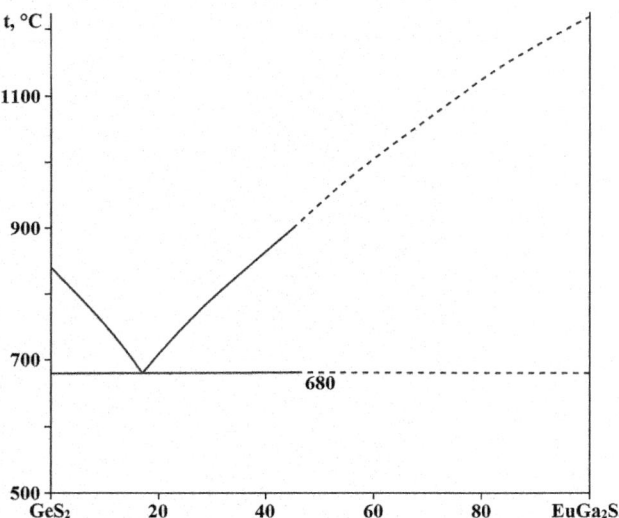

FIGURE 4.59 Phase diagram of the GeS₂–EuGa₂S₄ system. (From Barnier, S., et al., *Mater. Res. Bull.*, **15**(6), 689, 1980.)

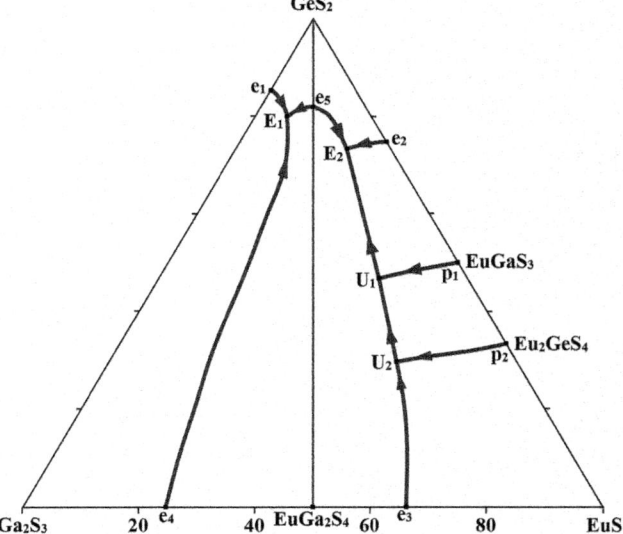

FIGURE 4.60 Scheme of the liquidus surface of the GeS₂–Ga₂S₃–EuS quasiternary system. (From Barnier, S., et al., *Mater. Res. Bull.*, **15**(6), 689, 1980.)

in 60 h and annealing it at that temperature for 100 h. This compound is stable in the air for more than 6 months.

4.215 Germanium–Gallium–Europium–Sulfur

GeS₂–EuGa₂S₄. The phase diagram of this system, constructed through DTA, is a eutectic type (Figure 4.59) (Barnier et al. 1980). The eutectic contains 17 mol% EuGa₂S₄ and crystallizes at 680°C.

GeS₂–Ga₂S₃–EuS. The scheme of the liquidus surface of this quasiternary system is presented in Figure 4.60 (Barnier et al. 1980). The ternary eutectic in the GeS₂–Ga₂S₃–EuGa₂S₄ subsystem crystallizes at 620°C. It was not possible to accurately determine the temperature and position of the ternary eutectic in the GeS₂–EuGa₂S₄–EuS subsystem. No quaternary compounds were found in the system.

The glass-forming region in the GeS₂–Ga₂S₃–EuS quasiternary system is given in Figure 4.61 (Barnier et al. 1980). The glass transition temperature of these glasses is within the interval from 410°C to 535°C.

According to the data of (Chen et al. 2011), the **Ga₂Eu₂GeS₇** quaternary compound, which crystallizes in the tetragonal structure with the lattice parameters $a = 963.76 \pm 0.06$, $c = 630.85 \pm 0.07$ pm, a calculated density of 4.196 g·cm⁻³, and an energy gap of 2.30 eV, is formed in this quasiternary system. The single orange crystals of this compound were initially discovered from the reaction of Ga, Eu, Ge, and S (molar ratio 2:2:1:7) heated to 850°C within 60 h, and kept at this temperature for 100 h, subsequently cooled to 300°C within 150 h, and then the furnace was turned off. Ga₂Eu₂GeS₇ is stable in the air for more than 6 months. All elements were stored inside an Ar-filled glove box (moisture and oxygen levels less than 0.1 ppm), and loading manipulations were carried out in the glove box.

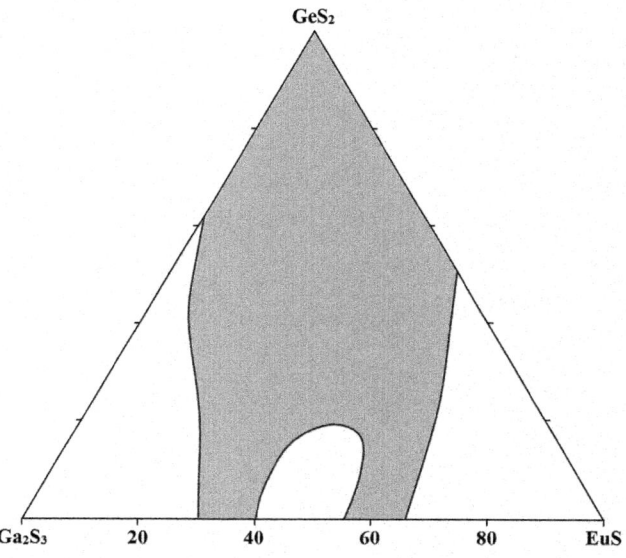

FIGURE 4.61 Glass-forming region in the GeS₂–Ga₂S₃–EuS quasiternary system. (From Barnier, S., et al., *Mater. Res. Bull.*, **15**(6), 689, 1980.)

4.216 Germanium–Gallium–Tin–Sulfur

The **Ga₂GeSnS₆** quaternary compound, which crystallizes in the orthorhombic structure with the lattice parameters $a = 4536.6 \pm 0.9$, $b = 722.88 \pm 0.14$, $c = 1160.7 \pm 0.2$ pm, and a calculated density of 3.651 g·cm⁻³, is formed in the Ge–Ga–Sn–S system (Lin et al. 2015). It has two optical transitions, one at about 1.12 eV and the other at about 2.04 eV. According to the studying using density functional theory, this compound shows a pseudo direct band gap of 1.74 eV (Yousaf et al. 2018).

To synthesize the title compound, the reaction mixture of SnS (0.5 mM), Ga₂S₃ (0.5 mM), and GeS₂ (0.5 mM) was ground and loaded into fused silica tube under an Ar atmosphere in a glove box (Lin et al. 2015). The tube was flame-sealed under a high vacuum (10^{-3} Pa) and then placed in a furnace. The sample was heated to 1000°C within 15 h, kept for 70 h, then slowly cooled to 300°C at a rate of 3°C·h^{-1}, and finally cooled to room temperature by switching off the furnace. The resulting dark-red air-stable crystals were hand-picked from the ampoule.

4.217 Germanium–Gallium–Lead–Sulfur

Two quaternary compounds, **Ga$_2$PbGeS$_6$** and **Ga$_4$Pb$_4$GeS$_{12}$**, are formed in the Ge–Ga–Pb–S system. First of them crystallizes in the orthorhombic structure with the lattice parameters $a = 4519.9 \pm 0.2$, $b = 728.38 \pm 0.02$, $c = 1160.19 \pm 0.04$ pm, a calculated density of 4.2537 ± 0.0004 g·cm^{-3}, and an energy gap of 2.37 eV (Fedorchuk et al. 2018) [$a = 4518.1 \pm 0.3$, $b = 728.15 \pm 0.05$, $c = 1159.28 \pm 0.08$ pm, a calculated density of 4.260 g·cm^{-3}, and an energy gap of 2.64 eV (Huang et al. 2018)]. Ga$_4$Pb$_4$GeS$_{12}$ crystallizes in the tetragonal structure with the lattice parameters $a = 1267.3 \pm 0.2$, $c = 612.8 \pm 0.2$ pm, a calculated density of 5.281 g·cm^{-3}, and an energy gap of ≈ 2.35 eV (Chen et al. 2013).

Ga$_2$PbGeS$_6$ was synthesized by melting stoichiometric amounts of the elements in an evacuated (10^{-2} Pa) quartz ampoule, which was subjected to local heating in the flame of an oxygen-gas burner under the visual observation of the reaction (Fedorchuk et al. 2018). The process was continued until the complete binding of elemental sulfur. The ampoule was placed in a rotating furnace to homogenize the alloy and heated to 900°C at a rate of 50°C·h^{-1}. The rotation at this temperature lasted for 24 h. Then, the furnace in the upright position was cooled at 30°C·h^{-1} to 400°C, and the alloy was annealed for 240 h. The synthesis was finished by cooling to ambient temperature in the inertial mode. The resulting alloy was a compact light brown ingot. This compound can also be prepared if PbS, Ga$_2$S$_3$, and GeS$_2$ are mixed in a molar ratio of 1:1:1 and sealed in an evacuated silica tube (Huang et al. 2018). The reactants were heated to 550°C over 10 h, then kept at that temperature for 100 h, cooled to 250°C over 60 h, and finally cooled to room temperature by shutting off the furnace. Light-yellow crystals were found at the bottom and inner wall of the tube, which were hand-picked.

The first synthesis attempt of Ga$_4$Pb$_4$GeS$_{12}$ started from reactants with a 3:2:1:8 molar ratio of Pb/Ga/Ge/S (Chen et al. 2013). The reagents were loaded in a silica tube and then flame-sealed under a high vacuum of 10^{-3} Pa. The mixture was heated at a rate of 24°C·h^{-1} to 980°C and annealed at that temperature for 100 h, then subsequently cooled at a rate of $\sim$6°C·h^{-1} to 350°C, before switching off the furnace. The green crystal products were stable in the air for a long time (more than 6 months). After the establishment of the formula, a stoichiometric reaction of Pb, Ga, Ge, and S in a silica crucible was carried out; this generated a pure phase of Pb$_4$Ga$_4$GeS$_{12}$. All acquired elements were stored inside an Ar-filled glove box with controlled moisture and oxygen levels, and all manipulations were carried out in the glove box.

4.218 Germanium–Gallium–Antimony–Sulfur

The glass-forming region in the Ge–Ga–Sb–Se system is represented in Figure 4.62 by two ternary systems Ge–Sb–Se with 5 and 10 at% at% Ga (Ma et al. 1999). It can be seen that the addition of the metallic element, Ga, destabilized the samples. With more than 15 at% Ga, it was difficult to obtain glass samples. The glasses were synthesized by using conventional methods: the appropriate quantities were weighed and

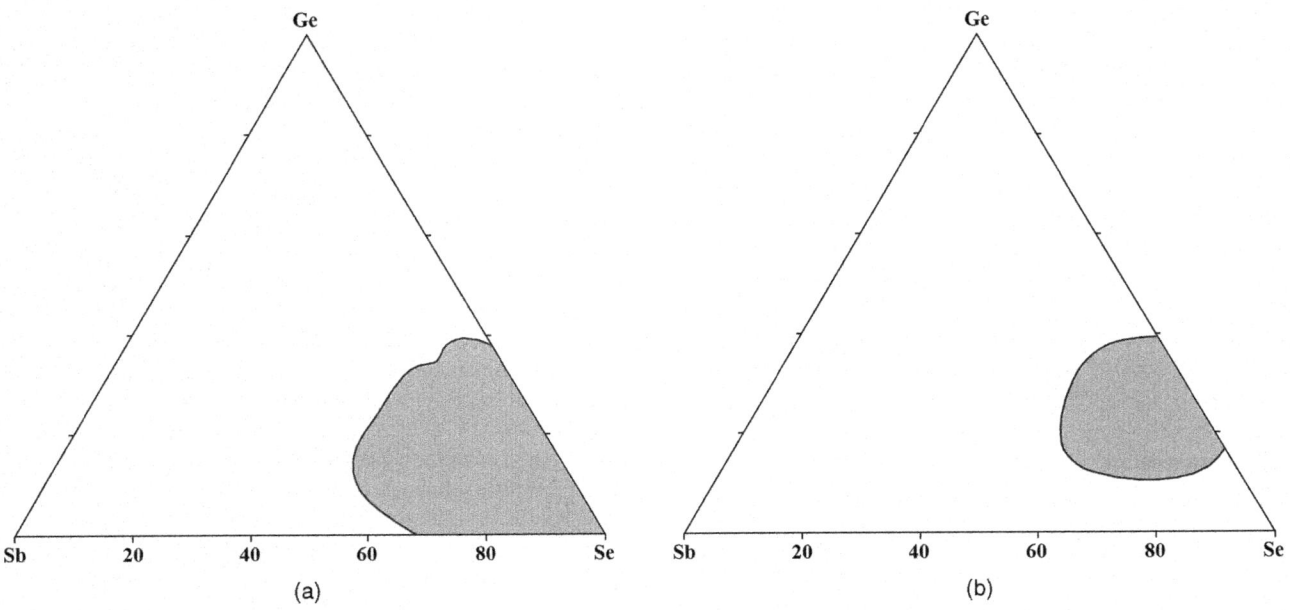

FIGURE 4.62 Glass-forming region in the Ge–Ga–Sb–Se system with (a) 5% and (b) 10% at% Ga. (From Ma, H., et al., *J. Non-Cryst. Solids*, **256–257**, 165, 1999.)

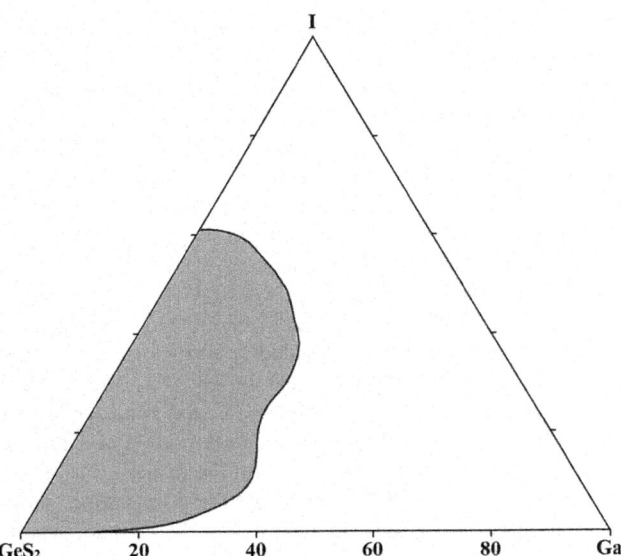

FIGURE 4.63 Glass-forming region in the GeS₂–Ga–I system. (From Ivanova, Z.G., et al., *J. Non-Cryst. Solids*, **162**(1–2), 123, 1993.)

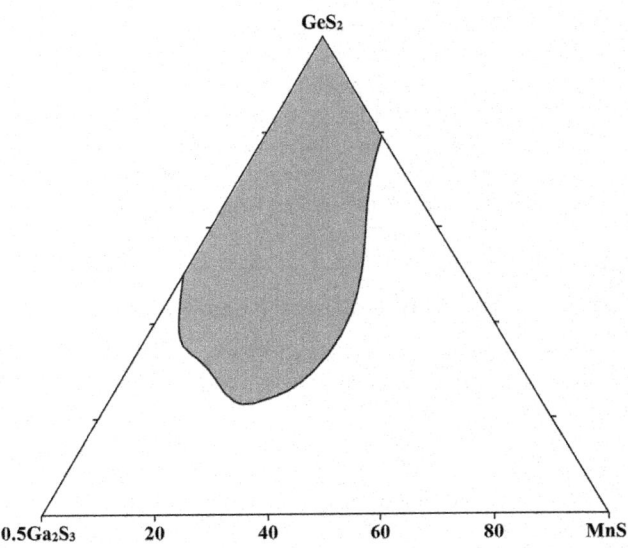

FIGURE 4.64 Glass-forming region in the GeS₂–Ga₂S₃–MnS system. (From Barnier, S., *Mater. Res. Bull.*, **19**(7), 837, 1984.)

introduced into a silica ampoule which was then sealed under vacuum. The mixture was heated and homogenized in a rocking furnace at about 800°C for 12 h. The ampoule was then cooled in the air. The glass transition temperature is within the interval from 186°C to 316°C.

4.219 Germanium–Gallium–Iodine–Sulfur

GeS₂–Ga–I. The glass-forming region in this system (Figure 4.63) lies between the GeS₂–Ga and GeS₂–I sides of the Gibbs triangle, as the gallium and iodine contents are limited to 32 and 62 at% , respectively (Ivanova et al. 1993). The glass transition temperature for the $(GeS_2)_{100-x}Ga_xI_{10}$ compositions increases with the addition of Ga from 342°C to 400°C. Glasses were prepared by placing appropriate amounts of the starting materials into quartz ampoules. The ampoules evacuated to 10^{-2} Pa with the mixtures were melted in a rotating furnace. The formation of homogeneous glassy phases requires 2–3 days, taking into account the melting points of the components. Iodine-enriched samples were heated first to 150°C and held for 2 h, then to 300°C for 2 h, and after holding at 500°C for 48 h the ampoules with the liquids were quenched in ice water. For GeS₂-enriched compositions, the synthesis was continued at 700°C for an additional 6 h and finally at 800°C for another 6 h, from which they were again water-cooled. The heating rate was approximately $2°C \cdot h^{-1}$.

4.220 Germanium–Gallium–Manganese–Sulfur

GeS₂–Ga₂S₃–MnS. Any new quaternary phase was observed in this quasiternary system (Barnier et al. 1984). Figure 4.64 shows the extent of the glass-forming region in the system. The glasses were homogeneous, and their color was dark red and increased with the Mn content. The glass-forming zone

extends from the GeS₂ top, mainly along the GeS₂–MnGa₂S₄ and GeS₂–Mn₂Ga₂S₄ axes, and it is possible to prepare homogeneous glassy samples in which one-third of the metal atoms are manganese. The ternary products were obtained from the mixture of constituent sulfides. The sulfide mixtures were heated in sealed silica ampoules under vacuum to ~1050°C. The heating was interrupted by quenching in water.

4.221 Germanium–Indium–Thallium–Sulfur

Three quaternary compounds, **InTlGeS₄**, **InTlGe₂S₆**, and **InTlGe₃S₈**, are formed in the Ge–In–Tl–S system. According to the data of Nakamura et al. (1984b), InTlGeS₄ decomposes at 320°C and crystallizes in the orthorhombic structure with the lattice parameters $a = 1170 \pm 1$, $b = 1694 \pm 1$, $c = 735.3 \pm 0.1$ pm, and the calculated and experimental densities of 4.74 and 4.73 $g \cdot cm^{-3}$, respectively.

InTlGe₂S₆ apparently has three polymorphic modifications. The first modification crystallizes in the trigonal structure with the lattice parameters $a = 977.10 \pm 0.04$, $c = 891.95 \pm 0.04$ pm, a calculated density of 4.4354 ± 0.0006 $g \cdot cm^{-3}$, and an energy gap of 2.52 eV (Khyzhun et al. 2018a), and the second one crystallizes in the cubic structure with the lattice parameter $a = 1299.7 \pm 0.6$ and a calculated density of 4.720 $g \cdot cm^{-3}$ (Yohannan and Vidyasagar 2016b). The third modification crystallizes in the orthorhombic structure with the lattice parameters $a = 1727.6 \pm 0.1$, $b = 734.04 \pm 0.05$, $c = 1166.71 \pm 0.08$ pm, a calculated density of 4.670 $g \cdot cm^{-3}$, and an energy gap of 2.3 eV (Yohannan and Vidyasagar 2016b).

InTlGe₃S₈ crystallizes in the monoclinic structure with the lattice parameters $a = 672.45 \pm 0.02$, $b = 3807.7 \pm 0.1$, $c = 679.22 \pm 0.02$ pm, $\beta = 90.616 \pm 0.002°$, and a calculated density of 3.7882 ± 0.0004 $g \cdot cm^{-3}$ (Khyzhun et al. 2018b).

Synthesis of InTlGeS₄ was carried out by the conventional evacuated silica tube method (Nakamura et al. 1984b). The stoichiometric mixture of the elements (Tl was used as TlS) was sealed in an evacuated silica glass tube and heated in a furnace at 800°C for 4 days. The melt was then shaken vigorously a number of times to ensure thorough mixing, then it was annealed at 500°C for 6 days, and slowly cooled to room temperature by turning off the furnace power. The title compound with reddish brown color was obtained.

Trigonal modification of $InTlGe_2S_6$ was prepared by comelting the stoichiometric amounts of high-purity elements (Khyzhun et al. 2018a). The reagents were placed in a quartz ampoule that was evacuated to 10^{-2} Pa and sealed. The ampoule was then put into a furnace and heated at rate of 30°C·h^{-1} up to 450°C, held there for a day for the binding of sulfur, and heated to 950°C at the rate of 20°C·h^{-1}. The melt was kept at the maximum temperature for 6 h, and then it was slowly cooled to 400°C, annealed for 240 h, and finally quenched into cold water. The as-prepared crystalline alloy was a compact ingot.

Cubic modification of $InTlGe_2S_6$ was synthesized by heating in a continuous stream of CS_2 vapor appropriate stoichiometric mixture of Tl, In_2O_3, and GeO_2 with one intermittent grinding (Yohannan and Vidyasagar 2016b). An alumina boat was loaded with the reactant mixture and placed in a ceramic tube in a horizontal tubular furnace. A continuous stream of N_2 was bubbled through CS_2 liquid and then passed through the ceramic tube to provide CS_2 vapor for the synthetic reaction. This modification also was obtained as block-shaped yellow single crystals by heating a stoichiometric reactant mixture of $TlInS_2$ (0.96 mM), Ge (0.96 mM), and S (1.9 mM) at 550°C for 2 days and then cooling to 450°C at 2°C·h^{-1} before the furnace was turned off.

Orthorhombic modification of $InTlGe_2S_6$ was prepared when a reactant mixture of Tl_2CO_3 (0.5 mM), In_2O_3 (0.5 mM), and a slight excess of GeO_2 (1.2 mM) was heated at 600°C for 1 day in CS_2 atmosphere, as in the synthesis of polycrystalline cubic modification (Yohannan and Vidyasagar 2016b). Single crystals were separated from the solidified melt.

To obtain $InTlGe_3S_8$, a batch of 2-g mass was composed of Tl, In, Ge, and S (Khyzhun et al. 2018b). The synthesis was performed in an evacuated to 10^{-2} Pa and sealed quartz ampoule. At the preliminary stage, the loaded ampoule was heated in the oxygen-gas burner flame to complete the bonding of S under the visual observation of the process. Then it was heated at the rate of 30°C·h^{-1} to 1000°C in a furnace. After keeping at this temperature for 3 h, the alloy was cooled at the rate of 20°C·h^{-1} to 400°C, annealed for 250 h, and cooled to room temperature with the furnace turned off. A plate-like single crystal was selected from the synthesized alloy.

4.222 Germanium–Indium–Lanthanum–Sulfur

Two quaternary compounds, $In_{0.33}La_3GeS_7$ and $InLa_3Ge_{0.5}S_7$, which crystallizes in the hexagonal structure with the lattice parameters $a = 1036.08 \pm 0.07$, $c = 583.71 \pm 0.07$ pm, and a calculated density of 4.604 g·cm^{-3} for the first compound and $a = 1027.9 \pm 0.3$, $c = 593.0 \pm 0.2$ pm, a calculated density of 4.849 g·cm^{-3}, and an energy gap of 2.61 eV for the second one, are formed in the Ge–In–La–S system (Shi et al. 2015).

Both compounds were synthesized by the next way (Shi et al. 2015). The mixture of the elements with corresponding ratios was loaded into a silica crucible, which was then put inside a longer silica jacket. This assembly was flame-sealed under a high vacuum of 10^{-3} Pa. The mixture was heated to 950°C in 35 h and annealed at that temperature for 120 h, then subsequently cooled at a rate of 5°C·h^{-1} to 300°C, before switching off the furnace. Yellow single crystals were obtained. After the establishment of the formula, $InLa_3Ge_{0.5}S_7$ could be prepared as a pure phase from the stoichiometric element mixtures at 950°C. Both compounds are stable in the air for more than 6 months. All acquired elements were stored inside an Ar-filled glove box (moisture and oxygen levels less than 0.1 ppm), and all manipulations were carried out in the glove box or under a vacuum.

4.223 Germanium–Indium–Samarium–Sulfur

The $In_{0.33}Sm_3GeS_7$ quaternary compound, which crystallizes in the hexagonal structure with the lattice parameters $a = 1002.9 \pm 0.2$, $c = 577.0 \pm 0.2$ pm, a calculated density of 5.197 g·cm^{-3}, and an energy gap of 2.36 eV, is formed in the Ge–In–Sm–S system (Shi et al. 2015). It was prepared in the same way as $In_{0.33}La_3GeS_7$ was synthesized.

4.224 Germanium–Indium–Gadolinium–Sulfur

The $In_{0.33}Gd_3GeS_7$ quaternary compound, which crystallizes in the hexagonal structure with the lattice parameters $a = 993.8 \pm 0.3$, $c = 574.7 \pm 0.2$ pm, and a calculated density of 5.453 g·cm^{-3}, is formed in the Ge–In–Gd–S system (Shi et al. 2015). It was prepared in the same way as $In_{0.33}La_3GeS_7$ was synthesized.

4.225 Germanium–Thallium–Lead–Sulfur

Tl_2GeS_3–PbS. The phase diagram of this system, constructed through DTA and XRD, is given in Figure 4.65 (Mozolyuk et al. 2012b). The eutectic contains 18 mol% PbS and crystallizes at 421°C. The Tl_2PbGeS_4 quaternary compound, which melts incongruently at 508°C (peritectic point at 48 mol% PbS) and has a polymorphic transformation at 395°C, is formed in this system. One of the modifications crystallizes in the monoclinic structure with the lattice parameters $a = 890.79 \pm 0.07$, $b = 909.51 \pm 0.07$, $c = 1047.72 \pm 0.07$ pm, and $\beta = 94.116 \pm 0.007°$ [$a = 890.7 \pm 0.2$, $b = 911.4 \pm 0.2$, $c = 1050.2 \pm 0.3$ pm, $\beta = 94.08 \pm 0.02°$, and a calculated density of 6.38 g·cm^{-3} (Eulenberger 1980)]. To prepare this modification, a stoichiometric mixture of GeS_2, Tl_2S, and PbS in an evacuated and sealed quartz ampoule was heated to 600°C for 3 days, then annealed at 465°C for 11 days, and finally cooled in the air to room temperature (Eulenberger 1980). A salmon-colored aggregate of acicular crystals was obtained.

GeS_2–Tl_2S–PbS. The isothermal sections of this quasiternary system at 250°C and 400°C are presented in Figure 4.66 (Mozolyuk et al. 2012b). Two quaternary compounds, Tl_2PbGeS_4 and $Tl_2PbGe_3S_8$, were found at 250°C. One more quaternary phase, **Tl$_{0.5}$Pb$_{1.75}$GeS$_4$**, which crystallizes in the cubic structure with the lattice parameter $a = 1420.082 \pm 0.006$ pm and a calculated density of 6.1747 ± 0.0001 g·cm^{-3}, was found in this system at 400°C. The alloys for investigations were annealed at 250°C for 250 h with the next quenching into cold water. A series of alloys in the region of Tl$_{0.5}$Pb$_{1.75}$GeS$_4$ existence were synthesized and annealed at 400°C.

4.226 Germanium–Thallium–Arsenic–Sulfur

The glass-forming region in the **GeS$_2$–Tl$_2$S–As$_2$S$_3$** quasiternary system is presented in Figure 4.67 (Imaoka and Jamadzaki 1967).

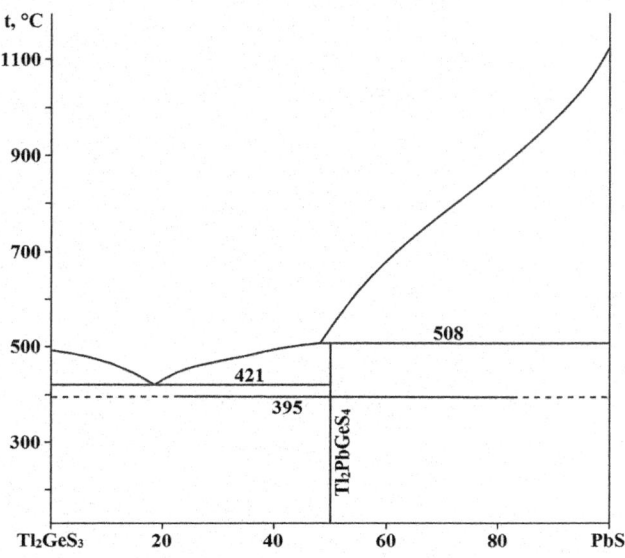

FIGURE 4.65 Phase diagram of the Tl$_2$GeS$_3$–PbS system. (From Mozolyuk, M.Yu., et al., *Chem. Met. Alloys*, **5**(1–2), 37, 2012.)

4.227 Germanium–Yttrium–Lanthanum–Sulfur

GeS$_2$–Y$_2$S$_3$–La$_2$S$_3$. The isothermal section of this quasiternary system at 500°C is shown in Figure 4.68 (Smitiukh et al. 2017c). The **Y$_{4x}$La$_{4-4x}$Ge$_3$S$_{12}$** solid solution ($x = 0$–0.75), which crystallizes in the tetragonal structure with the lattice parameters $a = 1925.87 \pm 0.09$, $c = 791.21 \pm 0.05$ pm, and a calculated density of 4.1409 ± 0.0007 g·cm^{-3} for $x = 0.5$, is formed in this system. There are also seven single-phase, eleven two-phase, and five three-phase fields in the system.

4.228 Germanium–Yttrium–Chromium–Sulfur

The **Y$_6$CrGe$_2$S$_{14}$** quaternary compound, which crystallizes in the hexagonal structure with the lattice parameters $a = 968 \pm 2$ and $c = 586 \pm 2$ pm, is formed in the Ge–Y–Cr–S system (Michelet and Flahaut 1969). This compound was synthesized in the same way as Y$_6$MnGe$_2$S$_{14}$ was prepared using CrS instead of MnS.

4.229 Germanium–Yttrium–Manganese–Sulfur

The **Y$_6$MnGe$_2$S$_{14}$** quaternary compound, which crystallizes in the hexagonal structure with the lattice parameters $a = 979.60 \pm 0.08$, $c = 574.84 \pm 0.07$ pm, and a calculated density of 4.110 g·cm^{-3} (Daszkiewicz et al. 2014a) [$a = 981 \pm 2$ and $c = 575 \pm 2$ pm (Michelet and Flahaut 1969)], is formed in the Ge–Y–Mn–S system.

A sample with the nominal composition Y$_6$MnGe$_2$S$_{14}$ was prepared by sintering the elemental constituents in an evacuated quartz tube (Daszkiewicz et al. 2014a). The synthesis was carried out in a tube resistance furnace. The ampoule was first

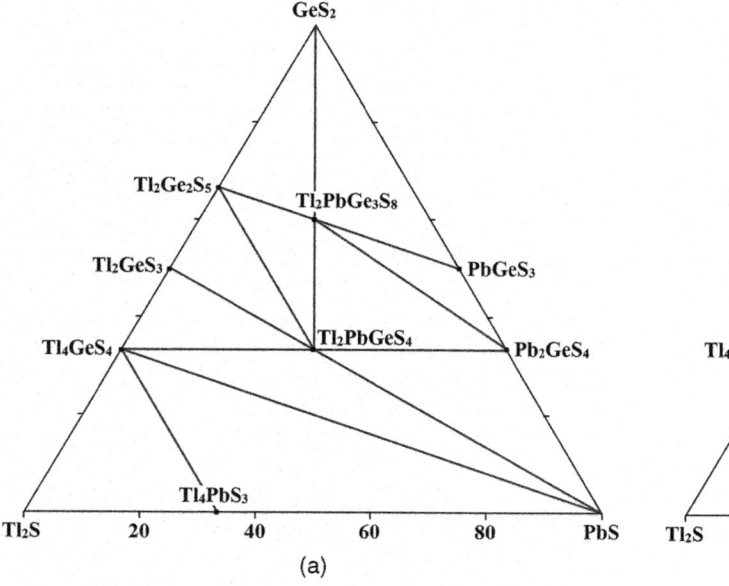

(a)

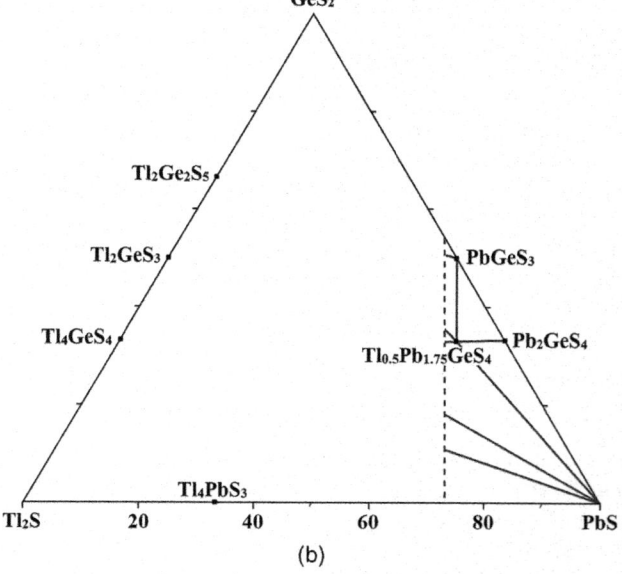

(b)

FIGURE 4.66 Isothermal section of the GeS$_2$–Tl$_2$S–PbS quasiternary system at (a) 250°C and (b) 400°C. (From Mozolyuk, M.Yu., et al., *Chem. Met. Alloys*, **5**(1–2), 37, 2012.)

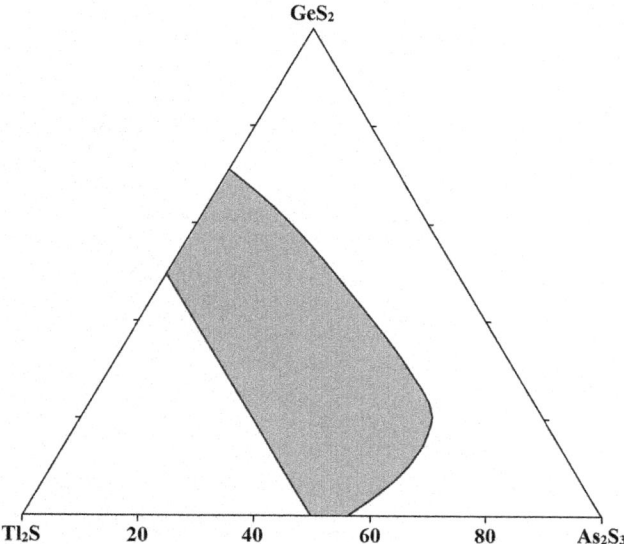

FIGURE 4.67 Glass-forming region in the GeS_2–Tl_2S–As_2S_3 quasiternary system. (From Imaoka, M. and Jamadzaki, T. [in Japanese], *Monthly J. Inst. Industr. Sci. Univ. Tokyo*, **19**(9), 261, 1967.) Open access

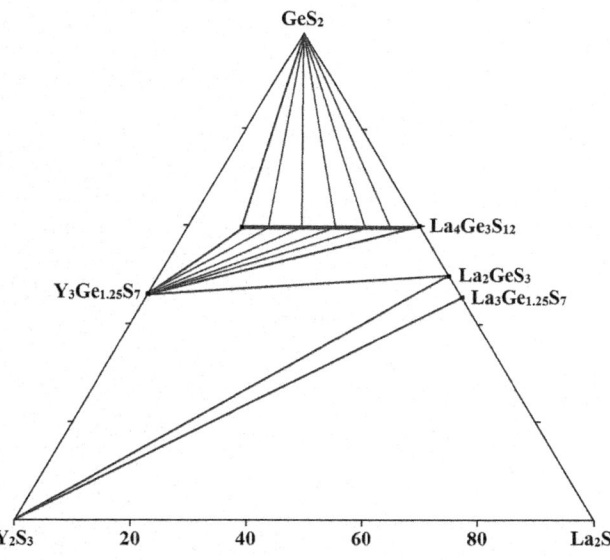

FIGURE 4.68 Isothermal section of the GeS_2–Y_2S_3–La_2S_3 quasiternary system at 500°C. (From Smitiukh, O.V., et al., *J. Alloys Compd.*, **698**, 739, 2017.)

heated with a rate of 30°C·h⁻¹ up to 1150°C, and then kept at this temperature for 3 h. Afterwards, the samples were cooled slowly (10°C·h⁻¹) down to 500°C, and annealed at this temperature for 720 h. Subsequently, the ampoule was quenched in cold water. Orange crystals of $Y_6MnGe_2S_{14}$ were obtained.

It can be also obtained in a sealed silica ampoule at around 1000°C by the interaction of MnS and Y_2S_3 with Ge and the desired amount of S (Michelet and Flahaut 1969). It appears as a polycrystalline mass. Sufficiently prolonged heating makes it possible to obtain small single crystals.

4.230 Germanium–Yttrium–Iron–Sulfur

The $Y_6FeGe_2S_{14}$ quaternary compound, which crystallizes in the hexagonal structure with the lattice parameters $a = 976.12 \pm 0.09$, $c = 577.58 \pm 0.06$ pm, and a calculated density of 4.123 g·cm⁻³ (Daszkiewicz et al. 2014b) [$a = 976 \pm 2$ and $c = 579 \pm 2$ pm (Michelet and Flahaut 1969)], is formed in the Ge–Y–Fe–S system. This compound was synthesized in the same way as $Y_6MnGe_2S_{14}$ was prepared. Dark-red crystals of the title compound were obtained.

4.231 Germanium–Yttrium–Cobalt–Sulfur

The $Y_6CoGe_2S_{14}$ quaternary compound, which crystallizes in the hexagonal structure with the lattice parameters $a = 972.9 \pm 0.1$, $c = 577.44 \pm 0.09$ pm, and a calculated density of 4.162 g·cm⁻³, is formed in the Ge–Y–Co–S system (Daszkiewicz et al. 2015). It was synthesized in the same way as $Y_6MnGe_2S_{14}$ was prepared.

4.232 Germanium–Yttrium–Nickel–Sulfur

The $Y_6NiGe_2S_{14}$ quaternary compound, which crystallizes in the hexagonal structure with the lattice parameters $a = 971.85 \pm 0.09$, $c = 578.32 \pm 0.07$ pm, and a calculated density of 4.164 g·cm⁻³, is formed in the Ge–Y–Ni–S system (Daszkiewicz et al. 2015). It was synthesized in the same way as $Y_6MnGe_2S_{14}$ was prepared.

4.233 Germanium–Lanthanum–Terbium–Sulfur

The $La_{2.02}Tb_{1.98}Ge_3S_{12}$ quaternary phase, which crystallizes in the trigonal structure with the lattice parameters $a = 1926.27 \pm 0.05$, $c = 792.63 \pm 0.02$ pm, and a calculated density of 4.685 g·cm⁻³, is formed in the Ge–La–Tb–S system (Daszkiewicz et al. 2018).

4.234 Germanium–Lanthanum–Dysprosium–Sulfur

The $La_{2.64}Dy_{1.36}Ge_3S_{12}$ quaternary phase, which crystallizes in the trigonal structure with the lattice parameters $a = 1928.68 \pm 0.05$, $c = 794.98 \pm 0.02$ pm, and a calculated density of 4.630 g·cm⁻³, is formed in the Ge–La–Dy–S system (Daszkiewicz et al. 2018).

4.235 Germanium–Lanthanum–Holmium–Sulfur

The $La_{2.25}Ho_{1.75}Ge_3S_{12}$ quaternary phase, which crystallizes in the trigonal structure with the lattice parameters $a = 1924.48 \pm 0.06$, $c = 790.96 \pm 0.04$ pm, and a calculated density of 4.727 g·cm⁻³, is formed in the Ge–La–Ho–S system (Daszkiewicz et al. 2018).

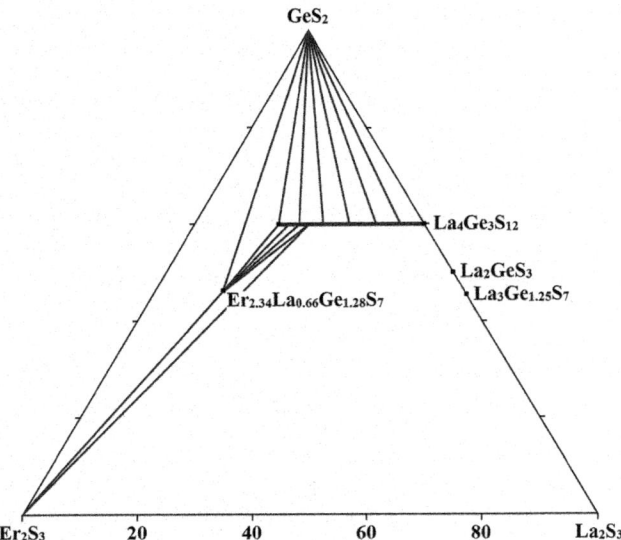

FIGURE 4.69 Isothermal section of the GeS$_2$–La$_2$S$_3$–Er$_2$S$_3$ quasiternary system at 500°C. (From Daszkiewicz, M., et al., *J. Alloys Compd.*, **738**, 263, 2018.)

4.236 Germanium–Lanthanum–Erbium–Sulfur

GeS$_2$–La$_2$S$_3$–Er$_2$S$_3$. The isothermal section of this quasiternary system at 500°C is shown in Figure 4.69 (Daszkiewicz et al. 2018). The **La$_{0.66}$Er$_{2.34}$Ge$_{1.28}$S$_7$** quaternary compound, which crystallizes in the hexagonal structure with the lattice parameters $a = 969.34 \pm 0.03$, $c = 586.80 \pm 0.02$ pm, and a calculated density of 5.572 g·cm^{-3}, and the **La$_{4-4x}$Er$_{4x}$Ge$_3$S$_{12}$** solid solution ($x = 0$–0.63), which crystallizes in the trigonal structure with the lattice parameters $a = 1921.65 \pm 0.04$, $c = 787.57 \pm 0.02$ pm, and a calculated density of 4.788 g·cm^{-3} for $x = 0.46$, are formed in this system. The alloys for the investigations were annealed at 500°C for 240 h and then quenched into cold water.

4.237 Germanium–Lanthanum–Lead–Sulfur

The **La$_2$Pb(GeS$_4$)$_2$** quaternary compound, which crystallizes in the trigonal structure with the lattice parameters $a = 906.1$ and $c = 2718.7$ pm, is formed in the Ge–La–Pb–S system (Marchuk 2019). This compound was synthesized by melting a stoichiometric mixture of the elements in evacuated (10^{-2} Pa) quartz ampoule. The ampoule was heated up to 1150°C (12°C·h^{-1}), held at this temperature for 4 h, cooled to 500°C (12°C·h^{-1}), annealed at 500°C for 500 h, and quenched in the air.

4.238 Germanium–Lanthanum–Chromium–Sulfur

The **La$_6$CrGe$_2$S$_{14}$** quaternary compound, which crystallizes in the hexagonal structure with the lattice parameters $a = 1034 \pm 2$ and $c = 584 \pm 2$ pm, is formed in the Ge–La–Cr–S system (Michelet and Flahaut 1969). This compound was synthesized in the same way as Y$_6$MnGe$_2$S$_{14}$ was prepared.

4.239 Germanium–Lanthanum–Chlorine–Sulfur

The **La$_3$Cl(GeS$_4$)$_2$** quaternary compound is formed in the Ge–La–Cl–S system (Choudhury and Dorhout 2015). Single crystals of this compound were obtained from the reaction of the anhydrous LaCl$_3$ and Na$_2$GeS$_3$. Typically, CeCl$_3$ (0.36 mM) and Na$_2$GeS$_3$ (0.28 mM) were mixed and ground in an agate mortar inside a N$_2$-filled glove box. The mixture was transferred to a graphite crucible, which was then placed inside a fused silica tube, evacuated, flame-sealed, placed vertically in a furnace, and heated at 825°C for 96 h.

4.240 Germanium–Lanthanum–Iodine–Sulfur

The **La$_3$Ge$_2$S$_8$I** quaternary compound, which crystallizes in the monoclinic structure with the lattice parameters $a = 1615.6 \pm 0.4$, $b = 797.76 \pm 0.18$, $c = 1101.8 \pm 0.3$ pm, $\beta = 93.192 \pm 0.005°$, a calculated density of 4.467 g·cm^{-3}, and an energy gap of 3.1 eV, is formed in the Ge–La–I–S system (Mumbaraddi et al. 2019). To prepare this compound, freshly La pieces, Ge powder, S flakes, and iodine crystals were used. The elements were combined in a stoichiometric ratio on a 0.5-g scale to La$_3$Ge$_2$S$_8$I, with a 10% excess (by mass) of I$_2$ added. The mixture was finely ground, cold-pressed into pellets, and loaded into fused silica tube, which was] evacuated to 0.1 Pa and sealed. To minimize volatilization losses of sulfur and iodine, and to avoid catastrophic failure of the tube, the sample was heated slowly at 2°C·min^{-1} to 300°C, held there for 2 days, heated at the same rate to 900°C, held there for 7 days, and then cooled to room temperature over 2 days. La$_3$Ge$_2$S$_8$I formed as light-yellow or nearly colorless powder, with excess I$_2$ deposited at the other end of the ampoule.

4.241 Germanium–Lanthanum–Manganese–Sulfur

The **La$_6$MnGe$_2$S$_{14}$** quaternary compound, which crystallizes in the hexagonal structure with the lattice parameters $a = 1035 \pm 2$ and $c = 578 \pm 2$ pm, is formed in the Ge–La–Mn–S system (Michelet and Flahaut 1969). This compound was synthesized in the same way as Y$_6$MnGe$_2$S$_{14}$ was prepared.

4.242 Germanium–Lanthanum–Iron–Sulfur

The **La$_6$FeGe$_2$S$_{14}$** quaternary compound, which crystallizes in the hexagonal structure with the lattice parameters $a = 1032.44 \pm 0.07$, $c = 581.32 \pm 0.05$ pm, and a calculated density of 4.590 g·cm^{-3} (Daszkiewicz et al. 2014b) [$a = 1031 \pm 2$ and $c = 585 \pm 2$ pm (Michelet and Flahaut 1969)], is formed in the Ge–La–Fe–S system. This compound was synthesized in the same way as Y$_6$MnGe$_2$S$_{14}$ was prepared. Black crystals of the title compound were obtained.

4.243 Germanium–Lanthanum–Cobalt–Sulfur

The **La$_6$CoGe$_2$S$_{14}$** quaternary compound, which crystallizes in the hexagonal structure with the lattice parameters $a = 1031.79 \pm 0.07$, $c = 581.07 \pm 0.04$ pm, and a calculated density of 4.607 g·cm^{-3} (Iyer et al. 2017b) [$a = 1031.81 \pm 0.07$, $c = 580.91 \pm 0.06$ pm, and a calculated density of 4.608 g·cm^{-3} (Daszkiewicz et al. 2015)], is formed in the Ge–La–Co–S system. It was synthesized in the same way as Y$_6$MnGe$_2$S$_{14}$ was prepared.

4.244 Germanium–Lanthanum–Nickel–Sulfur

The **La$_6$NiGe$_2$S$_{14}$** quaternary compound, which crystallizes in the hexagonal structure with the lattice parameters $a = 1029.83 \pm 0.04$, $c = 580.78 \pm 0.02$ pm, and a calculated density of 4.626 g·cm^{-3} (Iyer et al. 2017b) [$a = 1030.8 \pm 0.3$, $c = 581.4 \pm 0.2$ pm (Marchuk et al. 2015)], is formed in the Ge–La–Ni–S system. To prepare this compound, the stoichiometric mixture of the elements with a total mass of 0.2 g were pressed into pellets and loaded in a fused silica tube, which was evacuated and sealed (Iyer et al. 2017b). The tube was heated at 1050°C for 4 days, cooled to 600°C over 4 days, and then the furnace was turned off.

GeS$_2$–La$_2$S$_3$–NiS. The isothermal section of this quaternary system at 500°C is presented in Figure 4.70 (Marchuk et al. 2015). Twelve two-phase and six three-phase fields exist in this system. The solubility based on binary, ternary, and quaternary compounds is negligible (2–3 mol%). The alloys for the investigations were annealed at 500°C for 500 h.

4.245 Germanium–Lanthanum–Palladium–Sulfur

The **La$_6$PdGe$_2$S$_{14}$** quaternary compound, which crystallizes in the monoclinic structure with the lattice parameters $a = 1022.6 \pm 0.2$, $b = 583.77 \pm 0.12$, $c = 1032.0 \pm 0.2$ pm, $\beta = 119.495 \pm$

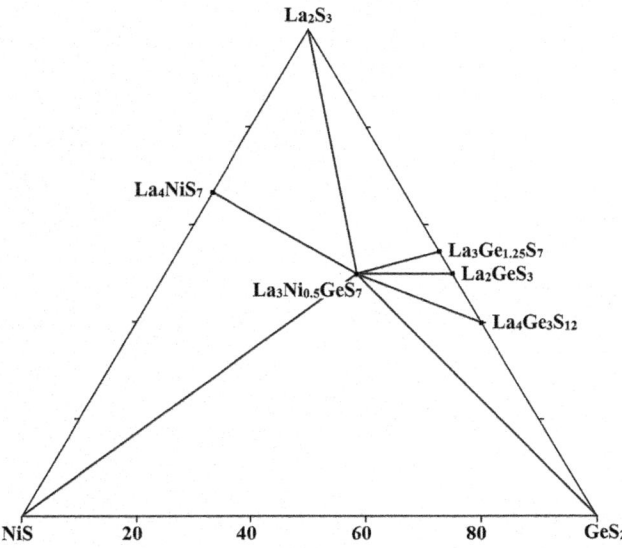

FIGURE 4.70 Isothermal section of the GeS$_2$–La$_2$S$_3$–NiS quasiternary system at 500°C. (From Marchuk, O.V., et al., *Nauk. notatky*, (50), 176, 2015.)

0.006°, a calculated density of 4.731 g·cm^{-3} at 173 K, and an energy gap of 2.57 eV, is formed in the Ge–La–Pd–S system (Akopov et al. 2022b). To obtain this compound, first the La–Pd–Ge precursor was prepared via arc-melting using pieces of La, Pd, and Ge. A sample with a total mass of 0.6–1 g was weighed out in a ratio of La/Pd/Ge = 6:1:2.05. Single crystals were grown using arc-melted precursor, S powder, and KI flux. The ratio of (precursor + S)/KI was kept at a 1:30 ratio by mass. The reactants and KI were added to a silica ampoule, which was sealed under a vacuum. The reaction products were washed with deionized water to remove KI flux, the sample was then vacuum filtered and allowed to dry. Bulk powder of the material was synthesized using the appropriate arc-melted precursor and sulfur powder. Sample was prepared by loading the germanide binary and sulfur (1:14 ratio) in a fused silica ampoule, which was then flame-sealed under vacuum. For both crystals and bulk powder, the samples were heated over 12 h to 1000–1050°C, dwelled for 72–120 h, and then cooled to room temperature over 8 h.

4.246 Germanium–Cerium–Erbium–Sulfur

The **Ce$_{0.66}$Er$_{2.34}$Ge$_{1.28}$S$_7$** quaternary compound, which crystallizes in the hexagonal structure with the lattice parameters $a = 973.86 \pm 0.05$, $c = 586.98 \pm 0.05$ pm, and a calculated density of 5.507 ± 0.001 g·cm^{-3}, is formed in the Ge–Ce–Er–S system (Smitiukh et al. 2018a). This compound was obtained by fusing a stoichiometric mixture of the elements in evacuated quartz ampoules according to the following scheme: heating to 1150°C (12°C·h^{-1}); exposure 4 h; cooling up to 500°C (12°C·h^{-1}); annealing at 500°C for 500 h; quenching in cold water.

4.247 Germanium–Cerium–Lead–Sulfur

The **Ce$_2$Pb(GeS$_4$)$_2$** quaternary compound, which crystallizes in the trigonal structure with the lattice parameters $a = 901.76 \pm 0.06$, $c = 2698.0 \pm 0.3$ pm, and a calculated density of 4.662 g·cm^{-3} (Daszkiewicz et al. 2012) [$a = 901.8$ and $c = 2698.0$ pm (Marchuk 2019)], is formed in the Ge–Ce–Pb–S system. This compound was synthesized in the same way as **La$_2$PbGeS$_4$** was prepared.

4.248 Germanium–Cerium–Iodine–Sulfur

The **Ce$_3$Ge$_2$S$_8$I** quaternary compound, which crystallizes in the monoclinic structure with the lattice parameters $a = 1605.4 \pm 0.2$, $b = 792.33 \pm 0.11$, $c = 1096.24 \pm 0.15$ pm, $\beta = 98.262 \pm 0.003$°, a calculated density of 4.568 g·cm^{-3}, and an energy gap of 2.7 eV, is formed in the Ge–Ce–I–S system (Mumbaraddi et al. 2019). This compound was prepared in the same way as La$_3$Ge$_2$S$_8$I was synthesized using Ce instead of La. Ce$_3$Ge$_2$S$_8$I formed as a yellow powder, with excess I$_2$ deposited at the other end of the ampoule.

4.249 Germanium–Cerium–Manganese–Sulfur

The $Ce_6MnGe_2S_{14}$ quaternary compound, which crystallizes in the hexagonal structure with the lattice parameters $a = 1026.36 \pm 0.06$, $c = 577.92 \pm 0.05$ pm, and a calculated density of 4.692 g·cm^{-3} (Daszkiewicz et al. 2014a) [$a = 1026 \pm 2$ and $c = 576 \pm 2$ pm (Michelet and Flahaut 1969)], is formed in the Ge–Ce–Mn–S system. Red crystals of $Ce_6MnGe_2S_{14}$ were obtained in the same way as $Y_6MnGe_2S_{14}$ was prepared.

4.250 Germanium–Cerium–Iron–Sulfur

The $Ce_6FeGe_2S_{14}$ quaternary compound, which crystallizes in the hexagonal structure with the lattice parameters $a = 1022.04 \pm 0.08$, $c = 577.54 \pm 0.06$ pm, and a calculated density of 4.738 g·cm^{-3}, is formed in the Ge–Ce–Fe–S system (Daszkiewicz et al. 2014b). This compound was synthesized in the same way as $Y_6MnGe_2S_{14}$ was prepared. Black crystals of the title compound were obtained.

4.251 Germanium–Cerium–Cobalt–Sulfur

The $Ce_6CoGe_2S_{14}$ quaternary compound, which crystallizes in the hexagonal structure with the lattice parameters $a = 1020.6 \pm 0.3$, $c = 579.06 \pm 0.17$ pm, and a calculated density of 4.748 g·cm^{-3} (Iyer et al. 2017b) [$a = 1021.68 \pm 0.07$, $c = 578.43 \pm 0.06$ pm and a calculated density of 4.743 g·cm^{-3} (Daszkiewicz et al. 2015)], is formed in the Ge–Ce–Co–S system. It was synthesized in the same way as $Y_6MnGe_2S_{14}$ was prepared.

4.252 Germanium–Cerium–Nickel–Sulfur

The $Ce_6NiGe_2S_{14}$ quaternary compound, which crystallizes in the hexagonal structure with the lattice parameters $a = 1020.52 \pm 0.13$, $c = 579.21 \pm 0.07$ pm, and a calculated density of 4.747 g·cm^{-3} (Iyer et al. 2017b) [$a = 1019.20 \pm 0.08$, $c = 578.17 \pm 0.06$ pm and a calculated density of 4.768 g·cm^{-3} (Daszkiewicz et al. 2015)], is formed in the Ge–Ce–Ni–S system. It was synthesized in the same way as $Y_6MnGe_2S_{14}$ was prepared.

4.253 Germanium–Praseodymium–Erbium–Sulfur

The $Pr_{0.66}Er_{2.34}Ge_{1.28}S_7$ quaternary compound, which crystallizes in the hexagonal structure with the lattice parameters $a = 972.81 \pm 0.05$, $c = 584.59 \pm 0.04$ pm, and a calculated density of 5.5218 ± 0.0009 g·cm^{-3}, is formed in the Ge–Pr–Er–S system (Smitiukh et al. 2018a). This compound was obtained in the same way as $Ce_{0.66}Er_{2.34}Ge_{1.28}S_7$ was prepared.

4.254 Germanium–Praseodymium–Lead–Sulfur

The $Pr_2Pb(GeS_4)_2$ quaternary compound, which crystallizes in the trigonal structure with the lattice parameters $a = 898.40 \pm 0.02$, $c = 2686.70 \pm 0.09$ pm, and a calculated density of 4.721 g·cm^{-3} (Daszkiewicz et al. 2012; Marchuk 2019)], is formed in the Ge–Pr–Pb–S system. This compound was synthesized in the same way as La_2PbGeS_4 was prepared.

4.255 Germanium–Praseodymium–Iodine–Sulfur

The $Pr_3Ge_2S_8I$ quaternary compound, which crystallizes in the monoclinic structure with the lattice parameters $a = 1597.60 \pm 0.09$, $b = 787.86 \pm 0.05$, $c = 1092.81 \pm 0.06$ pm, $\beta = 98.4525 \pm 0.0010°$, a calculated density of 4.644 g·cm^{-3}, and an energy gap of 2.9 eV, is formed in the Ge–Pr–I–S system (Mumbaraddi et al. 2019). This compound was prepared in the same way as $La_3Ge_2S_8I$ was synthesized using Ce instead of La. $Pr_3Ge_2S_8I$ formed as green powder, with excess I_2 deposited at the other end of the ampoule.

4.256 Germanium–Praseodymium–Manganese–Sulfur

The $Pr_6MnGe_2S_{14}$ quaternary compound, which crystallizes in the hexagonal structure with the lattice parameters $a = 1018.84 \pm 0.06$, $c = 576.28 \pm 0.05$ pm, and a calculated density of 4.790 g·cm^{-3} (Daszkiewicz et al. 2014a) [$a = 1019 \pm 2$ and $c = 576 \pm 2$ pm (Michelet and Flahaut 1969)], is formed in the Ge–Pr–Mn–S system. Red crystals of $Pr_6MnGe_2S_{14}$ were obtained in the same way as $Y_6MnGe_2S_{14}$ was prepared.

4.257 Germanium–Praseodymium–Iron–Sulfur

The $Pr_6FeGe_2S_{14}$ quaternary compound, which crystallizes in the hexagonal structure with the lattice parameters $a = 1015.49 \pm 0.02$, $c = 578.02 \pm 0.02$ pm, and a calculated density of 4.8097 g·cm^{-3}, is formed in the Ge–Pr–Fe–S system (Daszkiewicz et al. 2014b). This compound was synthesized in the same way as $Y_6MnGe_2S_{14}$ was prepared. Black crystals of the title compound were obtained.

4.258 Germanium–Praseodymium–Cobalt–Sulfur

The $Pr_6CoGe_2S_{14}$ quaternary compound, which crystallizes in the hexagonal structure with the lattice parameters $a = 1013.06 \pm 0.06$, $c = 577.85 \pm 0.04$ pm, and a calculated density of 4.845 g·cm^{-3} (Iyer et al. 2017b) [$a = 1014.74 \pm 0.07$, $c = 577.35 \pm 0.05$ pm and a calculated density of 4.833 g·cm^{-3} (Daszkiewicz et al. 2015)], is formed in the Ge–Pr–Co–S system. It was synthesized in the same way as $Y_6MnGe_2S_{14}$ was prepared.

4.259 Germanium–Praseodymium–Nickel–Sulfur

The $Pr_6NiGe_2S_{14}$ quaternary compound, which crystallizes in the hexagonal structure with the lattice parameters $a = 1013.03 \pm 0.04$, $c = 577.52 \pm 0.02$ pm, and a calculated density of 4.847 g·cm^{-3} (Iyer et al. 2017b), is formed in the Ge–Pr–Ni–S system. It was synthesized in the same way as $La_6NiGe_2S_{14}$ was prepared.

4.260 Germanium–Neodymium–Manganese–Sulfur

The $Nd_6MnGe_2S_{14}$ quaternary compound, which crystallizes in the hexagonal structure with the lattice parameters $a = 1013.60 \pm 0.06$, $c = 575.04 \pm 0.05$ pm, and a calculated density of 4.915 g·cm^{-3} (Daszkiewicz et al. 2014a) [$a = 1013 \pm 2$ and $c = 576 \pm 2$ pm (Michelet and Flahaut 1969)], is formed in the Ge–Nd–Mn–S system. Red crystals of $Nd_6MnGe_2S_{14}$ were obtained in the same way as $Y_6MnGe_2S_{14}$ was prepared.

4.261 Germanium–Neodymium–Cobalt–Sulfur

The $Nd_6CoGe_2S_{14}$ quaternary compound, which crystallizes in the hexagonal structure with the lattice parameters $a = 1007.75 \pm 0.06$, $c = 577.04 \pm 0.03$ pm, and a calculated density of 4.968 g·cm^{-3} (Iyer et al. 2017b) [$a = 1008.68 \pm 0.07$, $c = 576.06 \pm 0.05$ pm and a calculated density of 4.967 g·cm^{-3} (Daszkiewicz et al. 2015)], is formed in the Ge–Nd–Co–S system. It was synthesized in the same way as $Y_6MnGe_2S_{14}$ was prepared.

4.262 Germanium–Neodymium–Nickel–Sulfur

The $Nd_6NiGe_2S_{14}$ quaternary compound, which crystallizes in the hexagonal structure with the lattice parameters $a = 1007.60 \pm 0.03$, $c = 576.05 \pm 0.02$ pm, and a calculated density of 4.977 g·cm^{-3} (Iyer et al. 2017b), is formed in the Ge–Nd–Ni–S system. It was synthesized in the same way as $La_6NiGe_2S_{14}$ was prepared.

4.263 Germanium–Samarium–Manganese–Sulfur

The $Sm_6MnGe_2S_{14}$ quaternary compound, which crystallizes in the hexagonal structure with the lattice parameters $a = 997.66 \pm 0.07$, $c = 578.86 \pm 0.05$ pm, and a calculated density of 5.162 g·cm^{-3} (Daszkiewicz et al. 2014a) [$a = 1003 \pm 2$ and $c = 575 \pm 2$ pm (Michelet and Flahaut 1969)], is formed in the Ge–Sm–Mn–S system. Red crystals of $Sm_6MnGe_2S_{14}$ were obtained in the same way as $Y_6MnGe_2S_{14}$ was prepared.

4.264 Germanium–Samarium–Iron–Sulfur

The $Sm_6FeGe_2S_{14}$ quaternary compound, which crystallizes in the hexagonal structure with the lattice parameters $a = 996.2 \pm 0.1$, $c = 576.70 \pm 0.07$ pm, and a calculated density of 5.199 g·cm^{-3}, is formed in the Ge–Sm–Fe–S system (Daszkiewicz et al. 2014b). This compound was synthesized in the same way as $Y_6MnGe_2S_{14}$ was prepared. Black crystals of the title compound were obtained.

4.265 Germanium–Samarium–Cobalt–Sulfur

The $Sm_6CoGe_2S_{14}$ quaternary compound, which crystallizes in the hexagonal structure with the lattice parameters $a = 997.42 \pm 0.14$, $c = 573.93 \pm 0.08$ pm, and a calculated density of 5.222 g·cm^{-3} (Iyer et al. 2017b) [$a = 997.66 \pm 0.07$, $c = 574.69 \pm 0.05$ pm, and a calculated density of 5.213 g·cm^{-3} (Daszkiewicz et al. 2015)], is formed in the Ge–Sm–Co–S system. It was synthesized in the same way as $Y_6MnGe_2S_{14}$ was prepared.

4.266 Germanium–Samarium–Nickel–Sulfur

The $Sm_6NiGe_2S_{14}$ quaternary compound, which crystallizes in the hexagonal structure with the lattice parameters $a = 996.42 \pm 0.08$, $c = 574.36 \pm 0.05$ pm, and a calculated density of 5.228 g·cm^{-3} (Iyer et al. 2017b) [$a = 995.97 \pm 0.06$, $c = 574.30 \pm 0.05$ pm, and a calculated density of 5.233 g·cm^{-3} (Daszkiewicz et al. 2015)], is formed in the Ge–Sm–Ni–S system. It was synthesized in the same way as $Y_6MnGe_2S_{14}$ was prepared.

4.267 Germanium–Gadolinium–Manganese–Sulfur

The $Gd_6MnGe_2S_{14}$ quaternary compound, which crystallizes in the hexagonal structure with the lattice parameters $a = 992.36 \pm 0.09$, $c = 573.41 \pm 0.06$ pm, and a calculated density of 5.407 g·cm^{-3} (Daszkiewicz et al. 2014a) [$a = 994 \pm 2$ and $c = 573 \pm 2$ pm (Michelet and Flahaut 1969)], is formed in the Ge–Gd–Mn–S system. Black crystals of $Gd_6MnGe_2S_{14}$ were obtained in the same way as $Y_6MnGe_2S_{14}$ was prepared.

4.268 Germanium–Gadolinium–Iron–Sulfur

The $Gd_6FeGe_2S_{14}$ quaternary compound, which crystallizes in the hexagonal structure with the lattice parameters $a = 990.68 \pm 0.07$, $c = 573.93 \pm 0.05$ pm, and a calculated density of 5.424 g·cm^{-3}, is formed in the Ge–Gd–Fe–S system (Daszkiewicz et al. 2014b). This compound was synthesized in the same way as $Y_6MnGe_2S_{14}$ was prepared. Black crystals of the title compound were obtained.

4.269 Germanium–Gadolinium–Cobalt–Sulfur

The **Gd$_6$CoGe$_2$S$_{14}$** quaternary compound, which crystallizes in the hexagonal structure with the lattice parameters $a = 989.09 \pm 0.08$, $c = 574.06 \pm 0.06$ pm, and a calculated density of 5.451 g·cm^{-3}, is formed in the Ge–Gd–Co–S system (Daszkiewicz et al. 2015). It was synthesized in the same way as Y$_6$MnGe$_2$S$_{14}$ was prepared.

4.270 Germanium–Gadolinium–Nickel–Sulfur

The **Gd$_6$NiGe$_2$S$_{14}$** quaternary compound, which crystallizes in the hexagonal structure with the lattice parameters $a = 986.9 \pm 0.1$, $c = 573.21 \pm 0.07$ pm, and a calculated density of 5.482 g·cm^{-3}, is formed in the Ge–Gd–Ni–S system (Daszkiewicz et al. 2015). It was synthesized in the same way as Y$_6$MnGe$_2$S$_{14}$ was prepared.

4.271 Germanium–Terbium–Manganese–Sulfur

The **Tb$_6$MnGe$_2$S$_{14}$** quaternary compound, which crystallizes in the hexagonal structure with the lattice parameters $a = 987.61 \pm 0.08$, $c = 571.68 \pm 0.07$ pm, and a calculated density of 5.510 g·cm^{-3}, is formed in the Ge–Tb–Mn–S system (Daszkiewicz et al. 2014a). Orange crystals of Tb$_6$MnGe$_2$S$_{14}$ were obtained in the same way as Y$_6$MnGe$_2$S$_{14}$ was prepared.

4.272 Germanium–Terbium–Iron–Sulfur

The **Tb$_6$FeGe$_2$S$_{14}$** quaternary compound, which crystallizes in the hexagonal structure with the lattice parameters $a = 983.51 \pm 0.08$, $c = 574.81 \pm 0.05$ pm, and a calculated density of 5.529 g·cm^{-3}, is formed in the Ge–Tb–Fe–S system (Daszkiewicz et al. 2014b). This compound was synthesized in the same way as Y$_6$MnGe$_2$S$_{14}$ was prepared. Black crystals of the title compound were obtained.

4.273 Germanium–Terbium–Cobalt–Sulfur

The **Tb$_6$CoGe$_2$S$_{14}$** quaternary compound, which crystallizes in the hexagonal structure with the lattice parameters $a = 981.47 \pm 0.09$, $c = 573.76 \pm 0.07$ pm, and a calculated density of 5.573 g·cm^{-3}, is formed in the Ge–Tb–Co–S system (Daszkiewicz et al. 2015). It was synthesized in the same way as Y$_6$MnGe$_2$S$_{14}$ was prepared.

4.274 Germanium–Terbium–Nickel–Sulfur

The **Tb$_6$NiGe$_2$S$_{14}$** quaternary compound, which crystallizes in the hexagonal structure with the lattice parameters $a = 980.38 \pm 0.08$, $c = 573.73 \pm 0.06$ pm, and a calculated

density of 5.585 g·cm^{-3}, is formed in the Ge–Tb–Ni–S system (Daszkiewicz et al. 2015). It was synthesized in the same way as Y$_6$MnGe$_2$S$_{14}$ was prepared.

4.275 Germanium–Dysprosium–Manganese–Sulfur

The **Dy$_6$MnGe$_2$S$_{14}$** quaternary compound, which crystallizes in the hexagonal structure with the lattice parameters $a = 981.68 \pm 0.06$, $c = 574.09 \pm 0.05$ pm, and a calculated density of 5.628 g·cm^{-3} (Daszkiewicz et al. 2014a) [$a = 982 \pm 2$ and $c = 575 \pm 2$ pm (Michelet and Flahaut 1969)], is formed in the Ge–Dy–Mn–S system. Yellow crystals of Dy$_6$MnGe$_2$S$_{14}$ were obtained in the same way as Y$_6$MnGe$_2$S$_{14}$ was prepared.

4.276 Germanium–Dysprosium–Iron–Sulfur

The **Dy$_6$FeGe$_2$S$_{14}$** quaternary compound, which crystallizes in the hexagonal structure with the lattice parameters $a = 977.10 \pm 0.08$, $c = 576.90 \pm 0.06$ pm, and a calculated density of 5.657 g·cm^{-3}, is formed in the Ge–Dy–Fe–S system (Daszkiewicz et al. 2014b). This compound was synthesized in the same way as Y$_6$MnGe$_2$S$_{14}$ was prepared. Black crystals of the title compound were obtained.

4.277 Germanium–Dysprosium–Cobalt–Sulfur

The **Dy$_6$CoGe$_2$S$_{14}$** quaternary compound, which crystallizes in the hexagonal structure with the lattice parameters $a = 975.09 \pm 0.08$, $c = 576.18 \pm 0.06$ pm, and a calculated density of 5.698 g·cm^{-3}, is formed in the Ge–Dy–Co–S system (Daszkiewicz et al. 2015). It was synthesized in the same way as Y$_6$MnGe$_2$S$_{14}$ was prepared.

4.278 Germanium–Dysprosium–Nickel–Sulfur

The **Dy$_6$NiGe$_2$S$_{14}$** quaternary compound, which crystallizes in the hexagonal structure with the lattice parameters $a = 974.30 \pm 0.07$, $c = 577.27 \pm 0.05$ pm, and a calculated density of 5.696 g·cm^{-3}, is formed in the Ge–Dy–Ni–S system (Daszkiewicz et al. 2015). It was synthesized in the same way as Y$_6$MnGe$_2$S$_{14}$ was prepared.

4.279 Germanium–Holmium–Manganese–Sulfur

The **Ho$_6$MnGe$_2$S$_{14}$** quaternary compound, which crystallizes in the hexagonal structure with the lattice parameters $a = 975.41 \pm 0.07$, $c = 575.74 \pm 0.05$ pm, and a calculated density of 5.736 g·cm^{-3}, is formed in the Ge–Ho–Mn–S system (Daszkiewicz et al. 2014a). Yellow crystals of Ho$_6$MnGe$_2$S$_{14}$ were obtained in the same way as Y$_6$MnGe$_2$S$_{14}$ was prepared.

4.280 Germanium–Holmium–Iron–Sulfur

The $Ho_6FeGe_2S_{14}$ quaternary compound, which crystallizes in the hexagonal structure with the lattice parameters $a = 969.48 \pm 0.07$, $c = 579.25 \pm 0.06$ pm, and a calculated density of 5.774 $g \cdot cm^{-3}$, is formed in the Ge–Ho–Fe–S system (Daszkiewicz et al. 2014b). This compound was synthesized in the same way as $Y_6MnGe_2S_{14}$ was prepared. Black crystals of the title compound were obtained.

4.281 Germanium–Holmium–Cobalt–Sulfur

The $Ho_6CoGe_2S_{14}$ quaternary compound, which crystallizes in the hexagonal structure with the lattice parameters $a = 967.93 \pm 0.09$, $c = 580.06 \pm 0.06$ pm, and a calculated density of 5.795 $g \cdot cm^{-3}$, is formed in the Ge–Ho–Co–S system (Daszkiewicz et al. 2015). It was synthesized in the same way as $Y_6MnGe_2S_{14}$ was prepared.

4.282 Germanium–Holmium–Nickel–Sulfur

The $Ho_6NiGe_2S_{14}$ quaternary compound, which crystallizes in the hexagonal structure with the lattice parameters $a = 968.26 \pm 0.07$, $c = 581.82 \pm 0.06$ pm, and a calculated density of 5.773 $g \cdot cm^{-3}$, is formed in the Ge–Ho–Ni–S system (Daszkiewicz et al. 2015). It was synthesized in the same way as $Y_6MnGe_2S_{14}$ was prepared.

4.283 Germanium–Erbium–Manganese–Sulfur

The $Er_6MnGe_2S_{14}$ quaternary compound, which crystallizes in the hexagonal structure with the lattice parameters $a = 969.50 \pm 0.07$, $c = 580.33 \pm 0.05$ pm, and a calculated density of 5.809 $g \cdot cm^{-3}$ (Daszkiewicz et al. 2014a) [$a = 968 \pm 2$ and $c = 580 \pm 2$ pm (Michelet and Flahaut 1969)], is formed in the Ge–Er–Mn–S system. Orange crystals of $Er_6MnGe_2S_{14}$ were obtained in the same way as $Y_6MnGe_2S_{14}$ was prepared.

4.284 Germanium–Erbium–Iron–Sulfur

The $Er_6FeGe_2S_{14}$ quaternary compound, which crystallizes in the hexagonal structure with the lattice parameters $a = 964.28 \pm 0.06$, $c = 583.94 \pm 0.06$ pm, and a calculated density of 5.839 $g \cdot cm^{-3}$, is formed in the Ge–Er–Fe–S system (Daszkiewicz et al. 2014b). This compound was synthesized in the same way as $Y_6MnGe_2S_{14}$ was prepared. Black crystals of the title compound were obtained.

4.285 Germanium–Erbium–Cobalt–Sulfur

The $Er_6CoGe_2S_{14}$ quaternary compound, which crystallizes in the hexagonal structure with the lattice parameters $a = 960.84 \pm 0.09$, $c = 584.32 \pm 0.08$ pm, and a calculated density of 5.888 $g \cdot cm^{-3}$, is formed in the Ge–Er–Co–S system (Daszkiewicz et al. 2015). It was synthesized in the same way as $Y_6MnGe_2S_{14}$ was prepared.

4.286 Germanium–Erbium–Nickel–Sulfur

The $Er_6NiGe_2S_{14}$ quaternary compound, which crystallizes in the hexagonal structure with the lattice parameters $a = 958.96 \pm 0.07$, $c = 585.69 \pm 0.06$ pm, and a calculated density of 5.896 $g \cdot cm^{-3}$, is formed in the Ge–Er–Ni–S system (Daszkiewicz et al. 2015). It was synthesized in the same way as $Y_6MnGe_2S_{14}$ was prepared.

4.287 Germanium–Thulium–Manganese–Sulfur

The $Tm_6MnGe_2S_{14}$ quaternary compound, which crystallizes in the hexagonal structure with the lattice parameters $a = 965 \pm 2$ and $c = 584 \pm 2$ pm, is formed in the Ge–Tm–Mn–S system (Michelet and Flahaut 1969). This compound was synthesized in the same way as $Y_6MnGe_2S_{14}$ was prepared.

4.288 Germanium–Thulium–Iron–Sulfur

The $Tm_6FeGe_2S_{14}$ quaternary compound, which crystallizes in the hexagonal structure with the lattice parameters $a = 956.53 \pm 0.08$, $c = 587.52 \pm 0.07$ pm, and a calculated density of 5.933 $g \cdot cm^{-3}$, is formed in the Ge–Tm–Fe–S system (Daszkiewicz et al. 2014b). This compound was synthesized in the same way as $Y_6MnGe_2S_{14}$ was prepared. Black crystals of the title compound were obtained.

4.289 Germanium–Thulium–Cobalt–Sulfur

The $Tm_6CoGe_2S_{14}$ quaternary compound, which crystallizes in the hexagonal structure with the lattice parameters $a = 956.21 \pm 0.06$, $c = 588.38 \pm 0.05$ pm, and a calculated density of 5.940 $g \cdot cm^{-3}$, is formed in the Ge–Tm–Co–S system (Daszkiewicz et al. 2015). It was synthesized in the same way as $Y_6MnGe_2S_{14}$ was prepared.

4.290 Germanium–Thulium–Nickel–Sulfur

The $Tm_6NiGe_2S_{14}$ quaternary compound, which crystallizes in the hexagonal structure with the lattice parameters $a = 953.2 \pm 0.1$, $c = 588.6 \pm 0.1$ pm, and a calculated density of 5.974 $g \cdot cm^{-3}$, is formed in the Ge–Tm–Ni–S system (Daszkiewicz et al. 2015). It was synthesized in the same way as $Y_6MnGe_2S_{14}$ was prepared.

4.291 Germanium–Tin–Lead–Sulfur

The $GeSn_{0.5}Pb_{1.5}S_4$ quaternary phase, which crystallizes in the cubic structure with the lattice parameter $a = 1406.0 \pm 0.1$ pm, is formed in the Ge–Sn–Pb–S system (Poduska et al. 2002a).

4.292 Germanium–Lead–Selenium–Sulfur

GeS–PbSe. The solubility of GeS in PbSe is 27 mol% at 705°C and decreases to 7.5 mol% at 470°C (Krebs and Langner 1964; Nikolić 1969). The lattice parameter of the solid solution

changes linearly. The samples for the investigations were annealed at 705°C for 2 weeks (Nikolić 1969) and at 470°C for 70 h (Krebs and Langner 1964).

GeSe–PbS. The solubility of GeSe in PbS at 540°C is 8 mol%, and the solubility of PbS in GeSe at 500°C reaches 10 mol% (Krebs and Langner 1964). The samples for the investigations were annealed at 500°C and 540°C for 80 h and were studied through XRD (Krebs and Langner 1964).

GeS + PbSe ↔ GeSe + PbS. The region of the solid solutions in the GeSe–PbS–PbSe subsystem was determined by Krebs and Langner (1964) and is presented in Figure 4.71.

4.293 Germanium–Lead–Tellurium–Sulfur

GeS–PbTe. The solubility of GeS in PbTe is 5.7 mol% at 630°C (Nikolić 1969). The samples for the investigations were annealed at 630°C for 2 weeks.

4.294 Germanium–Lead–Iron–Sulfur

Two quaternary phases, **$(Pb,Fe)_3Ge_{1-x}S_4$** (mineral morozeviczite) and **$(Fe,Pb)_3(Ge,Fe)_{1-x}S_4$** (mineral polkovicite) ($x = 0.18$–0.69 for both phases), are formed in the Ge–Pb–Fe–S system (Fleischer and Pabst 1981). At $x = 0.47$ this phase crystallizes in the cubic structure with the lattice parameter $a = 1061$ pm and a calculated density of 6.62 $g \cdot cm^{-3}$. These two phases form a continuous solid solution series.

4.295 Germanium–Arsenic–Bromine–Sulfur

GeS_2–AsSBr. The equilibrium and non-equilibrium phase diagrams of this system have been constructed through DTA, XRD, and metallography and is presented in Figure 4.72 (Khiminets

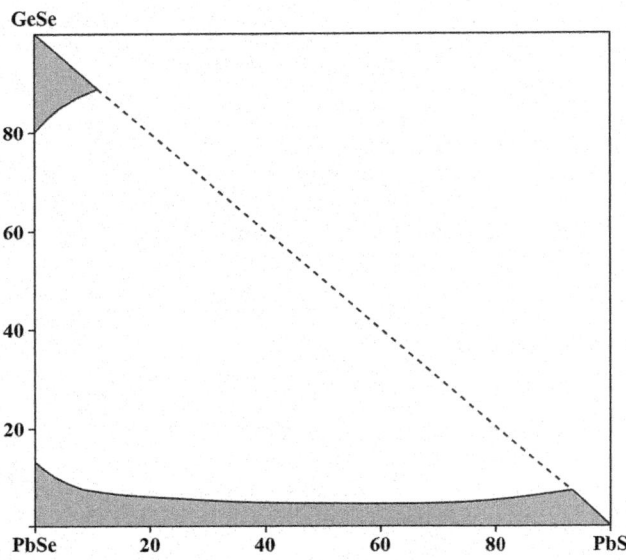

FIGURE 4.71 Region of the solid solutions in the GeSe–PbS–PbSe subsystem of the GeS + PbSe ↔ GeSe + PbS ternary mutual system. (From Krebs, H. and Langner, D., *Z. anorg. und allg. Chem.*, **334**(1–2), 37, 1964.)

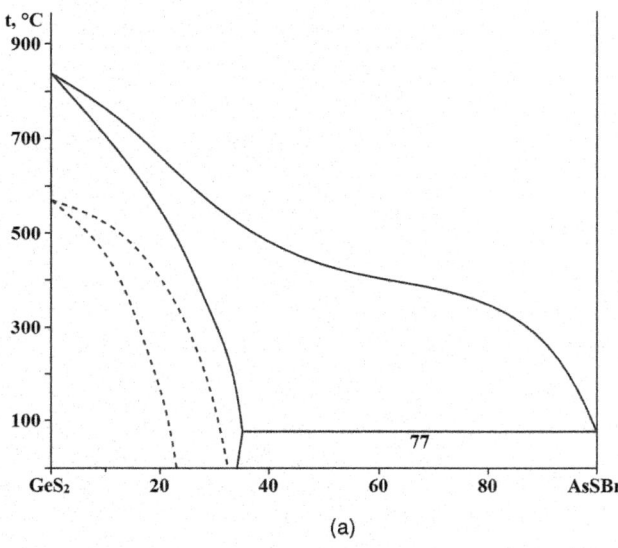

(a)

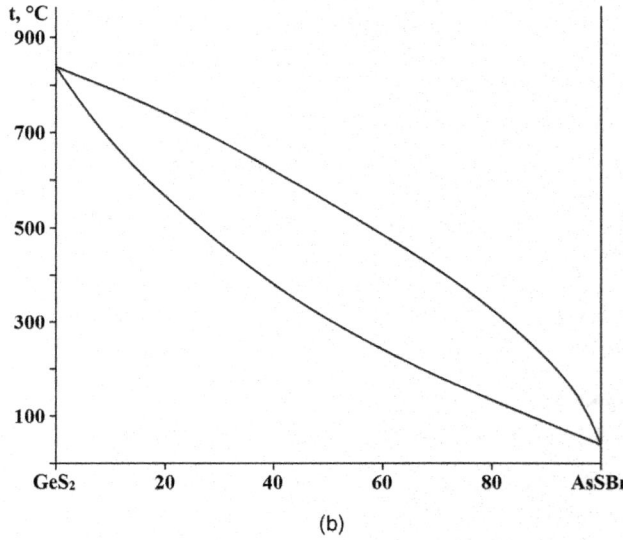

(b)

FIGURE 4.72 Equilibrium (a) and non-equilibrium (b) phase diagrams of the GeS₂–AsSBr system. (From Khiminets, et al., *Zhurn. neorgan. khimii*, **35**(11), 2945, 1990.)

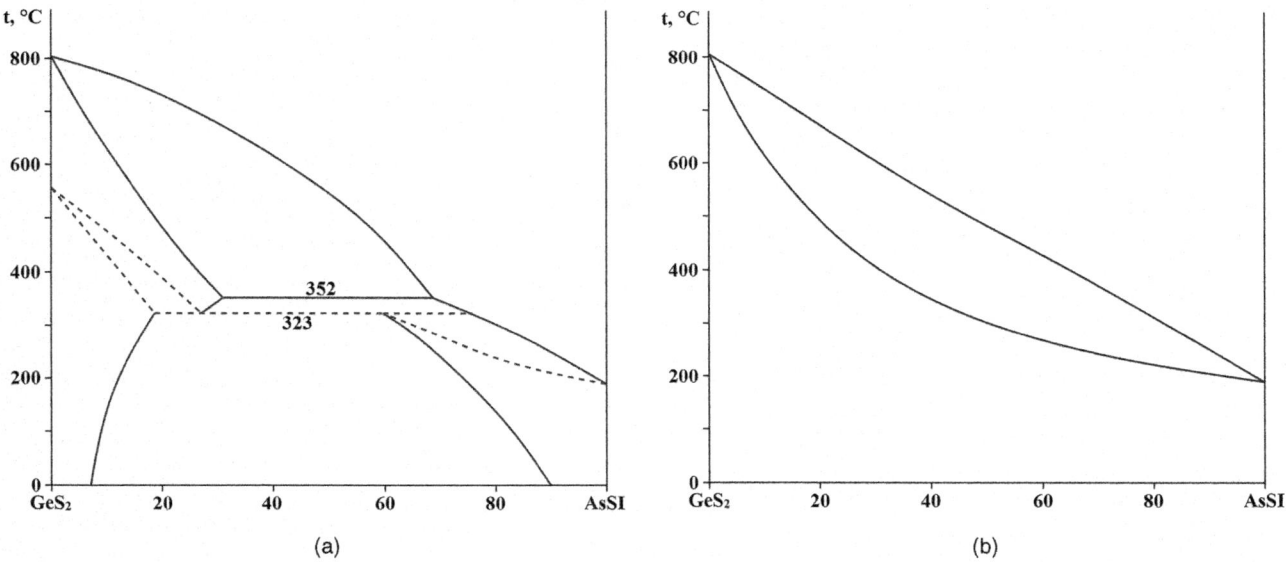

FIGURE 4.73 Equilibrium (a) and non-equilibrium (b) phase diagrams of the GeS$_2$–AsSI system. (From Khiminets, V.V., *Zhurn. neorgan. khimii*, **27**(8), 2094, 1982.)

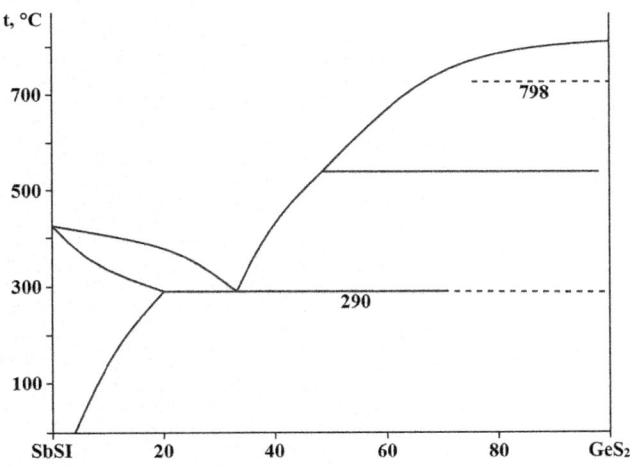

FIGURE 4.74 Phase diagram of the GeS$_2$–SbSI system. (From Bogdashevskaya, N.N., et al., *Ukr. khim. zhurn.*, **47**(7), 702, 1981.)

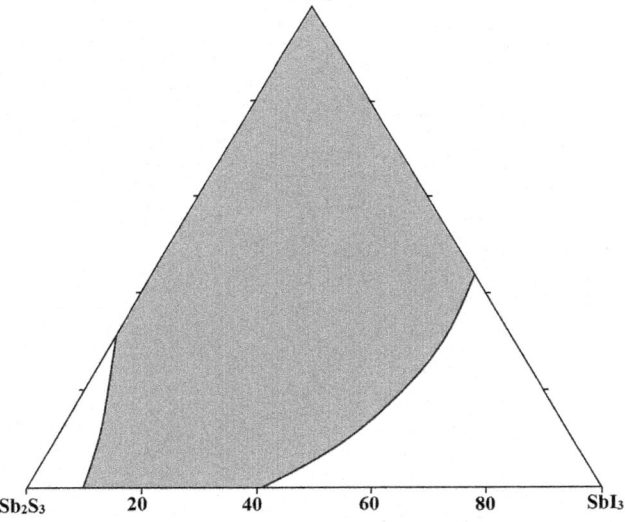

FIGURE 4.75 Glass-forming region in the GeS$_2$–Sb$_2$S$_3$–SbI$_3$ quasiternary system. (From Bogdashevskaya, N.N., et al., *Ukr. khim. zhurn.*, **47**(7), 702, 1981.)

et al. 1990). In the equilibrium state, the solubility of AsSBr in GeS$_2$ reaches 35 mol%, and in the non-equilibrium state, a continuous solid solution is formed. A degenerated eutectic from the AsSBr-side crystallizes at 77°C ± 20°C. The alloys containing up to 80 mol% AsSBr can be obtained in the glassy state already at the cooling rate of the melt ca. 1°C·min^{-1}. The crystallization ability of these glasses is low expressed. At the higher AsSBr content, for obtaining glassy alloys, the cooling rate must be higher than 50°C·sec^{-1}.

4.296 Germanium–Arsenic–Iodine–Sulfur

GeS$_2$–AsSI. The equilibrium and non-equilibrium phase diagrams of this system have been constructed through DTA, XRD, and metallography and are presented in Figure 4.73 (Khiminets 1982). The equilibrium phase diagram of the system is of a peritectic type with limited solubility in the solid state. The solubility of AsSI in GeS$_2$ and GeS$_2$ in AsSI at room temperature is 7 and 10 mol%, respectively. Peritectic interaction takes place at 356°C. An increase in the melt cooling rate makes it possible to significantly overcool a homogeneous liquid and obtain homogeneous samples. It was established that at a melt cooling rate of more than 50°C·sec^{-1}, all alloys are glassy.

4.297 Germanium–Antimony–Iodine–Sulfur

GeS$_2$–SbSI. The phase diagram of this system, constructed through DTA, XRD, and metallography, is a eutectic type (Figure 4.74) (Bogdashevskaya et al. 1981). The eutectic contains 33 mol% GeS$_2$ and crystallizes at 290°C. The solubility of GeS$_2$ in SbSI at the eutectic temperature is 20 mol%. For some samples with a content of more than 65 mol% GeS$_2$, a thermal effect was detected at 728°C, which, apparently, is associated with partial separation in the GeS$_2$–S system. When the GeS$_2$ content in the samples is more than 2–3 mol%, glass-formation is observed during their rapid cooling (Kopinets et al. 1975; Bogdashevskaya et al. 1981). The homogeneous and uniform in structure glasses can be obtained for samples in the range of compositions up to 53 mol% GeS$_2$.

The glass-forming region in the **GeS$_2$–Sb$_2$S$_3$–SbI$_3$** quasiternary system is shown in Figure 4.75 (Bogdashevskaya et al. 1981).

4.298 Germanium–Iron–Cobalt–Sulfur

The **Co$_{1-x}$Fe$_x$Ge$_{1.5}$S$_{1.5}$** solid solutions exist in the Ge–Fe–Co–S system (Vaqueiro et al. 2008). To obtain it, mixtures of the elements corresponding to their stoichiometry were ground in an agate mortar prior to sealing into evacuated (< 10^{-2} Pa) silica tubes. The inner walls of the silica tubes were coated with a thin layer of carbon by pyrolysis of acetone. The reaction mixtures were heated at 500–600°C for 2–4 days and then cooled to room temperature at 0.5°C·min^{-1} prior to removal from the furnace. Following regrinding, the material was sealed into a second silica tube, re-fired at 700°C for 4–5 days, and cooled to room temperature at 0.1°C·min^{-1}.

5

Systems Based on Germanium Selenides

5.1 Germanium–Hydrogen–Oxygen–Selenium

The **Ge(SeO$_3$)$_2$·H$_2$SeO$_3$** or **H$_2$[Ge(SeO$_3$)$_3$]** quaternary compound is formed in the Ge–H–O–Se system (Shokol et al. 1967; Gospodinov and Barkov 1982). This compound could be obtained at the interaction of GeO$_2$ and H$_2$SeO$_3$ at the components ratio of 6.6–13.0 M Se for 1 M of Ge and pH ≤ 2. It is a white, finely crystalline substance, poorly soluble in water. Thermal dissociation of Ge(SeO$_3$)$_2$·H$_2$SeO$_3$ proceeds according to the following scheme: at 180–200°C, Ge(SeO$_3$)$_2$·SeO$_2$ is formed, further heating to 360–380°C leads to the formation of Ge(SeO$_3$)$_2$, and at 500–510°C, GeO$_2$ is formed as a decomposition product.

5.2 Germanium–Lithium–Barium–Selenium

The **Li$_2$BaGeSe$_4$** quaternary compound, which crystallizes in the tetragonal structure with the lattice parameters $a = 697.9 \pm 0.5$, $c = 830.3 \pm 0.4$ pm, a calculated density of 4.431 g·cm^{-3}, and an energy gap of 2.40 eV, is formed in the Ge–Li–Ba–Se system (Wu et al. 2017c). To synthesize the title compound, a mixture of Li (2 mM), Ba (1 mM), Ge (1 mM), and Se (4 mM) was first loaded into a graphite crucible that avoids the reaction between Li and silica tube, then put it into the silica tube and flame-sealed under 10^{-3} Pa. The tube was heated to 800°C in 30 h and kept at this temperature for about 90 h, then cooled slowly down to ambient temperature for 5 days. The product was washed with dimethylformamide (DMF). Finally, many small deep-yellow transparent crystals of Li$_2$BaGeSe$_4$ were found, which are stable in the air after several months.

5.3 Germanium–Lithium–Zinc–Selenium

The **Li$_2$ZnGeSe$_4$** quaternary compound, which melts at 684°C and crystallizes in the monoclinic structure with the lattice parameters $a = 655.15 \pm 0.01$, $b = 685.89 \pm 0.01$, $c = 824.27 \pm 0.01$ pm, $\beta = 90.063 \pm 0.001°$, a calculated density of 4.193 g·cm^{-3}, and an energy gap of 1.86 eV [~2.5 eV (Weiland et al. 2017)], is formed in the Ge–Li–Zn–Se system (Zhang et al. 2015). This compound was prepared via traditional high-temperature, solid-state reactions. The starting materials were weighed in stoichiometric amounts of Li (138 mg), Zn (65.4 mg), Ge (72.6 mg), and Se (315.6 mg) in an Ar-filled glove box. The reactants were placed into a graphite tube inside a fused silica tube. The tube was sealed under a pressure

of approximately 0.1 Pa, placed into a furnace, and heated to 190°C in 2 h, held at 190°C for 10 h, and then heated up to 700°C over 5 h, held at 700°C for 100 h and slowly cooled to 350°C in 100 h. The sample was then radiatively cooled to room temperature by switching off the furnace. The reaction vessel was opened under ambient conditions. Li$_2$ZnGeSe$_4$ was prepared as orange-red rod-like polycrystals.

5.4 Germanium–Lithium–Cadmium–Selenium

The **Li$_2$CdGeSe$_4$** quaternary compound, which melts incongruently at 686°C and crystallizes in the orthorhombic structure with the lattice parameters $a = 1414.41 \pm 0.02$, $b = 830.46 \pm 0.01$, $c = 668.05 \pm 0.01$ pm, a calculated density of 4.357 g·cm^{-3}, and an energy gap of 2.5 eV, is formed in the Ge–Li–Cd–Se system (Zhang et al. 2017). Traditional high-temperature, solid-state reactions were carried out to prepare Li$_2$CdGeSe$_4$. The starting materials were weighed in stoichiometric portions, Li (15.8 mg), Cd (128.7 mg), Ge (83.1 mg), and Se (361.4 mg) in an Ar atmosphere in glove box. The mixture was inserted in graphite tube inside a fused silica tube. The tube was flame-sealed under a pressure of approximately 0.1 Pa and placed into a furnace. The reaction was heated to 190°C in 2 h, held at 190°C for 10 h, and then heated up to 675°C in 5 h, held at 675°C for 100 h, and slowly ramped down to 350°C in 100 h. The furnace was then shut off and the sample was allowed to cool to room temperature radiatively. The reaction vessel was opened under ambient conditions. Pink-red polyhedral-shaped crystals of the title compound were obtained.

According to the data of Guo et al. (2018), the composition of this compound is Li$_7$Cd$_{4.5}$Ge$_4$Se$_{16}$, it melts congruently at 712°C, and crystallizes in the orthorhombic structure with the lattice parameters $a = 1416.51 \pm 0.11$, $b = 829.90 \pm 0.09$, $c = 668.69 \pm 0.03$ pm, a calculated density of 4.453 g·cm^{-3} at 153 K, and an energy gap of 2.18 eV. The polycrystalline sample of this compound was synthesized by using a mixture of Li$_2$Se, CdSe, and GeSe$_2$ in a stoichiometric ratio. The sample was heated to 600°C for 20 h, kept at that temperature for 72 h, and then the furnace was turned off. To prepare single crystals, the analogous mixture was ground and loaded into a fused silica tube under an Ar atmosphere in a glove box, which was evacuated and sealed. The tube was heated to 1000°C for over 20 h and kept at that temperature for 48 h, slowly cooled to 300°C at a rate of 2.5°C·h^{-1}, and finally cooled to room temperature. The resultant red crystals were manually selected from the ampoule.

5.5 Germanium–Lithium–Gallium–Selenium

The $LiGaGe_2Se_6$ quaternary compound, which melts congruently at 710°C (Mei et al. 2012) and crystallizes in the orthorhombic structure with the lattice parameters $a = 1258.64 \pm 0.14$, $b = 2374.6 \pm 0.3$, $c = 713.04 \pm 0.07$ pm, a calculated density of 4.384 g·cm⁻³, and an energy gap of 2.35 eV (Yelisseyev et al. 2016) [$a = 1250.1 \pm 0.3$, $b = 2368.3 \pm 0.5$, $c = 711.96 \pm 0.14$ pm, a calculated density of 4.384 g·cm⁻³ at 93 K, and an energy gap of 2.64 eV (Mei et al. 2012)], is formed in the Ge–Li–Ga–Se system.

This compound was synthesized from elementary starting components, which were weighed and loaded inside a waterproof box filled with pure Ar (Yelisseyev et al. 2016). The total mass of the substance was within 10–15 g, whereas Ge and Se contents were preset above the stoichiometric one in order to provide a 3 to 10 mass% excess of $GeSe_2$. Starting components were loaded into a glassy-carbon crucible, which was placed inside a silica ampoule. This ampoule was joined to the vacuum system, evacuated to 0.1 Pa, and sealed. Synthesis was carried out in a vertical two-zone resistance furnace. The ampoule with the load was placed in an area with a temperature gradient of about 5–6°C·mm⁻¹ and heated to 750°C over 2 days. Then the furnace was switched off. The obtained fine-grained charge was displaced to the crystal growth ampoule. Single crystals of $LiGaGe_2Se_6$ were grown by the Bridgman–Stockbarger technique.

Polycrystalline samples of $LiGaGe_2Se_6$ were also synthesized by solid-state reaction technique from stoichiometric amounts of Li_2Se (1 mM), $GeSe_2$ (4 mM), and Ga_2Se_3 (1 mM) (Mei et al. 2012). The starting materials were well mixed and loaded into a fused silica tube under an Ar atmosphere in a glove box and then sealed under a 10⁻³ Pa atmosphere. The above sample was placed into a furnace, heated to 700°C in 30 h, kept at 700°C for 72 h, and then the furnace was turned off. To grow single crystals, the as-prepared yellow $LiGaGe_2Se_6$ powder was put into a fused silica tube, which was then flame-sealed under a high vacuum of 10⁻³ Pa and placed in a furnace. The sample was heated to 800°C in 20 h, kept at 800°C for 72 h, and then cooled at 2°C·h⁻¹ to 300°C, at which point the furnace was switched off. Many yellow block-shaped single crystals were produced. The crystals were air and moisture stable.

5.6 Germanium–Lithium–Indium–Selenium

The $Li_2In_2GeSe_6$ quaternary compound, which melts congruently at 722°C and crystallizes in the monoclinic structure with the lattice parameters $a = 1272.26 \pm 0.04$, $b = 745.27 \pm 0.04$, $c = 1266.98 \pm 0.05$ pm, $\beta = 109.826 \pm 0.004°$, a calculated density of 4.642 g·cm⁻³, and an energy gap of 2.30 eV, is formed in the Ge–Li–In–Se system (Yin et al. 2012a). The structural, electronic, and optical properties of this compound have been investigated by *ab initio* calculations based on the density functional theory by Wong et al. (2018) using the full potential linearized augmented plane wave method. The next values of the lattice parameters were determined: $a = 1271.83$, $b = 745.18$, $c = 1266.84$ pm (the value of β is not calculated).

To prepare this compound, the mixture of $LiInSe_2$ (1 mM) and $GeSe_2$ (0.5 mM) was ground, loaded into a fused silica tube under an Ar atmosphere in a glove box, then moved to a high-vacuum line, and flame-sealed under a high vacuum of 10⁻³ Pa (Yin et al. 2012a). The tube was then placed in a furnace and heated to 850°C for 12 h, left for 48 h, cooled to 200°C at a rate of 4°C·h⁻¹, and finally cooled to room temperature by switching off the furnace. Many orange block-shaped crystals, which are stable in the air, were found in the ampoule. Polycrystalline samples of $Li_2In_2GeSe_6$ were synthesized by solid-state reaction techniques. The mixtures of Li_2Se, $GeSe_2$, and In_2Se_3 (molar ratio 1:1:1) were heated to 700°C in 15 h, kept at that temperature for 72 h, and then the furnace was turned off.

5.7 Germanium–Sodium–Silver–Selenium

The $Na_{4.24}Ag_{1.76}Ge_2Se_7$ quaternary phase and the $Na_{8-x}Ag_xGe_4Se_{10}$ solid solution are formed in the Ge–Na–Ag–Se system (Choudhury et al. 2007b). $Na_{4.24}Ag_{1.76}Ge_2Se_7$ crystallizes in the monoclinic structure with the lattice parameters $a = 972.34 \pm 0.18$, $b = 1054.7 \pm 0.2$, $c = 1523.6 \pm 0.3$ pm, $\beta = 102.985 \pm 0.003°$, and a calculated density of 4.297 g·cm⁻³. $Na_{8-x}Ag_xGe_4Se_{10}$ also crystallizes in the monoclinic structure with the lattice parameters $a = 810.1 \pm 0.2$, $b = 2014.7 \pm 0.6$, $c = 1406.3 \pm 0.4$ pm, $\beta = 106.182 \pm 0.005°$, and a calculated density of 3.888 g·cm⁻³ for $x = 0.31$; $a = 810.4 \pm 0.2$, $b = 2016.3 \pm 0.3$, $c = 1399.8 \pm 0.2$ pm, $\beta = 105.861 \pm 0.003°$, and a calculated density of 3.988 g·cm⁻³ for $x = 0.67$; $a = 810.4 \pm 0.2$, $b = 2015.6 \pm 0.5$, $c = 1398.5 \pm 0.4$ pm, $\beta = 105.757 \pm 0.004°$, and a calculated density of 4.016 g·cm⁻³ for $x = 0.77$; $a = 810.30 \pm 0.09$, $b = 2015.5 \pm 0.3$, $c = 1396.65 \pm 0.09$ pm, $\beta = 105.699 \pm 0.002°$, and a calculated density of 4.046 g·cm⁻³ for $x = 0.87$; $a = 809.8 \pm 0.4$, $b = 2013.1 \pm 0.9$, $c = 1392.2 \pm 0.6$ pm, $\beta = 105.547 \pm 0.009°$, and a calculated density of 4.109 g·cm⁻³ for $x = 1.04$; and $a = 810.13 \pm 0.07$, $b = 2013.42 \pm 0.08$, $c = 1392.13 \pm 0.03$ pm, $\beta = 105.541 \pm 0.002°$, and a calculated density of 4.117 g·cm⁻³ for $x = 1.09$. All phases were prepared by the reaction of elemental Ge, Se, Ag and Na_2Se_2, or by solid-state metathesis reactions using preformed salts such as $Na_8Ge_4Se_{10}$. The mixtures of Ge, Se, Ag, and Na_2Se_2 were loaded into fused silica ampoules inside a N_2-filled glove box. The ampoules were flame-sealed under a vacuum and placed in a furnace. The furnace was ramped to 725°C at a rate of 35°C·h⁻¹, the temperature was held constant at 725°C for 150 h, and then slowly cooled to ambient temperature at a rate of 5°C·h⁻¹. After the reaction products cooled, the ampoules were opened in an inert atmosphere glove box, and the solid product was soaked in dry DMF and sonicated for 1 min to dissolve and wash away any remaining flux and loosen the product crystals. The color of the crystals varied from orange to red as the concentration of the Ag^+ ion increased. The phases were modestly stable in air; however, they decomposed gradually under longer exposure to air and light.

The solid solution $Na_{8-x}Ag_xGe_4Se_{10}$ with $x = 0.77$ was also prepared by solid-state metathesis reactions from the mixture of $Na_8Ge_4Se_{10}$ (0.06 mM) and AgCl (0.12 mM), which was loaded into fused silica ampoule inside a N_2-filled glove box, flame-sealed under vacuum, and placed in a furnace (Choudhury et al. 2007b). The furnace was ramped to 825°C at a rate of 35°C·h^{-1}, and the temperature was held at a constant 825°C for 96 h. The furnace was then slowly cooled to ambient temperature at a rate of 5°C·h^{-1}, and the orange crystals were recovered similarly as described previously.

5.8 Germanium–Sodium–Magnesium–Selenium

The **Na$_4$MgGe$_2$Se$_6$** quaternary compound, which crystallizes in the monoclinic structure with the lattice parameters $a = 722.24 \pm 0.14$, $b = 1199.4 \pm 0.2$, $c = 709.65 \pm 0.13$ pm, $\beta = 106.660 \pm 0.012°$, a calculated density of 3.878 g·cm^{-3}, and an energy gap of 2.53 eV, is formed in the Ge–Na–Mg–Se system (Wu et al. 2015b). It was synthesized using a mixture of Na, Mg, Ge, and Se at stoichiometric proportions. Firstly, the mixture was filled into a graphite crucible in a vacuum-sealed quartz tube, then put the tube into a furnace with the following procedure: heated to 700°C within 40 h and left at this temperature for about 50 h, then slowly cooled to 300°C with 5 days, finally rapidly down to room temperature. After spontaneous crystallization, many orange-yellow crystals were found in the tube. The pure phase of this compound was successfully achieved by repetitious grinding and sintering a mixture with a stoichiometric ratio of the elements at 600°C. All the preparation processes were completed in an Ar-filled glove box.

5.9 Germanium–Sodium–Barium–Selenium

The **Na$_2$BaGeSe$_4$** quaternary compound, which crystallizes in the trigonal structure with the lattice parameters $a = 2520 \pm 2$, $c = 734.3 \pm 1.3$ pm, a calculated density of 4.231 g·cm^{-3}, and an energy gap of 2.46 eV, is formed in the Ge–Na–Ba–S system (Wu et al. 2016a,b). The crystals of this compound were obtained using the mixture of Na, Ba, Ge, and Se (molar ratio 1:6:2:8), which was first loaded into a graphite crucible that avoids the reaction between Na and silica tube, then put into the silica tube and flame-sealed under vacuum (10^{-3} Pa). The tube was heated to 650°C for 30 h and kept there for about 90 h, then slowly cooled down to ambient temperature for 5 days. After the reaction and washing with DMF, deep-yellow crystals of Na$_2$BaGeSe$_4$ were found. They are stable in the air. Owing to Na and Ba metals being easily oxidized in the air, an Ar-filled glove box was used to complete the preparation processes.

5.10 Germanium–Sodium–Zinc–Selenium

The **Na$_2$ZnGe$_2$Se$_6$** quaternary compound, which crystallizes in the tetragonal structure with the lattice parameters $a = 775.68 \pm 0.12$, $c = 1873.4 \pm 0.4$ pm, a calculated density of 4.303 g·cm^{-3}, and an energy gap of 2.36 eV, is formed in the Ge–Na–Zn–S system (Zhou et al. 2016a). The traditional solid-state reaction technique can be applied to grow the single crystals of the title compound. The mixtures of K$_2$Se (0.5 mM), GeSe$_2$ (1.1 mM), and ZnSe (0.5 mM) were mixed and loaded into a fused silica

tube under an Ar atmosphere in a glove box, then sealed under a high vacuum (10^{-3} Pa). The tube was then placed in a furnace and heated to 1000°C within 15 h, left for 48 h, cooled to 350°C at a rate of 3°C·h^{-1}, and finally cooled to room temperature by switching off the furnace. Many prism-shaped crystals were found in the tube. The obtained crystals are stable in the air.

5.11 Germanium–Sodium–Cadmium–Selenium

The **Na$_2$CdGe$_2$Se$_6$** quaternary compound, which crystallizes in the monoclinic structure with the lattice parameters $a = 758.9 \pm 0.4$, $b = 1318.2 \pm 0.7$, $c = 1206.9 \pm 0.6$ pm, $\beta = 98.829 \pm 0.007°$, and a calculated density of 3.994 g·cm^{-3}, is formed in the Ge–Na–Cd–Se system (Li et al. 2017a). The crystals of this compound were synthesized by solid-state reaction. First, Na, CdS, Ge, and S (molar ratio 2:1:2:5) were placed into the graphite crucible, which was flame-sealed into a fused silica tube under a high vacuum (10^{-3} Pa). The tube was moved in a furnace, heated to 850°C in 60 h, dwelled there for 30 h, cooled to 450°C at a rate of 5°C·h^{-1}, then rapidly cooled to room temperature by switching off the furnace. After washing with DMF and deionized water, yellow crystals of Na$_2$CdGe$_2$Se$_6$ were obtained.

5.12 Germanium–Sodium–Gallium–Selenium

Two quaternary compounds, **NaGaGe$_3$Se$_8$** and **Na$_2$Ga$_2$GeSe$_6$**, are formed in the Ge–Na–Ga–Se system. NaGaGe$_3$Se$_8$ melts congruently at ~702°C and crystallizes in the monoclinic structure with the lattice parameters $a = 723.29 \pm 0.14$, $b = 1188.9 \pm 0.2$, $c = 1755.0 \pm 0.4$ pm, $\beta = 101.75 \pm 0.03°$, a calculated density of 4.235 g·cm^{-3} at 153 K, and an energy gap of 2.35 eV (Li et al. 2016c). Na$_2$Ga$_2$GeSe$_6$ also melts congruently at 571°C and crystallizes in the orthorhombic structure with the lattice parameters $a = 1298.5 \pm 0.4$, $b = 2388.0 \pm 0.8$, $c = 758.5 \pm 0.3$ pm, a calculated density of 4.133 g·cm^{-3}, and an energy gap of 1.61 eV (Li et al. 2018b).

A polycrystalline sample of NaGaGe$_3$Se$_8$ was synthesized using a solid-state reaction technique (Li et al. 2016c). A mixture of Na$_2$Se, Ga$_2$Se$_3$, and GeSe$_2$ according to the stoichiometric ratio, was ground and loaded into fused silica tube under an Ar atmosphere in a glove box, which was sealed under vacuum (10^{-3} Pa) and then placed in a furnace. The sample was heated to 800°C over 20 h, kept at that temperature for 72 h, and then the furnace was turned off. Single crystals of this compound were obtained from a mixture of Na$_2$Se (0.15 M), Ga$_2$Se$_3$ (0.15 M), and GeSe$_2$ (0.60 M), which was ground and loaded into fused silica tube under an Ar atmosphere in a glove box. The tube was sealed under a vacuum (10^{-3} Pa) and then placed in a furnace. The sample was heated to 1000°C over 20 h and kept at that temperature for 48 h, then cooled at a slow rate (2.5°C·h^{-1}) to 300°C, and finally cooled to room temperature. The resultant orange and light-yellow crystals were manually selected.

For the synthesis of Na$_2$Ga$_2$GeSe$_6$, a stoichiometric mixture of the starting materials Na$_2$Se, Ga, Ge, and Se (molar ratio 1:2:1:6) was loaded into a graphite crucible and placed in a quartz tube (Li et al. 2018b). The tube was flame-sealed under vacuum (~10^{-2} Pa) and placed in a furnace, heated from room temperature to 800°C over 40 h, kept at that temperature for 96 h, and then cooled to room temperature at 4°C·h^{-1}. The product was washed with degassed DMF and dried with ethanol. Yellow crystals, which are stable in the air and water, were obtained.

5.13 Germanium–Sodium–Indium–Selenium

The **$Na_2In_2GeSe_6$** quaternary compound, which melts congruently at 671°C and crystallizes in the monoclinic structure with the lattice parameters a = 1270.7 ± 0.3, b = 811.57 ± 0.16, c = 1276.0 ± 0.3 pm, β = 108.70 ± 0.03°, a calculated density of 4.380 g·cm^{-3}, and an energy gap of 2.47 eV, is formed in the Ge–Na–In–Se system (Zhou et al. 2016a). The traditional solid-state reaction technique can be applied to grow the single crystals of the title compound. The mixtures of Na_2Se (0.5 mM), $GeSe_2$ (0.5 mM), and In_2Se_3 (0.5 mM) were mixed and loaded into a fused silica tube under an Ar atmosphere in a glove box, then sealed under a high vacuum (10^{-3} Pa). The tube was then placed in a furnace and heated to 900°C within 15 h, left for 48 h, cooled to 300°C at a rate of 3°C·h^{-1}, and finally cooled to room temperature by switching off the furnace. Many chip-shaped crystals were found in the tube. The obtained crystals are stable in the air. A polycrystalline sample of $Na_2In_2GeSe_6$ was synthesized by mixtures of the corresponding binary materials in the stoichiometric ratio. The mixture was heated to 700°C in 15 h and kept at that temperature for 48 h, and finally, the furnace was turned off.

5.14 Germanium–Sodium–Lanthanum–Selenium

The **$Na_9La(Ge_2Se_6)_2$** quaternary compound, which crystallizes in the monoclinic structure with the lattice parameters a = 797.26 ± 0.08, b = 1233.49 ± 0.13, c = 711.21 ± 0.07 pm, β = 107.112 ± 0.002°, and a calculated density of 3.934 g·cm^{-3} (Martin et al. 2006) [a = 797.4 ± 0.1, b = 1233.7 ± 0.2, c = 711.4 ± 0.1 pm, and β = 107.101 ± 0.003° at 173 K (Martin et al. 2005)], is formed in the Ge–Na–La–Se system. Crystals of this compound were formed from a molten chalcogenide flux reaction of 19.2 mg La, 39.5 mg Ge, 110.2 mg Se, and 57.2 mg Na_2Se_2 (Martin et al. 2005). These reactants were combined, loaded into a carbon crucible, and then placed in a fused silica ampoule in an inert atmosphere glove box. The ampoule was sealed under a vacuum and heated to 725°C at a rate of 35°C·h^{-1}. After 150 h of heating, the ampoule was cooled at 4°C·h^{-1} to room temperature. Platy crystals were obtained after the product was washed with DMF to dissolve any remaining flux.

5.15 Germanium–Sodium–Samarium–Selenium

Three quaternary compounds, **NaSmGeSe₄**, **NaSmGeSe₄·0.25Na₂Se**, and **$Na_9Sm(Ge_2Se_6)_2$** are formed in the Ge–Na–Sm–Se system. First of them crystallizes in the triclinic structure with the lattice parameters a = 689.7 ± 0.2, b = 991.9 ± 0.2, c = 1118.3 ± 0.2 pm, α = 84.067 ± 0.004°, β = 88.105 ± 0.004°, and γ = 73.999 ± 0.004° (Martin and Dorhout 2004).

NaSmGeSe₄·0.25Na₂Se crystallizes in the monoclinic structure with the lattice parameters a = 2250.8 ± 0.3, b = 1103.07 ± 0.06, c = 682.61 ± 0.01 pm, β = 103.038 ± 0.002°, a calculated density of 4.771 g·cm^{-3} at 100 K, and an energy gap of 2.06 eV (Choudhury and Dorhout 2006). $Na_9Sm(Ge_2Se_6)$ quaternary compound also crystallizes in the monoclinic structure with the lattice parameters a = 791.58 ± 0.15, b = 1224.4 ± 0.2, c = 710.48 ± 0.13 pm, β = 106.993 ± 0.003°, and a calculated density of 4.022 g·cm^{-3} (Martin et al. 2006).

NaSmGeSe₄ was synthesized by combining Sm (0.264 mM), Ge (0.789 mM), Se (1.06 mM), and Na_2Se_2 (0.132 mM) (Martin and Dorhout 2004). The mixture was loaded into a fused silica ampoule inside an inert atmosphere glove box. The ampoule was flame-sealed under a vacuum and placed in a furnace. The furnace was ramped to 725°C at a rate of 35°C·h^{-1}, and the temperature was held constant at 725°C for 150 h. The furnace was then slowly cooled to ambient temperature at a rate of 4°C·h^{-1}. After the reaction products cooled, the ampoule was opened in an inert atmosphere glove box, and the solid product was soaked in DMF for 6 h in order to dissolve and wash away any remaining flux and loosen the product crystals. After treatment with DMF, a small number of orange-red plates of the title compound were observed. Although NaSmGeSe₄ is stable in the air for a period of days, the inert atmosphere prevented the precipitation of elemental selenium resulting from decomposition reactions of binary and ternary species during the washing process.

NaSmGeSe₄·0.25Na₂Se was synthesized by combining Sm (0.3 mM), $GeSe_2$ (0.1 mM), Se (0.5 mM), and Na_2Se (0.1 mM) (Choudhury and Dorhout 2006). Reactants were loaded into a fused silica ampoule inside a N_2-filled glove box. The ampoule was flame-sealed under a vacuum and placed in a furnace. The furnace was ramped to 825°C at a rate of 20°C·h^{-1}, and the temperature was held constant at 825°C for 72 h. The furnace was then slowly cooled (2°C·h^{-1}) to 550°C and then to ambient temperature at a rate of 4°C·h^{-1}. The ampoule was opened in the air, and the solid product contained red plate-like crystals and the little dark-colored product identified as SmSe$_{1.9}$. Although the title compound is stable in the air and dry DMF for a period of days, it decomposes immediately in the presence of a trace amount of water.

$Na_9Sm(Ge_2Se_6)$ was synthesized by combining 28.3 mg Sm, 83.3 mg Ge, 60.0 mg Se, and 159.8 mg Na_2Se_2 (Martin et al. 2006). Reactants were loaded into a fused silica ampoule inside an inert atmosphere glove box, which was flame-sealed under vacuum, and placed in a furnace. The mixture was heated to 725°C at 35°C·h^{-1}, where it remained for 150 h. The sample was cooled to room temperature at a rate of 4°C·h^{-1}. After the reaction was complete, the ampoule was opened, and the products were washed with DMF to remove unreacted flux. Ruby-red plates of $Na_9Sm(Ge_2Se_6)$ intergrown with yellow plates of $Na_8Ge_4Se_{10}$ and other ternary sodium germanium selenides were obtained.

5.16 Germanium–Sodium–Europium–Selenium

Three quaternary compounds, **$Na_{0.75}Eu_{1.625}GeSe_4$**, **$Na_2EuGeSe_4$**, and **$Na_8Eu_2(Ge_2Se_6)_2$**, are formed in the Ge–Na–Eu–Se system. First two compounds crystallize in the cubic structure with the lattice parameter $a = 1470.65 \pm 0.08$ and 734.66 ± 0.03 pm, a calculated density of 5.451 and 4.911 g·cm^{-3}, and an energy gap of 2.0 and 2.03 eV for the fist and the second compounds, respectively (Choudhury et al. 2007a). Na$_8$Eu(Ge$_2$Se$_6$)$_2$ crystallizes in the monoclinic structure with the lattice parameters $a = 713.88 \pm 0.16$, $b = 1233.4 \pm 0.3$, $c = 801.95 \pm 0.18$ pm, $\beta = 107.267 \pm 0.004°$, and a calculated density of 4.250 g·cm^{-3} (Martin et al. 2006) [$a = 713.7 \pm 0.2$, $b = 1233.0 \pm 0.4$, $c = 801.9 \pm 0.3$ pm, and $\beta = 107.279 \pm 0.005°$ (Knaust et al. 2005)].

Na$_{0.75}$Eu$_{1.625}$GeSe$_4$ and Na$_2$EuGeSe$_4$ were prepared by the well-known reactive flux method using the alkali metal binary polychalcogenides as well as by the reaction of metal halide with ternary sodium seleno-germanate (Choudhury et al. 2007a). Na$_{0.75}$Eu$_{1.625}$GeSe$_4$ was prepared by combining Eu (0.260 mM), Ge (0.655 mM), Se (0.98 mM), and Na$_2$Se$_2$ (0.262 mM) with an approximate compositional ratio of Na/Eu/Ge/Se = 4:2:5:11.5. The furnace was heated to 750°C at a rate of 35°C·h^{-1} and held constant at 750°C for 150 h and then slowly cooled (4°C·h^{-1}) to room temperature. After DMF treatment, orange-red crystals were obtained. This compound was also synthesized by combining EuI$_2$ (0.1 mM) and Na$_6$Ge$_2$Se$_6$ (0.1 mM). The furnace was ramped to 650°C at a rate of 35°C·h^{-1}, held constant at 650°C for 96 h, and then slowly cooled (5°C·h^{-1}) to room temperature. The product, after treatment with DMF, revealed the presence of red cubic crystals of the title compound.

Na$_2$EuGeSe$_4$ was synthesized by combining Eu (0.285 mm), Ge (0.717 mM), Se (0.79 mM), and Na$_2$Se$_2$ (0.573 mM) with an approximate compositional ratio of Na/Eu/Ge/Se = 8:2:5:13.5 (Choudhury et al. 2007a). The furnace was ramped to 800°C at a rate of 35°C·h^{-1}, held constant at 800°C for 150 h, and then slowly cooled (4°C·h^{-1}) to room temperature. After the product cooled, the ampoule was opened in an inert atmosphere glove box, and the solid product was soaked in DMF for 6 h in order to dissolve and wash away any remaining flux and loosen the product crystals. The majority of the product contained red crystals of Na$_2$EuGeSe$_4$. It was also synthesized by combining EuCl$_3$ (0.1 mM) and Na$_6$Ge$_2$Se$_6$ (0.2 mM). The furnace was ramped to 650°C at a rate of 35°C·h^{-1}, held constant at 650°C for 96 h and then slowly cooled (5°C·h^{-1}) to room temperature. The product, after treatment with DMF, revealed red cubes of Na$_2$EuGeSe$_4$, orange-red plates of Na$_8$Eu$_2$(Ge$_2$Se$_6$)$_2$, and yellow crystals of Na$_2$GeSe$_3$.

Crystals of Na$_8$Eu$_2$(Ge$_2$Se$_6$)$_2$ were formed from a molten chalcogenide flux reaction of Eu (40.7 mg), Ge (49.0 mg), Se (58.5 mg), and Na$_2$Se$_2$ (109.8 mg) (Knaust et al. 2005; Martin et al. 2006). The reactants were placed into a fused silica ampoule in an inert atmosphere glove box. The ampoule was sealed under vacuum, and heated to 750°C at a rate of 35°C·h^{-1}. After 150 h of annealing, the ampoule was cooled at 3°C·h^{-1} to room temperature. Orange-brown prismatic crystals were obtained after the product was washed with DMF to dissolve remaining flux.

5.17 Germanium–Sodium–Tin–Selenium

The **Na$_8$Sn$_2$(Ge$_2$Se$_6$)$_2$** quaternary compound is formed in the Ge–Na–Sn–Se system (Marking and Kanatzidis 1997).

5.18 Germanium–Sodium–Lead–Selenium

The **Na$_{0.5}$Pb$_{1.75}$GeSe$_4$** quaternary compound, which crystallizes in the cubic structure with the lattice parameter $a = 1466.2 \pm 0.4$ pm, a calculated density of 6.427 g·cm^{-3} at 173 K, and an energy gap of 1.60 eV, is formed in the Ge–Na–Pb–Se system (Iyer et al. 2004). It was the product of a reaction of Na$_2$Se, Pb, Ge, and Se (molar ratio 0.25:1.75:1:3.75). The reactants were loaded in a fused silica tube and flame-sealed under vacuum (< 0.1 Pa). The tube was then put in a furnace and heated to 500°C after an initial isotherm at 250°C. The tube was kept at 500°C for 72 h, followed by cooling to 200°C at a rate of 5°C·h^{-1}. It was then cooled to room temperature.

5.19 Germanium–Sodium–Antimony–Selenium

The **Na$_9$Sb(Ge$_2$Se$_6$)$_2$** quaternary compound, which crystallizes in the monoclinic structure with the lattice parameters $a = 895.0 \pm 1.0$, $b = 2433 \pm 3$, $c = 706.6 \pm 0.8$ pm, $\beta = 122.103 \pm 0.011°$, and a calculated density of 3.992 g·cm^{-3}, is formed in the Ge–Na–Sb–Se a system (Wu et al. 2016d). To synthesize this compound, a mixture of Na, Sb, Ge, and Se (molar ratio 9:1:4:12) was loaded into a sealed silica tube evacuated to 10^{-3} Pa. The mixture was first heated to 600°C in 30 h, left at this temperature for 72 h, then cooled to 300°C at a rate of 3°C·h^{-1}, and finally cooled rapidly to room temperature by switching off the furnace. After repeatedly washing with DMF, many deep-yellow crystals of Na$_8$Ge$_4$Se$_{10}$ and a few red crystals of Na$_9$Sb(Ge$_2$Se$_6$)$_2$ were found in the tube. As for the easily oxidized Na element, the preparation process was completed in a glove box with an Ar atmosphere.

5.20 Germanium–Sodium–Bromine–Selenium

The **Na$_3$GeSe$_3$Br** quaternary compound is formed in the Ge–Na–Br–Se system (Feltz and Pfaff 1980). To prepare this compound, 10 mL of bromine solution in CH$_3$OH (43.7 mg Br$_2$/mL) was added to a solution of Na$_6$Ge$_2$Se$_6$·4CH$_3$OH (2.43 g) in CH$_3$OH (50 mL) drop-wise while cooling at room temperature. Conversion occurs immediately, and a colorless, finely crystalline precipitate separates from the colorless solution after extensive concentration in vacuum. It is sensitive to hydrolysis and very readily soluble in methanol. Precipitation by drop-wise addition of CCl$_4$ increases the yield.

5.21 Germanium–Sodium–Manganese–Selenium

Two quaternary compounds, **Na$_2$MnGe$_2$Se$_6$** and **Na$_8$Mn$_2$(Ge$_2$Se$_6$)$_2$**, are formed in the Ge–Na–Mn–Se system. Na$_2$MnGe$_2$Se$_6$ crystallizes in the tetragonal structure with the lattice parameters $a = 777.18 \pm 0.16$, $c = 1907.7 \pm 0.7$ pm, a calculated density of 4.150 g·cm^{-3} at 153 K, and an energy gap of 1.93 eV (Zhou et al. 2016b). Polycrystalline samples of this compound were synthesized by the traditional solid-state reaction technique. Mixtures of Na$_2$Se, MnSe, and GeSe$_2$ according to the stoichiometric ratio were ground and loaded into fused silica tubes under an Ar atmosphere in a glove box, then flame-sealed under a high vacuum of 10^{-3} Pa and placed in a furnace. The samples were heated to 750°C in 20 h, kept at that temperature for 72 h, and then the furnace was turned off. Single crystals of Na$_2$MnGe$_2$Se$_6$ were achieved by the spontaneous crystallization method. Mixtures of Na$_2$Se, MnSe, and GeSe$_2$ (molar ratio 1:1:2) were ground and loaded into fused silica tubes under an Ar atmosphere in a glove box, then flame-sealed under a high vacuum of 10^{-3} Pa and placed in a furnace. The samples were heated to 950°C in 20 h and kept at that temperature for 96 h, then cooled at a slow rate of 3°C·h^{-1} to 300°C, and finally cooled to room temperature. The resultant brown crystals were manually selected from the product.

Na$_8$Mn$_2$(Ge$_2$Se$_6$)$_2$ crystallizes in the monoclinic structure with the lattice parameters $a = 772.7 \pm 0.4$, $b = 1194.3 \pm 0.7$, $c = 708.5 \pm 0.4$ pm, $\beta = 106.682 \pm 0.006°$, a calculated density of 4.061 g·cm^{-3} at 220 K, and an energy gap of 1.95 eV (Balijapelly et al. 2022). It was synthesized using the metathesis route by reacting Na$_6$Ge$_2$Se$_6$ and MnCl$_2$ (molar ratio 1:1), which were hand-ground together for 15 min, and the mixture was loaded into a carbon-coated quartz ampoule. The sealed quartz ampoule was heated to 750°C at a rate of 25°C·h^{-1} and held for 96 h, followed by cooling to room temperature at a rate of 35°C·h^{-1}. Orange crystals were recovered after breaking the ampoule in air, followed by washing with 10% water in DMF solution to remove the NaCl byproduct and to loosen the crystals. The title compound was also prepared by reacting a stoichiometric combination of elements; Na (4 mM), Mn (1 mM), Ge (2 mM), and Se (6 mM) were heated in a sealed quartz tube. The temperature of the furnace was raised to 600°C at a rate of 20°C·h^{-1} and held for 96 h followed by cooling to room temperature at a rate of 30°C·h^{-1}.

5.22 Germanium–Potassium–Gold–Selenium

The **K$_2$Au$_2$Ge$_2$Se$_6$** quaternary compound, which melts at 574°C and crystallizes in the monoclinic structure with the lattice parameters $a = 1063.3 \pm 0.2$, $b = 1112.7 \pm 0.2$, $c = 1130.3 \pm 0.2$ pm, $\beta = 115.37 \pm 0.03°$, a calculated density of 4.451 g·cm^{-3}, and an energy gap of 2.33 eV, is formed in the Ge–K–Au–Se a system (Löken and Tremel 1998, 1999). To prepare this compound, a mixture of Au (0.4 mM), Ge (0.4 mM), K$_2$Se (0.2 mM), and Se (1.6 mM) was mixed and sealed in a quartz ampoule under vacuum. It was heated to 650°C in 12 h, maintained at this temperature for 4 days, and then cooled slowly to room temperature at 5°C·h^{-1}. The resulting material was washed with absolute ethanol. The obtained dark-yellow plate-like crystals of K$_2$Au$_2$Ge$_2$Se$_6$ are stable in air and water.

5.23 Germanium–Potassium–Mercury–Selenium

The **K$_2$Hg$_3$Ge$_2$Se$_8$** quaternary compound, which crystallizes in the orthorhombic structure with the lattice parameters $a = 1985.1 \pm 0.2$, $b = 1000.40 \pm 0.07$, and $c = 860.97 \pm 0.07$ pm, and a calculated density 5.659 g·cm^{-3}, is formed in the Ge–K–Hg–Se system (Kanatzidis et al. 1997a, 1997b; Jin et al. 2002).

Dark-red granular crystals of this compound were isolated from a flux reaction containing K$_2$Se$_4$ (0.267 mM), HgSe (0.089 mM), Ge (0.134 mM), and Se (0.890 mM) (Jin et al. 2002). The reaction was carried out in a vacuum-sealed Pyrex tube at 500°C for 5 days. Subsequently, it was cooled slowly (4°C·h^{-1}) to 200°C, and then cooled naturally to room temperature. The product was washed with DMF and anhydrous ethanol to remove the excess flux, and dried with anhydrous diethyl ether.

5.24 Germanium–Potassium–Gallium–Selenium

Two quaternary compounds, **KGaGeSe$_4$** and **K$_3$Ga$_3$Ge$_7$Se$_{20}$**, are formed in the Ge–K–Ga–Se system. First of them crystallizes in the monoclinic structure with the lattice parameters $a = 735.52 \pm 0.03$, $b = 1241.51 \pm 0.03$, $c = 1762.13 \pm 0.04$ pm, $\beta = 97.026 \pm 0.002°$, a calculated density of 4.136 g·cm^{-3}, and an energy gap of 2.32 eV (Friedrich et al. 2020). K$_3$Ga$_3$Ge$_7$Se$_{20}$ melts congruently at 742°C and also crystallizes in the monoclinic structure with the lattice parameters $a = 705.20 \pm 0.04$, $b = 3903.3 \pm 0.2$, $c = 694.88 \pm 0.04$ pm, $\beta = 90.433 \pm 0.005°$, a calculated density of 4.191 g·cm^{-3} at 153 K, and an energy gap of 2.23 eV (Li et al. 2016c).

Yellow crystals of KGaGeSe$_4$ were prepared by reacting K$_2$Se with a stoichiometric mixture of Ga$_2$Se$_3$, Ge, and Se in a sealed evacuated quartz tube at 1000°C for 2 days and cooling to room temperature with a cooling rate of 15°C·min^{-1}. This compound decomposes within several minutes by releasing gaseous H$_2$Se.

A polycrystalline sample of K$_3$Ga$_3$Ge$_7$Se$_{20}$ was synthesized using a solid-state reaction technique (Li et al. 2016c). A mixture of K$_2$Se, Ga$_2$Se$_3$, and GeSe$_2$, according to the stoichiometric ratio, was ground and loaded into fused silica tube under an Ar atmosphere in a glove box, which was sealed under vacuum (10^{-3} Pa) and then placed in a furnace. The sample was heated to 800°C over 20 h, kept at that temperature for 72 h, and then the furnace was turned off. Single crystals of this compound were obtained from a mixture of K$_2$Se (0.15 M), Ga$_2$Se$_3$ (0.15 M), and GeSe$_2$ (0.60 M), which was ground and loaded into fused silica tube under an Ar atmosphere in a glove box. The tube was sealed under a vacuum (10^{-3} Pa) and then placed in a furnace. The sample was heated to 1000°C over 20 h and kept at that temperature for 48 h, then cooled at a slow rate (2.5°C·h^{-1}) to 300°C, and finally cooled to room temperature. The resultant orange and light-yellow crystals were manually selected.

5.25 Germanium–Potassium–Indium–Selenium

The **KInGeSe$_4$** quaternary compound, which crystallizes in the monoclinic structure with the lattice parameters $a = 761.08 \pm 0.03$, $b = 1244.73 \pm 0.06$, $c = 1808.96 \pm 0.08$ pm, $\beta = 97.238 \pm 0.004°$, a calculated density of 4.239 g·cm^{-3}, and an energy gap of 2.32 eV, is formed in the Ge–K–In–Se system (Friedrich et al. 2021). A polycrystalline sample of this compound was obtained by reacting K$_2$Se (0.92 mM), In (1.84 mM), Ge (1.84 mM), and Se (6.45 mM) in a sealed evacuated fused silica tube. The reaction mixture was heated to 1000°C with a heating rate of 3°C·min^{-1}. After annealing at 1000°C for 24 h the tube was cooled to room temperature with a cooling rate of 1°C·min^{-1}. The air-sensitive yellow polycrystalline product forms plate-like crystallites and is stored in a nitrogen glove box until further use.

5.26 Germanium–Potassium–Lanthanum–Selenium

The **KLaGeSe$_4$** quaternary compound, which crystallizes in the monoclinic structure with the lattice parameters $a = 687.5 \pm 0.3$, $b = 700.2 \pm 0.2$, $c = 894.2 \pm 0.4$ pm, and $\beta = 107.84 \pm 0.02°$ at 115 K, is formed in the Ge–K–La–Se system (Wu and Ibers 1993a). It was prepared by the reaction of Ge with K$_2$Se$_5$ and La$_2$Se$_3$. The starting materials were placed in a quartz tube that was subsequently evacuated to 10^{-3} Pa and sealed. The molar ratio of K/La/Ge/Se was 1:1:1:4. The quartz tube was heated gradually to 500°C, where it was kept for 24 h before being successively brought to 700°C for 24 h and to 1000°C for 150 h. The tube was then cooled (4°C·h^{-1}) to 300°C, and the furnace was shut off. Yellow-brown crystals of the title compound were obtained. KLaGeSe$_4$ appears to be modestly stable in the air but decomposes gradually in the presence of water. Bulk samples of this compound were prepared by reaction of stoichiometric amounts of starting materials at 1000°C for 10 days with intermittent grinding.

5.27 Germanium–Potassium–Cerium–Selenium

The **KCeGeSe$_4$** quaternary compound, which crystallizes in the monoclinic structure with the lattice parameters $a = 685.2 \pm 0.2$, $b = 702.4 \pm 0.3$, $c = 901.6 \pm 0.2$ pm, $\beta = 108.116 \pm 0.002°$, and a calculated density of 4.570 g·cm^{-3}, is formed in the Ge–K–Ce–S system (Choudhury et al. 2006). It was synthesized by combining CeCl$_3$ (0.2 mM) and K$_4$Ge$_4$Se$_{10}$ (0.2 mM) with an exact compositional ratio K/Ce/Ge/Se/Cl of 4:1:4:10:3. The mixture was first thoroughly ground in an agate mortar and then transferred to a graphite crucible, which was then placed inside a fused silica ampoule. The ampoule was flame-sealed under vacuum and placed in a furnace. The furnace was ramped to 650°C at a rate of 20°C·h^{-1}, and annealed at this temperature for 148 h. Then, the furnace was slowly cooled to ambient temperature at a rate of 3°C·h^{-1}. After the reaction product was cooled, the ampoule was opened in an inert atmosphere glove box, and the solid product was soaked in DMF for 6 h in order to dissolve and wash away any remaining flux and loosen the product crystals. The product, after treatment with DMF, revealed red plates of KCeGeSe$_4$ yellow crystals of GeSe$_2$ and colorless crystal of KCl. KCl can be completely removed by further washing with 10 % water in DMF.

5.28 Germanium–Potassium–Praseodymium–Selenium

The **KPrGeSe$_4$** quaternary compound, which crystallizes in the monoclinic structure with the lattice parameters $a = 682.80 \pm 0.08$, $b = 701.05 \pm 0.08$ $c = 898.7 \pm 0.1$ pm, and $\beta = 108.157 \pm 0.002°$, is formed in the Ge–K–Pr–Se system (Chan and Dorhout

2005). Crystals of this compound were formed from a molten chalcogenide flux reaction of Pr (51.5 mg), Ge (21.8 mg), Se (94.0 mg), and K$_2$Se$_2$ (40.6 mg). The reactants were combined in a fused silica ampoule in an inert atmosphere glove box, sealed under vacuum, and heated to 725°C at a rate of 35°C·h^{-1}. After 150 h of heating, the ampoule was cooled at 4°C·h^{-1} to room temperature. DMF was added to dissolve the remaining potassium selenide flux, resulting in well-formed plates of KPrGeSe$_4$.

5.29 Germanium–Potassium–Samarium–Selenium

The **KSmGeSe$_4$** quaternary compound, which crystallizes in the monoclinic structure with the lattice parameters $a = 677.4 \pm 0.1$, $b = 699.4 \pm 0.1$, $c = 896.0 \pm 0.2$ pm, $\beta = 108.225 \pm 0.003°$, a calculated density of 4.760 g·cm^{-3}, and an energy gap of 2.2 eV, is formed in the Ge–K–Sm–Se system (Martin and Dorhout 2004). KSmGeSe$_4$ was synthesized by combining Sm (0.504 mM), Ge (1.02 mM), Se (2.02 mM), and K$_2$Se$_2$ (0.342 mM). The mixture was loaded into a fused silica ampoule inside an inert atmosphere glove box. The ampoule was flame-sealed under a vacuum and placed in a furnace. The furnace was ramped to 725°C at a rate of 35°C·h^{-1}, and the temperature was held constant at 725°C for 150 h. The furnace was then slowly cooled to ambient temperature at a rate of 4°C·h^{-1}. After the reaction products cooled, the ampoule was opened in an inert atmosphere glove box, and the solid product was soaked in DMF for 6 h in order to dissolve and wash away any remaining flux and loosen the product crystals. After treatment with DMF, orange plates of the title compound were observed, along with orange hemispheres of amorphous potassium seleno-germanate. Although KSmGeSe$_4$ is stable in the air for a period of days, the inert atmosphere prevented the precipitation of elemental selenium resulting from the decomposition reactions of binary and ternary species during the washing process.

5.30 Germanium–Potassium–Europium–Selenium

The **K$_2$EuGeSe$_5$** quaternary compound, which crystallizes in the monoclinic structure with the lattice parameters $a = 1180.56 \pm 0.03$, $b = 996.30 \pm 0.01$ $c = 894.56 \pm 0.01$ pm, $\beta = 91.195 \pm 0.001°$, a calculated density of 4.404 g·cm^{-3} at 167 K, and an energy gap of 1.84 eV, is formed in the Ge–K–Eu–Se system (Evenson IV and Dorhout 2001). This compound was synthesized by combining Se (0.493 mM), K$_2$Se$_4$ (0.318 mM), Ge (0.164 mM), and Eu (0.164 mM) that were loaded into a fused silica ampoule inside an inert atmosphere glove box. The ampoule was flame-sealed under a vacuum and placed in a tube furnace. The temperature was ramped to 725°C where it remained for 150 h, and the furnace was allowed to cool back to room temperature at 4°C·h^{-1}. The product was washed with DMF, yielding red plates of the title compound, K$_2$GeSe$_4$ and EuSe.

5.31 Germanium–Potassium–Tin–Selenium

The **K$_8$Sn$_2$(Ge$_2$Se$_6$)$_2$** quaternary compound is formed in the Ge–K–Sn–Se system (Marking and Kanatzidis 1997).

5.32 Germanium–Potassium–Lead–Selenium

The **K$_8$Pb$_2$(Ge$_2$Se$_6$)$_2$** quaternary compound is formed in the Ge–K–Pb–Se system (Marking and Kanatzidis 1997).

5.33 Germanium–Potassium–Phosphorus–Selenium

The **K$_4$GeP$_4$Se$_{12}$** quaternary compound, which melts congruently at 415°C and crystallizes in the orthorhombic structure with the lattice parameters $a = 1394.65 \pm 0.04$, $b = 724.35 \pm 0.02$, $c = 2405.11 \pm 0.09$ pm, a calculated density of 3.555 g·cm^{-3}, and an energy gap of 2.0 eV, is formed in the Ge–K–P–Se system (Morris et al. 2012). It was synthesized by combining K$_2$Se (0.5 mM), Ge (0.25 mM), P (1.00 mM), and S (2.50 mM) in a fused silica tube inside a N$_2$-filled glove box. The tube was flame-sealed under a vacuum (10^{-2} Pa), heated to 500°C in 6 h, held there for 24 h, cooled to 200°C in 24 h, and then to room temperature in 3 h. The resulting product contained red, rod-like crystals of the target compound. If the mixture, after being heated at 500°C for 24 h, was quenched in a cool water bath, then a glassy ingot of this compound was obtained.

5.34 Germanium–Potassium–Manganese–Selenium

The **K$_2$MnGe$_2$Se$_6$** quaternary compound, which crystallizes in the tetragonal structure with the lattice parameters $a = 813.15 \pm 0.11$, $c = 1887.7 \pm 0.4$ pm, a calculated density 4.002 g·cm^{-3}, and an energy gap of 2.0 eV, is formed in the Ge–K–Mn–Se system (Li et al. 2020b). Polycrystalline sample of this compound was obtained by heating a mixture of K$_2$Se (0.157 g), MnSe (0.134 g), and GeSe$_2$ (0.448 g) (molar ratio 1:1:2). The starting materials were thoroughly ground and loaded into silica tube, which was then evacuated to 10^{-3} Pa and flame-sealed. The sample was then placed into a computer-controlled furnace and heated to 530°C over a 24 h period and then kept at that temperature for 120 h. Finally, the furnace was turned off and the sample was allowed to cool naturally. Single crystals of K$_2$MnGe$_2$Se$_6$ could be obtained by spontaneous crystallization. A mixture of K$_2$Se (0.031 g), MnSe (0.027 g), and GeSe$_2$ (0.090 g) (molar ratio 1:1:2) was thoroughly ground and put into a silica tube, which was then evacuated to 10^{-3} Pa and flame-sealed. The sample was gradually heated to 960°C in a horizontal furnace and kept at that temperature for 96 h. Then, the sample was slowly cooled at a rate of 3°C·h^{-1}. The product consists of hundreds of green crystals of the title compound.

5.35 Germanium–Potassium–Iron–Selenium

The **K$_2$FeGe$_3$Se$_8$** quaternary compound, which crystallizes in the monoclinic structure with the lattice parameters $a = 740.88 \pm 0.15$, $b = 1226.8 \pm 0.3$ $c = 3497.4 \pm 0.7$ pm, $\beta = 96.04 \pm 0.03°$, a calculated density of 4.133 g·cm^{-3} at 153 K, and an

energy gap of 1.95 eV, is formed in the Ge–K–Fe–Se system (Feng et al. 2013). To prepare this compound, the mixture of K$_2$Se, FeSe, and GeSe$_2$ (molar ratio 1:1:3) was loaded into a fused silica tube under an argon atmosphere in a glove box. The tube was sealed under 10^{-3} Pa atmosphere and then placed in a furnace. The sample was heated to 900°C in 24 h and kept for 48 h, then cooled at a slow rate of 4°C·h^{-1} to 400°C, and finally cooled to room temperature. The product consisted of thin dark-red plates of K$_2$FeGe$_3$Se$_8$. The crystals are stable in the air.

5.36 Germanium–Rubidium–Aluminum–Selenium

The **Rb$_3$(AlSe$_2$)$_3$(GeSe$_2$)$_7$** quaternary compound, which melts congruently at 730°C and crystallizes in the monoclinic structure with the lattice parameters $a = 705.80 \pm 0.05$, $b = 3941.9 \pm 0.2$, $c = 704.12 \pm 0.04$ pm, $\beta = 90.360 \pm 0.005°$, a calculated density of 4.111 g·cm^{-3}, and an energy gap of 2.4 eV, is formed in the Ge–Rb–Al–Se system (Rothenberger et al. 2010). The synthesis of this compound was achieved by heating stoichiometric amounts of Rb$_2$Se$_5$, Al, Ge, and Se in a sealed and evacuated fused silica tube at 850°C for 3 days and cooling to 400°C in 4 days.

5.37 Germanium–Rubidium–Gallium–Selenium

The **RbGaGeSe$_4$** quaternary compound, which crystallizes in the orthorhombic structure with the lattice parameters $a = 1757.50 \pm 0.05$, $b = 747.18 \pm 0.02$, $c = 1244.49 \pm 0.04$ pm, a calculated density of 4.419 g·cm^{-3}, and an energy gap of 2.18 eV, is formed in the Ge–Rb–Ga–Se system (Friedrich et al. 2020). RbGaGeSe$_4$ was prepared by reacting Rb$_2$Se, Ga$_2$Se$_3$, GeSe$_2$ or Ge, and Se in a sealed evacuated quartz tube. The reaction mixture was heated to 900°C with a heating rate of 50°C·h^{-1}, annealed at 900°C for 2 days and cooled to room temperature with a cooling rate of 20°C·min^{-1}. The resulting crude product was finely ground in an agate mortar under N$_2$ atmosphere and annealed at 900°C for 2 additional days. Orange crystals were obtained. This compound decomposes within several minutes by releasing gaseous H$_2$Se.

5.38 Germanium–Rubidium–Indium–Selenium

The **RbInGeSe$_4$** quaternary compound, which crystallizes in the orthorhombic structure with the lattice parameters $a = 1799.48 \pm 0.06$, $b = 769.08 \pm 0.03$, $c = 1260.03 \pm 0.06$ pm, a calculated density of 4.475 g·cm^{-3}, and an energy gap of 2.2 eV, is formed in the Ge–Rb–In–Se system (Friedrich et al. 2021). A polycrystalline sample of this compound was obtained by reacting Rb$_2$Se (0.85 mM), In (1.70 mM), Ge (1.70 mM), and Se (5.95 mM) in a sealed evacuated fused silica tube. The reaction mixture was heated to 1000°C with a heating rate of 3°C·min^{-1}. After annealing at 1000°C for 24 h the tube was cooled to room temperature with a cooling rate of 1°C·min^{-1}. The air-sensitive yellow polycrystalline product forms plate-like crystallites and is stored in a nitrogen glove box until further use.

5.39 Germanium–Rubidium–Samarium–Selenium

The **RbSmGeSe$_4$** quaternary compound, which crystallizes in the orthorhombic structure with the lattice parameters a = 673.47 ± 0.08, b = 701.85 ± 0.09, c = 1772.3 ± 0.2 pm, a calculated density of 4.950 g·cm^{-3}, and an energy gap of 2.2 eV, is formed in the Ge–Rb–Sm–Se system (Martin and Dorhout 2004). This compound was synthesized by combining Sm (0.483 mM), Ge (0.968 mM), Se (1.62 mM), and Rb$_2$Se$_3$ (0.322 mM). The mixture was loaded into a fused silica ampoule inside an inert atmosphere glove box. The ampoule was flame-sealed under vacuum and placed in a furnace. The furnace was ramped to 725°C at a rate of 35°C·h^{-1}, and the temperature was held constant at 725°C for 150 h. The furnace was then slowly cooled to ambient temperature at a rate of 4°C·h^{-1}. After the reaction products cooled, the ampoule was opened in an inert atmosphere glove box, and the solid product was soaked in DMF for 6 h in order to dissolve and wash away any remaining flux and loosen the product crystals. After treatment with DMF, orange plates of the title compound were observed, along with orange hemispheres of amorphous rubidium seleno-germanate. Although RbSmGeSe$_4$ is stable in the air for a period of days, the inert atmosphere prevented the precipitation of elemental selenium resulting from decomposition reactions of binary and ternary species during the washing process.

5.40 Germanium–Rubidium–Phosphorus–Selenium

The **Rb$_4$GeP$_4$Se$_{12}$** quaternary compound, which melts congruently at 416°C and crystallizes in the orthorhombic structure with the lattice parameters a = 1418.81 ± 0.05, b = 735.66 ± 0.03, c = 2429.14 ± 0.15 pm, a calculated density of 3.893 g·cm^{-3}, and an energy gap of 2.0 eV, is formed in the Ge–Rb–P–Se system (Morris et al. 2012). It was prepared in the same way as K$_4$GeP$_4$Se$_{12}$ was synthesized using Rb$_2$Se instead of K$_2$Se.

5.41 Germanium–Rubidium–Manganese–Selenium

The **Rb$_2$MnGe$_3$Se$_8$** quaternary compound, which crystallizes in the orthorhombic structure with the lattice parameters a = 758.09 ± 0.04, b = 1249.69 ± 0.08, c = 1758.04 ± 0.10 pm, a calculated density of 4.288 g·cm^{-3}, and an energy gap of 1.72 eV, is formed in the Ge–Rb–Mn–Se system (Hu et al. 2018). To synthesize this compound, a mixture of RbCl, Mn, Ge, and Se (molar ratio 4:2:3:8) was loaded into a silica tube inside a glove box. Subsequently, the assembly was flame-sealed under a high vacuum (10^{-6} Pa), then heated to 400°C in 20 h, dwelled for 10 h, then heated to 800°C in 20 h, maintained for 100 h, and finally cooled to 350°C at a rate of 3°C·h^{-1}. The product was washed with distilled water to remove the chloride. The red plate-like crystals of Rb$_2$MnGe$_3$S$_8$ were obtained.

5.42 Germanium–Cesium–Magnesium–Selenium

The **Cs$_2$MgGe$_3$Se$_8$** quaternary compound, which crystallizes in the orthorhombic structure with the lattice parameters a = 766.35 ± 0.02, b = 1272.42 ± 0.05, c = 1777.24 ± 0.06 pm, a calculated density of 4.368 g·cm^{-3}, and an energy gap of 2.04 eV, is formed in the Ge–Cs–Mg–Se system (Morris et al. 2013). To obtain this compound, a mixture of Cs$_2$Se$_2$ (0.25 mM), Mg (0.25 mM), Ge (0.75 mM), and Se (1.5 mM) was added to a fused silica tube inside a N$_2$-filled glove box. The tube was then evacuated, flame-sealed, and heated gently at first in a weak methane/oxygen flame to start the reaction. It was then heated more strongly until a glowing liquid was obtained for a total of ~2 min. The tube was allowed to cool in the air at a rate of ~200°C·min^{-1}. Reddish-orange plate-like crystals of Cs$_2$MgGe$_3$Se$_8$ were obtained from such a reaction. Glassy Cs$_2$MgGe$_3$Se$_8$ was prepared after quenching the melt into a cool water bath.

5.43 Germanium–Cesium–Zinc–Selenium

The **Cs$_2$ZnGe$_3$Se$_8$** quaternary compound, which crystallizes in the orthorhombic structure with the lattice parameters a = 757.56 ± 0.05, b = 1261.91 ± 0.10, c = 1765.43 ± 0.11 pm, a calculated density of 4.647 g·cm^{-3}, and an energy gap of 2.31 eV, is formed in the Ge–Cs–Zn–Se system (Morris et al. 2013). To obtain this compound, a mixture of Cs$_2$Se$_2$ (0.25 mM), Zn (0.25 mM), Ge (0.75 mM), and Se (1.5 mM) was added to a fused silica tube inside a N$_2$-filled glove box. The tube was then evacuated, flame-sealed, and heated gently at first in a weak methane/oxygen flame to start the reaction. It was then heated more strongly until a glowing liquid was obtained for a total of ~2 min. The tube was allowed to cool in the air at a rate of ~200°C·min^{-1}. Yellowish-orange plate-like crystals of Cs$_2$ZnGe$_3$Se$_8$ were obtained from such a reaction.

5.44 Germanium–Cesium–Cadmium–Selenium

The **Cs$_2$CdGe$_3$Se$_8$** quaternary compound, which crystallizes in the orthorhombic structure with the lattice parameters a = 770.48 ± 0.03, b = 1268.22 ± 0.06, c = 1782.08 ± 0.07 pm, a calculated density of 4.683 g·cm^{-3}, and an energy gap of 2.49 eV, is formed in the Ge–Cs–Cd–Se system (Morris et al. 2013). To obtain this compound, a mixture of Cs$_2$Se$_2$ (0.25 mM), Cd (0.25 mM), Ge (0.75 mM), and Se (1.5 mM) was added to a fused silica tube inside a N$_2$-filled glove box. The tube was then evacuated, flame-sealed, heated in a furnace to 750°C in 8 h, held there for 12 h, and cooled to room temperature in 24 h.

5.45 Germanium–Cesium–Gallium–Selenium

Two quaternary compounds, **CsGaGeSe$_4$** and **Cs$_2$Ga$_6$Ge$_3$Se$_{14}$**, are formed in the Ge–Cs–Ga–Se system. CsGaGeSe$_4$ crystallizes in the orthorhombic structure with the lattice parameters

$a = 1776.66 \pm 0.07$, $b = 751.71 \pm 0.03$, $c = 1263.83 \pm 0.05$ pm, a calculated density of 4.652 g·cm⁻³, and an energy gap of 2.14 eV (Friedrich et al. 2020). It was prepared by reacting Cs_2Se, Ga_2Se_3, $GeSe_2$ or Ge, and Se in a sealed evacuated quartz tube. The reaction mixture was heated to 900°C with a heating rate of 50°C·h⁻¹, annealed at 900°C for 2 days and cooled to room temperature with a cooling rate of 20°C·min⁻¹. The resulting crude product was finely ground in an agate mortar under a N_2 atmosphere and annealed at 900°C for 2 additional days. Orange-red crystals were obtained. This compound decomposes within several minutes by releasing gaseous H_2Se.

$Cs_2Ga_6Ge_3Se_{14}$ crystallizes in the trigonal structure with the lattice parameters $a = 763.96 \pm 0.03$, $c = 1358.66 \pm 0.06$ pm, a calculated density of 4.854 g·cm⁻³, and an energy gap of 2.41 eV (Ma et al. 2019). This compound was synthesized by a solid-state reaction of CsCl, Ge, Ga, and Se (molar ratio 4:4:6:14) with slightly extra Ge and CsCl as reactive fluxes. After flame-sealing inside a silica tube, the reactant assembly was heated to 900°C and kept at this temperature for 125 h, then slowly cooled to 300°C before shutting down the furnace. The lamellar dark-red crystals of $Cs_2Ge_3Ga_6Se_{14}$ are stable in the air and insoluble in water. Above 800°C, it decomposes to give Ga_2Se_3.

5.46 Germanium–Cesium–Indium–Selenium

Three quaternary compounds, **$CsInGeSe_4$**, **$Cs_2In_6Ge_3Se_{14}$**, and **$Cs_4In_8GeSe_{16}$**, are formed in the Ge–Cs–In–Se system. $CsInGeSe_4$ crystallizes in the orthorhombic structure with the lattice parameters $a = 1816.01 \pm 0.06$, $b = 772.08 \pm 0.02$, $c = 1263.90 \pm 0.04$ pm, and an energy gap of 2.32 eV (Ward et al. 2014). $Cs_2In_6Ge_3Se_{14}$ crystallizes in the trigonal structure with the lattice parameters $a = 799.51 \pm 0.06$, $c = 4172.6 \pm 0.4$ pm, a calculated density of 4.913 g·cm⁻³, and an energy gap of 2.08 eV (Wu et al. 2018). $Cs_4In_8GeSe_{16}$ crystallizes in the monoclinic structure with the lattice parameters $a = 1276.00 \pm 0.03$, $b = 957.71 \pm 0.02$, $c = 3071.85 \pm 0.06$ pm, $\beta = 99.561 \pm 0.001°$, and an energy gap of 2.20 eV (Ward et al. 2014).

$CsInGeSe_4$ was obtained from the reaction of In (0.126 mM), Ge (0.063 mM), Se (0.377 mM), and CsCl (0.445 mM) (Ward et al. 2014). The reactants were loaded into a carbon-coated fused silica tube under an Ar atmosphere and then evacuated to 10⁻² Pa and flame-sealed. It was loaded into a furnace and heated to 900°C within 12 h, held there for 6 h, cooled to 800°C over 12 h, and held at 800°C for 4 days. The reaction was cooled to 500°C at 5°C·h⁻¹ and then cooled to room temperature for a further 12 h. It yielded yellow blocks of the title compound that were washed with water and dried with acetone to remove excess flux.

$Cs_2In_6Ge_3Se_{14}$ was synthesized by the high-temperature solid-state reaction of a mixture of CsCl, Ge, In, and Se (molar ratio 5:3:6:14) with a gross mass of 500 mg; CsCl served as a flux to reduce the reaction temperature (Wu et al. 2018). The reactants were loaded into a silica tube that was flame-sealed under high vacuum. The mixture was heated to 900°C at 20°C·h⁻¹ and soaked for 100 h, followed by cooling to 200°C before turning off the furnace. The product was washed with distilled water in an ultrasonic cleaner and dried with ethanol; deep red crystals

of $Cs_2In_6Ge_3Se_{14}$ were obtained. Raw materials were stored in a glove box with a purified Ar-filled atmosphere.

$Cs_4In_8GeSe_{16}$ was initially prepared from the reaction of depleted uranium (0.126 mM), In (0.126 mM), GeI₂ (0.063 mM), Se (0.378 mM), and CsCl (0.445 mM) (Ward et al. 2014). There were a few orange crystals in the product. To achieve a rational synthesis of this compound, a second reaction was loaded with In (0.502 mM), Ge (0.063mM), Cs_2Se_3 (0.126 mM), and Se (0.628 mM). These two reactions were subjected to the same heated profile as for $CsInGeSe_4$. The major product comprised orange crystals. These were dried with acetone to remove excess flux. The yield of $Cs_4In_8GeSe_{16}$ was about 80 mass% based on Ge. The compound is stable in both air and moisture.

5.47 Germanium–Cesium–Samarium–Selenium

The **$CsSmGeSe_4$** quaternary compound, which crystallizes in the orthorhombic structure with the lattice parameters $a = 670.7 \pm 0.2$, $b = 706.7 \pm 0.2$, $c = 1833.4 \pm 0.6$ pm, a calculated density of 5.134 g·cm⁻³, and an energy gap of 2.2 eV, is formed in the Ge–Cs–Sm–Se system (Martin and Dorhout 2004). This compound was synthesized by combining Sm (0.501 mM), Ge (1.00 mM), Se (2.08 mM), and Ce_2Se_3 (0.341 mM). The mixture was loaded into a fused silica ampoule inside an inert atmosphere glove box. The ampoule was flame-sealed under vacuum and placed in a furnace. The furnace was ramped to 725°C at a rate of 35°C·h⁻¹, and the temperature was held constant at 725°C for 150 h. The furnace was then slowly cooled to ambient temperature at a rate of 4°C·h⁻¹. After the reaction products cooled, the ampoule was opened in an inert atmosphere glove box, and the solid product was soaked in DMF for 6 h in order to dissolve and wash away any remaining flux and loosen the product crystals. After treatment with DMF, orange plates of the title compound were observed, along with orange hemispheres of amorphous cesium seleno-germanate. Although $CsSmGeSe_4$ is stable in the air for a period of days, the inert atmosphere prevented the precipitation of elemental selenium resulting from the decomposition reactions of binary and ternary species during the washing process.

5.48 Germanium–Cesium–Phosphorus–Selenium

The **$Cs_4GeP_4Se_{12}$** quaternary compound, which melts congruently at 426°C and crystallizes in the orthorhombic structure with the lattice parameters $a = 1432.68 \pm 0.07$, $b = 762.02 \pm 0.03$, $c = 2463.01 \pm 0.14$ pm, a calculated density of 4.139 g·cm⁻³, and an energy gap of 2.0 eV, is formed in the Ge–Cs–P–Se system (Morris et al. 2012). It was prepared in the same way as $K_4GeP_4Se_{12}$ was synthesized using Cs_2Se_2 instead of K_2Se.

5.49 Germanium–Cesium–Arsenic–Selenium

The **$Cs_3AsGeSe_5$** quaternary compound, which crystallizes in the monoclinic structure with the lattice parameters $a = 763.0 \pm 0.2$, $b = 968.8 \pm 0.2$, $c = 1977.3 \pm 0.4$ pm, $\beta = 98.77 \pm 0.03°$, and a calculated density of 4.327 $g \cdot cm^{-3}$, is formed in the Ge–Cs–As–Se system (van Almsick and Sheldrick 2005). To synthesize $Cs_3AsGeSe_5$, Ge (1.0 mM), As (1.0 mM), Se (2.5 mM), and Cs_2CO_3 (1.0 mM) were heated in a sealed glass tube to 190°C in MeOH (0.8 mL). After 80 h, the contents were allowed to cool to room temperature at $2°C \cdot h^{-1}$ to afford orange crystals of $Cs_3AsGeSe_5$, red blocks of Cs_3AsSe_4, and yellow prisms of $Cs_4Ge_2Se_6$. The compounds were separated manually under a polarizing microscope.

5.50 Germanium–Cesium–Manganese–Selenium

The $Cs_2MnGe_3Se_8$ quaternary compound, which melts at about 708°C and crystallizes in the orthorhombic structure with the lattice parameters $a = 763.76 \pm 0.07$, $b = 1264.84 \pm 0.11$, $c = 1776.25 \pm 0.13$ pm, a calculated density of 4.530 $g \cdot cm^{-3}$, and an energy gap of 1.74 eV, is formed in the Ge–Cs–Mn–Se system (Hu et al. 2018). To synthesize this compound, a mixture of CsCl, Mn, Ge, and Se (molar ratio 4:2:3:8) was loaded into a silica tube inside a glove box. Subsequently, the assembly was flame-sealed under a high vacuum (10^{-6} Pa), and then heated to 400°C in 20 h, dwelled for 10 h, then heated to 800°C in 20 h, maintained for 100 h, and finally cooled to 350°C at a rate of $3°C \cdot h^{-1}$. The product was washed with distilled water to remove the chloride. The red plate-like crystals of $Cs_2MnGe_3Se_8$ were obtained.

5.51 Germanium–Copper–Silver–Selenium

Cu_2GeSe_3–Ag_2GeSe_3. This section, studied through differential thermal analysis (DTA) and X-ray difractometry (XRD), is a non-quasibinary section of the Ge–Cu–Ag–Se system (Abbasova 2019). Solid solution based on high-temperature modification of Ag_8GeSe_6 primarily crystallizes in the compositional range up to 55 mol% Cu_2GeSe_3, and solid solution based on Cu_2GeSe_3 crystallizes at 55–100 mol% Cu_2GeSe_3. The solubility of Ag_2GeSe_3 in Cu_2GeSe_3 is 20 mol% at 630°C and 10 mol% at room temperature.

Cu_8GeSe_6–Ag_8GeSe_6. Phase equilibria in the system were studied through the DTA, XRD, and electromotive force (EMF) method using the samples annealed at 630°C for about 500 h, and it was shown that the system is characterized by continuous series of solid solutions between high-temperature modifications of the starting compounds (Abbasova et al. 2017a; Moroz 1990). Formation of the solid solutions decreases the temperatures of the polymorphic transition of both compounds.

5.52 Germanium–Copper–Strontium–Selenium

The $Cu_2SrGeSe_4$ quaternary compound, which crystallizes in the orthorhombic structure with the lattice parameters $a = 1080.7 \pm 0.4$, $b = 1073.5 \pm 0.4$, $c = 654.1 \pm 0.2$ pm, a calculated density of 5.279 $g \cdot cm^{-3}$, and an energy gap of 0.7 eV [1.79–1.80 eV (Zhu et al. 2017)] (Tampier and Johrendt 2001), is formed in the Ge–Cu–Sr–S system. Dark-red crystals of this compound were synthesized by heating a stoichiometric

mixture of the elements in the Ar atmosphere. The mixture was heated to 700°C for 15 h and homogenized 24 h at 750°C.

5.53 Germanium–Copper–Barium–Selenium

The $Cu_2BaGeSe_4$ quaternary compound, which crystallizes in the trigonal structure with the lattice parameters $a = 650.14 \pm 0.06$, $c = 1626.6 \pm 0.3$ pm, a calculated density of 5.462 $g \cdot cm^{-3}$, and an energy gap of 1.88 eV (Nian et al. 2017) [$a = 649.0 \pm 0.1$, $c = 1635.5 \pm 0.3$ pm, a calculated density of 5.451 $g \cdot cm^{-3}$, and an energy gap of 0.5 eV (Tampier and Johrendt 2001); $a = 650.4 \pm 0.4$, $c = 1631 \pm 2$ pm and an energy gap of 1.60–1.61 eV (Zhu et al. 2017)], is formed in the Ge–Cu–Ba–S system.

A conventional high-temperature solid-state method was used to synthesize the title compound (Nian et al. 2017). A stoichiometric mixture of Cu, Ba, Ge, and Se (molar ratio 1:2:1:4) was loaded into the silica tube and sealed under a high vacuum of 10^{-3} Pa. The furnace was programmed by the following steps: heated from room temperature to 600°C in 30 h and kept at this temperature for 40 h; then heated to 1000°C in 20 h and left at this temperature for 100 h; finally, cooled to 400°C at a rate of $5°C \cdot h^{-1}$ and then the furnace was shut down to room temperature. After the reaction, the product was washed with DMF to remove the unreacted sulfur and other byproducts. Many red crystals of $Cu_2BaGeSe_4$ were gained that can be stable in the air for several months. Since the Ba metal is easily oxidized in the air, an Ar-filled glove box was used to avoid the effects of oxygen and moisture in the preparation processes.

$Cu_2BaGeSe_4$ can also be prepared as follows (Zhu et al. 2017). A stoichiometric mixture of BaSe, CuSe, Ge, and Se were ground and cold-pressed into pellets. The BaSe and CuSe starting materials were baked on a hotplate prior to use at 175°C, in a N_2-filled glove box, to remove SeO_2. These pellets were transferred to a quartz ampoule evacuated to $\sim10^{-4}$ Pa and flame-sealed. Before sealing, a small amount of excess Se powder (~2.5 mg for 167.3 mg of $Cu_2BaGeSe_4$) was added to create an overpressure of Se vapor and avoid oxide formation. The ampoule was then heated to 650°C in a box furnace ($100°C \cdot h^{-1}$) and held at this temperature for 24 h. The steps of grinding, homogenizing, cold-pressing, and annealing the pellets were repeated several times to ensure a homogeneous powder.

5.54 Germanium–Copper–Zinc–Selenium

Cu_2GeSe_3–ZnSe. The phase diagram of this system, constructed through DTA, XRD, metallography, and EPMA, using the samples annealed at 400°C for 500 h with the next quenching in ice water, is shown in Figure 5.1 (Parasyuk et al. 2001). The eutectic contains 3 mol% ZnSe and crystallizes at 775°C. The $Cu_2ZnGeSe_4$ quaternary compound, which melts incongruently (Caldera et al. 2014) at 890°C (Matsushita and Katsui 2005; Matsushita et al. 2000a, 2005; Parasyuk et al. 2001) [at 785°C (Schäfer and Nitsche 1977)] and has a polymorphous transformation at 798°C (Parasyuk et al. 2001), is formed in this system. First modification of this compound crystallizes in

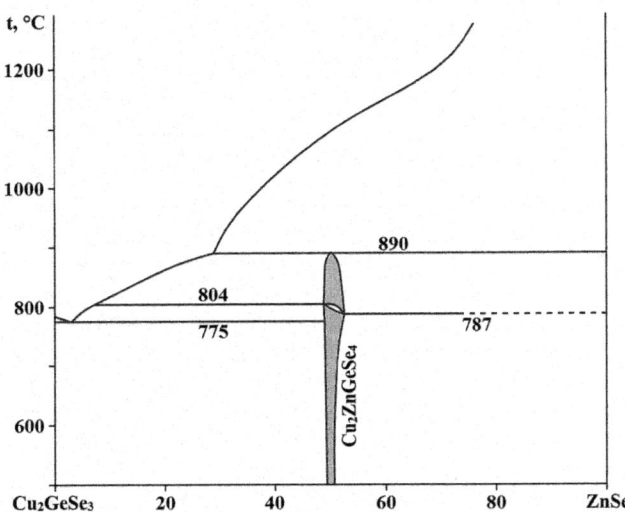

FIGURE 5.1 Phase diagram of the Cu₂GeSe₃–ZnSe system. (From Parasyuk, O.V., et al., *J. Alloys Compd.*, **329**(1–2), 202, 2001.)

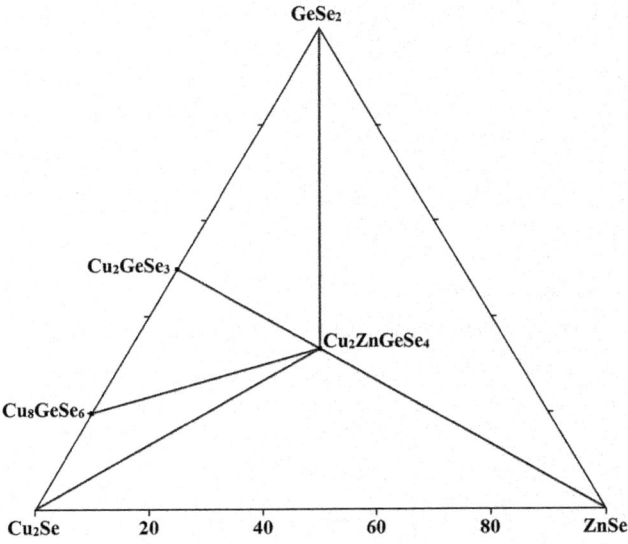

FIGURE 5.2 Isothermal section of the GeSe₂–Cu₂Se–ZnSe quasiternary system at 400°C. (From Romanyuk, Ya.E., and Parasyuk, O.V., *J. Alloys Compd.*, **348**(1–2), 195, 2003.)

the tetragonal structure with the lattice parameters $a = 561.12 \pm 0.01$ and $c = 1104.73 \pm 0.03$ pm (Khyzhun et al. 2015) [$a = 561$ and $c = 1102$ pm (Nitsche et al. 1967); $a = 562.2$, $c = 1106$ pm, the calculated and experimental densities of 5.52 and 5.50 g·cm⁻³, respectively (Hahn and Schulze 1965; Schäfer and Nitsche 1974; Guen et al. 1979; Guen and Glaunsinger 1980); $a = 561.3$ and $c = 1104.8$ pm (Schäfer and Nitsche 1977); $a = 561$ and $c = 1105$ pm (Schleich and Wold 1977); $a = 560.6$ and $c = 1104.2$ pm (Matsushita et al. 2000a, 2005; Matsushita and Katsui 2005); $a = 561.043 \pm 0.008$ and $c = 1104.57 \pm 0.03$ pm (Parasyuk et al. 2001)], and an energy gap of 1.38 eV (Sinagra III et al. 2021) [1.29 eV (Schleich and Wold 1977); 1.63 eV (Matsushita et al. 2000a, 2005)].

Because the energy between kesterite- and stannite-type structures is small, both structures can coexist in synthesized samples of Cu₂ZnGeSe₄. According to the first principles calculations, kesterite-type structure of this compound is characterized by the tetragonal structure with the lattice parameters $a = 560.2$ and $c = 1125.9$ pm and an energy gap of 1.60 eV, and stannite-type structure is also characterized by the tetragonal structure with the lattice parameters $a = 558.3$ and $c = 1132.5$ pm and an energy gap of 1.32 eV (Chen and Ravindra 2013).

Single crystals of Cu₂ZnGeSe₄ were obtained by the chemical transport reactions using I₂ as the transport agent (Nitsche et al. 1967; Schäfer and Nitsche 1974; Schleich and Wold 1977; Guen et al. 1979; Guen and Glaunsinger 1980; Yao et al. 1987; Parasyuk et al. 2001) or by the sintering of mixtures from binary compounds at 650–900°C (Hahn and Schulze 1965) or using the horizontal gradient method from respective melts (Matsushita et al. 2000a, 2005). They can also be obtained by employing the solution-fusion method (Khyzhun et al. 2015). The growth was performed in a two-zone furnace with a constant temperature gradient (4.2°C·mm⁻¹). The growth rate was set to be 2 mm per day. A comparatively long annealing of the crystal at 430°C for 150 h was made with the next cooling to room temperature over 100 h duration.

According to the data of Sinagra III et al. (2021), one more quaternary compound, **Cu₄ZnGe₂Se₇**, is formed in the Cu₂GeSe₃–ZnSe system. It crystallizes in the monoclinic structure with the lattice parameters $a = 1234.43 \pm 0.04$, $b = 561.95 \pm 0.02$, $c = 879.04 \pm 0.03$ pm, $\beta = 98.693 \pm 0.002°$, a calculated density of 5.606 g·cm⁻³, and an energy gap of 0.91 eV. To synthesize this compound, stoichiometric amounts of the elemental reagents, with the exception of Se that was added in 5% excess, were weighed and combined (not ground) in an Ar-filled glove box in order to make ca. 2 mM of product. The reagent mixtures were placed into a small fused silica tube, which was subsequently inserted into a longer fused silica tube. The tube was removed from the glove box, connected to a vacuum line, evacuated to a pressure of ~0.1 Pa, and sealed. The reaction vessel was inserted into a furnace, heated to 800°C at a rate of 65°C·h⁻¹, held at that temperature for 4 days, cooled to 400°C at a rate of 7.5°C·h⁻¹, and subsequently cooled to room temperature. After the reaction, the tube was split open, and the content was inspected under an optical microscope. The products were grayish-black microcrystalline powders. Needle-like and irregular polyhedral-shaped crystals were observed. Cu₄ZnGe₂Se₇ is air and moisture stable, thermally stable up to relatively high temperatures, and undergoes a phase transition.

Cu₈GeSe₆–ZnSe. This system is not a quasibinary section of the Ge–Cu–Zn–Se quaternary system and crosses two three-phase and one two-phase region (Romanyuk and Parasyuk 2003). There are two peritectic equilibria at 809°C and 818°C in this section. The polymorphous transformation of Cu₂ZnGeSe₄ was not observed in the experiments. This system was investigated through DTA, XRD, and metallography. The samples were annealed at 400°C during 500 h followed by quenching in cold water.

GeSe₂–Cu₂Se–ZnSe. Isothermal section of the GeSe₂–Cu₂Se–ZnSe quasiternary system at 400°C is given in Figure 5.2 (Romanyuk and Parasyuk 2003).

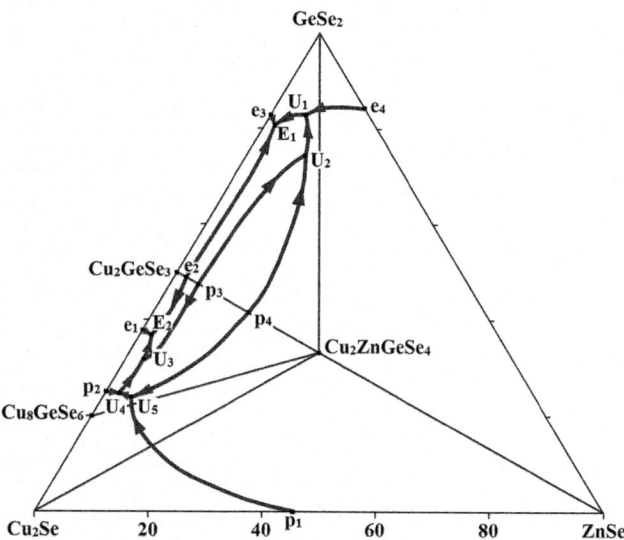

FIGURE 5.3 Projection of the liquidus surface of the GeSe₂–Cu₂Se–ZnSe quasiternary system. (From Romanyuk, Ya.E., and Parasyuk, O.V., *J. Alloys Compd.*, **348**(1–2), 195, 2003.)

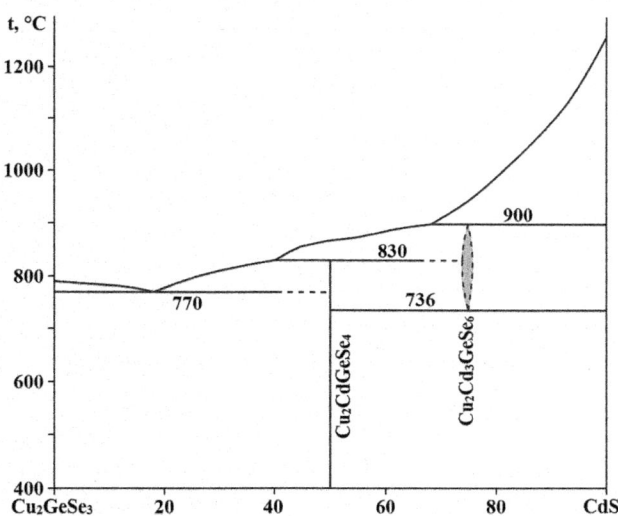

FIGURE 5.4 Phase diagram of the Cu₂GeSe₃–CdSe system. (From Piskach, L.V., et al., *J. Alloys Compd.*, **299**(1–2), 227, 2000.)

The projection of the liquidus surface of this system is shown in Figure 5.3 (Romanyuk and Parasyuk 2003). It consists of seven fields of primary crystallization. Sixteen monovariant lines divide the various fields of primary crystallization. Temperatures of the ternary invariant points in this system are the next: 697°C for E_1, 750°C for E_2, 710°C for U_1, 791°C for U_2, 795°C for U_3, 809°C for U_4, and 818°C for U_5.

5.55 Germanium–Copper–Cadmium–Selenium

Cu₂GeSe₃–CdSe. The phase diagram of this system, constructed through DTA, XRD, metallography, and measuring of the microhardness and density, is presented in Figure 5.4 (Piskach et al. 2000). Besides the quaternary compound

Cu₂CdGeSe₄, possessing a narrow homogeneity region and forming incongruently (Caldera et al. 2008) at 830°C (Piskach et al. 2000; Gulay et al. 2002d) [at 824°C (Schäfer and Nitsche 1977); at 852°C (Piskach et al. 1997); at 840°C°C (Zhukov et al. 1984; Matsushita et al. 2000a, 2005)], another quaternary phase with the approximate composition **Cu₂Cd₃GeSe₆** was revealed. It forms according to the peritectic reaction at 900°C and decomposes at 736°C according to the eutectoid reaction. The eutectic between Cu₂CdGeSe₄ and Cu₂GeSe₃ melts at 770°C and contains 18 mol% CdSe [at 652°C and 25 mol% CdSe (Piskach et al. 1997); at 660°C and 25 mol% CdSe (Zhukov et al. 1984)]. The solid solubility ranges of the components of this section do not exceed 2 mol% (Piskach et al. 2000). According to the data of Zhukov et al. (1984), the solubility of CdSe in Cu₂GeSe₃ reaches 12 mol% and the solubility of CdSe in Cu₂GeSe₃ is not higher than 7 mol%. The increasing of Cu₂GeSe₃ contents in the solid solutions based on CdSe leads to the monotonous changing of electrophysical properties of forming solid solutions (Dovletov et al. 1987).

Cu₂CdGeSe₄ has two polymorphic modifications. α-Cu₂CdGeSe₄ crystallizes in the tetragonal structure of the stannite type with the lattice parameters $a = 574.82 \pm 0.02$ and $c = 1105.33 \pm 0.03$ pm (Gulay et al. 2002d) [$a = 565.7 \pm 0.5$ and $c = 1098.8 \pm 0.1$ pm (Schäfer and Nitsche 1974; Parthé et al. 1969); $a = 574.7$ and $c = 1105.9$ pm (Schäfer and Nitsche 1977); $a = 565$, $c = 1096$ pm, and an experimental density of 5.45 g·cm⁻³ (Zhukov et al. 1984); $a = 574.89 \pm 0.05$ and $c = 1195.5 \pm 0.1$ pm (Piskach et al. 2000)] and an energy gap of 1.29 eV (Konstantinova et al. 1989) [1.20 eV (Mkrtchian et al. 1988a, Matsushita et al. 2000a, 2005)].

β-Cu₂CdGeSe₄ crystallizes in the orthorhombic structure with the lattice parameters $a = 809.68 \pm 0.09$, $b = 689.29 \pm 0.06$, and $c = 662.64 \pm 0.06$ pm (Gulay et al. 2002d) [$a = 808.8$, $b = 687.5$, and $c = 656.4$ pm (Matsushita and Katsui 2005; Matsushita et al. 2000a, 2005); $a = 806.2$, $b = 687.1$, and $c = 659.7$ pm (Schäfer and Nitsche 1977)].

The ingots for investigations were annealed at 400°C for 250 h, with the next quenching in cold water (Piskach et al. 2000) or at 800°C for 600 h (Zhukov et al. 1984). Cu₂CdGeSe₄ was obtained by heating at 1030–1080°C for 2 h and annealed at 530–580°C for 600 h (Mkrtchian et al. 1988b). β-Cu₂CdGeSe₄ was produced by the cooling of the respective melt without any annealing by directly quenching in water, while annealing of the alloy at 400°C leads to the obtaining of α-Cu₂CdGeSe₄ (Gulay et al. 2002d). Single crystals of this compound were grown by the chemical transport reactions (Parthé et al. 1969, Schäfer and Nitsche 1974, Matsushita et al. 2005), or by oriented crystallization (Konstantinova et al. 1989), or by using horizontal gradient freezing (Matsushita et al. 2000a, 2005).

Cu₈GeSe₆–CdSe. This section, investigated through DTA, XRD, and metallography using the alloys annealed at 400°C for 500 h with next quenching in cold water is not a quasibinary section of the Ge–Cu–Cd–Se quaternary system (Olekseyuk et al. 2000b). It crosses the fields of primary crystallization of the solid solutions based on Cu₂Se and CdSe. The horizontal at 650°C corresponds to the polymorphic transformation of Cu₈GeSe₆, and at 560°C corresponds to the decomposition of the Cu₂Cd₃GeSe₆ phase.

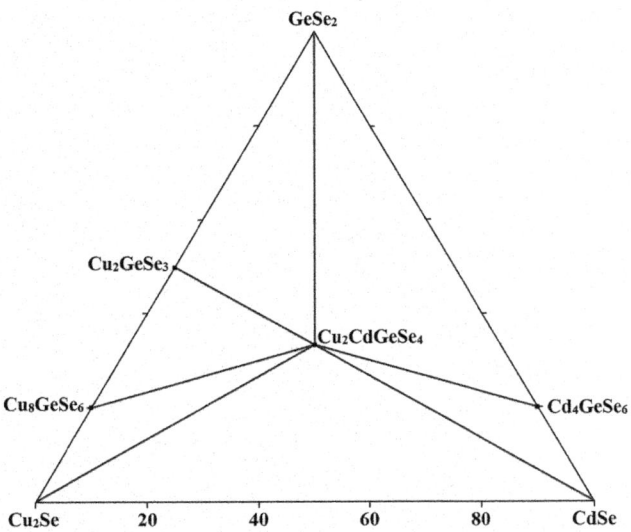

FIGURE 5.5 Isothermal section of the $GeSe_2$–Cu_2Se–$CdSe$ quasiternary system at 400°C. (From Olekseyuk, I.D., et al., *J. Alloys Compd.*, **298**(1–2), 203, 2000.)

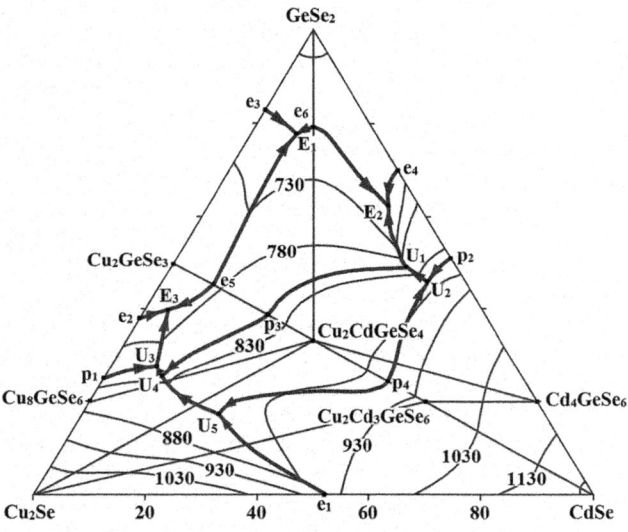

FIGURE 5.6 Liquidus surface of the $GeSe_2$–Cu_2Se–$CdSe$ quasiternary system. (From Olekseyuk, I.D., et al., *J. Alloys Compd.*, **298**(1–2), 203, 2000.)

$GeSe_2$–Cu_2Se–$CdSe$. There is only one quaternary compound, $Cu_2CdGeSe_4$, at 400°C presented in the $GeSe_2$–Cu_2Se–$CdSe$ quasiternary system (Figure 5.5) (Olekseyuk et al. 2000b). It is in equilibrium with all existing binary and ternary compound. The $Cu_2Cd_3GeSe_6$ phase decomposes at the cooling at higher temperatures and does not become apparent at this temperature.

The liquidus surface of this quasiternary system is shown in Figure 5.6 (Olekseyuk et al. 2000b). It consists of eight fields of primary crystallization of phases. Three of them belong to the crystallization of Cu_2Se, $CdSe$, and $GeSe_2$. Another three belong to the crystallization of the ternary phases Cu_8GeSe_6, Cu_2GeSe_3, and Cd_4GeSe_6, and two belong to the crystallization of the quaternary phases $Cu_2CdGeSe_4$ and $Cu_2Cd_3GeSe_6$. The

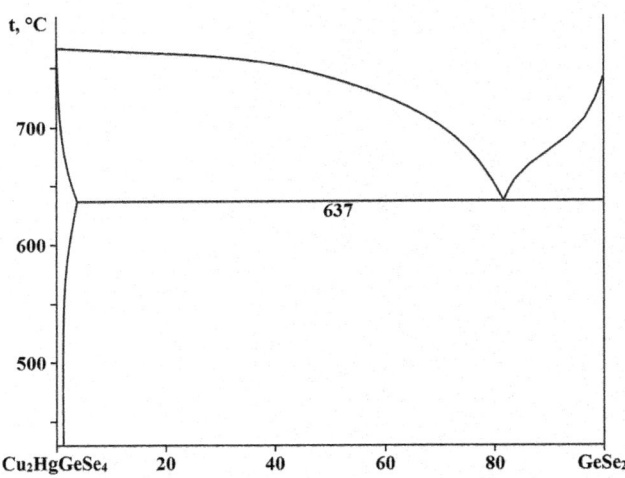

FIGURE 5.7 Phase diagram of the $GeSe_2$–$Cu_2HgGeSe_4$ system. (From Marchuk, O.V., et al., *J. Alloys Compd.*, **457**(1–2), 337, 2008.)

various fields of primary crystallization are divided by 19 monovariant lines and by 18 invariant points, 10 corresponding to binary invariant processes, and 8 corresponding to the ternary ones. There are three ternary eutectics and five transition reactions in this system: E_1 (692°C) – L ↔ $Cu_2CdGeSe_4$ + Cu_2GeSe_3 + $GeSe_2$; E_2 (686°C) – L ↔ $Cu_2CdGeSe_4$ + Cd_4GeSe_6 + $GeSe_2$; E_3 (754°C) – L ↔ $Cu_2CdGeSe_4$ + Cu_2GeSe_3 + β-Cu_8GeSe_6; U_1 (789°C) – L + $Cu_2Cd_3GeSe_6$ ↔ $Cu_2CdGeSe$ + Cd_4GeSe_6; U_2 (850°C) – L + $CdSe$ ↔ $Cu_2Cd_3GeSe_6$ + Cd_4GeSe_6; U_3 (798°C) – L + Cu_2Se ↔ β-Cu_8GeSe_6 + $Cu_2CdGeSe_4$; U_4 (805°C) – L + $Cu_2Cd_3GeSe_6$ ↔ $Cu_2CdGeSe_4$ + Cu_2Se; and U_5 (848°C) – L + $CdSe$ ↔ $Cu_2Cd_3GeSe_6$ + Cu_2Se.

5.56 Germanium–Copper–Mercury–Selenium

$GeSe_2$–$Cu_2HgGeSe_4$. The phase diagram of this system, constructed through DTA, XRD, and metallography using the ingots annealed at 400°C for 500 h, is a eutectic type (Figure 5.7) (Marchuk et al. 2008b). The eutectic contains 82 mol% $GeSe_2$ and crystallizes at 637°C.

Cu_2GeSe_3–$HgSe$. The phase diagram of this system, constructed through DTA, XRD, metallography, and measuring of the microhardness and density, is shown in Figure 5.8 (Parasyuk et al. 1997, 1998). The eutectic compositions and temperatures are 15 and 86 mol% HgSe and 753°C and 716°C [667°C and 597°C (Mkrtchian et al. 1988b)], respectively. The **$Cu_2HgGeSe_4$** quaternary compound is formed in this system, which melts congruently at 764°C (Parasyuk et al. 1997, 1998) [at 760°C (Hirai et al. 1967); at 754°C (Schäfer and Nitsche 1977); at 737°C (Mkrtchian et al. 1988b)]. The homogeneity region of this compound at 400°C is within the interval of 32–43 mol% HgSe (Parasyuk et al. 1997, 1998, Olekseyuk et al. 2005a) [35–42 mol% HgSe (Olekseyuk et al. 2001c)].

Low-temperature modification of this compound crystallizes in the tetragonal structure with the lattice parameters $a = 574.56 ± 0.02$, $c = 1108.34 ± 0.04$ pm, and a calculated density of 6.4993 g cm⁻³ (Olekseyuk et al. 2005a) [$a = 569.4$, $c = 1102$ pm, and the calculated and experimental densities of

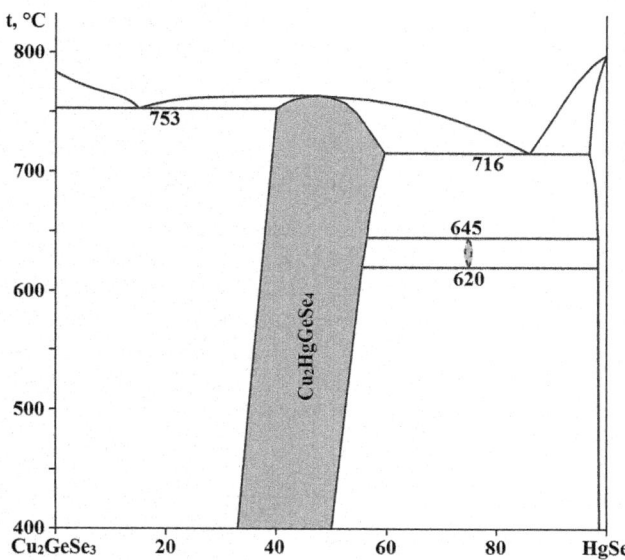

FIGURE 5.8 Phase diagram of the Cu$_2$GeSe$_3$–HgSe system. (From Parasyuk, O.V., *Ukr. khim. zhurn.*, **64**(9), 20, 1998.)

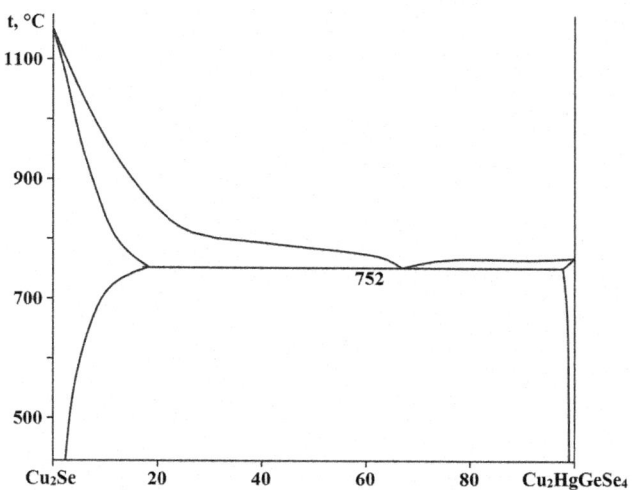

FIGURE 5.9 Phase diagram of the Cu$_2$HgGeSe$_4$–Cu$_2$Se system. (From Marchuk, O.V., et al., *J. Alloys Compd.*, **457**(1–2), 337, 2008.)

6.66 and 6.52 g·cm^{-3}, respectively (Hahn and Schulze 1965); a = 949 and c = 1796 pm (Hirai et al. 1967); a = 574.4 and c = 1110.3 pm (Schäfer and Nitsche 1977), a = 567, c = 1100 pm, an experimental density of 6.48 g·cm^{-3}, and an energy gap of 0.16 eV (Mkrtchian et al. 1988a,b)].

High-temperature modification of Cu$_2$HgGeSe$_4$ crystallizes in the orthorhombic structure with the lattice parameters a = 768.22 ± 0.03, b = 655.61 ± 0.02, c = 631.44 ± 0.03 pm, and a calculated density of 5.5185 g·cm^{-3} (Olekseyuk et al. 2005a).

A new phase is formed within the interval of 620–645°C, but its composition was not determined (a ratio of HgSe and Cu$_2$GeSe$_3$ in this phase is approximately 3/1) (Parasyuk et al. 1997, 1998).

The solubility of Cu$_2$GeSe$_3$ in HgSe is equal to 1.5 mol% [15 mol% at room temperature (Mkrtchian et al. 1988b)], and the solubility of HgSe in Cu$_2$GeSe$_3$ is not higher than 1 mol% (Parasyuk et al. 1997, 1998) [9 mol% (Mkrtchian et al. 1988b)].

The ingots for investigations were annealed at 400°C for 250 h [for 500 h (Olekseyuk et al. 2001c)] (Parasyuk et al. 1997, 1998; Olekseyuk et al. 2005a) [at 430–530°C for 500 h (Mkrtchian et al. 1988b)]. Single crystals of the Cu$_2$HgGeSe$_4$ compound were grown by the directional crystallization of the melt (Olekseyuk et al. 2001c).

Cu$_2$HgGeSe$_4$–Cu$_2$Se. The phase diagram of this system, constructed through DTA, XRD, and metallography, using the ingots annealed at 400°C for 500 h, is a eutectic type (Figure 5.9) (Marchuk et al. 2008b). The coordinates of the eutectic point are 33 mol% Cu$_2$Se and 752°C. The solubility of Cu$_2$HgGeSe$_4$ in Cu$_2$Se at 752°C is 18 mol% and decreases to 2 mol% at 400°C. The solubility of Cu$_2$Se in Cu$_2$HgGeSe$_4$ at 400°C does not exceed 2 mol%.

The **Cu$_2$HgGeSe$_4$–Cu$_8$GeSe$_6$** and **Cu$_2$HgGeSe$_4$–Hg$_2$GeSe$_4$** sections of the Ge–Cu–Hg–Se system are quasibinary only in the subsolidus regions (Marchuk et al. 2008b). The crystallization of all the alloys of the first section is completed at the temperature of 747°C. In the case of the second section, the

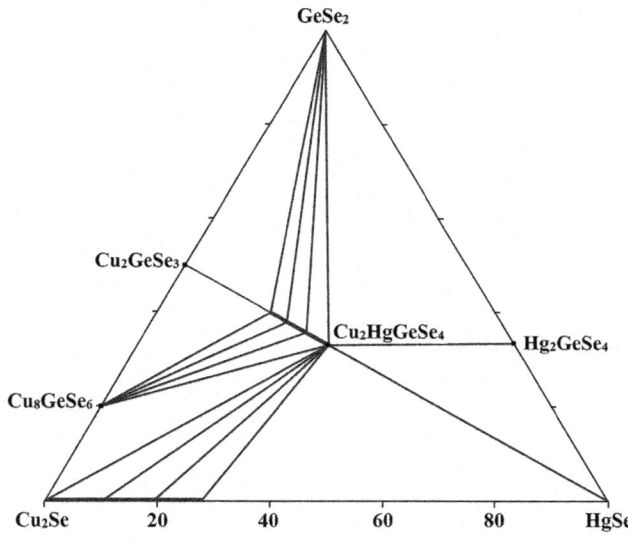

FIGURE 5.10 Isothermal section of the GeSe$_2$–Cu$_2$Se–HgSe quasiternary system at 400°C. (From Olekseyuk, I.D., et al., *J. Alloys Compd.*, **398**(1–2), 80, 2005.)

incongruent melting mode of Hg$_2$GeSe$_4$ is conditioned by the presence of two three-phase regions in the phase equilibria.

GeSe$_2$–Cu$_2$Se–HgSe. The isothermal section of the GeSe$_2$–Cu$_2$Se–HgSe quasiternary system at 400°C is shown in Figure 5.10 (Olekseyuk et al. 2005a). Seven single-phase regions, 12 two-phase regions, and 6 three-phase regions exist in the system at this temperature. The solubility of binary and ternary phases is less than 2–3 mol% of the respective component.

The liquidus surface of this quasiternary system (Figure 5.11) consists of seven fields of primary crystallization of the phases Cu$_2$Se, Cu$_8$GeSe$_6$, Cu$_2$GeSe$_3$, HgSe, Hg$_2$GeSe$_4$, GeSe$_2$, and Cu$_2$HgGeSe$_4$ (Marchuk et al. 2008b). The fields of primary crystallization are separated by 16 monovariant

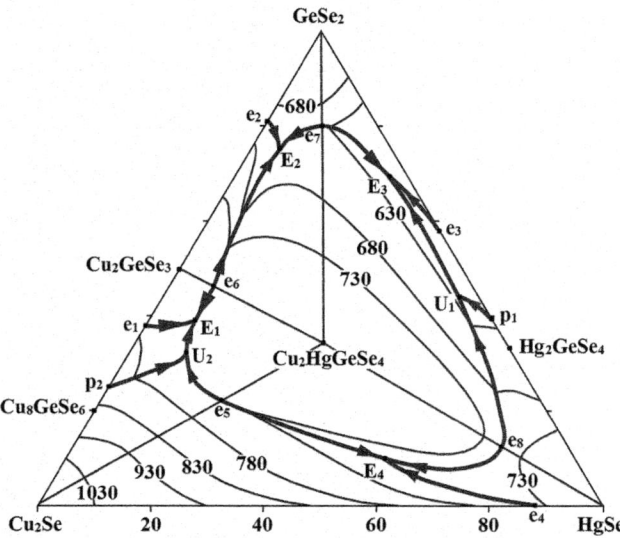

FIGURE 5.11 Liquidus surface of the GeSe₂–Cu₂Se–HgSe quasiternary system. (From Marchuk, O.V., et al., *J. Alloys Compd.*, **457**(1–2), 337, 2008.)

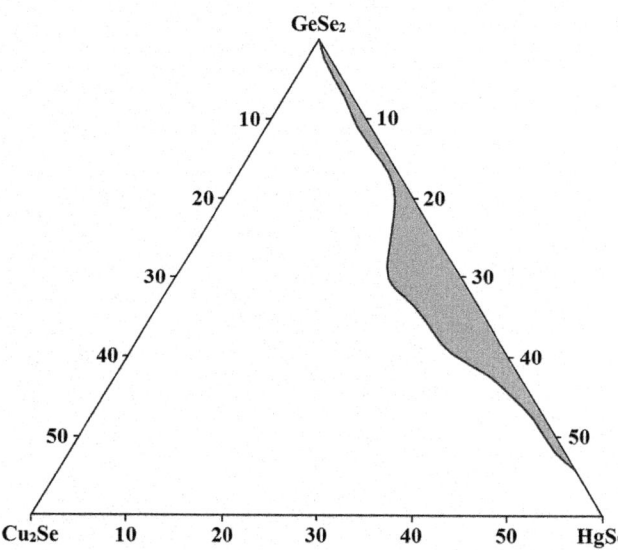

FIGURE 5.12 Glass-forming region in the GeSe₂–Cu₂Se–HgSe quasiternary system. (From Olekseyuk, I.D., et al., *Fiz. i khim. tv. tila*, **2**(1), 69, 2001.)

lines and 16 nonvariant points. Four ternary eutectics and two transition points exist in this system: E_1 (737°C) – L ↔ Cu_8GeSe_6 + $Cu_2HgGeSe_4$ + Cu_2GeSe_3; E_2 (622°C) – L ↔ $GeSe_2$ + $Cu_2HgGeSe_4$ + Cu_2GeSe_3; E_3 (575°C) – L ↔ $GeSe_2$ + $Cu_2HgGeSe_4$ + Hg_2GeSe_4; E_4 (680°C) – L ↔ β-Cu_2Se + $Cu_2HgGeSe_4$ + HgSe; U_1 (600°C) – L + HgSe ↔ $Cu_2HgGeSe_4$ + Hg_2GeSe_4; and U_2 (747°C) – L + β-Cu_2Se ↔ $Cu_2HgGeSe_4$ + Cu_8GeSe_6. The ingots for the investigations were annealed at 400°C for 500 h with the next quenching in 25% NaCl aqueous solution.

The glass-forming region in the HgSe–Cu₂Se–GeSe₂ quasiternary system is elongated along the GeSe₂–HgSe quasibinary system (Figure 5.12) (Olekseyuk et al. 2001d). The maximum

content of Cu_2Se that could be incorporated in the glasses is equal to 7 mol%. The glass transition temperature for such alloys is within the interval from 218°C to 316°C.

5.57 Germanium–Copper–Boron–Selenium

The **CuBGeSe₄** quaternary compound, which crystallizes in the tetragonal structure with the lattice parameters $a = 559.6$, $c = 547.8$ pm, and the calculated and experimental densities of 4.48 and 4.47 g·cm⁻³, respectively, is formed in the Ge–Cu–B–Se system (Hahn and Strick 1967b). This compound was obtained by long annealing of the elements' mixture or the mixture of binary components between 500°C and 1000°C. It has a green color.

5.58 Germanium–Copper–Aluminum–Selenium

The **CuAlGeSe₄** quaternary compound, which crystallizes in the tetragonal structure (Goodchild et al. 1981) with the lattice parameters $a = 557.5$, $c = 1068.2$ pm, and an energy gap of 2.34 eV (Hughes et al. 1980; Woolley et al. 1980) [$a = 556.6$, $c = 538.9$ pm, the calculated and experimental densities of 4.76 and 4.75 g·cm⁻³, respectively (Hahn and Strick 1967b), and an energy gap of 2.25 eV (Goodchild et al. 1982)], is formed in the Ge–Cu–Al–Se system.

This compound was obtained by the melt and anneals technique (Hahn and Strick 1967b; Goodchild et al. 1981, 1982; Hughes et al. 1980). To prepare it, the required masses of the elements were sealed under a vacuum in a quartz ampoule (Hughes et al. 1980). The sample so formed was slowly heated (20–40°C·h⁻¹) up to 1050–1100°C, left at that temperature for 5–10 h, slowly cooled (20–40°C·h⁻¹) down to ca. 500°C and then annealed at that temperature for 10–100 h before cooling to room temperature. The heating was carried out in a rocking furnace, and during the heating up and also during the reaction at high temperatures, the sample was rocked to ensure good mixing of the components. This compound has a brown color.

5.59 Germanium–Copper–Gallium–Selenium

GeSe₂–CuGaSe₂. The phase relations in this system, constructed through DTA, XRD, and metallography, are given in Figure 5.13 (Strok et al. 2000). CuGaSe₂ is formed according to the peritectic reaction, therefore, this section is non-quasibinary. γ-Ga₂Se₃, (CuGaSe₂), (CuGaGeSe₄), and (GeSe₂) primary crystallize in the system. The eutectic contains ~92 mol% GeSe₂ and crystallizes at 690°C. The maximum solubility of GeSe₂ in CuGaSe₂ is 30 mol% at 720°C and decreases abruptly to 7 mol% at 500°C. The solubility of CuGaSe₂ in GeSe₂ is not higher than 7 mol% at 690°C and decreases with temperature decreasing (< 2 mol% at 500°C).

In this system, a quaternary compound **CuGaGeSe₄** is formed, melting incongruently at 720°C [at 836°C (Matsushita et al. 2000a; Matsushita and Katsui 2005)] and apparently having two polymorphic modifications (Strok et al. 2000). It characterized by narrow temperature (690–720°C) and

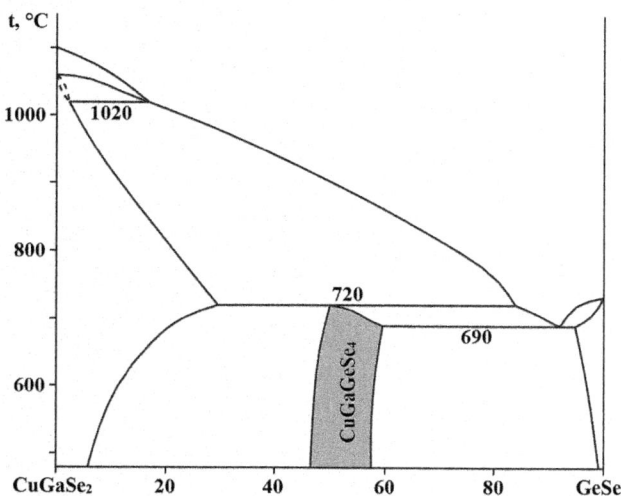

FIGURE 5.13 Phase equilibria in the GeSe$_2$–CuGaSe$_2$ system. (From Strok, O., et al., *Visnyk L'viv. Univ. Ser. khim.*, (39), 72, 2000.)

concentration (86–92 mol% GeSe$_2$) crystallization intervals with the homogeneity region 10 mol% (47–57 mol% GeSe$_2$) at 500°C.

The first modification of CuGaGeSe$_4$ crystallizes in the tetragonal structure (Goodchild et al. 1981) with the lattice parameters $a = 552.9 \pm 0.2$, $c = 1055.9 \pm 0.4$ pm, and a calculated density of 5.005 ± 0.006 g·cm^{-3} (Strok et al. 2000) [$a = 556.8$, $c = 1084.1$ pm, and an energy gap of 1.85 eV (Hughes et al. 1980; Woolley et al. 1980); $a = 557.6$, $c = 1087.8$ pm, and an energy gap of 1.38 eV (Matsushita et al. 2000a; Matsushita and Katsui 2005); an energy gap of 1.87 eV (Goodchild et al. 1982)].

This compound was obtained in the same way as CuAlGeSe$_4$ was prepared (Hahn and Strick 1967b; Goodchild et al. 1981, 1982; Hughes et al. 1980). The brown-green crystals of the title compound were synthesized. The ingots for the investigations were annealed at 500°C for 700 h (Strok et al. 2000). The ingots containing 50–98 mol% GeSe$_2$ were annealed additionally at 500°C for 350 h.

Cu$_8$GeSe$_6$–CuGaSe$_2$. As far as Cu$_8$GeSe$_6$ and CuGaSe$_2$ melt incongruently, this section is not a quasibinary (Olekseyuk et al. 2001f). Solid solution based on CuGaSe$_2$ reaches up to 5 mol% Cu$_8$GeSe$_6$ at 770°C and decreases with decreasing temperature to 2.5 mol% at 550°C. The solubility of CuGaSe$_2$ in Cu$_8$GeSe$_6$ reaches 5 mol% at 710°C.

GeSe$_2$–Cu$_2$Se–Ga$_2$Se$_3$. The isothermal section of this quasiternary system at 400°C is presented in Figure 5.14 (Olekseyuk et al. 2008; Strok et al. 2013b). The Cu$_{0.24}$Ga$_{1.61}$□$_{0.15}$Se$_2$ phase and CuGaGeSe$_4$ quaternary compound exist at this temperature. Eight single-phase regions were established at 400°C. Solid solutions of a large extent were observed on the base of Ga$_2$Se$_3$ up to 11 mol% Cu$_2$Se, most ternary and quaternary phases. Solid solutions based on CuGaSe$_2$ and Cu$_2$GeSe$_3$ have the greatest extent along the section. Solid solutions based on CuGaSe$_2$ extend up to 6 mol% GeSe$_2$, and solid solutions based on Cu$_2$GeSe$_3$ extend up to 29 mol% GeSe$_2$. The single-phase region based on CuGa$_3$Se$_5$ is essential in the system Cu$_2$Se–Ga$_2$Se$_3$ (73–85 mol% Ga$_2$Se$_3$) and extends in a plane of the

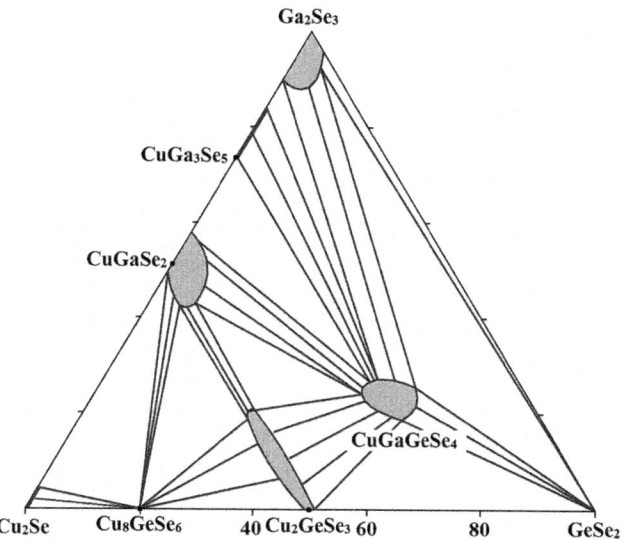

FIGURE 5.14 Isothermal section of the GeSe$_2$–Cu$_2$Se–Ga$_2$Se$_3$ quasiternary system at 400°C. (From Strok, O.M., et al., *J. Phase Equilib. Diffus.*, **34**(2), 94, 2013.)

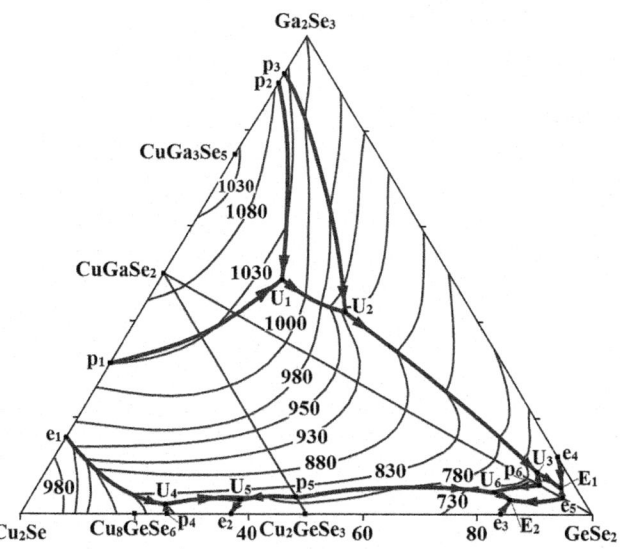

FIGURE 5.15 Liquidus surface of the GeSe$_2$–Cu$_2$Se–Ga$_2$Se$_3$ quasiternary system. (From Strok, O.M., et al., *J. Phase Equilib. Diffus.*, **34**(2), 94, 2013.)

triangle to 1 mol% GeSe$_2$. Solid solutions based on Cu$_2$Se, GeSe$_2$, and α-Cu$_8$GeSe$_6$ are insignificant. In addition, 14 two-phase and seven three-phase fields exist in the system at 400°C.

Some non-quasibinary sections and reaction schemes of the GeSe$_2$–Cu$_2$Se–Ga$_2$Se$_3$ quasiternary system were also constructed by Strok et al. (2013b).

The liquidus surface of this system (Figure 5.15) consists of nine fields of the primary crystallization (Strok et al. 2013b). The next invariant points exist in the system: U$_1$ (1020°C) – L + (Cu$_2$Ga$_4$Se$_7$) ↔ (CuGaSe$_2$) + (CuGa$_3$Se$_5$); U$_2$ (920°C) – L + (CuGa$_3$Se$_5$) ↔ (CuGaSe$_2$) + (Ga$_2$Se$_3$); U$_3$ (680°C) – L + (CuGaSe$_2$) ↔ (CuGaGeSe$_4$) + (Ga$_2$Se$_3$); U$_4$ (775°C) – L + (α-Cu$_2$Se) ↔ (γ-Cu$_8$GeSe$_6$) + (CuGaSe$_2$); U$_5$ (765°C) – L +

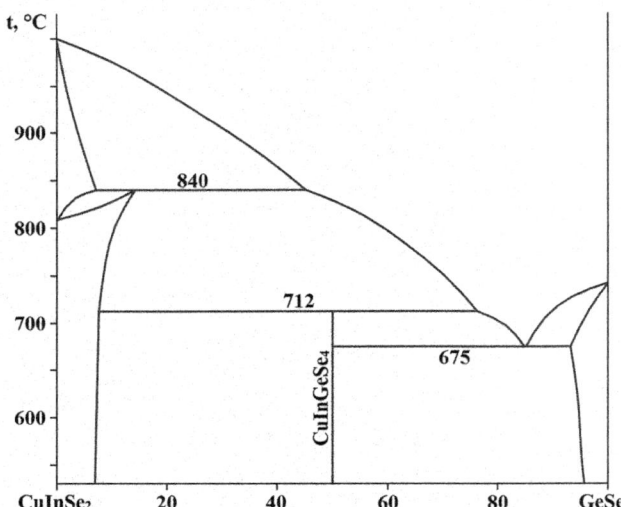

FIGURE 5.16 Phase equilibria in the GeSe₂–CuInSe₂ system. (From Vakulovich, and Olekseyuk I.D., *J. Alloys Compd.*, **367**(1–2), 47, 2004.)

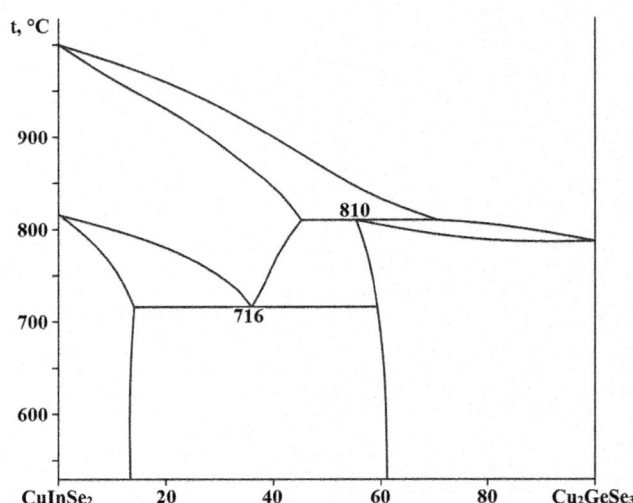

FIGURE 5.17 Phase equilibria in the Cu₂GeSe₃–CuInSe₂ system. (From Vakulovich, and Olekseyuk I.D., *J. Alloys Compd.*, **367**(1–2), 47, 2004.)

$(CuGaSe_2) \leftrightarrow (\gamma\text{-}Cu_8GeSe_6) + (Cu_2GeSe_3)$; U_6 (667°C) – L + $(CuGaSe_2) \leftrightarrow (CuGaGeSe_4) + (Cu_2GeSe_3)$; E_1 (650°C) – L $\leftrightarrow$ $(CuGaGeSe_4) + (Ga_2Se_3) + (GeSe_2)$; and E_2 (640°C) – L $\leftrightarrow$ $(Cu_2GeSe_3) + (CuGaGeSe_4) + (GeSe_2)$.

5.60 Germanium–Copper–Indium–Selenium

GeSe₂–CuInSe₂. The phase diagram of this system, constructed through DTA and metallography, is presented in Figure 5.16 (Vakulovich and Olekseyuk 2004; Sadykhova et al. 1988). The eutectic contains 85 mol% GeSe₂ and crystallizes at 675°C [90 mol% GeSe₂ and crystallizes at 690°C (Matsushita et al. 2000b)]. The **CuInGeSe₄**, which melts incongruently at 712°C [at 700°C (Sadykhova et al. 1988); at 810°C (Matsushita et al. 2000a; Matsushita and Katsui 2005)], is formed in this system. The existence of solid solutions based on both modifications of CuInSe₂ (9 mol% GeSe₂ in α-CuInSe₂ and 15 mol% GeSe₂ in β-CuInSe₂ at 840°C) was established. The solubility of GeSe₂ in α-CuInSe₂ at 530°C is 7 mol%. The ingots for the investigations were annealed at 500°C for over 300 h.

According to the data of Pamplin et al. (1977), the solid solutions are formed in the GeSe₂–CuInSe₂ system within the interval 0–50 mol% CuInSe₂.

CuInGeSe₄ crystallizes in the tetragonal structure (Goodchild et al. 1981) with the lattice parameters $a = 556.5 \pm 0.2$ and $c = 1111.6 \pm 0.3$ pm (Vakulovich and Olekseyuk 2004) [$a = 564.0$, $c = 1116.7$ pm, and an energy gap of 1.23 eV (Hughes et al. 1980; Woolley et al. 1980); $a = 556.2$, $c = 1111.4$ pm, and an energy gap of 1.30 eV (Matsushita et al. 2000a; Matsushita and Katsui 2005); an energy gap of 1.26 eV (Goodchild et al. 1982)]. This compound was obtained in the same way as CuAlGeSe₄ was prepared (Hahn and Strick 1967b; Goodchild et al. 1981, 1982; Hughes et al. 1980). The black crystals of the title compound were synthesized.

CuInSe₂–Ge. The phase diagram of this system, constructed through DTA, XRD, metallography, and measuring of the microhardness using the ingots annealing at 600°C for 200 h,

is a eutectic type (Allazova et al. 2020). The eutectic contains 50 mol% Ge and crystallizes at 790°C. There is an immiscibility area in this system in the region of 67–87 mol% Ge. The monotectic process occurs at 850°C.

Cu₂GeSe₃–CuInSe₂. The phase diagram of this system, constructed through DTA and metallography, is presented in Figure 5.17 (Vakulovich and Olekseyuk 2004). The peritectic reaction takes place in the concentration range of 44–70 mol% Cu₂GeSe₃ at 810°C. The solubility of Cu₂GeSe₃ in CuInSe₂ is 13 mol% at 530°C and 45 mol% at 810°C. The coordinates of the eutectoid point are 36 mol% Cu₂GeSe₃ and 716°C. The ingots for the investigations were annealed at 500°C for over 300 h.

Two other quaternary compounds, **Cu₃InGeSe₅** and **Cu₅InGe₂Se₈**, which melts (the melting temperature is only foe second compound) at 805°C, were found in the Ge–Cu–In–Se system (Hirai et al. 1967).

Ge–CuInSe₂–Se. Some vertical sections of this quasiternary system were constructed by Allazova et al. (2020). It was shown that the GeSe–CuInSe₂ system is non-quasibinary section as GeSe melts incongruently. The liquidus of this section consists of the primary crystallization curves of α-CuInSe₂ and Ge. The CuInGeSe₄–Ge and CuInGeSe₄–GeSe systems are also non-quasibinary sections as CuInGeSe₄ melts incongruently. The liquidus surface of the Ge–CuInSe₂–Se is divided by the GeSe₂–CuInSe₂ system into two subsystems. Seven fields of primary crystallization of α- and β-CuInSe₂, GeSe₂, CuInGeSe₄, GeSe, Ge, and Se, and two immiscibility regions exist on the liquidus surface. There are also one ternary eutectic and two transition points on the liquidus surface.

5.61 Germanium–Copper–Yttrium–Selenium

GeSe₂–Cu₂Se–Y₂Se₃. The isothermal section of this quasiternary system at 600°C is shown in Figure 5.18 (Lychmanyuk et al. 2006b). The formation of the **CuY₃GeSe₇** quaternary compound, which crystallizes in the hexagonal structure with the lattice parameters $a = 1023.68 \pm 0.08$, $c = 605.69 \pm 0.03$ pm,

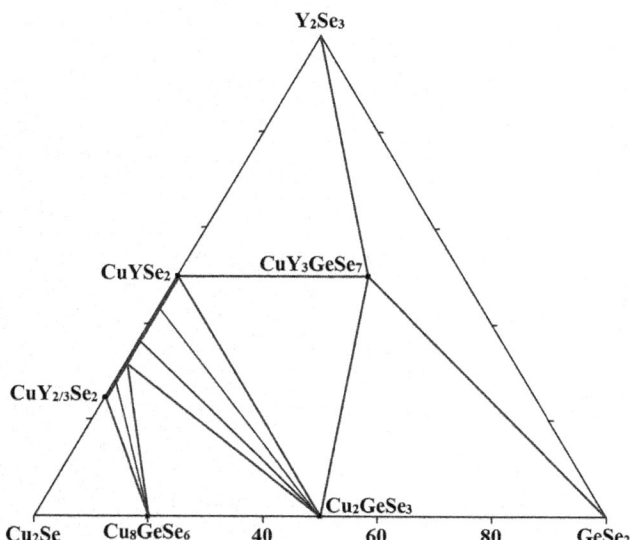

FIGURE 5.18 Isothermal section of the GeSe₂–Cu₂Se–Y₂Se₃ quasiternary system at 400°C. (From Lychmanyuk, O.S., et al., *Pol. J. Chem.*, **80**(3), 463, 2006.)

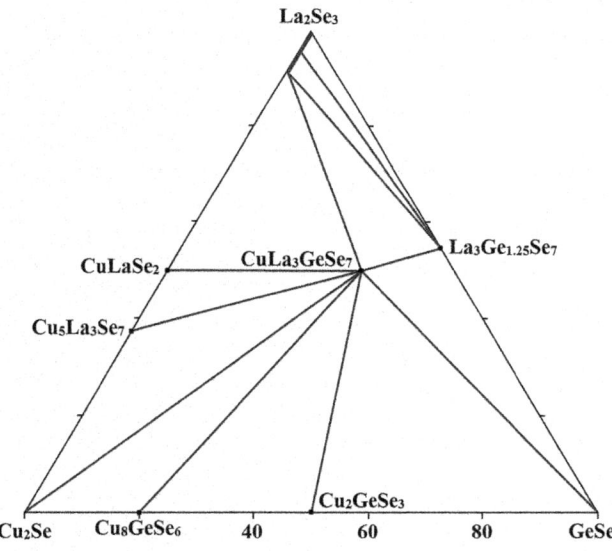

FIGURE 5.19 Isothermal section of the GeSe₂–Cu₂Se–La₂Se₃ quasiternary system at 600°C. (From Olekseyuk, I.D., et al., *Nauk. Visnyk Volyns'k. Derzh. Univ. im. Lesi Ukrainky. Ser. Khim. nauky*, (29), 13, 2009.)

and a calculated density of 5.773 g·cm⁻³, was established. The alloys for the investigation were annealed at 600°C for 240 h. After annealing, the ampoules with samples were quenched in cold water.

5.62 Germanium–Copper–Lanthanum–Selenium

GeSe₂–Cu₂Se–La₂Se₃. The isothermal section of this system at 600°C is presented in Figure 5.19 (Olekseyuk et al. 2009e). The existence of five ternary and one quaternary compound, **CuLa₃GeSe₇**, in this system has been confirmed. Solid

solutions of small extent have been established on the base of binary Cu₂Se, GeSe₂, and La₂Se₃, ternary and quaternary compounds. CuLa₃GeSe₇ forms equilibria with all binary and ternary compounds. Eight fields of three-phase equilibria exist in the system.

CuLa₃GeSe₇ crystallizes in the hexagonal structure with the lattice parameters $a = 1072.5 \pm 0.1$, $c = 613.3 \pm 0.1$ pm, and an energy gap of ~2 eV (Poduska et al. 2002b) [$a = 1062 \pm 1$ and $c = 614.0 \pm 0.5$ pm (Guittard et al. 1968; Guittard and Julien-Pouzol 1970)].

To prepare CuLa₃GeSe₇, GeSe₂ was mixed with La shavings, Cu turnings, and additional Se pellets in a stoichiometric ratio (Poduska et al. 2002b). This mixture was then sealed in an evacuated silica tube, heated for 48 h at 800°C, and cooling to room temperature over 12 h. Investigating other reaction conditions, other samples included iodine in the form of CuI powder in addition to a stoichiometric amount of the reactants; the iodine promoted single-crystal growth, and no iodine was incorporated into the CuLa₃GeSe₇ structure. Products from these reactions were dark orange-red ingots, with dark orange/silver needles on the wall of the silica reaction tube and embedded in the ingot.

CuLa₃GeSe₇ can also be prepared from a mixture of La₂Se₃, Cu₂Se, Ge, and Se, which was pressed and introduced into the silica ampoule and gradually heats up to 800–850°C in about a week (Guittard and Julien-Pouzol 1970). At the end of the preparation, it is necessary to heat the sample to 1200°C to obtain the pure hexagonal phase.

5.63 Germanium–Copper–Cerium–Selenium

The **CuCe₃GeSe₇** quaternary compound, which crystallizes in the hexagonal structure with the lattice parameters $a = 1064.3 \pm 0.1$, $c = 609.73 \pm 0.07$ pm, and a calculated density of 6.159 g·cm⁻³ (Gulay et al. 2006b) [$a = 1056$, $c = 612$ pm (Guittard and Julien-Pouzol 1970)], is formed in the Ge–Cu–Ce–Se system.

For the synthesis of this compound, the calculated amounts of the elements were sealed in an evacuated silica ampoule (Gulay et al. 2006b). The ampoule was heated with a heating rate of 30°C·h⁻¹ to the maximal temperature of 1150°C. It was kept at this temperature for 3 h. After that they were cooled slowly (10°C·h⁻¹) to 600°C and annealed at this temperature for 240 h. After annealing, the ampoule with the sample was quenched in cold water. CuCe₃GeSe₇ can also be prepared in the same way as CuLa₃GeSe₇ was synthesized using Ce₂Se₃ instead of La₂Se₃ (Guittard and Julien-Pouzol 1970).

5.64 Germanium–Copper–Praseodymium–Selenium

GeSe₂–Cu₂Se–Pr₂S₃. The isothermal section of this system at 600°C was constructed by Lychmanyuk et al. (2007a) using the alloys annealed at this temperature for 240 h and quenched in the air and is presented in Figure 5.20. The **CuPr₃GeSe₇** quaternary compound, which crystallizes in the hexagonal structure with the lattice parameters $a = 1055.9 \pm 0.1$, $c = 608.4$

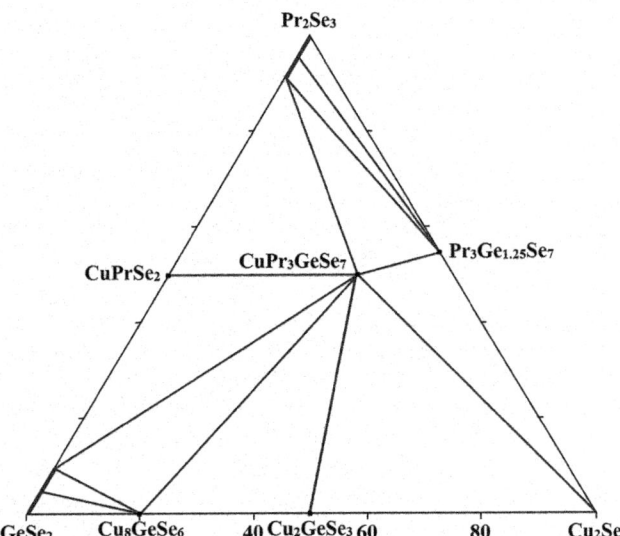

FIGURE 5.20 Isothermal section of the GeSe$_2$–Cu$_2$Se–Pr$_2$Se$_3$ quasiternary system at 600°C. (From Lychmanyuk, O.S., et al., *Nauk. Visnyk Volyns'k. Derzh. Univ. im. Lesi Ukrainky. Ser. Khim. nauky,* (15), 10, 2007.)

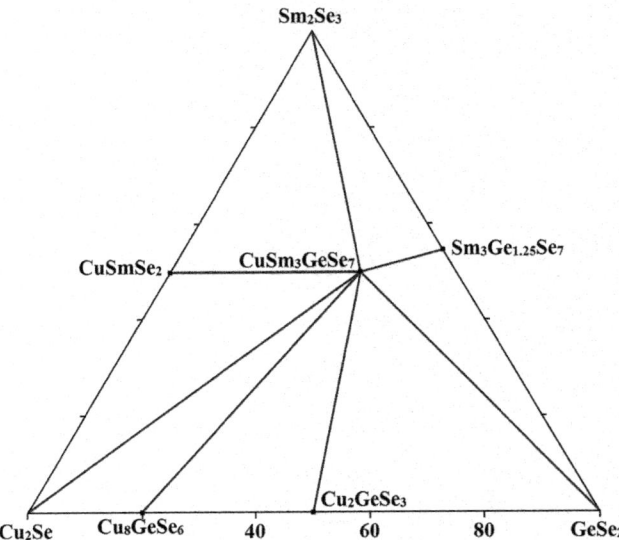

FIGURE 5.21 Isothermal section of the GeSe$_2$–Cu$_2$Se–Sm$_2$Se$_3$ quasiternary system at 600°C. (From Olekseyuk, I.D., et al., *Nauk. Visnyk Volyns'k. Nats. Univ. im. Lesi Ukrainky. Ser. Khim. nauky,* (16), 54, 2010.)

± 0.1 pm, and a calculated density of 6.284 g·cm^{-3} (Gulay et al. 2006b) [a = 1051, c = 610 pm (Guittard and Julien-Pouzol 1970)], is formed in the Ge–Cu–Pr–Se system. This compound was prepared in the same way as CuLa$_3$GeSe$_7$ was synthesized using Pr$_2$Se$_3$ instead of La$_2$Se$_3$ (Guittard and Julien-Pouzol 1970; Gulay et al. 2006b).

5.65 Germanium–Copper–Neodymium–Selenium

The **CuNd$_3$GeSe$_7$** quaternary compound, which crystallizes in the hexagonal structure with the lattice parameters a = 1051.9 ± 0.1, c = 607.07 ± 0.09 pm, and a calculated density of 6.403 g·cm^{-3} (Gulay et al. 2006b) [a = 1046, c = 608 pm (Guittard and Julien-Pouzol 1970)], is formed in the Ge–Cu–Nd–Se system. This compound was prepared in the same way as CuLa$_3$GeSe$_7$ was synthesized using Nd$_2$Se$_3$ instead of La$_2$Se$_3$ (Guittard and Julien-Pouzol 1970; Gulay et al. 2006b).

5.66 Germanium–Copper–Samarium–Selenium

GeSe$_2$–Cu$_2$Se–Sm$_2$Se$_3$. The isothermal section of this system at 600°C is presented in Figure 5.21 (Olekseyuk et al. 2010c*).* The existence of four ternary and one quaternary, **CuSm$_3$GeSe$_7$**, compounds have been confirmed in this system. CuSm$_3$GeSe$_7$ crystallizes in the hexagonal structure with the lattice parameters a = 1040.91 ± 0.09, c = 603.82 ± 0.08 pm, and a calculated density of 6.682 g·cm^{-3} (Strok et al. 2010) [a = 1039 c = 607 pm (Guittard and Julien-Pouzol 1970); a = 1042.16 ± 0.03, c = 604.47 ± 0.03 pm, and a calculated density of 6.657 g·cm^{-3} (Gulay et al. 2006b)]. This compound was prepared in the same way as CuLa$_3$GeSe$_7$ was synthesized using Sm$_2$Se$_3$ instead of La$_2$Se$_3$ (Guittard and Julien-Pouzol 1970; Gulay et al. 2006b; Strok et al. 2010).

5.67 Germanium–Copper–Europium–Selenium

The **Cu$_2$EuGeSe$_4$** quaternary compound, which crystallizes in the orthorhombic structure with the lattice parameters a = 1078.43 ± 0.07, b = 1066.06 ± 0.06, c = 653.60 ± 0.04 pm, a calculated density of 5.900 g·cm^{-3}, and an energy gap of 1.74 eV (Sun et al. 2019), is formed in the Ge–Cu–Eu–Se system. Single crystals of Cu$_2$EuGeSe$_4$ were obtained by high-temperature solid-state reactions with KI as a flux. The reagents were weighted by stoichiometry and the total reagents mass were 500 mg with additional 400 mg KI. Namely, 90.48 mg Cu, 125.28 mg Eu$_2$O$_3$, 224.86 mg Se, 51.68 mg Ge, and 7.70 mg B were used to obtain the title compound. The mixture was ground into a fine powder using an agate mortar and pressed into pellets. Then pellets were loaded into a quartz tube evacuated to 10^{-2} Pa and sealed using O$_2$/H$_2$ flame. The tube was put into a furnace and heated according to a profile as follows. It was heated from room temperature to 300°C by 5 h and kept for 5 h, then heated to 600°C by 5 h and kept for another 5 h, then heated to 800°C by 5 h and kept for 90 h, and finally cooled down slowly to 300°C after 5 days and shut down the furnace. Brown block crystals of Cu$_2$EuGeSe$_4$, which are air and water stable, were obtained. They were washed using hot distilled water and ethanol for several times under ultrasonic wave irradiation to eliminate KI and some impurities.

5.68 Germanium–Copper–Gadolinium–Selenium

The **CuGd$_3$GeSe$_7$** quaternary compound, which crystallizes in the hexagonal structure with the lattice parameters a = 1034.91 ± 0.08, c = 603.95 ± 0.07 pm, and a calculated density of 6.880 g·cm^{-3}, is formed in the Ge–Cu–Gd–Se system (Gulay et al. 2006b). This compound was prepared in the same way as CuLa$_3$GeSe$_7$ was synthesized using Gd$_2$Se$_3$ instead of La$_2$Se$_3$.

5.69 Germanium–Copper–Terbium–Selenium

The **CuTb₃GeSe₇** quaternary compound, which crystallizes in the hexagonal structure with the lattice parameters $a = 1029.40 \pm 0.05$, $c = 603.87 \pm 0.04$ pm, and a calculated density of 6.985 g·cm⁻³, is formed in the Ge–Cu–Tb–Se system (Gulay et al. 2006b). This compound was prepared in the same way as CuLa₃GeSe₇ was synthesized using Tb₂Se₃ instead of La₂Se₃.

5.70 Germanium–Copper–Dysprosium–Selenium

The **CuDy₃GeSe₇** quaternary compound, which crystallizes in the hexagonal structure with the lattice parameters $a = 1024.99 \pm 0.09$, $c = 603.22 \pm 0.08$ pm, and a calculated density of 7.118 g·cm⁻³, respectively, is formed in the Ge–Cu–Dy–Se system (Huang and Ibers 1999). Crystals of this compound were obtained from an initial mixture of Dy (1.0 mM), Ge (1.0 mM), Cu (1.0 mM), Se (4.0 mM), and DyCl₃ (2.5 mM) as a flux. The mixture was loaded under Ar, sealed under 10⁻² Pa in a fused silica tube, heated in a furnace to 850°C at 1°C·min⁻¹, kept at 850°C for 70 h, cooled at a rate of 0.05°C·min⁻¹ to 200°C, and finally cooled to room temperature. The reaction mixture was washed with water and acetone. The major component, CuDy₃GeSe₇, forms hexagonal brown needles.

5.71 Germanium–Copper–Holmium–Selenium

GeSe₂–Cu₂Se–Ho₂Se₃. The isothermal section of this system at 600°C was constructed by Lychmanyuk et al. (2007b) using the alloys annealed at this temperature for 240 h and quenched in cold water and is presented in Figure 5.22. The **CuHo₃GeSe₇** quaternary compound is formed in this system. It crystallizes in the hexagonal structure with the lattice parameters $a = 1019.78$

± 0.07 and $c = 606.12 \pm 0.06$ pm (Gulay et al. 2006b). This compound was prepared in the same way as CuLa₃GeSe₇ was synthesized using Ho₂Se₃ instead of La₂Se₃.

5.72 Germanium–Copper–Tin–Selenium

Cu₂GeSe₃–Cu₂SnSe₃. The phase diagram of this system, constructed through DTA and XRD using the alloys annealed at 480°C, is a peritectic type (Figure 5.23) (Mashadieva et al. 2022). The peritectic contains 60 mol% Cu₂SnSe₃ and crystallizes at 712°C. According to the data of Morihama et al. (2014), the Cu₂Ge$_x$Sn$_{1-x}$Se₃ solid solutions are formed in this quasibinary system. To prepare them, Cu, Sn, Ge, and Se were weighed to give a molar ratio of Cu/Sn/Ge/Se = 2:(1-x):x:3. They were put in a zirconia grinding jar with zirconia balls. A ball-to-powder mass ratio was maintained at 4:1. The milling was conducted in a planetary ball mill at 800 rpm for 20 min in a N₂ gas atmosphere. The mixed powders were post-heated at 600°C for 30 min also in a N₂ gas atmosphere.

GeSe₂–Cu₂Se–SnSe₂. The isothermal section of this quasiternary system at 480°C is shown in Figure 5.24 (Mashadieva et al. 2022). It characterizes by the existence of extensive solid solutions along the Cu₂GeSe₃–Cu₂SnSe₃ quasibinary system. Limited solid solutions based on GeSe₂, SnSe₂, and β-Cu₈GeSe₆ also exist in the system. The liquidus surface of this system (Figure 5.25) includes 6 primary crystallization fields of (β-Cu₂Se), (β-Cu₈GeSe₆), (Cu₂GeSe₃), (Cu₂SnSe₃), (GeSe₂), and (SnSe₂). There is one ternary eutectic and three transition points in this system: E (532°C) – L ↔ (GeSe₂) + (SnSe₂) + (Cu₂GeSe₃); U₁ (572°C) – L + (Cu₂SnSe₃) ↔ (SnSe₂) + (Cu₂GeSe₃); U₂ (692°C) – L + (Cu₂GeSe₃) ↔ (β-Cu₈GeSe₆) + (Cu₂SnSe₃); and U₃ (677°C) – L + (β-Cu₈GeSe₆) ↔ (β-Cu₂Se) + (Cu₂SnSe₃). Some vertical sections of the GeSe₂–Cu₂Se–SnSe₂ system were also constructed.

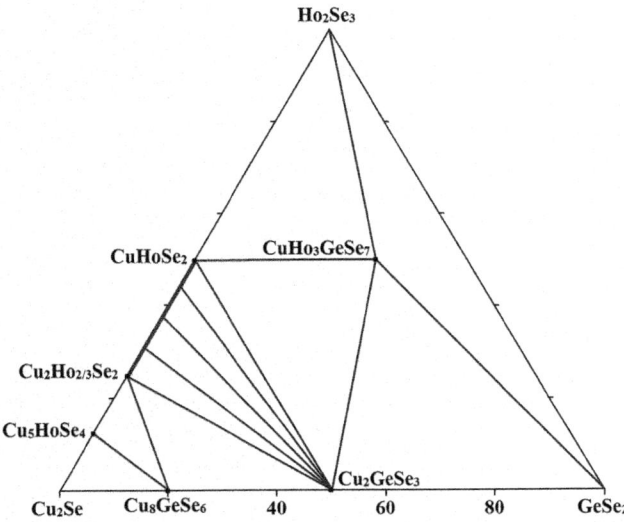

FIGURE 5.22 Isothermal section of the GeSe₂–Cu₂Se–Ho₂Se₃ quasiternary system at 600°C. (From Lychmanyuk, O.S., et al., *Pol. J. Chem.*, **81**(3), 353, 2007.)

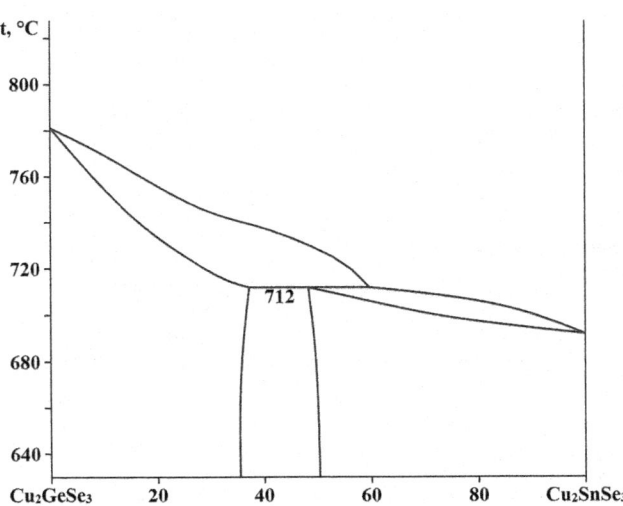

FIGURE 5.23 Phase diagram of the Cu₂GeSe₃–Cu₂SnSe₃ system. (From Mashadieva, L.F., et al., *Russ. J. Inorg. Chem.*, **67**(5), 670, 2022.)

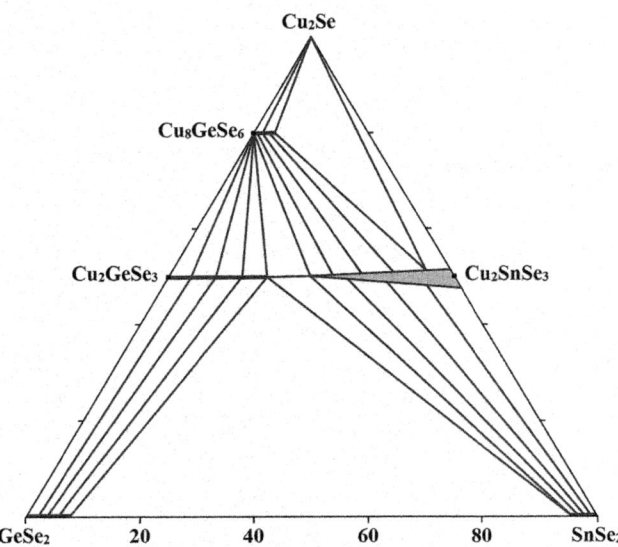

FIGURE 5.24 Isothermal section of the GeSe₂–Cu₂Se–SnSe₂ quasiternary system at 480°C. (From Mashadieva, L.F., et al., *Russ. J. Inorg. Chem.*, **67**(5), 670, 2022.)

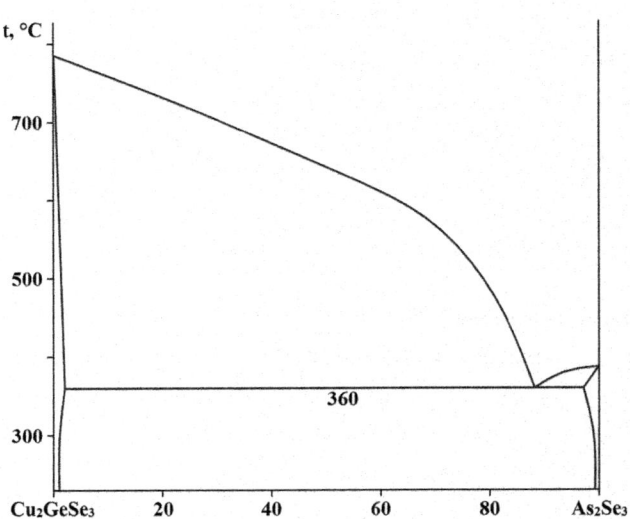

FIGURE 5.26 Phase equilibria in the Cu₂GeSe₃–As₂Se₃ system. (From Klymovych, O., et al., *J. Phase Equilib. Diffus.*, **41**(2), 157, 2020.)

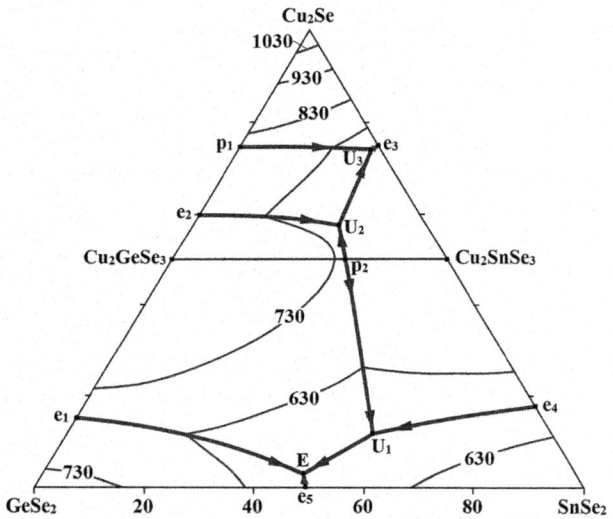

FIGURE 5.25 Liquidus surface of the GeSe₂–Cu₂Se–SnSe₂ quasiternary system. (From Mashadieva, L.F., et al., *Russ. J. Inorg. Chem.*, **67**(5), 670, 2022.)

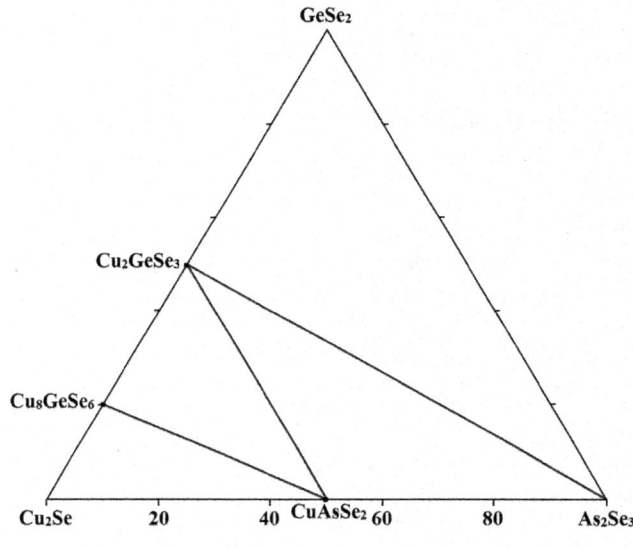

FIGURE 5.27 Isothermal section of the GeSe₂–Cu₂Se–As₂Se₃ quasiternary system at 240°C. (From Klymovych, O., et al., *J. Phase Equilib. Diffus.*, **41**(2), 157, 2020.)

5.73 Germanium–Copper–Phosphorus–Selenium

Cu₂GeSe₃–CuGe₂P₃. In this system, the existence of the solid solutions appears in the region of 9–25 mol% CuGe₂P₃ (Omar 1990).

5.74 Germanium–Copper–Arsenic–Selenium

GeSe–CuAsSe₂. The **CuGeAsS₃** quaternary compound, which characterized by the significant area of homogeneity and crystallizes in the tetragonal structure with the lattice parameter

$a = 394 \pm 1$ and $a = 548 \pm 1$ pm, is formed in this system (Mel'nikova et al. 1995).

Cu₂GeSe₃–As₂Se₃. The phase diagram of this system, constructed through DTA and metallography using the ingots annealed at 240°C for 600°C with the next quenching into 25% aqueous NaCl solution, is a eutectic type (Figure 5.26) (Klymovych et al. 2020a). The eutectic contains 11.7 mol% Cu₂GeSe₃ and crystallizes at 360°C.

GeSe₂–Cu₂Se–As₂Se₃. The isothermal section of this system at 240°C is shown in Figure 5.27 (Klymovych et al. 2020a). The quasiternary system is divided into 4 subsystems. No extensive solid solutions were found, and the solid solubility

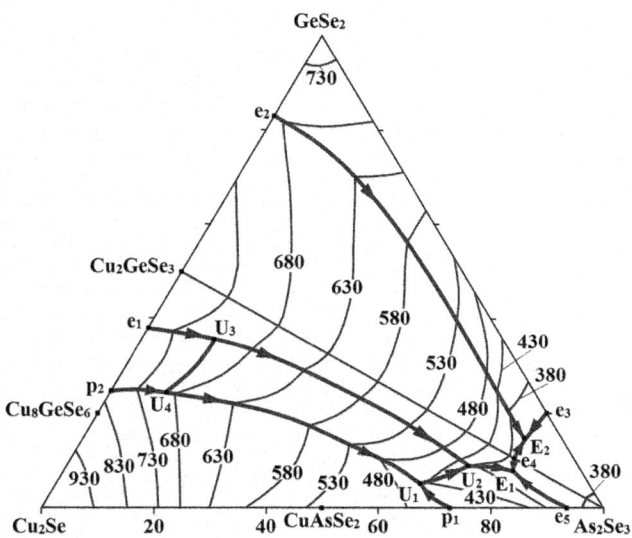

FIGURE 5.28 Liquidus surface of the $GeSe_2$–Cu_2Se–As_2Se_3 quasiternary system. (From Klymovych, O., et al., *J. Phase Equilib. Diffus.*, **41**(2), 157, 2020.)

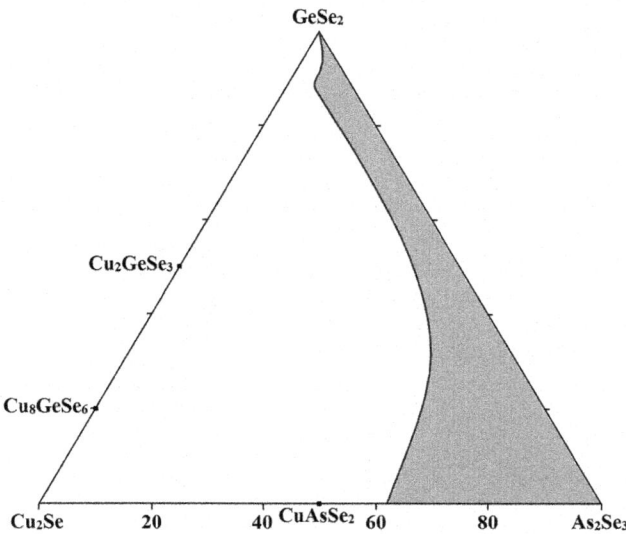

FIGURE 5.29 Glass-forming region in the $GeSe_2$–Cu_2Se–As_2Se_3 quasiternary system. (From Klymovych, O.S., et al., *Nauk. Visnyk Volyns'k. Derzh. Univ. im. Lesi Ukrainky. Ser. Khim. nauky*, (15), 14, 2007.)

based on binary and ternary compounds is less than 2 mol%. Four vertical sections of this system were also constructed.

The liquidus surface of this quasiternary system is given in Figure 5.28 (Klymovych et al. 2020a). It consists of the fields of primary crystallization of solid solutions based on Cu_2Se, $CuAsSe_2$ (the largest field), $GeSe_2$, As_2Se_3, and α- and β-Cu_8GeSe_6. These fields are separated by 11 monovariant curves and 13 nonvariant points between them two ternary eutectics, two transition points, and two ternary peritectics: U_1 (427°C) – L + (Cu_2Se) ↔ ($CuAsSe_2$) + α-Cu_8GeSe_6; U_2 (377°C) – L + α-Cu_8GeSe_6 ↔ ($CuAsSe_2$) + (Cu_2GeSe_3); P_1 (697°C) – L + α-Cu_8GeSe_6 + (Cu_2Se) ↔ β-Cu_8GeSe_6; P_2 (697°C) – L + α-Cu_8GeSe_6 + (Cu_2GeSe_3) ↔ β-Cu_8GeSe_6; E_1 (292°C) – L ↔ (As_2Se_3) + ($CuAsSe_2$) + (Cu_2GeSe_3); and E_2 (267°C) – L ↔ (As_2Se_3) + ($GeSe_2$) + (Cu_2GeSe_3).

The glass-forming region in the $GeSe_2$–Cu_2Se–As_2Se_3 quasiternary system is presented in Figure 5.29 (Klymovych et al. 2007). The tendency of alloys to glass-formation decreases with increasing Cu_2Se content. The glass transition temperature increases with increasing of the $GeSe_2$ content.

5.75 Germanium–Copper–Antimony–Selenium

Cu_2GeSe_3–Sb_2Se_3. The phase diagram of this system, constructed through DTA, XRD, and metallography, is a eutectic type (Figure 5.30) (Ostapyuk et al. 2009b; Ivashchenko et al. 2020). The eutectic contains 70 mol% Sb_2Se_3 and crystallizes at 507°C. The mutual solubility of the components is up to 5 mol%.

Cu_2GeSe_3–$CuSbSe_2$. The phase diagram of this system, constructed through DTA, XRD, and metallography, is also a eutectic type (Figure 5.31) (Ivashchenko et al. 2020). The eutectic contains 10 mol% Cu_2GeSe_3 and crystallizes at 477°C. The mutual solubility of the components is up to 5 mol%.

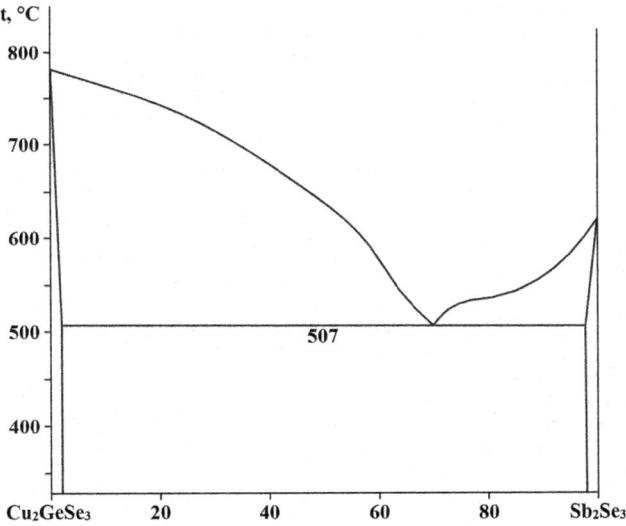

FIGURE 5.30 Phase equilibria in the Cu_2GeSe_3–Sb_2Se_3 system. (From Ivashchenko, I.A., et al., *J. Phase Equilib. Diffus.*, **41**(6), 827, 2020.)

$GeSe_2$–Cu_3SbSe_4. This system is a non-quasibinary section of the Ge–Cu–Sb–Se quaternary system (Ismayilova et al. 2021). Cu_3SbSe_4 dissolves up to 15 mol% $GeSe_2$. It was found that in the interval of 15–58 mol% $GeSe_2$ the alloys contain four phases: (Cu_3SbSe_4), Cu_2GeSe_3, Sb_2Se_3, and Se. The alloys containing 60 mol% $GeSe_2$ are three-phase: Cu_2GeSe_3, Sb_2Se_3, and Se, and the alloys containing more than 60 mol% $GeSe_2$ consist of four phases: Cu_2GeSe_3, Sb_2Se_3, $GeSe_2$, and Se. The ingots for the investigations were annealed at 380°C for 700 h.

$GeSe_2$–Cu_2Se–Sb_2Se_3. The isothermal section of this system at 350°C is shown in Figure 5.32 (Ostapyuk et al. 2009b; Ivashchenko et al. 2020). The quasiternary system is divided into five subsystems. Solid solubility in binary and ternary

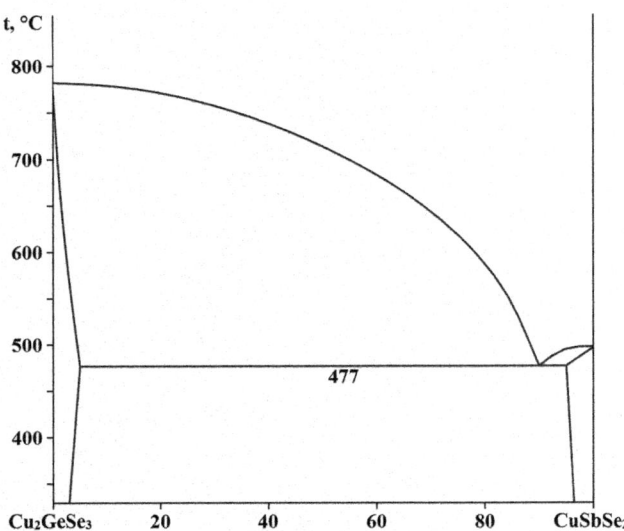

FIGURE 5.31 Phase equilibria in the Cu₂GeSe₃–CuSbSe₂ system. (From Ivashchenko, I.A., et al., *J. Phase Equilib. Diffus.*, **41**(6), 827, 2020.)

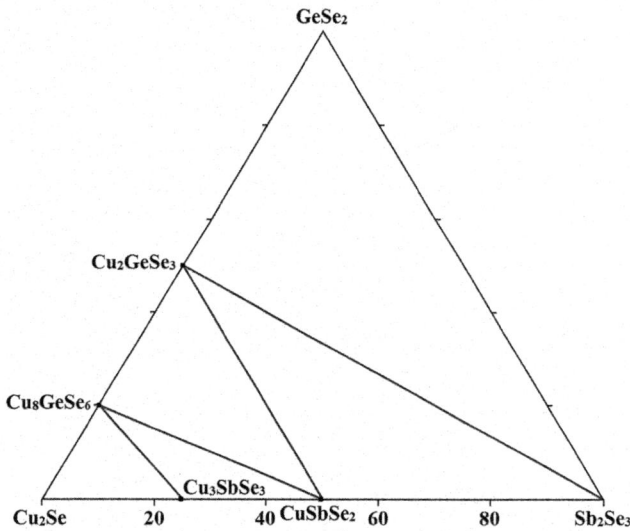

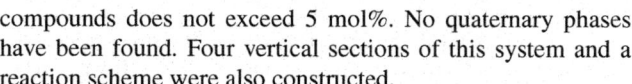

FIGURE 5.32 Isothermal section of the GeSe₂–Cu₂Se–Sb₂Se₃ quaternary system at 350°C. (From Ostapyuk, T.A., et al., *Nauk. Visnyk Volyns'k. Nats. Univ. im. Lesi Ukrainky. Ser. Khim. nauky*, (24), 23, 2009.)

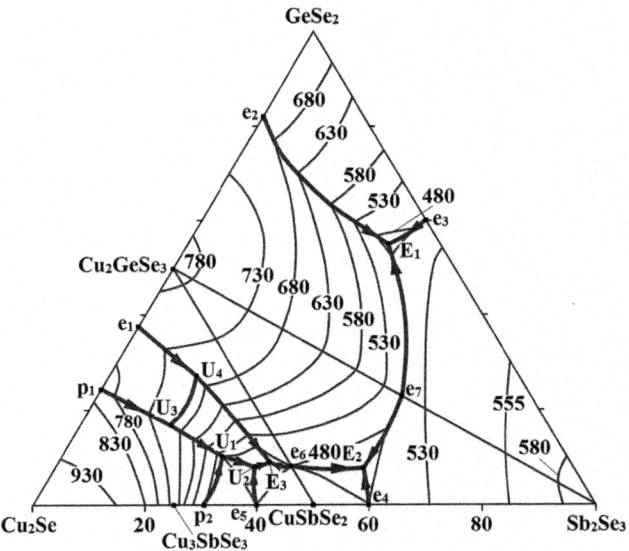

FIGURE 5.33 Liquidus surface of the GeSe₂–Cu₂Se–Sb₂Se₃ quaternary system. (From Ivashchenko, I.A., et al., *J. Phase Equilib. Diffus.*, **41**(6), 827, 2020.)

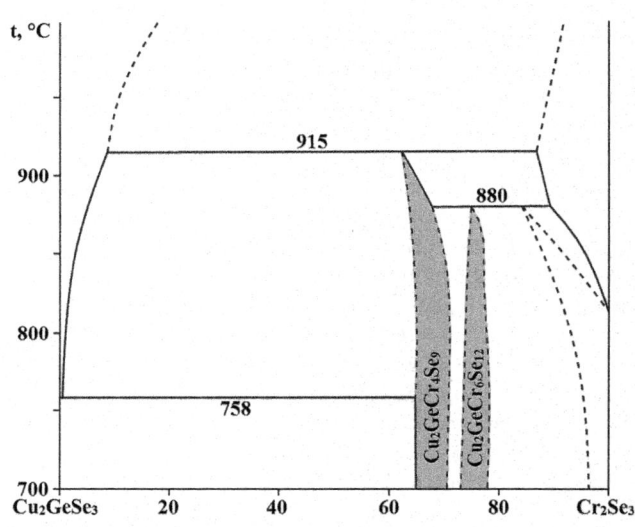

FIGURE 5.34 Phase diagram of the Cu₂GeSe₃–Cr₂Se₃ system. (From Shabunina, G.G., et al., *Russ. J. Inorg. Chem.*, **51**(1), 126, 2006.)

compounds does not exceed 5 mol%. No quaternary phases have been found. Four vertical sections of this system and a reaction scheme were also constructed.

The liquidus surface of this quasiternary system (Figure 5.33) is represented by the fields of the primary crystallization of the phases β-Cu₂Se, β-Cu₈GeSe₆, α-Cu₈GeSe₆, Cu₂GeSe₃, GeSe₂, Sb₂Se₃, CuSbSe₂, and Cu₃SbSe₃ (Ivashchenko et al. 2020). The fields are separated by nine nonvariant points corresponding to the processes in quasibinary systems and seven nonvariant points on the planes in the quasiternary systems. There are the next ternary eutectics, ternary peritectics, and transition points in the GeSe₂–Cu₂Se–Sb₂Se₃ quasiternary system: E₁ (472°C) – L ↔ (Cu₂GeSe₃) + (GeSe₂) + (Sb₂Se₃); E₂ (457°C)

– L ↔ (Cu₂GeSe₃) + (CuSbSe₂) + (Sb₂Se₃); E₃ (447°C) – L ↔ (Cu₂GeSe₃) + (CuSbSe₂) + α-Cu₈GeSe₆; U₁ (477°C) – L + (Cu₂Se) ↔ α-Cu₈GeSe₆ + (Cu₃SbSe₃); U₂ (457°C) – L + (Cu₃SbSe₃) ↔ α-Cu₈GeSe₆ + (CuSbSe₂); P₁ (697°C) – L + α-Cu₈GeSe₆ + (Cu₂Se) ↔ β-Cu₈GeSe₆; and P₂ (697°C) – L + α-Cu₈GeSe₆ + (Cu₂GeSe₃) ↔ β-Cu₈GeSe₆.

5.76 Germanium–Copper–Chromium–Selenium

Cu₂GeSe₃–Cr₂Se₃. The phase diagram of this system, constructed through DTA and XRD, is a eutectic type (Figure 5.34) (Shabunina et al. 2006). The eutectic degenerates from the

Cu$_2$GeSe$_3$ side. Three quaternary compounds, **Cu$_2$GeCr$_4$Se$_9$**, **Cu$_2$GeCr$_6$Se$_{12}$**, and **Cu$_4$Ge$_2$Cr$_6$Se$_{15}$**, are formed in the system. First of them crystallizes in the tetragonal structure with the lattice parameters $a = 1204.3$, $c = 918.0$ pm, and an experimental density of 5.57 g·cm^{-3} (Aminov et al. 2009). This compound crystallizes as a primary phase in the region up to 9 mol % Cr$_2$Se$_3$ in the temperature range 915–758°C. The homogeneity region of Cu$_2$GeCr$_4$Se$_9$ along Cu$_2$GeSe$_3$–Cr$_2$Se$_3$ section is 64.7–70 mol% Cr$_2$Se$_3$; the lattice parameters change from 1206.1 to 1204.1 pm for a, and from 920.9 to 917.9 pm for c at Cr$_2$Se$_3$ content increasing (Aminov et al. 2009) [from 1207.3 at 65 mol% Cr$_2$Se$_3$ to 1198.7 pm at 70 mol% Cr$_2$Se$_3$ for a while the parameter c decreases from 920.4 to 915.6 pm, respectively (Shabunina et al. 2006)].

Cu$_2$GeCr$_6$Se$_{12}$ is formed by the peritectoid reaction at 880°C, has a homogeneity region 73–79 mol% Cr$_2$Se$_3$, and also crystallizes in the tetragonal structure with the lattice parameters $a = 1204.3$, $c = 918.0$ pm, and an experimental density of 5.47 g·cm^{-3} (Shabunina et al. 2006; Aminov et al. 2009).

To prepare both compounds, appropriate powder mixtures of Cu, Ge, Cr, and Se were reacted at 850–900°C for 2 weeks and then at 750°C for a week in silica ampoules evacuated to 10^{-2} Pa (Aminov et al. 2009). Next, the ampoules were opened, and the sintered samples were thoroughly ground, sealed again in ampoules under vacuum, and equilibrated 100°C below the solidus temperature for 2 weeks.

Cu$_4$Ge$_2$Cr$_6$Se$_{15}$ decomposes at 978°C (Babitsyna and Novotortsev 1986). It was synthesized from the elements at 960–980°C for 100 h, with the next annealing at 900°C for 300 h.

5.77 Germanium–Copper–Tungsten–Selenium

The **Cu$_6$GeWSe$_8$** quaternary compound, which crystallizes in the hexagonal structure with the lattice parameters $a = 786.51 \pm 0.05$, $c = 1292.75 \pm 0.06$ pm, a calculated density of 6.087 g·cm^{-3}, and an energy gap of 1.586 eV, is formed in the Ge–Cu–W–Se system (Zhou et al. 2021). The polycrystalline sample of this compound was prepared by solid-state reaction from stoichiometric amounts of Cu, Ge, W, and Se powders. The thoroughly mixed ingredient was pelletized and sealed in an evacuated quartz tube with a vacuum of less than 10^{-4} Pa. The bulk sample for characterization was obtained by sintering at 600°C. The initial heating cycle was performed at 600°C for 3 days, and then the sample was furnace-cooled to room temperature. In order to improve the uniformity of the sample, the as-prepared sample was fully ground, pressed into pellets, and resealed in an evacuated quartz tube, which was annealed at 600°C for 10 days and finally cooled down to room temperature. This compound is stable at ambient temperature and pressure, and without any detectable degeneration by keeping it in the air for over 2 years.

5.78 Germanium–Copper–Iodine–Selenium

The **Cu$_{8-x}$GeSe$_{6-x}$I$_x$** quaternary phase, which melts incongruently at 820°C, has a polymorphic transformation at 567°C for $x = 1$ (Studenyak et al. 2007a), and crystallizes in the cubic structure with the lattice parameter $a = 1037.99 \pm 0.01$ and a calculated density of 6.218 g·cm^{-3} for $x = 0.48$ (Nilges and Pfitzner 2005) and $a = 1037.6 \pm 0.3$ for $x = 0.54$ (Kuhs et al. 1979), is formed in the Ge–Cu–I–Se system. The phase with $x = 0.48$ was obtained by the reaction of the stoichiometric mixture of elements in an evacuated, sealed silica ampoule at 1000°C (Nilges and Pfitzner 2005). After grinding, the phase was annealed at 690°C for 25 days. The phase with $x = 0.54$ was prepared by reacting stoichiometric amounts of the elements in an evacuated, sealed quartz ampoule for about 6 days at temperatures of 600°C to 700°C (Kuhs et al. 1979).

5.79 Germanium–Copper–Manganese–Selenium

The **Cu$_2$MnGeSe$_4$** quaternary compound, which melts incongruently (Caldera et al. 2014) at 812°C (Quintero et al. 2014b) [at 762°C (Schäfer and Nitsche 1977)] and crystallizes in the orthorhombic structure with the lattice parameters $a = 799.58$, $b = 685.74$, $c = 657.17$ pm, and a calculated density of 5.26 g·cm^{-3} (Quintero et al. 2014b) [$a = 797.9$, $b = 686.5$, $c = 655.7$ pm (Schäfer and Nitsche 1974, 1977); $a = 797.7$, $b = 685.4$, $c = 655.2$ pm, and the calculated and experimental densities of 5.29 and 5.27 g·cm^{-3}, respectively (Guen et al. 1979; Guen and Glaunsinger 1980); $a = 798.1$, $b = 687.4$, $c = 655.0$ pm (Lamarche et al. 1991)], is formed in the Ge–Cu–Mn–Se system.

Cu$_2$MnGeSe$_4$ was obtained by the chemical vapor reactions using I$_2$ as the transport agent (Schäfer and Nitsche 1974; Guen et al. 1979; Guen and Glaunsinger 1980) or by the melt and anneal technique using stoichiometric mixtures of the elements (Lamarche et al. 1991; Quintero et al. 2014b). Typically, the components of each 1.0-g sample were sealed under vacuum in a quartz capsule which had previously been coated with carbon in order to prevent reaction of the charge with the quartz (Lamarche et al. 1991). The capsule was then raised to a temperature of 1150°C for approximately an hour, held at this temperature for 0.5–1 h, and then cooled to room temperature. Samples were then annealed at 625°C for a week and quenched to room temperature.

5.80 Germanium–Copper–Iron–Selenium

The **Cu$_2$FeGeSe$_4$** quaternary compound, which melts congruently at 773°C (Quintero et al. 2014b) [at 780°C (Schäfer and Nitsche 1977); at 733°C (Quintero et al. 1998); at 739°C (Quintero et al. 1999)] and crystallizes in the tetragonal structure with the lattice parameters $a = 560.08 \pm 0.04$, $c = 1105.61 \pm 0.12$ pm, and a calculated density of 5.47 g·cm^{-3} (Quintero et al. 2000, 2014b) [$a = 559.0$ and $c = 1107.2$ pm (Schäfer and Nitsche 1974, 1977); $a = 559.06 \pm 1.53$ and $c = 1103.00 \pm 1.68$ pm (Quintero et al. 1998, 1999)], is formed in the Ge–Cu–Fe–Se system.

Cu$_2$FeGeSe$_4$ was produced by the melt and anneal technique (Schäfer and Nitsche 1974; Quintero et al. 1998, 1999, 2014b). The mixture of the elements of the 1-g sample was sealed under vacuum ($\approx$10^{-3} Pa) in a small quartz ampoule, and was heated up to 200°C and kept for about 1–2 h, then the temperature was raised to 500°C at a rate of 40°C·h⁻¹, and held at this temperature for 14 h. After that, the sample was heated from 500°C to 800°C at a rate of 30°C·h⁻¹ and kept at this temperature for another 14 h. Then, the temperature was raised to 1150°C at 60°C·h⁻¹, and the components were melted together at this temperature. The furnace temperature was brought slowly (4°C·h⁻¹) down to 600°C, and the sample was annealed at this temperature for one month. Then, the sample was slowly cooled to room temperature using a rate of about 2°C·h⁻¹.

5.81 Germanium–Copper–Cobalt–Selenium

The **Cu₂CoGeSe₄** quaternary compound, which melts incongruently at 781°C (Quintero et al. 2014b) and crystallizes in the orthorhombic structure with the lattice parameters $a = 558.58 \pm 0.06$, $b = 555.60 \pm 0.06$, $c = 549.35 \pm 0.05$ pm, and a calculated density of 5.594 g·cm⁻³ (Gulay et al. 2004a) [$a = 560.1$, $b = 556.1$, and $c = 550.0$ pm (Schäfer and Nitsche 1974); $a = 559.83$, $b = 555.48$, $c = 549.53$ pm, and a calculated density of 5.58 g·cm⁻³ (Quintero et al. 2014b); $a = 560.25 \pm 0.01$, $b = 557.75 \pm 0.02$, $c = 550.13 \pm 0.01$ pm, and a calculated density of 5.57 g·cm⁻³ (Quintero et al. 2014a)] or in the tetragonal structure with the lattice parameters $a = 556.1$, $c = 550.0$ pm, and the calculated and experimental densities of 5.57 and 5.55 g·cm⁻³ (Guen and Glaunsinger 1980), is formed in the Ge–Cu–Co–Se system.

Cu₂CoGeSe₄ was obtained by the melt and anneal technique using stoichiometric mixtures of the elements (Gulay et al. 2004a; Quintero et al. 2014a,b). Typically, calculated amounts of the elements were sealed in an evacuated quartz container (Gulay et al. 2004a). The synthesis was realized in a shaft furnace. The ampoule with the components was heated at a rate of 30°C·h⁻¹ to 1100°C and kept at this temperature for 4 h. After that, it was cooled slowly to 400°C with a rate of 10°C·h⁻¹. Further annealing at 400°C during 240 h was applied. After annealing, the ampoule with the sample was quenched in cold water. Single crystals of the title compound were grown through chemical transport reactions using iodine as a transport agent (Guen and Glaunsinger 1980).

5.82 Germanium–Copper–Nickel–Selenium

The **Cu₂NiGeSe₄** quaternary compound, which crystallizes in the tetragonal structure with the lattice parameters $a = 558.1$ and $c = 550.0$ pm, is formed in the Ge–Cu–Ni–Se system (Schäfer and Nitsche 1974). It was synthesized from a mixture of the elements in an evacuated quartz ampoule. To accelerate the reaction, about 30 mg of iodine was added. The ampoule was heated to 600°C within 2 h and kept at this temperature for 24 h. Then it was brought to 800–900°C and remained there for 72 h.

5.83 Germanium–Silver–Strontium–Selenium

The **Ag₂SrGeSe₄** quaternary compound, which crystallizes in the orthorhombic structure with the lattice parameters $a = 711.5$, $b = 738.9$, $c = 795.1$ pm, and an energy gap of 0.68 eV, is formed in the Ge–Ag–Sr–Se system (Zhu et al. 2017).

5.84 Germanium–Silver–Barium–Selenium

The **Ag₂BaGeSe₄** quaternary compound, which crystallizes in the orthorhombic structure with the lattice parameters $a = 711.7 \pm 1.6$, $b = 731.6 \pm 1.6$, $c = 831.6 \pm 1.8$ pm, a calculated density of 5.687 g·cm⁻³, and an energy gap of 1.57 eV (Nian et al. 2018) [$a = 705.8 \pm 0.2$, $b = 726.3 \pm 0.2$, $c = 825.3 \pm 0.2$ pm, a calculated density of 5.821 g·cm⁻³, and an energy gap of 0.35 eV (Tampier and Johrendt 2017); calculated value of an energy gap is 0.85 eV (Zhu et al. 2017)], is formed in the Ge–Ag–Ba–Se system.

Conventional high-temperature solid-state method was used to synthesize the title compound (Nian et al. 2018). A stoichiometric mixture of Ag, Ba, Ge, and Se (molar ratio 2:1:1:4) was loaded into a silica tube and sealed under a high vacuum of 10⁻³ Pa. After that, the tube was moved into a furnace. The furnace was programmed by the following steps: heated from room temperature to 600°C in 30 h and kept at this temperature for 40 h; then, heated to 1000°C in 20 h and left at this temperature for 100 h; finally, cooled to 400°C at a rate of 5°C·h⁻¹ and then cooled to room temperature with 10 h. After the reaction, the product was washed with DMF to remove the unreacted sulfur and other byproducts. Many red crystals of Ag₂BaGeSe₄ were gained that can be stable in the air for several months. Since the Ba metal is easily oxidized in the air, an Ar-filled glove box was used to avoid the effects of oxygen and moisture in the preparation processes.

5.85 Germanium–Silver–Zinc–Selenium

AgGeSe₂–ZnSe. According to the data of Parthé et al. (1969), the **AgZn₂GeSe₄** quaternary compound is formed in this system. It crystallizes in the hexagonal structure with the lattice parameters $a = 426.9 \pm 0.5$ and $c = 565.9 \pm 0.5$ pm, but this compound can be one of the solid solution compositions. This phase was obtained by heating the mixture of chemical elements at 600–800°C.

Ag₈GeSe₆–ZnSe. The phase diagram of this system, constructed through DTA, XRD, and metallography, is a eutectic type (Figure 5.35) (Piskach et al. 2006). The eutectic composition and temperature are 44 mol% ZnSe and 860°C, respectively. The solubility of ZnSe in Ag₈GeSe₆ at the eutectic temperature is 42 mol%.

GeSe₂–Ag₂Se–ZnSe. The glass-forming region in this quasiternary system is shown in Figure 5.36 (Olekseyuk et al. 2009b). Such a region in the Ag₂Se–GeSe₂ boundary system is limited to the range of 53–56 mol% GeSe₂. The maximum amount of ZnSe that can be introduced in the glass is 10 mol%. The maximum GeSe₂ content is 63 mol% at 4–6 mol% ZnSe. Characteristic temperatures of the glassy alloys, namely, the glass transition temperature (T_g), the crystallization temperature (T_c), and the melting temperature (T_m) of the crystallized alloys, have been measured. It was determined that T_g of the ZnSe containing glasses lies in a fairly narrow range (253°C ± 8°C), probably because the region of glass existence is rather small. This system was investigated through DTA and XRD.

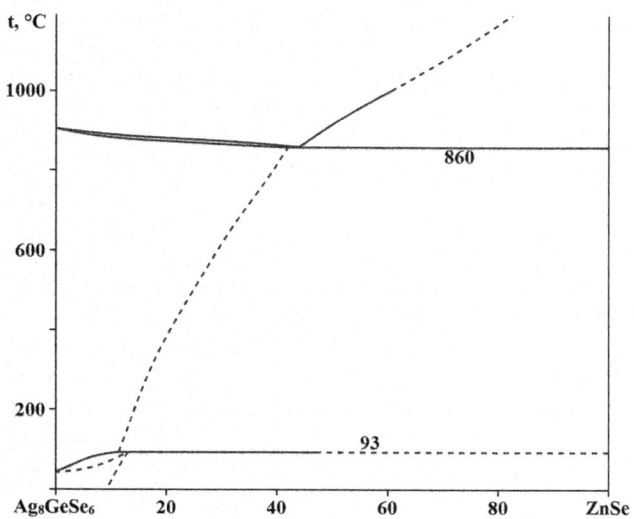

FIGURE 5.35 Phase diagram of the Ag$_8$GeSe$_6$–ZnSe system. (From Piskach, L.V., et al., *J. Alloys Compd.*, **421**(1–2), 98, 2006.)

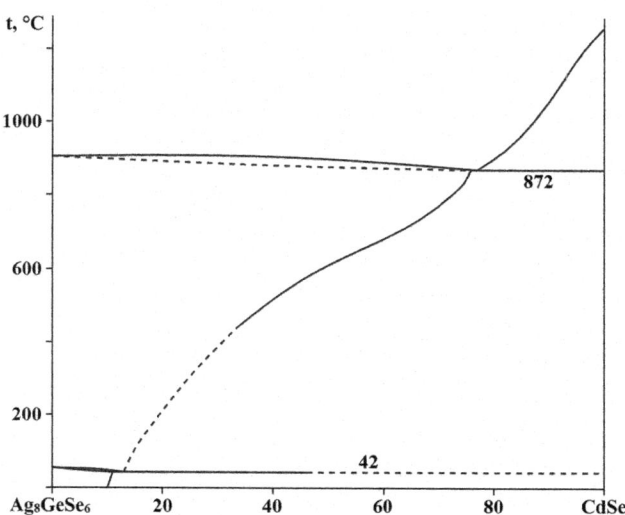

FIGURE 5.37 Phase diagram of the Ag$_8$GeSe$_6$–CdSe system. (From Piskach, L.V., et al., *J. Alloys Compd.*, **421**(1–2), 98, 2006.)

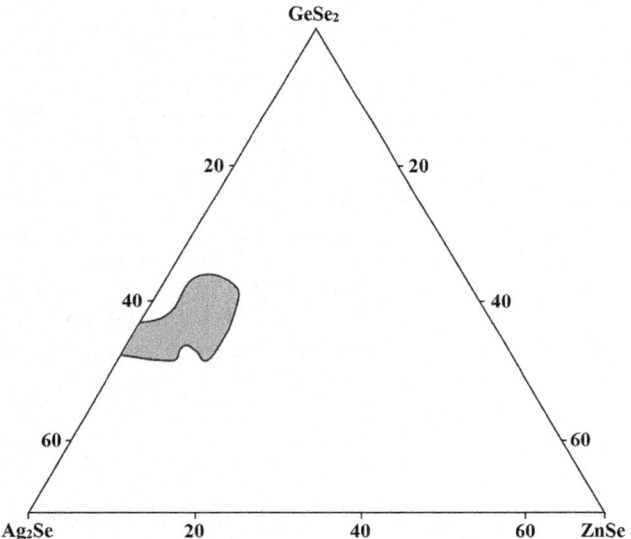

FIGURE 5.36 Glass-forming region in the GeSe$_2$–Ag$_2$Se–ZnSe quasiternary system. (From Olekseyuk, I.D., et al., *Chem. Met. Alloys*, **2**(3–4), 146, 2009.)

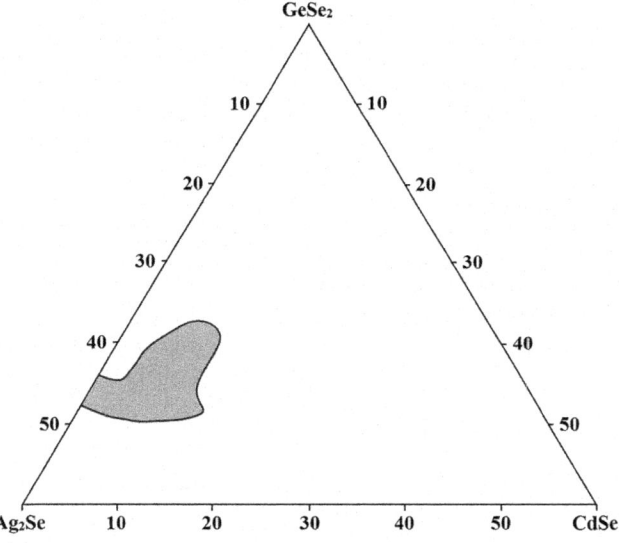

FIGURE 5.38 Glass-forming region in the GeSe$_2$–Ag$_2$Se–CdSe quasiternary system. (From Olekseyuk, I.D., et al., *Chem. Met. Alloys*, **2**(3–4), 146, 2009.)

According to the data of Vassilev et al. (1997), the glass-forming region is at the vertex of Gibbs' concentration triangle, corresponding to GeSe$_2$, and is located on the GeSe$_2$–Ag$_2$Se (from 54 to 100 mol% GeSe$_2$) and GeSe$_2$–ZnSe (from 78 to 100 mol% GeSe$_2$) sides. The maximum solubility of ZnSe in the glassy phases is about 22 mol% at a ratio GeSe$_2$/Ag$_2$Se = 9:1. To obtain the glasses, quartz ampoules with stoichiometric quantities of the starting components, evacuated in advance to a residual pressure of 0.1 Pa and sealed, were heated stepwise to the maximum temperature of the synthesis that does not exceed 1000°C. After vibrational agitation of the melt, the latter is quenched in a mixture of water and ice.

5.86 Germanium–Silver–Cadmium–Selenium

Ag$_8$GeSe$_6$–CdSe. The phase diagram of this system, constructed through DTA, XRD, and metallography, is a eutectic type (Figure 5.37) (Piskach et al. 2006). The eutectic composition and temperature are 77 mol% CdSe and 872°C, respectively. The solubility of CdSe in Ag$_8$GeSe$_6$ at the eutectic temperature is 76 mol%.

GeSe$_2$–Ag$_2$Se–CdSe. The glass-forming region in this system is shown in Figure 5.38 (Olekseyuk et al. 2009b). Such region in the Ag$_2$Se–GeSe$_2$ boundary system is limited to the range 53–56 mol% GeSe₂. The maximum amount of CdSe that can be introduced in the glass is 12 mol%. The maximum GeSe₂ content is 62 mol% at 8 mol% CdSe. Characteristic temperatures of the glassy alloys, namely, the glass transition temperature (T_g), the crystallization temperature (T_c), and the melting temperature (T_m) of the crystallized alloys, have been measured. It was determined that T_g of the CdSe-containing glasses lies in a fairly narrow range (253°C ± 8°C), probably because the region of glass existence is rather small. This system was investigated by DTA and XRD.

5.87 Germanium–Silver–Mercury–Selenium

GeSe₂–Ag₂Se–HgSe. The isothermal section of this quasiternary system at room temperature is shown in Figure 5.39 (Parasyuk et al. 2003a). The formation of five intermediate phases β (Ag₋₇.₁₂₋₆.₃₂Hg₋₀.₄₄₋₀.₈₂GeSe₆), γ (Ag₋₆.₀₈₋₄.₀₀Hg₋₀.₉₆₋₂.₀₀GeSe₆), δ (Ag₃.₄Hg₂.₃GeSe₆), ε (Ag₋₂.₂₄₋₂.₀₀Hg₋₂.₈₈₋₃.₀₀GeSe₆), and **~Ag₁.₄Hg₁.₃GeSe₆** was established.

β-Phase crystallizes in the cubic structure with the lattice parameter a = 1090.26 ± 0.04 pm and a calculated density of 7.3330 ± 0.0008 g·cm⁻³ for the composition Ag₆.₅₀₄Hg₀.₉₁₂GeSe₆ (Parasyuk et al. 2003a). γ-Phase crystallizes in the orthorhombic structure with the lattice parameters a = 770.65 ± 0.09, b = 770.73 ± 0.07, c = 1089.8 ± 0.1 pm, and a calculated density of 7.192 ± 0.003 g·cm⁻³ for the composition **Ag₆HgGeSe₆** (Gulay et al. 2002b); a = 769.61 ± 0.07, b = 769.58 ± 0.06, c = 1088.4 ± 0.1 pm, and a calculated density of 7.172 ± 0.002 g·cm⁻³ for the composition Ag₅.₆Hg₁.₂GeSe₆; a = 767.22 ± 0.08, b = 767.26 ± 0.07, c = 1085.0 ± 0.1 pm, and a calculated density of 7.241 ± 0.002 g·cm⁻³ for the composition Ag₄.₈Hg₁.₆GeSe₆; and a = 764.25 ± 0.07, b = 764.14 ± 0.07, c = 1080.6 ± 0.1 pm, and a calculated density of 7.285 ± 0.002 g·cm⁻³ for the composition **Ag₄Hg₂GeSe₆** (Parasyuk et al. 2003a). Ag₃.₄Hg₂.₃GeSe₆ (δ-phase) crystallizes in the cubic

structure with the lattice parameter a = 1077.67 ± 0.08 pm and a calculated density of 7.500 ± 0.002 g·cm⁻³. **Ag₂HgGeSe₄** that crystallizes in the cubic structure with the lattice parameter a = 1079.6 pm (Haeuseler and Himmrich 1989) was not observed at 25°C by Parasyuk et al. (2003a). Ag₂HgGeSe₄ was obtained by the interaction of chemical elements at 700°C (Haeuseler and Himmrich 1989).

According to the data of Li et al. (2019a), Ag₆HgGeSe₆ (γ phase) melts congruently at 915°C and its energy gap is 0.92 eV. Polycrystalline sample of this phase was obtained by traditional solid-state reactions in a stoichiometric mixture of Ag₂Se (1.5 mM), HgSe (0.5 mM), and GeSe₂ (0.5 mM). The mixture was loaded into a fused silica tube, which was then evacuated to 10⁻³ Pa and sealed. Afterward, the sample was heated in a furnace to 750°C over 15 h and kept at that temperature for 120 h. Then the furnace was shut off. All manipulations were performed in an Ar-filled glove box with H₂O and O₂ contents of less than 0.1 ppm.

The existence of one more quaternary compound, **Ag₂Hg₃GeSe₆**, was found by Moroz et al. (2017c). The linear dependence of the EMF of galvanic cells in the range of 152–182°C was used to calculate the standard thermodynamic values of this compound and ~Ag₁.₄Hg₁.₃GeSe₆: $-\Delta G^0_{f,\,298}$ = 139.4 ± 3.3 kJ·M⁻¹ and $-\Delta H^0_{f,\,298}$ = 247.8 ± 2.9 kJ·M⁻¹ for Ag₂Hg₃GeSe₆ and $-\Delta G^0_{f,\,298}$ = 90.2 ± 2.3 kJ·M⁻¹ and $-\Delta H^0_{f,\,298}$ = 173.1 ± 2.0 kJ·M⁻¹ for ~Ag₁.₄Hg₁.₃GeSe₆.

Glass-forming region in the GeSe₂–Ag₂Se–HgSe quasiternary system crosses the concentration triangle (Figure 5.40) (Olekseyuk et al. 2009b). The minimum content of the glass-forming component is 43 mol% GeSe₂. For GeSe₂ concentration over 80 mol%, the glass-forming region narrows along the HgSe–GeSe₂ side, and the content of the modifier Ag₂Se does not exceed 3 mol%. The characteristic temperatures of the glassy alloys, namely, the glass transition temperature, the crystallization temperature, and the melting temperature of the crystallized alloys, have been measured.

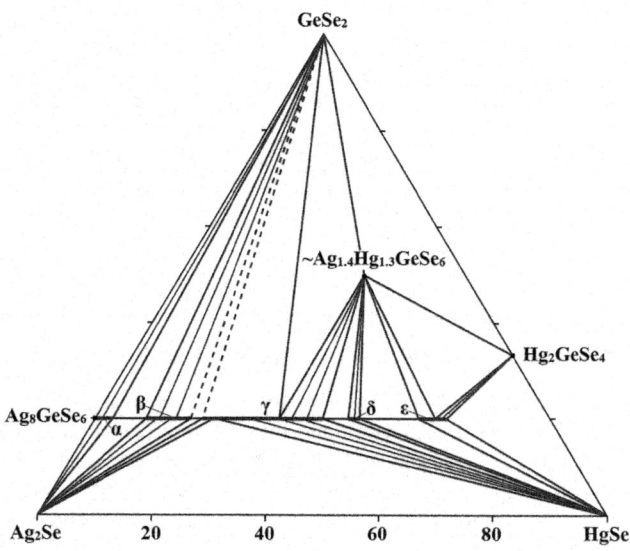

FIGURE 5.39 Isothermal section of the GeSe₂–Ag₂Se–HgSe quasiternary system at room temperature. (From Parasyuk, O.V. et al., *J. Alloys Compd.*, **351**, 135, 2003.)

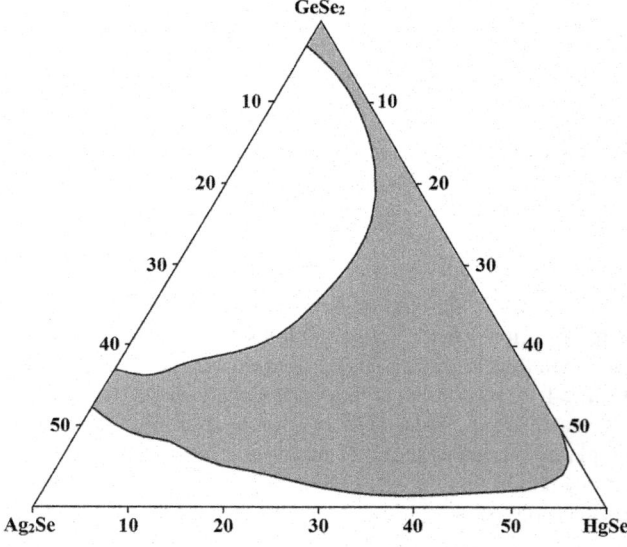

FIGURE 5.40 Glass-forming region in the GeSe₂–Ag₂Se–HgSe quasiternary system. (From Olekseyuk, I.D., et al., *Chem. Met. Alloys*, **2**(3–4), 146, 2009.)

This system was investigated by DTA, XRD, and metallography (Parasyuk et al. 2003a; Olekseyuk et al. 2009b). The alloys were annealed at 400°C for 500 h (Parasyuk et al. 2003a). After that, the ampoules were cooled to 100°C at the rate of 10°C·h⁻¹, and at this temperature, the furnace was turned off, and the ampoules with the samples were cooled to room temperature.

5.88 Germanium–Silver–Aluminum–Selenium

The **AgAlGeS₄** quaternary compound, which crystallizes in the tetragonal structure (Goodchild et al. 1981) with the lattice parameters $a = 587.1$, $c = 1030.4$ pm, and an energy gap of 1.89 eV [2.02 eV (Goodchild et al. 1981)] , is formed in the Ge–Ag–Al–Se system (Hughes et al. 1980; Woolley et al. 1980). This compound was obtained by the melt and anneals technique (Goodchild et al. 1981, 1982; Hughes et al. 1980). To prepare it, the required masses of the elements were sealed under a vacuum in a quartz ampoule (Hughes et al. 1980). The sample so formed was slowly heated (20–40°C·h⁻¹) up to 1050–1100°C, left at that temperature for 5–10 h, slowly cooled (20–40°C·h⁻¹) down to ca. 500°C and then annealed at that temperature for 10–100 h before cooling to room temperature. The heating was carried out in a rocking furnace during the heating up and also during the reaction at high temperature; the sample was rocked to ensure good mixing of the components.

5.89 Germanium–Silver–Gallium–Selenium

GeSe₂–AgGaSe₂. The phase diagram of this system, constructed through DTA, XRD, and metallography, is given in Figure 5.41 (Badikov et al. 1991; Olekseyuk et al. 1996a). According to the diagram given, solid solutions exist in the system at 500°C: a tetragonal solid solution based on AgGaSe₂ up to 53 mol% GeS₃ and a solid solution region of the monoclinic

germanium GeSe₂ up to 4 mol% AgGaSe₂ (Olekseyuk et al. 1996a). In the interval 56–90 mol% GeSe₂, the intermediate solid solution (γ-phase) exists. The compound **AgGaGeSe₄**, which crystallizes in the tetragonal structure and the existence of which was claimed in Hughes et al. (1980) and Woolley et al. (1980) ($a = 582.6$ and $c = 1038.2$ pm), and also in (Goodchild et al. 1981, 1982) ($E_g = 1.85$ eV), was not found (Olekseyuk et al. 1996a). The sample with 50 mol% GeSe₂ at 550°C corresponds to the one composition of the solid solution, which crystallizes in the tetragonal structure ($a = 581.6 \pm 0.2$ and $c = 1037.4 \pm 0.4$ pm). There is no glass-forming region in the GeSe₂–AgGaSe₂ system (Badikov et al. 1991).

The intermediate solid solution (γ-phase) could be expressed by the formula **Ag$_x$Ga$_x$Ge$_{1-x}$Se₂**. They crystallize in the orthorhombic structure with the next lattice parameters: $a = 1249.67 \pm 0.06$, $b = 2390.5 \pm 0.1$, $c = 714.20 \pm 0.03$ pm, a calculated density of 4.9592 ± 0.0007 g·cm⁻³ (Parasyuk et al. 2012; Reshak et al. 2013b) [$a = 1253$, $b = 2391$, and $c = 706$ pm (Panyutin et al. 2008)] for $x = 0.333$ or AgGaGe₂Se₆; $a = 1244.23 \pm 0.06$, $b = 2382.0 \pm 0.1$, $c = 714.03 \pm 0.03$ pm, a calculated density of 4.8349 ± 0.0007 g·cm⁻³ (Parasyuk et al. 2012; Reshak et al. 2013b) [$a = 1241$, $b = 2380$, and $c = 712$ pm (Panyutin et al. 2008)] for $x = 0.250$ or AgGaGe₃Se₈; $a = 1241.26 \pm 0.05$, $b = 2376.89 \pm 0.09$, $c = 713.84 \pm 0.03$ pm, a calculated density of 4.7589 ± 0.0006 g·cm⁻³ (Parasyuk et al. 2012; Reshak et al. 2013b) [$a = 1236$, $b = 2371$, and $c = 721$ pm (Panyutin et al. 2008)] for $x = 0.200$ or AgGaGe₄Se₁₀; $a = 1241.37 \pm 0.01$, $b = 2377.33 \pm 0.01$, $c = 714.38 \pm 0.00$ pm, a calculated density of 4.6886 ± 0.0007 g·cm⁻³ (Huang et al. 2021) [$a = 1232$, $b = 2364$, and $c = 726$ pm (Panyutin et al. 2008); $a = 1241.07 \pm 0.06$, $b = 2376.7 \pm 0.1$, $c = 713.64 \pm 0.03$ pm, a calculated density of 4.6946 ± 0.0007 g·cm⁻³ (Parasyuk et al. 2012; Reshak et al. 2013b)] for $x = 0.167$ or AgGaGe₅Se₁₂. The last phase melts at 707°C and has an energy gap of 2.131 eV (Huang et al. 2021).

Single crystals of the Ag$_x$Ga$_x$Ge$_{1-x}$Se₂ solid solutions have been grown by the two-zone Bridgman–Stockbarger method (Badikov et al. 1991; Panyutin et al. 2008; Parasyuk et al. 2012; Reshak et al. 2013b; Huang et al. 2021).

To prepare the polycrystalline samples of the solid solution with $x = 0.167$ or AgGaGe₅Se₁₂, stoichiometric amounts of liquid Ga, Ag shots, Ge chunks, and Se particles were loaded into a quartz ampoule (Huang et al. 2021). Then the quartz ampoule was evacuated to 10⁻⁴ Pa, sealed, and put into a two-zone tube furnace which was kept a little leaning. One side of the furnace was first heated to 650°C at 100°C·h⁻¹, equilibrated for 20 h, and then heated to 960°C at 90°C·h⁻¹, equilibrated for 24 h, while the other side was kept at 300°C all the time. When this period finished, the temperature of both sides would realize inversion, heating up to 1040°C and 1080°C, respectively. Then at the reaction side, a temperature oscillation process between 960°C to 1040°C was carried out, and a mechanical oscillation process was undergone through the furnace rotation. At last, the furnace was cooled down to room temperature at 30°C·h⁻¹, and the synthetic product was taken out of the quartz ampoule. This solid solution was also synthesized by vapor transport in sealed quartz ampoules from the elements, and single crystals were grown by the horizontal gradient freeze techniques in a transparent furnace (Schunemann et al. 2006).

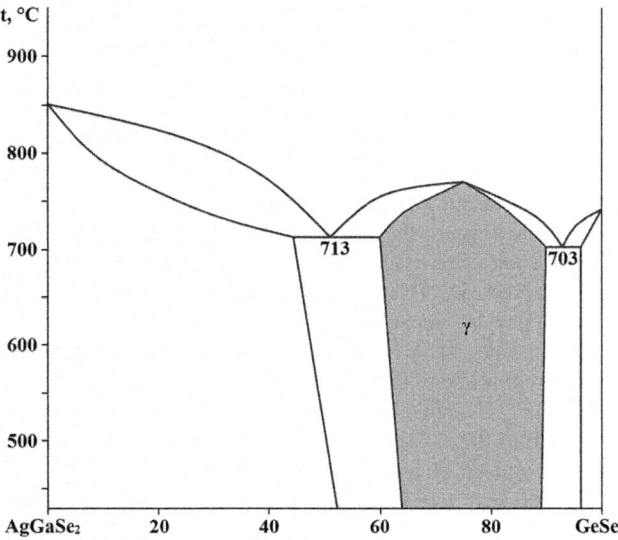

FIGURE 5.41 Phase diagram of the GeSe₂–AgGaSe₂ system. (From Olekseyuk, I.D., et al., *J. Alloys Compd.*, **241**(1–2), 187, 1996.)

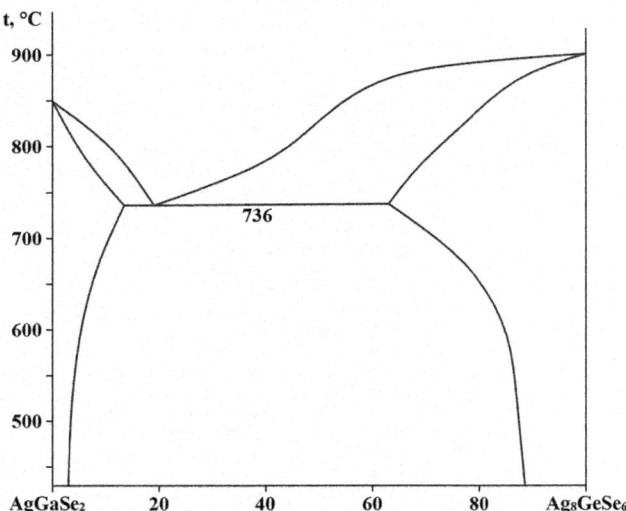

FIGURE 5.42 Phase diagram of the Ag₈GeSe₆–AgGaSe₂ system. (From Olekseyuk, I.D., et al., *J. Alloys Compd.*, **260**(1–2), 111, 1997.)

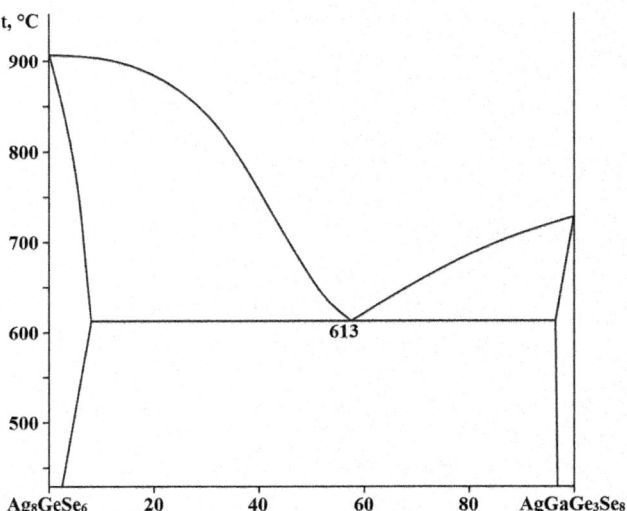

FIGURE 5.43 Phase diagram of the Ag₈GeSe₆–AgGaGe₃Se₈ system. (From Olekseyuk, I.D., et al., *J. Alloys Compd.*, **260**(1–2), 111, 1997.)

Ag₈GeSe₆–AgGaSe₂. The phase diagram of this system, constructed through DTA, XRD, metallography, and measuring of the microhardness, is a eutectic type (Figure 5.42) (Olekseyuk et al. 1997a). The eutectic point coordinates are 19 mol% Ag₈GeSe₆ and 736°C. AgGaSe₂ dissolves 14 mol% Ag₈GeSe₆ at the eutectic temperature and 3 mol% Ag₈GeSe₆ at 450°C. Solid solutions based on Ag₈GeSe₆ were found in the range 63–100 mol% Ag₈GeSe₆ at 736°C and 89–100 mol% Ag₈GeSe₆ at 450°C. The ingots for the investigations were annealed at 450°C for 250 h and quenched in cold water.

Ag₈GeSe₆–Ag₉GaSe₆. The phase diagram of this system, constructed through DTA, XRD, metallography, and measuring of the microhardness, is of type I according to Rozeboom's classification; a continuous solid solution range is formed in the system (Olekseyuk et al. 1997a). The ingots for the investigations were annealed at 450°C for 250 h and quenched in cold water.

Ag₈GeSe₆–AgGaGe₃Se₈. The phase diagram of this system, constructed through DTA, XRD, metallography, and measuring of the microhardness, is a eutectic type (Figure 5.43) (Olekseyuk et al. 1997a). The composition of the eutectic point is 57.5 mol% AgGaGe₃Se₈, and the eutectic temperature is 613°C. The solubility in both components is lower than 5 mol% at 450°C. The ingots for the investigations were annealed at 450°C for 250 h and quenched in cold water.

AgGaGe₃Se₈–Ga₂Se₃. The phase diagram of this system, constructed through DTA, XRD, metallography, and measuring of the microhardness, is a eutectic type (Figure 5.44) (Olekseyuk et al. 1997a). The eutectic point coordinates are 58 mol% AgGaGe₃Se₈ and 690°C. The solid solubility limit based on Ga₂Se₃ is 12 mol% AgGaGe₃Se₈ at the eutectic temperature. The solubility decreases with decreasing temperature to 2 mol% AgGaGe₃Se₈ at 450°C. In AgGaGe₃Se₈ the solubility is lower than 5 mol% Ga₂Se₃. The ingots for the investigations were annealed at 450°C for 250 h and quenched in cold water.

Olekseyuk et al. (1997a) constructed four more vertical sections, a reaction scheme, an isothermal section at 450°C, and a liquidus surface of the **GeSe₂–Ag₂Se–Ga₂Se₃** quasiternary

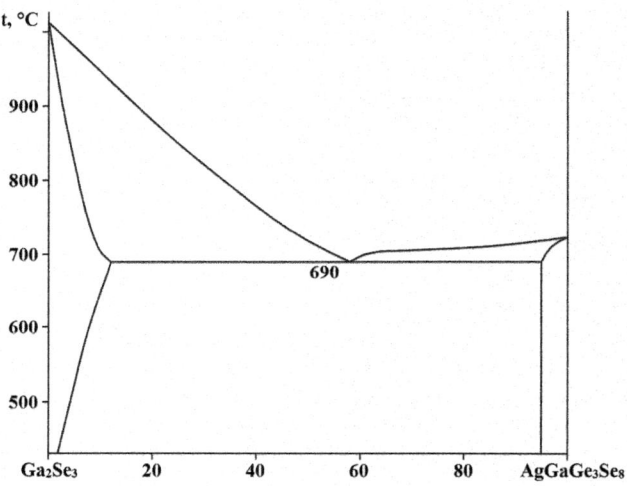

FIGURE 5.44 Phase diagram of the AgGaGe₃Se₈–Ga₂Se₃ system. (From Olekseyuk, I.D., et al., *J. Alloys Compd.*, **260**(1–2), 111, 1997.)

system. The isothermal section at 450°C reveals four ternary regions, which are narrow because of larger homogeneity regions of phases in the quasiternary system.

The liquidus surface of the GeSe₂–Ag₂Se–Ga₂Se₃ quasiternary system (Figure 5.45) contains six fields of the primary crystallization of (Ga₂Se₃), (Ag₈GeSe₆), (Ag₉GaSe₆), (Ag₂Se), (AgGaGe₃Se₈), and (GeSe₂) solid solutions (Olekseyuk et al. 2001b). They are separated by 14 monovariant lines and 15 invariant points, including four ternary eutectics. The glass regions were found in this system at the quenching of the melts from 1030°C. The first region extends inside the concentration triangle from the GeSe₂–Ga₂Se₃ system. Maximum quantities of Ga₂Se₃ and Ag₂Se reach up to 28 and 10 mol%, respectively. The second region is located near the eutectic of the GeSe₂–Ag₂Se system and contains up to 45 mol% Ag₂Se, 6 mol% Ga₂Se₃, and 56 mol% GeSe₂. The glass transition temperature is within the interval 340–385°C for the first interval and 220–240°C for the second one.

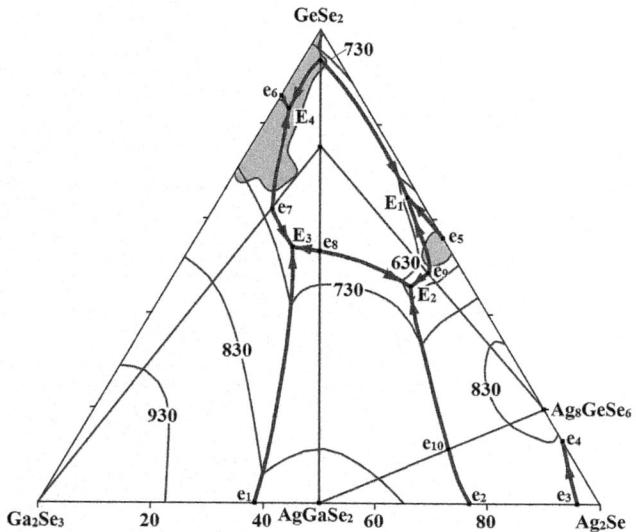

FIGURE 5.45 Liquidus surface and the glass-forming regions (shaded areas) of the $GeSe_2$–Ag_2Se–Ga_2Se_3 quasiternary system. (From Olekseyuk, I.D., et al., *Ukr. khim. zhurn.*, **67**(5–6), 68, 2001.)

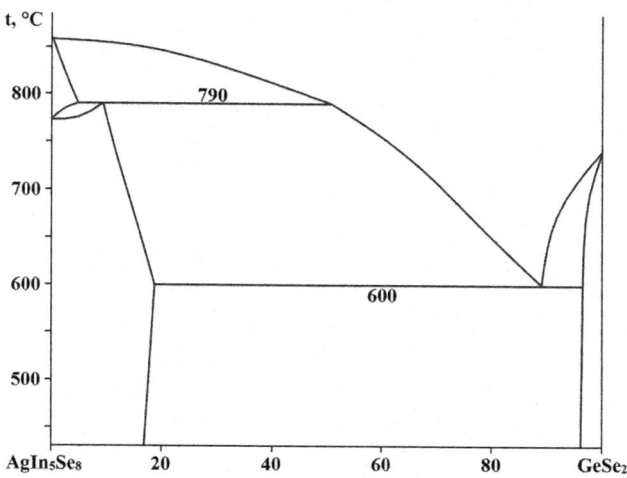

FIGURE 5.47 Phase diagram of the $GeSe_2$–$AgIn_5Se_8$ system. (From Olekseyuk, I.D., et al., *Zhurn. neorgan. khimii*, **42**(8), 1392, 1997.)

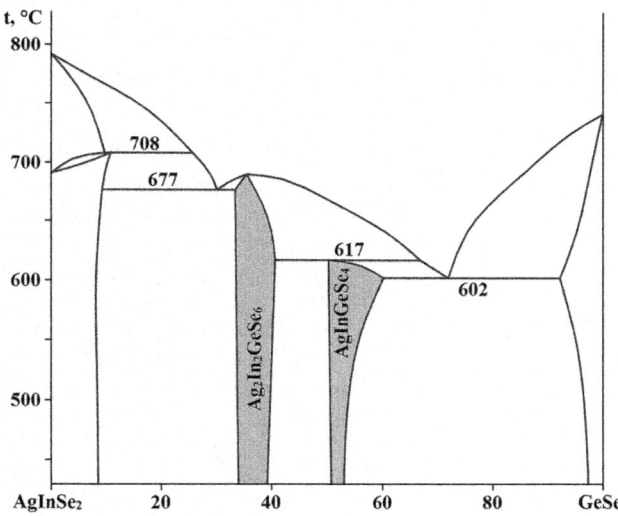

FIGURE 5.46 Phase diagram of the $GeSe_2$–$AgInSe_2$ system. (From Olekseyuk, I.D., and Kryrhovets', O.V., *Ukr. khim. zhurn.*, **67**(7–8), 81, 2001.)

5.90 Germanium–Silver–Indium–Selenium

$GeSe_2$–$AgInSe_2$. The phase diagram of this system, constructed through DTA, XRD, metallography, and measuring of the microhardness is presented in Figure 5.46 (Krykhovets et al. 2001; Olekseyuk and Krykhovets 2001a). The eutectics contain 30 and 72 mol% $GeSe_2$ and crystallize at 677°C and 602°C, respectively. At 480°C, the solubility of $GeSe_2$ in $AgInSe_2$ and $AgInSe_2$ in $GeSe_2$ reaches 8 and 5 mol%, respectively. Two quaternary compounds, **$AgInGeSe_4$** and **$Ag_2In_2GeSe_6$**, are formed in this system. The ingots for the investigations were annealed at 450°C for 250 h and quenched in cold water.

$AgInGeSe_4$ melts incongruently at 617°C and crystallizes in the orthorhombic structure with the lattice parameters $a = 1621.4 \pm 0.7$, $b = 863.9 \pm 0.2$, $c = 650.1 \pm 0.3$ pm (Krykhovets et al. 2001; Olekseyuk and Krykhovets 2001a) [$a = 1750.0$, $b = 635.3$, $c = 423.3$ pm (Olekseyuk et al. 1999a); in the tetragonal structure with the lattice parameters $a = 575.9$, $c = 1081.0$ pm (Hughes et al. 1980; Woolley et al. 1980), and an energy gap of 1.58 eV (Woolley et al. 1980; Goodchild et al. 1982); $a = 576.76 \pm 0.01$, $c = 1082.72 \pm 0.03$ pm, and a calculated density of 5.394 ± 0.001 g·cm^{-3} for $Ag_{0.735}InGeSe_4$ (Krykhovets et al. 2002)]. This compound was obtained by the melt and annealing technique (Hughes et al. 1980; Goodchild et al. 1981, 1982; Krykhovets et al. 2002).

$Ag_2In_2GeSe_6$ melts congruently at 690°C and crystallizes in the monoclinic structure with the lattice parameters $a = 1269.2 \pm 0.3$, $b = 749.21 \pm 0.01$, $c = 1264.4 \pm 0.3$ pm, $\beta = 109.50 \pm 0.03°$ (Krykhovets et al. 1999, 2001; Olekseyuk et al. 1999a; Olekseyuk and Krykhovets 2001a). The synthesis of the title compound was performed from elements in an evacuated quartz container by a direct one-temperature method (Krykhovets et al. 1999, 2002). The elements were mixed by vibration at a maximum temperature of 980°C. The sample was then cooled at 480°C and annealed at this temperature for 500 h.

$GeSe_2$–$AgIn_5Se_8$. The phase diagram of this system, constructed through DTA, XRD, and metallography, is given in Figure 5.47 (Olekseyuk et al. 1997b). The eutectic contains 90 mol% $GeSe_2$ and crystallizes at 600°C. The solubility of $GeSe_2$ in β-$AgIn_5Se_8$ is 7.5 mol% at 790°C, and the solubility of $GeSe_2$ in α $AgIn_5Se_8$ reaches 18 mol% at 600°C and practically does not change up to 430°C. The solubility of α-$AgIn_5Se_8$ in $GeSe_2$ is not higher than 5 mol%. The ingots for the investigations were annealed at 480°C for 250 h and quenched in a cold aqueous NaCl solution.

Ag_8GeSe_6–$Ag_2In_2GeSe_6$. The phase diagram of this system, constructed through DTA, XRD, metallography, and

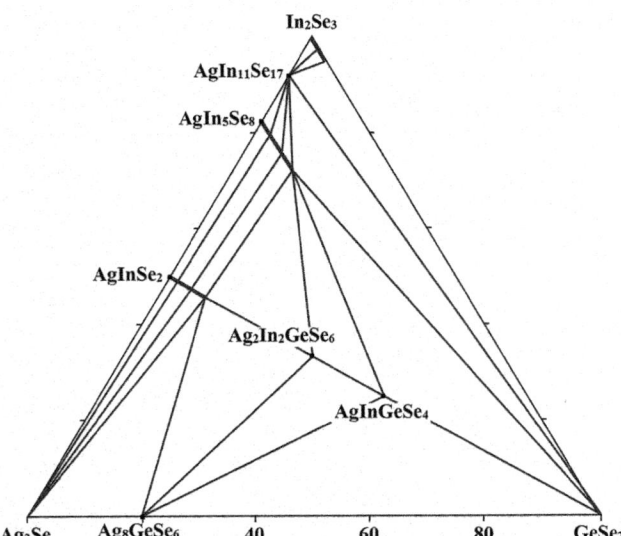

FIGURE 5.48 Isothermal section of the $GeSe_2$–Ag_2Se–In_2Se_3 quasiternary system at 480°C. (From Olekseyuk, I.D., et al., *Pol. J. Chem.*, **73**(3), 431, 1999.)

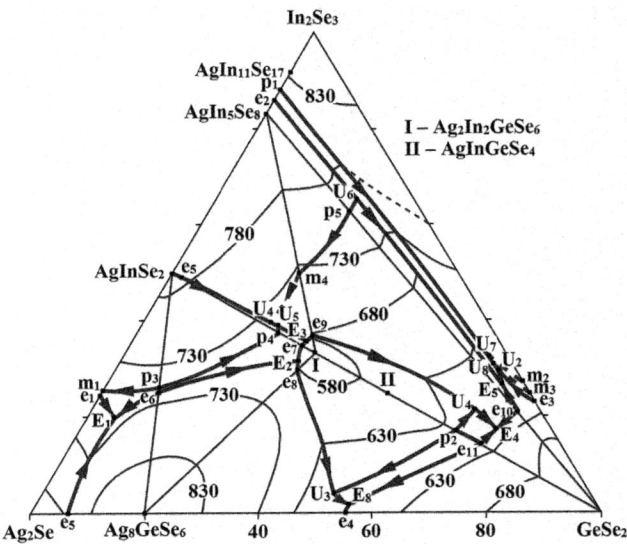

FIGURE 5.49 Liquidus surface of the $GeSe_2$–Ag_2Se–In_2Se_3 quasiternary system. (From Krykhovets, A.V., et al., *Zhurn. neorgan. khimii*, **46**(7), 1180, 2001.)

measuring of the microhardness, is a eutectic type (Krykhovets et al. 2001). The eutectic contains 7 mol% Ag_8GeSe_6 and crystallizes at 677°C. The solubility of $Ag_2In_2GeSe_6$ in β-Ag_8GeSe_6 is 84 mol% and decreases to 25 mol% at 480°C. The ingots for the investigations were annealed at 450°C for 250 h and quenched in cold water.

$Ag_2In_2GeSe_6$–$AgIn_5Se_8$. The phase diagram of this system, constructed through DTA, XRD, metallography, and measuring of the microhardness, is a eutectic type (Krykhovets et al. 2001). The eutectic contains 6 mol% $AgIn_5Se_8$ and crystallizes at 675°C. The maximum solubility of $Ag_2In_2GeSe_6$ in β-$AgIn_5Se_8$ reaches 28 mol% at 720°C. For α-$AgIn_5Se_8$, the solid solution region decreases from 15 mol% $Ag_2In_2GeSe_6$ at 720°C to 10 mol% $Ag_2In_2GeSe_6$ at 480°C. Polymorphic transformation of the solid solution based on $AgIn_5Se_8$ takes place at 720°C. The ingots for the investigations were annealed at 450°C for 250 h and quenched in cold water.

$GeSe_2$–Ag_2Se–In_2Se_3. The isothermal section of this quasiternary system at 480°C is shown in Figure 5.48 (Olekseyuk et al. 1999a; Krykhovets et al. 2001). Solid solutions based on $AgIn_5Se_8$ and In_2Se_3 reach 18 and 5 mol% $GeSe_2$, respectively.

The liquidus surface of this system (Figure 5.49) consists of 13 fields of primary crystallization, which are divided by 31 monovariant lines and 14 nonvariant equilibria, nine of which are four-phase (Krykhovets et al. 2001): E_1 (647°C) – L ↔ β-Ag_2Se + (α-$AgInSe_2$) + (β-Ag_8GeSe_6); E_2 (657°C) – L ↔ (α-$AgInSe_2$) + $Ag_2In_2GeSe_6$ + (β-Ag_8GeSe_6); E_3 (600°C) – L ↔ (α-$AgInSe_2$) + (α-$AgIn_5Se_8$) + $Ag_2In_2GeSe_6$; E_4 (570°C) – L ↔ (α-$AgIn_5Se_8$) + (AgInGeSe₄) + ($GeSe_2$); E_5 (580°C) – L ↔ (α-$AgIn_5Se_8$) + ($GeSe_2$) + $AgIn_{11}Se_{17}$; E_6 (550°C) – L ↔ (β-Ag_8GeSe_6) + (AgInGeSe₄) + ($GeSe_2$); U_1 (595°C) – L + $Ag_2In_2GeSe_6$ ↔ (AgInGeSe₄) + (α-$AgIn_5Se_8$); U_2 (602°C) – L + (β-In_2Se_3) ↔ $AgIn_{11}Se_{17}$ + ($GeSe_2$); and U_3 (560°C) – L + $Ag_2In_2GeSe_6$ ↔ (AgInGeSe₄) + (β-Ag_8GeSe_6).

5.91 Germanium–Silver–Lanthanum–Selenium

The $Ag_{0.94}La_3GeSe_7$ quaternary compound, which crystallizes in the hexagonal structure with the lattice parameters $a = 1080.6 \pm 0.1$ and $c = 611.1 \pm 0.1$ pm (Daszkiewicz et al. 2009b), is formed in the Ge–Ag–La–Se system. This compound was prepared by fusion of the elements (molar ratio La/Ag/Ge/Se = 3:1:1:7) into an evacuated silica ampoule in a tube. The ampoule was heated with a heating rate of 30°C·h⁻¹ to the maximal temperature, 1150°C, and was kept at this temperature for 3 h. Afterwards, it was cooled slowly (10°C·h⁻¹) to 600°C and annealed at this temperature for 240 h. After annealing the ampoule was quenched in cold water.

5.92 Germanium–Silver–Cerium–Selenium

The $Ag_{0.89}Ce_3GeSe_7$ quaternary compound, which crystallizes in the hexagonal structure with the lattice parameters $a = 1071.1 \pm 0.1$ and $c = 607.19 \pm 0.05$ pm (Daszkiewicz et al. 2009b), is formed in the Ge–Ag–Ce–Se system. This compound was prepared in the same way as $Ag_{0.94}La_3GeSe_7$ was synthesized using Ce instead of La.

5.93 Germanium–Silver–Praseodymium–Selenium

The $Ag_{0.89}Pr_3GeSe_7$ quaternary compound, which crystallizes in the hexagonal structure with the lattice parameters $a = 1061.9 \pm 0.1$ and $c = 607.0 \pm 0.1$ pm (Daszkiewicz et al. 2009b), is formed in the Ge–Ag–Pr–Se system. This compound was prepared in the same way as $Ag_{0.94}La_3GeSe_7$ was synthesized using Pr instead of La.

5.94 Germanium–Silver–Neodymium–Selenium

The **$Ag_{0.87}Nd_3GeSe_7$** quaternary compound, which crystallizes in the hexagonal structure with the lattice parameters a = 1062.1 ± 0.1 and c = 605.31 ± 0.06 pm (Daszkiewicz et al. 2009b), is formed in the Ge–Ag–Nd–Se system. This compound was prepared in the same way as $Ag_{0.94}La_3GeSe_7$ was synthesized using Nd instead of La.

5.95 Germanium–Silver–Samarium–Selenium

The **$Ag_{0.81}Sm_3GeSe_7$** quaternary compound, which crystallizes in the hexagonal structure with the lattice parameters a = 1047.9 ± 0.1 and c = 602.58 ± 0.08 pm (Daszkiewicz et al. 2009b), is formed in the Ge–Ag–Sm–Se system. This compound was prepared in the same way as $Ag_{0.94}La_3GeSe_7$ was synthesized using Sm instead of La.

5.96 Germanium–Silver–Gadolinium–Selenium

The **$Ag_{0.85}Gd_3GeSe_7$** quaternary compound, which crystallizes in the hexagonal structure with the lattice parameters a = 1042.8 ± 0.1 and c = 603.21 ± 0.05 pm (Daszkiewicz et al. 2009b), is formed in the Ge–Ag–Gd–Se system. This compound was prepared in the same way as $Ag_{0.94}La_3GeSe_7$ was synthesized using Gd instead of La.

5.97 Germanium–Silver–Terbium–Selenium

The **$Ag_{0.82}Tb_3GeSe_7$** quaternary compound, which crystallizes in the hexagonal structure with the lattice parameters a = 1036.4 ± 0.1 and c = 603.99 ± 0.06 pm (Daszkiewicz et al. 2009b), is formed in the Ge–Ag–Tb–Se system. This compound was prepared in the same way as $Ag_{0.94}La_3GeSe_7$ was synthesized using Tb instead of La.

5.98 Germanium–Silver–Dysprosium–Selenium

The **$Ag_{0.72}Dy_3GeSe_7$** quaternary compound, which crystallizes in the hexagonal structure with the lattice parameters a = 1027.1 ± 0.1 and c = 605.7 ± 0.1 pm (Daszkiewicz et al. 2009b), is formed in the Ge–Ag–Dy–Se system. This compound was prepared in the same way as $Ag_{0.94}La_3GeSe_7$ was synthesized using Dy instead of La.

5.99 Germanium–Silver–Tin–Selenium

Ag_8GeSe_6–Ag_8SnSe_6. The phase diagram of this system, constructed through DTA, XRD, and EMF, is presented in Figure 5.50 (Alieva et al. 2014). Continuous series of solid solutions are formed: low-temperature solid solution crystallizes in the orthorhombic structure, while the high-temperature one crystallizes in the cubic structure. The liquidus and solidus curves

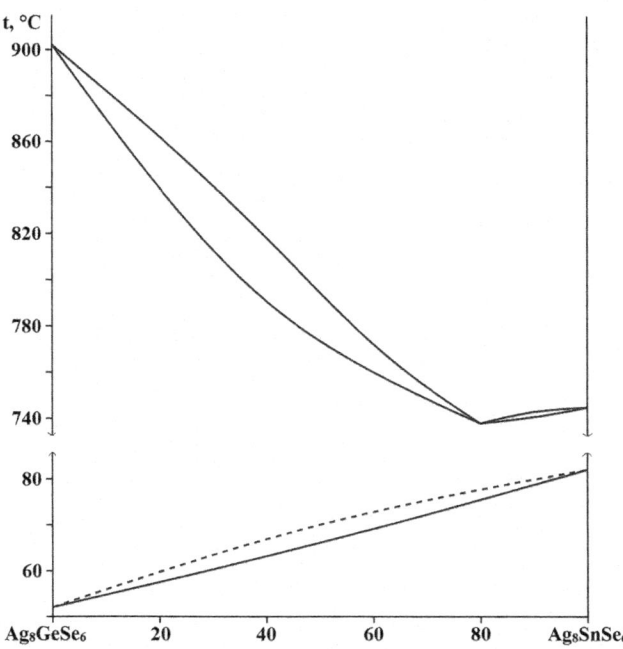

FIGURE 5.50 Phase diagram of the Ag_8GeS_6–Ag_8SnS_6 system. (From Alieva, Z.M., et al., *Inorg. Mater.*, **50**(10), 981, 2014.)

have a minimum at 737°C and 80 mol% Ag_8SnSe_6. The temperature of polymorphic transformation of solid solutions decreases monotonically with increasing Ag_8GeSe_6 content. The two vertical sections, an isothermal section at room temperature, and the liquidus surface of the Ag_8GeS_6–Ag_8SnSe_6–Ag_2Se quasiternary system were also constructed.

$GeSe_2$–Ag_2Se–$SnSe_2$. The isothermal sections of this quasiternary system at room temperature, 530°C, and 630°C are shown in Figure 5.51 (Alverdiev et al. 2017a). At room temperature, solid solutions between the low-temperature phases of Ag_8GeS_6 and Ag_8SnS_6 form two phase fields with Ag_2Se, ($GeSe_2$), and ($SnSe_2$). At 530°C, a part of the section is occupied by the liquid phase, and this part increases at 630°C. Some vertical sections, as well as the liquidus section of the $GeSe_2$–Ag_2Se–$SnSe_2$ system, were also constructed. The liquidus surface (Figure 5.52) consists of four fields of the primary crystallization of Ag_2Se, ($GeSe_2$), ($SnSe_2$), and $Ag_8Ge_xSn_{1-x}S_6$ solid solution. Ternary eutectic contains 21 mol% $GeSe_2$, 34 mol% Ag_2Se, and 45 mol% $SnSe_2$ and crystallizes at 412°C. Using EMF measurements, the thermodynamic properties of low- and high-temperature modifications of the $Ag_8Ge_xSn_{1-x}S_6$ solid solution were determined. The alloys for the investigations were annealed at 630°C for 500 h, followed by cooling in the furnace.

5.100 Germanium–Silver–Lead–Selenium

$PbSe$–Ag_8GeSe_6. The phase diagram of this system, constructed through DTA, XRD, and metallography, is a eutectic type (Figure 5.53) (Bilousov et al. 2006). The eutectic contains 75 mol% PbSe and crystallizes at 673°C. The eutectoid

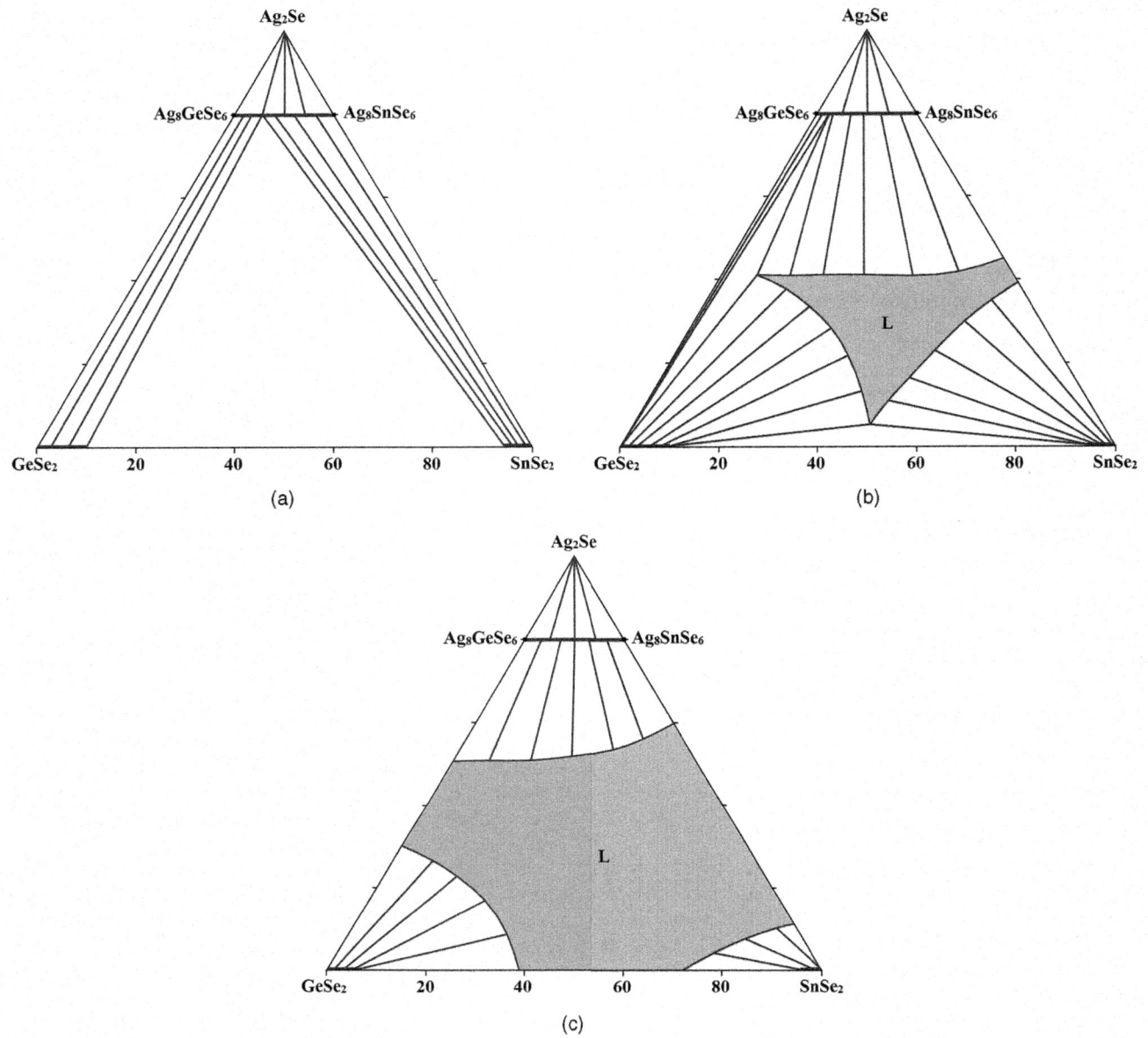

FIGURE 5.51 Isothermal section of the GeSe₂–Ag₂Se–SnSe₂ quasiternary system at (a) room temperature, (b) 530°C and (c) 630°C. (From Alverdiev, I.J., et al., *Inorg. Mater.*, **53**(8), 786, 2017.)

transformation γ-Ag₈GeSe₆ ↔ β-Ag₈GeSe₆ + PbSe takes place at 42°C. The solubility of PbSe in Ag₈GeSe₆ at the eutectic temperature is 8 mol% and <5 mol% at room temperature.

The **Ag₀.₅Pb₁.₇₅GeSe₄** quaternary compound, which crystallizes in the cubic structure with the lattice parameter $a = 1459.49 \pm 0.19$ pm, a calculated density of 6.879 g·cm⁻³, and an energy gap of 1.51 eV, is formed in the Ge–Ag–Pb–Se system (Iyer et al. 2004). Black, shiny crystals of this compound were prepared by a stoichiometric combination of the elements. The reactants were loaded in a fused silica tube and flame-sealed under vacuum (< 0.1 Pa). The tube was then put in a furnace, and the temperature was ramped up to 250°C where it was maintained for 6 h. The tube was then heated to 650°C, kept at that temperature for 48 h, and then cooled to 250°C at a rate of 5°C·h⁻¹ followed by rapid cooling to room temperature.

5.101 Germanium–Silver–Arsenic–Selenium

GeSe₂–AgAsSe₂. The phase diagram of this system, constructed through DTA, XRD, and metallography, is a eutectic type (Figure 5.54) (Ivashchenko et al. 2022). The eutectic contains 7.7 mol% GeSe₂ and crystallizes at 397°C. The eutectoid reaction β-AgAsSe₂ ↔ α-AgAsSe₂ + GeSe₂ takes place at 337°C. No significant solid solubility based on the starting compounds was observed.

Ag₈GeSe₆–AgAsSe₂. The phase diagram of this system, constructed through DTA, XRD, and metallography, is also a eutectic type (Figure 5.55) (Ivashchenko et al. 2022). The eutectic contains 4 mol% Ag₈GeSe₆ and crystallizes at 387°C. The eutectoid reaction β-AgAsSe₂ ↔ α-AgAsSe₂ + Ag₈GeSe₆

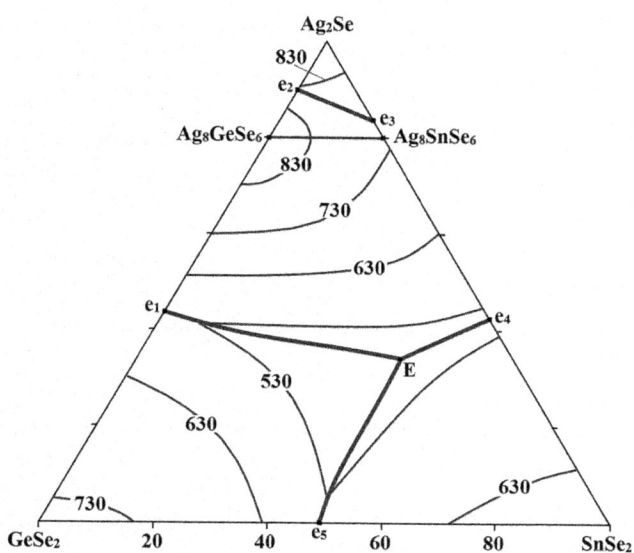

FIGURE 5.52 Liquidus surface of the GeSe$_2$–Ag$_2$Se–SnSe$_2$ quasiternary system. (From Alverdiev, I.J., et al., *Inorg. Mater.*, **53**(8), 786, 2017.)

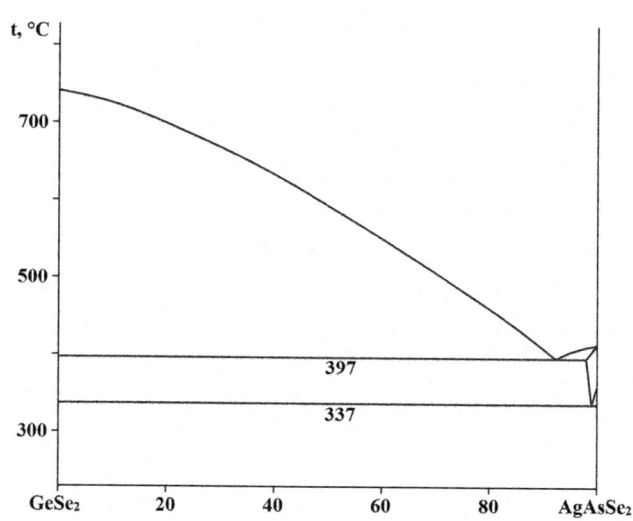

FIGURE 5.54 Phase diagram of the GeSe$_2$–AgAsSe$_2$ system. (From Ivashchenko, I.A., et al., *J. Phase Equilib. Diffus.*, **43**(4), 483, 2022.)

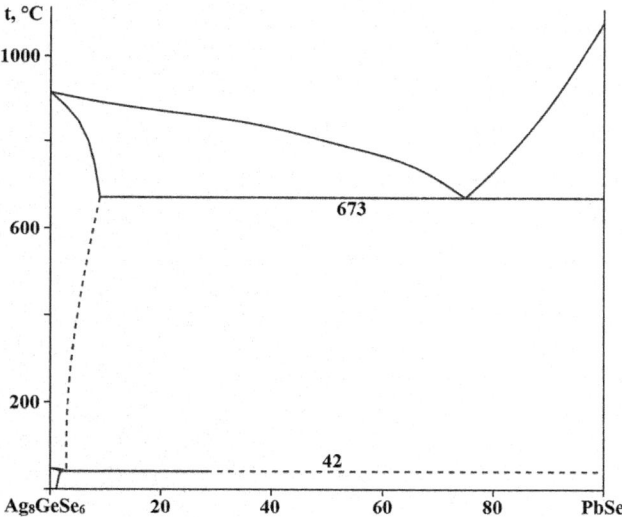

FIGURE 5.53 Phase diagram of the PbSe–Ag$_8$GeSe$_6$ system. (From Bilousov, O.V., et al., *Nauk. Visnyk Volyns'k. Derzh. Univ. im. Lesi Ukrainky. Ser. Khim. nauky*, (4), 128, 2006.)

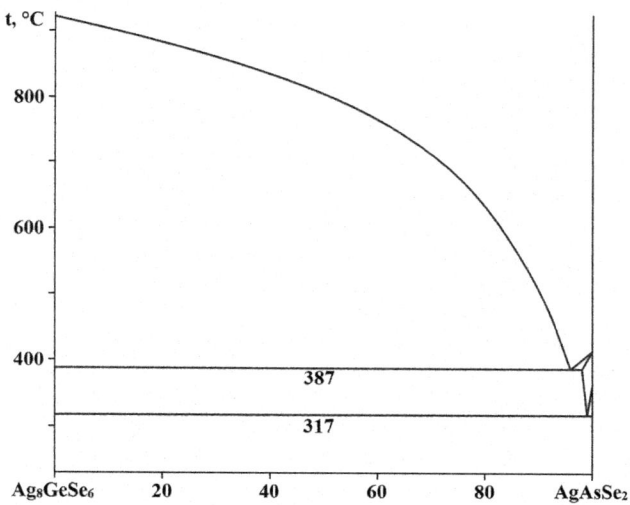

FIGURE 5.55 Phase diagram of the Ag$_8$GeSe$_6$–AgAsSe$_2$ system. (From Ivashchenko, I.A., et al., *J. Phase Equilib. Diffus.*, **43**(4), 483, 2022.)

takes place at 317°C. No significant solid solubility based on the starting compounds was observed.

GeSe$_2$–Ag$_2$Se–As$_2$Se$_3$. The isothermal section of this quasiternary system at 240°C is divided by the GeSe$_2$–AgAs$_3$Se$_5$, GeSe$_2$–AgAsSe$_2$, AgAsSe$_2$–Ag$_8$GeSe$_6$, and Ag$_8$GeSe$_6$–Ag$_3$AsSe$_3$ quasibinary systems into five subsystems (Ivashchenko et al. 2022). No significant solid solubility ranges were found for the binary compounds or intermediate ternary compounds. The system's liquidus surface (Figure 5.56) consists of the fields of the primary crystallization of Ag$_2$Se, Ag$_3$AsSe$_3$, β-AgAsSe$_2$, AgAs$_3$Se$_5$, As$_2$Se$_3$, Ag$_8$GeSe$_6$, and GeSe$_2$. The largest fields correspond to the primary crystallization of Ag$_8$GeSe$_6$ and GeSe$_2$. The fields of the primary

crystallization are separated by 13 monovariant curves of binary eutectic and ternary peritectic reactions and 14 invariant points. The system undergoes 5 quaternary invariant reactions: E$_1$ (367°C) – L ↔ Ag$_8$GeSe$_6$ + Ag$_3$AsSe$_3$ + β-AgAsSe$_2$; E$_2$ (357°C) – L ↔ Ag$_8$GeSe$_6$ + GeSe$_2$ + β-AgAsSe$_2$; E$_3$ (327°C) – L ↔ GeSe$_2$ + As$_2$Se$_3$ + AgAs$_3$Se$_5$; U$_1$ (377°C) – L + Ag$_2$Se ↔ Ag$_8$GeSe$_6$ + Ag$_3$AsSe$_3$; and U$_2$ (361°C) – L + β-AgAsSe$_2$ ↔ AgAs$_3$Se$_5$ + GeSe$_2$.

Some vertical sections of this quasiternary system were also constructed. The alloys for the investigations were annealed at 240°C for 600 h, after which they were quenched into 25 mass% aqueous NaCl solution (Ivashchenko et al. 2022).

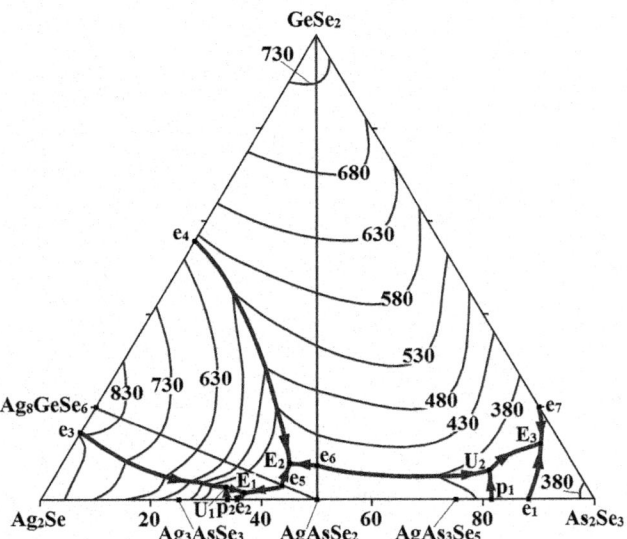

FIGURE 5.56 Liquidus surface of the GeSe₂–Ag₂Se–As₂Se₃ quasiternary system. (From Ivashchenko, I.A., et al., *J. Phase Equilib. Diffus.*, **43**(4), 483, 2022.)

5.102 Germanium–Silver–Antimony–Selenium

GeSe₂–Ag₂Se–Sb₂Se₃. The isothermal section of this quasiternary system at 300°C was constructed by Olekseyuk et al. (2009d). It was shown that two quasibinary systems, GeSe₂–AgSbSe₂ and Ag₈GeSe₆–AgSbSe₂, divide it for three three-phase regions: Ag₂Se–Ag₈GeSe₆–AgSbSe₂, Ag₈GeSe₆–AgSbSe₂–GeSe₂, and AgSbSe₂–GeSe₂–Sb₂Se₃. The ingots for the investigations were annealed at 300°C for 600 h, followed by quenching in cold water.

The glass-forming region in this quasiternary system is located on the GeSe₂–Sb₂Se₃ side of the Gibbs triangle in the interval 0–70 mol% Sb₂Se₃ (Figure 5.57) (Vassilev et al. 1998). The glass transition temperature varies in the range of 125–370°C. A decrease in the values with increasing Sb₂Se₃ concentration at a constant content of Ag₂Se was observed. The glasses were prepared by direct synthesis of the binary compounds up to 850–1000°C, depending on the Ag₂Se content. The mixtures were melted in evacuated (~10⁻³ Pa) quartz ampoules in a rotary furnace for several hours and then quenched in ice water.

5.103 Germanium–Silver–Tellurium–Selenium

Ag₈GeSe₆–Ag₈GeTe₆. According to the results of DTA and XRD, high- and low-temperature modifications of these compounds form a continuous series of solid solutions (Mikolaychuk and Moroz 1985b). The alloys, containing 5–20 and 40–100 mol% Ag₈GeTe₆, are homogeneous at room temperature. There is a small gentle minimum at 98 mol% Ag₈GeTe₆ and 640°C on the phase diagram.

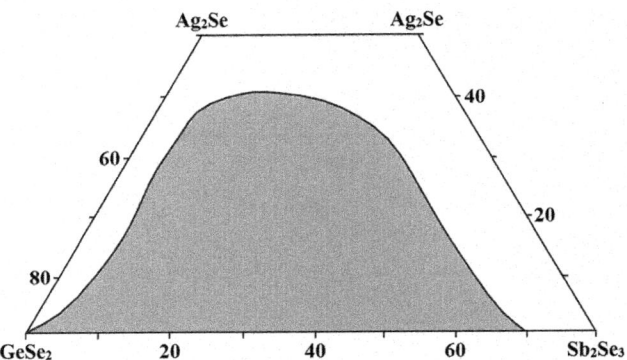

FIGURE 5.57 Glass-forming region in the GeSe₂–Ag₂Se–Sb₂Se₃ quasiternary system. (From Vassilev, V.S., et al., *J. Mater. Sci. Lett.*, **17**(23), 2007, 1998.)

5.104 Germanium–Silver–Bromine–Selenium

The **Ag₈₋ₓGeSe₆₋ₓBrₓ** quaternary phase, which crystallizes in the cubic structure with the lattice parameter $a = 1095.5 \pm 0.3$ pm for an unspecified x value, is formed in the Ge–Ag–Br–Se system (Kuhs et al. 1979). This phase was prepared by reacting stoichiometric amounts of the elements in an evacuated, sealed quartz ampoule for about 6 days at temperatures of 600°C to 700°C.

5.105 Germanium–Silver–Iodine–Selenium

The **Ag₈₋ₓGeSe₆₋ₓIₓ** quaternary phase, which melts at 870°C and crystallizes in the cubic structure with the lattice parameter $a = 1098.64 \pm 0.07$ pm and an experimental density of 6.740 g·cm⁻³ for $x = 1$ (Aldon et al. 2001) [$a = 1101.6 \pm 0.3$ pm for an unspecified x value (Kuhs et al. 1979); $a = 1100.19$ for $x = 1$ (Laqibi et al. 1987); $a = 1092.1 \pm 0.2$ pm and a calculated density of 6.5095 g·cm⁻³ at 173 K, and $a = 1097.2 \pm 0.3$ pm and the calculated and experimental densities of 6.4188 and 6.50 ± 0.04 g·cm⁻³ at room temperature for $x = 0.31$ (Belin et al. 2001)], is formed in the Ge–Ag–I–Se system. According to Albert et al. (2008), the lattice parameter of this compound with $x = 1$ has the following values depending on temperature: $a = 1095.58 \pm 0.05$ pm and a calculated density of 6.823 g·cm⁻³ at 15 K; $a = 1096.00 \pm 0.07$ pm and 6.815 g·cm⁻³ at 50 K; $a = 1096.83 \pm 0.10$ pm and 6.799 g·cm⁻³ at 100 K; $a = 1099.13 \pm 0.04$ pm and 6.757 g·cm⁻³ at 150 K; $a = 1099.55 \pm 0.07$ pm and 6.749 g·cm⁻³ at 200 K; $a = 1102.56 \pm 0.04$ pm and 6.694 g·cm⁻³ at 250 K; $a = 1103.4 \pm 0.5$ pm and 6.679 g·cm⁻³ at 325 K; $a = 1104.1 \pm 0.4$ pm and 6.666 g·cm⁻³ at 375 K; $a = 1105.2 \pm 1.0$ pm and 6.646 g·cm⁻³ at 425 K; and $a = 1106.3 \pm 0.5$ pm and 6.626 g·cm⁻³ at 475 K.

Depending on the pressure, this phase is characterized by the following values of the lattice parameter (Albert et al. 2009): $a = 1102.0 \pm 0.3$ pm at ambient pressure; $a = 1102.10 \pm 0.01$ pm, and a calculated density of 6.69 g·cm⁻³ at 0.1 MPa; $a = 1095.78 \pm 0.01$ pm and a calculated density of 6.82 g·cm⁻³ at 0.57 GPa; $a = 1089.44 \pm 0.01$ pm and a calculated density of 6.94 g·cm⁻³ at 1.1 GPa; $a = 1086.0 \pm 1.0$ pm at 1.42 GPa; $a = 1084.4 \pm 0.3$

pm at 1.68 GPa; $a = 1083.6 \pm 0.7$ pm at 1.72 GPa; $a = 1084.40 \pm 0.01$ pm and a calculated density of 7.04 g·cm^{-3} at 1.8 GPa; $a = 1079.6 \pm 0.3$ pm at 2.34 GPa; $a = 1073.9 \pm 0.9$ pm at 3.11 GPa; $a = 1065.6 \pm 1.3$ pm at 4.56 GPa; $a = 1060.9 \pm 0.4$ pm at 5.76 GPa; $a = 1051.5 \pm 0.8$ pm at 7.73 GPa; $a = 1050.2 \pm 0.6$ pm at 8.06 GPa; $a = 1047.9 \pm 0.5$ pm at 8.77 GPa; and $a = 1045.6 \pm 0.5$ pm at 10.54 GPa.

The Ag$_7$GeSe$_5$I samples were synthesized as follows (Zerouale et al. 1988; Belin et al. 2001; Albert et al. 2008, 2009): a mixture of Ag$_2$Se, GeSe$_2$, and AgI in the proportions $3:1:(1+\delta)$ (excess AgI insured a complete reaction between the components) was placed in an evacuated fused silica tube. The tube was then heated up to 700°C at the rate of 8°C·h^{-1} and maintained at this temperature for 1 day. It was then cooled down to 580°C, held at this temperature for a week, and finally slowly cooled down to room temperature. The excess AgI was removed by washing the obtained powder in a KI solution. The thermal stability of this phase extends up to 350°C.

This phase with $x = 1$ was also obtained by solid-state reactions as follows (Laqibi et al. 1987): 3Ag$_2$Se + GeSe$_2$ + AgI = Ag$_7$GeSe$_5$I. Suitable quantities of the sulfides and AgI were finely ground and mixed thoroughly and then treated in a quartz ampoule sealed under vacuum (10^{-2} Pa). The preparation was heated to 700°C over a period of 3 days, kept at this temperature for 48 h and then maintained at 580°C for a week.

This phase with unspecified x value was prepared by reacting stoichiometric amounts of the elements in an evacuated, sealed quartz ampoule for about 6 days at temperatures of 600°C to 700°C (Kuhs et al. 1979).

Single crystals of Ag$_{8-x}$GeSe$_{6-x}$I$_x$ were obtained by the iodine transport method (Aldon et al. 2001; Belin et al. 2001).

5.106 Germanium–Silver–Manganese–Selenium

The **Ag$_2$MnGeSe$_4$** quaternary compound, which crystallizes in the orthorhombic structure with the lattice parameters $a = 783.0$, $b = 672.1$, and $c = 641.7$ pm, is formed in the Ge–Ag–Mn–Se system (Lamarche et al. 1991). This compound was prepared by the melt and anneals technique using stoichiometric mixtures of the elements. The components were sealed under a vacuum in a quartz capsule, which had previously been coated with carbon in order to prevent the reaction of the charge with the quartz. The capsule was then raised to a temperature of 1150°C, held at 1150°C for 0.5–1 h, and then cooled to room temperature. The sample was then annealed at 625°C for a week and quenched to room temperature.

5.107 Germanium–Silver–Iron–Selenium

The **Ag$_2$FeGeSe$_4$** quaternary compound, which melts at 742°C and crystallizes in the orthorhombic structure with the lattice parameters $a = 765.8$, $b = 651.5$, $c = 643.4$ pm, is formed in the Ge–Ag–Fe–Se system (Quintero et al. 1999). It was prepared by the melt and anneals technique. The mixture was made from the appropriate amounts of the elements and was sealed under vacuum in a small quartz ampoule, which had previously been carbonized to prevent the interaction of the components with the quartz. The components were melted together at 1150°C for about an hour, annealed to equilibrium at 500°C, and then cooled to room temperature by leaving the ampoule in the switched-off furnace.

The samples of Ag$_2$FeGeSe$_4$ were also synthesized by two more methods (Moroz et al. 2021a): (1) Solid-state synthesis of the calculated mixture of elements in an evacuated (~10^{-2} Pa) quartz ampoule at 360°C for 5 days. Then the samples were cooled to room temperature at the rate of 2°C·min^{-1} and ground to a particle size of ~5.0 µm. Vacuum homogenization of the fine phase mixture was held at 310°C for 5 days; (2) Vacuum melting of the calculated mixture of the elements at 830°C followed by vacuum annealing of the finely dispersed mixture at 310°C for 5 days.

Linear dependences of the EMF of galvanic cells versus temperature in the range of 127–227°C were used for calculations of the standard Gibbs energies, enthalpies, and entropies of formations of Ag$_2$FeGeSe$_4$ (Moroz et al. 2021a): $-\Delta G^0_{f, 298} = 227.0 \pm 34.6$ and 225.5 ± 36.3 kJ·M^{-1}, $-\Delta H^0_{f, 298} = 228.7 \pm 27.5$ and 221.1 ± 31.2 kJ·M^{-1}, $S^0_{298} = 307.0 \pm 9.9$ and 327.3 ± 6.5 J·(M·K)$^{-1}$ in two different phase regions.

5.108 Germanium–Magnesium–Cerium–Selenium

The **Mg$_{0.5}$Ce$_3$GeSe$_7$** quaternary compound, which crystallizes in the hexagonal structure with the lattice parameters $a = 1065.79 \pm 0.07$, $c = 604.30 \pm 0.07$ pm, and a calculated density of 5.910 g·cm^{-3}, is formed in the Ge–Mg–Ce–Se system (Strok et al. 2015). This compound was obtained by the fusion of the elements in an evacuated silica ampoule. In order to exclude the reaction of magnesium with quartz, the walls of the ampoule were lined with graphite. The ampoule was gradually heated in a furnace at a rate of 30°C·h^{-1} up to a maximal temperature of 1150°C, and kept at this temperature for 3 h. After that, it was slowly cooled (10°C·h^{-1}) to 600°C, and annealed at this temperature for 720 h. Subsequently, the ampoule was quenched in the air.

5.109 Germanium–Magnesium–Praseodymium–Selenium

The **Mg$_{0.5}$Pr$_3$GeSe$_7$** quaternary compound, which crystallizes in the hexagonal structure with the lattice parameters $a = 1061.5 \pm 0.1$, $c = 603.29 \pm 0.07$ pm, and a calculated density of 5.980 g·cm^{-3}, is formed in the Ge–Mg–Pr–Se system (Strok et al. 2015). This compound was prepared in the same way as Mg$_{0.5}$Ce$_3$GeSe$_7$ was synthesized.

5.110 Germanium–Strontium–Cadmium–Selenium

The **SrCdGeSe$_4$** quaternary compound, which melts congruently at 860°C and crystallizes in the orthorhombic structure with the lattice parameters $a = 1082.45 \pm 0.08$, $b = 1069.12 \pm$

0.08, $c = 647.92 \pm 0.05$ pm, a calculated density of 5.027 g·cm^{-3}, and an energy gap of 1.9 eV, is formed in the Ge–Sr–Cd–Se system (Dou et al. 2019). Single crystals of this compound were grown from reactions involving slow cooling from the melt. A mixture of Sr (0.50 mM), CdSe (0.50 mM), Ge (0.50 mM), and Se (1.50 mM) was loaded into a fused silica tube, which was evacuated and sealed. The tube was heated to 850°C over 30 h, kept at that temperature for 72 h, and then slowly cooled to room temperature over 96 h. Red air-stable crystals were obtained. Polycrystalline samples were prepared from reactions of the same mixture of starting materials as before. The tube was heated to 750°C for 24 h, kept at that temperature for 96 h, and then cooled by shutting off the furnace. The sample was reground and subjected to repeated heat treatment.

5.111 Germanium–Strontium–Mercury–Selenium

The **SrHgGeSe$_4$** quaternary compound, which melts incongruently at 745°C and crystallizes in the orthorhombic structure with the lattice parameters $a = 1083.8 \pm 0.7$, $b = 1074.6 \pm 0.6$, $c = 663.8 \pm 0.4$ pm, a calculated density of 5.814 g·cm^{-3} at 153 K, and an energy gap of 2.42 eV, is formed in the Ge–Sr–Hg–Se system (Guo et al. 2019c). To prepare polycrystalline samples of this compound, an equimolar mixture of SrSe, HgSe, and GeSe$_2$ was ground under an Ar atmosphere in a glove box and placed within a sealed fused silica tube, which was evacuated to 10^{-3} Pa and sealed. The tube was heated to 700°C over 20 h, kept at that temperature for 100 h, and cooled to room temperature by turning off the furnace. Single crystals of SrHgGeSe$_4$, were synthesized using the same procedure as that described above except that the tube was heated to 900°C over 20 h, kept at that temperature for 72 h, cooled to 400°C at a rate of 3°C·h^{-1}, and then cooled to room temperature by shutting off the furnace. Yellow crystals of SrHgGeSe$_4$ were obtained.

5.112 Germanium–Strontium–Antimony–Selenium

The **Sr$_3$GeSb$_2$Se$_8$** quaternary compound, which crystallizes in the orthorhombic structure with the lattice parameters $a = 1263.3 \pm 0.4$, $b = 430.1 \pm 0.1$, $c = 2869.3 \pm 0.7$ pm, a calculated density of 5.158 g·cm^{-3}, and an energy gap of 0.75 eV, is formed in the Ge–Sr–Sb–Se system (Yu et al. 2008). It was synthesized by a solid-state method. In a typical reaction, the elements in stoichiometric proportions were mixed in an Ar-filled glove box (total mass ~0.5 g), placed in a silica tube, sealed under dynamic vacuum, and heated slowly to 750°C within 72 h. This temperature was maintained for 3 days, followed by slow cooling to 550°C over 7 days, and finally to room temperature on simply terminating the power.

5.113 Germanium–Strontium–Bismuth–Selenium

The **Sr$_8$Ge$_2$Bi$_8$Se$_{24}$** quaternary compound, which melts at 787°C and crystallizes in the orthorhombic structure with the lattice parameters $a = 1280.36 \pm 0.09$, $b = 2901.6 \pm 0.2$, $c = 1281.92 \pm$

0.09 pm, and an energy gap of 0.41 eV for the Sr$_{8.01}$Ge$_{2.04}$Bi$_{7.95}$Se$_{24}$ composition, is formed in the Ge–Sr–Bi–Se system (Chung and Lee 2012). To synthesize this compound, stoichiometric proportions of the pure elements were mixed in a N$_2$-filled glove box (total mass ~0.5 g), placed in a carbon-coated silica tube, sealed under a dynamic vacuum, and slowly heated to 750°C for 48 h. This temperature was maintained for one day, followed by slow cooling to 400°C at a rate of 15°C·h^{-1}, and finally to room temperature by simply terminating the power. The polycrystalline ingot with a metallic luster was obtained.

5.114 Germanium–Barium–Zinc–Selenium

The **BaZnGeSe$_4$** quaternary compound, which crystallizes in the orthorhombic structure with the lattice parameters $a = 1132.55 \pm 0.06$, $b = 1125.27 \pm 0.06$, $c = 629.17 \pm 0.03$ pm, a calculated density of 4.897 g·cm^{-3}, and an energy gap of 2.46 eV, is formed in the Ge–Ba–Zn–Se system (Yin et al. 2017). To prepare this compound, a mixture of BaSe (0.40 mM), ZnSe (0.40 mM), Ge (0.40 mM), and Se (0.80 mM) were combined (molar ratio 1:1:1:2), finely ground, and loaded into fused silica tube which was evacuated and sealed. The tube was heated to 900°C in 15 h, kept at that temperature for 60 h, slowly cooled to 400°C over 144 h, and then cooled to room temperature by shutting off the furnace. The products contained orange-yellow crystals of BaZnGeSe$_4$ which are air stable. Powder sample was prepared by reaction of BaSe, ZnSe, Ge, and Se (molar ratio 1:1:1:2) as before, except that two heat treatments were applied. The mixture was heated to 750°C in 24 h, kept at that temperature for 24 h, and then cooled by shutting off the furnace. The obtained sample was reground and the heat treatment was repeated.

5.115 Germanium–Barium–Cadmium–Selenium

The **BaCdGeSe$_4$** quaternary compound, which crystallizes in the orthorhombic structure with the lattice parameters $a = 2216.7 \pm 0.6$, $b = 2276.8 \pm 0.5$, $c = 1330.3 \pm 0.3$ pm, a calculated density of 5.051 g·cm^{-3}, and an energy gap of 1.77 eV, is formed in the Ge–Ba–Cd–Se system (Yuan et al. 2021). The crystals of this compound were synthesized by a solid-state reaction technique. A mixture of BaSe, CdSe, and GeSe$_2$ (molar ratio 1:1:1) was ground and loaded into a graphite crucible sealed in the silica tube evacuated up to 10^{-2} Pa. The reactant batch was heated to 300°C (holding for 5 h) within 5 h, then heated to 850°C (holding for 70 h) in 15 h, and then the sample was slowly cooled to 750°C (holding for 10 h) in 100 h and then to room temperature in 15 h. Finally, red crystals of BaCdGeSe$_4$ were obtained. The pure phase was produced with the use of a stoichiometric mixture of the BaSe, CdSe, and GeSe$_2$, which was heated to 800°C in 15 h, kept for 50 h, and then the furnace was turned off.

5.116 Germanium–Barium–Mercury–Selenium

The **BaHgGeSe$_4$** quaternary compound, which melts incongruently at 732°C and crystallizes in the orthorhombic structure with the lattice parameters $a = 1125.5 \pm 0.2$, $b = 1103.3 \pm 0.2$,

$c = 668.47 \pm 0.13$ pm, a calculated density of 5.812 g·cm^{-3} at 153 K, and an energy gap of 2.49 eV, is formed in the Ge–Ba–Hg–Se system (Guo et al. 2019c). Polycrystalline sample and yellow-single crystals of this compound were obtained in the same way as SrHgGeSe$_4$ was prepared using BaSe instead of SrSe.

5.117 Germanium–Barium–Gallium–Selenium

Two quaternary compounds, **BaGa$_2$GeSe$_6$** and **Ba$_4$Ga$_4$GeSe$_{12}$**, are formed in the Ge–Ba–Ga–Se system. First of them melts congruently at 880°C [at 877°C (Badikov et al. 2016)] and crystallizes in the trigonal structure with the lattice parameters $a = 1000.8 \pm 0.1$, $c = 909.0 \pm 0.2$ pm, a calculated density of 5.201 g·cm^{-3}, and an energy gap of 2.22 eV (Yin et al. 2012b; Badikov et al. 2016) [$a = 1004.38 \pm 0.13$, $c = 911.4 \pm 0.2$ pm, a calculated density of 3.433 g·cm^{-3}, and an energy gap of 2.81 eV (Lin et al. 2012)]. Ba$_4$Ga$_4$GeSe$_{12}$ crystallizes in the tetragonal structure with the lattice parameters $a = 1354.68 \pm 0.04$, $c = 649.15 \pm 0.02$ pm, a calculated density of 5.153 g·cm^{-3}, and an energy gap of 2.18 eV (Yin et al. 2016).

Crystals of BaGa$_2$GeSe$_6$ were initially obtained from a reaction between Ba$_2$GeSe$_4$ and Ga$_2$Se$_3$ (molar ratio 1:2) (Yin et al. 2012b). The mixture was ground and loaded into a fused silica tube under an Ar atmosphere in a glove box. The tube was flame-sealed under a high vacuum (10^{-3} Pa) and then placed in a furnace. The reaction mixture was heated to 1000°C in 15 h, kept at 1000°C for 48 h, followed by slow cooling to 320°C at a rate of 4°C·h^{-1}, and finally cooled to room temperature by switching off the furnace. Many orange block-shaped crystals were found in the ampoule. A pure polycrystalline sample of this compound was synthesized by the solid-state reaction technique. A mixture of BaSe (0.216 g), Ga$_2$Se$_3$ (0.376 g), and GeSe$_2$ (0.231 g) in a molar ratio of 1:1:1 was ground and loaded into a fused silica tube under an Ar atmosphere in a glove box. The tube was sealed under 10^{-3} Pa atmosphere and then placed in a furnace. The sample was heated to 850°C in 15 h, kept at that temperature for 48 h, and then the furnace was turned off. Crystal growth experiments were then tried again using the pure BaGa$_2$GeSe$_6$ polycrystalline sample. The as-prepared BaGa$_2$GeSe$_6$ powder was put into a fused silica tube, which was then flame-sealed under a high vacuum of 10^{-3} Pa and placed in a furnace. The sample was heated to 1000°C in 15 h, kept at 1000°C for 48 h, and then cooled at 3°C·h^{-1} to 320°C, at which point the furnace was switched off. The tube was full of orange BaGa$_2$GeSe$_6$ crystals after the reaction.

The synthesis of this compound can also be achieved using the solid-state reaction of BaSe, Ga, Ge, and Se (Lin et al. 2012). The stoichiometric mixture was evacuated under high vacuum, and the sealed ampoule slowly heated to 700°C, held for about 70 h, and then was sintered at 980°C for 3 days in a furnace before being slowly cooled to 500°C at a rate of 5°C·h^{-1}. Single crystals of BaGa$_2$GeSe$_6$ were grown by the Bridgman–Stockbarger techniques (Lin et al. 2012; Badikov et al. 2016).

To prepare Ba$_4$Ga$_4$GeSe$_{12}$, a mixture of BaSe (0.60 mM), Ga$_2$Se$_3$ (0.30 mM), and GeSe$_2$ (0.15 mM) (molar ratio 4:2:1) was finely ground and loaded into a fused silica tube, which was evacuated and flame-sealed (Yin et al. 2016). The tube was heated to 900°C over 30 h, kept at that temperature for 72 h, cooled to 200°C over 48 h, and then slowly cooled to room temperature by shutting off the furnace. Orange-yellow block-shaped air-stable crystals were obtained. A powder sample was prepared as before by stoichiometric reaction of BaSe, Ga$_2$Se$_3$, and GeSe$_2$ in a 4:2:1 molar ratio except that the dwell temperature was 800°C and no slow cooling step was included; the sample was reground, and the heat treatment was repeated.

5.118 Germanium–Barium–Oxygen–Selenium

The **BaGeOSe$_2$** quaternary compound, which melts congruently at 920°C and crystallizes in the orthorhombic structure with the lattice parameters $a = 479.8 \pm 0.2$, $b = 860.9 \pm 0.5$, $c = 1252.4 \pm 0.7$ pm, and a calculated density of 4.929 g·cm^{-3}, is formed in the Ge–Ba–O–Se system (Liu et al. 2015a). For synthesizing this compound, a mixture of Ba, GeO$_2$, Se, B, and KI as a flux (molar ratio 3:3:6:2:15) was loaded into quartz tubes and then flame-sealed under vacuum about 10^{-2} Pa. The samples were placed in a furnace, held at 300°C for 5 h, then heated to 700°C for 5 h and dwelled for 4 days, finally cooled down to 350°C before the furnace power was switched off. Colorless single crystals of the title compound were obtained after washing with distilled water. They are stable in air and moisture conditions for several months. Boron can act as a reducing reagent to extract oxygen from metal oxides due to its strong oxygen affinity in the reactive system. All starting reactants were handled inside an Ar-filled glove box with controlled oxygen and moisture levels below 0.1 ppm.

5.119 Germanium–Barium–Tellurium–Selenium

The **BaGeSe$_{4-\delta}$Te$_\delta$** ($\delta < 2.5$) quaternary compound, which crystallizes in the monoclinic structure with the lattice parameters $a = 699.58 \pm 0.04$, $b = 709.38 \pm 0.04$, $c = 917.38 \pm 0.06$ pm, $\beta = 109.135 \pm 0.001°$, a calculated density of 5.120 g·cm^{-3} for $\delta = 0$; $a = 732.12 \pm 0.08$, $b = 719.09 \pm 0.08$, $c = 928.5 \pm 0.1$ pm, $\beta = 108.433 \pm 0.002°$, a calculated density of 5.445 g·cm^{-3}, and an energy gap of 1.0 eV for $\delta = 2$; and $a = 745.87 \pm 0.06$, $b = 731.63 \pm 0.06$, $c = 937.15 \pm 0.08$ pm, $\beta = 108.429 \pm 0.002°$, a calculated density of 5.361 g·cm^{-3}, and an energy gap of 0.8–1.1 eV for $\delta = 2.47$, is formed in the Ge–Ba–Te–Se system (Assoud et al. 2004b). This compound was prepared by reacting the elements with the exclusion of air at 800°C followed by slow cooling to room temperature.

5.120 Germanium–Barium–Chromium–Selenium

The **Ba$_2$Cr$_4$GeSe$_{10}$** quaternary compound, which crystallizes in the triclinic structure with the lattice parameters $a = 713.0 \pm 0.4$, $b = 958.1 \pm 0.5$, $c = 1277.8 \pm 0.7$ pm, $\alpha = 71.477 \pm 0.007°$, $\beta = 81.940 \pm 0.009°$, $\gamma = 89.962 \pm 0.001°$, a calculated density of 5.456 g·cm^{-3}, and an energy gap of 0.64 eV, is formed in the Ge–Ba–Cr–Se system (Chen et al. 2018a). This compound is thermally stable in vacuum up to 1000°C. It was synthesized by a conventional high-temperature solid-state reaction method. For this, Ba, Cr, Ge, and Se were weighed (molar ratio 2:4:1:10) and loaded into a carbon-coated fused silica tube with an overall loading of about 0.5 g within the glove box. Then the tube was evacuated and sealed under a dynamic vacuum of approximately 10^{-3} Pa. Subsequently, the tube was heated to 1000°C within 72 h in a furnace, kept at that temperature for 100 h, after that, cooled to 400°C with a rate of 3°·h⁻¹ and then the furnace was switched off. Plate-shaped black single crystals were obtained, and they were stable in the air for more than 3 months. All elements were stored in an Ar-filled glove box with controlled oxygen and moisture levels below 0.1 ppm.

5.121 Germanium–Barium–Chlorine–Selenium

The $Ba_4Ge_3Se_9Cl_2$ quaternary compound, which crystallizes in the hexagonal structure with the lattice parameters $a = 1009.24 \pm 0.06$, $c = 1251.17 \pm 0.14$ pm, and an energy gap of 1.89 eV, is formed in the Ge–Ba–Cl–Se system (Liu et al. 2017). This compound was obtained from a mixture of $BaCl_2$, Ba, Ge, and Se (molar ratio 1.2:3:3:9) with a slight excess of $BaCl_2$ as reactive flux. The reactants were loaded into a fused silica tube under vacuum and then heated to 950°C within 50 h, kept at this temperature for 140 h, and subsequently cooled to 300°C at 3°C·h⁻¹ before switching off the furnace. The product was washed with distilled water to eliminate excess $BaCl_2$ and then dried with ethanol. After this treatment, the compound was obtained as pure phase with dark-red color. All manipulations were conducted in a glove box (moisture and oxygen levels less than 0.1 ppm) or under a vacuum.

5.122 Germanium–Barium–Bromine–Selenium

The $Ba_4Ge_3Se_9Br_2$ quaternary compound, which melts congruently at 731°C and crystallizes in the hexagonal structure with the lattice parameters $a = 1023.45 \pm 0.04$, $c = 1253.06 \pm 0.05$ pm, a calculated density of 4.785 g·cm⁻³, and an energy gap of 2.60 eV, is formed in the Ge–Ba–Br–Se system (Xing et al. 2020b). Single crystals of this compound were initially obtained from a reaction of BaSe, $GeSe_2$, and $BaBr_2$ (molar ratio 3:3:1). Mixture of finely ground reactants was loaded into a fused silica tube, which was evacuated and sealed. The tube was placed in a furnace where it was heated to 900°C in 20 h, kept at that temperature for 72 h, cooled to 200°C at a rate of 4°C·h⁻¹, and then cooled to room temperature by shutting off the furnace. The polycrystalline sample was prepared in the same manner except that the tube was heated at 750°C for 120 h. $Ba_4Ge_3Se_9Br_2$ was stable in air for more than one month. All materials were stored or manipulated in an Ag-filled glove box (with moisture and oxygen content of <0.1 ppm) or under a vacuum.

5.123 Germanium–Barium–Rhodium–Selenium

The $BaRh_2Ge_4Se_6$ quaternary compound, which crystallizes in the orthorhombic structure with the lattice parameters $a = 613.18 \pm 0.03$, $b = 607.00 \pm 0.03$, $c = 3031.44 \pm 0.09$ pm, a

calculated density of 5.36 g·cm⁻³, and an energy gap of 1.089 eV, is formed in the Ge–Ba–Rh–Se system (Lei et al. 2014). A polycrystalline sample of this compound was synthesized using high-pressure and high-temperature methods. BaSe was mixed with stoichiometric amounts of Rh, Ge, and Se, ground well, and then pelletized. The pellet was loaded into an h-BN capsule and then heated to 1200°C and 5 GPa for 2 h using a belt-type high-pressure apparatus. All starting materials and precursors for the synthesis were prepared in a glove box filled with purified Ar. Single crystals were grown by prolonging the annealing time (12 h) under high pressure.

5.124 Germanium–Barium–Iridium–Selenium

The $BaIr_2Ge_4Se_6$ quaternary compound, which crystallizes in the orthorhombic structure with the lattice parameters $a = 613.46 \pm 0.04$, $b = 608.70 \pm 0.04$, $c = 3027.9 \pm 0.1$ pm, and an energy gap of 1.332 eV, is formed in the Ge–Ba–Ir–Se system (Lei et al. 2014). Polycrystalline sample and single crystals of this compound were obtained in the same way as $BaRh_2Ge_4Se_6$ was prepared using Ir instead of Rh.

5.125 Germanium–Zinc–Gallium–Selenium

$GeSe_2$–$ZnSe$–Ga_2Se_3. The liquidus surface of this quasiternary system (Figure 5.58) consists of four fields of primary crystallization, which belong to the solid solutions based on ZnSe, Ga_2Se_3 and $GeSe_2$, and the $ZnGa_2Se_4$ ternary compound (Olekseyuk et al. 2003). No quaternary compounds were found. The solid solution based on ZnSe occupies the largest area of the concentration triangle. The fields of primary crystallization are separated by five monovariant lines and six invariant points, four of which correspond to binary reactions

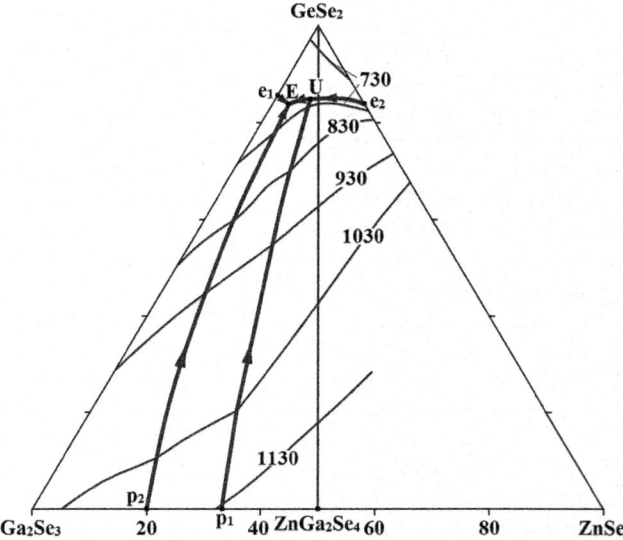

FIGURE 5.58 Liquidus surface of the $GeSe_2$–$ZnSe$–Ga_2Se_3 quasiternary system. (From Olekseyuk, I.D. et al., *J. Alloys Compd.*, **351**(1–2), 171, 2003.)

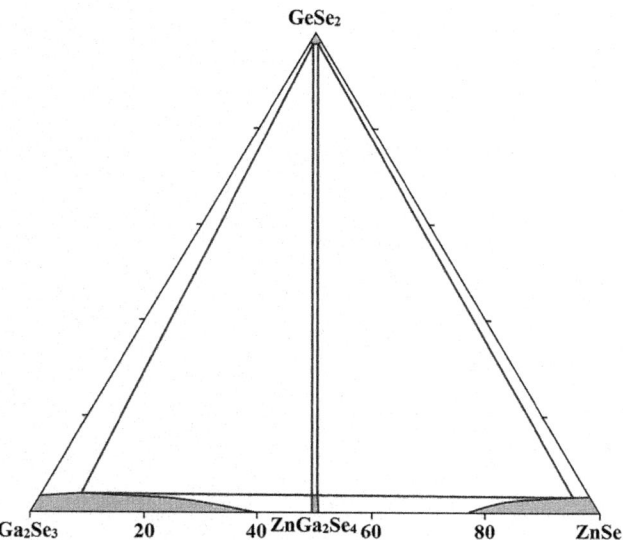

FIGURE 5.59 Isothermal section of the GeSe$_2$–ZnSe–Ga$_2$Se$_3$ quasiternary system at 400°C. (From Olekseyuk, I.D. et al., *J. Alloys Compd.*, **351**(1–2), 171, 2003.)

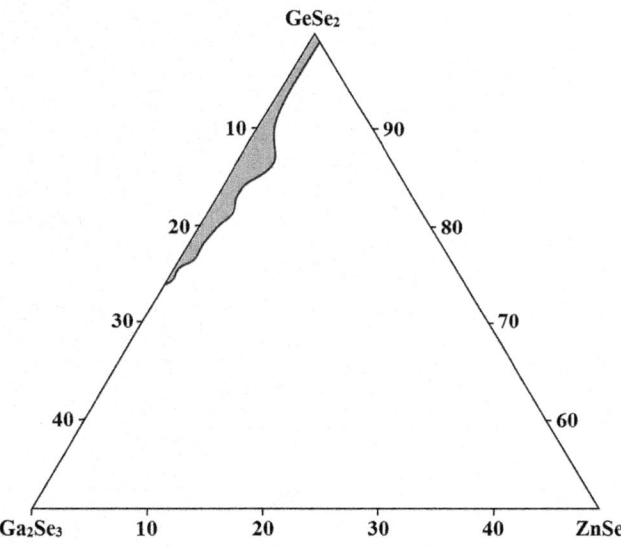

FIGURE 5.60 Glass-forming region in the GeSe$_2$–ZnSe–Ga$_2$Se$_3$ quasiternary system. (From Olekseyuk, I.D. et al., *Fizyka kondens. vysokomolek. system. Nauk. zap. Rivnens'kogo pedinstytutu*, (3), 148, 1997.)

and two to ternary ones. Ternary eutectic *E* (3 mol% ZnSe, 13 mol% Ga$_2$Se$_3$, and 84 mol% GeSe$_2$) crystallizes at 651°C, and the temperature of the ternary transition point *U* (6 mol% ZnSe, 9 mol% Ga$_2$Se$_3$, and 85 mol% GeSe$_2$) is equal to 693°C.

Isothermal section of this quasiternary system at 400°C is given in Figure 5.59 (Olekseyuk et al. 2003). The solid solution ranges of ZnSe and Ga$_2$Se$_3$ were found to be elongated along the ZnSe–Ga$_2$Se$_3$ quasibinary section. The GeSe$_2$ content in the solid solution based on ZnSe does not exceed 3 mol%, while in the solid solution based on Ga$_2$Se$_3$, it remains below 4 mol%.

The glass-forming region in this quasiternary system is shown in Figure 5.60 (Olekseyuk et al. 1997c). It is elongated along the GeSe$_2$–Ga$_2$Se$_3$ quasibinary system. The maximum

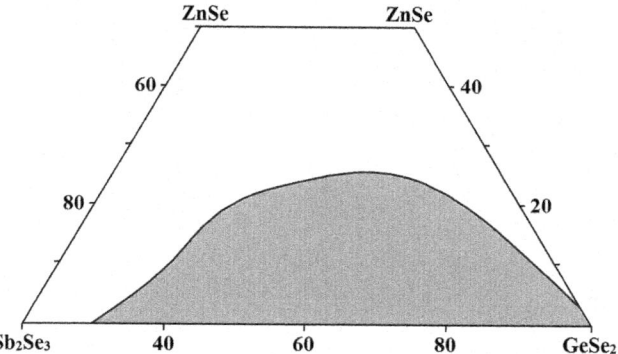

FIGURE 5.61 Glass-forming region in the GeSe$_2$–ZnSe–Sb$_2$Se$_3$ quasiternary system. (From Vassilev, V.S., et al., *J. Mater. Sci. Lett.*, **17**(23), 2007, 1998.)

amount of Ga$_2$Se$_3$ that can be introduced in the glass is 28 mol%, and the ZnSe content in this glass does not exceed 3 mol%.

This system was investigated through DTA, XRD, and metallography. All samples were annealed at 400°C for 500 h with subsequent quenching in cold water (Olekseyuk et al. 1997c, 2003).

5.126 Germanium–Zinc–Antimony–Selenium

GeSe$_2$–ZnSe–Sb$_2$Se$_3$. The glass-forming region in this quasiternary system is located on the GeSe$_2$–Sb$_2$Se$_3$ side of the Gibbs triangle in the interval 0–70 mol% Sb$_2$Se$_3$ (Figure 5.61) (Vassilev et al. 1998). The glass transition temperature varies in the range of 210–340°C. A decrease in the values with increasing Sb$_2$Se$_3$ concentration at a constant content of ZnSe was observed. The glasses were prepared by direct synthesis of the binary compounds up to 850–1000°C, depending on the ZnSe content. The mixtures were melted in evacuated (~10^{-3} Pa) quartz ampoules in a rotary furnace for several hours and then quenched in ice water.

5.127 Germanium–Zinc–Tellurium–Selenium

GeSe$_2$–ZnSe–ZnTe. The glass-forming region in this quasiternary system is presented in Figure 5.62 (Boycheva et al. 1999, 2002). To prepare the glasses, the required mass of the components were flame-sealed into a quartz ampoule under a vacuum of 10^{-3} Pa. The sealed ampoule was then placed in a rocking furnace and heated at 950°C. The final stage was water quenching with an average cooling rate of 2–5°C·s^{-1}.

5.128 Germanium–Cadmium–Gallium–Selenium

GeSe$_2$–CdGa$_2$Se$_4$. The phase diagram of this system, constructed through DTA, XRD, metallography, and measuring of the microhardness, is a eutectic type (Figure 5.63) (Olekseyuk and Parasyuk 1995). The eutectic contains 73 mol% GeSe$_2$ and crystallizes at 670°C. The solubility of GeSe$_2$ in α-CdGa$_2$Se$_4$

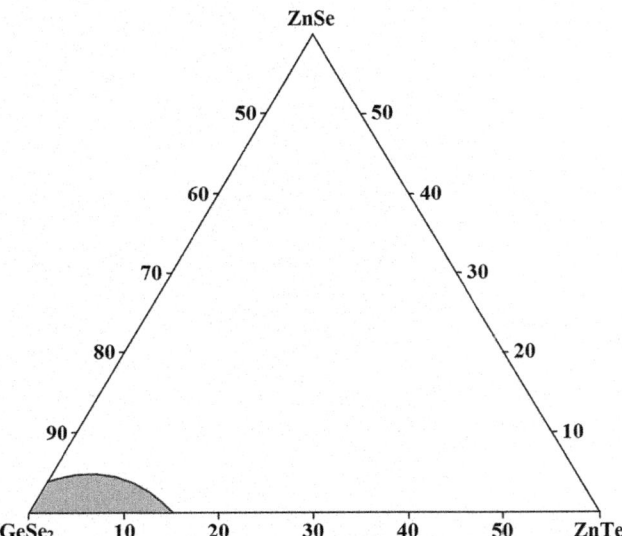

FIGURE 5.62 Glass-forming region in the GeSe$_2$–ZnSe–ZnTe quasiternary system. (From Boycheva, S.V., et al., *J. Appl. Electrochem.*, **32**(3) 281, 2002.)

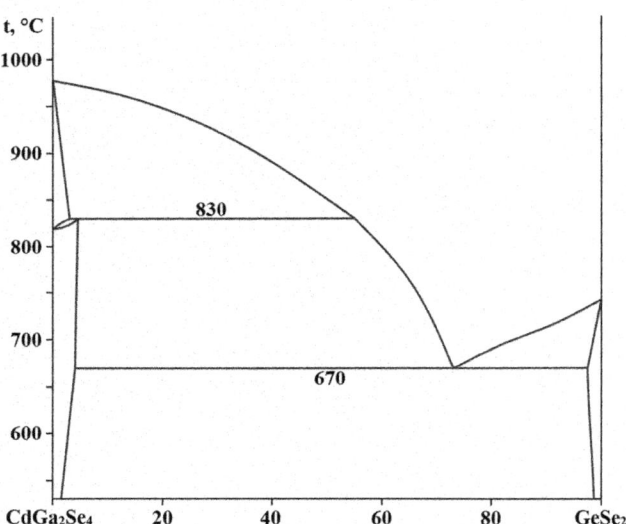

FIGURE 5.63 Phase diagram of the GeSe$_2$–CdGa$_2$Se$_4$ system. (From Olekseyuk, I.D., and Parasyuk, O.V., *Zhurn. neorgan. khimii*, **40**(2), 315, 1995.)

and α-CdGa$_2$Se$_4$ in GeSe$_2$ at the eutectic temperature is not higher than 4 and 3 mol%, respectively. The peritectic process L + (β-CdGa$_2$Se$_4$) ↔ (α-CdGa$_2$Se$_4$) proceeds in the system at 830°C within the range of 3–55 mol% GeSe$_2$. The ingots for the investigations were annealed at 600°C for 100 h.

GeSe$_2$–CdSe–Ga$_2$Se$_3$. Six fields of primary crystallization of the next phases, CdSe, Ga$_2$Se$_3$, GeSe$_2$, α- and β-CdGa$_2$Se$_4$, and Cd$_4$GeSe$_6$, exist on the liquidus surface of this system (Figure 5.64) (Olekseyuk and Parasyuk 1995). In the subsystem GeSe$_2$–CdGa$_2$Se$_4$–Ga$_2$Se$_3$, the crystallization ends in the ternary eutectic at 628°C. The crystallization in the GeSe$_2$–CdSe–CdGa$_2$Se$_4$ subsystem is more complex, and in the subsystem GeSe$_2$–CdGa$_2$Se$_4$–Cd$_4$GeSe$_6$, the crystallization

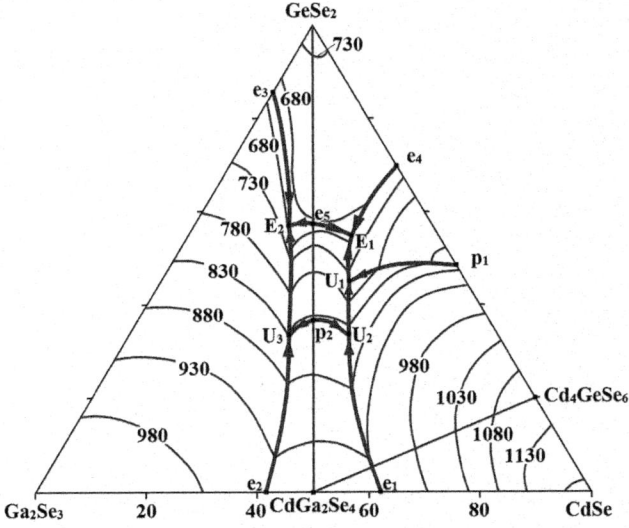

FIGURE 5.64 Liquidus surface of the GeSe$_2$–CdSe–Ga$_2$Se$_3$ quasiternary system. (From Olekseyuk, I.D., and Parasyuk, O.V., *Zhurn. neorgan. khimii*, **40**(2), 315, 1995.)

ends at 642°C. In the subsystem CdSe–CdGa$_2$Se$_4$–Cd$_4$GeSe$_6$, the crystallization ends in the transition point at 730°C. This reaction with liquid excess is representative for other alloys from β-CdGa$_2$Se$_4$–U_1–p_1–CdSe peritectic tetragon. Two ternary eutectics and three transition points exist in the CdSe–Ga$_2$Se$_3$–GeSe$_2$ system: E$_1$ (642°C) – L ↔ β-CdGa$_2$Se$_4$ + GeSe$_2$ + Cd$_4$GeSe$_6$; E$_2$ (628°C) – L ↔ GeSe$_2$ + Ga$_2$Se$_3$ + β-CdGa$_2$Se$_4$; U$_1$ (730°C) – L + CdSe ↔ Cd$_4$GeSe$_6$ + β-CdGa$_2$Se$_4$; U$_2$ (830°C) – L + CdSe ↔ α-CdGa$_2$Se$_4$ + β-CdGa$_2$Se$_4$; and U$_3$ (830°C) – L + Ga$_2$Se$_3$ ↔ α-CdGa$_2$Se$_4$ + β-CdGa$_2$Se$_4$.

Isothermal section of the GeSe$_2$–CdSe–Ga$_2$Se$_3$ quasiternary system at 600°C is shown in Figure 5.65 (Olekseyuk et al. 1997e). The alloys were annealed at this temperature for 250 h.

There are two glass-forming regions in this system (Figure 5.66) (Olekseyuk et al. 1997c). The first of them begins from GeSe$_2$ and is elongated toward E_2 ternary eutectic and the second is situated around E_1 ternary eutectic.

5.129 Germanium–Cadmium–Thallium–Selenium

GeSe$_2$–CdSe–Tl$_2$Se. The isothermal section of this quasiternary system at 300°C is shown in Figure 5.67 (Selezen et al. 2020). It is characterized by the formation of two new quaternary compounds, **Tl$_2$CdGeSe$_4$** at the Tl$_2$GeSe$_3$–CdSe section, and **Tl$_2$CdGe$_3$Se$_8$** at the Tl$_2$CdGeSe$_4$–GeSe$_2$ section. The section has 9 single-phase, 17 two-phase, and 9 three-phase fields in thermodynamic equilibrium. The solid solubility of CdSe at the Tl$_4$GeSe$_4$–CdSe and Tl$_2$GeSe$_3$–CdSe sections is less than 3 mol%.

Tl$_2$CdGeSe$_4$ crystallizes in the tetragonal structure with the lattice parameters $a = 801.45 \pm 0.09$, $c = 672.34 \pm 0.09$ pm, a calculated density of 6.995 ± 0.003 g·cm^{-3}, and an energy gap of 1.71 eV (Selezen et al. 2020) [$a = 845.02$, $c = 677.40$ pm (calculated values) and an energy gap of 1.52 eV (Vu et al. 2021)].

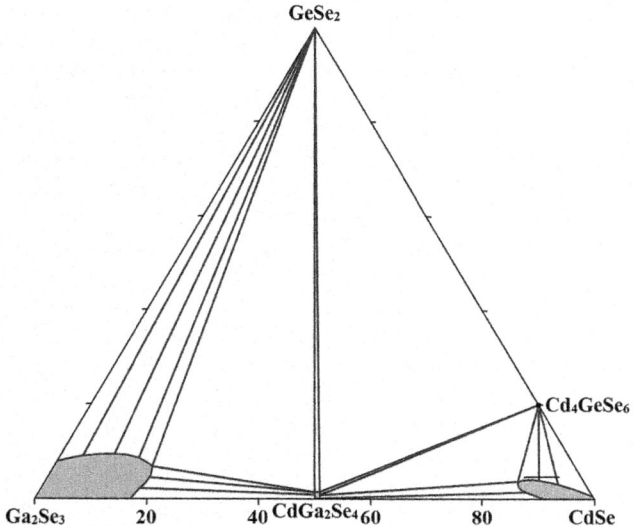

FIGURE 5.65 Isothermal section of the GeSe₂–CdSe–Ga₂Se₃ quasiternary system at 600°C. (From Olekseyuk, I.D. et al., *Pol. J. Chem.*, **71**(6), 701, 1997.)

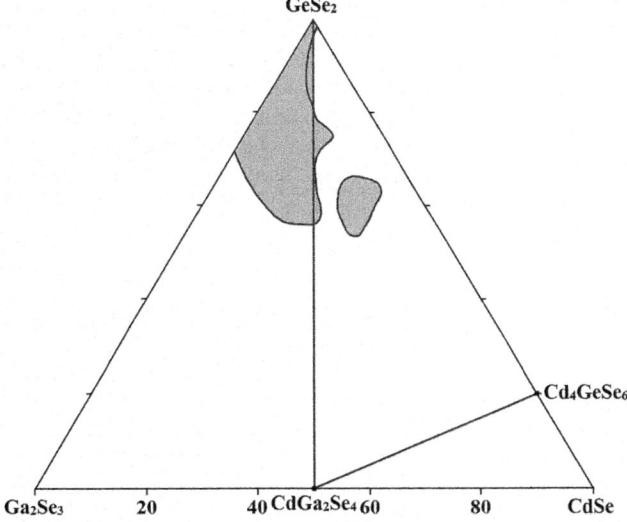

FIGURE 5.66 Glass-forming region in the GeSe₂–CdSe–Ga₂Se₃ quasiternary system. (From Olekseyuk, I.D. et al., *Fizyka kondens. vysokomolek. system. Nauk. zap. Rivnens'kogo pedinstytutu*, (3), 148, 1997.)

The quaternary compounds were synthesized from the mixtures of Tl, Cd, Ge, and Se in quartz ampoules that were evacuated to 0.1 Pa and sealed (Selezen et al. 2020). The ampoules were heated to 680°C at the rate of 20°C·h⁻¹, kept for 5 h, heated to 930°C at 10°C·h⁻¹ and held for 5 h. Then both series were cooled to 300°C at the rate of 10°C·h⁻¹ and annealed for 350 h. Finally, the samples were quenched into 20% NaCl aqueous solution.

5.130 Germanium–Cadmium–Europium–Selenium

The **EuCdGeSe₄** quaternary compound, which melts congruently at 882°C and crystallizes in the orthorhombic structure with the lattice parameters $a = 1075.25 \pm 0.08$, $b = 1062.75$

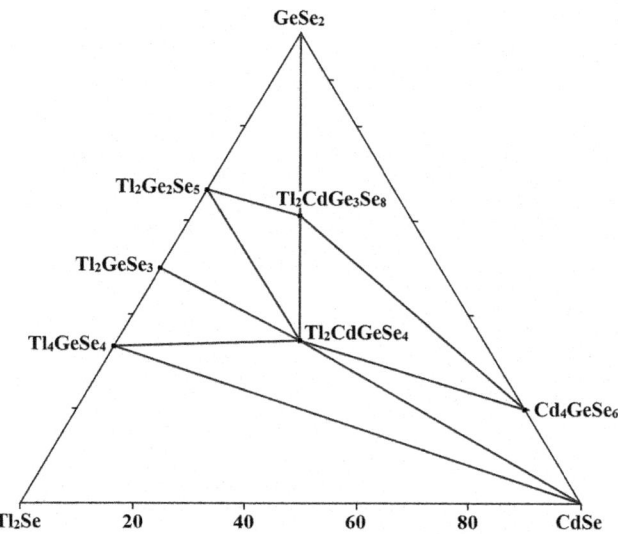

FIGURE 5.67 Isothermal section of the GeSe₂–CdSe–Tl₂Se quasiternary system at 300°C. (From Selezen, A.O., et al., *J. Solid State Chem.*, **289**, 121422, 2020.)

± 0.08, $c = 647.40 \pm 0.05$ pm, a calculated density of 5.685 g·cm⁻³, and an energy gap of 2.25 eV, is formed in the Ge–Cd–Eu–Se system (Xing et al. 2019). Polycrystalline sample of this compound was synthesized by a solid-state reaction using equimolar amounts of EuSe, CdSe, and GeSe₂, which were mixed and ground in an Ar-filled glove box. Then the mixture was loaded into quartz tube, sealed under vacuum and heated to 700°C in 15 h, maintained at this temperature for 72 h, and cooled down to room temperature by turning off the furnace. In order to enhance their crystallinity and purity, regrinding and a repeated heat treatment were necessary for the sample. Single crystals were grown using the same mixture of starting materials as before. The tube loading with starting materials was evacuated and heated in a furnace to 1000°C in 20 h and maintained at this temperature for 72 h. When the next procedure of cooling down to 400°C at a rate of 3°C·h⁻¹ was completed, the furnace was shut off. Red crystals of EuCdGeSe₄ were obtained.

5.131 Germanium–Cadmium–Arsenic–Selenium

GeSe₂–CdSe–As₂Se₃. A large glass-forming region exists in this quasiternary system (Figure 5.68) (Zhao et al. 2005, 2008). The addition of Cd increases the crystallization trend of the glasses. Besides, the Cd content accommodated by the system to form glasses is found to be lower in the As₂Se₃-rich region than in the GeSe₂-rich region. With the introduction of CdSe the density, microhardness, and glass transition temperature (T_g) of glasses increase, whereas the thermal stability decreases. The T_g for the obtained glasses are relatively high, varying from 240°C to 340°C. It was found that T_g increases with the content of Ge but decreases as As content increases. This system was investigated through DTA, DSC, XRD, measuring of the density and microhardness and using IR transmission spectra. The glasses were annealed at 300°C for 3 h.

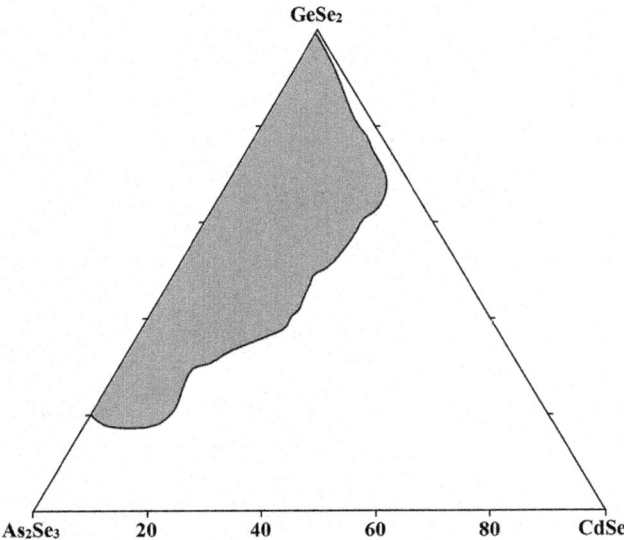

FIGURE 5.68 Glass-forming region in the GeSe₂–CdSe–As₂Se₃ quasiternary system. (From Zhao, D. et al., *J. Am. Ceram. Soc.*, **88**(11), 3143, 2005.)

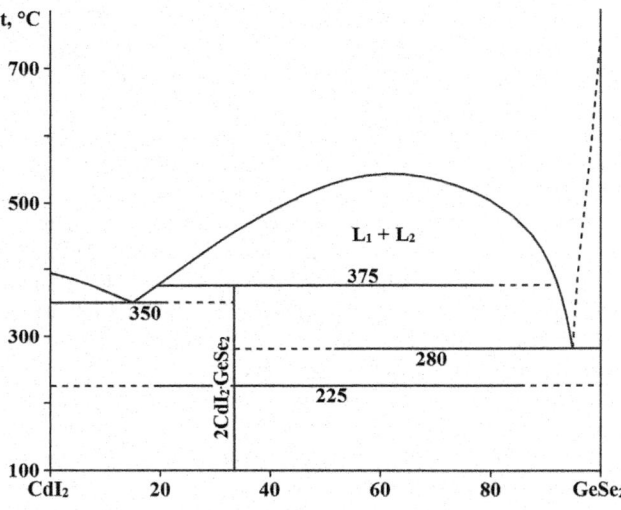

FIGURE 5.69 Phase diagram of the GeSe₂–CdI₂ system. (From Vassilev, V.S., et al., *J. Phys. Chem. Solids*, **63**(5), 815, 2002.)

5.132 Germanium–Cadmium–Tellurium–Selenium

GeSe₂–CdTe. The phase diagram of this system, constructed through DTA, XRD, metallography, and measuring of the microhardness, is a eutectic type (Asadov et al. 2012). The eutectic composition and temperature are 66 mol% GeSe₂ and 527°C, respectively. The **Cd₂GeSe₂Te₂** quaternary compound, which melts incongruently at 647°C and crystallizes in the hexagonal structure with the lattice parameters $a = 569$ and $c = 1132$ pm, is formed in this system (Asadov 2006; Asadov et al. 2012). The solubility of CdTe in GeSe₂ and GeSe₂ in CdTe at 527°C is equal to 16 and 22 mol%, respectively.

Asadov et al. (2016) studied the GeSe₂–CdSe–CdTe–GeTe system and noted that the CdSe–CdTe system is the stable diagonal of the abovementioned system. The authors designated the studied system as Cd,Ge ‖ Se,Te, but such designation can be used only for the GeSe + CdTe ↔ GeTe + CdSe ternary mutual system, which can be mapped in plane. The system studied by the authors is two ternary systems (GeTe–CdTe–CdSe and GeSe₂–CdSe–CdTe) connected by the CdSe–CdTe binary system, and their composition triangles do not lie in the same plane: they are at an angle to each other.

The homogeneous glasses containing up to 5 mol% CdTe have been found in this system (Vassilev et al. 2002a). Glass-forming region in the GeSe₂–CdTe system has been determined by visual observation, XRD, and EPMA.

5.133 Germanium–Cadmium–Iodine–Selenium

GeSe₂–CdI₂. The phase diagram of this system, constructed through DTA, XRD, and measuring of the density, is given in Figure 5.69 (Vassilev et al. 2002b). The eutectic compositions are at 15 and 95 mol% GeSe₂ with the eutectic temperatures 350°C and 280°C, respectively. The immiscibility region exists within the interval 20–90 mol% GeSe₂ with the syntectic temperature of 375°C. The **2CdI₂·GeSe₂** quaternary compound, which has a polymorphic transformation at 225°C and crystallizes in the triclinic structure with the lattice parameters $a = 750.7$, $b = 702.1$, $c = 662.8$ pm, $\alpha = 94.16°$, $\beta = 103.83°$, and $\gamma = 103.15°$, is formed in this system.

5.134 Germanium–Mercury–Gallium–Selenium

GeSe₂–HgGa₂Se₄. This section is a non-quasibinary one of the quasiternary GeSe₂–HgSe–Ga₂Se₃ system and crosses the fields of GeSe₂ and (Ga₂Se₃) primary crystallization (Olekseyuk et al. 1996b). Six other vertical sections were also constructed.

GeSe₂–HgSe–Ga₂Se₃. The liquidus surface of this system (Figure 5.70) consists of six regions of primary crystallization: (HgSe), GeSe₂, (Ga₂Se₃), α-HgGa₂Se₄, β-HgGa₂Se₄, and Hg₂GeSe₄ (Olekseyuk et al. 1996b). They are separated by 10 invariant lines and 5 invariant points. The next ternary invariant points exist in the system: E (572°C) – L ↔ β-HgGa₂Se₄ + Hg₂GeSe₄ + GeSe₂; U₁ (631°C) – L + (Ga₂Se₃) ↔ β-HgGa₂Se₄ + GeSe₂; U₂ (592°C) – L + (HgSe) ↔ β-HgGa₂Se₄ + Hg₂GeSe₄; U₃ (652°C) – L + (Ga₂Se₃) ↔ α-HgGa₂Se₄ + β-HgGa₂Se₄; and U₄ (652°C) – L + (HgSe) ↔ α-HgGa₂Se₄ + Hg₂GeSe₄.

The isothermal section of this quasiternary system at 500°C is given in Figure 5.71 (Olekseyuk et al. 1996b). Solid solutions based on HgSe and Ga₂Se₃ exist at this temperature. The solubility on the basis of other compounds is not significant. No quaternary phases were found in the system.

The glass-forming region in the GeSe₂–HgSe–Ga₂Se₃ system occupies almost all the field of GeSe₂ primary crystallization and is elongated along the GeSe₂–HgSe quasibinary

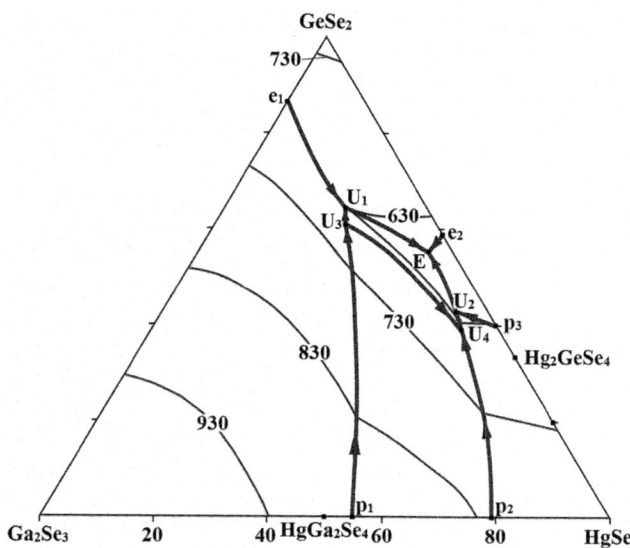

FIGURE 5.70 Liquidus surface of the GeSe₂–HgSe–Ga₂Se₃ quasiternary system. (From Olekseyuk, I.D. et al., *J. Alloys Compd.*, **238**(1–2), 141, 1996.)

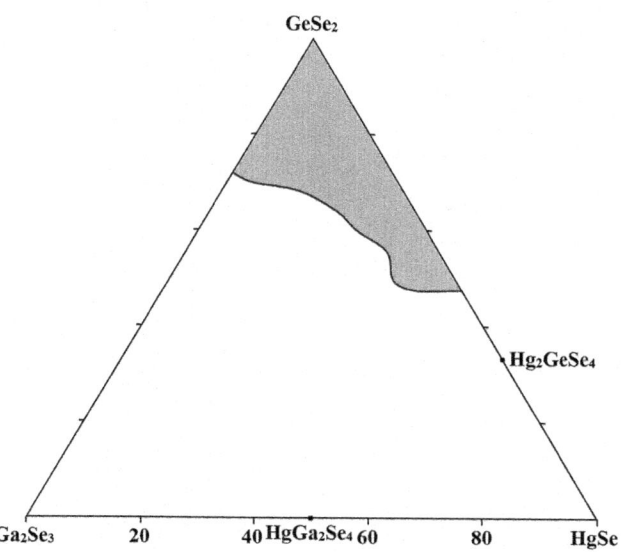

FIGURE 5.72 Glass-forming region in the GeSe₂–HgSe–Ga₂Se₃ quasiternary system. (From Olekseyuk, I.D. et al., *Fizyka kondens. vysokomolek. system. Nauk. zap. Rivnens'kogo pedinstytutu*, (3), 148, 1997.)

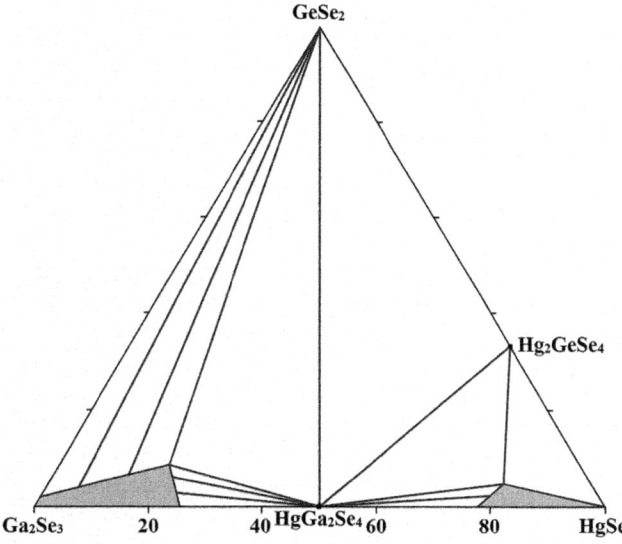

FIGURE 5.71 Isothermal section of the GeSe₂–HgSe–Ga₂Se₃ quasiternary system at 500°C. (From Olekseyuk, I.D. et al., *J. Alloys Compd.*, **238**(1–2), 141, 1996.)

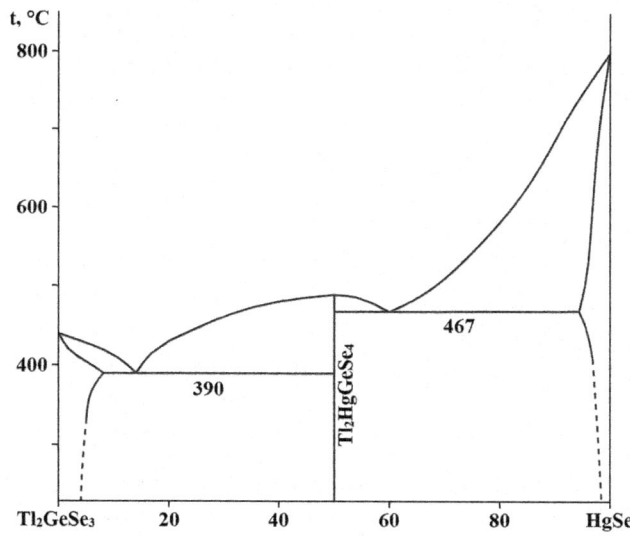

FIGURE 5.73 Phase diagram of the Tl₂GeSe₃–HgSe system. (From Mozolyuk, M.Yu., et al., *Chem. Met. Alloys*, **6**(1–2), 55, 2013.)

system (Figure 5.72) (Olekseyuk et al. 1997c). The characteristic temperatures of the glassy alloys, namely, the glass transition temperature (T_g), the crystallization temperature (T_c), and the melting temperature (T_m) of the crystallized alloys, have been measured.

This system was investigated by DTA, XRD, metallography, and the microhardness measurements, and the alloys were annealed at 500°C for 250°C and quenched in cold water (Olekseyuk et al. 1996b). The glass alloys were obtained by annealing at 1000°C for 6 h with the next quenching (Olekseyuk et al. 1997c).

5.135 Germanium–Mercury–Thallium–Selenium

Tl₂GeSe₃–HgSe. The phase diagram of this system (Figure 5.73), constructed through DTA and XRD, is a eutectic type with the formation of the **Tl₂HgGeSe₄** quaternary compound, which melts congruently at 489°C and crystallizes in the tetragonal structure with the lattice parameters $a = 799.47 \pm 0.04$, $c = 676.17 \pm 0.04$ pm, and a calculated density of 7.716 ± 0.001 g·cm⁻³ (Mozolyuk et al. 2013). The eutectics contain 14 and 60 mol% HgSe and crystallize at 390°C and 467°C, respectively.

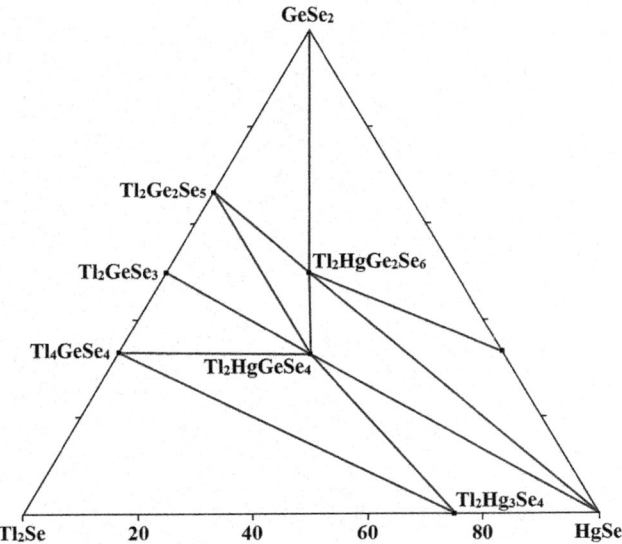

FIGURE 5.74 Isothermal section of the GeSe$_2$–HgSe–Tl$_2$Se quasiternary system at 500°C. (From Mozolyuk, M.Yu., et al., *Chem. Met. Alloys*, 6(1–2), 55, 2013.)

The solid solution ranges based on the system's component are smaller than 2 mol% Tl$_2$GeSe$_3$ and 5 mol% HgSe at 250°C.

GeSe$_2$–HgSe–Tl$_2$Se. The isothermal section of this quasiternary system at 500°C is presented in Figure 5.74 (Mozolyuk et al. 2013). One more quaternary compound, **Tl$_2$HgGe$_2$Se$_6$**, exists in the system at this temperature.

5.136 Germanium–Mercury–Europium–Selenium

The **HgEuGeSe$_4$** quaternary compound, which melts congruently at 764°C and crystallizes in the orthorhombic structure with the lattice parameters $a = 1065.30 \pm 0.07$, $b = 1059.90 \pm 0.06$, $c = 659.86 \pm 0.03$ pm, a calculated density of 6.606 g·cm^{-3}, and an energy gap of 1.97 eV, is formed in the Ge–Hg–Eu–Se system (Xing et al. 2020a). Single crystals of this compound were obtained by spontaneous crystallization from melts. EuSe (0.1 mM), HgSe (0.1 mM), GeSe$_2$ (0.1 mM) were fully milled in an Ar-filled glovebox. After that, the mixture of reagents was sealed into a quartz tube under a vacuum of 10^{-3} Pa and underwent a subsequent high-temperature solid-state reaction in a program-controlled muffle furnace with the following heating process: heated to 900°C at 50°C·h^{-1}, standing at this temperature for 3 days, cooled to 200°C at 4°C·h^{-1}, and then cooled naturally to ambient temperature by turning off the furnace. As a result, red crystals of HgEuGeSe$_4$ were obtained.

5.137 Germanium–Mercury–Tellurium–Selenium

GeSe$_2$–HgTe. The **Hg$_2$GeSe$_2$Te$_2$** quaternary compound, which melts incongruently at 477°C, is formed in this system (Asadov 2006). The solubility of HgTe in GeSe$_2$ and GeSe$_2$ in HgTe at 477°C is equal to 20 and 20 mol%, respectively. This system

was investigated through DTA, XRD, metallography, and measuring of the microhardness and EMF of concentrated chains.

The glass-forming regions in the **HgSe + GeTe ↔ HgTe + GeSe**, **GeSe$_2$–HgTe–GeSe**, and **GeSe$_2$–HgTe–GeTe** systems were determined by Feltz et al. (1976).

5.138 Germanium–Gallium–Thallium–Selenium

Two quaternary compounds, **TlGaGeSe$_4$** and **Tl$_2$Ga$_2$GeSe$_6$**, are formed in the Ge–Ga–Tl–Se system. The first compound has two polymorphic modifications, and one of them crystallizes in the orthorhombic structure with the lattice parameters $a = 1747.42 \pm 0.04$, $b = 741.05 \pm 0.02$, $c = 1194.06 \pm 0.03$ pm, a calculated density of 5.692 g·cm^{-3}, and an energy gap of 2.21 eV (Friedrich et al. 2020). The second modification of this compound crystallizes in the monoclinic structure with the lattice parameters $a = 1358.31 \pm 0.04$, $b = 740.15 \pm 0.02$, $c = 3074.10 \pm 0.07$ pm, $\beta = 96.066° \pm 0.002$, and a calculated density of 5.379 g·cm^{-3}. Orthorhombic modification of TlGaGeSe$_4$ was prepared by reacting Tl, Ga$_2$Se$_3$, Ge, and Se in a sealed evacuated quartz tube. The reaction mixture was heated to 900°C with a heating rate of 50°C·h^{-1}, annealed at 900°C for 2 days and cooled to room temperature with a cooling rate of 20°C·min^{-1}. The resulting crude product was finely ground in an agate mortar under a N$_2$ atmosphere and annealed at 900°C for 2 additional days. In the same batch, a few crystals of monoclinic modification were found. Both modifications decompose within several minutes by releasing gaseous H$_2$Se.

Tl$_2$Ga$_2$GeSe$_6$ crystallizes in the tetragonal structure with the lattice parameters $a = 807.70 \pm 0.04$, $c = 625.72 \pm 0.04$ pm, a calculated density of 5.91 g·cm^{-3}, and an energy gap of 2.05 eV (Babizhetskyy et al. 2020). This compound was prepared from the elements. Crystals were synthesized and grown in fused silica ampoules with a conical bottom. 37 mol% GeSe$_2$ and 63 mol% TlGaSe$_2$ (total mass 2 g) were placed in a fused silica tube, evacuated to a residual pressure of 0.1 Pa, and then sealed. The samples were synthesized in a furnace by heating to 900°C at the rate of 40°C·h^{-1}. The melt was kept at this temperature for 6 h with periodic vibration to assure homogeneity, cooled to 600°C at the rate of 20°C·h^{-1}, and annealed for 10 days. The process ended in cooling the ampoule to room temperature at 8.4°C·h^{-1}. Single crystals of Tl$_2$Ga$_2$GeSe$_6$ were grown using the Bridgman–Stockbarger method.

GeSe$_2$–Ga$_2$Se$_3$–Tl$_2$Se. The glass-forming region in this quasiternary system is presented in Figure 5.75 (Tsisar et al. 2017c). Glassy alloys with a maximum Tl$_2$Se content of 45 mol% were obtained in the Tl$_2$Se–GeSe$_2$ binary system. A continuous narrow strip of glass-formation from 13 to 33 mol% Ga$_2$Se$_3$ crosses the entire concentration triangle. A gap in the glass-formation area is observed on the section from GeSe$_2$ to the side Tl$_2$Se–Ga$_2$Se$_3$ at 75 mol% Tl$_2$Se. Up to 10 mol% Tl$_2$Se could be introduced into glasses at the GeSe$_2$–TlGaSe$_2$ section. Maximum content of 33 mol% Ga$_2$Se$_3$ was found in the glasses on the section from GeSe$_2$ to the Tl$_2$Se–Ga$_2$Se$_3$ binary system at 25 mol% Tl$_2$Se. The glass transition temperatures for these glasses are within the interval 188–304°C and the energy gap changes from 1.84 to 1.97 eV.

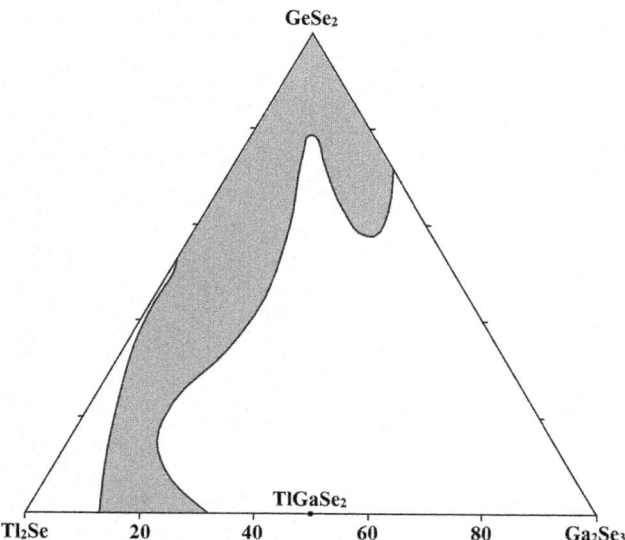

FIGURE 5.75 Glass-forming region in the GeSe$_2$–Tl$_2$Se–Ga$_2$Se$_3$ quasiternary system. (From Tsisar, O.V., et al., *Nauk. Visn. Uzhgorod. Univ. Ser. Khimiya*, [1(37)], 63, 2017.)

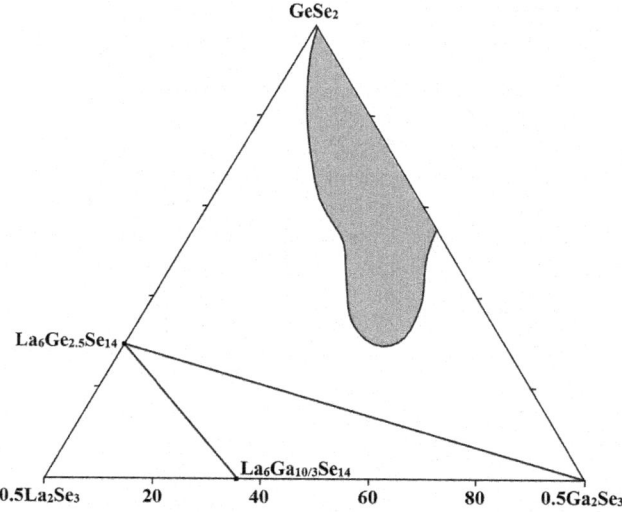

FIGURE 5.77 Glass-forming region in the GeSe$_2$–Ga$_2$Se$_3$–La$_2$Se$_3$ quasiternary system. (From Loireau-Lozac'h, A.-M., and Guittard, M., *Mater. Res. Bull.*, **12**(9), 887, 1977.)

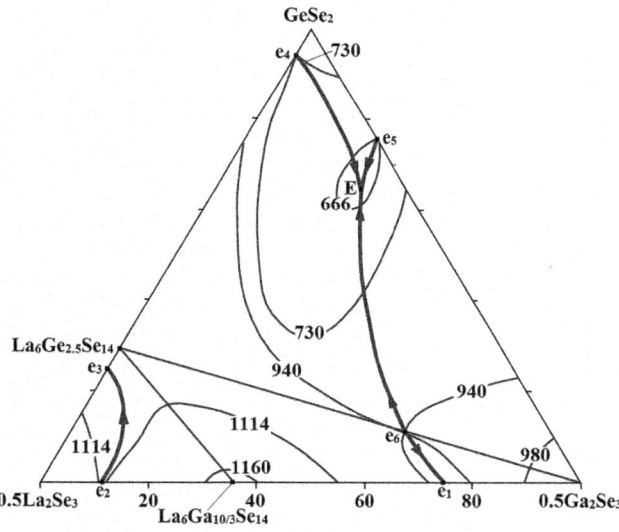

FIGURE 5.76 Liquidus surface of the GeSe$_2$–Ga$_2$Se$_3$–La$_2$Se$_3$ quasiternary system. (From Loireau-Lozac'h, A.-M., and Guittard, M., *Mater. Res. Bull.*, **12**(9), 887, 1977.)

5.139 Germanium–Gallium–Lanthanum–Selenium

GeSe$_2$–Ga$_2$Se$_3$–La$_2$Se$_3$. The liquidus surface of this quasiternary system is given in Figure 5.76 (Loireau-Lozac'h and Guittard 1977). The ternary eutectic *E* contain 5.5 mol% La$_2$Se$_3$, 16.7 mol% Ga$_2$Se$_3$, and 77.8 mol% GeSe$_2$ and crystallizes at 650°C. The primary crystallization field corresponding to the La$_2$Ge$_2$Se$_7$ compound was not found in the system. The glass-forming region in this system (Figure 5.77) is elongated along the eutectic line. The glass transition temperatures for these glasses are within the interval 385–449°C.

5.140 Germanium–Gallium–Praseodymium–Selenium

Pr$_3$Ge$_{1.25}$Se$_7$–Ga$_{1.67}$Pr$_3$Se$_7$. A continuous series of solid solutions are formed in this system (Blashko et al. 2020). The **GaPr$_3$Ge$_{0.5}$Se$_7$** quaternary phase crystallizes in the hexagonal structure with the lattice parameters $a = 1046.61 \pm 0.03$, $c = 620.67 \pm 0.03$ pm, and a calculated density of 6.0995 ± 0.0006 g·cm^{-3}.

5.141 Germanium–Gallium–Tin–Selenium

The **Ga$_2$GeSnSe$_6$** quaternary compound, which melts incongruently at 578°C and crystallizes in the orthorhombic structure with the lattice parameters $a = 4719.5 \pm 0.9$, $b = 752.13 \pm 0.15$, $c = 1218.3 \pm 0.2$ pm, and an energy gap of 1.98 eV, is formed in the Ge–Ga–Sn–Se system (Li et al. 2019b). A polycrystalline sample of this compound was gained by heating the stoichiometric mixture of SnSe, Ga$_2$Se$_3$, and GeSe$_2$ at 700°C for 6 days. Single crystals of Ga$_2$GeSnSe$_6$ were obtained through spontaneous crystallization. The reaction mixture of SnSe (0.5 mM), Ga$_2$Se$_3$ (0.5 mM), and GeSe$_2$ (0.5 mM) (a slight excess of GeSe$_2$ can be added as a flux) were ground and loaded into a fused silica tube under an Ar atmosphere in a glove box. The tube was flame-sealed under a high vacuum of 10^{-3} Pa and then placed in a furnace. The sample was heated to 1000°C within 15 h, kept for 70 h, then slowly cooled to 300°C at a rate of 3°C·h^{-1}, and finally cooled to room temperature by switching off the furnace. The resultant dark-red air stable crystals were hand-picked from the ampoule.

According to the studying using density functional theory, Ga$_2$GeSnSe$_6$ shows pseudo direct band gap of 1.24 eV (Yousaf et al. 2018).

5.142 Germanium–Gallium–Lead–Selenium

Two quaternary compounds, **Ga₂PbGeSe₆** and **Ga₄Pb₄GeSe₁₂**, are formed in the Ge–Ga–Pb–Se system. The first of them is thermally stable up to 600°C in a N_2 atmosphere and up to 520°C in the air and crystallizes in the orthorhombic structure with the lattice parameters $a = 4713.5 \pm 1.6$, $b = 757.8 \pm 0.3$, $c = 1216.1 \pm 0.4$ pm, a calculated density of 5.462 g·cm⁻³, and an energy gap of 1.96 eV (Luo et al. 2015). According to Badikov et al. (2019), this compound melts congruently at 720°C and has an energy gap of 2.25 eV.

Ga₂PbGeSe₆ was synthesized as follows (Luo et al. 2015). A mixture of PbSe (0.565 mM), Ga₂Se₃ (0.565 mM), Ge (0.565 mM), and Se (1.130 mM) was ground carefully, homogenized thoroughly with ethanol in an agate mortar, and pressed into sheets on the tablet machine. The sheets were put in a silica tube and flame-sealed under a 10⁻² Pa. Subsequently, the silica tube was placed in a muffle furnace and heated to 800°C in 48 h, dwelled at that temperature for 50 h, and then was cooled at a rate of 3°C·h⁻¹ to 300°C followed by rapid cooling to room temperature. A large number of bulk dark-red crystals of the title compound were obtained. The crystals are air stable for at least 5 months and stable under ethanol and water. A pure phase was produced by the stoichiometric mixtures of the Pb/Ga/Ge/Se molar ratio of 1:2:1:6. The mixture was heated to 600°C in 24 h, dwelled at that temperature for 50 h, and then cooled down to room temperature in 24 h. In order to improve the homogeneity and purity, the resulting product was reground by using a mortar and pestle and heated again at the same temperature.

This compound can be also obtained as follows (Badikov et al. 2019). The mixture of GeSe₂, Ga₂Se₃, and GeSe₂ (molar ratio 1:1:1) was placed in a graphitized quartz ampule, which was sealed off under vacuum (10⁻⁴ Pa) conditions by means of a gas burner and inserted into a horizontal furnace. After heating it to 770°C within 6 h, the melt was maintained at this temperature for additional 24 h, mixing it until complete homogenization.

Ga₄Pb₄GeS₁₂ crystallizes in the tetragonal structure with the lattice parameters $a = 1306.4 \pm 0.7$, $c = 631.0 \pm 0.5$ pm, a calculated density of 6.562 g·cm⁻³, and an energy gap of ≈ 1.91 eV (Chen et al. 2013).

To prepare Ga₄Pb₄GeSe₁₂, the stoichiometric mixture of Pb, Ga, Ge, and Se was loaded in a silica tube and then flame-sealed under a high vacuum of 10⁻³ Pa (Chen et al. 2013). The mixture was heated at a rate of 24°C·h⁻¹ to 800°C and annealed at that temperature for 100 h, then subsequently cooled at a rate of ~6°C·h⁻¹ to 350°C and the furnace was switched off. The dark-red crystals of this compound were obtained. All acquired elements were stored inside an Ar-filled glove box with controlled moisture and oxygen levels, and all manipulations were carried out in the glove box.

5.143 Germanium–Gallium–Tellurium–Selenium

GeTe–GaSe. The phase diagram of this system, constructed through DTA and XRD, is a eutectic type (Figure 5.78) (Magunov et al. 1979). The eutectic contains 50 mol% GeTe. Mutual solubility of GeTe and GaSe at the eutectic temperature

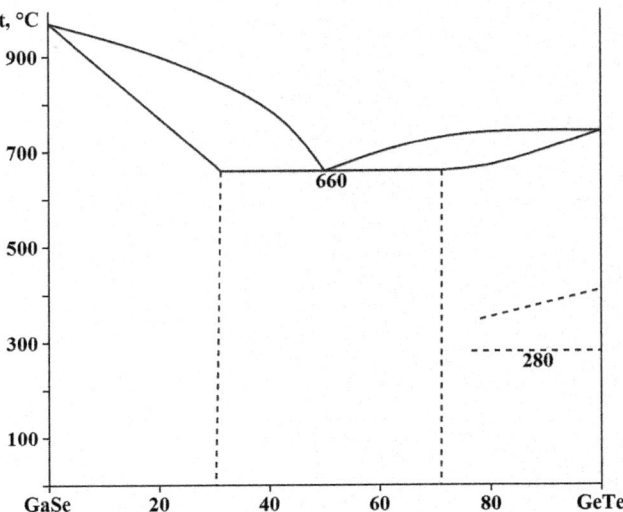

FIGURE 5.78   Phase diagram of the GeTe–GaSe system (From Magunov, R.L., et al., *Zhurn. neorgan. khimii*, **24**(11), 3133, 1979.)

reaches 30 mol%. Thermal effects in the solid state correspond to the polymorphic transformation of the solid solution based on GeTe.

5.144 Germanium–Indium–Thallium–Selenium

Two quaternary compounds, **TlInGeSe₄** and **TlInGe₂Se₆**, are formed in the Ge–In–Tl–Se system.

First of them crystallizes in the monoclinic structure with the lattice parameters $a = 758.6 \pm 0.1$, $b = 1213.4 \pm 0.1$, $c = 1802.3 \pm 0.1$ pm, $\beta = 96.037 \pm 0.006°$, a calculated density of 5.653 g·cm⁻³, and an energy gap of 1.93 eV (Friedrich et al. 2021). Polycrystalline sample of this compound was obtained by reacting Tl (1.41 mM), In (1.41 mM), Ge (1.41 mM), and Se (5.65 mM) in sealed evacuated fused silica tube. The reaction mixture was heated to 1000°C with a heating rate of 3°C·min⁻¹. After annealing at 1000°C for 24 h the tube was cooled to room temperature with a cooling rate of 1°C·min⁻¹. The air sensitive yellow polycrystalline product forms plate-like crystallites and is stored in a nitrogen glove box until further use.

TlInGe₂Se₆ quaternary compound crystallizes in the trigonal structure with the lattice parameters $a = 1017.98 \pm 0.02$, $c = 928.72 \pm 0.03$ pm, and an energy gap of 2.38 eV (Khyzhun et al. 2017). It was prepared by melting a stoichiometric mixture of the elements in an evacuated and sealed quartz ampoule. The ampoule was placed in a furnace and heated to 800°C at 20°C·h⁻¹. After holding at this temperature for 6 h, the furnace was cooled slowly down to 400°C and then annealed for 240 h at this temperature. A compact light-brown ingot of this compound was obtained.

GeSe₂–In₂Se₃–Tl₂Se. The isothermal section of this quasiternary system at 250°C is given in Figure 5.79 (Tsisar et al. 2018). The system is characterized by the formation of solid solutions of insignificant extent based on the initial binary compounds. The solid solution based on TlInSe₂ has the largest homogeneity region along the TlInSe₂–GeSe₂ section. One more quaternary compound, **TlInGeSe₄**, is formed in this

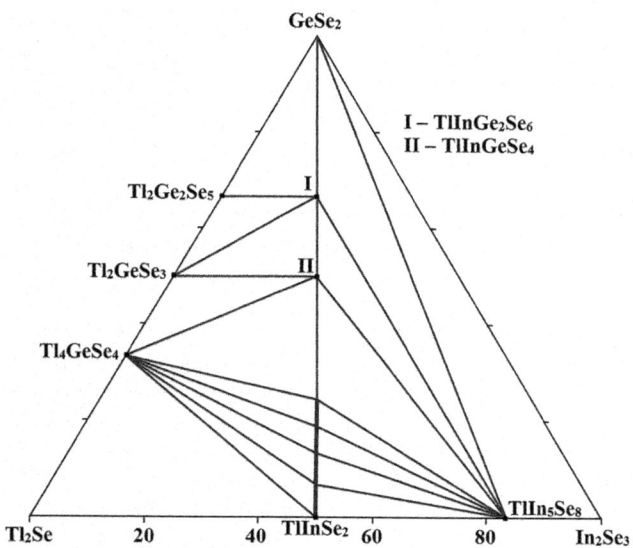

FIGURE 5.79 Isothermal section of the GeSe$_2$–In$_2$Se$_3$–Tl$_2$Se quasiternary system at 250°C. (From Tsisar, O., et al., *Visnyk L'viv. Univ. Ser. khim.*, (59), Pt. 1, 46, 2018.)

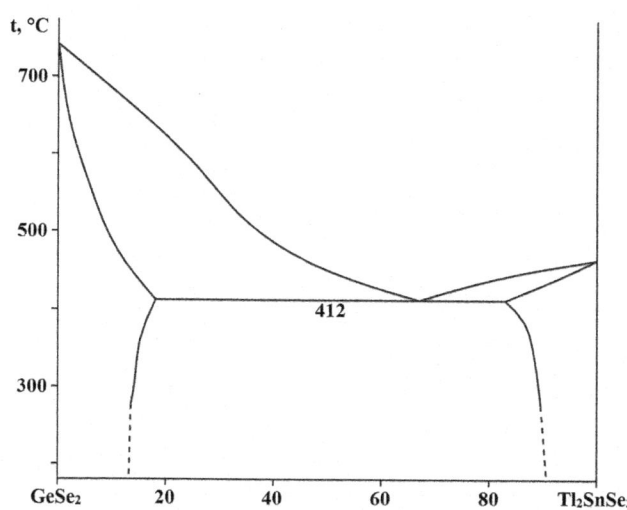

FIGURE 5.80 Phase diagram of the GeSe$_2$–Tl$_2$SnSe$_3$ system. (From Glukh, O.S., et al., *Nauk. visn. Uzhgorod. Univ., Ser. Khim.*, (19), 37, 2008.)

system. The extent of other solid solutions does not exceed 1 mol%. Ten single-phase regions, 19 two-phase regions, and 10 three-phase regions exist in the system at this temperature. The ingots for the investigations were annealed at 250°C for 240 h.

5.145 Germanium–Indium–Antimony–Selenium

The thermal studies carried out on **Ge$_{25-y}$In$_y$Sb$_{10}$Se$_{65}$** ($y = 0$–15) alloys reveal that their glass transition temperature decrease with increase in In content (Sharda et al. 2014). Bulk samples of these glasses were prepared using melt-quench technique.

5.146 Germanium–Thallium–Lanthanum– Selenium

The **TlLaGeSe$_4$** quaternary compound, which crystallizes in the orthorhombic structure with the lattice parameters $a = 1149.5 \pm 0.2$, $b = 713.18 \pm 0.14$, and $c = 1917.2 \pm 0.4$ pm, is formed in the Ge–Tl–La–Se system (McGuire et al. 2005a). This compound was synthesized from the elements. Since Tl will tarnish slowly in air it was handled in an argon glove box. Chunks of Tl were cut from the 0.5 inch diameter rod with wire cutters used only for this purpose, weighed inside the glove box and placed in capped glass vials. Appropriate amounts of the other elements were then weighed out and loaded into carbon-coated silica tube. The Ar-filled vials were removed from the glove box and the Tl was added to the silica tube. The tube was then quickly (to limit the exposure of Tl to air) attached to a vacuum line for sealing. It was heated over 24 h to 800°C, held at this temperature for 172 h and then cooled over 200 h to 300°C at which point the furnace was turned off and allowed to cool to room temperature. Crystals were extracted from the resulting polycrystalline mass for examination.

5.147 Germanium–Thallium–Tin–Selenium

GeSe$_2$–Tl$_2$SnSe$_3$. The phase diagram of this system, constructed through DTA and XRD using the ingots annealed at 150°C for 480 h, is a eutectic type (Figure 5.80) (Glukh et al. 2008). The eutectic contains 67 mol% Tl$_2$SnSe$_3$ and crystallizes at 412°C. The solubility of Tl$_2$SnSe$_3$ in GeSe$_2$ and GeSe$_2$ in Tl$_2$SnSe$_3$ at 412°C is 18 and 17 mol%, respectively, and decreases to 13 and 10 mol% at 150°C.

Tl$_2$GeSe$_3$–Tl$_2$SnSe$_3$. The phase diagram of this system, constructed through DTA and XRD using the ingots annealed at 150°C for 240 h, is a eutectic type (Barchiy et al. 2006). The eutectic contains 35 mol% Tl$_2$SnSe$_3$ and crystallizes at 372°C. The solubility of Tl$_2$SnSe$_3$ in Tl$_2$GeSe$_3$ at 372°C is ≈25 mol% and the solubility of Tl$_2$GeSe$_3$ in Tl$_2$SnSe$_3$ at this temperature is 55 mol%. At 150°C, the solid solutions contain 15 mol% Tl$_2$SnSe$_3$ and 35 mol% Tl$_2$GeSe$_3$, respectively.

Two parts of the **GeSe$_2$–Tl$_2$Se–GeSe$_2$** quasiternary system were investigated through DTA and XRD. The formation of quaternary compounds was not recorded in the **Tl$_4$GeSe$_4$– Tl$_4$SnSe$_4$–Tl$_2$Se** subsystem (Barchiy et al. 2005). The extent of solid solutions based on the initial components does not exceed 5 mol%. The liquidus surface consists of the fields of Tl$_4$Ge$_x$Sn$_{1-x}$Se$_4$ and (Tl$_2$Se) primary crystallization. The ingots for the investigations were annealed at 450°C for 480 h.

The formation of quaternary compounds was also not recorded in the **Tl$_2$GeSe$_3$–Tl$_4$GeSe$_4$–Tl$_4$SnSe$_4$–Tl$_2$SnSe$_3$** subsystem (Barchiy et al. 2006).The liquidus surface of this subsystem consists of the fields of Tl$_4$Ge$_x$Sn$_{1-x}$Se$_4$, (Tl$_2$GeSe$_3$), and (Tl$_2$SnSe$_3$) primary crystallization. The ternary eutectic contains 28 mol% GeSe$_2$, 15 mol% SnSe$_2$, and 57 mol% Tl$_2$Se and crystallizes at 336°C. The ingots for the investigations were annealed at 450°C for 240 h.

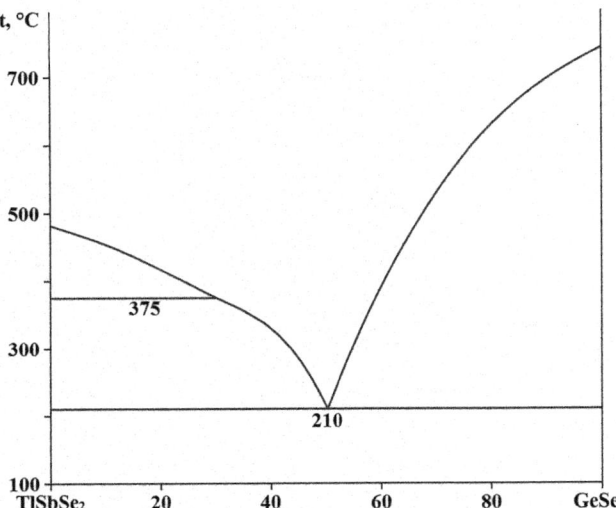

FIGURE 5.81 Phase diagram of the GeSe₂–TlSbSe₂ system. (From Makovskaya, Z.G., and Zhukov, E.G., *Zhurn. neorgan. khimii*, **28**(4), 1075, 1983.)

5.148 Germanium–Thallium–Antimony–Selenium

GeSe₂–TlSbSe₂. The phase diagram of this system, constructed through DTA, XRD, and metallography using the ingots annealed at 200°C for 7 days, is a eutectic type (Figure 5.81) (Makovskaya and Zhukov 1983b). The eutectic contains 50 mol% GeSe₂ and crystallizes at 210°C. TlSbSe₂ has a polymorphic transformation at 375°C. The region of stable glasses in this system lies in the range of 40–70 mol% GeSe₂. The glass transition temperature for such alloys is within the interval from 175°C to 190°C.

5.149 Germanium–Yttrium–Lead–Selenium

The **Y₁.₃₂Pb₁.₆₈Ge₁.₆₇Se₇** quaternary compound, which crystallizes in the hexagonal structure with the lattice parameters $a = 1039.4 \pm 0.1$, $c = 663.61 \pm 0.05$ pm, and a calculated density of 6.092 g·cm⁻³, is formed in the Ge–Y–Pb–Se system (Ruda et al. 2007; Marchuk 2018). This compound was obtained by melting the elements in evacuated (10⁻² Pa) quartz ampoule. The ampoule was heated to 1150°C (12°C·h⁻¹), kept at this temperature for 4 h, cooled to 500°C (12°C·h⁻¹), homogenized at this temperature for 500 h, and quenched in water.

5.150 Germanium–Lanthanum–Lead–Selenium

Two quaternary compounds, **La₁.₃₂Pb₁.₆₈Ge₁.₆₇Se₇** and **La₂Pb(GeSe₄)₂**, are formed in the Ge–La–Pb–Se system. First of them crystallizes in the hexagonal structure with the lattice parameters $a = 1059.0 \pm 0.5$, $c = 661.2 \pm 0.4$ pm, and a calculated density of 6.23 g·cm⁻³ (Ruda et al. 2007; Marchuk 2018).

La₂Pb(GeSe₄)₂ crystallizes in the trigonal structure with the lattice parameters $a = 939.8$, $c = 2827.3$ pm (Marchuk 2019). Both compounds were obtained in the same way as Y₁.₃₂Pb₁.₆₈Ge₁.₆₇Se₇ was prepared (Ruda et al. 2007; Marchuk 2018, 2019).

5.151 Germanium–Lanthanum–Antimony–Selenium

The **La₃GeSb₀.₃₁Se₇** quaternary compound, which crystallizes in the hexagonal structure with the lattice parameters $a = 1077.21 \pm 0.04$, $c = 609.11 \pm 0.05$ pm, and a calculated density of 5.86 g·cm⁻³, is formed in the Ge–La–Sb–Se system (Assoud et al. 2014). This compound was synthesized via a solid-state reaction method, starting with the elements in the stoichiometric ratios of La/Ge/Sb/Se = 3:1:0.33:7. The elements were placed into fused silica tube, which was then sealed under dynamic vacuum. This tube was heated in a furnace to 800°C and kept at that temperature for 4 days, followed by slow cooling to 300°C. Then the furnace was switched off to cool to room temperature. Dark-red crystals of La₃GeSb₀.₃₁Se₇ were obtained.

5.152 Germanium–Lanthanum–Iron–Selenium

The **La₃Fe₀.₅GeSe₇** quaternary compound, which crystallizes in the hexagonal structure with the lattice parameters $a = 1071.86 \pm 0.03$, $c = 607.97 \pm 0.03$ pm, and a calculated density of 4.7530 g·cm⁻³, is formed in the Ge–La–Fe–Se system (He et al. 2015). The crystals of the title compound were synthesized by traditional molten salt method with KI acting as a flux. The elements and KI were mixed with a molar ratio of La/Fe/Ge/Se/KI = 3:0.5:1:7:30. The mixture was ground and sealed in carbon-coated fused silica tube under vacuum (0.1 Pa). Then, the tube was slowly heated to 1000°C in a furnace followed by keeping at this temperature for 3 days. Afterwards, the tube was slowly cooled to 500°C at the rate of 2°C·h⁻¹, and then quenched in the air to room temperature. The final product was obtained by carefully breaking the tube, washed by distilled water several times, and dried by acetone to obtain black crystals of La₃Fe₀.₅GeSe₇. The direct combination reactions of the starting materials, without the flux (KI), were used to synthesize the powder samples. All operations were carried out in an Ar-protected glove box.

5.153 Germanium–Cerium–Lead–Selenium

The **Ce₁.₃₂Pb₁.₆₈Ge₁.₆₇Se₇** quaternary compound, which crystallizes in the hexagonal structure with the lattice parameters $a = 1054.2 \pm 0.6$, $c = 660.4 \pm 0.4$ pm, and a calculated density of 6.31 g·cm⁻³, is formed in the Ge–Ce–Pb–Se system (Ruda et al. 2007; Marchuk 2018). This compound was obtained in the same way as Y₁.₃₂Pb₁.₆₈Ge₁.₆₇Se₇ was prepared.

5.154 Germanium–Cerium–Cobalt–Selenium

Partially filled compounds **Ce$_x$Co$_4$Ge$_6$Se$_6$** were synthesized using two methods (Lin et al. 2007): (1) low-temperature interdiffusion and nucleation of ultrathin elemental layers and (2) a bulk high-temperature technique. The lattice parameters for the samples synthesized using the high-temperature technique were calculated to be 830 pm, while the samples synthesized using the elemental deposition technique showed lattice expansion proportional to the amount of incorporated Ce, with lattice parameters ranging from 830 pm for $x = 0$ up to 834 pm for $x = 0.2$ (cubic structure).

5.155 Germanium–Praseodymium–Lead–Selenium

The **Pr$_{1.32}$Pb$_{1.68}$Ge$_{1.67}$Se$_7$** quaternary compound, which crystallizes in the hexagonal structure with the lattice parameters $a = 1052.0 \pm 0.6$, $c = 662.3 \pm 0.4$ pm, and a calculated density of 6.32 g·cm^{-3}, is formed in the Ge–Pr–Pb–Se system (Ruda et al. 2007; Marchuk 2018). This compound was obtained in the same way as Y$_{1.32}$Pb$_{1.68}$Ge$_{1.67}$Se$_7$ was prepared.

5.156 Germanium–Neodymium–Lead–Selenium

The **Nd$_{1.32}$Pb$_{1.68}$Ge$_{1.67}$Se$_7$** quaternary compound, which crystallizes in the hexagonal structure with the lattice parameters $a = 1049.5 \pm 0.5$, $c = 664.0 \pm 0.4$ pm, and a calculated density of 6.35 g·cm^{-3}, is formed in the Ge–Nd–Pb–Se system (Ruda et al. 2007; Marchuk 2018). This compound was obtained in the same way as Y$_{1.32}$Pb$_{1.68}$Ge$_{1.67}$Se$_7$ was prepared.

5.157 Germanium–Samarium–Lead–Selenium

GeSe$_2$–Sm$_2$Se$_3$–PbSe. The isothermal section of this quasiternary system at 500°C is presented in Figure 5.82 (Olekseyuk et al. 2009a). The system is characterized by the formation of solid solutions of insignificant extent based on the initial binary compounds (1–2 mol%). The solid solution based on Sm$_2$Se$_3$ has the largest homogeneity region and is located along the Sm$_2$Se$_3$–PbSe section up to 50 mol% PbSe. Six single-phase regions, 10 two-phase regions, and 5 three-phase regions exist in the system at this temperature. The ingots for the investigations were annealed at 350°C for 240 h.

The **Sm$_{1.32}$Pb$_{1.68}$Ge$_{1.67}$Se$_7$** quaternary compound, which crystallizes in the hexagonal structure with the lattice parameters $a = 1044.2 \pm 0.5$, $c = 662.7 \pm 0.4$ pm, and a calculated density of 6.48 g·cm^{-3} (Ruda et al. 2007; Marchuk 2018) [$a = 1045.69 \pm 0.04$, $c = 662.22 \pm 0.05$ pm, and a calculated density of 6.462 ± 0.01 g·cm^{-3} (Olekseyuk et al. 2009a)], is formed in this system. This compound was obtained in the same way as Y$_{1.32}$Pb$_{1.68}$Ge$_{1.67}$Se$_7$ was prepared.

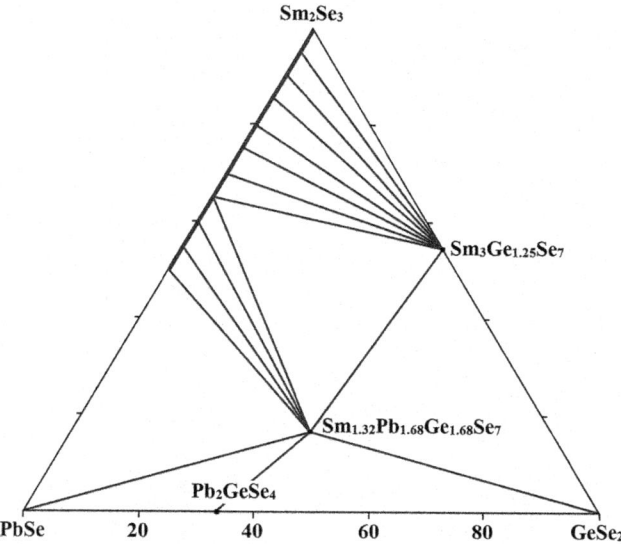

FIGURE 5.82 Isothermal section of the GeSe$_2$–Sm$_2$Se$_3$–PbSe quasiternary system at 500°C. (From Olekseyuk, I.D., et al., *Nauk. Visnyk Volyns'k. Nats. Univ. im. Lesi Ukrainky. Ser. Khim. nauky*, (24), 14, 2009.)

5.158 Germanium–Samarium–Antimony–Selenium

The crystal structure of **(Ge$_2$Sb$_2$Te$_5$)$_{100-x}$Sm$_x$** ($x = 0 - 1.2$) alloys was studied employing the XRD measurements which reveal the formation of both cubic and hexagonal phases (Kumar and Sharma 2021). There is an increase in the fraction of hexagonal phase up to $x = 0.4$ due to the substitution of vacancies by Sm leading to an increase in the tensile strain. The average crystallite size increases on Sm doping up to $x = 0.4$. Further increase in Sm content suppresses the growth of the hexagonal phase due to the grain refinement induced by the Sm addition with a decrease in average crystallite size and tensile strain. The bulk alloys were prepared by the conventional melt-quench technique. The mixtures of Ge, Sb, Te, and Sm were sealed according to their atomic percentage in evacuated (~10^{-3} Pa) quartz ampoules. The sealed ampoules were kept inside the furnace for 24 h in a multi-step heating program by maintaining the constant heating rate of 3°C·min^{-1} up to 1100°C temperature. There was continuous shaking of ampoules to ensure the homogeneity of the prepared samples. The ampoules were then immediately quenched into the ice-cold water. The quenched ampoules were kept inside the HF+H$_2$O$_2$ (1:3) solution for 24 h. The ampoules were then crushed gently to obtain the solid ingot of bulk alloys. The ingots were crushed into fine powder form.

5.159 Germanium–Samarium–Manganese–Selenium

The **Sm$_3$Mn$_{0.5}$GeSe$_7$** quaternary compound, which crystallizes in the hexagonal structure with the lattice parameters $a = 1039.58 \pm 0.08$, $c = 601.75 \pm 0.07$ pm, a calculated density

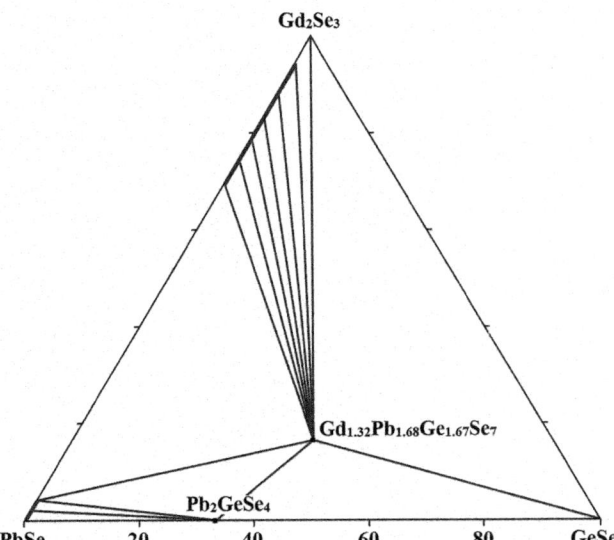

FIGURE 5.83 Isothermal section of the GeSe₂–Gd₂Se₃–PbSe quasiternary system at 500°C. (From Marchuk, O., et al., *Nauk. Visnyk Skhidnoyevrop. Nats. Univ. im. Lesi Ukrainky. Ser. Khim. nauky*, [20(297)], 29, 2014.)

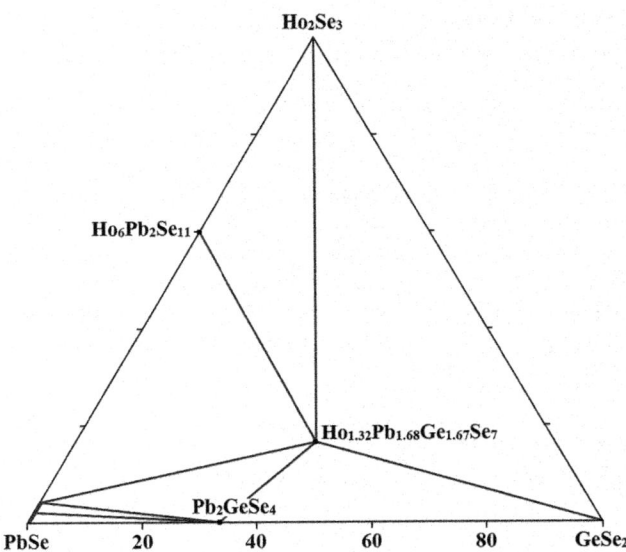

FIGURE 5.84 Isothermal section of the GeSe₂–Ho₂Se₃–PbSe quasiternary system at 500°C. (From Marchuk, O., et al., *Nauk. Visnyk Skhidnoyevrop. Nats. Univ. im. Lesi Ukrainky. Ser. Khim. nauky*, [20(297)], 29, 2014.)

of 4.2082 $g \cdot cm^{-3}$, and an energy gap of 1.37 eV, is formed in the Ge–Sm–Mn–Se system (He et al. 2015). The crystals of the title compound were synthesized in the same way as La₃Fe₀.₅GeSe₇ were prepared using Sm and Mn instead of La and Fe, respectively.

5.160 Germanium–Gadolinium–Lead–Selenium

GeSe₂–Gd₂Se₃–PbSe. The isothermal section of this quasiternary system at 500°C is presented in Figure 5.83 (Marchuk et al. 2014). The system is characterized by the formation of solid solutions of insignificant extent based on the initial binary compounds (1–2 mol%). The largest solid solution, $Gd_{2+2/3x}Pb_{1-x}Se_4$ ($x = 0.5–0.9$), is formed in the Gd₂Se₃–PbSe system. Six single-phase regions, 10 two-phase regions, and 5 three-phase regions exist in the system at this temperature. The ingots for the investigations were annealed at 500°C for 500 h.

The **Gd₁.₃₂Pb₁.₆₈Ge₁.₆₇Se₇** quaternary compound, which crystallizes in the hexagonal structure with the lattice parameters $a = 1042.8 \pm 0.2$, $c = 663.8 \pm 0.2$ pm, and a calculated density of 6.53 $g \cdot cm^{-3}$ (Ruda et al. 2007; Marchuk et al. 2014; Marchuk 2018), is formed in this system. This compound was obtained in the same way as Y₁.₃₂Pb₁.₆₈Ge₁.₆₇Se₇ was prepared.

5.161 Germanium–Terbium–Lead–Selenium

The **Tb₁.₃₂Pb₁.₆₈Ge₁.₆₇Se₇** quaternary compound, which crystallizes in the hexagonal structure with the lattice parameters $a = 1040.6 \pm 0.1$, $c = 663.84 \pm 0.09$ pm, and a calculated density of 6.57 $g \cdot cm^{-3}$, is formed in the Ge–Tb–Pb–Se system (Ruda et al. 2007; Marchuk 2018). This compound was obtained in the same way as Y₁.₃₂Pb₁.₆₈Ge₁.₆₇Se₇ was prepared.

5.162 Germanium–Dysprosium–Lead–Selenium

The **Dy₁.₃₂Pb₁.₆₈Ge₁.₆₇Se₇** quaternary compound, which crystallizes in the hexagonal structure with the lattice parameters $a = 1038.9 \pm 0.4$, $c = 664.7 \pm 0.3$ pm, and a calculated density of 6.61 $g \cdot cm^{-3}$, is formed in the Ge–Dy–Pb–Se system (Ruda et al. 2007; Marchuk 2018). This compound was obtained in the same way as Y₁.₃₂Pb₁.₆₈Ge₁.₆₇Se₇ was prepared.

5.163 Germanium–Holmium–Lead–Selenium

GeSe₂–Ho₂Se₃–PbSe. The isothermal section of this quasiternary system at 500°C is presented in Figure 5.84 (Marchuk et al. 2014). The system is characterized by the formation of solid solutions of insignificant extent based on the initial binary compounds (1–2 mol%). Six single-phase regions, 10 two-phase regions, and 5 three-phase regions exist in the system at this temperature. The ingots for the investigations were annealed at 500°C for 500 h.

The **Ho₁.₃₂Pb₁.₆₈Ge₁.₆₇Se₇** quaternary compound, which crystallizes in the hexagonal structure with the lattice parameters $a = 1038.1 \pm 0.1$, $c = 664.6 \pm 0.1$ pm, and a calculated density of 6.64 $g \cdot cm^{-3}$ (Ruda et al. 2007; Marchuk et al. 2014; Marchuk 2018), is formed in this system. This compound was obtained in the same way as Y₁.₃₂Pb₁.₆₈Ge₁.₆₇Se₇ was prepared.

5.164 Germanium–Erbium–Lead–Selenium

GeSe₂–Er₂Se₃–PbSe. The isothermal section of this quasiternary system at 500°C is constructed by Olekseyuk et al. (2009a). The system is characterized by the formation of solid solutions of insignificant extent based on the initial binary compounds

(1–2 mol%). It was shown that two quasibinary systems, $Pb_2GeSe_4–Er_2Se_3$ and $Pb_2GeSe_4–Er_2PbSe_4$, divided it for three 3-phase regions: $PbSe–Pb_2GeSe_4–Er_2PbSe_4$, $Er_2PbSe_4–Pb_2GeSe_4–Er_2Se_3$, and $Er_2Se_3–Pb_2GeSe_4–GeSe_2$. Besides these three-phase regions, five single-phase regions, and seven 2-phase regions exist in the system at this temperature. The ingots for the investigations were annealed at 350°C for 240 h.

5.165 Germanium–Tin–Lead–Selenium

The $GeSn_{0.5}Pb_{1.5}Se_4$ quaternary phase, which crystallizes in the cubic structure with the lattice parameter $a = 1458.0 \pm 0.2$ pm, is formed in the Ge–Sn–Pb–Se system (Poduska et al. 2002a).

5.166 Germanium–Tin–Arsenic–Selenium

GeSe–SnSe–As₂Se₃. The glass-forming region in this quasiternary system is presented in Figure 5.85 (Bakhtiyarov and Vasil'ev 1975; Vasil'ev et al. 1977). According to Mössbauer's studies, tin in the composition of glasses can be either Sn(II) or Sn(IV), which indicates that the system is not a quasiternary section of the Ge-Sn-As-Se quaternary system and explains the high glass-forming ability. Up to 40 mol% SnSe dissolves in the glasses of the GeSe–As₂Se₃ system. The criterion for glass-formation was X-ray amorphous, the absence of inclusions when viewed under a metal microscope, and the presence of a conchoidal fracture.

GeSe₂–SnSe–As₂Se₃. The glass-forming region in this quasiternary system is presented in Figure 5.86 (Pazin and Morozov 1978). According to electrophysical measurements, tin in the composition of glasses can be either Sn(II) or Sn(IV), which indicates that the system is not a quasiternary section of the Ge-Sn-As-Se quaternary system and explains the high glass-forming ability.

GeSe₂–SnSe₂–As₂Se₃. The glass-forming region in this quasiternary system adjoins the GeSe₂–As₂Se₃ side (Figure 5.85) (Bakhtiyarov and Vasil'ev 1975; Vasil'ev et al. 1977; Pazin et al. 1978b). Up to 40 mol% SnSe₂ [up to 50 mol% SnSe₂ (Bakhtiyarov and Vasil'ev 1975)] can be introduced into the composition of glassy alloys (Pazin et al. 1978b). According to the data of Pazin et al. (1978b), glasses containing up to 90 mol% GeSe₂ were obtained in this system.

5.167 Germanium–Tin–Antimony–Selenium

GeSe₂–SnSe–Sb₂Se₃. The glass-forming region in this quasiternary system is given in Figure 5.87 (Vassilev et al. 2009). Glasses were obtained in the GeSe₂-rich region and on the GeSe₂–Sb₂Se₃ side of the ternary system from 30 to 100 mol% GeSe₂. The glass transition temperature for such alloys is within the interval from 232°C to 323°C.

5.168 Germanium–Tin–Tellurium–Selenium

GeSe–SnTe. The solubility of GeSe in SnTe at 400°C and 600°C reaches 20 and 40 mol%, respectively, and the solubility of SnTe in GeSe at 430°C is 5 mol% (Krebs and Langner 1964; Nikolić 1965). The lattice parameters and an energy gap change linearly versus composition within the interval 0–40 mol% GeSe. The alloys containing more than 50 mol% GeSe were either two-phase or had an orthorhombic structure characteristic of γ-GeSe (Abrikosov et al. 1988).

The ingots for the investigations were annealed at 600°C for 24 days (10 days at the GeSe content less than 10 mol%) (Nikolić 1965) [at 400°C and 430°C for 80 and 120 h, respectively (Krebs and Langner 1964) and studied through XRD.

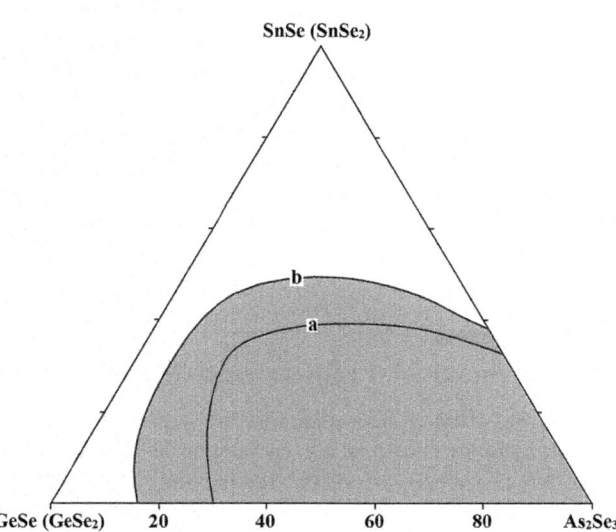

FIGURE 5.85 Glass-forming regions in the (a) GeSe–SnSe–As₂Se₃ and (b) GeSe₂–SnSe₂–As₂Se₃ quasiternary systems. (From Bakhtiyarov, and Vasil'ev, L.N., *Izv. AN SSSR. Neorgan. Mater.*, **11**(4), 741, 1975.)

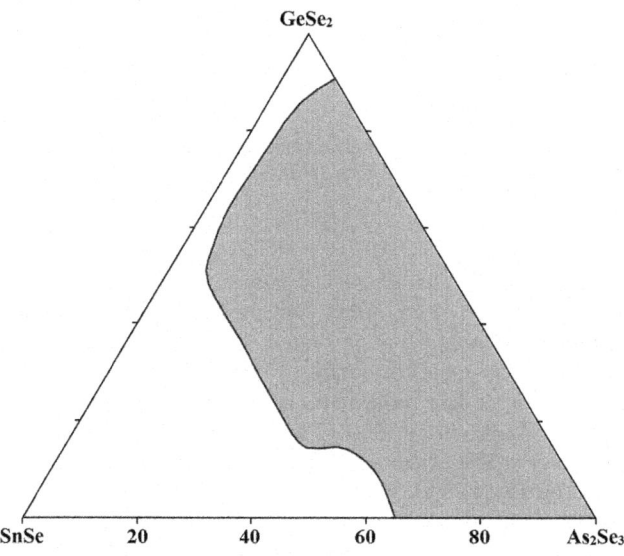

FIGURE 5.86 Glass-forming region in the GeSe₂–SnSe–As₂Se₃ quasiternary system. (From Pazin, A.V., and Morozov, V.A., *Izv. AN SSSR. Neorgan. mater.*, **14**(8), 1401, 1978.)

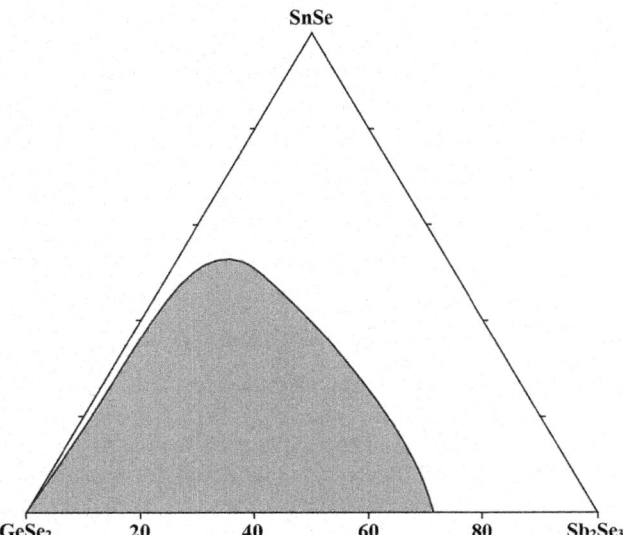

FIGURE 5.87 Glass-forming region in the GeSe₂–SnSe–Sb₂Se₃ quasiternary system. (From Vassilev, V., et al., *J. Alloys Compd.*, **485**(1–2), 569, 2009.)

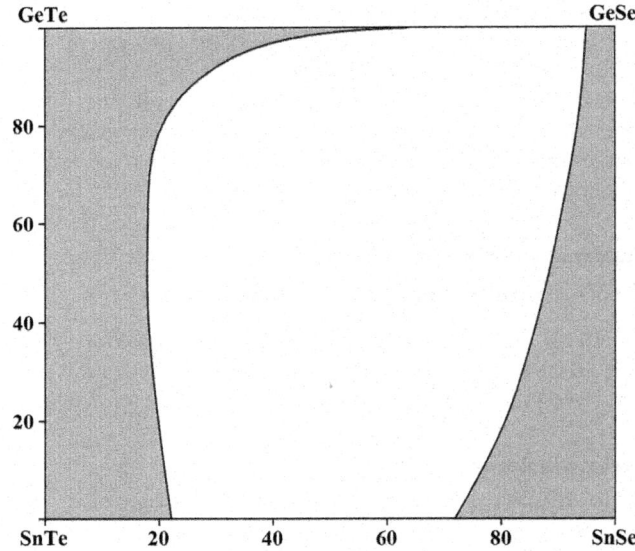

FIGURE 5.89 Region of the solid solutions in the GeSe + SnTe ↔ GeTe + SnSe ternary mutual system. (From Krebs, H. and Langner, D., *Z. anorg. und allg. Chem.*, **334**(1–2), 37, 1964.)

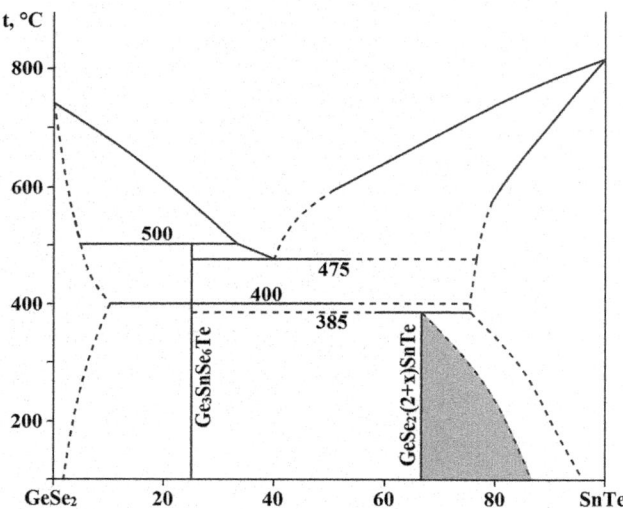

FIGURE 5.88 Phase diagram of the GeSe₂–SnTe system. (From Vassilev, V., et al., *J. Therm. Anal. Calorim.*, **76**(3), 727, 2004.)

GeSe₂–SnTe. The phase diagram of this system, constructed through DTA, XRD, and measuring of the microhardness and density, is presented Figure 5.88 (Vassilev et al. 2004). The eutectic contains 40 mol% SnTe and crystallizes at 475°C. The coordinates of peritectic and peritectoid are 33.3 mol% SnTe, 500°C and 75 mol% SnTe, 385°C, respectively. The existence of fields of limited solid solutions based on GeSe₂ and SnTe was observed. At elevated temperatures the solubility of SnTe in GeSe₂ and that of GeSe₂ in SnTe increase with maximum values: 10 mol% SnTe at 400°C and ≈25 mol% GeSe₂ at 385°C. At room temperature GeSe₂ (SnTe) does not dissolve SnTe (GeSe₂). Two quaternary phases were obtained in the system. **3GeSe₂·SnTe**

(**Ge₃SnSe₆Te**) melts incongruently at 500°C, undergoes a polymorphic transition at 400°C, and has a constant composition. α-3GeSe₂·SnTe crystallizes in the monoclinic structure with the lattice parameters $a = 795.5$, $b = 696.9$, $c = 606.4$ pm, and $\beta = 91.47°$. The other phase dissociates when heated through a peritectoidal reaction at 385°C, has a variable composition **GeSe₂·(2+x)SnTe** and crystallizes in the cubic structure with the lattice parameter $a = 606.3$ pm for $x = 0$. The maximum temperature and the duration of the sample's synthesis were 1050°C and 1 h, respectively. A vibrational stirring of the melt was applied. As a final step the melts were homogenized at 350°C for 3 h.

SnSe–GeTe. The solubility of SnSe in GeTe at 380–420°C reaches 40 mol% [20 mol% at 550°C (Krebs and Langner 1964)] and the solubility of GeTe in SnSe is 20 mol% (Krebs et al. 1961) [20 mol% at 480°C (Krebs and Langner 1964)]. The ingots for the investigations were annealed at 380–420°C for 90–100 h (Krebs et al. 1961) [at 480°C and 500°C for 100 and 50 h, respectively (Krebs and Langner 1964)] and were studied through XRD.

GeSe + SnTe ↔ GeTe + SnSe. The region of the solid solutions in this ternary mutual system was determined by Krebs and Langner (1964) and is presented in Figure 5.89.

5.169 Germanium–Lead–Arsenic–Selenium

Ge₁.₅As₀.₅Se₃–PbSe. Electron microscopy studies of growth of microstructure or crystal size as a function of heat treatment temperature have been used to determine immiscibility and liquidus temperatures in this quasibinary system (Mecholsky et al. 1973). The immiscibility dome was found to span nearly the entire pseudobinary composition region, in good agreement with the results of a previous heat capacity study of the system in the glass transition region.

5.170 Germanium–Lead–Oxygen–Selenium

The **Pb₂Ge(SeO₃)₄** quaternary compound, which is stable up to 345°C and crystallizes in the monoclinic structure with the lattice parameters $a = 897.1 \pm 0.5$, $b = 663.3 \pm 0.3$, $c = 925.6 \pm 0.5$ pm, $\beta = 96.376 \pm 0.009°$, a calculated density of 6.036 g·cm⁻³, and an energy gap of 4.10 eV, is formed in the Ge–Pb–O–Se system (Zhang et al. 2012b). To synthesize this compound, a mixture of PbO (0.2 mM), GeO₂ (0.1 mM), SeO₂ (2.0 mM) and H₂O (4 mL) was sealed in an autoclave equipped with a Teflon liner (23 mL), heated at 200°C for 4 days, and then cooled to room temperature at a rate of 6°C·h⁻¹. The final pH value of the reaction media is close to 0.5. The product was washed with water, and then dried in air. Colorless prism-shaped crystals of Pb₂Ge(SeO₃)₄ were obtained as a single phase.

5.171 Germanium–Lead–Tellurium–Selenium

GeSe–PbTe. The solubility of GeSe in PbTe at 600°C reaches 35 mol% (Krebs and Langner 1964). The lattice parameters and an energy gap change linearly versus composition within the interval 0–35 mol% GeSe. The alloys containing more GeSe were either two-phase or had an orthorhombic structure characteristic of γ-GeSe (Abrikosov et al. 1988). The ingots for the investigations were annealed at 600°C for 21 days and studied through XRD (Krebs and Langner 1964).

GeSe₂–PbTe. The phase diagram of this system, constructed through DTA, XRD, and measuring of the microhardness, is a eutectic type with peritectic transformation (Figure 5.90) (Rustamov et al. 1988; Vassilev et al. 2007b). The eutectic contains ≈53 mol% PbTe and crystallizes at 410°C [42 mol% PbTe and 430°C (Rustamov et al. 1988)]. The **GeSe₂·(2+x) PbTe** quaternary phase is formed in this system, which melts incongruently at 425°C, has a variable composition, and undergoes a polymorphic transition at temperatures from 320°C to 370°C depending on the composition (Vassilev et al. 2007b).

α-GeSe₂·(2+x)PbTe crystallizes in the triclinic structure with the lattice parameters $a = 620.0 \pm 0.2$, $b = 380.7 \pm 0.2$, $c = 364.5 \pm 0.2$ pm, $a = 106.47 \pm 0.02°$, $\beta = 99.85 \pm 0.02°$ and $\gamma = 87.22 \pm 0.02°$ at $x = 0$. The maximum temperature of the alloy synthesis was 970°C with a duration of 1.5 h, including vibrational stirring of the melt. During the synthesis of the phases rich in PbTe, the internal surface of the ampoules was preliminary covered with graphite. The synthesis ended with homogenizing temperature rise at 300°C throughout 1 month, with subsequent cooling of the melt when the furnace was switched off.

The alloys of the $(Ge_{0.97}Pb_{0.03})_{0.49}(Te_{1-x}Se_x)_{0.51}$ solid solutions ($0 \leq x \leq 0.20$) were studied through XRD, metallography, and dilatometry (Shelimova and Karpinskiy 1991). They were homogenized at 500°C for 500 h, and then the alloys were cooled to 300°C and annealed for 2000 h, followed by the cooling in the air. It was shown that at the increasing of the Se content, a concentration phase transition (γ-GeTe) → (α-GeTe) takes place. It was determined that this transition was realized at smaller x values compared with the $Ge_{0.49}(Te_{1-x}Se_x)_{0.51}$ alloys.

Ge₀.₉₈Te–PbSe. Continuous series of the solid solutions with a minimum at 647°C and 20 mol% PbSe (Figure 5.91) are formed in this system (Abrikosov et al. 1985a). This section is quasibinary only at temperatures close to the solidus temperatures. The forming solid solutions are stable within the interval 630–530°C. Below 530°C complex transformations are observed in the alloys associated with the decomposition of the solid solution, polymorphism of GeTe, and the formation of GeTe₀.₇₅Se₀.₂₅ compound (ε-phase). Two nonvariant reactions take place in the solid state: a transition reaction β + δ ↔ α + ε at ca. 360°C (α – solid solution based on α-GeTe, β – solid solution based on β-GeTe, δ – solid hsolution adjacent to the PbSe–PbTe system) and a peritectoid reaction β + γ' + δ ↔ ε (γ' – solid solution based on low-temperature modification of GeSe). The solubility of the components in each other decreases with decreasing temperature. The unit cell parameter in the region of solid solutions varies linearly (Nikolić 1969).

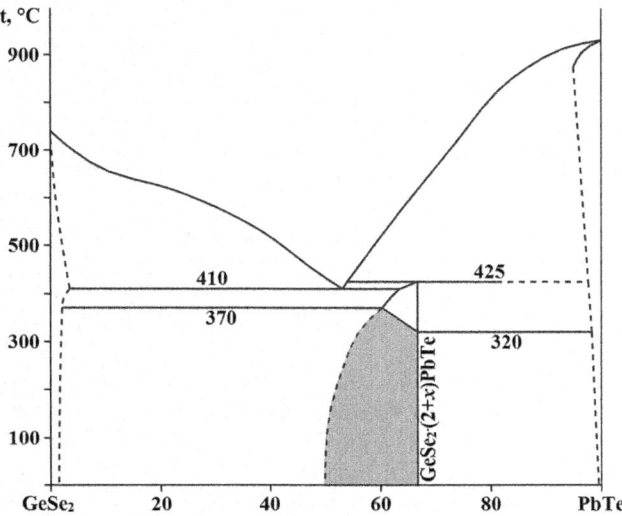

FIGURE 5.90 Phase diagram of the GeSe₂–PbTe system. (From Vassilev, V., et al., *Thermochim. Acta*, **459**(1–2), 12, 2007.)

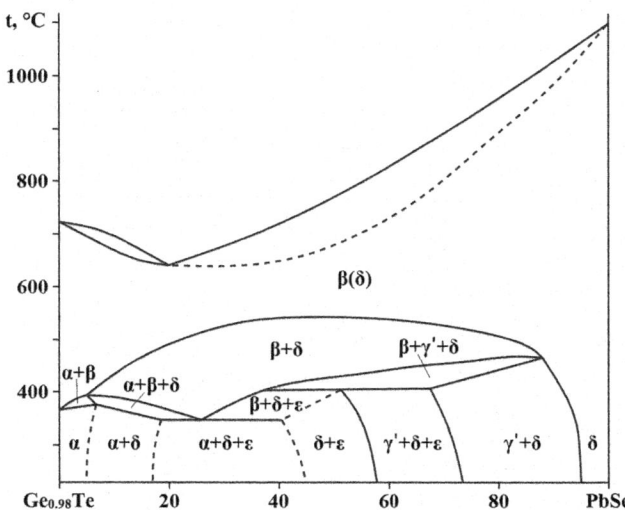

FIGURE 5.91 Phase equilibria in the Ge₀.₉₈Te–PbSe system. (From Abrikosov, N.H., et al., *Izv. AN SSSR. Neorgan. mater.*, **21**(10), 1664, 1985.)

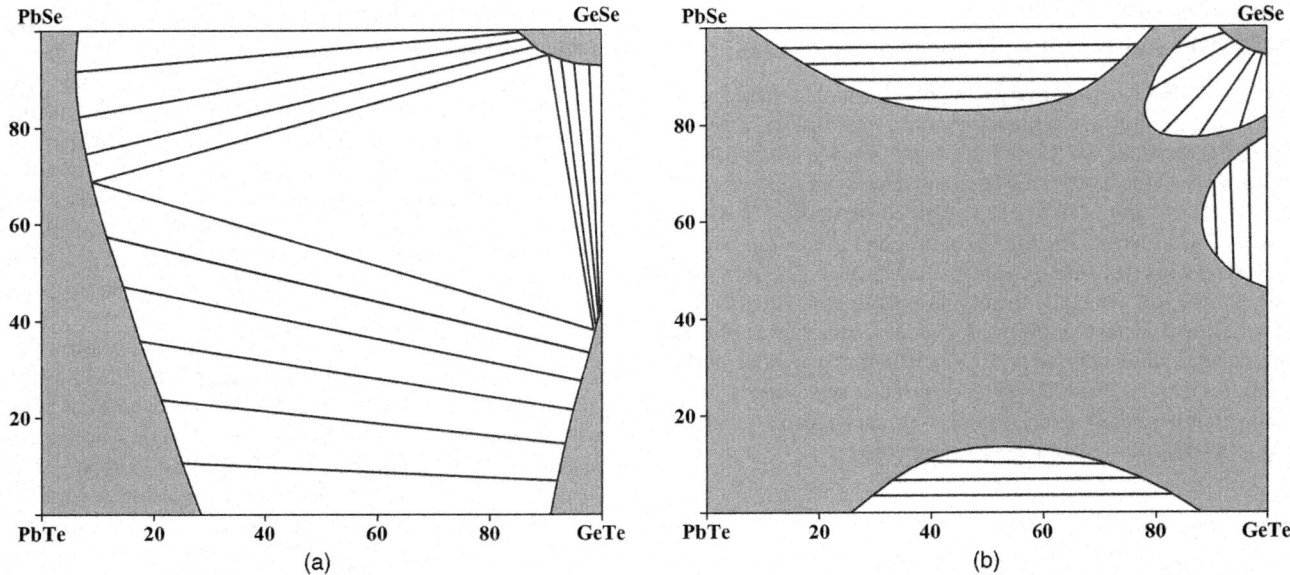

FIGURE 5.92 Isothermal sections of the GeSe + PbTe ↔ GeTe + PbSe ternary mutual system at (a) 410°C and (b) 540°C. (From Shelimova, L.E., and Abrikosov, N.H., *Izv. AN SSSR. Neorgan. mater.*, **4**(11), 1885, 1968.)

This system was studied through DTA, XRD, metallography, and dilatometry, and the ingots for the investigation were annealed first at 500°C for 500 h and then at 300°C for 2000 h (Abrikosov et al. 1985a) [at 700°C for 2 weeks (Nikolić 1969)].

GeSe + PbTe ↔ GeTe + PbSe. The isothermal sections of this ternary mutual system at 410°C and 540°C are presented in Figure 5.92 (Shelimova and Abrikosov 1968). Three regions of the solid solutions exist in the system at 410°C (Figure 5.92a): solid solution forming near the PbSe–PbTe quasibinary system, solid solution based on GeTe elongated along the GeSe–GeTe quasibinary system, and solid solution based on the GeSe low-temperature modification. Single-phase regions are divided by the two-phase ones, and a large three-phase region exists in the middle of the ternary mutual system. The isothermal section at 540°C (Figure 5.92b) is situated below the temperature of GeSe polymorphic transformation and above the temperature of GeTe polymorphic transformation. The main field of the ternary mutual system occupies the region of the $Ge_xPb_{1-x}Se_yTe_{1-y}$ solid solution, adjacent to the PbSe–PbTe quasibinary system. Near the GeSe-side there is a small region of the solid solution based on the GeSe low-temperature modification. There are four two-phase regions in this section: first of them is adjacent to the region of the solid solution based on the GeSe low-temperature modification, and three others are adjacent to the GeTe–PbTe, GeSe–PbSe, and GeSe–GeTe quasibinary systems.

The liquidus surface of this ternary mutual system (Figure 5.93) consisted of the fields of the primary crystallization of the $Ge_xPb_{1-x}Se_yTe_{1-y}$ solid solution that occupies the biggest part of the system and Ge near GeSe-corner (Shelimova and Abrikosov 1968). These fields are divided by the line of the monovariant equilibrium. On the liquidus surface, it is assumed that there is a region with the lowest crystallization

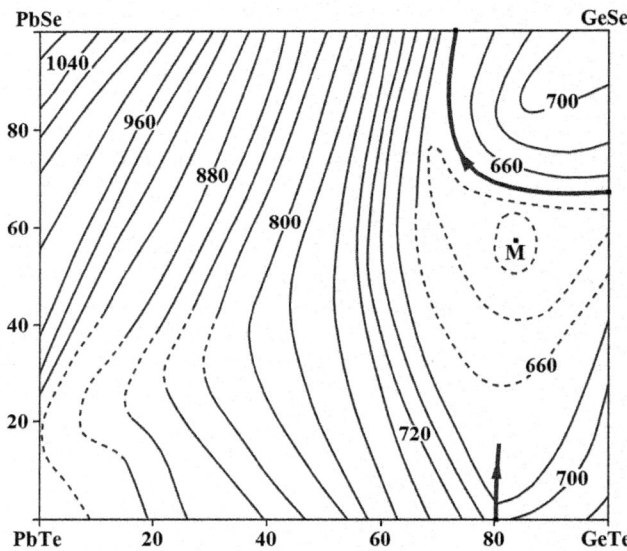

FIGURE 5.93 Liquidus surface of the GeSe + PbTe ↔ GeTe + PbSe ternary mutual system. (From Shelimova, L.E., and Abrikosov, N.H., *Izv. AN SSSR. Neorgan. mater.*, **4**(11), 1885, 1968.)

temperatures of the alloys. Point *M* denotes the supposed position of the minimum on the liquidus surface.

The thermal effects of heating and cooling of some samples of the GeTe–PbSe–PbTe subsystem were also determined by Alekseeva et al. (1969).

This system was studied through DTA, XRD, and metallography (Shelimova and Abrikosov 1968; Alekseeva et al. 1969), and the ingots for the investigation were annealed fist at 410°C and 540°C for 1800 and 1400 h, respectively (Shelimova and Abrikosov 1968) [at 600°C for 30 days (Alekseeva et al. 1969).

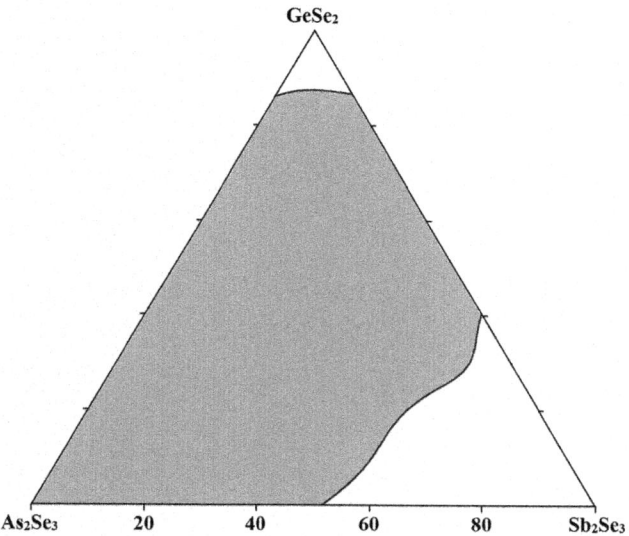

FIGURE 5.94 Glass-forming region in the $GeSe_2$–As_2Se_3–Sb_2Se_3 quasiternary system. (From Pazin, A.V., et al., *Izv. AN SSSR. Neorgan. mater.*, **14**(3), 417, 1978.)

5.172 Germanium–Arsenic–Antimony–Selenium

$GeSe_2$–As_2Se_3–Sb_2Se_3. The glass-forming region in this quasiternary system is given in Figure 5.94 (Pazin et al. 1978a). Up to 60 mol% Sb_2Se_3 and up to 80 mol% $GeSe_2$ can be introduced into the composition of glassy alloys.

5.173 Germanium–Arsenic–Tellurium–Selenium

Results of DSC of $Ge_xAs_{40-x}Se_{40}Te_{20}$ ($x = 0$–40) chalcogenide glasses were obtained by Shiryaev et al. (2004). The glass transition temperature with increasing the Ge content changes from 140°C to 320°C. The studied glasses with $x \leq 35$ have no exothermal peaks of crystallization, indicating their high glass-forming ability. The glass transition temperature in samples of the same composition increases with the increase in heating rate.

5.174 Germanium–Antimony–Tellurium–Selenium

$GeSe_2$–Sb_2Te_3. The phase diagram of this system, constructed through DTA, DSC, XRD, and metallography, is a eutectic type (Figure 5.95) (Makovskaya and Zhukov 1983a; Suriñach et al. 1984, 1985). The eutectic contains 24 mol% Sb_2Te_3 and crystallizes at 485°C (Suriñach et al. 1984, 1985) [42 mol% Sb_2Te_3 and 400°C (Makovskaya and Zhukov 1983a)]. The solubility

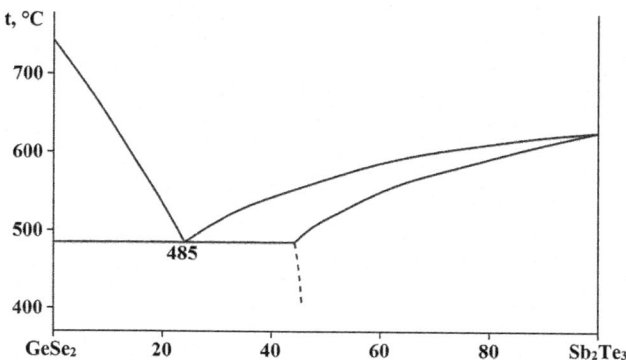

FIGURE 5.95 Phase diagram of the $GeSe_2$–Sb_2Te_3 system. (From Suriñach, S., et al., *J. Mater. Sci.*, **19**(9), 3005, 1984.)

of $GeSe_2$ in Sb_2Te_3 at the eutectic temperature is 57 mol%. By water quenching glasses were obtained from compositions in the range 5–30 mol% Sb_2Te_3. The glass transition temperature for these alloys is within the interval 207–340°C and decreases with the Sb_2Se_3 content increasing.

Bulk glasses of $(GeSe_2)_{60}(GeTe)_{40-x}(Sb_2Te_3)_x$ with $0 \leq x \leq 20$ were prepared by melting weighed amounts of elemental Ge, Sb, Se, and Te in evacuated and sealed quartz ampoules (Suriñach et al. 1983). The molten alloys were held at 1000°C for 12 h and constantly agitated to ensure homogeneity; subsequently, they were quenched in water at room temperature.

The glass-forming regions in the quaternary system Ge–Sb–Te–Se have been obtained for quenching in air and for a constant cooling rate of 5°C·min^{-1} by Bordas et al. (1979). The formation and stability of glasses were discussed in terms of the obtained glass-forming regions and the observed thermal behavior in DTA measurements.

5.175 Germanium–Bismuth–Tellurium–Selenium

GeTe–BiSe. The solubility of GeTe in BiSe at 450°C and 500°C is 20 and 23 mol%, respectively (Abrikosov et al. 1972). The system was studied through DTA, XRD, and measuring of the microhardness and the ingots for the investigations were annealed at 250°C and 450°C for 10 and 30 days, respectively.

5.176 Germanium–Tellurium–Iron–Selenium

GeTe–FeSe. The phase diagram of this system, constructed through DTA, XRD, metallography, and measuring of the microhardness, is constructed by Asadov et al. (2012). The eutectics crystallize at 600°C and 640°C. The **GeFeSeTe** quaternary compound, which melts congruently at 875°C and crystallizes in the cubic structure with the lattice parameter $a = 1073$ pm, is formed in this system.

6

Systems Based on Germanium Telluride

6.1 Germanium–Lithium–Antimony–Tellurium

The $(GeTe)_x(LiSbTe_2)_2$ solid solutions are formed in the Ge–Li–Sb–Te system (Schröder et al. 2013). They crystallize in the cubic structure with the lattice parameter $a = 606.904 \pm 0.007$ pm for $Li_2GeSb_2Te_5$ ($x = 1$), $a = 605.354 \pm 0.004$ pm for $Li_2Ge_2Sb_2Te_6$ ($x = 2$), $a = 603.207 \pm 0.006$ pm for $Li_2Ge_3Sb_2Te_7$ ($x = 3$), or in the tetragonal structure with the lattice parameters $a = 424.590 \pm 0.007$ and $c = 1043.82 \pm 0.02$ pm for $Li_2Ge_6Sb_2Te_{10}$ ($x = 6$), $a = 420.290 \pm 0.006$ and $c = 1053.60 \pm 0.02$ pm for $Li_2Ge_{11}Sb_2Te_{15}$ ($x = 11$). The solid solutions with $x = 6$ and 11 have a phase transition to a cubic high-temperature phase at approximately 280°C. The samples of these solid solutions were synthesized by heating stoichiometric mixtures of the elements to 700°C under an Ar atmosphere for 1 h. For $x = 1, 2, 3$, graphite crucibles in sealed silica ampoules were used; for lower Li contents, graphitized silica glass ampoules proved to be sufficient. The ampoules containing the resulting melts were quenched to room temperature in water. To obtain the amounts required for neutron diffraction, several samples with the same composition were combined, finely ground, annealed at 550°C for 12 h, and cooled to room temperature for 2 h to ensure homogeneity.

6.2 Germanium–Sodium–Antimony–Tellurium

The $(GeTe)_x(NaSbTe_2)_{1-x}$ solid solutions are formed in the Ge–Na–Sb–Te system (Schwarzmüller et al. 2019). They crystallize in the cubic structure with the lattice parameter $a = 632.73 \pm 0.04$, 630.41 ± 0.02, 628.24 ± 0.02, 625.69 ± 0.02, 621.52 ± 0.02, 617.61 ± 0.02 pm and a calculated density of 5.24, 5.30, 5.36, 5.42, 5.53, 5.64 $g \cdot cm^{-3}$ for respectively $x = 0, 0.1. 0.2, 0.3, 0.4, 0.5$. These solid solutions are not thermodynamically stable at room temperature, but can be obtained as metastable phases by quenching melts with $x \leq 0.5$. For $x > 0.5$ and in slowly cooled melts, samples are inhomogeneous. Despite the metastability, the lattice parameters show a Vegard-like behavior. Due to the exothermic reaction of Na and Te, NaSb was used as a precursor to the obtaining of these solid solutions. They were synthesized by fusing stoichiometric mixtures of NaSb, Ge, and Te at 950°C for 1 h in graphitized silica glass ampoules under Ar atmosphere. The ampoules were subsequently quenched in water. The samples are slightly sensitive to moisture and become covered with a black film when kept in air for several days; thus, they were stored under an argon atmosphere.

6.3 Germanium–Cesium–Zinc–Tellurium

The $Cs_2ZnGe_3Te_8$ quaternary compound, which crystallizes in the orthorhombic structure with the lattice parameters $a = 813.45 \pm 0.06$, $b = 1311.45 \pm 0.13$, and $c = 1882.63 \pm 0.19$ pm, a calculated density of 5.192 $g \cdot cm^{-3}$, and an energy gap of 1.07 eV, is formed in the Ge–Cs–Zn–Te system (Morris et al. 2013). To obtain this compound, a mixture of Cs_2Te_3, Zn, Ge, and Te in the appropriate ratios was added to a fused silica tube inside a N_2-filled glove box. The tube was evacuated, flame-sealed, and used in a flame-melting reaction. After cooling in air, a phase-pure ingot containing black plate-like crystals was obtained.

6.4 Germanium–Cesium–Gallium–Tellurium

The $Cs_2Ga_6Ge_3Te_{14}$ quaternary compound, which crystallizes in the trigonal structure with the lattice parameters $a = 824.75 \pm 0.02$, $c = 1427.34 \pm 0.08$ pm, and a calculated density of 5.309 $g \cdot cm^{-3}$, is formed in the Ge–Cs–Ga–Te system (Zhang et al. 2012a). This compound was synthesized from a mixture of Ga (0.6 mM), Ge (0.3 mM), and Te (1.4 mM), together with CsCl (1.2 mM) by solid-state reaction. CsCl worked as both a flux and a cesium source. Typically, the reactant mixture was loaded into a silica tube in an Ar-filled glove box, and then the silica tube was flame-sealed under a vacuum (10^{-3} Pa). Subsequently, the reaction assembly was placed in a furnace and heated to 820°C in 2 days, annealed at this temperature for 5 days, then slowly cooled at $3°C \cdot h^{-1}$ to 200°C before the furnace was turned off. The sample was cooled to ambient temperature in the furnace.

6.5 Germanium–Cesium–Indium–Tellurium

The $Cs_2In_6Ge_3Te_{14}$ quaternary compound, which crystallizes in the trigonal structure with the lattice parameters $a = 854.04 \pm 0.02$, $c = 1467.66 \pm 0.08$ pm, and a calculated density of 5.300 $g \cdot cm^{-3}$, is formed in the Ge–Cs–In–Te system (Zhang et al. 2012a). This compound was prepared in the same way as $Cs_2Ga_6Ge_3Te_{14}$ was synthesized using In instead of Ga.

DOI: 10.1201/9781003123484-6

6.6 Germanium–Copper–Barium–Tellurium

The **Cu$_{3.71}$Ba$_4$Ge$_2$Te$_9$** quaternary compound, which crystallizes in the orthorhombic structure with the lattice parameters $a = 864.64 \pm 0.02$, $b = 1353.05 \pm 0.04$, $c = 1008.10 \pm 0.03$ pm, a calculated density of 5.854 g·cm^{-3} and an energy gap of 1.0 eV, is formed in the Ge–Cu–Ba–Te system (Cui et al. 2009). It was prepared as follows. In the glove box, Ba, Cu, Ge, and Te (molar ratio 4:4:2:9) were loaded into a fused silica tube, which was then sealed under a vacuum of ca. 1 Pa. The reaction mixture was heated to 750°C within 48 h in a furnace, kept at that temperature for 2 h, and then cooled to 200°C at a rate of 3°C·h^{-1}, followed by switching off the furnace.

6.7 Germanium–Copper–Zinc–Tellurium

Cu$_2$GeTe$_3$–ZnTe. The phase relations in this system were studied by Parasyuk et al. (2005c). Since Cu$_2$GeTe$_3$ forms incongruently, there are phase fields related to phases that form upon decomposition of the ternary compound above the solidus. The liquidus of the system is represented by three fields of primary crystallization. The largest one belongs to the primary crystallization of ZnTe (~12–100 mol% ZnTe). The other two fields correspond to the primary crystallization of β_1-Cu$_{2-x}$Te and α-Cu$_2$ZnGeTe$_4$; they are small and located in the concentration range 0–12 mol% ZnTe. α-Cu$_2$ZnGeTe$_4$ melts incongruently at 550°C and crystallizes in the tetragonal structure with the lattice parameters $a = 595.40 \pm 0.04$ and $c = 1184.8 \pm 0.1$ pm (Parasyuk et al. 2005c) [$a = 599.9 \pm 0.2$ and $c = 1191.8 \pm 0.5$ pm (Haeuseler et al. 1991)]. The peritectic point is at ~12 mol% ZnTe. According to the data of Matsushita et al. (2005), Cu$_2$ZnGeTe$_4$ corresponds to a mixture of ZnTe and Cu$_2$GeTe$_3$.

Because the energy between kesterite- and stannite-type structures are small, both structures can coexist in synthesized samples of Cu$_2$ZnGeTe$_4$. According to the first-principles calculations, kesterite-type structure of this compound is characterized by the tetragonal structure with the lattice parameters $a = 610.2$ and $c = 1212.6$ pm and an energy gap of 0.81 eV, and stannite-type structure is also characterized by the tetragonal structure with the lattice parameters $a = 609.4$ and $c = 1222.0$ pm and an energy gap of 0.55 eV (Chen and Ravindra 2013).

This system was investigated through differential thermal analysis (DTA) and X-ray diffraction (XRD), and the ingots were annealed at 400°C for 700 h (Parasyuk et al. 2005c).

6.8 Germanium–Copper–Cadmium–Tellurium

Cu$_2$GeTe$_3$–CdTe. This section is a nonquasibinary one of the Ge–Cu–Cd–Te quaternary system since Cu$_2$Te and GeTe primarily crystallize from the Cu$_2$GeTe$_3$-rich side (Olekseyuk et al. 1996c; Piskach et al. 1997). The eutectic composition and temperature are 14 mol% CdTe and 482°C [500°C $\pm$ 5°C (Piskach et al. 1988)], respectively. The solubility of CdTe in Cu$_2$GeTe$_3$ and Cu$_2$GeTe$_3$ in CdTe at 430°C is equal to 13 and 8 mol%, respectively (Olekseyuk et al. 1996c). The

Cu$_2$CdGeTe$_4$ quaternary compound is formed in this system. It melts incongruently at 532°C [545°C $\pm$ 5°C (Piskach et al. 1988)] and crystallizes in the tetragonal structure with the lattice parameters $a = 611.4 \pm 0.3$ and $c = 1190.6 \pm 0.3$ pm [$a = 612.7 \pm 0.1$ and $c = 1191.9 \pm 0.3$ pm (Haeuseler et al. 1991); $a = 614$ and $c = 1196$ pm (Piskach et al. 1988)] and the calculated and experimental densities of 6.02 and 6.05 g·cm^{-3}, respectively (Olekseyuk et al. 1996c). Matsushita et al. (2005) noted that Cu$_2$CdGeTe$_4$ was not synthesized and had the mixed phases of CdTe and Cu$_2$GeTe$_3$.

This system was investigated through DTA, XRD, metallography, and measuring of the microhardness and density (Olekseyuk et al. 1996c; Piskach et al. 1997, 1988). The ingots were annealed at 430°C for 300–500 h (Olekseyuk et al. 1996c) [at 200°C for 100 h (Piskach et al. 1988)].

6.9 Germanium–Copper–Mercury–Tellurium

Cu$_2$GeTe$_3$–HgTe. The phase diagram of this quasibinary system was constructed by Hirai et al. (1967). The eutectic crystallizes at 512°C and contains 35 mol% Cu$_2$GeTe$_3$. The solid solution based on Cu$_2$GeTe$_3$ contains 12 mol% HgTe, and the solid solubility in HgTe at 400°C is 4 mol% Cu$_2$GeTe$_3$ (Parasyuk 1998). The **Cu$_2$HgGeTe$_4$** quaternary compound, which melts congruently at 532°C (Parasyuk 1998) [at 520°C (Hirai et al. 1967)] and crystallizes in the tetragonal structure with the lattice parameters $a = 612.940 \pm 0.005$, $c = 1193.74 \pm 0.02$ pm and a calculated density of 6.7430 $\pm$ 0.0002 g·cm^{-3} (Parasyuk et al. 2006a) [$a = 793$ and $c = 1715$ pm (Hirai et al. 1967); $a = 611.4 \pm 0.1$ and $c = 1192.8 \pm 0.3$ pm (Haeuseler et al. 1991)], is formed in this system. It was synthesized from the mixture of Cu, Ge, Te, and HgTe in an evacuated and sealed quartz ampoule (Parasyuk et al. 2006a). The mixture was heated to 750°C, dwelled at this temperature for 6 h, cooled at a rate of 10°C·h^{-1} to 400°C, and annealed at this temperature for 250 h.

6.10 Germanium–Copper–Indium–Tellurium

According to the data of Pamplin et al. (1977), the solubility based on CuInTe$_2$ along the "GeTe$_2$"–CuInTe$_2$ section is less than 25 mol%.

6.11 Germanium–Copper–Bismuth–Tellurium

GeTe–Cu$_2$Te–Bi$_2$Te$_3$. The liquidus surface of this quasiternary system (Figure 6.1) consists of six fields of the primary crystallization of Cu$_2$Te, GeTe, Bi$_2$Te$_3$, GeBi$_2$Te$_4$, GeBi$_4$Te$_7$, and Ge$_3$Bi$_2$Te$_6$ (Abrikosov et al. 1980c). The next reaction takes place in the transition point U_1: L + GeTe $\leftrightarrow$ Ge$_3$Bi$_2$Te$_6$ + Cu$_2$Te; and the nonvariant point E corresponds to the L $\leftrightarrow$ Cu$_2$Te + Bi$_2$Te$_3$ + GeBi$_4$Te$_7$ eutectic interaction.

The microstructural analysis of the (GeTe)$_{1-x}$[(Bi$_2$Te$_3$)$_{1-y}$(Cu$_2$Te)$_y$]$_x$ alloys showed that samples in the range of compositions $0 < x \leq 0.03$ at $y = 0.4$, 0.5, and 0.6 are single-phase (Abrikosov et al. 1980b). In the alloys with a higher content

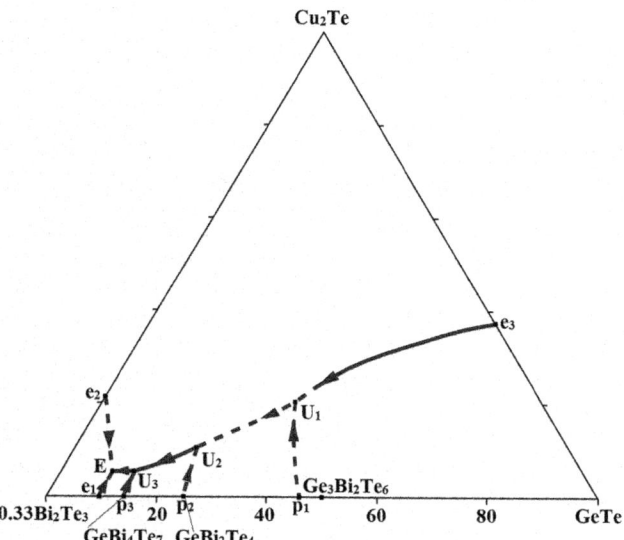

FIGURE 6.1 Liquidus surface of the GeTe–Cu$_2$Te–Bi$_2$Te$_3$ quasiternary system. (From Abrikosov, N.H., et al., *Izv. AN SSSR. Neorgan. mater.*, **16**(4), 648, 1980.)

of alloying additives ($x = 0.04$ and 0.05), traces of eutectoid decomposition were observed.

In the case of double doping of GeTe with Bi and Cu, while maintaining a constant content of GeTe and changing the ratio between the numbers of introduced Bi and Cu atoms in the direction of increasing the concentration of Cu, an increase in the rhombohedral distortion of the α-GeTe and the phase transition temperature was observed (Abrikosov et al. 1982). Bi was introduced into GeTe in the form of tellurides (Bi$_2$Te, BiTe, and Bi$_2$Te$_3$), and Cu – in the form of CuTe and Cu$_2$Te, as well as in the form of elemental copper.

This system was studied through DTA, XRD, and metallography and the ingots for the investigations were annealed at 400°C and 550°C for 600 h (Abrikosov et al. 1980b,c, 1982).

6.12 Germanium–Copper–Manganese–Tellurium

The **Cu$_2$MnGeTe$_4$** quaternary compound, which has two polymorphic modifications, is formed in the Ge–Cu–Mn–Te system (Lamarche et al. 1991). First modification crystallizes in the tetragonal structure with the lattice parameters $a = 598.9$ and $c = 1197.8$ pm, and the second one crystallizes in the cubic structure with the lattice parameter $a = 615$ pm. This compound was produced by the melt and anneals technique. To obtain it, the components of 1-g sample were sealed under vacuum in a quartz capsule that had previously been coated with carbon in order to prevent reaction of the charge with the quartz. The capsule was then raised to a temperature of 1150°C for approximately an hour to allow some reaction to occur and then cooled to room temperature. A sample annealed at 700°C and then brine-quenched showed only tetragonal plus very faint MnTe$_2$ lines. A sample similarly annealed at 700°C and then slowly cooled to room temperature (approximately 24 h) showed the tetragonal phase plus several other undetermined

phases and MnTe$_2$. However, a sample annealed at 600°C and air-cooled to room temperature had mainly tetragonal lines plus faint cubic and very faint MnTe$_2$ lines.

6.13 Germanium–Copper–Iron–Tellurium

The **Cu$_2$FeGeTe$_4$** quaternary compound, which melts congruently at 692°C (Quintero et al. 1998, 1999) and apparently has two polymorphic modifications, is formed in the Ge–Cu–Fe–Te system. First modification of this compound crystallizes in the orthorhombic structure with the lattice parameters $a = 765.6$, $b = 656.7$, and $c = 650.6$ pm (Quintero et al. 1999), and the second one crystallizes in the monoclinic structure with the lattice parameters $a = 1031.29 \pm 0.38$, $b = 403.52 \pm 0.12$, $c = 743.45 \pm 0.41$ pm, and $\beta = 89.261 \pm 0.025°$ (Quintero et al. 1998).

The samples of this compound were prepared by the melt and anneal technique (Quintero et al. 1998, 1999). The components of 1-g sample were made from the appropriate amounts of the elements and were sealed under vacuum in a small quartz ampoule, which had previously been carbonized to prevent interaction of the components with the quartz. The components were melted together at 1150°C for about an hour, annealed to equilibrium at 500°C, and then cooled to room temperature by leaving the ampoule in the switched-off furnace.

6.14 Germanium–Copper–Cobalt–Tellurium

The **Cu$_2$CoGeTe$_4$** quaternary compound, which crystallizes in the orthorhombic structure with the lattice parameters $a = 596.6 \pm 0.2$, $b = 592.2 \pm 0.3$, $c = 593.4 \pm 0.2$ pm, and a calculated density of 6.09 g·cm^{-3} (Quintero et al. 2014a), is formed in the Ge–Cu–Co–Te system. To synthesize this compound, the components of 1-g sample were sealed under vacuum ($\approx 10^{-3}$ Pa) in a quartz ampoule, which had previously been carbonized. The synthesis was realized inside a furnace. The ampoule with the components was heated up to 200°C and kept for about 1–2 h, and then the temperature was raised to 500°C using a rate of 40°C·h^{-1}, and held at this temperature for 14 h. After that, the sample was heated from 500°C to 800°C at a rate of 30°C·h^{-1} and kept at this temperature for another 14 h. Then it was raised to 1150°C at 60°C·h^{-1}, and the components were melted together at this temperature for about 2–3 h. The furnace temperature was brought slowly (4°C·h^{-1}) down to 600°C, and the sample was annealed at this temperature for 1 month. Then, the sample was slowly cooled to room temperature using a rate of about 2°C·h^{-1}.

6.15 Germanium–Silver–Barium–Tellurium

The **Ag$_{3.95}$Ba$_4$Ge$_2$Te$_9$** quaternary compound, which melts at 728°C and crystallizes in the orthorhombic structure with the lattice parameters $a = 868.35 \pm 0.03$, $b = 1364.21 \pm 0.04$, $c = 1026.12 \pm 0.03$ pm, a calculated density of 6.198 g·cm^{-3}, and an energy gap of 0.24 eV, is formed in the Ge–Ag–Ba–Te system (Cui et al. 2009). It was prepared in the following way. In the glove box, Ba, Ag, Ge, and Te (molar ratio 4:4:2:9) were loaded into a fused silica tube, which was then sealed under a vacuum of ca. 1 Pa. The reaction mixture was heated to 750°C within 48 h in a furnace, kept at that temperature for 2 h, and then cooled to 200°C at a rate of 3°C·h^{-1}, followed by switching off the furnace.

6.16   Germanium–Silver–Indium–Tellurium

A sample of the "GeTe₂"–AgInTe₂ vertical section, containing 50 mol% "GeTe₂", consists of three phases, the nature of which has not been established (Pamplin et al. 1977). The lattice parameters of one of the phases are close to the lattice parameters of AgInTe₂.

6.17   Germanium–Silver–Antimony–Tellurium

GeTe–"AgSbTe₂". This section is a nonquasibinary one of the Ge–Ag–Sb–Te quaternary system, as AgSbTe₂ melts incongruently at 360°C. (Plachkova et al. 1983; Plachkova and Odin 1983). Peritectic reaction at 536°C and eutectic reaction at 530°C take place in the system.

Solid solution based on GeTe contains up to 22 mol% "AgSbTe₂" and crystallizes in the rhombohedral structure (Borisova et al. 1971; Plachkova and Odin 1983). Polymorphic transformation exists in this region moreover, the temperature of the phase transition occurring practically without a change in volume decreases with an increase in the "AgSbTe₂" content (Abrikosov et al. 1984). Starting from 23 mol% "AgSbTe₂", the samples have a cubic crystal lattice of NaCl-type, and at a content of more than 30 mol% "AgSbTe₂", Ag₈Ge₁.₁Te₅.₉ is separated along with the main phase of β-GeTe (Borisova et al. 1971; Plachkova and Odin 1983). According to the date of Bushmarina et al. (1981), the $(GeTe)_{1-2x}(AgSbTe_2)_x$ ($0.05 \leq x \leq 0.15$) solid solution is a phase of variable composition in which the homogeneity region is shifted toward Te excess.

This system was studied through DTA, XRD, metallography, dilatometry, and measuring of the microhardness (Plachkova et al. 1983; Plachkova and Odin 1983; Abrikosov et al. 1984). The ingots for the investigations were annealed at 400°C for 2230 h [2000 h (Abrikosov et al. 1984)] and at 300°C for 2880 h (Plachkova and Odin 1983), or at 300°C and 400°C for 2160 h and at 500°C for 200 h (Plachkova et al. 1983), or at the temperatures that are two-thirds of the melting points for 300–500 h (Borisova et al. 1971).

GeTe–Ag₂Te–Sb₂Te₃. In the $(GeTe)_{1-x}[(Sb_2Te_3)_y(Ag_2Te)_{1-y}]$ alloys at $y = 1$, 0.75, 0.6, and 0.5, the phase transition temperature decreases at increasing of Sb₂Te₃ and Ag₂Te contents (Abrikosov et al. 1985b). At the same concentration of GeTe, the phase transition temperature increases when the ratio between Sb₂Te₃ and Ag₂Te changes toward an increase in the content of Ag₂Te. Increasing the Sb₂Te₃ and Ag₂Te content contributes to the stabilization of the cubic modification in the quenched alloys. The alloys for the investigations were annealed at 500°C for 1055 h, followed by quenching in ice water.

The regions of existence of phases in the part of the GeTe–Ag₂Te–Sb₂Te₃ system at 300°C were outlined in Plachkova and Odin (1983).

6.18   Germanium–Silver–Bismuth–Tellurium

GeTe–AgBiTe₂. The phase diagram of this system, constructed through DTA, XRD, metallography, and measuring of the microhardness, belongs to the type I according to Roseboom's classification (Figure 6.2) (Plachkova et al. 1980, 1984). High-temperature modification of GeTe forms continuous solid solutions with cubic structure with AgBiTe₂. In the range of 60–100 mol% GeTe, the decomposition of the solid solutions begins, apparently, from immiscibility in the solid state and subsequent precipitation of Ag₈Ge₁.₁Te₅.₉ phase. The decomposition of the solid solutions in the region 9–6 mol% GeTe starts with Ag₈Ge₁.₁Te₅.₉ phase precipitation, and in the region 5–0 mol% GeTe from precipitation of Bi₂Te₃. When alloys were quenched from 500°C, almost single-phase samples were obtained up to 75 mol% AgBiTe₂, and the solid solutions in the region of 80–100 mol% AgBiTe₂ are very unstable. According to the date of Bushmarina et al. (1981), the $(GeTe)_{1-2x}(AgBiTe_2)_x$ ($0.05 \leq x \leq 0.15$) solid solution is a phase of variable composition, in which the homogeneity region is shifted toward Te excess.

The $(GeTe)_{1-x}–[(Ag_2Te)_{0.4667}(Bi_2Te_3)_{0.5333}]$ ($0 < x < 0.35$) vertical section was constructed through DTA, XRD, and metallography by Plachkova et al. (1992). It was shown that at room temperature, the region of the solid solution based on α-GeTe is situated up to $x \leq 0.15$, and the region of the solid solution based on β-GeTe is very narrow ($0.16 < x < 0.20$).

The regions of existence of phases in the part of the GeTe–Ag₂Te–Bi₂Te₃ system at 300°C were outlined in Plachkova and Odin (1983).

The ingots for the investigations were annealed at 300°C and 400°C for 2160 h, and some ingots were annealed at 450–500°C for 200 h (Plachkova et al. 1984) (at 400°C and 500°C for 500 and 200 h, respectively (Plachkova et al. 1980). The $(GeTe)_{1-x}–[(Ag_2Te)_{0.4667}(Bi_2Te_3)_{0.5333}]$ alloys were annealed

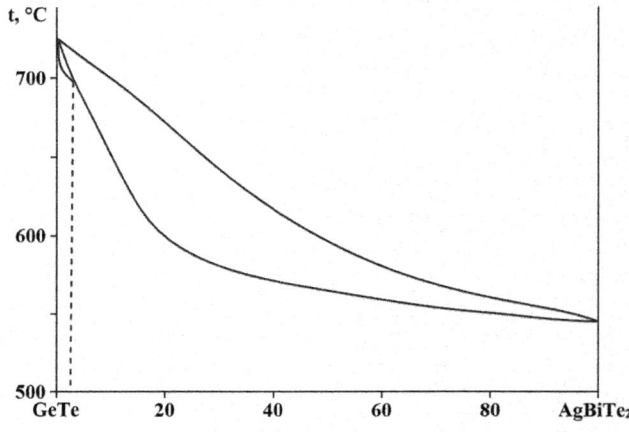

FIGURE 6.2 Phase diagram of the GeTe–AgBiTe₂ system. (From Plachkova, S.K., et al., *Izv. AN SSSR. Neorgan. mater.*, **20**(3), 403, 1984.)

at 500°C for 1000 h, followed by the quenching in ice water (Plachkova et al. 1992). After that, these alloys were additionally annealed at 300°C for 1720 h.

6.19 Germanium–Silver–Iodine–Tellurium

The $Ag_{8-x}GeTe_{6-x}I_x$ quaternary phase, which crystallizes in the cubic structure with the lattice parameter $a = 1156.1 \pm 0.3$ pm for an unspecified x value, is formed in the Ge–Ag–I–Te system (Kuhs et al. 1979). This phase was prepared by reacting stoichiometric amounts of the elements in an evacuated, sealed quartz ampoule for about 6 days at temperatures of 600°C to 700°C

6.20 Germanium–Silver–Manganese–Tellurium

The $Ag_2MnGeTe_4$ quaternary compound, which has two polymorphic modifications, is formed in the Ge–Ag–Mn–Te system (Lamarche et al. 1991). First modification crystallizes in the tetragonal structure with the lattice parameters $a = 579$ and $c = 1158$ pm, and the second one crystallizes in the cubic structure with the lattice parameter $a = 588.2$ pm. This compound was prepared in the same way as $Cu_2MnGeTe_4$ was synthesized. When the obtained samples were annealed at 600°C and either brine-quenched or air-cooled to room temperature, they showed mainly cubic lines plus other extra lines, including $MnTe_2$. Annealing at 550°C and 440°C and quenching showed similar results. However, a sample annealed at 400°C and quenched was tetragonal with a little inclusion of $MnTe_2$.

Polycrystalline samples of $Ag_2MnGeTe_4$ were also synthesized by Woolley et al. (1995). The appropriate quantities of the elements were sealed under a vacuum in quartz ampoules that had previously carbonized to prevent the interaction of the charge with the quartz. Each sample was melted at 1150°C and then slowly cooled to room temperature. The samples were then annealed for several days at 525°C and quenched to room temperature. Both the ordered rock-salt form and the ordered zinc-blend form were present in the sample obtained. These two phases crystallize in the orthorhombic structure with the lattice parameters $a = 1252.0$, $b = 418.2$, $c = 598.7$ pm and $a = 1230.7$, $b = 410.7$, $c = 575.2$ pm, respectively. Below approximately 350°C, this compound is unstable and decomposes, forming $MnTe_2$, Ag_8GeTe_6, and Ge_3MnTe_4.

6.21 Germanium–Silver–Iron–Tellurium

The $Ag_2FeGeTe_4$ quaternary compound, which melts congruently at 717°C and crystallizes in the orthorhombic structure with the lattice parameters $a = 804.8$, $b = 666.8$, and $c = 645.0$ pm, is formed in the Ge–Ag–Fe–Te system (Quintero et al. 1999). The samples of this compound were prepared by the melt and anneal technique. The components of 1-g sample were made from the appropriate amounts of the elements and were sealed under a vacuum in small quartz ampoules, which had previously been carbonized to prevent interaction of the components with

the quartz. The components were melted together at 1150°C for about an hour, annealed at 500°C, and then cooled to room temperature by leaving the ampoule in the switched-off furnace.

Equilibrium phase formations below 330°C in the Ag_8GeTe_6–GeTe–$FeTe_2$–$AgFeTe_2$ subsystem of the Ge–Ag–Fe–Te system were established by the EMF method (Moroz et al. 2021b). Linear dependences of the EMF of galvanic cells versus temperature in the range of 209–247°C were used for calculations of the standard Gibbs energies, enthalpies, and entropies of formations of $Ag_2FeGeTe_4$: $-\Delta G^0_{f, 298} = 167.04 \pm 2.3$ kJ·M^{-1}, $-\Delta H^0_{f, 298} = 166.6 \pm 2.6$ kJ·M^{-1}, $S^0_{298} = 344.3 \pm 4.9$ J·(M·K)$^{-1}$.

6.22 Germanium–Gold–Barium–Tellurium

The $Au_{3.69}Ba_4Ge_2Te_9$ quaternary compound, which crystallizes in the orthorhombic structure with the lattice parameters $a = 1358.06 \pm 0.06$, $b = 2070.33 \pm 0.09$, $c = 864.18 \pm 0.04$ pm, a calculated density of 7.025 g·cm^{-3} and an energy gap of 0.19 eV, is formed in the Ge–Au–Ba–Te system (Cui et al. 2010). A phase transition observed at 417°C indicates that the supercell is likely the low-temperature modification of the subcell. To prepare this compound, the starting materials, Ba, Au, Ge, and Te (molar ratio 4:4:2:9), were loaded into a fused silica tube, which was then sealed under a vacuum of approximately 1 Pa. The reaction mixture was heated to 750°C within 48 h in a furnace, kept at that temperature for 2 h, and then cooled to 200°C at a rate of 3°C·h^{-1}, followed by switching off the furnace. All chemical elements were stored in an argon-filled glove box with water and oxygen levels below 1 ppm.

6.23 Germanium–Gold–Tin–Tellurium

A thin film of $Au_{25}Ge_4Sn_{11}Te_{60}$ with a thickness of approximately 0.5 mkm was formed by sputtering on a glass disk (Matsunaga and Yamada 2002). The film was crystallized by means of laser irradiation and then scraped off with a spatula to create a powder. The obtained phase crystallizes in the cubic structure with the lattice parameter $a = 298.79 \pm 0.02$ pm at 100 K and $a = 299.49 \pm 0.05$ pm at 300 K.

6.24 Germanium–Barium–Antimony–Tellurium

The $Ba_4Ge_2Sb_2Te_{10}$ quaternary compound, which crystallizes in the monoclinic structure with the lattice parameters $a = 1398.4 \pm 0.3$, $b = 1347.2 \pm 0.3$, and $c = 1356.9 \pm 0.3$ pm, and $\beta = 90.16 \pm 0.03°$, is formed in the Ge–Ba–Sb–Te system (Jana et al. 2022a). It was synthesized at high temperatures via the reaction of the chemical elements.

6.25 Germanium–Cadmium–Thallium–Tellurium

The $CdTl_2GeTe_4$ quaternary compound, which is semiconductor and crystallizes in the tetragonal structure with the lattice parameters $a = 838.25 \pm 0.19$ and $c = 707.75 \pm 0.18$ pm, is formed in the Ge–Cd–Tl–Te system. (McGuire et al. 2005b). This compound was synthesized by reaction of Tl, Cd, Ge, and Te (molar ratio 2:1:1:4). At the syntheses, Tl metal was weighed in an Ar-filled glove box and placed into the capped vial. The other elements were weighed in air and loaded into a silica tube, which was coated with carbon to prevent a reaction between the silica and the sample. Thallium was then added to the tube, and the tube was quickly attached to a vacuum line for sealing. The evacuated, sealed silica tube was heated in a furnace. The sample was heated to 500°C for 4 days followed by cooling to room temperature over another 4 days. Several subsequent anneals for 4–7 days each at 250–300°C were required to obtain a single-phase sample. Single crystals were obtained by heating a stoichiometric mixture of the elements to 500°C in 12 h, held at this temperature for 24 h, and cooled to 30°C in 100 h by turning off the furnace.

6.26 Germanium–Mercury–Thallium–Tellurium

The $HgTl_2GeTe_4$ quaternary compound, which is semiconductor and crystallizes in the tetragonal structure with the lattice parameters $a = 835.71 \pm 0.11$ and $c = 706.84 \pm 0.14$ pm, is formed in the Ge–Hg–Tl–Te system (McGuire et al. 2005b). This compound was synthesized by reaction of Tl, HgTe, Ge, and Te (molar ratio 2:1:1:3). At the syntheses, Tl metal was weighed in an Ar-filled glove box and placed into capped vial. The other elements were weighed in air and loaded into silica tube which was coated with carbon to prevent reaction between the silica and the sample. Thallium was then added to the tube, and the tube was quickly attached to a vacuum line for sealing. The evacuated, sealed silica tubes were heated in a furnace. The sample was heated to 500°C over 36 h, held at this temperature for 5 days, at which point the furnace was turned off and allowed to cool naturally to room temperature. The product was then ground and pressed into pellets and annealed at 300°C for 7 days. The single-phase sample was prepared by the interaction of the stoichiometric amounts of Tl, HgTe, Ge, and Te at 800°C in 24 h, held at this temperature for 48 h, and cooled to 300°C in 200 h. Then the furnace was turned off and allowed to cool naturally to room temperature.

6.27 Germanium–Indium–Antimony–Tellurium

In_2GeTe–$InSb$. The specimens with compositions in the range of 0–50 mol% In_2GeTe were investigated by Woolley and Williams (1964). It was shown that for the quenched from 800°C materials, solid solubility extends out to about 33 mol% In_2GeTe, but this was reduced to 12 mol% for alloys annealed at 400°C for 14 days. Alloys containing 30 and 40 mol% In_2GeTe showed signs of partial melting when annealed at 450°C, indicating the presence of a eutectic or peritectic horizontal in the vicinity of 450°C. The energy gap of the solid solutions increases to a maximum value of 0.45 eV at 0.5 mol% In_2GeTe and then decreases with an increase in In_2GeTe content. According to the data by Aliev et al. (1981), the solubility of In_2GeTe in 2InSb reaches 12 mol% for the alloys annealed

at 400°C for 340 h. This system was investigated through DTA, XRD, and metallography (Woolley and Williams 1964; Aliev et al. 1981).

Several quaternary phases, $GeSbInTe_4$, $Ge_2SbInTe_5$, $Ge_3SbInTe_6$, and $Ge_{12}(Sb_xIn_{1-x})_2Te_5$, were synthesized in the Ge–In–Sb–Te system. The first three phases crystallize in the trigonal structure with the lattice parameters $a = 421.324 \pm 0.005$, $c = 4103.48 \pm 0.10$ pm, and a calculated density of 6.4722 ± 0.0002 g·cm⁻³ for $GeSbInTe_4$; $a = 420.204 \pm 0.006$, $c = 1720.76 \pm 0.04$ pm, and a calculated density of 6.4356 ± 0.0002 g·cm⁻³ for $Ge_2SbInTe_5$; and $a = 419.789 \pm 0.004$, $c = 6216.20 \pm 0.11$ pm, and a calculated density of 6.4064 ± 0.0002 g·cm⁻³ for $Ge_3SbInTe_6$ (Fahrnbauer et al. 2013). Bulk samples of these phases were obtained by quenching stoichiometric quaternary melts and subsequently annealing the ingots. Ge, Sb, In, and Te were melted in sealed silica glass ampoules under a dry argon atmosphere at 950°C (1, 2, and 18 h for $GeSbInTe_4$, $Ge_2SbInTe_5$, and $Ge_3SbInTe_6$, respectively), followed by quenching to room temperature in water. The samples were annealed in a tube furnace ($GeSbInTe_4$ and $Ge_2SbInTe_5$ at 450°C for 3 and 5 days, respectively, $Ge_3SbInTe_6$ at 300°C for 11 days) and subsequently quenched in water. Single crystals of $Ge_3SbInTe_6$ were grown via the gas phase by annealing 500 mg of the inhomogeneous quenched sample in the presence of 20 mg I_2 in an evacuated silica glass ampoule (450°C, 4 days).

$Ge_{12}(Sb_xIn_{1-x})_2Te_5$ crystallizes in the cubic structure with the lattice parameter $a = 596.03 \pm 0.01$ pm and a calculated density of 6.320 g·cm⁻³ for $x = 0.5$, and $a = 595.346 \pm 0.006$ pm and a calculated density of 6.334 g·cm⁻³ for $x = 0.75$ (Rosenthal et al. 2013). The samples of these phases were prepared by melting stoichiometric amounts of the elements in silica glass ampoules sealed under an Ar atmosphere at 950°C for 2 h. The obtained samples were quenched in water and then annealed at 590°C for 3 days, followed by quenching in water once again.

6.28 Germanium–Thallium–Bismuth–Tellurium

Tl_8GeTe_5–$TlBiTe_2$. The phase diagram, constructed through DTA and XRD using the alloys annealed at 430°C for 1300 h, shows that this system is practically quasibinary and is characterized by the eutectic and metatectic equilibria (Alakbarova 2022). The melting point of $TlBiTe_2$ is slightly shifted from the stoichiometric composition; therefore, a sample of stoichiometric composition melts in a narrow range (550–557°C) of temperatures. The eutectic contains 30 mol% $TlBiTe_2$ and crystallizes at 467°C. The homogeneity regions at the eutectic temperature cover the compositional area of 0–17 and 86–100 mol% $TlBiTe_2$. The formation of the solid solutions based on $TlBiTe_2$ is accompanied by a decrease in its polymorphic transformation temperature (512°C) and the formation of a metatectic equilibrium at 502°C.

Tl_8GeTe_5–Tl_9BiTe_6. The phase diagram of this system, constructed through DTA, XRD, and measuring of the microhardness and EMF of concentrated chains, belongs to type I according to Roseboom's classification (Alekperova et al. 2015). Continuous solid solutions with Tl_5Te_3-type structure are formed in the system. The ingots for the investigations were annealed at 430°C for 800 h.

Tl₂GeTe₂–TlBiTe₂. This system is a nonquasibinary section of the Ge–Tl–Bi–Te system as Tl_2GeTe_2 melts incongruently (Alakbarova 2022). There are wide solid solution regions based on both modifications of $TlBiTe_2$ in this section.

Tl₈GeTe₅–Tl₉BiTe₆–Tl₂Te. The phase equilibria in this compositional area of the Ge–Tl–Bi–Te quaternary system were studied through DTA and XRD (Alakbarova et al. 2017). Some vertical sections and isothermal sections at room temperature, 470°C, and 530°C, as well as the liquidus and solidus surfaces, were constructed. It was found that the homogeneity region of the solid solutions with Tl_5Te_3-type structure $[(Tl_8GeTe_5)_x(Tl_9BiTe_6)_{1-x}]$ occupied more than 80% of the concentration triangle. A narrow area of solid solutions based on Tl_2Te was detected along the Tl_2Te–Tl_9BiTe_6 system.

6.29 Germanium–Thallium–Manganese–Tellurium

The **MnTl₂GeTe₄** quaternary compound, which is semiconductor and crystallizes in the tetragonal structure with the lattice parameters $a = 839.9 \pm 0.3$ and $c = 696.3 \pm 0.3$ pm, is formed in the Ge–Tl–Mn–Te system (McGuire et al. 2005b). This compound was synthesized by reaction of Tl, Mn, Ge, and Te (molar ratio 2:1:1:4). At the syntheses, Tl metal was weighed in an Ar-filled glove box and placed into a capped vial. The other elements were weighed in air and loaded into a silica tube, which was coated with carbon to prevent a reaction between the silica and the sample. Thallium was then added to the tube, and the tube was quickly attached to a vacuum line for sealing. The evacuated, sealed silica tube was heated in a furnace. The sample was heated to 500°C in 12 h, held at this temperature for 24 h, and cooled to 30°C within 100 h by turning off the furnace. Attempt to make $MnTl_2GeTe_4$ as single-phase material failed. This compound is a semiconductor.

6.30 Germanium–Scandium–Lead–Tellurium

GeTe–ScTe–PbTe. The solubility region based on GeTe in this system is shown in Figure 6.3 (Shelimova et al. 1993b). The ingots for the investigations were annealed at 500°C for approximately 100 h, followed by quenching at a rate of 2°C·min⁻¹. At the quenching of annealed samples in water with ice, the boundaries of GeTe-based solid solutions do not change significantly, except for the alloys that are located near the GeTe–PbTe system, where the region of solid solutions increases.

6.31 Germanium–Lanthanum–Lead–Tellurium

GeTe–LaTe–PbTe. The solubility region based on GeTe in this system is shown in Figure 6.4 (Shelimova et al. 1993b). The ingots for the investigations were annealed at 500°C for approximately 100 h, followed by quenching at a rate of 2°C·min⁻¹. At the quenching of annealed samples in water with ice, the boundaries of GeTe-based solid solutions do not

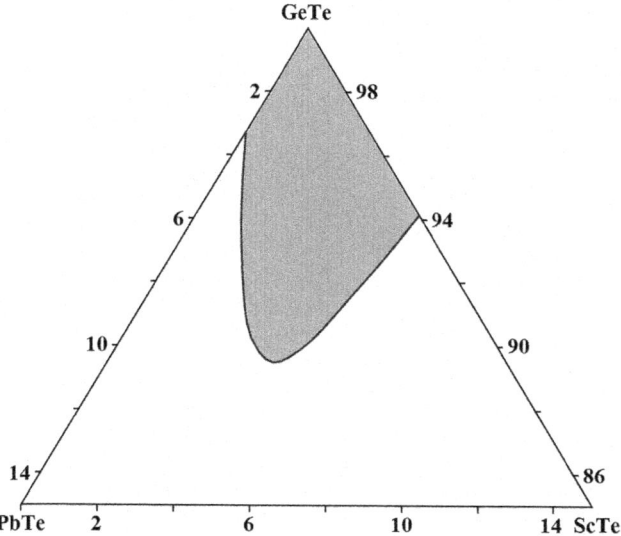

FIGURE 6.3 Solubility region based on GeTe in the GeTe–ScTe–PbTe system. (From Shelimova, L.E., et al., *Neorgan. mater.*, **29**(11), 1449, 1993.)

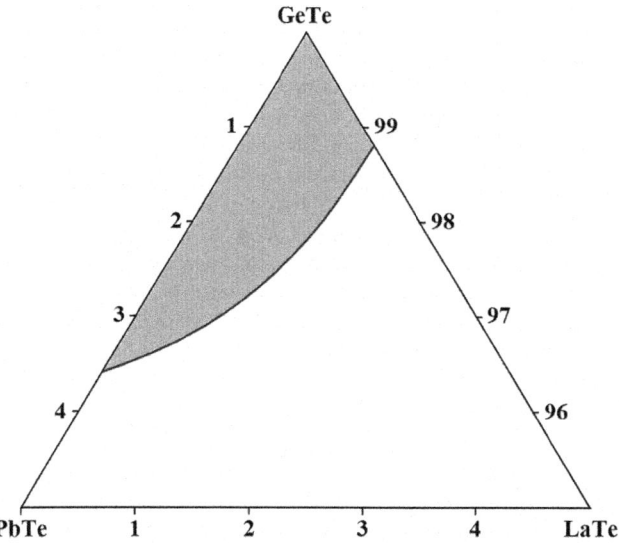

FIGURE 6.4 Solubility region based on GeTe in the GeTe–LaTe–PbTe system. (From Shelimova, L.E., et al., *Neorgan. mater.*, **29**(11), 1449, 1993.)

change significantly, except for the alloys that are located near the GeTe–PbTe system, where the region of solid solutions increases.

6.32 Germanium–Tin–Lead–Tellurium

GeTe–SnTe–PbTe. This quasiternary system was studied experimentally through DTA, XRD, EPMA (Yashina et al. 2006) and calculated using the interaction parameters of the quasibinary systems and taking into account ternary interactions (Volykhov et al. 2010). It was shown that at temperatures above 590°C, there is unlimited mutual solubility in both the liquid and the

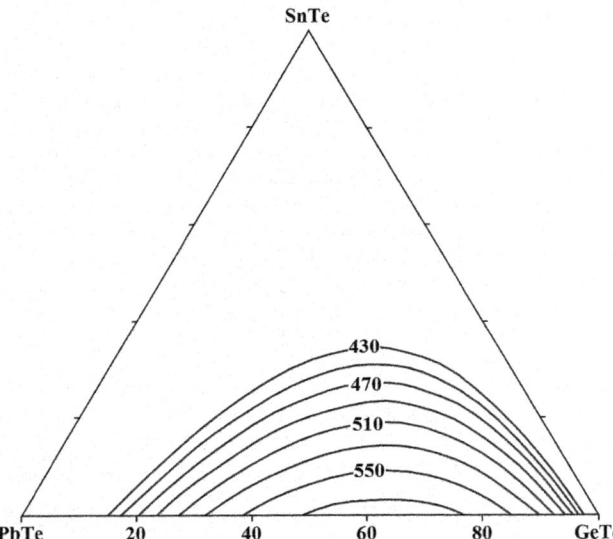

FIGURE 6.5 Isotherms of spinodal demixing in the GeTe–SnTe–PbTe quasiternary system. (From Yashina, L.V., et al., *J. Alloys Compd.*, **413**(1–2), 133, 2006.)

solid phase. Below this temperature, the solid solutions near the GeTe–PbTe boundary of the composition triangle decompose. Isotherms of spinodal demixing in this quasiternary system are presented in Figure 6.5 (Yashina et al. 2006; Volykhov et al. 2010). The comparison of these isotherms for different temperatures shows that with the decrease in temperature, the extension of the spinodal miscibility gap increases. In the whole temperature region, the absolute value of the slope of the tie lines relative to the PbTe–GeTe edge increases with the increase of SnTe mole fraction. The slope of these tie lines is mainly caused by the difference in the interaction parameters of the two quasibinary edge systems SnTe–PbTe and GeTe–SnTe. The observed slope of the tie lines is due to the different deviation from ideal behavior for the last two systems.

The isothermal sections of the GeTe–SnTe–PbTe quasiternary system 700°C, 710°C, and 740°C are shown in Figure 6.6, and the liquidus and solidus surfaces of this system are given in Figure 6.7 (Yashina et al. 2006; Volykhov et al. 2010.

6.33 Germanium–Tin–Antimony–Tellurium

GeTe–GeSnSb₄Te₈. The phase diagram of this system, constructed through DTA, XRD, metallography, and measuring of the microhardness and density, is a eutectic type (Figure 6.8) (Gurbanov and Adygezalova 2018). The eutectic contains 40 mol% GeTe and crystallizes at 427°C. The region of the solid solution based on GeSnSb₄Te₈ reaches 10 mol% GeTe, and that of the solid solution based on GeTe extends to 5 mol% GeSnSb₄Te₈ at room temperature. Single crystals of the $(GeTe)_x$ $(GeSnSb_4Te_8)_{1-x}$ solid solution at $x = 0.001-0.005$ were grown by directional crystallization.

GeSb₂Te₄–SnSb₂Te₄. A complete solid solution series is formed in this system (Welzmiller et al. 2014). The lattice parameters for the $(Ge_xSn_{1-x})Sb_2Te_4$ solid solution change linearly with composition.

Ge₂Sb₂Te₅–"Sn₂Sb₂Te₅". As "Sn₂Sb₂Te₅" does not exist, Sn can only partially replace Ge in Ge₂Sb₂Te₅; samples with more than 75 at% Sn are not homogeneous (Welzmiller et al. 2014). These solid solutions crystallize in the trigonal structure with the lattice parameters $a = 427.486 \pm 0.007$, $c = 1741.65 \pm 0.08$ pm, a calculated density of 6.46 g·cm⁻³ and $a = 425.792 \pm 0.011$, $c = 1736.57 \pm 0.14$ pm, a calculated density of 6.45 g·cm⁻³ for GeSnSb₂Te₅ and Ge₁.₃Sn₀.₇Sb₂Te₅ composition, respectively. Bulk samples of the solid solutions of both mentioned above systems were prepared by melting stoichiometric mixtures of Ge, Sn, Sb, and Te in sealed silica glass ampoules under an argon atmosphere at 950°C (for 2–24 h) and quenching in water. Subsequently, the samples were annealed for about 48 h at temperatures between 450°C and 590°C. Single crystals were grown by chemical transport reactions in sealed silica glass ampoules under a vacuum using I₂ or SbI₃ as transport agents.

SnTe–GeSnSb₄Te₈. The phase diagram of this system, constructed through DTA, XRD, metallography, and measuring of the microhardness and density, is a eutectic type (Figure 6.9) (Gurbanov and Adygezalova 2018). The eutectic contains 30 mol% SnTe and crystallizes at 477°C. The solubility of GeSnSb₄Te₈ in SnTe is 3 mol%, and the solubility of SnTe in GeSnSb₄Te₈ reaches 8 mol% at room temperature.

Four quaternary compounds, **GeSnSb₄Te₈**, **Ge₂Sn₂Sb₂Te₇**, **Ge₃SnSb₂Te₇**, and **Ge₃.₂₅Sn₁.₁₀Sb₁.₁₀Te₆**, and $(Ge_{1-x}Sn_xTe)_nSb_2Te_3$ ($n = 4, 7, 12; 0 \leq x \leq 1$) solid solutions are formed in the Ge–Sn–Sb–Te system. GeSnSb₄Te₈ melts congruently at 677°C and crystallizes in the orthorhombic structure with the lattice parameters $a = 492$, $b = 943$, and $c = 1805$ pm (Gurbanov and Adygezalova 2018). Single crystals of this compound were grown by means of chemical transport reactions.

Ge₃SnSb₂Te₇ and Ge₃.₂₅Sn₁.₁₀Sb₁.₁₀Te₆ crystallize in the trigonal structure with the lattice parameters $a = 424.990 \pm 0.004$, $c = 7346.77 \pm 0.09$ pm, a calculated density of 6.38 g·cm⁻³ and $a = 428.0 \pm 0.1$, $c = 2096.6 \pm 0.3$ pm, a calculated density of 6.32 g·cm⁻³, respectively (Rosenthal et al. 2014b). Samples of Ge₂Sn₂Sb₂Te₇, Ge₃SnSb₂Te₇, and Ge₃.₂₅Sn₁.₁₀Sb₁.₁₀Te₆ were prepared by melting stoichiometric mixtures of Ge, Sb, Sn, and Te in silica glass ampoules under argon atmosphere and subsequently quenching the ampoules in water. Samples with the nominal composition Ge₂Sn₂Sb₂Te₇ were melted at 950°C for 1.5 h, annealed at 590°C for 20 h, and subsequently quenched in water again. Crystals were grown from this product by chemical vapor transport with iodine as a transport agent. Ge₃SnSb₂Te₇ samples were prepared by melting a stoichiometric mixture at 950°C for 9 days, subsequently annealing at 380°C for 9 days, and cooling to room temperature by turning off the furnace (2 h to room temperature).

The $(Ge_{1-x}Sn_xTe)_nSb_2Te_3$ solid solutions crystallize in the cubic structure with the lattice parameters $a = 598.8 \pm 0.2$, 604.37 ± 0.03, 610.480 ± 0.017, 616.579 ± 0.007, 622.175 ± 0.008 pm and a calculated density of 6.307, 6.333, 6.337, 6.337, 6.349 g·cm⁻³ for $n = 4$ and $x = 0, 0.25, 0.50, 0.75$, and 1, respectively; $a = 599.160 \pm 0.006$, 605.668 ± 0.005, 612.377 ± 0.007, 619.206 ± 0.011, 625.520 ± 0.011 pm and a calculated density of 6.262, 6.303, 6.332, 6.350, 6.379 g·cm⁻³ for $n = 7$ and $x = 0, 0.25, 0.50, 0.75$, and 1, respectively; $a = 605.352$

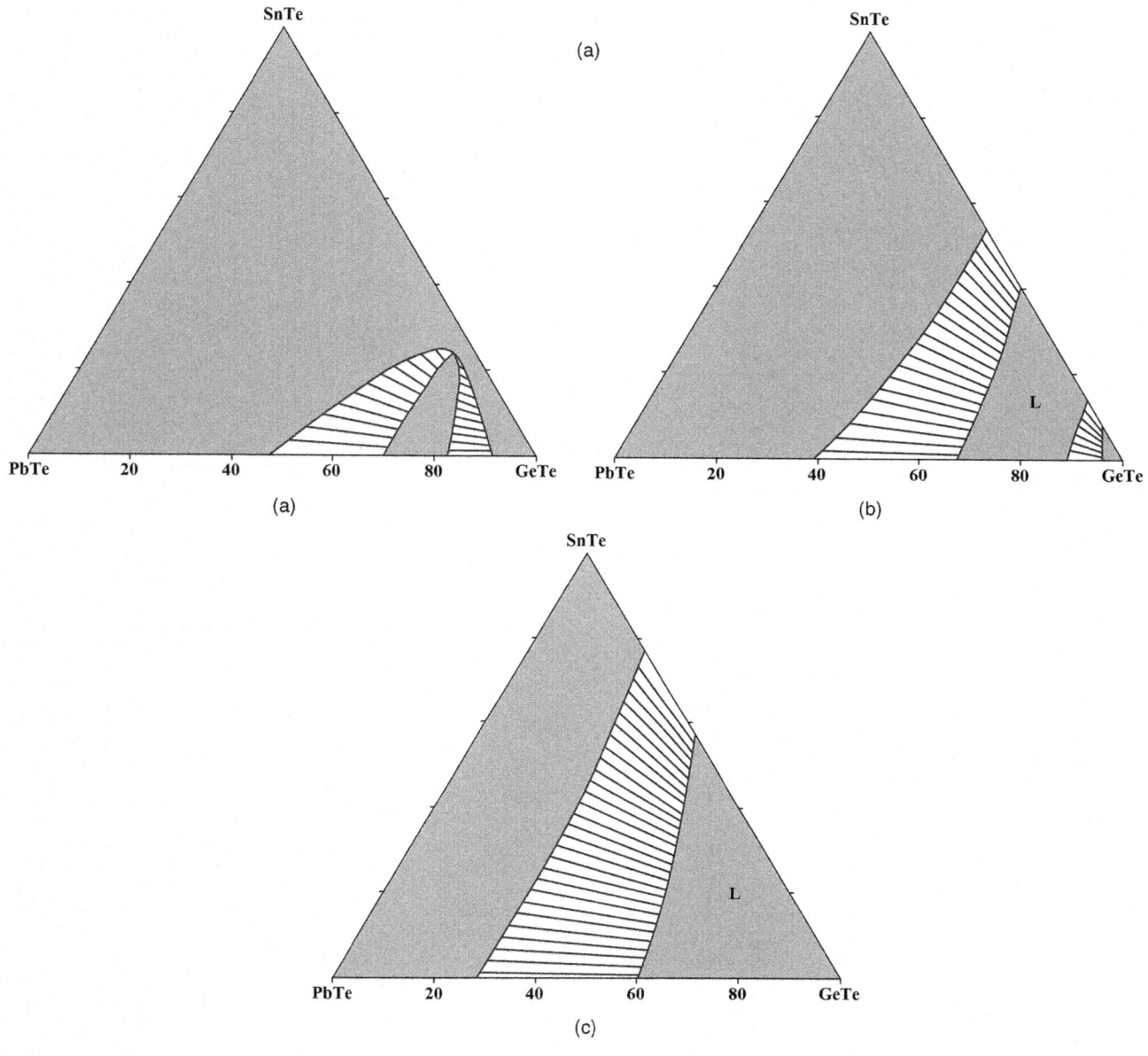

FIGURE 6.6 Isothermal sections of the GeTe–SnTe–PbTe quasiternary system at (a) 700°C, (b) 710°C, and (c) 740°C. (From Yashina, L.V., et al., *J. Alloys Compd.*, **413**(1–2), 133, 2006.)

± 0.004, 613.145 ± 0.006, 621.548 ± 0.012, 627.778 ± 0.009 pm and a calculated density of 6.322, 6.350, 6.350, 6.411 g·cm⁻³ for *n* = 12 and *x* = 0.25, 0.50, 0.75, and 1, respectively (Rosenthal 2014a). These solid solutions represent stable high-temperature phases and can be obtained as metastable compounds by quenching. The substitution with Sn significantly lowers the transition temperatures between the cubic high-temperature phase and the long-range ordered layered phases that are stable at ambient conditions. Their bulk samples were synthesized under Ar atmosphere in sealed silica glass ampoules. Stoichiometric mixtures of Ge, Sn, Sb, and Te were melted (minimum 1 h at 900–950°C) and quenched in water. The compact metallic gray ingots were annealed in the existing range of the cubic high-temperature phases (i.e. between 560°C and 610°C), for up to 10 days and quenched in water once again.

6.34 Germanium–Tin–Manganese–Tellurium

GeTe–SnTe–MnTe. The temperatures of the liquidus, solidus, and phase transition of some samples of this quasiternary system were determined by Abrikosov et al. (1978). The boundary of the solid solution regions based on GeTe along the GeTe–$Mn_{0.8}Sn_{0.2}Te$ and GeTe–$Mn_{0.5}Sn_{0.5}Te$ sections is near 80 mol% GeTe. Thermal effects corresponding to phase transitions in the solid state have not been observed for these alloys. They were detected by a decrease in the magnitude of the volume change during the phase transition in the process of adding tin and manganese tellurides to GeTe. This system was studied through DTA, metallography, and dilatometry. The ingots for the investigations were annealed at 500°C for 1100 h.

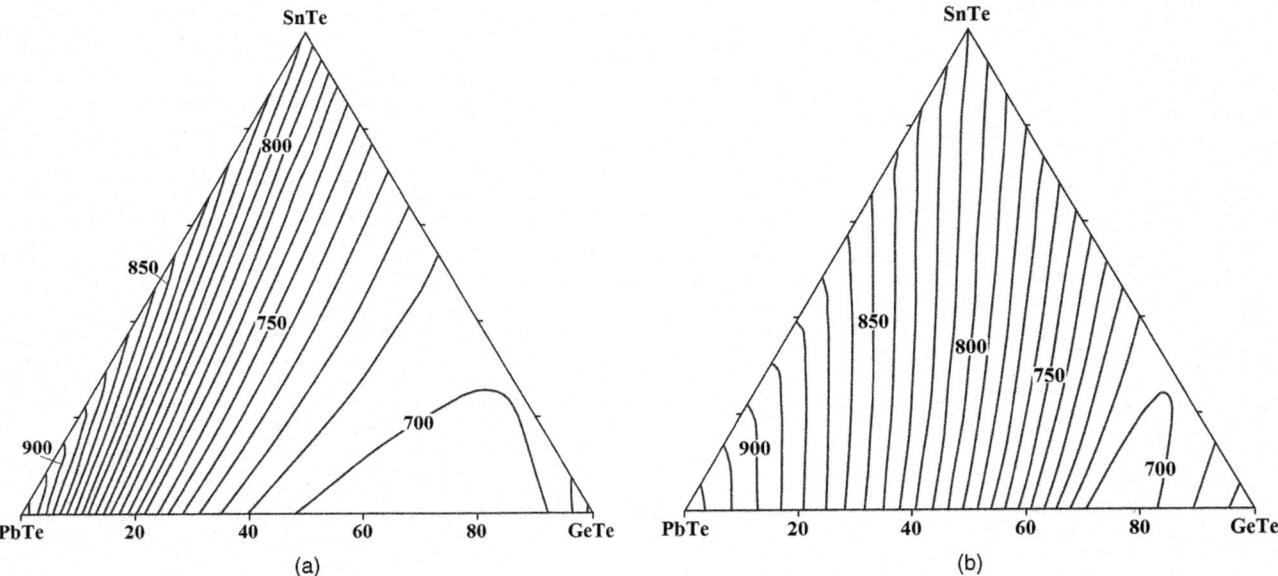

FIGURE 6.7 Solidus (a) and liquidus (b) surfaces of the GeTe–SnTe–PbTe quasiternary system. (From Yashina, L.V., et al., *J. Alloys Compd.*, **413**(1–2), 133, 2006.)

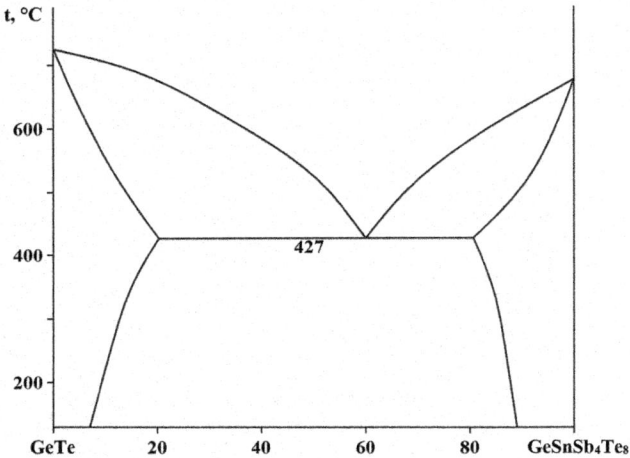

FIGURE 6.8 Phase diagram of the GeTe–GeSnSb₄Te₈ system. (From Gurbanov, G.R., et al., *Russ. J. Inorg. Chem.*, **63**(1), 111, 2018.)

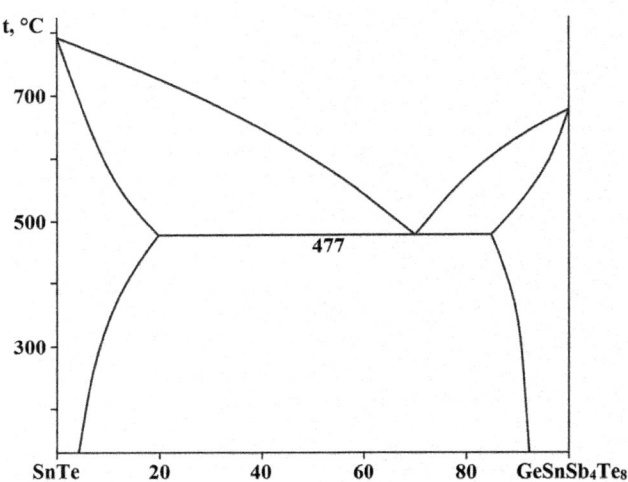

FIGURE 6.9 Phase diagram of the SnTe–GeSnSb₄Te₈ system. (From Gurbanov, G.R., et al., *Russ. J. Inorg. Chem.*, **63**(1), 111, 2018.)

6.35 Germanium–Lead–Manganese–Tellurium

GeTe–PbTe–MnTe. The solubility region based on GeTe in the GeTe–PbTe–MnTe quasiternary system is shown in Figure 6.10 (Shelimova et al. 1993a, b). The maximum extent of the homogeneity region at room temperature is observed along the GeTe–Pb$_x$Mn$_{1-x}$Te (x = 0.2 and 0.3) sections: the boundary of the homogeneity region runs between 20 and 25 mol % Pb$_x$Mn$_{1-x}$Te (Abrikosov et al. 1979; Shelimova 1993a, b).

The temperatures of the liquidus, solidus, and phase transition of some samples of this quasiternary system were determined by Abrikosov et al. (1979, 1980a). Thermal effects corresponding to phase transitions in the solid state have not

been observed for these alloys. However, in the region of the phase transition, a peak was clearly observed in the curve of the temperature dependence of the coefficient of linear thermal expansion. When lead and manganese tellurides are added to GeTe, the phase transition temperature decreases significantly.

This system was studied through DTA, metallography, and dilatometry. The ingots for the investigations were annealed at 500°C for 1000 h followed by the cooling with a rate of 2–3°C·min⁻¹ (Abrikosov et al. 1979; Shelimova 1993a, b). At the quenching of annealed samples in water with ice, the boundaries of GeTe-based solid solutions do not change significantly, except for alloys that are located near the GeTe–PbTe system, where the region of solid solutions increases.

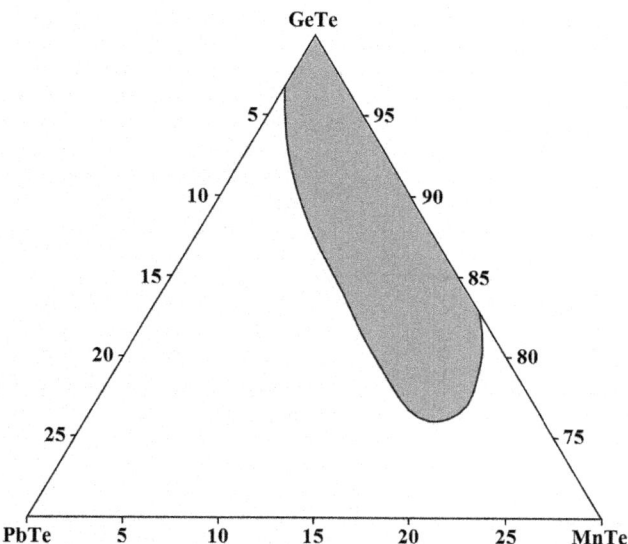

FIGURE 6.10 Solubility region based on GeTe in the GeTe–PbTe–MnTe quasiternary system. (From Shelimova, L.E., et al., *Neorgan. mater.*, **29**(8), 1089, 1993.)

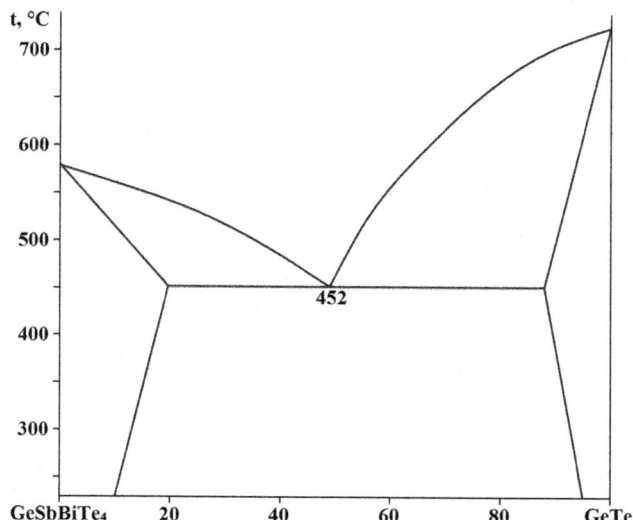

FIGURE 6.11 Phase diagram of the GeTe–GeSbBiTe₄ system. (From Gurbanov, G.R., *Russ. J. Inorg. Chem.*, **58**(1), 96, 2013.)

6.36 Germanium–Antimony–Bismuth–Tellurium

The **GeSbBiTe₄** quaternary compound, which melts congruently at 577°C and crystallizes in the trigonal structure with the lattice parameters $a = 627$ and $c = 3840$ pm, is formed in the Ge–Sb–Bi–Te system (Agaguseinova et al. 2012; Gurbanov 2013). Single crystals of this compound were grown by chemical transport reactions using iodine as a transport agent. The energy gap of this compound is 0.197 eV and 0.22 eV for $GeSb_{3.91}Bi_{0.03}Te_{6.91}$ composition (Gurbanov 2017).

GeTe–GeSbBiTe₄. The phase diagram of this system, constructed through DTA, XRD, MSA, and measuring of the microhardness and density, is a eutectic type (Figure 6.11) (Gurbanov 2013). The eutectic contains 49 mol% GeTe and crystallizes at 452°C. The solubility of alloys based on GeSbBiTe₄ is 10 mol% GeTe at room temperature and 20 mol% at 452°C. The solubility of GeSbBiTe₄ in GeTe is 3 mol% at room temperature.

GeSbBiTe₄–Bi₂Te₃. The phase diagram of this system, constructed through DTA, XRD, MSA, and measuring of the microhardness and density, is a eutectic type (Gurbanov 2013). The eutectic contains 60 mol% Bi₂Te₃ and crystallizes at 427°C. Solubilities based on GeSbBiTe₄ and Bi₂Te₃ are 12 and up to 7 mol% at room temperature, respectively.

GeSbBiTe₄–Sb₂Te₃. The phase diagram of this system, constructed through DTA, XRD, MSA, and measuring of the microhardness and density, is also a eutectic type with limited solubility of alloys based on the initial tellurides (Gurbanov 2013). The eutectic contains 40 mol% Sb₂Te₃ and crystallizes at 477°C.

The **GeSb₂Te₄–GeBi₂Te₄**, **Ge₂Sb₂Te₅–GeSbBiTe₄**, and **GeBi₄Te₇–GeSbBiTe₄** vertical sections were also constructed by Gurbanov (2013).

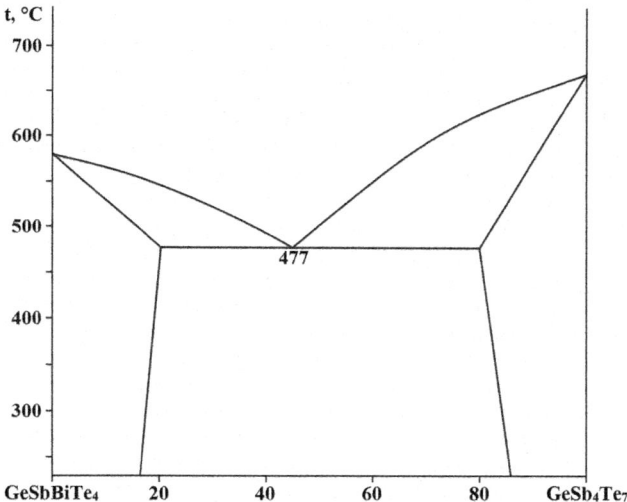

FIGURE 6.12 Phase diagram of the GeSb₄Te₇–GeSbBiTe₄ system. (From Gurbanov, G.R., *Russ. J. Inorg. Chem.*, **58**(1), 96, 2013.)

GeSb₄Te₇–GeSbBiTe₄. The phase diagram of this system, constructed through DTA, XRD, metallography, and measuring of the microhardness and density, is a eutectic type (Figure 6.12) (Gurbanov 2013). The eutectic contains 45 mol% GeSb₄Te₇ and crystallizes at 477°C.

The alloys for the construction of the phase diagram and vertical sections were annealed at 330°C for 250 h (Gurbanov 2013).

GeTe–Sb₂Te₃–Bi₂Te₃. On the isothermal section of this quasiternary system at 500°C (Figure 6.13), there are five single-phase fields separated by four two-phase regions (Abrikosov and Sokolova 1977). The solid solution based on GeTe does

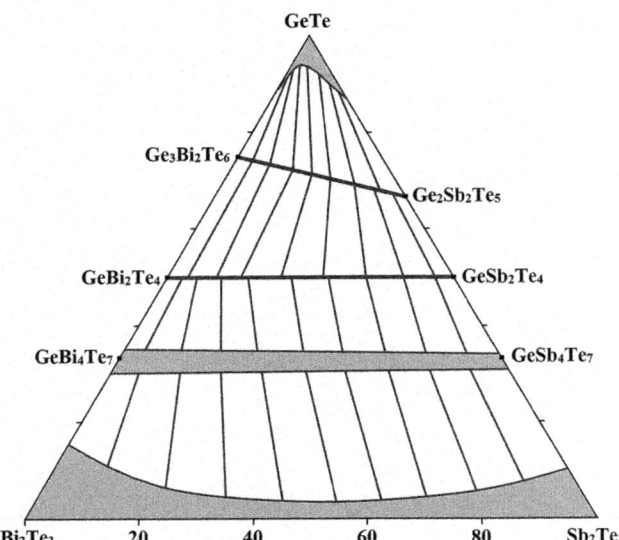

FIGURE 6.13 Isothermal section of the GeTe–Sb$_2$Te$_3$–Bi$_2$Te$_3$ quasiternary system at 500°C. (From Abrikosov, N.H., and Sokolova, I.F., *Zhurn. neorgan. khimii*, **22**(6), 1651, 1977.)

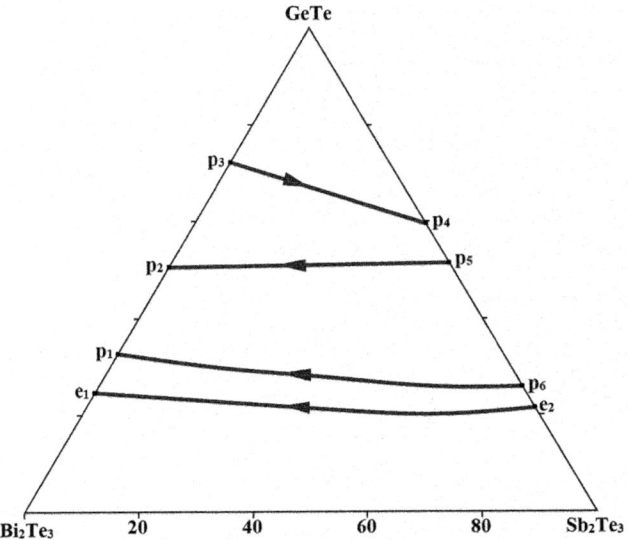

FIGURE 6.14 Liquidus surface of the GeTe–Sb$_2$Te$_3$–Bi$_2$Te$_3$ quasiternary system. (From Abrikosov, N.H., and Sokolova, I.F., *Zhurn. neorgan. khimii*, **22**(6), 1651, 1977.)

not exceed 5 mol % Bi$_x$Sb$_{2-x}$Te$_3$, and the solid solutions based on the Sb$_2$Te$_3$–Bi$_2$Te$_3$ system have a width that is within 4–8 mol % GeTe. The liquidus surface of the system (Figure 6.14) consists of five fields of primary crystallization. The system was studied by DTA, metallography, and the microhardness measurements, and the research alloys were annealed at 500°C for 1500 h.

6.37 Germanium–Antimony–Nickel–Tellurium

GeTe–NiSb. This system is a nonquasibinary section of the Ge–Ni–Sb–Te quaternary system (Zargarova et al. 1975c). Homogeneity regions based on GeTe and NiSb do not exceed 1 mol%.

7

Systems Based on Tin Sulfides

7.1 Tin–Hydrogen–Nitrogen–Sulfur

In liquid NH_3, SnS_2 reacts with $(NH_4)_2S$ to form $(NH_4)_2(SnS_3)$ and $(NH_4)_4(SnS_4)$, both of which decompose above 100°C forming SnS_2 (Behrens and Ostermeier 1962).

7.2 Tin–Hydrogen–Oxygen–Sulfur

Two quaternary compounds, $Sn_3(OH)_2OSO_4$ and $Sn_7(OH)_{12}(SO_4)_2$, are formed in the Sn–H–O–S system. First of them decomposes at 230°C with the formation of $Sn_3O_2SO_4$ and crystallizes in the orthorhombic structure with the lattice parameters $a = 498.3 \pm 0.2$, $b = 1312.8 \pm 0.5$, $c = 1221.4 \pm 0.5$ pm, and the calculated and experimental densities of 4.18 and 4.12 g·cm^{-3}, respectively (Davies and Donaldson 1967; Davies et al. 1975) [$a = 1302.03 \pm 0.34$, $b = 494.51 \pm 0.13$, and $c = 1207.83 \pm 0.35$ pm (Grimvall 1973); $a = 1304.5 \pm 0.2$, $b = 493.83 \pm 0.14$, and $c = 1214.0 \pm 0.2$ pm (Grimvall 1975)]. The standard Gibbs free energy of $Sn_3(OH)_2OSO_4$ formation is -1563.1 ± 0.9 kJ·M^{-1} (Edwards et al. 1996). To prepare this compound, $SnSO_4$ (5 g) was dissolved in a minimum of water containing a few drops of concentrated sulfuric acid (Davies and Donaldson 1967; Edwards et al. 1996). The pH of the clear solution was increased to approximately 1.8 by adding 0.5 M solution of NaOH. The resulting white precipitate was collected at the pump, washed with water and acetone, and dried in a vacuum over silica gel or KOH pellets. During storage, the resulting crystals are stable for at least 12 months. All preparations were carried out in degassed water under an atmosphere of N_2. Crystals of $Sn_3(OH)_2OSO_4$ were also obtained by slow hydrolysis of tin(II) sulphamate, $Sn(SO_3NH_2)_2$.

$Sn_7(OH)_{12}(SO_4)_2$ also crystallizes in the orthorhombic structure with the lattice parameters $a = 1247.2 \pm 0.1$, $b = 1264.9 \pm 0.3$, and $c = 1267.6 \pm 0.2$ pm (Grimvall 1982). Its crystals were prepared together with $Sn_3(OH)_2OSO_4$, according to Davies and Donaldson (1967), but with air access in order to permit oxidation Sn(II) to Sn(IV) by atmospheric oxygen.

7.3 Tin–Hydrogen–Manganese–Sulfur

The $H_{2x}Mn_xSn_{3-x}S_6$ quaternary phases with $x = 0.11-0.25$, which crystallize in the hexagonal structure with the lattice parameters $a = (359.1 \pm 0.2)$-(359.9 ± 0.5), $c = (583.8 \pm 0.5)$-(587.7 ± 0.1) pm, and the calculated and experimental densities of 4.69 and 3.65 g·cm^{-3}, respectively, are formed in the Sn–H–Mn–S system (Manos et al. 2009). The density measurements revealed that these phases may contain vacancies created by the removal of Mn^{2+} ions. $H_{0.4}Mn_{0.2}Sn_{2.8}S_6$ displays a band

gap energy of ~1.0 eV. To synthesize $H_{2x}Mn_xSn_{3-x}S_6$ with $x = 0.2-0.25$, $K_{1.9}Mn_{0.95}Sn_{2.05}S_6 \cdot 2H_2O$ (0.35 mM), prepared with solid-state synthesis, was added as a solid in 0.6 M solution of HCl (20 mL). The mixture was kept under magnetic stirring for ≈12 h. This procedure was performed twice in order to ensure the complete conversion of $K_{1.9}Mn_{0.95}Sn_{2.05}S_6$ to the proton-containing material. Then, the gray solid was isolated by filtration, washed several times with water, acetone, and ether, and dried in the air. These phases at $x = 0.11-0.15$ were prepared with a similar procedure with a difference that $H_{2x}Mn_xSn_{3-x}S_6$ used was prepared with hydrothermal synthesis (the pristine $K_{1.9}Mn_{0.95}Sn_{2.05}S_6$ used for the preparation of $H_{0.4}Mn_{0.2}Sn_{2.8}S_6$, was prepared with solid-state synthesis). Pristine materials prepared with hydrothermal synthesis are usually finer powders than those prepared with solid-state synthesis. Therefore, more Mn is leached out during the acid treatment of $H_{2x}Mn_xSn_{3-x}S_6$ and hydrothermally synthesized, likely due to its finer powder form.

7.4 Tin–Lithium–Strontium–Sulfur

The Li_2SrSnS_4 quaternary compound, which crystallizes in the tetragonal structure with the lattice parameters $a = 665.9 \pm 0.5$, $c = 791.8 \pm 1.2$ pm, a calculated density of 3.296 g·cm^{-3}, and an energy gap of 3.10 eV, is formed in the Sn–Li–Sr–S system (Wu et al. 2019). To prepare this compound, a mixture of Li (2 mM), SrS (1 mM), Sn (1 mM), and S (3 mM) was first loaded into a graphite crucible that avoids the reaction between alkali metal and silica tube, then put it into the silica tube and flame-sealed under 10^{-3} Pa. The tube was heated to ~750°C in 30 h and kept at this temperature for about 100 h, then cooled slowly down to ambient temperature for 5 days. After carefully washing with dimethylformamide (DMF), pale-yellow crystals of the title compound appeared, which were stable in the air. The whole preparation process is completed in an Ar-filled glove box.

7.5 Tin–Lithium–Barium–Sulfur

Two quaternary compounds, Li_2BaSnS_4 and $Li_2Ba_7Sn_4S_{16}$, are formed in the Sn–Li–Ba–S system. First of them crystallizes in the tetragonal structure with the lattice parameters $a = 677.4 \pm 0.7$, $c = 818.5 \pm 1.6$ pm, a calculated density of 3.520 g·cm^{-3}, and an energy gap of 3.07 eV (Wu et al. 2017c). To synthesize the title compound, a mixture of Li (2 mM), Ba (1 mM), Sn (1 mM), and S (4 mM) was first loaded into a graphite crucible that avoids the reaction between Li and silica tube, then put it into the silica tube and flame-sealed under 10^{-3} Pa. The tube was heated to 800°C in 30 h and kept at this temperature for about 90 h, then cooled slowly down to ambient temperature for 5 days. The product was washed with DMF. Finally, many small pale-yellow transparent crystals of Li_2BaSnS_4 were found, which are stable in the air.

$Li_2Ba_7Sn_4S_{16}$ crystallizes in the cubic structure with the lattice parameter $a = 1457.7 \pm 0.2$ pm, a calculated density of 4.209 g·cm⁻³, and an energy gap of 2.30 eV (Abudurusuli et al. 2018) [$a = 1460.34 \pm 0.07$ pm, a calculated density of 3.988 g·cm⁻³, and an energy gap of 2.23 eV for the $Li_{2.67}Ba_6Sn_{4.33}S_{16}$ composition (Duan et al. 2018)]. To prepare this compound, Li, BaS, Sn, and S (molar ratio 1:3.5:2:8) were weighed, well-mixed, and loaded into the graphite crucible (Abudurusuli et al. 2018). Then, the crucible was embedded in a silica tube that was sealed under a vacuum of 10⁻³ Pa. The tube was put into a furnace, and the setting temperature process was as follows: firstly, heated to 870°C in 30 h and kept at this temperature for 50 h to make the mixture completely melted, and then slowly cooled to 300°C at the rate of 5°C·h⁻¹ before switching off the furnace. The obtained product was carefully washed with DMF to remove the unreacted S powder. Finally, air-stable crystals of $Li_2Ba_7Sn_4S_{16}$ of a light-orange color appeared.

$Li_{2.67}Ba_6Sn_{4.33}S_{16}$ was synthesized using the mixture of Ba, Li_2S, Sn, and S (molar ratio 6:1.335:4.33:8.665) (Duan et al. 2018). The mixture was put in a graphite crucible, and under a vacuum of 10⁻³ Pa, the crucible was sealed into a silicon tube, which was loaded in a heater. Then the heater was warmed up to 930°C in 50 h and kept at this temperature for 100 h. Next, the heater was cooled to 350°C over 100 h before being closed. Finally, the red compound was successfully synthesized. It shows excellent thermal stability before 683°C, is very stable, and can be stored in the air for at least one year.

7.6 Tin–Lithium–Zinc–Sulfur

The Li_2ZnSnS_4 quaternary compound, which melts at 919°C and crystallizes in the orthorhombic structure with the lattice parameters $a = 637.28 \pm 0.13$, $b = 672.86 \pm 0.13$, $c = 796.21 \pm 0.16$ pm, a calculated density of 3.173 g·cm⁻³, and an energy gap of 2.87 eV, is formed in the Sn–Li–Zn–S system (Lekse et al. 2008). This compound was synthesized using Li_2S (1.2 mM), Zn (1 mM), Sn (1 mM), and S (3 mM), which was ground in an agate mortar and pestle in an Ar-filled glove box. The ground mixture was placed into a graphite crucible inside a fused silica tube, which was sealed under a vacuum of approximately 0.1 Pa, heated to 700°C in 12h, and then held at 700°C for 96 h. The sample was then slowly cooled to room temperature for 12 h. The graphite crucible is necessary to prevent lithium from reacting with the fused silica. The tube was opened under ambient conditions, and the yellow-green, microcrystalline powder was obtained. Single crystals of Li_2ZnSnS_4 were obtained in the same way, but the mixture was heated to 825°C in 9 h and then held at this temperature for 72 h. After that, the sample was cooled slowly to 500°C for 32.5 h and then rapidly to room temperature in 3 h. Yellow-green, single crystals trapezoidal plates were obtained.

7.7 Tin–Lithium–Cadmium–Sulfur

Two quaternary compounds, Li_2CdSnS_4 and $Li_4CdSn_2S_7$, are formed in the Sn–Li–Cd–S system. First of them crystallizes in the orthorhombic structure with the lattice parameters $a = 795.55 \pm 0.03$, $b = 696.84 \pm 0.03$, $c = 648.86 \pm 0.03$ pm, and an energy gap of 3.26 eV (Lekse et al. 2009) [$a = 785.6$, $b = 679.9$, $c = 639.4$ pm and an energy gap of 2.50 eV (Li et al. 2011)].

This compound has been synthesized by the reaction of CdS, SnS, Li_2S, and S in a stoichiometric ratio in an evacuated, sealed quartz tube (Devi and Vidyasagar 2002). A mixture of CdS (0.027 mM), SnS (0.027 mM), Li_2S (0.218 mM), and S (1.088 mM), taken in an evacuated, sealed quartz tube, was heated at 560°C for 4 days and then cooled to room temperature over a period of another 4 days. The entire product content of the reaction tube was washed thoroughly with water to isolate the water-insoluble, homogeneous phase of Li_2CdSnS_4 in the form of orange, needle crystals. Both single crystals and polycrystalline samples of this compound were also synthesized using polychalcogenide flux at 750°C and 650°C, respectively (Lekse et al. 2009).

$Li_4CdSn_2S_7$ crystallizes in the monoclinic structure with the lattice parameters $a = 1744.30 \pm 0.02$, $b = 693.532 \pm 0.007$, $c = 1032.71 \pm 0.01$ pm, $\beta = 93.9042 \pm 0.0005°$, a calculated density of 3.208 g·cm⁻³, and an energy gap of 2.59 eV (Zhang et al. 2020b). This compound was prepared via high-temperature solid-state synthesis. Stoichiometric quantities of Li_2S, Cd, Sn, and S were weighed and combined in an Ar-filled glove box. The mixture was loaded into a graphite tube that was loosely capped and subsequently inserted into a fused silica tube, which was then flame-sealed under vacuum (~0.1 Pa). Next, the tube was placed into a furnace. The reaction was heated to 800°C in 12 h, held at that temperature for 2 days, cooled at a rate of 2°C·h⁻¹ to 700°C, held at 700°C for 7 days, and finally cooled to room temperature for 24 h. Pale-yellow crystals were obtained.

7.8 Tin–Lithium–Mercury–Sulfur

The Li_2HgSnS_4 quaternary compound, which crystallizes in the orthorhombic structure with the lattice parameters $a = 794.00 \pm 0.17$, $b = 693.10 \pm 0.15$, $c = 651.22 \pm 0.14$ pm, a calculated density of 4.276 g·cm⁻³, and an energy gap of 2.32 eV (Wu and Pan 2017) [$a = 780.94*/793.22**$, $b = 684.14*/689.73**$ and $c = 645.18*/652.02**$ pm according to the calculations with (*) and without (**) including the spin-orbit coupling and an energy gap of 3.195 eV (Alnujaim et al. 2020)], is formed in the Sn–Li–Hg–S system. The target compound was prepared with a mixture of Li, HgS, Sn, and S (molar ratio 2:1:1:3) (Wu and Pan 2017). In the preparation process, a graphite crucible was added to the vacuum-sealed silica tube to avoid the reaction between Li and the silica tube at a high temperature. The temperature process was set as follows: first, it was heated to 700°C in 2 days, kept at this temperature for about 4 days, then slowly cooled down to 300°C within 4 days, and finally quickly cooled to room temperature by turning off the furnace. The obtained product was washed by DMF to remove the other by-products. Orange-red crystals of Li_2HgSnS_4 were obtained, and they remained stable in the air for over half a year.

7.9 Tin–Lithium–Indium–Sulfur

The **LiInSnS$_4$** quaternary compound, which crystallizes in the cubic structure with the lattice parameter $a = 1062.9 \pm 0.4$ pm, a calculated density of 4.079 g·cm^{-3}, and an energy gap of 2.3 eV, is formed in the Sn–Li–In–S system (Yohannan and Vidyasagar 2016b). It was synthesized by heating in a continuous stream of CS$_2$ vapor appropriate stoichiometric mixture of Li$_2$CO$_3$, In$_2$O$_3$, and SnO$_2$ with one intermittent grinding. An alumina boat was loaded with the reactant mixture and placed in a ceramic tube in a horizontal tubular furnace. A continuous stream of N$_2$ was bubbled through CS$_2$ liquid and then passed through the ceramic tube to provide CS$_2$ vapor for the synthetic reaction. The gray octahedral-shaped single crystals were grown using 0.25 g of the prepared polycrystalline sample, which was sealed in an evacuated silica tube with a small quantity of iodine. This tube was heated for 3 days in a two-zone furnace such that the sample was at 750°C and the other end of the tube was at a colder zone, which was about 50°C less. Single crystals were found, along with the polycrystalline sample, at the cold end of the quartz tube.

7.10 Tin–Lithium–Lanthanum–Sulfur

The **LiLa$_3$SnS$_7$** quaternary compound, which crystallizes in the hexagonal structure, is formed in the Sn–Li–La–S system (Craig et al. 2022). It was prepared via a direct combination of the elements or binary sulfides. This compound is a semiconductor, melts in the vicinity of 740°C, and shows high thermal stability (>1000°C).

7.11 Tin–Lithium–Cerium–Sulfur

The **LiCe$_3$SnS$_7$** quaternary compound, which crystallizes in the hexagonal structure, is formed in the Sn–Li–Ce–S system (Craig et al. 2022). It was prepared via direct combination of the elements or binary sulfides. This compound is a semiconductor and melts in the vicinity of 740°C.

7.12 Tin–Lithium–Praseodymium–Sulfur

The **LiPr$_3$SnS$_7$** quaternary compound, which crystallizes in the hexagonal structure, is formed in the Sn–Li–Pr–S system (Craig et al. 2022). It was prepared via direct combination of the elements or binary sulfides. This compound is a semiconductor and melts in the vicinity of 740°C.

7.13 Tin–Lithium–Neodymium–Sulfur

The **LiNd$_3$SnS$_7$** quaternary compound, which crystallizes in the hexagonal structure, is formed in the Sn–Li–Nd–S system (Craig et al. 2022). It was prepared via direct combination of the elements or binary sulfides. This compound is a semiconductor and melts in the vicinity of 740°C.

7.14 Tin–Lithium–Phosphor–Sulfur

Li$_4$SnS$_4$–Li$_3$PS$_4$. The **Li$_{4-x}$Sn$_{1-x}$P$_x$S$_4$** solid solutions are formed in the range of $0.67 \leq x \leq 0.75$ ($-0.25 < \delta < -0.01$ in Li$_{10+\delta}$Sn$_{1+\delta}$P$_{2-\delta}$S$_{12}$) (Hori et al. 2014). These solid solutions crystallize in the tetragonal structure with the lattice parameters $a = 870.054 \pm 0.003$ and $c = 1268.610 \pm 0.008$ pm at 12 K, $a = 873.764 \pm 0.003$ and $c = 1271.660 \pm 0.007$ pm at 300 K, $a = 881.049 \pm 0.003$ and $c = 1279.398 \pm 0.007$ pm at 800 K for Li$_{9.81}$Sn$_{0.81}$P$_{2.19}$S$_{12}$ composition (Hori et al. 2015b) [$a = 870.57 \pm 0.04$, $c = 1273.89 \pm 0.06$ pm, and a calculated density of 2.183 g·cm^{-3} at 100 K for Li$_{10}$SnP$_2$S$_{12}$ composition (Bron et al. 2013); $a = 875.06 \pm 0.18$ and $c = 1279.93 \pm 0.06$ pm at room temperature for Li$_{10}$SnP$_2$S$_{12}$ composition (Kuhn et al. 2014)]. Li$_{10}$SnP$_2$S$_{12}$ shows a high decomposition temperature above 700°C (Bron et al. 2013). Below this temperature, no phase transition was observed.

All samples of the solid solutions were synthesized using solid-state reactions. To obtain Li$_{10}$SnP$_2$S$_{12}$, stoichiometric quantities of Li$_4$SnS$_4$, Li$_2$S, and P$_2$S$_5$ were mixed and ground in an agate mortar, sealed in an evacuated quartz tube, and heated up to 600°C at a rate of 30°C·h^{-1} (Bron et al. 2013). The sample was kept for 48 h at 600°C and slowly cooled down to room temperature at a rate of 1°C·h^{-1}. This composition was also synthesized as follows (Kuhn et al. 2014). Li$_2$S, P, Sn, and S were first mechanically treated in a ball mill under Ar for 2 days. Pellets were pressed from the obtained amorphous precursor and heated in an evacuated quartz ampoule according to the following temperature program: 30°C·h^{-1} → 340°C (10 h) → 30°C·h^{-1} → 380°C (10 h) → 30°C·h^{-1} → 450°C (2 days). In the first step at 340°C, the material crystallizes in the orthorhombic modification, which then transforms to the desired tetragonal structure at 380°C. The subsequent sintering step at 450°C improves the crystallinity of the sample.

To prepare Li$_{10+\delta}$Sn$_{1+\delta}$P$_{2-\delta}$S$_{12}$, the Li$_2$S, P$_2$S$_5$, and SnS$_2$ powders were weighed in appropriate molar ratios in an Ar-filled glove box and then mechanically mixed by planetary ball-milling for 40 h (Hori et al. 2014, 2015b). The specimens were then pressed into pellets, sealed in a carbon-coated quartz ampoule at 10 Pa, and heated at a reaction temperature of 650°C for 3 days. After heating, the furnace was slowly cooled to room temperature.

7.15 Tin–Lithium–Niobium–Sulfur

The **Li$_x$SnNbS$_3$** quaternary intercalation compound is formed in the Sn–Li–Nb–S system (Hernán et al. 1991). It was prepared by lithium intercalation of SnNbS$_3$ using *n*-buthyllithium under an inert atmosphere. Li$_x$SnNbS$_3$ crystallizes in the orthorhombic structure with the lattice parameter $c = 1180 \pm 2$, 1208 ± 4, and 1218 ± 2 pm at $x = 0.13$, 0.50, and 1.51, respectively.

7.16 Tin–Lithium–Manganese–Sulfur

The **Li$_2$MnSnS$_4$** quaternary compound, which has two polymorphic modifications, is formed in the Sn–Li–Mn–S system (Devlin et al. 2015). α-Li$_2$MnSnS$_4$ crystallizes in the orthorhombic structure with the lattice parameters $a = 1370.36 \pm 0.03$, $b = 800.23 \pm 0.02$, $c = 641.55 \pm 0.02$ pm, and a calculated density of 2.981 g·cm^{-3} and β-Li$_2$MnSnS$_4$ crystallizes in the monoclinic structure with the lattice parameters $a = 641.43 \pm 0.06$, $b = 684.75 \pm 0.06$, $c = 800.78 \pm 0.07$ pm, $\beta = 89.980 \pm 0.006°$, and a calculated density of 2.981 g·cm^{-3}. The optical band gaps of both polymorphs were estimated as ~2.6–3.0 eV.

The polycrystalline samples of α- and β-Li$_2$MnSnS$_4$ were obtained by heating the reactants to 750°C at a rate of 50°C·h^{-1}, holding the reaction at 750°C for 144 h followed by slow cooling at a rate of 1°C·h^{-1} to 550°C before cooling to room temperature radiatively (Devlin et al. 2015). Li$_2$MnSnS$_5$ single crystals were synthesized by weighing stoichiometric amounts of Li$_2$S, Mn, Sn, and S in an Ar-filled glove box. The reagents were combined and ground using an agate mortar and pestle. The starting materials were then transferred to a graphite crucible and placed into a fused silica tube. The tube was flame-sealed under a vacuum of ~0.1 Pa. The sealed reaction tube was then placed into a furnace and heated. The single crystals of α- and β-Li$_2$MnSnS$_4$ were selected from a reaction that was heated from room temperature to 800°C over 11.3 h and held at that temperature for 30 h. The sample was allowed to cool to ambient temperature radiatively. The entire sample was rinsed with methanol in order to remove the Li$_2$S$_x$ flux. The obtained crystals were yellow-orange irregular polyhedra.

7.17 Tin–Lithium–Iron–Sulfur

Two quaternary compounds, **Li$_{0.5}$Fe$_{0.5}$[FeSn]S$_4$** and **Li$_2$FeSnS$_4$**, are formed in the Sn–Li–Fe–S system. First of them crystallizes in the cubic structure with the lattice parameter $a = 1040$ pm (Harada 1973). It was prepared by the usual ceramic techniques. The appropriate stoichiometric mixture of Fe, Li$_2$S, SnS$_2$, and S was fired using the following two-step process. First, the mixed powder was heated at 250–700°C for 3 days in either an evacuated Pyrex or quartz tube, and the prefired polysulfides thus obtained were ground and pressed into a pellet. The material was then refired in an evacuated quartz tube for about a week at 580–600°C. When necessary, the refiring procedure was repeated until single-phase products were obtained.

Li$_2$FeSnS$_4$ crystallizes in the monoclinic structure with the lattice parameters $a = 637.419 \pm 0.007$, $b = 678.396 \pm 0.007$, $c = 793.071 \pm 0.009$ pm, and $\beta = 90.2639 \pm 0.0009°$ at room temperature and $a = 637.3 \pm 0.2$, $b = 678.14 \pm 0.02$, $c = 791.83 \pm 0.03$ pm, and $\beta = 90.279 \pm 0.003°$ at 50 K (Brant et al. 2014b) [$a = 637.27 \pm 0.03$, $b = 677.76 \pm 0.03$, $c = 791.13 \pm 0.02$ pm, $\beta = 90.207 \pm 0.003°$, and a calculated density of 3.078 g·cm^{-3} at 173 K (Brunetta et al. 2013)]. This compound was prepared by grinding a mixture of Li$_2$S (1.2 mM), Fe (1 mM), Sn (1 mM), and S (3 mM) using an agate mortar and pestle in an Ar-filled glove box (Brunetta et al. 2013; Brant et al. 2014b). The reactants were placed into a graphite crucible inside a fused silica tube. The tube was sealed under a pressure of approximately 0.1 Pa. The sample was heated at 50°C·h^{-1} to 600°C, held at 600°C or 750°C for 96 h, slowly cooled to 350°C or 500°C over 50 h, and then allowed to cool to room temperature radiatively. The tube was opened under ambient conditions. The

excess Li$_2$S$_x$ flux was rinsed with DMF and hexane. Dark-red plate-like crystals of Li$_2$FeSnS$_4$ were obtained.

7.18 Tin–Lithium–Cobalt–Sulfur

The **Li$_2$CoSnS$_4$** quaternary compound, which crystallizes in the monoclinic structure with the lattice parameters $a = 634.32 \pm 0.02$, $b = 671.84 \pm 0.02$, $c = 794.04 \pm 0.03$ pm, $\beta = 89.988 \pm 0.002°$, a calculated density of 3.139 g·cm^{-3}, and an energy gap of 2.421 eV, is formed in the Sn–Li–Co–S system (Brant et al. 2015a). This compound was obtained by grinding in an agate mortar and pestle stoichiometric amounts of Co chips, Sn pieces, and S powder plus a 20% excess of Li$_2$S that can act as a molten Li$_2$S$_x$ flux at elevated temperatures. The mixture was placed into a graphite crucible inside a fused silica tube that was sealed under vacuum (10^{-2} Pa). The reaction vessel was heated at 650°C for 144 h, slowly cooled to 550°C for 1000 h, and then allowed to cool to room temperature naturally. The reaction vessel was opened under ambient conditions, and the product was rinsed with methanol to remove the excess Li$_2$S$_x$ flux. A bright-green polycrystalline powder of Li$_2$MnSnS$_4$ was obtained.

7.19 Tin–Sodium–Potassium–Sulfur

The **K$_2$Na$_2$Sn$_3$S$_8$** quaternary compound, which crystallizes in the monoclinic structure with the lattice parameters $a = 1123.96 \pm 0.14$, $b = 754.83 \pm 0.08$, $c = 1826.3 \pm 0.2$ pm, $\beta = 97.899 \pm 0.004°$, a calculated density of 3.188 g·cm^{-3}, and an energy gap of 2.38 eV, is formed in the Sn–Na–K–S system (Ji et al. 2021b). The crystals of this compound were prepared with a mixture of Na, KBr, Sn, and S (molar ratio 1:2:1:3). The mixture was loaded into a fused silica tube under a vacuum of 10^{-3} Pa. Next, the fused silica tube was put in the furnace and annealed at 850°C for 50 h and then kept at this temperature for 100 h. This was followed by slowly cooling at 5°C·h^{-1} to 200°C, and then the furnace was turned off. The anhydrous ethanol and DMF were used to wash the reaction product for removing the other by-products. Finally, yellow crystals were obtained, which were stable in the air for more than 3 months.

7.20 Tin–Sodium–Copper–Sulfur

The **Na$_2$Cu$_2$Sn$_2$S$_6$** quaternary compound is formed in the Sn–Na–Cu–S system (Liao and Kanatzidis 1993). To prepare this compound, a mixture of Sn, Cu, Na$_2$S, and S (molar ratio 1:1–2:4:16) was loaded into a Pyrex tube in a N$_2$ glove box. The tube was evacuated and sealed at a pressure of ~0.1 Pa. The mixture was heated slowly from room temperature to 400°C over 12 h in a furnace. The temperature was kept at 400°C for 4 days, and then was cooled slowly to room temperature at 4°C·h^{-1}. Black plate-like crystals were formed. The product was washed with degassed DMF to remove excess Na$_2$S$_x$ flux using a standard Schlenk technique and dried with acetone and ether. The obtained crystals are stable in the air.

7.21 Tin–Sodium–Strontium–Sulfur

The **Na₂SrSnS₄** quaternary compound, which crystallizes in the trigonal structure with the lattice parameters $a = 2383.8 \pm 0.4$, $c = 711.5 \pm 0.2$ pm, a calculated density of 3.248 g·cm^{-3}, and an energy gap of 3.12 eV, is formed in the Sn–Na–Sr–S system (Wu et al. 2019). To synthesize this compound, a mixture of Na (2 mM), SrS (1 mM), Sn (1 mM), and S (3 mM) was first loaded into a graphite crucible that avoids the reaction between an alkali metal and silica tube, then put it into the silica tube and flame-sealed under 10^{-3} Pa. The tube was heated to ~750°C in 30 h and kept at this temperature for about 100 h, then cooled slowly down to ambient temperature for 5 days. After carefully washing with DMF, pale-yellow crystals of the title compound appeared, which are stable in the air after. The whole preparation process is completed in an Ar-filled glove box.

7.22 Tin–Sodium–Barium–Sulfur

Two quaternary compounds, **Na₂BaSnS₄** and **Na₂Ba₇Sn₄S₁₆**, are formed in the Sn–Na–Ba–S system. First of them crystallizes in the tetragonal structure with the lattice parameters $a = 1010.3 \pm 0.4$, $c = 800.5 \pm 0.7$ pm, a calculated density of 3.498 g·cm^{-3}, and an energy gap of 3.27 eV (Wu et al. 2016a,b). This compound was prepared as follows. A mixture of Na (2 mM), Ba (1 mM), Sn (1 mM), and S (4 mM) was firstly loaded into a graphite crucible that avoids the reaction between Na and silica tube, then put into the silica tube and flame-sealed under 10^{-3} Pa. The tube was heated to 500°C in 30 h and kept at this temperature for about 90 h, then slowly cooled down to ambient temperature for 5 days. DMF was used to wash the products. Finally, many pale-yellow crystals were found. They are stable in the air after several months. Owing to Na and Ba metals being easily oxidized in the air, an Ar-filled glove box was used to complete the preparation processes.

Na₂Ba₇Sn₄S₁₆ crystallizes in the cubic structure with the lattice parameter $a = 1480.7 \pm 0.4$, a calculated density of 4.082 g·cm^{-3}, and an energy gap of 2.50 eV (Abudurusuli et al. 2018). To obtain this compound, Na, BaS, Sn, and S (molar ratio 1:3.5:2:8) were weighed, well mixed, and loaded into the graphite crucible. Then, the crucible was embedded in a silica tube that was sealed under a vacuum of 10^{-3} Pa. The tube was put into a furnace, and the setting temperature process was as follows: firstly heated to 870°C in 30 h and kept at this temperature for 50 h to make the mixture completely melted, and then slowly cooled to 300°C at the rate of 5°C·h^{-1} before switching off the furnace. The obtained product was carefully washed with DMF to remove the unreacted S powder. Finally, air-stable crystals of Na₂Ba₇Sn₄S₁₆ with light orange color appeared.

7.23 Tin–Sodium–Zinc–Sulfur

Two quaternary compounds, **Na₂ZnSnS₄** and **Na₂ZnSn₂S₆**, are formed in the Sn–Na–Zn–S system. The first crystallizes in the tetragonal structure with the lattice parameters $a = 648.35 \pm 0.06$, $c = 913.4 \pm 0.1$ pm, and an energy gap of 3.1 eV (He et al. 2018a). Single crystals of this compound were obtained using the reactive flux method. Na₂S₂ (4 mM), Zn (1 mM), Sn (1 mM), and S (2 mM) were mixed uniformly and pressed into pellets, followed by flame sealing in a silica tube under vacuum. The tube was slowly heated to 600°C in a furnace and held for 50 h. Afterward, the tube was cooled to 300°C over a period of 17 h and finally to room temperature by turning off the furnace. The products were washed and sonicated with DMF, distilled water, and acetone in sequence several times to remove extra Na₂S₂ and Na₂SnS₃ impurities. The obtained crystals were dried in a vacuum oven at 60°C for 1 h. Light-yellow crystals of Na₂ZnSnS₄ were obtained. Powder samples were prepared through a simple solid-state reaction method. Na₂S₂, Zn, Sn, and S in a stoichiometric ratio were repeatedly ground and annealed at 600°C for 20 h twice.

Na₂ZnSn₂S₆ melts congruently at 604°C and crystallizes in the orthorhombic structure with the lattice parameters $a = 1295.12$, $b = 2358.01$, $c = 728.51$ pm, and a calculated density of 3.231 g·cm^{-3} (Yohannan and Vidyasagar 2016a) [$a = 1294.5 \pm 0.3$, $b = 2355.6 \pm 0.5$, $c = 727.74 \pm 0.15$ pm, a calculated density of 3.239 g·cm^{-3}, and an energy gap of 2.71 eV (Li et al. 2017b)].

A polycrystalline sample of this compound was synthesized by heating a stoichiometric mixture of Na₂CO₃ (0.92 mM), ZnS (0.92 mM), and SnS (1.85 mM), in a continuous stream of CS₂ vapor, at 550°C for 24 h with one intermittent grinding (Yohannan and Vidyasagar 2016a). An alumina boat was loaded with the reactant mixture and placed in a ceramic tube in a horizontal tubular furnace. A continuous stream of N₂ gas was bubbled through CS₂ liquid and then passed through the ceramic tube to provide CS₂ vapor for the synthetic reaction. Thus N₂ was used as a carrier gas for CS₂ vapor. Yellow block-shaped single crystals of Na₂ZnSn₂S₆ were obtained by the molten Na₂S₂O₃ flux method, from a mixture of Na₂CO₃ (1.39 mM), Sn (1.85 mM), Zn (0.92 mM), and S (6.46 mM). A quartz tube containing the reactant mixture was initially heated at 350°C under a dynamic vacuum for 10 min to expel the CO₂ generated and then sealed under vacuum. This sealed quartz tube was heated at 750°C for 3 days and then cooled to 550°C at 4°C·h^{-1}, and the furnace was finally turned off. The solid products were washed with distilled water to dissolve away Na₂S₂O₃ formed and to isolate the single crystals of Na₂ZnSn₂S₆. After that, colorless single crystals were obtained.

The crystal and pure phase samples of the title compound were also synthesized by solid-state reaction of Na, Zn, Sn, S (molar ratio 2:1:2:6) (Li et al. 2017b). The mixture of the elements was weighted and placed into the graphite crucible in the Ar atmosphere, and then sealed into a fused silica tube under a vacuum of 10^{-3} Pa. The tube was moved into a furnace, heated to 750°C in 50 h, dwelled there for 30 h, cooled to 650°C at a rate of 5°C·h^{-3}, then cooled to room temperature at a rate of 10°C·h^{-3}. This compound is stable in moisture conditions and air for several months.

7.24 Tin–Sodium–Cadmium–Sulfur

Three quaternary compounds **Na₂CdSnS₄**, **Na₂CdSn₃S₈**, and **Na₆CdSn₄S₁₂**, are formed in the Sn–Na–Cd–S system. Na₂CdSnS₄ crystallizes in the monoclinic structure with the lattice parameters $a = 928.2 \pm 0.1$, $b = 942.1 \pm 0.3$, $c = 659.3$

± 0.9 pm, and β = 134.83° ± 0.09° and an energy gap of 1.52 eV. Na$_6$CdSn$_4$S$_{12}$ also crystallizes in the monoclinic structure with the lattice parameters a = 662.2 ± 0.4, b = 1148.9 ± 0.8, c = 699.9 ± 0.2 pm, and β = 108.56° ± 0.04° (Devi and Vidyasagar 2002).

A mixture of CdS, SnS, and Na$_2$S$_5$ heated at 780°C gives a biphasic mixture of Na$_2$CdSnS$_4$ and Na$_6$CdSn$_4$S$_{12}$. These compounds could not be prepared by conventional high-temperature solid-state reaction from a stoichiometric mixture of reactants. The attempts to prepare compound Na$_2$CdSnS$_4$ or Na$_6$CdSn$_4$S$_{12}$ as a single phase have always resulted in a mixture of both items.

Na$_2$CdSn$_3$S$_8$ has an energy gap of 2.14 eV (Pogu and Vidyasagar 2020). Its polycrystalline sample was synthesized by heating stoichiometric mixtures of Na$_2$CO$_3$, CdS, and SnO$_2$ in a continuous stream of CS$_2$ vapor at 625°C for 2 days, with one intermittent grinding. An alumina boat was loaded with the solid reactant mixture and placed in a ceramic tube, which was kept in a horizontal tubular furnace. A continuous stream of N$_2$ gas was bubbled, at a rate of two bubbles per second, through CS$_2$ liquid, and then passed through the ceramic tube to provide CS$_2$ vapor for the synthetic reaction. The outgoing gaseous mixture of N$_2$, CS$_2$, and CO$_2$ from the ceramic tube was passed through paraffin oil and then into a dilute aqueous solution of KOH. The furnace was turned off after heating and allowed to cool naturally to room temperature for over 12 h. Single crystals of Na$_2$CdSn$_3$S$_8$ were obtained by the thiosulfate flux method. It involved heating the solid reactant mixture of Na$_2$CO$_3$, S, Cd and Sn, or CdS and SnS$_2$ in a silica tube under dynamic vacuum first at about 300°C for about 15 min and then sealing the quartz tube along with solid content under vacuum. The sealed quartz ampoule was heated at a rate of 3°C·min^{-1} to 600°C for the synthesis of CdS and SnS$_2$ and to 625°C when Cd and Sn were used, held at that temperatures for 3 days and then cooled to 350°C over a period of 2 days. The ampoule was broken open after the reaction, and the solid contents were washed with water to dissolve away K$_2$S$_2$O$_3$ flux and isolate the desired solid product. The Na$^+$ ion of the layered Na$_2$CdSn$_3$S$_8$ compound undergoes facile ion-exchange, in an aqueous medium at room temperature, with monovalent NH$_4^+$, Rb$^+$, Cs$^+$, Tl$^+$ ions, divalent Sr^{2+}, Ca^{2+}, Ba^{2+} ions, and trivalent lanthanide Ln^{3+} (Ln = La, Ce, Pr, Sm, Eu, Gd, Tb, Dy, Ho, Er, Tm) ions.

7.25 Tin–Sodium–Mercury–Sulfur

The **Na$_2$Hg$_3$Sn$_2$S$_8$** quaternary compound, which crystallizes in the tetragonal structure with the lattice parameters a = 907.7 ± 0.4, c = 919.5 ± 0.7 pm, a calculated density of 5.005 g·cm^{-3}, and an energy gap of 2.45 eV, is formed in the Sn–Na–Hg–S system (Wu et al. 2016c). To prepare this compound, an Ar-filled glove box was used. A mixture of Na (2 mM), HgS (3 mM), Sn (2 mM), and excess sulfur (8 mM) were loaded into a graphite crucible, and then put into a tidy silica tube that was flame-sealed under 10^{-3} Pa. The tube was heated to about 600°C in 30 h and kept at this temperature for about 60 h, then slowly cooled down to 300°C for 4 days, and finally quickly cooled to ambient temperature. The product was carefully

washed with DMF and then dried at 100°C. Yellow crystals of Na$_2$Hg$_3$Sn$_2$S$_8$ were obtained. This compound is stable in the air.

7.26 Tin–Sodium–Gallium–Sulfur

Two quaternary compounds, **NaGaSnS$_4$** and **Na$_2$Ga$_2$SnS$_6$**, are formed in the Sn–Na–Ga–S system. NaGaSnS$_4$ has two polymorphic modifications (Kumari and Vidyasagar 2007). The first modification crystallizes in the cubic structure with the lattice parameter a = 1301.61 ± 0.01 pm, a calculated density of 3.069 g·cm^{-3}, and an energy gap of 1.75 eV and the second one crystallizes in the orthorhombic structure with the lattice parameters a = 1252.9 ± 0.2, b = 2336.0 ± 0.4, and c = 702.9 ± 0.3 pm. Polycrystalline samples of NaGaSnS$_4$ were prepared by heating stoichiometric mixtures of appropriate reactants at 850°C for 4 days and then cooling to room temperature at the rate of 2°C·h^{-1}. A cubic modification was synthesized from a reactant mixture of NaGaS$_2$, SnS, and S, whereas its orthorhombic modification was obtained from Na, Ga, SnS, and S. Single crystals of cubic modification were grown by chemical transport technique using I$_2$ as the transporting agent. Orthorhombic NaGaSnS$_4$ transforms to cubic modification upon quenching of its melt at 850°C or during vapor transport by iodine. However, it decomposes in molten NaI flux into single crystals of Na$_{1.263}$Ga$_{1.263}$Sn$_{0.737}$S$_4$.

Na$_2$Ga$_2$SnS$_6$ melts congruently at 775°C and also has two polymorphic modifications (Yohannan and Vidyasagar 2016a): α-Na$_2$Ga$_2$SnS$_6$ crystallizes in the orthorhombic structure with the lattice parameters a = 1273.8 ± 0.1, b = 2314.2 ± 0.2, c = 728.2 ± 0.1 pm, a calculated density of 3.073 g·cm^{-3}, and an energy gap of 2.29 eV [a = 1277.1 ± 0.6, b = 2318.1 ± 1.1, c = 729.9 ± 0.4 pm, a calculated density of 3.052 g·cm^{-3}, and an energy gap of 2.74 eV (Li et al. 2018c); a = 1272.62 ± 0.05, b = 2311.45 ± 0.09, c = 725.85 ± 0.03 pm, and a calculated density of 3.170 g·cm^{-3} for Na$_{1.263}$Ga$_{1.263}$Sn$_{0.737}$S$_4$ or Na$_{1.895}$Ga$_{1.895}$Sn$_{1.106}$S$_6$ (Kumari and Vidyasagar 2007)]; and β-Na$_2$Ga$_2$SnS$_6$ crystallizes in the monoclinic structure with the lattice parameters a = 854.7 ± 0.3, b = 1164.9 ± 0.3, c = 1249.4 ± 0.4 pm, β = 109.82° ± 0.01°, a calculated density of 2.818 g·cm^{-3}, and an energy gap of 2.20 eV (Yohannan and Vidyasagar 2016a). The structural transition of α-Na$_2$Ga$_2$SnS$_6$ to β-Na$_2$Ga$_2$SnS$_6$ was not observed.

α-Na$_2$Ga$_2$SnS$_6$ was obtained as a gray polycrystalline sample by heating a stoichiometric mixture of Na$_2$CO$_3$ (1 mM), Ga$_2$S$_3$ (1 mM), SnO$_2$ (1 mM) in a continuous stream of CS$_2$ vapor, at 650°C for 24 h with one intermittent grinding (Yohannan and Vidyasagar 2016a). Yellow block-shaped crystals were obtained by heating a stoichiometric mixture of NaGaS$_2$ (2 mM) and SnS$_2$ (1 mM) in an evacuated sealed tube at 750°C for 2 days and then cooling to 450°C at 4°C·h^{-1}. α-Na$_2$Ga$_2$SnS$_6$ could also be synthesized when a mixture of Na$_2$S (0.605 mM), Ga (1.209 mM), Sn (0.605 mM), and S (3.022 mM) was sealed in the evacuated glass ampoule, heated at 800°C for 4 days, and then cooled to 350°C at 4°C·h^{-1} (Li et al. 2018c). After washing with degassed DMF, light-yellow crystals were obtained.

Polycrystalline β-Na$_2$Ga$_2$SnS$_6$ was obtained by heating a stoichiometric mixture of NaGaS$_2$ (2 mM) and SnS$_2$ (1 mM) in an evacuated sealed tube at 850°C for 4 days and then cooling to 550°C at 4°C·h^{-1} (Yohannan and Vidyasagar 2016a). Gray single crystals of this modification were obtained from molten $Na_2S_2O_3$ flux by heating a mixture of Na_2CO_3 (1.5 mM), Ga_2S_3 (1 mM), Sn (1 mM), and S (4 mM) at 850°C for 2 days and then cooling to 750°C at 2°C·h^{-1}. This compound is stable in the open air for several months.

Crystals of $Na_{1.263}Ga_{1.263}Sn_{0.737}S_4$ formed inadvertently in crystal growth attempts for orthorhombic $NaGaSnS_4$ (Kumari and Vidyasagar 2007). A mixture of 0.20 g orthorhombic $NaGaSnS_4$ and 2.0 g of NaI was heated at 700°C for 1 day and then cooled to 600°C over a period of 3 days, followed by further cooling to room temperature for 1 day. The entire contents were then washed thoroughly with water to isolate the water-insoluble plate-like yellow-orange crystals of $Na_{1.263}Ga_{1.263}Sn_{0.737}S_4$ and a small quantity of gray powder of presumably SnS.

7.27 Tin–Sodium–Indium–Sulfur

The **$NaInSnS_4$** quaternary compound, which crystallizes in the hexagonal structure with the lattice parameters $a = 372.73 \pm 0.06$, $c = 856.0 \pm 0.2$ pm, and a calculated density of 3.102 g·cm^{-3} [the powder X-ray diffraction (XRD) of the polycrystalline sample was indexed in the hexagonal unit cell with $a = 371.9 \pm 0.6$, $c = 2566.7 \pm 0.4$ pm] and an energy gap of 2.4 eV, is formed in the Sn–Na–In–S system (Yohannan and Vidyasagar 2016b). It was synthesized by heating in a continuous stream of CS_2 vapor appropriate stoichiometric mixture of Na_2CO_3, In_2O_3, and SnO_2 with one intermittent grinding. An alumina boat was loaded with the reactant mixture and placed in a ceramic tube in a horizontal tubular furnace. A continuous stream of N_2 was bubbled through CS_2 liquid and then passed through the ceramic tube to provide CS_2 vapor for the synthetic reaction. Single crystals of this compound were obtained by heating a stoichiometric mixture of $NaInS_2$ (1.3 mM) and SnS_2 (1.3 mM), in an evacuated silica tube, at 700°C for 1 day and then cooling to 600°C at 1.5°C·h^{-1}. Yellow plates of $NaInSnS_4$ were obtained. This compound undergoes facile ion-exchange reactions at room temperature.

7.28 Tin–Sodium–Manganese–Sulfur

The **$NaSn_{0.5}Mn_{0.5}S_2$** quaternary compound, which crystallizes in the trigonal structure with the lattice parameters $a = 371.1 \pm 0.5$, $c = 2000 \pm 3$ pm, and an energy gap of 0.7 eV, is formed in the Sn–Na–Mn–S system (He et al. 2018b). The powder sample was obtained by solid-state reaction. A stoichiometric combination of starting materials with 5% extra Na_2S_2 and 10% extra S was flame-sealed in a quartz tube. The tube was heated to 600°C and kept at 600°C for 8 h. The tube was then quenched in air, and excess sulfur was condensed on the walls of the silica tube. For growing the single crystals of this compound, Na_2S_2 powder was used as a reactive flux. Na_2S_2 (4 mM), Mn (1 mM), Sn (1 mM), and S (2 mM) were mixed and ground uniformly and loaded into a silica tube. Then, the tube was flame-sealed under a vacuum (0.1 Pa). The reaction was carried out as follows: this tube was heated to 600°C, kept at this temperature for 50 h, and slowly cooled down to 300°C

over a period of 16.7 h. Finally, the furnace was turned off, and the tube was cooled to room temperature with the furnace. To remove extra flux, the melts were washed by DMF several times. The obtained crystals were dried with acetone. Black crystals of $NaSn_{0.5}Mn_{0.5}S_2$ were obtained. All operations were carried out in an Ar-protected glove box.

7.29 Tin–Sodium–Iron–Sulfur

The **$NaSn_{0.5}Fe_{0.5}S_2$** quaternary compound, which crystallizes in the trigonal structure with the lattice parameters $a = 376.30 \pm 0.05$, $c = 1989.6 \pm 0.3$ pm, and an energy gap of 1.5 eV, is formed in the Sn–Na–Fe–S system (He et al. 2018b). A powder sample of this compound was obtained in the same way as $NaSn_{0.5}Mn_{0.5}S_2$ was prepared. Trials to obtain the single crystals of this compound using the same method as for $NaSn_{0.5}Mn_{0.5}S_2$ were unsuccessful.

7.30 Tin–Potassium–Copper–Sulfur

The **$K_2Cu_2Sn_2S_6$** quaternary compound, which is characterized by an energy gap of 1.47 eV, is formed in the Sn–K–Cu–S system (Liao and Kanatzidis 1993). To prepare this compound, a mixture of Sn, Cu, K_2S, and S (molar ratio 1:1–2:4:16) was loaded into a Pyrex tube in a N_2 glove box. The tube was evacuated and sealed at a pressure of ~0.1 Pa. The mixture was heated slowly from room temperature to 400°C in 12 h in a furnace. The temperature was kept at 400°C for 4 days, and then was cooled slowly to room temperature at 4°C·h^{-1}. Black plate-like crystals were formed. The product was washed with degassed DMF to remove excess K_2S_x flux using a standard Schlenk technique and dried with acetone and ether. The obtained crystals are stable in the air.

7.31 Tin–Potassium–Silver–Sulfur

The **$K_2Ag_6Sn_3S_{10}$** quaternary compound, which crystallizes in the orthorhombic structure with the lattice parameters $a = 2402.01 \pm 0.02$, $b = 640.17 \pm 0.03$, $c = 1330.56 \pm 0.04$ pm, a calculated density of 4.552 g·cm^{-3}, and an energy gap of 1.8 eV, is formed in the Sn–K–Ag–S system (Baiyin et al. 2004). The synthesis of this compound was as follows: Sn (10 mg), $AgNO_3$ (45 mg), K_2CO_3 (89 mg), and S (12 mg) were put into a glass tube, to which 0.4 mL of an ethanol/$HSCH_2CH(SH)CH_2OH$ mixed solvent with volume ratio 3:1 was added. The glass tube was sealed (reagents filled about 10% of the tube), placed into a Teflon-lined stainless-steel autoclave, and heated at 120°C for 5 days. The products were washed with ethanol and water, respectively, and dark-red needle-like crystals were obtained.

7.32 Tin–Potassium–Gold–Sulfur

Two quaternary compounds, **$K_2Au_2SnS_4$** and **$K_2Au_2Sn_2S_6$**, are formed in the Sn–K–Au–S system (Liao and Kanatzidis 1993). The first compound crystallizes in the triclinic structure with the lattice parameters $a = 821.2 \pm 0.4$, $b = 911.0 \pm 0.4$, $c = 731.4 \pm 0.4$ pm, $\alpha = 97.82 \pm 0.03°$, $\beta = 111.72 \pm 0.02°$, $\gamma = 72.00 \pm 0.03°$, a calculated density of 4.941 g·cm⁻³, and an energy gap of 2.75 eV. $K_2Au_2Sn_2S_6$ crystallizes in the tetragonal structure with the lattice parameters $a = 796.8 \pm 0.2$, $c = 1920.0 \pm 0.6$ pm, a calculated density of 4.914 g·cm⁻³, and an energy gap of 2.30 eV. To prepare these compounds, a mixture of Sn, Au, K_2S, and S in molar ratios of 1:2:4:16 for $K_2Au_2SnS_4$ and 1:1.5:2:16 for $K_2Au_2Sn_2S_6$ was loaded into a Pyrex tube in a dry N_2 glove box. The tube was evacuated and flame-sealed under a vacuum of 0.1 Pa. The mixture was heated slowly from room temperature to 350°C in 12 h in a furnace. The temperature was kept at 350°C for 4 days and then was cooled slowly to room temperature at 4°C·h⁻¹. Yellow long parallelepiped crystals of $K_2Au_2SnS_4$ and chunky orange crystals of $K_2Au_2Sn_2S_6$ were obtained by removing the excess K_2S_x flux with degassed DMF under a N_2 atmosphere. The final products were washed and dried with acetone and ether. The crystals of both compounds are stable in the air.

7.33 Tin–Potassium–Magnesium–Sulfur

The $K_{2x}Mg_xSn_{3-x}S_6$ ($x = 0.5–1$) quaternary phase, which crystallizes in the hexagonal structure with the lattice parameters $a = 367.49 \pm 0.08$, $c = 1682.7 \pm 0.4$ pm, and an energy gap of 2.38 eV for $x = 0.69$, is formed in the Sn–K–Mg–S system (Mertz et al. 2013). Two methods were used to prepare this phase: (1) A mixture of Sn (17.7 mM), Mg (8.9 mM), K_2S (9.8 mM), and S (53.3 mM) was sealed under vacuum (10^{-2} Pa) in a fused silica tube and heated (25°C·h⁻¹) to 550°C for 48 h, followed by cooling to room temperature at 100°C·h⁻¹. The material was washed with H_2O or DMF to remove any unreacted material, revealing a bright yellow polycrystalline product. (2) When heated, the temperature can be raised at a rate of 500°C·h⁻¹ to 800°C, holding at this temperature for 6 h, followed by cooling (100°C·h⁻¹) to room temperature.

7.34 Tin–Potassium–Barium–Sulfur

The K_2BaSnS_4 quaternary compound, which melts incongruently and crystallizes in the trigonal structure with the lattice parameters $a = 2541.9 \pm 0.4$, $c = 749.74 \pm 0.15$ pm, a calculated density of 3.295 g·cm⁻³ at 153 K, and an energy gap of 3.09 eV, is formed in the Sn–K–Ba–S system (Luo et al. 2019). The powder of this compound was synthesized by the stoichiometric high-temperature solid-state reaction of K_2S, BaS, and SnS_2. The tube loaded with the above mixture was sealed and heated to 500°C for 12 h with a duration of 48 h, and then the furnace was shut off. Single crystals of K_2BaSnS_4 were achieved directly via a spontaneous crystallization method. First, K_2S (0.132 g), BaS (0.169 g), and SnS_2 (0.182 g) were mixed according to a molar ratio of 1.2:1:1 (with a little excess of K_2S serving as a flux) and then ground and transferred into fused silica tube in a glove box. The tube was placed in a heating furnace after it was sealed under a high vacuum. The heating temperature

was raised to 900°C in 20 h for 24 h, and then the furnace was slowly cooled to 200°C and at 4°C·h⁻¹ to room temperature.

7.35 Tin–Potassium–Zinc–Sulfur

Three quaternary compounds, $K_2ZnSn_3S_8$, $K_6Zn_4Sn_5S_{17}$, and $K_{10}Zn_4Sn_4S_{17}$, are formed in the Sn–K–Zn–S system. The first of them has two polymorphic modifications. α-$K_2ZnSn_3S_8$ crystallizes in the triclinic structure with the lattice parameters $a = 736.56 \pm 0.15$, $b = 799.96 \pm 0.16$, $c = 1531.9 \pm 0.3$ pm, $\alpha = 95.04 \pm 0.03°$, $\beta = 91.64 \pm 0.03°$, $\gamma = 115.76 \pm 0.03°$, a calculated density of 3.110 g·cm⁻³, and an energy gap of 2.30 eV; β-$K_2ZnSn_3S_8$ crystallizes in the cubic structure with the lattice parameter $a = 1312.6 \pm 0.2$ pm, a calculated density of 3.331 g·cm⁻³, and an energy gap of 2.55 eV (Fard and Kanatzidis 2012). $K_2ZnSn_3S_8$ melts at 445°C and its glassy form is characterized by an energy gap of 2.15 eV. α-$K_2ZnSn_3S_8$ was prepared from a mixture of K_2CO_3, S, Zn, and Sn (molar ratio 1:6.32:1:2), which was ball-milled at a rate of 250 rpm for 30 min. Then 3 g of this mixture was placed in a quartz tube under a N_2 atmosphere. A secured balloon was attached at the end of the reaction tube in order to absorb the created pressure of the CO_2 evolution. The mixture was heated gradually to 200°C where it was kept for 5 h before being successfully brought to 700°C for 8 h. Well-formed yellow plate-shaped crystals of α-$K_2ZnSn_3S_8$ were obtained by cooling at a rate of ~1°C·min⁻¹ to room temperature. However, water quenching of the same melt led to glassy $K_2ZnSn_3S_8$, which upon annealing below its melting point crystallizes to metastable single-phase β-$K_2ZnSn_3S_8$. Glassy $K_2ZnSn_3S_8$ shows a glass transition at ~220°C, followed by a sharp, single-step crystallization at 398°C and melting at 468°C. β-$K_2ZnSn_3S_8$ as well-formed orange block-shaped crystals was formed by annealing the amorphous $K_2ZnSn_3S_8$ at 420°C for 1 h followed by water quenching. It can also be obtained by annealing α-$K_2ZnSn_3S_8$ at 445°C for 24 h.

$K_6Zn_4Sn_5S_{17}$ crystallizes in the tetragonal structure with the lattice parameters $a = 1374.25 \pm 0.07$, $c = 972.72 \pm 0.05$ pm, and an energy gap of ≈2.87 eV (Manos et al. 2005) [$a = 1379.6 \pm 0.2$, $c = 958.0 \pm 0.2$ pm, a calculated density of 2.977 g·cm⁻³, and an energy gap of 2.94 eV (Kanatzidis et al. 1997a, 1997b)]. A mixture of Sn, Zn, K_2S, and S was sealed under vacuum in a silica tube, heated (≈40°C/h) to 400°C for 92–96 h, and then cooled to 25°C at a rate of 4–6°C/h to obtain yellowish-white crystals of this compound (Kanatzidis et al. 1997a, 1997b, Manos et al. 2005). The product must be washed with degassed DMF and dried with acetone and ether.

$K_{10}Zn_4Sn_4S_{17}$ melts congruently at 820°C and crystallizes in the cubic structure with the lattice parameter $a = 1992.3 \pm 0.3$ pm, a calculated density of 2.809 g·cm⁻³, and an energy gap of 3.1–3.2 eV (Palchik et al. 2003, 2004). This compound was obtained by reaction of Sn (0.5 mM), Zn (0.5 mM), K_2S (1.5 mM), and S (6.0 mM). The mixture was sealed under vacuum (< 0.1 Pa) in a silica tube and heated at a rate of 50°C·h⁻¹ to 500°C for 60 h, followed by cooling to room temperature at 5°C·h⁻¹. The excess flux (K_2S_x) was removed with MeOH to reveal yellowish-white cubic crystals, which are moderately air stable (for a few days) and need to be stored under an inert atmosphere. This compound could also be prepared by direct stoichiometric reaction.

7.36 Tin–Potassium–Cadmium–Sulfur

Three quaternary compounds, K_2CdSnS_4, $K_2CdSn_2S_6$, and $K_2CdSn_3S_8$, are formed in the Sn–K–Cd–S system. First of them crystallizes in the monoclinic structure with the lattice parameters $a = 1102.1 \pm 0.5$, $b = 1103.0 \pm 0.5$, $c = 1515.1 \pm 1.0$ pm, $\beta = 100.416 \pm 0.012°$, a calculated density of 3.209 g·cm^{-3}, and an energy gap of 2.2 eV (Baiyin et al. 2014). This compound was prepared under solvothermal conditions.

$K_2CdSn_2S_6$ has two polymorphic modifications: α-$K_2CdSn_2S_6$ is stable up to 750°C and has an energy gap of 2.14 eV, while β-$K_2CdSn_2S_6$ is not thermally stable and two values of energy gap were determined for this modification: 2.32 and 2.86 eV (Pogu et al. 2019). β-$K_2CdSn_2S_6$ crystallizes in the monoclinic structure with the lattice parameters $a = 751.56 \pm 0.04$, $b = 1614.46 \pm 0.10$, $c = 1022.38 \pm 0.06$ pm, $\beta = 91.940 \pm 0.002°$, and a calculated density of 3.324 g·cm^{-3}. Polycrystalline samples of α- and β-$K_2CdSn_2S_6$ were prepared by heating stoichiometric mixtures of appropriate solid reactants [$K_2C_2O_4·H_2O$ (1.61 mM), CdS (1.61 mM), and SnO_2 (3.22 mM)] in a continuous stream of CS_2 vapor at 700°C (α-$K_2CdSn_2S_6$) or 600°C (β-$K_2CdSn_2S_6$) for 24 h with one intermittent grinding. An alumina boat was loaded with the mixture and placed in a ceramic tube in a horizontal tubular furnace. A continuous stream of N_2 was bubbled through CS_2 liquid and then passed through the ceramic tube to provide CS_2 vapor for the synthetic reaction. Yellow samples of α-$K_2CdSn_2S_6$ and gray sample of β-$K_2CdSn_2S_6$ were obtained. The light-yellow single crystals of both modifications were isolated from the *in-situ* molten K_2SO_3 flux. A quartz tube containing an appropriate reactant mixture [K_2CO_3 (0.805 mM), SnS_2 (1.07 mM), CdS (0.537 mM), and S (1.07 mM) for β-$K_2CdSn_2S_6$ or K_2CO_3 (0.805 mM), Sn (1.07 mM), Cd (0.537 mM), and S (3.76 mM) for α- and β-$K_2CdSn_2S_6$] was initially heated at 350°C under dynamic vacuum for 10 min to expel the CO generated and then sealed under vacuum. This sealed quartz tube was heated at 700°C for α-$K_2CdSn_2S_6$ (at 650°C for β-$K_2CdSn_2S_6$) for 3 days and then cooled to 300°C at 4°C·h^{-1}, and the furnace was finally turned off. The solid product contents were washed with distilled water to dissolve away $K_2S_2O_3$ formed *in situ*, and the single crystals of α- and β-$K_2CdSn_2S_6$ were isolated.

$K_2CdSn_3S_8$ has an energy gap of 2.08 eV (Pogu and Vidyasagar 2020). Its polycrystalline sample was synthesized by heating stoichiometric mixtures of K_2CO_3, CdS, and SnO_2 in a continuous stream of CS_2 vapor at 575°C for 2 days, with one intermittent grinding. An alumina boat was loaded with the solid reactant mixture and placed in a ceramic tube, which was kept in a horizontal tubular furnace. A continuous stream of N_2 gas was bubbled, at a rate of two bubbles per second, through CS_2 liquid and then passed through the ceramic tube to provide CS_2 vapor for the synthetic reaction. The outgoing gaseous mixture of N_2, CS_2, and CO_2 from the ceramic tube was passed through paraffin oil and then into a dilute aqueous solution of KOH. The furnace was turned off after heating and allowed to cool naturally to room temperature for over 12 h. Single crystals of $K_2CdSn_3S_8$ were obtained by the thiosulfate flux method. It involved heating the solid reactant mixture of K_2CO_3, S, Cd and Sn or, CdS and SnS_2 in a silica tube under dynamic vacuum first at about 300°C for about 15 min and then sealing the quartz tube along with solid content under vacuum. The sealed quartz ampoule was heated at a rate of 3°C·min^{-1} to 750°C at the use for the synthesis of CdS and SnS_2 and to 625°C when Cd and Sn were used, held at that temperature for 3 days and then cooled to 350°C over a period of 2 days. The ampoule was broken open after the reaction, and the solid contents were washed with water to dissolve away $K_2S_2O_3$ flux and isolate the desired solid product. The K$^+$ ion of the layered $K_2CdSn_3S_8$ compound undergoes facile ion exchange in an aqueous medium at room temperature, with monovalent NH_4^+, Rb^+, Cs^+, Tl^+ ions, divalent Sr^{2+}, Ca^{2+}, Ba^{2+} ions, and trivalent lanthanide Ln^{3+} (Ln = La, Ce, Pr, Sm, Eu, Gd, Tb, Dy, Ho, Er, Tm) ions.

7.37 Tin–Potassium–Mercury–Sulfur

The $K_2Hg_3Sn_2S_8$ quaternary compound, melts incongruently at ~545°C, recrystallizes at ~430°C, and has two polymorphic modifications, is formed in the Sn–K–Hg–S system (Kanatzidis et al. 1997a, 1997b; Liao et al. 2003). α-$K_2Hg_3Sn_2S_8$ crystallizes in the orthorhombic structure with the lattice parameters $a = 1956.3 \pm 0.2$, $b = 985.3 \pm 0.1$, $c = 846.7 \pm 0.1$ pm, a calculated density of 9.555 g·cm^{-3}, and an energy gap of 2.40 eV (Liao et al. 2003) [$a = 1952.2 \pm 0.5$, $b = 983.5 \pm 0.3$, $c = 843.1 \pm 0.2$ pm, a calculated density of 4.816 g·cm^{-3}, and an energy gap of 2.39 eV (Kanatzidis et al. 1997a, 1997b)]. β-$K_2Hg_3Sn_2S_8$ crystallizes in the monoclinic structure and has an energy gap of 2.50 eV (2.78 eV for β-$K_{0.8}Hg_{3.6}Sn_2S_8$) (Liao et al. 2003).

α-$K_2Hg_3Sn_2S_8$ was obtained by the reaction of the mixture of Sn, HgS, K_2S, and S, taking in stoichiometric ratio, at 400°C for 96 h upon a cooling rate of 4°C·h^{-1} (Kanatzidis et al. 1997a, 1997b, Liao et al. 2003). The product was washed with degassed DMF and dried with acetone and ether. Yellow crystals were obtained.

β-$K_2Hg_3Sn_2S_8$ was synthesized by the reaction of Sn (0.25 mM), HgS (0.38 mM), K_2S (0.50 mM), and S (4.06 mM) in an evacuated glass ampoule at 400°C for 96 h with the next cooling to 160°C at a rate of 80°C·h^{-1} (Liao et al. 2003). The product was washed with degassed methanol and deionized water and then dried with acetone and ether. Large yellow crystals of β-$K_2Hg_3Sn_2S_8$ were obtained (α-$K_2Hg_3Sn_2S_8$ was also present).

7.38 Tin–Potassium–Gallium–Sulfur

The $KGaSnS_4$ quaternary compound, which has two polymorphic modifications, is formed in the Sn–K–Ga–S system. The first modification crystallizes in the cubic structure with the lattice parameter $a = 1305.0 \pm 0.5$ pm, a calculated density of 3.190 g·cm^{-3}, and an energy gap of 2.10 eV (Kumari and

Vidyasagar 2007). Single crystals of this modification were obtained from a molten flux of K_2S_2. A mixture of Ga (0.502 mM), Sn (0.50 mM), S (9.9 mM), and K_2S_2 (4.98 mM) was heated at 450°C for 4 days and then cooled to room temperature over a period of 1 day. The reaction product was washed thoroughly with water to isolate the water-insoluble, homogeneous phase in the form of reddish-brown, block-shaped crystals.

Second modification of $KGaSnS_4$ crystallizes in the triclinic structure with the lattice parameters $a = 715.2 \pm 0.2$, $b = 784.7 \pm 0.2$, $c = 1516.3 \pm 0.3$ pm, $\alpha = 83.78 \pm 0.02°$, $\beta = 87.21 \pm 0.02°$, and $\gamma = 63.83 \pm 0.02°$ at 153 K (Wu et al. 1992). It was prepared by the reaction of K_2S_5 with Ga, Sn, and S (molar ratio 1:2:2:3) in a quartz tube. The quartz tube was evacuated ($\sim 10^{-3}$ Pa), sealed, and then heated gradually to 500°C, where it was kept for 24 h before being successively brought to 700°C in 24 h and to 900°C over 100 h. Next, the tube was cooled at a rate of 4°C·h^{-1} to 300°C, and then the furnace was shut off. Colorless plate-like crystals formed in the tube. They are stable in the air and do not decompose significantly in water.

7.39 Tin–Potassium–Indium–Sulfur

The $KInSnS_4$ quaternary compound, which has two polymorphic modifications, and one of them crystallizes in the hexagonal structure with the lattice parameters $a = 374.5 \pm 0.2$, $c = 843.4 \pm 0.5$ pm, a calculated density of 3.249 g·cm^{-3}, and an energy gap of 2.3 eV (the powder XRD of the polycrystalline sample was indexed in the hexagonal unit cell with $a = 372.0 \pm 0.6$ and $c = 2508.6 \pm 0.2$ pm), is formed in the Sn–K–In–S system (Yohannan and Vidyasagar 2016b). The second modification of this compound crystallizes in the trigonal structure with the lattice parameters $a = 369.8 \pm 0.6$, $c = 2452 \pm 3$ pm, and a calculated density of 3.4399 g·cm^{-3}.

$KInSnS_4$ was synthesized by heating in a continuous stream of CS_2 vapor appropriate stoichiometric mixture of K_2CO_3 or KCH_3COO, In_2O_3, and SnO_2 with one intermittent grinding (Yohannan and Vidyasagar 2016b). An alumina boat was loaded with the reactant mixture and placed in a ceramic tube in a horizontal tubular furnace. A continuous stream of N_2 was bubbled through CS_2 liquid and then passed through the ceramic tube to provide CS_2 vapor for the synthetic reaction. Single crystals of this compound were obtained by employing $K_2S_2O_3$ flux. A stoichiometric mixture of K_2CO_3 (0.93 mM), In (1.2 mM), Sn (1.2 mM), and S (5.6 mM) was heated at 750°C for 2 days, then cooled to 550°C at 3°C·h^{-1} and the furnace was turned off. The solid product was washed with distilled water to dissolve away the flux formed and isolate yellow plates of $KInSnS_4$. This compound undergoes facile ion-exchange reactions at room temperature.

7.40 Tin–Potassium–Yttrium–Sulfur

The $K_2Y_4Sn_2S_{11}$ quaternary compound, which crystallizes in the tetragonal structure with the lattice parameters $a = 858.7 \pm 0.1$ and $c = 2789.2 \pm 0.4$ pm at 115 K, is formed in the Sn–K–Y–S system (Wu and Ibers 1994). It was prepared by the reaction of K_2S_5 with Y_2S_3 and Sn (molar ratio 1:1:2) in a quartz

ampoule that was evacuated to 10^{-3} Pa. The mixture was heated gradually to 700°C for 24 h and to 1000°C for 150 h. The tube was then cooled at a rate of 4°C·h^{-1} to 300°C, and then the furnace was shut off. Small light-yellow plates had grown in the ampoule, usually with a mixture of other binary and ternary sulfides.

7.41 Tin–Potassium–Phosphorus–Sulfur

The $KSnPS_4$ quaternary compound, which melts congruently at 530°C and crystallizes in the monoclinic structure with the lattice parameters $a = 667.63 \pm 0.08$, $b = 1198.06 \pm 0.11$, $c = 1099.70 \pm 0.14$ pm, $\beta = 127.347 \pm 0.008°$, and a calculated density of 3.011 g·cm^{-3} at 100 K, is formed in the Sn–K–P–S system (Banerjee et al. 2012). To prepare this compound, in a N_2-filled glove box K_2S, Sn, red P, and S (molar ratio 1:2.5:2:9) were loaded in a fused silica tube, which was then sealed under vacuum ($<10^{-2}$ Pa). The mixture was heated at 900°C for 72 h, followed by cooling to 300°C for 120 h and finally to room temperature in 3 h. The solidified yellow-orange colored melt was then taken out of the tube and washed with degassed DMF. Yellow-orange colored polyhedral crystals of $KSnPS_4$ were isolated after drying with ether. This compound forms glass upon melt quenching.

7.42 Tin–Potassium–Arsenic–Sulfur

Two quaternary compounds, $KSnAsS_4$ and $K_2SnAs_2S_6$, are formed in the Sn–K–As–S system (Iyer and Kanatzidis 2002). First of them crystallizes in the orthorhombic structure with the lattice parameters $a = 813.6 \pm 0.2$, $b = 1378.4 \pm 0.4$, $c = 742.8 \pm 0.2$ pm, a calculated density of 3.134 g·cm^{-3}, and an energy gap of 2.11 eV. It was synthesized from a mixture of K_2S (0.3 mM), Sn (0.3 mM), As_2S_3 (0.3 mM), and S (3 mM). The reactant mixture was sealed under vacuum ($\sim 10^{-3}$ Pa) in a fused silica tube and heated to 500°C for 60 h, followed by slow cooling to 250°C at 5°C·h^{-1} and rapid cooling to room temperature. Reddish-brown, air- and moisture-stable, rectangular block-like crystals were isolated after dissolving the flux away using DMF. $KSnAsS_4$ discovered a glass-forming property. The material melts at 400°C and, when cooled to room temperature, forms a red glass. However, the melt slowly generates SnS_2 nanocrystals, which were found, embedded in the glassy matrix.

$K_2SnAs_2S_6$ crystallizes in the trigonal structure with the lattice parameters $a = 671.71 \pm 0.05$, $c = 720.4 \pm 0.8$ pm, a calculated density of 3.18 g·cm^{-3}, and an energy gap of 1.89 eV (Iyer and Kanatzidis 2002). This compound was initially prepared in a low yield from a mixture of K_2S (0.9 mM), Sn (0.3 mM), As_2S_3 (0.6 mM), and S (3 mM) under vacuum in a silica tube at 500°C for 60 h. A clean synthesis with yields of over 90% was achieved from a ratio of 4:1:3:10 of the same reactants. The product, consisting of air and moisture-stable dark-red hexagonal crystals, was isolated by washing the excess flux, under nitrogen, with DMF and washed with ether. The material melts at 490°C and crystallizes at ~406°C without apparent decomposition

7.43 Tin–Potassium–Antimony–Sulfur

The $K_2Sb_2Sn_3S_{10}$ quaternary compound, which crystallizes in the monoclinic structure with the lattice parameters $a = 675.36 \pm 0.02$, $b = 1325.28 \pm 0.03$, $c = 1043.78 \pm 0.02$ pm, $\beta = 97.462 \pm 0.001°$, a calculated density of 3.579 g·cm^{-3}, and an energy gap of 2.30 eV, is formed in the Ge–K–Sb–S system (Yohannan and Vidyasagar 2015). It was obtained as a homogeneous phase, in the form of orange-yellow block-shaped single crystals, from a mixture of K_2CO_3 (0.75 mM), Sn (1.50 mM), Sb (1 mM), and S (5.50 mM). A quartz tube containing the reactant mixture was initially heated at 350°C under dynamic vacuum for 10 min to expel the CO_2 generated and then sealed under vacuum. This sealed quartz tube was heated at 500°C for 2 days and then cooled to 300°C at 2°C·h^{-1} and the furnace was finally turned off. The solid product contents were washed with distilled water to dissolve away any $K_2S_2O_3$ formed and isolate single crystals $K_2Sb_2Sn_3S_{10}$.

7.44 Tin–Potassium–Chromium–Sulfur

The $KCrSnS_4$ quaternary compound, which crystallizes in the trigonal structure with the lattice parameters $a = 357.86 \pm 0.05$, $c = 2483.2 \pm 0.5$ pm, and a calculated density of 3.06 g·cm^{-3} at 115 K, is formed in the Sn–K–Cr–S system (Wu and Ibers 1993b). It was prepared by the reaction of K_2S_5 with Cr and Sn powders and elemental S in a quartz tube. The reaction was carried out at 1000°C for 6 days; the tube was then cooled at a rate of 4°C·h^{-1}. Hexagonal and trigonal phases were present in the $KCrSnS_4$ samples as prepared.

7.45 Tin–Potassium–Manganese–Sulfur

Three quaternary phases, K_2MnSnS_4, $K_{2x}Mn_xSn_{3-x}S_6$ ($x = 0.5–0.95$), and $K_{10}Mn_4Sn_4S_{17}$, are formed in the Sn–K–Mn–S system. Second compound crystallizes in the trigonal structure with the lattice parameters $a = 369.69 \pm 0.05$, $c = 2540.3 \pm 0.5$ pm, and a calculated density of 3.105 g·cm^{-3} for $x = 0.95$ system (Manos et al. 2008; Manos and Kanatzidis 2009; Mertz et al. 2013). $K_{2x}Mn_xSn_{3-x}S_6$ can be synthesized by solid-state reaction of Sn (1.9 mM), Mn (1.1 mM), K_2S (2 mM), and S (16 mM). The mixture was sealed under vacuum (10^{-2} Pa) in a silica tube and heated at a rate of 50°C·h^{-1} to 400–500°C for 60 h, followed by cooling to room temperature at 50°C·h^{-1}. The excess flux was removed with DMF to reveal dark-brown polycrystalline material.

$K_{10}Mn_4Sn_4S_{17}$ crystallizes in the cubic structure with the lattice parameter $a = 1006.4 \pm 0.2$ pm, a calculated density of 2.656 g·cm^{-3}, and an energy gap of 2.25–2.3 eV (Palchik et al. 2003, 2004). This compound was obtained by reaction of Sn (0.5 mM), Mn (0.5 mM), K_2S (2.0 mM), and S (6.0 mM). The mixture was sealed under vacuum (< 0.1 Pa) in a silica tube and heated at a rate of 50°C·h^{-1} to 650°C for 96 h, followed by cooling to room temperature at 5°C·h^{-1}. The excess flux (K_2S_x) was removed with MeOH to reveal the mixture of red cubic crystals of $K_{10}Mn_4Sn_4S_{17}$ and orange-reddish crystals of K_2MnSnS_4.

This compound is stable in the air for approximately 5 h, after which some deterioration could be observed.

7.46 Tin–Potassium–Iron–Sulfur

The $K_{10}Fe_4Sn_4S_{17}$ quaternary compound, which decomposes at 661°C and crystallizes in the cubic structure with the lattice parameter $a = 997.56 \pm 0.12$ pm, a calculated density of 2.734 g·cm^{-3}, and an energy gap of 1.75–2.2 eV, is formed in the Sn–K–Fe–S system (Palchik et al. 2003, 2004). This compound was obtained by reaction of Sn (2.0 mM), Fe (2.0 mM), K_2S (2.5 mM), and S (6.0 mM). The mixture was sealed under vacuum (< 0.1 Pa) in a silica tube and heated at a rate of 50°C·h^{-1} to 650°C. It was kept at this temperature for 60 h, followed by cooling to room temperature at 5°C·h^{-1}. The excess flux (K_2S_x) was removed with MeOH to reveal black crystals of $K_{10}Fe_4Sn_4S_{17}$. This compound is moderately stable in the air (for a few days).

7.47 Tin–Potassium–Cobalt–Sulfur

The $K_{10}Co_4Sn_4S_{17}$ quaternary compound, which decomposes at 704°C and crystallizes in the cubic structure with the lattice parameter $a = 993.32 \pm 0.13$ pm, a calculated density of 2.790 g·cm^{-3}, and an energy gap of 1.31–1.8 eV, is formed in the Sn–K–Co–S system (Palchik et al. 2003, 2004). This compound was obtained in the same way as $K_{10}Fe_4Sn_4S_{17}$ was synthesized using Co instead of Fe. The excess flux (K_2S_x) was removed with MeOH to reveal dark (almost black) crystals of $K_{10}Co_4Sn_4S_{17}$. This compound is moderately stable in the air (for a few days).

7.48 Tin–Rubidium–Copper–Sulfur

Two quaternary compounds, $Rb_2Cu_2SnS_4$ and $Rb_2Cu_2Sn_2S_6$, are formed in the Sn–Rb–Cu–S system (Liao and Kanatzidis 1993). First of them crystallizes in the orthorhombic structure with the lattice parameters $a = 552.8 \pm 0.4$, $b = 1141.8 \pm 0.6$, $c = 1370.0 \pm 0.6$ pm, a calculated density of 4.185 g·cm^{-3}, and an energy gap 2.08 eV. $Rb_2Cu_2Sn_2S_6$ crystallizes in the monoclinic structure with the lattice parameters $a = 1102.6 \pm 0.2$, $b = 1101.9 \pm 0.3$, $c = 2029.9 \pm 0.4$ pm, $\beta = 97.79 \pm 0.02°$, a calculated density of 3.956 g·cm^{-3}, and an energy gap of 1.44 eV. To prepare these compounds, a mixture of Sn, Cu, Rb_2S, and S (molar ratios 1:4:4:16 for $Rb_2Cu_2SnS_4$ and 1:1–2:4:16 for $Rb_2Cu_2Sn_2S_6$) was loaded into a Pyrex tube in a N_2 glove box. The tube was evacuated and sealed under a vacuum of ~0.1 Pa. The mixture was heated slowly from room temperature to 400°C in 12 h in a furnace. The temperature was kept at 400°C for 4 days, and then was cooled slowly to room temperature at 4°C·h^{-1}. Orange plate-like crystals were formed. The product was washed with degassed DMF to remove excess Rb_2S_x flux using a standard Schlenk technique and dried with acetone and ether. The obtained orange crystals of $Rb_2Cu_2SnS_4$ decompose slowly in the air, and black crystals of $Rb_2Cu_2Sn_2S_6$ are stable in the air.

7.49 Tin–Rubidium–Zinc–Sulfur

Three quaternary compounds, **Rb$_2$ZnSn$_2$S$_6$**, **Rb$_2$ZnSn$_3$S$_8$**, and **Rb$_{10}$Zn$_4$Sn$_4$S$_{17}$**, are formed in the Sn–Rb–Zn–S system. First of them crystallizes in the monoclinic structure with the lattice parameters $a = 687.3 \pm 0.2$, $b = 1345.6 \pm 0.4$, $c = 728.5 \pm 0.2$ pm, $\beta = 113.12° \pm 0.02°$, a calculated density of 3.569 g·cm^{-3} at 183 K, and an energy gap of 3.00 eV (Kanatzidis et al. 1997a, 1997b). The reaction of Sn, Zn, Rb$_2$S, and S at 400°C for 4 days upon a cooling rate of 4°C·h^{-1} afforded brownish-orange crystals of Rb$_2$ZnSn$_2$S$_6$. The product must be washed with degassed DMF and dried with acetone and ether.

Rb$_2$ZnSn$_3$S$_8$ has two polymorphic modifications (Pogu and Vidyasagar 2020). α-Rb$_2$ZnSn$_3$S$_8$ crystallizes in the triclinic structure with the lattice parameters $a = 735.17 \pm 0.04$, $b = 797.54 \pm 0.06$, $c = 1514.14 \pm 0.08$ pm, $\alpha = 94.563 \pm 0.004°$, $\beta = 93.053 \pm 0.004°$, $\gamma = 113.753 \pm 0.004°$, a calculated density of 3.496 g·cm^{-3}, and an energy gap of 2.94 eV, and β-Rb$_2$ZnSn$_3$S$_8$ crystallizes in the cubic structure with the lattice parameter $a = 1334.06 \pm 0.04$ pm, a calculated density of 3.562 g·cm^{-3}, and an energy gap of 2.55 eV. Polycrystalline samples of both modifications were synthesized by heating stoichiometric mixtures of Rb$_2$CO$_3$, ZnS, and SnO$_2$ in a continuous stream of CS$_2$ vapor at 750°C for α-Rb$_2$ZnSn$_3$S$_8$, and at 600°C for β-Rb$_2$ZnSn$_3$S$_8$ for 2 days, with one intermittent grinding. An alumina boat was loaded with the solid reactant mixture and placed in a ceramic tube, which was kept in a horizontal tubular furnace. A continuous stream of N$_2$ gas was bubbled, at a rate of two bubbles per second, through CS$_2$ liquid and then passed through the ceramic tube to provide CS$_2$ vapor for the synthetic reaction. The outgoing gaseous mixture of N$_2$, CS$_2$, and CO$_2$ from the ceramic tube was passed through paraffin oil and then into a dilute aqueous solution of KOH. The furnace was turned off after heating and allowed to cool naturally to room temperature over 12 h. Single crystals of Rb$_2$ZnSn$_3$S$_8$ were obtained by the thiosulfate flux method. It involved heating the solid reactant mixture of Rb$_2$CO$_3$, S, Zn and Sn, or ZnS and SnS$_2$ in silica tube under dynamic vacuum first at about 300°C for about 15 min and then sealing the quartz tube along with solid content under vacuum. The sealed quartz ampoule was heated at a rate of 3°C·min^{-1} to 850°C for α-Rb$_2$ZnSn$_3$S$_8$ and to 650°C for β-Rb$_2$ZnSn$_3$S$_8$, held at that temperatures for 3 days and then cooled to 350°C over a period of 2 days. The ampoule was broken open after the reaction and the solid contents were washed with water to dissolve away Rb$_2$S$_2$O$_3$ flux and isolate the desired solid product.

Rb$_{10}$Zn$_4$Sn$_4$S$_{17}$ crystallizes in the cubic structure with the lattice parameter $a = 1018.56 \pm 0.01$ pm, a calculated density of 3.356 g·cm^{-3}, and an energy gap of 3.21 eV (Zhao et al. 2019). For the solid-state synthesis of this compound, Rb$_2$S$_3$, Zn powder, Sn powder, and sublimed S powder were used as raw materials and prepared in a glove box filled with Ar (both water and oxygen were less than 0.1 ppm). The raw materials were fully ground in a certain proportion in the mortar and then sealed in a quartz tube under vacuum. The quartz tube was placed in a muffle furnace, heated up to 750°C in 48 h, kept at this temperature for 24 h, cooled at a rate of about 8°C·h^{-1} to 300°C, and finally cooled naturally to room temperature. After the

solid phase synthesis sample was ground into powder, the powder was mixed in a 1:1 ratio with KI in an Ar protective glove box, and the mixture was filled into a graphite crucible. The graphite crucible was then placed in a quartz tube, sealed under vacuum, and transferred to a muffle furnace. Subsequently, the quartz tube was heated up to 750°C in 2 days and kept at the current temperature for 2 another days, then cooled to 300°C at a slow rate of 8°C·h^{-1}, and finally the sample was cooled to room temperature. The product was cleaned with deionized water to obtain stable crystals in the air.

7.50 Tin–Rubidium–Cadmium–Sulfur

Two quaternary compounds, **Rb$_2$CdSn$_2$S$_6$** and **Rb$_2$CdSn$_3$S$_8$**, are formed in the Sn–Rb–Cd–S system. The first melts congruently at 604°C, is thermally stable up to 900°C and crystallizes in the monoclinic structure with the lattice parameters $a = 702.91 \pm 0.02$, $b = 1369.85 \pm 0.04$, $c = 740.84 \pm 0.02$ pm, $\beta = 113.888° \pm 0.002°$, a calculated density of 3.631 g·cm^{-3}, and an energy gap of 2.93 eV (Pogu et al. 2019). A polycrystalline sample of this compound was prepared by heating stoichiometric mixtures of appropriate solid reactants [Rb$_2$CO$_3$ (1.40 mM), CdS (1.40 mM), and SnO$_2$ (2.80 mM)] in a continuous stream of CS$_2$ vapor at 650°C for 24 h with one intermittent grinding. An alumina boat was loaded with the mixture and placed in a ceramic tube in a horizontal tubular furnace. A continuous stream of N$_2$ was bubbled through CS$_2$ liquid and then passed through the ceramic tube to provide CS$_2$ vapor for the synthetic reaction. Pale-yellow sample of Rb$_2$CdSn$_2$S$_6$ was obtained. The light-yellow single crystals were isolated from the *in-situ* molten Rb$_2$S$_2$O$_3$ flux. A quartz tube containing an appropriate reactant mixture [Rb$_2$CO$_3$ (0.70 mM), SnS$_2$ (0.93 mM), CdS (0.46 mM), and S (0.93 mM) or Rb$_2$CO$_3$ (0.70 mM), Sn (0.93 mM), Cd (0.46 mM), and S (3.25 mM)] was initially heated at 350°C under dynamic vacuum for 10 min to expel the CO$_2$ generated and then sealed under vacuum. This sealed quartz tube was heated at 650°C for 3 days and then cooled to 300°C at 4°C·h^{-1}, and the furnace was finally turned off. The solid product contents were washed with distilled water to dissolve away Rb$_2$S$_2$O$_3$ formed *in situ*, and the single crystals of Rb$_2$CdSn$_2$S$_6$ were isolated.

Rb$_2$CdSn$_3$S$_8$ apparently has two polymorphic modifications (Pogu and Vidyasagar 2020). β-Rb$_2$CdSn$_3$S$_8$ crystallizes in the cubic structure with the lattice parameter $a = 1344.24 \pm 0.07$ pm, a calculated density of 3.675 g·cm^{-3}, and an energy gap of 2.59 eV. Polycrystalline samples and single crystals of this modification were prepared in the same way as for β-Rb$_2$ZnSn$_3$S$_8$ using CdS instead of ZnS.

7.51 Tin–Rubidium–Mercury–Sulfur

The **Rb$_2$Hg$_3$Sn$_2$S$_8$** quaternary compound, which melts incongruently (Liao et al. 2003) and crystallizes in the monoclinic structure with the lattice parameters $a = 1013.2 \pm 0.2$, $b = 654.0 \pm 0.2$, $c = 1343.4 \pm 0.2$ pm, $\beta = 97.93 \pm 0.01°$, a calculated density of 4.770 g·cm^{-3} (Marking et al. 1998), and an energy gap of 2.40–2.48 eV (Kanatzidis et al. 1997a, 1997b; Liao et al. 2003), is formed in the Sn–Rb–Hg–S. It was prepared with a slight excess of HgS powder in the mixture of Sn, HgS, Rb_2S, and S by heating at 500°C for 96 h and cooling to 200°C at 4 $C\cdot h^{-1}$ (Liao et al. 2003) or by interaction of HgS with Sn in the Rb_2S_x melt at 350°C (Marking et al. 1998). This compound is stable in the air.

7.52 Tin–Rubidium–Gallium–Sulfur

The **$RbGaSnS_4$** quaternary compound, which apparently has two polymorphic modifications, is formed in the Sn–Rb–Ga–S. α-$RbGaSnS_4$ crystallizes in the cubic structure with the lattice parameter $a = 1317.4 \pm 0.5$ pm, a calculated density of 3.505 $g\cdot cm^{-3}$, and an energy gap of 2.70 eV (Kumari and Vidyasagar 2007). β-$RbGaSnS_4$ crystallizes in the monoclinic structure with the lattice parameters $a = 725.22 \pm 0.12$, $b = 1237.7 \pm 0.2$, $c = 1757.0 \pm 0.3$ pm, $\beta = 96.554 \pm 0.006°$, a calculated density of 3.410 $g\cdot cm^{-3}$, and an energy gap of 2.96 eV (Liu et al. 2020a).

Polycrystalline samples of α-$RbGaSnS_4$ were prepared by heating stoichiometric mixtures of $RbGaS_2$, SnS or Sn, and S at 850°C for 4 days and then cooling to room temperature at the rate of 2°C·h^{-1} (Kumari and Vidyasagar 2007). Single crystals were grown by chemical transport technique using I_2 as the transporting agent.

The crystals of β-$RbGaSnS_4$ were obtained from a mixture of RbCl, Ga_2S_3, Sn, and S with the molar ratio of 2:1:1:4, which was loaded into a graphite crucible that was put into outer silica tubing, and finally the whole assembly was sealed under a vacuum of 10^{-3} Pa (Liu et al. 2020a). The sealed assembly was heated to 780°C in 40 h, and maintained for 60 h, followed by a slow cooling to 300°C before turning off the furnace. The product was washed with distilled water and CS_2 in an ultrasonic cleaner to remove the by-products. The air-stable light-yellow crystals were obtained.

7.53 Tin–Rubidium–Indium–Sulfur

The **$RbInSnS_4$** quaternary compound, which melts at 700°C and has two polymorphic modifications, is formed in the Sn–Rb–In–S system (Yohannan and Vidyasagar 2016b). α-$RbInSnS_4$ crystallizes in the hexagonal structure with the lattice parameters $a = 385.8 \pm 0.4$, $c = 885.3 \pm 0.3$ pm, and an energy gap of 2.4 eV, and β-$RbInSnS_4$ crystallizes in the cubic structure with the lattice parameter $a = 1347.29 \pm 0.05$ pm, a calculated density of 3.644 $g\cdot cm^{-3}$, and an energy gap of 2.7 eV. α-$RbInSnS_4$ do not undergo structural transformation to cubic β-$RbInSnS_4$ when heated up to 900°C.

Orange-yellow plate crystals of α-$RbInSnS_4$ were obtained, when a mixture of In (1.1 mM), Sn (1.1 mM), S (4.5 mM), RbCl (5.9 mM), and RbI (8.4 mM) was heated at 575°C in evacuated sealed silica tube (Yohannan and Vidyasagar 2016b). The product was isolated by washing off the eutectic flux with water and then dried in the open air at room temperature. Gray polycrystalline powder of β-$RbInSnS_4$ was obtained by

heating, at 650 °C in an evacuated sealed silica tube, a mixture of $RbInS_2$ (1.1 mM), SnS_2 (1.1 mM), RbCl (5.9 mM), and RbI (8.4 mM). This modification was obtained as single crystals by heating a mixture of $RbInS_2$ (1.1 mM) and SnS_2 (1.1 mM) at 800°C for 4 days and then cooled to 600°C at a rate of 2°C·h^{-1}.

7.54 Tin–Rubidium–Phosphorus–Sulfur

The **$RbSnPS_4$** quaternary compound, which melts congruently at 560°C and crystallizes in the monoclinic structure, is formed in the Sn–Rb–P–S system (Banerjee et al. 2012). To prepare this compound, in a N_2-filled glove box Rb_2S, Sn, red P, and S (molar ratio 1:2.5:2:9) were loaded in a fused silica tube, which was then sealed under vacuum (<10^{-2} Pa). The mixture was heated at 900°C for 72 h followed by cooling to 300°C for 120 h and finally to room temperature in 3 h. The solidified yellow-orange colored melt was then taken out of the tube and washed with degassed DMF. Yellow-orange colored polyhedral crystals of $RbSnPS_4$ were isolated after drying with ether. This compound forms glass upon melt quenching.

7.55 Tin–Rubidium–Antimony–Sulfur

The **$Rb_2Sb_2Sn_3S_{10}$** quaternary compound, which crystallizes in the monoclinic structure with the lattice parameters $a = 686.8 \pm 0.3$, $b = 1339.9 \pm 0.5$, $c = 1034.2 \pm 0.4$ pm, $\beta = 95.66 \pm 0.02°$, a calculated density of 3.826 $g\cdot cm^{-3}$, and an energy gap of 2.33 eV, is formed in the Sn–Rb–Sb–S system (Yohannan and Vidyasagar 2015). It was obtained as a homogeneous phase, in the form of orange-yellow block-shaped single crystals, from a mixture of Rb_2CO_3 (0.69 mM), Sn (1.37 mM), Sb (0.92 mM), and S (5 mM). A quartz tube containing the reactant mixture was initially heated at 350°C under a dynamic vacuum for 10 min to expel the CO_2 generated and then sealed under vacuum. This sealed quartz tube was heated at 500°C for 2 days and then cooled to 300°C at 2°C·h^{-1}, and the furnace was finally turned off. The solid product contents were washed with distilled water to dissolve away $Rb_2S_2O_3$ formed and isolate single crystals $Rb_2Sb_2Sn_3S_{10}$. A polycrystalline orange-yellow sample of this compound was also synthesized by heating, in a continuous stream of CS_2 vapor, stoichiometric mixtures of Rb_2CO_3, Sb_2S_3, and SnO_2 at 400°C for 24 h with one intermittent grinding. An alumina boat was loaded with the reactant mixture and placed in a ceramic tube in a horizontal tubular furnace. A continuous stream of N_2 was bubbled through the CS_2 liquid and then passed through the ceramic tube to provide CS_2 vapor for the synthetic reaction. In this procedure, single crystals of $Rb_2Sb_2Sn_3S_{10}$ were obtained along with a polycrystalline sample, when a slight excess of Rb_2CO_3 and Sb_2S_3 were employed.

7.56 Tin–Rubidium–Manganese–Sulfur

The **$Rb_{1.57}Mn_{0.95}Sn_{2.05}S_6$** quaternary compound, which crystallizes in the trigonal structure with the lattice parameters $a = 368.27 \pm 0.04$, $c = 2593.4 \pm 0.6$ pm, and a calculated density of 3.391 g·cm^{-3}, is formed in the Sn–Rb–Mn–S system (Manos and Kanatzidis 2009). To prepare this compound, single crystals of K$_{1.90}$Mn$_{0.95}$Sn$_{2.05}$S$_6$ (0.018 mM) were placed in an aqueous solution (10 mL) of RbI (0.24 mM). The mixture was then heated at 70°C and left undisturbed at this temperature for 1 week. The crystals were then isolated with filtration and washed several times with water, acetone, and diethyl ether.

7.57 Tin–Cesium–Copper–Sulfur

The **Cs$_2$Cu$_2$Sn$_2$S$_6$** quaternary compound is formed in the Sn–Cs–Cu–S system (Liao and Kanatzidis 1993). To prepare this compound, a mixture of Sn, Cu, Cs$_2$S, and S (molar ratio 1:1–2:4:16) was loaded into a Pyrex tube in a N$_2$ glove box. The tube was evacuated and sealed at a pressure of ~0.1 Pa. The mixture was heated slowly from room temperature to 400°C for 12 h in a furnace. The temperature was kept at 400°C for 4 days, and then was cooled slowly to room temperature at a rate of 4°C·h^{-1}. Black plate-like crystals were formed. The product was washed with degassed DMF to remove excess Cs$_2$S$_x$ flux using a standard Schlenk technique and dried with acetone and ether. The obtained crystals are stable in the air.

7.58 Tin–Cesium–Gold–Sulfur

The **Cs$_2$Au$_2$Sn$_2$S$_6$** quaternary compound, which melts at 421°C and crystallizes in the orthorhombic structure with the lattice parameters $a = 614.3 \pm 0.1$, $b = 1429.6 \pm 0.3$, $c = 2457.8 \pm 0.5$ pm, a calculated density of 5.58 g·cm^{-3}, and an energy gap of 2.28 eV, is formed in the Sn–Cs–Au–S a system (Löken and Tremel 1998, 1999). To prepare this compound, a mixture of Au (0.8 mM), Sn (0.4 mM), Cs$_2$S (0.4 mM), and S (1.2 mM) was mixed and sealed in a quartz ampoule under vacuum. It was heated to 400°C in 6 h, kept at this temperature for 4 days, and then cooled slowly to room temperature at a rate of 4°C·h^{-1}. Yellow plate-like crystals of Cs$_2$Au$_2$Sn$_2$S$_6$ were isolated by washing with absolute ethanol. The obtained crystals are stable in air and water.

7.59 Tin–Cesium–Zinc–Sulfur

The **Cs$_2$ZnSn$_3$S$_8$** quaternary compound, which has two polymorphic modifications, is formed in the Sn–Cs–Zn–S system (Pogu and Vidyasagar 2020). α-Cs$_2$ZnSn$_3$S$_8$ crystallizes in the monoclinic structure with the lattice parameters $a = 752.94 \pm 0.11$, $b = 1768.8 \pm 0.2$, $c = 1247.71 \pm 0.16$ pm, $\beta = 94.796 \pm 0.004°$, a calculated density of 3.786 g·cm^{-3}, and an energy gap of 2.93 eV [2.82 eV (Tian et al. 2021)], and β-Cs$_2$ZnSn$_3$S$_8$ crystallizes in the cubic structure with the lattice parameter $a = 1353.23 \pm 0.06$ pm, a calculated density of 3.795 g·cm^{-3}, and an energy gap of 2.53 eV. Polycrystalline samples of both modifications were synthesized by heating stoichiometric mixtures of Cs$_2$CO$_3$, ZnS, and SnO$_2$ in a continuous stream of CS$_2$ vapor at 750°C for α-Cs$_2$ZnSn$_3$S$_8$ and at 600°C for β-Cs$_2$ZnSn$_3$S$_8$ for 2 days, with one intermittent grinding. An alumina boat was

loaded with the solid reactant mixture and placed in a ceramic tube, which was kept in a horizontal tubular furnace. A continuous stream of N$_2$ gas was bubbled, at a rate of two bubbles per second, through CS$_2$ liquid and then passed through the ceramic tube to provide CS$_2$ vapor for the synthetic reaction. The outgoing gaseous mixture of N$_2$, CS$_2$, and CO$_2$ from the ceramic tube was passed through paraffin oil and then into a dilute aqueous solution of KOH. The furnace was turned off after heating and allowed to cool naturally to room temperature over 12 h. Single crystals of Cs$_2$ZnSn$_3$S$_8$ were obtained by the thiosulfate flux method. It involved heating the solid reactant mixture of Cs$_2$CO$_3$, S, Zn and Sn, or ZnS and SnS$_2$ in a silica tube under dynamic vacuum first at ≈300°C for about 15 min and then sealing the quartz tube along with solid content under vacuum. The sealed quartz ampoule was heated at a rate of 3°C·min^{-1} to 850°C for α-Cs$_2$ZnSn$_3$S$_8$ and to 650°C for β-Cs$_2$ZnSn$_3$S$_8$, held at that temperatures for 3 days and then cooled to 350°C over a period of 2 days. The ampoule was broken open after the reaction and the solid contents were washed with water to dissolve away Cs$_2$S$_2$O$_3$ flux and isolate the desired solid product.

7.60 Tin–Cesium–Cadmium–Sulfur

Three quaternary compounds, **Cs$_2$CdSn$_2$S$_6$**, **Cs$_2$CdSn$_3$S$_8$**, and **Cs$_{10}$Cd$_4$Sn$_4$S$_{17}$**, are formed in the Sn–Cs–Cd–S system. The first of them melts congruently at 614°C, is thermally stable up to 900°C, and crystallizes in the monoclinic structure with the lattice parameters $a = 717.810 \pm 0.010$, $b = 1385.20 \pm 0.03$, $c = 748.81 \pm 0.02$ pm, $\beta = 112.8830° \pm 0.0010°$, and a calculated density of 3.912 g·cm^{-3}, and an energy gap of 2.89 eV (Pogu et al. 2019). Polycrystalline sample of this compound was prepared by heating stoichiometric mixtures of appropriate solid reactants [Cs$_2$CO$_3$ (1.23 mM), CdS (1.23 mM), and SnO$_2$ (2.46 mM)] in a continuous stream of CS$_2$ vapor at 650°C for 24 h with one intermittent grinding. An alumina boat was loaded with the mixture and placed in a ceramic tube in a horizontal tubular furnace. A continuous stream of N$_2$ was bubbled through CS$_2$ liquid and then passed through the ceramic tube to provide CS$_2$ vapor for the synthetic reaction. Pale-yellow samples of Cs$_2$CdSn$_2$S$_6$ were obtained. The light-yellow single crystals were isolated from the *in-situ* molten Cs$_2$S$_2$O$_3$ flux. A quartz tube containing an appropriate reactant mixture [Cs$_2$CO$_3$ (0.62 mM), SnS$_2$ (0.82 mM), CdS (0.412 mM), and S (0.82 mM) or Cs$_2$CO$_3$ (0.62 mM), Sn (0.82 mM), Cd (0.41 mM), and S (2.87 mM)] was initially heated at 350°C under dynamic vacuum for 10 min to expel the CO$_2$ generated and then sealed under vacuum. This sealed quartz tube was heated at 650°C for 3 days and then cooled to 300°C at 4°C·h^{-1} and the furnace was finally turned off. The solid product contents were washed with distilled water to dissolve away Cs$_2$S$_2$O$_3$ formed *in situ* and the single crystals of Cs$_2$CdSn$_2$S$_6$ were isolated.

Cs$_2$CdSn$_3$S$_8$ has two polymorphic modifications (Pogu and Vidyasagar 2020). α-Cs$_2$CdSn$_3$S$_8$ crystallizes in the orthorhombic structure with the lattice parameters $a = 768.98 \pm 0.07$, $b = 1243.22 \pm 0.12$, $c = 1789.75 \pm 0.15$ pm, a calculated density of 3.846 g·cm^{-3}, and an energy gap of 2.63 eV, and β-Cs$_2$CdSn$_3$S$_8$ crystallizes in the cubic structure with the lattice parameter $a = 1367.62 \pm 0.05$ pm, a calculated density of 3.859 $g\cdot cm^{-3}$, and an energy gap of 2.56 eV. Polycrystalline samples of both modifications were synthesized by heating stoichiometric mixtures of Cs_2CO_3, CdS, and SnO_2 in a continuous stream of CS_2 vapor at 750°C for α-$Cs_2CdSn_3S_8$ and at 600°C for β-$Cs_2CdSn_3S_8$ for 2 days, with one intermittent grinding. An alumina boat was loaded with the solid reactant mixture and placed in a ceramic tube, which was kept in a horizontal tubular furnace. A continuous stream of N_2 gas was bubbled, at a rate of two bubbles per second, through CS_2 liquid, and then passed through the ceramic tube to provide CS_2 vapor for the synthetic reaction. The outgoing gaseous mixture of N_2, CS_2 and CO_2 from the ceramic tube was passed through paraffin oil and then into a dilute aqueous solution of KOH. The furnace was turned off after heating and allowed to cool naturally to room temperature over 12 h. Single crystals of $Cs_2CdSn_3S_8$ were obtained by the thiosulfate flux method. It involved heating the solid reactant mixture of Cs_2CO_3, S, Cd and Sn, or CdS and SnS_2 in silica tube under dynamic vacuum first at about 300°C for about 15 min and then sealing the quartz tube along with solid content under vacuum. The sealed quartz ampoule was heated at a rate of $3°C\cdot min^{-1}$ to 900°C for α-$Cs_2CdSn_3S_8$ and to 650°C for β-$Cs_2CdSn_3S_8$, held at that temperatures for 3 days and then cooled to 350°C over a period of 2 days. The ampoule was broken open after the reaction, and the solid contents were washed with water to dissolve away $Cs_2S_2O_3$ flux and isolate the desired solid product.

$Cs_{10}Cd_4Sn_4S_{17}$ decomposes at 714°C and crystallizes in the tetragonal structure with the lattice parameters $a = 1504.6 \pm 0.3$, $c = 1053.4 \pm 0.3$ pm, a calculated density of 3.897 $g\cdot cm^{-3}$, and an energy gap 3.16 eV (Palchik et al. 2004). This compound was obtained by reaction of Sn (0.3 mM), Cd (0.3 mM), Cs_2S (1.5 mM), and S (3.6 mM). The mixture was sealed under vacuum (< 0.1 Pa) in a silica tube and heated at a rate of $50°C\cdot h^{-1}$ to 500°C for 96 h, followed by cooling to room temperature at $5°C\cdot h^{-1}$. The excess flux (Cs_2S_x) was removed with MeOH to reveal yellowish-white transparent rectangular crystals, which were air stable.

7.61 Tin–Cesium–Mercury–Sulfur

Two quaternary compounds, $Cs_2HgSn_3S_8$ and $Cs_2Hg_3Sn_2S_8$, are formed in the Sn–Cs–Hg–S system. $Cs_2HgSn_3S_8$ crystallizes in the orthorhombic structure with the lattice parameters $a = 766.72 \pm 0.04$, $b = 1234.05 \pm 0.10$, $c = 1790.11 \pm 0.11$ pm, a calculated density of 4.231 $g\cdot cm^{-3}$, and an energy gap of 2.72 eV (Morris et al. 2013). This compound was obtained by the interaction of Cs_2S, Hg, Sn, and S in a stoichiometric ratio, which was placed in a fused silica tube and heated in a furnace to 750°C in 8 h, held there for 12 h, and cooled to room temperature in 24 h.

$Cs_2Hg_3Sn_2S_8$ crystallizes in the triclinic structure with the lattice parameters $a = 787.8 \pm 0.2$, $b = 915.7 \pm 0.2$, $c = 680.3 \pm 0.2$ pm, $\alpha = 92.96 \pm 0.02°$, $\beta = 109.45 \pm 0.02°$, $\gamma = 107.81 \pm 0.02°$, and a calculated density of 5.207 $g\cdot cm^{-3}$ (Marking et al. 1998). This compound was obtained by the interaction of HgS with Sn in the Cs_2S_x melt at 520°C. This compound was

taken to its glassy state by heating it at 550 °C in a hermetically closed system.

7.62 Tin–Cesium–Gallium–Sulfur

The $CsGaSnS_4$ quaternary compound, which has two polymorphic modifications, is formed in the Sn–Cs–Ga–S system. α-$CsGaSnS_4$ crystallizes in the orthorhombic structure with the lattice parameters $a = 1757.08 \pm 0.05$, $b = 738.46 \pm 0.02$, $c = 1243.43 \pm 0.02$ pm, a calculated density of 3.699 $g\cdot cm^{-3}$, and an energy gap of 3.07 eV [$a = 1757.07 \pm 0.11$, $b = 737.88 \pm 0.04$, $c = 1243.38 \pm 0.08$ pm, a calculated density of 3.705 $g\cdot cm^{-3}$, and an energy gap of 3.02 eV (Liu et al. 2020a)], and β-$CsGaSnS_4$ crystallizes in the cubic structure with the lattice parameter $a = 1337.9 \pm 0.1$ pm, a calculated density of 3.747 $g\cdot cm^{-3}$, and an energy gap of 3.00 eV (Friedrich et al. 2020) [$a = 1335.7 \pm 0.7$ pm and an energy gap of 2.75 eV (Kumari and Vidyasagar 2007); $a = 1337.29 \pm 0.02$ pm, a calculated density of 3.746 $g\cdot cm^{-3}$, and an energy gap of 2.92 eV (Liu et al. 2020a)].

Polycrystalline samples of α-$CsGaSnS_4$ were obtained by melting Cs_2S with a stoichiometric mixture of Ga_2S_3, Sn, and S in a sealed quartz ampoule at 1000°C for 2 days and cooling to room temperature with a rate of $15°C\cdot min^{-1}$ (Friedrich et al. 2020). Polycrystalline batches of white β-$CsGaSnS_4$ were obtained by annealing Cs_2S with stoichiometric mixtures of Ga_2S_3, Sn, and S in a sealed evacuated quartz tube at 500°C for 7 days. After the initial annealing, the crude product was thoroughly grounded in an agate mortar under a N_2 atmosphere and pressed to a pellet. The pellet was annealed in an evacuated quartz tube for an additional week at 500°C.

To prepare the crystals of α-$CsGaSnS_4$, a mixture of CsCl, Ga_2S_3, Sn, and S (molar ratio 2:1:1:4) was loaded into a graphite crucible that was put into an outer silica tubing, and finally the whole assembly was sealed under a vacuum of 10^{-3} Pa (Liu et al. 2020a). The sealed assembly was heated to 810°C in 40 h, and maintained for 60 h, followed by a slow cooling to 300°C before turning off the furnace. The product was washed with distilled water and CS_2 in an ultrasonic cleaner to remove the by-products; the air-stable light-yellow crystals were obtained. Similarly, the crystals of β-$CsGaSnS_4$ were obtained at 700°C.

7.63 Tin–Cesium–Indium–Sulfur

The $CsInSnS_4$ quaternary compound, which apparently has two polymorphic modifications, is formed in the Sn–Cs–In–S system. α-$CsInSnS_4$ crystallizes in the orthorhombic structure with the lattice parameters $a = 1801.38 \pm 0.04$, $b = 762.38 \pm 0.04$, $c = 1243.32 \pm 0.03$ pm, a calculated density of 3.848 $g\cdot cm^{-3}$, and an energy gap of 2.92 eV (Friedrich et al. 2021), and β-$CsInSnS_4$ crystallizes in the cubic structure with the lattice parameter $a = 1368.08 \pm 0.01$ pm, a calculated density of 3.850 $g\cdot cm^{-3}$, and an energy gap of 2.9 eV (Yohannan and Vidyasagar 2016b).

Polycrystalline white α-$CsInSnS_4$ was obtained by reacting Cs_2S (1.01 mM), In (2.02 mM), Sn (2.0 mM), and S (7.08 mM) in a sealed evacuated fused silica tube (Friedrich et al. 2021).

The reaction mixture was heated to 1000°C with a heating rate of 3°C·min^{-1}. After annealing at 1000°C for 24 h, the tube was cooled to room temperature with a cooling rate of 3°C·min^{-1}. The white polycrystalline product is stable in moist air and forms plate-like crystallites upon.

β-CsInSnS$_4$ was synthesized by heating in a continuous stream of CS$_2$ vapor appropriate stoichiometric mixture of Cs$_2$CO$_3$, In$_2$O$_3$, and SnO$_2$ with one intermittent grinding (Yohannan and Vidyasagar 2016b). An alumina boat was loaded with the reactant mixture and placed in a ceramic tube in a horizontal tubular furnace. A continuous stream of N$_2$ was bubbled through the CS$_2$ liquid and then passed through the ceramic tube to provide CS$_2$ vapor for the synthetic reaction. Single crystals of this modification were obtained by heating a stoichiometric mixture of CsInS$_2$ (1 mM) and SnS$_2$ (1 mM), in an evacuated silica tube, at 800°C for 4 days and then cooling to 600°C at 2°C·h^{-1}.

7.64 Tin–Cesium–Phosphorus–Sulfur

The **CsSnPS$_4$** quaternary compound, which melts congruently at 536°C and crystallizes in the monoclinic structure with the lattice parameters $a = 1804.77 \pm 0.14$, $b = 620.21 \pm 0.05$, $c = 684.15 \pm 0.05$ pm, $\beta = 90°$, and a calculated density of 3.5617 g·cm^{-3} at 100 K, is formed in the Sn–Cs–P–S system (Banerjee et al. 2012). To prepare this compound, in a N$_2$-filled glove box Cs$_2$S, Sn, red P, and S (molar ratio 1:2.5:2:9) were loaded in a fused silica tube, which was then sealed under vacuum (<10^{-2} Pa). The mixture was heated at 900°C for 72 h, followed by cooling to 300°C for 120 h and finally to room temperature for 3 h. The solidified yellow-orange colored melt was then taken out of the tube and washed with degassed DMF. Yellow-orange colored polyhedral crystals of KSnPS$_4$ were isolated after drying with ether. This compound forms glass upon melt quenching.

7.65 Tin–Cesium–Arsenic–Sulfur

The **Cs$_2$SnAs$_2$S$_9$** quaternary compounds, which crystallize in the orthorhombic structure with the lattice parameters $a = 738.6 \pm 0.3$, $b = 1461.4 \pm 0.5$, $c = 1441.7 \pm 0.5$ pm, a calculated density of 3.512 g·cm^{-3}, and an energy gap of 1.98 eV, is formed in the Sn–Cs–As–S system (Iyer et al. 2003). This compound melts at 350°C with no recrystallization observed on cooling, instead giving a glass product. The glass has the same color as the crystalline compound and recrystallizes exothermically at 268°C upon subsequent heating.

It was synthesized from a mixture of Cs$_2$S (0.3 mM), Sn (0.15 mM), As$_2$S$_3$ (0.15 mM), and S (1.5 mM). The reagents were mixed, sealed in an evacuated silica tube, heated at 500°C for 3 days, and then cooled at a rate of 5°C·h^{-1} to 250°C followed by rapid cooling to room temperature. The solid products were washed with DMF and ether to remove the flux. Red plate crystals (60% yields) were obtained. A single phase of Cs$_2$SnAs$_2$S$_9$ with over 95% yields was achieved by using a stoichiometric ratio of Cs$_2$S/Sn/As/S = 1:1:2:8 at 550°C.

7.66 Tin–Cesium–Antimony–Sulfur

The **Cs$_2$Sb$_2$Sn$_3$S$_{10}$** quaternary compound, which crystallizes in the monoclinic structure with the lattice parameters $a = 700.36 \pm 0.03$, $b = 1345.90 \pm 0.05$, $c = 1040.13 \pm 0.04$ pm, $\beta = 94.631 \pm 0.002°$, a calculated density of 4.030 g·cm^{-3}, and an energy gap of 2.34 eV, is formed in the Sn–Cs–Sb–S system (Yohannan and Vidyasagar 2015). It was obtained as a homogeneous phase, in the form of orange-yellow block-shaped single crystals, from a mixture of Cs$_2$CO$_3$ (0.63 mM), Sn (1.26 mM), Sb (0.84 mM), and S (4.64 mM). A quartz tube containing the reactant mixture was initially heated at 350°C under a dynamic vacuum for 10 min to expel the CO$_2$ generated and then sealed under vacuum. This sealed quartz tube was heated at 500°C for 2 days and then cooled to 300°C at 2°C·h^{-1}, and the furnace was finally turned off. The solid product contents were washed with distilled water to dissolve away Cs$_2$S$_2$O$_3$ formed and isolate single crystals Cs$_2$Sb$_2$Sn$_3$S$_{10}$. Polycrystalline orange-yellow sample of this compound was also synthesized by heating, in a continuous stream of CS$_2$ vapor, stoichiometric mixtures of Cs$_2$CO$_3$, Sb$_2$S$_3$, and SnS at 450°C for 24 h with one intermittent grinding. An alumina boat was loaded with the reactant mixture and placed in a ceramic tube in a horizontal tubular furnace. A continuous stream of N$_2$ was bubbled through CS$_2$ liquid and then passed through the ceramic tube to provide CS$_2$ vapor for the synthetic reaction. Single crystals of Cs$_2$Sb$_2$Sn$_3$S$_{10}$ were obtained.

7.67 Tin–Cesium–Magnesium–Sulfur

The **CsMn$_{0.95}$Sn$_{2.05}$S$_6$** quaternary compound, which crystallizes in the trigonal structure with the lattice parameters $a = 316.139 \pm 0.008$, $c = 2686.7 \pm 1.2$ pm, and a calculated density of 3.392 g·cm^{-3}, is formed in the Sn–Cs–Mn–S system (Manos and Kanatzidis 2009). To prepare this compound, single crystals of K$_{1.9}$Mn$_{0.95}$Sn$_{2.05}$S$_6$ (0.018 mM) were placed in an aqueous solution (10 mL) of CsCl (0.24 mM). The mixture was then heated at 70°C and left undisturbed at this temperature for 1 week. The crystals were then isolated with filtration and washed several times with water, acetone, and diethyl ether. For the solid synthesis of **Cs$_{2x}$Mn$_x$Sn$_{3-x}$S$_6$** ($x = 0.5–0.95$), a mixture of Sn (1.9 mM), Mn (1.1 mM), Cs$_2$S (2 mM), and S (16 mM) was sealed under vacuum (10^{-2} Pa) in a silica tube and heated (50°C·h^{-1}) to 400–500°C for 60 h, followed by cooling to room temperature at 50°C·h^{-1}. The excess flux was removed with H$_2$O or DMF to reveal dark-brown polycrystalline material.

7.68 Tin–Copper–Silver–Sulfur

Cu$_2$SnS$_3$–Ag$_8$SnS$_6$. The phase diagram of this system, constructed through differential thermal analysis (DTA), XRD, metallography, and measuring of the microhardness is a eutectic type (Rzaguluyev et al. 2019). The eutectic contains 50 mol% Cu$_2$SnS$_3$ and crystallizes at 627°C. The solubility of Cu$_2$SnS$_3$ in Ag$_8$SnS$_6$ at room temperature is 5 mol% and increases up to 20 mol% at the eutectic temperature, and the solubility of

Ag_8SnS_6 in Cu_2SnS_3 is 10 and 28 mol% at room temperature and 627°C, respectively. A eutectoid decomposition of the solid solution based on Ag_8SnS_6 takes place at 137°C.

$Cu_2Sn_4S_9$–Ag_2SnS_3. This system is a nonquasibinary section of the Sn–Cu–Ag–S quaternary system (Rzaguluyev et al. 2019). The eutectic contains 50 mol% $Cu_2Sn_4S_9$ and crystallizes at 577°C. The component-based solubility regions are narrow: Ag_2SnS_3 and $Cu_2Sn_4S_9$ dissolve up to 10 mol% $Cu_2Sn_4S_9$ and 2.5 mol% Ag_2SnS_3, respectively. The alloys for the investigation of both systems were annealed at 580°C for 240 h.

7.69 Tin–Copper–Magnesium–Sulfur

The **$Cu_2MgSn_3S_8$** quaternary compound, which crystallizes in the cubic structure with the lattice parameter $a = 1041.73 \pm 0.02$ pm, a calculated density of 4.49 g·cm^{-3}, and an energy gap of 1.65 eV, is formed in the Sn–Cu–Mg–S system (Heppke et al. 2020). This compound was synthesized in a high-energy planetary mill with a subsequent annealing step in a tube furnace. Stoichiometric amounts of CuS, MgS, SnS, and S were filled in a 45 mL zirconia grinding beaker with six zirconia balls and ground at a rotational speed of 450 rpm for 4 h. In order to obtain highly crystalline powders, the product was annealed at 550°C for 3 h under flowing H_2S gas after milling.

7.70 Tin–Copper–Strontium–Sulfur

Two quaternary compounds, **Cu_2SrSnS_4** and **$Cu_4Sr_6Sn_4S_{16}$**, are formed in the Sn–Cu–Sr–S system. The first of them has two polymorphic modifications. α-Cu_2SrSnS_4 crystallizes in the trigonal structure with the lattice parameters $a = 628.87 \pm 0.07$, $c = 1557.5 \pm 0.4$ pm, a calculated density of 4.311 g·cm^{-3}, and an energy gap of 2.01 eV (Yang et al. 2020c) [$a = 629.0$, $c = 1557.8$ pm, and the calculated and experimental densities of 4.310 and 4.327 g·cm^{-3}, respectively (Teske 1976); $a = 626.8 \pm 0.2$, $c = 1553.4 \pm 0.4$ pm, and an energy gap of 2.1 eV (Llanos et al. 2003); an energy gap of 1.73–1.75 eV (Zhu et al. 2017)]. β-Cu_2SrSnS_4 crystallizes in the orthorhombic structure with the lattice parameters $a = 1051.4 \pm 0.3$, $b = 1045.6 \pm 0.3$, $c = 642.5 \pm 0.2$ pm, a calculated density of 4.341 g·cm^{-3}, and an energy gap of 2.06 eV (Yang et al. 2020c).

To prepare both modifications of this compound, the spontaneous crystallization processes were completed in the vacuum-sealed silica tubes (Yang et al. 2020c). The crystals were prepared under different sinter temperatures (800°C for α-Cu_2SrSnS_4 and 1000°C for β-Cu_2SrSnS_4) and ratio of raw materials (SrS/Cu_2S/SnS_2 = 1:1:1 for α-Cu_2SrSnS_4 and SrS/Cu/SnS_2/S = 1:2:1:1 for β-Cu_2SrSnS_4). Their high yields (~95%) were achieved after repeated grinding and calcination. Finally, many high-quality crystals with red color were found in the tubes after washing with DMF, and they were stable in the air within several months. The whole preparation process is completed in an Ar-filled glove box because of the instability of SrS in the air. Polycrystalline α-Cu_2SrSnS_4 was also synthesized from a thoroughly ground mixture of SrS, Cu, Sn, and S in stoichiometric proportions (Llanos et al. 2003). The mixture

was placed in an alumina boat and heated in a CS_2/Ar stream at 450°C for about 36 h, with two intermittent grindings. The title compound was obtained, reddish-brown in color and stable in the air.

$Cu_4Sr_6Sn_4S_{16}$ crystallizes in the cubic structure with the lattice parameter $a = 1398.2 \pm 0.3$ pm, a calculated density of 4.295 g·cm^{-3}, and an energy gap of 2.07 eV (Yang et al. 2020a). The whole preparation process of this compound was completed in an Ar-filled glove box. The targeted product was prepared under the stoichiometric ratio of SrS, Cu_2S, and SnS_2 in the vacuum-sealed silica tube. Firstly, the tube was heated to 600°C in 30 h and kept at this temperature for about 100 h, then cooled quickly down to room temperature. The obtained product was ground and then loaded into the vacuum-sealed silica tube. It was further sintered again at 900°C for 10 days and cooled slowly down to room temperature at 5°C·h^{-1}. Finally, many millimeter-level high-quality crystals with red color were found in the tube after washing with DMF. They are stable in the air within several weeks.

7.71 Tin–Copper–Barium–Sulfur

Two quaternary compounds, **Cu_2BaSnS_4** and **$Cu_2Ba_3Sn_2S_8$**, are formed in the Sn–Cu–Ba–S system. The first crystallizes in the trigonal structure with the lattice parameters $a = 635.3 \pm 0.3$, $c = 1578.8 \pm 1.4$ pm, a calculated density of 4.616 g·cm^{-3}, and an energy gap of 1.96 eV (Nian et al. 2017) [$a = 636.7$, $c = 1583.3$ pm, and the calculated and experimental densities of 4.582 and 4.532 g·cm^{-3}, respectively (Teske and Vetter 1976b); an energy gap of 1.74–1.75eV (Zhu et al. 2017)], The crystals of Cu_2BaSnS_4 were synthesized from a mixture of BaS, Cu, Sn, and S (molar ratio 1:2:1:3.5) (Nian et al. 2017). First, the reaction mixture was loaded into a silica tube and flame-sealed under a high vacuum (10^{-3} Pa). Second, the furnace was programmed by the following steps: heated from room temperature to 600°C in 30 h and kept at this temperature for 40 h; then heated to 1050°C in 20 h and left at this temperature for 100 h; finally, cooled to 400°C at a rate of 5°C·h^{-1} and then the furnace was shut down to room temperature. Third, the obtained products were washed with DMF to remove the unreacted sulfur and other by-products. Finally, red crystals of Cu_2BaSnS_4 were obtained after drying in the air. They are stable in the air for several months. Since Ba is easily oxidized in the air, an Ar-filled glove box was used to avoid the effects of oxygen and moisture in the preparation processes.

According to Márquez et al. (2020), Cu_2BaSnS_4 is stable at high sulfur pressure and decomposes at high temperatures into Cu_4BaS_3 and $Cu_2Ba_3Sn_2S_8$ if the synthesis is performed under low partial pressure of sulfur. **$Cu_2Ba_3Sn_2S_8$**, crystallizes in the cubic structure with the lattice parameter $a = 1453 \pm 1$ pm and an energy gap of 2.19 eV. For the synthesis of yellow powder of this compound, Cu_2S, BaS, SnS, and S were combined in a 1:3:2:2 molar ratio, ground with mortar and pestle, and cold-pressed, all in a N_2-filled glovebox. Pellets were sealed under dynamic vacuum (10^{-4} Pa) in fused silica ampoules. Ampoules were heated at 550°C for 96 h and rapidly quenched to room temperature. The resulting material was ground, pressed, and annealed under the same conditions three additional times. $Cu_2Ba_3Sn_2S_8$ is stable at high temperatures (560°C) under low sulfur partial pressure conditions.

7.72 Tin–Copper–Zinc–Sulfur

Cu_2SnS_3–ZnS. This system is a quasibinary section of the SnS_2–Cu_2S–ZnS quasiternary system (Figure 7.1) (Olekseyuk et al. 2000a). The eutectic, which melts at 850°C and contains 5 mol% ZnS, is form between Cu_2SnS_3 and **Cu_2ZnSnS_4** compounds. Solid solubility based on the system components at 400°C is insignificant (less than 1 mol% based on Cu_2SnS_3 and less than 2 mol% based on ZnS). Solid solution based on sphalerite modification of ZnS tends to increase with increasing temperature. The highest solubility (10 mol% Cu_2SnS_3) takes place at the peritectic temperature. Its transition to wurtzite modification is accompanied with the peritectic process at 1080°C. The existence of the **$Cu_4ZnSn_2S_7$** quaternary compound was not confirmed. The alloys were annealed at 400°C for 250 h and this system was investigated through DTA, metallography, and XRD.

According to the data of Moh (1975), in the Cu_2SnS_3–ZnS system α-Cu_2SnS_3 forms a complete solid solution extending beyond Cu_2ZnSnS_4 toward ZnS. In contrast, ZnS takes measurable amounts of Cu_2ZnSnS_4 in solid solution only above ~600°C, reaching a maximum of ≈12 mol% (≈38 mass%) at 972°C. At higher temperatures, the solubility decreases to ≈3.1 mol% (≈12.5 mass%) at 1050°C where sphalerite inverts to wurtzite, which takes only ≈1.4 mol% (≈6 mass.%) Cu_2ZnSnS_4 in the solid solution. Cu_2ZnSnS_4 can coexist with liquid sulfur, the ternary liquid of the Cu–Sn–S system, and all other binary and ternary sulfides, but not with the metals or their alloys.

Cu_2ZnSnS_4 melts incongruently at 980°C [990°C (Ichikawa et al. 2000; Matsushita et al. 2005; Matsushita and Katsui 2005); 982°C (Schäfer and Nitsche 1977)], possesses narrow homogeneity region (Olekseyuk et al. 2000a), and crystallizes in the tetragonal structure with the lattice parameters given in Table 7.1. Using the first-principles density functional method, it was shown that the low-energy crystal structure of this compound is a kesterite-type structure (Chen et al. 2009). However, the stannite or partially disordered kesterite structure can also exist in synthesized samples due to the small energy cost. According to the first-principles calculations, kesterite- and stannite-type structures of this compound are characterized by the tetragonal structure (Bensalem et al. 2014).

Cu_2ZnSnS_4 was obtained by the chemical vapor reactions using iodine as a transport agent (Nitsche et al. 1967; Schäfer and Nitsche 1974; Guen et al. 1979; Guen and Glaunsinger 1980), or by sintering of binary sulfide or constituent elements' mixtures (Hahn and Schulze 1965; Kissin and Owens 1979; Kheraj et al. 2013), or by the horizontal gradient freezing (Ichikawa et al. 2000; Matsushita et al. 2005). This compound could also be obtained if an aqueous solution (150 mL) of $Zn(NO_3)_2·6H_2O$ with 10–15% excess amounts (about 0.029 M/L) and $SnCl_4·5H_2O$ (0.025 M/L) was purged with N_2 with the next addition of CuCl into the mixture solution (Tsuji et al. 2010). Cu_2ZnSnS_4 is precipitated by bubbling an H_2S gas into

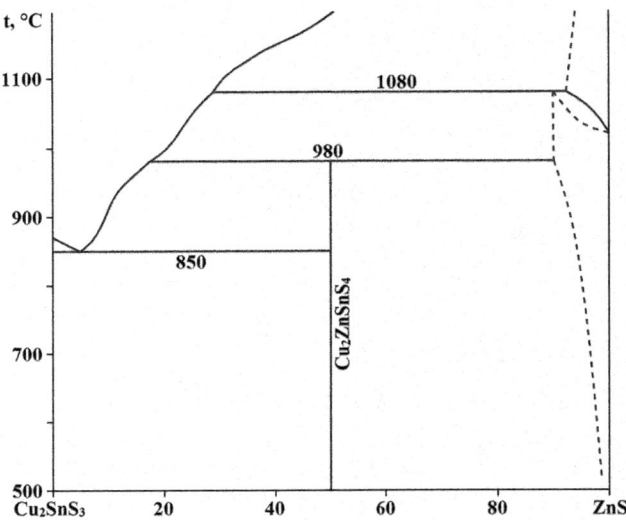

FIGURE 7.1 Phase diagram of the Cu_2SnS_3–ZnS system. (From Olekseyuk, I., et al., *Visnyk L'viv. Univ. Ser. khim.*, (39), 48, 2000.)

the mixed solution containing $Zn(NO_3)_2$, $SnCl_4$, and CuCl. Single crystals of the title compound have been produced within sealed quartz ampoules via the chemical vapor transport technique using I_2 as the transport agent (Colombara et al. 2013) or by the traveling heater method, where the Sn solvent was used (Nagaoka et al. 2011, 2012, 2013).

In-situ diffraction experiments using high-energy synchrotron X-rays revealed the existence of a structural phase transition in Cu_2ZnSnS_4 from the tetragonal kesterite type to the cubic sphalerite type structure (Schorr and Gonzalez-Aviles 2009; Schorr 2011). The transition occurs between 866–883°C, where a tetragonal and a cubic phase coexists. Outside the phase transition region, the lattice parameter of the tetragonal phase increase nearly linearly with increasing temperature, whereas the lattice parameter of the cubic phase show first a negative thermal expansion changing at ~930°C to a linear increase of a_{cub}.

Orthorhombic Cu_2ZnSnS_4 was synthesized *via* a hydrothermal method with the assistance of ethylenediamine (Jiang et al. 2012). In a typical experiment, $ZnCl_2$ (0.5 mM), $SnCl_2·2H_2O$ (0.5 mM), $CuCl_2·2H_2O$ (1 mM), and thiocarbamide (4 mM) were dissolved in a 40 mL mixed solvent of ultrapure water and ethylenediamine (volume ratio 1:1) with magnetic stirring. Then the precursors were transferred into a 55 mL Teflon-lined stainless-steel autoclave and kept at 200°C for 24 h. Afterwards it was cooled to room temperature, and a black precipitate was collected, which was washed with ethanol several times and dried under vacuum. Ethylenediamine plays a key role in the synthesis of the wurtzite-derived structure. Orthorhombic Cu_2ZnSnS_4 particles transform into the kesterite (tetragonal) structure when they are annealed at 500°C. It shows that the orthorhombic structure is a metastable phase for this compound. The band gap of as-synthesized Cu_2ZnSnS_4 is about 1.45 eV.

Wurtzite Cu_2ZnSnS_4 nanocrystals (hexagonal structure, a = 383.87, c = 633.88 pm, and an energy gap of 1.4 eV) were synthesized through a hot-injection method (Lu et al. 2011).

TABLE 7.1

Crystallographic Data and an Energy Gap for Cu_2ZnSnS_4

a, pm	c, pm	$d_{calc.}$	$d_{meas.}$	E_g, eV	References
		$g \cdot cm^{-3}$			
543.6	1085	4.55	4.49	—	Hahn and Schulze (1965)
543	1083	—	—	—	Nitsche et al. (1967)
542.7	1084.8	—	—	—	Schäfer and Nitsche (1974, 1977)
543.39 ± 0.09	1086.79 ± 0.18	—	—	—	Kissin and Owens (1979)
543.5	1084.3	4.56–4.57	4.60–4.61	—	Guen et al. (1979); Guen and Glaunsinger (1980)
543	1080–1083	—	—	1.38	Ichikawa et al. (2000)
541.98 ± 0.03	1089.2 ± 0.5				Olekseyuk et al. (2000a)
543	1081	—	—	1.39	Matsushita et al. (2005); Matsushita and Katsui (2005)
—	—	—	—	1.50	Chen et al. (2009);
532.6	1066.3	—	—	1.50	Persson (2010)[a]
532.5	1062.9	—	—		
—	—	—	—	1.4	Tsuji et al. (2010)
545.5	1088.0	—	—	—	Nagaoka et al. (2011, 2012)
543.440 ± 0.015	1083.82 ± 0.06	—	—	—	Choubrac et al. (2012)
543.006 ± 0.005	1082.22 ± 0.02	—	—	—	Choubrac et al. (2012)[b]
543.44 ± 0.010	1083.37 ± 0.02	—	—	—	Choubrac et al. 2012[c]
542.936 ± 0.008	1083.91 ± 0.02	—	—	—	Lafond et al. (2012)[d]
542.1 ± 0.1	1081.9 ± 0.2	—	—	—	Colombara et al. (2013)
542.90 ± 0.02	1083.40 ± 0.05	—	—	—	
542.7	1085.4	—	—	1.45	Kheraj et al. (2013)
539.3	1079.7	4.6484	—	—	Bensalem et al. (2014)[a]
539.5	1078.4	4.6494			
543.53 ± 0.01	1084.64 ± 0.03	—	—	—	Lafond et al. (2014)

Note:

[a] theoretical values

[b] for $Cu_{1.71}Zn_{1.18}Sn_{0.99}S_4$ composition

[c] for $Cu_{1.879}Zn_{1.061}SnS_4$ composition

[d] for $Cu_{2.12}Zn_{0.82}Sn_{1.06}S_4$ composition

For this, $CuCl_2 \cdot 2H_2O$ (0.10 mM), $ZnCl_2$ (0.05 mM), and $SnCl_4 \cdot 5H_2O$ (0.05 mM) were first dissolved in dodecanethiol (1 mL) under ca. 120°C to produce a homogenous precursor solution of metal thiolates. Then the hot solution was injected into dodecanethiol (2 mL) and oleylamine (2 mL) or into oleic acid (4 mL) in a three-neck flask at 240°C. The mixture soon became brownish and was maintained for 1 h to allow the growth of nanocrystals. Finally, the products were precipitated by ethanol and redispersed in cyclohexane.

$Cu_2Sn_4S_9$–ZnS. The phase relations in this system were studied by Olekseyuk et al. (2004). The liquidus consists of lines that correspond to the fields of primary crystallization of the solid solutions based on α- and β-ZnS, SnS_2, and Cu_2ZnSnS_4. The secondary crystallizations L + Cu_2SnS_3 + (SnS_2) and L + (SnS_2) + (Cu_2ZnSnS_4) end at 762°C in a ternary eutectic reaction. The secondary crystallization of the binary peritectic L + (β-ZnS) + (Cu_2ZnSnS_4) ends at the temperature of the ternary peritectic reaction (790°C). In the subsolidus region of the diagram, the change of the phase equilibria

occurs, which is caused by the formation of $Cu_2Sn_4S_9$ on the side and of **$Cu_2ZnSn_3S_8$** within the concentration triangle. This system was investigated through DTA, metallography, and XRD.

SnS_2–Cu_2S–ZnS. The isothermal section of this system at 400°C is given in Figure 7.2 (Olekseyuk et al. 2004). Cu_2ZnSnS_4 and $Cu_2ZnSn_3S_8$ quaternary compounds are formed in this system. $Cu_2ZnSn_3S_8$ melts incongruently at 700°C and crystallizes in the tetragonal structure with the lattice parameters $a = 543.5 \pm 0.1$ and $c = 1082.5 \pm 0.6$ pm.

The liquidus surface projection of the SnS_2–Cu_2S–ZnS quasiternary system consists of seven fields of primary crystallization (Figure 7.3) (Olekseyuk et al. 2004). ZnS has the largest area of the primary crystallization field. This causes the elongated form of the primary crystallization fields of Cu_2SnS_3, SnS_2, and Cu_2ZnSnS_4 along the Cu_2S–SnS_2 side. The fields of the primary crystallization are separated by 16 monovariant lines and 15 invariants points, of which 8 correspond to the binary and 7 to the ternary invariant reactions.

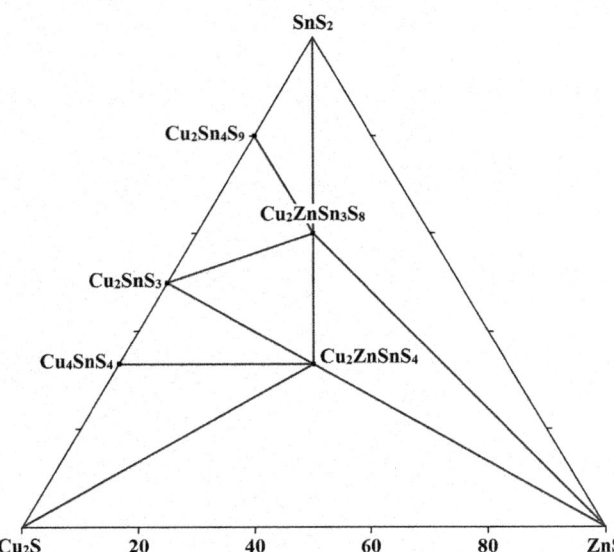

FIGURE 7.2 Isothermal section of the SnS_2–Cu_2S–ZnS quasiternary system at 400°C. (From Olekseyuk, I.D., et al., *J. Alloys Compd.*, **368**(1–2), 135, 2004.)

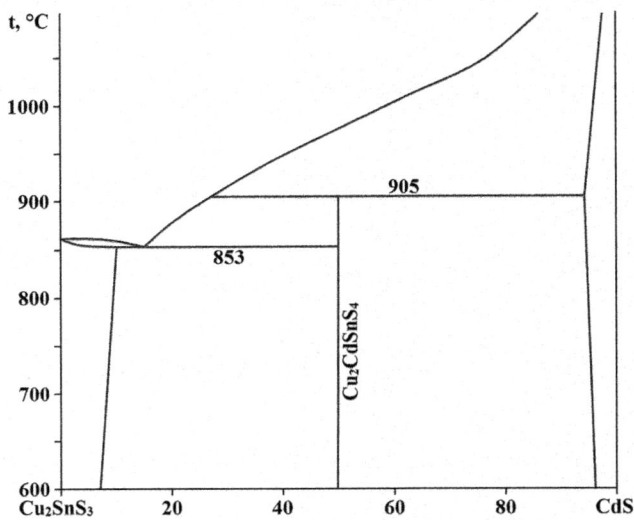

FIGURE 7.4 Phase diagram of the Cu_2SnS_3–CdS system. (From Olekseyuk, I.D., and Piskach, L.V., *Zhurn. neorgan. khimii*, **42**(2), 331, 1997.)

reaches 6 mol% and decreases with temperature decreasing. The **Cu_2CdSnS_4** quaternary compound is formed in this system with peritectic composition and temperature 27 mol% CdS and 905°C [926°C (Ichikawa et al. 2000; Matsushita et al. 2005; Matsushita and Katsui 2005); 920°C (Schäfer and Nitsche 1977)], respectively. This compound crystallizes in the tetragonal structure with the lattice parameters given in Table 7.2.

The Cu_2CdSnS_4 compound was obtained by the horizontal gradient freezing (Ichikawa et al. 2000; Matsushita et al. 2005), or by the chemical transport reactions using iodine as a transport agent (Nitsche et al. 1967; Schäfer and Nitsche 1974; Davydyuk et al. 2005), or by the heating of binary compounds at 650–900°C (Hahn and Schulze 1965). For example, this compound was prepared via high-temperature solid-state synthesis as follows (Rosmus et al. 2014). The starting materials were weighed in stoichiometric amounts and ground in an agate mortar and pestle for 30 min in an Ar-filled glove box. The mixture was placed into a graphite crucible, which was then inserted into a fused silica tube. The tube was flame-sealed under a vacuum of 0.1 Pa, heated to 800°C in 12 h, and held at that temperature for 125 h. The sample was then cooled to 500°C over 50 h, at a rate of 6°C·h⁻¹, and then cooled to ambient temperature.

Cu_4SnS_4–CdS. This section is a nonquasibinary one and crosses the fields of primary crystallization of the solid solutions based on Cu_2S, Cu_2SnS_3, and CdS (Piskach et al. 1998). In the subsolidus part, this section crosses one-phase fields of Cu_4SnS_4 and solid solution based on CdS, two-phase field (Cu_2S) + (CdS), and two ternary regions Cu_4SnS_4 + (Cu_2S) + Cu_2CdSnS_4 and (CdS) + (Cu_2S) + Cu_2CdSnS_4. The two ternary regions are separated by the secondary quasibinary system Cu_2S–Cu_2CdSnS_4. This system was investigated through DTA, metallography, X-ray diffraction (XRD), and scanning electron microscopy (SEM) with energy dispersive X-ray analysis (EDAX), and the alloys were annealed at 500°C for 500 h with subsequent quenching in cold water.

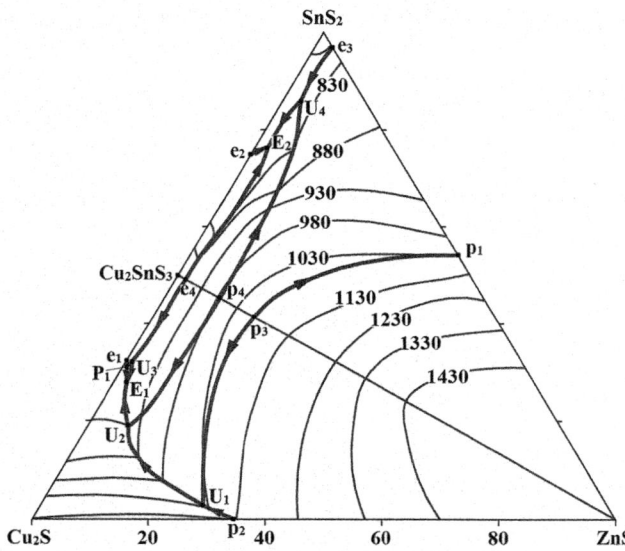

FIGURE 7.3 Liquidus surface of the SnS_2–Cu_2S–ZnS quasiternary system. (From Olekseyuk, I.D., et al., *J. Alloys Compd.*, **368**(1–2), 135, 2004.)

7.73 Tin–Copper–Cadmium–Sulfur

Cu_2SnS_3–CdS. The phase diagram of this system, constructed through DTA, metallography, XRD, and measuring of the microhardness and density, is shown in Figure 7.4 (Olekseyuk and Piskach 1997; Piskach et al. 1997). The eutectic composition and temperature are 15 mol% CdS and 853°C, respectively. The solubility of CdS in Cu_2SnS_3 is equal to 10 mol% at the eutectic temperature and decreases to 8 mol% at 550°C, and the solubility of Cu_2SnS_3 in CdS at the peritectic temperature

TABLE 7.2

Crystallographic Data and an Energy Gap for Cu_2CdSnS_4

a, pm	*c*, pm	$d_{calc.}$	$d_{meas.}$	E_g, eV	References
		g·cm⁻³			
558.2	1086	4.77	4.75	—	Hahn and Schulze (1965)
558	1082	—	—	—	Nitsche et al. (1967)
558.6	1083.4	—	—	—	Schäfer and Nitsche (1974, 1977)
533.30 ± 0.27	1082.66 ± 0.79	4.776	—	—	Kissin et al. (1978); Fleischer et al.
548.71 ± 0.15	1084.54 ± 0.55	4.618	—	—	(1979b)[a]
548.7 ± 0.2	1084.8 ± 0.3	4.62	—	—	Szymański (1978)[a]
558.09 ± 0.01	1082.6 ± 0.2	4.77	4.75	—	Olekseyuk and Piskach (1997)
558–559	1080–1082	—	—	1.38	Ichikawa et al. (2000)
—	—	—	—	1.4	Davydyuk et al. (2005)
559	1084	—	—	1.39	Matsushita et al. (2005); Matsushita and Katsui (2005)
559.2 ± 0.1	1084.0 ± 0.2	4.766	—	0.92	Rosmus et al. (2014)

Note:

[a] for mineral černýite

SnS_2–Cu_2S–CdS. The isothermal sections of this system at 500°C and 700°C are shown in Figure 7.5 (Piskach et al. 1998). Two quaternary compounds Cu_2CdSnS_4 and **$Cu_2CdSn_3S_8$** exist at these temperatures. The liquidus surface of this quasiternary system is shown in Figure 7.6 (Piskach et al. 1998). It consists of seven fields of primary crystallization of the solid solutions based on Cu_2S, CdS, SnS_2, and Cu_2SnS_3 and phases $Cu_2CdSn_3S_8$, Cu_2CdSnS_4, and Cu_4SnS_4. They are separated by 18 monovariant lines and 16 invariant points. The next ternary invariant points exist in the system: E_1 (770°C) – L ↔ Cu_4SnS_4 + Cu_2CdSnS_4 + (Cu_2SnS_3); E_2 (782°C) – L ↔ Cu_2CdSnS_4 + (Cu_2SnS_3) + (SnS_2); E_3 (762°C) – L ↔ (SnS_2) + (CdS) + $Cu_2CdSn_3S_8$; P (810°C) – L + (Cu_2S) + (Cu_2SnS_3) ↔ Cu_4SnS_4; U_1 (761°C) – L + (CdS) ↔ Cu_2CdSnS_4 + (Cu_2S); U_2 (798°C) – L + (Cu_2S) ↔ Cu_4SnS_4 + Cu_2CdSnS_4; U_3 (805°C) – L + (CdS) ↔ Cu_2CdSnS_4 + $Cu_2CdSn_3S_8$; and U_4 (795°C) – L + Cu_2CdSnS_4 ↔ $Cu_2CdSn_3S_8$ + (Cu_2SnS_3).

$Cu_2CdSn_3S_8$ melts incongruently at 818°C (Piskach et al. 1998) and crystallizes in the tetragonal structure with the lattice parameters $a = 738.27 ± 0.04$ and $c = 1041.52 ± 0.09$ pm (Chykhriy et al. 2000). The samples of this compound were prepared from elemental components, which were brought into evacuated quartz containers under residual Ar pressure of 0.1 Pa. Synthesis was carried out in two steps. The first stage of interaction took place

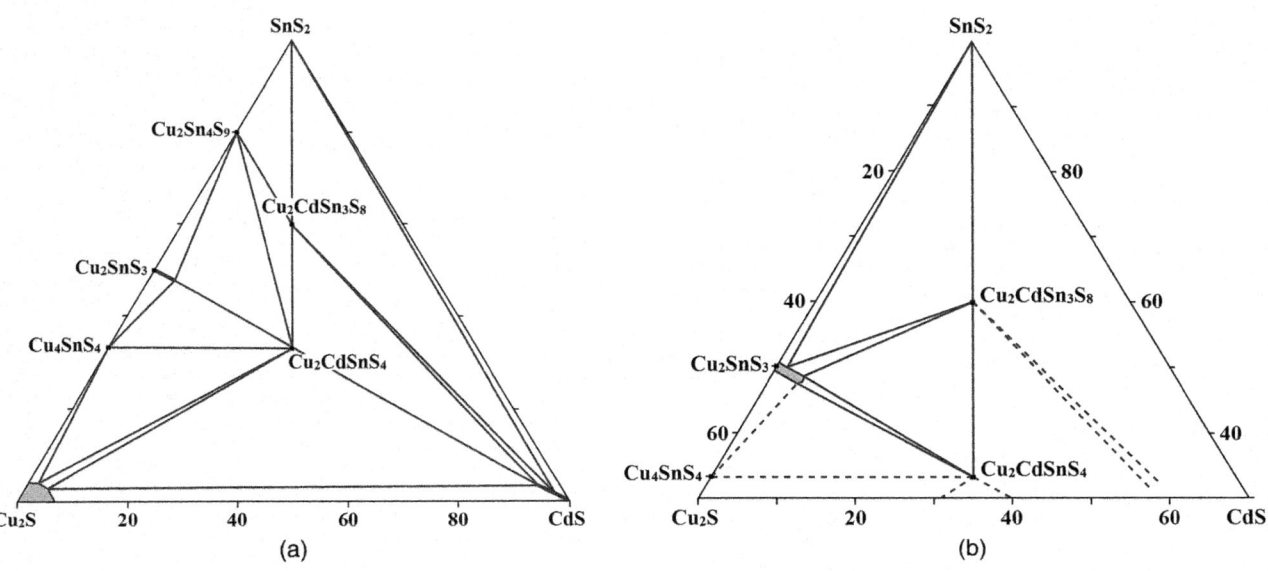

FIGURE 7.5 Isothermal section of the SnS_2–Cu_2S–CdS quasiternary system at (a) 500°C and (b) 700°C. (From Piskach, L.V., et al., J. Alloys Compd., **279**(2), 142, 1998.)

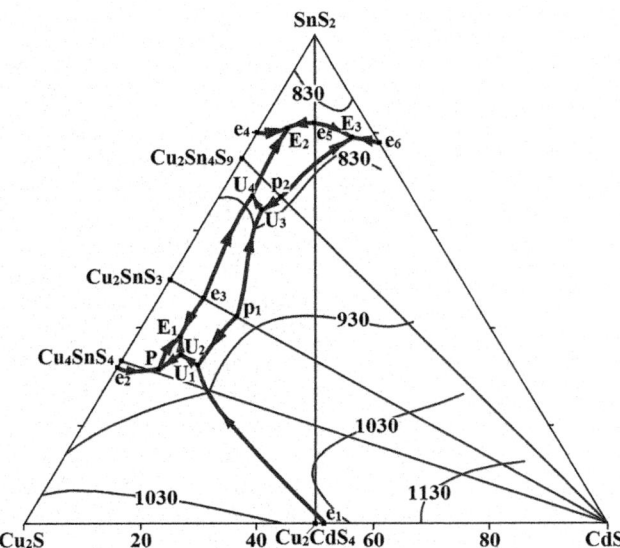

FIGURE 7.6 Liquidus surface of the SnS₂–Cu₂S–CdS quasiternary system. (From Piskach, L.V., et al., *J. Alloys Compd.*, **279**(2), 142, 1998.)

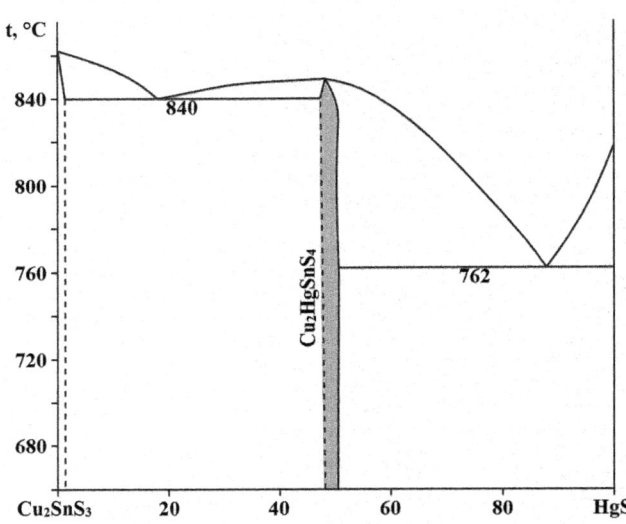

FIGURE 7.7 Phase diagram of the Cu₂SnS₃–HgS system. (From Olekseyuk, I., et al., *Visnyk L'viv. Univ. Ser. khim.*, (39), 48, 2000.)

in an O₂/CH₄ flame with visual control of the reaction, the aim of this stage being a partial reaction of pure sulfur. The second stage of the synthesis process was carried out in the furnace by gradually increasing the temperature (40–50°C·h⁻¹) up to 950°C. The samples then were heat treated at this temperature for 6 h and cooled to 400°C at a rate of 10°C·min⁻¹. At 818°C (melting point of Cu₂CdSn₃S₈), the samples were additionally treated during 10 h and then annealed at 400°C for 250 h, followed by quenching in cold water without breaking the ampoules.

7.74 Tin–Copper–Mercury–Sulfur

Cu₂SnS₃–HgS. The phase diagram of this system, constructed through DTA, XRD, and metallography using the alloys annealed at 400°C during 250 h, is shown in Figure 7.7 (Olekseyuk et al. 2000a; Marchuk et al. 2021). Two eutectics exist in the system, which melt at 840°C (18 mol% HgS) and

at 762°C (88 mol% HgS). The solid solubility based on system components at the annealing temperature is less than 2 mol%. **Cu₂HgSnS₄** quaternary compound is formed in this system. The melting maximum of this quaternary phase is somewhat shifted from stoichiometric composition, and it takes place at 47–49 mol% HgS. It melts congruently at 849°C [at 845°C (Schäfer and Nitsche 1977)] and crystallizes in the tetragonal structure with the lattice parameters given in Table 7.3.

SnS₂–Cu₂HgSnS₄. The phase diagram of this system, constructed through DTA, XRD, and metallography using the alloys annealed at 400°C during 500 h, is given in Figure 7.8 (Marchuk et al. 2021). The eutectic contains 83 mol% SnS₂ and crystallizes at 748°C. The solid solubility based on system components is negligible.

Cu₂HgSnS₄–Cu₂S. The phase diagram of this system, constructed through DTA, XRD, and metallography using the alloys annealed at 400°C during 500 h, is a eutectic type (Marchuk et al. 2021). The eutectic contains 73 mol% Cu₂S and crystallizes at 787°C. The solid solubility in high-temperature modification of Cu₂S at 787°C does not exceed 18 mol%

TABLE 7.3

Crystallographic Data for Cu₂HgSnS₄

a, pm	*c*, pm	$d_{calc.}$	$d_{meas.}$	References
		g·cm⁻³		
556.8	1088	5.66	5.56	Hahn and Schulze (1965)
554.2 ± 0.3	1090.8 ± 0.7	5.59	5.48	Kaplunnik et al. (1977)[a]
557.5	1084.4	—	—	Schäfer and Nitsche (1977)
555.4 ± 0.3	1091.1 ± 0.8	5.27	5.59	Gruzdev et al. (1988); Jambor and Grew (1990)[a]
556.0 ± 0.5	1090.5 ± 1.0	5.537–5.450	5.45–5.59	Gruzdev et al. (1997); Jambor et al. (1998a)[a]
557.49 ± 0.06	1088.2 ± 0.1	—	—	Kabalov et al. (1998)
556.0	1090.5	—	—	Mandarino (1998)[a]
558.0 ± 0.2	1089.5 ± 0.3	—	—	Olekseyuk et al. (2000); Marchuk et al. (2021)

Note:

[a] for mineral velikite; according to Kaplunnik et al. (1977), the composition of this mineral is Cu₃.₇₅Hg₁.₇₅Sn₂S₈

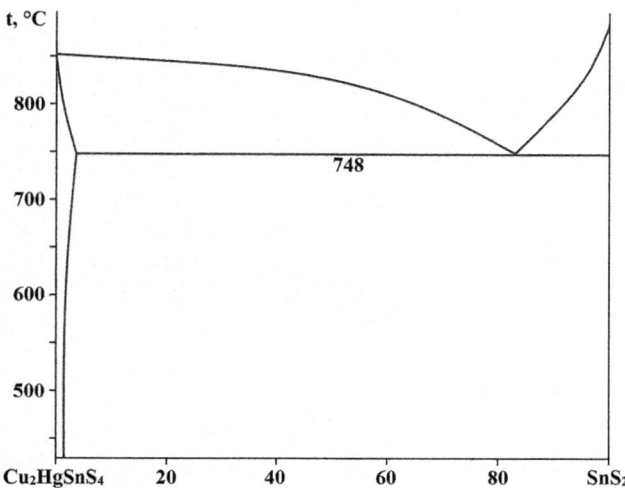

FIGURE 7.8 Phase diagram of the SnS_2–Cu_2HgSnS_4 system. (From Marchuk, O.V., et al., *J. Phase Equilib. Diffus.*, **42**(2), 245, 2021.)

Cu_2HgSnS_4 and decreases with decreasing temperature: at 400°C the solubility is less than 3 mol% Cu_2HgSnS_4. The solid solubility based on Cu_2HgSnS_4 is not higher than 2 mol% Cu_2S.

SnS_2–Cu_2S–HgS. Some vertical sections, the isothermal section at 400°C (Figure 7.9) and liquidus surface (Figure 7.10) of this quasiternary system were also constructed by Marchuk et al. (2021). Six three-phase regions exist in this system at 400°C, and Cu_2HgSnS_4 is in equilibrium with Cu_2S, HgS, SnS_2, as well as Cu_4SnS_4, Cu_2SnS_3, and $Cu_2Sn_4S_9$. The liquidus surface consists of 6 fields of the primary crystallization of Cu_2S, HgS, SnS_2, Cu_2SnS_3, Cu_4SnS_4, and Cu_2HgSnS_4. They are separated by 14 monovariant lines and 14 invariant points, of which eight correspond to binary and six to ternary invariant processes. There are four ternary eutectics, one ternary peritectic, and one transition point in the system: E_1 (742°C) – L

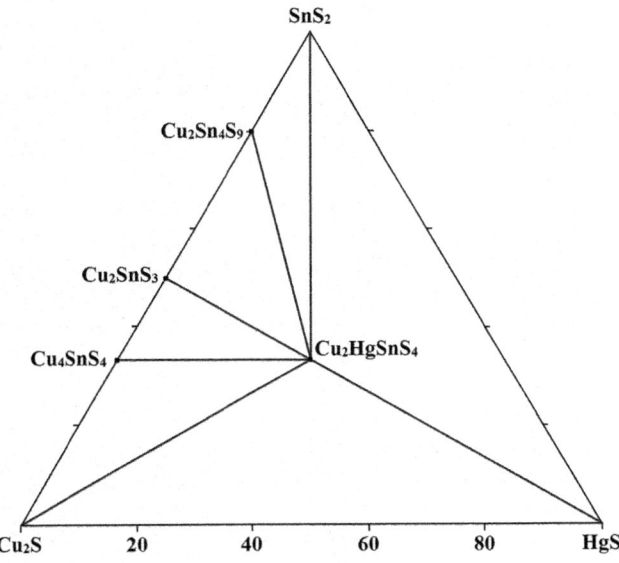

FIGURE 7.9 Isothermal section of the SnS_2–Cu_2S–HgS quasiternary system at 400°C. (From Marchuk, O.V., et al., *J. Phase Equilib. Diffus.*, **42**(2), 245, 2021.)

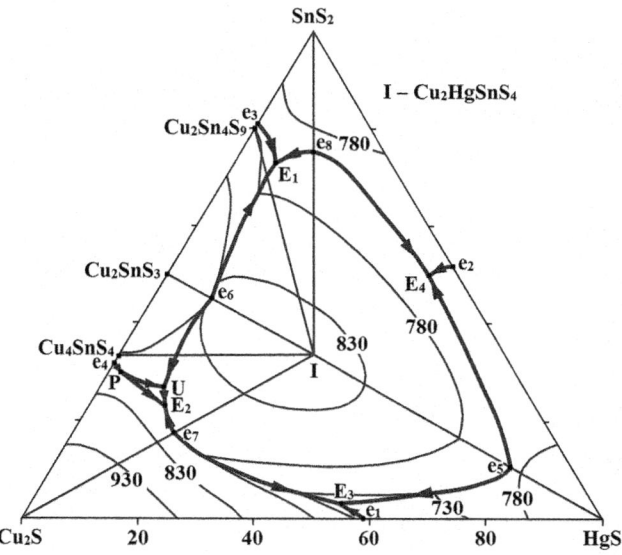

FIGURE 7.10 Liquidus surface of the SnS_2–Cu_2S–HgS quasiternary system. (From Marchuk, O.V., et al., *J. Phase Equilib. Diffus.*, **42**(2), 245, 2021.)

$\leftrightarrow Cu_2SnS_3 + SnS_2 + Cu_2HgSnS_4$; E_2 (772°C) – L $\leftrightarrow$ (γ-Cu_2S) + Cu_2HgSnS_4 + Cu_4SnS_4; E_3 (692°C) – L $\leftrightarrow$ (γ-Cu_2S) + Cu_2HgSnS_4 + (HgS); E_4 (692°C) – L $\leftrightarrow$ SnS_2 + Cu_2HgSnS_4 + (HgS); P (810°C) – L + Cu_2SnS_3 + (γ-Cu_2S) $\leftrightarrow$ Cu_4SnS_4; and U (777°C) – L + Cu_2SnS_3 $\leftrightarrow$ Cu_4SnS_4 + Cu_2HgSnS_4.

7.75 Tin–Copper–Aluminum–Sulfur

The **CuAlSnS₄** quaternary compound, which decomposes at 800°C forming $CuAlS_2$ and SnS_2 and has two polymorphic modifications, is formed in the Sn–Cu–Al–S system (Garbato et al. 1991). α-$CuAlSnS_4$ crystallizes in the cubic structure with the lattice parameter $a = 1022.25 \pm 0.03$ pm (Garbato et al. 1991) [$a = 1024$ pm and the calculated and experimental densities of 4.171 and 4.08 $g \cdot cm^{-3}$, respectively (Hahn and Strick 1967a; Strick et al. 1968); $a = 1021.3$ pm and the calculated and experimental densities of 4.27 and 4.08 $g \cdot cm^{-3}$ for $Cu_{0.82}Al_{0.82}Sn_{1.18}S_4$, respectively (Mähl et al. 1984b)]. β-$CuAlSnS_4$ crystallizes in the orthorhombic structure with the lattice parameters $a = 711 \pm 1$, $b = 745 \pm 1$, and $c = 1014.2 \pm 0.4$ pm (Garbato et al. 1991). Above 700°C, α-$CuAlSnS_4$ transforms into β-$CuAlSnS_4$. Polycrystalline samples of this compound were prepared by a direct reaction from the elements or binary compounds (Hahn and Strick 1967a; Strick et al. 1968; Garbato et al. 1991). Alumina crucibles were employed in order to minimize a chemical reaction with the holder. Quartz ampoules, sealed under vacuum (10^{-2} Pa), were slowly heated up to 1000°C, kept at this temperature for a few days, and quenched in water. Small single crystals of α-$CuAlSnS_4$ were obtained using a solid-vapor growth technique, but no crystals were grown for β-$CuAlSnS_4$ (Garbato et al. 1991). Single crystals of $Cu_{0.82}Al_{0.82}Sn_{1.18}S_4$ were grown by chemical transport reactions using $AlCl_3/I_2$ mixtures as transporting agent (Mähl et al. 1984b).

7.76 Tin–Copper–Gallium–Sulfur

The **CuGaSnS₄** quaternary compound, which decomposes at 750°C forming CuGaS₂ and SnS₂ and crystallizes in the orthorhombic structure with the lattice parameters $a = 1395.4 \pm 0.2$, $b = 888.6 \pm 0.3$, and $c = 647.6 \pm 0.3$ pm, is formed in the Sn–Cu–Ga–S system (Garbato et al. 1991). Polycrystalline samples of this compound were prepared in the same way as in the case of CuAlSnS₄ without using alumina crucibles. No crystals of CuGaSnS₄ were grown.

7.77 Tin–Copper–Indium–Sulfur

SnS₂–CuInS₂. The phase diagram of this system, constructed through DTA and XRD, is presented in Figure 7.11 (Shabunina and Aminov 1999). The eutectics contain 46 and 87 mol% SnS₂ and melt at 908°C and 718°C, respectively. Thermal effects at 945°C and 1040°C correspond to polymorphic transformations of CuInS₂. The solubility of SnS₂ in CuInS₂ at 908°C reaches 5 mol% [is not higher than 25 mol% (Pamplin et al. 1977)]. The ingots for the investigation were annealed at 700°C for 12 days, followed by the cooling to 500°C over 3–4 h and new annealing at this temperature for 7 days.

The **CuInSnS₄** quaternary compound, which melts congruently at 923°C (Shabunina and Aminov 1999) [at 916°C (Matsushita and Katsui 2005); according to the data of Garbato et al. (1991), this compound decomposes at 680°C forming CuInS₂ and SnS₂]. It crystallizes in the cubic structure with the lattice parameter $a = 1050.65 \pm 0.03$ pm (Garbato et al. 1991) [$a = 1050$ pm and the calculated and experimental densities of 4.906 and 4.86 g·cm⁻³, respectively (Hahn and Strick 1967a; Strick et al. 1968); $a = 1049.38 \pm 0.02$ pm (Ohachi and Pamplin 1977a); $a = 1048.7$ pm and the calculated and experimental densities of 4.80 and 4.70 g·cm⁻³ for Cu₀.₈₇₅InSnS₄, respectively (Mähl et al. 1982); $a = 1062$ pm (Matsushita and

Katsui 2005)]. Polycrystalline samples of this compound were prepared in the same way as in the case of CuAlSnS₄ without using alumina crucibles (Hahn and Strick 1967a; Strick et al. 1968; Garbato et al. 1991). CuInSnS₄ could also be prepared by the interaction of metal alloys with sulfur vapor at 680°C for some days (Ohachi and Pamplin 1977a). Small single crystals of CuInSnS₄ were obtained using a solid-vapor growth technique (Garbato et al. 1991; Ohachi and Pamplin 1977a). Single crystals of Cu₀.₈₇₅InSnS₄ were grown by chemical transport reactions using TeCl₄ or AlCl₃/I₂ mixtures as transporting agents (Mähl et al. 1984b).

7.78 Tin–Copper–Thallium–Sulfur

The **Cu₂Tl₂SnS₄** quaternary compound, which crystallizes in the orthorhombic structure with the lattice parameters $a = 537.94 \pm 0.11$, $b = 1126.8 \pm 0.2$, $c = 1341.3 \pm 0.3$ pm, and an energy gap of 1.4 eV, is formed in the Sn–Cu–Tl–S system (McGuire et al. 2005a). This compound was synthesized from the elements. Since Tl will tarnish slowly in the air, it was handled in an Ar glove box. Chunks of Tl were cut from the 0.5-inch diameter rod with wire cutters used only for this purpose, weighed inside the glove box, and placed in capped glass vials. Appropriate amounts of the other elements were then weighed out and loaded into a carbon-coated silica tube. The Ar-filled vials were removed from the glove box, and the Tl was added to the silica tube. The tube was then quickly (to limit the exposure of Tl to air) attached to a vacuum line for sealing. It was heated over 24 h to 800°C, held at this temperature for 172 h, and then cooled over 200 h to 300°C, at which point the furnace was turned off and allowed to cool to room temperature. Crystals were extracted from the resulting polycrystalline mass for examination. Cu₂Tl₂SnS₄ was made as a single-phase polycrystalline powder by reaction of the elements at 700°C for 2 days, annealing for 7 days at 300°C, and then grinding, pressing a pellet, and annealing at 400°C for 4 days.

7.79 Tin–Copper–Yttrium–Sulfur

SnS–Cu₂S–Y₂S₃. The isothermal section of this quasiternary system at 450°C was constructed by Shemet et al. (2006b). No quaternary compounds were found, and SnS is in the equilibria with YCuS₂ ternary compound and (Y₀.₆₆Cu₂S₂)ₓ(Y₀.₈₄Cu₁.₄₈S₂)₁₋ₓ solid solution. Therefore the section could be divided into 3 three-phase regions. The ingots for the investigations were annealed at 450°C for 240 h, followed by quenching in cold water.

SnS₂–Cu₂S–Y₂S₃. The isothermal section of this quasiternary system at 600°C is presented in Figure 7.12 (Shemet et al. 2006d). The formation of the **Cu₁₋₄ₓY₃Sn₁₊ₓS₇** ($0 \leq x \leq 0.09$) solid solution and the **Cu₀.₂₀Y₂Sn₀.₉₅S₅** quaternary compound was established.

The solid solution crystallizes in the hexagonal structure with the lattice parameters $a = 969.4 \pm 0.1$, $c = 616.8 \pm 0.1$ pm, and a calculated density of 4.374 g·cm⁻³ for $x = 0.09$ (Shemet et al. 2006d) [$a = 967$ and $c = 601$ pm for $x = 0$ (Collin and Flahaut 1972); $a = 967.66 \pm 0.01$, $c = 617.17 \pm 0.01$ pm, and

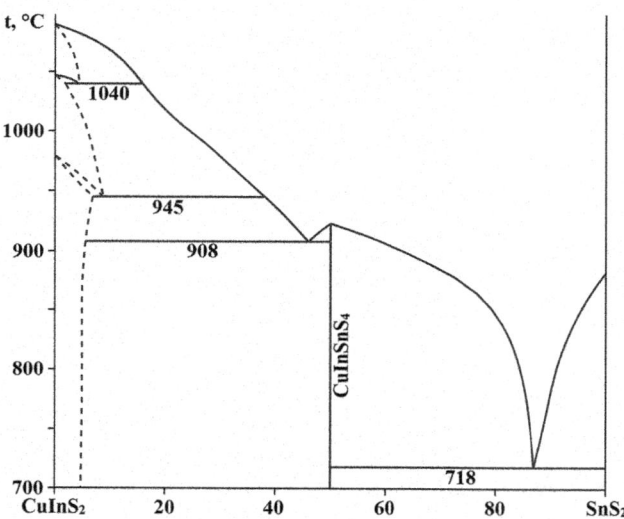

FIGURE 7.11 Phase diagram of the SnS₂–CuInS₂ system. (From Shabunina, G.G., and Aminov, T.G., *Zhurn. neorgan. khimii*, **44**(5), 859, 1999.)

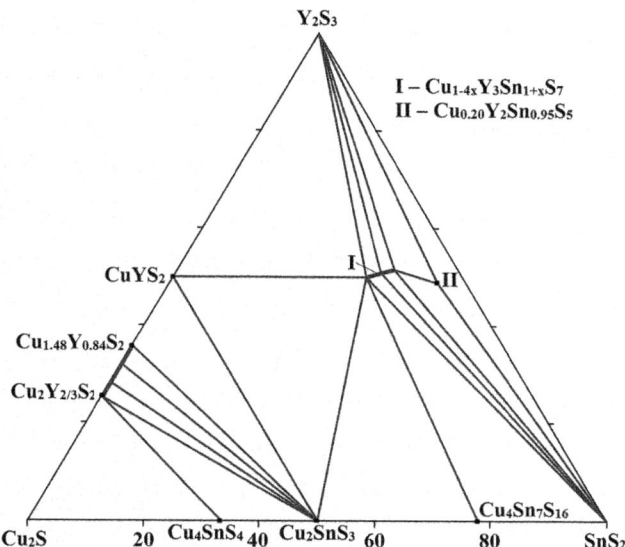

FIGURE 7.12 Isothermal section of the SnS_2–Cu_2S–Y_2S_3 quasiternary system at 600°C. (From Shemet, V.Ya., et al., *Pol. J. Chem.*, **80**(6), 943, 2006.)

a calculated density of 4.4680 g·cm^{-3} for $x = 0$ (Gulay et al. 2005f)]. This phase was prepared by the fusion of the elements in an evacuated quartz ampoule (Gulay et al. 2005f). The synthesis was realized in a shaft furnace with a heating rate of 20°C·h^{-1}. The ampoule with the sample was heated to a maximal temperature of 1150°C and kept at this temperature for 4 h. After that, it was cooled slowly to 600°C with a rate of 10°C·h^{-1} and annealed at 600°C for 240 h. After the annealing, the ampoule with the sample was quenched in cold water.

$Cu_{0.20}Y_2Sn_{0.95}S_5$ crystallizes in the orthorhombic structure with the lattice parameters $a = 1124.91 \pm 0.04$, $b = 769.94 \pm 0.03$, $c = 378.50 \pm 0.02$ pm, and a calculated density of 4.696 ± 0.003 g·cm^{-3} (Shemet et al. 2006d).

7.80 Tin–Copper–Scandium–Sulfur

The **$Cu_{0.54}Sc_{1.02}Sn_{1.1}S_4$** quaternary compound, which crystallizes in the cubic structure with the lattice parameter $a = 1041.76 \pm 0.06$ pm and a calculated density of 3.983 g·cm^{-3}, is formed in the Sn–Cu–Sc–S system (Shemet et al. 2006a). It was prepared by fusing high-purity elements in an evacuated silica ampoule. The synthesis was realized in a tube furnace with a heating rate of 20°C·h^{-1}. The ampoule with the sample was heated to 1250°C and kept at this temperature for 3 h. After that, it was cooled slowly to 600°C with a rate of 10°C·h^{-1} and annealed at this temperature for 240 h. After annealing, the ampoule and the sample were quenched in cold water.

7.81 Tin–Copper–Lanthanum–Sulfur

The **$CuLa_3SnS_7$** quaternary compound, which crystallizes in the hexagonal structure with the lattice parameters $a = 1031.7 \pm 0.1$, $c = 602.74 \pm 0.07$ pm, and a calculated density of 4.922 g·cm^{-3}

(Gulay et al. 2005e) [$a = 1027$, $c = 601$ pm (Collin and Flahaut 1972)], is formed in the Sn–Cu–La–S system. It was prepared by fusing high-purity elements in an evacuated silica ampoule in a tube furnace (Gulay et al. 2005e). The ampoule was heated to 1250°C with a rate of 30°C·h^{-1} and kept at this temperature for 4 h. After that, it was cooled slowly to 600°C with a rate of 10°C·h^{-1} and annealed at this temperature for 240 h. After annealing, the ampoule and the sample were quenched in cold water. The products were brown-colored compact buttons.

7.82 Tin–Copper–Cerium–Sulfur

The **$CuCe_3SnS_7$** quaternary compound, which crystallizes in the hexagonal structure with the lattice parameters $a = 1020.9 \pm 0.1$, $c = 601.50 \pm 0.07$ pm, and a calculated density of 5.059 g·cm^{-3} (Gulay et al. 2005e) [$a = 1017$, $c = 601$ pm (Collin and Flahaut 1972)], is formed in the Sn–Cu–Ce–S system. It was prepared in the same way as $CuLa_3SnS_7$ was synthesized.

7.83 Tin–Copper–Praseodymium–Sulfur

The **$CuPr_3SnS_7$** quaternary compound, which crystallizes in the hexagonal structure with the lattice parameters $a = 1014.17 \pm 0.02$, $c = 602.10 \pm 0.02$ pm, and a calculated density of 5.14 g·cm^{-3} (Gulay et al. 2005e) [$a = 1013$, $c = 601$ pm (Collin and Flahaut 1972)], is formed in the Sn–Cu–Pr–S system. It was prepared in the same way as $CuLa_3SnS_7$ was synthesized.

7.84 Tin–Copper–Neodymium–Sulfur

The **$CuNd_3SnS_7$** quaternary compound, which crystallizes in the hexagonal structure with the lattice parameters $a = 1007.5 \pm 0.1$, $c = 602.12 \pm 0.06$ pm, and a calculated density of 5.267 g·cm^{-3} (Gulay et al. 2005e) [$a = 1004$, $c = 601$ pm (Collin and Flahaut 1972)], is formed in the Sn–Cu–Nd–S system. It was prepared in the same way as $CuLa_3SnS_7$ was synthesized.

7.85 Tin–Copper–Samarium–Sulfur

The **$CuSm_3SnS_7$** quaternary compound, which crystallizes in the hexagonal structure with the lattice parameters $a = 992.88 \pm 0.02$, $c = 608.00 \pm 0.06$ pm, and a calculated density of 5.49 g·cm^{-3} (Gulay et al. 2005e) [$a = 989$, $c = 601$ pm (Collin and Flahaut 1972)], is formed in the Sn–Cu–Sm–S system. It was prepared in the same way as $CuLa_3SnS_7$ was synthesized.

7.86 Tin–Copper–Europium–Sulfur

The **Cu_2EuSnS_4** quaternary compound melting at approximately 762°C (Aitken et al. 2009) and, apparently, having two polymorphic modifications, is formed in the Sn–Cu–Eu–S system. α-Cu₂EuSnS₄ crystallizes in the trigonal structure with the lattice parameters $a = 624.9 \pm 0.8$, $c = 1548 \pm 7$ pm, and an energy gap of 2.2 eV (Llanos et al. 2003). β-Cu₂EuSnS₄ crystallizes in the orthorhombic structure with the lattice parameters $a = 1047.93 \pm 0.01$, $b = 1036.10 \pm 0.02$, $c = 640.15 \pm 0.01$ pm, a calculated density of 5.026 g·cm⁻³, and an energy gap of 1.85 eV (Aitken et al. 2009).

Polycrystalline α-Cu₂EuSnS₄ was synthesized from a thoroughly ground mixture of EuS, Cu, Sn, and S in stoichiometric proportions (Llanos et al. 2003). The mixture was placed in an alumina boat and heated in a CS₂/Ar stream at 450°C for about 36 h, with two intermittent grindings. The title compound was obtained, reddish-brown in color and stable in the air.

β-Cu₂EuSnS₄ was synthesized from a stoichiometric mixture of the elements (Aitken et al. 2009). The starting materials were ground for 15 min in an Ar-filled glove box and placed into a graphite crucible. The crucible was then sealed under a vacuum of 0.1 Pa in a fused silica tube. The tube was placed into a furnace and heated to 700°C in 12 h. The sample was held at this temperature for 125 h and then quenched in an ice-water bath. The tube was opened and the sample removed from the crucible. The dark-red microcrystalline air-stable powder obtained. Single crystals of this modification were obtained using the same procedure as that described above, with the exception of the heating profile. This reaction was heated to 800°C in 12 h, held at 800°C for 125 h, slow cooled to 500°C for 50 h, and then rapidly cooled to room temperature. The product consisted of dark-red block-like crystals.

7.87 Tin–Copper–Gadolinium–Sulfur

The **CuGd₃SnS₇** quaternary compound, which crystallizes in the hexagonal structure with the lattice parameters $a = 980.87 \pm 0.03$, $c = 615.11 \pm 0.03$ pm, and a calculated density of 5.69 g·cm⁻³ (Gulay et al. 2005e) [$a = 980$, $c = 601$ pm (Collin and Flahaut 1972)], is formed in the Sn–Cu–Gd–S system. It was prepared in the same way as CuLa₃SnS₇ was synthesized.

7.88 Tin–Copper–Terbium–Sulfur

The **CuTb₃SnS₇** quaternary compound, which crystallizes in the hexagonal structure with the lattice parameters $a = 975.12 \pm 0.02$, $c = 615.73 \pm 0.02$ pm, and a calculated density of 5.79 g·cm⁻³ (Gulay et al. 2005e), is formed in the Sn–Cu–Tb–S system. It was prepared in the same way as CuLa₃SnS₇ was synthesized.

7.89 Tin–Copper–Dysprosium–Sulfur

SnS₂–Cu₂S–Dy₂S₃. The isothermal section of this quasiternary system at 600°C is presented in Figure 7.13 (Gulay and Shemet 2012). The **CuDy₃SnS₇** quaternary compound was found, and $(CuDy_{2,0.66}S_2)_x(Cu_{1.45}Dy_{0.85}S_2)_{1-x}$ solid solution existed in the CuDy₂S–Dy₂S₃ system. The ingots for the investigations were annealed at 600°C for 240 h.

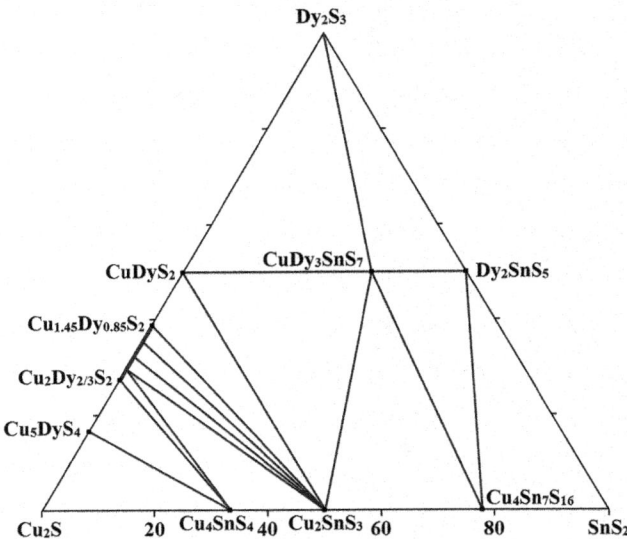

FIGURE 7.13 Isothermal section of the SnS₂–Cu₂S–Dy₂S₃ quasiternary system at 600°C. (From Gulay, L.D., and Shemet, V.Ya., *Nauk. Visnyk Volyns'k. Nats. Univ. im. Lesi Ukrainky. Ser. Khim. nauky*, (17), 69, 2012.)

CuDy₃SnS₇ quaternary compound crystallizes in the hexagonal structure with the lattice parameters $a = 970.24 \pm 0.02$, $c = 616.94 \pm 0.02$ pm, and a calculated density of 5.90 g·cm⁻³ (Gulay et al. 2005e) [$a = 968$, $c = 601$ pm (Collin and Flahaut 1972)]. It was prepared in the same way as CuLa₃SnS₇ was synthesized.

7.90 Tin–Copper–Holmium–Sulfur

The **CuHo₃SnS₇** quaternary compound, which crystallizes in the hexagonal structure with the lattice parameters $a = 965.27 \pm 0.02$, $c = 617.31 \pm 0.02$ pm, and a calculated density of 6.01 g·cm⁻³ (Gulay et al. 2005e), is formed in the Sn–Cu–Ho–S system. It was prepared in the same way as CuLaSnS₇ was synthesized.

7.91 Tin–Copper–Erbium–Sulfur

The **CuEr₃SnS₇** quaternary compound, which crystallizes in the hexagonal structure with the lattice parameters $a = 960$ and $c = 601$ pm (Collin and Flahaut 1972), is formed in the Sn–Cu–Er–S system. This compound was prepared by prolonged heating a stoichiometric mixture of Er₂S₃, Cu, SnS, and S at 800–850°C in a quartz ampoule sealed under vacuum.

7.92 Tin–Copper–Lead–Sulfur

SnS₂–Cu₂S–PbS. The isothermal section of this quasiternary system at 400°C is shown in Figure 7.14 (Kogut et al. 2006). No ternary compounds were found in the system. The ingots for the investigations were annealed at 400°C for 250 h.

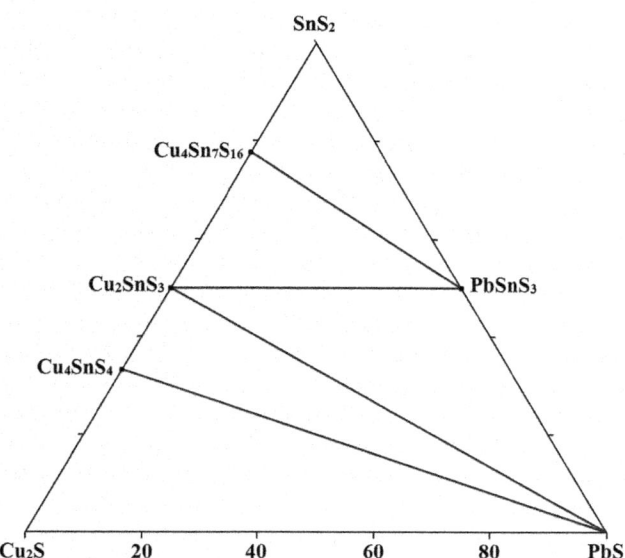

FIGURE 7.14 Isothermal section of the SnS$_2$–Cu$_2$S–PbS quasiternary system at 400°C. (From Kogut, Yu.M., et al., *Nauk. Visnyk Volyns'k. Derzh. Univ. im. Lesi Ukrainky. Ser. Khim. nauky*, (4), 63, 2006.)

7.93 Tin–Copper–Phosphorus–Sulfur

SnS$_2$–Cu$_2$S–P$_2$S$_5$. There are two glass-forming regions in the SnS$_2$–Cu$_2$S–P$_2$S$_5$ quasiternary system; the first is in the region of 5–15 mol% P$_2$S$_5$, which can include up to 5 mol% Cu$_2$S, and the second one is in the region of 35–65 mol% P$_2$S$_5$ with the maximum content of 15 mol% Cu$_2$S (Bereznyuk and Petrus' 2020).

7.94 Tin–Copper–Antimony–Sulfur

The **Cu$_{12-x}$Sn$_x$Sb$_4$S$_{13}$** ($x = 0.75-0.8$) solid solution, which crystallizes in the cubic structure, is formed in the Sn–Cu–Sb–Te quaternary system (Nasonova et al. 2019). It was prepared by a standard ampoule technique using the mixtures of Sn, Sb, CuO, and S, which were sealed in evacuated quartz ampoules under a vacuum of 0.2 Pa. The ampoules were heated to 700°C, annealed at this temperature for 3 h, slowly cooled down to 550°C in 30 h, and finally cooled down to room temperature in a switched-off furnace. The reaction products were finely ground and pressed into pellets at a pressure of 8–10 mPa at room temperature. The pellets were sealed in evacuated quartz ampoules and annealed at 450°C for 25 h, followed by switching off the furnace and cooling down to room temperature. The next lattice parameters of obtained solid solutions were obtained: $a = 1034.74 \pm 0.01, 1036.472 \pm 0.005, 1039.041 \pm 0.002, 1039.140 \pm 0.002$ pm and a calculated density of 5.0364, 5.0524, 5.0558, 5.0544 g·cm^{-3} for $x = 0.25, 0.5, 0.75,$ and 1.0, respectively.

7.95 Tin–Copper–Bismuth–Sulfur

SnS–Cu$_2$S–Bi$_2$S$_3$. The isothermal section of this quasiternary system at 400°C was constructed by Olekseyuk and Zhbankov (2005). It characterized by the equilibria between SnS and

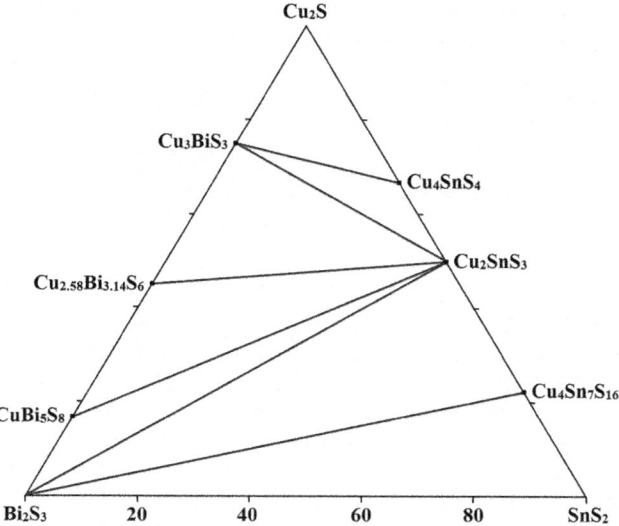

FIGURE 7.15 Isothermal section of the SnS$_2$–Cu$_2$S–Bi$_2$S$_3$ quasiternary system at 400°C. (From Olekseyuk, I., and Zhbankov, O., *Visnyk L'viv. Univ. Ser. khim.*, (46), 53, 2005.)

CuBi$_5$S$_6$, Cu$_{2.58}$Bi$_{3.14}$S$_6$ and Cu$_3$BiS$_3$ ternary compounds. No quaternary significant solid solutions were found.

SnS$_2$–Cu$_2$S–Bi$_2$S$_3$. The isothermal section of this quasiternary system at 400°C is presented in Figure 7.15 (Olekseyuk and Zhbankov 2005). Seven three-phase regions divided by six quasibinary equilibria exist in this system. No quaternary compounds and significant solid solutions were found. The alloys for the investigations of both isothermal sections were annealed at 400°C for 300 h with the next quenching in cold water.

7.96 Tin–Copper–Vanadium–Sulfur

The **Cu$_{26}$V$_2$Sn$_6$S$_{32}$** quaternary compound, which crystallizes in the cubic structure, is formed in the Sn–Cu–V–S system. To prepare it, Cu, V, Sn, and S were mixed in the stoichiometric ratio (Kikuchi et al. 2016). The mixture of approximately 3 g was loaded into a fused quartz tube, which was evacuated to a pressure of ~5 × 10^{-3} Pa and then flame-sealed. The mixture was heated to 1050°C in 60 h, held at 1050°C for 12 h, cooled to 400°C in 19 h, and then to room temperature over 5 h. The obtained ingot was hand-ground into powders, which were placed into graphite die for hot-pressing. The graphite die was inserted into the hot-press furnace. Sintering was performed at 750°C for 1 h under 70 MPa uniaxial pressures in an Ar gas flow atmosphere. The heating and cooling rates were 10°C·min^{-1} and 20°C·min^{-1}, respectively. The ingots melted in evacuated and sealed quartz tubes were found to be composed of two phases with slightly different lattice parameters. The successive hot-pressing converted the two-phase samples to single-phase ones of high density. The lattice parameter a for as-prepared ingot are 1079.8 and 1075.9 pm, for sintered compact is 1081.2 pm and the calculated and experimental densities of 4.59 and 4.73 g·cm^{-3}, respectively (Kikuchi et al. 2016) [$a = 1081.1 \pm 0.5$ and 1084.4 ± 0.5 pm (Suekuni et al. 2014)].

7.97 Tin–Copper–Niobium–Sulfur

The **Cu$_{26}$Nb$_2$Sn$_6$S$_{32}$** quaternary compound, which crystallizes in the cubic structure, is formed in the Sn–Cu–Nb–S system. It was prepared in the same way as Cu$_{26}$V$_2$Sn$_6$S$_{32}$ was obtained using Nb instead of V (Kikuchi et al. 2016; Bouyrie et al. 2017). The ingots melted in evacuated and sealed quartz tubes were found to be composed of two phases with slightly different lattice parameters. The successive hot-pressing converted the two-phase samples to single-phase ones of high density. The lattice parameter *a* for as-prepared ingot are 1085.2 and 1081.0 pm, for sintered compact is 1087.7 pm and the calculated and experimental densities of 4.61 and 4.80 g·cm^{-3}, respectively (Kikuchi et al. 2016) [*a* = 1083.2 for as-prepared ingot and 1087.2 pm for sintered compact (Bouyrie et al. 2017)].

7.98 Tin–Copper–Tantalum–Sulfur

The **Cu$_{26}$Ta$_2$Sn$_6$S$_{32}$** quaternary compound, which crystallizes in the cubic structure, is formed in the Sn–Cu–Ta–S system. It was prepared in the same way as Cu$_{26}$V$_2$Sn$_6$S$_{32}$ was obtained using Ta instead of V (Kikuchi et al. 2016; Bouyrie et al. 2017). The ingots melted in evacuated and sealed quartz tubes were found to be composed of two phases with slightly different lattice parameters. The successive hot-pressing converted the two-phase samples to single-phase ones of high density. The lattice parameter *a* for as-prepared ingot are 1086.5 and 1082.7 pm, for sintered compact is 1088.5 pm and the calculated and experimental densities of 4.83 and 5.03 g·cm^{-3}, respectively (Kikuchi et al. 2016) [*a* = 1082.6 for as-prepared ingot and 1089.5 pm for sintered compact (Bouyrie et al. 2017)].

7.99 Tin–Copper–Selenium–Sulfur

The **Cu$_2$Sn(S$_{1-x}$Se$_x$)$_3$** ($0 \leq x \leq 1.0$) and **Cu$_4$Sn(S$_{1-x}$Se$_x$)$_4$** ($x < 0.30 \pm 0.01$) solid solutions are formed in the Sn–Cu–Se–S quaternary. To prepare Cu$_2$Sn(S$_{1-x}$Se$_x$)$_3$, elemental powders of Cu, Sn, S, and Se were weighed to give a molar ratio of Cu/Sn/S/Se = 2:1:3(1-*x*):3*x* and put in a zirconia grinding pot with zirconia balls (Nomura et al. 2013a, b). The ball-to-powder weight ratio was maintained at 2:1. Milling was conducted in a planetary ball mill at 800 rpm for 20 min in a N$_2$ gas atmosphere. The solid solutions could not be synthesized only by ball milling. Therefore, the mixed powders were post-annealed at 600°C for 30 min in a N$_2$ gas atmosphere. The refined lattice parameters *a*, *b*, and *c* of these solid solutions monotonically increased with increasing Se content, and band gap decreased from 0.87 eV (*x* = 0.0) to 0.67 eV (*x* = 0.60) with increasing *x*.

The Cu$_4$Sn(S$_{1-x}$Se$_x$)$_4$ solid solutions also crystallize in the orthorhombic structure and the lattice parameters *a*, *b*, and *c* also increased with *x* (Anzai et al. 1990).

7.100 Tin–Copper–Chromium–Sulfur

Two quaternary compounds, **CuSnCrS$_4$** and **Cu$_4$Sn$_2$Cr$_6$S$_{15}$**, are formed in the Sn–Cu–Cr–S system. CuSnCrS$_4$ crystallizes in the cubic structure with the lattice parameter *a* = 1017.4 ± 0.6

pm and the calculated and experimental densities of 4.58 and 4.49 g·cm^{-3}, respectively, for Cu$_{0.875}$SnCrS$_4$ composition (Mähl et al. 1982) [*a* = 1016 pm and the calculated and experimental densities of 4.587 and 4.51 g·cm^{-3}, respectively (Hahn and Strick 1967a; Strick et al. 1968); *a* = 1017.5 pm and an energy gap of 0.9 eV (Riedel and Morlock 1978); *a* = 1010.1 ± 0.2 pm for Cu$_{1.10}$Sn$_{0.70}$Cr$_{1.30}$S$_{3.90}$ composition (Danot et al. 1985)].

Polycrystalline samples of this compound were prepared by a direct reaction from the elements or binary compounds by heating at 850°C for 4 days (Hahn and Strick 1967a; Strick et al. 1968). Single crystals of Cu$_{0.875}$InSnS$_4$ were grown by chemical transport reactions using TeCl$_4$ or AlCl$_3$/I$_2$ mixtures as transporting agents (Mähl et al. 1982).

Cu$_4$Sn$_2$Cr$_6$S$_{15}$ decomposes at temperatures of more than 1100°C (Babitsyna and Novotortsev 1986). It was synthesized from the elements at 960–980°C for 100 h, with the next annealing at 900°C for 300 h.

7.101 Tin–Copper–Molybdenum–Sulfur

The **Cu$_6$MoSnS$_8$** quaternary compound (mineral hemusite), which crystallizes in the cubic structure with the lattice parameter *a* = 1080.6 pm and a calculated density of 4.55 g·cm^{-3} (Shimizu et al. 1988) [*a* = 1082 pm (Terziev 1971)], is formed in the Sn–Cu–Mo–S system.

7.102 Tin–Copper–Tungsten–Sulfur

The **Cu$_6$WSnS$_8$** quaternary compound, which crystallizes in the cubic structure with the lattice parameter *a* = 1082.99 ± 0.07 pm and the calculated and experimental densities of 4.917 and 3.996 g·cm^{-3}, respectively (Zhou and Dong 2019) [*a* = 1085.6 ± 0.2 pm and a calculated density of 4.880 g·cm^{-3} for mineral kiddcreekite (Harris et al. 1984; Dunn et al. 1985b); *a* = 1081.78 ± 0.03 pm and a calculated density of 4.934 g·cm^{-3} for mineral kiddcreekite (Wenyuan et al. 2014)], is formed in the Sn–Cu–W–S system.

The polycrystalline sample of this compound was synthesized by solid-state reaction (Zhou and Dong 2019). Stoichiometric amounts of Cu, W, Sn, and S powders were thoroughly mixed, then pressed into pellets, and sealed in an evacuated quartz tube. The first heating cycle was performed at 600°C for 3 days. The sample was furnace-cooled to room temperature and then ground, and pressed into pellets which were resealed in another evacuated quartz tube. The sample was annealed at 600°C for 1 week, and finally furnace-cooled to room temperature.

7.103 Tin–Copper–Manganese–Sulfur

Four quaternary compounds are formed in the Sn–Cu–Mn–S system. **Cu$_{0.5}$Mn$_{0.5}$[MnSn]S$_4$** and **Cu$_{0.5}$Mn$_{0.5}$[Mn$_{0.75}$Sn$_{1.25}$]S$_4$** crystallize in the cubic structure with the lattice parameter *a* = 1044 and 1043 pm, respectively (Harada 1973). They were prepared with the usual ceramic techniques. The appropriate stoichiometric mixtures of Cu, Mn, SnS$_2$, and S were fired using the following two-step process. First, the mixed powder was heated at 250–700°C for three days in either an evacuated Pyrex or quartz tube, and the prefired polysulfides thus obtained were ground and pressed into a pellet. These materials were then refired in an evacuated quartz tube for about a week at 750°C and 800°C for the first and the second compound, respectively. When necessary, the refiring procedure was repeated until single-phase products were obtained.

Cu_2MnSnS_4 melts incongruently at 904°C (Quintero et al. 2014b) [melts congruently at 909°C (Schäfer and Nitsche 1977)] and crystallizes in the tetragonal structure with the lattice parameters given in Table 7.4. This compound was produced by the melt and anneal technique (Lamarche et al. 1991; Bernert and Pfitzner 2005; Sachanyuk et al. 2006b; Quintero et al. 2014a,b, 2017). In a typical synthesis of Cu_2MnSnS_4, the constituents elements of 1-g sample were sealed under vacuum ($\approx 10^{-3}$ Pa) in a small quartz ampoule, and then the components were heated up to 200°C and kept for about 1–2 h (Quintero et al. 2017). Then the temperature was raised to 500°C using a rate of 40°C·h^{-1}, and held at this temperature for 14 h. After that, the sample was heated from 500°C to 800°C at a rate of 30°C·h^{-1} and kept at this temperature for another 14 h. Then it was raised to 1150°C at 60°C·h^{-1}, and the components were melted together at this temperature. The furnace temperature was brought slowly (4°C·h^{-1}) down to 600°C, and the sample was annealed at this temperature for 1 month. Then, the sample was slowly cooled to room temperature using a rate of about 2°C·h^{-1}. Single crystals of this compound were obtained by the chemical vapor reactions using iodine as a transport agent (Allemand and Winterberger 1970; Guen et al. 1979; Guen and Glaunsinger 1980).

$Cu_2MnSn_3S_8$ crystallizes in the cubic structure with the lattice parameter $a = 1041.6 \pm 0.7$ pm (Lavela et al. 1996) [$a = 1041.45 \pm 0.07$ pm and a calculated density of 4.611 g·cm^{-3} for $Cu_{7.38\pm0.11}Mn_4Sn_{12}S_{32}$ composition (Garg et al. 2002)]. It was synthesized as follows (Lavela et al. 1996; Garg et al. 2002). Stoichiometric amounts of the chemical elements were mixed and sealed in an evacuated (10^{-3} Pa) quartz tube. The mixture was heated at 250–300°C for 24 h, after which the temperature was increased to achieve 680–750°C and annealed for 8 days. After that, the mixture was cooled by switching off the furnace. After the reaction, black cuboidal crystals of the title compound were found as a major phase.

7.104 Tin–Copper–Iron–Sulfur

Some quaternary compounds are formed in the Sn–Cu–Fe–S system. $Cu_{0.5}Fe_{0.5}[FeSn]S_4$ and $Cu_{0.5}Fe_{0.5}[Fe_{0.75}Sn_{1.25}]S_4$ crystallize in the cubic structure with the lattice parameter $a = 1033$ and 1035 pm, respectively (Harada 1973). They were prepared in the same way as $Cu_{0.5}Mn_{0.5}[MnSn]S_4$ and $Cu_{0.5}Mn_{0.5}[Mn_{0.75}Sn_{1.25}]S_4$ were synthesized using Fe instead of Mn.

Cu_2FeSnS_4 melts congruently at 855°C (Ohtsuki et al. 1980) [at 864°C and has polymorphic transformation at 706°C (Moh 1975); at 860°C (Schäfer and Nitsche 1977); melts incongruently at 849–880°C and has polymorphic transformation at 709°C (Nenasheva and Kalinina 1986)]. According to the data of Ohtsuki et al. (1980), a low-temperature cubic form of this compound is stable in the temperature range from 350°C to 565°C at least, and sluggishly inverts to a high-temperature tetragonal form. The high-temperature tetragonal form also transforms to a high-temperature cubic form at 710°C. The lattice parameters for the tetragonal modification are given in Table 7.5.

The second modification of Cu_2FeSnS_4 crystallizes in the cubic structure with the lattice parameter $a = 1083.8 \pm 0.1$ pm (Ohtsuki et al. 1980) [$a = 547.4 \pm 0.5$ pm at 800°C (Moh 1975); $a = 541.79 \pm 0.05$ pm for $Cu_{1.86}Fe_{0.80}Sn_{0.99}S_4$ composition (Evstigneeva and Kabalov 2001)]. According to the data of Rincón et al. (2011), this compound could crystallize in the trigonal structure with the lattice parameter $a = 543.29 \pm 0.03$, $c = 541.04 \pm 0.02$ pm, and a calculated density of 4.47 g·cm^{-3}.

To prepare the tetragonal form of this compound, the stoichiometric proportions of the pure elements were heated at 800°C for 20 days in a sealed, evacuated silica ampoule (Kissin and Owens 1979) or at 1050–1080°C for 18–20 h (Eibschütz et al. 1967). The ampoule was then quenched in ice water, and the reaction products were ground in a mortar (Kissin and Owens 1979). The products of the initial reaction were then placed in a vertical furnace in a sealed, evacuated silica tube with a sufficient flux of NaCl and KCl (molar ratio 1:1) to completely immerse them and heated to 800°C for 4 days. This modification was also prepared by heating a stoichiometric mixture of Fe, Cu_2S, SnS, and S (Nenasheva and Kalinina 1986). Its crystals were obtained by solid-state reaction of SnS

TABLE 7.4

Crystallographic Data for Cu_2MnSnS_4

a, pm	*c*, pm	$d_{calc.}$	$d_{meas.}$	References
		g·cm^{-3}		
549	1072	4.41	4.38	Allemand and Winterberger (1970); Guen et al. (1979); Guen and Glaunsinger (1980)
551.8	1082.1	—	—	Schäfer and Nitsche (1977)
551.1	1082.0	—	—	Lamarche et al. (1991)
554.8 ± 0.1	1084.4 ± 0.2	4.268	—	Bernert and Pfitzner (2005)
550.63 ± 0.02	1081.78 ± 0.05	4.3430	—	Sachanyuk et al. (2006b)
550.9 ± 0.2	1079.6 ± 0.1	—	—	López-Vergara et al. (2014)
552.02 ± 0.02	1081.24 ± 0.04	4.32	—	Quintero et al. (2014a)
551.80	1080.70	4.33	—	Quintero et al. (2014b, 2017)

TABLE 7.5

Crystallographic Data for the Tetragonal Modification of Cu$_2$FeSnS$_4$

		$d_{calc.}$	$d_{meas.}$	
		gcm^{-3}		
a, pm	c, pm			References
546	1072.5	4.44	—	Brockwey (1934)[a]
546.6 ± 0.5	1076 ± 1	—	—	Eibschütz et al. (1967)
544	1074	—	—	Nitsche et al. (1967)
545.4 ± 0.3	1072.0 ± 0.4	—	—	Moh (1975)
544.3	1072.6	—	—	Schäfer and Nitsche (1977)
544.9 ± 0.2	1075.7 ± 0.3	—	—	Hall et al. (1978)[a]
544.32 ± 0.11	1072.99 ± 0.51	—	—	Kissin and Owens (1979)
545.4 ± 0.1	1073.5 ± 0.3	—	—	Ohtsuki et al. (1980)
538.59	1077.18	—	—	Ohtsuki et al. (1981)[b]
543.3 ± 3.6	1088.4 ± 8.9	4.490	—	Kissin and Owens (1989); Miyawaki et al. (2019c,d)
541.4 ± 0.3	541.4 ± 0.3	5.416	—	Llanos et al. (2000)
543.4 ± 0.4	1073.4 ± 0.1	—	—	López-Vergara et al. (2014)
543.29	541.04	4.47	—	Quintero et al. (2014b)

Note:

[a] for mineral stannite

[b] for Cu$_{3.2}$Fe$_{0.9}$Sn$_{1.0}$S$_{4.9}$ composition, which is stable up to 875°C

and CuFeS$_2$ (Llanos et al. 2000). The mixture was heated in a tightly sealed graphite crucible at 1050°C in a vertical furnace for 24 h under and Ar atmosphere. The sample was then allowed to cool to room temperature over a period of 50 h.

The synthesis of cubic modification of Cu$_2$FeSnS$_4$ was also performed in a quartz tube (Evstigneeva and Kabalov 2001). Weighed sample of crushed and thoroughly mixed chemical elements (the total mass was 1 g) were placed into the tube, evacuated to 10^{-4} Pa, and sealed. Then, the tube was first heated to 900°C (at a step of 200°C and annealed for 0.5 h at each step), kept at this temperature for 6 h, and finally cooled to 400°C and annealed at this temperature for 10 days. Upon annealing, the tube with the sample was cooled in the furnace to ~80°C, taken away from the furnace, and opened. The products obtained were ground, and the synthesis procedure was repeated with a week-long annealing at 400°C.

Trigonal modification of Cu$_2$FeSnS$_4$ was grown from the melt (Rincón et al. 2011). The components of the 1.0-g sample, sealed under vacuum in a quartz ampoule, were heated up to 200°C and kept at this temperature for 2 h. The temperature was then increased to 500°C using a rate of 40°C·h^{-1} and held this temperature for 14 h. After that, the sample was heated to 800°C at a rate of 30°C·h^{-1} and kept at this temperature for another 14 h. Then, they were raised at 60°C·h^{-1} to 1300°C and maintained at this temperature for 2 h. The furnace temperature was brought slowly (4°C·h^{-1}) down to 600°C, and the sample was annealed for 1 month. Finally, it was slowly cooled to room temperature at a rate of about 2°C·h^{-1}.

Cu$_2$FeSn$_3$S$_8$ (mineral rhodostannite) is stable below 715°C, at which temperature it inverts to an unquenchable high-temperature form. It, in turn, melts congruently at 850°C (Ohtsuki et al. 1980) [at 831°C (Quintero et al. 2014b)]. This compound crystallizes in the tetragonal structure with the lattice parameters a = 730.92 ± 0.03 and c = 1033.9 ± 0.2 pm (Ohtsuki et al.

1980) [a = 729 and c = 1031 pm (Wang 1975); a = 730.5 ± 0.2, c = 1033.0 ± 0.5 pm, and a calculated density of 4.79 gcm^{-3} (Jumas et al. 1979); in the hexagonal structure with the lattice parameters a = 727, c = 1807 pm, and a calculated density of 4.79 gcm^{-3} (Springer 1968); Fleischer 1969)]. This compound was prepared by the reaction of a stoichiometric mixture of FeS, SnS$_2$, Cu, and S in the solid state at 700°C in an evacuated and sealed ampoule (Jumas et al. 1979). The microcrystalline powder of Cu$_2$FeSn$_3$S$_8$ was found in the ampoule. Single crystals of the title compound were grown through the chemical transport reactions using I$_2$ as a transport agent.

According to the data of Lavela et al. (1996), Cu$_2$FeSn$_3$S$_8$ crystallizes in the cubic structure with the lattice parameter a = 1031.2 ± 0.5 pm. It was synthesized as follows. Stoichiometric amounts of the constituent elements were mixed and sealed in an evacuated quartz tube. The mixture was heated at 300°C for 24 h, after which the temperature was increased to achieve a final constant value of 750°C for 8 days. The product obtained was reground and kept under an Ar atmosphere in a dry box. In order to extract Cu$^+$ ions from the structure, Cu$_2$FeSn$_3$S$_8$ was suspended in an excess of 0.1 M solution of I$_2$ in CH$_3$CN at 50°C with magnetic stirring for different periods of time (1, 2, and 3 weeks) and **Cu$_{1.8}$FeSn$_3$S$_8$** quaternary phase was obtained.

The existence of two more quaternary phases, **Cu$_{4.06}$Fe$_{0.63}$Sn$_{0.31}$S$_{5.00}$** and **Cu$_{4.07}$Fe$_{0.38}$Sn$_{0.55}$S$_{5.00}$**, was determined by Ohtsuki et al. (1981). These phases decompose at 525°C and 528°C, respectively, and the first phase crystallizes in the trigonal structure with the lattice parameters a = 378.98 and c = 4305.0 pm. These phases, as well as **Cu$_{3.2}$Fe$_{0.9}$Sn$_{1.0}$S$_{4.9}$**, were synthesized by the interaction of the elements at temperatures of up to 600°C, or by hydrothermal method at temperatures of up to 400°C and pressures up to 78.5 MPa, or by the interaction of binary sulfides at the temperatures up to 500°C following by the growth of single crystals using KCl/LiCl flux.

$Cu_{5.47}Fe_{2.9}Sn_{13.1}S_{32}$ crystallizes in the cubic structure with the lattice parameter $a = 1033.62 \pm 0.02$ pm (Garg et al. 2001b). To prepare this compound, stoichiometric amounts of Cu, Fe, Sn, and S were loaded in the composition $Cu_4Fe_4Sn_{12}S_{32}$ in a N_2-filled glove box. This was sealed in a silica ampoule under high vacuum (10^{-3} Pa) and heated at 250°C for 1 day followed by annealing at 680°C for 8 days and then cooled by switching off the furnace. After the reaction, two phases was observed: the major phase was in the form of black cuboidal crystals (the title compound) and the minor phase of orange-colored crystals.

$Cu_6Fe_2SnS_8$ (mineral mawsonite) is stable at temperatures below 715°C, at which temperature it decomposes, and crystallizes in the tetragonal structure with the lattice parameters $a = 759.7 \pm 0.1$ and $c = 535.4 \pm 0.1$ pm (Ohtsuki et al. 1980) [$a = 760.3 \pm 0.2$ and $c = 535.8 \pm 0.1$ pm (Szymański 1976; Kissin and Owens 1979); $a = 1074.5 \pm 0.1$ and $c = 1071.1 \pm 0.6$ pm (Yamanaka and Kato 1976); in the cubic structure with the lattice parameter $a = 1074 \pm 1$ (Markham and Lawrence 1965); $a = 761 \pm 1$ and $c = 537.3 \pm 0.5$ pm for mineral chatkalite of the same composition (Kovalenker et al. 1981; Fleischer et al. 1982a)].

Cu_2FeSnS_4–$CuFeS_2$. This system is quasibinary with a peritectic configuration and limited solid solutions on both sides (Moh 1975). At the addition of $CuFeS_2$, the temperature of Cu_2FeSnS_4 phase transition decreases from 706°C to 610°C, and at the addition of Cu_2FeSnS_4 the temperature of the $CuFeS_2$ phase transition decreases from 550°C to 462°C.

Cu_2SnS_3 forms a complete solid solution with Cu_2FeSnS_4, and Cu_2SnS_3 and $CuFeS_2$ form a limited mutual solid solution series (Moh 1975).

SnS–Cu_2S–FeS. The vapor pressures of SnS in the molten SnS–Cu_2S–FeS system have been measured by means of the transportation method (Koike and Yazawa). The data obtained was used to calculate the activities, and slight negative deviations of $a(Cu_2S)$, $a(FeS)$, and $a(SnS)$ from ideality were observed in this molten system (Figure 7.16). Also,

the distribution behavior of tin between the matte and silica-saturated slag was investigated under the controlled partial pressures of SO_2 at 1250°C. From the results obtained it was found that tin was concentrated in the matte phase in general, but with increasing matte grade, tin became concentrated in the slag phase. The activity coefficient of SnO in the slag equilibrating with the matte was calculated by thermodynamic analysis.

7.105 Tin–Copper–Cobalt–Sulfur

Four quaternary compounds are formed in the Sn–Cu–Co–S system. $Cu_{0.5}Co_{0.5}[CoSn]S_4$ and $Cu_{0.5}Co_{0.5}[Co_{0.75}Sn_{1.25}]S_4$ crystallize in the cubic structure with the lattice parameter $a = 1021$ for both compounds (Harada 1973). They were prepared in the same way as $Cu_{0.5}Mn_{0.5}[MnSn]S_4$ and $Cu_{0.5}Mn_{0.5}[Mn_{0.75}Sn_{1.25}]S_4$ were synthesized using Co instead of Mn.

Cu_2CoSnS_4 melts congruently at 918°C (Quintero et al. 2014b) [at 909°C (Schäfer and Nitsche 1977)] and crystallizes in the tetragonal structure with the lattice parameters $a = 539.56 \pm 0.08$, $c = 1078.9 \pm 0.3$ pm, and a calculated density of 4.577 g·cm^{-3} (Gulay et al. 2004a; López-Vergara et al. 2013) [$a = 540.2$ and $c = 1080.5$ pm (Schäfer and Nitsche 1974, 1977); $a = 539.92$, $c = 1081.80$ pm, and a calculated density of 4.56 g·cm^{-3} (Quintero et al. 2014b)].

Cu_2CoSnS_4 was obtained by the chemical vapor reactions using I_2 as the transport agent (Schäfer and Nitsche 1974) or by the melt and anneals technique using stoichiometric mixtures of the elements (Gulay et al. 2004a; Quintero et al. 2014b). Typically, calculated amounts of the elements were sealed in an evacuated quartz container. The synthesis was realized in a shaft furnace. The ampoule with the components was heated at a rate of 30°C·h^{-1} to 1100°C. The sample was kept at this temperature for 4 h. After that, it was cooled slowly to 400°C at a rate of 10°C·h^{-1}. Further annealing at 400°C during 240 h was applied. After annealing, the ampoule with the sample was quenched in cold water.

$Cu_2CoSn_3S_8$ crystallizes in the cubic structure with the lattice parameter $a = 1027.2 \pm 0.5$ pm (Lavela et al. 1996). It was synthesized in the same way as $Cu_2FeSn_3S_8$ was prepared. In order to extract Cu^+ ions from the structure, $Cu_2CoSn_3S_8$ was suspended in an excess of 0.1 M solution of I_2 in CH_3CN at 50°C with magnetic stirring for different periods of time (1, 2, and 3 weeks) and $Cu_{1.1}CoSn_3S_8$ quaternary phase was obtained.

7.106 Tin–Copper–Nickel–Sulfur

Four quaternary compounds are formed in the Sn–Cu–Ni–S system. $Cu_{0.5}Ni_{0.5}[NiSn]S_4$ and $Cu_{0.5}Ni_{0.5}[Ni_{0.75}Sn_{1.25}]S_4$ crystallize in the cubic structure with the lattice parameter $a = 1030$ and 1031 pm, respectively (Harada 1973). They were prepared in the same way as $Cu_{0.5}Mn_{0.5}[MnSn]S_4$ and $Cu_{0.5}Mn_{0.5}[Mn_{0.75}Sn_{1.25}]S_4$ were synthesized using Ni instead of Mn.

Cu_2NiSnS_4 crystallizes in the cubic structure with lattice parameter $a = 542.5$ pm (Schäfer and Nitsche 1974). Synthesis

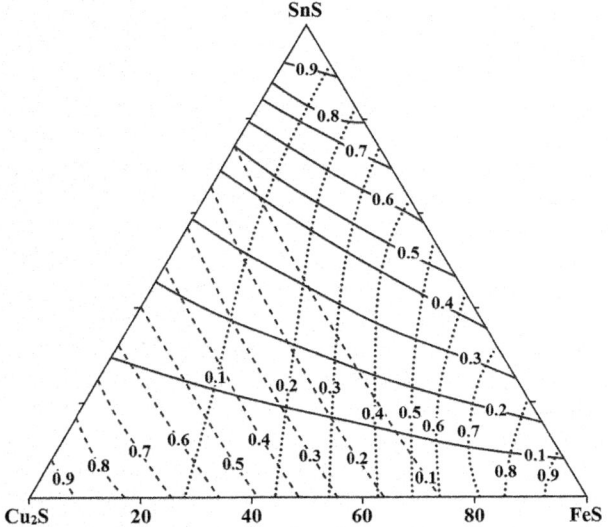

FIGURE 7.16 Isoactivity curves of SnS (solid lines), Cu_2S (dashed lines) and FeS (dotted lines) in the Cu_2S–FeS–SnS quasiternary system at 1100°C. (From Koike, K., and Yazawa, A., *Shigen-to-Sozai*, **110**(1), 43, 1994.) Open access

of this compound from elements was performed in an evacuated quartz ampoule. To accelerate the reaction, about 30 mg of iodine was added. The ampoule was heated to 600°C within 2 h and kept at this temperature for 24 h. Then it was brought to the final reaction temperature of 800–900°C and remained there for 72 h.

Cu₂NiSn₃S₈ also crystallizes in the cubic structure with lattice parameter $a = 1030.22 \pm 0.06$ pm, a calculated density of 4.850 g·cm⁻³ and an energy gap of 0.028 eV in the temperature range of 80–120 K and 0.046 eV in the temperature range of 120–300 K (Garg et al. 2001a) [$a = 1027.6 \pm 0.5$ pm (Lavela et al. 1996); $a = 1030.5 \pm 0.1$ pm and a calculated density of 4.764 g·cm⁻³ for Cu₇.₀₇Ni₄Sn₁₂S₃₂ composition (Garg et al. 2002)]. This compound was obtained by the melt and anneals technique using stoichiometric mixtures of the elements (Lavela et al. 1996; Garg et al. 2001a, 2002). Typically, the elements were mixed in a N₂-filled glove box and sealed in silica ampoule under high vacuum. The ampoule was heated at 750°C for 8 days and then cooled to room temperature after switching off the furnace. This led to single-phase title compound in the form of block cuboidal crystals.

7.107 Tin–Silver–Magnesium–Sulfur

The **Ag₂MgSn₃S₈** quaternary compound, which crystallizes in the cubic structure with the lattice parameter $a = 1069.38 \pm 0.02$ pm, a calculated density of 4.63 g·cm⁻³, and an energy gap of 1.65 eV, is formed in the Sn–Ag–Mg–S system (Heppke et al. 2020). This compound was synthesized in a high energy planetary mill with a subsequent annealing step in a tube furnace. Stoichiometric amounts of Ag₂S, MgS, SnS, and S were filled in a 45 mL zirconia grinding beaker with six zirconia balls and ground at a rotational speed of 450 rpm for 4 h. In order to obtain highly crystalline powders, the product was annealed at 550°C for 3 h under flowing H₂S gas after milling.

7.108 Tin–Silver–Strontium–Sulfur

Two quaternary compounds, **Ag₂SrSnS₄** and **Ag₂Sr₃Sn₂S₈**, are formed in the Sn–Ag–Sr–S system. The first of them crystallizes in the orthorhombic structure with the lattice parameters $a = 691.0$, $b = 721.0$, $c = 783.1$ pm, and an energy gap of 1.08 eV (Zhu et al. 2017).

Ag₂Sr₃Sn₂S₈ crystallizes in the cubic structure with the lattice parameters $a = 1433.20 \pm 0.01$ pm, and a calculated density of 4.430 g·cm⁻³ at 80 K and $a = 1437.04 \pm 0.01$ pm, a calculated density of 4.353 g·cm⁻³, and an energy gap of 2.66 eV at room temperature (McKeown Wessler et al. 2022) [$a = 1422.19 \pm 0.04$ pm, a calculated density of 4.491 g·cm⁻³ and an energy gap of 1.94 eV (Yang et al. 2020a); $a = 1436.37 \pm 0.04$ pm, a calculated density of 4.127 g·cm⁻³ and an energy gap of 1.38 eV for the Ag₂.₇Sr₆Sn₄.₃S₁₆ composition (Haynes et al. 2017)].

The whole preparation process of Ag₂Sr₃Sn₂S₈ was completed in an Ar-filled glove box. Targeted product was prepared under stoichiometric ratio of SrS, Ag₂S, and SnS₂ in the vacuum-sealed silica tube. Firstly, the tube was heated to 600°C in 30 h and kept at this temperature for about 100 h, then cooled quickly down to room temperature. The obtained product was ground and then loaded into the vacuum-sealed silica tube. It

was further sintered again at 900°C for 10 days and cooled slowly down to room temperature with 5°C·h⁻¹. Finally, many millimeter-level high-quality crystals with red color were found in the tube after washed with DMF. They are stable in the air within several weeks.

To prepare Ag₂.₇Sr₆Sn₄.₃S₁₆, a mixture of Sr (6.0 mM), Ag (3.8 mM), Sn (4.1 mM), S (16.0 mM), and KBr (4.2 mM) were loaded into a fused silica tube in a dry N₂ atmosphere in a glove box and then sealed under vacuum of <0.01 Pa. The reaction was heated to 300°C for 4 h, raised to 800°C and held for 48 h, then cooled to 300°C at a rate of 5°C·h⁻¹. The experiment was finished by turning off the power. Deionized water was used to remove the KBr flux in an ultrasonic cleaner. Dark-red and chunk-shaped crystals of the title compound were observed. They are air and water stable.

7.109 Tin–Silver–Barium–Sulfur

Two quaternary compounds, **Ag₂BaSnS₄** and **Ag₂Ba₃Sn₂S₈**, are formed in the Sn–Ag–Ba–S system. The first of them is thermally stable up to 833°C (Chen et al. 2018b) and crystallizes in the orthorhombic structure with the lattice parameters $a = 688.0 \pm 0.7$, $b = 712.9 \pm 0.7$, $c = 812.2 \pm 0.8$ pm, a calculated density of 5.003 g·cm⁻³, and an energy gap of 1.88 eV [1.26 eV (Zhu et al. 2017)] (Liu et al. 2020b) [$a = 712.7$, $b = 811.7$, $c = 685.9$ pm, and the calculated and experimental densities of 5.022 and 4.994 g·cm⁻³, respectively (Teske and Vetter 1976a,b); $a = 688.5 \pm 0.6$, $b = 712.2 \pm 0.4$, $c = 813.6 \pm 0.6$ pm, a calculated density of 4.99 g·cm⁻³, and an energy gap of 1.77 eV (Chen et al. 2018b)].

Dark-red irregular bulk crystals of this compound can be prepared from a mixture of Ba (1.0 mM), Ag (2.0 mM), Sn (1.0 mM), S (4.0 mM), and BaCl₂ (3.1 mM) (Chen et al. 2018b). The chemicals were loaded in a fused silica tube and sealed under vacuum (<10⁻³ Pa). The tube was heated at 300°C for 20 h, raised to 900°C and stayed for 100 h, and cooled to 200°C at 3°C·h⁻¹ before turning off the furnace. The product was washed with deionized water to remove the BaCl₂ flux and soluble byproducts, and dried by ethanol in the air.

The next method also can be used to prepare Ag₂BaSnS₄ (Liu et al. 2020b). The mixture of AgNO₃ (12 mg), Sn (5 mg), S (37 mg), Ba(OH)₂·8H₂O (13 mg), 1.3-propanediamine (about 330 mg), and H₂O (60 mL) was sealed in a Pyrex glass tube (about 10% filling volume of the tube) at air atmosphere, and then the glass tube was placed in a stainless-steel autoclave (about 80% filling volume of water to balance the pressure), and finally heated in the furnace at 170°C for 10 days. After being cooled to ambient temperature naturally, the product was washed with water and ethanol, respectively, and orange-red block crystals were obtained.

Ag₂Ba₃Sn₂S₈, with a homogeneity region that can be expressed as Ag₂.₆₇₊₄δBa₆Sn₄.₃₃₋₈S₁₆, crystallizes in the cubic structure with the lattice parameters $a = 1474.32 \pm 0.02$ pm, a calculated density of 4.433 g·cm⁻³, and an energy gap of 1.58 eV for $\delta = 0.03$ (Haynes et al. 2017); $a = 1474.32 \pm 0.02$ pm, a calculated density of 4.510 g·cm⁻³, and an energy gap of 1.58 eV for $\delta = 0.13$; $a = 1529.83 \pm 0.03$ pm and a calculated density of 5.499 g·cm⁻³ for $\delta = 0.24$; $a = 1474.22 \pm 0.07$ pm, a calculated density of 4.639 g·cm⁻³, and an energy gap of 1.85 eV for $\delta = 0.31$; $a = 1474.82$

± 0.03 pm, a calculated density of 4.645 g·cm⁻³, and an energy gap of 2.37 eV for δ = 0.33 (Lai et al. 2015) [*a* = 1471.32 ± 0.13 pm, a calculated density of 4.678 g·cm⁻³, and an energy gap of 2.02 eV for δ = 0.33 (Liu et al. 2020b)].

To prepare this compound with δ = 0.03, a mixture of Ba (6.0 mM), Ag (3.2 mM), Sn (4.2 mM), S (16.0 mM), and KBr (4.2 mM) were loaded into a fused silica tube in a dry N₂ atmosphere in a glove box and then sealed under vacuum of <0.01 Pa (Haynes et al. 2017). The reaction was heated to 300°C for 4 h, raised to 800°C and held for 48 h, then cooled to 300°C at a rate of 5°C·h⁻¹. The experiment was finished by turning off the power. Deionized water was used to remove the KBr flux in an ultrasonic cleaner. Dark-red and chunk-shaped crystals of the title compound were observed. They are air and water stable.

This compound with δ = 0.13, 0.31, and 0.33 was synthesized using a mixture of Ba (6.00 mM), Ag (3.21 mM), Sn (4.20 mM), S (16.00 m), and KBr (4.20 mM), which was heated at 300°C for 4 h, raised to 800°C and kept there for 48 h, cooled to 300°C within 100 h, and finished by turning off the power (Lai et al. 2015). Dark-red irregular-shaped crystals with δ = 0.13 were obtained in the final product. Deionized water was used to remove the KBr flux in an ultrasonic cleaner. On the basis of the stoichiometric composition of Ag₂.₆₇₊₄₈Ba₆Sn₄.₃₃₋₈S₁₆, slight adjustment of the Ag/Sn molar ratio produced crystals appearing with various colors. Dark-red crystals at δ = 0.13 were obtained as a single product with the range of 0.0 < δ < 0.24. Orange crystals at δ = 0.31 grew gradually with crystals at δ = 0.13 in the product with increasing values of δ. The major product at δ = 0.31 was obtained using the range of 0.30 < δ < 0.40 in the reactions. However, several large crystals at δ = 0.13 coexisted with the crystals at δ = 0.31 in the final product. The materials can be separated with forceps under an optical microscope. The continued increase of δ values caused yellow crystals at δ = 0.33 to precipitate with the crystals at δ = 0.13 and 0.31, or both in the final products. All these crystals are stable in air and water. For preparing the crystals with δ = 0.24, a mixture of Ba (6.00 mM), Ag (3.63 mM), Sn (4.09 mM), S (16 mM), and KBr (4.20 mM) was reacted in the same heating profile as that operated for the crystals with δ = 0.13. Dark-red crystals with a yield of about 66% were obtained as a single product. DMF was used to remove the KBr flux since the obtained crystals dissolved completely in water giving black solutions. The surface of the crystals became cloudy after a few days when exposed to external moisture.

This compound with δ = 0.33 can also be prepared by using Ag₂S (9 mg), Sn power (8 mg), S power (35 mg), Ba(OH)₂ (26 mg), 1,3-propanediamine (about 330 mg) with the next heating at 170°C for 7 days (Liu et al. 2020b). After washing with ethanol, yellow block crystals were obtained. A small amount of Ag₂BaSnS₄ also exists in this experiment.

7.110 Tin–Silver–Zinc–Sulfur

Ag₂SnS₃–ZnS. The phase diagram of this system, constructed through DTA, is given in Figure 7.17 (Nagaoka et al. 2021). The eutectic contains ~10 mol% ZnS and crystallizes at 630°C and the peritectic point is at ~20 mol% ZnS. The **Ag₂ZnSnS₄** quaternary compound, which crystallizes in the tetragonal

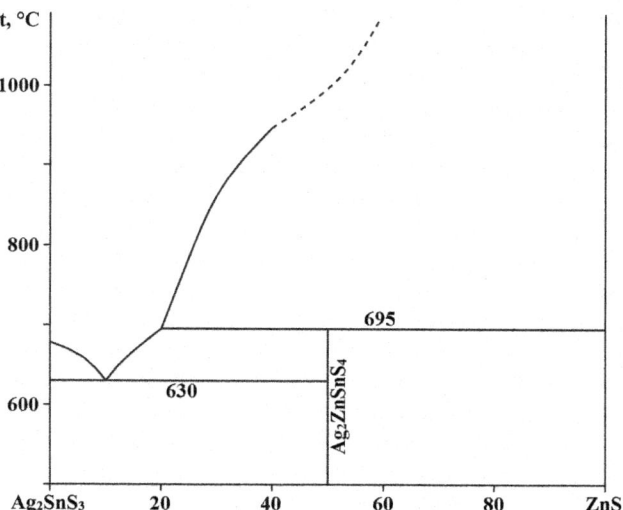

FIGURE 7.17 Phase diagram of the Ag₂SnS₃–ZnS system. (From Nagaoka, A., et al., *J. Cryst. Growth*, **555**, 125967, 2021.)

structure with the lattice parameters *a* = 581.365 ± 0.008 and *c* = 1078.247 ± 0.001 pm (Pietak et al. 2020) [*a* = 578.6 ± 0.4, *c* = 1082.9 ± 0.6 pm, and a calculated density of 4.822 g·cm⁻³ for mineral pirquitasite (Johan and Picot 1982; Dunn et al. 1983); *a* = 577.57 ± 0.12, *c* = 1087.0 ± 0.2 pm, and a calculated density of 4.765 g·cm⁻³ for mineral pirquitasite (Schumer et al. 2013); *a* = 577.6, *c* = 1086.9 pm (Nagaoka et al. 2021), and an energy gap of 2.0 eV (Tsuji et al. 2010)], is formed in this system.

To obtain this compound, a mixed aqueous solution (150 mL) of Zn(NO₃)₂·6H₂O with 10–15% excess amounts (about 0.029 M/L) and SnCl₄·5H₂O (0.025 M/L) was bubbled with an H₂S gas (Tsuji et al. 2010). Then the aqueous solution containing 7.5 mM of AgNO₃ was slowly added to the mixed solution. It was also grown by the chemical transport reactions using Ag, Zn, Sn, and S and ZnCl₂ as transport agents (Pietak et al. 2020).

Ag₈SnS₆–ZnS. The phase diagram of this system, constructed through DTA, XRD, and metallography, is a eutectic type (Figure 7.18) (Dudchak and Piskach 2001a, Piskach et al. 2006). The eutectic composition and temperature are 32 mol% ZnS and 814°C, respectively. The solubility of ZnS in Ag₈SnS₆ at the eutectic temperature is 24 mol%. The crystallization of the solid solutions occurs in a narrow temperature range that does not exceed 20°C. These solid solutions are characterized by the polymorph transformation at 169°C and the solid solutions based on ZnS have the transformation sphalerite-wurtzite at 1039°C. The solubility of ZnS in Ag₈SnS₆ is not higher than 5 mol%. The alloys for the investigations were annealed at 400°C for 250 h followed by the quenching in the 25% aqueous solution of NaCl.

7.111 Tin–Silver–Cadmium–Sulfur

Ag₂SnS₃–CdS. The phase diagram of this system, constructed through DTA, XRD, and metallography, is a eutectic type (Figure 7.19) (Parasyuk and Piskach 1998). The eutectic

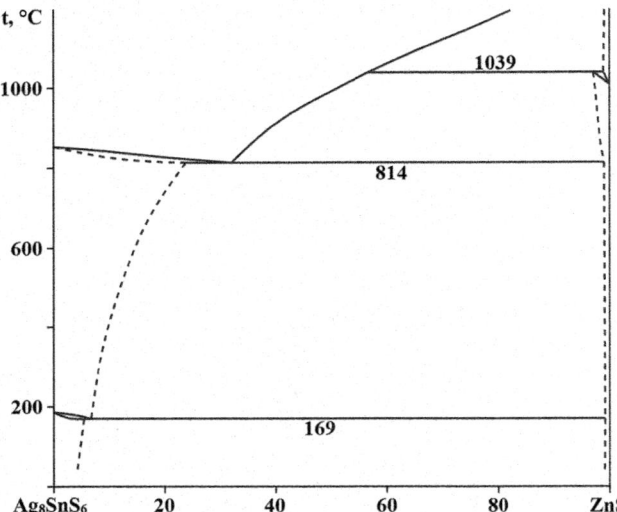

FIGURE 7.18 Phase diagram of the Ag$_8$SnS$_6$–ZnS system. (From Dudchak, I.V., and Piskach L.V., *Nauk. Visnyk Volyns'k. Derzh. Univ. im. Lesi Ukrainky. Ser. Khim. nauky*, (6), 59, 2001.)

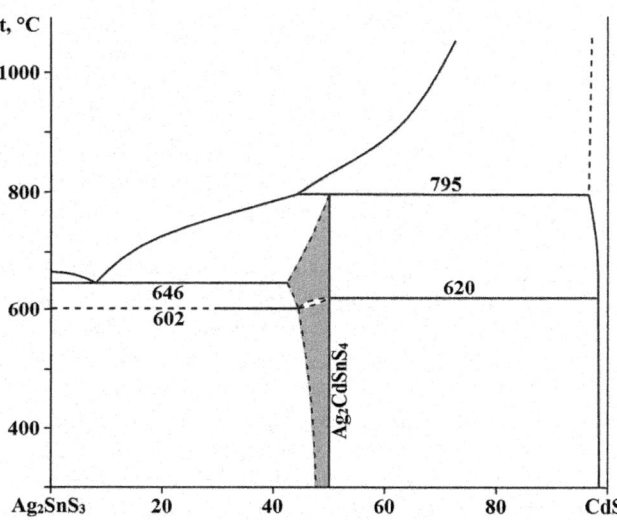

FIGURE 7.19 Phase diagram of the Ag$_2$SnS$_3$–CdS system. (From Parasyuk, O.V., and Piskach L.V., *Pol. J. Chem.*, **72**(5), 966, 1998.)

crystallizes at 646°C and contains 8 mol% CdS. The solubility of Ag$_2$SnS$_3$ in CdS and of CdS in Ag$_2$SnS$_3$ is less than 2 mol% at 400°C. The alloys were annealed at 400°C for 250 h.

The quaternary compound **Ag$_2$CdSnS$_4$** is formed in this system. It melts incongruently at 795°C and has two polymorphic modifications. The homogeneity region of Ag$_2$CdSnS$_4$ at 400°C is less than 5 mol% CdS. It crystallizes in the orthorhombic structure with the lattice parameters for stoichiometric composition $a = 410.15 \pm 0.03$, $b = 702.24 \pm 0.04$, and $c = 669.46 \pm 0.04$ pm and a calculated density of 4.9521 ± 0.0009 g·cm^{-3} (Parasyuk et al. 2005b) [$a = 411.1 \pm 0.5$, $b = 703.8 \pm 0.5$, and $c = 668.5 \pm 0.5$ pm (Parthé et al. 1969)]. According to the data of Parasyuk and Piskach (1998), XRD of low-temperature Ag$_2$CdSnS$_4$ modification (the alloy of stoichiometric composition) confirms the existence of the orthorhombic structure with

two possible variants of space groups. The lattice parameters for *Cmc2$_1$* space group are $a = 1215.8 \pm 0.1$, $b = 704.8 \pm 0.1$, and $c = 665.9 \pm 0.1$ pm and for *Pnm2$_1$*, $a = 818.2 \pm 0.5$, $b = 701.5 \pm 0.3$, and $c = 666.6 \pm 0.5$ pm.

The thermodynamic functions of Ag$_2$CdSnS$_4$ are the next: $\Delta G^0_{f,298} = -375.5 \pm 0.4$ kJ·M^{-1}, $\Delta H^0_{f,298} = -383.3 \pm 0.5$ kJ·M^{-1} and $S^0_{298} = 294.4 \pm 0.4$ J·(M·K)$^{-1}$ (Moroz et al. 2018a). It was obtained through solid-state synthesis from a finely dispersed mixture of Ag$_2$SnS$_3$ and CdS (molar ratio 1:1) at 680°C for 200 h. The high-temperature annealing was performed at 530°C for 240 h.

Ag$_8$SnS$_6$–CdS. The phase diagram of this system, constructed through DTA, XRD, and metallography, is a eutectic type (Figure 7.20) (Parasyuk et al. 2005b, Piskach et al. 2006). The eutectic point coordinates are ~47 mol% CdS and 809°C. The solubility of CdS in β-Ag$_8$SnS$_6$ exceeds 40 mol% CdS at the eutectic temperature. As the temperature decreases, the homogeneity region of the solid solution based on Ag$_8$SnS$_6$ narrows, and at 160°C, it undergoes eutectoid decomposition. At room temperature, the solubility of CdS in α-Ag$_8$SnS$_6$ determined only by the results of metallography lies between 10 and 20 mol%.

SnS$_2$–Ag$_2$S–CdS. The isothermal section of this quasiternary system at room temperature is presented in Figure 7.21 (Parasyuk et al. 2005b). At this temperature, the existence of the Ag$_2$CdSnS$_4$ and **Ag$_2$CdSn$_3$S$_8$** quaternary compounds is established. Ag$_2$CdSn$_3$S$_8$ melts incongruently at 732°C, undergoes a polymorphic transformation at 599–616°C due to the existence of a homogeneity region (65–67 mol% CdS) (Moroz et al. 2018a) and crystallizes in the cubic structure of the chalcospinel type ($a = 1076.35 \pm 0.02$ pm and a calculated density of 5.0102 ± 0.0003 g·cm^{-3}) or in the tetragonal structure of the rhodostannine type ($a = 761.63 \pm 0.06$ and $c = 1077.1 \pm 0.2$ pm and a calculated density of 5.000 ± 0.002 g·cm^{-3}) (Parasyuk et al. 2005b).

The thermodynamic functions of Ag$_2$CdSn$_3$S$_8$ are the next: $\Delta G^0_{f,298} = -672.7 \pm 0.2$ kJ·M^{-1}, $\Delta H^0_{f,298} = -699.8 \pm 0.2$ kJ·M^{-1}

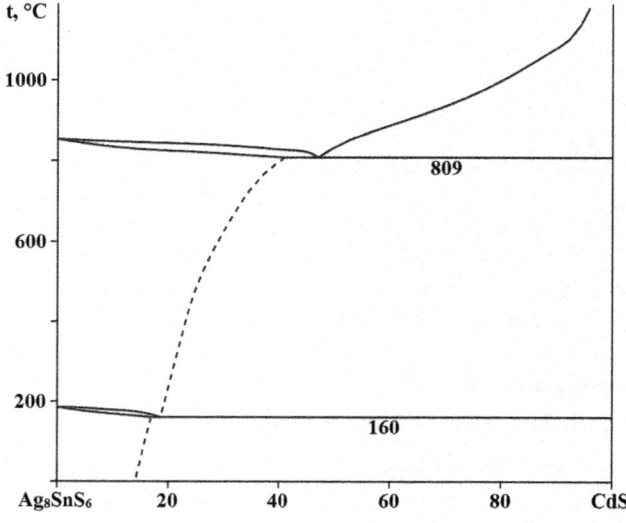

FIGURE 7.20 The phase diagram of the Ag$_8$SnS$_6$–CdS system. (From Parasyuk, O.V. et al., *J. Alloys Compd.*, **399**(1–2), 173, 2005.)

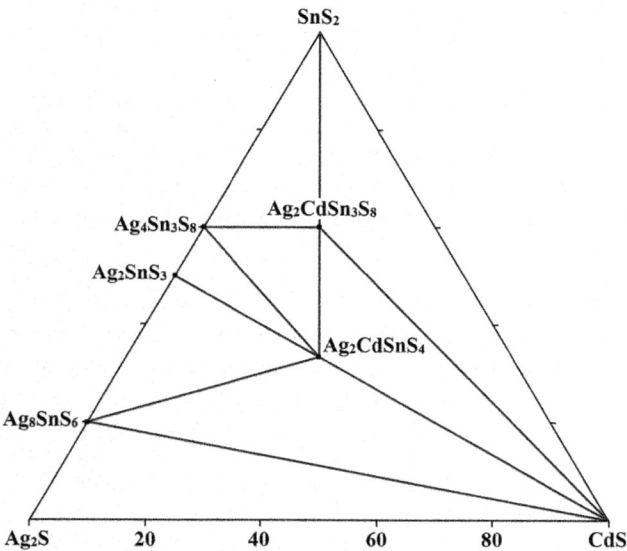

FIGURE 7.21 Isothermal section of the SnS$_2$–Ag$_2$S–CdS quasiternary system at room temperature (From Parasyuk, O.V. et al., *J. Alloys Compd.*, **399**(1–2), 173, 2005.)

and $S^0_{298} = 456.3 \pm 0.2$ J·(M·K)$^{-1}$ (Moroz et al. 2018a). It was obtained through solid-state synthesis from finely dispersed mixture of Ag$_2$SnS$_3$, CdS, and SnS$_2$ (molar ratio 1:1:2) at 680°C for 200 h. The high-temperature annealing was performed at 530°C for 240 h.

7.112 Tin–Silver–Mercury–Sulfur

Ag$_2$SnS$_3$–HgS. The phase diagram of this system, constructed through DTA, XRD, and metallography, is shown in Figure 7.22 (Parasyuk 1999). The coordinates of the eutectic points are 23 mol% HgS and 623°C and 77 mol% HgS and 639°C. The ingots for the investigations were annealed at 400°C for 250 h with further quenching in cold water.

Ag$_2$HgSnS$_4$ quaternary compound is formed in this system. It melts congruently at 660°C and crystallizes in the orthorhombic structure with the lattice parameters $a = 820.74 \pm 0.04$, $b = 703.30 \pm 0.03$, $c = 671.80 \pm 0.03$ pm, a calculated density of 5.7740 ± 0.0008 g·cm^{-3}, and an energy gap of 1.38 eV at room temperature and 1.45 eV at liquid N$_2$ temperature (Parasyuk et al. 2005a). [$a = 814.9 \pm 0.2$, $b = 700.8 \pm 0.2$, and $c = 670.4 \pm 0.3$ pm (Parasyuk 1999); $a = 820.3$, $b = 702.6$, and $c = 671.0$ pm (Haeuseler and Himmrich 1989)]. The highest point of melting is found for Ag$_2$HgSnS$_4$ situated in the interval 50–52.5 mol% HgS. The homogeneity region does not exceed 3 mol% from the Ag$_2$SnS$_3$ side and 6 mol% from the HgS side and it decreases with decreasing temperature (Parasyuk 1999). This compound was obtained by the interaction of the chemical elements at 700°C (Haeuseler and Himmrich 1989). Single crystals of this compound were grown using the Bridgman–Stockbarger technique (Parasyuk et al. 2005a).

Ag$_8$SnS$_6$–HgS. The phase diagram of this system, constructed through DTA and XRD, is a eutectic type (Figure 7.23) (Parasyuk et al. 2005a). The eutectic contains 89 mol% HgS and crystallizes at 608°C. Solid solution range based on β-Ag$_8$SnS$_6$ extends to ~57 mol% HgS and significantly decreases with decreasing temperature. A eutectoid decomposition of this solid solution with the formation of the solid solution based on α-Ag$_8$SnS$_6$ and HgS occurs at 157°C. The horizontal line at 344°C corresponds to the phase transition of HgS. The ingots for the investigations were annealed at 400°C during 250 h.

SnS$_2$–Ag$_2$S–HgS. Isothermal section of this quasiternary system is shown in Figure 7.24 (Parasyuk et al. 2005a). Only one quaternary compound, Ag$_2$HgSnS$_4$, exists in the system at this temperature.

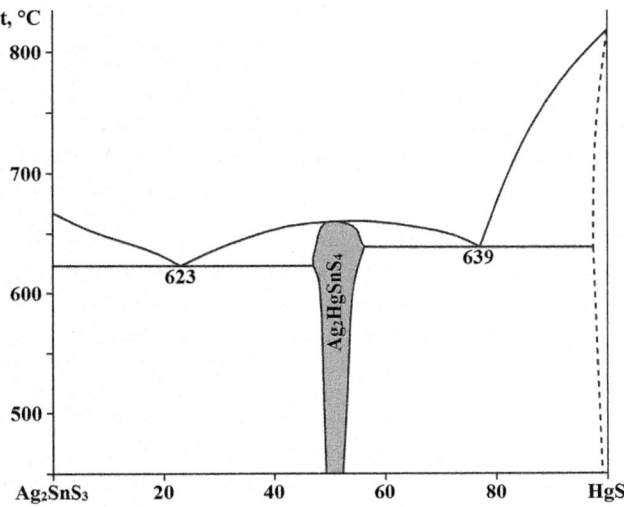

FIGURE 7.22 Phase diagram of the Ag$_2$SnS$_3$–HgS system. (From Parasyuk, O.V., *J. Alloys Compd.*, **291**(1–2), 215, 1999.)

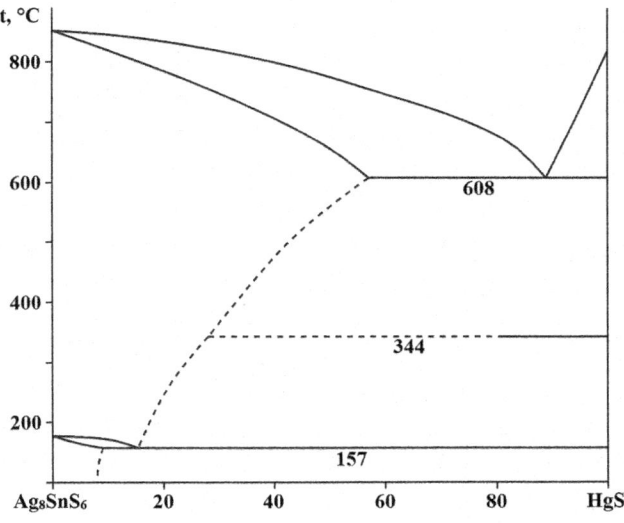

FIGURE 7.23 Phase diagram of the Ag$_8$SnS$_6$–HgS system. (From Parasyuk, O.V. et al., *J. Alloys Compd.*, **399**(1–2), 32, 2005.)

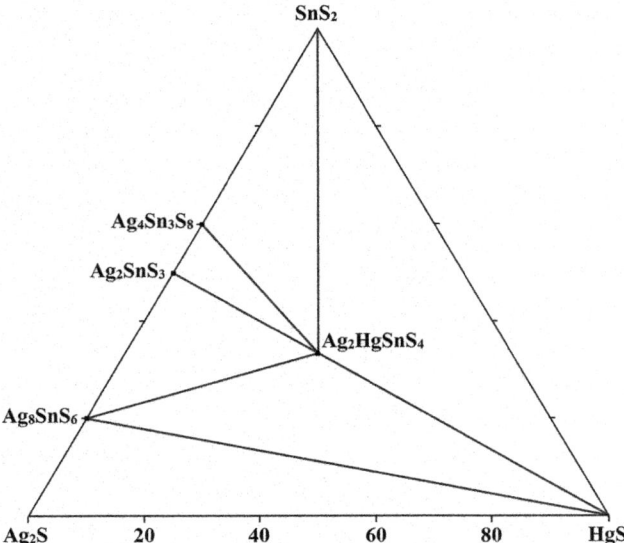

FIGURE 7.24 Isothermal section of the SnS$_2$–Ag$_2$S–HgS quasiternary system at room temperature (From Parasyuk, O.V. et al., *J. Alloys Compd.*, **399**(1–2), 32, 2005.)

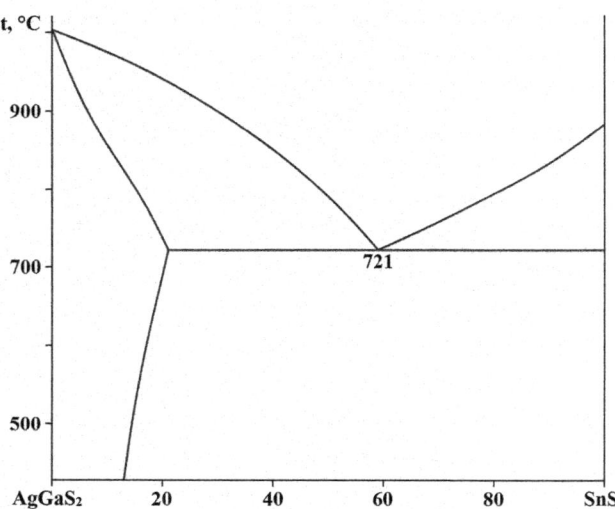

FIGURE 7.25 Phase diagram of the SnS$_2$–AgGaS$_2$ system. (From Shevchuk, M.V., and Olekseyuk, I.D., *J. Alloys Compd.*, **433**, 171, 2007.)

7.113 Tin–Silver–Gallium–Sulfur

SnS$_2$–AgGaS$_2$. The phase diagram of this system, constructed through DTA, XRD, and metallography, is a eutectic type (Figure 7.25) (Shevchuk and Olekseyuk 2007). The eutectic contains 59 mol% SnS$_2$ and crystallizes at 721°C. The wide solid solution range based on AgGaS$_2$ extends at the eutectic temperature to 21 mol% SnS$_2$ and narrows to 13 mol% at 450°C. The solid solubility based on SnS$_2$ is practically absent. The ingots for the investigations were annealed at 450°C for 480 h followed by quenching into cold water.

7.114 Tin–Silver–Indium–Sulfur

Four quaternary compounds, **AgInSnS$_4$**, **Ag$_2$In$_6$SnS$_{12}$**, **Ag$_2$In$_4$Sn$_4$S$_{11}$**, and **Ag$_7$In$_7$Sn$_{10}$S$_{34}$**, which crystallizes in the cubic structure with the lattice parameters correspondingly $a = 1075.88 \pm 0.05$ pm [$a = 1074$ pm and the calculated and experimental densities of 4.923 and 4.91 g·cm^{-3}, respectively (Hahn and Strick 1967a; Strick et al. 1968)], 1076 ± 1 pm, 1081 ± 1 pm, and 1073.8 ± 1.5 pm (Ohachi and Pamplin 1977b), are formed in the Sn–Ag–In–S system. Polycrystalline samples of these compounds of dark-brown color were prepared by a direct reaction from the elements or binary compounds at 700°C for 5 days. AgInSnS$_4$ melts at 720°C (Ohachi and Pamplin 1977b). All these compounds were also prepared by the interaction of metal alloys with sulfur vapor, and AgInSnS$_4$ can be synthesized by heating AgInS$_2$ with SnS$_2$ at 680°C.

SnS$_2$–AgInS$_2$. The phase diagram of this system is apparently a eutectic type, and the eutectic crystallizes at 785°C (Pamplin et al. 1977).

7.115 Tin–Silver–Lanthanum–Sulfur

The **AgLa$_3$SnS$_7$** quaternary compound, which crystallizes in the hexagonal structure with the lattice parameters $a = 1037.28 \pm 0.15$ and $c = 600.29 \pm 0.12$ pm for Ag$_{0.82}$La$_3$SnS$_7$ composition (Daszkiewicz and Gulay 2012) [$a = 1039.9 \pm 0.1$, 1040.4 ± 0.1, and 1040.8 ± 0.1 pm, $c = 601.6 \pm 0.1$, 603.9 ± 0.1, and 605.0 ± 0.1, and a calculated density of 4.999, 4.976, and 4.966 g·cm^{-3} for Ag$_{0.82}$La$_3$SnS$_7$ composition at room temperature, 180°C, and 260°C, respectively (Daszkiewicz et al. 2007); $a = 1037.80 \pm 0.15$, $c = 599.00 \pm 0.12$, and a calculated density of 5.158 g·cm^{-3} (Zeng et al. 2008)], is formed in the Sn–Ag–La–S system. The compressibility of the unit cell is anisotropic since the dimension of the c axis decreases by ~1.6% while contraction along a and b axes is ~2.5% each at the applied pressure up to 4.5 gPa (Daszkiewicz and Gulay 2012). The phase transition up to 4.5 GPa was not found. Zero-pressure bulk modulus is $B_0 = 61.74$ GPa and its pressure derivative has a value of B.' = 4.02.

Dark-red crystals of Ag$_{0.82}$La$_3$SnS$_7$ were prepared by sintering the elemental constituents with the atomic ratio La/Ag/Sn/S = 3:1:1:7 in an evacuated silica ampoule in a tube furnace (Daszkiewicz et al. 2007; Daszkiewicz and Gulay 2012). The ampoule was heated at a rate of 30°C·h^{-1} to a maximum temperature of 1150°C and kept at this temperature for 3 h. Afterwards the ampoule was cooled slowly (10°C·h^{-1}) to 600°C, annealed at this temperature for 240 h and then quenched in cold water.

AgLa$_3$SnS$_7$ was obtained as follows (Zeng et al. 2008). The mixture of La$_2$S$_3$, Ag$_2$S, and SnS$_2$ (molar ratio 0.360:0.121:0.241) was mixed together and ground thoroughly within a N$_2$-filled glove box, pressed into pellets, and then sealed in evacuated quartz ampoules before being heated in a resistant furnace to 750°C in 72 h, held at this temperature for 24 h, then heated to 1000°C in 20 h, annealed at this temperature for 240 h, and the furnace was turned off. The precursor obtained from the first-step solid-state reaction was mixed with KBr (0.80 g). Upon regrinding, repelleting, and resealing, the precursor/flux mixture was heated to 850°C for 54 h, held at this temperature for 360 h, then slowly cooled to 600°C, annealed at this temperature for 166.67 h, and finally to room temperature by switching off the furnace powers. Transparent red crystals of La$_3$AgSnS$_7$ were isolated manually from the residue after the flux was removed, by washing with distilled water.

7.116 Tin–Silver–Cerium–Sulfur

The **Ag$_{0.81}$Ce$_3$SnS$_7$** quaternary compound, which crystallizes in the hexagonal structure with the lattice parameters $a = 1030.2 \pm 0.1$, 1030.0 ± 0.1, 1029.2 ± 0.1, and 1029.9 ± 0.1 pm, $c = 596.9$ ± 0.1, 600.2 ± 0.1, 603.3 ± 0.1, and 604.1 ± 0.1, and a calculated density of 5.153, 5.126, 5.112, and 5.095 g·cm^{-3} at 12, 298, 450, and 530 K, respectively, are formed in the Sn–Ag–La–S system (Daszkiewicz et al. 2007). This compound was prepared in the same way as Ag$_{0.82}$La$_3$SnS$_7$ was synthesized.

7.117 Tin–Silver–Lead–Sulfur

SnS$_2$–Ag$_2$S–PbS. The isothermal section of this quasiternary system at 400°C is shown in Figure 7.26 (Kogut et al. 2006). No ternary compounds were found in the system. The solubility based on Ag$_8$SnS$_6$ is less than 2 mol%. The ingots for the investigations were annealed at 400°C for 250 h.

7.118 Tin–Silver–Phosphorus–Sulfur

Ag$_8$SnS$_6$–Ag$_7$PS$_6$. The phase diagram of this quasibinary system is a eutectic type (Bereznyuk et al. 2020). The eutectic contains 35 mol% Ag$_8$SnS$_6$ and crystallizes at 737°C. The existence of Ag$_{8-x}$Sn$_{1-x}$P$_x$S$_6$ ($x = 0$–0.21) and Ag$_{7+x}$Sn$_x$P$_{1-x}$S$_6$

($x = 0$–0.31) solid solutions was determined in the system. The Ag$_{7.8}$Sn$_{0.8}$P$_{0.2}$S$_6$ phase crystallizes in the orthorhombic structure with the lattice parameters $a = 1520.5 \pm 0.3$,.$b = 752.7 \pm 0.1$, $c = 1066.3 \pm 0.2$ pm and a calculated density of 6.189 ± 0.003 g·cm^{-3} and Ag$_{7.31}$Sn$_{0.31}$P$_{0.69}$S$_6$ phase crystallizes in the cubic structure with the lattice parameter $a = 1052.6 \pm 0.1$ pm and a calculated density of 4.1409 ± 0.0007 g·cm^{-3}.

SnS$_2$–Ag$_2$S–P$_2$S$_5$. The isothermal section of this quasiternary system at 150°C (Figure 7.27) includes 10 single-phase, 17 two-phase, and 8 three-phase regions (Bereznyuk et al. 2020). The alloys for the investigations were annealed at 150°C for 500 h.

There are two glass-forming regions in this quasiternary system (Bereznyuk and Petrus 2020). Both regions are adjacent to the SnS$_2$–P$_2$S$_5$ system; the first in the range of 5–15 mol. % P$_2$S$_5$ and the second in the range of 35–65 mol. % P$_2$S$_5$.

7.119 Tin–Silver–Antimony–Sulfur

Ag$_2$SnS$_3$–AgSbS$_2$. The phase diagram of this system, constructed through DTA, XRD, metallography, and measuring of the microhardness and density, is a eutectic type (Figure 7.28) (Mammadov et al. 2020). The eutectic contains 30 mol% Ag$_2$SnS$_3$ and crystallizes at 427°C. The solubility of AgSbS$_2$ in Ag$_2$SnS$_3$ is 12 mol% at room temperature and 19 mol% at the eutectic temperature, and the solubility of Ag$_2$SnS$_3$ in AgSbS$_2$ is 5 and 18 mol% at room temperature and 427°C, respectively. The ingots for the investigations were annealed at 230–430°C for 300 h.

SnS–Ag$_2$S–Sb$_2$S$_3$. The isothermal section of this quasiternary system at 400°C is presented in Figure 7.29 (Chang 1987). The system is divided into two halves by the join SnS–AgSbS$_2$, along which mutual solid solubility exists. AgSbS$_2$ at 400°C has the NaCl-type structure, and its solid solution extends

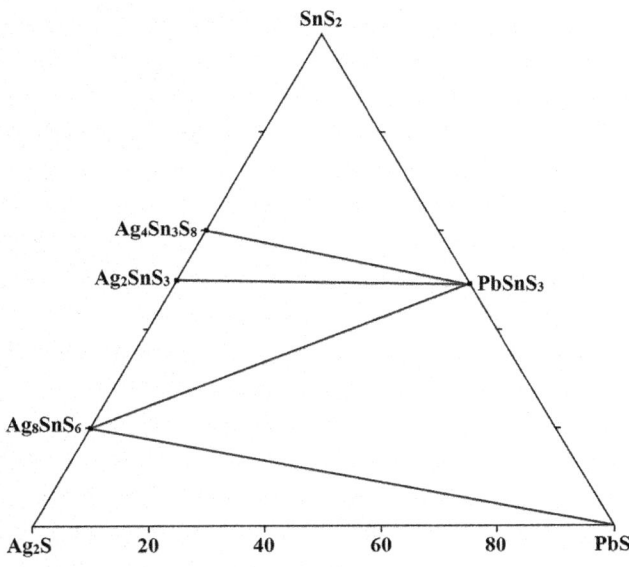

FIGURE 7.26 Isothermal section of the SnS$_2$–Ag$_2$S–PbS quasiternary system at 400°C. (From Kogut, Yu.M., et al., *Nauk. Visnyk Volyns'k. Derzh. Univ. im. Lesi Ukrainky. Ser. Khim. nauky*, (4), 63, 2006.)

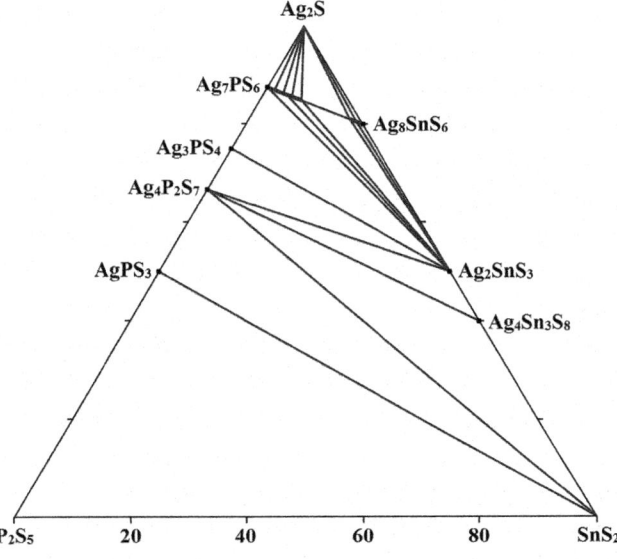

FIGURE 7.27 Isothermal section of the SnS$_2$–Ag$_2$S–P$_2$S$_5$ quasiternary system at 150°C. (From Bereznyuk, O.P., et al., *Visnyk Odes'k. Nats. Univ.*, **25**[4(76)], 32, 2020.) Open access

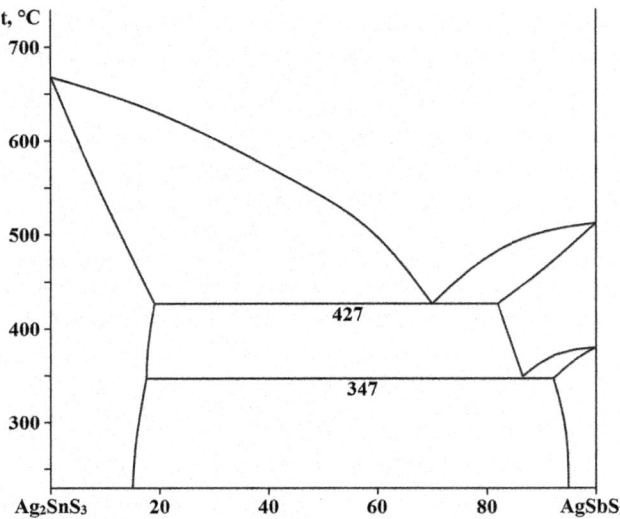

FIGURE 7.28 Phase diagram of the Ag_2SnS_3–$AgSbS_2$ system. (From Mammadov, Sh.G., et al., *Russ. J. Inorg. Chem.*, **65**(2), 217, 2020.)

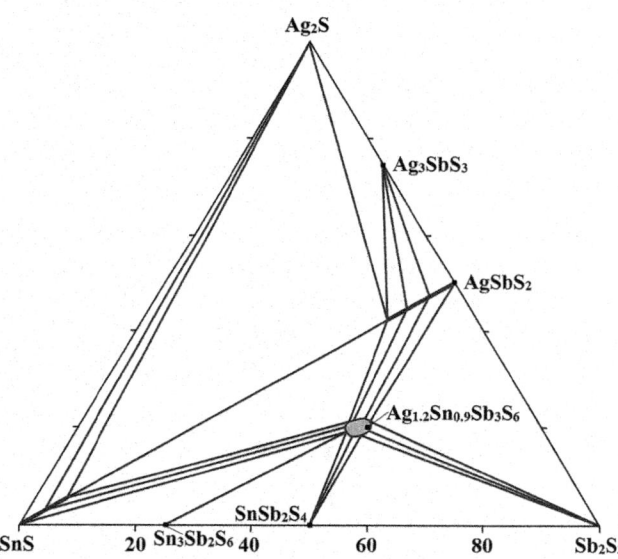

FIGURE 7.29 Isothermal section of the SnS–Ag_2S–Sb_2S_3 quasiternary system at 400°C. (From Chang, L.L.Y, *Mineralog. Mag.*, **51**(363), 741, 1987.)

to $Ag_{0.9}Sn_{0.2}Sb_{0.9}S_2$. The solid solution of SnS has a range of $Ag_{0.1}Sn_{0.8}Sb_{0.1}S_2$. The $Ag_{1.2}Sn_{0.9}Sb_3S_6$ quaternary compound is stable in the system and forms the equilibria with SnS, $Sn_3Sb_2S_6$, $SnSb_2S_4$, Sb_2S_3, and $AgSbS_2$. Its composition falls on the $SnSb_2S_4$–$AgSbS_2$ section. It melts congruently at 433°C and has eutectic relations with $AgSbS_2$ and $SnSb_2S_4$. The eutectic temperatures determined in both relations are 427°C. The compositions at the eutectic points are 45 mol% $AgSbS_2$ and 47 mol% $SnSb_2S_4$. This compound crystallizes in the orthorhombic structure with the lattice parameters $a = 1289$, $b = 1915$, and $c = 422$ pm.

7.120 Tin–Silver–Selenium–Sulfur

Ag_8SnS_6–Ag_8SnSe_6. This system is characterized by formation of a continuous series of solid solutions between the high-temperature modifications of the starting compounds (Figure 7.30) (Bagheri et al. 2014b). No maximum or minimum were observed on the liquidus and solidus curves. Polymorphic transformations in solid solutions are accompanied by peritectoid reactions. The peritectoid point has the coordinates 40 mol% Ag_8SnSe_6 and 82°C. The maximum solubility of α-Ag_8SnSe_6 in α-Ag_8SnS_6 is 22 mol%, and the solubility of α-Ag_8SnS_6 in α-Ag_8SnSe_6 reaches 63 mol%. The concentration dependence of the lattice parameters for high-temperature solid solutions is practically linear. The phase diagram was constructed through DTA, XRD, and metallography.

$6Ag_2S + Ag_8SnSe_6 \leftrightarrow 6Ag_2Se + Ag_8SnS_6$. This system is a reversible mutual one and is characterized by the formation of continuous solid solutions at 530°C, and at room temperature, the system consists of 3 two-phase and 2 three-phase regions (Figure 7.31) (Alverdiyev 2019). The ingots for the investigations were annealed at 530°C for 500 h.

The $Ag_{7.98}Sn_{1.02}(S_{4.19}Se_{1.81})$ quaternary phase (Se-rich mineral canfieldite) crystallizes in the cubic structure with the lattice parameter $a = 1081.45 \pm 0.08$ pm (Zhai et al. 2019).

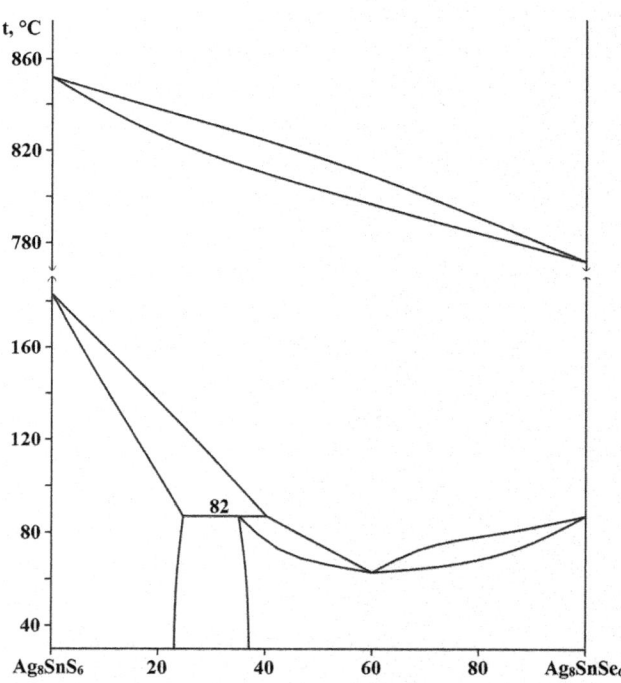

FIGURE 7.30 Phase diagram of the Ag_8SnS_6–Ag_8SnSe_6 system. (From Bagheri, S.M., et al., *Int. J. Adv. Sci. Tech. Res.*, **2**(4), 291, 2014.) Open access

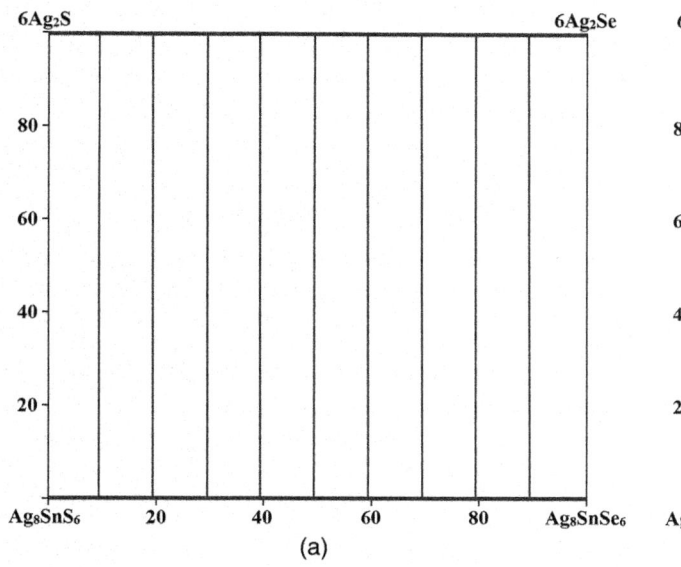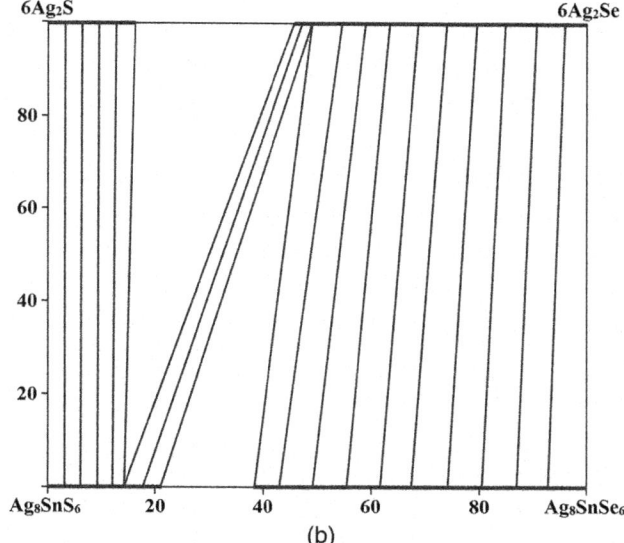

FIGURE 7.31 Isothermal section of the $6Ag_2S + Ag_8SnSe_6 \leftrightarrow 6Ag_2Se + Ag_8SnS_6$ ternary mutual system (a) at 530°C and (b) at room temperature. (From Alverdiyev, I.J., *Azerb. Chem. J.*, (4), 70, 2019.)

7.121 Tin–Silver–Tellurium–Sulfur

The $Ag_8Sn(S,Te)_6$ quaternary phase (Te-rich mineral canfieldite) crystallizes in the cubic structure with the lattice parameter $a = 1100.03 \pm 0.06$ pm (Bindi et al. 2012) [$a = 2170$ pm (Harris and Owens 1971)].

7.122 Tin–Silver–Chromium–Sulfur

The $AgCrSnS_4$ quaternary compound, which crystallizes in the cubic structure with the lattice parameter $a = 1041.42 \pm 0.03$ pm for $Ag_{1.41}Cr_{1.47}S_{2.52}S_8$ composition (Garg et al. 2005) [$a = 1044$ pm and the calculated and experimental densities of 4.752 and 4.61 g·cm^{-3}, respectively (Hahn and Strick 1967a; Strick et al. 1968); $a = 1043.7 \pm 0.2$ pm and the calculated and experimental densities of 4.72 and 4.55 g·cm^{-3}, respectively for $Ag_{0.926}Cr_{0.926}Sn_{1.074}S_4$ composition (Mähl et al. 1984b)], is formed in the Sn–Ag–Cr–S system.

Polycrystalline samples of $AgCrSnS_4$ were prepared by the direct reaction from the elements or binary compounds by heating at 750°C for 4 days (Hahn and Strick 1967a; Strick et al. 1968). To prepare $Ag_{1.41}Cr_{1.47}S_{2.52}S_8$, stoichiometric amounts of Ag, Cr, Sn, and S were mixed in a N_2-filled glove box and sealed in silica ampoule under high vacuum (10^{-3} Pa) (Garg et al. 2005). In a typical reaction, about 250 mg of reactant mixture was loaded in the stoichiometric ratio. The ampoule was heated in a furnace at 750°C for 8 days and then cooled to room temperature after switching off the furnace. The single crystals of the desired material were grown from the melt during the cooling of the furnace. It may be noted that initially, the stoichiometry attempted was corresponding to $Ag_2CrSn_3S_8$. However, a monophasic composition was obtained from the nominal composition of $Ag_{1.63}CrSn_3S_8$. Single crystals of $Ag_{0.926}Cr_{0.926}Sn_{1.074}S_4$ were grown by chemical transport

reactions using $AlCl_3/I_2$ mixtures as transporting agents (Mähl et al. 1984b).

7.123 Tin–Silver–Chlorine–Sulfur

The $Ag_{8-x}SnS_{6-x}Cl_x$ quaternary phase (x value is not specified) was obtained in the Sn–Ag–Cl–S system by reacting stoichiometric amounts of high-purity elements in an evacuated quartz ampoule at temperatures of 600°C to 700°C for about 6 days (Kuhs et al. 1979).

7.124 Tin–Silver–Bromine–Sulfur

$Ag_2SnS_3–AgBr$. The phase diagram of this system, constructed through DTA, XRD, and metallography, is a eutectic type (Figure 7.32) (Mikolaichuk et al. 2010). The eutectic contains 5.3 mol% Ag_2SnS_3 and crystallizes at 388°C.

$Ag_8SnS_6–AgBr$. The phase diagram of this system, constructed through DTA, XRD, and metallography, is a eutectic type with a eutectoid decomposition of solid solution based on Ag_8SnS_6 (Figure 7.33) (Mikolaichuk et al. 2010). The eutectic degenerates from the AgBr side, contains 1.0 mol% Ag_8SnS_6, and crystallizes at 404°C. The coordinates of the eutectoid point are 1.8 mol % AgBr and 134°C.

$Ag_2SnS_3–Ag_8SnS_6–AgBr$. The liquidus surface of this system includes the fields of primary crystallizations of Ag_2SnS_3, Ag_8SnS_6, which occupies a significant part of the liquidus surface, $Ag_6SnS_4Br_2$, and AgBr (Mikolaichuk et al. 2010). $Ag_6SnS_4Br_2$ melts incongruently at 439°C, has a homogeneity region and a polymorphic transformation at −38°C. It crystallizes in the orthorhombic structure with the lattice parameters $a = 667.050 \pm 0.010$, $b = 782.095 \pm 0.009$, $c = 2314.04 \pm 0.03$ pm, and a calculated density of 5.737 g·cm^{-3}.

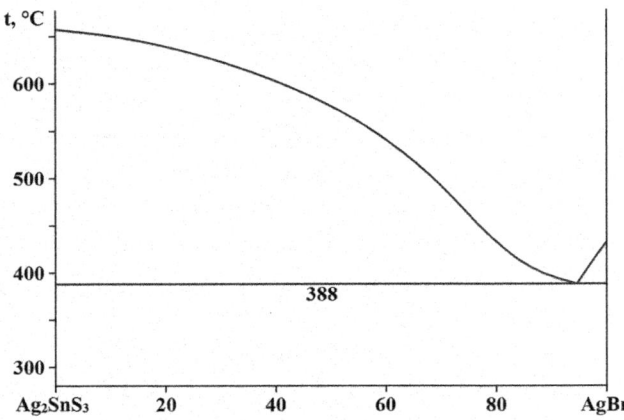

FIGURE 7.32 Phase diagram of the Ag_2SnS_3–AgBr system. (From Mikolaichuk, O.G., et al., *Inorg. Mater.*, **46**(6), 590, 2010.)

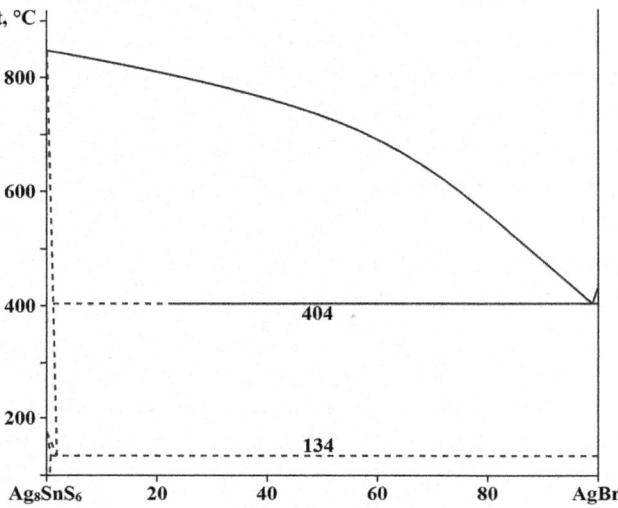

FIGURE 7.33 Phase diagram of the Ag_8SnS_6–AgBr system. (From Mikolaichuk, O.G., et al., *Inorg. Mater.*, **46**(6), 590, 2010.)

The $\mathbf{Ag_{8-x}SnS_{6-x}Br_x}$ quaternary phase, which melts at 825°C (Laqibi et al. 1987) and crystallizes in the cubic structure with the lattice parameter $a = 1077.6 \pm 0.2$ pm (x value is not specified) (Kuhs et al. 1979) [$a = 1077.8 \pm 0.3$ pm for $x = 1$ (Nagel and Range 1978); $a = 1076.76 \pm 0.01$ pm for $x = 1$ (Laqibi et al. 1987)], is also formed in the Sn–Ag–Br–S system. It was obtained by a solid-state reaction using suitable quantities of Ag_2S, SnS_2, and AgBr, which were finely ground and mixed thoroughly and then treated in a quartz ampoule sealed under vacuum (Laqibi et al. 1987). The mixture was heated to 700°C over a period of 3 days, kept at this temperature for 48 h, and then maintained at 580°C for 1 week. It is necessary to note that Ag_7SnS_5Br was not detected by Mikolaichuk et al. (2010).

7.125 Tin–Silver–Iodine–Sulfur

The $\mathbf{Ag_{8-x}SnS_{6-x}I_x}$ quaternary phase, which melts at 816°C and crystallizes in the cubic structure with the lattice parameter $a = 1082.81 \pm 0.01$ pm for $x = 1$ (Laqibi et al. 1987) [$a = 1085.2 \pm$

0.4 pm (x value is not specified) (Kuhs et al. 1979); $a = 1082.3 \pm 0.4$ pm for $x = 1$ (Nagel and Range 1978)], is formed in the Sn–Ag–I–S system. This phase was obtained in the same way as $Ag_{8-x}SnS_{1-x}Br_x$ was prepared (Laqibi et al. 1987).

7.126 Tin–Silver–Manganese–Sulfur

Two quaternary compounds, $\mathbf{Ag_2MnSnS_4}$ and $\mathbf{Ag_2MnSn_3S_8}$, are formed in the Sn–Ag–Mn–S system. Ag_2MnSnS_4 melts incongruently at 790°C (Delgado et al. 2018) and apparently has some polymorphic modifications. According to the data of Friedrich et al. (2018), it is stable up to 700°C and apparently undergoes a reversible phase transition above 400°C.

First modification crystallizes in the orthorhombic structure with the lattice parameters $a = 817.05 \pm 0.05$, $b = 694.13 \pm 0.05$, $c = 665.32 \pm 0.05$ pm, and a calculated density of 4.55 g·cm⁻³ (Delgado et al. 2018) [$a = 801.9$, $b = 696.4$, and $c = 652.7$ pm (Lamarche et al. 1991); $a = 663.2 \pm 0.2$, $b = 692.2 \pm 0.2$, $c = 815.6 \pm 0.2$ pm, and a calculated density of 4.574 g·cm⁻³ for mineral agmantinite (Keutsch et al. 2015, 2019)]. The second modification crystallizes in the cubic structure with the lattice parameter $a = 523.5$ pm (Lamarche et al. 1991), and the third modification crystallizes in the monoclinic structure with the lattice parameters $a = 665.1 \pm 0.1$, $b = 694.3 \pm 0.1$, $c = 1053.6 \pm 0.2$ pm, $\beta = 129.15 \pm 0.01°$, a calculated density of 4.56 g·cm⁻³, and an energy gap of 2.0 eV (Friedrich et al. 2018) [$a = 669.6 \pm 0.1$, $b = 699.1 \pm 0.1$, $c = 822.2 \pm 0.2$ pm, and $\beta = 90.00 \pm 0.03°$ (Greil and Pfitzner 2012)].

Orthorhombic Ag_2MnSnS_4 was synthesized by the melt and anneals technique (Delgado et al. 2018). Ag, Mn, Sn, and S of 1-g sample were sealed under vacuum (~10⁻³ Pa) in a quartz ampoule, and then the mixture was heated up to 200°C and kept at this temperature for about 1–2 h. The temperature was then raised to 500°C at a rate of 40°C·h⁻¹, and held at this temperature for 14 h. After that, the sample was heated from 500°C to 800°C at a rate of 30°C·h⁻¹ and kept at 800°C for another 14 h. Then the temperature was raised to 1150°C at 60°C·h⁻¹, and the mixture was melted at this temperature. The furnace temperature was brought slowly (4°C·h⁻¹) down to 600°C, and the sample was annealed at this temperature for 1 month. Finally, the sample was slowly cooled to room temperature using a rate of about 2°C·h⁻¹.

Cubic Ag_2MnSnS_4 was prepared by the melt and anneals technique using stoichiometric mixtures of the elements (Lamarche et al. 1991). The components were sealed under a vacuum in a quartz capsule which had previously been coated with carbon in order to prevent the reaction of the charge with the quartz. The capsule was then raised to a temperature of 1150°C, held at 1150°C for 0.5–1 h, and then cooled to room temperature. The sample was then annealed 625°C for a week and quenched to room temperature. Annealing at 550°C and quenching provided a sample showing a single-phase orthorhombic form.

Monoclinic Ag_2MnSnS_4 was obtained using a high-temperature solid-state technique using a mixture of stoichiometric amounts of Ag, Mn, Sn, and S in an evacuated quartz ampoule (Greil and Pfitzner 2012; Friedrich et al. 2018). After an initial heating process of 1 day at 400°C, which is necessary to avoid a too-high sulfur pressure, the ampoule was then heated to 800°C for 1 day to ensure a complete reaction of the elements. Finally, the reaction mixture was annealed at 500°C for 8 days, homogenized by grinding, and subsequently annealed at 500°C for several days.

$Ag_2MnSn_3S_8$ crystallizes in the cubic structure with the lattice parameter $a = 1069.84 \pm 0.02$ pm (Garg et al. 2003b). For preparing this compound, stoichiometric amounts of Ag, Mn, Sn, and S were mixed in a N_2-filled glove box and sealed in a silica ampoule under vacuum (10^{-3} Pa). The ampoule was heated in a furnace at 680°C for 8 days and then the furnace was cooled to room temperature. This led to the formation of a single-phase $Ag_2MnSn_3S_8$ in the form of black cuboidal crystals.

7.127 Tin–Silver–Iron–Sulfur

Ag_2SnS_3–FeS. The phase diagram of this system, constructed through DTA, is shown in Figure 7.34 (Moroz et al. 2018b). The eutectic contains 14 mol% FeS and crystallizes at 642°C. The immiscibility region was found to be at 708°C within the interval ~67–79 mol% Ag_2SnS_3. The **Ag_2FeSnS_4** quaternary compound, which melts incongruently at 649°C and has two polymorphic transitions (at 547°C and 603°C), is formed in the system. This compound is thermally unstable. It was obtained through solid-state synthesis from a finely dispersed mixture of Ag_2SnS_3 and FeS (molar ratio 1:1) at 600–680°C for 200 h (Moroz et al. 2018b, c). The high-temperature annealing of the compound was performed at 410–480°C for 240 h. The standard thermodynamic properties of this compound, calculated using the measured temperature dependence of EMF values are the next: $\Delta G^0_{f,298} = -330.1 \pm 6.6$ kJ·M^{-1}, $\Delta H^0_{f,298} = -322.1 \pm 7.4$ kJ·M^{-1}, $S^0_{f,298} = 319.0 \pm 11.3$ J·(M·K)$^{-1}$ (Moroz et al. 2018b) [$\Delta G^0_{f,298} = -332.14 \pm 14.12$ kJ·M^{-1}, $\Delta H^0_{f,298} = -330.18 \pm 12.17$ kJ·M^{-1}, $S^0_{f,298} = 285.51 \pm 23.99$ J·(M·K)$^{-1}$ (Moroz et al. 2018c); $\Delta G^0_{f,298} = -339.6 \pm 10.5$ kJ·M^{-1}, $\Delta H^0_{f,298} = -345.3 \pm 15.4$ kJ·M^{-1}, $S^0_{f,298} = 272.8 \pm 10.1$ J·(M·K)$^{-1}$ (Moroz et al. 2020b).

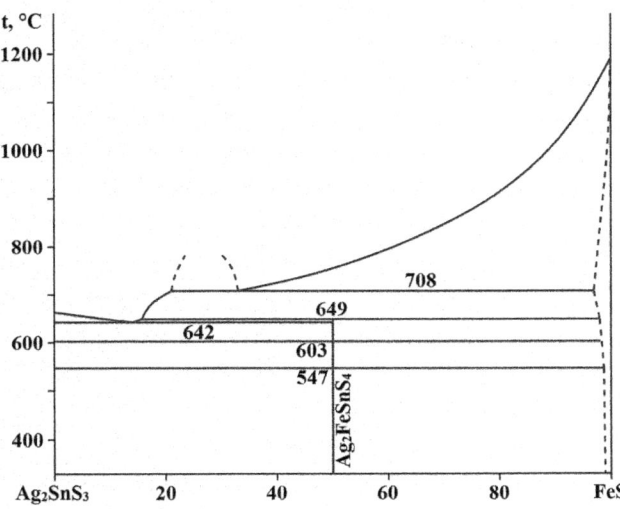

FIGURE 7.34 Phase diagram of the Ag_2SnS_3–FeS system. (From Moroz, M., et al., *J. Electron. Mater.*, **47**(9), 5433, 2018.)

The enthalpy of the phase transition of Ag_2FeSnS_4 is 1.60 ± 0.02 kJ·M^{-1} at 547°C (Moroz et al. 2018b).

Two more quaternary compounds, **$Ag_{0.5}Fe_{0.5}[FeSn]S_4$** and **$Ag_2FeSn_3S_8$**, are formed in the Sn–Ag–Fe–S system. First of them crystallizes in the cubic structure with the lattice parameter $a = 1053$ pm (Harada 1973). It was prepared by the usual ceramic techniques. The appropriate stoichiometric mixture of Ag, Fe, SnS_2, and S was fired using the following two-step process. First, the mixed powder was heated at 250–700°C for 3 days in either an evacuated Pyrex or quartz tube, and the prefired polysulfides thus obtained were ground and pressed into a pellet. The material was then refired in an evacuated quartz tube for about a week at 700°C. when necessary, the refiring procedure was repeated until single-phase products were obtained.

$Ag_2FeSn_3S_8$ melts incongruently at 758°C (Moroz et al. 2018b). It undergoes a polymorphic transformation at 545–558°C due to the existence of a homogeneity region (65–67 mol% FeS). One of the modifications of this compound crystallizes in the cubic structure with the lattice parameter $a = 1057.23 \pm 0.04$ pm and a calculated density of 4.970 g·cm^{-3} (Garg et al. 2003a), and the second modification (mineral toyohaite) crystallizes in the tetragonal structure with the lattice parameters $a = 746.4 \pm 0.3$, $c = 1080 \pm 2$ pm, and a calculated density of 4.94 g·cm^{-3} (Yajima et al. 1991; Jambor and Puziewicz 1992). This compound is also thermally unstable (Moroz et al. 2018b). It was obtained through solid-state synthesis from finely dispersed mixture of Ag_2SnS_3, FeS, and SnS_2 (molar ratio 1:1:2) at 600–680°C for 200 h (Moroz et al. 2018b,c). The high-temperature annealing of the compound was performed at 410–480°C for 240 h. $Ag_2FeSn_3S_8$ was also synthesized using a stoichiometric amounts of Ag, Fe, Sn, and S, which were mixed in a N_2-filled glove box and sealed in silica ampoule under vacuum (10^{-3} Pa) (Garg et al. 2003a). The ampoule was heated at 480°C for 8 days and then cooled to room temperature after switching off the furnace. The product was obtained in the form of black crystals.

The standard thermodynamic properties of $Ag_2FeSn_3S_8$, calculated using the measured temperature dependence of EMF values are the next: $\Delta G^0_{f,298} = -640.4 \pm 12.1$ kJ·M^{-1}, $\Delta H^0_{f,298} = -665.8 \pm 12.8$ kJ·M^{-1}, $S^0_{f,298} = 437.6 \pm 14.1$ J·(M·K)$^{-1}$ (Moroz et al. 2018b) [$\Delta G^0_{f,298} = -647.87 \pm 28.10$ kJ·M^{-1}, $\Delta H^0_{f,298} = -694.44 \pm 24.22$ kJ·M^{-1}, $S^0_{f,298} = 678.88 \pm 47.55$ J·(M·K)$^{-1}$ (Moroz et al. 2018c); $\Delta G^0_{f,298} = -633.4 \pm 18.2$ kJ·M^{-1}, $\Delta H^0_{f,298} = -655.6 \pm 23.7$ kJ·M^{-1}, $S^0_{f,298} = 448.8 \pm 15.5$ J·(M·K)$^{-1}$ (Moroz et al. 2020b). The enthalpy of the phase transition of this compound is 3.40 ± 0.04 kJ·M^{-1} at 547°C (Moroz et al. 2018b).

7.128 Tin–Gold–Barium–Sulfur

The **Au_2BaSnS_4** quaternary compound, which crystallizes in the orthorhombic structure with the lattice parameters $a = 1098.2 \pm 0.4$, $b = 1109.3 \pm 0.4$, $c = 665.2 \pm 0.4$ pm, and the calculated and experimental densities of 6.34 and 6.78 g·cm^{-3}, respectively (Teske 1978). This compound was obtained by a solid-state reaction from the mixture of binary sulfides at 400–540°C in a sealed quartz ampoule with sintered corundum crucible, excess S, and under argon-reduced pressure. In the usual way of working, with repeated interruptions to the annealing process for renewed mixing, the reaction is complete after a total annealing time of 20–30 h. The light red microcrystalline sample always contains some metallic gold as a minor impurity. After an annealing period of 10 weeks, with a slow increase in temperature from 400°C to a maximum of 560°C without interrupting the test, some samples contained dark-red, shiny small crystal needles that could be seen under the microscope. Au_2BaSnS_4 has low thermal stability.

7.129 Tin–Calcium–Lanthanum–Sulfur

The $Ca_3La_2Sn_3S_{12}$ quaternary compound, which crystallizes in the hexagonal structure with the lattice parameters $a = 1159.97 \pm 0.06$, $c = 394.07 \pm 0.04$ pm, a calculated density of 4.118 g·cm^{-3}, and an energy gap of 1.45 eV, is formed in the Sn–Ca–La–S system (Zeng et al. 2012). Single crystals of this compound were obtained via a precursor/flux method. La, CaS, Sn, and S (molar ratio 1:1.5:1.5:4.52; mixture A) or La_2S_3, CaS, Sn, and S (molar ratio 1:2:3:6.05; mixture B) were mixed together and ground thoroughly within a N_2-filled glove box, pressed into pellets, and then sealed in an evacuated fused silica ampoule. The ampoule was then put in a furnace and heated slowly to 1000°C, kept at this temperature for 110 h with one intermediate grinding and re-pressing of the materials. The product obtained from this procedure was mixed with KBr according to the two next schemes. (1) The mixture A (0.298 g) and KBr (2.920 g) was heated to 900°C in 60 h, held at this temperature for 240 h, cooled to 630°C in 90 h, annealed at 630°C for 90 h, cooled to 360°C in 90 h, and then the furnace was switch off; (2) The mixture B (0.880 g) and KBr (2.148 g) was heated to 600°C in 9 h, held at this temperature for 15 h, heated to 850°C in 4.5 h, annealed at this temperature for 120 h, cooled to 700°C in 50 h, heated to 860°C in 3 h, cooled to 560°C in 150 h, and then the furnace was switch off. Stable in air, black crystals of $Ca_3La_2Sn_3S_{12}$, as a major byproduct, were isolated manually from the residues after the fluxes were removed by washing with distilled water.

7.130 Tin–Calcium–Erbium–Sulfur

The $Ca_3Er_2Sn_3S_{12}$ quaternary compound, which crystallizes in the hexagonal structure with the lattice parameters $a = 1144.8 \pm 0.3$, $c = 387.80 \pm 0.11$ pm, and a calculated density of 4.510 g·cm^{-3}, is formed in the Sn–Ca–Er–S system (Zeng et al. 2012). Single crystals of this compound were obtained via a precursor/flux method. Er_2S_3, CaS, and SnS_2 (molar ratio 3.54:5.31:7.08; mixture A) or Er_2S_3, CaS, Sn, and S (molar ratio 0.710:1.070:1.420:2.844; mixture B) were mixed together and ground thoroughly within a N_2-filled glove box, pressed into pellets, and then sealed in an evacuated fused silica ampoule. The ampoule was then put in a furnace and heated slowly to 1000°C or 950°C, kept at this temperature for 316 or 120 h for mixture A and mixture B, respectively, with one intermediate grinding and re-pressing of the materials. The product obtained from this procedure was mixed with KBr according to the two

next schemes. (2) The mixture A (0.256 g) and KBr (2.208 g) was heated to 850°C in 20 h, held at this temperature for 216 h, cooled to 700°C in 45 h, heated to 860°C for 0.5 h, cooled to 600°C in 144 h, held at 600°C for 15 h, and then the furnace was switched off; (2) (1) The mixture B (0.324 g) and KBr (1.872 g) was heated to 600°C in 20 h, held at this temperature for 15 h, heated to 850°C in 4.5 h, annealed at this temperature for 240 h, cooled to 700°C in 50 h, heated to 860°C in 3 h, cooled to 560°C in 150 h, and then the furnace was switched off. Stable in air, black crystals of $Ca_3Er_2Sn_3S_{12}$ as a minor phase were isolated manually from the residues after the fluxes were removed by washing with distilled water.

7.131 Tin–Strontium–Zinc–Sulfur

The $SrZnSnS_4$ quaternary compound, which crystallizes in the orthorhombic structure with the lattice parameters $a = 1263.36 \pm 0.04$, $b = 2092.42 \pm 0.08$, $c = 2085.91 \pm 0.09$ pm, a calculated density of 3.854 g·cm^{-3}, and an energy gap of 2.83 eV, is formed in the Sn–Sr–Zn–S system (Zhang et al. 2019). The pure phase sample of this compound was synthesized through high-temperature solid-state reaction. SrS, Zn, Sn, and S (molar radio o 1:1:1:3) were ground, then loaded into a silica tube coated with carbon in a sealed under vacuum (<10^{-3} Pa) ampoule, and finally, moved into a furnace with the following controlled warming procedure: heated to 800°C within 30 h, dwelled there for 30 h, and cooled slowly (6°C·h^{-1}) until ambient temperature. At last, the powder of $SrZnSnS_4$ was obtained. The prepared powder was mixed with KI flux (molar ratio 1:1), and then the mixture was loaded into the graphitic crucible and placed in the silica tube. Next, the tube was sealed under vacuum (<10^{-3} Pa) and put into a muffle furnace with the following controlled heating ramp: heated to 800°C within 30 h, kept at this temperature for 2 days, and slowly cooled to ambient temperature with a rate of 3°C·h^{-1}. Finally, the product was washed with water to remove excess KI, and yellow air-stable crystals were obtained. The starting reagents were stored in an Ar-filled glove box without oxygen and moisture.

7.132 Tin–Strontium–Cadmium–Sulfur

Two quaternary compounds, $SrCdSnS_4$ and $SrCdSn_2S_6$, are formed in the Sn–Sr–Cd–S system. The first of them melts congruently at 889°C and crystallizes in the orthorhombic structure with the lattice parameters $a = 2090.17 \pm 0.14$, $b = 2106.26 \pm 0.12$, $c = 1306.76 \pm 0.07$ pm, a calculated density of 4.128 g·cm^{-3}, and an energy gap of 2.05 eV (Lin et al. 2019a). Single crystals of this compound were obtained through solid-state reactions. Dark-red crystals were crystallized from a reaction mixture containing Sr (0.66 mM), Cd (0.66 mM), Sn (0.66 mM), and S (2.65 mM). The mixture was loaded into a quartz tube and flame-sealed under a vacuum (~10^{-2} Pa). The tube was placed into a furnace, firstly heated from room temperature to 200°C in 4 h, kept at that temperature for 2 h, and then heated to 400°C in 5 h and kept for another 5 h, subsequently heated to 850°C in 22.5 h, kept for 5 days, and then cooled to room temperature in 6 days before switching off the furnace. The obtained crystals are stable in the air and moisture conditions.

$SrCdSn_2S_6$ was prepared using an ion-exchange reaction, which was carried out at room temperature by stirring, for 10–12 h, 0.25 g of polycrystalline samples of α-$K_2CdSn_2S_6$ with 20 mL aqueous solution of $SrCl_2·6H_2O$ (molar ratio 1:2.5) (Pogu et al. 2019). The ion-exchanged solid product was filtered, washed with deionized water, and dried at room temperature in the open air.

7.133 Tin–Strontium–Mercury–Sulfur

The **$SrHgSnS_4$** quaternary compound, which is stable up to 710°C and crystallizes in the orthorhombic structure with the lattice parameters $a = 1040.72 \pm 0.13$, $b = 1048.73 \pm 0.10$, $c = 655.78 \pm 0.06$ pm, a calculated density of 4.966 g·cm^{-3}, and an energy gap of 2.72 eV, is formed in the Sn–Sr–Hg–S system (Guo et al. 2019a). To prepare single crystals of this compound, equimolar mixtures of SrS, HgS, and SnS_2 were ground under an Ar atmosphere in a glove box. Then the mixture of reagents was placed into a quartz tube and sealed under a vacuum of 10^{-3} Pa. The tube was heated to 1000°C in 20 h, maintained at that temperature for 72 h, cooled to 400°C in 200 h, and then the furnace was shut off. Orange crystals of $SrHgSnS_4$ were found, which are air and moisture stable.

7.134 Tin–Strontium–Yttrium–Sulfur

The **$Sr_2Y_{2.67}Sn_3S_{12}$** quaternary compound, which crystallizes in the hexagonal structure with the lattice parameters $a = 1160.0 \pm 0.7$, $c = 394.8 \pm 0.3$ pm, and a calculated density of 4.163 g·cm^{-3}, is formed in the Sn–Sr–Y–S system (Zeng et al. 2012). Single crystals of this compound were obtained via a precursor/flux method. Y, SrS, Sn, and S (molar ratio 0.200:0.300:0.300:0.901) were mixed together and ground thoroughly within a N_2-filled glove box, pressed into pellets, and then sealed in an evacuated fused silica ampoule. The ampoule was then put in a furnace and heated slowly to 1000°C, kept at this temperature for 120 h with one intermediate grinding and re-pressing of the materials. The product obtained from this procedure (0.336 g) was mixed with KBr (2.460 g), heated to 600°C in 9 h, held at this temperature for 15 h, heated to 850°C in 4.5 h, annealed at this temperature for 120 h, cooled to 700°C in 50 h, heated to 860°C in 3 h, cooled to 560°C in 150 h, and then the furnace was switch off. Stable in air black crystals of $Ca_3Er_2Sn_3S_{12}$ as a minor phase were isolated manually from the residues after the fluxes were removed by washing with distilled water.

7.135 Tin–Strontium–Fluorine–Sulfur

The **$Sr_2SnS_3F_2$** quaternary compound, which crystallizes in the orthorhombic structure with the lattice parameters $a = 593.65 \pm 0.09$, $b = 1884.2 \pm 0.2$, $c = 594.90 \pm 0.09$ pm, and

an energy gap of 3.06 eV, is formed in the Sn–Sr–F–S system (Kabbour et al. 2006). It was synthesized using a high-temperature ceramic method. A stoichiometric proportion of SrF_2, SrS, SnS, and S was weighted and ground in a glove box under an Ar atmosphere. This mixture was subsequently pressed into pellets and sealed under a vacuum in a silica tube. The tube was then heated to 700°C (at 50°C·h^{-1}) for 12 h.

7.136 Tin–Strontium–Manganese–Sulfur

The **$Sr_3MnSn_2S_8$** quaternary compound, which crystallizes in the cubic structure with the lattice parameter $a = 1422.87 \pm 0.06$ pm, a calculated density of 3.743 g·cm^{-3}, and an energy gap 3.02 eV, is formed in the Sn–Sr–Mn–S system (Liu et al., 2019a). To synthesize this compound, SrS, Mn, Sn, and S were mixed (molar radio 3:1:2:5). Next, the mixed reagents were loaded into the graphite crucible. Then, they were put into the silica tube sealed under vacuum (<10^{-3} Pa) and transferred to the furnace. The furnace was heated to 900°C in one day, and then there was a 20 h process of heat preservation. Finally, the sample was cooled to room temperature. To grow single crystals of $Sr_3MnSn_2S_8$, the prepared powder was mixed with flux KI at the ratio of 1:1, and then the mixture was loaded into the graphite crucible. Next, the graphite crucible was put into the silica tube and then sealed under vacuum, and transferred to a programmed muffle furnace. Afterwards, the programmed temperature was heated to 900°C in one day and maintained at the current temperature for 2 days, subsequently, the temperature was cooled to 300°C at the slow rate of 2°C·h^{-1}, and finally, the sample was cooled to ambient temperature. The product was cleaned with deionized water to obtain yellow-green crystals which are stable in air.

7.137 Tin–Barium–Zinc–Sulfur

The **$BaZnSnS_4$** quaternary compound, which crystallizes in the orthorhombic structure with the lattice parameters $a = 2196.4$, $b = 2150.4$, $c = 1270.1$ pm and the calculated and experimental densities of 3.99 and 3.84 g·cm^{-3}, respectively, is formed in the Sn–Ba–Zn–S system (Teske 1980a). This compound was obtained by the solid-state reactions at 700–840°C. It is stable in the air.

7.138 Tin–Barium–Cadmium–Sulfur

Two quaternary compounds, **$BaCdSnS_4$** and **$Ba_3CdSn_2S_8$**, are formed in the Sn–Ba–Cd–S system. $BaCdSnS_4$ crystallizes in the orthorhombic structure with the lattice parameters $a = 2156.6 \pm 1.3$, $b = 2176.0 \pm 1.3$, $c = 1311.0 \pm 0.8$ pm, a calculated density of 4.290 g·cm^{-3}, and an energy gap of 2.30 eV (Zhen 2016b) [$a = 2186 \pm 2$, $b = 2169 \pm 1$, $c = 1318.0 \pm 0.5$ pm, and the calculated and experimental densities of 4.23 and 4.11 g·cm^{-3}, respectively (Teske 1980b)]. The crystals of the title compound were prepared through the conventional high-temperature solid-state method in the vacuum-sealed silica tube (Zhen et al. 2016b). A stoichiometric mixture of Ba (1 mM), CdS (1 mM), Sn (1 mM), and S (3 mM) was loaded into a graphite crucible embedded within a silica tube, and then the tube was flame-sealed under a high vacuum of 10^{-3} Pa. The tube was put into a muffle furnace. In order to avoid latent explosion of silica tube derived from the high vapor pressure of S at 440°C, one critical step is chosen: held at 400°C for 10 h to achieve the reaction of partial S. Subsequently, the furnace was heated to 950°C in 30 h, annealed at this temperature for 3 days, and cooled to 400°C at a rate of 4°C·h⁻¹, and finally cooled to room temperature in 40 h. The product was washed with DMF to remove the other by-products. Finally, yellow crystals of $BaCdSnS_4$ were obtained after dried in air and found that they are stable in air for several months.

$Ba_3CdSn_2S_8$ crystallizes in the cubic structure with the lattice parameter $a = 1467.9 \pm 0.3$ pm, a calculated density of 4.277 g·cm⁻³, and an energy gap of 2.75 eV (Zhen et al. 2016b) [$a = 1472.35 \pm 0.07$ pm and the calculated and experimental densities of 4.24 and 4.21 ± 0.08 g·cm⁻³, respectively (Teske 1985)]. To prepare this compound, a stoichiometric mixture of Ba (3 mM), CdS (1 mM), Sn (2 mM), and S (7 mM) was loaded into a silica tube, and the preparation process is similar to that of $BaCdSnS_4$. After the reaction, the product was washed with DMF and dried in 100°C. Finally, many yellow stable in the air crystals of this compound were obtained. For the synthesis of both compounds, all the raw reactants were operated under a dry Ar atmosphere in a glove box that is used to avoid the effect of oxygen and moisture.

7.139 Tin–Barium–Mercury–Sulfur

The **BaHgSnS₄** quaternary compound, which has a high thermal stability and a phase transition, is formed in the Sn–Ba–Hg–S system. β-$BaHgSnS_4$ crystallizes in the orthorhombic structure with the lattice parameters $a = 1084.50 \pm 0.07$, $b = 1080.42 \pm 0.06$, $c = 661.78 \pm 0.04$ pm, a calculated density of 5.010 g·cm⁻³, and an energy gap of 2.77 eV (Guo et al. 2019a) [$a = 1080.4$, $b = 1084.0$, and $c = 661.3$ pm the calculated and the experimental densities of 4.91 and 4.96 g·cm⁻³, respectively (Teske 1980c)]. To prepare single crystals of this compound, equimolar mixtures of BaS, HgS, and SnS₂ were ground under an Ar atmosphere in a glove box. Then the mixture of reagents was placed into a quartz tube and sealed under a vacuum of 10^{-3} Pa. The tube was heated to 1000°C in 20 h, maintained at that temperature for 72 h, cooled to 400°C in 200 h, and then the furnace was shut off. Orange crystals of β-$BaHgSnS_4$ were found, which are air and moisture stable.

7.140 Tin–Barium–Gallium–Sulfur

The **Ba₄Ga₄SnS₁₂** quaternary compound, which is thermally stable up to 1073°C and crystallizes in the tetragonal structure with the lattice parameters $a = 1306.65 \pm 0.06$, $c = 630.40 \pm 0.05$ pm, a calculated density of 4.109 g·cm⁻³, and an energy gap of 2.90 eV, is formed in the Sn–Ba–Ga–S system (Duan et al. 2017). To synthesize this compound, a total 0.3 g of Ba, Ga, Sn and S reactants with the molar ratio of 4:4:1:12 were

mixed in a silica crucible in the glove box. The silica crucible was embedded within another larger silica tube, and then this silica tube was flame-sealed under a vacuum of 10^{-3} Pa. The tube was heated to 900°C in 72 h and holding 100 h at this temperature. Subsequently, it was cooled to 300°C in 100 h before the furnace was turned off. The light-yellow crystals of this compound, which are stable in the air for at least 10 months at room temperature, were obtained.

7.141 Tin–Barium–Cerium–Sulfur

The **BaCeSn₂S₆** quaternary compound, which crystallizes in the orthorhombic structure with the lattice parameters $a = 406.65 \pm 0.08$, $b = 1985.9 \pm 0.4$, $c = 1187.3 \pm 0.2$ pm, and a calculated density of 4.899 g·cm⁻³, is formed in the Sn–Ba–Ce–S system (Feng et al. 2014). To prepare this compound, the mixture of BaS (1 mM), SnS₂ (2 mM), Ce (1 mM), and S (1 mM) were ground and loaded into a fused silica tube under an Ar atmosphere in a glove box. The tube was flame-sealed under a high vacuum (10^{-3} Pa) and then placed in a furnace. It was heated to 1100°C in 24 h, left for 48 h, cooled to 420°C at a rate of 3°C·h⁻¹, and finally cooled to room temperature by switching off the furnace. Dark-red air-stable crystals of the title compound were found.

7.142 Tin–Barium–Praseodymium–Sulfur

The **BaPrSn₂S₆** quaternary compound, which crystallizes in the orthorhombic structure with the lattice parameters $a = 404.78 \pm 0.02$, $b = 1989.14 \pm 0.07$, $c = 1193.03 \pm 0.05$ pm, and a calculated density of 4.896 g·cm⁻³, is formed in the Sn–Ba–Pr–S system (Feng et al. 2014). It was synthesized in the same way as $BaCeSn_2S_6$ was prepared using Pr instead of Ce. Dark-red air-stable crystals of the title compound were found.

7.143 Tin–Barium–Neodymium–Sulfur

The **BaNdSn₂S₆** quaternary compound, which crystallizes in the orthorhombic structure with the lattice parameters $a = 400.98 \pm 0.08$, $b = 1976.1 \pm 0.4$, $c = 1184.1 \pm 0.2$ pm, and a calculated density of 5.036 g·cm⁻³, is formed in the Sn–Ba–Nd–S system (Feng et al. 2014). It was synthesized in the same way as $BaCeSn_2S_6$ was prepared using Nd instead of Ce. Dark-red air-stable crystals of the title compound were found.

7.144 Tin–Barium–Lead–Sulfur

The **Ba₅Pb₂Sn₃S₁₃** quaternary compound, which crystallizes in the orthorhombic structure with the lattice parameters $a = 1220.8 \pm 0.5$, $b = 2349.9 \pm 1.0$, $c = 883.8 \pm 0.4$ pm, a calculated density of 4.863 g·cm⁻³, and an energy gap of ≈2.11 eV, is formed in the Sn–Ba–Pb–S system (Abudurusuli et al. 2017). This compound was synthesized as follows. The mixtures of BaS, PbS, Sn, and S (molar ratio of 1:1:1:2) were planned to obtain the **BaPbSnS₄** crystal, which were ground and loaded into a fused silica tube under Ar atmosphere in a glove box, then this tube was flame-sealed under a high vacuum of 10^{-3} Pa. The tube was then placed in a furnace and heated to 850°C in 30 h, left for 100 h to ensure the mixture completely melted, cooled to 300°C at a rate of 3°C·h^{-1}, and finally cooled to room temperature by switching off the furnace. This compound can also be prepared using the stoichiometric mixture of the precursors. Obtained products were washed by DMF to remove the other by-products. Many block-shaped crystals with bright red color were found in the ampoule. These crystals are stable in air and moisture conditions.

7.145 Tin–Barium–Antimony–Sulfur

The $Ba_{2.77}Sb_{2.16}SnS_8$ quaternary compound, which crystallizes in the orthorhombic structure with the lattice parameters $a = 1171.25 \pm 0.18$, $b = 438.70 \pm 0.09$, $c = 2977.8 \pm 0.5$ pm, a calculated density of 4.418 g·cm^{-3}, and an energy gap of 1.50 eV, is formed in the Sn–Ba–Sb–S system (Zhang 2018). It was synthesized from the Ba, Sn, Sb, S mixture (300 mg) in a stoichiometric ratio, which was loaded into a graphite crucible sealed in an evacuated fused silica tube. The sample was heated to 950°C in 24 h and kept at that temperature for 5 days. After that, it was cooled (5°C·h^{-1}) to 320°C; then, the furnace was turned off. The black block crystals of the title compound were obtained. They are stable in air.

7.146 Tin–Barium–Bismuth–Sulfur

The $Ba_{3.14}Bi_{2.39}Sn_{0.61}S_8$ quaternary compound, which crystallizes in the orthorhombic structure with the lattice parameters $a = 1244.79 \pm 0.07$, $b = 431.40 \pm 0.02$, $c = 2878.60 \pm 0.17$ pm, a calculated density of 5.411 g·cm^{-3}, and an energy gap of 1.4 eV, is formed in the Sn–Ba–Bi–S system (Jana et al. 2022b). Black-colored block-shaped crystals of the title compound were obtained from the reaction of Ba, Sn, Bi, and S with the loaded composition of $Ba_{3.2}Bi_{2.4}Sn_{0.6}S_8$. The reactants were first loaded into a carbon-coated fused silica tube inside the Ar-filled glove box. The reaction tube was evacuated to ~10^{-2} Pa and sealed. The sealed tube was placed in a muffle furnace, which temperature was initially increased to 300°C in 12 h, and held constant for 6 h before increasing the temperature to 900°C in 48 h. The reaction mixture was annealed at 900°C for 72 h, and the temperature of the furnace was then slowly cooled down to 700°C with a cooling rate of 3°C·h^{-1} and annealed for 48 h at this temperature before cooling to 300°C in 100 h. Finally, the furnace was switched off, and the reaction mixture was allowed to cool down to room temperature. A black homogeneous-looking lump was obtained from the reaction.

7.147 Tin–Barium–Fluorine–Sulfur

The $Ba_2SnS_3F_2$ quaternary compound, which crystallizes in the orthorhombic structure with the lattice parameters $a = 620.37 \pm 0.04$, $b = 1929.50 \pm 0.11$, $c = 632.36 \pm 0.01$ pm,

and an energy gap of 3.21 eV, is formed in the Sn–Ba–F–S system (Kabbour et al. 2006). It was synthesized using a high-temperature ceramic method. A stoichiometric proportion of BaF_2, BaS, SnS, and S was weighted and ground in a glove box under Ar atmosphere. This mixture was subsequently pressed into pellets and sealed under a vacuum in a silica tube. The tube was then heated to 700°C (at 50°C·h^{-1}) for 12 h. White powder of the title compound was obtained.

7.148 Tin–Barium–Manganese–Sulfur

The $BaMnSnS_4$ quaternary compound, which melts congruently at 829°C and crystallizes in the orthorhombic structure with the lattice parameters $a = 2154.7 \pm 0.5$, $b = 2179.6 \pm 0.3$, $c = 1298.1 \pm 0.3$ pm, a calculated density of 3.828 g·cm^{-3}, and an energy gap of 1.90 eV (Lin et al. 2019b) [$a = 2173.1$, $b = 2144.9$, $c = 1295.1$ pm and the calculated and experimental densities of 3.81 and 3.74 g·cm^{-3}, respectively], is formed in the Sn–Ba–Mn–S system (Teske 1980a). Dark-red crystals of this compound were obtained from a reaction mixture containing Ba (0.66 mM), Mn (0.66 mM), Sn (0.66 mM), and S (2.65 mM), which was loaded into a quartz tube and flame-sealed under vacuum (~10^{-2} Pa). The tube was placed into a furnace, heated up to 250°C and held at that temperature for 2 h, then heated up to 875°C in 24 h and kept for 96 h, and then cooled down to 300°C at 4°C·h^{-1} before switching off the furnace. They are stable in the air and moisture conditions.

7.149 Tin–Zinc–Tellurium–Sulfur

SnTe–ZnS. The phase diagram of this system, constructed through DTA, XRD, metallography, and measuring of the microhardness, is a eutectic type (Figure 7.35) (Dubrovin et al.

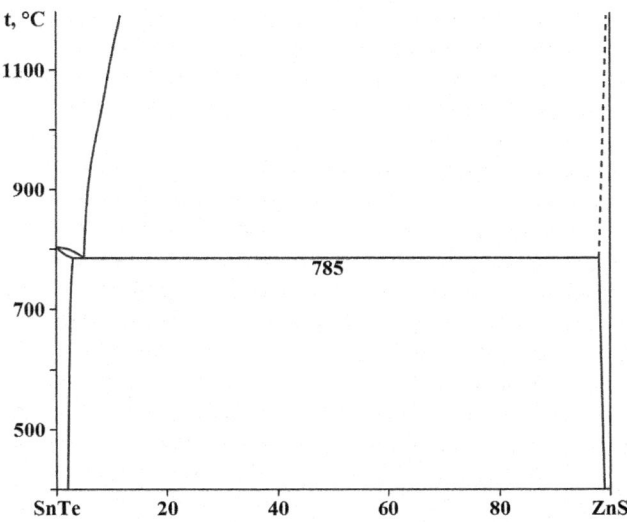

FIGURE 7.35 Phase diagram of the SnTe–ZnS system. (From Dubrovin, I.V. et al., *Izv. AN SSSR. Neorgan. mater.*, **19**(11), 1816, 1983.)

1983). The eutectic composition and temperature are 5 mol% ZnS and 785°C, respectively. The solubility of ZnS in SnTe at the eutectic temperature is equal to 3 and 2 mol% at 400°C, and the solubility of SnTe in ZnS is not higher than 1 mol%. The ingots for investigations were annealed at 700°C and 400°C for 120 and 250 h, respectively.

7.150 Tin–Zinc–Iron–Sulfur

The $Zn_{0.5}Fe_{0.5}[FeSn]S_4$ quaternary compound, which crystallizes in the cubic structure with the lattice parameter $a = 1034$ pm, is formed in the Sn–Zn–Fe–S system (Harada 1973). It was prepared by the usual ceramic techniques. The appropriate stoichiometric mixture of ZnS, Fe, SnS_2, and S were fired using the following two-step process. First, the mixed powder was heated at 250–700°C for 3 days in either evacuated Pyrex or quartz tube, and the prefired polysulfides thus obtained were ground and pressed into a pellet. The material was then refired in an evacuated quartz tube for about a week at 800°C. When necessary, the refiring procedure was repeated until single-phase products were obtained.

7.151 Tin–Cadmium–Mercury–Sulfur

SnS_2–CdS–HgS. Using co-precipitation of CdS with HgS and SnS_2 by the hydrogen sulfide and then heating up to 300–400°C, any quaternary compounds were obtained in this system (Kislinskaya 1974).

7.152 Tin–Cadmium–Cerium–Sulfur

The $CdCe_{2/3}Sn_2S_6$ quaternary compound is formed in the Sn–Cd–Ce–S system (Pogu et al. 2019). It was prepared using an ion-exchange reaction, which was carried out at room temperature by stirring, for 10–12 h, 0.25 g of polycrystalline samples of α-$K_2CdSn_2S_6$ with 20 mL aqueous solution of $CeCl_3 \cdot 7H_2O$ (molar ratio 1:1.5). The ion-exchanged solid product was filtered, washed with deionized water, and dried at room temperature in the open air.

7.153 Tin–Cadmium–Tellurium–Sulfur

CdS–SnTe. This section is a nonquasibinary section of the Sn–Cd–Te–S quaternary system (Dubrovin et al. 1985). The solubility of CdS in SnTe at 725°C is equal to 12 mol% and decreases to 2 mol% at 400°C. The solubility of SnTe in CdS is not higher than 1 mol%. Thermal effects at 605°C correspond to the polymorphous transformation of the solid solutions based on SnS forming at the exchange interaction of CdS and SnTe. This system was investigated through DTA, XRD, and metallography. The ingots for investigations were annealed at 400°C, 500°C, and 670°C for 250, 200, and 100 h, respectively.

7.154 Tin–Cadmium–Iron–Sulfur

The $Cd_{1-x}Fe_x[FeSn]S_4$ quaternary solid solution is formed in the Sn–Cd–Fe–S system (Harada 1973). It crystallizes in the cubic structure with the lattice parameter decreasing from $a = 1061$ pm for $x = 0$ ($Cd[FeSn]S_4$) to $a = 1035$ pm for $x = 0.9$. The solid solution was prepared by the usual ceramic techniques. The appropriate stoichiometric mixture of Fe, CdS, SnS_2, and S was fired using the following two-step process. First, the mixed powder was heated at 250–700°C for 3 days in either an evacuated Pyrex or quartz tube, and the prefired polysulfides thus obtained were ground and pressed into a pellet. The material was then refired in an evacuated quartz tube for about a week at 450–650°C (the firing temperature decreases with x increasing). When necessary, the refiring procedure was repeated until single-phase products were obtained. This solid solution is p-type semiconductor.

7.155 Tin–Mercury–Thallium–Sulfur

SnS_2–HgS–Tl_2S. The isothermal section of this system at 250°C is shown in Figure 7.36 (Olekseyuk et al. 2010a; Piskach et al. 2017). The $HgTl_2SnS_4$ quaternary compound, which has a narrow homogeneity region, exists in this quasiternary system at this temperature. It crystallizes in the tetragonal structure with the lattice parameters $a = 785.71 \pm 0.06$, $c = 669.89 \pm 0.07$ pm, and a calculated density of 6.876 g·cm^{-3} (Piskach et al. 2017) [$a = 785.86 \pm 0.03$ and $c = 670.05 \pm 0.03$ pm (Olekseyuk et al. 2010a)]. No solid solubility was found for

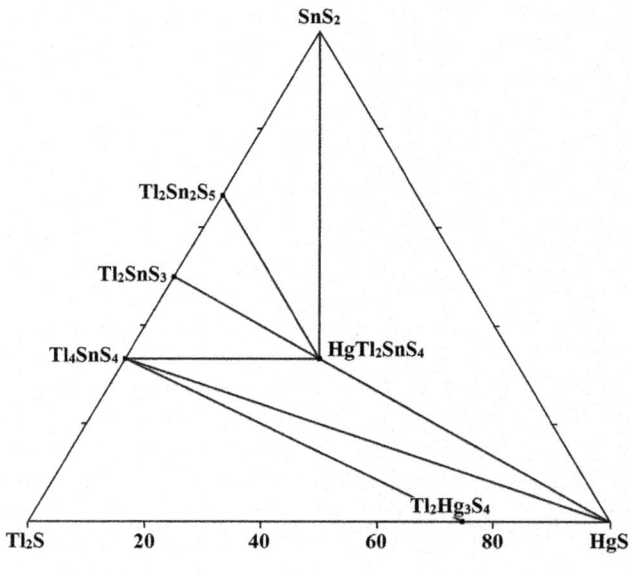

FIGURE 7.36 Isothermal section of the SnS_2–HgS–Tl_2S quasiternary system at 250°C. (From Olekseyuk, I.D. et al., *Nauk. Vsnyk Volyns'k. Nats. Univ. im. Lesi Ukrainky. Khim. nauky*, (30), 19, 2010.)

the system's binary components or the ternary compounds. The ingots for the investigations were annealed at 250°C for 250 h.

7.156 Tin–Mercury–Europium–Sulfur

The **HgEuSnS$_4$** quaternary compound, which decomposes at temperatures above 780°C and crystallizes in the orthorhombic structure with the lattice parameters $a = 1037.30 \pm 0.08$, $b = 1043.80 \pm 0.08$, $c = 656.80 \pm 0.04$ pm, a calculated density of 5.599 g·cm^{-3} and an energy gap of 2.14 eV, is formed in the Sn–Hg–Eu–S system (Xing et al. 2020a). Single crystals of this compound were obtained by spontaneous crystallization from melts. EuS (0.1 mM), HgS (0.1 mM), SnS$_2$ (0.1 mM) were fully milled in an Ar-filled glovebox. After that, the mixture of reagents was sealed into a quartz tube under a vacuum of 10^{-3} Pa and underwent a subsequent high-temperature solid-state reaction in a program-controlled muffle furnace with the following heating process: heated to 900°C at 50°C·h^{-1}, standing at this temperature for 3 days, cooled to 200°C at 4°C·h^{-1}, and then cooled naturally to ambient temperature by turning off the furnace. As a result, red crystals of HgEuSnS$_4$ were obtained.

7.157 Tin–Mercury–Bromine–Sulfur

SnBr$_2$–HgS. There is a metastable eutectic in this system, which melts at 202°C (Blachnik et al. 1996). **Hg$_2$SnS$_2$Br$_2$** quaternary compound is formed in this system. It melts incongruently at 292°C and crystallizes in the monoclinic structure with the lattice parameters $a = 935.6 \pm 0.1$, $b = 802.8 \pm 0.1$, $c = 1063.0 \pm 0.2$ pm, $\beta = 103.06 \pm 0.1°$, and a calculated density of 6.532 g·cm^{-3} (Blachnik et al. 1996) [$a = 950.3$, $b = 815.2$, $c = 1071.8$ pm, and $\beta = 102.68°$ according to *ab initio* calculation (Ruiz and Payne 1998)]. This compound was prepared by solid-state reaction of HgS and SnBr$_2$, which were mixed in stoichiometric amounts and loaded into a dry silica glass tube that was evacuated and then sealed under vacuum. The ampoule was placed in a furnace and annealed at 220°C for 4 weeks. With this procedure, transparent trapezoidal single crystals of yellow color were obtained; larger crystals with a diameter of more than 2 mm were red.

According to the data of Ruiz and Payne (1998), α-HgS reacts with SnBr$_2$ at 220°C to form the 1D intercalation 2HgS·SnBr$_2$ compound. In this compound, the helical structure of the HgS chains remains, but the turns are modified to give tetragonal symmetry. The formation of 2HgS·SnBr$_2$ can be understood as a process of miscibility between the two solid phases.

7.158 Tin–Mercury–Iodine–Sulfur

The **Hg$_2$SnS$_2$I$_2$** quaternary compound, which crystallizes in the orthorhombic structure with the lattice parameters $a = 928.2 \pm 0.3$, $b = 956.5 \pm 0.2$, and $c = 1346.0 \pm 0.5$ pm, is formed in the Sn–Hg–I–S system (Blachnik et al. 1986).

7.159 Tin–Mercury–Iron–Sulfur

The **Hg$_{0.5}$Fe$_{0.5}$[FeSn]S$_4$** quaternary compound, which crystallizes in the cubic structure with the lattice parameter $a = 1056$ pm, is formed in the Sn–Hg–Fe–S system (Harada 1973). It was prepared by the usual ceramic techniques. The appropriate stoichiometric mixture of HgS, Fe, SnS$_2$, and S was fired using the following two-step process. First, the mixed powder was heated at 250–700°C for 3 days in either an evacuated Pyrex or quartz tube, and the prefired polysulfides thus obtained were ground and pressed into a pellet. The material was then refired in an evacuated quartz tube for about a week at 650°C. When necessary, the refiring procedure was repeated until single-phase products were obtained.

7.160 Tin–Gallium–Indium–Sulfur

SnS$_2$ Ga$_2$S$_3$–In$_2$S$_3$. A study of SnS$_2$, Ga$_2$S$_3$ and In$_2$S$_3$ co-precipitation processes shows that a certain dependence of the degree of precipitation of gallium and indium on their atomic ratio to tin is observed in the system, and that the precipitation of gallium and indium is practically independent of their ratio in solution (Kislinskaya 1974).

7.161 Tin–Gallium–Thallium–Sulfur

The **GaTlSnS$_4$** quaternary compound, which has two polymorphic modifications, is formed in the Sn–Ga–Tl–S system. α-GaTlSnS$_4$ crystallizes in the monoclinic structure with the lattice parameters $a = 725.16 \pm 0.02$, $b = 1166.29 \pm 0.04$, $c = 1750.44 \pm 0.05$ pm, $\beta = 95.267 \pm 0.003°$, a calculated density of 4.695 g·cm^{-3}, and an energy gap of 2.49 eV, and β-CsGaSnS$_4$ crystallizes in the cubic structure with the lattice parameter $a = 1299.38 \pm 0.04$ pm, a calculated density of 4.732 g·cm^{-3}, and an energy gap of 2.50 eV (Friedrich et al. 2020) [$a = 1324 \pm 1$ pm and an energy gap of 1.90 eV (Kumari and Vidyasagar 2007)].

Polycrystalline samples of yellow α-GaTlSnS$_4$ were obtained by melting Tl with a stoichiometric mixture of Ga$_2$S$_3$, Sn, and S in a sealed quartz ampoule at 1000°C for 2 days and cooling to room temperature with a rate of 15°C·h^{-1} (Friedrich et al. 2020). Polycrystalline batches of yellow β-GaTlSnS$_4$ were obtained by annealing Tl with stoichiometric mixtures of Ga$_2$S$_3$, Sn, and S in a sealed evacuated quartz tube at 500°C for 7 days. After the initial annealing, the crude product was thoroughly ground in an agate mortar under a N$_2$ atmosphere and pressed into a pellet. The pellet was annealed in an evacuated quartz tube for an additional week at 500°C.

7.162 Tin–Gallium–Lanthanum–Sulfur

The **Ga$_3$LaSnS$_7$** quaternary compound, which melts incongruently at 807°C and crystallizes in the monoclinic structure with the lattice parameters $a = 1135.55 \pm 0.09$, $b = 961.77 \pm 0.05$,

$c = 1151.37 \pm 0.08$ pm, $\beta = 115.693 \pm 0.009°$, a calculated density of 4.052 g·cm^{-3}, and an energy gap of 2.39 eV, is formed in the Sn–Ga–La–S system (Tang et al. 2020). Crystals of this compound were initially identified from a reaction of La, SnS, Ga$_2$S$_3$, and S in the molar ratio of 2:2:3:3. The mixture with a total mass of ~0.3 g was pressed into a pellet and loaded into a silica tube, and then this tube was flame-sealed under a high vacuum of 10^{-3} Pa and placed in a furnace. The tube was heated to 900°C over 15 h, kept at that temperature for 48 h, cooled to 300°C over 120 h, and then slowly cooled to room temperature by switching off the furnace. Many small block-shaped crystals of Ga$_3$LaSnS$_7$ with yellow color were found in the tube. These crystals are stable in the air for months. A polycrystalline sample of Ga$_3$LaSnS$_7$ was obtained in an optimized synthesis using a slightly different heat treatment. As before, a mixture of La (80 mg), SnS (87 mg), Ga$_2$S$_3$ (205 mg), and S (28 mg) was placed in a silica tube, and then this tube was flame-sealed under a high vacuum of 10^{-3} Pa. The tube was heated to 700°C over 10 h and kept there for 96 h, after which the furnace was turned off. The sample was finely ground, loaded into a new tube, and reheated at 700°C for another 96 h.

7.163 Tin–Indium–Thallium–Sulfur

The **InTlSnS$_4$** quaternary compound, which crystallizes in the hexagonal structure with the lattice parameters $a = 367.6 \pm 0.5$, $c = 1517.2 \pm 0.2$, and an energy gap of 1.8 eV, is formed in the Sn–In–Tl–S system (Yohannan and Vidyasagar 2016b). It was synthesized by heating in a continuous stream of CS$_2$ vapor appropriate stoichiometric mixture of Tl, In$_2$O$_3$, and SnO$_2$ with one intermittent grinding. An alumina boat was loaded with the reactant mixture and placed in a ceramic tube in a horizontal tubular furnace. A continuous stream of N$_2$ was bubbled through the CS$_2$ liquid and then passed through the ceramic tube to provide CS$_2$ vapor for the synthetic reaction.

7.164 Tin–Indium–Lanthanum–Sulfur

The **InLa$_3$Sn$_{0.5}$S$_7$** quaternary compound, which crystallizes in the hexagonal structure with the lattice parameters $a = 1029.93 \pm 0.11$, $c = 609.21 \pm 0.06$, a calculated density of 4.838 g·cm^{-3}, and an energy gap of 1.45 eV, is formed in the Sn–In–La–S system (Iyer et al. 2017a). To synthesize this compound, a stoichiometric mixture of the elements with a total mass of 0.2 g was pressed into a pellet and loaded into a fused silica tube, which was evacuated and sealed. The tube was heated at 1050°C for 4 days and cooled to 600°C over another 4 days, and then the furnace was turned off.

7.165 Tin–Indium–Selenium–Sulfur

The **~In$_{19}$Sn$_{12}$(Se,S)$_{41}$** quaternary phase, which crystallizes in the monoclinic structure with the lattice parameters $a = 5629 \pm 4$, $b = 392.4 \pm 0.3$, $c = 1591.6 \pm 1.0$ pm, $\beta = 102.56 \pm 0.01°$, and a calculated density of 5.85 g·cm^{-3}, is formed in the Sn–In–Se–S system (Topa and Makovicky 2012). It was synthesized in the

dry condensed phase system Sn–In–Sb–S–Fe–S at 600°C from a mixture of pure elements. It occurs as small needle-like crystals together with Fe$_9$S$_{10}$, FeIn$_2$(S,Se)$_4$, and an intimate aggregate of ~**SbSn$_9$(Se,S)$_{10}$** and needle-like ~**Sn$_4$Sb$_3$In$_2$(S,Se)$_{11}$** that may have resulted from eutectic crystallization.

7.166 Tin–Indium–Cobalt–Sulfur

The **In$_x$Sn$_{2-x}$Co$_3$S$_2$** solid solution ($0 \leq x \leq 2$) is formed in the Sn–In–Co–S quaternary system. It crystallizes in the rhombohedral structure with the lattice parameters $a = 531.922 \pm 0.004$ and $c = 1347.99 \pm 0.02$ pm in hexagonal setting for $x = 1$ (Corps et al. 2015) [$a = 530$ and $c = 1346$ pm in hexagonal setting for $x = 1$ (Natarajan et al. 1988); melts at 890°C and crystallizes in the rhombohedral structure with the lattice parameters $a = 531.41 \pm 0.03$ and $c = 1348.3 \pm 0.2$ pm in hexagonal setting or $a = 544.17$ pm and $\alpha = 58.5°$ for $x = 1$ (Rothballer et al. 2013); $a = 531.24 \pm 0.06$ and $c = 1347.8 \pm 0.2$ pm in hexagonal setting or $a = 760.35 \pm 0.05$ pm and $\alpha = 88.642 \pm 0.001°$ for $x = 1$ (Rothballer et al. 2014); $a = 531.986 \pm 0.006$ and $c = 1350.451 \pm 0.012$ pm in hexagonal setting for $x = 1$ (Kassem et al. 2015)]. The concentration dependences of a and c are anisotropic and do not obey Vegard's law (Weihrich and Anusca 2006; Corps et al. 2013, 2015; Kassem et al. 2015): c increases monotonically with x, whereas a shows decrease until $x = 0.8–1.0$ and almost remains constant for a higher In concentration region.

This solid solution has been synthesized by the direct combination of stoichiometric quantities of high-purity elements in evacuated (10^{-3} Pa) and sealed quartz tubes (Natarajan et al. 1988; Weihrich and Anusca 2006; Corps et al. 2013, 2015; Rothballer et al. 2014). In a typical experiment, the powders were ground in an agate mortar and pestle and sealed into evacuated (<10^{-2} Pa) fused silica tubes. The mixtures were heated to 850°C and then slowly cooled down to 700°C within 12 h. After 4 days, the samples were quenched in H$_2$O. Single crystals were found within the grayish polycrystalline samples of metallic luster. If needed, the intermediate regrinding was used to homogenize the samples between firings.

Single crystals of the In$_x$Sn$_{2-x}$Co$_3$S$_2$ solid solution up to $x \approx 0.35$ were grown from the initial composition of molar ratio Co/S/(Sn+In) = 8:6:86 with the stoichiometric molar ratio of binary Sn-In (Kassem et al. 2015). The solid solution crystals with $x > 0.35$ were grown by using a mixture flux of Sn, In, and Pb with an initial composition of Co/S/Sn/In/Pb = 6:6:(36–18x_{nom}):18x_{nom}:52, where x_{nom} is the starting composition of In. Two Al$_2$O$_3$ crucibles were used to provide the growth environment. The constituent elements were placed in one crucible, which was used as a growth crucible, while the other one filled with quartz wool was put on the growth crucible upside down to reduce the flow-out of the evaporated sulfur and to filter and immobilize the excess flux at the end of the growth as explained below. Both were sealed in an evacuated quartz ampoule, which was put in a furnace and heated according to the temperature sequence as follows. The mixture was initially heated to 400°C over 2 h and was held for an extra 2 h to avoid evaporation of sulfur by making it reacts with Sn. Then it was heated up to 1050°C, higher than the melting temperature of the mixture, for over 6 h. After keeping at 1050°C for 6 h, the melt was cooled slowly to 700°C over 70 h. The ampoule was removed from the furnace at 700°C to remove the flux via rapid decanting and subsequent spinning of the ampoule in a centrifuge. Sheet-shaped crystals were obtained.

7.167 Tin–Thallium–Lead–Sulfur

SnS–Tl₂S–PbS. The isothermal section of this quasiternary system at 250°C is presented in Figure 7.37 (Filep et al. 2014). The region of the solid solution based on Tl_2S does not exceed 3 mol%, and those for the solid solutions based on the ternary compounds are 10 mol%. The homogeneity regions are the largest for the phases based on PbS and SnS. The liquidus surface of this system (Figure 7.38) is formed by seven

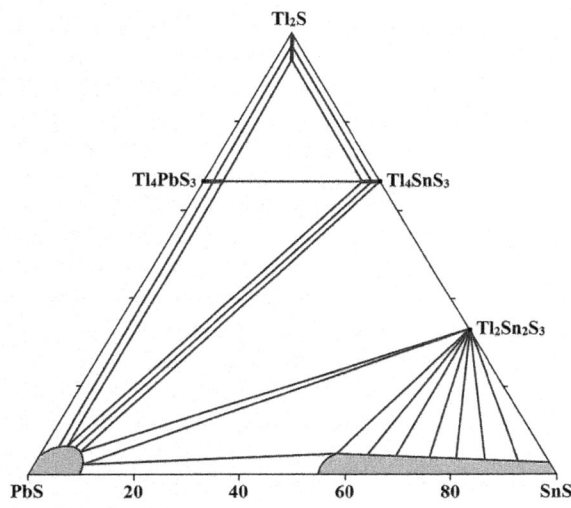

FIGURE 7.37 Isothermal section of the SnS–Tl₂S–PbS quasiternary system at 250°C. (From Filep, M.Y., et al., *Russ. J. Inorg. Chem.*, **59**(9), 1026, 2014.)

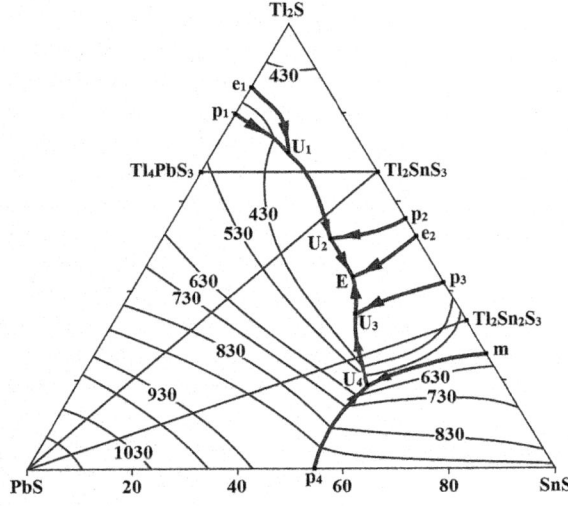

FIGURE 7.38 Liquidus surface of the SnS–Tl₂S–PbS quasiternary system. (From Filep, M.Y., et al., *Russ. J. Inorg. Chem.*, **59**(9), 1026, 2014.)

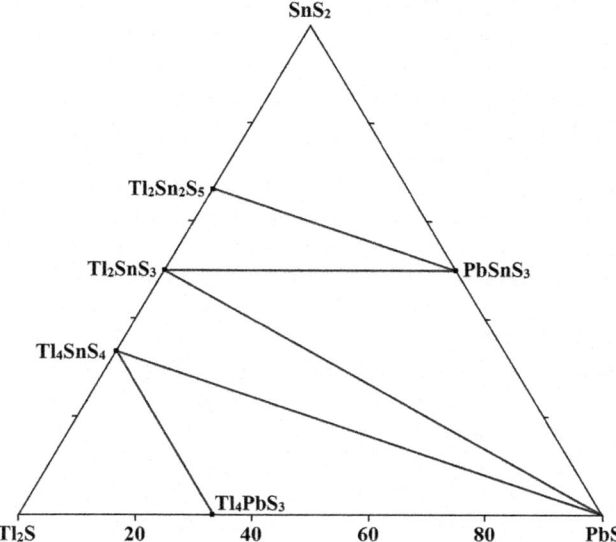

FIGURE 7.39 Isothermal section of the SnS₂–Tl₂S–PbS quasiternary system at 250°C. (From Olekseyuk, I.D., et al., *Nauk. Visnyk Volyns'k. Nats. Univ. im. Lesi Ukrainky. Ser. Khim. nauky*, (14), 40, 2011.)

fields of the primary crystallization of phases. No quaternary compounds were found. One ternary eutectic and 4 transition points exist in the system: E (300°C) – L ↔ (β-Tl₄SnS₃) + (Tl₂Sn₂S₃) + (PbS); U₁ (382°C) – L + (Tl₂S) ↔ (β-Tl₄SnS₃) + (PbS); U₂ (316°C) – L + (PbS) ↔ (β-Tl₄PbS₃) + (Tl₂S); U₃ (358°C) – L + (α-SnS) ↔ (Tl₂Sn₂S₃) + (Tl₂S); and U₄ (588°C) – L + (β-SnS) ↔ (α-SnS) + (PbS). The ingots for the investigations were annealed at 250°C for 168 h.

SnS₂–Tl₂S–PbS. The isothermal section of this quasiternary system at 250°C is given in Figure 7.39 (Olekseyuk et al. 2011). No quaternary phases were found in this system. The ingots for the investigations were annealed at 250°C for 250 h.

7.168 Tin–Thallium–Phosphor–Sulfur

The **TlSnPS₄** quaternary compound, which melts at 575°C and crystallizes in the orthorhombic structure with the lattice parameters $a = 1175.8 \pm 0.5$, $b = 890.1 \pm 0.4$, $c = 663.3 \pm 0.4$ pm, and a calculated density of 4.61 g·cm⁻³, is formed in the Sn–Tl–P–S system (Becker et al. 1987). It was prepared from stoichiometric amounts of the elements in an evacuated quartz ampoule by slowly heating them to 800°C, annealing at this temperature for 10 h, and slowly cooling them down to room temperature. This compound is extremely brittle (it can be pulverized by shaking the ampoule). During the gas transport of TlSnPS₄ in a temperature gradient of 550°C→350°C, orange-red prismatic crystals were formed. This compound is not moisture sensitive.

7.169 Tin–Thallium–Arsenic–Sulfur

The **Tl₂SnAs₂S₆** quaternary compound (mineral erniggliite), which crystallizes in the trigonal structure with the lattice parameters $a = 668.0 \pm 0.3$, $c = 716.4 \pm 0.9$ pm, and a calculated density of 5.24 g·cm^{-3}, is formed in the Sn–Tl–As–S system (Graeser et al. 1992; Jambor and Burke 1993).

7.170 Tin–Thallium–Selenium–Sulfur

SnS+Tl$_2$Se ↔ **SnSe+Tl$_2$S**. The **SnS–Tl$_4$SnSe$_3$** and **Tl$_4$SnSe$_3$–Tl$_2$S** systems are the quasibinary sections of this ternary mutual system, and the Tl$_4$SnS$_3$–Tl$_4$SnSe$_3$, Tl$_2$Sn$_2$S$_3$–Tl$_2$Sn$_2$Se$_3$, and SnS–Tl$_2$Sn$_2$Se$_3$ are partly quasibinary ones (Filep et al. 2011).

7.171 Tin–Thallium–Tellurium–Sulfur

SnS–Tl$_4$SnTe$_3$. The phase diagram of this system, constructed through DTA, XRD, and metallography, is a eutectic type (Figure 7.40) (Filep et al. 2013). The eutectic contains 50 mol% SnS and crystallizes at 433°C. The solid solutions based on SnS and Tl$_4$SnTe$_3$ do not exceed 5 mol%. The phase transformation of the solid solution based on SnS takes place at 593°C.

Tl$_4$SnTe$_3$–Tl$_2$S. The phase diagram of this system, constructed through DTA, XRD, and metallography, is a eutectic type (Figure 7.41) (Filep et al. 2013). The eutectic contains 82 mol% Tl$_2$S and crystallizes at 332°C. At 250°C, Tl$_4$SnTe$_3$ dissolves 65 mol% Tl$_2$S and Tl$_2$S dissolves 10 mol% Tl$_4$SnTe$_3$.

SnS+Tl$_2$Te ↔ **SnTe+Tl$_2$S**. Two vertical sections, an isothermal section at 250°C, and the liquidus surface (Figure 7.42) of this ternary mutual system were constructed by Filep et al. (2013). The liquidus surface consists of seven fields of primary crystallization of (Tl$_2$S), (Tl$_2$Te), (Tl$_4$SnTe$_3$), (β-Tl$_4$SnS$_3$), (α-SnS), (β-SnS), and (SnTe). The fields of the primary crystallization are divided by 14 monovariant lines.

7.172 Tin–Yttrium–Lead–Sulfur

SnS$_2$–Y$_2$S$_3$–PbS. The isothermal section of this quasiternary system at 500°C is shown in Figure 7.43 (Marchuk et al. 2007). The **Y$_2$Pb$_3$Sn$_3$S$_{12}$** quaternary compound, which crystallizes

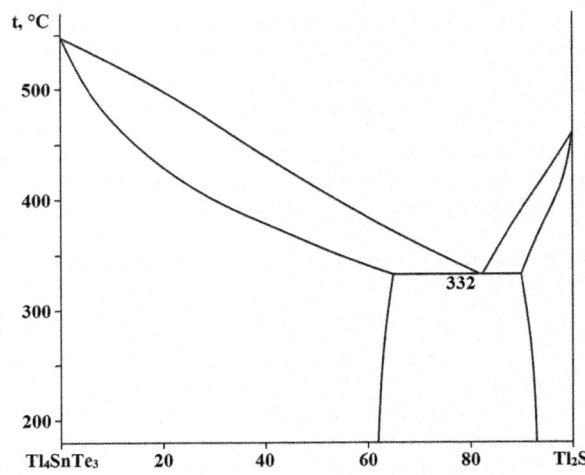

FIGURE 7.41 Phase diagram of the Tl$_4$SnTe$_3$–Tl$_2$S system. (From Filep, M.Y., et al., *Nauk. visn. Uzhgorod. Univ., Ser. Khim.*, [1(25)], 7, 2011.)

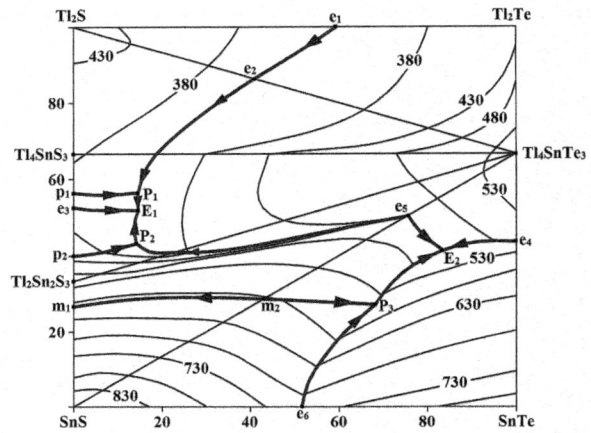

FIGURE 7.42 Liquidus surface of the SnS+Tl$_2$Te ↔ SnTe+Tl$_2$S ternary mutual system. (From Filep, M.Y., et al., *Nauk. visn. Uzhgorod. Univ., Ser. Khim.*, [1(25)], 7, 2011.)

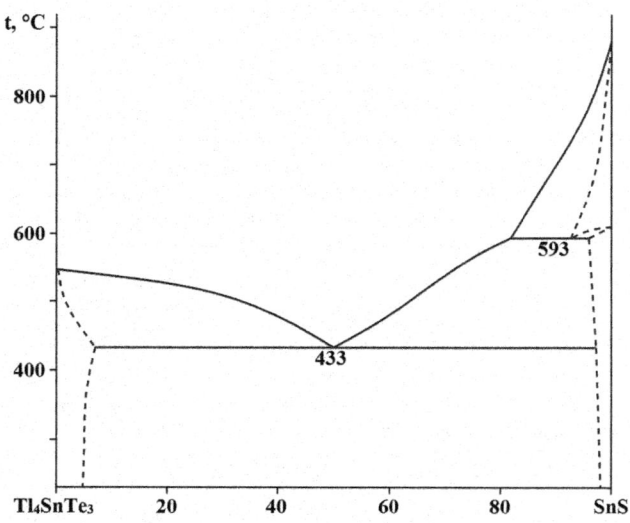

FIGURE 7.40 Phase diagram of the SnS–Tl$_4$SnTe$_3$ system. (From Filep, M.Y., et al., *Nauk. visn. Uzhgorod. Univ., Ser. Khim.*, [1(25)], 7, 2011.)

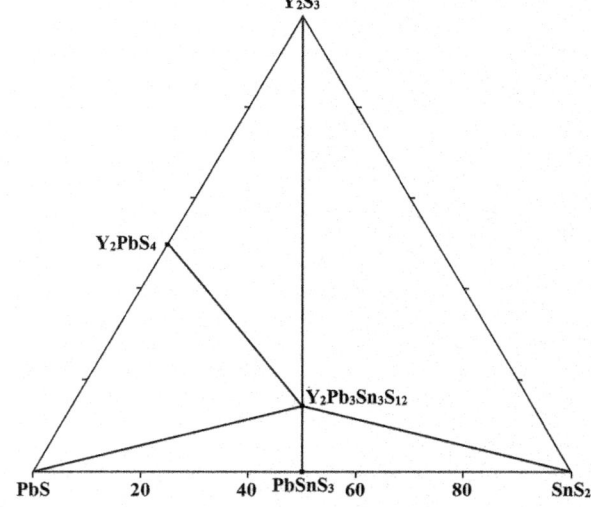

FIGURE 7.43 Isothermal section of the SnS$_2$–Y$_2$S$_3$–PbS quasiternary system at 250°C. (From Marchuk, O.V., et al., *Pol. J. Chem.*, **81**(3), 425, 2007.)

in the orthorhombic structure with the lattice parameters $a = 390.21 \pm 0.02$, $b = 2010.03 \pm 0.08$, $c = 1151.69 \pm 0.05$ pm, and a calculated density of 5.6622 $g \cdot cm^{-3}$, is formed in this system. At 500°C, this compound is in equilibrium with all binary and ternary compounds in the ternary system. The samples for the investigations were annealed at 500°C for 500 h, followed by quenching in cold water.

7.173 Tin–Yttrium–Iron–Sulfur

The $Y_3Fe_{0.5}SnS_7$ quaternary compound, which crystallizes in the hexagonal structure with the lattice parameters $a = 959.48 \pm 0.10$, $c = 618.88 \pm 0.08$ pm, and a calculated density of 4.293 $g \cdot cm^{-3}$, is formed in the Sn–Y–Fe–S system (Melnychuk et al. 2020). It was prepared by solid-state syntheses carried out in a furnace. The calculated amounts of the elemental constituents were sealed in an evacuated quartz tube. The ampoule was first heated at a rate of 30°C·h⁻¹ up to 1150°C and then kept at this temperature for 3 h. Afterwards, the sample was cooled slowly (10°C·h⁻¹) down to 500°C and annealed at this temperature for 720 h. Subsequently, the ampoule was quenched in cold water.

7.174 Tin–Yttrium–Cobalt–Sulfur

The $Y_3Co_{0.5}SnS_7$ quaternary compound, which crystallizes in the hexagonal structure with the lattice parameters $a = 960.675 \pm 0.013$, $c = 618.970 \pm 0.011$ pm, and a calculated density of 4.292 $g \cdot cm^{-3}$, is formed in the Sn–Y–Co–S system (Melnychuk et al. 2020). This compound was prepared in the same way as $Y_3Fe_{0.5}SnS_7$ was obtained.

7.175 Tin–Yttrium–Nickel–Sulfur

The $Y_3Ni_{0.5}SnS_7$ quaternary compound, which crystallizes in the hexagonal structure with the lattice parameters $a = 959.589 \pm 0.017$, $c = 619.632 \pm 0.011$ pm, and a calculated density of 4.296 $g \cdot cm^{-3}$, is formed in the Sn–Y–Ni–S system (Melnychuk et al. 2020). This compound was prepared in the same way as $Y_3Fe_{0.5}SnS_7$ was obtained.

7.176 Tin–Lanthanum–Lead–Sulfur

The $La_2Pb_3Sn_3S_{12}$ quaternary compound, which crystallizes in the orthorhombic structure with the lattice parameters $a = 396.97 \pm 0.07$, $b = 2032.9 \pm 0.6$, $c = 1160.6 \pm 0.3$ pm, and a calculated density of 5.821 $g \cdot cm^{-3}$, is formed in the Sn–La–Pb–S system (Gulay et al. 2008b). This compound was prepared by melting the mixture of the high-purity elements in an evacuated silica ampoule. The synthesis was realized in a tube furnace. The ampoule was heated with a heating rate of 30°C·h⁻¹ to a maximal temperature of 1100°C. The sample was kept at maximal temperature for 4 h. After that, it was cooled slowly (10°C·h⁻¹) to 500°C and annealed at this temperature for 500 h. After annealing, the ampoule with the sample was quenched in cold water.

7.177 Tin–Lanthanum–Oxygen–Sulfur

The $(LaO)_4Sn_2S_6$ quaternary compound, which crystallizes in the orthorhombic structure with the lattice parameters $a = 584.1 \pm 0.3$, $b = 585.1 \pm 0.2$, $c = 1900.3 \pm 0.3$ pm, and a calculated density of 5.37 $g \cdot cm^{-3}$, is formed in the Sn–La–O–S system (Bénazeth et al. 1985). This compound could not be obtained by melting in a vacuum-sealed ampoule, because its melting point exceeds 1200°C. A study of the phase diagram shows that in the SnS_2–La_2O_2S system, at 800°C, a eutectic exists whose composition is very close to SnS_2. Therefore, a sample with an intermediate composition between that of the eutectic and that of the research compound ($7SnS_2 + 1.5La_2O_2S$) was prepared. After fusion at 1000°C and slow cooling, $(LaO)_4Sn_2S_6$ was obtained in the form of single yellow crystals mixed with SnS_2.

7.178 Tin–Lanthanum–Iron–Sulfur

The $La_3Fe_{0.5}SnS_7$ quaternary compound, which crystallizes in the hexagonal structure with the lattice parameters $a = 1029.119 \pm 0.019$, $c = 600.220 \pm 0.012$ pm, and a calculated density of 4.752 $g \cdot cm^{-3}$, is formed in the Sn–La–Fe–S system (Melnychuk et al. 2020). This compound was prepared in the same way as $Y_3Fe_{0.5}SnS_7$ was obtained.

7.179 Tin–Lanthanum–Cobalt–Sulfur

The $La_3Co_{0.5}SnS_7$ quaternary compound, which crystallizes in the hexagonal structure with the lattice parameters $a = 1027.33 \pm 0.11$, $c = 599.80 \pm 0.07$ pm, and a calculated density of 4.782 $g \cdot cm^{-3}$, is formed in the Sn–La–Co–S system (Melnychuk et al. 2020). This compound was prepared in the same way as $Y_3Fe_{0.5}SnS_7$ was obtained.

7.180 Tin–Lanthanum–Nickel–Sulfur

The $La_3Ni_{0.5}SnS_7$ quaternary compound, which crystallizes in the hexagonal structure with the lattice parameters $a = 1024.869 \pm 0.019$, $c = 604.474 \pm 0.011$ pm, and a calculated density of 4.767 $g \cdot cm^{-3}$, is formed in the Sn–La–Ni–S system (Melnychuk et al. 2020). This compound was prepared in the same way as $Y_3Fe_{0.5}SnS_7$ was obtained.

7.181 Tin–Cerium–Lead–Sulfur

The $Ce_2Pb_3Sn_3S_{12}$ quaternary compound, which crystallizes in the orthorhombic structure with the lattice parameters $a = 395.75 \pm 0.09$, $b = 2027.5 \pm 0.7$, $c = 1159.0 \pm 0.4$ pm, and a calculated density of 5.873 $g \cdot cm^{-3}$, is formed in the Sn–Ce–Pb–S system (Gulay et al. 2008b). This compound was synthesized in the same way as $La_2Pb_3Sn_3S_{12}$ was obtained.

7.182 Tin–Cerium–Iron–Sulfur

The **Ce$_3$Fe$_{0.5}$SnS$_7$** quaternary compound, which crystallizes in the hexagonal structure with the lattice parameters $a = 1017.333 \pm 0.012$, $c = 601.116 \pm 0.008$ pm, and a calculated density of 4.878 g·cm^{-3}, is formed in the Sn–Ce–Fe–S system (Melnychuk et al. 2020). This compound was prepared in the same way as Y$_3$Fe$_{0.5}$SnS$_7$ was obtained.

7.183 Tin–Cerium–Cobalt–Sulfur

The **Ce$_3$Co$_{0.5}$SnS$_7$** quaternary compound, which crystallizes in the hexagonal structure with the lattice parameters $a = 1017.08 \pm 0.08$, $c = 598.37 \pm 0.06$ pm, and a calculated density of 4.913 g·cm^{-3}, is formed in the Sn–Ce–Co–S system (Melnychuk et al. 2020). This compound was prepared in the same way as Y$_3$Fe$_{0.5}$SnS$_7$ was obtained.

7.184 Tin–Cerium–Nickel–Sulfur

The **Ce$_3$Ni$_{0.5}$SnS$_7$** quaternary compound, which crystallizes in the hexagonal structure with the lattice parameters $a = 1013.62 \pm 0.03$, $c = 603.855 \pm 0.016$ pm, and a calculated density of 4.901 g·cm^{-3}, is formed in the Sn–Ce–Ni–S system (Melnychuk et al. 2020). This compound was prepared in the same way as Y$_3$Fe$_{0.5}$SnS$_7$ was obtained.

7.185 Tin–Praseodymium–Lead–Sulfur

The **Pr$_2$Pb$_3$Sn$_3$S$_{12}$** quaternary compound, which crystallizes in the orthorhombic structure with the lattice parameters $a = 394.48 \pm 0.05$, $b = 2007.1 \pm 0.3$, $c = 1170.2 \pm 0.2$ pm, and a calculated density of 5.894 g·cm^{-3}, is formed in the Sn–Pr–Pb–S system (Gulay et al. 2008b). This compound was synthesized in the same way as La$_2$Pb$_3$Sn$_3$S$_{12}$ was obtained.

7.186 Tin–Praseodymium–Iron–Sulfur

The **Pr$_3$Fe$_{0.5}$SnS$_7$** quaternary compound, which crystallizes in the hexagonal structure with the lattice parameters $a = 1009.43 \pm 0.02$, $c = 600.764 \pm 0.013$ pm, and a calculated density of 4.973 g·cm^{-3}, is formed in the Sn–Pr–Fe–S system (Melnychuk et al. 2020). This compound was prepared in the same way as Y$_3$Fe$_{0.5}$SnS$_7$ was obtained.

7.187 Tin–Praseodymium–Cobalt–Sulfur

The **Pr$_3$Co$_{0.5}$SnS$_7$** quaternary compound, which crystallizes in the hexagonal structure with the lattice parameters $a = 1008.19 \pm 0.03$, $c = 601.07 \pm 0.02$ pm, and a calculated density of 4.992 g·cm^{-3}, is formed in the Sn–Pr–Co–S system (Melnychuk et al. 2020). This compound was prepared in the same way as Y$_3$Fe$_{0.5}$SnS$_7$ was obtained.

7.188 Tin–Praseodymium–Nickel–Sulfur

The **Pr$_3$Ni$_{0.5}$SnS$_7$** quaternary compound, which crystallizes in the hexagonal structure with the lattice parameters $a = 1001.36 \pm 0.03$, $c = 609.21 \pm 0.02$ pm, and a calculated density of 4.992 g·cm^{-3}, is formed in the Sn–Pr–Ni–S system (Melnychuk et al. 2020). This compound was prepared in the same way as Y$_3$Fe$_{0.5}$SnS$_7$ was obtained.

7.189 Tin–Neodymium–Lead–Sulfur

The **Nd$_2$Pb$_3$Sn$_3$S$_{12}$** quaternary compound, which crystallizes in the orthorhombic structure with the lattice parameters $a = 393.61 \pm 0.06$, $b = 2004.9 \pm 0.3$, $c = 1168.0 \pm 0.2$ pm, and a calculated density of 5.952 g·cm^{-3}, is formed in the Sn–Nd–Pb–S system (Gulay et al. 2008b). This compound was synthesized in the same way as La$_2$Pb$_3$Sn$_3$S$_{12}$ was obtained.

7.190 Tin–Neodymium–Iron–Sulfur

The **Nd$_3$Fe$_{0.5}$SnS$_7$** quaternary compound, which crystallizes in the hexagonal structure with the lattice parameters $a = 999.378 \pm 0.019$, $c = 605.675 \pm 0.012$ pm, and a calculated density of 5.095 g·cm^{-3}, is formed in the Sn–Nd–Fe–S system (Melnychuk et al. 2020). This compound was prepared in the same way as Y$_3$Fe$_{0.5}$SnS$_7$ was obtained.

7.191 Tin–Neodymium–Cobalt–Sulfur

The **Nd$_3$Co$_{0.5}$SnS$_7$** quaternary compound, which crystallizes in the hexagonal structure with the lattice parameters $a = 995.18 \pm 0.08$, $c = 609.86 \pm 0.06$ pm, and a calculated density of 5.113 g·cm^{-3}, is formed in the Sn–Nd–Co–S system (Melnychuk et al. 2020). This compound was prepared in the same way as Y$_3$Fe$_{0.5}$SnS$_7$ was obtained.

7.192 Tin–Neodymium–Nickel–Sulfur

The **Nd$_3$Ni$_{0.5}$SnS$_7$** quaternary compound, which crystallizes in the hexagonal structure with the lattice parameters $a = 995.32 \pm 0.02$, $c = 610.245 \pm 0.016$ pm, and a calculated density of 5.108 g·cm^{-3}, is formed in the Sn–Nd–Ni–S system (Melnychuk et al. 2020). This compound was prepared in the same way as Y$_3$Fe$_{0.5}$SnS$_7$ was obtained.

7.193 Tin–Samarium–Lead–Sulfur

SnS$_2$–Sm$_2$S$_3$–PbS. The isothermal section of this quasiternary system at 500°C is presented in Figure 7.44 (Gulay and Marchuk 2010). There are 7 single-phase, 11 two-phase, and 6 three-phase fields in the system at this temperature. The solid solutions based on the initial binary components are 1–2 mol%. The highest solubility is observed on the basis of the Sm$_2$PbS$_4$

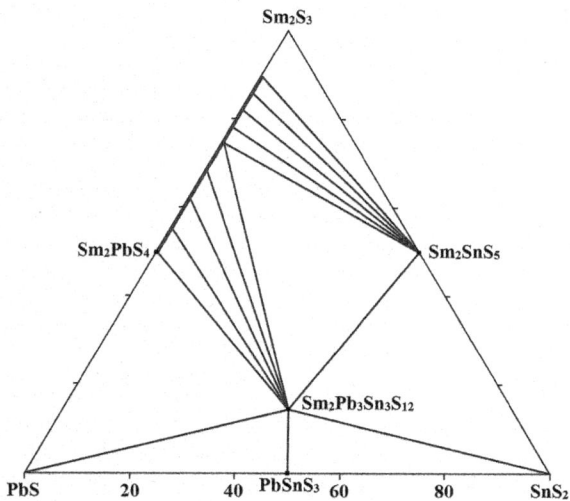

FIGURE 7.44 Isothermal section of the SnS_2–Sm_2S_3–PbS quasiternary system at 500°C. (From Gulay, L.D., and Marchuk, O.V., *Nauk. Visnyk Volyns'k. Nats. Univ. im. Lesi Ukrainky. Ser. Khim. nauky*, (16), 50, 2010.)

compound: the $Sm_{2+2/3x}Pb_{1-x}S_4$ solid solution is localized along the PbS–Sm_2S_3 system. The **$Sm_2Pb_3Sn_3S_{12}$** quaternary compound, which crystallizes in the orthorhombic structure with the lattice parameters $a = 392.30 \pm 0.03$, $b = 2011.9 \pm 0.3$, $c = 1161.1 \pm 0.2$ pm, and a calculated density of 6.031 g·cm⁻³, is formed in this system (Gulay et al. 2008b). This compound was synthesized in the same way as $La_2Pb_3Sn_3S_{12}$ was obtained.

7.194 Tin–Samarium–Iron–Sulfur

The **$Sm_3Fe_{0.5}SnS_7$** quaternary compound, which crystallizes in the hexagonal structure with the lattice parameters $a = 982.15 \pm 0.09$, $c = 613.01 \pm 0.06$ pm, and a calculated density of 5.331 g·cm⁻³, is formed in the Sn–Sm–Fe–S system (Melnychuk et al. 2020). This compound was prepared in the same way as $Y_3Fe_{0.5}SnS_7$ was obtained.

7.195 Tin–Samarium–Cobalt–Sulfur

The **$Sm_3Co_{0.5}SnS_7$** quaternary compound, which crystallizes in the hexagonal structure with the lattice parameters $a = 982.09 \pm 0.07$, $c = 612.58 \pm 0.07$ pm, and a calculated density of 5.346 g·cm⁻³, is formed in the Sn–Sm–Co–S system (Melnychuk et al. 2020). This compound was prepared in the same way as $Y_3Fe_{0.5}SnS_7$ was obtained.

7.196 Tin–Samarium–Nickel–Sulfur

The **$Sm_3Ni_{0.5}SnS_7$** quaternary compound, which crystallizes in the hexagonal structure with the lattice parameters $a = 979.01 \pm 0.08$, $c = 614.60 \pm 0.06$ pm, and a calculated density of 5.361 g·cm⁻³, is formed in the Sn–Sm–Ni–S system (Melnychuk et al. 2020). This compound was prepared in the same way as $Y_3Fe_{0.5}SnS_7$ was obtained.

7.197 Tin–Europium–Lutetium–Sulfur

The **$Eu_4LuSn_3S_{12}$** quaternary compound, which crystallizes in the orthorhombic structure with the lattice parameters $a = 392.0 \pm 0.1$, $b = 2013.2 \pm 0.4$, $c = 1145.9 \pm 0.2$ pm, and a calculated density of 5.595 g·cm⁻³ (Jakubcová et al. 2007) This compound was synthesized by heating the elements or binary sulfides in a stoichiometric ratio in an alumina crucible and sealed in silica ampoule under an Ar atmosphere. At first, the temperature was raised slowly (50°C·h⁻¹) to 870°C and kept there for 24 h. After cooling, the sample was homogenized and heated to 900°C for 24 h. This procedure was repeated once more and resulted in a homogeneous dark powder, which was not sensitive to air.

7.198 Tin–Gadolinium–Lead–Sulfur

The **$Gd_2Pb_3Sn_3S_{12}$** quaternary compound, which crystallizes in the orthorhombic structure with the lattice parameters $a = 391.53 \pm 0.05$, $b = 2020.6 \pm 0.3$, $c = 1155.6 \pm 0.2$ pm, and a calculated density of 6.091 g·cm⁻³, is formed in the Sn–Gd–Pb–S system (Gulay et al. 2008b). This compound was synthesized in the same way as $La_2Pb_3Sn_3S_{12}$ was obtained.

7.199 Tin–Gadolinium–Iron–Sulfur

The **$Gd_3Fe_{0.5}SnS_7$** quaternary compound, which crystallizes in the hexagonal structure with the lattice parameters $a = 972.2 \pm 0.2$ and $c = 617.6 \pm 0.2$ pm, is formed in the Sn–Gd–Fe–S system (Melnychuk et al. 2020). This compound was prepared in the same way as $Y_3Fe_{0.5}SnS_7$ was obtained.

7.200 Tin–Gadolinium–Cobalt–Sulfur

The **$Gd_3Co_{0.5}SnS_7$** quaternary compound, which crystallizes in the hexagonal structure with the lattice parameters $a = 972.2 \pm 0.1$ and $c = 618.2 \pm 0.8$ pm, is formed in the Sn–Gd–Co–S system (Melnychuk et al. 2020). This compound was prepared in the same way as $Y_3Fe_{0.5}SnS_7$ was obtained.

7.201 Tin–Gadolinium–Nickel–Sulfur

The **$Gd_3Ni_{0.5}SnS_7$** quaternary compound, which crystallizes in the hexagonal structure with the lattice parameters $a = 970.1 \pm 0.2$ and $c = 617.8 \pm 0.2$ pm, is formed in the Sn–Gd–Ni–S system (Melnychuk et al. 2020). This compound was prepared in the same way as $Y_3Fe_{0.5}SnS_7$ was obtained.

7.202 Tin–Terbium–Lead–Sulfur

The **$Tb_2Pb_3Sn_3S_{12}$** quaternary compound, which crystallizes in the orthorhombic structure with the lattice parameters $a = 390.76 \pm 0.04$, $b = 2017.4 \pm 0.2$, $c = 1153.2 \pm 0.1$ pm, and a calculated density of 6.143 g·cm⁻³, is formed in the Sn–Tb–Pb–S system (Gulay et al. 2008b). This compound was synthesized in the same way as $La_2Pb_3Sn_3S_{12}$ was obtained.

7.203 Tin–Terbium–Iron–Sulfur

The $Tb_3Fe_{0.5}SnS_7$ quaternary compound, which crystallizes in the hexagonal structure with the lattice parameters $a = 966.51 \pm 0.08$, $c = 616.74 \pm 0.06$ pm, and a calculated density of 5.643 g·cm⁻³, is formed in the Sn–Tb–Fe–S system (Melnychuk et al. 2020). This compound was prepared in the same way as $Y_3Fe_{0.5}SnS_7$ was obtained.

7.204 Tin–Terbium–Cobalt–Sulfur

The $Tb_3Co_{0.5}SnS_7$ quaternary compound, which crystallizes in the hexagonal structure with the lattice parameters $a = 966.3 \pm 0.2$ and $c = 617.9 \pm 0.2$ pm, is formed in the Sn–Tb–Co–S system (Melnychuk et al. 2020). This compound was prepared in the same way as $Y_3Fe_{0.5}SnS_7$ was obtained.

7.205 Tin–Terbium–Nickel–Sulfur

The $Tb_3Ni_{0.5}SnS_7$ quaternary compound, which crystallizes in the hexagonal structure with the lattice parameters $a = 964.3 \pm 0.2$ and $c = 618.7 \pm 0.2$ pm, is formed in the Sn–Tb–Ni–S system (Melnychuk et al. 2020). This compound was prepared in the same way as $Y_3Fe_{0.5}SnS_7$ was obtained.

7.206 Tin–Dysprosium–Lead–Sulfur

The $Dy_2Pb_3Sn_3S_{12}$ quaternary compound, which crystallizes in the orthorhombic structure with the lattice parameters $a = 390.00 \pm 0.04$, $b = 2015.3 \pm 0.2$, $c = 1152.4 \pm 0.1$ pm, and a calculated density of 6.194 g·cm⁻³, is formed in the Sn–Dy–Pb–S system (Gulay et al. 2008b). This compound was synthesized in the same way as $La_2Pb_3Sn_3S_{12}$ was obtained.

7.207 Tin–Dysprosium–Oxygen–Sulfur

The $Dy_8SnS_{13.61}O_{39}$ quaternary compound, which crystallizes in the orthorhombic structure with the lattice parameters $a = 378.22 \pm 0.08$, $b = 2362.0 \pm 0.5$, $c = 2127.1 \pm 0.4$ pm, and a calculated density of 6.506 g·cm⁻³, is formed in the Sn–Dy–O–S system (Daszkiewicz et al. 2008b). Single crystals of the title compound were obtained unintentionally by the fusion of Dy, Sn, and S (molar ratio 2:1:5) in an evacuated silica ampoule. The ampoule was heated in a tube furnace with a heating rate of 30°C·h⁻¹ to 1150°C and kept at this temperature for 4 h. It was then cooled down slowly (10°C·h⁻¹) to 600°C and annealed at this temperature for a further 240 h, and finally quenched in cold water. The product was a brown-colored compact alloy containing red crystals with a prismatic habit.

7.208 Tin–Dysprosium–Iron–Sulfur

The $Dy_3Fe_{0.5}SnS_7$ quaternary compound, which crystallizes in the hexagonal structure with the lattice parameters $a = 960.93 \pm 0.09$, $c = 618.15 \pm 0.06$ pm, and a calculated density of 5.768 g·cm⁻³, is formed in the Sn–Dy–Fe–S system (Melnychuk et al. 2020). This compound was prepared in the same way as $Y_3Fe_{0.5}SnS_7$ was obtained.

7.209 Tin–Dysprosium–Cobalt–Sulfur

The $Dy_3Co_{0.5}SnS_7$ quaternary compound, which crystallizes in the hexagonal structure with the lattice parameters $a = 961.6 \pm 0.2$ and $c = 618.9 \pm 0.1$ pm, is formed in the Sn–Dy–Co–S system (Melnychuk et al. 2020). This compound was prepared in the same way as $Y_3Fe_{0.5}SnS_7$ was obtained.

7.210 Tin–Dysprosium–Nickel–Sulfur

The $Dy_3Ni_{0.5}SnS_7$ quaternary compound, which crystallizes in the hexagonal structure with the lattice parameters $a = 959.1 \pm 0.2$ and $c = 619.2 \pm 0.2$ pm, is formed in the Sn–Dy–Ni–S system (Melnychuk et al. 2020). This compound was prepared in the same way as $Y_3Fe_{0.5}SnS_7$ was obtained.

7.211 Tin–Holmium–Lead–Sulfur

SnS_2–Sm_2S_3–PbS. The isothermal section of this quasiternary system at 500°C is presented in Figure 7.45 (Gulay and Marchuk 2010). There are 5 single-phase, 10 two-phase, and 5 three-phase fields in the system at this temperature. The solid solutions based on initial binary components are 1–2 mol%.

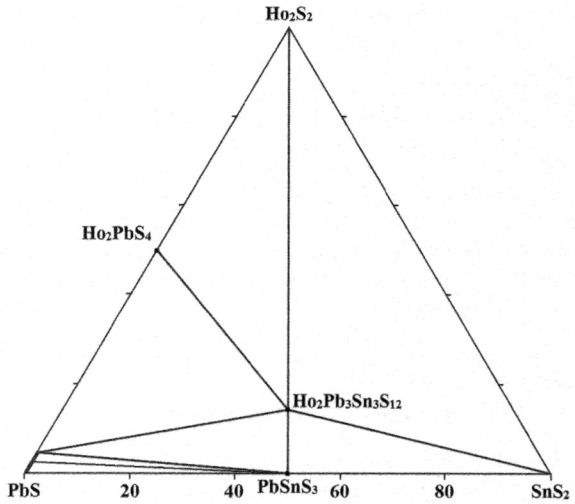

FIGURE 7.45 Isothermal section of the SnS_2–Ho_2S_3–PbS quasiternary system at 500°C. (From Gulay, L.D., and Marchuk, O.V., *Nauk. Visnyk Volyns'k. Nats. Univ. im. Lesi Ukrainky. Ser. Khim. nauky*, (16), 50, 2010.)

The $Ho_2Pb_3Sn_3S_{12}$ quaternary compound, which crystallizes in the orthorhombic structure with the lattice parameters $a = 389.92 \pm 0.01$, $b = 2011.75 \pm 0.07$, $c = 1151.40 \pm 0.04$ pm, and a calculated density of 6.222 g·cm⁻³, is formed in this system (Gulay et al. 2008b). This compound was synthesized in the same way as $La_2Pb_3Sn_3S_{12}$ was obtained.

7.212 Tin–Holmium–Iron–Sulfur

The $Ho_3Fe_{0.5}SnS_7$ quaternary compound, which crystallizes in the hexagonal structure with the lattice parameters $a = 957.9 \pm 0.2$ and $c = 618.7 \pm 0.2$ pm, is formed in the Sn–Ho–Fe–S system (Melnychuk et al. 2020). This compound was prepared in the same way as $Y_3Fe_{0.5}SnS_7$ was obtained.

7.213 Tin–Holmium–Cobalt–Sulfur

The $Ho_3Co_{0.5}SnS_7$ quaternary compound, which crystallizes in the hexagonal structure with the lattice parameters $a = 957.6 \pm 0.1$ and $c = 619.1 \pm 0.8$ pm, is formed in the Sn–Ho–Co–S system (Melnychuk et al. 2020). This compound was prepared in the same way as $Y_3Fe_{0.5}SnS_7$ was obtained.

7.214 Tin–Holmium–Nickel–Sulfur

The $Ho_3Ni_{0.5}SnS_7$ quaternary compound, which crystallizes in the hexagonal structure with the lattice parameters $a = 954.1 \pm 0.2$ and $c = 618.4 \pm 0.2$ pm, is formed in the Sn–Ho–Ni–S system (Melnychuk et al. 2020). This compound was prepared in the same way as $Y_3Fe_{0.5}SnS_7$ was obtained.

7.215 Tin–Erbium–Lead–Sulfur

The $Er_2Pb_3Sn_3S_{12}$ quaternary compound, which crystallizes in the orthorhombic structure with the lattice parameters $a = 390.06 \pm 0.05$, $b = 2002.9 \pm 0.3$, $c = 1151.3 \pm 0.3$ pm, and a calculated density of 6.262 g·cm⁻³, is formed in the Sn–Er–Pb–S system (Gulay et al. 2008b). This compound was synthesized in the same way as $La_2Pb_3Sn_3S_{12}$ was obtained.

7.216 Tin–Thulium–Lead–Sulfur

The $Tm_2Pb_3Sn_3S_{12}$ quaternary compound, which crystallizes in the orthorhombic structure with the lattice parameters $a = 389.36 \pm 0.07$, $b = 2005.6 \pm 0.4$, $c = 1150.1 \pm 0.3$ pm, and a calculated density of 6.293 g·cm⁻³, is formed in the Sn–Tm–Pb–S system (Gulay et al. 2008b). This compound was synthesized in the same way as $La_2Pb_3Sn_3S_{12}$ was obtained.

7.217 Tin–Lead–Phosphor–Sulfur

$Sn_2P_2S_6+Pb_2P_2Se_6 \leftrightarrow Sn_2P_2Se_6+Pb_2P_2S_6$. Continuous series of the solid solutions are formed in the $Sn_2P_2S_6–Sn_2P_2Se_6$, $Pb_2P_2S_6–Pb_2P_2Se_6$, $Sn_2P_2S_6–Pb_2P_2S_6$, $Sn_2P_2Se_6–Pb_2P_2Se_6$,

and $Sn_2P_2Se_6–Pb_2P_2S_6$ sections of this ternary mutual system (Potoriy et al. 1988). The liquidus and solidus surfaces (Figure 7.46) represent a continuous field of primary crystallization of solid solutions in the full concentration range.

7.218 Tin–Lead–Antimony–Sulfur

$PbSnS_2–PbSb_2S_4$. The phase diagram of this system, constructed through DTA, XRD, metallography, and measuring of the microhardness using the ingots annealing at 430°C for 450 h, is given in Figure 7.47 (Mamedov et al. 2010). The eutectics contain 20 and 72 mol% $PbSb_2S_4$ and crystallize at 577°C and 542°C, respectively. The solubility of $PbSb_2S_4$ in $PbSnS_2$ is 6 mol% and the solubility of $PbSnS_2$ in $PbSb_2S_4$ reaches 12 mol%. The $Pb_2SnSb_2S_6$ quaternary compound, which melts congruently at 677°C and crystallizes in the orthorhombic

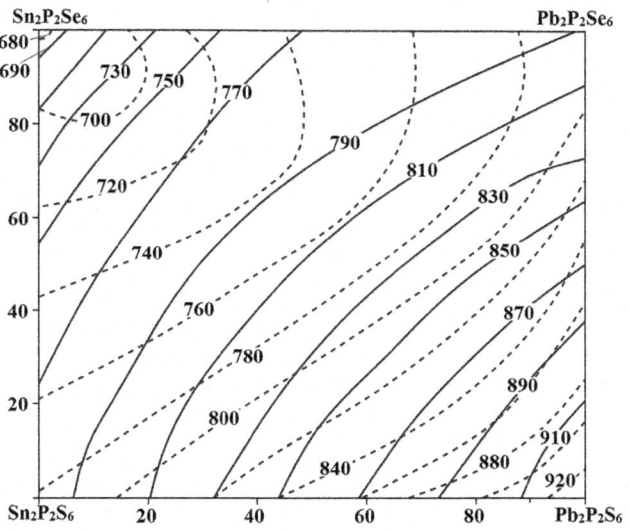

FIGURE 7.46 Liquidus (solid lines) and solidus (dashed lines) surfaces of the $Sn_2P_2S_6+Pb_2P_2Se_6 \leftrightarrow Sn_2P_2Se_6+Pb_2P_2S_6$ ternary mutual system. (From Potoriy, M.V., et al., *ChemChemTech [Izv. Vyssh. Uchebn. Zaved. Khim. Khim. Tekhnol.]*, **31**(8), 21, 1988.)

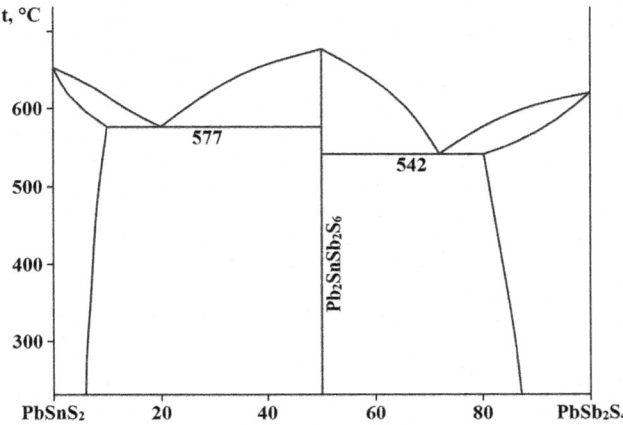

FIGURE 7.47 Phase diagram of the $PbSnS_2–PbSb_2S_4$ system. (From Mamedov, Sh.G., et al., *Russ. J. Inorg. Chem.*, **55**(4), 626, 2010.)

structure with the lattice parameters $a = 1522$, $b = 1068$, and $c = 390$ pm, is formed in this system (Mamedov 2010; Mamedov et al. 2010). The thermodynamic functions of this compound are the next: $\Delta G^0_{f,298} = -596.36$ kJ·M^{-1}, $\Delta H^0_{f,298} = -600.5$ kJ·M^{-1} and $S^0_{298} = 442.2$ J·(M·K)$^{-1}$ (Mammadov et al. 2014). Its single crystals were grown by chemical vapor reactions using iodine as a transport agent (Mamedov et al. 2010).

SnS–Pb₂SnSb₂S₆. The phase diagram of this system, constructed through DTA, XRD, metallography, and measuring of the microhardness using the ingots annealing at 430°C for 300 h, is given in Figure 7.48 (Mamedov 2010). The eutectic contains 25 mol% SnS and crystallizes at 502°C. The extents of the solid solutions based on the initial components were determined to be 10 mol% SnS and 5 mol% Pb₂SnSb₂S₆.

According to the data of Bakhtiyarly et al. (2009) and Mammadov et al. (2014), another quaternary compound, **PbSnSb₄S₈**, is also formed in the Sn–Pb–Sb–S system. The thermodynamic functions of this compound are the next: $\Delta G^0_{f,298} = -689.5$ kJ·M^{-1}, $\Delta H^0_{f,298} = -694.3$ kJ·M^{-1} and $S^0_{298} = 531.1$ J·(M·K)$^{-1}$ (Mammadov et al. 2014).

SnS–PbSnSb₄S₈. The phase diagram of this system, constructed through DTA, XRD, metallography, and measuring of the microhardness using the ingots annealing at 380°C for 240 h, is a eutectic type (Figure 7.49) (Bakhtiyarly et al. 2009). The eutectic contains 35 mol% 4SnS and crystallizes at 452°C. The line at 552°C corresponds to the polymorphic transformation of SnS. The solid solution based on both initial components reaches 8 mol%.

SnS–PbS–Sb₂S₃. In this system, the liquid is in a predominant phase and forms equilibrium assemblages with Sb₂S₃, PbSb₂S₄, and Sn₃Sb₂S₄ at 500°C (Sachdev and Chang 1975). Neither solid solution nor ternary phases were found at this temperature.

According to the data of Sachdev and Chang (1975), the **Pb₀.₆₀Sn⁺⁴₀.₂₃Sb₀.₃₄S₁.₅₇** quaternary phase, which melts incongruently at 617°C and has a composition in the ranges limited by 57.0 to 64.0 mol% PbS, 18.0 to 28.0 mol% SnS₂, and 13.5 to 21.0 mol% Sb₂S₃, was found in the Sn–Pb–Sb–S system. This

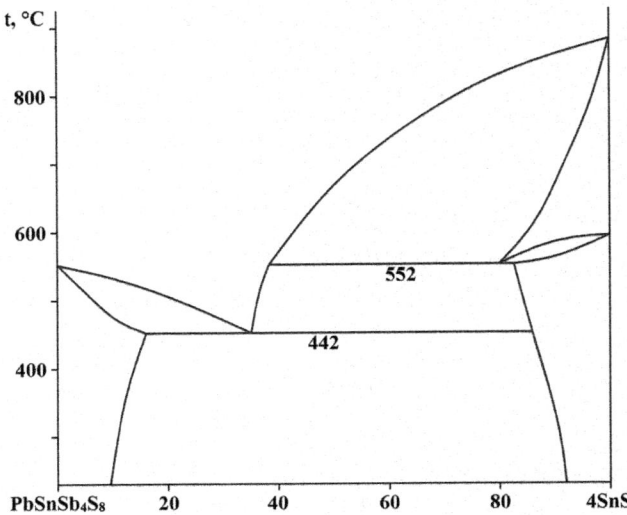

FIGURE 7.49 Phase diagram of the 4SnS–PbSnSb₄S₈ system. (From Bakhtiyarly, I.B., et al., *ChemChemTech* [*Izv. Vyssh. Uchebn. Zaved. Khim. Khim. Tekhnol.*], **52**(4), 120, 2009.)

compound forms a complete series of the solid solution with **Sn⁺²₀.₇₀Sn⁺⁴₀.₁₅Sb₀.₃₀S₁.₄₅**. The solid-solution series was found to be stable at 400°C; however, at 300°C, an immiscibility gap develops in the PbS-rich portion.

7.219 Tin–Lead–Bismuth–Sulfur

SnS–PbSnBi₄S₈. The phase diagram of this system, constructed through DTA, XRD, metallography, and measuring of the microhardness using the ingots annealing at 480°C for 120 h, is a eutectic type (Gurbanov 2012a). The eutectic contains 45 mol% SnS and crystallizes at 477°C. PbSnBi₄S₈ dissolves 10 mol% SnS and SnS dissolves 4 mol% PbSnBi₄S₈.

SnS–Pb₂SnBi₂S₆. The phase diagram of this system, constructed through DTA, XRD, metallography, and measuring of the microhardness using the ingots annealing at 480°C for 120 h, is a eutectic type (Gurbanov 2012a). The eutectic contains 40 mol% SnS and crystallizes at 427°C. Pb₂SnBi₂S₆ dissolves 12 mol% SnS, and SnS dissolves up to 4 mol% Pb₂SnBi₂S₆.

SnBi₂S₄–PbBi₂S₄. This system is nonquasibinary section of the SnS–PbS–Bi₂S₃ quasiternary system as PbBi₂S₄ melts incongruently (Gurbanov 2012a). The **PbSnBi₄S₈** quaternary compound, which melts congruently at 677°C and crystallizes in the orthorhombic structure with the lattice parameters $a = 2178$, $b = 752$, and $c = 420$ pm, is formed in this system. This compound forms a eutectic with SnBi₂S₄ at 30 mol% PbBi₂S₄ and 527°C and is in equilibrium with Pb₂SnBi₂S₆, SnS, and PbBi₄S₇. The solubility based on SnBi₂S₄ is 14 and 8 mol% PbBi₂S₄ at the eutectic and room temperature, respectively.

SnBi₂S₄–PbBi₄S₇. The **PbSnBi₆S₁₁** quaternary compound, which melts congruently at 607°C, has a homogeneity region in the range of 48–57 mol% SnBi₂S₄, and crystallizes in the orthorhombic structure with the lattice parameters $a = 1118$, $b = 412$, and $c = 1154$ pm, is formed in this system (Gurbanov 2012a). This compound divides the system into two eutectic subsytems with the eutectics at 35 mol% SnBi₂S₄ and 523°C and 25 mol%

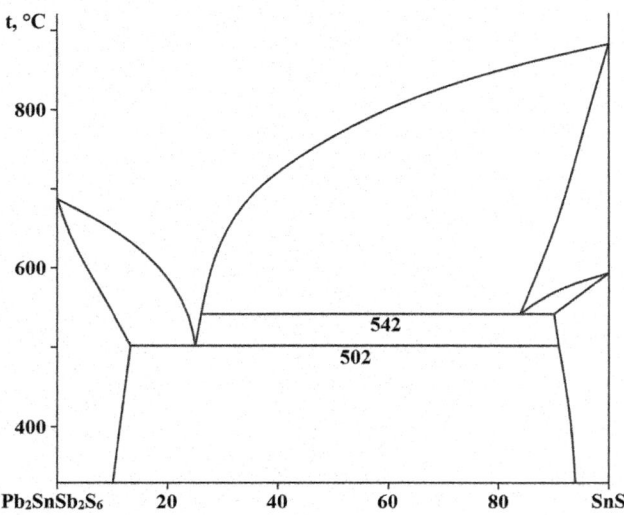

FIGURE 7.48 Phase diagram of the SnS–Pb₂SnSb₂S₆ system. (From Mamedov, Sh.G., *Russ. J. Inorg. Chem.*, **55**(8), 1292, 2010.)

PbBi$_4$S$_7$ and 447°C. The solubility based on SnBi$_2$S$_4$ is 7 mol% PbBi$_4$S$_7$ and based on PbBi$_4$S$_7$ – 10 mol% SnBi$_2$S$_4$.

PbSnS$_2$–PbBi$_2$S$_4$. This system is a nonquasibinary section of the Sn–Pb–Bi–S quaternary system as PbBi$_2$S$_4$ melts incongruently (Gurbanov 2012a; Gurbanov and Mamedov 2016). The **Pb$_2$SnBi$_2$S$_6$** quaternary compound, which melts congruently at 727°C and crystallizes in the orthorhombic structure with the lattice parameters $a = 1560$, $b = 780$, and $c = 426$ pm, is formed in this system. This compound divides the PbSnS$_2$–PbBi$_2$S$_4$ system into two parts, and one part, Pb$_2$SnBi$_2$S$_6$–PbSnS$_2$ subsystem, is a eutectic type: the eutectic contains 20 mol% PbBi$_2$S$_4$ and crystallizes at 577°C. The solubility based on PbSnS$_2$ is 5 mol% PbBi$_2$S$_4$. This system was investigated through DTA, XRD, metallography, and measuring of the microhardness. The crystals of the quaternary compound were grown by a chemical transport reaction using iodine as a transport agent.

PbSnBi$_4$S$_8$–PbBi$_4$S$_7$. The phase diagram of this system, constructed through DTA, XRD, metallography, and measuring of the microhardness using the ingots annealing at 480°C for 120 h, is a eutectic type (Gurbanov 2012a). The eutectic contains 55 mol% PbSnBi$_4$S$_8$ and crystallizes at 497°C. PbSnBi$_4$S$_8$ dissolves 10 mol% PbBi$_4$S$_7$ and PbBi$_4$S$_7$ dissolves 15 mol% PbSnBi$_4$S$_8$.

PbSnBi$_4$S$_8$–Pb$_2$SnBi$_2$S$_6$. The phase diagram of this system, constructed through DTA, XRD, metallography, and measuring of the microhardness using the ingots annealing at 480°C for 120 h, is a eutectic type (Gurbanov 2012a). The eutectic contains 45 mol% Pb$_2$SnBi$_2$S$_6$ and crystallizes at 567°C. PbSnBi$_4$S$_8$ dissolves 15 mol% Pb$_2$SnBi$_2$S$_6$ and Pb$_2$SnBi$_2$S$_6$ dissolves 14 mol% PbSnBi$_4$S$_8$.

PbSnBi$_6$S$_{11}$–Bi$_2$S$_3$. The phase diagram of this system, constructed through DTA, XRD, metallography, and measuring of the microhardness using the ingots annealing at 480°C for 120 h, is a eutectic type (Gurbanov 2012a). The eutectic contains 40 mol% Bi$_2$S$_3$ and crystallizes at 427°C. PbSnBi$_6$S$_{11}$ dissolves 7 mol% Bi$_2$S$_3$ and Bi$_2$S$_3$ dissolves 6 mol% PbSnBi$_6$S$_{11}$.

Pb$_2$SnBi$_2$S$_6$–PbS. The phase diagram of this system, constructed through DTA, XRD, metallography, and measuring of the microhardness using the ingots annealing at 480°C for 120 h, is a eutectic type (Gurbanov 2012a). The eutectic contains 33 mol% PbS and crystallizes at 407°C. Pb$_2$SnBi$_2$S$_6$ dissolves 5 mol% PbS, and PbS dissolves 6 mol% Pb$_2$SnBi$_2$S$_6$.

Pb$_2$SnBi$_2$S$_6$–PbBi$_4$S$_7$. The phase diagram of this system, constructed through DTA, XRD, metallography, and measuring of the microhardness using the ingots annealing at 480°C for 120 h, is a eutectic type (Gurbanov 2012a). The eutectic contains 44 mol% Pb$_2$SnBi$_2$S$_6$ and crystallizes at 582°C. Pb$_2$SnBi$_2$S$_6$ dissolves 15 mol% PbBi$_4$S$_7$ and PbBi$_4$S$_7$ dissolves 10 mol% Pb$_2$SnBi$_2$S$_6$.

Pb$_2$SnBi$_2$S$_6$–Pb$_3$Bi$_2$S$_6$. This system is a nonquasibinary section of the SnS–PbS–Bi$_2$S$_3$ quasiternary system as Pb$_3$Bi$_2$S$_6$ melts incongruently (Gurbanov 2012a). Pb$_2$SnBi$_2$S$_6$ dissolves 6 mol% Pb$_3$Bi$_2$S$_6$.

Single crystals of the PbSnBi$_4$S$_8$, PbSnBi$_6$S$_{11}$, and Pb$_2$SnBi$_2$S$_6$ were grown through chemical transport reactions (Gurbanov 2012a). The standard thermodynamic properties of these quaternary compounds were calculated: $\Delta G^0_{f,298} = -782 \pm 20$ kJ·M^{-1}, $\Delta H^0_{f,298} = -783 \pm 20$ kJ·M^{-1}, $\Delta S^0_{f,298} = 0.20$

± 0.05 J·(M·K)$^{-1}$ and $S^0_{f,298} = 586 \pm 25$ J·(M·K)$^{-1}$ for PbSnBi$_4$S$_8$; $\Delta G^0_{f,298} = -1128 \pm 30$ kJ·M^{-1}, $\Delta H^0_{f,298} = -1129 \pm 30$ kJ·M^{-1}, $\Delta S^0_{f,298} = -1.1 \pm 0.2$ J·(M·K)$^{-1}$ and $S^0_{f,298} = 795 \pm 30$ J·(M·K)$^{-1}$ for PbSnBi$_6$S$_{11}$; and $\Delta G^0_{f,298} = -641 \pm 20$ kJ·M^{-1}, $\Delta H^0_{f,298} = -645 \pm 20$ kJ·M^{-1}, $S^0_{f,298} = -3.4 \pm 0.4$ J·(M·K)$^{-1}$ and $\Delta S^0_{f,298} = 470 \pm 20$ J·(M·K)$^{-1}$ for Pb$_2$SnBi$_2$S$_6$.

SnS–PbS–Bi$_2$S$_3$. This quasiternary system is divided by the given above quasibinary sections into 9 secondary subsystems (Gurbanov, 2012a).

7.220 Tin–Lead–Selenium–Sulfur

SnS+PbSe ↔ SnSe+PbS. The decomposition region in this system was studied systematically by Leute et al. (1993). Using XRD and X-ray microanalysis data, the isothermal sections at 555°C, 630°C, and 705°C of this ternary mutual system were constructed. No thermodynamic modeling of this system has been reported. Krebs and Langner (1964) determined the mutual solubility of the end-members in the SnSe–PbS and SnS–PbSe joins. Their results agree within experimental uncertainty with the data reported by Leute et al. (1993). The isothermal sections constructed earlier were used to evaluate the phase diagram of the system under consideration. Given that the existence of two orthorhombic phases was shown in this study to be unlikely, all of the X-ray microanalysis data were interpreted as boundaries of the decomposition region between the cubic and orthorhombic phases. Using the modeling results for solid-state equilibria and the interaction parameters for the liquid phase in the quasibinary systems, the solidus and liquidus surfaces as well as some isothermal sections of this ternary mutual system were calculated by Volykhov et al. (2009). The calculated isothermal sections at 555°C, 830°C, and 1030°C are presented in Figure 7.50.

The **PbSnSeS$_2$** quaternary compound, which melts incongruently at 740°C and crystallizes in the orthorhombic structure with the lattice parameters $a = 890.2 \pm 0.3$, $b = 384.0 \pm 0.3$, $c = 1419.7 \pm 0.3$ pm, and the calculated and experimental densities of 3.43 and 6.41 g·cm^{-3}, respectively, is formed in the Sn–Pb–Se–S system (Jumas et al. 1972).

7.221 Tin–Lead–Tellurium–Sulfur

SnS+PbTe ↔ SnTe+PbS. This ternary mutual system was studied experimentally at 630°C by Leute et al. (1995) using XRD and X-ray microanalysis. They developed a thermodynamic model for the solid phases in this system using the interaction parameters for the binary systems PbS–PbTe and SnS–PbS and the hypothetical parameters of the transitions of their components (except for PbTe) known from previous reports. In addition, the SnTe–PbS and PbTe–SnS systems were investigated by Layner and Samedov (1971), Matyas (1985), and Matyas and Borisenko (1990). There are significant differences between the solubility data: according to Matyas (1985), SnS solubility in PbTe exceeds 20 mol% even at 550°C, and PbTe solubility in SnS is above 7 mol%, whereas according to Leute et al. (1995), SnS solubility in PbTe is below 15 mol% at 630°C, and PbTe solubility in SnS is 3 mol%.

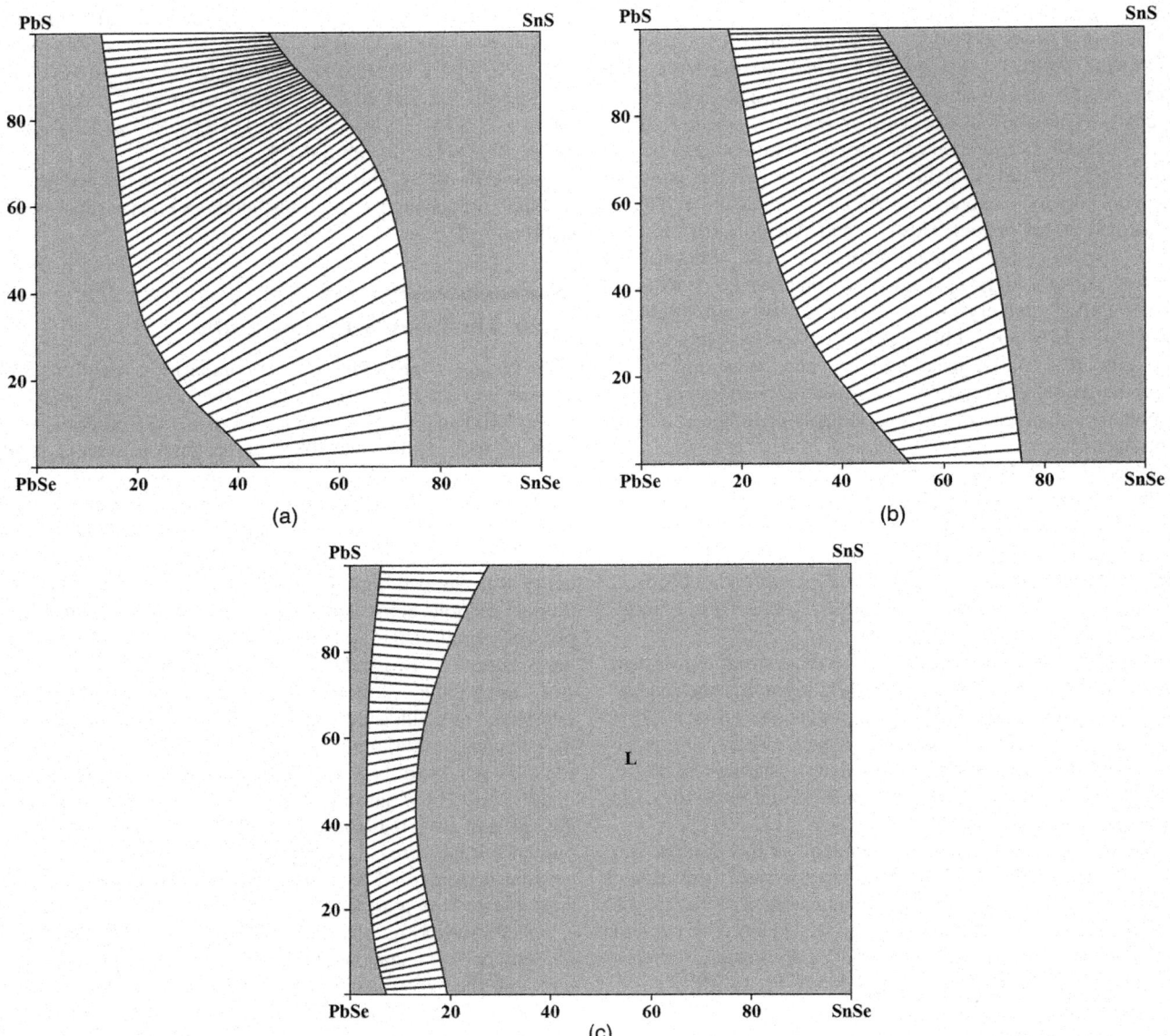

FIGURE 7.50 Isothermal section of the SnS+PbSe $\leftrightarrow$ SnSe+PbS ternary mutual system at (a) 555°C, (b) 830°C, and (c) 1030°C. (From Volykhov, A.A., et al., *Inorg. Mater.*, **45**(9), 968, 2009.)

In modeling this system, the experimental data reported by Leute et al. (1995), Matyas (1985) and Matyas and Borisenko (1990) were left out of consideration as considerably less accurate (Volykhov et al. 2009). The assumption that there are two solid-solution regions with an orthorhombic structure was not taken into account. As a result, the agreement with experimental data was substantially better than in the case of the calculations performed by Leute et al. (1995), which may be due to the more accurate determination of the interaction parameters in the constituent binary systems. Using the modeling results for solid-state equilibria and the interaction parameters for the liquid phase in the quasibinary systems, the solidus and liquidus surfaces, as well as some isothermal sections of this ternary mutual system, were calculated by Volykhov et al. (2009). The calculated isothermal sections at 630°C, 730°C, 830°C, and 930°C are presented in Figure 7.51.

Samples of $Sn_xPb_{1-x}Te_{1-y}S_y$ solid solution were investigated in the 4–200 K temperature range using electrical and X-ray methods by Lebedev and Sluchinskaya (1994). The region where low-temperature phase transition takes place was established (Figure 7.52).

7.222 Tin–Lead–Chlorine–Sulfur

SnS–PbCl₂. The phase diagram of this system, constructed through DTA, is a eutectic type (Figure 7.53) (Morozov and Li 1963). The eutectic contains 17.4 mol% SnS and crystallizes at 440°C.

SnS+PbCl₂ $\leftrightarrow$ SnCl₂+PbS. The liquidus surface of this ternary mutual system consists from the fields of primary crystallization of the $Sn_xPb_{1-x}S$ and $Sn_xPb_{1-x}Cl_2$ solid solutions,

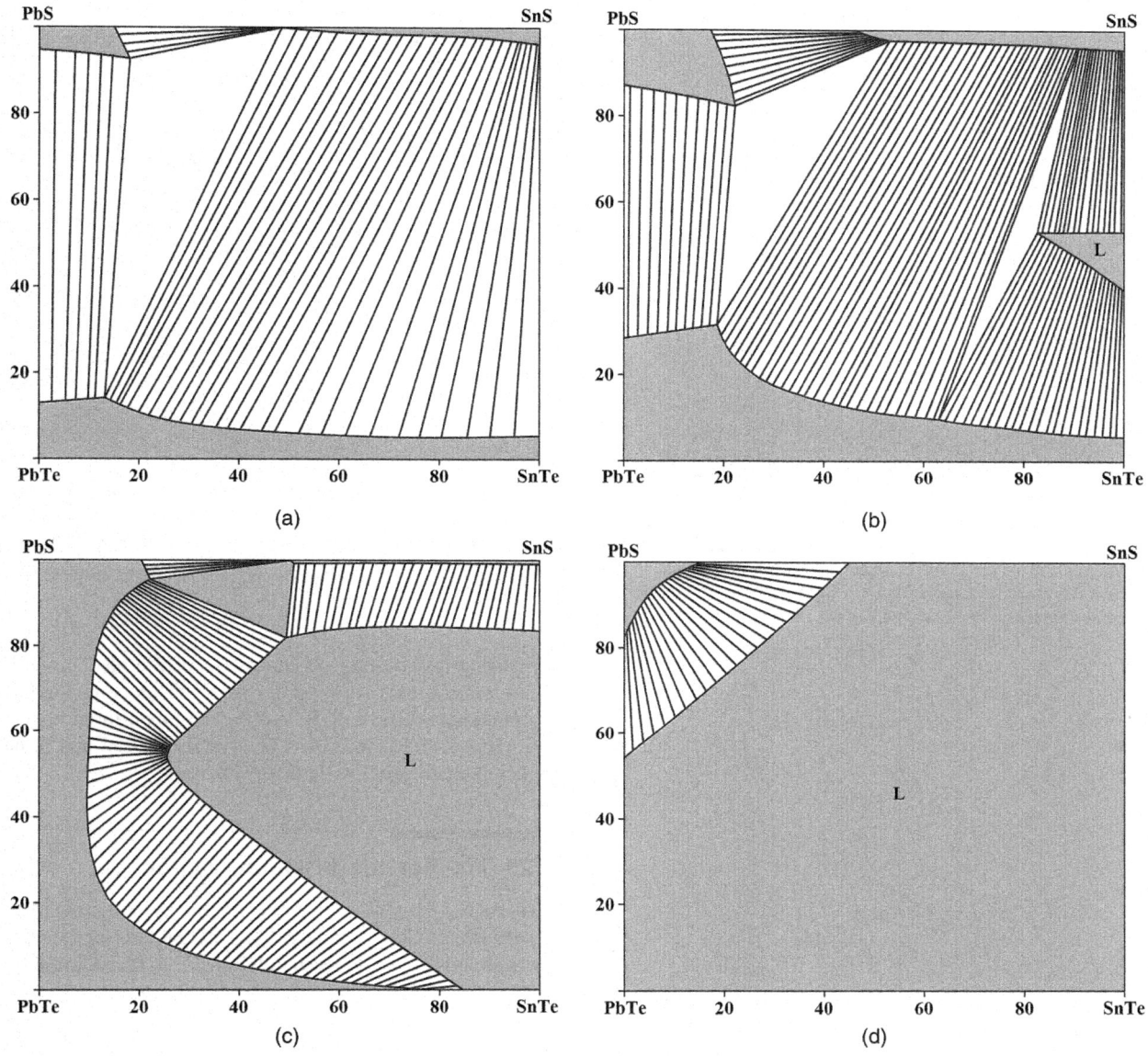

FIGURE 7.51 Isothermal section of the SnS+PbTe ↔ SnTe+PbS ternary mutual system at (a) 630°C, (b) 730°C, (c) 830°C, and (d) 930°C. (From Volykhov, A.A., et al., *Inorg. Mater.*, **45**(9), 968, 2009.)

the first of which occupies most of the system (Figure 7.54) (Morozov and Li 1963). All liquidus isotherms crossing the SnS–PbCl₂ stable diagonal split into two branches, which indicates the singularity of the ternary mutual system, i.e. about a sharp shift of the equilibrium toward SnS–PbCl₂. The PbS–SnCl₂ system is nonquasibinary section of this ternary mutual system.

7.223 Tin–Nitrogen–Chlorine–Sulfur

The **SnCl₄·2S₂N₄** quaternary compound, which crystallizes in the orthorhombic structure with the lattice parameters $a = 1146.7 \pm 0.5$, $b = 1199.5 \pm 1.5$, $c = 1237.4 \pm 0.5$ pm, and a calculated density of 2.45 g·cm⁻³, is formed in the Sn–N–Cl–S system (Martan and Weiss 1984). To obtain this compound,

a 100-mL two-necked with two 50 mL-dropping funnels was filled halfway up the dropping funnels with dry, N₂-saturated CHCl₃. With the taps closed, S₄N₄ (200 mg) and freshly distilled SnCl₄ (400 mg) were added to the dropping funnels. S₄N₄ was dissolved by shaking the entire arrangement, and then the stopcocks of both funnels were opened a little. The two solutions slowly flowed into the flask and the solution in the flask turned red. Red crystals of the title compound were grown from this solution over the course of several weeks.

7.224 Tin–Phosphorus–Selenium–Sulfur

The **SnPS₂.₈₆Se₀.₁₄** quaternary phase, which crystallizes in the monoclinic structure with the lattice parameters $a = 654.3 \pm 0.3$, $b = 749.8 \pm 0.3$, $c = 938.4 \pm 0.4$ pm, $\beta = 91.16 \pm 0.10°$, a

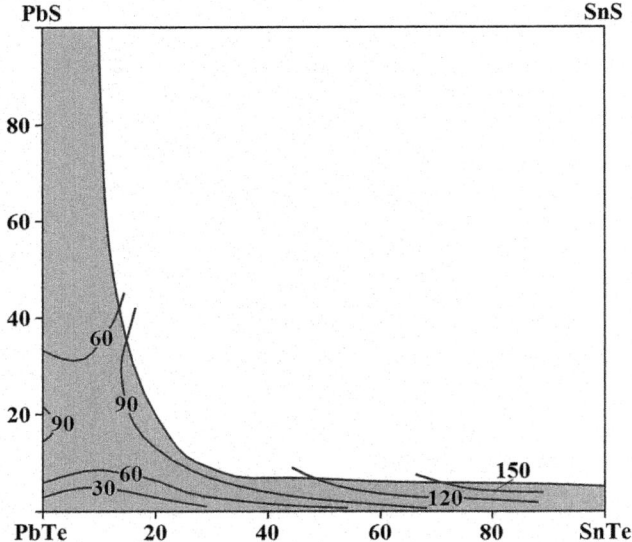

FIGURE 7.52 The dependence of low-temperature phase transition (in K) on composition for SnS+PbTe ↔ SnTe+PbS ternary mutual system. The dotted line shows the stability limits for a solid solution. (From Lebedev, A.I., and Sluchinskaya. I.A., *J. Alloys Compd.*, **203**(1–2), 51, 1994.)

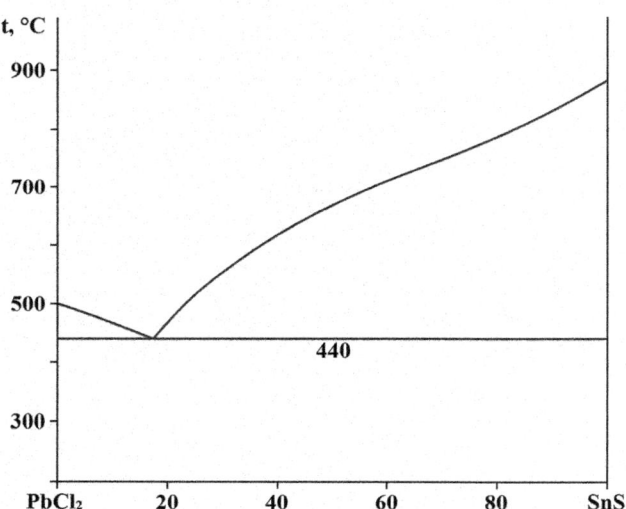

FIGURE 7.53 Phase diagram of the SnS–PbCl₂ system. (From Morozov, I.S., and Li, C.-F., *Zhurn. neorgan. khimii*, **9**(7), 1688, 1963.)

calculated density of 3.642 g·cm⁻³, and an energy gap of 2.17 eV, is formed in the Sn–P–Se–S system (Shi et al. 2021). Single crystals were synthesized by the high-temperature solid-state method. The starting reagents Sn (1.1 mM), P₂S₅ (0.56 mM), and Se (0.56 mM) were mixed homogeneously with KI flux. The mixture was loaded into a fused silica tube under an Ar atmosphere in a glove box, which was sealed under 10⁻² Pa and then placed into a furnace. The reactants were heated to 950°C within 25 h, kept at that temperature for 48 h, and then cooled to 400°C at a rate of 5°C·h⁻¹ before the furnace was switched

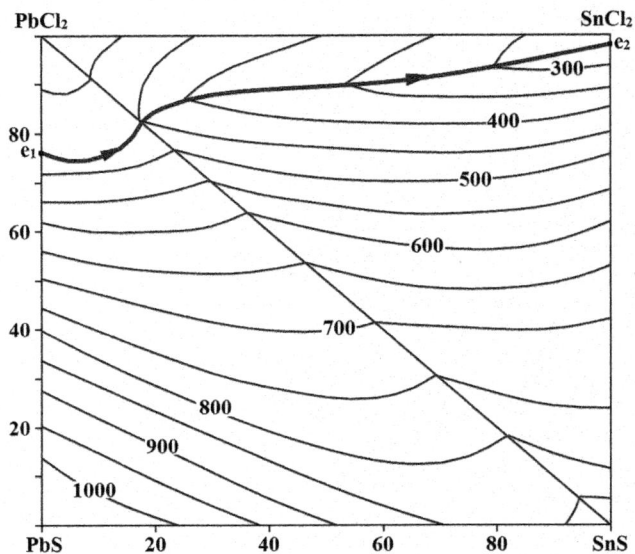

FIGURE 7.54 Liquidus surface of the SnS+PbCl₂ ↔ SnCl₂+PbS ternary mutual system. (From Morozov, I.S., and Li, C.-F., *Zhurn. neorgan. khimii*, **9**(7), 1688, 1963.)

off. During this heating process, several intermediate equilibrated temperatures were set for safety. This compound was also obtained with Sn, P, S, and Se combined in an approximate ratio of 1:1:2.86:0.14. The crystals are red and remain stable in air and water without any changes.

7.225 Tin–Arsenic–Selenium–Sulfur

SnSe–As₂S₃. The phase diagram of this system, constructed through DTA, XRD, metallography, and measuring the microhardness and density, is given in Figure 7.55 (Gurshumov et al.

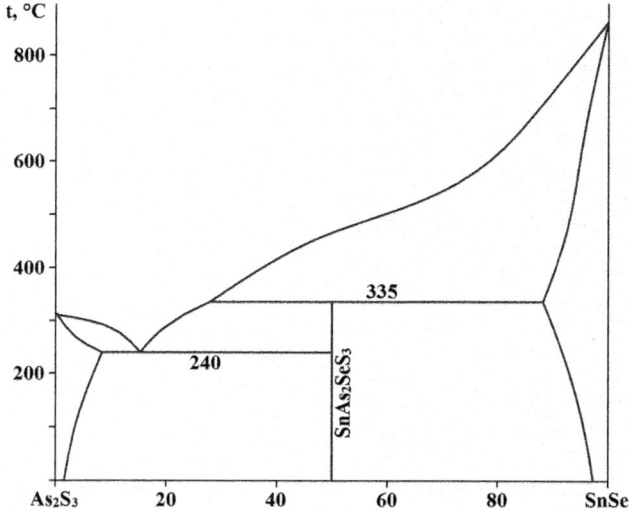

FIGURE 7.55 Phase diagram of the SnSe–As₂S₃ system. (From Gurshumov, A.P., et al., *Zhurn. neorgan. khimii*, **31**(1), 266, 1986.)

1986b). The eutectic contains 15 mol% SnSe and crystallizes at 240°C. The **SnAs$_2$SeS$_3$** quaternary compound, which melts incongruently at 325°C, is formed in the system. The solubility of SnSe in As$_2$S$_3$ and As$_2$S$_3$ in SnSe at room temperature is 1.3 and 2.7 mol%, respectively. The ingots for the investigations were annealed at 220°C for 560 h.

7.226 Tin–Arsenic–Tellurium–Sulfur

SnTe–As$_2$S$_3$. The phase diagram of this system, constructed through DTA, XRD, metallography, and measuring the microhardness and density, is presented in Figure 7.56 (Kuliev et al. 1982). The eutectic contains 25 mol% SnTe and crystallizes at 240°C. The **SnAs$_2$TeS$_3$** quaternary compound, which melts incongruently at 335°C, is formed in the system. The solubility of As$_2$S$_3$ in SnTe at 100°C, 200°C, and 250°C is 5, 8, and 10 mol%, respectively. A solid solution based on As$_2$S$_3$ was not deetermined during slow cooling of the alloys, the region of glass formation based on As$_3$S$_3$ extends to 15 mol% SnTe. The ingots for the investigations were annealed at 200°C for 780 h.

7.227 Tin–Antimony–Bismuth–Sulfur

Two quaternary compounds, **SnSbBiS$_4$** and **Sn$_2$Sb$_{2-x}$Bi$_x$S$_5$**, are formed in the Sn–Sb–Bi–S system. First of them melts incongruently at 627°C and crystallizes in the hexagonal structure with the lattice parameters $a = 1552$ and $c = 381$ pm (Kurbanov 2010). Single crystals of this compound were grown by chemical transport reactions, and I$_2$ was used as a transport agent. They are stable in air.

Sn$_2$Sb$_{2-x}$Bi$_x$S$_5$ ($0.4 < x < 0.2$) crystallizes in the orthorhombic structure with the lattice parameters $a = 395 \pm 2$, $b = 1126 \pm 4$, $c = 1949 \pm 6$ pm (Kupčik and Wendschuh 1982).

SnS–SnSbBiS$_4$. The phase diagram of this system, constructed through DTA, XRD, metallography, and measuring the microhardness, is shown in Figure 7.57 (Agaguseinova et al. 2011). The eutectic contains 30 mol% SnS and crystallizes at 427°C. The isotherm at 527°C corresponds to the polymorphic transformation in SnS. Solubility boundaries were determined at 8 and 3 mol% for the alloys based on SnSbBiS$_4$ and SnS, respectively. The alloys for the investigations were annealed at 230–380°C for 240–250 h.

SnSbBiS$_4$–Sn$_2$Sb$_6$S$_{11}$. The phase diagram of this system, constructed through DTA, XRD, metallography, and measuring the microhardness and density, is a eutectic type (Figure 7.58) (Agaguseinova et al. 2011). The eutectic contains 35 mol% Sn$_2$Sb$_6$S$_{11}$ and crystallizes at 327°C. The boundaries of

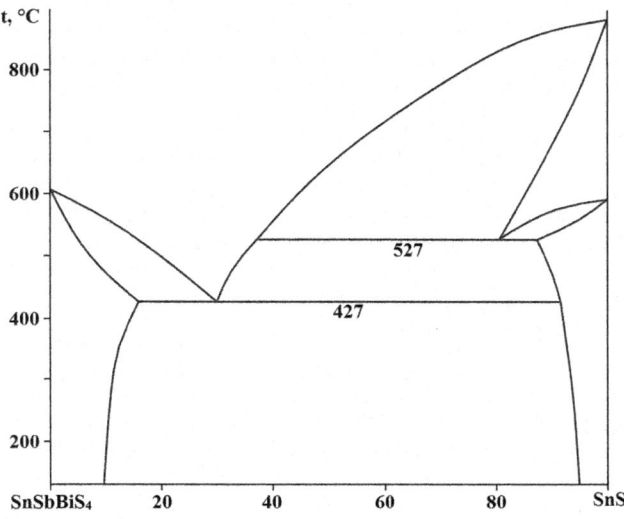

FIGURE 7.57 Phase diagram of the SnS–SnSbBiS$_4$ system. (From Agaguseinova, M.M., et al., *Russ. J. Inorg. Chem.*, **56**(8), 1331, 2011.)

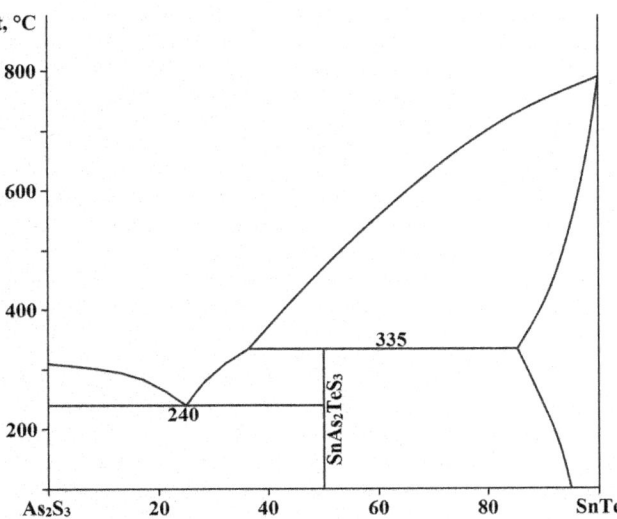

FIGURE 7.56 Phase diagram of the SnTe–As$_2$S$_3$ system. (From Kuliev, et al., *Izv. AN SSSR. Neorgan. mater.*, **18**(5), 738, 1982.)

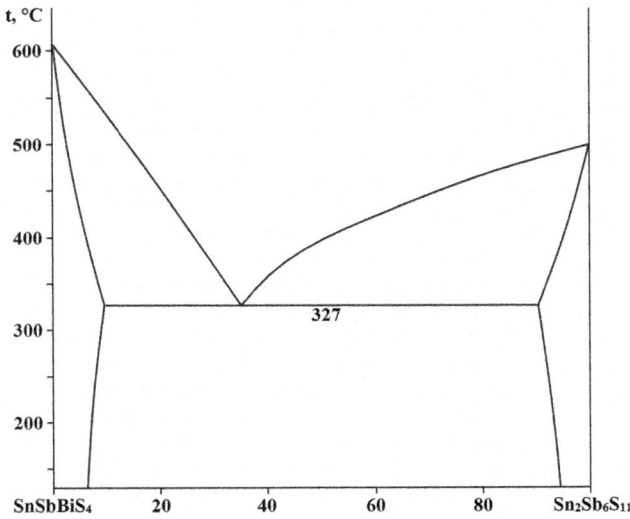

FIGURE 7.58 Phase diagram of the SnSbBiS$_4$–Sn$_2$Sb$_6$S$_{11}$ system. (From Agaguseinova, M.M., et al., *Russ. J. Inorg. Chem.*, **56**(8), 1331, 2011.)

the solid-solution regions based on SnSbBiS$_4$ and Sn$_2$Sb$_6$S$_{11}$ reach 3 mol% Sn$_2$Sb$_6$S$_{11}$ and ~2 mol% SnSbBiS$_4$, respectively, at room temperature.

SnSbBiS$_4$–Bi$_2$S$_3$. The phase diagram of this system, constructed through DTA, XRD, metallography, and measuring the microhardness and density, is a eutectic type (Gurbanov 2012d). The eutectic contains 30 mol% Bi$_2$S$_3$ and crystallizes at 427°C. SnSbBiS$_4$ dissolves 7 mol% Bi$_2$S$_3$ and solid solution based on Bi$_2$S$_3$ reaches 3 mol%.

Sn$_2$Sb$_2$S$_5$–SnSbBiS$_4$. This system is nonquasibinary section of the Sn–Sb–Bi–S system as Sn$_2$Sb$_2$S$_5$ melts incongruently, and SnS crystallizes primarily from the melts from the side of this compound (Gurbanov 2010).

7.228 Tin–Antimony–Selenium–Sulfur

SnS–Sb$_2$Se$_3$ and SnSe–Sb$_2$S$_3$. These systems are nonquasibinary sections of the SnS+Sb$_2$Se$_3$ ↔ SnSe–Sb$_2$S$_3$ ternary mutual system (Gospodinov et al. 1972, 1974a,b). The Sb$_2$S$_x$Se$_{1-x}$, Sn$_2$Sb$_6$S$_x$Se$_{1-x}$, and SnS$_x$Se$_{1-x}$ solid solutions primary crystallize from the melt (a solid solution based on Sn$_2$Sb$_2$S$_5$ also primarily crystallizes in the case of SnSe–Sb$_2$S$_3$ system). SnS$_x$Se$_{1-x}$ solid solutions have a polymorphic transformation within the interval from 510°C to 540°C and contain up to 9 mol% Sb$_2$Se$_3$. The Sb$_2$S$_x$Se$_{1-x}$ solid solution contains negligible quantity of SnS. The system was investigated through DTA, XRD, metallography, and measuring of the microhardness. The ingots for the investigations were annealed at 400°C for 500 h.

SnS+Sb$_2$Se$_3$ ↔ SnSe–Sb$_2$S$_3$. The liquidus surface of this ternary mutual system is presented in Figure 7.59 (Gospodinov et al. 1974b).

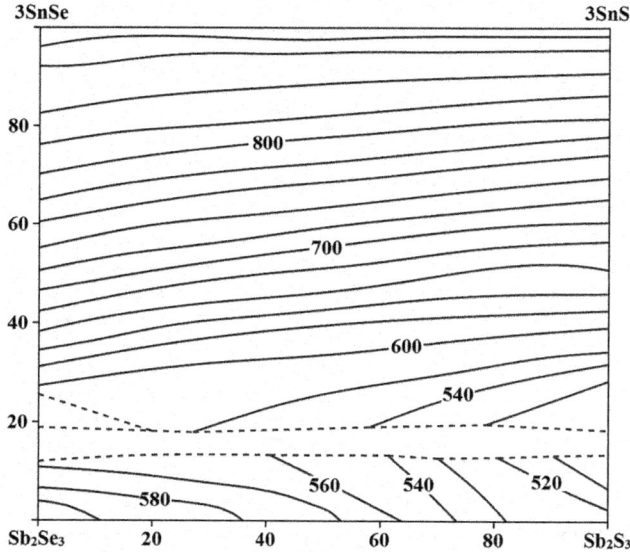

FIGURE 7.59 Liquidus surface of the SnS+Sb$_2$Se$_3$ ↔ SnSe–Sb$_2$S$_3$ ternary mutual system. (From Gospodinov, G.G., et al., *Zhurn. neorgan. khimii*, **19**(6), 1644, 1974.)

7.229 Tin–Antimony–Iodine–Sulfur

The **Sn$_2$SbS$_2$I$_3$** quaternary compound, which melts congruently at 415°C (Gorak and Starosta 1991) [at 400°C (Dolgikh 1985b)] and crystallizes in the orthorhombic structure with the lattice parameters $a = 472.5 \pm 0.1$, $b = 1405.9 \pm 0.3$, $c = 1646.5 \pm 0.3$ pm, a calculated density of 5.44 g·cm^{-3} (Olivier-Fourcade et al. 1980) and an energy gap of 1.5 eV (Gorak and Starosta 1991) [1.65 eV (Dolgikh 1985b); 1.77 eV (Ibañez et al. 1984)] is formed in the Sn–Sb–I–S system. The melting enthalpy of this compound is 53.4 kJ·M^{-1} (Gorak and Starosta 1991).

Sn$_2$SbS$_2$I$_3$ was obtained by the interaction of SnS$_2$ and SbI$_3$ at 650°C for 5 days (Olivier-Fourcade et al. 1985). It was also synthesized by heating in an evacuated quartz ampoule, a stoichiometric mixture of the elements, using a rocking furnace (Gorak and Starosta 1991). The mixture was heated to 650°C, kept at this temperature for 24–36 h, and cooled by switching off the furnace. Its single crystals were grown by recrystallization through the vapor phase (Dolgikh 1985b) or using chemical transport reactions (Olivier-Fourcade et al. 1985; Ibañez et al. 1984). This compound is stable in the air.

7.230 Tin–Bismuth–Iodine–Sulfur

The **SnBi$_{12}$S$_{18}$I$_2$** quaternary compound, which crystallizes in the hexagonal structure with the lattice parameters $a = 1564.8 \pm 0.1$ and $a = 402.9 \pm 0.1$ pm, is formed in the Sn–Bi–I–S system (Guseynov et al. 1989). Single crystals of this compound were grown through chemical transport reactions using a mixture of SnS and Bi$_2$S$_3$ with I$_2$ as a transport agent.

7.231 Tin–Selenium–Tellurium–Sulfur

SnS–SnSe–SnTe. This ternary system contains two solid-solution regions: based on the cubic phase SnTe and between the orthorhombic phases SnS and SnSe. The fragmentary data on the solid-solution regions in this system were reported by Liu and Chang (1992). The boundaries of the single-phase regions and the slope of the tie lines corresponding to the equilibrium between the cubic and orthorhombic phases were determined through X-ray microanalysis (Volykhov et al. 2010). Figure 7.60 presents isothermal sections of this quasiternary system at 430–730°C. From the experimental data, the solid-state interaction parameter was estimated at −20 kJ·M^{-1} for the orthorhombic phase and zero for the cubic phase. It is necessary to note that the homogeneity region of the cubic phase is very small, which makes it difficult to accurately determine the ternary interaction parameter of this phase. The positions of the solidus and liquidus surfaces were determined using DTA. The results show a systematic positive deviation from the values obtained under the assumption that there is no ternary interaction in the liquid phase. A liquid-phase ternary interaction parameter of −3 kJ·M^{-1} ensures a better description of the system. Modeling was based mainly on liquidus data because the solidus was poorly defined in DTA curves and could not be accurately located. The liquidus and solidus surfaces are shown in Figure 7.61.

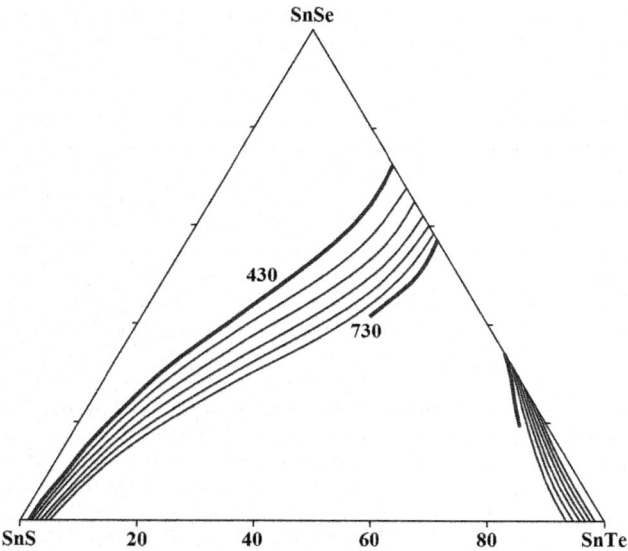

FIGURE 7.60 Isothermal sections of the SnS–SnSe–SnTe quasiternary system at 430–830°C. (From Volykhov, A.A., et al., *Inorg. Mater.*, **46**(5), 464, 2010.)

7.232 Tin–Selenium–Iodine–Sulfur

Two quaternary phases, **SnI₄·(SeS₇)** and **SnI₄·(Se₈₋ₙSₙ)**, are formed in the Sn–Se–I–S system. First of them melts at 98.0°C and crystallizes in the orthorhombic structure with the lattice parameters $a = 2094$, $b = 2220$, $c = 1151$ pm and a calculated density of 3.05 g·cm⁻³ (Hawes 1963). It was prepared by precipitation using a minimal amount of SnI₄ with the mother liquor from the Se₂S₆ separation. SnI₄·(Se₈₋ₙSₙ) melts at 96–98°C and crystallizes in the triclinic structure with the lattice parameters

$a = 1141.3 \pm 0.6$, $b = 1613.7 \pm 1.4$, $c = 809.8 \pm 0.6$ pm, $\alpha = 92.34 \pm 0.07°$, $\beta = 110.47 \pm 0.05°$, and $\gamma = 68.90 \pm 0.05°$ (n value is not specified) (Laitinen et al. 1980). It was prepared by melting S and Se (molar ratio 3:1). After quenching, the mixture was extracted in boiling toluene and SnI₄ was added to the extract. By cooling to −20°C two kinds of crystals appeared: orange needles of Se₈₋ₙSₙ molecules and a few red plate-like crystals of the title compound.

7.233 Tin–Selenium–Cobalt–Sulfur

The **Sn₂Co₃S₂₋ₓSeₓ** quaternary solid solution is formed in the Sn–Se–Co–S system (Weihrich et al. 2015). It crystallizes in the rhombohedral structure with the lattice parameters $a = 537.7 \pm 0.1$ and $c = 1324.8 \pm 0.1$ pm in hexagonal setting or $a = 539.8 \pm 0.1$ pm and $\alpha = 59.74 \pm 0.02°$ for $x = 0.2$; $a = 538.3 \pm 0.1$ and $c = 1330.9 \pm 0.1$ pm in hexagonal setting or $a = 541.7 \pm 0.1$ pm and $\alpha = 59.59 \pm 0.03°$ for $x = 0.4$; $a = 538.9 \pm 0.1$ and $c = 1337.4 \pm 0.1$ pm in hexagonal setting or $a = 543.7 \pm 0.1$ pm and $\alpha = 59.43 \pm 0.02°$ for $x = 0.6$; $a = 539.7 \pm 0.1$ and $c = 1344.1 \pm 0.1$ pm in hexagonal setting or $a = 545.7 \pm 0.1$ pm and $\alpha = 59.27 \pm 0.03°$ for $x = 0.8$; and $a = 540.0 \pm 0.1$ and $c = 1349.4 \pm 0.1$ pm in hexagonal setting or $a = 547.3 \pm 0.1$ pm and $\alpha = 59.13 \pm 0.03°$ and an energy gap of 0.30 eV for $x = 1$. According to the data of calculations, carried out within the framework of density functional theory with generalized gradient approximation the lattice parameters for $x = 1$ are the next: $a = 539.90$ and $c = 1347.88$ pm in hexagonal setting or $a = 546.83$ pm and $\alpha = 59.16°$ and a calculated density of 7.74 g·cm⁻³. The samples of this solid solution were prepared by high-temperature synthesis from the corresponding elements in sealed quartz ampoules. They were heated to 550°C and cooled down slowly to room temperature after 5 days.

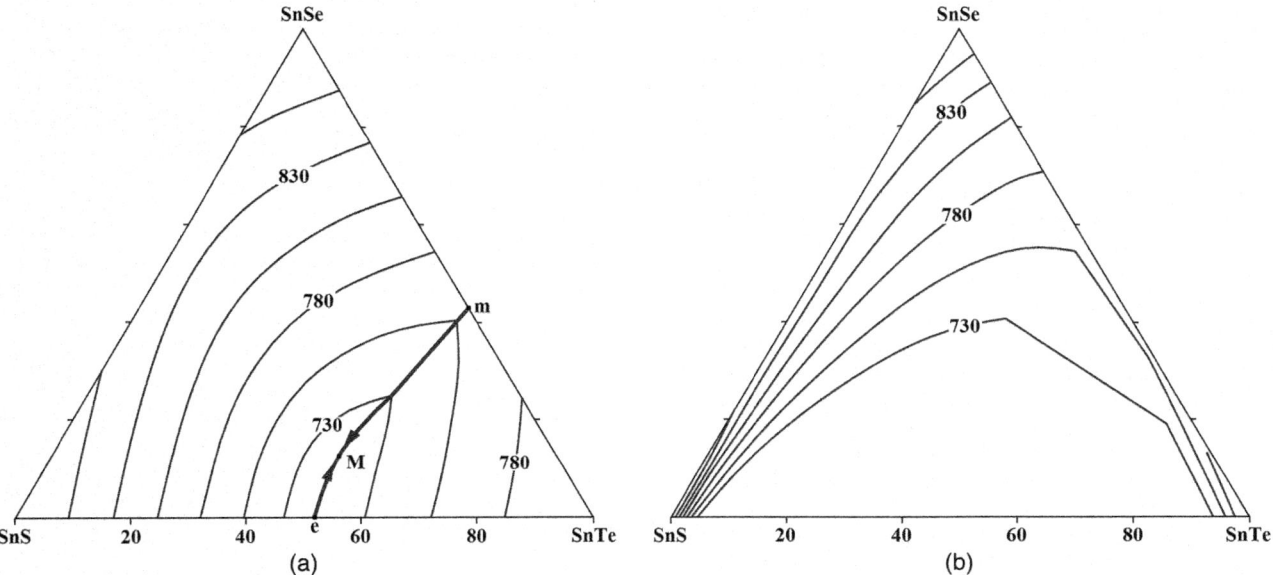

FIGURE 7.61 Liquidus (a) and solidus (b) surfaces of the SnS–SnSe–SnTe quasiternary system. (From Volykhov, A.A., et al., *Inorg. Mater.*, **46**(5), 464, 2010.)

7.234 Tin–Tellurium–Iodine–Sulfur

The **SnI$_4$[(TeS$_7$)(S$_8$)$_2$]** quaternary compound, which has an experimental density of 2.97 g·cm^{-3}, is form in the Sn–Te–I–S system (Hawes 1963).

7.235 Tin–Chromium–Iron–Sulfur

SnS$_2$–Cr$_2$S$_3$–FeS. XRD and nuclear gamma resonance were used to study the part of the system adjacent to FeCr$_2$S$_4$ (Filimonov et al. 1996). A wide ranges of cation-deficient solid solutions Fe$_{1-x}$[Cr$_{2-2x-2y}$Fe$_y$Sn$_{2x+y}$]S$_4$ (0 ≤ x ≤ 0.33; 0 ≤ y ≤ 0.33; y ≤ x) with structure of FeCr$_2$S$_4$, as well as regions of solid solutions Fe$_{1.63}$Cr$_{0.38}$Sn$_{0.91}$S$_4$–Fe$_{1.75}$Cr$_{0.12}$Sn$_{1.04}$S$_4$ (B) and Fe$_{1.03}$Cr$_{1.81}$Sn$_{0.13}$S$_4$–Fe$_{1.62}$Cr$_{0.63}$Sn$_{0.72}$S$_4$ (C), cation-deficient with respect to stoichiometry of FeCr$_2$S$_4$, were found. Based on the results of the studies, an isothermal section at 730°C was

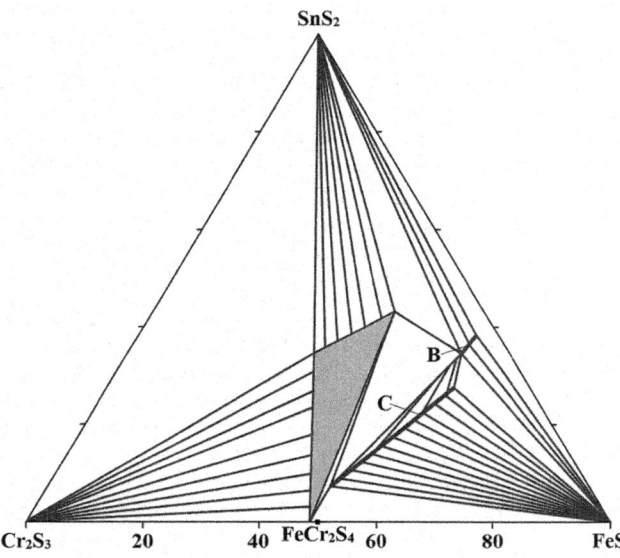

FIGURE 7.62 Isothermal section of the SnS$_2$–Cr$_2$S$_3$–FeS quasiternary system at 730°C. (From Filimonov, D.S., et al., *Neorgan. mater.*, **32**(8), 930, 1996.)

constructed (Figure 7.62), reflecting the possible phase equilibria in the system.

7.236 Tin–Iron–Cobalt–Sulfur

The **Co$_{3-x}$Fe$_x$Sn$_2$S$_2$** quaternary solid solution is formed in the Sn–Fe–Co–S system. It crystallizes in the trigonal structure with the lattice parameters a = 536.81 ± 0.04 and c = 1318.758 ± 0.009 pm for x = 0.14; a = 536.72 ± 0.04 and c = 1319.47 ± 0.09 pm for x = 0.29; a = 536.71 ± 0.02 and c = 1319.50 ± 0.05 pm for x = 0.32; a = 536.70 ± 0.07 and c = 1319.9 ± 0.2 pm for x = 0.39; a = 536.67 ± 0.03 and c = 1320.65 ± 0.06 for x = 0.49; and a = 536.65 ± 0.06 and c = 1320.6 ± 0.2 pm for x = 0.53 (Kassem et al. 2016). To prepare this solid solution, lumps of Co and Fe, grains of Sn and grinded crystals of S were mixed in an initial mixture of molar ratios (Co+Fe)/S/Sn = 8:6:86. The constituent was placed in a 5 mL Al$_2$O$_3$ growth crucible that was covered by another upside-down crucible filled with quartz wool. Both were sealed in an evacuated quartz ampoule, which was put in a furnace. After heated to 400°C over 2 h and held for extra 2 h, the mixture was heated up to 1050°C over 6 h and kept for another 6 h to wait for the constituent melts homogeneously. The melt was cooled slowly to 700°C over 70 h. At 700°C the ampoule was removed from the furnace and the flux was removed via rapid decanting and subsequent spinning in a centrifuge. Pb flux was also employed in an initial composition of (Co+Fe)/S/Sn/Pb = 6:6:36:52 in a trial to extend the solubility of Co$_{3-x}$Fe$_x$Sn$_2$S$_2$. Single crystals were successfully grown out of Sn self-flux with the starting Fe concentration up to x = 0.7 in the molten.

Polycrystalline Co$_{3-x}$Fe$_x$Sn$_2$S$_2$ was also synthesized from a stoichiometric mixture of Sn, Co, Fe, and S (Sakai et al. 2015). The mixture was sealed in an evacuated quartz tube, heated at 600°C for 40 h, and then spontaneously cooled to room temperature. Polycrystalline samples with nominal x ≥ 0.6 were used as starting materials for high-pressure and high-temperature experiments. *In-situ* energy dispersive XRD measurements were carried out in the pressure range of 0–11 GPa and the temperature range from room temperature to 1400°C. The Fe solubility limit x_{max} rises from below 0.6 up to 1.0 following high-pressure and high-temperature heat treatments.

8

Systems Based on Tin Selenides

8.1 Tin–Lithium–Barium–Selenium

Two quaternary compounds, **Li$_2$BaSnSe$_4$** and **Li$_2$Ba$_7$Sn$_4$Se$_{16}$**, are formed in the Sn–Li–Ba–Se system. First of them crystallizes in the tetragonal structure with the lattice parameters $a = 712.0 \pm 0.1$, $c = 845 \pm 2$ pm, a calculated density of 4.451 g·cm^{-3}, and an energy gap of 2.18 eV (Wu et al. 2017c). To synthesize the title compound, a mixture of Li (2 mM), Ba (1 mM), Sn (1 mM), and Se (4 mM) were first loaded into a graphite crucible that avoids the reaction between Li and silica tube, then put it into the silica tube and flame-sealed under 10^{-3} Pa. The tube was heated to 800°C in 30 h and kept at this temperature for about 90 h, then cooled slowly down to ambient temperature for 5 days. The product was washed with dimethylformamide (DMF). Finally, many small light-red transparent crystals of Li$_2$BaSnSe$_4$ were found, which are stable in the air.

Li$_2$Ba$_7$Sn$_4$Se$_{16}$ crystallizes in the cubic structure with the lattice parameter $a = 1516.28 \pm 0.05$ pm, a calculated density of 5.170 g·cm^{-3}, and an energy gap of 1.75 eV (Abudurusuli et al. 2018). To prepare this compound, Li, Ba, Sn, and Se (molar ratio 2:3:2:8) were weighed, well mixed, and loaded into the graphite crucible. Then, the crucible was embedded in a silica tube that was sealed under a vacuum of 10^{-3} Pa. The tube was put into a furnace and the setting temperature process was as follows: firstly heated to 870°C in 30 h and kept at this temperature for 50 h to make the mixture completely melted, and then slowly cooled to 300°C at a rate of 5°C·h^{-1} before switching off the furnace. The obtained product was carefully washed with DMF to remove the unreacted S powder. Finally, air-stable crystals of Li$_2$Ba$_7$Sn$_4$Se$_{16}$ of red color appeared.

8.2 Tin–Lithium–Zinc–Selenium

The **Li$_2$ZnSnSe$_4$** quaternary compound, which melts at 736°C and crystallizes in the monoclinic structure with the lattice parameters $a = 668.48 \pm 0.01$, $b = 703.03 \pm 0.01$, $c = 832.34 \pm 0.01$ pm, $\beta = 90.050 \pm 0.001°$, a calculated density of 4.362 g·cm^{-3}, and an energy gap of 1.87 eV [2.0 eV (Weiland et al. 2017)], is formed in the Sn–Li–Zn–Se system (Zhang et al. 2015). This compound was prepared via traditional high-temperature solid-state reactions. The starting materials were weighed in stoichiometric amounts of Li (132 mg), Zn (62.8 mg), Sn (114.0 mg), and Se (303.0 mg) in an Ar-filled glove box. The reactants were placed into a graphite tube inside a fused silica tube. The tube was sealed under a pressure of approximately 0.1 Pa, placed into a furnace, and heated to 190°C in 2 h, held at 190°C for 10 h, and then heated up to 700°C in 5 h, held at 700°C for 100 h and slowly cooled to 350°C in 100 h. The sample was then radiatively cooled to

room temperature by switching off the furnace. The reaction vessel was opened under ambient conditions. Li$_2$ZnSnSe$_4$ was prepared as orange-red irregular polyhedral-shaped crystals.

8.3 Tin–Lithium–Cadmium–Selenium

The **Li$_2$CdSnSe$_4$** quaternary compound, which crystallizes in the orthorhombic structure with the lattice parameters $a = 1445.08 \pm 0.03$, $b = 839.86 \pm 0.02$, $c = 680.32 \pm 0.01$ pm, a calculated density of 4.511 g·cm^{-3}, and an energy gap of 2.2 eV, is formed in the Sn–Li–Cd–Se system (Zhang et al. 2017). Traditional high-temperature solid-state reactions were carried out to prepare this compound. The starting materials were weighed in stoichiometric portions, Li (13.8 mg), Cd (112.4 mg), Sn (118.7 mg), and Se (315.6 mg), in an Ar atmosphere in a glove box. The mixture was inserted into a graphite tube inside a fused silica tube. The tube was flame-sealed under a pressure of approximately 0.1 Pa and placed into a furnace. The reaction was heated to 190°C in 2 h, held at 190°C for 10 h, and then heated up to 675°C in 5 h, held at 675°C for 100 h, and slowly cooled down to 350°C for 100 h. The furnace was then shut off, and the sample was allowed to cool to room temperature radiatively. The reaction vessel was opened under ambient conditions. Orange irregularly shaped crystals of the title compound were obtained.

According to the data of Guo et al. (2018), the composition of this compound is Li$_{6.4}$Cd$_{4.8}$Sn$_4$Se$_{16}$, it melts congruently at 787°C, and crystallizes in the orthorhombic structure with the lattice parameters $a = 1440.0 \pm 0.3$, $b = 839.90 \pm 0.17$, $c = 678.75 \pm 0.03$ pm, a calculated density of 4.697 g·cm^{-3} at 153 K, and an energy gap of 1.95 eV. The polycrystalline sample of this compound was synthesized by using a mixture of Li$_2$Se, CdSe, and SnSe$_2$ in a stoichiometric ratio. The sample was heated to 600°C in 20 h, kept at that temperature for 72 h, and then the furnace was turned off. To prepare single crystals, the analogous mixture was ground and loaded into a fused silica tube under an Ar atmosphere in a glove box, which was evacuated and sealed. The tube was heated to 1000°C in 20 h and kept at that temperature for 48 h, slowly cooled to 300°C at a rate of 2.5°C·h^{-1}, and finally cooled to room temperature. The resultant red crystals were manually selected from the ampoule.

8.4 Tin–Lithium–Mercury–Selenium

The **Li$_4$HgSn$_2$Se$_7$** quaternary compound, which melts at 661°C and crystallizes in the monoclinic structure with the lattice parameters $a = 1828.6 \pm 0.4$, $b = 719.72 \pm 0.14$, $c = 1080.9 \pm$

0.2 pm, $\beta = 93.44 \pm 0.03°$, a calculated density of 4.764 g·cm⁻³ at 153 K, and an energy gap of 2.10 eV, is formed in the Sn–Li–Hg–Se system (Guo et al. 2019b). Polycrystalline samples of this compound were synthesized by solid-state reaction technique. The mixture of Li, HgSe, Sn, and Se (molar ratio 4:1:2:6) was ground and loaded into a fused silica tube. Then the tube was evacuated to a high vacuum of 10⁻³ Pa and sealed. The sample was heated to 600°C in 10 h in a furnace and stayed at that temperature for 100 h. Then it was cooled to room temperature by switching off the furnace. The single crystals were prepared by a spontaneous crystallization method using pure single-phase powder. The powder was loaded into quartz tube in a glove box, which was sealed under a high vacuum of 10⁻³ Pa and then placed in a furnace. The sample was heated to 700°C in 15 h and kept at that temperature for 72 h. Then the sample was cooled at a rate of 3°C·h⁻¹ to 200°C, and finally cooled to room temperature. The red crystals were found in the tube.

8.5 Tin–Lithium–Vanadium–Selenium

The $Li_{0.5}SnVSe_3$ quaternary intercalated compound with the c-lattice parameter of 2442 pm is formed in the Sn–Li–V–Se system (Hernán et al. 1992). It was prepared by lithium intercalation of SnVSe₃ using the n-buthyllithium technique in a hexane solution (1.6 M·L⁻¹) at room temperature under an inert atmosphere. Unreacted n-buthyllithium was removed by washing off the product with dry hexane with suction filtering.

8.6 Tin–Lithium–Niobium–Selenium

Two quaternary intercalated compounds, $Li_{0.5}SnNbSe_3$ and $Li_{1.7}SnNbSe_3$, with the c-lattice parameter of 3728 and 3787 pm, respectively, are formed in the Sn–Li–Nb–Se system (Hernán et al. 1992). They were obtained in the same way as $Li_{0.5}SnVSe_3$ was prepared.

8.7 Tin–Lithium–Manganese–Selenium

Two quaternary compounds, $Li_2MnSnSe_4$ and $Li_4MnSn_2Se_7$, are formed in the Sn–Li–Mn–Se system. The first melts congruently at ~673°C and crystallizes in the orthorhombic structure with the lattice parameters $a = 838.29 \pm 0.17$, $b = 715.96 \pm 0.14$, $c = 675.11 \pm 0.14$ pm, a calculated density of 4.126 g·cm⁻³, and an energy gap of 2.03 eV (Li et al. 2017c). Polycrystalline sample of $Li_2MnSnSe_4$ was synthesized by solid-state reaction technique. Mixtures of Li₂Se, MnSe, and SnSe₂ according to the stoichiometric ratio were ground and loaded into a fused silica tube under an Ar atmosphere in a glove box, which were sealed under vacuum (10⁻³ Pa) and then placed in a furnace. The sample was heated to 550°C in 20 h, kept at that temperature for 72 h, and then the furnace was turned off. To prepare single crystals of the title compound, a mixture of Li₂Se, MnSe, and SnSe₂ (molar ratio 1:1:1) was

ground and loaded into quartz tubes under an Ar atmosphere in a glove box, which was sealed under vacuum (10⁻³ Pa), and then placed in a furnace. The sample was heated to 800°C in 20 h and kept at that temperature for 48 h, then cooled at a rate of 2.5°C·h⁻¹ to 300°C, and finally cooled to room temperature. Red crystals were obtained. The as-obtained samples are not stable in the air: when the samples were stored in the air for about a week, their color changed from red to black.

$Li_4MnSn_2Se_7$ crystallizes in the monoclinic structure with the lattice parameters $a = 1812.6 \pm 0.4$, $b = 722.09 \pm 0.14$, $c = 1074.0 \pm 0.2$ pm, and $\beta = 93.43 \pm 0.03°$ and a calculated density of 4.132 g·cm⁻³ at 100 K (Kaib et al. 2013). To prepare orange block crystals of the title compound, a stoichiometric mixture of Li₂Se (15.36 mM), Sn (7.39 mM), Se (29.57 mM), and Mn (1.85 mM) was melted together in a furnace with a definite temperature program: heating up to 650°C with a rate of 18°C·h⁻¹, keeping for 24 h at 650°C, and cooling down to room temperature with a rate of 6°C·h⁻¹. To obtain the phase-pure product, the prepared mixture was washed with water to remove the soluble by-products, and the surplus of Se was removed at 500°C and reduced pressure of ≈ 10⁻⁶ Pa. All reaction steps were performed with the strong exclusion of air and external moisture under the Ar atmosphere at a high-vacuum double-manifold Schlenk line or under N₂ atmosphere in a glove box.

8.8 Tin–Sodium–Potassium–Selenium

The $NaK_3Sn_3Se_8$ quaternary compound, which melts at about 468°C and crystallizes in the tetragonal structure with the lattice parameters $a = 812.1 \pm 0.1$, $c = 1367.2 \pm 0.3$ pm, a calculated density of 4.155 g·cm⁻³, and an energy gap of 2.2 eV, is formed in the Sn–Na–K–Se system (Chen et al. 2000b). Direct reaction of a stoichiometric mixture of K₂Se, Na₂Se, Sn, and Se at 450°C for 1 week yielded an almost single-phase polycrystalline sample of the title compound with small amounts of Se and SnSe₂ impurities.

8.9 Tin–Sodium–Rubidium–Selenium

The $NaRb_3Sn_3Se_8$ quaternary compound, which crystallizes in the tetragonal structure with the lattice parameters $a = 836.97 \pm 0.03$, $c = 1376.77 \pm 0.05$ pm, a calculated density of 4.363 g·cm⁻³, and an energy gap of 1.71 eV, is formed in the Sn–Na–Rb–Se system (Ji et al. 2021b). The crystals of this compound were prepared with a mixture of Na, RbCl, Sn, and Se (molar ratio 2:1:1:3). The mixture was loaded into a fused silica tube under a vacuum of 10⁻³ Pa. Next, the fused silica tube was put in the furnace and annealed at 850°C in 50 h and then kept at this temperature for 100 h. This was followed by slowly cooling at 5°C·h⁻¹ to 200°C, and then the furnace was turned off. The anhydrous ethanol and DMF were used to wash the reaction product for removing the other by-products. Finally, black crystals were obtained, which were stable in the air for more than 3 months.

8.10 Tin–Sodium–Barium–Selenium

Two quaternary compounds, **Na$_2$BaSnSe$_4$** and **Na$_2$Ba$_7$Sn$_4$Se$_{16}$**, are formed in the Sn–Na–Ba–Se system. First of them crystallizes in the tetragonal structure with the lattice parameters $a = 2416.4 \pm 0.8$, $c = 703.7 \pm 0.5$ pm, a calculated density of 3.227 g·cm^{-3}, and an energy gap of 2.25 eV (Wu et al. 2016a,b). This compound was prepared as follows. A mixture of Na (2 mM), Ba (1 mM), Sn (1 mM), and Se (4 mM) were firstly loaded into a graphite crucible that avoids the reaction between Na and silica tube, then put it into the silica tube and flame-sealed under 10^{-3} Pa. The tube was heated to 650°C in 30 h and kept at this temperature for about 90 h, then slowly cooled down to ambient temperature in 5 days. DMF was used to wash the products. Finally, many deep-yellow crystals were found. They are stable in the air after several months. Owing to Na and Ba metals being easily oxidized in the air, an Ar-filled glove box was used to complete the preparation processes.

Na$_2$Ba$_7$Sn$_4$Se$_{16}$ crystallizes in the cubic structure with the lattice parameter $a = 1541.63 \pm 0.13$ pm, a calculated density of 4.977 g·cm^{-3}, and an energy gap of 2.10 eV (Abudurusuli et al. 2018). To obtain this compound, Na, Ba, Sn, and Se (molar ratio 2:3:2:8) were weighed, well mixed, and loaded into the graphite crucible. Then, the crucible was embedded in a silica tube that was sealed under a vacuum of 10^{-3} Pa. The tube was put into a furnace, and the setting temperature process was as follows: firstly heated to 870°C in 30 h and kept at this temperature for 50 h to make the mixture completely melted, and then slowly cooled to 300°C at a rate of 5°C·h^{-1} before switching off the furnace. The obtained product was carefully washed with DMF to remove the unreacted Se powder. Finally, air-stable crystals of Na$_2$Ba$_7$Sn$_4$S$_{16}$ of light-red color appeared.

8.11 Tin–Sodium–Mercury–Selenium

The **Na$_2$HgSn$_2$Se$_6$** quaternary compound, which crystallizes in the tetragonal structure with the lattice parameters $a = 785.71 \pm 0.04$, $c = 2021.38 \pm 0.12$ pm, a calculated density 5.098 g·cm^{-3}, and an energy gap of 1.83 eV, is formed in the Sn–Na–Hg–Se system (Li et al. 2020b). A polycrystalline sample of this compound was obtained by heating a mixture of Na$_2$Se$_3$ (0.283 g), HgSe (0.279 g), and SnSe (0.395 g) (molar ratio 1:1:2). The starting materials were thoroughly ground and loaded into silica tube, which was then evacuated to 10^{-3} Pa and flame-sealed. The sample was then placed into a computer-controlled furnace and heated to 500°C over a 24 h period and then kept at that temperature for 120 h. Finally, the furnace was turned off, and the sample was allowed to cool naturally. Single crystals of Na$_2$HgSn$_2$Se$_6$ could be obtained by spontaneous crystallization. A mixture of Na$_2$Se (0.025 g), HgSe (0.056 g), and SnSe$_2$ (0.111 g) (molar ratio 1:1:2) was thoroughly ground and put into a silica tube, which was then evacuated to 10^{-3} Pa and flame-sealed. The sample was gradually heated to 960°C in a horizontal furnace and kept at that temperature for 96 h. Then, the sample was slowly cooled at a rate of 3°C·h^{-1}. The product consists of hundreds of red crystals of the title compound.

8.12 Tin–Sodium–Gallium–Selenium

The **Na$_{2-x}$Ga$_{2-x}$Sn$_{1+x}$Se$_6$** quaternary compound, which apparently has a homogeneity region and polymorphic transformation, is formed in the Sn–Na–Ga–Se system. First modification melts incongruently at 709°C, is kinetically stable phase, and crystallizes in the monoclinic structure with the lattice parameters $a = 1330.8 \pm 0.3$, $b = 759.4 \pm 0.2$, $c = 1384.2 \pm 0.3$ pm, and $\beta = 118.730 \pm 0.004°$, a calculated density of 4.246 g·cm^{-3}, and an energy gap of 1.50 eV for Na$_{1.84}$Ga$_{1.81}$Sn$_{1.20}$Se$_6$ composition (Hwang et al. 2004a). Its red crystals were synthesized by the stoichiometric reaction of Na$_2$Se, Ga, Sn, and Se with the molar ratio of Na/Ga/Sn/Se =1:1:1:4. The starting materials were loaded into a fused silica tube and sealed under vacuum. They were thoroughly melted in a torch flame, and then the liquid was allowed to cool rapidly in the air. This corresponds to ~200°C·min^{-1} cooling rate. Transparent plate crystals were isolated in ~50% yield.

The second modification melts congruently at 689°C and crystallizes in the orthorhombic structure with the lattice parameters $a = 1332.9 \pm 0.3$, $b = 2429.1 \pm 0.7$, $c = 762.1 \pm 0.2$ pm, a calculated density of 4.188 g·cm^{-3}, and an energy gap of 1.73 eV (Li et al. 2018b). For the obtaining of this modification, a stoichiometric mixture of Na$_2$Se, Ga, Sn, and Se (molar ratio 1:2:1:6) was loaded into a graphite crucible and placed in a quartz tube. The tube was flame-sealed under vacuum (~10^{-2} Pa) and placed in a furnace, heated from room temperature to 800°C in 40 h, kept at that temperature for 96 h, and then cooled to room temperature at 4°C·h^{-1}. The product was washed with degassed DMF and dried with ethanol. Red crystals, which are stable in air and water, were obtained.

8.13 Tin–Potassium–Copper–Selenium

The **K$_2$Cu$_2$Sn$_2$Se$_6$** quaternary compound, which is characterized by an energy gap of 1.04 eV, is formed in the Sn–K–Cu–Se system (Liao and Kanatzidis 1993). To prepare this compound, a mixture of Sn, Cu, K$_2$Se, and Se (molar ratio 1:1–2:4:16) was loaded into a Pyrex tube in a N$_2$ glove box. The tube was evacuated and sealed at a pressure of ~0.1 Pa. The mixture was heated slowly from room temperature to 400°C over 12 h in a furnace. The temperature was kept at 400°C for 4 days, and then was cooled slowly to room temperature at 4°C·h^{-1}. Black plate-like crystals were formed. The product was washed with degassed DMF to remove excess K$_2$Se$_x$ flux using a standard Schlenk technique and dried with acetone and ether. The obtained crystals are stable in air.

8.14 Tin–Potassium–Silver–Selenium

Three quaternary compounds, **K$_2$Ag$_2$SnSe$_4$**, **K$_2$Ag$_2$Sn$_2$Se$_6$**, and **K$_3$AgSn$_3$Se$_8$**, are formed in the Sn–K–Ag–Se system. First of them melts congruently at about 510°C and crystallizes in the monoclinic structure with the lattice parameters $a = 757.5 \pm 0.2$, $b = 592.0 \pm 0.1$, $c = 1214.8 \pm 0.2$ pm, and $\beta = 113.56 \pm$

0.03°, a calculated density of 4.845 g·cm^{-3}, and an energy gap of 1.8 eV (Chen et al. 2000a). For the preparation of this compound, K$_2$Se (0.5 mM), Ag (0.5 mM), Sn (0.25 mM), and Se (2 mM) were weighed in a glove box under an atmosphere of Ar. The mixture was introduced into a thin-walled Pyrex tube and sealed at a pressure of ~0.1 Pa. The tube was gradually heated to 500°C, where it was kept for 1 week, then cooled at a rate of 4°C·h^{-1} to 150°C. The red plate-like crystals were obtained after the reaction product was washed with DMF and anhydrous ethanol and dried with anhydrous diethyl ether. This compound is relatively stable in air and water.

K$_2$Ag$_2$Sn$_2$Se$_6$ crystallizes in the tetragonal structure with the lattice parameters $a = 817.30 \pm 0.10$, $c = 2027.8 \pm 0.4$ pm, and a calculated density of 4.929 g·cm^{-3} (Guo et al. 2001). To prepare this compound, a mixture of K$_2$Se (0.417 mM), Ag (0.417 mM), Sn (0.417 mM), and Se (1.264 mM) was loaded into a Pyrex tube in a glove box under an Ar atmosphere and then sealed under vacuum (about 0.1 Pa). The tube was gradually heated to 500°C and kept at that temperature for 5 days. It was then cooled at a rate of 4°C·h^{-1} to 200°C, followed by natural cooling to room temperature. Orange-red block-like crystals were isolated from the reaction product, washed with DMF and ethanol, and finally dried with anhydrous ether.

K$_3$AgSn$_3$Se$_8$ also crystallizes in the tetragonal structure with the lattice parameters $a = 809.6 \pm 0.6$, $c = 1338.4 \pm 1.2$ pm, and a calculated density of 4.592 g·cm^{-3} (Ji et al. 2007). This compound was synthesized using a mixture of Ag, Sn, and Se powders and K$_2$CO$_3$, which was put into a Pyrex glass tube, to which 0.6 mL of pyridine/glycol = 2:1 was added. The glass tube was sealed with a 10% filling, placed into a Teflon-lined stainless-steel autoclave, and heated at 180°C for 7 days. The product was washed with ethanol and water, respectively, and pure crystals of the title compound were obtained.

8.15 Tin–Potassium–Gold–Selenium

The **K$_2$Au$_2$Sn$_2$Se$_6$** quaternary compound, which melts at about 483°C and crystallizes in the tetragonal structure with the lattice parameters $a = 825.1 \pm 0.1$, $c = 1996.1 \pm 0.4$ pm, a calculated density of 5.784 g·cm^{-3} at 198 K, and an energy gap of 1.88 eV, is formed in the Sn–K–Au–Se system (Löken and Tremel 1998, 1999). It was prepared by combining Au powder (0.5 mM), Sn (0.5 mM), K$_2$Se (0.25 mM), and Se (1.25 mM) in a quartz tube and sealed under a vacuum. The mixture was heated up to 550°C in 6 h and kept at this temperature for 4 days before being cooled to room temperature at a rate of 5°C·h^{-1}. The products were washed with absolute ethanol. The sample contained red crystals and black amorphous powder. The crystals were separated manually from the charge, they are air and water stable over several hours.

8.16 Tin–Potassium–Barium–Selenium

The **K$_2$BaSnSe$_4$** quaternary compound, which crystallizes in the trigonal structure with the lattice parameters $a = 2644.4 \pm 0.4$, $c = 774.52 \pm 0.15$ pm, and a calculated density of

4.142 g·cm^{-3} at 153 K, is formed in the Sn–K–Ba–Se system (Luo et al. 2019). The powder of this compound was synthesized by the stoichiometric high-temperature solid-state reaction of K$_2$Se, BaSe, and SnSe$_2$. The tube loaded with the above mixture was sealed and heated to 500°C in 12 h with a duration of 48 h, and then the furnace was shut off. Single crystals of K$_2$BaSnSe$_4$ were achieved directly via a spontaneous crystallization method. First, K$_2$Se (0.188 g), BaSe (0.216 g), and SnSe$_2$ (0.276 g) were mixed according to a molar ratio of 1.2:1:1 (with a little excess of K$_2$Se serving as a flux) and then ground and transferred into fused silica tube in a glove box. The tube was placed in a heating furnace after it was sealed under a high vacuum. The heating temperature was raised to 900°C in 20 h with a duration of 24 h, and then the furnace was slowly cooled to 200°C and at 4°C·h^{-1} to room temperature.

8.17 Tin–Potassium–Zinc–Selenium

Three quaternary compounds, **K$_2$ZnSn$_3$Se$_8$**, **K$_2$Zn$_2$SnSe$_6$**, and **K$_6$Zn$_4$Sn$_5$Se$_{17}$**, are formed in the Sn–K–Zn–Se system. First of them crystallizes in the orthorhombic structure with the lattice parameters $a = 763.87 \pm 0.01$, $b = 1240.0 \pm 0.3$, $c = 1838.4 \pm 0.4$ pm, a calculated density of 4.359 g·cm^{-3}, and an energy gap of 2.10 eV (Zhou et al. 2017b). Single crystals of this compound were achieved by the spontaneous crystallization method. The reaction mixture of K$_2$Se, ZnSe, and SnSe$_2$ (molar ratio 1.5:1:3) was mixed and loaded into a fused silica tube under an Ar atmosphere in a glove box, then flame-sealed under a high vacuum of 10^{-3} Pa. The tube was then gradually heated to 900°C in a furnace, kept for 72 h, and then slowly cooled at a rate of 3°C·h^{-1}, finally cooled to room temperature by switching off the furnace.

K$_2$ZnSn$_2$Se$_6$ crystallizes in the tetragonal structure with the lattice parameters $a = 815.11 \pm 0.12$, $c = 1957.0 \pm 0.4$ pm, a calculated density of 4.366 g·cm^{-3}, and an energy gap of 1.71 eV (Zhou et al. 2016a). Traditional solid-state reaction techniques can be applied to grow the single crystals of the title compound. The mixtures of K$_2$Se (0.5 mM), SnSe$_2$ (1.1 mM), and ZnSe (0.5 mM) were mixed and loaded into a fused silica tube under an Ar atmosphere in a glove box, then sealed under a high vacuum (10^{-3} Pa). The tube was then placed in a furnace and heated to 1000°C within 15 h, left for 48 h, cooled to 350°C at a rate of 3°C·h^{-1}, and finally cooled to room temperature by switching off the furnace. Many prism-shaped crystals were found in the tube. The obtained crystals are stable in the air.

The reaction of Sn, Zn, K$_2$Se, and Se at 400°C for 4 days upon a cooling rate of 4°C/h afforded yellowish-white crystals of K$_6$Zn$_4$Sn$_5$Se$_{17}$ (Kanatzidis et al. 1997a, 1997b). The product must be washed with degassed DMF and dried with acetone and ether.

8.18 Tin–Potassium–Cadmium–Selenium

Three quaternary compounds, **K$_2$CdSnSe$_4$**, **K$_6$Cd$_4$Sn$_3$Se$_{13}$**, and **K$_{14}$Cd$_{15}$Sn$_{12}$Se$_{46}$** are formed in this system. K$_6$Cd$_4$Sn$_3$Se$_{13}$ crystallizes in the trigonal structure with the lattice parameters

$a = 1503.38 \pm 0.11$, $c = 1651.2 \pm 0.2$ pm, a calculated density of 3.186 g·cm^{-3}, and an energy gap of 2.33 eV (Ding et al. 2004). This compound decomposes at 280°C to K$_2$CdSnSe$_4$ and CdSe and is accessible only through hydrothermal synthesis. The synthesis of this compound was achieved in two steps: (1) K$_2$Se (1.2 mM), Cd (1.6 mM), Sn (1.2 mM), and Se (4 mM) were combined and melted at ~900°C. The product was loaded in a Pyrex tube along with 0.3 mL of deionized water, which was then evacuated to < 3×10^{-3} Pa and flame-sealed. The tube was kept at ~115°C for 3.5 days. The products were filtered off and washed with water, ethanol, and ether. They were in the form of yellow/orange powder and square-shaped orange crystals. This compound appears to be stable in the air for days.

K$_{14}$Cd$_{15}$Sn$_{12}$Se$_{46}$ crystallizes in the cubic structure with the lattice parameter $a = 2309.0 \pm 0.3$ pm and calculation density 3.933 g·cm^{-3} (Ding and Kanatzidis 2006a,b). To prepare this compound, K$_2$Se (1.2 mM), Cd (2.6 mM), Sn (2.0 mM), and Se (6.6 mM) were combined in an evacuated and flame-sealed fused silica tube and melted at 900°C. The obtained product was loaded into a Pyrex tube with deionized water (0.3 mL), and the tube was evacuated, flame-sealed, heated up to 180°C, and kept at this temperature for 6 days. Dark orange polyhedral crystals of the title compound were isolated.

8.19 Tin–Potassium–Mercury–Selenium

Two quaternary compounds, **K$_2$HgSnSe$_4$** and **K$_2$Hg$_3$Sn$_2$Se$_8$**, are formed in the Sn–K–Hg–Se system (Kanatzidis et al. 1997b; Brandmayer et al. 2004). First of them crystallizes in the tetragonal structure with the lattice parameters $a = 806.81 \pm 0.11$, $c = 694.97 \pm 0.14$ pm, and a calculated density of 5.237 g·cm^{-3} at 203 K (Brandmayer et al. 2004). To obtain this compound, K$_4$SnSe$_4$·1.5MeOH (0.15 mM) was suspended in MeOH (5 mL) and added to a solution of Hg(CH$_3$COO)$_2$ (0.15 mM) in a mixture of MeOH (4.5 mL) and H$_2$O (0.5 mL). A black, insoluble precipitate formed immediately. After stirring overnight, the precipitate was removed by filtration, and toluene (10 mL) was allowed to flow under the filtrate. After one week, large amounts of [K$_4$(MeOH)$_4$(Sn$_2$Se$_6$)] had formed; after 9 weeks, crystallization of small ruby-red rhombuses of K$_2$HgSnSe$_4$ started.

8.20 Tin–Potassium–Gallium–Selenium

The **KGaSnSe$_4$** quaternary compound, which melts congruently at 746°C and has polymorphic transformations, is formed in the Sn–K–Ga–Se system. α-KGaSnSe$_4$ crystallizes in the tetragonal structure with the lattice parameters $a = 818.6 \pm 0.5$, $c = 640.3 \pm 0.5$ pm, and a calculated density of 8.411 g·cm^{-3} and β-KGaSnSe$_4$ crystallizes in the monoclinic structure with the lattice parameters $a = 749.0 \pm 0.2$, $b = 1257.8 \pm 0.3$, $c = 1830.6 \pm 0.5$ pm, $\beta = 98.653 \pm 0.005°$, a calculated density of 4.218 g·cm^{-3}, and an energy gap of 1.73 eV for K$_{0.996}$Ga$_{1.038}$Sn$_{0.963}$Se$_4$ composition (Hwang et al. 2004a). Yellowish-brown needles of α-KGaSnSe$_4$ and reddish-orange crystals of β-KGaSnSe$_4$ were synthesized by the stoichiometric reaction of K$_2$Se, Ga, Sn, and

Se with the molar ratio of K/Ga/Sn/Se =1:1:1:4. The starting materials were loaded into a fused silica tube and sealed under vacuum. They were thoroughly melted in a torch flame and then the liquid was allowed to rapidly cool in air. This corresponds to ~200°C·min^{-1} cooling rate.

On cooling α-KGaSnSe$_4$ to room temperature, it converted to the cubic γ-form (Hwang et al. 2004a) with the lattice parameter $a = 1355.55 \pm 0.03$ pm, a calculated density of 4.347 g·cm^{-3}, and an energy gap of 2.10 eV (Friedrich et al. 2020). Polycrystalline batches of red γ-KGaSnSe$_4$ were obtained by annealing K$_2$Se with stoichiometric mixtures of Ga$_2$Se$_3$, Sn, and Se in a sealed evacuated quartz tube at 500°C for 7 days. After the initial annealing, the crude product was thoroughly grounded in an agate mortar under N$_2$ atmosphere and pressed to a pellet. The pellet was annealed in an evacuated quartz tube for an additional week at 500°C. This compound decomposes within several minutes by releasing gaseous H$_2$Se.

8.21 Tin–Potassium–Indium–Selenium

The **KInSnSe$_4$** quaternary compound, which melts at 700–710°C and has polymorphic transformations, is formed in the Sn–K–In–Se system (Hwang et al. 2004b). α-KInSnSe$_4$ crystallizes in the tetragonal structure with the lattice parameters $a = 819.3 \pm 0.2$, $c = 668.8 \pm 0.2$ pm, a calculated density of 4.353 g·cm^{-3}, and an energy gap of 2.40 eV; β-KInSnSe$_4$ crystallizes in the monoclinic structure with the lattice parameters $a = 777.2 \pm 0.3$, $b = 1245.7 \pm 0.4$, $c = 1865.8 \pm 0.6$ pm, and $\beta = 98.149 \pm 0.006°$, a calculated density of 4.371 g·cm^{-3}, and an energy gap of 1.80 eV; and γ-KInSnSe$_4$ crystallizes in the cubic structure with the lattice parameter $a = 1378.49 \pm 0.11$ pm, a calculated density of 4.476 g·cm^{-3}, and an energy gap of 1.49 eV.

α-KInSnSe$_4$ was prepared from a stoichiometric mixture of K$_2$Se, In, Sn, and Se that was sealed under vacuum (~0.02 Pa) in a fused silica tube (Hwang et al. 2004b). Well-formed yellow needle-shaped crystals were obtained by forming a melt at 850°C and cooling at 10°C·min^{-1} to room temperature. β-KInSnSe$_4$ was synthesized by the stoichiometric reaction of K$_2$Se, In, Sn, and Se. The starting materials were vacuum-sealed in a fused silica tube, thoroughly melted in a torch flame, and allowed to rapidly cool in air. The cooling rate from the melt temperature to ~200°C was estimated at approximately 600°C·min^{-1}. At room temperature, reddish plate crystals could be isolated. γ-KInSnSe$_4$ was obtained at the heating of α- or β-KInSnSe$_4$ in the solid state at 650°C for 24 h, followed by quenching in air.

8.22 Tin–Potassium–Phosphor–Selenium

Two quaternary compounds, **K$_5$Sn(PSe$_5$)$_3$** and **K$_{10}$Sn$_3$(P$_2$Se$_6$)$_4$**, are formed in the Sn–K–P–Se system. The first of them was synthesized from a mixture of Sn (0.3 mM), P$_2$Se$_5$ (0.6 mM), K$_2$Se (0.90 mM), and Se (3 mM) sealed under vacuum in a Pyrex tube and heated to 440°C for 4 days followed by cooling to 150°C at a rate of 4°C·h^{-1} (Chondroudis and Kanatzidis

1996). The excess $K_xP_ySe_z$ flux was removed with DMF to reveal black plate-like crystals, which are air- and water-sensitive.

$K_{10}Sn_3(P_2Se_6)_4$ melts congruently at 538°C and crystallizes in the trigonal structure with the lattice parameters $a = 2411.84 \pm 0.07$, $c = 1764.82 \pm 0.02$ pm, and a calculated density of 3.736 g·cm⁻³ at 100 K, and an energy gap of 1.82 eV (Chung and Kanatzidis 2011). The synthesis of this compound was achieved by reacting a mixture of K_2Se, Sn, P, and Se (molar ratio 5:3:8:19) under a vacuum in a carbon-coated fused silica tube at 850°C for 3 days, followed by cooling at a rate of 2°C·h⁻¹ to 300°C. The excess flux was dissolved with degassed DMF under a N_2 atmosphere to reveal red, stable-in-air plate crystals of the title compound with the by-product of $Sn_2P_2Se_6$. The pure compound could be obtained by reacting a stoichiometric mixture of the same reagents in a fused silica tube at 800°C for 2 h, followed by quenching in air.

8.23 Tin–Potassium–Bismuth–Selenium

Some quaternary compounds are formed in the Sn–K–Bi–Se system. $K_xSn_{6-2x}Bi_{2+x}Se_9$ crystallizes in the orthorhombic structure with the lattice parameters $a = 420.96 \pm 0.04$, $b = 1400.6 \pm 0.1$, $c = 3245.1 \pm 0.3$ pm, and a calculated density of 6.428 g·cm⁻³ for $x = 0.54$ (Mrotzek and Kanatzidis 2003). To synthesize this compound, a mixture of K_2Se (1 mM), Sn (8 mM), Se (8 mM), and Bi_2Se_3 (3 mM) was loaded in a carbon-coated quartz tube and sealed at a residual pressure of <0.01 Pa. The starting materials were heated within 24 h to 800°C and kept there for 24 h, followed by slow cooling to 50°C at a rate of 0.5°C·min⁻¹. A silver-gray polycrystalline ingot of the title compound was obtained. All manipulations were carried out under a dry nitrogen atmosphere in a glove box.

$KSn_5Bi_5Se_{13}$ melts congruently at 672°C and also crystallizes in the monoclinic structure with the lattice parameters $a = 1387.9 \pm 0.4$, $b = 420.5 \pm 0.1$, $c = 2336.3 \pm 0.6$ pm, $\beta = 99.012 \pm 0.004°$, and a calculated density of 6.668 g·cm⁻³ (Mrotzek and Kanatzidis 2003). To obtain this compound, a mixture of K_2Se (1 mM), Sn (10 mM), Se (10 mM), and Bi_2Se_3 (5 mM) was loaded in a quartz tube and sealed at a residual pressure of <0.01 Pa. To prevent an explosion of the tube during the following procedure, the length of the tube was longer than usual. The tube was carefully placed under the flame of a natural gas-oxygen torch until the mixture melted, and then, the tube was removed from the flame and let to solidify. A silver-gray ingot of $KSn_5Bi_5Se_{13}$ was obtained.

$K_{1+x}Sn_{3-2x}Bi_{7+x}Se_{14}$ melts congruently at 689°C and crystallizes in the monoclinic structure with the lattice parameters $a = 1740.2 \pm 0.2$, $b = 420.54 \pm 0.06$, $c = 2122.7 \pm 0.3$ pm, $\beta = 109.524 \pm 0.004°$, and a calculated density of 6.732 g·cm⁻³ for $K_{1.40}Sn_{2.20}Bi_{7.40}Se_{14}$ composition (Mrotzek et al. 2001c). To prepare this compound, a mixture of K_2Se (1 mM), Sn (7 mM), Se (7 mM), and Bi_2Se_3 (7 mM) was loaded in a carbon-coated quartz tube and sealed at a residual pressure of <0.01 Pa. Then it was heated within 24 h to 800°C and kept there for 24 h, followed by slow cooling to 50°C at a rate of 0.5°C·min⁻¹. A silver, shiny, polycrystalline ingot of the title compound was obtained after washing any impurities with DMF, methanol, and diethyl ether. All manipulations were carried out under dry N_2 atmosphere.

$K_{1.46}Sn_{3.09}Bi_{7.45}Se_{15}$ melts at 662°C and also crystallizes in the monoclinic structure with the lattice parameters $a = 1745.4 \pm 0.5$, $b = 420.1 \pm 0.1$, $c = 2175.0 \pm 0.6$ pm, $\beta = 98.550 \pm 0.005°$, a calculated density of 6.665 g·cm⁻³, and an energy gap of ~0.39 eV (Choi et al. 2001). This compound was discovered from the reaction of K_2Se (1 mM), Sn (4 mM), Bi_2Se_3 (3 mM), and Se (4 mM). They were thoroughly mixed, sealed in a carbon-coated silica tube and heated at 900°C for 6 h, followed by cooling to room temperature at a rate of 100°C·h⁻¹. A shiny silver ingot of the pure phase was obtained after isolation in DMF and washing with methanol and diethyl ether. Single crystals of $K_{1.46}Sn_{3.09}Bi_{7.45}Se_{15}$ were prepared when the reaction mixture was heated at 800°C for 120 h and slowly cooled at a rate of 10°C·h⁻¹. The silver ribbon-like crystals were obtained by isolation in degassed DMF and washed with methanol and ether. The crystals are air and water stable. After the stoichiometry was determined, this compound could be prepared from the stoichiometric proportions of $K_2Se/Sn/Bi_2Se_3/Se$, that is, 0.75:3:3.75:3, respectively.

$K_{1-x}Sn_{3-x}Bi_{11+x}Se_{20}$ melts congruently at 676°C and also crystallizes in the monoclinic structure with the lattice parameters $a = 1613.9 \pm 0.2$, $b = 414.70 \pm 0.05$, $c = 1700.9 \pm 0.3$ pm, $\beta = 116.881 \pm 0.002°$, a calculated density of 7.013 g·cm⁻³, and an energy gap of 0.30 eV for $K_{0.70}Sn_{2.70}Bi_{11.30}Se_{20}$ composition (Mrotzek et al. 2001c). To obtain this compound, a mixture of K_2Se (1 mM), Sn (3 mM), Se (3 mM), and Bi_2Se_3 (5.5 mM) was loaded in a carbon-coated quartz tube and sealed at a residual pressure of <0.01 Pa. The mixture was heated within 12 h to 800°C and kept there for 24 h, followed by slow cooling to 400°C at a rate of 0.1°C·min⁻¹ and then to 50°C in 10 h. A silver, shiny, polycrystalline ingot of the title compound was obtained after washing any impurities with DMF, methanol, and diethyl ether.

$K_xSn_{4-x}Bi_{11+x}Se_{21}$ melts at 673°C and also crystallizes in the monoclinic structure with the lattice parameters $a = 3191.0 \pm 0.8$, $b = 414.6 \pm 0.1$, $c = 1731.1 \pm 0.4$ pm, $\beta = 112.064 \pm 0.004°$, and a calculated density of 7.030 g·cm⁻³ for $K_{0.54}Sn_{3.54}Bi_{11.46}Se_{21}$ composition (Mrotzek et al. 2001a). This compound was prepared as follows. A mixture of K_2Se (0.5 mM), Sn (4 mM), Se (4 mM), and Bi_2Se_3 (5.5 mM) was loaded in a carbon-coated quartz tube and sealed at a residual pressure of <0.01 Pa. The starting materials were heated over 24 h to 800°C and kept at this temperature for a further 24 h, followed by slow cooling to 400°C at a rate of 0.1°C·min⁻¹ and then to 50°C over a period of 10 h. A silver, shiny polycrystalline ingot of pure title compound was obtained after washing away any impurities with DMF, methanol, and diethyl ether. All manipulations were carried out under a dry nitrogen atmosphere in a glove box.

$K_{0.66}Sn_{4.82}Bi_{11.18}Se_{22}$ melts congruently at 680°C and also crystallizes in the monoclinic structure with the lattice parameters $a = 1577.7 \pm 0.2$, $b = 416.69 \pm 0.06$, $c = 1735.8 \pm 0.2$ pm, $\beta = 99.249 \pm 0.002°$, and an energy gap of ≈0.12 eV (Mrotzek et al. 2000). This compound was prepared using the next two methods. (1) A mixture of K_2Se (0.5 mM), Sn (3 mM), Se (2 mM), and Bi_2Se_3 (6.5 mM) was loaded in a carbon-coated silica tube and sealed at a residual pressure of <0.01 Pa. The starting materials were heated within 24 h to 800°C and kept there for 24 h, followed by slow cooling to 400°C at a rate of 0.1°C·min⁻¹ and to 50°C in 10 h. A shiny silver ingot of pure this compound was obtained after washing any impurities with DMF, methanol, and diethyl ether. (2) Needle-like single crystals were obtained in a mixture of the starting materials in the molar ratio $K_2Se/Sn/Bi_2Se_3/Se = 1:4:4:4$. They were heated within 24 h to 800°C and kept there for 72 h, followed by slowly cooling to 400°C at a rate of 5°C·h⁻¹, then to 100°C at a rate of 15°C·h⁻¹ and to 50°C in 1 h. All manipulations were carried out under a dry nitrogen atmosphere in a glove box.

$K_{1.1}Sn_{9.5}Bi_{11.6}Se_{26}$ also crystallizes in the monoclinic structure with the lattice parameters $a = 1728.1 \pm 0.4$, $b = 416.1 \pm 0.1$, $c = 1924.9 \pm 0.5$ pm, $\beta = 106.582 \pm 0.004°$, a calculated density of 6.849 g·cm⁻³, and an energy gap of ~0.50 eV (Mrotzek and Kanatzidis 2002). It was prepared according to the predicted composition by combining K_2Se, Bi_2Se_3, Sn, and Se (molar ratio 1:11:18:18) in a sealed evacuated carbon-coated quartz ampoule. The tube was placed under the flame of a natural gas-oxygen torch until the mixture melted and then removed from the flame to solidify. A silver-gray ingot of plates was obtained after annealing the reaction mixture at 800°C for 12 h.

8.24 Tin–Potassium–Manganese–Selenium

Two quaternary compounds, $K_2MnSnSe_4$ and $K_2MnSn_2Se_6$, are formed in the Sn–K–Mn–Se system (Chen et al. 2000a). Both compounds crystallize in the tetragonal structure with the lattice parameters $a = 1057.4 \pm 0.1$, $c = 830.1 \pm 0.2$ pm, a calculated density of 4.063 g·cm⁻³, and an energy gap of 1.7 eV for the first compound, and $a = 816.7 \pm 0.1$, $c = 1972.4 \pm 0.4$ pm, a calculated density of 4.262 g·cm⁻³, and an energy gap of 2.0 eV for the second one. $K_2MnSnSe_4$ was synthesized from a reaction containing K_2Se (0.34 mM), Mn (0.18 mM), Sn (0.17 mM), and Se (1.36 mM). The mixture was slowly heated to and maintained at 400°C for 1 week, followed by a slow cooling (4°C·h⁻¹) to room temperature. After the excess flux was removed with DMF, red polyhedral-shaped crystals were isolated. This compound is relatively stable in air and water.

For the preparation of $K_2MnSn_2Se_6$, K_2Se (0.5 mM), Mn (0.5 mM), SnSe (1.0 mM), and Se (1.5 mM) were weighed in a glove box under an atmosphere of Ar (Chen et al. 2000a). The mixture was introduced into a thin-walled Pyrex tube and sealed at a pressure of 0.1 Pa. The tube was gradually heated to 520°C, where it was kept for 4 days, then cooled at a rate of 4°C·h⁻¹ to 150°C. The orange-red rectangular plate-like crystals were collected after the reaction product was washed with DMF and anhydrous ethanol and dried with anhydrous diethyl ether. Direct reaction of a stoichiometric mixture of elements at 500°C for 1 week yielded a single-phase polycrystalline sample. This compound is relatively stable in air and water and thermally stable up to 590°C.

8.25 Tin–Rubidium–Copper–Selenium

The $Rb_2Cu_2Sn_2Se_6$ quaternary compound, which is characterized by an energy gap of 1.04 eV, is formed in the Sn–Rb–Cu–Se system (Liao and Kanatzidis 1993). To prepare this compound, a mixture of Sn, Cu, Rb_2Se, and Se (molar ratio

1:1–2:4:16) was loaded into a Pyrex tube in a N_2 glove box. The tube was evacuated and sealed at a pressure of ~0.1 Pa. The mixture was heated slowly from room temperature to 400°C in 12 h in a furnace. The temperature was kept at 400°C for 4 days, and then was cooled slowly to room temperature at 4°C·h⁻¹. Black plate-like crystals were formed. The product was washed with degassed DMF to remove excess Rb_2Se_x flux using a standard Schlenk technique and dried with acetone and ether. The obtained crystals are stable in air.

8.26 Tin–Rubidium–Silver–Selenium

The $Rb_3AgSn_3Se_8$ quaternary compound, which crystallizes in the tetragonal structure with the lattice parameters $a = 834.5 \pm 0.3$, $c = 1345.5 \pm 0.8$ pm, a calculated density of 4.792 g·cm⁻³ and an energy gap of 1.8 eV, is formed in the Sn–Rb–Ag–Se system (Ji et al. 2007). The synthesis of this compound was as follows: Ag (10 mg), Sn (60 mg), and Se powders (60 mg) and Rb_2CO_3 (120 mg) were put into a Pyrex glass tube, to which 0.6 mL of pyridine/glycol = 2:1 was added. The glass tube was sealed with a 10% filling, placed into a Teflon-lined stainless-steel autoclave, and heated at 180°C for 7 days. The product was washed with ethanol and water, respectively, and pure red crystals of the title compound were obtained.

8.27 Tin–Rubidium–Zinc–Selenium

The $Rb_2ZnSn_2Se_6$ quaternary compound is formed in the Sn–Rb–Zn–Se system (Kanatzidis et al. 1997a, 1997b). The reaction of Sn, Zn, Rb_2Se, and Se at 400°C for 4 days upon a cooling rate of 4°C·h⁻¹ afforded brownish orange crystals of the title compound. The product must be washed with degassed DMF and dried with acetone and ether.

8.28 Tin–Rubidium–Cadmium–Selenium

The $Rb_{14}Cd_{15}Sn_{12}Se_{46}$ quaternary compound is formed in the Sn–Rb–Cd–Se system (Ding and Kanatzidis 2006a,b). This compound could be obtained if the Rb⁺ ions replace the K⁺ ions in the $K_{14}Cd_{15}Sn_{12}Se_{46}$.

8.29 Tin–Rubidium–Mercury–Selenium

The $Rb_2Hg_3Sn_2Se_8$ quaternary compound is formed in the Sn–Rb–Hg–Se system (Kanatzidis et al. 1997b).

8.30 Tin–Rubidium–Gallium–Selenium

Two quaternary compounds, $RbGaSnSe_4$ and $RbGaSn_2Se_6$, are formed in the Sn–Rb–Ga–Se system. $RbGaSnSe_4$ melts congruently at 734°C (Hwang et al. 2004a) and has two polymorphic modifications. First modification crystallizes in the monoclinic structure with the lattice parameters $a = 756.7 \pm 0.2$, $b = 1265.6 \pm 0.3$, $c = 1827.7 \pm 0.4$ pm, and $\beta = 95.924$

± 0.004°, a calculated density of 4.529 g·cm^{-3}, and an energy gap of 1.88 eV for RbGa$_{0.92}$Sn$_{1.08}$Se$_4$ composition (Hwang et al. 2004a). Reddish-orange crystals of this compound were synthesized by the stoichiometric reaction of Rb$_2$Se, Ga, Sn, and Se with the molar ratio of Rb/Ga/Sn/Se =1:1:1:4. The starting materials were loaded into a fused silica tube and sealed under vacuum. They were thoroughly melted in a torch flame, and then the liquid was allowed to cool rapidly in air. This corresponds to ~200°C·min^{-1} cooling rate. Pure phase of this compound was obtained. On cooling to room temperature, it converted to the cubic form.

The second modification crystallizes in the cubic structure with the lattice parameter a = 1372.00 ± 0.01 pm and a calculated density of 4.550 g·cm^{-3} (Friedrich et al. 2020). Polycrystalline batches of cubic orange RbGaSnSe$_4$ were obtained by annealing Rb$_2$Se with stoichiometric mixtures of Ga$_2$Se$_3$, Sn, and Se in a sealed evacuated quartz tube at 500°C for 7 days. After the initial annealing, the crude product was thoroughly grounded in an agate mortar under N$_2$ atmosphere and pressed to a pellet. The pellet was annealed in an evacuated quartz tube for an additional week at 500°C. This compound decomposes in moist air within several minutes by releasing gaseous H$_2$Se.

RbGaSn$_2$Se$_6$ crystallizes in the trigonal structure with the lattice parameters a = 1046.97 ± 0.02, c = 947.6 ± 0.2 pm, a calculated density of 4.798 g·cm^{-3}, and an energy gap of 1.8 eV (Lin et al. 2017a). It was synthesized by a solid-state reaction technique. The optimal synthesis route was determined as follows: the mixture of RbCl, Ga, Sn, and Se (molar ratio 2:1.5:1.625:6) was placed into a fused silica tube under vacuum, annealed at 700°C for 50 h, and then kept at this temperature for 100 h, followed by slow cooling at 3°C·h^{-1} to 200°C, at which point the furnace was turned off. The raw product was washed with distilled water and then dried with ethanol. Deep-red crystals, which are thermally stable up to 693°C, were obtained.

8.31 Tin–Rubidium–Indium–Selenium

Two quaternary compounds, **RbInSnSe$_4$** and **RbInSn$_2$Se$_6$**, are formed in the Sn–Rb–In–Se system.

RbInSnSe$_4$ has two polymorphic modifications, first of them crystallizes in the monoclinic structure with the lattice parameters a = 781.04 ± 0.02, b = 1265.33 ± 0.04, c = 1870.62 ± 0.04 pm, β = 96.297 ± 0.003°, a calculated density of 4.589 g·cm^{-3}, and an energy gap of 2.18 eV and the second one crystallizes in the cubic structure with the lattice parameter a = 1397.28 ± 0.01, a calculated density of 4.637 g·cm^{-3}, and an energy gap of 2.10 eV (Friedrich et al. 2021). Polycrystalline monoclinic RbInSnSe$_4$ was obtained by reacting Rb$_2$Se (0.73 mM), In (1.47 mM), Sn (1.47 mM), and Se (5.13 mM) in a sealed evacuated fused silica tube. The reaction mixture was heated to 1000°C with a heating rate of 3°C·min^{-1}. After annealing at 1000°C for 24 h the tube was cooled to room temperature with a cooling rate of 1°C·min^{-1}. The air sensitive orange polycrystalline product forms plate-like crystallites and is stored in a nitrogen glove box until further use. Polycrystalline red cubic RbInSnSe$_4$ was obtained by reacting Rb$_2$Se (0.79 mM),

In (0.158 mM), Sn (0.158 mM), and Se (5.51 mM) in sealed evacuated fused silica tube. The reaction mixture was heated to 500°C with a heating rate of 2°C·min^{-1} and annealed at 550°C for 7 days. After this initial annealing step, the crude products were thoroughly grounded in an agate mortar under N$_2$ atmosphere, pressed to a pellet using a pellet press, and annealed at 550°C for an additional week. The air sensitive red polycrystalline product was stored in a nitrogen glove box until further use.

RbInSn$_2$Se$_6$ crystallizes in the trigonal structure with the lattice parameters a = 1060.44 ± 0.08, c = 966.0 ± 0.2 pm, a calculated density of 4.832 g·cm^{-3} and an energy gap of 1.92 eV (Lin et al. 2017a). It was synthesized in the same way as RbGaSn$_2$Se$_6$ was prepared using In instead of Ga. Deep-red crystals, which are thermally stable up to 678°C, were obtained.

8.32 Tin–Rubidium–Phosphor–Selenium

Some quaternary compounds are formed in the Sn–Rb–P–Se system. **Rb$_3$Sn(PSe$_5$)(P$_2$Se$_6$)** melts incongruently at 426°C and crystallizes in the monoclinic structure with the lattice parameters a = 1401.3 ± 0.2, b = 734.36 ± 0.08, c = 2198.3 ± 0.4 pm, β = 106.61 ± 0.01°, a calculated density of 4.095 g·cm^{-3}, and an energy gap of 1.51 eV (Chondroudis and Kanatzidis 1998). To prepare this compound, a mixture of Sn (0.3 mM), P$_2$Se$_5$ (0.3 mM), Rb$_2$Se (0.3 mM), and Se (2.0 mM) was sealed under vacuum in a Pyrex tube and heated to 495°C for 4 days followed by cooling to 150°C at 4°C·h^{-1}. The excess Rb$_x$P$_y$Se$_z$ flux was removed by washing with DMF under a N$_2$ atmosphere. The product was then washed with tri-n-butylphosphine to remove residual elemental Se and then with ether to reveal a mixture of black plates of the title compound and black, irregular crystals of **Rb$_4$Sn$_5$(P$_2$Se$_6$)$_3$Se$_2$**. The plate crystals of Rb$_3$Sn(PSe$_5$)(P$_2$Se$_6$) were manually separated, and they are air and water stable.

Rb$_4$Sn$_5$P$_4$Se$_{20}$ melts congruently at 517°C and crystallizes in the trigonal structure with the lattice parameters a = 761.63 ± 0.04, c = 1869.0 ± 0.1 pm, and the calculated and experimental densities of 4.666 and 4.64 g·cm^{-3}, respectively (Chung et al. 2011). It was synthesized by the reaction of Sn, Rb$_2$Se, P$_2$Se$_5$, Se (molar ratio 1:1:1:5) at 490°C for 4 days. However, the complicated Lewis acid-base equilibria in the flux also yielded Rb$_3$Sn(PSe$_5$)(P$_2$Se$_6$) as a by-product. Pure phase of Rb$_4$Sn$_5$P$_4$Se$_{20}$ could be obtained only by direct combination reactions of Sn, Rb$_2$Se, P, and Se at 850°C.

Rb$_5$Sn(PSe$_5$)$_3$ melts congruently at 406°C, has an energy gap of 1.61 eV and two polymorphic modifications, both of which crystallize in the triclinic structure with the lattice parameters a = 1174.5 ± 0.3, b = 1923.0 ± 0.5, c = 727.8 ± 0.3 pm, α = 99.97 ± 0.03°, β = 107.03 ± 0.02°, γ = 87.16 ± 0.02°, and a calculated density of 3.932 g·cm^{-3} at 26°C for α-Rb$_5$Sn(PSe$_5$)$_3$ and a = 1356.6 ± 0.3, b = 1624.4 ± 0.4, c = 730.3 ± 0.2 pm, α = 102.04 ± 0.02°, β = 94.22 ± 0.02°, γ = 76.91 ± 0.02°, and a calculated density of 3.951 g·cm^{-3} at 23°C for β-Rb$_5$Sn(PSe$_5$)$_3$ (Chondroudis and Kanatzidis 1996). This compound was synthesized from a mixture of Sn (0.3 mM), P$_2$Se$_5$ (0.6 mM), Rb$_2$Se (0.90 mM), and Se (3 mM) sealed under vacuum in a

Pyrex tube and heated to 440°C for 4 days followed by cooling to 150°C at a rate of 4°C·h^{-1}. The excess $Rb_xP_ySe_z$ flux was removed with DMF to reveal black plate-like crystals, which are air- and water-sensitive. β-$Rb_5Sn(PSe_5)_3$ is formed at a reaction temperature 20°C higher than α-$Rb_5Sn(PSe_5)_3$.

$Rb_6Sn_2Se_4(PSe_5)_2$ was synthesized from a mixture of Sn (0.3 mM), P_2Se_5 (0.45 mM), Rb_2Se (1.20 mM), and Se (3 mM) by heating at 495°C for 4 days and isolated as in the case of $Rb_5Sn(PSe_5)_3$ to reveal orange rod-like crystals, which are air- and water-sensitive (Chondroudis and Kanatzidis 1996).

8.33 Tin–Rubidium–Bismuth–Selenium

Three quaternary compounds, $Rb_{0.36}Sn_{2.36}Bi_{11.64}Se_{20}$, $Rb_{1.9}Sn_{4.8}Bi_{10.3}Se_{21}$ and $Rb_{0.9}Sn_{9.4}Bi_{10.9}Se_{26}$, are formed in the Sn–Rb–Bi–Se system. $Rb_{0.36}Sn_{2.36}Bi_{11.64}Se_{20}$ melts congruently at 675°C and crystallizes in the monoclinic structure with the lattice parameters $a = 1603.8 \pm 0.2$, $b = 415.40 \pm 0.06$, $c = 1702.8 \pm 0.3$ pm, and $β = 116.517 \pm 0.002°$, a calculated density of 7.071 g·cm^{-3} (Mrotzek et al. 2001c). Initially, this compound was synthesized in the following way. A mixture of Rb_2Se (0.160 mM), Bi_2Se_3 (1.118 mM), Sn (0.320 mM), and Se (0.317 mM) was transferred into a carbon-coated silica tube, which was flame-sealed under vacuum. The tube was heated to 760°C in 24 h, kept at 760°C for 3 days, cooled to 450°C for 60 h, and further cooled to 50°C in 11 h. The product consisted of a silvery chunk with needles growing across its surface. $Rb_{0.36}Sn_{2.36}Bi_{11.64}Se_{20}$ is the major phase with a second minor unidentified phase. A pure compound can be made as follows. A mixture of Rb_2Se (0.056 mM), Bi_2Se_3 (0.763 mM), Sn (0.219 mM), and Se (0.215 mM) was transferred into a carbon-coated silica tube. The tube was placed under the flame of a natural gas-oxygen torch until the mixture melted, and then the tube was removed from the flame and allowed to solidify.

$Rb_{1.9}Sn_{4.8}Bi_{10.3}Se_{21}$ melts at 679°C and is characterized by an energy gap of 0.50 eV (Mrotzek et al. 2001a). This compound was prepared as follows. A mixture of Rb_2Se (0.5 mM), Sn (4 mM), Se (4 mM), and Bi_2Se_3 (5.5 mM) was loaded in a carbon-coated quartz tube and sealed at a residual pressure of <0.01 Pa. The starting materials were heated over 24 h to 800°C and kept at this temperature for a further 24 h, followed by slow cooling to 400°C at a rate of 0.1°C·min^{-1} and then to 50°C over a period of 10 h. A silver, shiny polycrystalline ingot of the pure title compound was obtained after washing away any impurities with DMF, methanol, and diethyl ether. All manipulations were carried out under a dry nitrogen atmosphere in a glove box.

$Rb_{0.9}Sn_{9.4}Bi_{10.9}Se_{26}$ was prepared according to the predicted composition by combining Rb_2Se, Bi_2Se_3, Sn, and Se (molar ratio 1:11:18:18) in a sealed evacuated carbon-coated quartz ampoule (Mrotzek and Kanatzidis 2002). The tube was placed under the flame of a natural gas-oxygen torch until the mixture melted and then removed from the flame to solidify. A silver-gray ingot of plates was obtained after annealing the reaction mixture at 800°C for 12 h.

8.34 Tin–Rubidium–Manganese–Selenium

The $Rb_2MnSn_3Se_8$ quaternary compound, which crystallizes in the monoclinic structure with the lattice parameters $a = 778.64 \pm 0.04$, $b = 1263.43 \pm 0.06$, $c = 1848.78 \pm 0.09$ pm, $β = 95.8099 \pm 0.0018°$, a calculated density of 4.455 g·cm^{-3}, and an energy gap of 1.83 eV, is formed in the Sn–Rb–Mn–Se system (Hu et al. 2018). To synthesize this compound, a mixture of RbCl, Mn, Sn, and Se (molar ratio 4:2:3:8) was loaded into a silica tube inside a glove box. Subsequently, the assembly was flame-sealed under a high vacuum (10^{-6} Pa), and then heated to 400°C over 20 h, dwelled for 10 h, then heated to 800°C in 20 h, maintained for 100 h, and finally cooled to 350°C at a rate of 3°C·h^{-1}. The product was washed with distilled water to remove the chloride. The red plate-like crystals of $Rb_2MnSn_3S_8$ were obtained.

8.35 Tin–Cesium–Magnesium–Selenium

The $Cs_2MgSn_3Se_8$ quaternary compound, which crystallizes in the orthorhombic structure with the lattice parameters $a = 790.55 \pm 0.03$, $b = 1276.61 \pm 0.05$, $c = 1847.40 \pm 0.06$ pm, a calculated density of 4.553 g·cm^{-3}, and an energy gap of 1.92 eV, is formed in the Sn–Cs–Mg–Se system (Morris et al. 2013). To obtain this compound, a mixture of Cs_2Se_2 (0.25 mM), Mg (0.25 mM), Sn (0.75 mM), and Se (1.5 mM) was added to a fused silica tube inside a N_2-filled glove box. The tube was then evacuated, flame-sealed, and heated gently at first in a weak methane/oxygen flame to start the reaction. It was then heated more strongly until a glowing liquid was obtained for a total of ~2 min. The tube was allowed to cool in the air at a rate of ~200°C·min^{-1}. Red plate-like crystals of $Cs_2MgSn_3Se_8$ were obtained from such a reaction.

8.36 Tin–Cesium–Zinc–Selenium

The $Cs_2ZnSn_3Se_8$ quaternary compound, which crystallizes in the orthorhombic structure with the lattice parameters $a = 781.46 \pm 0.03$, $b = 1265.34 \pm 0.04$, and $c = 1837.47 \pm 0.05$ pm, a calculated density of 4.822 g·cm^{-3}, and an energy gap of 2.12 eV, is formed in the Sn–Cs–Zn–Se system (Morris et al. 2013). It was obtained in the same way as $Cs_2MgSn_3Se_8$ was prepared using Zn instead of Mg. After cooling in air, a phase-pure ingot containing reddish-orange plate-like crystals was obtained.

8.37 Tin–Cesium–Cadmium–Selenium

The $Cs_2CdSn_3Se_8$ quaternary compound, which crystallizes in the orthorhombic structure with the lattice parameters $a = 793.35 \pm 0.04$, $b = 1273.36 \pm 0.09$, and $c = 1853.87 \pm 0.10$ pm, a calculated density of 4.845 g·cm^{-3}, and energy gap 2.16 eV, is formed in the Sn–Cs–Cd–Se system (Morris et al. 2013). To obtain $Cs_2CdSn_3Se_8$, a mixture of Cs_2Se_2 (0.25 mM), Cd (0.25 mM),

Sn (0.75 mM), and Se (1.5 mM) was added to a fused silica tube inside a nitrogen-filled glove box. The tube was evacuated, flame-sealed, placed in a furnace, heated to 750°C in 8 h, held there for 12 h, and cooled to room temperature for 12 h. After cooling in air, a phase-pure ingot containing orange plate-like crystals was obtained.

8.38 Tin–Cesium–Mercury–Selenium

The $Cs_2HgSn_3Se_8$ quaternary compound, which melts congruently at 595°C and crystallizes in the orthorhombic structure with the lattice parameters $a = 791.96 \pm 0.03$, $b = 1266.57 \pm 0.07$, and $c = 1853.79 \pm 0.09$ pm, a calculated density of 5.194 g·cm^{-3}, and an energy gap of 1.96 eV, is formed in the Sn–Cs–Hg–Se system (Morris et al. 2013). It was obtained in the same way as $Cs_2CdSn_3Se_8$ was prepared using HgSe instead of Cd. After cooling in air, a phase-pure ingot containing dark red plate-like crystals was obtained.

8.39 Tin–Cesium–Gallium–Selenium

Two quaternary compounds, $CsGaSnSe_4$ and $CsGaSn_2Se_6$, are formed in the Sn–Cs–Ga–Se system. First of them melts congruently at 759°C and crystallizes in the orthorhombic structure with the lattice parameters $a = 767.9 \pm 0.2$, $b = 1265.5 \pm 0.3$, $c = 1827.8 \pm 0.5$ pm, a calculated density of 4.735 g·cm^{-3}, and an energy gap of 1.97 eV (Hwang et al. 2004a). Reddish-orange crystals of this compound were synthesized by the stoichiometric reaction of Cs_2Se, Ga, Sn, and Se with the molar ratio of Cs/Ga/Sn/Se =1:1:1:4. The starting materials were loaded into a fused silica tube and sealed under vacuum. They were thoroughly melted in a torch flame, and then the liquid was allowed to cool rapidly in the air. This corresponds to ~200°C·min^{-1} cooling rate.

$CsGaSn_2Se_6$ is thermally stable up to 684°C and crystallizes in the trigonal structure with the lattice parameters $a = 1051.8 \pm 0.8$, $c = 953.9 \pm 0.2$ pm, a calculated density of 4.98 g·cm^{-3}, and an energy gap of 1.87 eV (Lin et al. 2017b). It was synthesized by solid-state reaction technique. The optimal synthesis route is as follows: the mixture of Ga, Sn, Se, and CsCl (molar ratio 1.3:1.775:6:2) was placed into a fused silica tube under vacuum, annealing at 800°C for 40 h and then kept at this temperature for 100 h, followed by slow cooling at 3°C·h^{-1} to 200°C and then the furnace was turned off. The raw products were washed with distilled water and then dried with ethanol. Deep-red crystals, which are stable in air for more than month, were obtained.

8.40 Tin–Cesium–Indium–Selenium

Two quaternary compounds, $CsInSnSe_4$ and $CsInSn_2Se_6$, are formed in the Sn–Cs–In–Se system. $CsInSnSe_4$ has two polymorphic modifications, first of them crystallizes in the orthorhombic structure with the lattice parameters $a = 1866.50 \pm 0.05$, $b = 789.76 \pm 0.02$, $c = 1273.40 \pm 0.04$ pm, a calculated

density of 4.828 g·cm^{-3} [$a = 1871.1 \pm 0.2$, $b = 790.7 \pm 0.6$, $c = 1276.6 \pm 0.9$ pm (Lin et al. 2017b)], and an energy gap of 2.15 eV and the second one crystallizes in the cubic structure with the lattice parameter $a = 1419.32 \pm 0.03$ pm, a calculated density of 4.755 g·cm^{-3}, and an energy gap of 2.02 eV (Friedrich et al. 2021). Polycrystalline orthorhombic $CsInSnSe_4$ was obtained by reacting Cs_2Se (0.73 mM), In (1.47 mM), Sn (1.47 mM), and Se (5.13 mM) in a sealed evacuated fused silica tube. The reaction mixture was heated to 1000°C with a heating rate of 3°C·min^{-1}. After annealing at 1000°C for 24 h the tube was cooled to room temperature with a cooling rate of 1°C·min^{-1}. The air-sensitive yellow polycrystalline product forms plate-like crystallites and is stored in a nitrogen glove box until further use. Polycrystalline red cubic $CsInSnSe_4$ was obtained by reacting Cs_2Se (0.73 mM), In (1.47 mM), Sn (1.47 mM), and Se (5.13 mM) in a sealed evacuated fused silica tube. The reaction mixtures were heated to 500°C with a heating rate of 2°C·min^{-1} and annealed at 550°C for 7 days. After this initial annealing step, the crude products were thoroughly grounded in an agate mortar under N_2 atmosphere, pressed to a pellet using a pellet press, and annealed at 550°C for an additional week. The air-sensitive red polycrystalline product was stored in a nitrogen glove box.

$CsInSn_2Se_6$ is thermally stable up to 671°C and crystallizes in the trigonal structure with the lattice parameters $a = 1065.2 \pm 0.5$, $c = 968.8 \pm 0.8$ pm, a calculated density of 5.02 g·cm^{-3}, and an energy gap of 1.78 eV (Lin et al. 2017b). It was synthesized in the same way as $CsGaSn_2Se_6$ was prepared using In instead of Ga. Deep-red crystals, which are stable in air for more than a month, were obtained.

8.41 Tin–Cesium–Phosphor–Selenium

Two quaternary compounds, $Cs_2SnP_2Se_6$ and $Cs_6Sn_2Se_4(PSe_5)_2$, are formed in the Sn–Cs–P–Se system. The first of them melts congruently at 573°C and crystallizes in the monoclinic structure with the lattice parameters $a = 1011.60 \pm 0.04$, $b = 1278.67 \pm 0.05$, $c = 1108.28 \pm 0.05$ pm, $\beta = 98.463 \pm 0.003°$, a calculated density of 4.277 g·cm^{-3} at 100 K, and an energy gap of 2.06 eV (Chung and Kanatzidis 2011). The synthesis of this compound was achieved by reacting a mixture of Cs_2Se, Sn, P, and Se (molar ratio 5:3:8:19) under a vacuum in a carbon-coated fused silica tube at 850°C for 3 days, followed by cooling at a rate of 2°C·h^{-1} to 300°C. The excess flux was dissolved with degassed DMF under a N_2 atmosphere to reveal orange thick rod, stable in air crystals of the title compound with the by-product of $Sn_2P_2Se_6$. The pure compound could be obtained by reacting a stoichiometric mixture of the same reagents in a fused silica tube at 800°C for 2 h, followed by quenching in air.

$Cs_6Sn_2Se_4(PSe_5)_2$ melts congruently at 456°C and crystallizes in the triclinic structure with the lattice parameters $a = 989.9 \pm 0.1$, $b = 1241.6 \pm 0.2$, $c = 749.7 \pm 0.1$ pm, $\alpha = 91.62 \pm 0.01°$, $\beta = 110.19 \pm 0.01°$, $\gamma = 79.94 \pm 0.01°$, a calculated density of 4.297 g·cm^{-3}, and an energy gap of 2.09 eV (Chondroudis and Kanatzidis 1996). It was synthesized from a mixture of Sn (0.3 mM), P_2Se_5 (0.45 mM), Cs_2Se (1.20 mM), and Se (3 mM) by heating at 495°C for 4 days followed by cooling to 150°C at a rate of 4°C·h⁻¹. The excess $Cs_xP_ySe_z$ flux was removed with DMF to reveal orange rod-like crystals, which are air and water sensitive.

8.42 Tin–Cesium–Arsenic–Selenium

The $Cs_2SnAs_2Se_9$ quaternary compound, which melts congruently at ~258°C and crystallizes in the monoclinic structure with the lattice parameters $a = 717.5 \pm 0.5$, $b = 1755.5 \pm 1.2$, $c = 766.3 \pm 0.5$ pm, $\beta = 115.857 \pm 0.011°$, a calculated density of 4.761 g·cm⁻³, and an energy gap of 1.45 eV, is formed in the Sn–Cs–As–Se system (Iyer et al. 2003). The title compound was synthesized from a mixture of Cs_2Se (0.5 mM), Sn (0.5 mM), As (1.5 mM), and Se (4.75 mM). The reagents were mixed, sealed in an evacuated silica tube, and heated at 550°C for 4 days. The tube was then cooled at a rate of 5°C·h⁻¹ to 250°C followed by rapid cooling to room temperature. The solid products were washed with DMF and ether. Product recovery gave black platy single crystals of $Cs_2SnAs_2Se_9$ together with an unidentified glassy phase. Quantitative synthesis of pure $Cs_2SnAs_2Se_9$ was achieved by a reaction of the stoichiometric amounts of Cs_2Se, Sn, As, and Se at 500°C. This compound was stable in air and water.

8.43 Tin–Cesium–Bismuth–Selenium

Some compounds are formed in the Sn–Cs–Bi–Se quaternary system. $Cs_{1.1}Sn_{1.9}Bi_{7.9}Se_{14}$ melts congruently at 669°C and has an energy gap of 0.40 eV (Mrotzek et al. 2001c). To prepare this compound, a mixture of Cs_2Se (1 mM), Sn (7 mM), Se (7 mM), and Bi_2Se_3 (7 mM) was loaded in a carbon-coated quartz tube and sealed at a residual pressure of <0.01 Pa. Then it was heated within 24 h to 800°C and kept there for 24 h, followed by slow cooling to 50°C at a rate of 0.5°C·min⁻¹. A silver, shiny, polycrystalline ingot of the title compound was obtained after washing any impurities with DMF, methanol, and diethyl ether.

$Cs_{1.1}Sn_{0.9}Bi_{9.3}Se_{15}$ crystallizes in the monoclinic structure with the lattice parameters $a = 2728.7 \pm 0.8$, $b = 411.6 \pm 0.1$, $c = 1400.4 \pm 0.4$ pm, $\beta = 103.346 \pm 0.005°$, a calculated density of 7.168 g·cm⁻³, and an energy gap of 0.53 eV (Mrotzek et al. 2001b). It was prepared involving Cs_2Se, Bi_2Se_3, Sn, and Se (molar ratio 1:2:11:2) in a sealed evacuated carbon-coated quartz ampoule. The tube was heated within 24 h to 800°C and kept at this temperature for 24 h, followed by slow cooling to 400°C at a rate of 0.1°C·min⁻¹ and then to 50°C in 10 h resulting in a silver, shiny, polycrystalline ingot of the title compound as the major phase with $Cs_{1-x}Sn_{3+x}Bi_{11+x}Se_{20}$ as impurity.

$Cs_{1-x}Sn_{3-x}Bi_{11+x}Se_{20}$ melts congruently at 668°C and also crystallizes in the monoclinic structure with the lattice parameters $a = 1610.0 \pm 0.3$, $b = 413.60 \pm 0.08$, $c = 1707.7 \pm 0.3$ pm, and $\beta = 116.79 \pm 0.03°$, a calculated density of 7.106 g·cm⁻³, and an energy gap of 0.40 eV for $Cs_{0.46}Sn_{2.46}Bi_{11.54}Se_{20}$ composition (Mrotzek et al. 2001c). To obtain this compound, a mixture of Cs_2Se (1 mM), Sn (3 mM), Se (3 mM), and Bi_2Se_3 (5.5 mM) was loaded in a carbon-coated quartz tube and sealed at a residual pressure of <0.01 Pa. The mixture was heated within 12 h to 800°C and kept there for 24 h, followed by slow cooling

to 400°C at a rate of 0.1°C·min⁻¹ and then to 50°C in 10 h. A silver, shiny, polycrystalline ingot of the title compound was obtained after washing any impurities with DMF, methanol, and diethyl ether.

$Cs_{0.8}Sn_{4.3}Bi_{10.6}Se_{21}$ melts at 690°C and is characterized by an energy gap of 0.52 eV (Mrotzek et al. 2001a). This compound was prepared as follows. A mixture of Cs_2Se (0.5 mM), Sn (4 mM), Se (4 mM), and Bi_2Se_3 (5.5 mM) was loaded in a carbon-coated quartz tube and sealed at a residual pressure of <0.01 Pa. The starting materials were heated over 24 h to 800°C and kept at this temperature for a further 24 h, followed by slow cooling to 400°C at a rate of 0.1°C·min⁻¹ and then to 50°C over a period of 10 h. A silver, shiny polycrystalline ingot of the title compound in a mixture with another quaternary compound was obtained after washing away any impurities with DMF, methanol, and diethyl ether. All manipulations were carried out under a dry nitrogen atmosphere in a glove box.

$Cs_{1.3}Sn_{8.9}Bi_{11.3}Se_{26}$ was prepared according to the predicted composition by combining Cs_2Se, Bi_2Se_3, Sn, and Se (molar ratio 1:11:18:18) in a sealed evacuated carbon-coated quartz ampoule (Mrotzek and Kanatzidis 2002). The tube was placed under the flame of a natural gas-oxygen torch until the mixture melted and then removed from the flame to solidify. A silver-gray ingot of plates was obtained after annealing the reaction mixture at 800°C for 12 h.

8.44 Tin–Cesium–Manganese–Selenium

The $Cs_2MnSn_3Se_8$ quaternary compound, which crystallizes in the orthorhombic structure with the lattice parameters $a = 787.98 \pm 0.03$, $b = 1270.40 \pm 0.05$, $c = 1845.56 \pm 0.08$ pm, a calculated density of 4.704 g·cm⁻³, and an energy gap of 1.86 eV, is formed in the Sn–Cs–Mn–Se system (Hu et al. 2018). To synthesized this compound, a mixture of CsCl, Mn, Sn, and Se (molar ratio 4:2:3:8) was loaded into a silica tube inside a glove box. Subsequently, the assembly was flame-sealed under high vacuum (10⁻⁶ Pa), and then heated to 400°C in 20 h, dwell for 10 h, then heated to 800°C in 20 h, maintained for 100 h, and finally cooled to 350°C at a rate of 3°C·h⁻¹. The product was washed by distilled water to remove the chloride. The red plate-like crystals of $Cs_2MnSn_3S_8$ were obtained.

8.45 Tin–Copper–Silver–Selenium

$Cu_2SnSe_3–Ag_2Se$. The phase diagram of this system, constructed through differential thermal analysis (DTA), X-ray diffraction (XRD), metallography, and the measuring of microhardness and density is a eutectic type (Figure 8.1) (Rzaguliev et al. 2020). No quaternary compounds were detected. The eutectic contains 40 mol% Ag_2Se and crystallizes at 637°C. Cu_2SnSe_3 dissolves up to 10 mol% Ag_2Se. The alloys for the investigation were annealed at 580°C for 250–740 h.

$Cu_2SnSe_3–Ag_8SnSe_6$. The phase diagram of this system, constructed through DTA, XRD, metallography and the measuring of the microhardness and density is a eutectic type (Figure 8.2) (Rzaguliev et al. 2020). No quaternary compounds were detected.

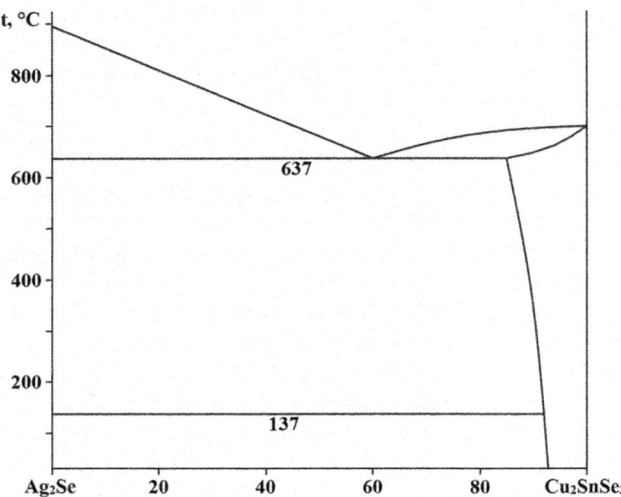

FIGURE 8.1 Phase diagram of the Cu₂SnSe₃–Ag₂Se system. (From Rzaguliev, V.A., et al., *Russ. J. Inorg. Chem.*, **65**(12), 1899, 2020.)

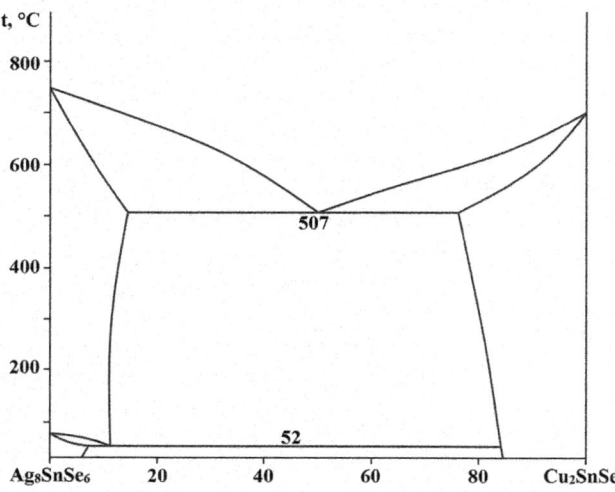

FIGURE 8.2 Phase diagram of the Cu₂SnSe₃–Ag₈SnSe₆ system. (From Rzaguliev, V.A., et al., *Russ. J. Inorg. Chem.*, **65**(12), 1899, 2020.)

The eutectic contains 50 mol% Ag₈SnSe₆ and crystallizes at 507°C. Cu₂SnSe₃ dissolves up to 15 mol% Ag₈SnSe₆. The alloys for the investigation were annealed at 470°C for 250–740 h.

8.46 Tin–Copper–Magnesium–Selenium

The **Cu₂MgSnSe₄** quaternary compound, which crystallizes in the tetragonal structure with the lattice parameters $a = 572.1 \pm 0.3$, $c = 1143.5 \pm 0.5$ pm, and the calculated and experimental densities of 5.197 and 5.188 ± 0.7 g·cm⁻³, is formed in the Sn–Cu–Mg–Se system (Odin et al. 2021). For the Cu₂₋ₓMg₁₋ₓSnSe₄ solid solution, the lattice parameters decrease with increasing x up to 0.15 ($a = 570.9 \pm 0.3$ and $c = 1141.5 \pm 0.5$ pm for $x = 0.15$). The lattice parameters for the copper doped compositions Cu₂₊ₓMg₁₋ₓSnSe for $x = 0.1$ are the next: $a = 571.1$, $c = 1139.9$ pm, and an energy gap of 1.7 eV (Kumar et al. 2015).

Polycrystalline samples of Cu₂MgSnSe₄ and Cu₂₋ₓMgSnSe₄ were synthesized in two steps (Odin et al. 2021). In the first step, appropriate amounts of Cu, Mg, Sn, and Se were fired at 750°C for 48 h in graphitized quartz ampoules sealed under vacuum (0.3 Pa). To eliminate the risk of explosion during the synthesis process, the samples were heated to the synthesis temperature at a rate of 1°C·min⁻¹. After the ampoules were opened, the resultant materials were ground in an agate mortar. Next, the mixtures were again sealed in quartz ampoules under vacuum and homogenized by annealing at a temperature of 650°C for 600 h.

Powders of the Cu₂₋ₓMg₁₋ₓSnSe₄ solid solutions were also prepared from a stoichiometric mixture of the elements, which was loaded in carbon-coated evacuated fused quartz ampoules (Kumar et al. 2015). The ampoules were then slowly heated up to 800°C with a rate of 2°C·min⁻¹, held at this temperature for 48 h, and the furnace was cooled to room temperature. The harvested powders were ground, sealed, and annealed for 96 h at 800°C. The final powders were densified in a spark plasma sintering furnace in vacuum at 550°C for 5 min in a graphite die under a pressure of 50 mPa.

8.47 Tin–Copper–Strontium–Selenium

The **Cu₂SrSnSe₄** quaternary compound, which has an energy gap of 1.46 eV, is formed in the Sn–Cu–Sr–Se system (Zhu et al. 2017).

8.48 Tin–Copper–Barium–Selenium

Two quaternary compounds, **Cu₂BaSnSe₄** and **Cu₂Ba₃Sn₃Se₁₀**, are formed in the Sn–Cu–Ba–Se system. First of them crystallizes in the orthorhombic structure with the lattice parameters $a = 1110.1 \pm 1.0$, $b = 1118.9 \pm 1.0$, $c = 672.8 \pm 0.6$ pm, a calculated density of 5.555 g·cm⁻³, and an energy gap of 1.72 eV (Nian et al. 2017) [$a = 1112.15 \pm 0.13$, $b = 1123.73 \pm 0.13$, $c = 675.31 \pm 0.08$ pm, a calculated density of 5.501 g·cm⁻³, and an energy gap of 0.48 eV (Assoud et al. 2005); 1.50 eV (Zhu et al. 2017)].

The conventional high-temperature solid-state method was used to synthesize the title compound (Nian et al. 2017). A stoichiometric mixture of Cu, Ba, Sn, and Se (molar ratio 1:2:1:4) was loaded into the silica tube and sealed under a high vacuum of 10⁻³ Pa. The furnace was programmed by the following steps: heated from room temperature to 600°C in 30 h and kept at this temperature for 40 h; then heated to 1000°C in 20 h and left at this temperature for 100 h; finally, cooled to 400°C at a rate of 5°C·h⁻¹ and then the furnace was shut down to room temperature. After the reaction, the product was washed with DMF to remove the unreacted sulfur and other by-products. Many red crystals of Cu₂BaSnSe₄ were gained that can be stable in the air for several months. Since the Ba metal is easily oxidized in the air, an Ar-filled glove box was used to avoid the effects of oxygen and moisture in the preparation processes.

Cu₂BaSnSe₄ can also be obtained if the stoichiometric mixture of the elements was placed into a silica tube, which was then sealed under vacuum (Assoud et al. 2005). The tube was put into a muffle furnace and the mixture was heated to 800°C within 24 h, kept at 800°C for 6 h, slowly cooled (within 100 h) to 650°C, and then annealed at 650°C for 100 h. Thereafter, the furnace was switched off, to allow for rapid cooling to room temperature. The temperature profile was chosen to initially exceed the expected melting points (at about 700°C), to cool slowly through the melting points to obtain crystals, and then to anneal below the melting points to allow homogenization.

$Cu_2Ba_3Sn_3Se_{10}$ crystallizes in the monoclinic structure with the lattice parameters $a = 662.46 \pm 0.07$, $b = 1360.12 \pm 0.14$, $c = 2309.7 \pm 0.2$ pm, $\beta = 94.723 \pm 0.002°$, and a calculated density of 5.384 g·cm⁻³ (Assoud et al. 2005). The phase-pure compound was obtained by using the elements in a stoichiometric ratio. The mixture was placed into a silica tube, which was then sealed under vacuum and put into a muffle furnace. The temperature profile applied was heat to 800°C within 1 day, keep at 800°C for 6 h, cool to 600°C within 6 h, heat at 600°C for 200 h, and finally finish by switching off the furnace.

8.49 Tin–Copper–Zinc–Selenium

Cu_2SnSe_3–ZnSe. The phase diagram of this system, constructed through DTA, XRD, metallography, and EPMA, is shown in Figure 8.3 (Dudchak and Piskach 2001b). The eutectic contains 2.5 mol% ZnSe and crystallizes at 694°C. The **$Cu_2ZnSnSe_4$** quaternary compound, which melts incongruently at 790°C (Dudchak and Piskach 2001b) [at 805°C (Matsushita et al. 2000a, 2005); at 796°C (Schäfer and Nitsche 1977)], is formed in this system. The composition of the peritectic point is 14 mol% ZnSe. This compound has a homogeneity range and undergoes a polymorphic transformation in the temperature interval 616–619°C [Matsushita et al. (2000a) indicated that this compound has three polymorphic modifications]. Solubility of Cu_2SnSe_3 in ZnSe at the peritectic temperature reaches up to 4 mol%.

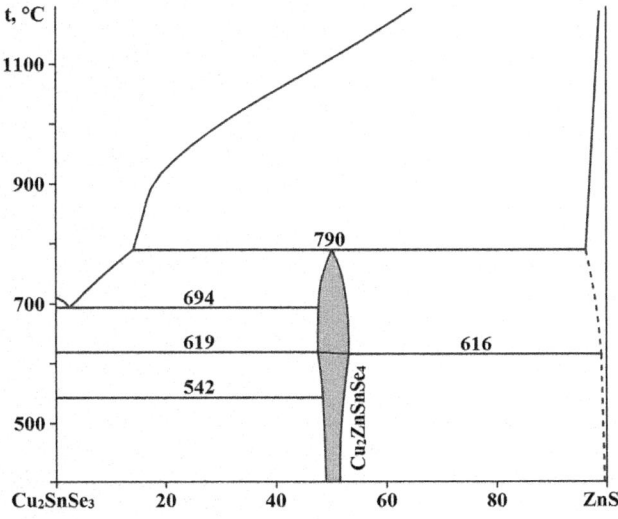

FIGURE 8.3 Phase diagram of the Cu_2SnSe_3–ZnSe system. (From Dudchak, I., and Piskach, L., *Visnyk L'viv. Univ. Ser. khim.*, (40), 73, 2001.)

$Cu_2ZnSnSe_4$ crystallizes in the tetragonal structure with the lattice parameters $a = 568.82 \pm 0.02$, $c = 1133.78 \pm 0.09$ pm, and experimental densities of 5.689 ± 0.003 g·cm⁻³ (Olekseyuk et al. 2002b) [$a = 568.1$, $c = 1134$ pm, and the calculated and experimental densities of 5.69 and 5.68 g·cm⁻³ (Hahn and Schulze 1965; Guen et al. 1979; Guen and Glaunsinger 1980); $a = 569.4$ and $c = 1134.7$ pm (Schäfer and Nitsche 1977); $a = 565.5 \pm 0.1$ and $c = 1137.9 \pm 0.3$ pm (Dudchak and Piskach 2001b; Olekseyuk et al. 2001a); $a = 569.3$ and $c = 1133.3$ pm (Matsushita et al. 2000a, 2005); $a = 568.2$ and $c = 1134.2$ pm (Stolyarova et al. 2019)], and an energy gap of 1.44 eV (Matsushita et al. 2000a, 2005) [0.96 eV (Chen et al. 2009); ≈1.0 eV (Persson 2010)].

The ingots for investigations were annealed at 400°C during 250 h (Dudchak and Piskach 2001b). Single crystals of $Cu_2ZnSnSe_4$ were obtained by the chemical transport reactions (Schäfer and Nitsche 1974; Guen et al. 1979; Guen and Glaunsinger 1980), or by the sintering of mixtures from binary compounds at 650–900°C (Hahn and Schulze 1965), or using the horizontal gradient method from respective melts (Matsushita et al. 2000a, 2005), or by crystallization from the solution in the melt (Olekseyuk et al. 2002b).

Using the first-principles density functional method, it was shown that the low-energy crystal structure of $Cu_2ZnSnSe_4$ is a kesterite-type structure (Chen et al. 2009). However, the stannite or partially disordered kesterite structure can also exist in synthesized samples due to the small energy cost. According to the first-principles calculations, kesterite-type structure of this compound is characterized by the tetragonal structure with the lattice parameters $a = 564.2$ and $c = 1130.3$ pm [$a = 560.5$ and $c = 1120.0$ pm (Persson 2010)] and a calculated density of 5.7871 g·cm⁻³, and stannite-type structure is characterized also by the tetragonal structure with the lattice parameters $a = 565.0$ and $c = 1127.0$ pm [$a = 560.4$ and $c = 1120.8$ pm (Persson 2010)] and a calculated density of 5.7882 g·cm⁻³ (Bensalem et al. 2014).

The standard enthalpy of formation of $Cu_2ZnSnSe_4$ was determined by measuring the heat of its formation from the elements in a calorimeter (Stolyarova et al. 2019): $\Delta H^0_{f,298} = -391.91 \pm 2.34$ kJ·M⁻¹.

$Cu_{2+x}Zn_{1-x}SnSe_4$ ($x \leq 0.2$) solid solutions were prepared by direct reaction of the high-purity elements in stoichiometric ratios (Dong et al. 2014). The mixtures were loaded in the ampoules, which were sealed in quartz tubes, heated to 700°C and subsequently held at this temperature for 2 days. The furnace was turned off and the reaction tubes were quenched in air to room temperature. After annealing the products were again grounds into fine powders, cold pressed into pellets and annealed at 600°C for one week. The products were then ground into fine powders inside a glove box and loaded into graphite dies for hot pressing. Densification was accomplished by hot pressing at 600°C and 150 MPa for 3 h under N_2 flow. They crystallize in the tetragonal structure with the lattice parameters $a = 570.15 \pm 0.02$, 569.19 ± 0.02, 568.98 ± 0.02, $c = 1135.23 \pm 0.05$, 1135.00 ± 0.07, 1134.87 ± 0.07 pm, and a calculated density of 5.643, 5.660, 5.664 g·cm⁻³ for $x = 0$, 0.15, and 0.2, respectively.

$SnSe_2$–Cu_2Se–ZnSe. The isothermal section of the $SnSe_2$–Cu_2Se–ZnSe quasiternary system at 400°C is shown in

Figure 8.4 (Olekseyuk et al. 2001a; Dudchak and Piskach 2003). The solubility in Cu$_2$Se is lower than 2 mol% and is elongated along the SnSe$_2$–Cu$_2$Se boundary side; in ZnSe and SnSe$_2$, it is lower than 1 mol%; and in Cu$_2$SnSe$_3$, it is lower than 0.5 mol%. The miscibility gap of Cu$_2$ZnSnSe$_4$ was found to be 3 mol% along the triangulated section Cu$_2$SnSe$_3$–ZnSe. The ingots were annealed at 400°C during 250 h.

The liquidus surface of this system (Figure 8.5) consists of five fields of primary crystallization of Cu$_2$SnSe$_3$ and solid solutions based on Cu$_2$Se, ZnSe, SnSe$_2$, and Cu$_2$ZnSnSe$_4$, respectively (Olekseyuk et al. 2001a; Dudchak and Piskach 2003). There are 10 monovariant lines and 10 invariant points

(2 binary peritectics, 4 binary eutectics, 2 ternary transition points, and 2 ternary eutectics) in the liquidus surface. The field of primary crystallization of ZnSe occupies the largest area of the concentration triangle as the most refractory component. The field of primary crystallization of Cu$_2$ZnSnSe$_4$ is elongated along the SnSe$_2$–Cu$_2$Se system with the temperature maximum in the quasibinary section Cu$_2$SnSe$_3$–ZnSe. The polymorphic transformation of the solid solutions based on Cu$_2$ZnSnSe$_4$ occurs at 627°C in the quasiternary system ZnSe–Cu$_2$SnSe$_3$–Cu$_2$Se and at 583°C for the ZnSe–Cu$_2$SnSe$_3$–SnSe$_2$ quasiternary subsystem.

8.50 Tin–Copper–Cadmium–Selenium

Cu$_2$SnSe$_3$–CdSe. The phase diagram of this system, constructed through DTA, XRD, metallography, and measuring of the microhardness and density, is shown in Figure 8.6 (Olekseyuk and Piskach 1997; Piskach et al. 1997). The eutectic contains 23 mol% CdSe and crystallizes at 657°C [640°C (Zhukov et al. 1982)]. At 530°C, the solubility of CdSe in Cu$_2$SnSe$_3$ reaches 20 mol% and the solubility of Cu$_2$SnSe$_3$ in CdSe is not higher than 7 mol% (Olekseyuk and Piskach 1997). According to the data of Zhukov et al. (1982), at room temperature, the solubility of CdSe in Cu$_2$SnSe$_3$ is equal to 15 mol% and Cu$_2$SnSe$_3$ in CdSe is not higher than 12 mol%. The increasing of Cu$_2$SnSe$_3$ contents in the solid solutions based on CdSe leads to the monotonic changing of electrophysical properties of forming solid solutions (Dovletov et al. 1987).

The **Cu$_2$CdSnSe$_4$** quaternary compound is formed in this system. It melts incongruently at 782°C (Olekseyuk and Piskach 1997) [780°C (Matsushita et al. 2000a, 2005); 795°C (Zhukov et al. 1982); 768°C (Schäfer and Nitsche 1977)] and crystallizes in the tetragonal structure with the lattice parameters $a = 583.37 \pm 0.02$, $c = 1140.39 \pm 0.04$ pm and the calculated density of 5.78 ± 0.02 g·cm^{-3} (Olekseyuk et al. 2002b) [$a = 581.4$,

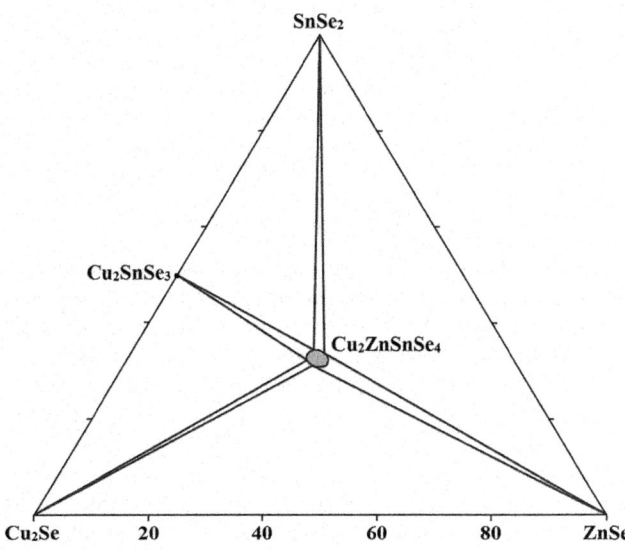

FIGURE 8.4 Isothermal section of the SnSe$_2$–Cu$_2$Se–ZnSe quasiternary system at 400°C. (From Dudchak, I.V., and Piskach, L.V., *J. Alloys Compd.*, **351**, 145, 2003.)

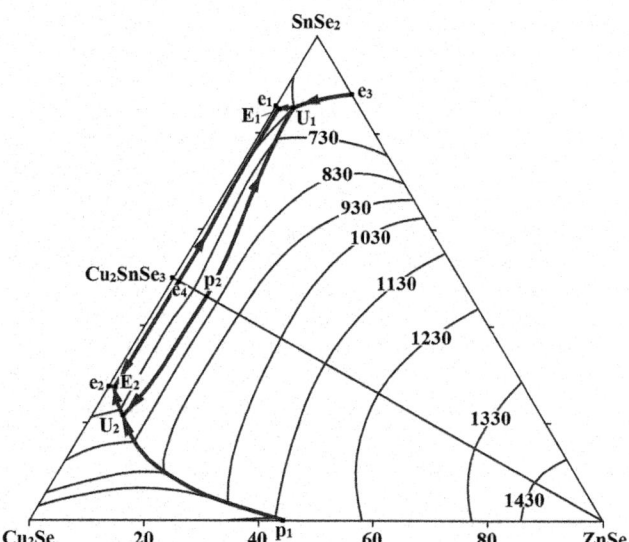

FIGURE 8.5 Liquidus surface of the SnSe$_2$–Cu$_2$Se–ZnSe quasiternary system. (From Dudchak, I.V., and Piskach, L.V., *J. Alloys Compd.*, **351**, 145, 2003.)

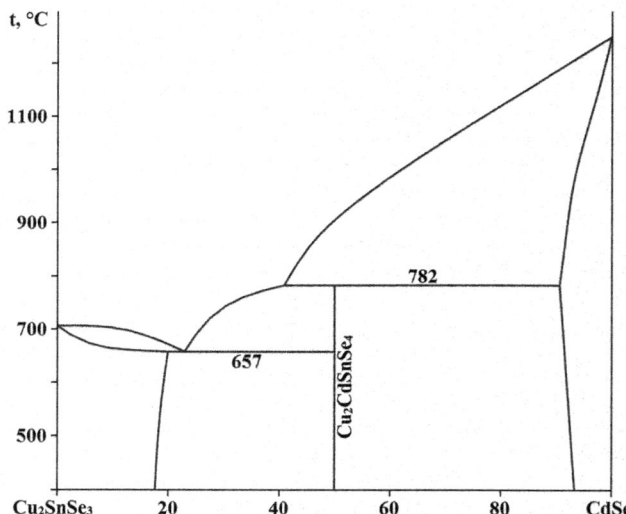

FIGURE 8.6 Phase diagram of the Cu$_2$SnSe$_3$–CdSe system. (From Piskach, L.V., et al., *Fiz. kondens. vysokomolek. system. Nauk. zap. Rivnens'kogo pedinstytutu*, (3), 153, 1997.)

$c = 1147$ pm, and the calculated and experimental densities of 5.77 and 5.75 g·cm^{-3} (Hahn and Schulze 1965); $a = 582.6$ and $c = 1139.0$ pm (Schäfer and Nitsche 1977); $a = 572$, $c = 1112$ pm, and an experimental density of 5.77 g·cm^{-3} (Zhukov et al. 1982); $a = 581.88 \pm 0.06$, $c = 1113.83 \pm 0.02$ pm, and an experimental density of 5.76 g·cm^{-3} (Olekseyuk and Piskach 1997); $a = 583.2$ and $c = 1138.9$ pm (Matsushita et al. 2000a, 2005)], and an energy gap of 1.30 eV (Mkrtchian et al. 1988a) [0.96 eV (Matsushita et al. 2000a, 2005); 0.89 eV (Konstantinova et al. 1989)].

$Cu_2CdSnSe_4$ was obtained by heating the binary constituents at 1030–1080°C for 2 h and annealed at 530–580°C for 600 h (Mkrtchian et al. 1988a). Its single crystals were grown by the oriented crystallization (Konstantinova et al. 1989), or using a solution-fusion method (Olekseyuk et al. 2002a), or using the horizontal gradient method from the melt (Matsushita et al. 2000a).

$SnSe_2$–Cu_2Se–$CdSe$. Isothermal section of this quasiternary system at 550°C is shown in Figure 8.7 (Parasyuk et al. 1999b). The ingots for the investigations were annealed at this temperature for 250 h.

The Liquidus surface of this quasiternary system is shown in Figure 8.8 (Parasyuk et al. 1999b). It consists of five fields of primary crystallization of phases: $CdSe$, Cu_2Se, $SnSe_2$, Cu_2SnSe_3, and $Cu_2CdSnSe_4$. There are two ternary eutectics and two transition points in the system: E_1 (635°C) – $L \leftrightarrow Cu_2SnSe_3 + Cu_2Se + Cu_2SnSnSe_4$; E_2 (575°C) – $L \leftrightarrow Cu_2SnSe_3 + Cu_2CdSnSe_4 + SnSe_2$; U_1 (715°C) – $L + CdSe \leftrightarrow Cu_2Se + Cu_2CdSnSe_4$; and U_2 (605°C) – $L + CdSe \leftrightarrow Cu_2CdSnSe_4 + SnSe_2$.

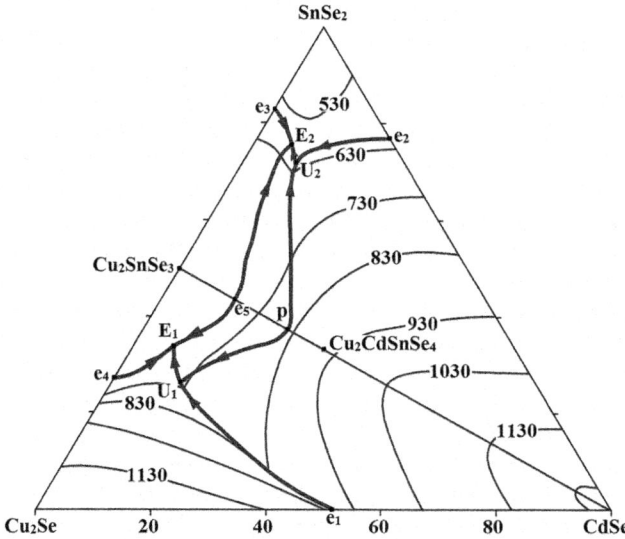

FIGURE 8.8 Liquidus surface of the $SnSe_2$–Cu_2Se–$CdSe$ quasiternary system. (From Parasyuk,O.V., et al., *Zhurn. neorgan. khimii*, **44**(8), 1363, 1999.)

8.51 Tin–Copper–Mercury–Selenium

Cu_2SnSe_3–$HgSe$. The phase diagram of this system, constructed through DTA, XRD, metallography, and measuring of the microhardness and density, is presented in Figure 8.9 (Parasyuk et al. 1997, 1998). The eutectics contain 11 and 84 mol% HgSe and crystallize at 682°C and 677°C [at 627°C and 652°C (Mkrtchian et al. 1988b)], respectively. The solubility of Cu_2SnSe_3 in HgSe at 400°C is equal to 9 mol% [15 mol% at room temperature (Mkrtchian et al. 1988b)], and the solubility of HgSe in Cu_2SnSe_3 is insignificant (Parasyuk et al. 1997, 1998) [17 mol% (Mkrtchian et al. 1988b)]. The ingots for the investigations were annealed at 400°C for 250 h and quenched

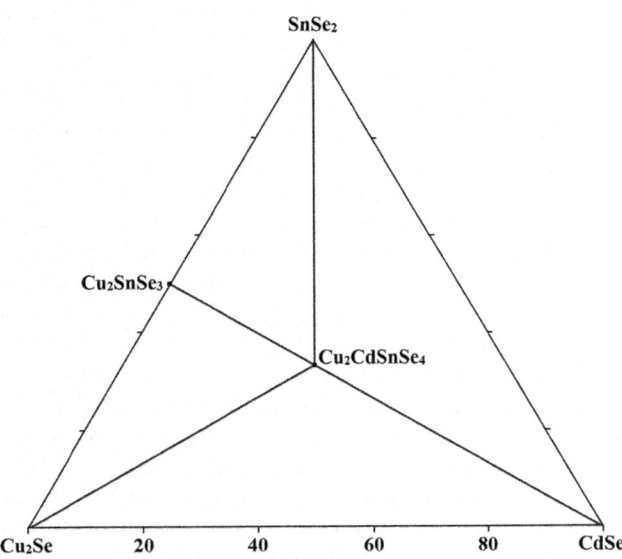

FIGURE 8.7 Isothermal section of the $SnSe_2$–Cu_2Se–$CdSe$ quasiternary system at 550°C. (From Parasyuk,O.V., et al., *Zhurn. neorgan. khimii*, **44**(8), 1363, 1999.)

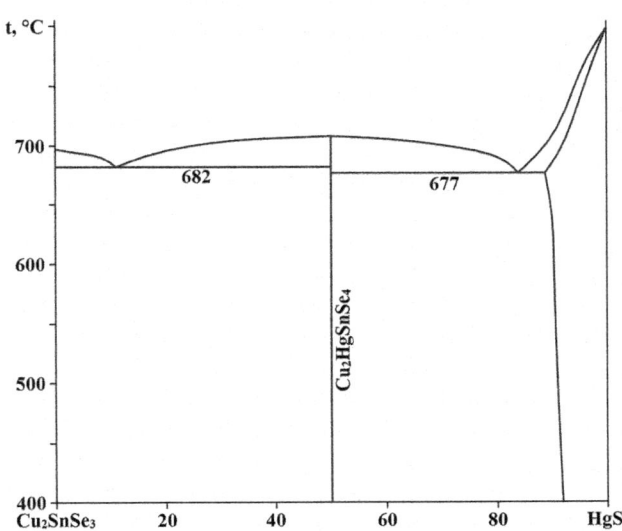

FIGURE 8.9 Phase diagram of the Cu_2SnSe_3–$HgSe$ system. (From Parasyuk, O.V. et al., *Ukr. khim. zhurn.*, **64**(9), 20, 1998.)

in cold water (Parasyuk et al. 1998) [at 430–530°C for 500 h (Mkrtchian et al. 1988b)].

The **Cu$_2$HgSnSe$_4$** quaternary compound is formed in this system. It melts congruently at 708°C (Parasyuk et al. 1997, 1998; Schäfer and Nitsche 1977) [at 712°C (Parasyuk et al. 1999a); at 710°C (Hirai et al. 1967); at 697°C (Mkrtchian et al. 1988b)] and crystallizes in the tetragonal structure with the lattice parameters $a = 582.88 \pm 0.01$, $c = 1141.79 \pm 0.02$ pm, and a calculated density of 6.514 g·cm^{-3} (Olekseyuk et al. 2002b) [$a = 581.8$, $c = 1148$ pm, and the calculated and experimental densities of 6.51 and 6.47 g·cm^{-3}, respectively (Hahn and Schulze 1965); $a = 949$ and $c = 1715$ pm (Hirai et al. 1967); $a = 582.5$ and $c = 1141.3$ pm (Schäfer and Nitsche 1977); $a = 579$, $c = 1145$ pm, and an experimental density of 6.45 g·cm^{-3} (Mkrtchian et al. 1988b); $a = 579.2 \pm 0.3$ and $c = 1138.9 \pm 0.5$ (Parasyuk et al. 1997, 1998)], and an energy gap of 0.17 eV (Mkrtchian et al. 1988a). This compound has a homogeneity region along the Cu$_2$SnSe$_3$–HgSe section within the interval 35–42 mol% HgSe (Olekseyuk et al. 2001c).

The single crystals of Cu$_2$HgSnSe$_4$ were grown by the directional crystallization of the melt (Olekseyuk et al. 2001c) or using a solution-fusion method (Olekseyuk et al. 2002b).

SnSe$_2$–Cu$_2$HgSnSe$_4$. The phase diagram of this system, constructed through DTA, XRD, metallography, and measuring of the microhardness, is a eutectic type (Figure 8.10) (Parasyuk et al. 1999a). The eutectic contains 87 mol% SnSe$_2$ and crystallizes at 598°C. Cu$_2$HgSnSe$_4$ dissolves up to 5 mol% SnSe$_2$ and SnSe$_2$ dissolves 2 mol% Cu$_2$HgSnSe$_4$.

Cu$_2$HgSnSe$_4$–Cu$_2$Se. The phase diagram of this system, constructed through DTA, XRD, metallography, and measuring of the microhardness, is also a eutectic type (Parasyuk et al. 1999a). The eutectic contains 33 mol% Cu$_2$HgSnSe$_4$ and crystallizes at 671°C. The solid solubility based on Cu$_2$Se exceeds 20 mol% Cu$_2$HgSnSe$_4$ and decreases with decreasing temperature. The solid solubility based on Cu$_2$HgSnSe$_4$ is rather low.

SnSe$_2$–Cu$_2$Se–HgSe. Some vertical sections, the isothermal section at 400°C (Figure 8.11) and the liquidus surface (Figure 8.12) of this quasiternary system were constructed by Parasyuk et al. (1999b). Cu$_2$HgSnSe$_4$ quaternary compound is in thermal equilibrium with all phases of this system at 400°C.

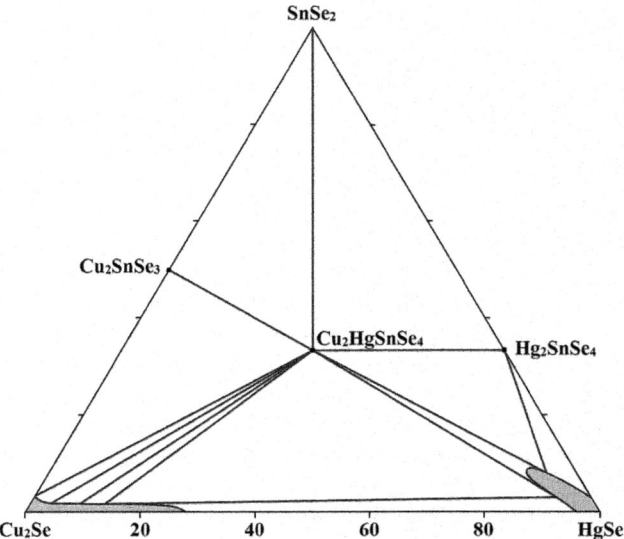

FIGURE 8.11 Isothermal section of the SnSe$_2$–Cu$_2$Se–HgSe quasiternary system at 400°C. (From Parasyuk, O.V., et al., *J. Alloys Compd.*, **287**(1–2), 197, 1999.)

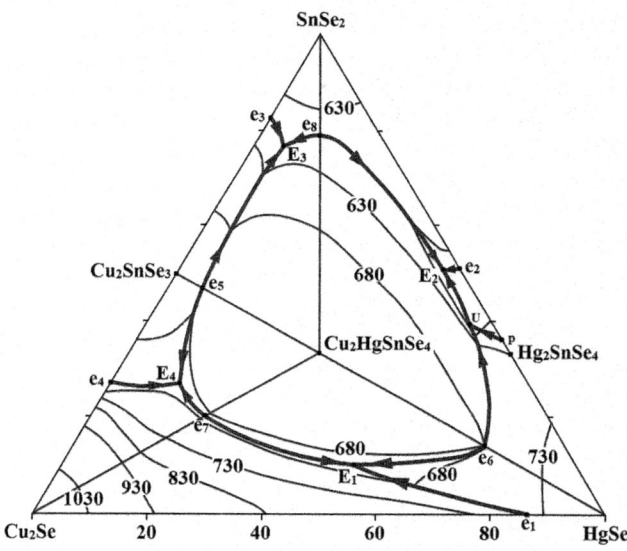

FIGURE 8.12 Liquidus surface of the SnSe$_2$–Cu$_2$Se–HgSe quasiternary system. (From Parasyuk, O.V., et al., *J. Alloys Compd.*, **287**(1–2), 197, 1999.)

The liquidus surface of this system consists of six surfaces of primary crystallization of Cu$_2$Se, Cu$_2$SnSe$_3$, HgSe, SnSe$_2$, β-Hg$_2$SnSe$_4$, and Cu$_2$HgSnSe$_4$. There are four ternary eutectics and one transition point in the system: E$_1$ (645°C) – L ↔ Cu$_2$Se + HgSe + Cu$_2$HgSnSe$_4$; E$_2$ (546°C) – L ↔ Cu$_2$HgSnSe$_4$ + β-Hg$_2$SnSe$_4$ + SnSe$_2$; E$_3$ (915°C) – L ↔ Cu$_2$HgSnSe$_4$ + Cu$_2$SnSe$_3$ + SnSe$_2$; E$_4$ (680°C) – L ↔ Cu$_2$Se + Cu$_2$SnSe$_3$ + Cu$_2$HgSnSe$_4$; and U (667°C) – L + HgSe ↔ Cu$_2$HgSnSe$_4$ + β-Hg$_2$SnSe$_4$. The alloys for the investigations were annealed at 400°C for 250 h and quenched in cold water.

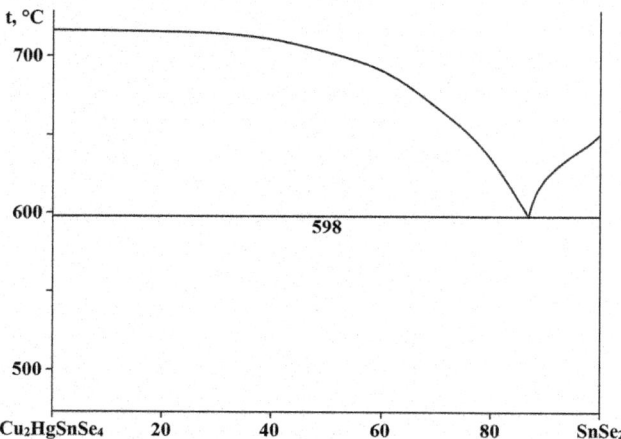

FIGURE 8.10 Phase diagram of the SnSe$_2$–Cu$_2$HgSnSe$_4$ system. (From Parasyuk, O.V., et al., *J. Alloys Compd.*, **287**(1–2), 197, 1999.)

8.52 Tin–Copper–Aluminum–Selenium

The **CuAlSnSe$_4$** quaternary compound, which crystallizes in the tetragonal structure (Goodchild et al. 1981) with the lattice parameters $a = 560.4$, $c = 1095.0$ pm, and an energy gap of 2.34 eV (Hughes et al. 1980; Woolley et al. 1980) [E_g = 1.90 eV (Goodchild et al. 1982)], is formed in the Sn–Cu–Al–Se system.

This compound was obtained by the melt and anneals technique (Goodchild et al. 1981, 1982; Hughes et al. 1980). To prepare it, the required masses of the elements were sealed under vacuum in a quartz ampoule (Hughes et al. 1980). The sample so formed was slowly heated (20–40°C·h^{-1}) up to 1050–1100°C, left at that temperature for 5–10 h, slowly cooled (20–40°C·h^{-1}) down to ca. 500°C and then annealed at that temperature for 10–100 h before cooling to room temperature. The heating was carried out in a rocking furnace, and during the heating up and also during the reaction at high temperature, the sample was rocked to ensure good mixing of the components.

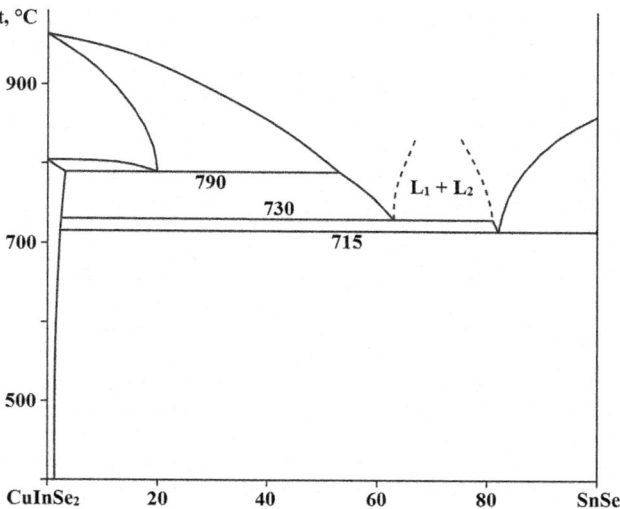

FIGURE 8.13 Phase diagram of the SnSe–CuInSe$_2$ system. (From Allazova, N.M., et al., *Russ. J. Inorg. Chem.*, **56**(10), 1634, 2011.)

8.53 Tin–Copper–Gallium–Selenium

The **CuGaSnSe$_4$** quaternary compound, which melts at 936°C, decomposes at 590°C (López-Rivera et al. 1983), and crystallizes in the tetragonal structure (Goodchild et al. 1981) with the lattice parameters $a = 561.1$, $c = 1098.6$ pm, and an energy gap of 1.42 eV (Hughes et al. 1980; Woolley et al. 1980; Goodchild et al. 1982) [$a = 562.2$, $c = 550.3$ pm, and the calculated and experimental densities of 5.42 and 5.49 g·cm^{-3}, respectively (Hahn and Strick 1967b)], is formed in the Sn–Cu–Ga–Se system. The lattice parameters increase linearly with temperature up to 590°C (López-Rivera et al. 1983). This compound was obtained in the same way as CuAlSnSe$_4$ was prepared (Goodchild et al. 1981, 1982; Hughes et al. 1980). Iodine transport has been used to grow single crystals of CuGaSnSe$_4$ (Woolley et al. 1980).

8.54 Tin–Copper–Indium–Selenium

SnSe–CuInSe$_2$. The phase diagram of this system, constructed through DTA, XRD, metallography, and measuring of the microhardness, is a eutectic type with an immiscibility region (Figure 8.13) (Allazova et al. 2011). The eutectic contains 82 mol% SnSe and crystallizes at 715°C. The immiscibility region takes place within the region of 63–81 mol% SnSe at 730°C, and at 790°C the phase transformation of CuInSe$_2$ in the alloys occurs.

SnSe$_2$–CuInSe$_2$. The phase diagram of this system, constructed through DTA, XRD, metallography, and measuring of the microhardness, is also a eutectic type (Figure 8.14) (Zargarova et al. 1995; Allazova et al. 2011). The eutectic contains 85 mol% SnSe$_2$ and crystallizes at 615°C [87 mol% SnSe$_2$ and crystallizes at 580°C (Zargarova et al. 1995)]. The solubility of SnSe$_2$ in β-CuInSe$_2$ at 790°C is 10 mol% and in α-CuInSe$_2$ is 3 mol% at 790°C and 0.2 mol% at room temperature (Allazova et al. 2011). At 790°C [at 780°C (Zargarova et al. 1995)] the phase transformation of CuInSe$_2$ in the alloys

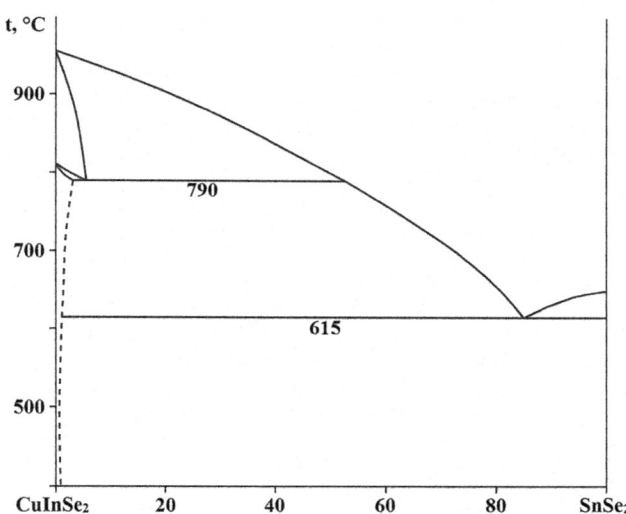

FIGURE 8.14 Phase diagram of the SnSe$_2$–CuInSe$_2$ system. (From Allazova, N.M., et al., *Russ. J. Inorg. Chem.*, **56**(10), 1634, 2011.)

occurs. According to the data of Pamplin et al. (1977), the solubility of SnSe$_2$ in CuInSe$_2$ is not higher than 25 mol%.

CuInSe$_2$–Sn. The phase diagram of this system, constructed through DTA, XRD, metallography, and measuring of the microhardness, is also a eutectic type with an immiscibility region (Allazova et al. 2011). The eutectic contains 96 mol% Sn and crystallizes at 200°C. The immiscibility region takes place at 650°C within the region of 55–95 mol% Sn. **CuInSnSe$_2$** quaternary compound is formed in this system; it melts incongruently at 610°C and exists only within a certain temperature range.

CuInSe$_2$–Se. The phase diagram of this system, constructed through DTA, XRD, metallography, and measuring of the microhardness, is also a eutectic type with an immiscibility region (Allazova et al. 2011). The eutectic contains 99 mol% Se and crystallizes at 210°C. The immiscibility region takes

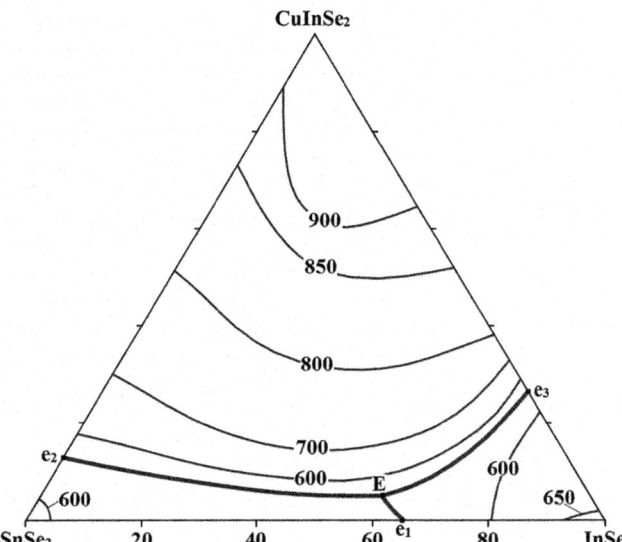

FIGURE 8.15 Liquidus surface of the SnSe₂–CuInSe₂–InSe quasiternary system. (From Azhdarova, D.S., et al., *Neorgan. mater.*, **35**(8), 923, 1999.)

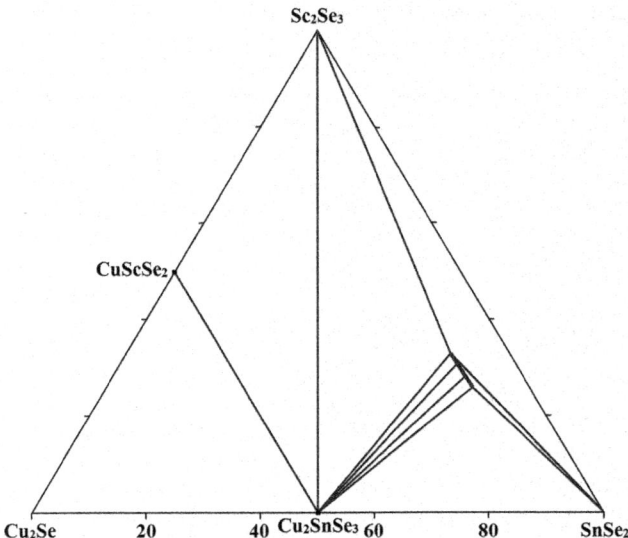

FIGURE 8.16 Isothermal section of the SnSe₂–Cu₂Se–Sc₂Se₃ quasiternary system at 600°C. (From Gulay, L.D., et al., *Pol. J. Chem.*, **82**(5), 1001, 2008.)

place at 800°C within the region of 72–96 mol% Se. At 790°C, the phase transformation of CuInSe₂ in the alloys occurs. The selenium solubility in CuInSe₂ is about 1 mol%.

The equilibrium state of the alloys for the investigations of all mentioned above systems was reached by their annealing at 600°C for 600 h with subsequent slow cooling to room temperature (Allazova et al. 2011).

SnSe₂–CuInSe₂–InSe. Three vertical sections and the liquidus surface (Figure 8.15) of this quasiternary system were constructed by Azhdarova et al. (1999) using the alloys annealed at 400°C for 300 h. Three fields of the primary crystallization exist in the system. CuInSe₂ is characterized by the largest field of primary crystallization. The solubility of SnSe and InSe in CuInSe₂ is ≤ 0.2 mol%. The ternary eutectic contains 36 mol% SnSe₂, 4.5 mol% CuInSe₂, and 59.5 mol% InSe.

CuInSe₂–Sn–Se. The projection of the liquidus surface of this subsystem was constructed by Allazova et al. (2011). The largest crystallization region belongs to CuInSe₂, and the primary crystallization field of Se is degenerated. There are three immiscibility regions on the liquidus surface.

The **CuInSnSe₄** quaternary compound, which crystallizes in the tetragonal structure (Goodchild et al. 1981) with the lattice parameters $a = 570.0$, $c = 1140$ pm, and an energy gap of 0.71 eV (Hughes et al. 1980; Woolley et al. 1980; Goodchild et al. 1982) [$a = 567.0$, $c = 1134$ pm, and the calculated and experimental densities of 5.58 and 5.70 g·cm⁻³, respectively (Hahn and Strick 1967b)], is formed in the Sn–Cu–In–Se system. This compound was obtained in the same way as CuAlSnSe₄ was prepared (Goodchild et al. 1981, 1982; Hughes et al. 1980).

8.55 Tin–Copper–Scandium–Selenium

SnSe₂–Cu₂Se–Sc₂Se₃. The isothermal section of this quasiternary system at 600°C was constructed using the alloys annealed at this temperature for 240 h and is presented in Figure 8.16

(Gulay et al. 2008a). The existence of the $Cu_{0.37}Sc_{0.96-1.18}Sn_{1.19-1.02}Se_4$ solid solution, which crystallizes in the cubic structure with the lattice parameters from $a = 1082.2 ± 0.2$ pm to $a = 1093.5 ± 0.2$ pm, was established. **CuSc₃Sn₃Se₁₁**, which is one of these solid solution compositions, crystallizes in the cubic structure with the lattice parameters from $a = 1088.27 ± 0.4$ pm and a calculated density of 5.3461 g·cm⁻³ (Gulay et al. 2005g). The synthesis of CuSc₃Sn₃Se₁₁ was realized in a shaft furnace with a heating rate of 20°C·h⁻¹. The ampoule with the sample was heated to a maximal temperature of 1150°C and kept at this temperature for 4 h. After that, it was cooled slowly to 600°C with a rate of 10°C·h⁻¹ and annealed at this temperature for 240 h. After annealing, the ampoule with the sample was quenched in cold water.

8.56 Tin–Copper–Yttrium–Selenium

SnSe–Cu₂Se–Y₂Se₃. The isothermal section of this system at 600°C was constructed using the alloys annealed at this temperature for 240 h and is given in Figure 8.17 (Shemet et al. 2005). No quaternary compounds were found.

SnSe₂–Cu₂Se–Y₂Se₃. The isothermal section of this system at 600°C was constructed using the alloys annealed at this temperature for 240 h and is shown in Figure 8.18 (Shemet et al. 2006d). The **CuY₃SnSe₇** quaternary compound, which crystallizes in the hexagonal structure with the lattice parameters $a = 1012.70 ± 0.04$, $c = 635.91 ± 0.01$, and a calculated density of 5.8564 g·cm⁻³ (Gulay et al. 2004c) is formed in this system. This compound was prepared by fusion of the high-purity elements (Gulay et al. 2004c; Shemet et al. 2006d). The calculated amounts of the components were sealed in evacuated quartz ampoule. The synthesis was realized in a shaft furnace with a heating rate of 30°C·h⁻¹. The maximal temperature of the synthesis was about 1150°C. The sample was kept at this temperature for 4 h. After that, it was cooled slowly to 600°C with

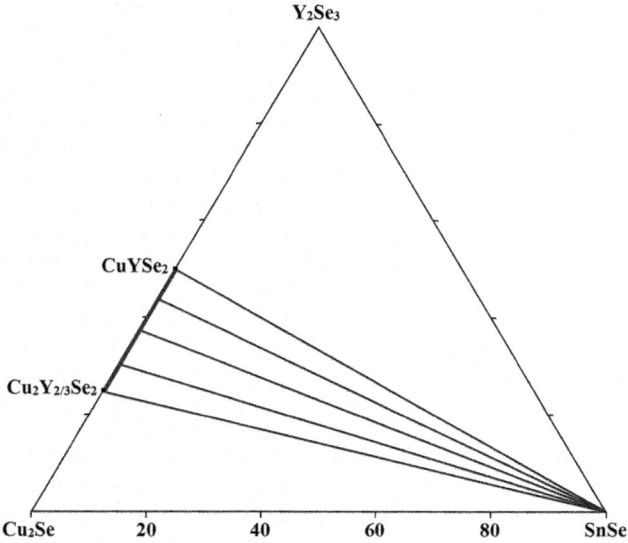

FIGURE 8.17 Isothermal section of the $SnSe–Cu_2Se–Y_2Se_3$ quasiternary system at 600°C. (From Shemet, V.Ya., et al., *Pol. J. Chem.*, **79**(8), 1315, 2005.)

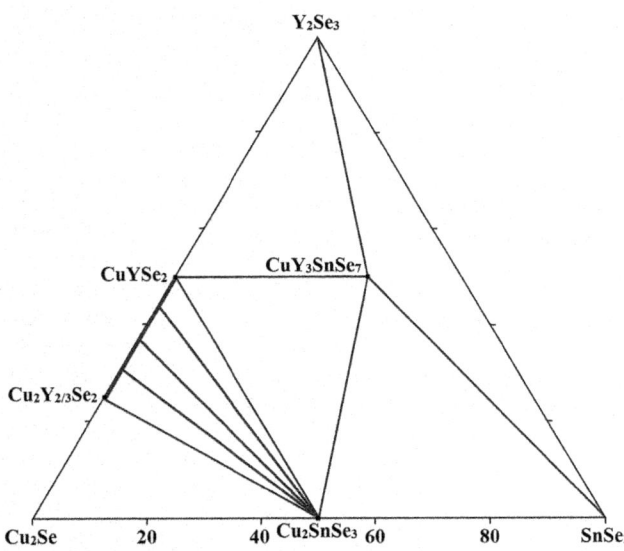

FIGURE 8.18 Isothermal section of the $SnSe_2–Cu_2Se–Y_2Se_3$ quasiternary system at 600°C. (From Shemet, V.Ya., et al., *Pol. J. Chem.*, **80**(6), 943, 2006.)

a rate of 10°C·h⁻¹. The alloys were annealed at this temperature for 240 h. After annealing, the ampoule with the sample was quenched in cold water.

8.57 Tin–Copper–Lanthanum–Selenium

The **CuLa₃SnSe₇** quaternary compound, which crystallizes in the hexagonal structure with the lattice parameters $a = 1072.11 \pm 0.01$, $c = 627.70 \pm 0.01$ pm, and a calculated density of 6.1208 g·cm⁻³, is formed in the Sn–Cu–La–Se system (Gulay and Olekseyuk 2005c). The samples of this compound were prepared by melting high-purity elements. Calculated amounts

of the components were sealed in evacuated quartz ampoules. The synthesis was realized in a shaft furnace. The ampoules with the components were heated with a heating rate of 30°C·h⁻¹ to the maximal temperature, 1150°C. They were kept at this temperature for 4 h. After that, the furnace with the samples was cooled slowly (10°C·h⁻¹) to 600°C. Homogeneous annealing was applied at this temperature for 240 h. After annealing, the ampoules with the samples were quenched in cold water.

8.58 Tin–Copper–Cerium–Selenium

The **CuCe₃SnSe₇** quaternary compound, which crystallizes in the hexagonal structure with the lattice parameters $a = 1063.10 \pm 0.02$, $c = 625.62 \pm 0.01$ pm, and a calculated density of 6.2654 g·cm⁻³ (Gulay and Olekseyuk 2005c) [$a = 1063.7 \pm 0.1$, $c = 625.4 \pm 0.1$, and a calculated density of 6.261 g·cm⁻³ (Gulay et al. 2005a)], is formed in the Sn–Cu–Ce–Se system. This compound was synthesized in the same way as CuLa₃SnSe₇ was prepared.

8.59 Tin–Copper–Praseodymium–Selenium

$SnSe_2–Cu_2Se–Pr_2Se_3$. The isothermal section of this system at 600°C is shown in Figure 8.19 (Strok 2012). The **CuPr₃SnSe₇** quaternary compound, which crystallizes in the hexagonal structure with the lattice parameters $a = 1056.13 \pm 0.04$, $c = 625.32 \pm 0.03$ pm, and a calculated density of 6.3645 g·cm⁻³ (Gulay and Olekseyuk 2005c), is formed in this system. This compound was synthesized in the same way as CuLa₃SnSe₇ was prepared.

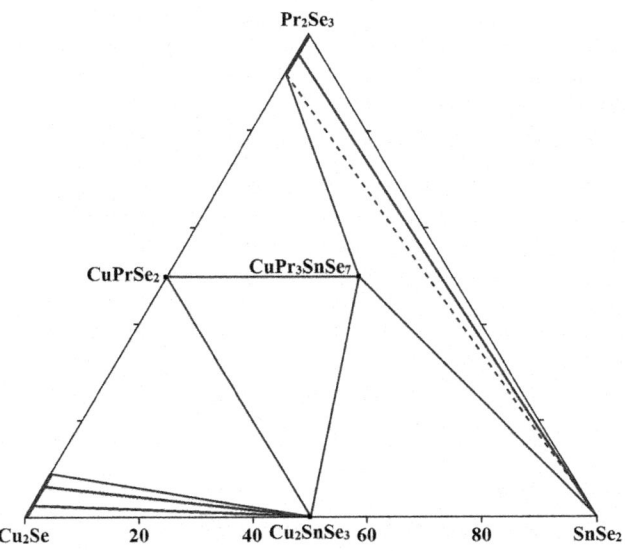

FIGURE 8.19 Isothermal section of the $SnSe_2–Cu_2Se–Pr_2Se_3$ quasiternary system at 600°C. (From Strok, O.M., *Nauk. Visnyk Volyns'k. Nats. Univ. im. Lesi Ukrainky. Ser. Khim. nauky*, (17), 100, 2012.)

8.60 Tin–Copper–Neodymium–Selenium

The **CuNd$_3$SnSe$_7$** quaternary compound, which crystallizes in the hexagonal structure with the lattice parameters $a = 1050.02 \pm 0.04$, $c = 625.23 \pm 0.02$ pm, and a calculated density of 6.4953 g·cm^{-3}, is formed in the Sn–Cu–Nd–Se system (Gulay and Olekseyuk 2005c). This compound was synthesized in the same way as CuLa$_3$SnSe$_7$ was prepared.

8.61 Tin–Copper–Samarium–Selenium

The **CuSm$_3$SnSe$_7$** quaternary compound, which crystallizes in the hexagonal structure with the lattice parameters $a = 1038.09 \pm 0.02$, $c = 628.48 \pm 0.02$ pm, and a calculated density of 6.7148 g·cm^{-3}, is formed in the Sn–Cu–Sm–Se system (Gulay and Olekseyuk 2005c). This compound was synthesized in the same way as CuLa$_3$SnSe$_7$ was prepared.

8.62 Tin–Copper–Gadolinium–Selenium

The **CuGd$_3$SnSe$_7$** quaternary compound, which crystallizes in the hexagonal structure with the lattice parameters $a = 1024.35 \pm 0.02$, $c = 634.09 \pm 0.02$ pm, and a calculated density of 6.9545 g·cm^{-3}, is formed in the Sn–Cu–Gd–Se system (Gulay and Olekseyuk 2005c). This compound was synthesized in the same way as CuLa$_3$SnSe$_7$ was prepared.

8.63 Tin–Copper–Terbium–Selenium

The **CuTb$_3$SnSe$_7$** quaternary compound, which crystallizes in the hexagonal structure with the lattice parameters $a = 1018.94 \pm 0.02$, $c = 636.42 \pm 0.02$ pm, and a calculated density of 7.0319 g·cm^{-3}, is formed in the Sn–Cu–Tb–Se system (Gulay and Olekseyuk 2005c). This compound was synthesized in the same way as CuLa$_3$SnSe$_7$ was prepared.

8.64 Tin–Copper–Dysprosium–Selenium

SnSe$_2$–Cu$_2$Se–Dy$_2$Se$_3$. The isothermal section of this system at 600°C was constructed using the alloy annealed at this temperature for 240 h and is presented in Figure 8.20 (Gulay and Shemet 2012). The **CuDy$_3$SnSe$_7$** quaternary compound, which crystallizes in the hexagonal structure with the lattice parameters $a = 1013.59 \pm 0.03$, $c = 638.09 \pm 0.03$ pm, and a calculated density of 7.1505 g·cm^{-3} (Gulay and Olekseyuk 2005c), is formed in this system. This compound was synthesized in the same way as CuLa$_3$SnSe$_7$ was prepared.

8.65 Tin–Copper–Lead–Selenium

SnSe$_2$–Cu$_2$Se–PbSe. The isothermal section of this system at room temperature was constructed using the alloy annealed at 400°C for 250 h (Kogut et al. 2006). No quaternary compounds

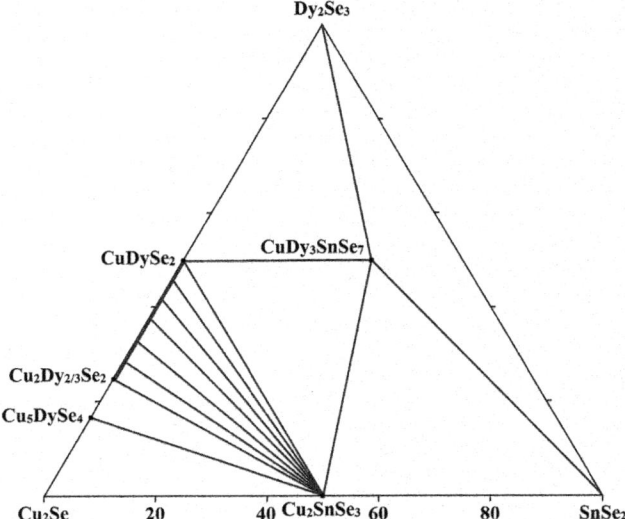

FIGURE 8.20 Isothermal section of the SnSe$_2$–Cu$_2$Se–Dy$_2$Se$_3$ quasiternary system at 600°C. (From Gulay, L.D., and Shemet, V.Ya., *Nauk. Visnyk Volyns'k. Nats. Univ. im. Lesi Ukrainky. Ser. Khim. nauky*, (17), 69, 2012.)

were found in the system. The system is divided into two subsystems by the Cu$_2$SnSe$_3$–PbSe section. Alloys adjacent directly to SnSe$_2$ contain not 3, but 4 phases. This means that the tetrahedration of the Sn–Cu–Pb–Se system does not pass through SnSe$_2$, but through SnSe.

8.66 Tin–Copper–Arsenic–Selenium

Cu$_2$SnSe$_3$–As$_2$Se$_3$. The phase diagram of this system, constructed through DTA and XRD using the alloys annealed at 240°C for 600 h with the next quenching in the 25 mass% NaCl aqueous solution, is a eutectic type Figure 8.21 (Klymovych et al. 2020b). The eutectic contains 4 mol%

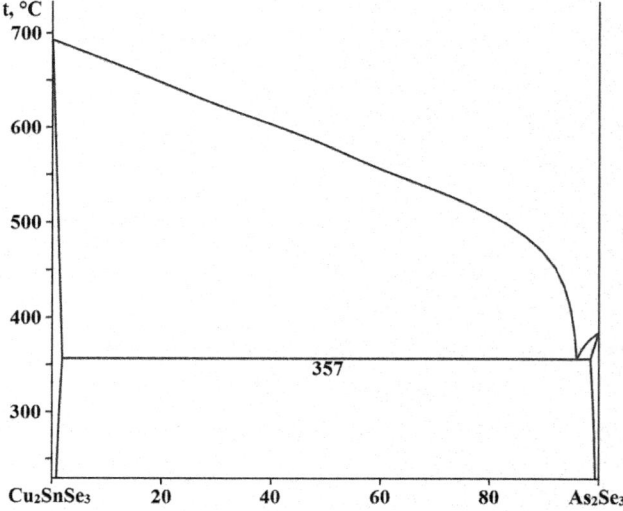

FIGURE 8.21 Phase diagram of the Cu$_2$SnSe$_3$–As$_2$Se$_3$ system. (From Klymovych, O., et al., *Visnyk Odes'k. Nats. Univ.*, **25**[1(73)], 31, 2020.)

As$_2$Se$_3$ and crystallizes at 357°C. The mutual solubility of starting components is negligible.

SnSe$_2$–Cu$_2$Se–As$_2$Se$_3$. The isothermal section of this system at 240°C was constructed using the alloy annealed at this temperature for 600 h with the next quenching in water (Klymovych et al. 2013, 2020b). No quaternary compounds were found in the system. The system is divided into three subsystems by the Cu$_2$SnSe$_3$–As$_2$Se$_3$ and Cu$_2$SnSe$_3$–CuAsSe$_2$ sections.

Four vertical sections and the liquidus surface (Figure 8.22) of the SnSe$_2$–Cu$_2$Se–As$_2$Se$_3$ were constructed by Klymovych et al. (2020b). Five fields of the primary crystallization of Cu$_2$Se, Cu$_2$SnSe$_3$, SnSe$_2$, As$_2$Se$_3$, and CuAsSe$_2$ exist on the liquidus surface. These fields are divided by 8 monovariant curves and 9 nonvariant points, three of which are ternary: E$_1$ (327°C) – L ↔ As$_2$Se$_3$ + CuAsSe$_2$ + Cu$_2$SnSe$_3$; E$_2$ (337°C) – L ↔ As$_2$Se$_3$ + Cu$_2$SnSe$_3$+ SnSe$_2$; and U (427°C) – L + Cu$_2$Se ↔ CuAsSe$_2$ + Cu$_2$SnSe$_3$.

The glass-forming region in this system is presented in Figure 8.23 (Klymovych et al. 2013). The glass transition temperature for such alloys is within the interval from 182°C to 231°C.

8.67 Tin–Copper–Antimony–Selenium

SnSe–CuSbSe$_2$. The phase diagram of this system, constructed through DTA and XRD, is a eutectic type (Figure 8.24) (Ismayilova et al. 2018, 2022). The eutectic contains 63 mol% CuSbSe$_2$ and crystallizes at 427°C. The solubility based on SnSe and CuSbSe$_2$ reaches ~22 and 3 mol% at room temperature and 25 and 5 mol% at the eutectic temperature, respectively. The **CuSnSbSe$_3$** quaternary compound, which melts incongruently at 450°C and exist in a very narrow temperature range (427–450°C), is formed in this system. The ingots for the investigations were annealed at 400°C for 500 h.

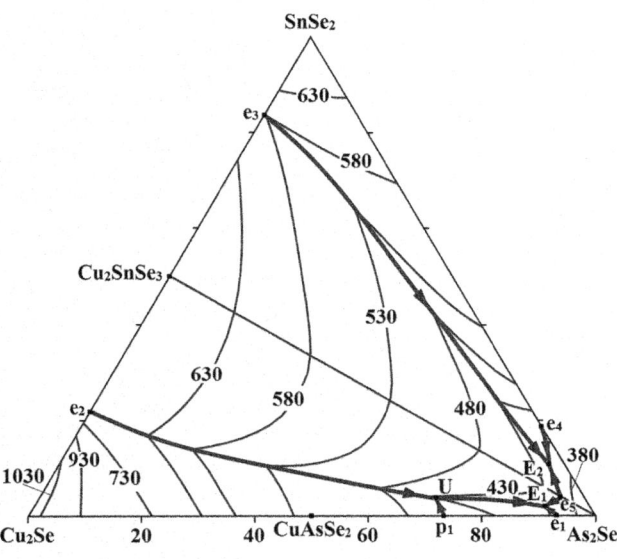

FIGURE 8.22 Liquidus surface of the SnSe$_2$–Cu$_2$Se–As$_2$Se$_3$ quasiternary system. (From Klymovych, O., et al., *Visnyk Odes'k. Nats. Univ.*, **25**[1(73)], 31, 2020.)

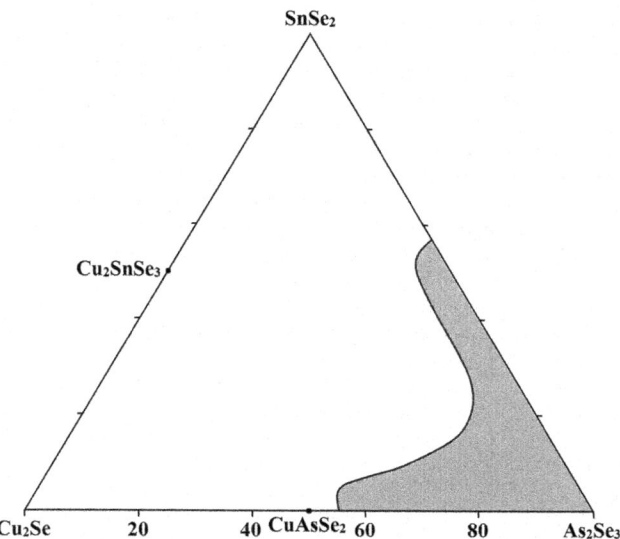

FIGURE 8.23 Glass-forming region in the SnSe$_2$–Cu$_2$Se–As$_2$Se$_3$ system. (From Klymovych, O., et al., *Nauk. Visnyk Skhidnoyevrop. Nats. Univ. im. Lesi Ukrainky. Ser. Khim. nauky*, [23(272)], 89, 2013.)

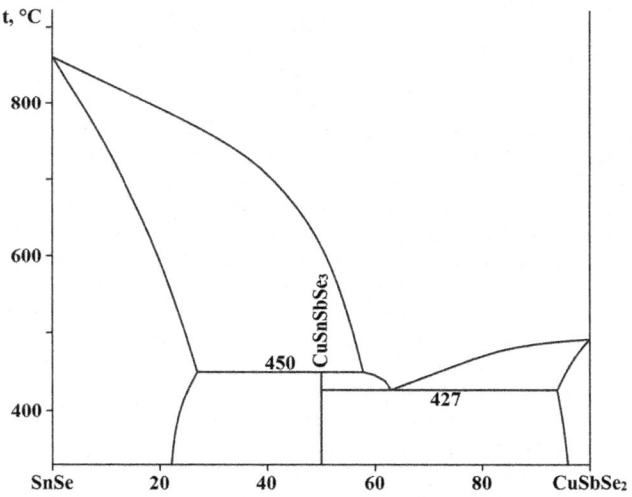

FIGURE 8.24 Phase diagram of the SnSe–CuSbSe$_2$ system. (From Ismayilova, E.N., et al., *Azerb. Chem. J.*, (1), 73, 2022.)

SnSe–Cu$_3$SbSe$_3$. This system is a nonquasibinary section as Cu$_3$SbSe$_3$ melts incongruently (Ismayilova and Mashadieva 2018). SnSe dissolves ~3 mol% Cu$_3$SbSe$_3$. This section is stable below the solidus temperatures.

Cu$_2$SnSe$_3$–CuSbSe$_2$. The phase diagram of this system, constructed through DTA, XRD, and metallography, is a eutectic type (Figure 8.25) (Ostapyuk et al. 2009a). The eutectic contains 93 mol% CuSbSe$_2$ and crystallizes at 457°C. The ingots for the investigations were annealed at 350°C for 600 h.

Cu$_2$SnSe$_3$–Sb$_2$Se$_3$. The phase diagram of this system, constructed through DTA, XRD, and metallography, is also a eutectic type (Figure 8.26) (Ostapyuk et al. 2009a). The eutectic contains 72 mol% Sb$_2$Se$_3$ and crystallizes at 496°C. The ingots for the investigations were annealed at 350°C for 600 h.

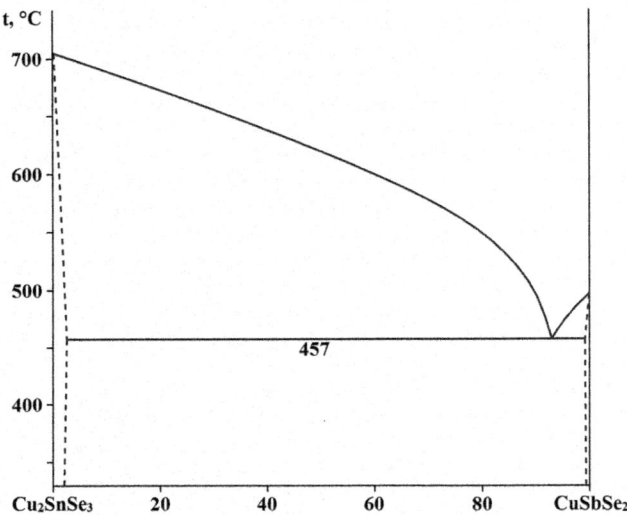

FIGURE 8.25 Phase diagram of the Cu$_2$SnSe$_3$–CuSbSe$_2$ system. (From Ostapyuk, T.A., et al., *Chem. Met. Alloys*, **2**(3–4), 164, 2009.)

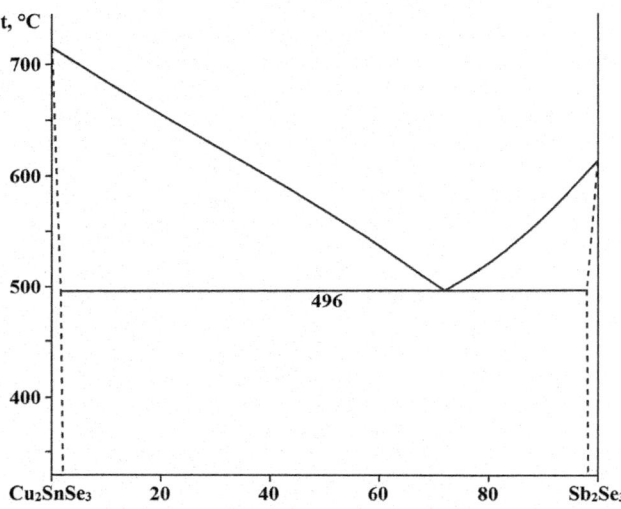

FIGURE 8.26 Phase diagram of the Cu$_2$SnSe$_3$–Sb$_2$Se$_3$ system. (From Ostapyuk, T.A., et al., *Chem. Met. Alloys*, **2**(3–4), 164, 2009.)

SnSe–Cu$_2$Se–CuSbSe$_2$. The isothermal section of this system at room temperature is presented in Figure 8.27 (Ismayilova et al. 2019, 2022). The concentration triangle includes 7 two-phase and 5 three-phase regions. There are seven two-phase and five three-phase regions on the isothermal section. The liquidus surface (Figure 8.28) consists of the fields of the (SnSe), (Cu$_2$Se), (Sb$_2$Se$_3$) Cu$_3$SbSe$_3$, (CuSbSe$_2$) Sn$_2$Sb$_2$Se$_5$, ε-phase, and CuSnSbSe$_3$ primary crystallization. The largest are the fields of (Cu$_2$Se), (Cu$_2$Se), and (Sb$_2$Se$_3$) primary crystallization. There are two ternary eutectics and five transition points in the system: E$_1$ (402°C) – L ↔ CuSnSbSe$_3$ + (Cu$_2$Se) + (CuSbSe$_2$); E$_2$ (407°C) – L ↔ Cu$_3$SbSe$_3$ + CuSnSbSe$_3$ + (CuSbSe$_2$); U$_1$ (442°C) – L + (SnSe) ↔ Sn$_2$Sb$_2$Se$_5$ + CuSnSbSe$_3$; U$_2$ (414°C) – L + Sn$_2$Sb$_2$Se$_5$ ↔ ε + CuSnSbSe$_3$; U$_3$ (462°C) – L + (Sb$_2$Se$_3$) ↔ ε + (CuSbSe$_2$); U$_4$ (454°C) – L + (Cu$_2$Se) ↔ (SnSe) + Cu$_3$SbSe$_3$; and U$_5$ (432°C) – L + (SnSe) ↔ Cu$_3$SbSe$_3$ + CuSnSbSe$_3$. Some

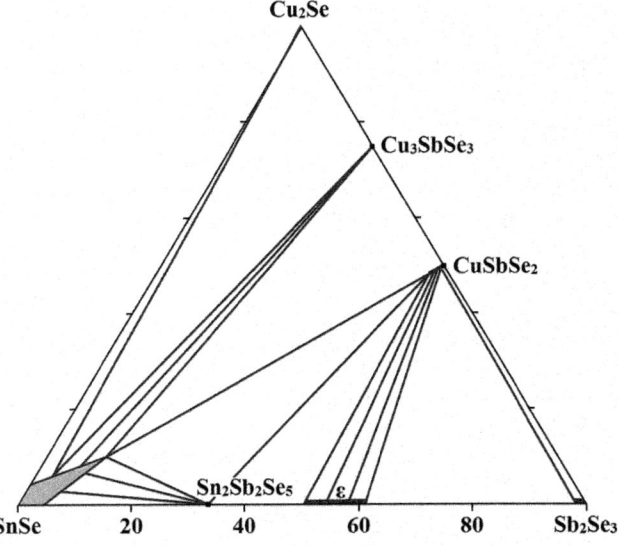

FIGURE 8.27 Isothermal section of the SnSe–Cu$_2$Se–Sb$_2$Se$_3$ quasiternary system at room temperature. (From Ismayilova, E.N., et al., *Azerb. Chem. J.*, (1), 73, 2022.)

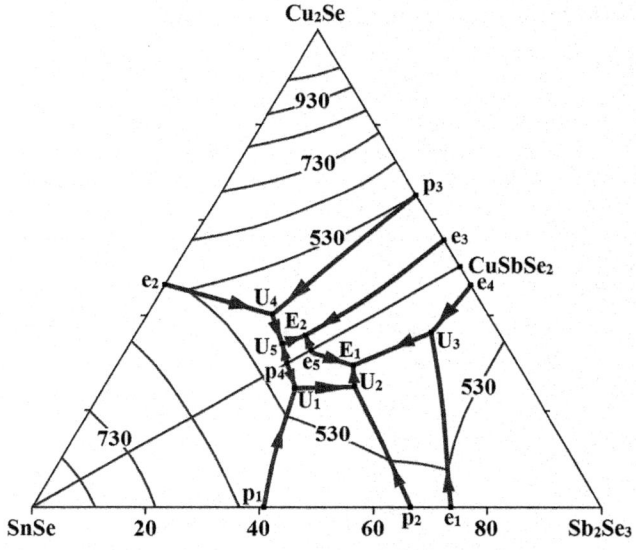

FIGURE 8.28 Liquidus surface of the SnSe–Cu$_2$Se–Sb$_2$Se$_3$ quasiternary system. (From Ismayilova, E.N., et al., *Azerb. Chem. J.*, (1), 73, 2022.)

vertical sections of this quasiternary system were also constructed. Large regions of the solid solutions based on SnSe and CuSbSe$_2$ were revealed in the system. The alloys for the investigations were annealed at 380°C for 500–800 h.

SnSe$_2$–Cu$_2$Se–Sb$_2$Se$_3$. The isothermal section of this system at 350°C is given in Figure 8.29 (Ostapyuk et al. 2009a). No quaternary compounds were found in the system. The extent of solid solutions based on binary and ternary compounds is negligible. There are 3 binary equilibria in the system, which divide it on 4 regions of three-phase equilibria. The liquidus surface (Figure 8.30) consists of the fields of Cu$_2$Se, Cu$_2$SnSe$_3$, SnSe$_2$, Sb$_2$Se$_3$, CuSbSe$_2$, and Cu$_3$SbSe$_3$ primary crystallization. The largest is the field of Cu$_2$SnSe$_3$ primary crystallization.

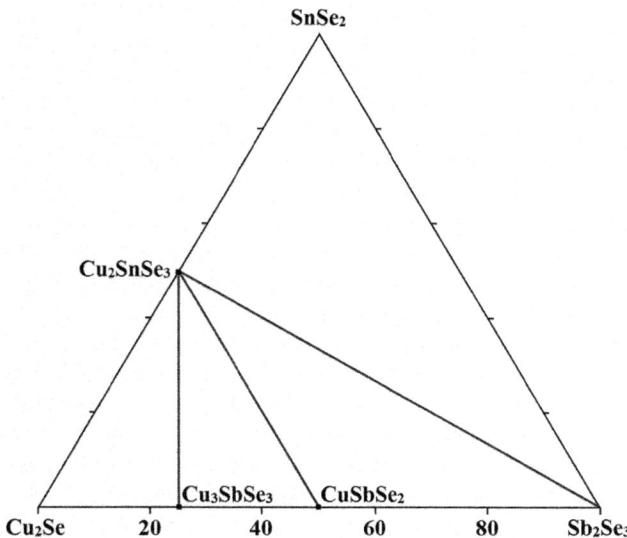

FIGURE 8.29 Isothermal section of the $SnSe_2-Cu_2Se-Sb_2Se_3$ quasiternary system at 600°C. (From Ostapyuk, T.A., et al., *Chem. Met. Alloys*, 2(3–4), 164, 2009.)

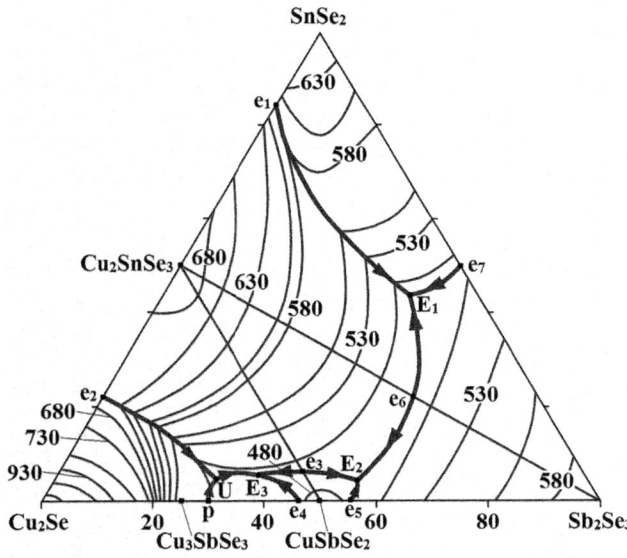

FIGURE 8.30 Liquidus surface of the $SnSe_2-Cu_2Se-Sb_2Se_3$ quasiternary system. (From Ostapyuk, T.A., et al., *Chem. Met. Alloys*, 2(3–4), 164, 2009.)

There are three ternary eutectics and one transition point in this quasiternary system: E_1 (472°C) – L $\leftrightarrow$ $SnSe_2$ + Cu_2SnSe_3 + Sb_2Se_3; E_2 (438°C) – L $\leftrightarrow$ Cu_2SnSe_3 + $CuSbSe_2$ + Sb_2Se_3; E_3 (442°C) – L $\leftrightarrow$ Cu_2SnSe_3 + $CuSbSe_2$ + Cu_3SbSe_3; and U (460°C) – L + Cu_2Se $\leftrightarrow$ Cu_2SnSe_3 + Cu_3SbSe_3.

8.68 Tin–Copper–Chromium–Selenium

$Cu_2SnSe_3-Cr_2Se_3$. The phase diagram of this system, constructed through DTA, XRD, metallography, and measuring of the microhardness, is presented in Figure 8.31 (Babitsyna and

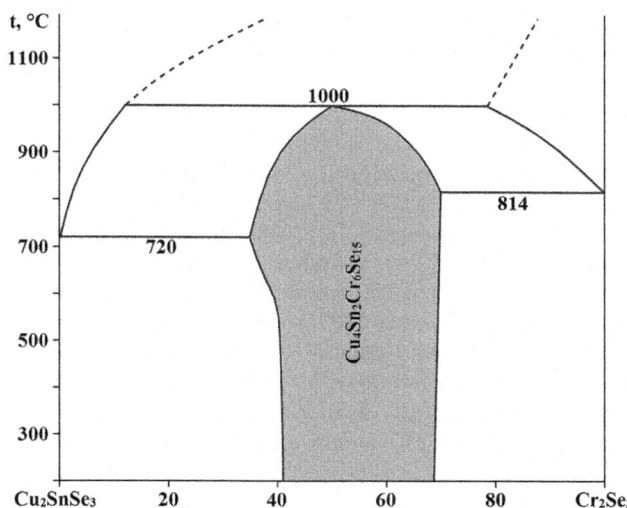

FIGURE 8.31 Phase diagram of the $Cu_2SnSe_3-Cr_2Se_3$ system. (From Babitsyna, A.A., and Novotortsev V.M., *Zhurn. neorgan. khimii*, 31(7), 1825, 1986.)

Novotortsev 1986). The $Cu_4Sn_2Cr_6Se_{15}$ quaternary compound, which melts incongruently at 1000°C, has a wide homogeneity region, and crystallizes in the cubic structure with the lattice parameter $a = 1124$ pm, is formed in this system. There is an enough wide solid solution region based on β-Cr_2Se_3. Based on α-Cr_2Se_3, no noticeable solubility was found. The alloys for the investigations were annealed at 630°C for 300 h.

One more quaternary compound, $CuCrSnSe_4$, is formed in the Sn–Cu–Cr–Se system. It crystallizes in the cubic structure with the lattice parameter $a = 1067.2$ pm and the calculated and experimental densities of 6.01 and 5.85 g·cm^{-3}, respectively (Mähl et al. 1984b) [$a = 1070$ pm and the calculated and experimental densities of 5.962 and 5.94 g·cm^{-3}, respectively (Hahn and Strick 1967a; Strick et al. 1968); $a = 1068.5$ pm and an energy gap of 0.5 eV (Riedel and Morlock 1978)]. It was synthesized from the elements at 800°C for 4 days (Hahn and Strick 1967a; Strick et al. 1968). Single crystals of $CuCrSnSe_4$ were grown by chemical transport using mixtures of $AlCl_3+I_2$ as transporting agents (Mähl et al. 1984b).

8.69 Tin–Copper–Manganese–Selenium

The $Cu_2MnSnSe_4$ quaternary compound and $Cu_{2-x}Mn_xSnSe_3$ solid solution are formed in the Sn–Cu–Mn–Se system. $Cu_2MnSnSe_4$ melts congruently at 694°C (Quintero et al. 2014b) [at 725°C (Schäfer and Nitsche 1977)] and crystallizes in the tetragonal structure with the lattice parameters $a = 576.40 \pm 0.02$ and $c = 1136.47 \pm 0.04$ pm (Song et al. 2016) [$a = 574.4$ and $c = 1142.3$ pm (Schäfer and Nitsche 1974, 1977); $a = 574.4$, $c = 1137.1$ pm, and the calculated and experimental densities of 5.43 and 5.42 g·cm^{-3}, respectively (Guen et al. 1979; Guen and Glaunsinger 1980); $a = 577.4$ and $c = 1142.3$ pm (Lamarche et al. 1991); $a = 576.23$, $c = 1136.82$ pm, and a calculated density of 5.42 g·cm^{-3} (Quintero et al. 2014b)].

This compound was produced by the melt and anneals technique (Lamarche et al. 1991; Quintero et al. 2010, 2014b; Song et al. 2016). In a typical synthesis of $Cu_2MnSnSe_4$, the constituents elements were sealed under vacuum ($\approx 10^{-3}$ Pa) in a small quartz ampoule, and then they were heated up to 200°C and kept for about 1–2 h (Quintero et al. 2010, 2014b). After that, the temperature was raised to 500°C using a rate of 40°C·h^{-1} and held at this temperature for 14 h. Then the sample was heated from 500°C to 800°C at a rate of 30°C·h^{-1} and kept at this temperature for another 14 h. The temperature was raised to 1150°C at 60°C·h^{-1}, and the mixture was melted at this temperature for about 1 h. The furnace temperature was brought slowly (4°C·h^{-1}) down to 600°C, and the sample was annealed at this temperature for 1 month. Then, the sample was slowly cooled to room temperature with a rate of about 2°C·h^{-1}.

According to the data of Song et al. (2018), $Cu_2MnSnSe_4$ can include the excess Cu atoms, which enter into the Mn-sites without forming the secondary phases. The obtained solid solution $Cu_{2+\delta}Mn_{1-\delta}SnSe_4$ ($\delta \leq 0.1$) also crystallizes in the tetragonal structure with the lattice parameters $a = 576.4 \pm 0.8$ and $c = 1136.6 \pm 0.2$ pm for $\delta = 0$, $a = 576.1 \pm 0.5$ and $c = 1136.6 \pm 0.1$ pm for $\delta = 0.025$, $a = 576.1 \pm 0.2$ and $c = 1136.7 \pm 0.1$ pm for $\delta = 0.05$, $a = 576.0 \pm 0.8$ and $c = 1136.6 \pm 0.5$ pm for $\delta = 0.075$, and $a = 576.0 \pm 0.9$ and $c = 1137.0 \pm 0.5$ pm for $\delta = 0.1$. To synthesize this phase, Cu, Mn, Sn, and Se were weighed out in the needed atomic ratio and then sealed in evacuated quartz ampoules. The sealed ampoules were heated up to 600°C at a rate of 50°C·h^{-1} and then kept at 600°C for 14 h. Subsequently, they were heated to 1150°C at a rate of 100°C·h^{-1} and remained at this temperature for 72 h. Then the sealed ampoules were naturally cooled to room temperature. The obtained ingots were manually ground into fine powders and then cold pressed into disks. These disks were sealed in evacuated quartz ampoules again and annealed at 600°C for 7 days. The final products were manually ground into fine powders and then sintered by spark plasma sintering at 500°C for 10 min under an axis pressure of 60 MPa.

The $Cu_{2-x}Mn_xSnSe_3$ solid solution crystallizes in the cubic structure with the lattice parameters $a = 568.7 \pm 0.6$, 569.5 ± 0.4, 568.9 ± 0.4, 569.7 ± 0.3, and 569.5 ± 0.4 pm for $x = 0$, 0.05, 0.10, 0.15, and 0.20 (Gurukrishna et al. 2022). Polycrystalline bulk samples of $Cu_{2-x}Mn_xSnSe_3$ were synthesized via the conventional solid-state reaction route. The powders of Cu, Mn, Sn, and Se were mixed according to the nominal stoichiometry and ground well in an agate mortar for 2 h. The powder was consolidated into pellets using uniaxial hydraulic press applying a load of about 60 kPa. The obtained pellets were sealed under a high vacuum (10^{-4} Pa) in quartz ampoules. Sintering of pellets was carried out at 500°C for 72 h and allowed to cool naturally. The process of grinding, pelletizing, and vacuum sealing of the powder was repeated, and samples were annealed at 350°C for 12 h to promote homogeneity and densified microstructure.

8.70 Tin–Copper–Iron–Selenium

Two quaternary compounds, $Cu_2FeSnSe_4$ and $Cu_6Fe_4Sn_{12}Se_{32}$, are formed in the Sn–Cu–Fe–Se system. $Cu_2FeSnSe_4$ melts incongruently at 772°C (Quintero et al. 2014b) [at 686°C (Schäfer and

Nitsche 1977); at 678.3°C (Quintero et al. 1999)] and crystallizes in the tetragonal structure with the lattice parameters $a = 570.40 \pm 0.01$ and $c = 1128.31 \pm 0.04$ pm (Song et al. 2016) [$a = 566.4$, $c = 1133$ pm, and the calculated and experimental densities of 5.63 and 5.60 g·cm^{-3}, respectively (Hahn and Schulze 1965; $a = 569.3$ and $c = 1133.8$ pm (Schäfer and Nitsche 1974, 1977); $a = 569.4 \pm 0.2$ and $c = 1128.6 \pm 0.4$ pm (Infante et al. 1997); $a = 572.0$ and $c = 1129.2$ pm (Quintero et al. 1999); $a = 570.54$, $c = 1127.10$ pm, and a calculated density of 5.59 g·cm^{-3} (Quintero et al. 2014b); $a = 568.98 \pm 0.01$, $c = 1131.97 \pm 0.01$ pm, a calculated density of 5.60 g·cm^{-3}, and an energy gap of 0.18 eV for $Cu_{2.1}Fe_{0.9}SnSe_4$ composition, and $a = 569.27 \pm 0.01$, $c = 1129.81 \pm 0.01$ pm, a calculated density of 5.61 g·cm^{-3}, and an energy gap of 0.25 eV for $Cu_{2.2}Fe_{0.8}SnSe_4$ composition (Dong et al. 2015b)]. This compound was obtained in the same way as $Cu_2MnSnSe_4$ was prepared (Infante et al. 1997; Quintero et al. 2010, 2014b; Song et al. 2016). Single crystals of $Cu_2FeSnSe_4$ were grown by chemical transport reactions using iodine as the transport agent (Infante et al. 1997).

$Cu_6Fe_4Sn_{12}Se_{32}$ crystallizes in the cubic structure with the lattice parameter $a = 1075.84 \pm 0.14$ pm and a calculated density of 6.079 g·cm^{-3} for $Cu_{6.16}Fe_{4.13}Sn_{11.87}Se_{32}$ composition (Suekuni et al. 2013). A single crystal of this compound was grown using a melt-growth method. The pure elements were mixed in an atomic ratio of Cu/Fe/Sn/Se = 6:4:12:32. The mixture was sealed in an evacuated quartz tube, heated at 650°C, held for 3 h, and cooled slowly to 550°C for 100 h. Then it was cooled naturally to room temperature after switching off the furnace.

8.71 Tin–Copper–Cobalt–Selenium

The $Cu_2CoSnSe_4$ quaternary compound, which melts incongruently at 845°C (Quintero et al. 2014b) and has apparently two polymorphic modifications, is formed in the Sn–Cu–Co–Se system. First modification of this compound crystallizes in the tetragonal structure with the lattice parameters $a = 566.65 \pm 0.03$ and $c = 1132.93 \pm 0.01$ pm (Song et al. 2016) [$a = 566.76 \pm 0.02$, $c = 1131.46 \pm 0.09$ pm, and a calculated density of 5.6699 g·cm^{-3} (Gulay et al. 2004a); $a = 567.11 \pm 0.01$, $c = 1132.98 \pm 0.05$ pm, and a calculated density of 5.66 g·cm^{-3} (Quintero et al. 2014a); $a = 567.28$, $c = 1132.20$ pm, and a calculated density of 5.66 g·cm^{-3} (Quintero et al. 2014b)]. Second modification of $Cu_2CoSnSe_4$ crystallizes in the cubic structure with the lattice parameter $a = 569.7$ pm (Schäfer and Nitsche 1974). This compound was synthesizes in the same way as $Cu_2MnSnSe_4$ was prepared (Gulay et al. 2004a; Quintero et al. 2014a,b; Song et al. 2016).

8.72 Tin–Copper–Nickel–Selenium

The $Cu_2NiSnSe_4$ quaternary compound, which crystallizes in the cubic structure with the lattice parameter $a = 570.5$ pm, is formed in the Sn–Cu–Ni–Se system (Schäfer and Nitsche 1974). Single crystals of the title compound were grown through chemical transport reactions using iodine as a transport agent.

8.73 Tin–Silver–Strontium–Selenium

The **Ag₂SrSnSe₄** quaternary compound, which crystallizes in the orthorhombic structure with the lattice parameters $a = 719.3$, $b = 765.7$, $c = 803.4$ pm, and an energy gap of 0.66 eV, is formed in the Sn–Ag–Sr–Se system (Zhu et al. 2017).

8.74 Tin–Silver–Barium–Selenium

The **Ag₂BaSnSe₄** quaternary compound, which crystallizes in the orthorhombic structure with the lattice parameters $a = 710.1 \pm 1.0$, $b = 746.9 \pm 1.0$, $c = 830.2 \pm 1.1$ pm, a calculated density of 5.941 g·cm⁻³, and an energy gap of 1.42 eV (Nian et al. 2018) [$a = 711.54 \pm 0.07$, $b = 749.94 \pm 0.07$, $c = 833.75 \pm 0.08$ pm, a calculated density of 5.88 g·cm⁻³, and an energy gap of 0.24 eV (Assoud et al. 2005); an energy gap of 0.77 eV (Zhu et al. 2017)], is formed in the Sn–Ag–Ba–Se system.

The conventional high-temperature solid-state method was used to synthesize the title compound (Nian et al. 2018). A stoichiometric mixture of Ag, Ba, Sn, and Se (molar ratio 2:1:1:4) was loaded into a silica tube and sealed under a high vacuum of 10⁻³ Pa. After that, the tube was moved into a furnace. The furnace was programmed by the following steps: heated from room temperature to 600°C in 30 h and kept at this temperature for 40 h; then, heated to 1000°C in 20 h and left at this temperature for 100 h; finally, cooled to 400°C at a rate of 5°C·h⁻¹ and then cooled to room temperature in 10 h. After the reaction, the product was washed with DMF to remove the unreacted sulfur and other by-products. Many red crystals of Ag₂BaSnSe₄ were gained that can be stable in the air for several months. Since the Ba metal is easily oxidized in the air, an Ar-filled glove box was used to avoid the effects of oxygen and moisture in the preparation processes.

8.75 Tin–Silver–Zinc–Selenium

Ag₈SnSe₆–ZnSe. The phase diagram of this system, constructed through DTA, XRD, and metallography, is a eutectic type (Figure 8.32) (Piskach et al. 2006). The eutectic contains 37 mol% ZnSe and crystallizes at 715°C. Solubility of ZnSe in Ag₈SiSe₆ at the eutectic temperature is 33 mol%.

8.76 Tin–Silver–Cadmium–Selenium

"Ag₂SnSe₃"–CdSe. This section is a nonquasibinary one of the Sn–Ag–Cd–Se quaternary system (Olekseyuk et al. 1997f). Ag₂CdSnSe₄ quaternary compound is formed in this section. It melts incongruently at 635°C and crystallizes in the orthorhombic structure with the lattice parameters $a = 426.43 \pm 0.03$, $b = 731.75 \pm 0.04$, $c = 698.48 \pm 0.05$ pm, and a calculated density of 5.821 ± 0.03 g·cm⁻³ (Parasyuk et al. 2002c) [$a = 426.2 \pm 0.5$, $b = 731.4 \pm 0.5$, and $c = 697.9 \pm 0.5$ pm (Parthé

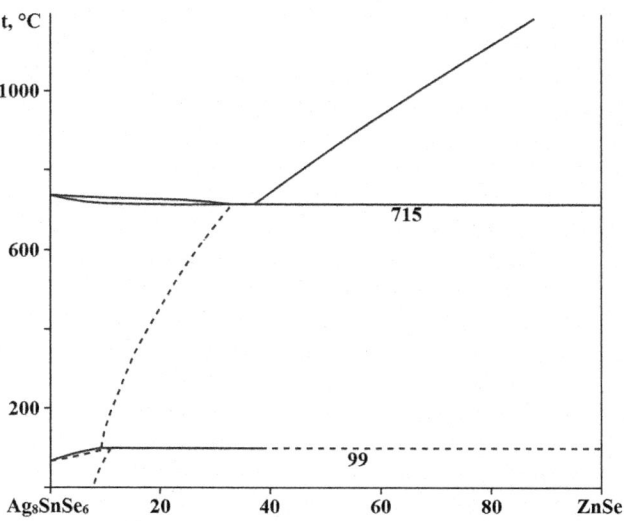

FIGURE 8.32 Phase diagram of the Ag₈SnSe₆–ZnSe system. (From Piskach, et al., *J. Alloys Compd.*, **421**(1–2), 98, 2006.)

et al. 1969) $a = 426.3 \pm 0.3$, $b = 737.2 \pm 0.2$, and $c = 681.1 \pm 0.2$ pm (Olekseyuk et al. 1997f)]. The values of the standard thermodynamic functions of Ag₂CdSnSe₄ were calculated by Moroz and Prokhorenko (2015) using EMF measurements: $\Delta G^0_{f,298} = -310.9 \pm 2.1$ kJ·M⁻¹, $\Delta H^0_{f,298} = -300.8 \pm 1.4$ kJ·M⁻¹ and $\Delta S^0_{f,298} = 33.9 \pm 2.9$ J·(M·K)⁻¹.

The homogeneity region of Ag₂CdSnSe₄ at 480°C is within the interval from 43.5 to 50 mol% CdSe. The solid solutions based on CdSe do not exceed 2 mol% "Ag₂SnSe₃" at 480°C. Alloys in the region 0–43.5 mol% CdSe are three-phase and contain Ag₂CdSnSe₄, Ag₈SnSe₆, and SnSe₂. In the region of 50–98 mol% CdSe, only two phases (CdSe and Ag₂CdSnSe₄) are in equilibrium (Olekseyuk et al. 1997f). This system was investigated through DTA, XRD, and metallography, and the alloys were annealed at 480°C for 400 h.

Ag₈SnSe₆–CdSe. The phase diagram of this system, constructed through DTA, XRD, and metallography, is a eutectic type (Figure 8.33) (Parasyuk et al. 2002c, Piskach et al. 2006). The eutectic composition and temperature are 65 mol% CdSe and 703°C, respectively. The solubility of CdSe in Ag₈SnSe₆ at the eutectic temperature is 42 mol%. The alloys for the investigations were annealed at 400°C for 250 h, with the next quenching in cold water.

SnSe₂–Ag₂Se–CdSe. Isothermal section of this quasiternary system at 400°C is shown in Figure 8.34 (Parasyuk et al. 2002c). The Ag₂CdSnSe₄ compound is only quaternary intermediate phase in this system. The limit of the solid solution region of Ag₈SnSe₆ is localized between 5 and 10 mol% CdSe and the solid solution range of CdSe is less than 2.5 mol%.

8.77 Tin–Silver–Mercury–Selenium

"Ag₂SnSe₃"–HgSe. This system is a nonquasibinary section of the Sn–Ag–Hg–Se quaternary system (Parasyuk 1999). The **Ag₂HgSnSe₄** quaternary compound is formed in the system. It melts congruently at 557°C, is characterized by a homogeneity region within 47–52 mol% HgSe at 400°C (Parasyuk

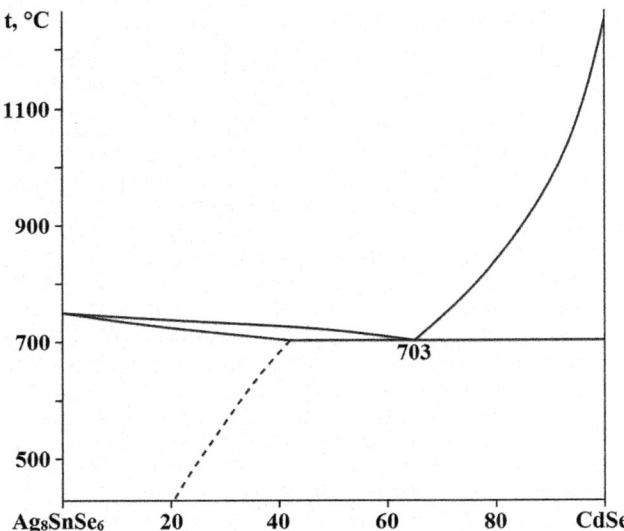

FIGURE 8.33 Phase diagram of the Ag₈SnSe₆–CdSe system. (From Parasyuk, O.V., et al., *J. Alloys Compd.*, **335**(1–2), 176, 2002.)

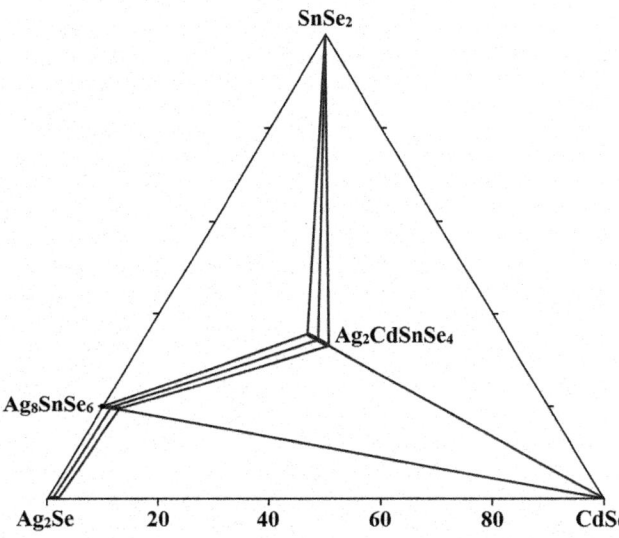

FIGURE 8.34 Isothermal section of the SnSe₂–Ag₂Se–CdSe quasiternary system at 400°C. (From Parasyuk, O.V., et al., *J. Alloys Compd.*, **335**(1–2), 176, 2002.)

1999), and crystallizes in the orthorhombic structure with the lattice parameters $a = 846.1 \pm 0.1$, $b = 734.0 \pm 0.1$, $c = 699.01 \pm 0.06$ pm, and a calculated density of 6.523 ± 0.002 g·cm⁻³ (Parasyuk et al. 2002b) [$a = 849.4$, $b = 731.4$, and $c = 697.2$ pm (Haeuseler and Himmrich 1989)]. In the homogeneity region, the lattice parameters decrease from $a = 850.2 \pm 0.3$, $b = 736.2 \pm 0.3$, and $c = 700.7 \pm 0.3$ pm to $a = 848.0 \pm 0.3$, $b = 732.3 \pm 0.3$, and $c = 699.6 \pm 0.4$ pm (Parasyuk 1999).

A new quaternary phase *X* with composition lying in the interval 58–60 mol% HgSe was found in Parasyuk (1999). This phase melts incongruently at 566°C and crystallizes in the tetragonal structure. A eutectic with a melting temperature of 554°C and a composition of 53 mol% HgSe is situated between quaternary compounds. HgSe forms a solid solution

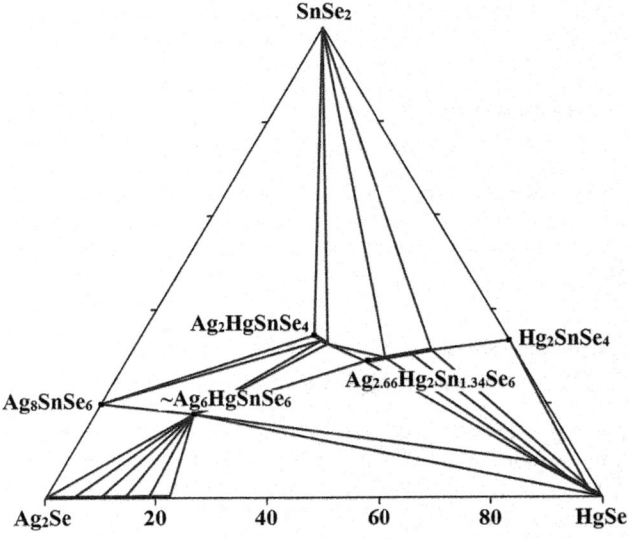

FIGURE 8.35 Isothermal section of the SnSe₂–Ag₂Se–HgSe quasiternary system at 400°C. (From Parasyuk, O.V. et al., *J. Alloys Compd.*, **339**(1–2), 140, 2002.)

that exceeds 10 mol% at the peritectic temperature and equals 8.5 mol% at 400°C.

This system was investigated through DTA, XRD, and metallography, and the ingots for the investigations were annealed at 250°C during 500 h with further quenching in cold water (Parasyuk 1999). The Ag₂HgSnSe₄ compound was obtained by the interaction of chemical elements at 700°C (Haeuseler and Himmrich 1989).

SnSe₂–Ag₂Se–HgSe. The isothermal section of this quasiternary system at 400°C is shown in Figure 8.35 (Parasyuk et al. 2002b). Three intermediate phases, Ag₂HgSnSe₄, **Ag₂.₆₆Hg₂Sn₁.₃₄Se₆**, and **~Ag₆HgSnSe₆**, exist at this temperature. Ag₂.₆₆Hg₂Sn₁.₃₄Se₆ crystallizes in the orthorhombic structure with the lattice parameters $a = 1279.5 \pm 0.2$, $b = 426.31 \pm 0.06$, $c = 582.07 \pm 0.04$ pm, and a calculated density of 6.914 ± 0.003 g·cm⁻³ (Gulay and Parasyuk 2002).

This system was investigated through XRD and metallography, and the ingots for the investigations were annealed at 400°C during 250–500 h with the next quenching in air (Gulay and Parasyuk 2002; Parasyuk et al. 2002b).

8.78 Tin–Silver–Aluminum–Selenium

The **AgAlSnSe₄** quaternary compound, which crystallizes in the tetragonal structure (Goodchild et al. 1981) with the lattice parameters $a = 588.2$, $c = 1071.1$ pm, and an energy gap of 1.56 eV (Hughes et al. 1980; Woolley et al. 1980) [$E_g = 1.85$ eV (Goodchild et al. 1982)], is formed in the Sn–Ag–Al–Se system.

This compound was obtained by the melt and anneals technique (Goodchild et al. 1981, 1982; Hughes et al. 1980). To prepare it, the required masses of the elements were sealed under vacuum in a quartz ampoule (Hughes et al. 1980). The sample so formed was slowly heated (20–40°C·h⁻¹) up to 1050–1100°C, left at that temperature for 5–10 h, slowly cooled (20–40°C·h⁻¹) down to ca. 500°C and then annealed at that temperature for 10–100 h before cooling to room temperature. The heating was carried out in a rocking furnace, and during the heating up and also during the reaction at high temperatures, the sample was rocked to ensure good mixing of the components.

8.79 Tin–Silver–Gallium–Selenium

The **AgGaSnSe₄** quaternary compound, which crystallizes in the tetragonal structure (Goodchild et al. 1981) with the lattice parameters $a = 585.3 \pm 0.2$, $c = 1082.0 \pm 0.4$ pm, and a calculated density of 5.419 g·cm⁻³ (Gil de et al. 1985) [$a = 586.9$, $c = 1081.7$ pm, and an energy gap of 1.70 eV (Hughes et al. 1980; Woolley et al. 1980; Goodchild et al. 1982)], is formed in the Sn–Ag–Ga–Se system. This compound was prepared in the same way as AgAlSnSe₄ (Goodchild et al. 1981, 1982; Hughes et al. 1980). Iodine transport has been used to grow single crystals of AgGaSnSe₄ (Woolley et al. 1980; Gilde et al. 1985).

SnSe₂–AgGaSe₂. The phase diagram of this system, constructed through DTA, XRD, and metallography, is a eutectic type (Figure 8.36) (Shevchuk and Olekseyuk 2001, 2007). The eutectic contains 71 mol% SnSe₂ and crystallizes at 570°C. The solid solution range based on AgGaSe₂ extends at the eutectic temperature to 29 mol% SnSe₂ and decreases to 26 mol% at 450°C. The solid solubility based on SnSe₂ is practically absent. The ingots for the investigations were annealed at 450°C for 480 h followed by quenching into cold water. The existence of the AgGaSnSe₄ compound did not confirm (Shevchuk and Olekseyuk 2007). According to the data of Shevchuk and Olekseyuk (2001), the **Ag₃Ga₃SnSe₈** quaternary compound, which melts incongruently at 656°C, is formed in this system.

8.80 Tin–Silver–Indium–Selenium

SnSe₂–AgInSe₂. The phase diagram of this system, constructed through DTA, XRD, and metallography, is shown in Figure 8.37 (Olekseyuk et al. 1998a; Olekseyuk and Krykhovets 2001a,b). The eutectic contains 78 mol% SnSe₂ and crystallizes at 557°C. At 667°C, in the concentration range of 17–43 mol% SnSe₂, a polymorphic transformation of solid solutions based on AgInSe₂ is observed. On the basis of α-AgInSe₂, solid solutions are formed containing up to 17 mol% SnSe₂, and on the basis of β-AgInSe₂ – up to 20 mol% SnSe₂. The solubility of AgInSe₂ in SnSe₂ is 5 mol%. According to the data of Pamplin et al. (1977), the solubility of SnSe₂ in AgInSe₂ is not higher than 25 mol%.

The **AgInSnSe₄** quaternary compound, which melts incongruently at 587°C (Olekseyuk et al. 1998a) and crystallizes in the tetragonal structure (Goodchild et al. 1981) with the lattice parameters $a = 469.7 \pm 0.3$ and $c = 1\,209.5 \pm 0.3$ pm (Olekseyuk et al. 1999a) [$a = 587.7$, $c = 1128.4$ pm, and an energy gap of 0.94 eV (Hughes et al. 1980; Woolley et al. 1980; Goodchild et al. 1982)], is formed in the SnSe₂–AgInSe₂ system. This compound was prepared in the same way as AgAlSnSe₄ was synthesized (Goodchild et al. 1981, 1982; Hughes et al. 1980).

SnSe₂–AgIn₅Se₈. The phase diagram of this system, constructed through DTA, XRD, and metallography, is shown in Figure 8.38 (Olekseyuk et al. 1998a; Olekseyuk and Krykhovets 2001b). The eutectic contains 90 mol% SnSe₂ and crystallizes at 557°C. At 747°C, a polymorphic transformation of solid solutions based on AgIn₅Se₈ is observed. On the basis of β-AgIn₅Se₈, solid solutions are formed containing up to 8 mol% SnSe₂ at 747°C, and on the basis of α-AgIn₅Se₈ – up to 18 mol% SnSe₂ at 430–560°C. The solubility of α-AgIn₅Se₈ in SnSe₂ reaches 5 mol%.

The ingots for the investigations of both systems were annealed at 580°C for 250 h with the next quenching in

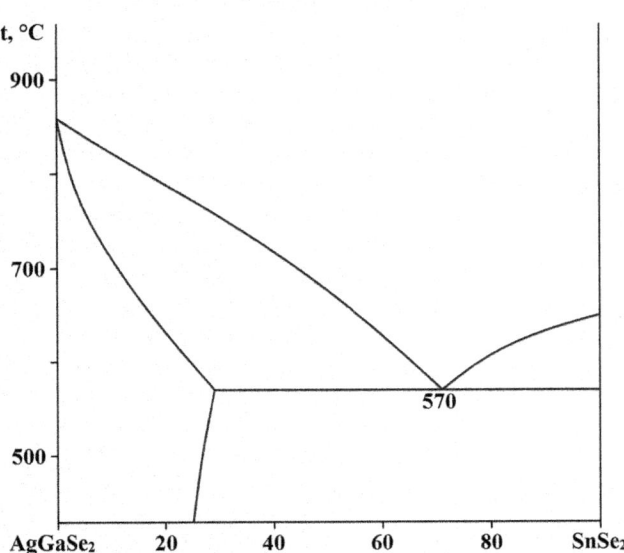

FIGURE 8.36 Phase diagram of the SnSe₂–AgGaSe₂ system. (From Shevchuk, M.V., and Olekseyuk, I.D., *J. Alloys Compd.*, **433**, 171, 2007.)

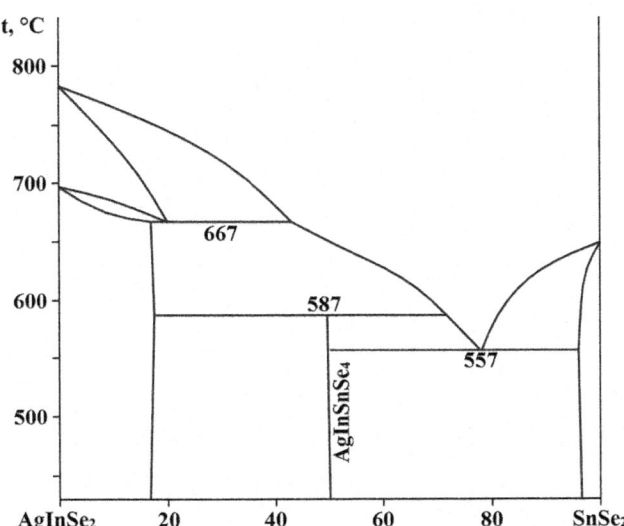

FIGURE 8.37 Phase diagram of the SnSe₂–AgInSe₂ system. (From Olekseyuk, I.D., et al., *Zhurn. neorgan. khimii*, **43**(12), 2084, 1998.)

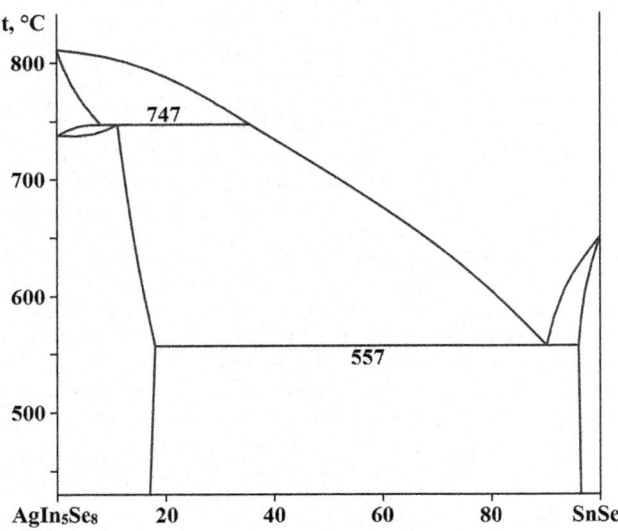

FIGURE 8.38 Phase diagram of the SnSe₂–AgIn₅Se₈ system. (From Olekseyuk, I.D., et al., *Zhurn. neorgan. khimii*, **43**(12), 2084, 1998.)

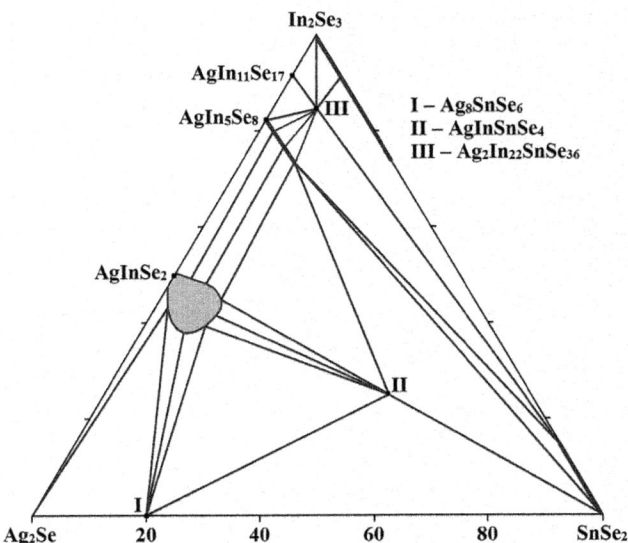

FIGURE 8.39 Isothermal section of the SnSe₂–Ag₂Se–In₂Se₃ quasiternary system at 480°C. (From Olekseyuk, I.D., and Kryrhovets, O.V., *J. Alloys Compd.*, **316**(1–2), 193, 2001.)

saturated aqueous solution of NaCl (Olekseyuk et al. 1998a; Olekseyuk and Krykhovets 2001a).

SnSe₂–AgIn₁₁Se₁₇. This system is nonquasibinary section of the SnSe₂–Ag₂Se–In₂Se₃ system as AgIn₁₁Se₁₇ melts incongruently (Olekseyuk and Krykhovets 2001b). The **Ag₂In₂₂SnSe₃₆** quaternary compound is formed in this system.

Ag₈SnSe₆–AgInSe₂. The phase diagram of this system, constructed through DTA, XRD, and metallography, is a eutectic type (Olekseyuk and Krykhovets 2001b). The eutectic contains 32 mol% Ag₈SnSe₆ and crystallizes at 650°C. The invariant reactions at 735°C and 87°C are caused by the polymorphous transformation of AgInSe₂ and Ag₈SnSe₆, respectively. The solid solution ranges based on γ-AgInSe₂ reach 8 mol% Ag₈SnSe₆ at the eutectic temperature and 2 mol% at 450°C. The solubility of AgInSe₂ in Ag₈SnSe₆ reaches its maximum at 750°C (45 mol%) and decreases with decreasing temperature.

SnSe₂–Ag₂Se–In₂Se₃. The isothermal section of this quasiternary system at 450°C is shown in Figure 8.39 (Olekseyuk and Krykhovets 2001b). Homogeneity regions form for all binary compounds. In In₂Se₃ the solubility reaches up to 25 mol% SnSe₂, and in SnSe₂ can dissolve 20 mol% In₂Se₃. AgInSe₂ and AgIn₅Se₈ have solid solution ranges extending up to 18 mol% SnSe₂. Two quaternary compounds, AgInSnSe₄ and Ag₂In₂₂SnSe₃₆, exist at this temperature in the quasiternary system. The liquidus surface of this system is given in Figure 8.40. It is seen that this system is divided by quasibinary sections into four subsystems. Some vertical sections and the reaction scheme were also constructed.

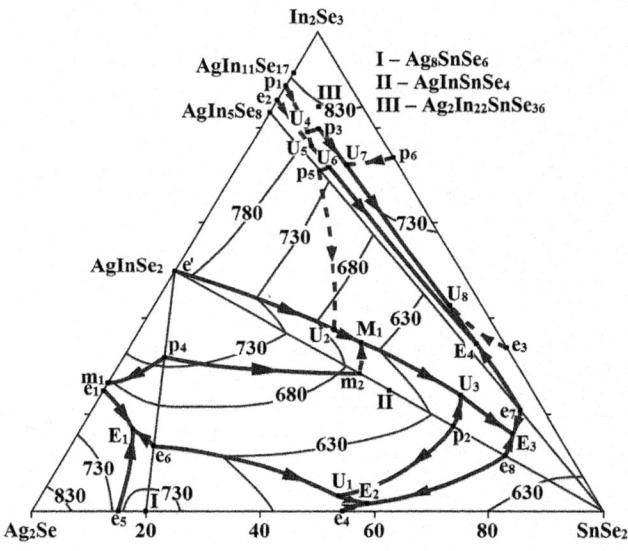

FIGURE 8.40 Liquidus surface of the SnSe₂–Ag₂Se–In₂Se₃ quasiternary system. (From Olekseyuk, I.D., and Kryrhovets, O.V., *J. Alloys Compd.*, **316**(1–2), 193, 2001.)

8.81 Tin–Silver–Lanthanum–Selenium

The **AgLa₃SnSe₇** quaternary compound, which crystallizes in the hexagonal structure with the lattice parameters $a = 1080.5 \pm 0.4$, $c = 624.6 \pm 0.1$ pm and a calculated density of 6.290 g·cm⁻³, is formed in the Sn–Ag–La–Se system (Daszkiewicz et al. 2007). Dark red crystals of this compound were prepared

by sintering the elemental constituents with the atomic ratio La/Ag/Sn/Se = 3:1:1:7 in an evacuated silica ampoule in a tube furnace. The ampoule was heated at a rate of 30°C·h⁻¹ to a maximum temperature of 1150°C and kept at this temperature for 3 h. Afterwards the ampoule was cooled slowly (10°C·h⁻¹) to 600°C, annealed at this temperature for 240 h and then quenched in cold water.

8.82 Tin–Silver–Lead–Selenium

SnSe₂–Ag₂Se–PbSe. The isothermal section of this system at room temperature was constructed using the alloy annealed at 400°C for 250 h (Kogut et al. 2006). No quaternary compounds were found in the system. The system is divided into two subsystems by the Ag₈SnSe₆–PbSe section. Alloys adjacent directly to SnSe₂ contain not 3, but 4 phases. This means that the tetrahedration of the Sn–Ag–Pb–Se system does not pass through SnSe₂, but through SnSe.

8.83 Tin–Silver–Arsenic–Selenium

SnSe₂–Ag₂Se–As₂Se₃. The isothermal section of this quasiternary system at 250°C is presented in Figure 8.41 (Zmiy et al. 2008). The formation of **Ag₂SnAs₆Se₁₂** quaternary compound was established. It melts congruently at 417°C and crystallizes in the trigonal structure with the lattice parameters $a = 381.18 \pm 0.02$, $c = 397.24 \pm 0.03$ pm and a calculated density of 5.751 g·cm⁻³. Two-phase equilibria exist between the quaternary Ag₂SnAs₆Se₁₂ and the ternary Ag₈SnSe₆, AgAsSe₂, and AgAs₃Se₅, and the binary SnSe₂ and As₂Se₃ compounds. Two-phase Ag₈SnSe₆+Ag₃AsSe₃ and Ag₈SnSe₆+AgAsSe₂ equilibria were also observed. Seven three-phase fields exist in this quasiternary system at 250°C. No significant solubility of the third (fourth) component in the binary (ternary) compounds was detected (<1–2 mol%).

8.84 Tin–Silver–Antimony–Selenium

SnSe–AgSbSe₂. The phase diagram of this system, constructed through DTA, XRD, and metallography, is a eutectic type Figure 8.42 (Olekseyuk et al. 2012b). The eutectic contains

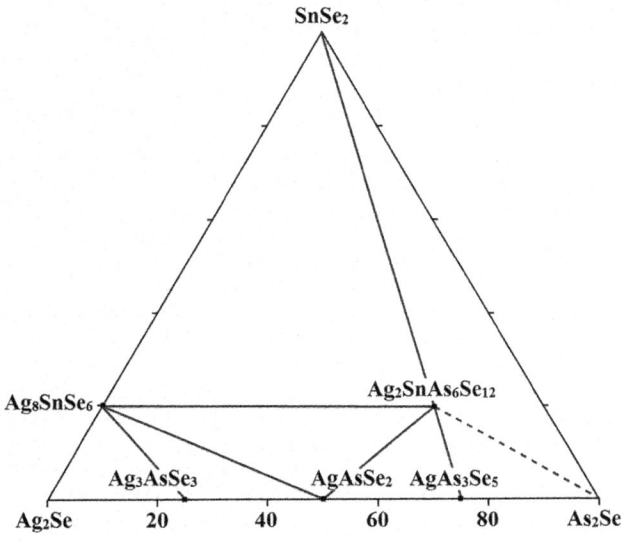

FIGURE 8.41 Isothermal section of the SnSe₂–Ag₂Se–As₂Se₃ quasiternary system at 250°C. (From Zmiy, O.F., et al., *Chem. Met. Alloys*, 1(2), 115, 2008.)

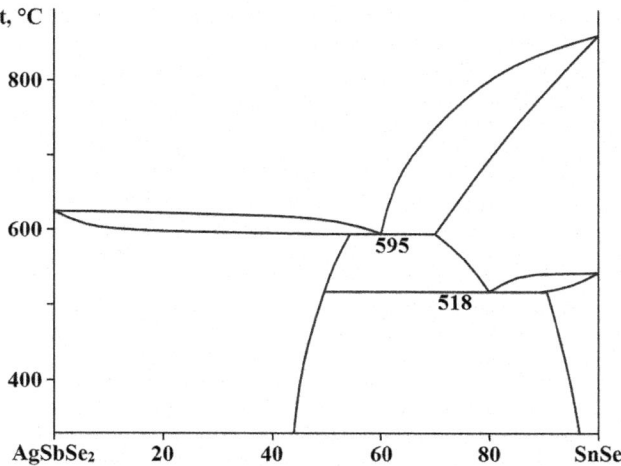

FIGURE 8.42 Phase diagram of the SnSe–AgSbSe₂ system. (From Olekseyuk, I.D., et al., *Nauk. Visnyk Volyns'k. Nats. Univ. im. Lesi Ukrainky. Ser. Khim. nauky*, (17), 105, 2012.)

40 mol% AgSbSe₂ and crystallizes at 595°C. A polymorphic transformation of solid solutions based on SnSe takes place at 518°C. At 595°C, the region of solid solutions based on AgSbSe₂ is 50 mol% β-SnSe, and the solubility of AgSbSe₂ in β-SnSe is 30 mol%. The solubility of AgSbSe₂ in α-SnSe is not higher than 5 mol%.

SnSe₂–Ag₂Se–Sb₂Se₃. The isothermal section of this quasiternary system at 300°C was constructed by Olekseyuk et al. (2009d). It was shown that two quasibinary systems, SnSe₂–AgSbSe₂ and Ag₈SnSe₆–AgSbSe₂, divide it for three three-phase regions: Ag₂Se–Ag₈SnSe₆–AgSbSe₂, Ag₈SnSe₆–AgSbSe₂–SnSe₂, and AgSbSe₂–SnSe₂–Sb₂Se₃. The ingots for the investigations were annealed at 300°C for 600 h followed by quenching in cold water.

8.85 Tin–Silver–Chromium–Selenium

The **AgCrSnSe₄** quaternary compound, which crystallizes in the cubic structure with the lattice parameter $a = 1097$ pm and the calculated and experimental densities of 5.978 and 5.96 g·cm⁻³, respectively is formed in the Sn–Ag–Cr–Se system (Hahn and Strick 1967a; Strick et al. 1968). It was synthesized from the elements at 750°C for 4 days.

8.86 Tin–Silver–Bromine–Selenium

Tetrahedration of the Sn–Ag–Br–Se quaternary system in the part of SnSe–SnSe₂–Ag₂Se–AgBr was carried out by Moroz et al. (2017b) using DTA and XRD. The linear dependencies of the EMF of cells on temperature in the range 210–250°C have been used to calculate the standard thermodynamic values of Ag₈SnSe₆ saturated with bromine and **Ag₈₋ₓSnSe₆₋ₓBrₓ** at $x = 1$: $\Delta G^0_{f,298} = -347.6 \pm 27.2$ kJ·M⁻¹, $\Delta H^0_{f,298} = -336.2 \pm 19.2$ kJ·M⁻¹ and $\Delta S^0_{f,298} = 38.3 \pm 36.2$ J·(M·K)⁻¹ for Ag₈SnSe₆ saturated with bromine, and $\Delta G^0_{f,298} = -410.2 \pm 17.6$ kJ·M⁻¹,

$\Delta H^0_{f,298} = -404.2 \pm 15.2$ kJ·M^{-1} and $\Delta S^0_{f,298} = 20.6 \pm 28.4$ J·(M·K)$^{-1}$ for Ag$_7$SnSe$_5$Br.

The Ag$_{8-x}$SnSe$_{6-x}$Br$_x$ quaternary phase (x value is not specified) was obtained by reacting stoichiometric amounts of high-purity elements in an evacuated quartz ampoule at temperatures of 600°C to 700°C for about 6 days (Kuhs et al. 1979).

8.87 Tin–Silver–Iodine–Selenium

The **Ag$_{8-x}$SnSe$_{6-x}$I$_x$** quaternary phase, which crystallizes in the cubic structure with the lattice parameter $a = 1113.7 \pm 0.4$ (x value is not specified), is formed in the Sn–Ag–I–Se system (Kuhs et al. 1979). It was obtained by reacting stoichiometric amounts of high-purity elements in an evacuated quartz ampoule at temperatures of 600°C to 700°C for about 6 days.

8.88 Tin–Silver–Manganese–Selenium

The **Ag$_2$MnSnSe$_4$** quaternary compound, which has apparently two polymorphic modifications, is formed in the Sn–Ag–Mn–Se system (Lamarche et al. 1991). First modification crystallizes in the orthorhombic structure with the lattice parameters $a = 842.4$, $b = 734.8$, and $c = 698.6$ pm, and the second one in the cubic structure with the lattice parameter $a = 549$ pm. This compound was obtained by the melt and anneals technique using stoichiometric mixtures of the elements. Typically, the components of each 1.0-g sample were sealed under vacuum in a quartz capsule that had previously been coated with carbon in order to prevent reaction of the charge with the quartz. The capsule was then raised to a temperature of 1150°C for approximately an hour, held at this temperature for 0.5–1 h, and then cooled to room temperature. Samples were then annealed at 600°C for a week and quenched to room temperature.

8.89 Tin–Silver–Iron–Selenium

The **Ag$_2$FeSnSe$_4$** quaternary compound, which melts at 369°C and crystallizes in the orthorhombic structure with the lattice parameters $a = 739.8$, $b = 699.3$, and $c = 640.1$ pm, is formed in the Sn–Ag–Fe–Se system (Quintero et al. 1999). It was prepared by the melt and anneals technique. The compound of 1-g was made from the appropriate amounts of the elements that were sealed under vacuum in a small quartz ampoule, which had previously been carbonized to prevent interaction of the components with the quartz. The mixture was melted together at 1150°C for about an hour, annealed at 500°C, and then cooled to room temperature by leaving the ampoule in the switched-off furnace.

The samples of Ag$_2$FeSnSe$_4$ were also synthesized by two more methods (Moroz et al. 2021a): (1) Solid-state synthesis of the calculated mixture of elements in an evacuated (~10^{-2} Pa) quartz ampoule at 360°C for 5 days. Then the samples were cooled to room temperature at the rate of 2°C·min^{-1} and ground to a particle size of ~ 5.0 μm. Vacuum homogenization of the fine phase mixture was held at 310°C for 5 days; (2) Vacuum

melting of the calculated mixture of the elements at 830°C followed by vacuum annealing of the finely dispersed mixture at 310°C for 5 days.

Linear dependences of the EMF of galvanic cells versus temperature in the range of 127–227°C were used for calculations of the standard Gibbs energies, enthalpies, and entropies of formations of Ag$_2$FeSnSe$_4$ (Moroz et al. 2021a): $-\Delta G^0_{f,298} = 250.1 \pm 38.6$ and 253.0 ± 40.9 kJ·M^{-1}, $-\Delta H^0_{f,298} = 251.8 \pm 30.9$ and 252.4 ± 36.3 kJ·M^{-1}, $S^0_{298} = 327.0 \pm 10.4$ and 334.9 ± 6.0 J·(M·K)$^{-1}$ in two different phase regions.

8.90 Tin–Gold–Thallium–Selenium

The **Tl$_2$Au$_2$Sn$_2$Se$_6$** quaternary compound, which crystallizes in the tetragonal structure with the lattice parameters $a = 1145.13 \pm 0.07$, $c = 2007.1 \pm 0.3$ pm, and an energy gap of 1.2 eV, is formed in the Sn–Au–Tl–Se system (McGuire et al. 2005a). This compound was synthesized from the elements. Since Tl will tarnish slowly in air, it was handled in an argon glove box. Chunks of Tl were cut from the 0.5-inch diameter rod with wire cutters used only for this purpose, weighed inside the glove box and placed in capped glass vials. Appropriate amounts of the other elements were then weighed out and loaded into a carbon-coated silica tube. The Ar-filled vials were removed from the glove box and the Tl was added to the silica tube. The tube was then quickly (to limit the exposure of Tl to air) attached to a vacuum line for sealing. It was heated over 24 h to 800°C, held at this temperature for 172 h, and then cooled over 200 h to 300°C, at which point the furnace was turned off and allowed to cool to room temperature. Crystals were extracted from the resulting polycrystalline mass for examination.

8.91 Tin–Magnesium–Strontium–Selenium

The **MgSrSnSe$_4$** quaternary compound, which crystallizes in the orthorhombic structure with the lattice parameters $a = 2176.01 \pm 0.01$, $b = 2185.9 \pm 0.2$, $c = 1340.60 \pm 0.09$ pm, a calculated density of 4.554 g·cm^{-3}, and an energy gap of 1.8 eV, is formed in the Sn–Mg–Sr–Se system (Assoud et al. 2004a). This compound was synthesized from a stoichiometric mixture of the elements, which was heated to 850°C within 24 h, kept at 850°C for 12 h, and then cooled to 700°C within 2 h. After sintering at 700°C for 300 h, the furnace was slowly cooled to 150°C within 200 h. Thereafter, the furnace was switched off to reach room temperature within a few hours. The obtained product appeared to be red.

8.92 Tin–Strontium–Zinc–Selenium

The **SrZnSnSe$_4$** quaternary compound, which crystallizes in the orthorhombic structure with the lattice parameters $a = 2169.8$, $b = 2187.3$, $c = 1311$ pm, and an energy gap of 2.14 eV, is formed in the Sn–Sr–Zn–Se system (Pang et al. 2021). This compound was prepared by the traditional high-temperature solid-state technique. The powders of SrSe (0.1418 g),

Zn (0.0557 g), SnSe (0.1682 g), and Se (0.1344 g) were mixed stoichiometrically and loaded into a silica tube. After flame sealing under a high vacuum of 10^{-3} Pa, the tube was heated to 650°C at 5°C·min^{-1}, kept at 650°C for 1.5 days, and then cooled down by switching off the furnace. The as-prepared SrZnSnSe$_4$ is stable in air for at least 9 months. The decomposition of this compound began at 544°C, and most of Se and SnSe were taken away by Ar gas. The main components of the residue after heat treatment were ZnSe and SrSe.

8.93 Tin–Strontium–Cadmium–Selenium

The **SrCdSnSe$_4$** quaternary compound, which melts congruently at 790°C and crystallizes in the orthorhombic structure with the lattice parameters $a = 2189.08 \pm 0.10$, $b = 2189.45 \pm 0.09$, $c = 1354.07 \pm 0.10$ pm, a calculated density of 5.196 g·cm^{-3}, and an energy gap of 1.54 eV, is formed in the Sn–Sr–Cd–Se system (Lin et al. 2019a). Single crystals of this compound were obtained through solid-state reactions. Dark green crystals were crystallized from a reaction mixture containing Sr (0.66 mM), Cd (0.66 mM), Sn (0.66 mM), and Se (2.65 mM). The mixture was loaded into a quartz tube and flame-sealed under vacuum (~10^{-2} Pa). The tube was placed into a furnace, firstly heated from room temperature to 200°C in 4 h, kept at that temperature for 2 h, and then heated to 400°C in 5 h and kept for another 5 h, subsequently heated to 850°C in 22.5 h, kept for 5 days, and then cooled to room temperature in 6 days before switching off the furnace. The obtained crystals are stable in the air and moisture conditions.

8.94 Tin–Strontium–Mercury–Selenium

The **SrHgSnSe$_4$** quaternary compound, which is stable up to 694°C and crystallizes in the orthorhombic structure with the lattice parameters $a = 2190.27 \pm 0.08$, $b = 2190.59 \pm 0.07$, $c = 1350.10 \pm 0.08$ pm, a calculated density of 5.928 g·cm^{-3}, and an energy gap of 2.07 eV, is formed in the Sn–Sr–Hg–Se system (Guo et al. 2019a). To prepare single crystals of this compound, equimolar mixtures of SrSe, HgSe, and SnSe$_2$ were ground under an Ar atmosphere in a glove box. Then the mixture of reagents was placed into a quartz tube and sealed under a vacuum of 10^{-3} Pa. The tube was heated to 900°C in 20 h, maintained at that temperature for 72 h, cooled to 400°C in 200 tb, and then the furnace was shut off. Red crystals of SrHgSnSe$_4$ were found, which are air and moisture stable.

8.95 Tin–Strontium–Bismuth–Selenium

The **Sr$_3$Sn$_2$Bi$_2$Se$_8$** quaternary compound, which crystallizes in the orthorhombic structure with the lattice parameters $a = 1297.45 \pm 0.07$, $b = 423.52 \pm 0.02$, and $c = 2909.2 \pm 0.2$ pm for the Sr$_3$Sn$_{0.96}$Bi$_{2.04}$Se$_8$ composition, is formed in the Sn–Sr–Bi–Se system (Chung and Lee 2012). To synthesize this compound, stoichiometric proportions of the pure elements were mixed in a N$_2$-filled glove box (total mass ~0.5 g), placed in a carbon-coated silica tube, sealed under dynamic vacuum, and slowly heated to 750°C for 48 h. This temperature was maintained for one day, followed by slow cooling to 400°C at a rate of 15°C·h^{-1}, and finally to room temperature by simply terminating the power. The polycrystalline ingot with a metallic luster was obtained.

8.96 Tin–Strontium–Fluorine–Selenium

The **Sr$_2$SnSe$_3$F$_2$** quaternary compound, which crystallizes in the orthorhombic structure with the lattice parameters $a = 610.78 \pm 0.04$, $b = 1945.3 \pm 0.7$, $c = 610.51 \pm 0.05$ pm, and an energy gap of 2.24 eV, is formed in the Sn–Sr–F–Se system (Kabbour et al. 2006). It was synthesized using a high-temperature ceramic method. A stoichiometric proportion of SrF$_2$, SnSe, and Se was weighted and ground in glove box under Ar atmosphere. This mixture was subsequently pressed into pellets and sealed under vacuum in a silica tube. The tube was first heated at 220°C for 12 h, and then heated to 700°C (at 50°C·h^{-1}) for another 12 h, followed by regrinding and annealing at the same temperature. Orange powder of the title compound was obtained.

8.97 Tin–Barium–Cadmium–Selenium

The **BaCdSnSe$_4$** quaternary compound, which crystallizes in the orthorhombic structure with the lattice parameters $a = 2238.1 \pm 1.4$, $b = 2271.1 \pm 1.4$, $c = 1358.8 \pm 0.9$ pm, and an energy gap of 1.79 eV, is formed in the Sn–Ba–Cd–Se system (Wu et al. 2015a). This compound was prepared by a molten flux method with Cd as a flux. It can also be obtained by a stoichiometric combination of elements and was successfully prepared by solid-state method at 800°C for 100 h.

8.98 Tin–Barium–Mercury–Selenium

The **BaHgSnSe$_4$** quaternary compound, which melts congruently at 712°C and crystallizes in the orthorhombic structure with the lattice parameters $a = 2244.1 \pm 0.5$, $b = 2276.0 \pm 0.5$, $c = 1357.9 \pm 0.3$ pm, a calculated density of 5.918 g·cm^{-3}, and an energy gap of 1.98 eV, is formed in the Sn–Ba–Hg–Se system (Guo et al. 2019a). To prepare single crystals of this compound, equimolar mixtures of BaSe, HgSe, and SnSe$_2$ were ground under an Ar atmosphere in a glove box. Then the mixture of reagents was placed into a quartz tube and sealed under a vacuum of 10^{-3} Pa. The tube was heated to 900°C in 20 h, maintained at that temperature for 72 h, cooled to 400°C in 200 h, and then the furnace was shut off. Red crystals of BaHgSnSe$_4$ were found, which are air and moisture stable.

8.99 Tin–Barium–Gallium–Selenium

Three quaternary compounds, **BaGa$_2$SnSe$_6$**, **Ba$_4$Ga$_4$SnSe$_{12}$**, and **Ba$_6$Ga$_2$SnSe$_{11}$**, are formed in the Sn–Ba–Ga–Se system. First of them melts incongruently and crystallizes in the trigonal structure with the lattice parameters $a = 1014.49 \pm 0.14$,

c = 924.90 ± 0.18 pm, a calculated density of 5.253 g·cm^{-3} at 153 K, and an energy gap of 1.95 eV (Li et al. 2015). To prepare this compound, the mixture of BaSe, Ga$_2$Se$_3$, and SnSe$_2$ (molar ratio 2:1:6) was ground and loaded into a fused silica tube under an Ar atmosphere in a glove box. The tube was flame-sealed under a high vacuum (10^{-3} Pa) and then placed in a furnace. The sample was heated to 1000°C in 24 h and kept at that temperature for 48 h, then cooled at a slow rate of 3°C·h^{-1} to 350°C and finally cooled to room temperature naturally. Red crystals, which are stable in air for months, were obtained.

Ba$_4$Ga$_4$SnSe$_{12}$ crystallizes in the tetragonal structure with the lattice parameters a = 1360.7 ± 0.2, c = 651.0 ± 0.1 pm, a calculated density of 5.221 g·cm^{-3}, and an energy gap of 2.16 eV [2.14 eV (Azam et al. 2015)], and Ba$_6$Ga$_2$SnSe$_{11}$ crystallizes in the monoclinic structure with the lattice parameters a = 1871.5 ± 0.4, b = 710.9 ± 0.1, c = 1916.5 ± 0.4 pm, β = 103.29 ± 0.03°, a calculated density of 5.221 g·cm^{-3}, and an energy gap of 1.99 eV (Yin et al. 2015).

Crystals of Ba$_4$Ga$_4$SnSe$_{12}$ were initially obtained from a reaction between BaSe and SnGa$_4$Se$_7$ in the molar ratio of 4:1 (Yin et al. 2015). A mixture of 112 mg BaSe and 124 mg SnGa$_4$Se$_7$ was ground and loaded into a fused silica tube under an Ar atmosphere in a glove box. The tube was flame-sealed under a high vacuum of 10^{-3} Pa and then placed in a furnace. The reaction mixture was heated to 950°C within 15 h and kept at 950°C for 48 h, followed by slow cooling to 320°C at a rate of 3°C·h^{-1}, and finally cooled to room temperature by switching off the furnace. Many orange block-shaped crystals of Ba$_4$Ga$_4$SnSe$_{12}$ were found in the ampoule.

Crystals of Ba$_6$Ga$_2$SnSe$_{11}$ were initially obtained from a reaction between BaSe (108 mg) and SnGa$_4$Se$_7$ (95 mg) in the molar ratio of 5:1 (Yin et al. 2015). The preparation method is the same as for Ba$_4$Ga$_4$SnSe$_{12}$. Many orange block-shaped crystals were obtained.

Then the synthesis of both compounds was carried out by the next way (Yin et al. 2015). The mixtures of BaSe, Ga$_2$Se$_3$, and SnSe$_2$ (molar ratios of 4:2:1 for Ba$_4$Ga$_4$SnSe$_{12}$ or 6:1:1 for Ba$_6$Ga$_2$SnSe$_{11}$ were ground and loaded into fused silica tubes under vacuum (10^{-3} Pa) and then placed in a furnace. The samples were heated to 800°C within 15 h and kept at that temperature for 48 h, and then the furnace was turned off.

8.100 Tin–Barium–Cerium–Selenium

The **BaCeSn$_2$Se$_6$** quaternary compound, which crystallizes in the orthorhombic structure with the lattice parameters a = 419.08 ± 0.08, b = 2074.9 ± 0.4, c = 1240.6 ± 0.3 pm, and a calculated density of 6.087 g·cm^{-3}, is formed in the Sn–Ba–Ce–Se system (Feng et al. 2014). To prepare this compound, the mixture of BaSe (1 mM), SnSe$_2$ (2 mM), Ce (1 mM), and Se (1 mM) was ground and loaded into a fused silica tube under an Ar atmosphere in a glove box. The tube was flame-sealed under a high vacuum (10^{-3} Pa) and then placed in a furnace. It was heated to 1100°C in 24 h, left for 48 h, cooled to 420°C at a rate of 3°C·h^{-1}, and finally cooled to room temperature by switching off the furnace. Dark-red air-stable crystals of the title compound were found.

8.101 Tin–Barium–Lead–Selenium

The **Ba$_6$Sn$_3$PbSe$_{13}$** quaternary compound, which crystallizes in the orthorhombic structure with the lattice parameters a = 1273 ± 3, b = 2465 ± 5, c = 924 ± 2 pm, a calculated density of 5.528 g·cm^{-3}, and an energy gap of 1.89 eV, is formed in the Sn–Ba–Pb–Se system (Abudurusuli et al. 2017). This compound was synthesized as follows. The mixtures of BaSe, PbSe, Sn, and Se (molar ratio of 1:1:1:2) were ground and loaded into a fused silica tube under Ar atmosphere in a glove box, then this tube was flame-sealed under a high vacuum of 10^{-3} Pa. The tube was then placed in a furnace and heated to 800°C in 30 h, left for 100 h to ensure the mixture completely melted, cooled to 200°C at a rate of 5°C·h^{-1}, and finally cooled to room temperature within a short time. Ba$_6$PbSn$_3$Se$_{13}$ can also be prepared using the stoichiometric mixture of the precursors. After repeatedly washing with DMF, the crystals of the title compound with the crimson color and other unknown phases were found in the ampoule. These crystals are stable in the air.

8.102 Tin–Barium–Antimony–Selenium

The **Ba$_3$Sn$_2$Sb$_2$Se$_8$** quaternary compound, which melts at 697°C and crystallizes in the orthorhombic structure with the lattice parameters a = 1271.5 ± 0.3, b = 455.09 ± 0.09, c = 2966.0 ± 0.6 pm and an energy gap of 0.51 eV (melts at 857°C and crystallizes in the orthorhombic structure with the lattice parameters a = 1312.5 ± 0.3, b = 440.12 ± 0.09, c = 2989.3 ± 0.6 pm and an energy gap of 0.67 eV for Ba$_3$Sn$_{0.87}$Bi$_{2.13}$Se$_8$ composition), is formed in the Sn–Ba–Sb–Se system (Chung and Lee 2012). To synthesize this compound, stoichiometric proportions of the pure elements were mixed in a N$_2$-filled glove box (total mass ~0.5 g), placed in a carbon-coated silica tube, sealed under dynamic vacuum, and slowly heated to 750°C in 48 h. This temperature was maintained for one day, followed by slow cooling to 400°C at a rate of 15°C·h^{-1}, and finally to room temperature by simply terminating the power. The polycrystalline ingot with a metallic luster was obtained.

8.103 Tin–Barium–Bismuth–Selenium

The **Ba$_3$SnBi$_6$Se$_{13}$** quaternary compound, which crystallizes in the monoclinic structure with the lattice parameters a = 1724.98 ± 0.05, b = 429.12 ± 0.01, c = 1845.90 ± 0.05 pm, β = 90.679 ± 0.001°, and a calculated density of 6.833 g·cm^{-3}, is formed in the Sn–Ba–Bi–Se system (Wang and DiSalvo 2000). This compound was synthesized from a stoichiometric mixture of the elements, which was put in a vitreous carbon crucible and sealed in an evacuated quartz tube. Then it was gradually heated to 710°C in 48 h, kept at that temperature for 100 h, and slowly cooled to 510°C at 2°C·h^{-1}. Black ingot of Ba$_3$SnBi$_6$Se$_{13}$ was obtained. Due to the air sensitivity of barium, all work was carried out under an argon atmosphere in a vacuum-dry box.

8.104 Tin–Barium–Tellurium–Selenium

The $Ba_7Sn_3Se_{13-x}Te_x$ quaternary solid solution, which crystallizes in the orthorhombic structure with the lattice parameters $a = 1274.8 \pm 0.1$, $b = 2489.1 \pm 0.3$, $c = 921.2 \pm 0.1$ pm, and a calculated density of 5.326 g·cm⁻³ for $x = 0$; $a = 1284.5 \pm 0.3$, $b = 2509.8 \pm 0.6$, $c = 924.2 \pm 0.2$ pm, and a calculated density of 5.344 g·cm⁻³ for $x = 1.10$; $a = 1292.44 \pm 0.06$, $b = 2518.3 \pm 0.1$, $c = 927.46 \pm 0.04$ pm, and a calculated density of 5.345 g·cm⁻³ for $x = 1.76$; and $a = 1295.96 \pm 0.06$, $b = 2524.2 \pm 0.1$, $c = 929.20 \pm 0.04$ pm, and a calculated density of 5.361 g·cm⁻³ for $x = 2.24$, is formed in the Sn–Ba–Te–Se system (Assoud and Kleinke 2005). To prepare this solid solution, the reaction was started from the stoichiometric mixture of the elements with a total sample mass of around 500 mg. The mixture was put into an evacuated silica tube and heated to 900°C, followed by slow cooling to 200°C within 5 days.

8.105 Tin–Barium–Fluorine–Selenium

The $Ba_2SnSe_3F_2$ quaternary compound, which crystallizes in the orthorhombic structure with the lattice parameters $a = 638.44 \pm 0.01$, $b = 1975.79 \pm 0.05$, $c = 644.92 \pm 0.01$ pm, and an energy gap of 2.49 eV, is formed in the Sn–Ba–F–Se system (Kabbour et al. 2006). It was synthesized using a high-temperature ceramic method. A stoichiometric proportion of BaF_2, SnSe, and Se was weighted and ground in a glove box under Ar atmosphere. This mixture was subsequently pressed into pellets and sealed under vacuum in a silica tube. The tube was first heated at 220°C for 12 h, and then heated to 700°C (at 50°C·h⁻¹) for another 12 h, followed by regrinding and annealing at the same temperature. Green powder of the title compound was obtained.

8.106 Tin–Barium–Palladium–Selenium

The $Ba_8PdSn_4Se_{18.25}$ or $Ba_8Pd(SnSe_4)_{3.75}(SnSe_5)_{0.25}(Se_2)$ quaternary compound, which crystallizes in the triclinic structure with the lattice parameters $a = 1260.4 \pm 0.1$, $b = 1261.4 \pm 0.1$, $c = 1478.2 \pm 0.2$ pm, $\alpha = 67.85 \pm 0.01°$, $\beta = 66.02 \pm 0.01°$, $\gamma = 74.08 \pm 0.01°$, and a calculated density of 5.266 g·cm⁻³, is formed in the Sn–Ba–Pd–Se system (Johrendt 2002). The title compound was synthesized by heating mixtures of the elements with the nominal composition $Ba_8PdSn_4Se_{18.25}$ in a corundum crucible, sealed in a quartz ampoule under an argon atmosphere. The sample was first heated slowly (20°C·h⁻¹) to 700°C for 15 h and then cooled to room temperature. The inhomogeneous product was ground under argon and heated again to 800°C for 2 days. This yielded a black crystalline powder, which was stable in air for several weeks. Single crystals appeared deep red when crushed into transparent pieces.

8.107 Tin–Zinc–Gallium–Selenium

$SnSe_2$–$ZnSe$–Ga_2Se_3. Isothermal section of this quasiternary system at 400°C is shown in Figure 8.43 (Parasyuk et al. 2004). ZnSe and Ga_2Se_3 form solid solution ranges, which are

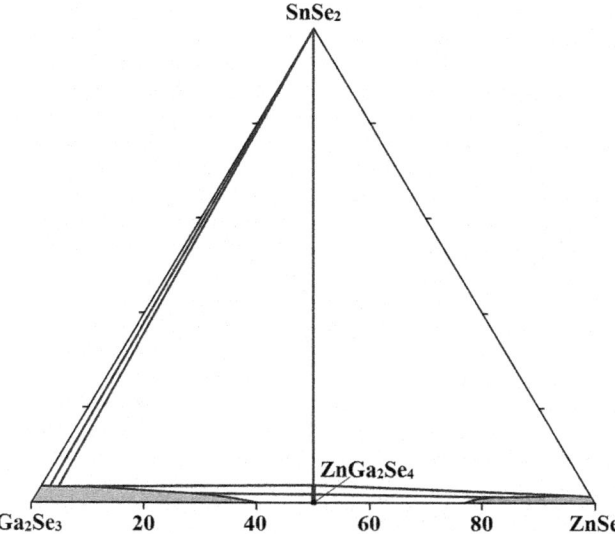

FIGURE 8.43 Isothermal section of the $SnSe_2$–$ZnSe$–Ga_2Se_3 quasiternary system at 400°C. (From Parasyuk, O.V. et al., *J. Alloys Compd.*, **379**(1–2), 143, 2004.)

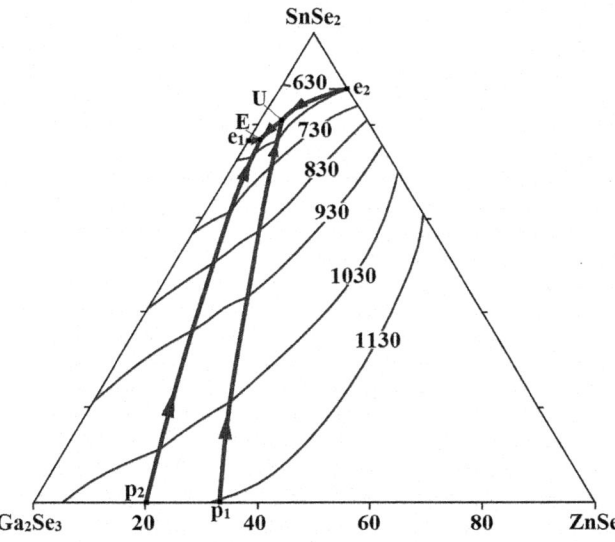

FIGURE 8.44 Liquidus surface of the $SnSe_2$–$ZnSe$–Ga_2Se_3 quasiternary system. (From Parasyuk, O.V. et al., *J. Alloys Compd.*, **379**(1–2), 143, 2004.)

elongated along the quasibinary section ZnSe–Ga_2Se_3, and the maximum $SnSe_2$ content in them equals 2 and 3 mol%, respectively. The liquidus surface of the system is formed by four fields of primary crystallization (Figure 8.44). Three of them correspond to the solid solutions based on ZnSe, Ga_2Se_3, and $SnSe_2$. The other one belongs to the solid solution based on $ZnGa_2Se_4$ ternary compound. No quaternary compounds were found. The fields of primary crystallization are separated by 5 monovariant lines and 6 invariant points, 4 of which correspond to binary reactions and 2 to ternary ones. Ternary eutectic E (2 mol% ZnSe, 21 mol% Ga_2Se_3, and 77 mol% $SnSe_2$) crystallizes at 578°C, and the temperature of the ternary transition

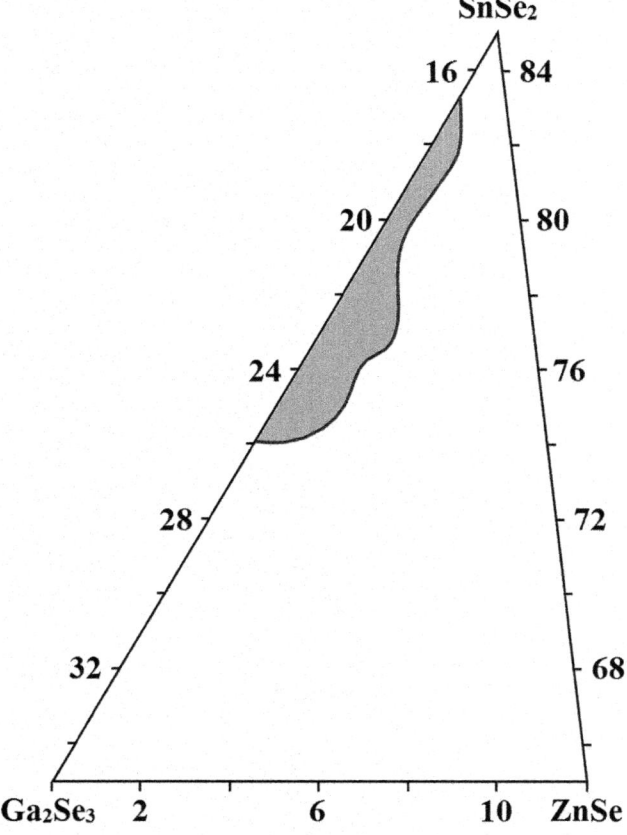

FIGURE 8.45 Glass-forming region in the SnSe₂–ZnSe–Ga₂Se₃ quasiternary system. (From Olekseyuk, I.D. et al., *Funct. Mater.*, **6**(3), 474, 1999.)

point *U* (4 mol% ZnSe, 15 mol% Ga₂Se₃and 81 mol% SnSe₂) is equal to 591°C.

The glass-forming region in this quasiternary system is shown in Figure 8.45 (Olekseyuk et al. 1999b). Glass transition temperature, the crystallization temperature, and the melting temperature were determined for obtained glasses. The glasses have considerable tendency to crystallization and they can be obtained only in the case of the rigid hardening.

This system was investigated through DTA, XRD, and metallography, and all samples were annealed at 400°C for 500 h and then quenched in cold water (Olekseyuk et al. 1999b, Parasyuk et al. 2004).

8.108 Tin–Zinc–Tellurium–Selenium

SnTe–ZnSe. The phase diagram of this system, constructed through DTA, XRD, metallography, and measuring of the microhardness, is a eutectic type (Figure 8.46) (Dubrovin et al. 1984). The eutectic composition and temperature are 6 mol% ZnSe and 780°C, respectively. Solubility of ZnSe in SnTe at the eutectic temperature is equal to 5 mol% and decreases to 4 mol% at 400°C and solubility of SnTe in ZnSe is not higher than 1 mol%. The ingots for the investigations were annealed at 700°C and 400°C for 120 and 250 h, respectively.

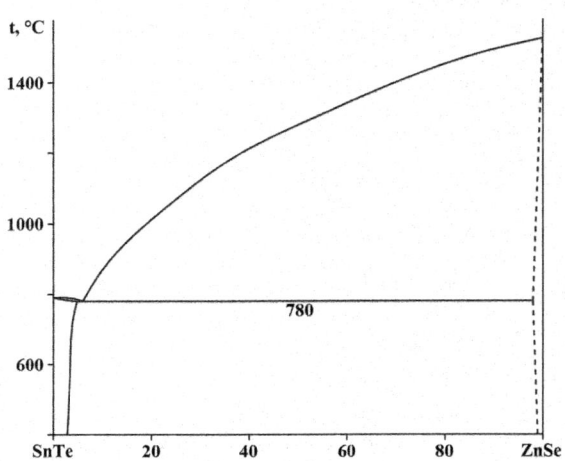

FIGURE 8.46 Phase diagram of the SnTe–ZnSe system. (From Dubrovin, I.V. et al., *Izv. AN SSSR. Neorgan. mater.*, **20**(4), 571, 1984.)

8.109 Tin–Cadmium–Gallium–Selenium

SnSe₂–CdSe–Ga₂Se₃. Isothermal section of this quasiternary system at 400°C is shown in Figure 8.47 (the alloys were annealed at this temperature for 250 h), and the liquidus surface of the system is given in Figure 8.48 (Piskach et al. 2002). The liquidus surface consists of five fields of primary crystallization of the phases CdSe, Ga₂Se₃, SnSe₂, α- and β-CdGa₂Se₄. There are two ternary eutectics and two transition points in the system: E₁ (585°C) – L ↔ CdSe + SnSe₂ + α-CdGa₂Se₄; E₂ (570°C) – L ↔ Ga₂Se₃ + SnSe₂ + α-CdGa₂Se₄; U₁ (812°C) – L + β-CdGa₂Se₄ ↔ CdSe + α-CdGa₂Se₄; and U₂ (812°C) – L + β-CdGa₂Se₄ ↔ Ga₂Se₃ + α-CdGa₂Se₄.

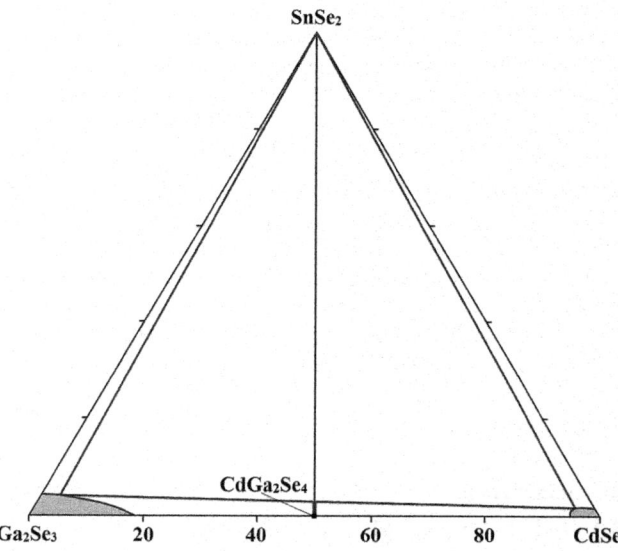

FIGURE 8.47 Isothermal section of the SnSe₂–CdSe–Ga₂Se₃ quasiternary system at 400°C system. (From Piskach, L.V. et al., *Fiz. i khim. tv. tila*, **3**(1), 25, 2002.)

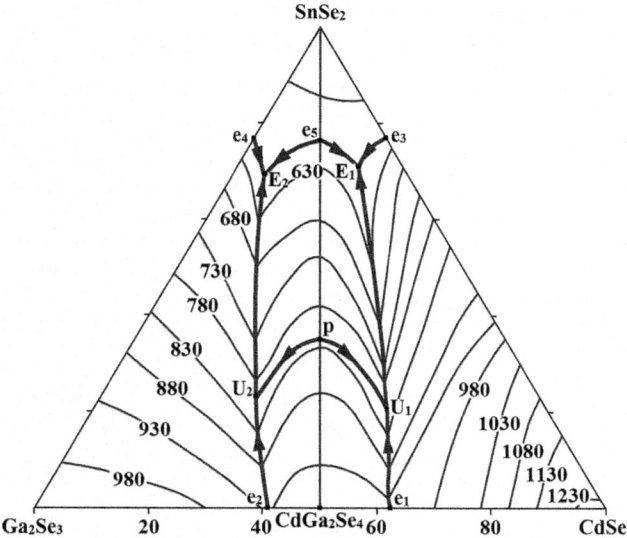

FIGURE 8.48 Liquidus surface of the SnSe$_2$–CdSe–Ga$_2$Se$_3$ quasiternary system. (From Piskach, L.V. et al., *Fiz. i khim. tv. tila*, **3**(1), 25, 2002.)

The glass-forming region in this quasiternary system is shown in Figure 8.49 (Olekseyuk et al. 1999b). The glass transition temperature, the crystallization temperature, and the melting temperature were determined for obtained glasses. The

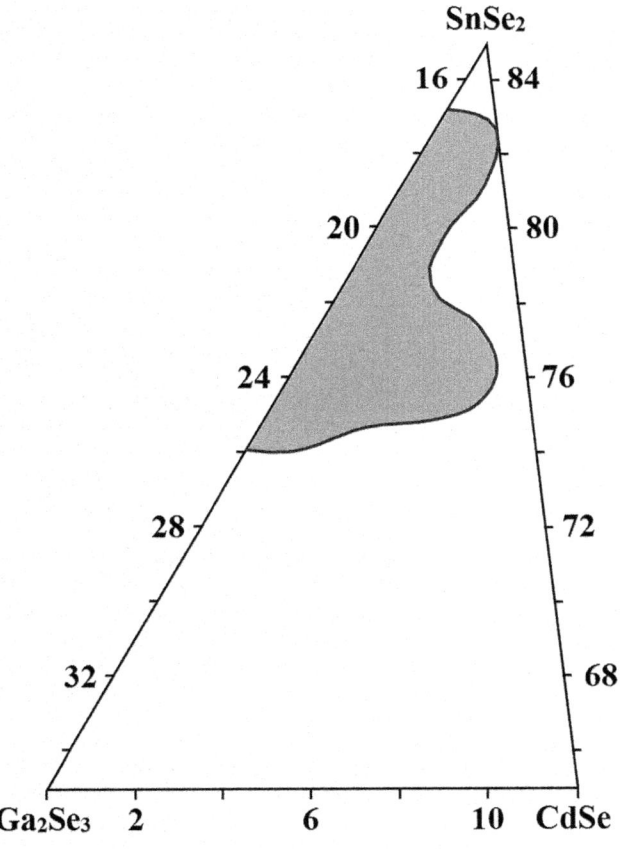

FIGURE 8.49 Glass-forming region in the SnSe$_2$–CdSe–Ga$_2$Se$_3$ quasiternary system. (From Olekseyuk, I.D. et al., *Funct. Mater.*, **6**(3), 474, 1999.)

glasses have considerable tendency for crystallization and they can be obtained only in the case of rigid hardening.

8.110 Tin–Cadmium–Thallium–Selenium

Tl$_2$SnSe$_3$–CdSe. The phase diagram of this system, constructed through DTA and XRD, is shown in Figure 8.50 (Selezen et al. 2019). The eutectic contains 11 mol% CdSe and crystallizes at 447°C. The mutual solubility of the components is negligible. The **CdTl$_2$SnSe$_4$** quaternary compound, which melts incongruently at 587°C and has a polymorphic transformation at 405°C, is formed in this system. The peritectic point composition is 25 mol% CdSe. α-CdTl$_2$SnSe$_4$ crystallizes in the tetragonal structure with the lattice parameters $a = 804.90 \pm 0.06$, $c = 685.73 \pm 0.08$ pm, a calculated density of 7.144 ± 0.002 g·cm^{-3}, and an energy gap of 1.39 eV (Selezen et al. 2019, 2020). CdTl$_2$SnSe$_4$ was synthesized from the mixtures of Tl, Cd, Sn, and Se in quartz ampoule that was evacuated to 0.1 Pa and sealed. The ampoule was heated to 680°C at the rate of 20°C·h^{-1}, kept for 5 h, heated to 930°C at 10°C·h^{-1} and held for 5 h. Then it was cooled to 300°C at a rate of 10°C·h^{-1} and annealed for 350 h. Finally, the samples were quenched into 20% NaCl aqueous solution.

SnSe$_2$–CdSe–Tl$_2$Se. The isothermal section of this quasiternary system at 300°C is shown in Figure 8.51 (Selezen et al. 2020). It is characterized by the formation of the quaternary compounds, CdTl$_2$SnSe$_4$ at the Tl$_2$SnSe$_3$–CdSe section, and CdTl$_2$Sn$_3$Se$_8$ at the CdTl$_2$SnSe$_4$–GeSe$_2$ section. The section contains 6 single-phase, 10 two-phase, and 5 three-phase fields in thermodynamic equilibrium. The solid solubility in Tl$_4$SnSe$_4$ is 6 mol% CdSe along the Tl$_4$SnSe$_4$–CdSe section, for CdSe it is 3 mol% along the Tl$_2$Se–CdSe, Tl$_4$SnSe$_4$–CdSe, and Tl$_2$SnSe$_3$–CdSe sections and for Tl$_2$Se it is 3 mol% along the Tl$_2$Se–CdSe section.

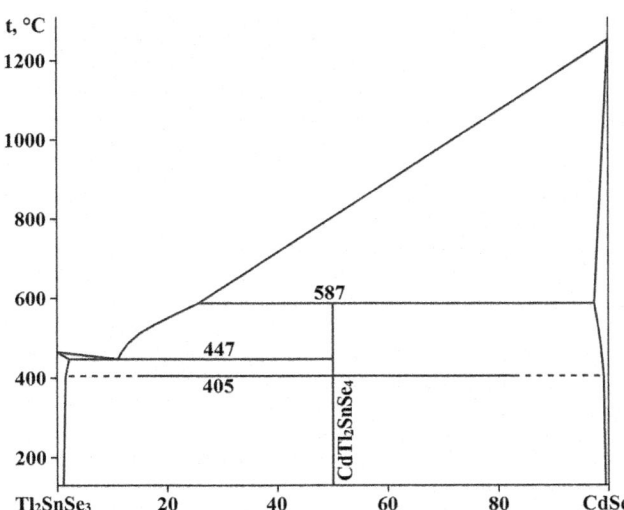

FIGURE 8.50 Phase diagram of the Tl$_2$SnSe$_3$–CdSe system. (From Selezen, A.O., et al., *J. Phase Equilib. Diffus.*, **40**(6), 797, 2019.)

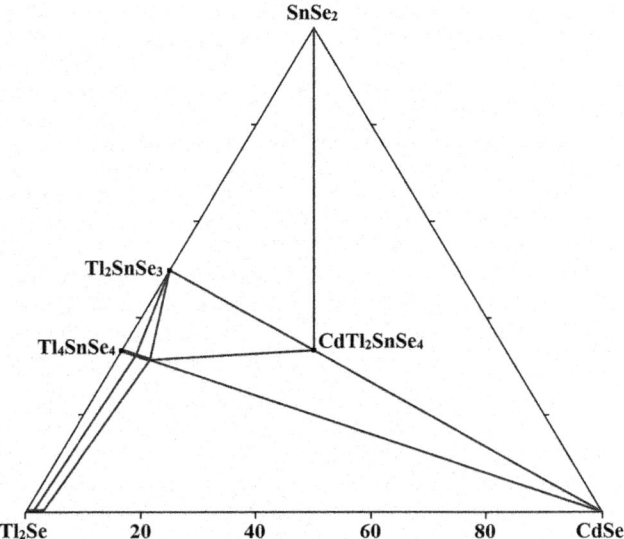

FIGURE 8.51 Isothermal section of the SnSe₂–CdSe–Tl₂Se quasiternary system at 300°C. (From Selezen, A.O., et al., *J. Solid State Chem.*, **289**, 121422, 2020.)

8.111 Tin–Cadmium–Arsenic–Selenium

CdSnAs₂–CdSe. The solubility of CdSnAs₂ in 2CdSe is 6 mol% (Dovletmuradov et al. 1971, 1972). This system was investigated through DTA, XRD, metallography, and measuring of the microhardness.

8.112 Tin–Cadmium–Tellurium–Selenium

SnTe–CdSe. This section is nonquasibinary section of the Sn–Cd–Te–Se quaternary system (Dubrovin et al. 1986). The solubility of CdSe in SnTe at 780°C is equal to 14 mol% and decreases to 3 mol% at 400°C. The solubility of SnTe in CdSe is not higher than 1 mol%. Solid solutions based on CdTe with the sphalerite structure and based on SnSe with the orthorhombic structure are formed at the interaction of CdSe and SnTe. This system was investigated through DTA, XRD, and metallography. The ingots were annealed at 400°C, 700°C, and 750°C for 250, 120, and 100 h, respectively.

SnTe + CdSe ↔ SnSe + CdTe. The isothermal sections of this ternary mutual system at 730°C and 530°C (Figure 8.52) have been constructed using EPMA and XRD (Leute and Menge 1992). The tie line fields for the structural miscibility gaps and the three-phase triangles were calculated from data, which mainly have been determined on the quasibinary edge systems. At an intermediate temperature between 530°C and 730°C, a region must exist, where four solid phases with different structures coexist in thermodynamic equilibrium. The ingots were equilibrated for 30 days at 730°C and for 70 days at 530°C.

8.113 Tin–Mercury–Gallium–Selenium

SnSe₂–HgSe–Ga₂Se₃. The liquidus surface of this system includes 6 fields of primary crystallizations (Figure 8.53) (Olekseyuk and Parasyuk 1997). The field of the Ga₂Se₃ primary crystallization occupies the most part of this surface.

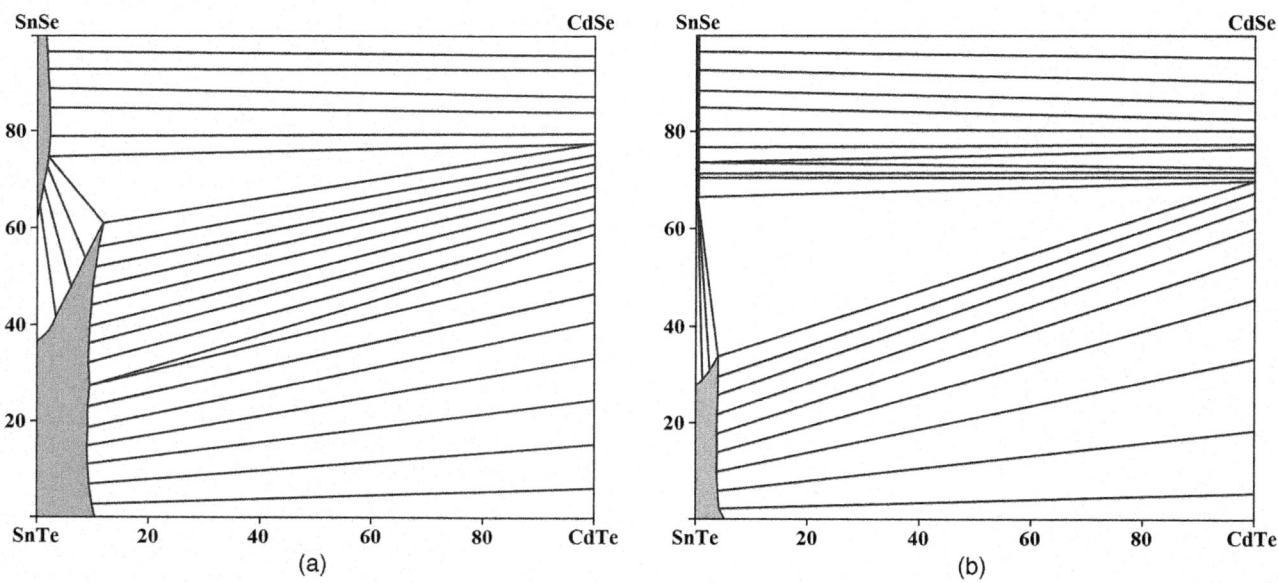

FIGURE 8.52 Isothermal sections of the SnTe + CdSe ↔ SnSe + CdTe ternary mutual system at (a) 730°C and (b) 530°C. (From Leute, V., and Menge, D., *Z. Phys. Chem. (Munchen)*, **176**(1), 65, 1992.)

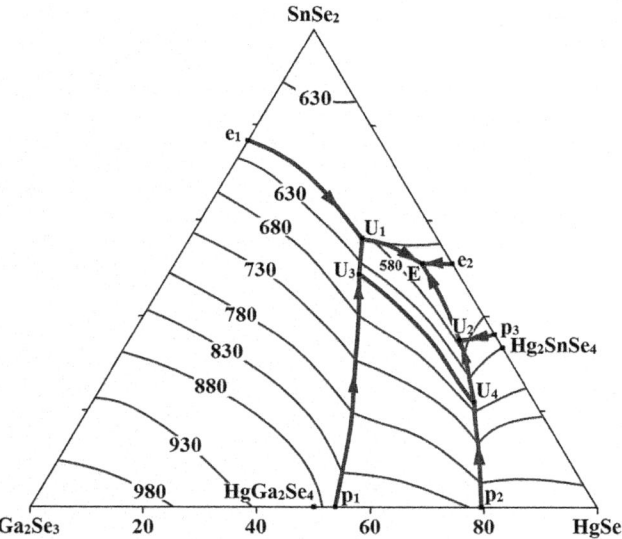

FIGURE 8.53 Liquidus surface of the SnSe₂–HgSe–Ga₂Se₃ quasiternary system. (From Olekseyuk, I.D. and Parasyuk, O.V., *Zhurn. neorgan. khimii*, **42**(5), 838, 1997.)

There are one ternary eutectic and four transition points in the system: E (552°C) – L ↔ β-HgGa₂Se₄+ Hg₂SnSe₄ + SnSe₂; U₁ (582°C) – L + Ga₂Se₃ ↔ β-HgGa₂Se₄ + SnSe₂; U₂ (567°C) – L + HgSe ↔ β-HgGa₂Se₄ + Hg₂SnSe₄; U₃ (652°C) – L + Ga₂Se₃ ↔ α-HgGa₂Se₄ + β-HgGa₂Se₄; and U₄ (652°C) – L + HgSe ↔ α-HgGa₂Se₄ + β-HgGa₂Se₄. This system was investigated by DTA, XRD, and metallography, and measuring of the microhardness. The ingots were annealed at 450°C for 250 h.

The glass-forming region in this quasiternary system is shown in Figure 8.54 (Olekseyuk et al. 1999b). The glass transition temperature, the crystallization temperature, and the melting temperature were determined for obtained glasses.

The glasses have considerable tendency to crystallization, and they can be obtained only in the case of the rigid hardening.

8.114 Tin–Mercury–Thallium–Selenium

SnSe₂–HgSe–Tl₂Se. The isothermal section of this system at 250°C is shown in Figure 8.55 (Olekseyuk et al. 2010a, Mozolyuk et al. 2012c). The **HgTl₂SnSe₄** quaternary compound, which crystallizes in the tetragonal structure with the lattice parameters $a = 804.07 \pm 0.01$, $c = 688.52 \pm 0.02$ pm, and a calculated density of 7.7871 ± 0.0005 g·cm⁻³, is formed in this system. The ingots for the investigations were annealed at 250°C for 250 h.

8.115 Tin–Gallium–Thallium–Selenium

Three quaternary compounds, **TlGaSnSe₄**, **TlGaSn₂Se₆**, and **Tl₂Ga₂SnSe₆**, are formed in the Sn–Ga–Tl–Se system. TlGaSnSe₄ has two polymorphic modifications: first of them crystallizes in the monoclinic structure with the lattice parameters $a = 750.1 \pm 0.1$, $b = 1217.5 \pm 0.1$, $c = 1820.3 \pm 0.1$ pm, $\beta = 97.164 \pm 0.003°$, a calculated density of 5.646 g·cm⁻³, and an energy gap of 1.95 eV and the second modification crystallizes in the cubic structure with the lattice parameter $a = 1347.55 \pm 0.02$ pm, a calculated density of 5.770 g·cm⁻³, and an energy gap of 1.84 eV (Friedrich et al. 2020). Polycrystalline samples of red α-GaTlSnSe₄ were obtained by melting Tl with stoichiometric mixture of Ga₂Se₃, Sn, and Se in sealed quartz ampoule at 1000°C for 2 days and cooling to room temperature with a rate of 15°C·h⁻¹. Polycrystalline batches of red β-GaTlSnSe₄ were obtained by annealing Tl with stoichiometric mixtures of Ga₂Se₃, Sn, and

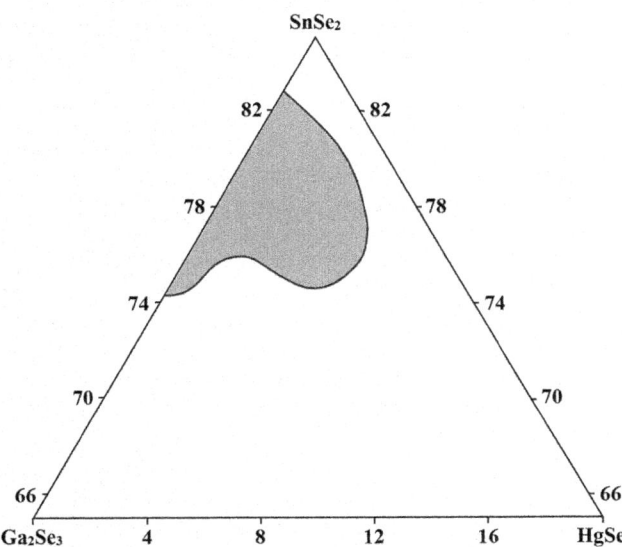

FIGURE 8.54 Glass-forming region in the SnSe₂–HgSe–Ga₂Se₃ quasiternary system. (From Olekseyuk, I.D. et al., *Funct. Mater.*, **6**(3), 474, 1999.)

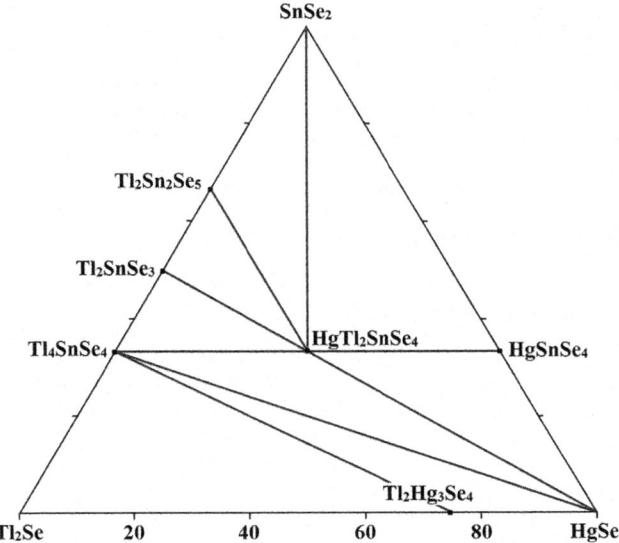

FIGURE 8.55 Isothermal section of the SnSe₂–HgSe–Tl₂Se quasiternary system at 250°C. (From Mozolyuk, M.Yu. et al., *Mater. Res. Bull.*, **47**(11), 3830, 2012.)

Se in a sealed evacuated quartz tube at 500°C for 7 days. After the initial annealing the crude product was thoroughly grounded in an agate mortar under an N_2 atmosphere and pressed to a pellet. The pellet was annealed in an evacuated quartz tube for an additional week at 500°C. This compound decomposes in moist air within several minutes by releasing gaseous H_2Se.

$TlGaSn_2Se_6$ crystallizes in the trigonal structure with the lattice parameters $a = 1032.89 \pm 0.02$, $c = 943.40 \pm 0.04$ pm, and a calculated density of 5.6301 ± 0.004 g·cm⁻³, and an energy gap of 1.86 eV (Parasyuk et al. 2017). Single crystals of this compound were grown by melting the batches of Tl, Ga, Sn, and Se. They were obtained in evacuated quartz ampoule in a furnace by heating to the maximum temperature of 800°C at a rate of 20°C·h⁻¹, exposure at this temperature for 6 h, cooling to 400°C at a rate of 10–20°C·h⁻¹, annealing at 400°C for 240 h, followed by quenching in air.

$Tl_2Ga_2SnSe_6$ crystallizes in the tetragonal structure with the lattice parameters $a = 809.5 \pm 0.1$, $c = 640.2 \pm 0.1$ pm, and a calculated density of 5.99 g·cm⁻³, and an energy gap of 2.15 eV (Babizhetskyy et al. 2020). This compound was prepared from the elements. Crystals were synthesized and grown in fused silica ampoules with a conical bottom. 40 mol% $SnSe_2$ and 60 mol% $TlGaSe_2$ (total mass 2 g) were placed in a fused silica tube, evacuated to a residual pressure of 0.1 Pa, and then sealed. The samples were synthesized in a furnace by heating to 900°C at a rate of 40°C·h⁻¹. The melt was kept at this temperature for 6 h with periodic vibration to assure homogeneity, cooled to 600°C at a rate of 20°C·h⁻¹, and annealed for 10 days. The process ended in cooling the ampoule to room temperature at 8.4°C·h⁻¹. Single crystals of $Tl_2Ga_2SnSe_6$ were grown using the Bridgman-Stockbarger method.

8.116 Tin–Indium–Thallium–Selenium

$SnSe_2$–$TlInSe_2$. The phase diagram system of this system, constructed through DTA and XRD, is a eutectic type (Figure 8.56) (Mozolyuk et al. 2011). The eutectic contains 63 mol% $SnSe_2$ and crystallizes at 515°C. The solubility of $SnSe_2$ in $TlInSe_2$ at 400°C is 28 mol%. The homogenizing annealing was held at 400°C for 250 h, followed by quenching the alloys into cold water.

8.117 Tin–Indium–Lanthanum–Selenium

The $InLa_3Sn_{0.5}Se_7$ quaternary compound, which crystallizes in the hexagonal structure with the lattice parameters $a = 1065.33 \pm 0.07$, $c = 624.45 \pm 0.04$ pm, and a calculated density of 6.015 g·cm⁻³, is formed in the Sn–In–La–Se system (Iyer et al. 2017a). To synthesize this compound, a stoichiometric mixture of the elements with a total mass of 0.2 g was pressed into a pellet and loaded into a fused silica tube, which was evacuated and sealed. The tube was heated at 950°C for 4 days and cooled to 600°C over another 4 days, and then the furnace was turned off.

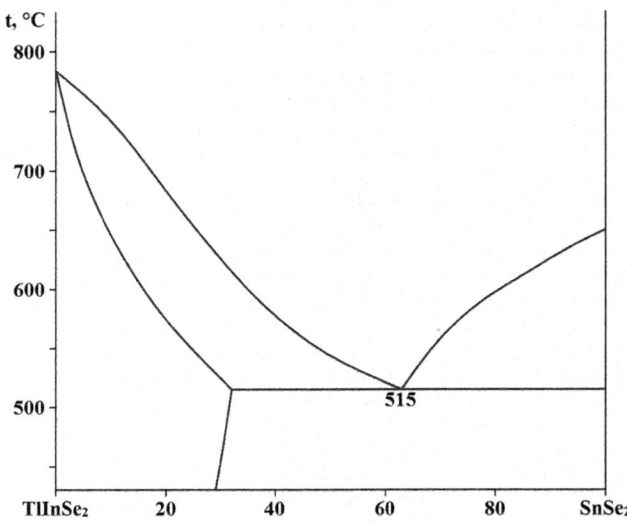

FIGURE 8.56 Phase diagram of the $SnSe_2$–$TlInSe_2$ system. (From Mozolyuk, M.Yu., *J. Alloys Compd.*, **509**(6), 2693, 2011.)

8.118 Tin–Indium–Lead–Selenium

The $In_4Pb_{0.01}Sn_xSe_3$ solid solution ($x = 0.01–0.05$), which formed in the Sn–In–Pb–Se quaternary system, was synthesized by solid-state reaction (Lin et al. 2013). Appropriate amounts of starting materials, In, Pb, Sn, and Se were weighed and mixed in a silica crucible in an Ar-filled glove box. The crucible was transferred into a larger silica jacket, which was flame-sealed under a high vacuum. The assembly was slowly heated to 950°C over 15 h and dwelled there for 1 day. The assembly was then cold-water quenched. The obtained product was ground into a fine powder and pressed into a pellet in an argon-filled glove box. The pellet was annealed at 480°C for 5 days, then ground into a fine powder and subsequently sintered by spark plasma sintering at 430°C for 5 min under uniaxial pressure of 70 MPa under vacuum (10⁻³ Pa). The experimental relative densities of sample pellets were all above 95%.

8.119 Tin–Indium–Antimony–Selenium

In_2SnSe–$InSb$. Specimens with compositions in the range of 0–20 mol% In_2SnSe were investigated by Woolley and Williams (1964). It was shown that for the specimens quenched from 800°C, the solid solubility extends out to about 5 mol% In_2SnSe. The energy gap of the solid solutions increases to a maximum value of 0.42 eV at 1.5 mol% In_2SnSe and then decreases with increasing In_2SnSe content.

Two quaternary compounds, $In_5Sn_4Sb_9Se_{25}$ and $In_5Sn_8Sb_{13}Se_{35}$, which crystallizes in the monoclinic structure with the lattice parameters $a = 3182.0 \pm 0.2$, $b = 402.41 \pm 0.03$, $c = 1970.3 \pm 0.2$ pm, $\beta = 114.246 \pm 0.005°$, a calculated density of 5.946 g·cm⁻³, and an energy gap of 0.66 eV for the first compound and $a = 3170.6 \pm 0.8$, $b = 404.99 \pm 0.09$, $c = 2643.4 \pm 0.6$ pm, $\beta = 105.538 \pm 0.004°$, and a calculated density of 5.961 g·cm^{-3} for the second one, are formed in the Sn–In–Sb–Se system (Chen et al. 2022a).

To prepare $In_5Sn_4Sb_9Se_{25}$, a mixture of Sn, In, Sb, and Se (molar ratio 12:7:15:45) were sealed in a silica tube under vacuum (~2 Pa), heated from room temperature to 750°C in 8 h, held at 750°C for 48 h, cooled to 400°C in 24 h, followed by cooling naturally to room temperature (Chen et al. 2022a). The products were then finely ground into powder and sealed in a silica tube again. The sealed tube was further annealed at 550°C for 24 h to obtain the pure product. The as-synthesized $In_5Sn_4Sb_9Se_{25}$ is stable under air.

A single-crystal sample of $In_5Sn_8Sb_{13}Se_{35}$ was first found as one of the by-products at the obtaining of $In_5Sn_4Sb_9Se_{25}$ (Chen et al. 2022a). The pure phase of this compound was obtained with the following procedures. The mixture of Sn, In, Sb, and Se was initially heated from room temperature to 800°C for 8 h; the temperature was maintained for 48 h, then cooled to 400°C over 24 h, and finally cooled down naturally by turning off the furnace. An additional annealing process was carried out at 580°C for 24 h to improve the crystallinity. $In_5Sn_8Sb_{13}Se_{35}$ is stable in the air under ambient conditions. All operations at the obtaining of both compounds were performed in a glove box with a dry N_2 atmosphere.

8.120 Tin–Indium–Bismuth–Selenium

Two quaternary compounds, $In_{0.2}Sn_6Bi_{1.8}Se_9$ and $InSn_2Bi_3Se_8$, are formed in the Sn–In–Bi–Se system (Wang et al. 2009b). First compound decomposes at ~760°C and crystallizes in the orthorhombic structure with the lattice parameters $a = 418.10 \pm 0.08$, $b = 1379.9 \pm 0.3$, and $c = 3195.3 \pm 0.6$ pm and the second one melts at 650–680°C, decomposes at ~720°C, crystallizes in the monoclinic structure with the lattice parameters $a = 1355.7 \pm 0.3$, $b = 412.99 \pm 0.08$, $c = 1525.2 \pm 0.3$ pm, $\beta = 115.73 \pm 0.03°$, and a calculated density of 6.954 g·cm^{-3}.

$In_{0.2}Sn_6Bi_{1.8}Se_9$ was first observed on annealing a pressed pellet sample of $InSn_2Bi_3Se_8$ at 800°C (Wang et al. 2009b). Needle-shaped crystals were observed on the cool side of the silica ampoule. After the structure and composition were confirmed, this compound was synthesized as a pure phase using a mixture of In, Sn, Bi, and Se (molar ratio 0.63:5.57:1.8:9), which was heated to 650°C over 12 h and held there for 24 h, followed by cooling to room temperature naturally in a gradient furnace. The crystals of the title compound were observed in the cool part of the silica ampoule, with SnSe, and Se attached on the tube wall.

$InSn_2Bi_3Se_8$ was synthesized from a mixture of In, Sn, Bi, and Se (molar ratio 1:2:3:8) (Wang et al. 2009b). The reaction mixture in a quartz evacuated and sealed ampoule was heated from room temperature to 800°C over 8 h; the latter temperature was maintained for 24 h followed by cooling to room temperature on simply terminating the power. This compound is stable under ambient conditions. All operations at the obtaining of both compounds were performed in a glove box with a dry Ar atmosphere.

8.121 Tin–Indium–Tellurium–Selenium

SnTe–InSe. The phase diagram of this system, constructed through DTA, XRD, metallography, and measuring of the microhardness, is presented in Figure 8.57 (Gurshumov et al. 1985; Ashirov et al. 1986). The eutectic contains 70 mol% InSe and crystallizes at 477°C. The **SnInTeSe** quaternary compound, which melts incongruently at 497°C, is formed in this system. At 500°C, the solubility of InSe in SnTe and SnTe in InSe is 20 and 15 mol%, respectively. At room temperature, the mutual solubility of SnTe and InSe is not higher than 1.5–2 mol%. The ingots for the investigations were annealed at 450°C for 240–250 h.

The $In_xSn_5Te_5Se_{90-x}$ glasses at $x \leq 9$ are formed in the Sn–In–Te–Se quaternary system (Kumar and Singh 2011, 2012). The crystallization kinetics in these glasses has been studied under nonisothermal conditions using different scanning calorimetry (DSC) technique. The alloys with $x = 9$ has better glass-forming ability than other glasses. The glass transition temperature is dependent on composition and heating rates. Their thermal stability increases with increasing x, and the glass with $x = 9$ is the better thermally stable in this system. To prepare these glasses, Se, Te, Sn, and In were weighed in appropriate atomic ratio. The mixtures were evacuated in quartz ampoules, which were sealed under vacuum of 10–3 Pa to remove possibility of any reaction of alloys with oxygen at high temperature. The ampoules were heated in a furnace at the rate of 3–4°C.min–1 up to 825°C for 12 h. They were frequently rocked to ensure homogeneity of the samples. The molten samples were then rapidly quenched in ice-cooled water to obtain the alloys in their glassy state.

8.122 Tin–Thallium–Lead–Selenium

Tl_4SnSe_3–Tl_4PbSe_3. The phase diagram of this system, constructed through DTA and XRD, is a peritectic type with the peritectic at 62 mol% Tl_4PbSe_3 and 481°C (Figure 8.58) (Filep et al. 2012b). At 250°C, the solubility of Tl_4PbSe_3 in Tl_4SnSe_3

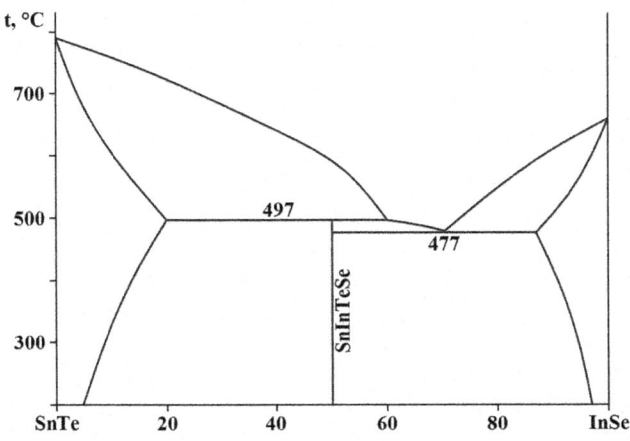

FIGURE 8.57 Phase diagram of the SnTe–InSe system. (From Ashirov, A., et al., *Zhurn. neorgan. khimii*, **31**(5), 1282, 1986.)

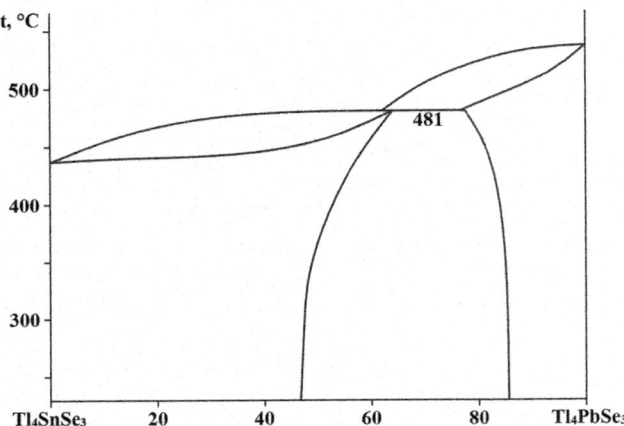

FIGURE 8.58 Phase diagram of the Tl₄SnSe₃–Tl₄PbSe₃ system. (From Filep, M.Y., et al., *Chem. Met. Alloys*, **5**(3–4), 118, 2012.)

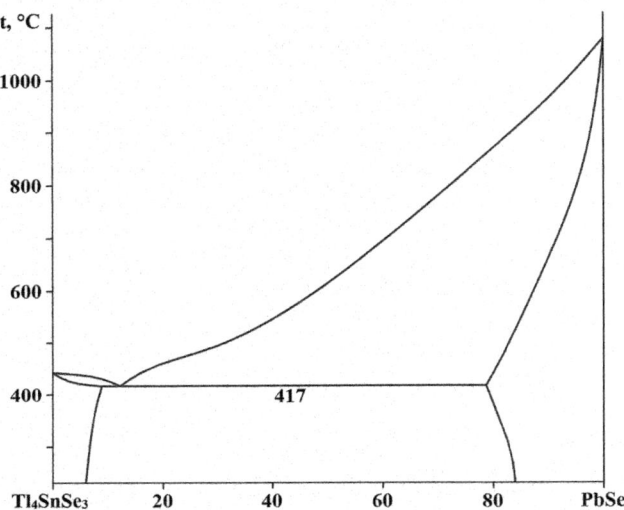

FIGURE 8.59 Phase diagram of the PbSe–Tl₄SnSe₃ system. (From Filep, M.Y., et al., *Chem. Met. Alloys*, **5**(3–4), 118, 2012.)

is 48 mol%, and the solubility of Tl₄SnSe₃ in Tl₄PbSe₃ does not exceed 15 mol%.

PbSe–Tl₄SnSe₃. The phase diagram of this system, constructed through DTA and XRD, is a eutectic type (Figure 8.59) (Filep et al. 2012b). The eutectic contains 12 mol% PbSe and crystallizes at 417°C. At 250°C, the solubility of PbSe in Tl₄SnSe₃ is 7 mol% and the solubility of Tl₄SnSe₃ in PbSe reaches 15 mol%.

SnSe–Tl₂Se–PbSe. The isothermal section of this quaternary system at 250°C is presented in Figure 8.60 (Filep et al. 2012b). No quaternary compounds were observed in the system. The liquidus surface of the system (Figure 8.61) consists of 6 fields of primary crystallization, which are delimited by 11 monovariant lines. There are two ternary eutectics and two transition points on the liquidus surface: E₁ (384°C) – L ↔ (Tl₄SnSe₃) + (PbSe) + (Tl₄PbSe₃); E₂ (374°C) – L ↔ (Tl₄SnSe₃) + (PbSe) + (Tl₂Sn₂Se₃); U₁ (452°C) – L + (β-SnSe) ↔ (PbSe) + (α-SnSe); and U₂ (400°C) – L + (α-SnSe) ↔ (Tl₂Sn₂Se₃) + (PbSe). The

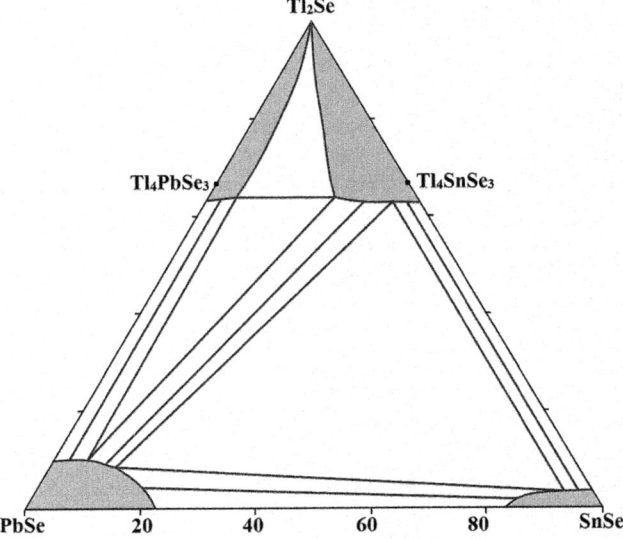

FIGURE 8.60 Isothermal section of the SnSe–Tl₂Se–PbSe quaternary system at 250°C. (From Filep, M.Y., et al., *Chem. Met. Alloys*, **5**(3–4), 118, 2012.)

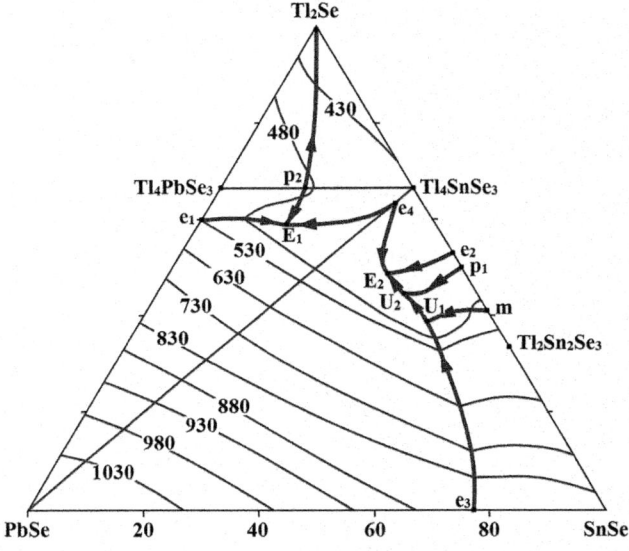

FIGURE 8.61 Liquidus surface of the SnSe–Tl₂Se–PbSe quaternary system. (From Filep, M.Y., et al., *Chem. Met. Alloys*, **5**(3–4), 118, 2012.)

alloys for the investigation of these systems were annealed at 250°C for 168 h, with the next quenching in cold water.

The isothermal section of the SnSe–Tl₂Se–PbSe quaternary system at 230°C was also constructed by Guseynov et al. (2001). However, the authors did not take into consideration the formation in the Sn(Pb)Se–Tl₂Se–system of the ternary compounds Tl₄SnSe₃ and Tl₄PbSe₃, which melt congruently.

SnSe₂–Tl₂Se–PbSe. The isothermal section of this system at room temperature was constructed using the alloy annealed at 400°C for 250 h (Olekseyuk et al. 2011). The system is divided into five subsystems by the Tl₂Sn₂Se₃–PbSe, Tl₂SnSe₃–PbSe, Tl₄SnSe₄–PbSe, and Tl₄SnSe₄–Tl₄PbSe₃ sections. Alloys adjacent directly to SnSe₂–PbSe section contain not 3, but 4 phases as this section is nonquasibinary. This means that the tetrahedration of the Sn–Tl–Pb–Se system does not pass through $SnSe_2$, but through SnSe.

8.123 Tin–Thallium–Arsenic–Selenium

SnSe–Tl₂Se–As₂Se₃. The glass-forming region in this system is presented in Figure 8.62 (Vasil'ev and Bakhtiyarov 1975). According to the data of nuclear gamma resonance, Sn can be both two- and four-valent. In the region of glass-formation, tin is mainly tetravalent, and divalent tin is present only at the boundary of the region in glasses. The maximal content of Sn in glasses reaches up to 40 mol%.

SnSe₂–Tl₂Se–As₂Se₃. The glass-forming region in this system is shown in Figure 8.63 (Bakhtiyarov and Vasil'ev 1974). In the glasses of the system, the structural units of tin are close to its structural units in the $SnSe_2$, Tl_2SnSe_3, and Tl_4SnSe_4 compounds.

8.124 Tin–Thallium–Antimony–Selenium

SnSe–TlSbSe₂. In this system, a continuous series of solid solutions is formed (Figure 8.64) (Medzhidov et al. 1993). The phase diagram does not reflect the phase transition of SnSe, which requires additional studies. It was studied through DTA and XRD using the alloys annealed at 320°C and 360°C for 240 h.

SnSe₂–Tl₂Se–Sb₂Se₃. This quasiternary system is divided onto 5 subsystems by the $SnSe_2$–$TlSbSe_2$, Tl_2SnSe_3–$TlSbSe_2$, Tl_4SnSe_4–$TlSbSe_2$, and Tl_4SnSe_4–Tl_9SbSe_6 (Tats'kar et al. 2013).

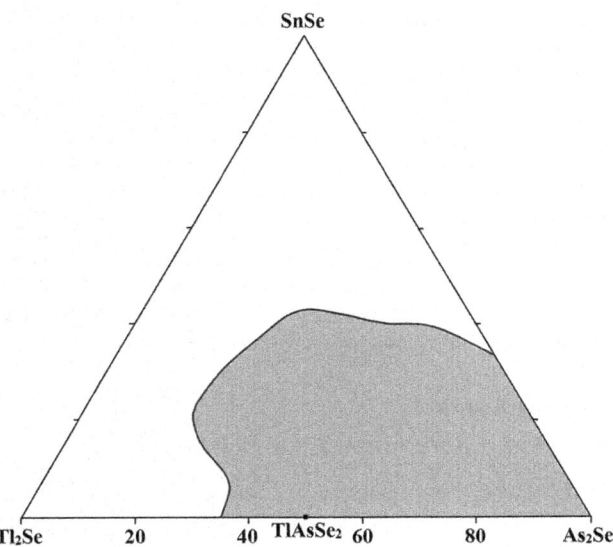

FIGURE 8.62 Glass-forming region in the SnSe–Tl₂Se–As₂Se₃ system. (From Vasil'ev, L.N., and Bahtiyarov, A.Sh., *Izv. AN SSSR. Neorgan. mater.*, **11**(11), 2074, 1975.)

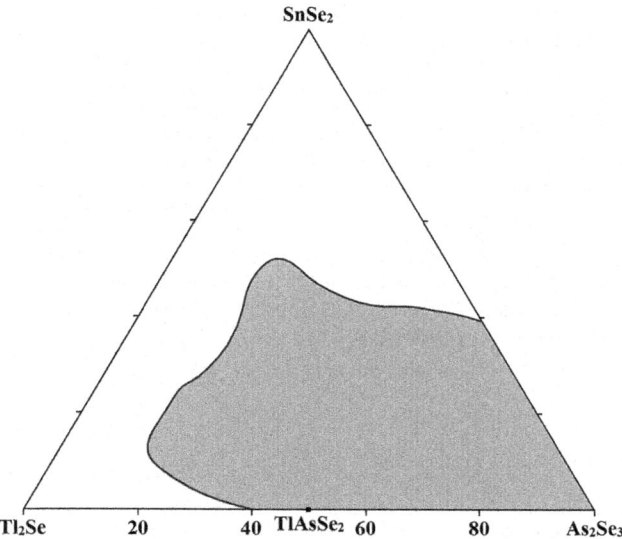

FIGURE 8.63 Glass-forming region in the SnSe₂–Tl₂Se–As₂Se₃ system. (From Bakhtiyarov, A.Sh., and Vasil'ev, L.N., *Izv. AN SSSR. Neorgan. mater.*, **10**(11), 2079, 1974.)

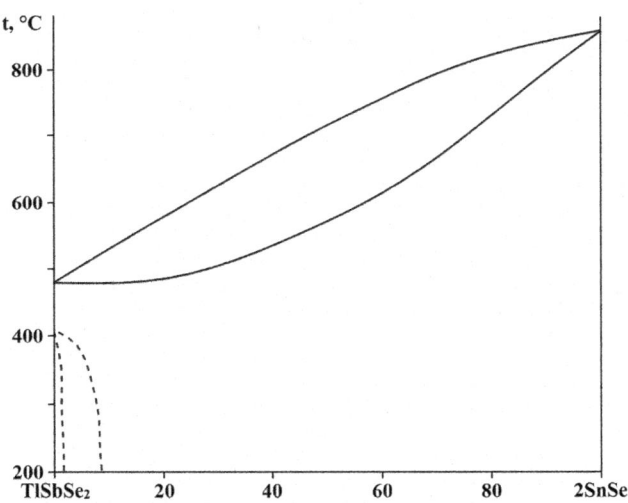

FIGURE 8.64 Phase diagram of the 2SnSe–TlSbSe₂ system. (From Medzhidov, G.A., et al., *Neorgan. mater.*, **29**(9), 1304, 1993.)

8.125 Tin–Thallium–Bismuth–Selenium

SnSe–Tl₂Se–Bi₂Se₃. This quasiternary system has three quasibinary sections: SnSe–TlBiSe₂, Tl₄SnSe₃–Tl₉BiSe₆, and Tl₄SnSe₃–TlBiSe₂, which divide it into four subsystems (Masalovych et al. 2011a).

SnSe₂–TlBiSe₂. The phase diagram of this system, constructed through DTA and XRD, is a eutectic type (Figure 8.65 (Koz'ma et al. 2008). The eutectic contains 40 mol% TlBiSe₂ and crystallizes at 454°C. The solubility of TlBiSe₂ in $SnSe_2$ and $SnSe_2$ in TlBiSe₂ at the eutectic temperature is 10 and 25 mol%, respectively. The solubility of the components in each other decreases with decreasing temperature. The ingots for the investigations were annealed at 150°C for 336 h with the next quenching in water.

SnSe₂–TlBiSe₂–Bi₂Se₃. The liquidus surface of this quasiternary system is given in Figure 8.66) (Koz'ma et al. 2009). The ternary eutectic contains 49 mol% $SnSe_2$, 24 mol% $TlBiSe_2$, and 27 mol% Bi_2Se_3 and crystallizes at 400°C.

Tl₄SnSe₄–Tl₂Se. The phase diagram of this system, constructed through DTA and XRD, is a eutectic type and the eutectic crystallizes at 355°C (Koz'ma 2013). The system was studied through DTA and XRD and the ingots for the investigations were annealed at 150°C for 290 h.

Tl₄SnSe₄–Tl₂Se–Tl₉BiSe₆. The liquidus surface of this quasiternary system is presented in Figure 8.67 (Koz'ma 2013).

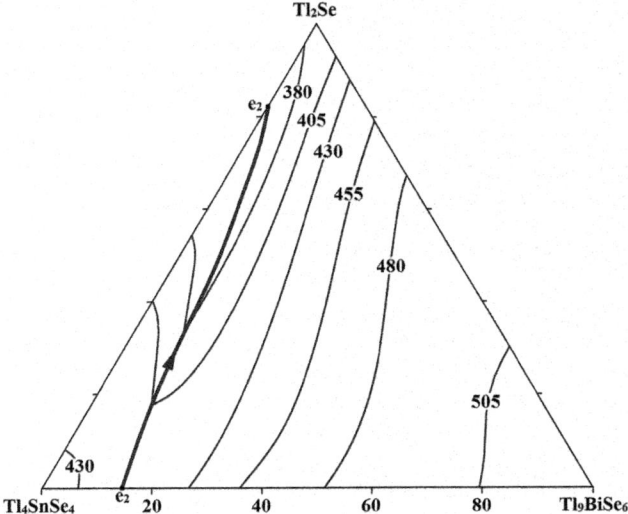

FIGURE 8.67 Liquidus surface of the Tl_4SnSe_4–Tl_2Se–Tl_9BiSe_6 quasiternary system. (From Koz'ma A.A., *Nauk. Visn. Uzhgorod. Univ. Ser. Khimiya*, [2(30)], 15, 2013.)

Solid solutions based on Tl_4SnSe_4 and $(Tl_2Se)_x(Tl_9BiSe_6)_{1-x}$ solid solutions primary crystallize in the system. The system is characterized by the monovariant eutectic process within the temperature interval of 414–355°C.

8.126 Tin–Thallium–Tellurium–Selenium

SnSe + Tl₂Te ↔ SnTe + Tl₂Se. There three quasibinary sections, $SnSe$–Tl_4SnTe_3, Tl_2Se–Tl_4SnTe_3, and Tl_4SnSe_3–Tl_4SnTe_3, in this ternary mutual system (Filep et al. 2019). The existence of **Tl₂SnSeTe** quaternary compound has not been confirmed. This system was studied through DTA, XRD, and metallography and the ingots for the investigations were annealed at 200°C for 72 h.

According to the data of Ashirov et al. (1986), the phase diagram of the **SnTe–Tl₂Se** is a eutectic type with peritectic transformation. The eutectic contains 86 mol% Tl_2Se and crystallizes at 350°C. The solubility of Tl_2Se in SnTe is 5 mol% at 400°C and decreases to 2 mol% at room temperature. The solubility of SnTe in Tl_2Se reaches 5 mol% at the eutectic temperature and is 1 mol% at room temperature. The $Tl_2SnSeTe$ quaternary compound, which melts incongruently at 450°C, is formed in the SnTe–Tl₂Se system. The system was studied through DTA, XRD, metallography, and measuring of the microhardness.

8.127 Tin–Lanthanum–Iron–Selenium

The **La₃SnFe₀.₆₁Se₇** quaternary compound, which crystallizes in the hexagonal structure with the lattice parameters $a = 1068.04 \pm 0.06$, $c = 630.40 \pm 0.06$ pm, and a calculated density of 5.99 g·cm⁻³, is formed in the Sn–La–Fe–Se system (Assoud et al. 2014). This compound was synthesized via a solid-state reaction method, starting with the elements in the

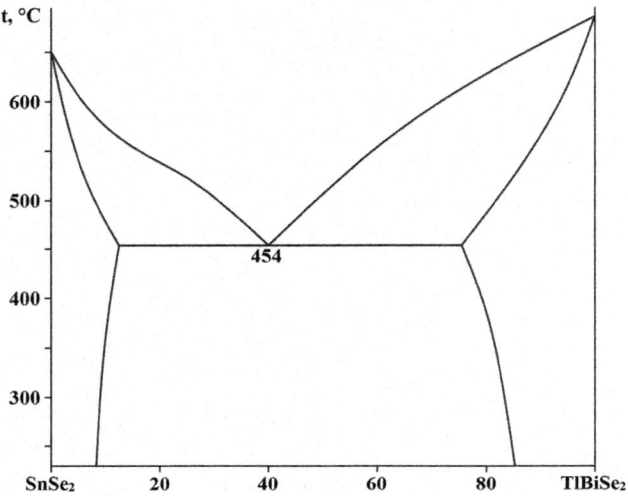

FIGURE 8.65 Phase diagram of the $SnSe_2$–$TlBiSe_2$ system. (From Koz'ma, A.A., et al., *Nauk. Visn. Uzhgorod. Univ. Ser. Khimiya*, (19–20), 89, 2008.)

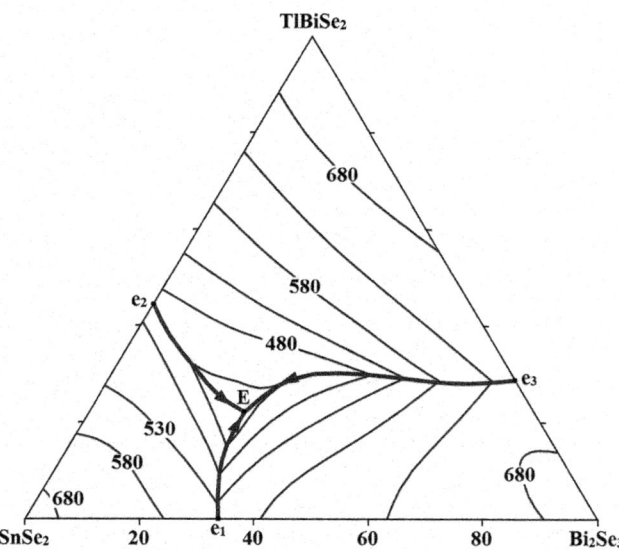

FIGURE 8.66 Liquidus surface of the $SnSe_2$–$TlBiSe_2$–Bi_2Se_3 quasiternary system. (From Koz'ma, A.A., et al., *Nauk. Visn. Uzhgorod. Univ. Ser. Khimiya*, (21), 6, 2009.)

stoichiometric ratios of La/Sn/Fe/Se = 3:1:0.5:7. The elements were placed into a fused silica tube, which was then sealed under a dynamic vacuum. This tube was heated in a furnace to 800°C and kept at that temperature for 4 days, followed by slow cooling to 300°C. Then the furnace was switched off to cool to room temperature. Black crystals of $La_3SnFe_{0.61}Se_7$ were obtained.

8.128 Tin–Lead–Bismuth–Selenium

Two quaternary phases, $Sn_{3-x}Pb_xBi_2Se_6$ ($x \leq 0.7$) and $Sn_{6-x}Pb_xBi_2Se_9$ ($x \leq 4.36$), and two quaternary compounds, $Sn_2Pb_5Bi_4Se_{13}$ and $Sn_{8.65}Pb_{0.35}Bi_4Se_{15}$, are formed in the Sn–Pb–Bi–Se system. $Sn_{3-x}Pb_xBi_2Se_6$ crystallizes in the orthorhombic structure with the lattice parameters $a = 2121.3 \pm 0.6$, $b = 416.2 \pm 0.1$, $c = 1364.0 \pm 0.4$ pm, and a calculated density of 6.883 g·cm⁻³ for $x = 0$; $a = 2137.3 \pm 0.4$, $b = 418.98 \pm 0.08$, $c = 1369.0 \pm 0.2$ pm, and a calculated density of 6.808 g·cm⁻³ for $x = 0.1$; $a = 2136.5 \pm 0.6$, $b = 419.0 \pm 0.1$, $c = 1374.1 \pm 0.4$ pm, and a calculated density of 6.881 g·cm⁻³ for $x = 0.3$; $a = 2139.9 \pm 0.8$, $b = 419.0 \pm 0.2$, $c = 1379.7 \pm 0.5$ pm, and a calculated density of 6.923 g·cm⁻³ for $x = 0.5$; and $a = 2123.5 \pm 0.8$, $b = 419.8 \pm 0.2$, $c = 1384.6 \pm 0.6$ pm, and a calculated density of 7.048 g·cm⁻³ for $x = 0.7$ (Chen and Lee 2010a). These phases were obtained by heating stoichiometric mixtures of Sn, Pb, Bi, and Se in evacuated quartz tubes to 800°C over 12 h, holding at this temperature for 18 h, slow cooling to 500°C for 12 h, and then cooling to room temperature naturally. All operations were performed in a glove box with a dry nitrogen atmosphere.

$Sn_{6-x}Pb_xBi_2Se_9$ also crystallizes in the orthorhombic structure with the lattice parameters $a = 420.6 \pm 0.2$, $b = 1390.3 \pm 0.6$, $c = 3212 \pm 1$ pm, and a calculated density of 6.510 g·cm⁻³ for $x = 0$; $a = 421.05 \pm 0.08$, $b = 1394.5 \pm 0.3$, $c = 3217.4 \pm 0.5$ pm, and a calculated density of 7.026 g·cm⁻³ for $x = 1.78$; and $a = 424.69 \pm 0.07$, $b = 1407.3 \pm 0.2$, $c = 3238.3 \pm 0.5$ pm, and a calculated density of 7.641 g·cm⁻³ for $x = 4.36$ (Chen and Lee 2010b). To prepare these phases, mixtures of Sn, Pb, Bi, and Se in stoichiometric ratios were placed in evacuated fused silica tubes and put in a furnace. They were heated from room temperature to 750°C over 8 h. The latter temperature was maintained for 16 h before natural cooling to room temperature. The cuboid-shaped crystals were obtained. All operations were performed in a glove box with a dry nitrogen atmosphere.

$Sn_2Pb_5Bi_4Se_{13}$ and $Sn_{8.65}Pb_{0.35}Bi_4Se_{15}$ crystallize in the monoclinic structure with the lattice parameters $a = 1400.1 \pm 0.6$, $b = 423.4 \pm 0.2$, $c = 2347.1 \pm 0.8$ pm, $\beta = 98.46 \pm 0.01°$, and a calculated density of 7.552 g·cm⁻³ for the first compound and $a = 1387.2 \pm 0.3$, $b = 420.21 \pm 0.08$, $c = 2685.5 \pm 0.5$ pm, $\beta = 95.92 \pm 0.03°$, and a calculated density of 6.7 g·cm⁻³ for the second one (Chen and Lee 2009). Both compounds were obtained in a glove box under a dry nitrogen atmosphere. Stoichiometric mixtures of the chemical elements were placed in evacuated silica glass tubes, which were put in a furnace and heated slowly to 800°C over 12 h. This temperature was maintained for 18 h, followed by slow cooling to 500°C over 12 h, and finally to room temperature on simply terminating the power.

8.129 Tin–Lead–Oxygen–Selenium

Using the method of diagrams of coexisting phases, the process of oxidation of $Pb_{1-x}Sn_xSe$ solid solution was analyzed by Medvedev et al. (1987). A good agreement with the results of experimental studies of the natural aging process has been obtained.

8.130 Tin–Lead–Tellurium–Selenium

SnTe–PbSe. The phase diagram of this system, constructed through DTA, XRD, and metallography, belongs to the type I according to Roseboom's classification (Figure 8.68) (Borovikova et al. 1972; Abrikosov and Goncharova 1974). In this system, a continuous series of solid solutions is formed, which crystallize in a cubic structure of the NaCl type (Krebs and Langner 1964; Nikolić 1967; Borovikova et al. 1972; Abrikosov and Goncharova 1974). The lattice parameter of the solid solutions changes with a slight positive deviation from the Vegard's law (Nikolić 1967). Alloys containing 80 and 90 mol% PbSe turned out to be nonequilibrium even after addition annealing at 750°C for 100 h (Borovikova et al. 1972). The PbSe, PbTe, SnSe, and SnTe molecules were found in the vapor phase, which indicates on the existing of the mutual chemical interaction in such conditions (Kalashnikov et al. 1981).

The ingots for the investigations were annealed at 630–650°C for 100–200 h (Borovikova et al. 1972) [at 700–800°C for 350 h (Abrikosov and Goncharova 1974); at 700°C for 21 days (Nikolić 1967); at 520°C and 660°C for 70 h (Krebs and Langner 1964).

SnSe–PbTe. This system is nonquasibinary section of the SnSe+PbTe ↔ SnTe+PbSe ternary mutual system (Abrikosov et al. 1973). The point of intersection of the monovariant line of the eutectic crystallization of two solid solutions by this section corresponds to 78 mol% SnSe and a temperature of

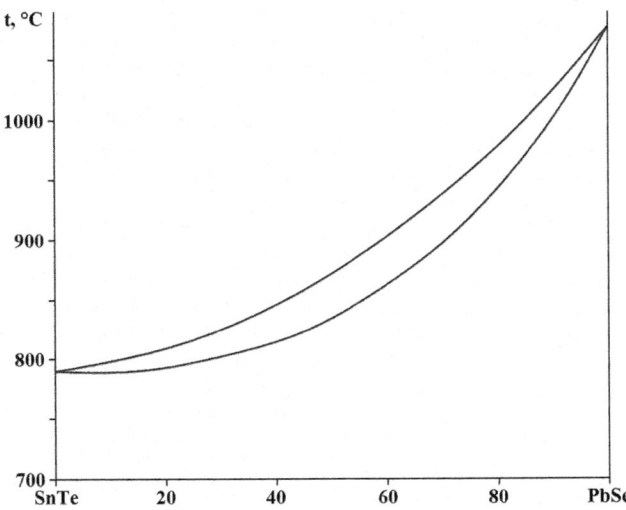

FIGURE 8.68 Phase diagram of the SnTe–PbSe system. (From Borovikova, R.P., et al., *Izv. AN SSSR. Neorgan. mater.*, **8**(10), 1762, 1972.)

820°C. The solubility of SnSe in PbTe at 620°C reaches 62.5 mol% (Abrikosov et al. 1973) [70 mol% at 700°C (Nikolić 1965)] and decreases to 39 mol% at 520°C (Krebs and Langner 1964), and the solubility of PbTe in SnSe is 10 mol% at 620°C (Abrikosov et al. 1973) [at 500°C (Krebs and Langner 1964)]. The lattice parameter and energy gap decrease linearly with composition in the field of the solid solution based on PbTe (Nikolić 1965; Abrikosov et al. 1973). The ingots for the investigations were annealed at 620°C for 500 h (Abrikosov et al. 1973) [at 500–520°C for 80 h (Krebs and Langner 1964); at 700°C for 14 days (Nikolić 1965)].

SnSe+PbTe ↔ SnTe+PbSe. Using the modeling results for solid-state equilibria and the interaction parameters for the liquid phase in the quasibinary systems, the solidus and liquidus surfaces as well as some isothermal sections of this ternary mutual system were calculated by Volykhov et al. (2009). The calculated isothermal sections at 630°C, 730°C, 830°C, and 930°C are presented in Figure 8.69.

The liquidus and solidus surfaces of the SnTe–PbSe–PbTe subsystem are presented in Figure 8.70 (Alekseeva et al. 1969). The continuous solid solutions are formed in this subsystem (Krebs and Langner 1964; Alekseeva et al. 1969; Abrikosov and Goncharova 1973), but according to the data of Krebs and Langner (1964), three-phase region exists along the SnSe–PbTe section.

The equilibrium compositions of the liquid and solid phases in the Sn–Pb–Te–Se system were calculated by Davarashvili et al. (1992) using the associated regular solution model for the liquid phase and regular solution model for the solid phase.

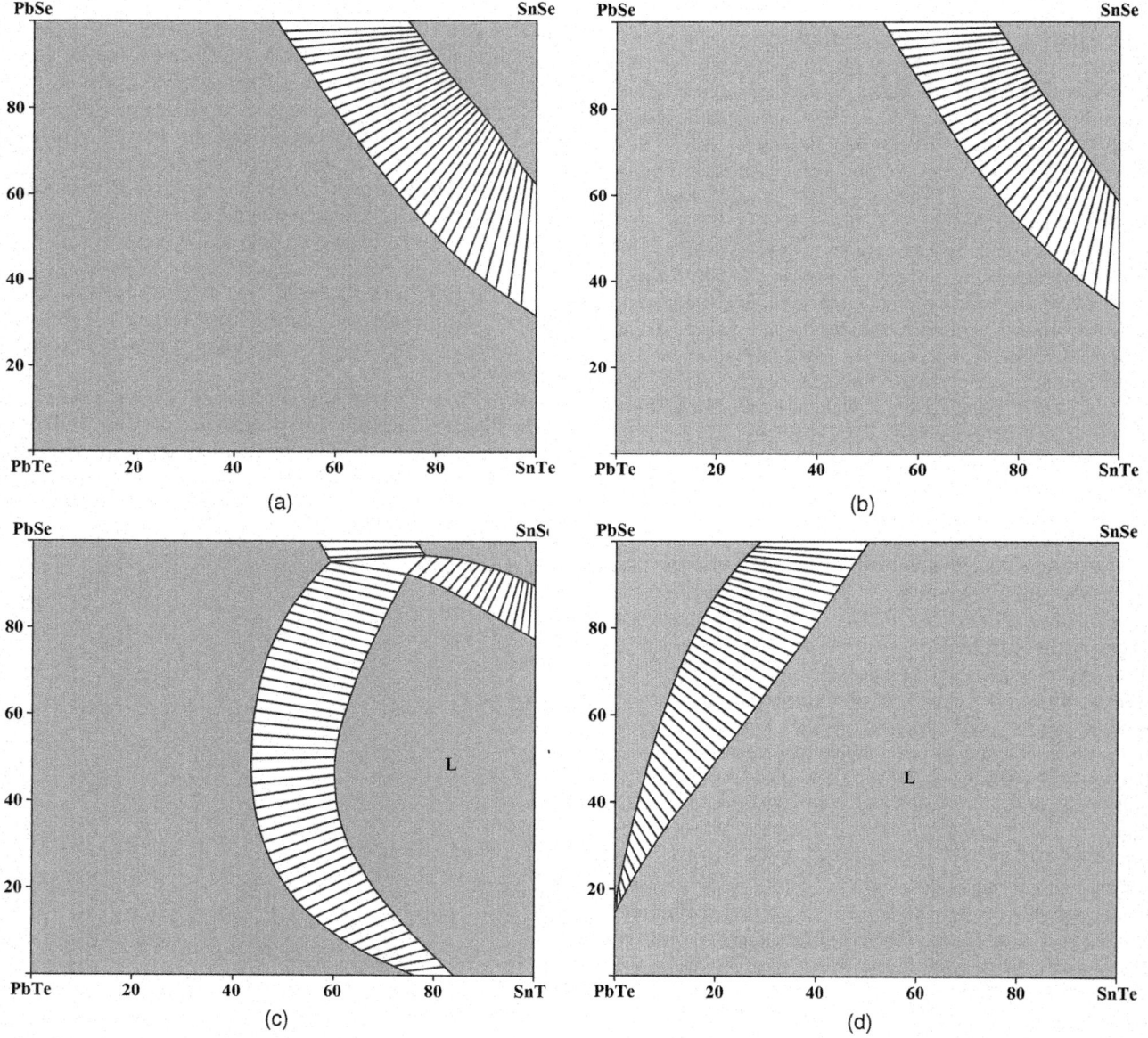

FIGURE 8.69 Isothermal section of the SnSe+PbTe ↔ SnTe+PbSe ternary mutual system at (a) 630°C, (b) 730°C, (c) 830°C, and (d) 930°C. (From Volykhov, A.A., et al., *Inorg. Mater.*, **45**(9), 968, 2009.)

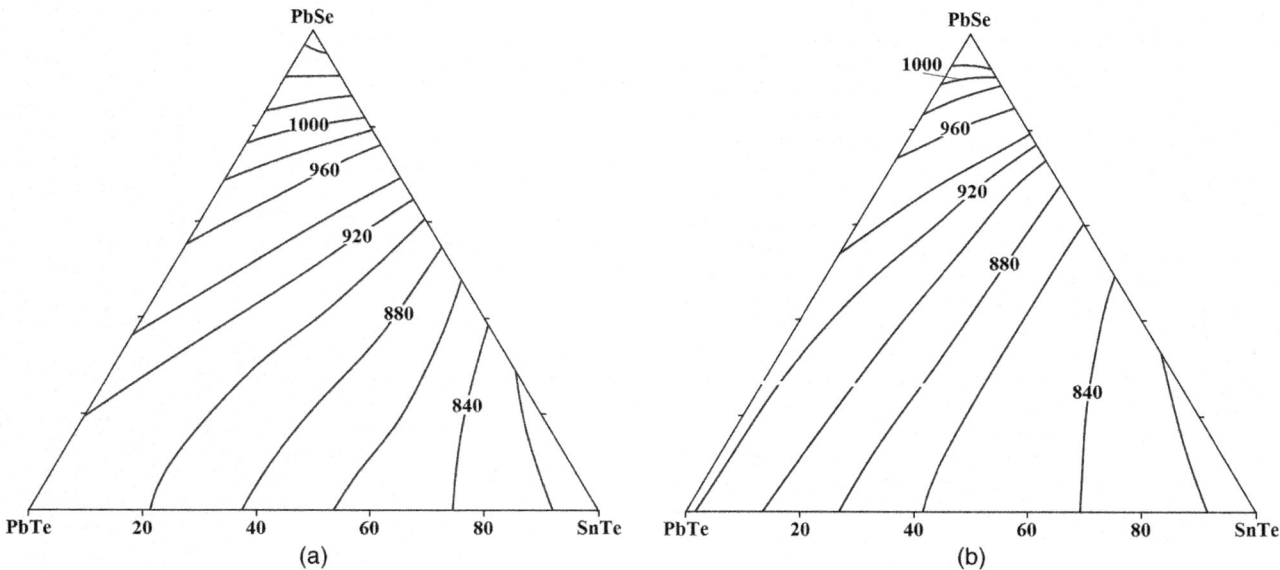

FIGURE 8.70 Liquidus (a) and solidus (b) surfaces of the SnTe–PbSe–PbTe subsystem of the SnSe+PbTe ↔ SnTe+PbSe ternary mutual system. (From Alekseeva, G.T., et al., *Izv. AN SSSR. Neorgan. mater.*, **5**(12), 2105, 1969.)

The liquidus temperatures of this system were determined by Kuznetsov et al. (1983) during cooling of alloys at different rates by extrapolating them to zero cooling rate. Using stepwise regression analysis, the liquidus surface coordinates of the Sn–Pb–Te–Se system in the composition region of $(Pb_{1-x}Sn_x)_{1-z}(Te_{1-y}Se_y)_z$ at $0.005 \leq z \leq 0.06$ were calculated. A characteristic feature of the liquidus surface is its wavy folds when x changes from 0 to 1.

Samples of $Sn_xPb_{1-x}Te_{1-y}Se_y$ solid solution were investigated in the 4–200 K temperature range using electrical and X-ray methods by Lebedev and Sluchinskaya (1994). The region where the low-temperature phase transition takes place was established (Figure 8.71). Single crystals of these solid solutions can be grown using the Bridgman method or by recrystallization through the vapor phase (Starik et al. 1979).

8.131 Tin–Lead–Iodine–Selenium

SnSe–PbI₂. According to the data of Odin et al. (2004), a metastable phase μ, which crystallizes in the rhombohedral structure, is formed in this system within the interval of 13–24 mol% SnSe. This phase is kept at room temperature for months.

SnSe+PbI₂ ↔ PbSe+SnI₂. The isothermal section of this ternary mutual system is given in Figure 8.72 (Odin et al. 2003a). It was shown, that 2*H*-PbTe and 6*R*-PbTe polytypic modification crystallize from the melt and take part in complex metastable equilibria at the liquid presence. On the liquidus surface (Figure 8.73), the fields of the primary crystallization of solid solutions based on SnSe and PbSe occupy the biggest part of the concentration square. The ternary eutectics crystallize at 371°C (E_1) and 291°C (E_2) and the transition point U_1 exists at 307°C. This system was investigated through DTA, XRD, and metallography.

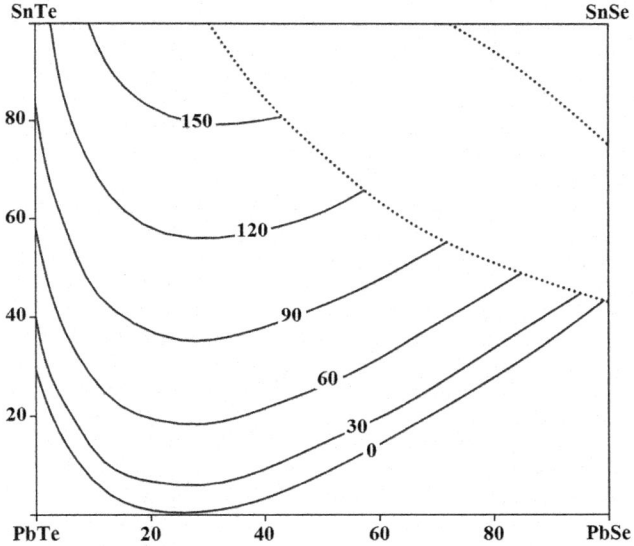

FIGURE 8.71 The dependence of low-temperature phase transition (in K) on composition for SnSe+PbTe ↔ SnTe+PbSe ternary mutual system. The dotted line shows the stability limits for a solid solution. (From Lebedev, A.I., and Sluchinskaya. I.A., *J. Alloys Compd.*, **203**(1–2), 51, 1994.)

8.132 Tin–Arsenic–Tellurium–Selenium

SnTe–As₂Se₃–As₂Te₃. The glass-forming region in this system is presented in Figure 8.74 (Vassilev et al. 2007a). The glasses in this system were produced via direct single temperature synthesis from mixtures of Sn, As, Te, and Se in evacuated (10^{-3} Pa) and sealed quartz ampoules. The maximum temperature of the glasses synthesis was 850°C where, in the course

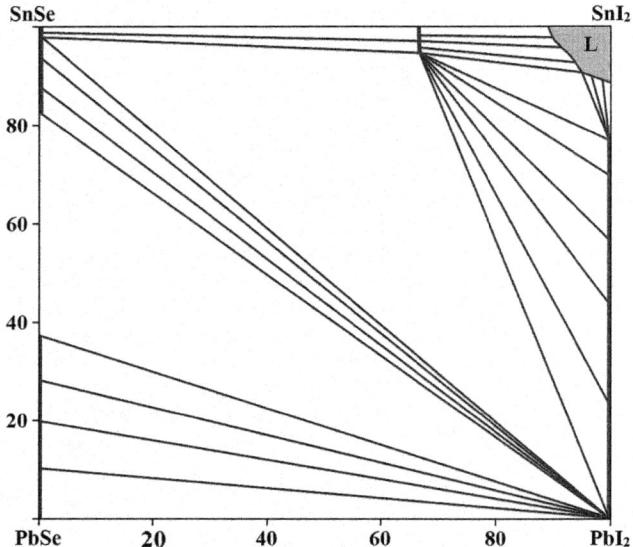

FIGURE 8.72 Isothermal section of the SnSe+PbI₂ ↔ PbTe+SnI₂ ternary mutual system at 324°C. (From Odin I.N., et al., *Zhurn. neorgan. khimii*, **48**(5), 842, 2003.)

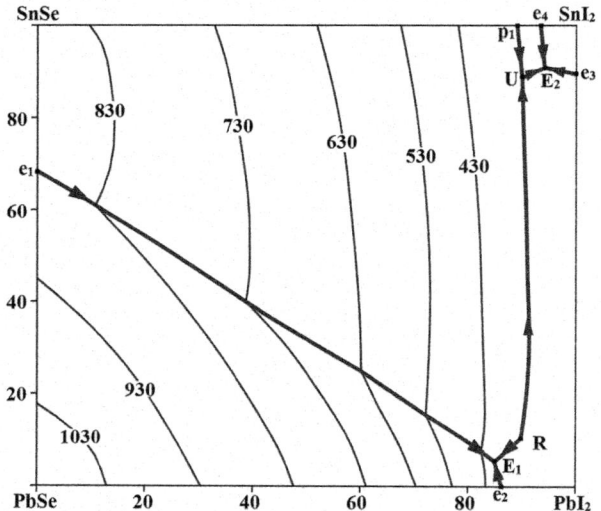

FIGURE 8.73 Liquidus surface of the SnSe+PbI₂ ↔ PbTe+SnI₂ ternary mutual system. (From Odin I.N., et al., *Zhurn. neorgan. khimii*, **48**(5), 842, 2003.)

of 2 h a vibration agitation of the melt was included. The last has been tempered at a temperature of 800°C and quenched in a mixture of water and ice with a cooling rate of 10–15°C·s⁻¹. The glass transition temperatures of these glasses are within the temperature interval from 145°C and 193°C.

8.133 Tin–Antimony–Iodine–Selenium

Two quaternary compounds, **Sn₂SbSe₂I₃** and **Sn₃SbSe₂I₅**, which have an energy gap of 1.22 and 1.66 eV, respectively, are formed in the Sn–Sb–I–Se system (Ibañez et al. 1984). They

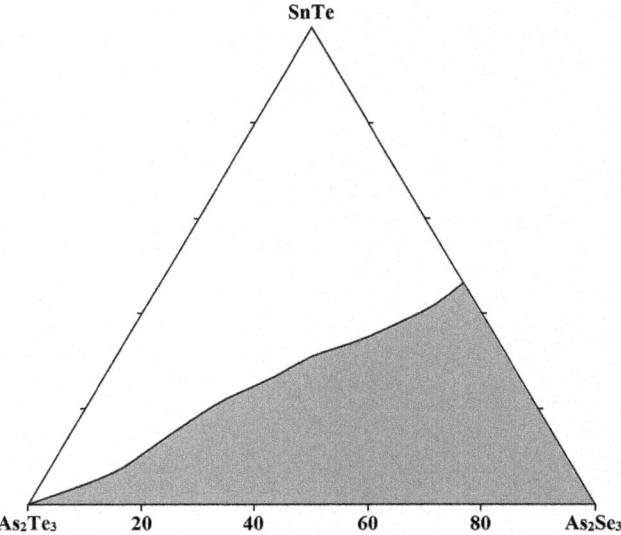

FIGURE 8.74 Glass-forming region in the SnTe–As₂Se₃–As₂Te₃ system. (From Vassilev, V., et al., *Mater. Let.*, **61**(17), 3676, 2007.)

were obtained using the chemical transport reactions from the mixtures of SnSe, SnI₂, and SbI₃.

8.134 Tin–Antimony–Iron–Selenium

The **Sn₃₁.₅₂Sb₆.₂₃Fe₃.₁₂Se₅₉.₁₂** or **Sn₁₀Sb₂FeSe₁₉** quaternary compound was synthesized in the Sn–Sb–Fe–Se system from pure elements at 500°C (Makovicky et al. 2008). This compound has a triclinic crystal structure composed of two alternating layer types, both with a pronounced one-dimensional modulation, and with noncommensurate layer match in two dimensions. The pseudo-tetragonal layer has the lattice parameters $a = 596.9 ± 0.2$, $b = 600.36 ± 0.03$, $c = 1223.82 ± 0.05$ pm, $\alpha = 87.98 ± 0.04°$, $\beta = 83.14 ± 0.03°$, and $\gamma = 90.01 ± 0.04°$. The pseudo-hexagonal layer has the lattice parameters $a = 383.08 ± 0.12$, $b = 658.0 ± 0.3$, $c = 1215.1 ± 0.5$ pm, $\alpha = 87.79 ± 0.04°$, $\beta = 90.59 ± 0.03°$, and $\gamma = 89.99 ± 0.03°$; the a and b vectors of the subsystems are parallel, the c vector diverge.

8.135 Tin–Bismuth–Vanadium–Selenium

The $([Sn_{1-x}Bi_xSe]_{1+\delta})_1(VSe_2)_1$ misfit layer phases with $0 \le x \le 0.66$, which crystallizes in the hexagonal structure, are formed in the Sn–Bi–V–Se quaternary system (Falmbigl et al. 2015). The next values of a-lattice parameters of $Sn_{1-x}Bi_xSe$ and VSe_2 sublattices, c-lattice parameters and x were obtained: $a(Sn_{1-x}Bi_xSe) = 593.5 ± 0.2$, $602.2 ± 0.3$, $599.9 ± 0.2$, $599.7 ± 0.3$, and $599.6 ± 0.2$ pm; $a(VSe_2) = 341.4 ± 0.2$, $343.4 ± 0.3$, $345.6 ± 0.2$, $346.6 ± 0.3$, and $347.5 ± 0.2$ pm; $c = 1203 ± 1$, $1200.8 ± 0.2$, $1194.6 ± 0.1$, $1188.9 ± 0.1$, $1185.6 ± 0.1$ and $x = 0$, $0.06 ± 0.03$, $0.23 ± 0.02$, $0.42 ± 0.03$, $0.57 ± 0.03$ for $\delta = 0.143$, 0.127, 0.150, 0.157, and 0.164, respectively. Physical vapor deposition was utilized to form the thin film samples in a vacuum deposition chamber evacuated to a base pressure of 10^{-5} Pa before deposition on (100) oriented silicon wafers. An effusion cell was used to evaporate Se and three Thermionics electron beam guns were used to evaporate Sn, V, and Bi.

8.136 Tin–Bismuth–Tellurium–Selenium

SnTe–BiSe. The solubility of SnTe in BiSe at 250°C and 500°C is 19 and 22 mol%, respectively (Abrikosov et al. 1972). The XRD, metallography, and measuring of the microhardness were used to investigate this system and the ingots for the investigations were annealed at 250°C and 500°C for 30 and 10 days, respectively.

8.137 Tin–Bismuth–Manganese–Selenium

The $Mn_{1.34}Sn_{6.66}Bi_8Se_{20}$ quaternary compound, which melts congruently at 688°C, is thermally stable up to 800°C and crystallize in the monoclinic structure with the lattice parameters $a = 1364.8 \pm 0.3$, $b = 417.5 \pm 0.1$, $c = 1746.3 \pm 0.4$ pm, $\beta = 93.42 \pm 0.03°$, and an energy gap of ≈ 0.29 eV, is formed in the Sn–Bi–Mn–Se system (Anglin et al. 2010). Single crystals of this compound were first obtained as by-product of a reaction designed to prepare polycrystalline powder of $Mn_2Sn_2Bi_8Se_{16}$. Polycrystalline single-phase powder of $Mn_{1.34}Sn_{6.66}Bi_8Se_{20}$ was later synthesized using Mn, Sn, and Se powders and Bi pieces as starting materials. All four components weighed in the desired ratio (total mass was 5 g) under Ar atmosphere in a dry glove box, were thoroughly mixed in an agate mortar with pestle and transferred into a fused silica tube. The tube was flame-sealed under residual pressure of $\sim 10^{-2}$ Pa. The sealed tube was then inserted into a mullite casing and isolated using ceramic fibers. This precaution is necessary to minimize the temperature gradient inside the quartz tube during the reaction. The protected tube was then placed into a furnace, and the temperature was raised to 350°C in 12 h, dwelled for 24 h and finally increased to 500°C. 72 h dwelling at 500°C was necessary to complete the reaction. The furnace was then slowly cooled to room temperature for over 80 h. The resulting product was dark gray polycrystalline powder containing small black needle-shaped crystals of the title compound.

8.138 Tin–Niobium–Molybdenum–Selenium

Metastable ferecrystal compounds $[(SnSe)_{1.16-1.09}]_1[(Nb_xMo_{1-x})Se_2]_1$ with $x = 0$, 0.26, 0.49, 0.83, and 1 were formed from designed modulated precursors prepared using the modulated elemental technique in a custom built physical vapor deposition vacuum technique (Westover et al. 2014). The c-lattice parameter of the ferecrystal alloys decreased from 1253 ± 2 pm for $x = 0$ to 1227 ± 2 for $x = 1$. The a-lattice parameter of the dichalcogenide constituent increased from 332.9 ± 0.8 for $x = 0$ to 346.1 ± 0.4 for $x = 1$. The a-lattice parameter of the rock-salt

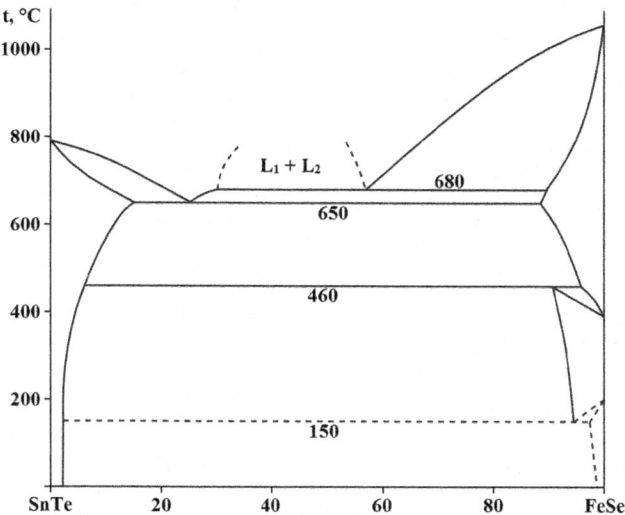

FIGURE 8.75 Phase diagram of the SnTe–FeSe system. (From Murguzov, M.I., et al., *Zhurn. neorgan. khimii*, **29**(11), 2968, 1984.)

constituent (SnSe) remained constant resulting in a linear increase in misfit parameter with increased Nb constituent.

8.139 Tin–Tellurium–Iron–Selenium

SnTe–FeSe. The phase diagram of this system, constructed through DTA, metallography, and measuring of the microhardness, is a eutectic type with monotectic transformation (Figure 8.75) (Murguzov et al. 1984). The eutectic contains 25 mol% FeSe and crystallizes at 650°C. The immiscibility region exists in this system within the interval 30–57 mol% FeSe at 680°C. The mutual solubility of SnTe and FeSe at the eutectic temperature is 10 mol%. The eutectoid and peritectoid transformations of the solid solution based on FeSe take place at 150°C and 460°C, respectively. The ingots for the investigations were annealed at 300–400°C.

8.140 Tin–Tellurium–Cobalt–Selenium

SnTe–CoSe. The phase diagram of this system, constructed through DTA, XRD, metallography, and measuring of the microhardness, is given in Figure 8.76 (Murguzov et al. 1985). The eutectic contains 15.5 mol% CoSe and crystallizes at 760°C. The **SnCoTeSe** quaternary compound, which melts incongruently at 815°C, is formed in this system. The ingots for the investigations were annealed at temperatures below solidus for 300–500 h.

8.141 Tin–Tellurium–Nickel–Selenium

SnTe–NiSe. The phase diagram of this system, constructed through DTA, XRD, metallography, and measuring of the microhardness, is presented in Figure 8.77 (Murguzov et al.

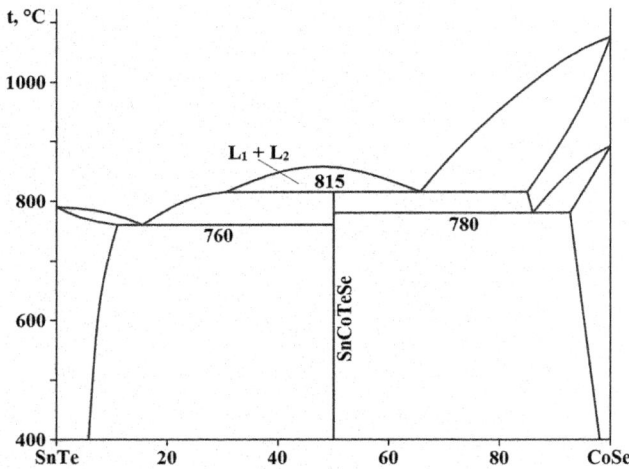

FIGURE 8.76 Phase diagram of the SnTe–CoSe system. (From Murguzov, M.I., *Zhurn. neorgan. khimii*, **30**(1), 186, 1985.)

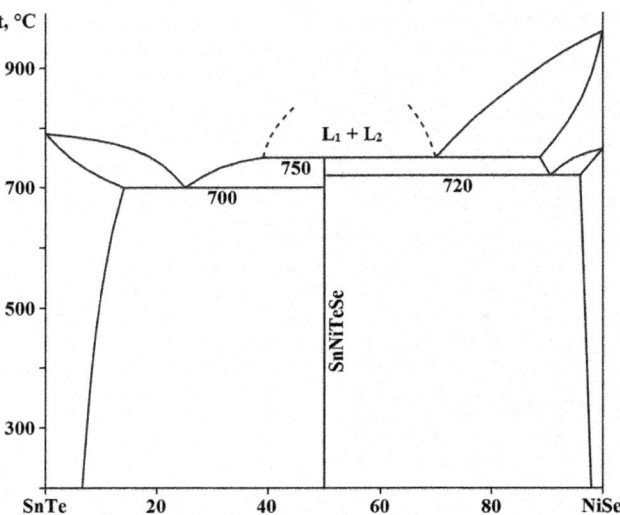

FIGURE 8.77 Phase diagram of the SnTe–NiSe system. (From Murguzov, M.I., *Zhurn. neorgan. khimii*, **30**(1), 186, 1985.)

1985). The eutectic contains 25 mol% NiSe and crystallizes at 700°C. The immiscibility region exists in this system within the interval 39–70 mol% FeSe. The **SnNiTeSe** quaternary compound, which melts incongruently at 750°C, is formed in this system. The ingots for the investigations were annealed at temperatures below solidus for 300–500 h.

The solid solutions $(SnTe)_{1-x}(NiSe)_x$ containing up to 5 mol% NiSe were obtain in the Sn–Te–Ni–Se system as a result of the alloys annealing at 700°C for 600 h (Alidzhanov et al. 1997). As the content of NiSe increases, the band gap also increases.

9

Systems Based on Tin Telluride

9.1 Tin–Potassium–Silver–Tellurium

The **K$_2$Ag$_2$SnTe$_4$** quaternary compound, which crystallizes in the tetragonal structure with the lattice parameters a = 837.2 ± 0.5, c = 742.5 ± 0.3 pm, and a calculated density of 5.417 g·cm^{-3}, is formed in the Sn–K–Ag–Te system (Li et al. 1995). The synthesis of this compound was carried out in a K$_2$Te+Te flux. Black, thin column-like crystals were isolated from a reaction mixture containing K$_2$Te (1 mM), Ag (1 mM), Sn (0.5 mM), and Te (2.5 mM) as starting materials. The reaction was carried out in a sealed Pyrex tube at 450°C for 4 days. After the heating process, it was then cooled slowly to 120°C at a rate of 4°C·h^{-1}. The excess flux was removed from the final product with DMF.

9.2 Tin–Potassium–Barium–Tellurium

The **K$_2$BaSnTe$_4$** quaternary compound, which crystallizes in the cubic structure with the lattice parameter a = 839.3 ± 0.1 pm and a calculated density of 4.745 g·cm^{-3}, is formed in the Sn–K–Ba–Te system (Li et al. 1995). This compound was crystallized in a K$_2$Te+BaTe+Te flux. The heating scheme was the same as described for K$_2$Ag$_2$SnTe$_4$. The reaction of K$_2$Te (1.5 mM), BaTe (0.5 mM), Sn (0.5 mM), and Te (2.5 mM) yielded black, nearly cubic-shaped crystals. The excess flux was also removed from the final product with DMF.

9.3 Tin–Potassium–Mercury–Tellurium

The **K$_2$HgSnTe$_4$** quaternary compound, which crystallizes in the tetragonal structure with the lattice parameters a = 858.0 ± 0.2 and c = 735.8 ± 0.4 pm, is formed in the Sn–K–Hg–Te system (Dhingra and Haushalter 1994). To obtain this compound, a mixture of K$_4$SnTe$_4$ (0.255 mM) and HgCl$_2$ (0.254 mM) was treated with 0.3 mL of ethylenediamine and sealed in a quartz tube under vacuum. After heating for 48 h at 100°C, the solution was cooled, and the small, metallic gray, cubic-shaped crystals were filtered and washed with ethylenediamine. K$_2$HgSnTe$_4$ can also be prepared at high temperature by the reaction of K$_2$Te (0.972 mM), Hg (0.972 mM), Sn (0.969 mM), and Te (2.915 mM) at 500°C for 10 h in an evacuated quartz tube. Single crystals of this compound were prepared by the treatment of K$_4$SnTe$_4$, which was prepared from the reaction of KSn and Te, with 1 equivalent HgCl$_2$ in ethylenediamine

at 100°C. All manipulations were performed under an atmosphere of oxygen-free helium.

9.4 Tin–Cesium–Bismuth–Tellurium

Two quaternary compounds, **CsSnBi$_3$Te$_6$** and **CsSn$_2$Bi$_3$Te$_7$**, are formed in the Sn–Cs–Bi–Te system (Hsu et al. 2002a). Both compounds crystallize in the orthorhombic structure with the lattice parameters a = 626.13 ± 0.07, b = 2847.9 ± 0.3, and c= 432.07 ± 0.05 pm for the first compound (Cs$_{0.84}$Sn$_{0.84}$Bi$_{3.16}$Te$_6$ composition) and a = 438.81 ± 0.06, b = 3313.9 ± 0.4, and c = 1260.8 ± 0.2 pm for the second one (Cs$_{0.88}$Sn$_{1.88}$Bi$_{3.12}$Te$_7$ composition). CsSnBi$_3$Te$_6$ was obtained by reacting CsBi$_4$Te$_6$ (0.173 mM), Sn (0.173 mM), and Te (0.173 mM) at 750°C for 1 h and slow cooling to 50°C. CsSn$_2$Bi$_3$Te$_7$ was prepared by reacting CsBi$_4$Te$_6$ (0.174 mM), Sn (0.522 mM), and Te (0.523 mM) at the same heating profile as above for CsSnBi$_3$Te$_6$. Both compounds can be prepared by the stoichiometric reactions of Cs$_2$Te, Sn, Bi, and Te reagents.

9.5 Tin–Cesium–Manganese–Tellurium

The **Cs$_2$MnSnTe$_4$** quaternary compound, which crystallizes in the orthorhombic structure with the lattice parameters a = 681.89 ± 0.14, b = 1506.3 ± 0.3, c = 2596.7 ± 0.5 pm, and a calculated density of 1.233 g·cm^{-3} at 203 K, is formed in the Sn–Cs–Mn–Te system (Zimmermann and Dehnen 2003). This compound was synthesized by adding MnCl$_2$·4H$_2$O (0.41 mM) to a solution of Cs$_4$SnTe$_4$·2MeOH (0.41 mM) in H$_2$O (10 mL). After stirring for 10 h, the black metallic shiny residue was filtered from the orange-red reaction solution and covered with tetrahydrofuran (10 mL). After a week, Cs$_2$MnSnTe$_4$ crystallized in the form of very small black cubes.

9.6 Tin–Copper–Zinc–Tellurium

The **Cu$_2$ZnSnTe$_4$** quaternary compound, which crystallizes in the tetragonal structure with the lattice parameters a = 608.8 ± 0.1 and c = 1218.0 ± 0.4 pm (Haeuseler et al. 1991) [a = 607.5 and c = 1201 pm and the calculated and experimental densities of 6.15 and 6.01 g·cm^{-3}, respectively (Hahn and Schulze 1965)], is formed in the Sn–Cu–Zn–Te system. According to the data of Matsushita et al. (2005), Cu$_2$ZnSnTe$_4$ is the mixture of ZnTe, SnTe, and Cu$_2$GeTe$_3$.

DOI: 10.1201/9781003123484-9

9.7 Tin–Copper–Cadmium–Tellurium

Cu₂SnTe₃–CdTe. This section is a nonquasibinary section of the Sn–Cu–Cd–Te quaternary system since Cu_2SnTe_3 melts incongruently (Olekseyuk and Piskach 1997; Piskach et al. 1997). The eutectic composition and temperature are 24 mol% CdTe and 405°C, respectively. SnTe primarily crystallizes within the interval of 0–7 mol% CdTe. The **Cu₂CdSnTe₄** quaternary compound is formed in this system. It melts incongruently at 470°C [(the peritectic composition is 17 mol% CdTe (Olekseyuk and Piskach 1997)] and has apparently two polymorphic modifications. First modification crystallizes in the monoclinic structure with the lattice parameters $a = 425.6 \pm 0.3$, $b = 427.7 \pm 0.3$, $c = 1042.1 \pm 0.4$ pm, and $\beta = 119.35 \pm 0.05°$ and an experimental density of 6.14 g·cm⁻³ (Olekseyuk and Piskach 1997). The second modification crystallizes in the tetragonal structure with the lattice parameters $a = 619.60 \pm 0.01$, $c = 1225.31 \pm 0.04$ pm, and a calculated density of 6.1318 ± 0.0004 g·cm⁻³ (Parasyuk et al. 2006b) [$a = 619.8 \pm 0.1$ and $c = 1225.6 \pm 0.3$ pm (Haeuseler et al. 1991)]. $Cu_2CdSnTe_4$ was synthesized from the mixture of elements in an evacuated and sealed ampoule (Parasyuk et al. 2006b). The mixture was heated to 930°C at a rate of 30°C·h⁻¹, dwelled at this temperature for 6 h, cooled at a rate of 10°C·h⁻¹ to 400°C and annealed at this temperature for 250 h.

The solubility of CdTe in Cu_2SnTe_3 and Cu_2SnTe_3 in CdTe at 350°C is equal to 12 and 4 mol%, respectively (Olekseyuk and Piskach 1997). Matsushita et al. (2005) noted that $Cu_2CdSnTe_4$ was not synthesized and it is a mixture of CdTe and Cu_2SnTe_3.

This system was investigated through differential thermal analysis (DTA), X-ray diffraction (XRD), metallography, and measuring of the microhardness and density (Olekseyuk and Piskach 1997; Piskach et al. 1997).

9.8 Tin–Copper–Mercury–Tellurium

Cu₂SnTe₃–HgTe. The **Cu₂HgSnTe₄** quaternary compound, which crystallizes in the tetragonal structure with lattice parameters $a = 619.44 \pm 0.01$, $c = 1225.99 \pm 0.01$ pm, and a calculated density of 6.7539 ± 0.0005 g·cm⁻³ (Parasyuk et al. 2006a) [$a = 616.2$, $c = 1228$ pm, and the calculated and experimental densities of 6.81 and 6.66 g·cm⁻³, respectively (Hahn and Schulze 1965); $a = 619.1 \pm 0.1$ and $c = 1226.3 \pm 0.3$ pm (Haeuseler et al. 1991)], is formed in this system. According to the data of Parasyuk (1998), this compound is one component of the solid solution based on Cu_2SnTe_3.

$Cu_2HgSnTe_4$ was synthesized from the mixture of Cu, Sn, Te, and HgTe in an evacuated and sealed quartz ampoule (Parasyuk et al. 2006a). The mixture was heated to 750°C, dwelled at this temperature for 6 h, cooled at a rate of 10°C·h⁻¹ to 400°C, and annealed at this temperature for 250 h.

9.9 Tin–Copper–Indium–Tellurium

SnTe–CuInTe₂–Te. The "SnTe₂"–CuInTe₂ section of this system is nonquasibinary (Pamplin et al. 1977). At 25 mol% "SnTe₂", the samples are two-phase and on the cooling curves

thermal effects at 700°C, 640°C, 590°C, and 380°C take place. At 50 mol% "SnTe₂", the samples are three-phase, and on the cooling curves thermal effects were found at 670°C, 635°C, 600°C, and 380°C. The nature of the formed phases and thermal effects has not been established. The system was studied through DTA, XRD, and metallography.

9.10 Tin–Copper–Thallium–Tellurium

The **Cu₂Tl₂SnTe₄** quaternary compound, which crystallizes in the tetragonal structure with the lattice parameters $a = 838.2 \pm 0.3$ and $c = 711.8 \pm 0.3$ pm, is formed in the Sn–Cu–Tl–Te system (McGuire et al. 2005a). This compound was synthesized from the elements. Since Tl will tarnish slowly in air it was handled in an Ar glove box. Chunks of Tl were cut from the 0.5 inch diameter rod with wire cutters used only for this purpose, weighed inside the glove box and placed in capped glass vials. Appropriate amounts of the other elements were then weighed out and loaded into carbon coated silica tube. The Ar-filled vials were removed from the glove box and the Tl was added to the silica tube. The tube was then quickly (to limit the exposure of Tl to air) attached to a vacuum line for sealing. It was heated over 24 h to 800°C, held at this temperature for 172 h and then cooled over 200 h to 300°C at which point the furnace was turned off and allowed to cool to room temperature. Crystals were extracted from the resulting polycrystalline mass for examination. $Cu_2Tl_2SnTe_4$ was made as a single-phase polycrystalline powder by reaction of the elements at 500°C for 2 days, annealing for 7 days at 300°C, and then grinding and annealing a pressed pellet at 300°C for 4 days.

9.11 Tin–Copper–Yttrium–Tellurium

SnTe–Cu₂Te–Y₂Te₃. The isothermal section of this quasiternary system at 600°C (Figure 9.1) was constructed by Shemet et al. (2006b). No quaternary compounds were found, and SnTe is in the equilibria with $Cu_3Y_7Te_{12}$ ternary compound and $(CuYTe_2)_x(Cu_2Y_{0.66}Te_3)_{1-x}$ solid solution. Therefore the section could be divided into three three-phase regions. The ingots for the investigations were annealed at 450°C for 240 h followed by quenching in cold water.

9.12 Tin–Copper–Manganese–Tellurium

The **Cu₂MnSnTe₄** quaternary compound, which has two polymorphic modifications, is formed in the Sn–Cu–Mn–Te system (Lamarche et al. 1991). First modification crystallizes in the tetragonal structure with the lattice parameters $a = 604.9$ and $c = 1209.8$ pm, and the second one crystallizes in the cubic structure with the lattice parameter $a = 628.6$ pm. This compound was produced by the melt and anneals technique. To obtain it, the components of 1-g sample were sealed under vacuum in a quartz capsule which had previously been coated with carbon in order to prevent reaction of the charge with the quartz. The capsule was then raised to a temperature of 1150°C in approximately an hour to allow some reaction to occur and

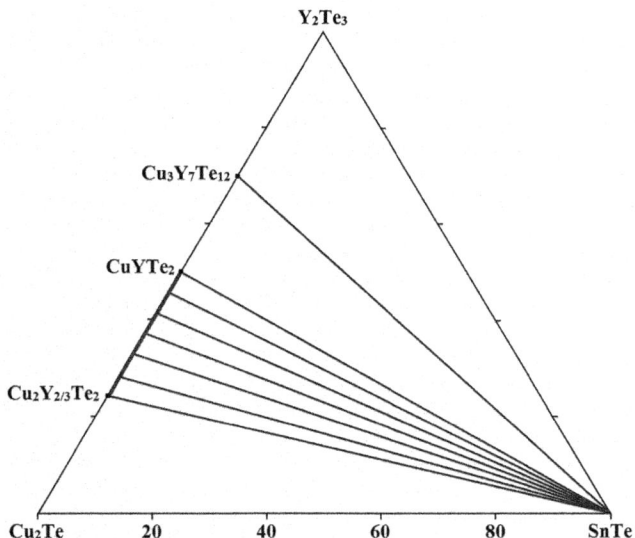

FIGURE 9.1 Isothermal section of the SnTe–Cu$_2$Te–Y$_2$Te$_3$ quasiternary system at 600°C. (From Shemet, V.Ya., et al., *Nauk. Visnyk Volyns'k. Derzh. Univ. im. Lesi Ukrainky. Ser. Khim. nauky*, (4), 124, 2006.)

then cooled to room temperature. A sample annealed at 525°C and then brine-quenched showed only tetragonal plus very faint MnTe$_2$ lines. However, when annealed at 450°C and then brine-quenched, the sample showed strong cubic lines, very faint MnTe$_2$ lines and no tetragonal lines.

9.13 Tin–Copper–Iron–Tellurium

The **Cu$_2$FeSnTe$_4$** quaternary compound, which melts congruently at 662.4°C and crystallizes in the orthorhombic structure with the lattice parameters $a = 768.7$, $b = 674.1$, and $c = 657.9$ pm (Quintero et al. 1999), is formed in the Sn–Cu–Fe–Te system. The samples of this compound were prepared by the melt and anneals technique. The components of 1-g sample were made from the appropriate amounts of the elements and were sealed under vacuum in a small quartz ampoule, which had previously been carbonized to prevent interaction of the components with the quartz. The components were melted together at 1150°C for about an hour, annealed to equilibrium at 500°C, and then cooled to room temperature by leaving the ampoule in the switched-off furnace.

9.14 Tin–Silver–Mercury–Tellurium

"Ag$_2$SnTe$_3$"–HgTe. This system is a nonquasibinary section of the Sn–Ag–Hg–Te quaternary system (Parasyuk 1999). Quaternary phases were not discovered in this section. The liquidus consists of the two fields corresponding to the primary crystallization of HgTe and SnTe. Under the solidus, four phases, HgTe, SnTe, AgSnTe$_2$, and Te, are present. This system was studied through DTA, XRD, and metallography, and the ingots for the investigations were annealed at 250°C for 500 h with further quenching in cold water.

9.15 Tin–Silver–Indium–Tellurium

SnTe–AgInTe$_2$–Te. The "SnTe$_2$"–AgInTe$_2$ section of this system is nonquasibinary (Pamplin et al. 1977). At 25 and 50 mol% "SnTe$_2$", the samples contain three phases, one of which corresponds to SnTe, while the nature of the other phases has not been established, and the lattice parameters of one of them are close to the lattice parameters of AgInTe$_2$. On the cooling curves of samples containing 25 mol% "SnTe$_2$", thermal effects were found at 570°C, 550°C, and 373°C. The system was studied through DTA, XRD, and metallography.

9.16 Tin–Silver–Thallium–Tellurium

The **Ag$_2$Tl$_2$SnTe$_4$** quaternary compound, which crystallizes in the tetragonal structure with the lattice parameters $a = 853.9 \pm 0.3$ and $c = 739.1 \pm 0.3$ pm, is formed in the Sn–Ag–Tl–Te system (McGuire et al. 2005a). This compound was synthesized in the same way as Cu$_2$Tl$_2$SnTe$_4$ was prepared. Ag$_2$Tl$_2$SnTe$_4$ was prepared as a single-phase polycrystalline phase by first interaction of the elements at 700°C with the next grinding and annealing a pressed pellet at 400°C for 4 days.

9.17 Tin–Silver–Antimony–Tellurium

SnTe–Ag$_2$Te–Sb$_2$Te$_3$. Phase equilibria in this quasiternary system were experimentally studied through DTA, XRD, and measuring of the microhardness and EMF of the concentration chain by Mashadieva et al. (2017a, c). The isothermal sections at 130°C and 430°C (Figure 9.2), the liquidus surface (Figure 9.3) as well as four vertical sections of the phase diagram were constructed. This system is characterized by the formation of wide continuous high-temperature solid solutions with a cubic structure along the SnTe–"AgSbTe$_2$" section (Fleischmann et al. 1959; Mashadieva et al. 2017a, c). The lattice parameters of these solid solutions change linearly with composition. There are five fields of γ-Ag$_2$Te, β-Ag$_2$Te, SnTe, SnSb$_2$Te$_4$, and Sb$_2$Te$_3$ primary crystallization on the liquidus surface of this system. The partial molar thermodynamic functions of Ag in alloys and standard integral thermodynamic functions of the (SnTe)$_{1-x}$(AgSbTe$_2$)$_x$ solid solutions were calculated based on the results of EMF measurements.

9.18 Tin–Silver–Bismuth–Tellurium

SnTe–AgBiTe$_2$. This system is a partially quasibinary (above 442°C) and is characterized by formation continuous series of high-temperature solid solutions, which crystallize in the cubic structure (Figure 9.4) (Fleischmann et al. 1959; Aliyev et al. 2013; Mashadieva et al. 2017b). The lattice parameters of these solid solutions change linearly with composition. Complex interaction exists in the system below 442°C on the AgBiTe$_2$-side (Aliyev et al. 2013; Mashadieva et al. 2017b). The partial molar thermodynamic functions of Ag in alloys and standard integral thermodynamic functions of the

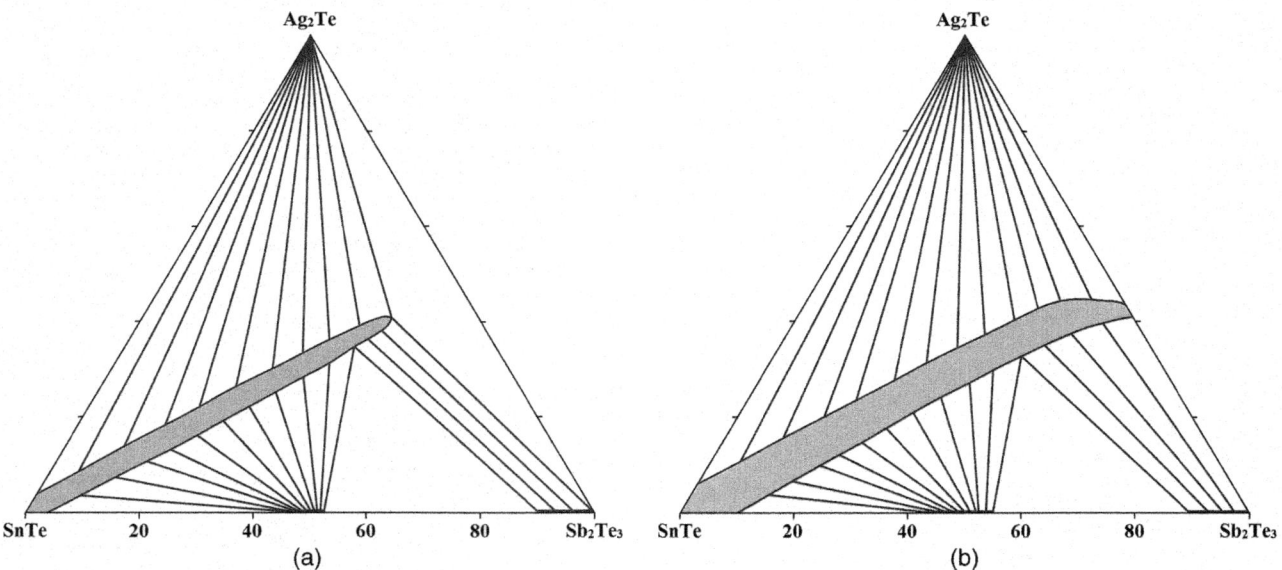

FIGURE 9.2 Isothermal section of the SnTe–Ag₂Te–Sb₂Te₃ quasiternary system at (a) 130°C and (b) 430°C. (From Mashadieva, L.F., et al., *J. Phase Equilib. Diffus.*, **38**(5), 603, 2017.)

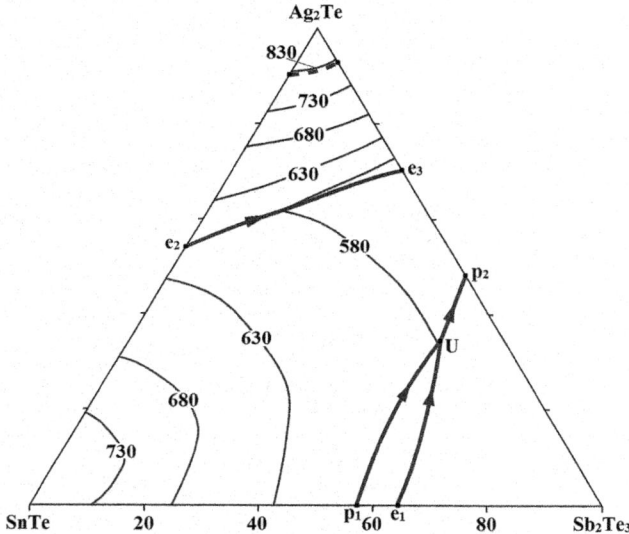

FIGURE 9.3 Liquidus surface of the SnTe–Ag₂Te–Sb₂Te₃ quasiternary system. (From Mashadieva, L.F., et al., *J. Phase Equilib. Diffus.*, **38**(5), 603, 2017.)

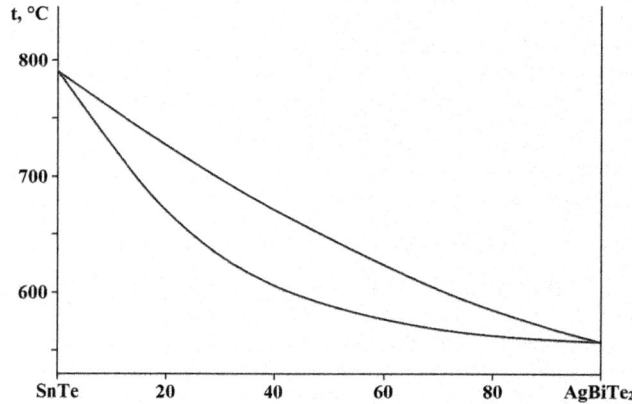

FIGURE 9.4 Phase diagram of the SnTe–AgBiTe₂ system. (From Mashadieva, L.F., et al., *J. Alloys Compd.*, **724**, 641, 2017.)

crystallization belongs to the solid solutions based on SnTe–AgBiTe₂ system. The nature of the various nonvariant and monovariant equilibria was also identified. The alloys were annealing 530°C for 800 h and then quenched in cold water. A series of selected alloys has been further annealed at 330°C for 300 h and then slowly cooled down to room temperature. This quasiternary system was experimentally studied through DTA, XRD, and measuring the microhardness and EMF of the concentration chain.

$(SnTe)_{1-x}(AgBiTe_2)_x$ solid solutions were calculated based on the results of EMF measurements.

SnTe–Ag₂Te–Bi₂Te₃. The isothermal sections at room temperature and 530°C (Figure 9.5), the liquidus surface (Figure 9.6) of this system as well as four vertical sections of the phase diagram were constructed (Mashadieva et al. 2017b). The system is characterized by the formation of wide continuous high-temperature solid solutions along the SnTe–AgSbTe₂ section. Liquidus surface consists of three fields of primary crystallization of SnTe, Ag₂Te, and Bi₂Te₃, as well as the SnBi₂Te₄ and SnBi₄Te₇ ternary compounds. The largest field of primary

9.19 Tin–Silver–Manganese–Tellurium

The **Ag₂MnSnTe₄** quaternary compound, which has two polymorphic modifications, is formed in the Sn–Ag–Mn–Te system (Lamarche et al. 1991). The first modification crystallizes in the

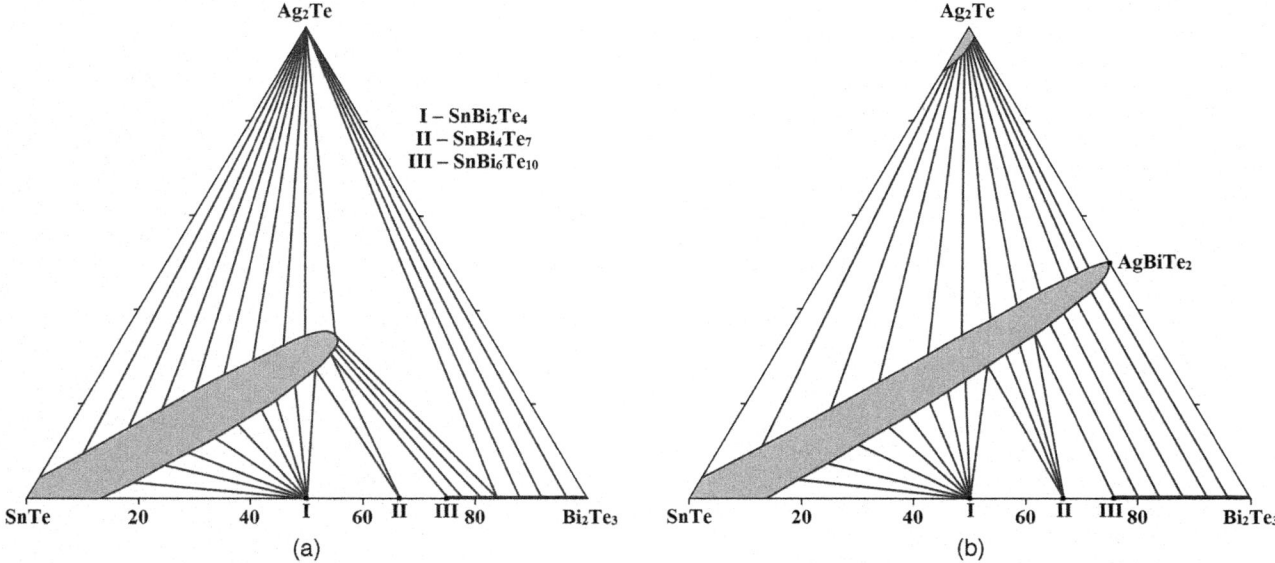

FIGURE 9.5 Isothermal section of the SnTe–Ag$_2$Te–Bi$_2$Te$_3$ quasiternary system at (a) room temperature and (b) 530°C. (From Mashadieva, L.F., et al., *J. Alloys Compd.*, **724**, 641, 2017.)

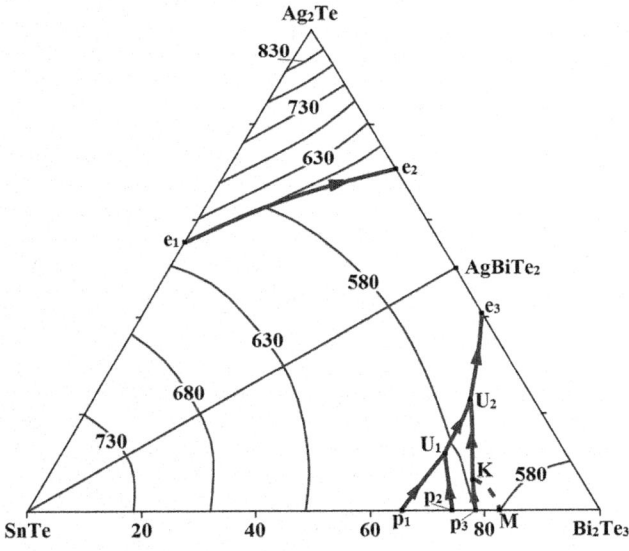

FIGURE 9.6 Liquidus surface of the SnTe–Ag$_2$Te–Bi$_2$Te$_3$ quasiternary system. (From Mashadieva, L.F., et al., *J. Alloys Compd.*, **724**, 641, 2017.)

tetragonal structure with the lattice parameters $a = 585$ and $c = 1170$ pm, and the second one crystallizes in the cubic structure with the lattice parameter $a = 600.5$ pm. This compound was synthesized in the same way as Cu$_2$MnSnTe$_4$ was prepared. A sample annealed at 600°C and air-cooled was cubic with very faint tetragonal and MnTe$_2$ lines present. When it was annealed at 400°C and brine-quenched, the sample showed both tetragonal and cubic lines and fainter MnTe$_2$ lines.

9.20 Tin–Silver–Iron–Tellurium

The **Ag$_2$FeSnTe$_4$** quaternary compound, which melts congruently at 647°C and crystallizes in the orthorhombic structure with the lattice parameters $a = 809.8$, $b = 678.5$, and $c = 633.5$ pm (Quintero et al. 1999), is formed in the Sn–Cu–Fe–Te system. The samples of this compound were prepared in the same way as Cu$_2$FeSnTe$_4$ were synthesized.

9.21 Tin–Cadmium–Thallium–Tellurium

The **CdTl$_2$SnTe$_4$** quaternary compound, which is semiconductor and crystallizes in the tetragonal structure with the lattice parameters $a = 842.50 \pm 0.04$ and $c = 721.71 \pm 0.05$ pm, is formed in the Sn–Cd–Tl–Te system (McGuire et al. 2005b). This compound was synthesized by reaction of Tl, Cd, Sn, and Te (molar ratio 2:1:1:4). At the syntheses, Tl metal was weighed in an Ar-filled glove box and placed into a capped vial. The other elements were weighed in air and loaded into a silica tube, which was coated with carbon to prevent reaction between the silica and the sample. Thallium was then added to the tube, and the tube was quickly attached to a vacuum line for sealing. The evacuated, sealed silica tube was heated in a furnace. The sample was heated to 500°C for 4 days, followed by cooling to room temperature over another 4 days. Several subsequent anneals for 4–7 days each at 250°C–300°C were required to obtain a single-phase sample. The single crystals were obtained by heating the mixture of the elements to 800°C for 24 h, held at this temperature for 100 h, cooled to 200°C for 300 h, and then to room temperature by turning off the furnace.

9.22 Tin–Cadmium–Arsenic–Tellurium

CdSnAs₂–CdTe. The solubility of 2CdTe in CdSnAs₂ is equal to 6 mol% (Dovletmuradov et al. 1971, 1972). This system was investigated through DTA, XRD, metallography, and measuring of the microhardness.

9.23 Tin–Mercury–Thallium–Tellurium

The **HgTl₂SnTe₄** quaternary compound, which is semiconductor and crystallizes in the tetragonal structure with the lattice parameters $a = 839.7 \pm 0.4$ and $c = 715.7 \pm 0.6$ pm, is formed in the Sn–Tl–Mn–Te system (McGuire et al. 2005b). This compound was synthesized by reaction of Tl, HgTe, Sn, and Te (molar ratio 2:1:1:3). At the syntheses, Tl metal was weighed in an Ar-filled glove box and placed into a capped vial. The other elements were weighed in air and loaded into a silica tube which was coated with carbon to prevent reaction between the silica and the sample. Thallium was then added to the tube, and the tube was quickly attached to a vacuum line for sealing. The evacuated, sealed silica tube was heated in a furnace. The sample was heated to 500°C in 36 h and held at this temperature for 5 days, at which point the furnace was turned off and allowed to cool naturally to room temperature. The product was then ground and pressed into pellets and annealed at 300°C for 7 days. The single-phase sample was prepared by the interaction of the stoichiometric amounts of Tl, HgTe, Sn, and Te at 500°C in 24 h, held at this temperature for 100 h, and cooled to 100°C over 100 h. Then the furnace was turned off and allowed to cool naturally to room temperature.

9.24 Tin–Gallium–Lead–Tellurium

SnTe–GaTe–PbTe. The solubility of GaTe in the Sn$_x$Pb$_{1-x}$Te solid solution is less than 1.5 mol% (Bushmarina et al. 1980). The ingots were annealed at 600°C and studied through XRD and metallography.

9.25 Tin–Indium–Antimony–Tellurium

In₂SnTe–InSb. Specimens with compositions in the range of 0–50 mol% In₂SnTe were investigated by Woolley and Williams (1964). It was shown that for the quenched from 800°C materials, solid solubility extends out to about 26 mol% In₂SnTe, but this was reduced to 7 mol% for alloys annealed at 400°C for 6 weeks, with a similar value for annealing temperatures up to 450°C. Alloys containing 30 and 40 mol% In₂SnTe were found to melt at 500°C and showed signs of melting when annealed at 450°C. The energy gap of the solid solutions increases to a maximum value of 0.47 eV at 0.5 mol% In₂SnTe, and then it remains constant. This system was investigated through XRD and metallography.

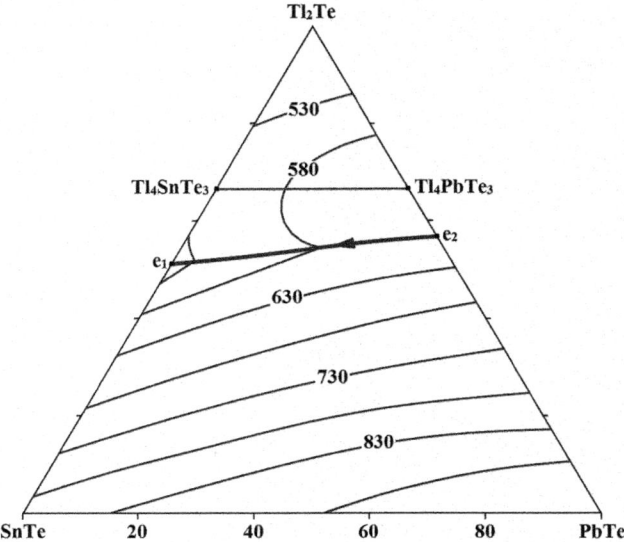

FIGURE 9.7 Liquidus surface of the SnTe–Tl₂Te–PbTe quasiternary system. (From Guseynov, Z.A., et al., *Zhurn. neorgan. khimii*, **46**(11), 1927, 2001.)

9.26 Tin–Thallium–Lead–Tellurium

Tl₄SnTe₃–Tl₄PbTe₃. A continuous solid solution series are formed in this system (Babanly et al. 1979; Guseynov et al. 2001; Filep and Sabov 2008).

SnTe–Tl₂Te–PbTe. The liquidus surface of this quasiternary system (Figure 9.7) consists of the field of primary crystallization of the Sn$_x$Pb$_{1-x}$Te and Tl₄Sn$_x$Pb$_{1-x}$Te₃ solid solutions (Guseynov et al. 2001). There are no nonvariant equilibria in the system. At 230°C, the wide region of the Tl₄Sn$_x$Pb$_{1-x}$Te solid solution is formed of ~2 mol % wide along the SnTe–PbTe system. The alloys for the investigations were annealed at 330°C (at 430°C for the alloys containing <85 mol% Tl₂Te) for 400–500 h.

Tl₄SnTe₃–Tl₄PbTe₃–Tl₅Te₃. The limiting quasibinary systems are characterized by the unlimited solubility of the components in the solid and liquid states and are characterized by the phase diagrams without nonvariant equilibrium (Babanly et al. 1979). The single-phase solid solution is formed in this system for any composition, and the liquidus and solidus surfaces do not have any extreme. The alloys for the investigations were annealed 400°C–550°C for 500 h with the next quenching in water.

9.27 Tin–Thallium–Zirconium–Tellurium

The **Tl₃Zr₁.₄Sn₁.₆Te₆** quaternary compound, which crystallizes in the trigonal structure with the lattice parameters $a = 434.4 \pm 0.5$ and $c = 2343 \pm 6$ pm, is formed in the Sn–Tl–Zr–Te system (McGuire et al. 2005a). Black needles of this compound were synthesized from the elements. Since Tl will tarnish slowly in air it was handled in an Ar glove box. Chunks of Tl were cut from the 0.5 inch diameter rod with wire cutters used only

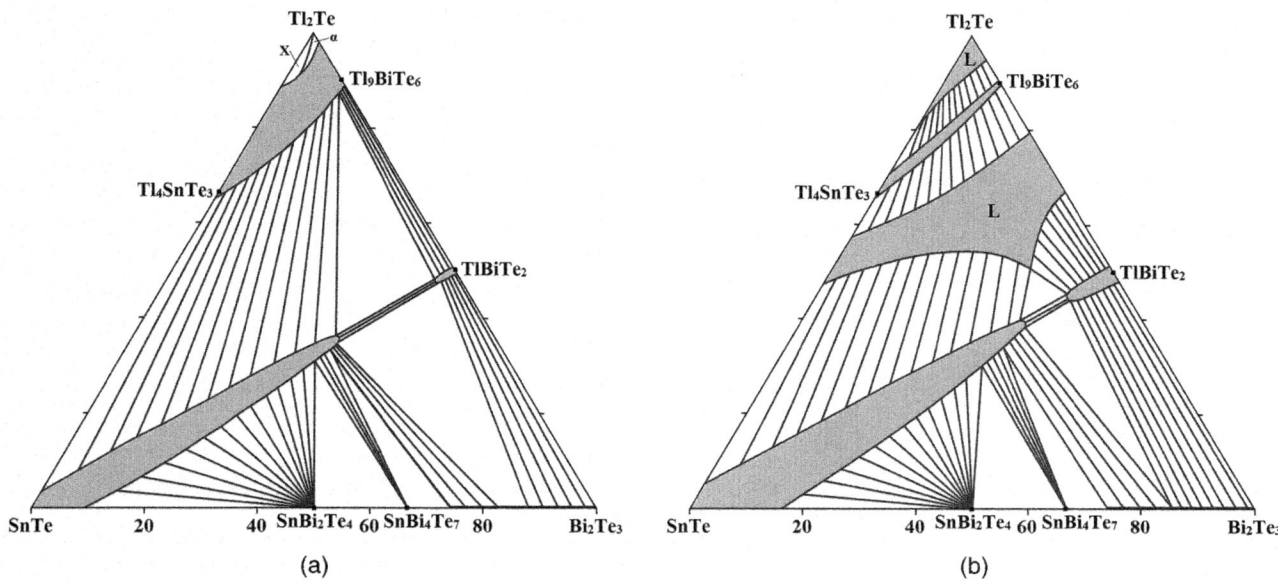

FIGURE 9.8 Isothermal section of the SnTe–Tl$_2$Te–Bi$_2$Te$_3$ quasiternary system at (a) 330°C and (b) 530°C. (From Babanly, D.M., et al., *Russ. J. Inorg. Chem.*, **56**(12), 1981, 2011.)

for this purpose, weighed inside the glove box and placed in capped glass vials. Appropriate amounts of the other elements were then weighed out and loaded into carbon coated silica tube. The Ar-filled vials were removed from the glove box and the Tl was added to the silica tube. The tube was then quickly (to limit the exposure of Tl to air) attached to a vacuum line for sealing. It was heated over 24 h to 800°C, held at this temperature for 100 h and then cooled to 400°C for over 100 h, at which point the furnace was turned off and allowed to cool to room temperature. Crystals were extracted from the resulting polycrystalline mass for examination.

9.28 Tin–Thallium–Antimony–Tellurium

SnTe–TlSbTe$_2$. The phase diagram of this system, constructed through DTA and XRD, is a eutectic type (Mazelsky and Lubell 1962). The eutectic contains 10 mol% SnTe (the system is viewed as SnTe–Tl$_{0.5}$Sb$_{0.5}$Te) and crystallizes at 465°C. The nature of the thermal effects at 404°C was not determined. The Sn$_{1-x}$Tl$_{x/2}$Sb$_{x/2}$Te solid solutions crystallizes in the cubic structure within the interval $0 < x < 0.42$ and in the rhombohedral structure at $0.75 < x < 1.0$.

9.29 Tin–Thallium–Bismuth–Tellurium

SnTe–TlBiTe$_2$. The phase diagram of this system, constructed through DTA, XRD, and measuring of the microhardness, is a eutectic type with eutectoid transformation of the solid solution based on TlBiTe$_2$ (Babanly et al. 2008). The eutectic contains ≈85 mol% TlBiTe$_2$ and crystallizes at 542°C, and the eutectoid contains ≈90 mol% TlBiTe$_2$ and takes place at 477°C. The solubility of SnTe in β-TlBiTe$_2$ reaches 10 mol% at 530°C

and that in α-TlBiTe$_2$ is ≈5 mol% at 330°C. The alloys for the investigations were annealed for 500 h at 330°C with the next furnace cooling and at 530°C with the next quenching in cold water. According to the data of Mazelsky and Lubell (1962), solid solutions in this system crystallize in the cubic structure at the SnTe content of 44–100 mol% and in the rhombohedral structure at 0–21 mol% SnTe.

SnTe–Tl$_2$Te–Bi$_2$Te$_3$. Some vertical sections, the isothermal section at 330°C and 530°C, the liquidus surface of the SnTe–Tl$_2$Te–TlBiTe$_2$ subsystem and all quasiternary system were constructed by Babanly et al. (2008) and Babanly et al. (2011b), respectively. The isothermal section at 330°C is presented in Figure 9.8a. It is characterized by the existing of three solid solutions regions based on Tl$_4$SnTe$_3$–Tl$_9$BiTe$_6$ system and SnTe and TlBiTe$_2$ compounds. Two regions near Tl$_2$Te, α and X, have been identified. Alloys in the α region consist of a monoclinic phase with the Tl$_2$Te structure. An $\alpha + \delta$ two-phase region was not detected. Equilibrium alloys in the region X contain not only the α- and δ-phases, but also a metallic phase based on thallium. The isothermal section at 530°C is shown in Figure 9.8b and the liquidus surface system of this system is given Figure 9.9. The alloys for the investigations were annealed at 330°C and 530°C for 1000 h and then quenched in cold water.

9.30 Tin–Thallium–Manganese–Tellurium

The **MnTl$_2$SnTe$_4$** quaternary compound, which is semiconductor and crystallizes in the tetragonal structure with the lattice parameters $a = 845.03 \pm 0.13$ and $c = 710.78 \pm 0.15$ pm, is formed in the Sn–Tl–Mn–Te system (McGuire et al. 2005b). This compound was synthesized by reaction of Tl, Mn, Sn, and Te (molar ratio 2:1:1:4). At the syntheses, Tl metal was weighed in an Ar-filled glove box and placed into the capped

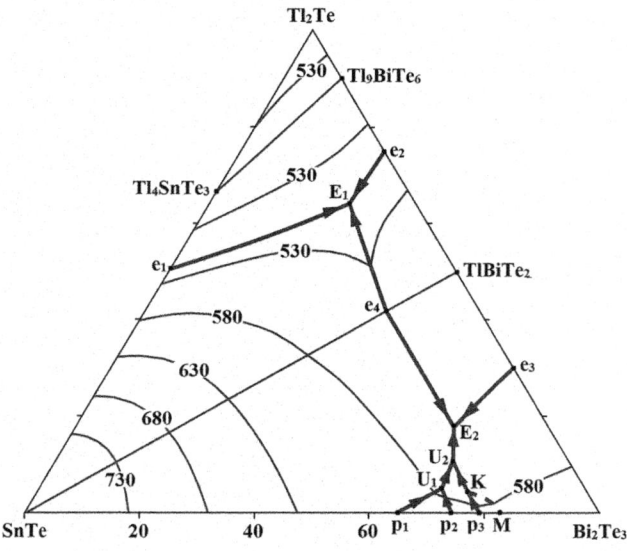

FIGURE 9.9 Liquidus surface of the SnTe–Tl₂Te–Bi₂Te₃ quasiternary system. (From Babanly, D.M., et al., *Russ. J. Inorg. Chem.*, **56**(12), 1981, 2011.)

vial. The other elements were weighed in air and loaded into a silica tube, which was coated with carbon to prevent a reaction between the silica and the sample. Thallium was then added to the tube, and the tube was quickly attached to a vacuum line for sealing. The evacuated, sealed silica tube was heated in a furnace. The sample was heated at 500°C for 4 days and then cooled to room temperature over 4 days. An attempt to make MnTl₂SnTe₄ as single-phase material failed.

9.31 Tin–Ytterbium–Lead–Tellurium

SnTe–YbTe–PbTe. *T-x-y*-relations for the solidus surface and the spinodal decomposition of solid solution in this quasiternary system were found by Mamedov et al. (2016).

9.32 Tin–Ytterbium–Antimony–Tellurium

SnSb₂Te₄–YbTe. This system is nonquasibinary section of the SnTe–YbTe–Sb₂Te₃ system as SnSb₂Te₄ melts incongruently (Ibadova et al. 2013). The mixture of starting components exists in the system below the solidus line. The solubility of YbTe in SnSb₂Te₄ at room temperature is less than 1 mol%. The system was studied through DTA, XRD, and metallography using the alloys annealed at 530°C for 1000 h.

SnTe–YbTe–Sb₂Te₃. The isothermal sections of this system at room temperature, 430°C, 630°C, and 930°C, the liquidus surface, as well as three vertical sections, were constructed by Aliev et al. (2014). No quaternary compounds were detected in this system. Four different solid solution fields were found at room temperature and at 430°C (Figure 9.10). Subsequently,

the border between the two regions completely disappeared and only one field exists in the system. The liquidus surface of the system is presented in Figure 9.11. It consists of four fields of primary crystallization. The critical point *C* exists in the system at ~680°C, the ternary eutectic (*E*) L ↔ (YbTe) + (Sb₂Te₃) + (SnSb₂Te₄) crystallizes at 577°C and the transition point (*U*) L + (SnTe) (YbTe) + (SnSb₂Te₄) takes place at 600°C. DTA, XRD, and SEM with EDS techniques were employed to check the purity of the synthesized starting compounds and analyze the samples. All alloys were heated up to 1030°C and held at this temperature for about 10–12 h and were then annealed at 430°C for ~1200 h. In most cases alloys were then quenched in cold water.

9.33 Tin–Ytterbium–Bismuth–Tellurium

SnTe–YbTe–Bi₂Te₃. The isothermal sections of this system at room temperature, 530°C, 630°C, and 930°C, the liquidus surface, as well as four vertical sections, were constructed by Aliev et al. (2018). No quaternary compounds were detected in this system. Three different solid solution fields were found at room temperature, at 530°C, and 630°C (Figure 9.12). Subsequently, the border between the two regions completely disappeared, and only one field exists in the system at 930°C. The ternary solubility based on YbTe dos not exceeds 2–3 mol% at room temperature and at 530°C. Areas of different solid phases that exist at room temperature and at 530°C completely disappear with increasing temperature and are replaced by areas of liquid phase or coexistence of solid and liquid phases at 630°C. Ultimately, seven different phase regions (3 single-phase, 3 two-phase, and 1 three-phase regions) have been detected at 630°C. The liquid phase areas continuously expand toward SnTe upon increasing temperature, and therefore, several phases that exist at 630°C completely disappear at 730°C. Consequently, only one solid phase, one liquid phase, and one two-phase region exist at this temperature. The liquidus surface of this system consists of five fields of primary crystallization of (SnTe), (YbTe), (Bi₂Te₃), SnBi₂Te₄, and SnBi₄Te₇ (Figure 9.13).

DTA, XRD, and SEM with EDS techniques were employed to check the purity of the synthesized starting compounds and analyze the samples. All alloys were melted at 1030°C for about 10–12 h and were then annealed at 530°C–540°C for ~1000 h. Most of the alloys were then slowly cooled down to room temperature with the furnace, but in some cases alloys were quenched at 530°C in cold water after annealing to keep the equilibrium state.

9.34 Tin–Lead–Bismuth–Tellurium

The **SnBi₂Te₄–PbBi₂Te₄** and **SnBi₄Te₇–PbBi₄Te₇** systems are nonquasibinary sections of the SnTe–PbTe–Bi₂Te₃ quasiternary system as all ternary compounds melt incongruently (Aghazade 2022). They are characterized by the formation of a continuous series of solid solutions with a tetradymite-like

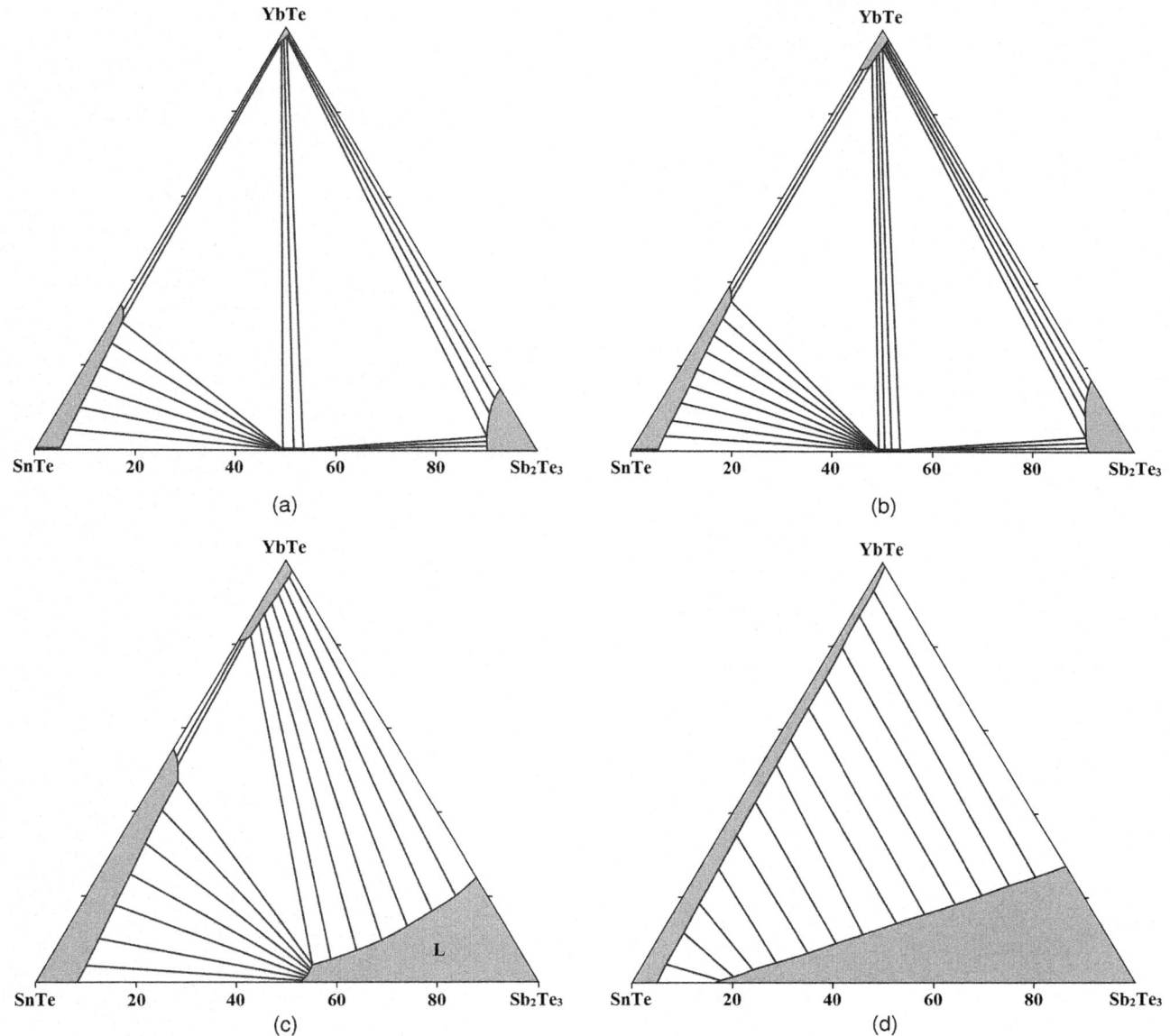

FIGURE 9.10 Isothermal section of the SnTe–YbTe–Sb$_2$Te$_3$ quasiternary system (a) at room temperature, (b) at 430°C, (c) at 630°C, and (d) at 730°C. (From Aliev, Z.S., et al., *J. Alloys Compd.*, **602**, 248, 2014.)

layered structure. The lattice parameters of solid solutions are linear functions of composition. The systems were studied through DTA and XRD using the alloys annealed at 450°C for 1000 h.

9.35 Tin–Lead–Oxygen–Tellurium

The **Pb$_2$(Sn$_{1.5}$Te$_{0.5}$)O$_{6.5}$** quaternary compound, which crystallizes in the cubic structure with the lattice parameter $a = 1059.62 \pm 0.02$ pm and a calculated density of 8.49 g·cm^{-3}, is formed in the Sn–Pb–O–Te system (Alonso et al. 1986). It was obtained from stoichiometric mixtures of PbO, SnO$_2$, and a light excess (15%) of TeO$_2$ to offset its partial sublimation. The mixture was ground and heated in air at 500°C for 3 days, at 650°C and 750°C for 24

h each, and 980°C also for 24 h. After each thermal treatment, the materials were quenched, weighed, and ground. This compound was obtained as an orange-yellow powder.

Using the method of diagrams of coexisting phases, the process of oxidation of Pb$_{1-x}$Sn$_x$Te solid solution was analyzed by Medvedev et al. (1987). A good agreement with the results of experimental studies of the natural aging process has been obtained.

9.36 Tin–Antimony–Bismuth–Tellurium

SnSb$_2$Te$_4$–SnBi$_2$Te$_4$ and **SnSb$_4$Te$_7$–SnBi$_4$Te$_7$**. Phase equilibria in these systems were experimentally investigated through DTA and XRD using the alloys annealed at 450°C for 500 h

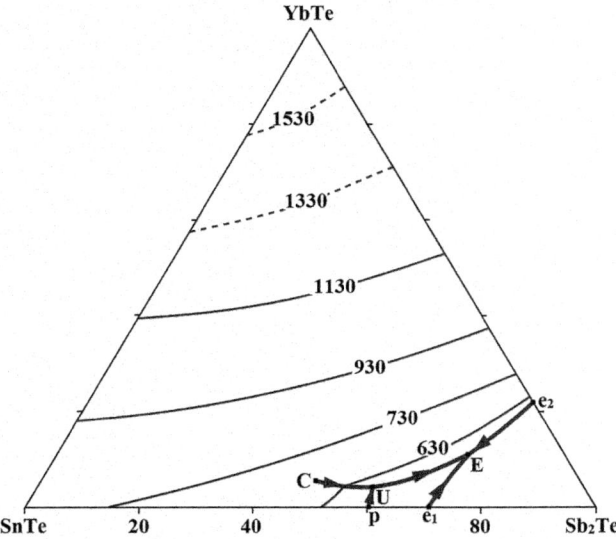

FIGURE 9.11 Liquidus surface of the SnTe–YbTe–Sb₂Te₃ quasiternary system. (From Aliev, Z.S., et al., *J. Alloys Compd.*, **602**, 248, 2014.)

(Seidzade and Babanly 2019; Seidzade 2019). It was shown that these systems are nonquasibinary due to the incongruent melting of the starting compounds, but stable in the subsolidus. The system is characterized by the formation of a continuous series of solid solutions with the hexagonal structure. The lattice parameters of the solid solutions change linearly from $a = 429.57 \pm 0.03$, $c = 4154.2 \pm 0.4$ pm for SnSb₂Te₄ to $a = 430.35 \pm 0.04$, $c = 4162.3 \pm 0.5$ pm for SnBi₂Te₄ and $a = 431.28 \pm 0.04$, $c = 2379.6 \pm 0.3$ pm for SnSb₄Te₇ to $a = 439.26 \pm 0.04$, $c = 2398.2 \pm 0.3$ pm for SnBi₄Te₇.

SnSbBiTe₄–Bi₂Te₃. The phase diagram of this system, constructed through DTA, XRD, metallography, and measuring the microhardness and density using the alloys annealed at 330°C for 200 h, is a eutectic type (Figure 9.14) (Gurbanov et al. 2010b). The eutectic contains 65 mol% 2Bi₂Te₃ and crystallizes at 402°C. The mutual solubility of the components is ~20 mol%.

SnSbBiTe₄–SnBi₄Te₇. The phase diagram of this system, constructed through DTA, XRD, metallography, and measuring the microhardness and density using the alloys annealed at 330°C for 200 h, is also a eutectic type (Gurbanov et al.

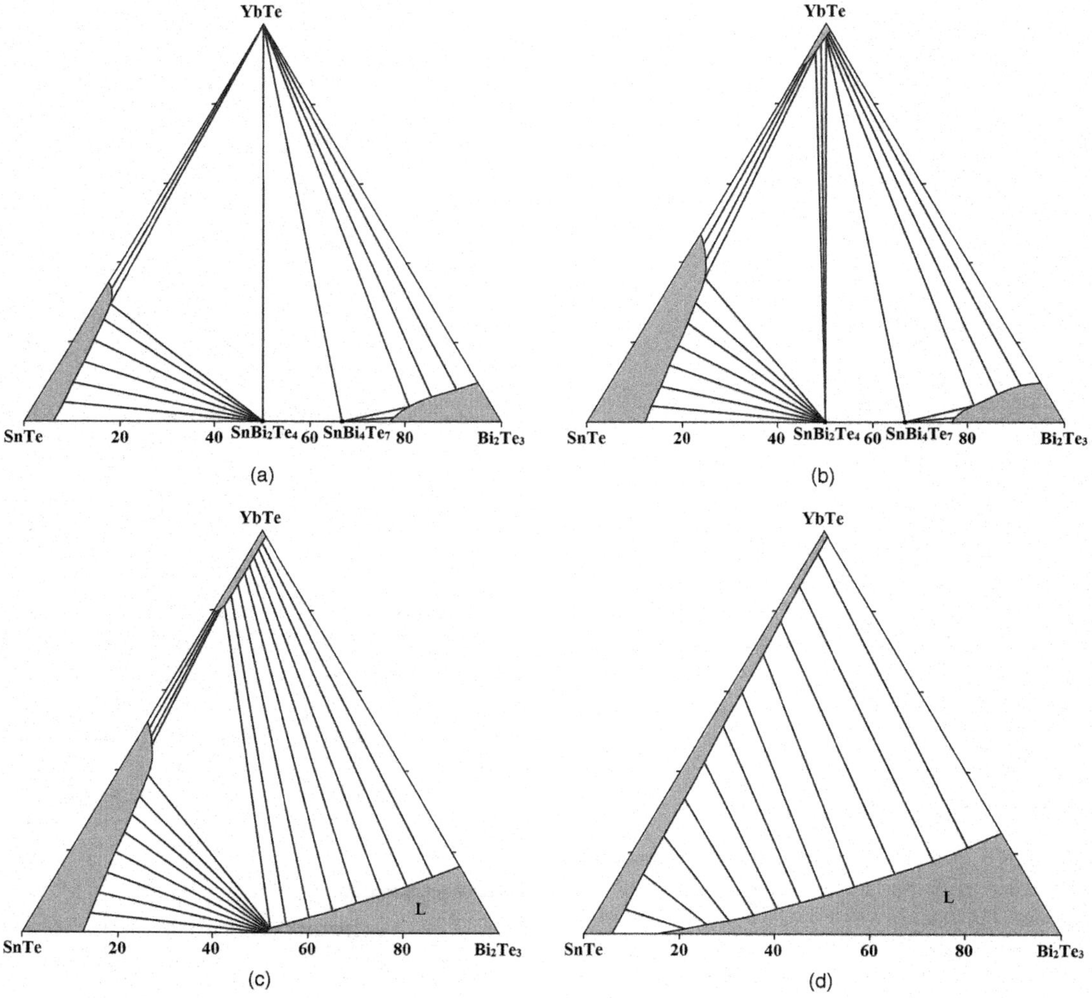

FIGURE 9.12 Isothermal sections of the SnTe–YbTe–Bi₂Te₃ system (a) at room temperature, (b) at 530°C, (c) at 630°C, and (d) at 730°C. (From Aliev, Z.S., et al., *J. Alloys Compd.*, **750**, 887, 2018.)

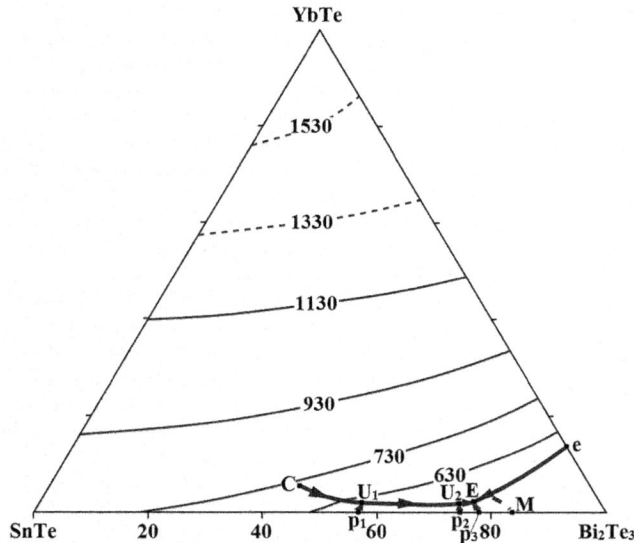

FIGURE 9.13 Liquidus surface of the SnTe–YbTe–Bi₂Te₃ quasiternary system. (From Aliev, Z.S., et al., *J. Alloys Compd.*, **750**, 887, 2018.)

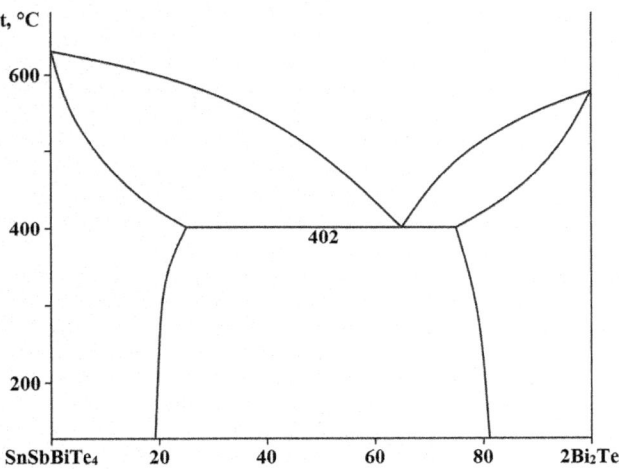

FIGURE 9.14 Phase diagram of the SnSbBiTe₄–2Bi₂Te₃ system. (From Gurbanov, G.R., et al, *Russ. J. Inorg. Chem.*, 55(7), 1149, 2010.)

2010b). The eutectic contains 35 mol% $SnBi_4Te_7$ and crystallizes at 477°C. Solid solutions are formed based on the starting components in the system.

9.37 Tin–Antimony–Manganese–Tellurium

$SnSb_2Te_4$–$MnSb_2Te_4$. Phase equilibria in these systems were experimentally investigated through DTA, XRD, and SEM with EDS using the alloys annealed at 450°C for 45 days with the next quenching in ice water (Orujlu 2020b). It was shown that this system is nonquasibinary due to the incongruent melting of the starting compounds, but stable in the subsolidus. The system is characterized by the formation of a continuous series of solid solutions with a trigonal structure. The lattice parameters of the solid solutions change linearly from $a = 429.58$

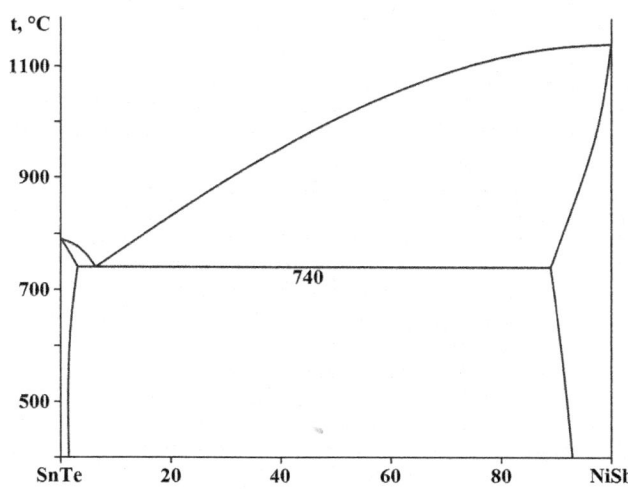

FIGURE 9.15 Phase diagram of the SnTe–NiSb system. (From Rustamov, P.G., *Azerb. khim. zhurn.*, (1), 116, 1975.)

± 0.03, $c = 4154.6 ± 0.4$ pm for $SnSb_2Te_4$ to $a = 424.51 ± 0.04$, $c = 4087.2 ± 0.3$ pm for $MnSb_2Te_4$.

$SnTe$–Sb_2Te_3–$MnTe$. An analytical method was used for $3D$ modeling of the crystallization surfaces of this quasiternary system based on the existing data of the boundary systems and a small number of experimental measurements through DTA (Orujlu et al. 2021). The obtained analytical dependencies of the liquidus and solidus temperatures on the composition allowed visualizing the crystallization surfaces of the phases of the system in a separate form as well as in one graph.

9.38 Tin–Antimony–Nickel–Tellurium

$SnTe$–$NiSb$. The phase diagram of this system, constructed through DTA, XRD, metallography, and measuring of the microhardness using the alloys annealed at 600°C for 500 h, is a eutectic type (Figure 9.15) (Rustamov et al. 1975). The eutectic contains 6 mol% NiSb and crystallizes at 740°C. The solubility of SnTe in NiSb reaches 7 mol% and the solubility of NiSb in SnTe is negligible. The solid solution based on NiSb is p-type semiconductor. According to the data of Kuliev et al. (1974), SnTe begins to interact chemically with NiSb at 760°C.

9.39 Tin–Bismuth–Manganese–Tellurium

$SnBi_2Te_4$–$MnBi_2Te_4$. Phase equilibria in this system were experimentally investigated through DTA and XRD using the alloys annealed at 450°C for ~500 h (Orujlu 2020a). It was shown that it is nonquasibinary due to the incongruent melting of the starting compounds, but stable in the subsolidus. The formation of a continuous series of solid solutions was observed. The lattice parameters of the $Sn_xMn_{1-x}Bi_2Te_4$ solid solutions increase linearly with the increasing x.

10

Systems Based on Lead Sulfide

10.1 Lead–Hydrogen–Nitrogen–Sulfur

The $PbN_2S_2 \cdot NH_3$ quaternary compound, which crystallizes in the monoclinic structure with the lattice parameters $a = 567.1 \pm 0.1$, $b = 1612.3 \pm 0.6$, $c = 610.2 \pm 0.2$ pm, $\beta = 95.11 \pm 0.02°$, and a calculated density of 3.78 g·cm^{-3} (Martan and Weiss 1984) [($a = 564$, $b = 1630$, $c = 613$ pm, $\beta = 94.4°$, and a calculated density of 3.72 g·cm^{-3} (Weiss and Neubauer 1958); $a = 565$, $b = 1621$, $c = 613$ pm, and $\beta = 94.3°$ (Weiss 1966)], is formed in the Pb–H–N–S system. This compound was prepared by interaction of PbI_2 or $Pb(NO_3)_2$ with S_4N_4 in liquid ammonia (Ruff and Geisel 1904; Goehring et al. 1955). Stable in air green crystals of the title compound were obtained.

10.2 Lead–Hydrogen–Oxygen–Sulfur

Some quaternary compounds are formed in the Pb–H–O–S system. $PbS_2O_6 \cdot 4H_2O$ crystallizes in the hexagonal structure with the lattice parameters $a = 634.13 \pm 0.09$, $c = 646.22 \pm 0.09$ pm, and a calculated density of 3.22 g·cm^{-3} (de Matos Gomes 1991). It was prepared by the addition of $PbCO_3$ to a solution of ditionic acid ($H_2S_2O_6$). Slow evaporation at room temperature yielded prismatic colorless crystals.

$3PbO \cdot PbSO_4 \cdot H_2O$ or $Pb_4O_2(OH)_2(SO_4)$ (mineral kennygayite) crystallizes in the triclinic structure with the lattice parameters $a = 637.8 \pm 0.1$, $b = 745.4 \pm 0.2$, $c = 1030.8 \pm 0.2$ pm, $\alpha = 75.26 \pm 0.03°$, $\beta = 79.37 \pm 0.03°$, $\gamma = 88.16 \pm 0.03°$, and a calculated density of 7.065 g·cm^{-3} (Steele et al. 1997) [$a = 637.85 \pm 0.05$, $b = 745.19 \pm 0.06$, $c = 1031.12 \pm 0.08$ pm, $\alpha = 75.234 \pm 0.005°$, $\beta = 79.388 \pm 0.006°$, and $\gamma = 88.175 \pm 0.006°$ for mineral kennygayite (Kampf et al. 2022i,k)].

This compound was synthesized from PbO and the stoichiometric quantity of H_2SO_4 (Burbank 1966). Powdered PbO (1 M) was mixed in 1 L of distilled water, and the required amount of H_2SO_4 ($d = 1.250$ g·cm^{-3}) was added dropwise with stirring, which was continued for 4 h after the addition of the acid. The mixture was digested for 4 days at room temperature, the solid collected on a filter, and air dried. The experimental study of the stability fields of the phases occurring in the $PbO \cdot PbSO_4$–H_2O system allows the construction of a solubility-pH-diagram and shows the temperature stability of various phases (Bode and Voss 1959). It was shown that $3PbO \cdot PbSO_4 \cdot H_2O$ is stable as a solid in equilibrium with the solution at low temperatures only. At the heating, it is stable only up to 150°C and decomposes at 350–400°C with the formation of a mixture of $4PbO \cdot PbSO_4$ and $PbO \cdot PbSO_4$ (Lander 1949; Bode and Voss 1959).

$Pb_4(S_2O_3)O_2(OH)_2$, mineral sidpietersite, crystallizes in the triclinic structure with the lattice parameters $a = 744.7 \pm 0.4$, $b = 650.2 \pm 0.4$, $c = 1120.6 \pm 0.4$ pm, $\alpha = 114.30 \pm 0.03°$, $\beta = 89.51 \pm 0.04°$, $\gamma = 89.04 \pm 0.06°$, and a calculated density of 6.765 g·cm^{-3} (Roberts et al. 1999; Jambor et al. 2000b) [$a = 745.5 \pm 0.2$, $b = 649.6 \pm 0.2$, $c = 1120.7 \pm 0.4$ pm, $\alpha = 114.33 \pm 0.02°$, $\alpha = 89.65 \pm 0.02°$, $\alpha = 88.69 \pm 0.02°$, and a calculated density of 6.765 g·cm^{-3} (Cooper and Hawthorne 1999)].

$Pb_4(OH)_4(S_2O_3)_2$, mineral finescreekite, crystallizes in the orthorhombic structure with the lattice parameters $a = 823.30 \pm 0.09$, $b = 1073.1 \pm 0.1$, and $c = 1482.6 \pm 0.3$ pm (Kampf et al. 2022e,f).

$[Pb_4(OH)_4][Pb(S_2O_3)_3]$ has two polymorphic modifications: first of them, mineral cubothioplumbite, crystallizes in the cubic structure with the lattice parameter $a = 1491.8 \pm 0.1$ pm (Kampf et al. 2022c,d) and the second, mineral hexathioplumbite, crystallizes in the hexagonal structure with the lattice parameters $a = 1072.1 \pm 0.1$, $c = 865.41 \pm 0.06$ pm (Kampf et al. 2022g,h).

$Pb_8O_4(OH)_2(S_2O_3)_3$, mineral boojumite, crystallizes in the orthorhombic structure with the lattice parameters $a = 1401.03 \pm 0.08$, $b = 2055.3 \pm 0.1$, and $c = 726.68 \pm 0.05$ pm (Kampf et al. 2022a,b).

10.3 Lead–Lithium–Titanium–Sulfur

The Li_xPbTiS_3 and $Li_xPbTi_2S_5$ quaternary intercalation compounds are formed in the Pb–Li–Ti–S system (Hernán et al. 1991). They were prepared by lithium intercalation of $PbTiS_3$ or $PbTi_2S_5$ using n-buthyllithium under an inert atmosphere. First of them crystallizes in the monoclinic structure with the lattice parameters $a = 581 \pm 3$, $b = 574 \pm 3$, $c = 2380 \pm 9$ pm, and $\beta = 98.5 \pm 0.4°$ at $x = 0.16$ and $a = 582 \pm 3$, $b = 575 \pm 3$, $c = 2384 \pm 9$ pm and $\beta = 98.5 \pm 0.4°$ at $x = 0.24$. This phase becomes amorphous at $x = 2.80$. $Li_xPbTi_2S_5$ crystallizes in the orthorhombic structure with the lattice parameters $a = 2292 \pm 9$, $b = 592 \pm 2$, $c = 3581 \pm 3$ pm at $x = 1.1$ and $a = 2326 \pm 3$, $b = 588 \pm 4$, $c = 3590 \pm 2$ pm at $x = 1.6$. This phase becomes amorphous at $x = 6.4$.

10.4 Lead–Lithium–Vanadium–Sulfur

The Li_xPbVS_3 quaternary intercalation compound is formed in the Pb–Li–V–S system (Hernán et al. 1991). It was prepared by lithium intercalation of $PbVS_3$ using n-buthyllithium under an inert atmosphere. Li_xPbVS_3 crystallizes in the monoclinic

DOI: 10.1201/9781003123484-10

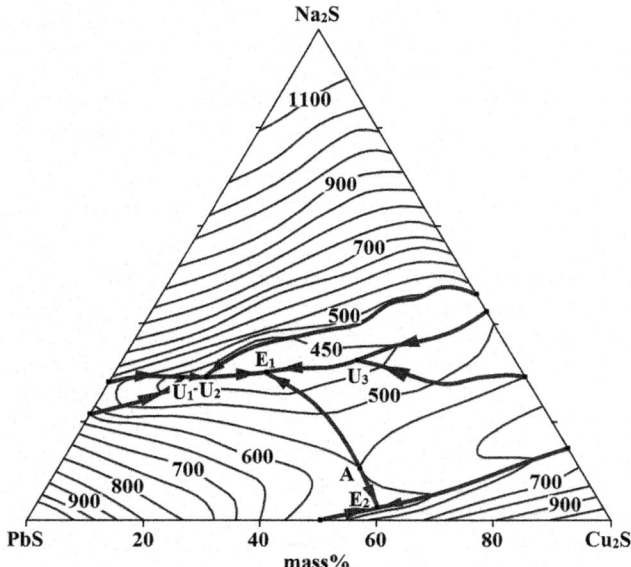

FIGURE 10.1 Liquidus surface of the PbS–Na₂S–Cu₂S quasiternary system. (From Bibenina, G.A., and Smirnov, M.P., *Zhurn. neorgan. khimii*, 11(9), 2133, 1966.)

structure with the lattice parameters $a = 572 \pm 3$, $b = 574 \pm 7$, $c = 2390 \pm 10$ pm and $\beta = 98.8 \pm 0.5°$ at $x = 0.89$. This phase becomes amorphous at $x = 2.40$.

10.5 Lead–Sodium–Copper–Sulfur

PbS–Na₂S–Cu₂S. The liquidus surface of this quasiternary system (Figure 10.1), constructed using 9 vertical sections, consists of 7 fields of PbS, 3PbS·Na₂S, Na₂S, Cu₂S·9Na₂S, 2Cu₂S·Na₂S, 5Cu₂S·2Na₂S, and Cu₂S primary crystallization (Bibenina and Smirnov 1966; Bibenina et al. 1967). There are two ternary eutectics (E_1 and E_2) in the system, which crystallize at 410°C and 500°C and contain 44.6 mass% PbS, 25 mass% Cu₂S, 30.4 mass% Na₂S and 39.1 mass% PbS, 58 mass% Cu₂S, 2.1 mass% Na₂S, respectively. On the E_1E_2 curve there is a saddle point (A), at which the temperature is maximal and is 550°C. There are also three transition points (U_1, U_2, and U_3) on the liquidus surface. The metallography revealed the existence of an unidentified phase in the system, the existence of which is not fixed by the differential thermal analysis (DTA) method. This phase occupies a large area in the region of PbS crystallization.

10.6 Lead–Sodium–Phosphorus–Sulfur

The **Na₆Pb₃(PS₄)₄** quaternary compound, which has two polymorphic modifications, is formed in the Pb–Na–P–S system (Aitken and Kanatzidis 2001). α-Na₆Pb₃(PS₄)₄ crystallizes in the trigonal structure with the lattice parameters $a = 1945.27 \pm 0.02$, $c = 617.20 \pm 0.01$ pm, a calculated density of 3.439 g·cm⁻³ at 173 K, and an energy gap of ≈ 2.7 eV. β-Na₆Pb₃(PS₄)₄ melts at ca. 580°C and crystallizes in the cubic structure with the lattice parameters $a = 1373.0 \pm 0.2$, a calculated density of 3.584 g·cm⁻³ at 173 K, and an energy gap of 2.57 eV.

α-Na₆Pb₃(PS₄)₄ was obtained from the reaction of Pb, P₂S₅, Na₂S, and S (molar ratio 1:2:3:8) that was sealed under vacuum (~0.02 Pa) in a Pyrex tube and heated to 500°C for 4 days followed by cooling to 250°C at 2°C·h⁻¹. The excess Na$_x$P$_y$S$_z$ flux was removed with DMF to reveal yellow needles that decompose slowly in air and H₂O. It was also prepared as a crystalline powder with PbS contamination from a stoichiometric mixture of Pb, P₂S₅, Na₂S, and S (molar ratio 3:2:2:3) that was sealed under vacuum and heated to 680°C for 12 h followed by cooling to 250°C at 10°C·h⁻¹. β-Na₆Pb₃(PS4)₄ was prepared as a yellow crystalline powder with PbS contamination by rapid quenching of a stoichiometric mixture of Pb, P₂S₅, Na₂S, and S. A few small, single crystals, which decomposed slowly in air and H₂O, were found. β-Na₆Pb₃(PS₄)₄ transforms to α-Na₆Pb₃(PS₄)₄ on cooling. Annealing of β-Na₆Pb₃(PS₄)₄ at 375°C for 2 days also converts it to α-Na₆Pb₃(PS₄)₄.

10.7 Lead–Sodium–Oxygen–Sulfur

PbS–Na₂SO₄. The phase diagram of this system, constructed through DTA, X-ray diffraction (XRD), and metallography, is a eutectic type with monotectic transformation (Smirnov and Kudryashova 1956). The eutectic contains 99.7 mol% Na₂SO₄ and crystallizes at 875°C. Immiscibility region exists in the system within the interval 1.7–97 mol% Na₂SO₄. The nature of the thermal effects at 985°C and below 875°C has not been elucidated.

10.8 Lead–Sodium–Chlorine–Sulfur

PbS–NaCl. The phase diagram of this system, constructed through DTA and by measuring of the PbS dissolution in NaCl melt, is a eutectic type (Pelton and Flengas 1970). The eutectic is degenerated from the NaCl side. The immiscibility region exists in the system with a monotectic temperature of 1096.2°C.

10.9 Lead–Sodium–Iron–Sulfur

PbS–Na₆Fe₄S₇. The phase diagram of this system, constructed through DTA and metallography, is a eutectic type (Figure 10.2) (Kopylov 1967). The eutectic contains 38 mol% Na₆Fe₄S₇ and crystallizes at 700°C.

PbS–Na₂S–FeS. The liquidus surface of this quasiternary system (Figure 10.3), constructed using 19 vertical sections, consists from the fields of PbS, FeS, Na₂S, Na₂Pb₃S₄, Na₆Fe₄S₇, and Na₁₀Fe₂S₇ primary crystallization (Kopylov 1967). The field of Na₁₀Fe₂S₇ primary crystallization occupies the biggest part of the liquidus surface. There are three ternary eutectics (E_1, E_2 and E_3) and one transition point (U) in the system, which crystallize at 640°C, 485°C, 600°C, and 515°C, respectively. The E_1E_2 and E_2E_3 lines, passing through the maximum, form two saddle points with temperatures of 700°C and 640°C, respectively.

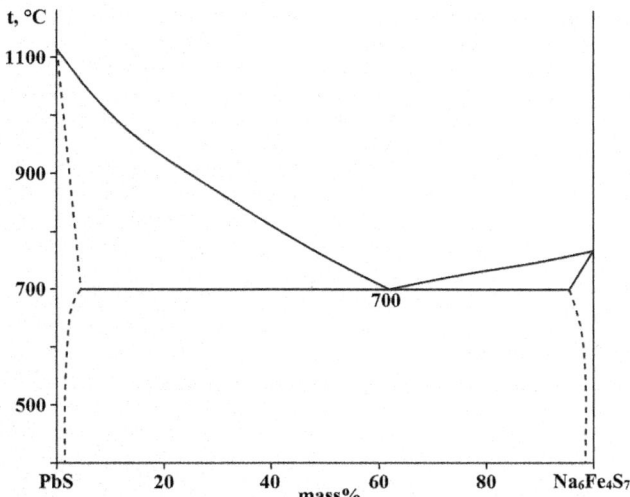

FIGURE 10.2 Phase diagram of the PbS–Na₆Fe₄S₇ system. (From Kopylov, N.I., *Zhurn. neorgan. khimii*, **12**(10), 2832, 1967.)

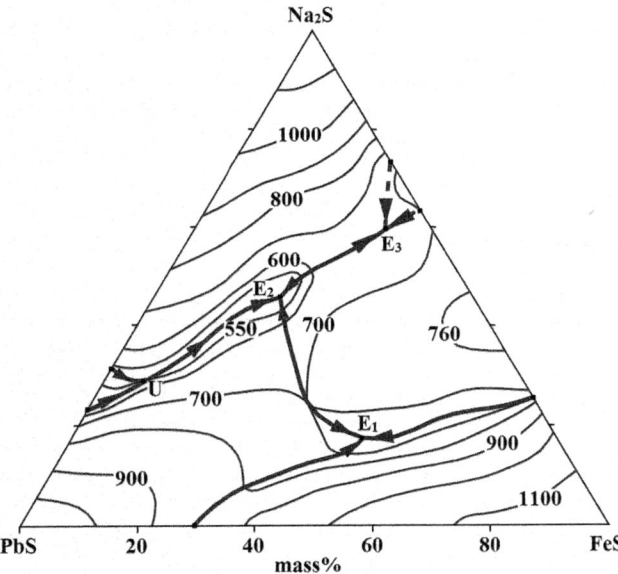

FIGURE 10.3 Liquidus surface of the PbS–Na₂S–FeS quasiternary system. (From Kopylov, N.I., *Zhurn. neorgan. khimii*, **12**(10), 2832, 1967.)

10.10 Lead–Potassium–Phosphorus–Sulfur

The **KPbPS₄** quaternary compound, which crystallizes in the orthorhombic structure with the lattice parameters $a = 1704.5 \pm 0.3$, $b = 666.0 \pm 0.1$, $c = 647.3 \pm 0.1$ pm, and a calculated density of 3.665 g·cm⁻³, is formed in the Pb–K–P–S system (Belkyal et al. 2006b). It was prepared by reacting a mixture of K₂S₃, Pb, P₂S₅, and S (molar ratio 1:1:1:6). The reaction mixture was thoroughly mixed in a N₂-filled glove box and loaded into a glass ampoule. After evacuation to 0.1 Pa the ampoule was flame-sealed, placed in a furnace, and heated to 500°C with 20°C·h⁻¹. After 6 days, the product was cooled to 100°C with 2°C·h⁻¹ and the furnace was turned off. The residual

K$_x$P$_y$S$_z$ flux was removed with dry DMF and acetone. After washing with ether transparent yellow platelets were obtained. This compound is slightly air and moisture sensitive.

10.11 Lead–Potassium–Oxygen–Sulfur

The **K₂Pb(SO₄)₂** quaternary compound, which crystallizes in the trigonal structure with the lattice parameters $a = 549.7 \pm 0.1$, $c = 2086.4 \pm 0.2$ pm, a calculated density of 4.36 ± 0.05 g·cm⁻³ (Tissot et al. 2001) [$a = 549 \pm 1$ and $c = 2083 \pm 5$ pm (Møller 1954); $a = 550.0$, $c = 2086.3$ pm in hexagonal setting or $a = 764.5$ pm, $\alpha = 42°10'$, and the calculated and experimental densities of 4.352 and 4.33 g·cm⁻³, respectively (Schwarz 1966)], is formed in the Pb–K–O–S system.

This compound was prepared via a chemical synthesis method (Tissot et al. 2001). The Pb(NO₃)₂ solution containing Pb (41.92 mM) was slowly dropped over 37 min into the K₂SO₄ solution containing potassium (175.8 mM), and held at 80.5–84°C. Upon completion of the lead nitrate addition, the precipitate was stirred for 2 h with a temperature ranging from 84°C to 86°C. After the 2-h hold, the pH of the slurry product was ~5.2 when taken at ~84°C. The pH of the K₂SO₄ solution was ~6.1. The precipitate was isolated by vacuum filtration at the elevated temperature and washed with deionized water, water + methanol (volume ratio 1:1), and methanol. The vacuum filtration was very slow, so the precipitate was transferred to a glass frit funnel for a final wash with methanol. The powder was dried at ~84°C overnight.

K₂Pb(SO₄)₂ was also synthesized in the microcrystalline state by stirring a precipitate of PbSO₄ with a sufficiently concentrated solution of K₂SO₄, filtering, and washing with dilute alcohol (Møller 1954) or by annealing PbSO₄ and K₂SO₄ at 700°C (2 times × 12 h, air) and from the melt at 800–850°C by quenching (Schwarz 1966). The title compound was also obtained from the solution in the following way (Schwarz 1966). Adding dropwise to a boiling solution of K₂SO₄ (10 g) in H₂O (100 mL), a solution of Pb(CH₃COO)₂·3H₂O (2.5 g) in 10 mL of H₂O. The suspension was stirred at the boiling point for 8 days at the reflux. Then it was filtered while hot, the crystal slurry was quickly pressed onto clay and air dried for 24 h.

10.12 Lead–Potassium–Chlorine–Sulfur

PbS–KCl. The phase diagram of this system, constructed through DTA and by measuring of the PbS dissolution in KCl melt, is a eutectic type (Pelton and Flengas 1970). The eutectic is degenerated, and its temperature practically coincides with the melting temperature of KCl. The immiscibility region exists in the system with a monotectic temperature of 1096°C.

10.13 Lead–Rubidium–Phosphorus–Sulfur

The **RbPbPS₄** quaternary compound, which has reversible structural phase transition at 182–184 K and crystallizes in the orthorhombic structure with the lattice parameters $a = 1748.6$

± 0.1, $b = 671.27 \pm 0.05$, and $c = 641.91 \pm 0.05$ pm at room temperature (Belkyal et al. 2009) [$a = 639.87 \pm 0.07$, $b = 668.99 \pm 0.07$, $c = 1729.75 \pm 0.19$ pm, and a calculated density of 4.053 g·cm⁻³ at 153 K (Yao and Ibers 2004)], is formed in the Pb–Rb–P–S system.

The title compound was prepared by reacting a mixture of Rb_2S_3 with Pb, P_2S_5, and additional sulfur (molar ratio 1:1:1:6) (Belkyal et al. 2009). The reagents were thoroughly mixed in a N_2-filled glove box and loaded into a glass ampoule, which was evacuated and flame-sealed. The sample was heated to 500°C, kept at this temperature for 6 days and afterward cooled down to 100°C at a rate of 2°C·h⁻¹, whereupon the furnace was turned off. The resulting melt was washed with DMF and acetone, and transparent yellow platelets were obtained. The compound is stable in the air.

$RbPbPS_4$ was also obtained from a solid-state reaction of Rb_2S_3 (0.5 mM), Pb (1.0 mM), P_2S_5 (0.5 mM), and S (1.0 mM) (Yao and Ibers 2004). The mixture was loaded into a fused silica tube under an Ar atmosphere in a glove box. The tube was sealed under a vacuum (10⁻² Pa) and then placed in a furnace. The sample was heated to 800°C over a period of 20 h, kept at 800°C for 84 h, cooled at 6°C·h⁻¹ to 100°C and then cooled rapidly to room temperature.

10.14 Lead–Rubidium–Bismuth–Sulfur

The $RbPbBi_3S_6$ quaternary compound, which melts at 739°C and crystallizes in the hexagonal structure with the lattice parameters $a = 409.9 \pm 0.3$, $c = 2373 \pm 1$ pm, and an energy gap of 0.89 eV, is formed in the Pb–Rb–Bi–S system (Chung et al. 1999). This compound is n-type semiconductor and did not show any sign of decomposition or phase transformation in DTA. To prepare this compound, a mixture of Rb_2S (0.197 mM), Pb (0.395 mM), Bi (1.177 mM), and S (2.183 mM) was transferred to a carbon-coated quartz tube and flame-sealed under high vacuum (<10⁻² Pa). The mixture was heated for 5 days at 800°C and cooled to 500°C at 10°C·h⁻¹ and then to 50°C in 10 h. A big chunk made up of big shiny gray-metallic plates was isolated by washing with water, methanol, and ether. All manipulations were carried out under a dry N_2 atmosphere in a glove box.

10.15 Lead–Rubidium–Oxygen–Sulfur

The $Rb_2Pb(SO_4)_2$ quaternary compound, which crystallizes in the trigonal structure with the lattice parameters $a = 559.4$, $c = 2159.2$ pm in hexagonal setting or $a = 788.9$ pm, $\alpha = 41°32'$, and the calculated and experimental densities of 4.855 and 4.81 g·cm⁻³, respectively (Schwarz 1966) [$a = 560 \pm 1$ and $c = 2158 \pm 5$ pm (Møller 1954)], is formed in the Pb–Rb–O–S system.

$Rb_2Pb(SO_4)_2$ was synthesized in the microcrystalline state by stirring a precipitate of $PbSO_4$ with a sufficiently concentration solution of Rb_2SO_4, filtering, and washing with dilute alcohol (Møller 1954) or by annealing $PbSO_4$ and K_2SO_4 at 700°C (2 times, 17 h + 21 h, air) (Schwarz 1966).

10.16 Lead–Rubidium–Chlorine–Sulfur

$PbS–RbCl$. The phase diagram of this system, constructed through DTA and by measuring of the PbS dissolution in RbCl melt, is a eutectic type (Pelton and Flengas 1970). The eutectic is degenerated and its temperature practically coincides with the melting temperature of RbCl. The immiscibility region exists in the system with a monotectic temperature of 1095.5°C.

10.17 Lead–Cesium–Phosphorus–Sulfur

The $CsPbPS_4$ quaternary compound, which crystallizes in the orthorhombic structure with the lattice parameters $a = 1818.5 \pm 0.1$, $b = 680.85 \pm 0.05$, and $c = 635.18 \pm 0.05$ pm, is formed in the Pb–Cs–P–S system (Belkyal et al. 2005). It was synthesized from a mixture of Cs_2S_3, Pb, P_2S_5 and S (molar ratio 1:1:1:6). The reagents were thoroughly mixed in a N_2-filled glove box and loaded into a glass ampoule, which was evacuated and flame-sealed. The sample was heated to 500°C, kept at this temperature for 6 days and afterwards cooled down to 100°C at a rate of 2°C·h⁻¹, and then turned off the furnace. To remove unreacted $Cs_xP_yS_z$, the solidified melt was washed with DMF and acetone. The product was consisted of transparent yellow crystals which are stable in air and moisture.

10.18 Lead–Cesium–Bismuth–Sulfur

The $CsPbBi_3S_6$ quaternary compound, which melts at 726°C and crystallizes in the hexagonal structure with the lattice parameters $a = 407.4 \pm 0.2$, $c = 2454.9 \pm 0.5$ pm, a calculated density of 5.457 g·cm⁻³, and an energy gap of 0.89 eV, is formed in the Pb–Rb–Bi–S system (Chung et al. 1999). This compound is n-type semiconductor and did not show any sign of decomposition or phase transformation in DTA. To prepare this compound, a mixture of Cs_2S (0.282 mM), Pb (0.560 mM), Bi_2S_3 (0.842 mM), and S (0.623 mM) was transferred to a carbon-coated quartz tube and flame-sealed under high vacuum (< 10⁻² Pa). The mixture was heated for 6 days at 850°C and cooled to 500°C at 5°C·h⁻¹ and then to 50°C in 10 h. A big chunk made up of big shiny gray-metallic plates was isolated by washing with water, methanol, and ether. All manipulations were carried out under a dry N_2 atmosphere in a glove box.

10.19 Lead–Cesium–Chlorine–Sulfur

$PbS–CsCl$. The phase diagram of this system, constructed through DTA and by measuring of the PbS dissolution in CsCl melt, is a eutectic type (Pelton and Flengas 1970). The eutectic is degenerated and its temperature practically coincides with the melting temperature of CsCl. The immiscibility region exists in the system with a monotectic temperature of 1095.3°C.

10.20 Lead–Copper–Silver–Sulfur

PbS–Cu₂S–Ag₂S. The isothermal section of this quasiternary system at 500°C was constructed through DTA, metallography and EPMA by Wu (1987a) and Chang et al. (1988). 0.5 g samples of appropriate composition were weighed and sealed in Pyrex tubes which were then heated in a horizontally placed muffle furnace at 500°C for about 50 days. In the middle of heating, the samples were taken out of the tubes and reground and then resealed into Pyrex tubes to ensure the reaction to be complete and homogeneous. The characteristic feature of phase relations in this system is the joins between PbS and each of the sulfides in the Cu₂S–Ag₂S boundary system, which divide the system into a number of the subsystems. There is also a narrow liquid field near the PbS–Cu₂S boundary system. This liquid will completely crystallize at about 480°C.

The **(Cu,Ag)₆PbS₄** quaternary compound (mineral furutobeite), which decomposes at 100°C and crystallizes in the monoclinic structure with the lattice parameters $a = 2002.5 \pm 1.3$, $b = 396.3 \pm 0.2$, $c = 970.5 \pm 0.4$ pm, $\beta = 101.57 \pm 0.04°$, and a calculated density of 6.74 g·cm⁻³, is formed in the Pb–Cu–Ag–S system (Sugaki et al. 1981; Fleischer et al. 1982b).

10.21 Lead–Copper–Zinc–Sulfur

PbS–Cu₁.₈S–ZnS. At 1200°C in this system, the solubility of ZnS in the liquid phase is around 10 mol% ZnS in the low composition range of PbS (Figure 10.4) (Surapunt et al. 1995, 1996). The Raoultian activity coefficients of ZnS at this temperature, which were determined from the obtained phase relation data, are very large at about 10.

The liquidus surface of this quasiternary system (Figure 10.5) was constructed according to three polythermal sections, which are traced from the ZnS corner (Kopylov et al. 1976b). It includes three fields of primary crystallization of binary sulfides, and the field of ZnS primary crystallization occupies the most part of the liquidus surface. The phase diagram of this system refers to diagrams with a four-phase transitional equilibrium preceded by one eutectic and one peritectic three-phase equilibrium. The border lines characterized the eutectic crystallization of ZnS and PbS, and the peritectic interaction L + ZnS ↔ (Cu₁.₈S) is situated near the Cu₂₋ₓS–PbS quasibinary system and converges in the *U* transition point. The crystallization in this system ends by the three-phase eutectic equilibrium L ↔ (Cu₁.₈S) + PbS. The most fusible part of the ZnS–Cu₁.₈S–PbS quasiternary system adjoins to the region of the binary eutectic in the Cu₂₋ₓS–PbS system at 560°C. Quaternary compounds are not formed in this system (Strohfeldt 1936, Kopylov et al. 1976b). This system was investigated through DTA and metallography.

In the Zn–Cu–S ternary system, a two-phase region consisting of the L_1 alloy phase rich in Cu and the L_2 matte phase exists in a considerably wide composition range from 0 to about 20 mol% Zn (Surapunt et al. 1995, 1996). When 0.8, 1.5, or 2.4 mol% Pb is added, the range of the two-liquid region does not change with the lead content.

10.22 Lead–Copper–Gallium–Sulfur

PbGa₂S₄–CuGaS₂. The phase diagram of this system, constructed through DTA, XRD, metallography, and measuring of the microhardness and density, is a eutectic type (Figure 10.6) (Jahangirova et al. 2019). The eutectic contains 60 mol% PbGa₂S₄ and crystallizes at 742°C. Solid solution based on CuGaS₂ contains up to 29 mol% PbGa₂S₄ at the eutectic

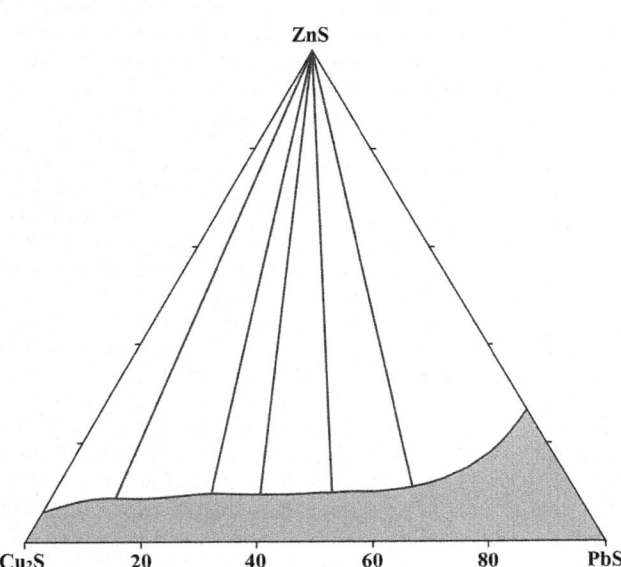

FIGURE 10.4 Isothermal section of the PbS–Cu₂S–ZnS quasiternary system at 1200°C. (From Surapunt, S., et al., *Shigen-to-Sozai*, **112**(1), 56, 1996.) Open access

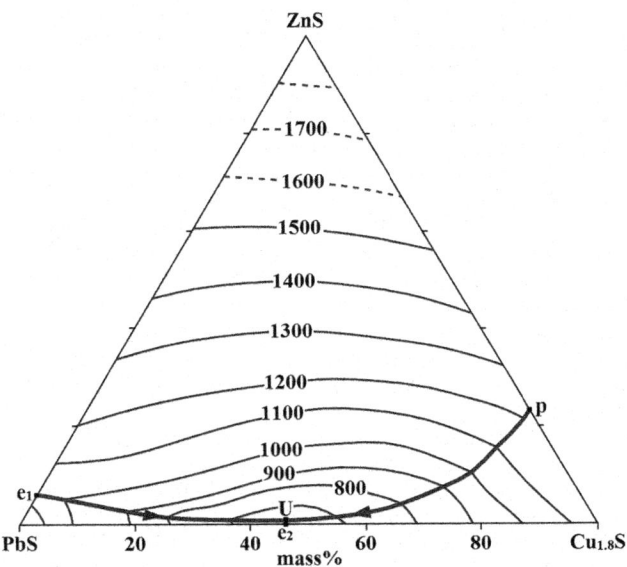

FIGURE 10.5 Liquidus surface of the PbS–Cu₁.₈S–ZnS quasiternary system. (From Kopylov, N.I. et al., *Izv. AN SSSR. Ser. Metally*, (6), 80, 1976.)

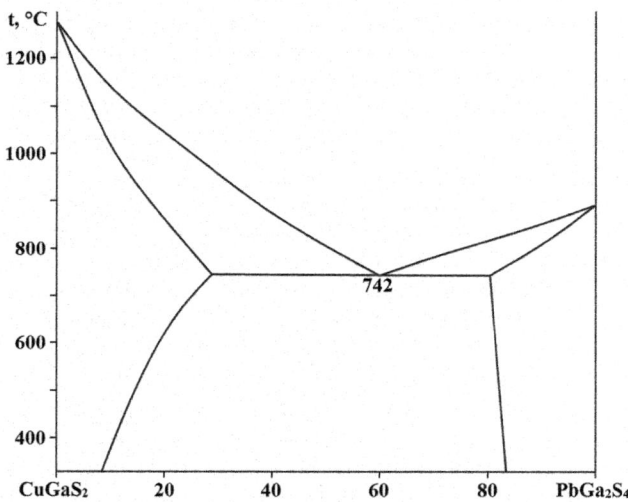

FIGURE 10.6 Phase diagram of the $PbGa_2S_4$–$CuGaS_2$ system. (From Jahangirova, S.K., et al., *Azerb. Chem. J.*, (1), 46, 2019.)

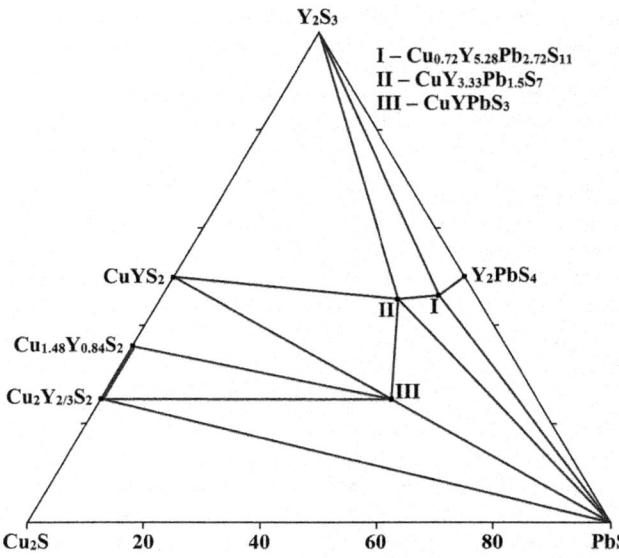

I – $Cu_{0.72}Y_{5.28}Pb_{2.72}S_{11}$
II – $CuY_{3.33}Pb_{1.5}S_7$
III – $CuYPbS_3$

FIGURE 10.7 Isothermal section of the PbS–Cu_2S–Y_2S_3 quasiternary system at 600°C. (From Gulay, L.D., et al., *J. Alloys Compd.*, **431**(1–2), 77, 2007.)

temperature and decreases to 7 mol% at 230°C. $PbGa_2S_4$ dissolves 20 and 17 mol% $CuGaS_2$ at 742°C and 230°C, respectively. The ingots for the investigations were annealed at 630–680°C for 20 days.

10.23 Lead–Copper–Indium–Sulfur

The $CuPb_3In_{17}S_{34}$ quaternary compounds, which crystallizes in the monoclinic structure with the lattice parameters $a = 1266.4 \pm 0.3$, $b = 388.16 \pm 0.10$, $c = 2732.9 \pm 0.7$ pm, $\beta = 94.377 \pm 0.006°$, and an energy gap of ~1.3 eV, is formed in the Pb–Cu–In–S system (Wang and Lee 2006). To synthesize this compound, the mixture of the elements in stoichiometric proportions was heated in an evacuated silica ampoule (~10^{-2} Pa) at 850°C for 24 h, kept at 850°C for 24 h, and then cooled slowly to 700°C at a rate 10°C·h⁻¹; finally, the furnace was turned off to allow cooling to room temperature. The product exhibits a black color with bar-shaped crystals and is stable in the air for at least 14 days. All materials were handled in an Ar-filled glove box in which the concentration of oxygen was less than 10 ppm.

10.24 Lead–Copper–Yttrium–Sulfur

PbS–Cu₂S–Y₂S₃. The isothermal section of this quasiternary system at 600°C is presented in Figure 10.7 (Gulay et al. 2007c). The alloys were annealed at 600°C for 240 h, with the next quenching in cold water. Three quaternary compounds, $CuYPbS_3$, $CuY_{3.33}Pb_{1.5}S_7$, and $Cu_{1-x}Y_{5+x}Pb_{3-x}S_{11}$, are formed in this system. The first crystallizes in the orthorhombic structure with the lattice parameters $a = 1019.57 \pm 0.07$, $b = 393.70 \pm 0.02$, $c = 1296.25 \pm 0.06$ pm, and a calculated density of 5.818 g·cm⁻³ (Gulay et al. 2005f). It was prepared by fusion of

the elements in an evacuated quartz ampoule. The synthesis was realized in a shaft furnace with a heating rate of 20°C·h⁻¹. The ampoule with the sample was heated to a maximal temperature of 1150°C and kept at this temperature for 4 h. After that, it was cooled slowly to 600°C with a rate of 10°C·h⁻¹ and annealed at 600°C for 240 h. After the annealing, the ampoule with the sample was quenched in cold water.

$CuY_{3.33}Pb_{1.5}S_7$ crystallizes in the monoclinic structure with the lattice parameters $a = 1302.46 \pm 0.05$, $b = 394.21 \pm 0.02$, $c = 1206.71 \pm 0.04$ pm, $\beta = 104.953 \pm 0.003°$ and a calculated density of 4.966 g·cm⁻³ (Gulay et al. 2005h). The compound was prepared by fusion of the high-purity constituent elements in an evacuated quartz ampoule. The synthesis was realized in a shaft furnace with a heating rate of 20°C·h⁻³. The ampoule with the sample was heated to a maximal temperature of 1150°C. The sample was kept at the maximal temperature for 4 h. After that, it was cooled slowly to 600°C with a rate of 10°C·h⁻³ and annealed at this temperature for 240 h. After annealing, the ampoule with the sample was quenched in cold water.

$Cu_{1-x}Y_{5+x}Pb_{3-x}S_{11}$ crystallizes in the orthorhombic structure with the lattice parameters $a = 394.23 \pm 0.07$, $b = 1300.7 \pm 0.1$, $c = 363.86 \pm 0.05$ pm for $x = 0.28$ (Gulay et al. 2007c).

10.25 Lead–Copper–Lanthanum–Sulfur

The $CuLaPbS_3$ quaternary compounds, which crystallize in the orthorhombic structure with the lattice parameters $a = 809.1 \pm 0.3$, $b = 409.3 \pm 0.1$, $c = 1599.6 \pm 0.5$ pm, a calculated density of 6.341 g·cm⁻³, and an energy gap of ~1.5 eV (Brennan and Ibers 1992) [$a = 826.0$, $b = 884.0$, $c = 796.0$ pm, and the calculated and experimental densities of 5.80 and 5.76 g·cm⁻³, respectively (Bayramova et al. 2010; Bairamova et al. 2011b)], is formed in the Pb–Cu–La–S system. This compound was synthesized by melting stoichiometric elemental mixture in an evacuated silica tube mounted in a single-zone furnace (Bairamova et al. 2011b). The highest synthesis temperature was 930–980°C. Cooling at a rate of 300°C·h⁻¹ led to the formation of polycrystalline ingot with a large amount of cracking. Single crystals of this compound were isolated from the reaction of a mixture of La$_2$S$_3$, Pb, Cu, and S powders (molar ratio 1:4:6:9) (Brennan and Ibers 1992). The mixture was heated in an evacuated quartz tube at 500°C, 700°C, 800°C, and 1000°C for 1 day each, and then the furnace was shut off and allowed to cool. Long deep-red transparent prisms of the title compound were obtained.

The standard thermodynamic functions of CuLaPbS$_3$ were calculated: $S^0_{298} = 235.4 \pm 5.0$ J·M⁻¹K⁻¹, $\Delta S^0_{298} = -15.7 \pm 3.0$ J·M⁻¹K⁻¹, $\Delta H^0_{298} = -802.6 \pm 30$ kJ·M⁻¹, and $\Delta G^0_{298} = -797.5 \pm 30$ kJ·M⁻¹ (Bayramova et al. 2010).

10.26 Lead–Copper–Praseodymium–Sulfur

PbS–Cu$_2$S–Pr$_2$S$_3$. The isothermal section of this quasiternary system at 600°C is shown in Figure 10.8 (Marchuk et al. 2006a). The ingots for the investigations were annealed for 24 h. On the basis of Pr$_2$PbS$_4$, a solid solution is formed, the limiting composition of which corresponds to Pr$_{2.36}$Pb$_{0.46}$S$_4$. The solubility on the basis of the starting compounds is negligible (< 2 mol%).

10.27 Lead–Copper–Neodymium–Sulfur

PbS–CuNdS$_2$. This system is nonquasibinary section of the Pb–Cu–Nd–S system as CuNdS$_2$ melts incongruently (Bayramova et al. 2010). The **CuNdPbS$_3$** quaternary compound, which melts incongruently at 1002°C and crystallizes in the orthorhombic structure with the lattice parameters

$a = 820.0$, $b = 880.0$, $c = 792.0$ pm, and the calculated and experimental densities of 5.96 and 5.90 g·cm⁻³, respectively, is formed in the system (Bayramova et al. 2010; Bairamova et al. 2011b). The eutectic contains 70 mol% PbS and crystallizes at 937°C. The solubility of CuNdS$_2$ in PbS at room temperature is 1.5 mol%.

CuNdPbS$_3$ was synthesized by melting a stoichiometric elemental mixture in an evacuated silica tube mounted in a single-zone furnace (Bairamova et al. 2011b). The highest synthesis temperature was 930–980°C. Cooling at a rate of 300°C·h⁻¹ led to the formation of a polycrystalline ingot with a large amount of cracking. The system was studied through DTA, XRD, and measuring of the microhardness using the ingots annealed at 780°C for 2 weeks.

The standard thermodynamic functions of CuNdPbS$_3$ were calculated: $S^0_{298} = 245.6 \pm 5.0$ J·M⁻¹K⁻¹, $\Delta S^0_{298} = -19.8 \pm 4.0$ J·M⁻¹K⁻¹, $\Delta H^0_{298} = -775.5 \pm 30$ kJ·M⁻¹, and $\Delta G^0_{298} = -756.1 \pm 30$ kJ·M⁻¹ (Bayramova et al. 2010).

10.28 Lead–Copper–Samarium–Sulfur

PbS–CuSmS$_2$. This system is nonquasibinary section of the Pb–Cu–Sm–S system as CuSmS$_2$ melts incongruently (Bayramova et al. 2010). The **CuSmPbS$_3$** quaternary compound, which crystallizes in the orthorhombic structure with the lattice parameters $a = 390$, $b = 1328$, $c = 1030$ pm, and a calculated density of 6.42 g·cm⁻³, is formed in the system. The system was studied through DTA, XRD, and measuring of the microhardness using the ingots annealed at 780°C for 2 weeks.

The standard thermodynamic functions of CuSmPbS$_3$ were calculated: $S^0_{298} = 253.2 \pm 5.0$ J·M⁻¹K⁻¹, $\Delta S^0_{298} = -12.2 \pm 3.0$ J·M⁻¹K⁻¹, $\Delta H^0_{298} = -809.1 \pm 30$ kJ·M⁻¹, and $\Delta G^0_{298} = -805.3 \pm 30$ kJ·M⁻¹ (Bayramova et al. 2010).

10.29 Lead–Copper–Gadolinium–Sulfur

PbS–CuGdS$_2$. The phase equilibria in this system, obtained through DTA, XRD, metallography, and measuring of the microhardness, are given in Figure 10.9 (Bayramova et al. 2010, 2015). This system is a nonquasibinary section as CuGdS$_2$ melts incongruently. The eutectic contains 70 mol% PbS and crystallizes at 952°C. The mutual solubility of the starting components in each other was not found. The **CuGdPbS$_3$** quaternary compound, which melts incongruently at 1217°C and crystallizes in the orthorhombic structure with the lattice parameters $a = 386$, $b = 1324$, $c = 1026$ pm, and a calculated density of 6.65 g·cm⁻³, is formed in this system. The system was studied through DTA, XRD, and measuring of the microhardness using the ingots annealed at 780°C for 2 weeks.

The standard thermodynamic functions of CuGdPbS$_3$ were calculated: $S^0_{298} = 252.4 \pm 5.0$ J·M⁻¹K⁻¹, $\Delta S^0_{298} = -9.8 \pm 2.0$ J·M⁻¹K⁻¹, $\Delta H^0_{298} = -815.5 \pm 30$ kJ·M⁻¹, and $\Delta G^0_{298} = -812.2 \pm 30$ kJ·M⁻¹ (Bayramova et al. 2010).

Cu$_2$S–CuGdPbS$_3$. The phase diagram of this system, constructed through DTA, XRD, metallography, and measuring of the microhardness, is a eutectic type (Bayramova et al. 2015).

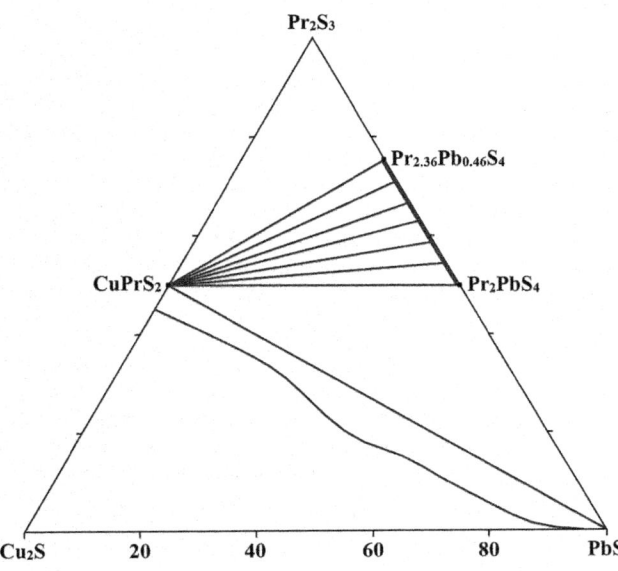

FIGURE 10.8 Isothermal section of the PbS–Cu$_2$S–Pr$_2$S$_3$ quasiternary system at 600°C. (From Marchuk, O.V., et al., *Nauk. Visnyk Volyns'k. Derzh. Univ. im. Lesi Ukrainky. Ser. Khim. nauky*, (4), 96, 2006.)

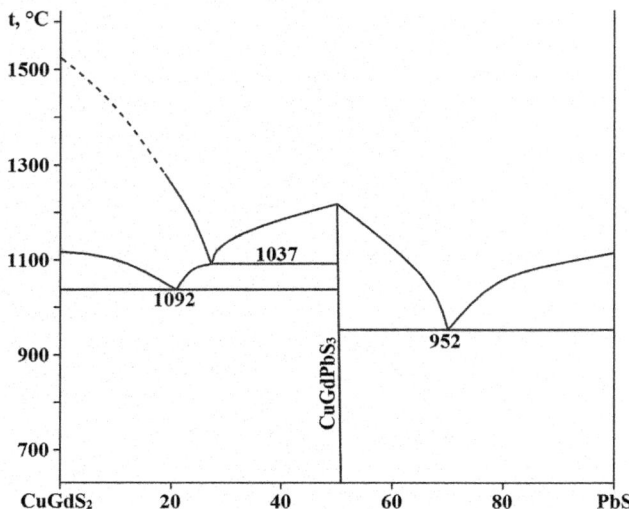

FIGURE 10.9 Phase equilibria in the system PbS–GdCuS₂ system. (From Bayramova, S.T., et al., *Kimya Problem.*, (4), 424, 2015.)

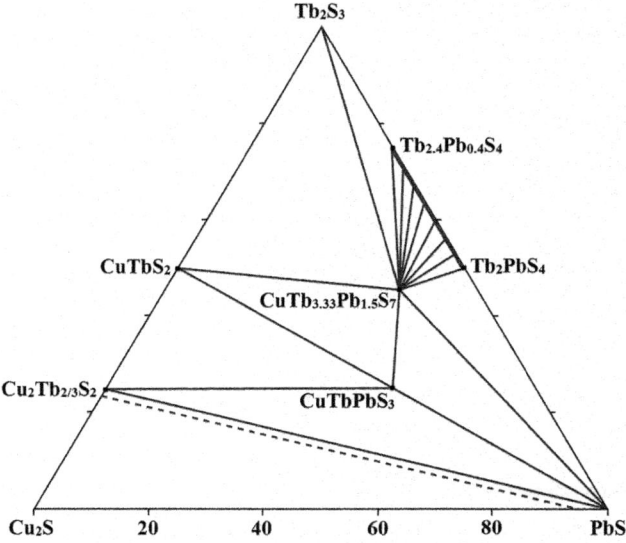

FIGURE 10.10 Isothermal section of the PbS–Cu₂S–Tb₂S₃ quasiternary system at 600°C. (From Khvaleba, N.V., et al., *Nauk. Visnyk Volyns'k. Derzh. Univ. im. Lesi Ukrainky. Ser. Khim. nauky*, (4), 112, 2006.)

The eutectic contains 40 mol% CuGdPbS₃ and crystallizes at 882°C. The solubility of CuGdPbS₃ in Cu₂S at 882°C is 10 mol% and the solid solution based on CuGdPbS₃ was not found.

10.30 Lead–Copper–Terbium–Sulfur

PbS–Cu₂S–Tb₂S₃. The isothermal section of this quasiternary system at 600°C is presented in Figure 10.10 (Khvaleba et al. 2006). The alloys were annealed at 600°C for 240 h, with the next quenching in cold water. Two quaternary compounds, **CuTbPbS₃** and **CuTb₃.₃₃Pb₁.₅S₇**, are formed in this system. The first of them crystallizes in the orthorhombic structure with the lattice parameters $a = 394.13 \pm 0.03$, $b = 1288.5 \pm 0.1$, $c = 1026.2 \pm 0.1$ pm, and a calculated density of 6.701 g·cm⁻³ (Gulay et al. 2005d). CuTb₃.₃₃Pb₁.₅S₇ crystallizes in the monoclinic structure with the lattice parameters $a = 1312.9 \pm 0.2$, $b = 397.91 \pm 0.06$, $c = 1217.6 \pm 0.2$ pm, and $\beta = 104.96 \pm 0.01°$ (Gulay and Olekseyuk 2006a). Both compounds were prepared by fusion of the high-purity elements in evacuated silica ampoules. The synthesis was performed in a tube furnace. The ampoules were heated with a heating rate of 30°C·h⁻¹ to maximal temperature of 1150°C. The samples were kept at maximal temperature for 4 h. After that, they were cooled slowly (10°C·h⁻¹) to 600°C and annealed at this temperature for 240 h. After annealing, the ampoules were quenched in cold water. The prepared products were gray-colored compact alloys.

10.31 Lead–Copper–Dysprosium–Sulfur

PbS–Cu₂S–Dy₂S₃. The isothermal section of this quasiternary system at 600°C is presented in Figure 10.11 (Gulay et al. 2007c). The alloys were annealed at 600°C for 240 h with the next quenching in cold water. Three quaternary compounds,

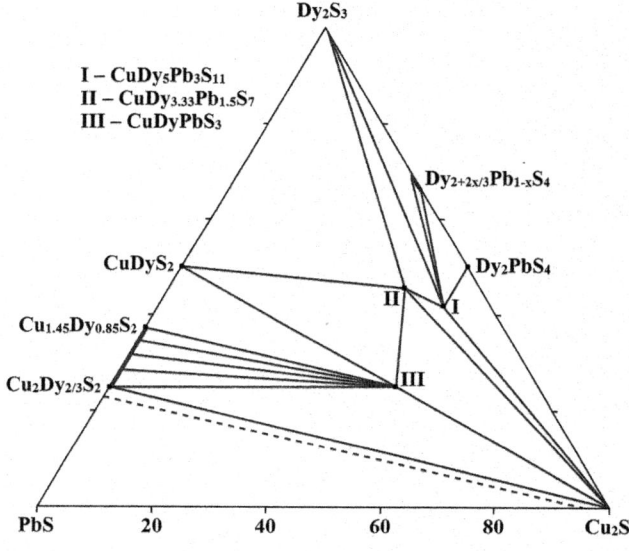

FIGURE 10.11 Isothermal section of the PbS–Cu₂S–Dy₂S₃ quasiternary system at 600°C. (From Gulay, L.D., et al., *J. Alloys Compd.*, **431**(1–2), 77, 2007.)

CuDyPbS₃, CuDy₃.₃₃Pb₁.₅S₇, and **CuDy₅Pb₃S₁₁,** are formed in this system. The first crystallizes in the orthorhombic structure with the lattice parameters $a = 392.97 \pm 0.03$, $b = 1289.1 \pm 0.1$, $c = 1020.5 \pm 0.1$ pm, and a calculated density of 6.802 g·cm⁻³ (Gulay et al. 2005d). CuDy₃.₃₃Pb₁.₅S₇ crystallizes in the monoclinic structure with the lattice parameters $a = 1307.9 \pm 0.2$, $b = 395.91 \pm 0.06$, $c = 1211.9 \pm 0.2$ pm, and $\beta = 104.996 \pm 0.008°$ (Gulay and Olekseyuk 2006a). These compounds were synthesized in the same way as CuTbPbS₃ and CuTb₃.₃₃Pb₁.₅S₇ were prepared.

CuDy₅Pb₃S₁₁ crystallizes in the orthorhombic structure with the lattice parameters $a = 396.04 \pm 0.07$, $b = 1306.8 \pm 0.2$, $c = 366.53 \pm 0.07$ pm (Gulay et al. 2007c).

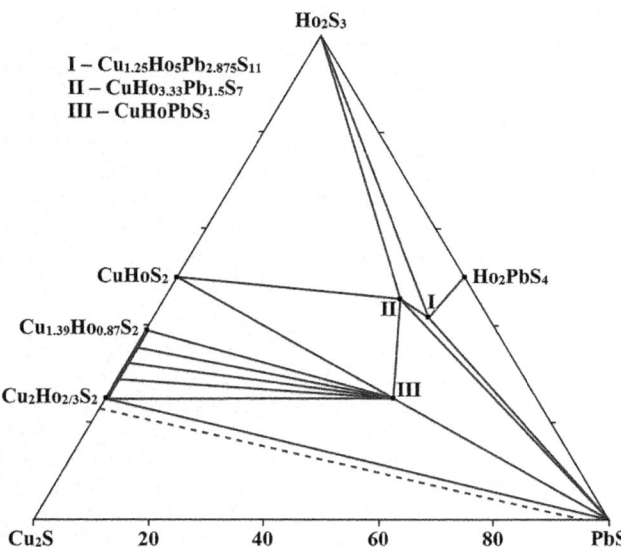

FIGURE 10.12 Isothermal section of the PbS–Cu$_2$S–Ho$_2$S$_3$ quasiternary system at 600°C. (From Gulay, L.D., et al., *J. Alloys Compd.*, **431**(1–2), 77, 2007.)

10.32 Lead–Copper–Holmium–Sulfur

PbS–Cu$_2$S–Ho$_2$S$_3$. The isothermal section of this quasiternary system at 600°C is presented in Figure 10.12 (Gulay et al. 2007c). The alloys were annealed at 600°C for 240 h, with the next quenching in cold water. Three quaternary compounds, **CuHoPbS$_3$**, **CuHo$_{3.33}$Pb$_{1.5}$S$_7$**, and **Cu$_{1+x}$Ho$_5$Pb$_{3−x/2}$S$_{11}$**, are formed in this system. The first crystallizes in the orthorhombic structure with the lattice parameters $a = 392.5 \pm 0.1$, $b = 1291.6 \pm 0.4$, $c = 1016.1 \pm 0.3$ pm, and a calculated density of 6.858 g·cm^{-3} (Gulay et al. 2005d). CuHo$_{3.33}$Pb$_{1.5}$S$_7$ crystallizes in the monoclinic structure with the lattice parameters $a = 1304.7 \pm 0.2$, $b = 394.62 \pm 0.05$, $c = 1209.0 \pm 0.2$ pm, $\beta = 104.885 \pm 0.007°$, and a calculated density of 6.340 g·cm^{-3} (Gulay and Olekseyuk 2006a). These compounds were synthesized in the same way as CuTbPbS$_3$ and CuTb$_{3.33}$Pb$_{1.5}$S$_{70}$ were prepared.

Cu$_{1+x}$Ho$_5$Pb$_{3−x/2}$S$_{11}$ crystallizes in the orthorhombic structure with the lattice parameters $a = 394.00 \pm 0.08$, $b = 1302.2 \pm 0.2$, $c = 367.07 \pm 0.07$ pm for $x = 0.25$ (Gulay et al. 2007c).

10.33 Lead–Copper–Erbium–Sulfur

PbS–CuErS$_2$. This system is nonquasibinary section of the Pb–Cu–Er–S system as CuErS$_2$ melts incongruently (Bayramova et al. 2010). The **CuErPbS$_3$** quaternary compound, which crystallizes in the orthorhombic structure with the lattice parameters $a = 382$, $b = 1320$, $c = 1018$ pm, and a calculated density of 6.90 g·cm^{-3}, is formed in the system. The standard thermodynamic functions of this compound were calculated: $S^0_{298} = 255.2 \pm 5.0$ J·M^{-1}K^{-1}, $\Delta S^0_{298} = −12.2 \pm 3.0$ J·M^{-1}K^{-1}, $\Delta H^0_{298} = −830.1 \pm 30$ kJ·M^{-1}, and $\Delta G^0_{298} = −826.2 \pm 30$ kJ·M^{-1}.

The system was studied through DTA, XRD, and measuring of the microhardness using the ingots annealed at 780°C for 2 weeks.

PbS–Cu$_2$S–Er$_2$S$_3$. The isothermal section of this quasiternary system at 600°C is analogous as one for PbS–Cu$_2$S–Ho$_2$S$_3$ system (Gulay et al. 2007c). The alloys were annealed at 600°C for 240 h, with the next quenching in cold water. Three quaternary compounds, **CuErPbS$_3$**, **CuEr$_{3.33}$Pb$_{1.5}$S$_7$**, and **Cu$_{1+x}$Er$_5$Pb$_{3−x/2}$S$_{11}$**, are formed in this system. The first crystallizes in the orthorhombic structure with the lattice parameters $a = 391.6 \pm 0.1$, $b = 1293.4 \pm 0.3$, $c = 1010.6 \pm 0.2$ pm, and a calculated density of 6.932 g·cm^{-3} (Gulay et al. 2005d). CuEr$_{3.33}$Pb$_{1.5}$S$_7$ crystallizes in the monoclinic structure with the lattice parameters $a = 1301.1 \pm 0.1$, $b = 393.13 \pm 0.03$, $c = 1206.06 \pm 0.09$ pm, $\beta = 104.884 \pm 0.004°$, and a calculated density of 6.440 g·cm^{-3} (Gulay and Olekseyuk 2006a). These compounds were synthesized in the same way as CuTbPbS$_3$ and CuTb$_{3.33}$Pb$_{1.5}$S$_7$ were prepared.

CuEr$_5$Pb$_3$S$_{11}$ crystallizes in the orthorhombic structure with the lattice parameters $a = 393.29 \pm 0.04$, $b = 1300.4 \pm 0.1$, $c = 366.64 \pm 0.03$ pm (Gulay et al. 2007c).

According to the data of Aliev et al. (2009), one more quaternary compound, **CuEr$_3$PbS$_6$**, which crystallizes in the orthorhombic structure with the lattice parameters $a = 1130$, $b = 1155$, and $c = 403.6$ pm, is formed in this system. It was prepared by reacting the elements in an evacuated (0.133 Pa) silica tube with vibration stirring. The highest reaction temperature was 930°C. The synthesized compound was homogenized at 600°C for a month. As a result, a dense, homogeneous dark gray ingot was obtained.

10.34 Lead–Copper–Thulium–Sulfur

The **CuTmPbS$_3$** quaternary compound, which crystallizes in the orthorhombic structure with the lattice parameters $a = 390.96 \pm 0.03$, $b = 1295.39 \pm 0.09$, $c = 1007.8 \pm 0.1$ pm, and a calculated density of 6.973 g·cm^{-3}, is formed in the Pb–Cu–Tm–S system (Gulay et al. 2005d). It was synthesized in the same way as CuTbPbS$_3$ was prepared.

10.35 Lead–Copper–Ytterbium–Sulfur

The **CuYbPbS$_3$** quaternary compound, which crystallizes in the orthorhombic structure with the lattice parameters $a = 391.1 \pm 0.1$, $b = 1295.6 \pm 0.3$, $c = 1006.4 \pm 0.3$ pm, and a calculated density of 7.033 g·cm^{-3}, is formed in the Pb–Cu–Yb–S system (Gulay et al. 2005d). It was synthesized in the same way as CuTbPbS$_3$ was prepared.

10.36 Lead–Copper–Lutetium–Sulfur

Two quaternary compounds, **CuLuPbS$_3$** and **CuLu$_{3.33}$Pb$_{1.5}$S$_7$**, are formed in the Pb–Cu–Lu–S system. First of them crystallizes in the orthorhombic structure with the lattice parameters $a = 388.9 \pm 0.1$, $b = 1292.0 \pm 0.4$, $c = 1003.0 \pm 0.3$ pm, and a calculated density of 7.124 g·cm^{-3} (Gulay et al. 2005d). CuLu$_{3.33}$Pb$_{1.5}$S$_7$ crystallizes in the monoclinic structure with the lattice parameters $a = 1292.62 \pm 0.09$, $b = 389.48 \pm 0.02$,

$c = 1199.16 \pm 0.08$ pm, $\beta = 104.794 \pm 0.004°$, and a calculated density of 6.724 g·cm⁻³ (Gulay and Olekseyuk 2006a). These compounds were synthesized in the same way as $CuTbPbS_3$ and $CuTb_{3.33}Pb_{1.5}S_7$ were prepared.

10.37 Lead–Copper–Arsenic–Sulfur

PbS–Cu₃As. This system is a nonquasibinary section of the Pb–Cu–As–S quaternary system (Kopylov et al. 1976c). The system has a wide immiscibility region in the liquid state, and the monotectic reaction is not isothermal and proceeds in the temperature range of 640–765°C. The maximum on the binodal curve corresponds to 1020°C. The immiscibility is accompanied by the formation of the sulfide layer and arsenide layer. The main phases of the arsenide layer are Cu_3As and Cu_5As_2, as well as minor precipitates of Pb, PbS, and the PbS+Cu₂S eutectic. The basis of the sulfide layer is the PbS+Cu₂S eutectic, a phase close to Cu_2As, and Pb. Copper sulfide is present in all alloys only as a part of the PbS+Cu₂S binary eutectic. The maximum values of thermal effects of crystallization of the ternary eutectic (530°C) and complex eutectic based on lead (290°C) correspond to compositions of 67.3–69.3 mol% PbS. Small thermal effects at 412°C and 350°C for some alloys apparently correspond to a more complex interaction in the Cu,Pb ‖ As,S ternary mutual system. The system was studied through DTA, metallography, and by using local microprobes.

PbS–CuAsS₂. The phase diagram of this system, constructed through DTA, XRD, metallography, and measuring of the microhardness, is given in Figure 10.13 (Bairamova et al. 2011a). The eutectics contain 20 and 65 mol% PbS and crystallize at 467°C and 622°C, respectively. The CuAsS₂-based solid solution extends to 7 mol% PbS and the solubility of $CuAsS_2$ in PbS is negligible. The **CuPbAsS₃** quaternary compound, which melts congruently at 772°C [melts incongruently at 460°C (Kala et al. 1971)] and crystallizes in the orthorhombic

structure with the lattice parameters $a = 812.5$, $b = 874.8$, $c = 764.4$ pm, and an energy gap of 1.31 eV, is formed in this system (Kala et al. 1971; Bairamova et al. 2011a). This compound was synthesized by melting the mixture of $CuAsS_2$ and PbS (molar ratio 1:1) in sealed silica ampoule. The highest synthesis temperature was 730–880°C. After the reaction reached completion, the temperature was slowly (60–70°C·h⁻¹) lowered to 330–380°C, and the alloys were homogenized by annealing at this temperature for a week. Single crystals of $CuPbAsS_3$ were grown by chemical-vapor transport using iodine as a transport agent.

Mineral seligmannite ($CuPbAsS_3$) also crystallizes in the orthorhombic structure with the lattice parameters $a = 804 \pm 5$, $b = 866 \pm 5$, $c = 756 \pm 5$ pm, and the calculated and experimental densities of 5.54 and 5.38 g·cm⁻³, respectively (Frondel 1941) [$a = 808.1$, $b = 874.7$, and $c = 763.6$ pm (Hellner and Leineweber 1956); $a = 807.6 \pm 0.2$, $b = 873.7 \pm 0.5$, $c = 763.4 \pm 0.3$ pm and a calculated density of 5.41 g·cm⁻³ (Edenharter and Nowacki 1970); $a = 808$ and $c = 764$ pm (Bakakin and Godovikov 1980)].

PbS–Cu₂S–Cu₃As. The liquidus surface of this system, constructed using five vertical sections, consists from the fields of primary crystallization of Cu_3As or Cu_5As_2, Cu_2S, PbS, and the immiscibility region (Kopylov et al. 1976a, 1982). The ternary eutectic in the system is degenerated. Since Cu_5As_2, Cu_2As, Pb, and As are formed in the system, it is not quasiternary, but is a phase plane onto which the elements that characterize transformations for a wide concentration region of the Pb–Cu–As–S quaternary system are projected. At 400–450°C, rhombic crystals of arsenic precipitate from the lead melt formed as a result of the interaction of Cu_3As and PbS. The crystallization ends at 290°C with the solidification of a complex eutectic based on lead, including small amounts of copper, PbS, and arsenic. The system was studied through DTA, metallography, and using X-ray spectral analysis.

10.38 Lead–Copper–Antimony–Sulfur

PbS–CuSbS₂. The phase diagram of this system, constructed through DTA, XRD, metallography, and measuring of the microhardness using the alloys annealed at 630°C for 400 h, is presented in Figure 10.14 (Alieva et al. 2010; Aliyev et al. 2016). The eutectics contain 20 and 70 mol% PbS and crystallize at 402°C and 677°C, respectively. There is a solid solution based on $CuSbS_2$ in this system. The **CuPbSbS₃** quaternary compound, which melts congruently at 852°C [at 512°C (Dong et al. 2015a)], and has no homogeneity region (Alieva et al. 2010), is formed in the system. According to the data of Aliyev et al. (2016), this compound is a phase of changing composition and its field of homogeneity is within the interval from 46 to 52 mol% PbS. $CuPbSbS_3$ crystallizes in the orthorhombic structure with the lattice parameters given in Table 10.1. The standard thermodynamic functions of $CuPbSbS_3$ were calculated: $S^0_{298} = 243.7 \pm 5.0$ J·M⁻¹K⁻¹, $\Delta S^0_{298} = 3.8 \pm 0.5$ J·M⁻¹K⁻¹, $\Delta H^0_{298} = -291.2 \pm 10$ kJ·M⁻¹, and $\Delta G^0_{298} = -292.5 \pm 10$ kJ·M⁻¹ (Bayramova et al. 2010).

This compound was synthesized by conventional solid-state reactions from PbS, Cu_2S, and Sb_2S_3, in the stoichiometric ratio

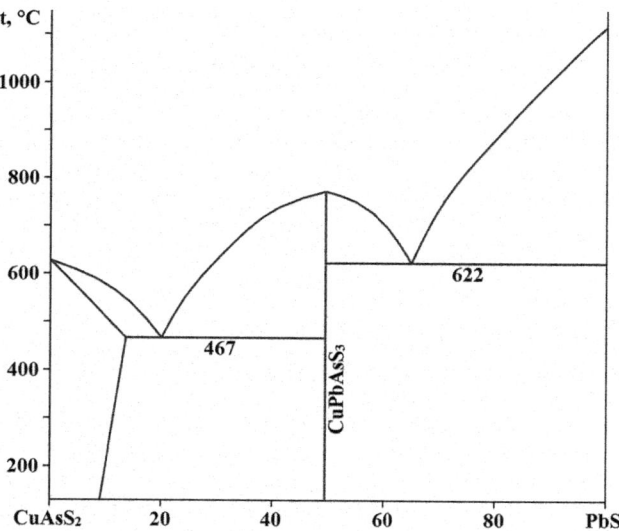

FIGURE 10.13 Phase diagram of the PbS–CuAsS₂ system. (From Bairamova, S.T., et al., *Inorg. Mater.*, **47**(3), 231, 2011.)

before being loaded into a quartz ampoule. This ampoule was sealed in a quartz tube, heated to 570°C and subsequently held at this temperature for 4 days before the furnace was turned off and the reaction tube was cooled to room temperature (Dong et al. 2015a) [heated at 450°C over 56 days with four intermediate grindings (Pruseth et al. 2001)]. The product was then ground into a fine powder, cold pressed into a pellet, and annealed at 400°C for one month. CuPbSbS₃ was also synthesized by melting a stoichiometric elemental mixture in an evacuated silica tube mounted in a single-zone furnace (Bairamova et al. 2011b). The highest synthesis temperature was 930–980°C. Cooling at a rate of 300°C·h⁻¹ led to the formation of a polycrystalline ingot with a large amount of cracking.

Cu₂S–CuPbSbS₃. The phase diagram of this system, constructed through DTA, XRD, metallography, and measuring of the microhardness and density, is a eutectic type (Figure 10.15) (Aliyev et al. 2018). The eutectic contains 40 mol% Cu₂S and crystallizes at 662°C. The constituent sulfides form a limited solid solution series. Similar phase relations were observed in the Sb_2S_3–$CuPbSbS_3$, $PbSb_2S_4$–$CuPbSbS_3$, and $Pb_2Sb_4S_{11}$–$CuPbSbS_3$ systems: the eutectics contain 63 mol% Sb_2S_3, 55 mol% $PbSb_2S_4$, and 50 mol% $Pb_2Sb_4S_{11}$ and crystallize at 412°C, 477°C, and 567°C, respectively. All these systems have limited solubility of the constituent sulfides in each other. The alloys for the investigations were homogenized by annealing for a week at temperatures 50–60°C below their solidus temperatures.

PbS–Cu₂S–Sb₂S₃. The isothermal sections of this quasiternary system at 300°C, 400°C, and 500°C were constructed by Hoda and Chang (1975a). At 500°C, there is a small region of Cu₂S-enriched liquid in the system, which is in the range 63–70 mol% Cu₂S, 18–20 mol% PbS, and 10–17 mol% Bi₂S₃. This liquid crystallizes completely at 477°C. The second liquid region exists near Sb₂S₃-side and is situated in the concentration interval 5–25 mol% Cu₂S, 0–47 mol% PbS, and 45–82 mol% Bi₂S₃. Complete crystallization of this liquid occurs at 461°C. From the PbS–Sb₂S₃ side of the ternary system, there is a region of solid solutions based on $CuPb_{13}Sb_7S_{24}$ (mineral meneghinite), which is in equilibrium with all lead sulfoantimonides, PbS and CuPbSbS₃. At 400°C, solid solutions based on $CuPb_{13}Sb_7S_{24}$ are separated by a two-phase region, which increases at 300°C. The ingots for the investigations were annealed at 300°C, 400°C, and 500°C at 300, 72, and 42 days, respectively.

The extent of solid solution in $Pb_9Sb_{22}S_{42}$, $Pb_4Sb_6S_{13}$ (mineral robinsonite), and $CuPb_{13}Sb_7S_{24}$ has been determined from EPMA established compositions of their synthetic analogs in pertinent assemblages in the course of a phase-equilibrium study of this quasiternary system (Pruseth et al. 1998). All three solid solution fields are elongated more or less parallel to the PbS–Sb₂S₃ binary system, and are relatively broader at their PbS-rich ends. The positions of the PbS-rich ends are practically insensitive to variations in temperature. Robinsonite and meneghinite are not stable at 300°C.

Evacuated silica tube experiments [+ halide flux; an eutectic mixture of KCl–LiCl (molar ratio 42:58) was chosen as the flux] were conducted in portions of this quasiternary system at 300°C, 440°C, and 500°C using two-pyrrhotite indicator method to measure the sulfur fugacity [*log* f(S₂)] (Mishra and Pruseth 1994; Pruseth et al. 1997). Product phases were identified by optical

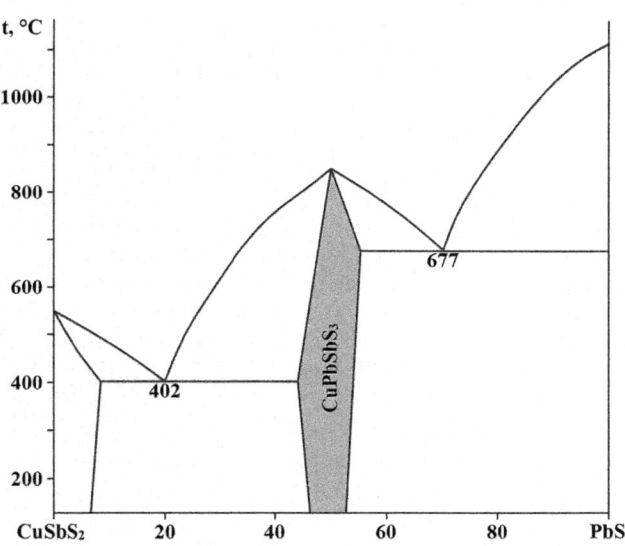

FIGURE 10.14 Phase diagram of the PbS–CuSbS₂ system. (From Aliyev, O.M., et al., *Azerb. khim. zhurn.*, (2), 51, 2016.)

TABLE 10.1

Crystallographic Data for CuPbSbS₃

a, pm	b, pm	c, pm	$d_{calc.}$ g·cm⁻³	$d_{meas.}$	References
810 ± 5	865 ± 5	775 ± 5	5.83	—	Oftedal (1932)[a]
816.2	871.05	781.05	—	—	Hellner and Leineweber (1956)[a]
815.3 ± 0.3	869.2 ± 0.3	779.3 ± 0.2	5.84	—	Edenharter and Nowacki (1970)[a]
817	869	782			Hoda and Chang (1975b)
816	—	781	—	—	Bakakin and Godovikov (1980)[a]
816.2	871	781	—	5.86	Alieva et al. (2010); Bayramova et al. (2010)
817.6	866.0	779.6	5.49	5.40	Bairamova et al. (2011b)
781.01 ± 0.02	815.04 ± 0.02	870.09 ± 0.03	5.860	—	Dong et al. (2015a)

Note:

[a] for mineral bournonite.

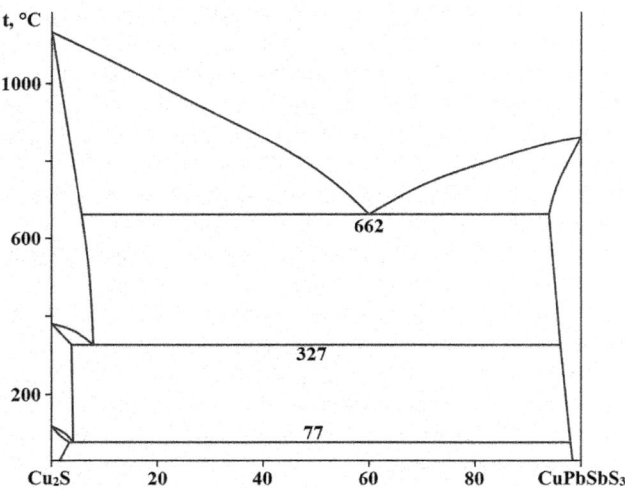

FIGURE 10.15 Phase diagram of the Cu₂S–CuPbSbS₃ system. (From Aliyev, O.M., et al., *Inorg. Mater.*, **54**(12), 1999, 2018.)

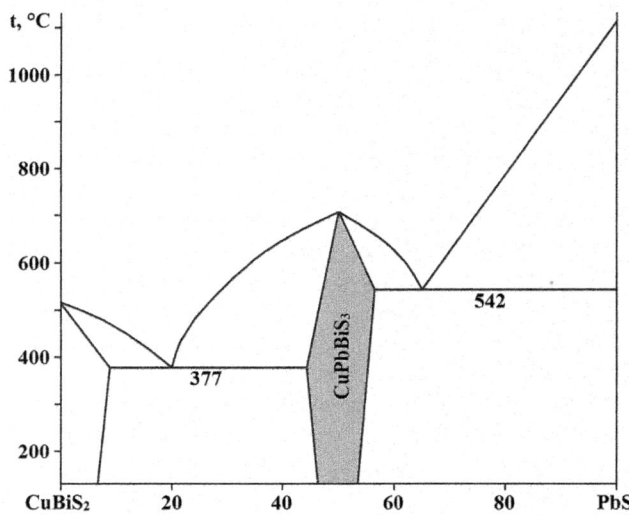

FIGURE 10.16 Phase diagram of the PbS–CuBiS₂ system. (From Aliev, O.M., et al., *Kondens. sredy i mezhfaz. granitsy*, **22**(2), 182, 2020.)

and XRD methods supplemented with EPMA. Microprobe data indicate that almost all sulfosalts depart from stoichiometry. The Cu₂S-rich melt appeared at 459°C and the Sb₂S₃-rich melt was observed to appear at 457°C. The values of *log* f(S₂) vary from −6.37 to −10.53 at 500°C, from −8.07 to −11.87 at 440°C, and from −11.02 to −12.86 at 310°C.

Two quaternary compounds, **Cu₀.₇Pb₉.₇Sb₂₁.₃S₄₂** and CuPb₁₃Sb₇S₂₄, are formed in the PbS–Cu₂S–Sb₂S₃ quasiternary system. Fist of them (Cu-bearing zinkenite) has a hexagonal subcell with the lattice parameters $a = 2212.19 \pm 0.11$ and $c = 432.07 \pm 0.03$ pm (Biagioni et al. 2018). The occurrence of weak superstructure reflections points to a triclinic unit cell with the lattice parameters $a = 3827.1 \pm 0.2$, $b = 2212.19 \pm 0.13$, $c = 864.75 \pm 0.05$ pm, $\alpha = 89.931 \pm 0.003°$, $\beta = 90.030 \pm 0.003°$, $\gamma = 89.957 \pm 0.003°$.

CuPb₁₃Sb₇S₂₄ crystallizes in the orthorhombic structure with the lattice parameters $a = 2405.49 \pm 0.03$, $b = 412.91 \pm 0.06$, and $c = 1133.61 \pm 0.16$ pm (Bindi et al. 2017) [$a = 1129$, $b = 2378$, $c = 412$ pm, and a calculated density of 6.391 g·cm⁻³ (Palache et al. 1938); $a = 1136.3 \pm 0.3$, $b = 2405.7 \pm 0.6$, and $c = 9908 \pm 2$ (or 24 × 412.82) pm (Euler and Hellner 1960)].

10.39 Lead–Copper–Bismuth–Sulfur

PbS–CuBiS₂. The phase diagram of this system, constructed through DTA, XRD, metallography, and measuring of the microhardness and density, is presented in Figure 10.16 (Aliev et al. 2020). The eutectics contain 20 and 65 mol% PbS and crystallize at 377°C and 542°C, respectively. There is a solid solution based on CuSbS₂ in this system. The **CuPbBiS₃** quaternary compound, which melts congruently at 707°C [at 525°C (Guseynov et al. 1972)] and has the homogeneity region (10 and 7 mol% PbS at 380°C and room temperature, respectively), is formed in the system. This compound crystallizes in the orthorhombic structure with the lattice parameters $a = 1163.2$, $b = 1166$, $c = 401.7$ pm and an energy gap of 0.84 eV

(Aliev et al. 2020) [$a = 1133$, $b = 1156$, and $c = 407$ pm for mineral aikinite with the same composition (Povilaytis et al. 1969; Fleischer 1970a); $a = 1163.8 \pm 0.3$, $b = 403.9 \pm 0.1$, and $c = 1131.9 \pm 0.2$ pm for mineral aikinite with the same composition (Ohmasa and Nowacki 1970); $a = 1160.83 \pm 0.10$, $b = 402.79 \pm 0.03$, $c = 1127.54 \pm 0.17$ pm and the calculated and experimental densities of 7.25 and 7.07 ± 0.1 g·cm⁻³, respectively, also for mineral aikinite (Kohatsu and Wuensch 1971); $a = 1142.4$, $b = 1148.4$, $c = 402$ pm, the calculated and experimental densities of 7.24 and 7.20 g·cm⁻³, respectively, and an energy gap of 0.12 eV (Guseynov et al. 1972); $a = 1129.6 \pm 0.5$, $b = 1171.6 \pm 0.5$, and $c = 403.9 \pm 0.2$ pm also for mineral aikinite (Johan et al. 1987)]. The standard thermodynamic functions of the CuPbBiS₃ compound were calculated by Aliev et al. (2020): $S°_{298} = 253.1 \pm 5.0$ J·M⁻¹K⁻¹, $\Delta S°_{298} = −14.1 \pm 3.0$ J·M⁻¹K⁻¹, $\Delta H°_{298} = −270.2 \pm 10.0$ kJ·M⁻¹, and $\Delta G°_{298} = −266.3 \pm 10.0$ kJ·M⁻¹.

The single crystals of this compound were obtained by directional crystallization of a stoichiometric melt in quartz ampoules using a vertical furnace (Aliev et al. 2020).

CuPbBiS₃–Cu₂S. The phase diagram of this system, constructed through DTA, XRD, metallography, and measuring of the microhardness and density, is a eutectic type (Aliev et al. 2020). The eutectic contains 40 mol% Cu₂S and crystallizes at 577°C. The solubility of CuPbBiS₂ in Cu₂S is 2 mol% at room temperature. The phase transitions (α-Cu₂S) ↔ (β-Cu₂S) ↔ (γ-Cu₂S) have a eutectoid nature and take place at 102°C and 307°C, respectively.

CuPbBiS₃–Bi₂S₃. The phase diagram of this system, constructed through DTA, XRD, metallography, and measuring of the microhardness and density, is a eutectic type (Aliev et al. 2020). The eutectic contains 50 mol% Bi₂S₃ and crystallizes at 527°C. The solubility based on Bi₂S₃ reaches 5 mol%, and one on CuPbBiS₃ is 7 mol%. According to the data of Chang and Hoda (1977), Kupčik et al. (1978), and Wu (1987b), the complete solid solution exists between Bi₂S₃ and CuPbBiS₃ above 300°C.

PbBi$_4$S$_7$–CuPbBiS$_3$. The phase diagram of this system, constructed through DTA, XRD, metallography, and measuring of the microhardness and density, is a eutectic type (Aliev et al. 2020). The eutectic contains 55 mol% PbBi$_4$S$_7$ and crystallizes at 527°C. The solubility of PbBi$_4$S$_7$ in CuPbBiS$_3$ is 18 mol% at the eutectic temperature and decreases to 10 mol% at room temperature.

PbBi$_2$S$_4$–CuPbBiS$_3$. This system is nonquasibinary section of the PbS–Cu$_2$S–Bi$_2$S$_3$ quasiternary system as PbBi$_2$S$_4$ melts incongruently (Aliev et al. 2020). The eutectic contains 40 mol% CuPbBiS$_3$ and crystallizes at 552°C. The solubility of PbBi$_2$S$_4$ in CuPbBiS$_3$ is 15 mol% at the eutectic temperature and decreases to 8 mol% at room temperature.

PbS–Cu$_2$S–Bi$_2$S$_3$. The isothermal section at 400°C and 500°C of this quasiternary system was constructed by Chang and Hoda (1977), Chang et al. (1988), and Wu (1987b). It was shown that the compositional range of the solid solution based on CuBi$_3$S$_5$ is extremely sensitive to temperature change. It narrows down to a range between 73 and 77 mol% Bi$_2$S$_3$ along the Cu$_2$S–Bi$_2$S$_3$ section and is limited between 0 and 17 mol% PbS and 23 and 27 mol% Cu$_2$S in the ternary system. CuBiS$_2$ shows a small range of solid solution along the Cu$_2$S–Bi$_2$S$_3$ section with a maximum of 2 mol% excess Bi$_2$S$_3$ or Cu$_2$S. At 500°C, a narrow liquid field exists, trending parallel with CuPbBiS$_3$–Cu$_2$S join, midway between 10 and 30 mol% PbS and 55 and 80 mol% Cu$_2$S. Complete solidification occurs at 492°C. In addition, CuPbBiS$_3$ shows a range of solid solution elongated along Pb$_3$Bi$_2$S$_6$–Cu$_3$BiS$_3$ join between 42 and 55 mol% PbS. CuBi$_3$S$_5$ has a range of solid solution between 60 and 77 mol% Bi$_2$S$_3$ in the Cu$_2$S–Bi$_2$S$_3$ system and can take a maximum of 26 mol% PbS into solid solution in the ternary system. Cu$_3$BiS$_3$ was found to have a range of solid solution between 73 and 75 mol% Cu$_2$S along the Cu$_2$S–Bi$_2$S$_3$ system.

According to the data of Wu (1987b), a narrow liquid field exists between Cu$_3$BiS$_3$, PbS, and high-temperature modification of Cu$_2$S.

The liquidus surface of the PbS–Cu$_2$S–Bi$_2$S$_3$ quasiternary system was constructed by Godovikov and Fedorova (1969) and Godovikov et al. (1976).

Besides CuPbBiS$_3$, some other quaternary compounds are formed in the PbS–Cu$_2$S–Bi$_2$S$_3$ quasiternary system. **CuPbBi$_3$S$_6$** (mineral krupkaite) crystallizes in the orthorhombic structure with the lattice parameters $a = 401.45 \pm 0.04$, $b = 1120.2 \pm 0.1$, $c = 1156.0 \pm 0.1$ pm, and a calculated density of 6.963 g·cm^{-3} (Topa et al. 2002) [$a = 1115$, $b = 1151$, and $c = 401$ pm (Fleischer et al. 1975); $a = 400.6 \pm 0.3$, $b = 1120 \pm 1$, $c = 1156 \pm 1$ pm (Large and Mumme 1975); $a = 400.3 \pm 0.3$, $b = 1120.0 \pm 0.9$, $c = 1156.0 \pm 0.9$ pm (Mumme 1975); $a = 401.7 \pm 0.2$, $b = 1121.0 \pm 0.3$, $c = 1156.4 \pm 0.4$ pm (Kaplunnik et al. 1975); $a = 1120$, $b = 1156$, and $c = 400$ pm (Aliev et al. 2009)].

CuPbBi$_5$S$_9$ (mineral gladite) also crystallizes in the orthorhombic structure with the lattice parameters $a = 400.44 \pm 0.04$, $b = 3357.5 \pm 0.3$, $c = 1148.0 \pm 0.1$ pm, and a calculated density of 6.903 g·cm^{-3} (Topa et al. 2002) [$a = 3353.1 \pm 0.6$, $b = 1148.6 \pm 0.2$, $c = 400.3 \pm 0.2$ pm, and the calculated and experimental densities of 6.91 and 6.96 g·cm^{-3}, respectively (Kohatsu and Wuensch 1973, 1976); $a = 1149.8 \pm 0.2$, $b = 3354.6 \pm 0.2$, $c = 400.1 \pm 0.1$ pm (Mumme and Watts 1976a,b)].

CuPbBi$_7$S$_{12}$ (unnamed mineral species) also crystallizes in the orthorhombic structure with the lattice parameters $a = 1119$, $b = 1149$, $c = 402$ pm, and a calculated density of 6.9 g·cm^{-3} (Dunn et al. 1984a).

CuPbBi$_{11}$S$_{18}$ also crystallizes in the orthorhombic structure with the lattice parameters $a = 1132.2 \pm 0.2$, $b = 3350.4 \pm 0.2$, and $c = 398.7 \pm 0.1$ pm (Mumme and Watts 1976a,b).

Cu$_9$Pb$_9$Bi$_{12}$S$_{28}$ also crystallizes in the orthorhombic structure with the lattice parameters $a = 5467$, $b = 403$, and $c = 2256$ pm (Kupčik 1983).

Cu$_{1.6}$Pb$_{1.6}$Bi$_{6.4}$S$_{12}$ (mineral salzburgite) also crystallizes in the orthorhombic structure with the lattice parameters $a = 400.74 \pm 0.09$, $b = 4481 \pm 1$, $c = 1151.3 \pm 0.3$ pm, and a calculated density of 6.904 g·cm^{-3} (Topa et al. 2000a, 2005; Jambor and Roberts 2001; Locock et al. 2006b).

Cu$_{1.7}$Pb$_{1.7}$Bi$_{6.3}$S$_{12}$ (mineral paarite) also crystallizes in the orthorhombic structure with the lattice parameters $a = 400.70 \pm 0.06$, $b = 5599.8 \pm 0.8$, $c = 1151.2 \pm 0.2$ pm, and a calculated density of 6.944 g·cm^{-3} (Makovicky et al. 2001; Jambor and Roberts 2004; Topa et al. 2005; Locock et al. 2006b).

Cu$_2$Pb$_2$Bi$_4$S$_9$ (mineral hammarite) also crystallizes in the orthorhombic structure with the lattice parameters $a = 3377.26 \pm 0.08$, $b = 1158.57 \pm 0.03$, $c = 401$ pm, and a calculated density of 7.05 g·cm^{-3} (Horiuchi and Wuensch 1976) [$a = 1118$, $b = 1144$, $c = 400$ pm, and a calculated density of 6.734 g·cm^{-3} (Povilaytis et al. 1969; Fleischer 1970a); $a = 1158$, $b = 3345$, $c = 401$ pm (Mumme et al. 1976)]. Synthetic Cu$_2$Pb$_2$Bi$_4$S$_9$ was prepared by reacting stoichiometric proportions of PbS, Cu$_2$S, and Bi$_2$S$_3$ in a sealed, evacuated silica tube at 500°C for 25 days (Pring 1995). The sample was quenched, reground, and heated for an additional 25 days at 500°C to obtain a more homogeneous product.

Cu$_2$Pb$_6$Bi$_8$S$_{19}$ (mineral felbertalite) crystallizes in the monoclinic structure with the lattice parameters $a = 2763.7 \pm 0.4$, $b = 404.99 \pm 0.06$, $c = 2074.1 \pm 0.3$ pm, $\beta = 131.258 \pm 0.002°$, and a calculated density of 6.948 g·cm^{-3} (Topa et al. 2000b, 2001; Jambor and Roberts 2001).

Cu$_{2.68}$Pb$_{2.68}$Bi$_{5.32}$S$_{12}$ (mineral emilite) crystallizes in the orthorhombic structure with the lattice parameters $a = 402.85 \pm 0.08$, $b = 4498.6 \pm 0.9$, $c = 1159.9 \pm 0.2$ pm, and a calculated density of 7.025 g·cm^{-3} (Balić-Žunić et al. 2002; Jambor and Roberts 2004; Locock et al. 2006a; Topa et al. 2006).

Cu$_3$Pb$_3$Bi$_7$S$_{15}$ (mineral lindströmite) also crystallizes in the orthorhombic structure with the lattice parameters $a = 5599 \pm 3$, $b = 1154.9 \pm 0.3$, and $c = 401.0 \pm 0.1$ pm (Pring et al. 1998) [$a = 1157$, $b = 5607$, and $c = 401$ pm (Mumme et al. 1976); $a = 5611.5 \pm 0.4$, $b = 1156.95 \pm 0.08$, $c = 400.1 \pm 0.5$ pm, and a calculated density of 7.03 g·cm^{-3} (Horiuchi and Wuensch 1977)].

Cu$_5$Pb$_5$Bi$_7$S$_{18}$ (mineral friedrichite) also crystallizes in the orthorhombic structure with the lattice parameters $a = 3389.7 \pm 1.5$, $b = 1164.0 \pm 0.5$, and $c = 403.4 \pm 0.2$ pm (Shimizu et al. 1998a) [$a = 3384$, $b = 1165$, $c = 401$ pm and the calculated and experimental densities of 7.06 and 6.98 g·cm^{-3}, respectively (Chen et al. 1978; Fleischer et al. 1979b)].

Cu$_7$Pb$_{27}$Bi$_{25}$S$_{68}$ (mineral cuproneyite) crystallizes in the monoclinic structure with the lattice parameters $a = 3743.2 \pm 0.8$, $b = 405.29 \pm 0.09$, $c = 4354.5 \pm 0.9$ pm, $\beta = 108.800 \pm 0.005°$, and a calculated density of 7.113 g·cm^{-3} (Ilinca et al. 2012).

According to the data of Chang and Hoda (1977), the ~$Cu_{12.9}Pb_{2.4}Bi_{2.3}S_{12.3}$ phase is formed in this system at 400°C.

10.40 Lead–Copper–Tellurium–Sulfur

$PbS+Cu_2Te \leftrightarrow PbTe+Cu_2S$. With the help of physicochemical analysis and thermodynamic calculations, it has been established that the exchange interaction in this ternary mutual system is characterized by a slight shift of equilibrium toward the formation of PbS and Cu_2Te (Grytsiv et al. 1994). In the system, PbS_xTe_{1-x} and $Cu_2S_xTe_{1-x}$ solid solutions initially crystallize, and along the $PbTe–Cu_2S$ vertical section, crystallization of PbS_xTe_{1-x} proceeds under equilibrium conditions, and $Cu_2S_xTe_{1-x}$ – in non-equilibrium. The eutectic point on the $PbTe–Cu_2S$ diagonal section corresponds to 571°C and 67.5 mol% Cu_2S.

10.41 Lead–Copper–Iron–Sulfur

The **$Cu_{10}(Fe,Pb)S_6$** quaternary compound (mineral betekhtinite), which crystallizes in the orthorhombic structure with the lattice parameters $a = 386 \pm 1$, $b = 1467 \pm 3$, $c = 2280 \pm 3$ pm and a calculated density of 5.73 g·cm^{-3} (Dornberger-Schiff and Höhne 1959) [$a = 385$, $b = 1467$, $c = 2280$ pm and an experimental density of 6.14 g·cm^{-3} (Fleischer 1956a)], is formed in the Pb–Cu–Bi–S system.

$PbS–Cu_2S–FeS$. The liquidus surface of this quaternary system consists of the fields of PbS, FeS, Cu_2S, and $CuFeS_2$ primary crystallization (Kopylov 1969). The vapor pressure of PbS gradually decreases with an increase in the Cu_2S/FeS ratios (Shendyapin et al. 1965; Nesterov et al. 1969). The activity and activity coefficients of PbS in alloys decrease with increasing temperature and increase with increasing PbS concentration in the melt.

The dew-point method was used to determine the vapor pressures of PbS over liquid sulfides of this system (Eriç and Timuçin 1981). From the PbS activity data, activities of Cu_2S and FeS were evaluated by Gibbs-Duhem calculations. The isoactivity curves of PbS, Cu_2S, and FeS in ternary sulfide melts at 1200°C are presented in Figures 10.17 and 10.18.

10.42 Lead–Silver–Gallium–Sulfur

$PbS–AgGaS_2$. The phase diagram of this system, constructed through DTA and XRD, is a eutectic type (Figure 10.19) (Sashital and Gentile 1984). The eutectic contains 43.5 mol% $AgGaS_2$ and crystallizes at 665°C.

10.43 Lead–Silver–Indium–Sulfur

The **$AgPb_3In_{17}S_{34}$** quaternary compounds, which crystallize in the monoclinic structure with the lattice parameters $a = 1265.7 \pm 0.3$, $b = 388.10 \pm 0.08$, $c = 2745.9 \pm 0.6$ pm, $\beta = 94.36 \pm 0.03°$, and an energy gap of ~1.3 eV, is formed in

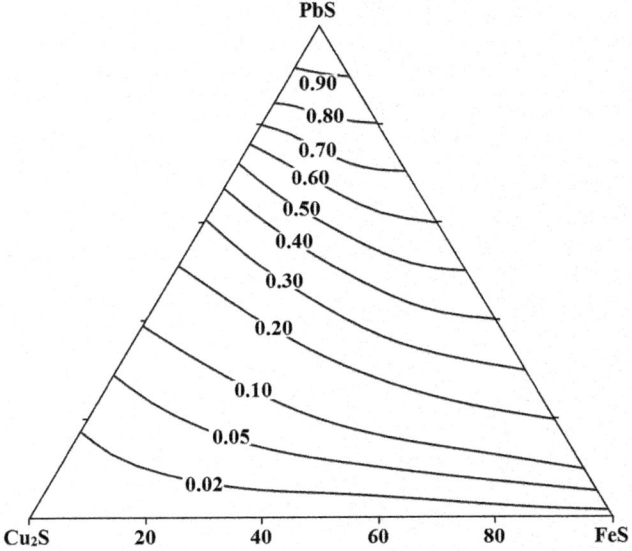

FIGURE 10.17 Isoactivity curves for PbS in ternary sulfide melts of the $PbS–Cu_2S–FeS$ quasiternary system at 1200°C. (From Eriç, H., and Timuçin, M., *Metall. Trans. B*, **12**(3), 493, 1981.)

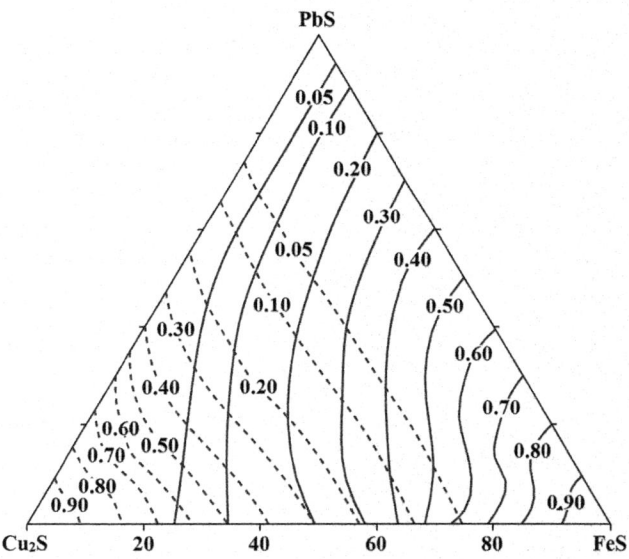

FIGURE 10.18 Isoactivity curves for FeS (solid lines) and Cu_2S (dashed lines) in ternary sulfide melts of the $PbS–Cu_2S–FeS$ quasiternary system at 1200°C. (From Eriç, H., and Timuçin, M., *Metall. Trans. B*, **12**(3), 493, 1981.)

the Pb–Ag–In–S system (Wang and Lee 2006). To synthesize this compound, the mixture of the elements in stoichiometric proportions was heated in an evacuated silica ampoule (~10^{-2} Pa) at 850°C for 24 h, kept at 850°C for 24 h, and then cooled slowly to 700°C at a rate 10°C·h^{-1}; finally, the furnace was turned off to allow cooling to room temperature. The product exhibits a black color with bar-shaped crystals and is stable in air for at least 14 days. All materials were handled in an Ar-filled glove box in which the concentration of oxygen was less than 10 ppm.

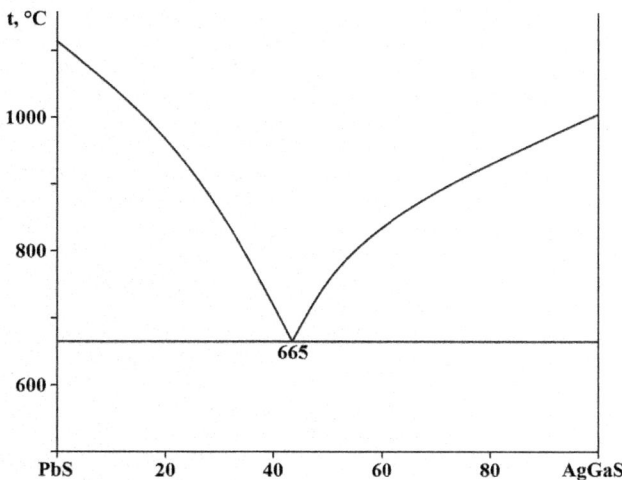

FIGURE 10.19 Phase diagram of the PbS–AgGaS$_2$ system. (From Sashital, S.R., and Gentile, A.L., *J. Cryst. Growth*, 69(2–3), 379, 1984.)

10.44 Lead–Silver–Arsenic–Sulfur

Some quaternary compounds are formed in the Pb–Ag–As–S system. AgPbAsS$_3$ (mineral marrite) crystallizes in the monoclinic structure with the lattice parameters $a = 727.05 \pm 0.06$, $b = 1263.19 \pm 0.04$, $c = 598.53 \pm 0.03$ pm, and $\beta = 91°13.7' \pm 0.2'$ (Wuensch and Nowacki 1967) [$a = 729.1 \pm 0.8$, $b = 1268.5 \pm 3.4$, $c = 599.8 \pm 0.5$ pm, $\beta = 91°13' \pm 2'$, and a calculated density of 5.822 g·cm^{-3} (Fleischer 1965)].

Ag$_{1.5}$Pb$_{22}$As$_{33.5}$S$_{72}$ (mineral argentobaumhauerite) crystallizes in the triclinic structure with the lattice parameters $a = 790.53 \pm 0.10$, $b = 846.80 \pm 0.10$, $c = 4441.02 \pm 0.53$ pm, $\alpha = 84.614 \pm 0.002°$, $\beta = 86.469 \pm 0.002°$, $\gamma = 89.810 \pm 0.002°$, and a calculated density of 5.29 g·cm^{-3} (Topa and Makovicky 2016) [crystallizes in the monoclinic structure with the lattice parameters $a = 4475$, $b = 848$, $c = 789$ pm, and $\beta = 93.63°$ (Pring and Graeser 1994)]. The name "argentobaumhauerite" replaces the preliminary name "baumhauerite-2a" (Hålenius et al. 2015; Topa and Makovicky 2016).

Ag$_3$Pb$_{38.1}$As$_{51.9}$S$_{96}$ (unnamed mineral species) crystallizes in the orthorhombic structure with the lattice parameters $a = 6830$, $b = 848$, and $c = 789$ pm (Pring and Graeser 1994).

Ag$_3$Pb$_{26}$As$_{35}$S$_{80}$ (mineral argentodufrénoysite) crystallizes in the triclinic structure with the lattice parameters $a = 787.7 \pm 0.3$, $b = 841.8 \pm 0.3$, $c = 4943.9 \pm 1.9$ pm, $\alpha = 89.338 \pm 0.007°$, $\beta = 90.012 \pm 0.007°$, and $\gamma = 89.993 \pm 0.006°$ (Topa et al. 2016b).

Ag$_x$Pb$_{40-2x}$As$_{48+x}$S$_{112}$ ($3 < x < 4$) (mineral argentoliveingite) also crystallizes in the triclinic structure with the lattice parameters $a = 790.5 \pm 0.2$, $b = 846.9 \pm 0.2$, $c = 1379.6 \pm 0.4$ pm, $\alpha = 89.592 \pm 0.002°$, $\beta = 88.969 \pm 0.002°$, and $\gamma = 89.893 \pm 0.002°$ (Topa et al. 2016a).

10.45 Lead–Silver–Antimony–Sulfur

PbS–AgSbS$_2$. The phase diagram of this system, constructed through DTA, XRD, metallography, and measuring of the microhardness and density, belongs to the type I according to the Roseboom's classification (Wernick 1960; Godovikov and Nenasheva 1969a,b; Hoda and Chang 1975b; Godovikov et al. 1976). At temperatures above 480°C, a continuous series of solid solutions with a cubic structure is formed in the system. The dependence of the unit cell parameter, microhardness, and density on the composition is characterized by a stepped character, which is inconsistent with the form of the phase diagram. At 400°C, in the range of 65–90 mol% PbS, the solid solution decomposes in the system, and at 300°C, a wide two-phase region and small areas of solid solutions based on AgPbSbS$_3$ and in the range of 25–40 mol% PbS based on Ag$_3$Pb$_2$Sb$_3$S$_8$ are observed. According to data from Amcoff (1978), the region of solid solutions based on PbS reaches 4 mol% AgSbS$_2$ at 300°C, 2 mol% at 250°C, and less than 2 mol% at 200°C.

The calculated phase diagram of the PbS–AgSbS$_2$ system is in agreement with experimental results (Chutas et al. 2008). The calculated miscibility gap in this system is at 441.7°C.

PbS–Ag$_2$S–Sb$_2$S$_3$. The isothermal sections at 300°C, 400°C, and 500°C of this quasiternary system were constructed by (Hoda and Chang 1975b), and two of them are presented in Figure 10.20. At 500°C, PbSb$_2$S$_4$, Pb$_5$Sb$_4$S$_{11}$, and Pb$_6$Sb$_{10}$S$_{21}$ solid phases, as well as the region of solid solutions, are stable in the system. At 400°C, solid solutions decompose in the system along PbS–AgSbS$_2$ section in the range of 65–90 mol% PbS, the liquid field disappears, and the Ag$_6$Sb$_2$S$_6$ phase appears. The solid solution region based on AgPbSbS$_3$S$_6$ significantly decreases at 300°C. The ingots for the investigations were annealed at 300°C, 400°C, and 500°C for 147, 52, and 42 days, respectively.

The following quaternary compounds have been found in the Pb–Ag–Sb–S system. AgPbSbS$_3$ (mineral freieslebenite) crystallizes in the monoclinic structure with the lattice parameters $a = 751.8 \pm 0.1$, $b = 1280.9 \pm 0.4$, $c = 594.0 \pm 0.1$ pm, $\beta = 92.25 \pm 0.1°$, and the calculated and experimental densities of 6.194 and 6.20 g·cm^{-3}, respectively (Ito and Nowacki 1974a) [$a = 753$, $b = 1279$, $c = 588$ pm, $\beta = 92.14'$, and the calculated and experimental densities of 6.27 and 6.20–6.23 g·cm^{-3}, respectively (Palache et al. 1938); $a = 766$, $b = 1269$, $c = 583$ pm, and $\beta = 90.75°$ (Hoda and Chang 1975b)]. At decreasing temperatures below 340–350°C, this compound transforms into the cubic modification with the lattice parameter $a = 578.7$ pm (Hoda and Chang 1975b).

AgPbSb$_3$S$_6$ (mineral andorite VI *or* senandorite) crystallizes in the orthorhombic structure with the lattice parameters $a = 1300.5$, $b = 1915.5$, $c = 2562.2$ pm, and a calculated density of 5.456 g·cm^{-3} (Sawada et al. 1987) [$a = 1300 \pm 1$, $b = 1917 \pm 1$, and $c = 424.6 \pm 0.2$ pm (Organova et al. 1982); $a = 1302$, $b = 1918$, and $c = 2548$ pm (Dunn et al. 1985a); $a = 1304$, $b = 1917$, and $c = 429$ pm (Chang 1987)].

AgPb$_4$Sb$_4$S$_{10}$ (mineral zoubekite) also crystallizes in the orthorhombic structure with the lattice parameters $a = 1869.8 \pm 0.5$, $b = 649.2 \pm 0.3$, $c = 457.7 \pm 0.1$ pm, and a calculated density of 5.15 g·cm^{-3} (Hawthorne et al. 1987).

AgPb$_7$Sb$_5$S$_{15}$ (mineral chukotkaite) crystallizes in the monoclinic structure with the lattice parameters $a = 405.75 \pm 0.03$, $b = 3595.02 \pm 0.11$, $c = 1922.15 \pm 0.19$ pm, $\beta = 90.525 \pm 0.008°$, and a calculated density of 6.255 g·cm^{-3} for AgPb$_{6.6}$Sb$_{5.4}$S$_{15}$ composition (Kasatkin et al. 2020a,b,c).

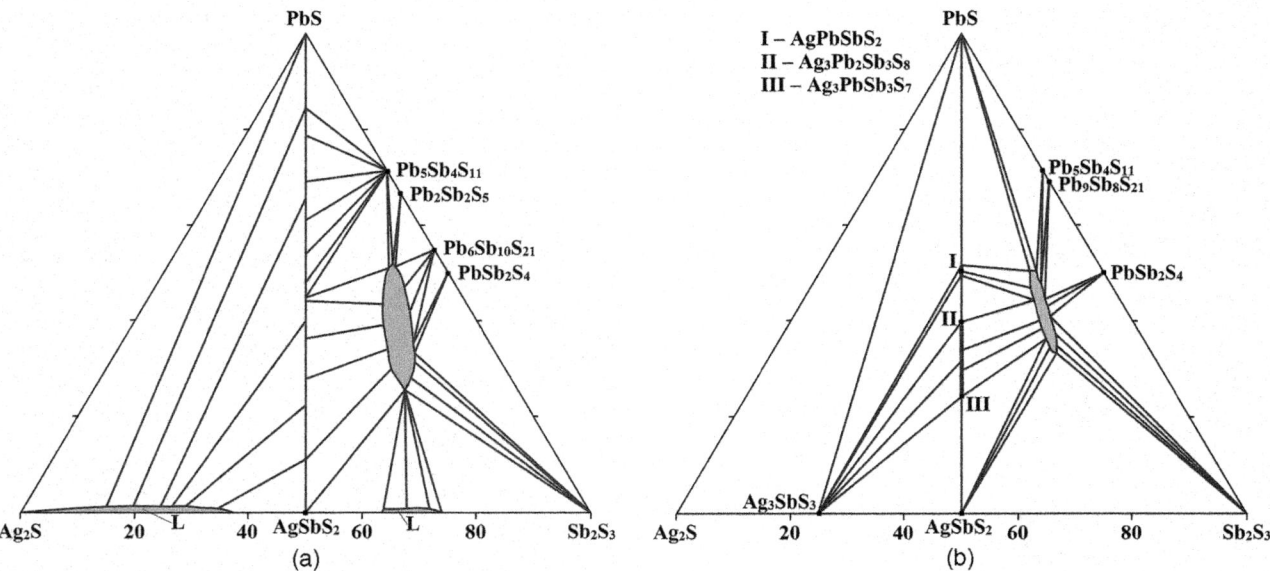

FIGURE 10.20 Isothermal sections of the PbS–Ag₂S–Sb₂S₃ quasiternary system (a) at 500°C and (b) at 300°C. (From Hoda, S.N., and Chang, L.L.Y., *Amer. Mineralog.*, **60**(7–8), 621, 1975.) Open access

$Ag_2Pb_{22}Sb_{20}S_{53}$ (mineral tubulite) also crystallizes in the monoclinic structure with the lattice parameters a = 413.2 ± 0.02, b = 4310 ± 20, c = 2740 ± 10 pm, β = 93.2 ± 0.3°, and a calculated density of 6.05 g·cm⁻³ (Moëlo et al. 2012, 2013; Belakovskiy et al. 2014).

$Ag_3Pb_2Sb_3S_8$ (mineral diaphorite) also crystallizes in the monoclinic structure with the lattice parameters a = 1785.20 ± 0.36, b = 588.70 ± 0.12, c = 1580.90 ± 0.32 pm, β = 116.165 ± 0.030°, and a calculated density of 6.06 g·cm⁻³ (Armbruster et al. 2003) [a = 1584.9 ± 0.4, b = 1791.4 ± 0.4, c = 590.1 ± 0.1 pm, and β = 116°25.5 ± 2′ (Hellner 1958); a = 1586 ± 2, b = 1795 ± 2, c = 589 ± 1 pm, and β = 116.67 ± 0.09° (Mozgova et al. 1989); crystallizes in the orthorhombic structure with the lattice parameters a = 1583, b = 3223, c = 589 pm, and the calculated and experimental densities of 5.97 and 5.90–6.04 g·cm⁻³, respectively (Palache et al. 1938); a = 1586, b = 3199, c = 595 pm (Hoda and Chang 1975b)]. At decreasing temperatures below 340–350°C, this compound transforms into the cubic modification with the lattice parameter a = 576.0 pm (Hoda and Chang 1975b).

$Ag_3Pb_6Sb_{11}S_{24}$ (mineral ramdohrite) also crystallizes in the monoclinic structure with the lattice parameters a = 873.48 ± 0.03, b = 1305.43 ± 0.04, c = 1931.17 ± 0.06 pm, β = 90.179 ± 0.002°, and a calculated density of 5.543 g·cm⁻³ (Makovicky et al. 2013) [a = 1924, b = 1308, c = 873 pm, and β = 90.28° (Dunn et al. 1984a, 1985a)].

$Ag_3Pb_{10}Sb_{11}S_{28}$ (mineral owyheeite) also crystallizes in the monoclinic structure with the lattice parameters a = 410.35 ± 0.01, b = 2731.44 ± 0.03, c = 2293.66 ± 0.03 pm, β = 90.359 ± 0.001°, and a calculated density of 5.86 g·cm⁻³ (Laufek et al. 2007a) [crystallizes in the orthorhombic structure with the lattice parameters a = 2282, b = 2720, c = 819 pm, and the calculated and experimental densities of 6.43 and 6.25 ± 0.02 g·cm⁻³, respectively (Robinson 1949); a = 2258, b = 2668, c = 408 or 816 pm, and the calculated and experimental densities of 6.65

and 6.25 g·cm⁻³, respectively (Timofeevskiy 1967)]. According to the data of Makovicky and Olsen (2015), owyheeite, monoclinic, and pseudo-orthorhombic, is invariably intimately twinned, and its structure displays an alternation of two types of unit order-disorder layers. The solid solution field of this mineral may be described by the formula $Ag_{3+x}Pb_{10-2x}Sb_{11+x}S_{28}$ (−0.13 ≤ x ≤ +0.20) (Moëlo et al. 1984; Dunn et al. 1985b).

$Ag_4Pb_4Sb_{12}S_{24}$ (mineral andorite IV or quatrandorite) also crystallizes in the monoclinic structure with the lattice parameters a = 1916.86 ± 0.19, b = 1716.0 ± 0.3, c = 1304.2 ± 0.2 pm, β = 90.008 ± 0.012°, and a calculated density of 5.403 g·cm⁻³ (Nespolo et al. 2012) [a = 1304, b = 1918, c = 1707 pm, β = 90° (Dunn et al. 1985a)].

$Ag_5Pb_{14}Sb_{21}S_{48}$ (mineral fizélyite) also crystallizes in the monoclinic structure with the lattice parameters a = 1927.67 ± 0.06, b = 1323.45 ± 0.04, c = 872.30 ± 0.03 pm, β = 90.401 ± 0.002°, and a calculated density of 5.644 g·cm⁻³ for $Ag_{5.94}Pb_{13.74}Sb_{20.84}S_{48}$ (Yang et al. 2009) [a = 1321, b = 1927, c = 868 pm, β = 90.03° (Dunn et al. 1985a)].

$Ag_{15}Pb_{18}Sb_{47}S_{96}$ crystallizes in the orthorhombic structure with the lattice parameters a = 1277.8 ± 0.6, b = 1924.4 ± 0.8, c = 424.8 ± 0.3 pm (Johan and Mantienne 2000).

$Ag_{19}Pb_{10}Sb_{51}S_{96}$ (mineral roshchinite) crystallizes in the orthorhombic structure with the lattice parameters a = 1290 ± 3, b = 1904 ± 6, c = 423.3 ± 0.9 pm, and the calculated and experimental densities of 5.263 and 5.265 ± 0.001 g·cm⁻³, respectively (Spiridonov et al. 1990a,b; Jambor and Vanko 1992).

There are the data in the literature about two more compounds formed in the Pb–Ag–Sb–S system: $Ag_3PbSb_3S_7$, which crystallizes in the orthorhombic structure with the lattice parameters a = 1586, b = 3154, c = 590 pm the Pb–Ag–Sb–S system (Hoda and Chang 1975b), and $Ag_2PbSb_2S_5$ (mineral brogniardite), which crystallizes in the monoclinic structure with the lattice parameters a = 1591 ± 1, b = 1794 ± 1,

$c = 590.3 \pm 0.6$ pm, $\beta = 116.79 \pm 0.05°$ (Mozgova et al. 1989). Comparing the available data on the lattice parameters of these compounds, it can be assumed that both of them correspond to the $Ag_3Pb_2Sb_3S_8$ (mineral diaphorite).

10.46 Lead–Silver–Bismuth–Sulfur

PbS–AgBiS₂. The phase diagram of this system, constructed through DTA, XRD, and metallography, belongs to the type I according to the Roseboom's classification (Van Hook 1960; Wernick 1960; Craig 1967; Hoda and Chang 1975b; Godovikov et al. 1976). At temperatures above 220°C, a continuous series of solid solutions with a cubic structure is formed in the system. Significant negative deviations from the Vegard's law are observed in the concentration dependence of the unit cell parameter. Below 220°C, the decomposition of solid solutions is observed in the system. The solubility of $AgBiS_2$ in PbS at 170°C is 10 mol% and PbS in $AgBiS_2$ is 2 mol%. The **AgPbBiS₃** quaternary compound [melting temperature 850°C and an energy gap of 1.2 eV (Guseynov et al. 1972)] was not found in the system. The ingots for the investigations were annealed at 300°C, 400°C, and 500°C for 147, 52, and 42 days, respectively (Hoda and Chang 1975b).

PbS–Ag₂S–Bi₂S₃. The isothermal sections at 400°C, and 500°C of this quasiternary system were constructed by Hoda and Chang (1975b), Wu (1987b), and Chang et al. (1988) and the last is presented in Figure 10.21. At 500°C, there are wide regions of solid solutions in the system along the section (73 mol% PbS + 27 mol% Bi_2S_3)–(25 mol% PbS + 75 mol% Bi_2S_3) with a discontinuity in solubility in the range of 25–46 mol% PbS. At 670°C, complete solubility is observed along this section. As the temperature drops to 400°C, the regions of solid solutions decrease. The ingots for the investigations of both

systems were annealed at 300°C, 400°C, and 500°C for 147, 52, and 42 days, respectively (Hoda and Chang 1975b).

The part of the liquidus surface of this quasiternary system was constructed by Van Hook (1960). It was found that the fields of Ag_2S and $(PbS)_x(AgBiS_2)_{1-x}$ solid solution primary crystallization are divided by the line of secondary crystallization with a minimum at 592°C containing 81 mol% Ag_2S, 5 mol% Bi_2S_3, and 14 mol% PbS. The region of solid solutions in the system narrows from the Ag_2S–Bi_2S_3 system to the PbS–Bi_2S_3 system (Van Hook 1960; Craig 1967; Hoda and Chang 1975b).

Some quaternary compounds are formed in the Pb–Ag–Bi–S system. **Ag₀.₄Pb₀.₂Bi₀.₄S** (mineral schapbachite) has a homogeneity range from $Ag_{0.41}Pb_{0.18}Bi_{0.4}S_{1.01}$ to $Ag_{0.49}Pb_{0.07}Bi_{0.45}S_{1.00}$ (Staude et al. 2010).

Despite previous research (see above), the $AgPbBiS_3$ quaternary compound was synthesized by combining the chemical elements in stoichiometric ratios under static vacuum at 900°C for 3 days (Sportouch et al. 1998). It do not melt below 900°C and crystallizes in the cubic structure with the lattice parameter $a = 586.5 \pm 0.1$ pm and an energy gap of 0.54 eV.

AgPbBi₃S₆ (mineral gustavite) has apparently two polymorphic modifications. One of them crystallizes in the monoclinic structure with the lattice parameters $a = 705.67 \pm 0.14$, $b = 1969.05 \pm 0.39$, $c = 822.19 \pm 0.16$ pm, $\beta = 106.961 \pm 0.003°$, and a calculated density of 6.789 g·cm⁻³ (Makovicky and Topa 2011) [$a = 707.7$, $b = 1956.6$, $c = 827.2$ pm, $\beta = 107.18°$ (Harris and Chen 1975); $a = 707.7 \pm 0.9$, $b = 1973.8 \pm 0.5$, $c = 821 \pm 7$ pm, $\beta = 107.1°$ (Bente and Anton 1995)]. Second modification crystallizes in the orthorhombic structure with the lattice parameters $a = 1347.7 \pm 0.9$, $b = 1987.8 \pm 0.5$, and $c = 407.7 \pm 0.7$ pm (Bente and Anton 1995) [$a = 1354.8$, $b = 1944.9$, and $c = 410.5$ pm (Karup-Møller 1970; Fleischer 1971); $a = 1334.9 \pm 0.5$, $b = 1966.2 \pm 0.8$, $c = 407.0 \pm 0.8$ pm and $a = 1351.0 \pm 0.8$, $b = 1962.9 \pm 1.0$, $c = 409.2 \pm 0.8$ pm for two various samples (Karup-Møller and Makovicky 1979); $a = 407.7 \pm 0.2$, $b = 1347.7 \pm 0.7$, and $c = 1988 \pm 2$ pm (Bente et al. 1993)]. The orthorhombic modification was prepared by recrystallization in silica tubes at 500°C (Bente et al. 1993).

The composition of the mineral schirmerite extends from **Ag₃Pb₃Bi₉S₁₈** to **Ag₃Pb₆Bi₇S₁₈** (Fleischer et al. 1979a). It crystallizes in the orthorhombic structure with the lattice parameters $a = 1344.8 \pm 2.6$, $b = 4438.6 \pm 10.0$, $c = 402.2 \pm 1.2$ pm, and a calculated density of 7.58 g·cm⁻³ (Karup-Møller 1973; Fleischer and Mandarino 1974).

Ag₃Pb₄Bi₅S₁₃ (mineral ourayite) crystallizes in the orthorhombic structure with the lattice parameters $a = 1349 \pm 2$, $b = 4417 \pm 4$, and $c = 405 \pm 2$ pm and $a = 1315 \pm 2$, $b = 4417 \pm 4$, and $c = 405 \pm 2$ pm for $Pb_{3.6}Ag_{2.8}Bi_{5.6}S_{13}$ composition (Makovicky and Karup-Møller 1984) [$a = 1345.7 \pm 1.4$, $b = 4404.2 \pm 4.0$, $c = 410.0 \pm 1.0$ pm, and a calculated density of 7.24 g·cm⁻³ (Karup-Møller 1977; Fleischer et al. 1979a)].

According to the data of Godovikov et al. (1976), the **Ag₄PbBi₄S₉** quaternary compound is formed in the Pb–Ag–Bi–S system.

Ag₅Pb₈Bi₁₃S₃₀ (mineral vikingite) crystallizes in the monoclinic structure with the lattice parameters $a = 1360.3 \pm 0.6$, $b = 2524.8 \pm 0.7$, $c = 411.2 \pm 0.4$ pm, and $\beta = 95.55 \pm 0.03°$ (Karup-Møller 1977; Fleischer et al. 1979a).

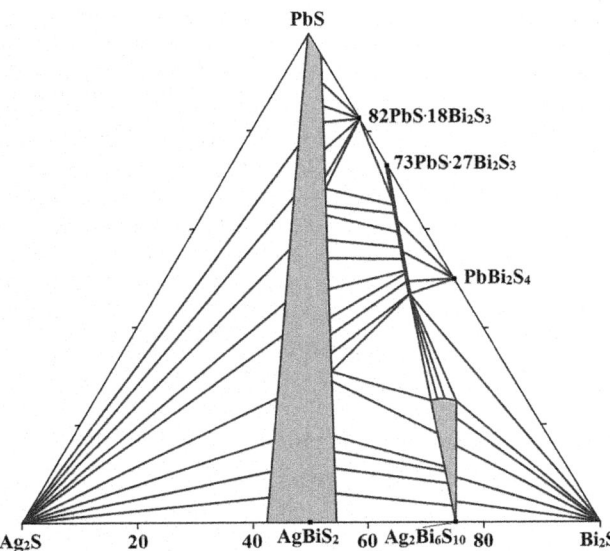

FIGURE 10.21 Isothermal sections of the PbS–Ag₂S–Bi₂S₃ quasiternary system at 500°C. (From Hoda, S.N., and Chang, L.L.Y., *Amer. Mineralog.*, **60**(7–8), 621, 1975.) Open access

Ag₇Pb₆Bi₁₅S₃₂ (mineral treasurite) also crystallizes in the monoclinic structure with the lattice parameters $a = 1334.9 \pm 1.0$, $b = 2653.8 \pm 2.0$, $c = 409.2 \pm 0.7$ pm, $\beta = 92.77 \pm 0.07°$, and a calculated density of 7.25 g·cm⁻³ (Karup-Møller 1977; Fleischer et al. 1979a).

Ag₇Pb₁₀Bi₁₅S₃₆ (mineral eskimoite) also crystallizes in the monoclinic structure with the lattice parameters $a = 1345.9 \pm 0.5$, $b = 3019.4 \pm 0.8$, $c = 410.0 \pm 0.5$ pm, $\beta = 93.35 \pm 0.05°$, and a calculated density of 7.12 g·cm⁻³ (Karup-Møller 1977; Fleischer et al. 1979a).

Ag₈Pb₁₂Bi₁₆S₄₀ (mineral erzwiesite) crystallizes in the orthorhombic structure with the lattice parameters $a = 408.5 \pm 0.5$, $b = 1346.2 \pm 1.5$, and $c = 3392 \pm 4$ pm (Topa et al. 2013b).

10.47 Lead–Silver–Chlorine–Sulfur

PbS + 2AgCl ↔ PbCl₂ + Ag₂S. The liquidus surface of this ternary mutual system, constructed through DTA and metallography, consists of the fields of PbS, PbCl₂, Ag₂S, and AgCl primary crystallization and the immiscibility region (Urazov and Sokolova 1941). The ternary eutectics contain 54 mol% PbCl₂, 43.6 mol% 2AgCl, 2.4 mol% Ag₂S and 20.5 mol% PbS, 69.4 mol% PbCl₂, 10.1 mol% Ag₂S, respectively.

The nonquasibinary section of this system is PbS–2AgCl. The area of immiscibility in this section is in the range of 33–62.5 mol % PbS. The primary crystallization temperature of Ag₂S reaches its maximum value (660°C) at 47.5 mol % PbS and is shifted from the Ag₂S–PbCl₂ stable diagonal section toward AgCl. In this section, the binary eutectics Ag₂S+AgCl and PbS+Ag₂S crystallize at 355°C and 565°C, respectively, while the ternary eutectics AgCl+Ag₂S+PbCl₂ and PbS+Ag₂S+PbCl₂ crystallize at 310°C and 420°C, respectively. Thermal effects at 455°C are due to the peculiarities of alloys crystallization in the region of PbS primary precipitation.

10.48 Lead–Silver–Iron–Sulfur

Pb–Ag–FeS. The isothermal section of this subsystem of the Pb–Ag–Fe–S quaternary system at 1200°C is shown in Figure 10.22 (Rybkin et al. 2006; Raghavan 2009a). The tie-lines between the metallic liquid and the sulfide liquid are shown. FeS dissolves 26.9 mass% Pb and 23.6 mass% Ag. The silver distribution coefficient, defined as the ratio of the mass fractions in the coexisting metal and sulfide liquids, decreases from 4.7 to 2.0, as the Pb content of the metallic phase increases from 13.2 to 87.2 mass%. The alloys for the investigations were equilibrated at 1200°C for 60 min.

10.49 Lead–Silver–Palladium–Sulfur

The **Ag₂Pb₂Pd₉S₄** quaternary compound, which crystallizes in the tetragonal structure with the lattice parameters $a = 797.3 \pm 0.3$, $c = 913.9 \pm 0.3$ pm, and a calculated density of 9.81 g·cm⁻³ (Vymazalová et al. 2021) [$a = 798$ and $c = 914$ pm for mineral panskyite with the same composition (Vymazalová et al. 2020a, b)], is formed in the Pb–Ag–Pd–S system. Synthetic

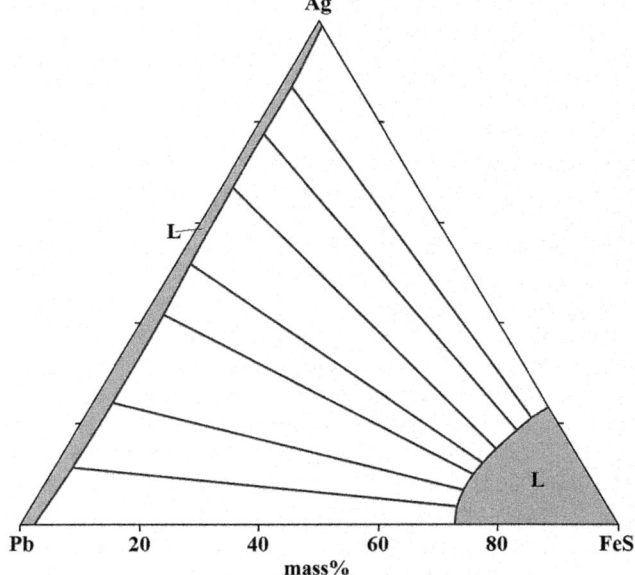

FIGURE 10.22 Isothermal section of the Pb–Ag–FeS subsystem at 1200°C. (From Rybkin, S.G., et al., *Russ. J. Inorg. Chem.*, **51**(3), 470, 2006.)

Ag₂Pb₂Pd₉S₄ phase was prepared using an evacuated silica-glass tube method (Vymazalová et al. 2021). Pb, Ag, Pd, and S were used as starting materials for synthesis. The evacuated tube with its charge was sealed and annealed. After cooling in a cold-water bath, the charge was ground into powder in acetone using an agate mortar, and thoroughly mixed to homogenize. The pulverized charge was sealed in an evacuated silica-glass tube again, and heated at 400°C for 7 months. The experimental product was quenched rapidly in cold water.

10.50 Lead–Gold–Indium–Sulfur

The **AuPb₃In₁₇S₃₄** quaternary compounds, which crystallize in the monoclinic structure with the lattice parameters $a = 1270.1 \pm 0.3$, $b = 387.94 \pm 0.09$, $c = 2735.9 \pm 0.6$ pm, $\beta = 94.347 \pm 0.004°$, and an energy gap of ~1.3 eV, is formed in the Pb–Au–In–S system (Wang and Lee 2006). To synthesize this compound, the mixture of the elements in stoichiometric proportions was heated in an evacuated silica ampoule (~10⁻² Pa) to 850°C within 24 h, kept at 850°C for 24 h, and then cooled slowly to 700°C at a rate 10°C·h⁻¹; finally, the furnace was turned off to allow cooling to room temperature. The product exhibits a black color with bar-shaped crystals and is stable in air for at least 14 days. All materials were handled in an Ar-filled glove box in which the concentration of oxygen was less than 10 ppm.

10.51 Lead–Gold–Bismuth–Sulfur

Two quaternary compounds are formed in the Pb–Au–Bi–S system. **[(Bi,Pb)₃S₃][AuS₂]** (mineral jaszczakite) crystallizes in the orthorhombic structure with the lattice parameters $a = 385.8 \pm 0.1$, $b = 1255.2 \pm 0.3$, and $c = 928.9 \pm 0.2$ pm, and a calculated density of 7.327 g·cm⁻³ (Bindi and Paar 2016, 2017).

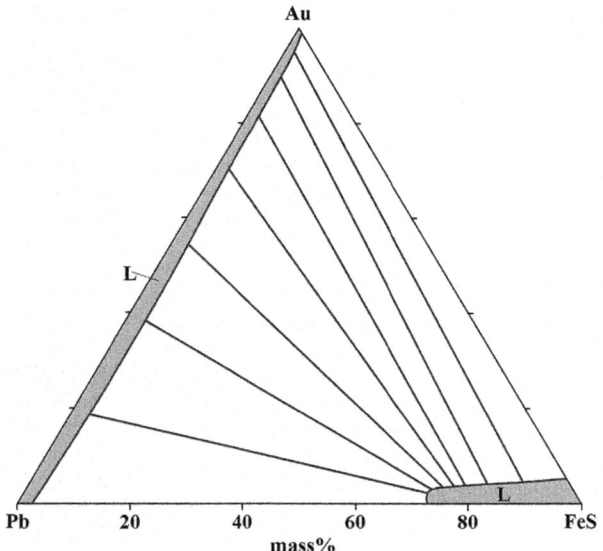

FIGURE 10.23 Isothermal section of the Pb–Au–FeS subsystem at 1200°C. (From Rybkin, S.G., et al., *Russ. J. Inorg. Chem.*, **51**(3), 470, 2006.)

Au(Bi,Pb)$_5$S$_4$ (mineral jonassonite) crystallizes in the in the monoclinic structure with the lattice parameters $a = 1832.9 \pm 2.9$, $b = 410.8 \pm 0.4$, $c = 1397.4 \pm 1.6$ pm, $\beta = 100.90 \pm 0.10°$, and a calculated density of 8.64 g·cm^{-3} (Paar et al. 2006; Vasiliev et al. 2011).

10.52 Lead–Gold–Iron–Sulfur

Pb–Au–FeS. The isothermal section of this subsystem of the Pb–Au–Fe–S quaternary system at 1200°C is shown in Figure 10.23 (Rybkin et al. 2006; Raghavan 2009b). The tie-lines between the metallic liquid and the sulfide liquid are shown. FeS dissolves 26.9 mass% Pb and 4.4 mass% Au. The FeS solubility in Au is 1.4 mass% The gold distribution coefficient, defined as the ratio of the mass fractions in the coexisting metal and sulfide liquids, decreases from 32.8 to 12.7, as the Pb mass fraction in the metallic phase increases from 10.2 to 78.3 mass%. The alloys for the investigations were equilibrated at 1200°C for 60 min.

10.53 Lead–Barium–Antimony–Sulfur

The **Ba$_{2.62}$Pb$_{1.38}$Sb$_4$S$_{10}$** quaternary compound, which melts congruently at 749°C and crystallizes in the monoclinic structure with the lattice parameters $a = 884.02 \pm 0.02$, $b = 820.38 \pm 0.02$, $c = 2676.23 \pm 0.06$ pm, $\beta = 99.488 \pm 0.001°$, a calculated density of 5.045 g·cm^{-3}, and an energy gap of 1.64 eV, is formed in the Pb–Ba–Sb–S system (Choi and Kanatzidis 2000). This compound was prepared from a mixture of BaS (0.6 mM), Pb (0.6 mM), Sb (1.2 mM), and S (3.3 mM). The reagents were thoroughly mixed, flame-sealed in an evacuated silica tube, and heated at 800°C for 5 days. The reaction was then cooled to 200°C at a rate of 2°C·h^{-1}. Pure silvery-red, needle-like crystals were obtained by isolation in degassed DMF and water.

10.54 Lead–Barium–Bismuth–Sulfur

The **Ba$_2$Pb$_2$Bi$_6$S$_{13}$** quaternary compound, which crystallizes in the monoclinic structure, is formed in the Pb–Ba–Bi–S system (Kanatzidis et al. 1998).

10.55 Lead–Barium–Chromium–Sulfur

The **Ba$_{0.5}$Pb$_{0.5}$Cr$_2$S$_4$** quaternary compound, which crystallizes in the hexagonal structure with the lattice parameters $a = 2173.7 \pm 0.6$ and $c = 344.0 \pm 0.2$ pm, is formed in the Pb–Ba–Cr–S system (Omloo et al. 1971). This compound was prepared by heating a stoichiometric mixture of the four elements or a mixture of PbS, BaS, and Cr$_2$S$_3$ in evacuated quartz tubes at temperatures between 1000°C and 1200°C for 1 week. After cooling to room temperature the sample was ground to powder. This powder was annealed at the same temperature for 1 to several weeks, slowly cooled to 750°C for 3 weeks, and quenched or cooled for 1 day to room temperature.

10.56 Lead–Barium–Chlorine–Sulfur

PbS–BaCl$_2$. The liquidus curve of this quasibinary system was determined by DTA from the PbS side up to 4 mol% BaCl$_2$ by Bell and Flengas (1966).

10.57 Lead–Zinc–Gallium–Sulfur

The **ZnGa$_6$Pb$_5$S$_{15}$** quaternary compound, which crystallizes in the orthorhombic structure with the lattice parameters $a = 2223.2 \pm 0.9$, $b = 1793.3 \pm 0.7$, $c = 622.5 \pm 0.2$ pm, a calculated density of 5.36 g·cm^{-3}, and an energy gap of 2.32 eV, is formed in the Pb–Zn–Ga–S system (Duan et al. 2016). To synthesize this compound, the mixture of PbS, Zn, Ga, and S in the stoichiometric ratio was heated to 900°C at 24°C·h^{-1} and held at this temperature for 100 h. Then it was cooled to 300°C through 100 h and shot out the furnace. Some earthy yellow crystals of the title compound were obtained with some byproducts of black PbS and yellow ZnGa$_2$S$_4$. All operations were carried out in the glove box.

10.58 Lead–Zinc–Oxygen–Sulfur

PbSO$_4$–ZnS. Research carried out on the interaction between ZnS and PbSO$_4$ showed that the process starts at 530°C (Malinowski 1992; Malinowski et al. 1996). It was found that a few reactions occur at the same time, including the transformation of ZnS into ZnO and PbS formation. The secondary PbSO$_4$ containing the sulfur from ZnS also appears. The PbS is derived primarily from PbSO$_4$, and some of it contains the sulfur from ZnS also. The presence of PbO·PbSO$_4$ in the products is possible in the case of a low ZnS content. The dependence of the preparation mass reduction and the degree of change of ZnS into ZnO have been determined. This system was investigated through DTA, XRD, thermogravimetry and differential thermogravimetry, and chemical analysis.

PbO–ZnS. Using DTA, XRD, and chemical analysis, it was determined that annealing of ZnS + PbO mixtures leads to the formation of ZnO and Pb (Suleimanov et al. 1974a, 1974b). This indicates that ZnS, PbO, and forming products interact according to the next reactions: ZnS + PbO = ZnO + PbS and PbS + 2PbO = 3Pb + SO_2.

10.59 Lead–Zinc–Selenium–Sulfur

PbSe–ZnS. The phase diagram of this system, constructed through DTA, XRD, and metallography, is a eutectic type (Figure 10.24) (Oleinik et al. 1982). The eutectic composition and temperature are 18 mol% ZnS and 1019°C, respectively. Mutual solubility of ZnS and PbSe is not higher than 1 mol%.

10.60 Lead–Zinc–Iron–Sulfur

PbS–ZnS–Fe₀.₉₂S. A part of the liquidus surface of this quasiternary system (Figure 10.25) was constructed according to 7 polythermal sections (Dutrizac 1980). It includes three fields

of primary crystallization. The ternary eutectic composition and temperature are 41 mol% PbS + 13 mol% ZnS + 46 mol% $Fe_{0.92}S$ and 850°C, respectively [40.9 mol% PbS + 4.2 mol% ZnS + 54.9 mol% FeS and 820°C, respectively (Avetisian and Gnatyshenko 1956)]. The liquidus surface fall rather gradually from either PbS and $Fe_{0.92}S$, but rise steeply on the ZnS-rich side. This system was investigated through DTA, metallography, and chemical analysis.

10.61 Lead–Cadmium–Bismuth–Sulfur

The **(Cd,Pb)Bi₂S₄** quaternary compound (mineral kudriavite), which crystallizes in the monoclinic structure with the lattice parameters $a = 1309.5 \pm 0.1$, $b = 400.32 \pm 0.03$, $c = 1471.1 \pm 0.1$ pm, $\beta = 115.59 \pm 0.01°$, and a calculated density of 6.578 g·cm⁻³, is formed in the Pb–Cd–Bi–S system (Chaplygin et al. 2005; Piilonen et al. 2005b; Balić-Žunić et al. 2007). The crystal-chemical analysis suggests that the observed ratio Pb/Cd = 1:1 in natural kudriavite represents the upper limit for the substitution of Pb in the structure.

10.62 Lead–Cadmium–Selenium–Sulfur

PbSe–CdS. The phase diagram of this system, constructed through DTA, XRD, metallography, and measuring of the microhardness using the ingots annealed at 750°C and 800°C for 100 h, is a eutectic type (Figure 10.26) (Tomashik et al. 1981). The eutectic composition and temperature are 42 mol% CdS and 1000°C, respectively. Solubility of CdS in PbSe at the eutectic temperature is 33 mol% [30 mol% (Leute and Böttner 1978)] and decreases to 11 mol% at 800°C and 9 mol% at 750°C, and the solubility of PbSe in CdS is not higher than 1 mol%.

PbS–CdSe. This section is a nonquasibinary section of the Pb–Cd–Se–S quaternary system since at the interaction of CdSe and PbS, lead selenide and CdS were determined (Leute and Böttner 1978).

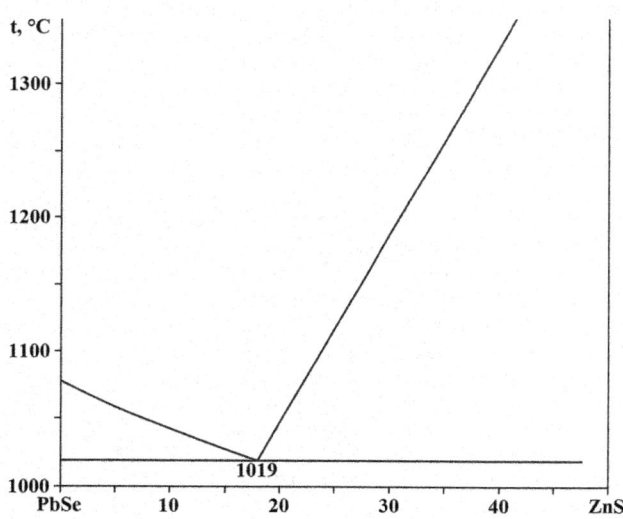

FIGURE 10.24 Phase diagram of the PbSe–ZnS system. (From Oleinik, G.S., et al., *Izv. AN SSSR. Neorgan. mater.*, **18**(5), 873, 1982.)

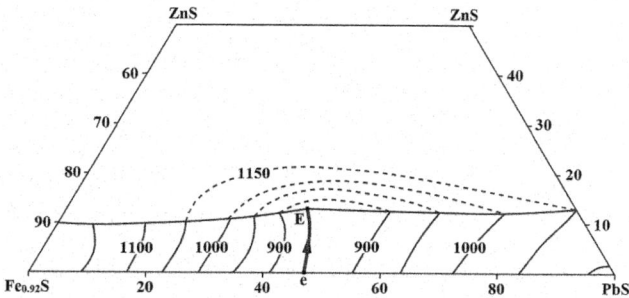

FIGURE 10.25 Part of the liquidus surface of the PbS–ZnS–Fe₀.₉₂S quasiternary system. (From Dutrizac, J.E., *Can. J. Chem.*, **58**(7), 739, 1980.)

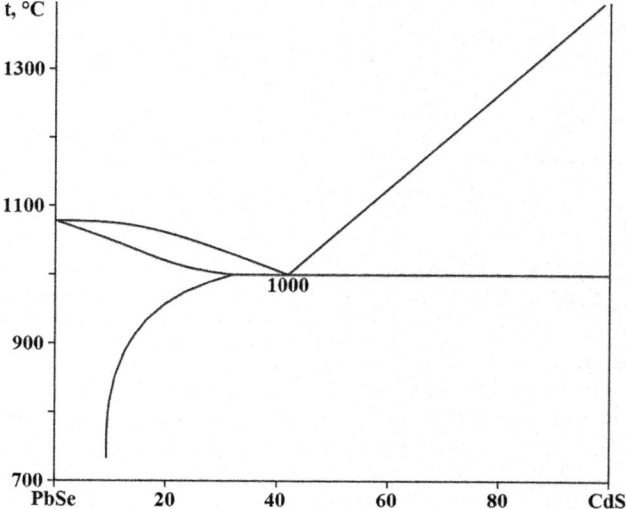

FIGURE 10.26 Phase diagram of the PbSe–CdS system. (From Tomashik, Z.F., et al., *Izv. AN SSSR. Neorgan. mater.*, **17**(12), 2155, 1981.)

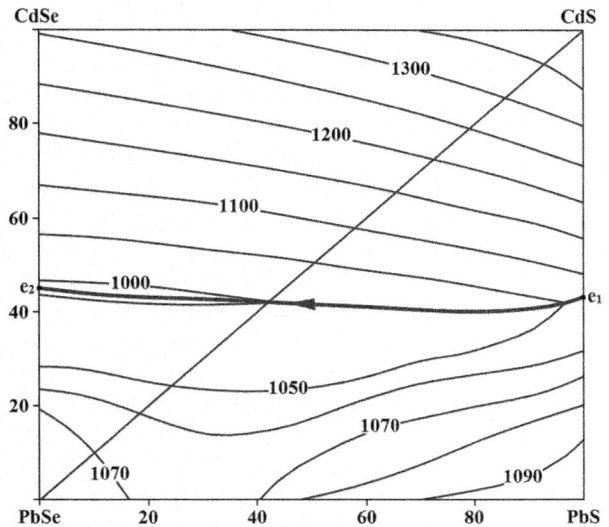

FIGURE 10.27 Liquidus surface of the PbSe + CdS ↔ PbS + CdSe ternary mutual system. (From Tomashik, Z.F. and Tomashik, V.N. et al., *Izv. AN SSSR. Neorgan. mater.*, **20**(4), 568, 1984.)

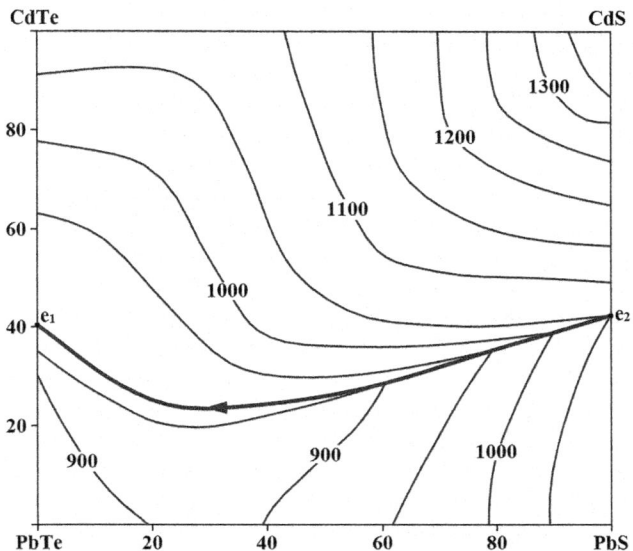

FIGURE 10.29 Liquidus surface of the PbTe + CdS ↔ PbS+ CdTe ternary mutual system. (From Tomashik, Z.F. and Tomashik, V.N., *Izv. AN SSSR. Neorgan. mater.*, **23**(12), 1981, 1987.)

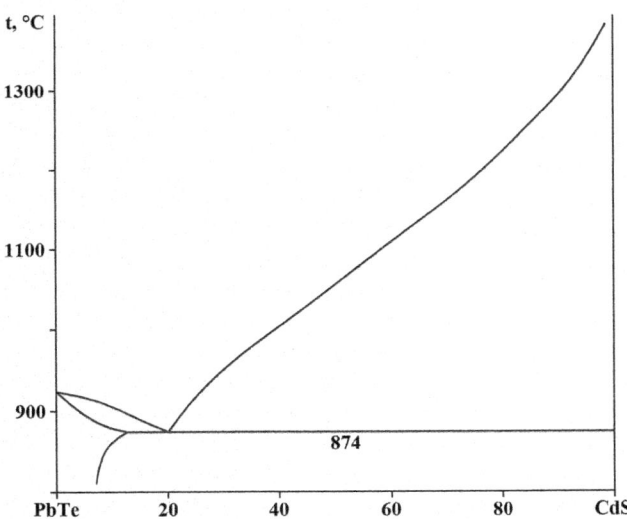

FIGURE 10.28 Phase diagram of the PbTe–CdS system. (From Tomashik, Z.F. and Tomashik, V.N., *Izv. AN SSSR. Neorgan. mater.*, **23**(12), 1981, 1987.)

PbSe + CdS ↔ PbS + CdSe. The fields of primary crystallization of CdS_xSe_{1-x} and PbS_xSe_{1-x} solid solutions exist on the liquidus surface of this ternary mutual system (Figure 10.27) (Tomashik and Tomashik 1984). This system was investigated through DTA and using mathematical simulation of the experiment.

10.63 Lead–Cadmium–Tellurium–Sulfur

PbTe–CdS. The phase diagram of this system, constructed through DTA and metallography, is a eutectic type (Figure 10.28) (Tomashik and Tomashik 1987). The eutectic composition and temperature are 20 mol% CdS and 874°C, respectively. The solubility of CdS in PbTe at the eutectic temperature is 13 mol% and the solubility of PbTe in CdS is not higher than 1 mol%.

PbTe + CdS ↔ PbS + CdTe. The fields of primary crystallization of CdS_xTe_{1-x} and PbS_xTe_{1-x} solid solutions exist on the liquidus surface of this ternary mutual system (Figure 10.29) (Tomashik and Tomashik 1987). This system was investigated through DTA, metallography, and using mathematical simulation of the experiment.

10.64 Lead–Cadmium–Iodine–Sulfur

PbI$_2$–CdS. The phase diagram of this system, constructed through DTA and metallography using the ingots annealing for 1000 h, is a eutectic type (Figure 10.30) (Odin 2001). The eutectic composition and temperature are 23 mol% CdS and 366°C, respectively. The mutual solubility of CdS and PbI$_2$ is negligible: the solubility of PbI$_2$ in CdS is 0.15 mol%. The crystallization of the melts from the PbI$_2$ side leads to the formation of two polytypic forms of this compound: $6R$-PbI$_2$ and $2H$-PbI$_2$ (Odin 2001; Odin and Chukichev 2001).

PbS +CdI$_2$ ↔ PbI$_2$ + CdS. The isothermal section of this ternary mutual system at 350°C is shown in Figure 10.31 (Odin 2001).

The fields of primary crystallization of (CdS), (PbS), $Pb_5S_2I_6$, and $Cd_xPb_{1-x}I_2$ solid solutions exist on the liquidus surface of this ternary mutual system (Figure 10.32) (Odin 2001). The reaction $L + PbS ↔ Pb_5S_2I_6 + CdS$ takes place in the transition point U at 409°C. The ternary eutectic E at 357°C is formed by the solid solutions based on CdS, $Pb_5S_2I_6$ and $Cd_xPb_{1-x}I_2$ solid solutions. A minimum m exists on the e_2e_4 line of the secondary crystallization. Metastable phases can form at the crystallization of the melts (Odin and Chukichev 2001).

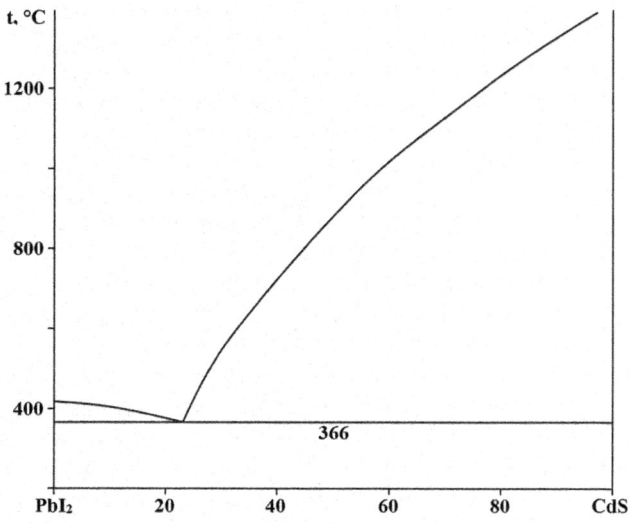

FIGURE 10.30 Phase diagram of the PbI$_2$–CdS system. (From Odin I.N., *Zhurn. neorgan. khimii*, **46**(10), 1733, 2001.)

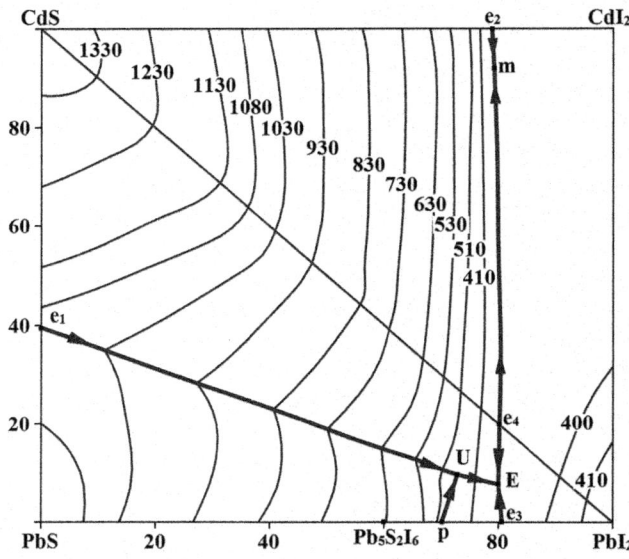

FIGURE 10.32 Liquidus surface of the PbS + CdI$_2$ ↔ PbI$_2$ + CdS ternary mutual system. (From Odin, I.N., *Zhurn. neorgan. khimii*, **46**(10), 1733, 2001.)

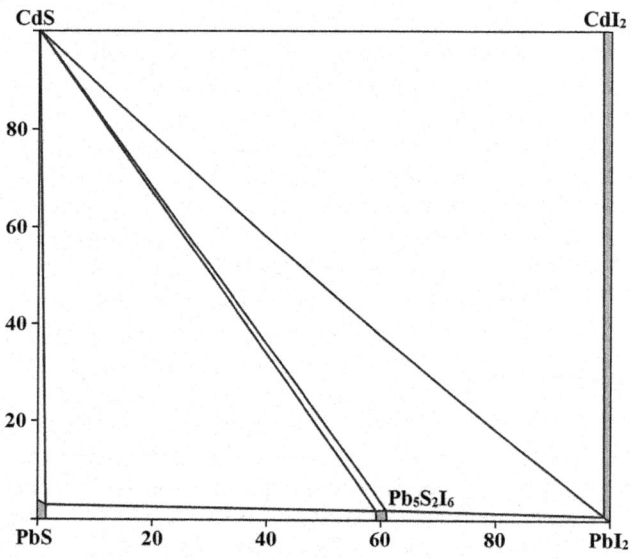

FIGURE 10.31 Isothermal section of the PbS + CdI$_2$ ↔ PbI$_2$ + CdS ternary mutual system at 350°C. (From Odin, I.N., *Zhurn. neorgan. khimii*, **46**(10), 1733, 2001.)

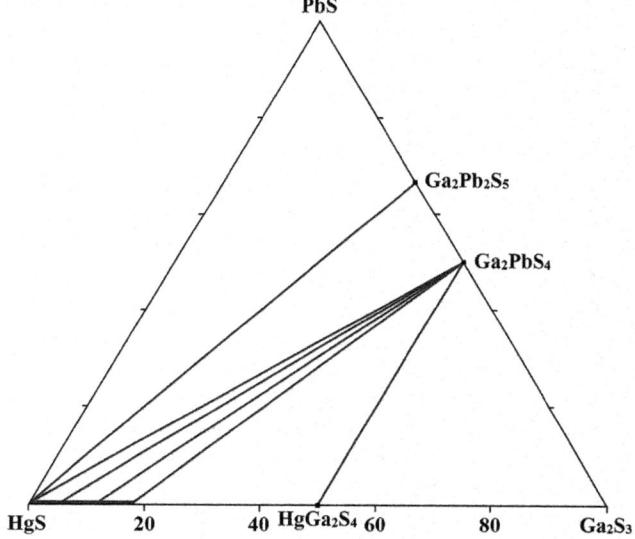

FIGURE 10.33 Isothermal section of the PbS–HgS–Ga$_2$S$_3$ quasiternary system at 400°C. (From Filyuk, T.O., et al., *Nauk. Visnyk Volyns'k. Derzh. Univ. im. Lesi Ukrainky. Ser. Khim. nauky*, (13), 12, 2007.)

10.65 Lead–Mercury–Gallium–Sulfur

PbS–HgS–Ga$_2$S$_3$. The isothermal section of this quasiternary system at 400°C is shown in Figure 10.33 (Filyuk et al. 2007). It is divided by the three quasibinary sections into four subsystems. No quaternary compounds were found. At this temperature, the solubility of Ga$_2$S$_3$ in HgS reaches 18 mol%.

10.66 Lead–Mercury–Arsenic–Sulfur

The **HgPbAs$_2$S$_6$** quaternary compound (mineral daliranite), which crystallizes in the monoclinic structure with the lattice parameters $a = 1911.3 \pm 0.5$, $b = 423.3 \pm 0.2$, $c = 2295.8 \pm 0.8$ pm, $\beta = 114.78 \pm 0.05°$, and a calculated density of 5.93 g·cm^{-3}, is formed in the Pb–Hg–As–S system (Paar et al. 2009; Piilonen and Poirier 2010). Reinvestigation by Lanza et al. (2019) demonstrates that the mineral is orthorhombic with

$a = 2124.6 \pm 0.5$, $b = 428.97 \pm 0.09$, $c = 952.57 \pm 0.12$ pm, and incommensurately modulated.

10.67 Lead–Mercury–Antimony–Sulfur

The $Hg_3Pb_{16}Sb_{18}S_{46}$ quaternary compound (mineral marrucciite), which crystallizes in the monoclinic structure with the lattice parameters $a = 4807.7 \pm 0.8$, $b = 410.68 \pm 0.06$, $c = 2397.6 \pm 0.3$ pm, $\beta = 118.752 \pm 0.008°$, and a calculated density of 6.00 g·cm^{-3} (Sejkora et al. 2011b) [$a = 4812.4 \pm 1.1$, $b = 410.83 \pm 0.02$, $c = 2399.0 \pm 0.5$ pm, $\beta = 118.76 \pm 0.02°$, and a calculated density of 6.055 g·cm^{-3} (Laufek et al. 2007b); $a = 4832 \pm 1$, $b = 411.70 \pm 0.08$, $c = 2405.6 \pm 0.5$ pm, $\beta = 118.84 \pm 0.03°$, and a calculated density of 6.013 g·cm^{-3} (Orlandi et al. 2007; Tait et al. 2008)], is formed in the Pb–Hg–Sb–S system.

10.68 Lead–Mercury–Bromine–Sulfur

$PbBr_2$–HgS. The $Hg_2PbS_2Br_2$ quaternary compound, which crystallizes in the monoclinic structure with the lattice parameters $a = 971.8$, $b = 813.1$, $c = 1088.8$ pm, and $\beta = 105.60°$ according to *ab initio* calculation, is formed in this system (Ruiz and Payne 1998). α-HgS reacts with $PbBr_2$ to form the *1D* intercalation compound 2HgS·PbBr$_2$. In this compound, the helical structure of the HgS chains remains but the turns are modified to give tetragonal symmetry. The formation of 2HgS·PbBr$_2$ can be understood as a process of miscibility between the two solid phases.

10.69 Lead–Mercury–Iodine–Sulfur

PbI_2–HgS. The $Hg_2PbS_2I_2$ quaternary compound, which melts congruently at 395°C and crystallizes in the tetragonal structure with the lattice parameters $a = 1350.1 \pm 0.1$, $c = 459.3 \pm 0.1$ pm and a calculated density of 7.349 g·cm^{-3}, is formed in this system (Blachnik et al. 1986). This compound was prepared by mixing the proper amounts of PbI$_2$ and HgS. The mixture, enclosed in an evacuated quartz ampoule, was heated to 730°C, kept at this temperature for several hours, slowly cooled to 280°C, and held at this temperature for one month. The sample was then finely ground and treated with a pressure of 0.735 GPa. The specimen was again annealed at 380°C for 6 months. Orange crystals in the form of rectangular prisms were obtained after this period.

10.70 Lead–Boron–Oxygen–Sulfur

Two quaternary compounds, $Pb[B_2(SO_4)_4]$ and $Pb_4(BO_3)_2SO_4$, are formed in the Pb–B–O–S system. First of them crystallizes in the orthorhombic structure with the lattice parameters $a = 1251.5 \pm 0.1$, $b = 1252.3 \pm 0.1$, $c = 730.2 \pm 0.1$ pm, and a calculated density of 3.558 g·cm^{-3} at 173 K (Schönegger

et al. 2018). To prepare this compound, a mixture of PbO$_2$ (50 mg), H$_3$BO$_3$ (100 mg), and 1 mL of oleum (20 mass% SO$_3$) were loaded into a thick-walled glass ampoule. Afterwards, the ampoule was torch sealed under reduced pressure, placed into a tube furnace and heated up to 120°C, kept there for 24 h, and finally reduced to room temperature with a cooling rate of 0.1°C·min^{-1}. A large number of colorless crystals were obtained. The product was handled under strictly inert conditions due to the sensitivity toward moisture.

$Pb_4(BO_3)_2SO_4$ melts congruently at ~689°C, is thermally stable up to about 680°C, and crystallizes in the monoclinic structure with the lattice parameters $a = 693.63 \pm 0.04$, $b = 1587.84 \pm 0.08$, $c = 960.53 \pm 0.05$ pm, $\beta = 110.770 \pm 0.005°$, and a calculated density of 7.000 g·cm^{-3}, and an energy gap of 4.03 eV (Ruan et al. 2018). Polycrystalline samples of this compound were synthesized by the solid-state reaction of the stoichiometric mixture of PbO (8 mM), H$_3$BO$_3$ (4 mM), and (NH$_4$)$_2$SO$_4$ (2 mM). The mixture was ground thoroughly in an agate mortar and then heated at 620°C for 3 days with several intermittent grindings. Crystals of Pb$_4$(BO$_3$)$_2$SO$_4$ were obtained through the high-temperature melt method. Polycrystalline samples placed in a platinum crucible were heated at 700°C for 2 h in muffle furnace, forming homogeneous transparent melt, and then cooled down to 550°C at a rate of 3°C·h^{-1}, followed by cooling to room temperature at a rate of 20°C·h^{-1}. Finally, transparent crystals were obtained.

10.71 Lead–Gallium–Yttrium–Sulfur

PbS–Ga_2S_3–Y_2S_3. The isothermal section of this quasiternary system at 500°C is shown in Figure 10.34 (Filyuk et al. 2008). There are 11 two-phase and 5 three-phase regions in the system. The solubility of Y$_2$S$_3$ in PbS does not exceed 3 mol%,

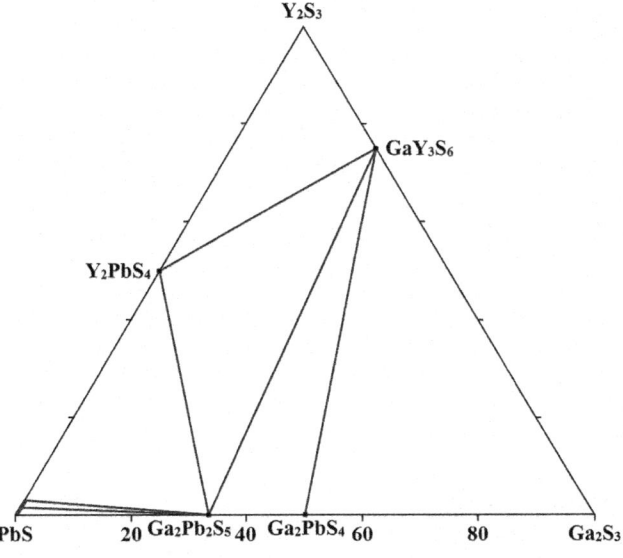

FIGURE 10.34 Isothermal section of the PbS–Ga$_2$S$_3$–Y$_2$S$_3$ quasiternary system at 500°C. (From Filyuk, T.O., et al., *Nauk. Visnyk Volyns'k. Nats. Univ. im. Lesi Ukrainky. Ser. Khim. nauky*, (16), 51, 2008.)

while the solubility based on other binary and ternary compounds is negligible. The ingots for the investigations were annealed at 500°C for 250 h, followed by quenching in air.

10.72 Lead–Gallium–Manganese–Sulfur

The **Pb$_{0.65}$Mn$_{2.85}$Ga$_3$S$_8$** quaternary compound, which crystallizes in the hexagonal structure with the lattice parameters a = 1680.4 ± 0.2, c = 371.67 ± 0.07 pm, a calculated density of 4.149 g·cm^{-3}, and two values of the energy gap – 1.68 and 2.25 eV, is formed in the Pb–Ga–Mn–S system (Zhou et al. 2017a). It was obtained from a reacting mixture of PbS (0.65 mM), MnS (2.85 mM), and Ga$_2$S$_3$ (1.5 mM). The raw materials were carefully ground and loaded into a fused silica tube, which was evacuated to a high vacuum of 10^{-3} Pa and flame-sealed. Afterward, the tube was heated from room temperature to 800°C within 15 h in a furnace, kept at that temperature for 100 h, and then the furnace was turned off. Single crystals were obtained by the spontaneous crystallization method from the solid-state reactions. The mixture of PbS, MnS, and Ga$_2$S$_3$ (molar ratio 1:2:1) was blended and loaded into a fused silica tube in a glove box. Then, the tube was evacuated to a high vacuum of 10^{-3} Pa and flame-sealed. Subsequently, the tube was gradually heated to 950°C, kept for 72 h, then slowly cooled at a rate of 3°C·h^{-1}, and finally cooled to room temperature by turning off the furnace. All operations were conducted in an argon-filled glove box with controlled oxygen and moisture levels of <0.1 ppm.

10.73 Lead–Gallium–Iron–Sulfur

PbGa$_2$S$_4$–FeS. The phase diagram of this system, constructed through DTA, XRD, and metallography, is a eutectic type (Hasanova et al. 2019). The eutectic contains 35 mol% FeS and crystallizes at 602°C. PbGa$_2$S$_4$ dissolves up to 15 mol% FeS, and the solubility of PbGa$_2$S$_4$ in FeS is negligible.

PbGa$_2$S$_4$–FeGa$_2$S$_4$. The phase diagram of this system, constructed through DTA, XRD, and metallography, is a eutectic type (Hasanova et al. 2019). The eutectic contains 45 mol% FeGa$_2$S$_4$ and crystallizes at 677°C. α-FeGa$_2$S$_4$ dissolves 5 mol% PbGa$_2$S$_4$ and PbGa$_2$S$_4$ dissolves 10 mol% α-FeGa$_2$S$_4$.

Pb$_2$Ga$_2$S$_5$–FeS. The phase diagram of this system, constructed through DTA, XRD, and metallography, is a eutectic type (Hasanova et al. 2019). The eutectic contains 40 mol% FeS and crystallizes at 577°C. Pb$_2$Ga$_2$S$_5$ dissolves up to 20 mol% FeS and the solubility of Pb$_2$Ga$_2$S$_5$ in FeS is negligible.

PbS–Ga$_2$S$_3$–FeS. At room temperature, this quasiternary system is divided by the sections indicated above into four subsystems (Hasanova et al. 2019).

10.74 Lead–Indium–Bismuth–Sulfur

Three quaternary compounds are formed in the Pb–In–Bi–S system. **In$_2$Pb$_4$Bi$_4$S$_{13}$** crystallizes in the orthorhombic structure with the lattice parameters a = 2134.4 ± 0.4, b = 400.2 ± 0.1, c = 2649.4 ± 0.5 pm, and a calculated density of 6.783 g·cm^{-3}

(Krämer 1986b) [a = 2133.1 ± 0.4, b = 2643.5 ± 0.5, and c = 400.6 ± 0.1 pm for mineral znamenskyite with the same composition (Chaplygin et al. 2014)]. Needle-shaped crystals of this compound were grown by chemical-vapor transport with iodine in a two-zone furnace adjusted at 550°C/500°C (Krämer 1986b). Black crystals with high metallic luster were obtained.

In$_3$Pb$_4$Bi$_7$S$_{18}$ crystallizes in the monoclinic structure with the lattice parameters a = 2102.1 ± 0.5, b = 401.4 ± 0.2, c = 1889.8 ± 0.5 pm, β = 97.07 ± 0.02°, and a calculated density of 6.744 g·cm^{-3} (Krämer 1986a). Needle-shaped crystals of this compound were grown by chemical-vapor transport with iodine in a two-zone furnace adjusted at 640°C/570°C. Black crystals with high metallic luster were obtained.

In$_8$Pb$_{1.6}$Bi$_4$S$_{19}$ also crystallizes in the monoclinic structure with the lattice parameters a = 2916.7 ± 0.5, b = 387.2 ± 0.2, c = 1555.4 ± 0.5 pm, β = 121.6 ± 0.1°, and a calculated density of 5.983 g·cm^{-3} (Krämer 1983). Needle-shaped crystals of this compound were grown by chemical-vapor transport with iodine in a two-zone furnace adjusted at 650°C/600°C. Black crystals with high metallic luster were obtained.

10.75 Lead–Indium–Iron–Sulfur

The **In$_{10}$Pb$_{5.5}$Fe$_{1.5}$S$_{22}$** quaternary compound, which melts congruently at 825°C and crystallizes in the monoclinic structure with the lattice parameters a = 1455.8 ± 0.1, b = 386.56 ± 0.03, c = 1555.8 ± 0.1 pm, β = 96.876 ± 0.001°, a calculated density of 5.893 g·cm^{-3}, and an energy gap of 0.95 eV, is formed in the Pb–In–Fe–S system (Matsushita and Ueda 2006). This compound was synthesized by a solid-state reaction using FeS, Pb, In$_2$S$_3$, and S (molar ratio 1:3:2.5:3). These starting materials were roughly mixed, and the mixture was sealed into an evacuated silica tube (~10^{-3} Pa). The mixture was slowly heated to 850°C over 12 h, annealed at this temperature for 3 days, cooled to 400°C over 100 h, and then cooled to room temperature within 10 h. Bar-shaped shiny metallic silver crystals were obtained. Both ball-shaped PbS impurities and chunk PbIn$_2$S$_4$ impurities were observed in the products.

10.76 Lead–Indium–Rhodium–Sulfur

The **InPbRh$_3$S$_2$** quaternary compound, which crystallizes in the hexagonal structure with the lattice parameters a = 562 and c = 1369 pm, is formed in the Pb–In–Rh–S system (Natarajan et al. 1988). It was synthesized by the direct combination of stoichiometric quantities of high-purity elements in evacuated (~10^{-3} Pa) and sealed quartz tube. The quartz tube containing the reactants was heated initially at 400–500°C for 2–3 days until all the sulfur was reacted. The temperature was then raised to 800–1000°C and kept for 24 h. In most cases, a molten mass was noted inside the quartz tube after this treatment. The compounds were recovered, ground thoroughly in an agate pestle and mortar, and pressed in the form of disks with a WC-lined stainless steel die, and sintered at 600°C for 1–2 days in an evacuated quartz tube. Single-phase material has thus been obtained. This compound oxidizes when heated in air, but is stable when heated to temperatures higher than 900°C in an evacuated and sealed quartz ampoule and sinters well.

10.77 Lead–Thallium–Phosphor–Sulfur

The $TlPbPS_4$ quaternary compound, which crystallizes in the orthorhombic structure with the lattice parameters $a = 1209.76 \pm 0.09$, $b = 658.16 \pm 0.05$, $c = 880.93 \pm 0.06$ pm, and a calculated density of 5.405 g·cm^{-3}, is formed in the Pb–Tl–P–S system (Belkyal et al. 2006a). It was prepared from a stoichiometric mixture of Pb, P_2S_5, S, and Tl_2S. The reaction mixture was thoroughly mixed in a N_2-filled glove box and loaded into a quartz ampoule. After evacuation to 0.1 Pa, the ampoule was flame-sealed and placed in a furnace. The sample was heated to 870°C, kept at this temperature for 3 days, and then cooled to 135°C at a rate of 4.5°C·h^{-1}; the furnace was then turned off. After washing with diethyl ether, transparent orange crystals were obtained. $TlPbPS_4$ is slightly air and moisture sensitive.

10.78 Lead–Thallium–Arsenic–Sulfur

Some quaternary compounds are formed in the Pb–Tl–As–S system. $TlPbAsS_3$ (mineral richardsollyite) crystallizes in the monoclinic structure with the lattice parameters $a = 889.25 \pm 0.02$, $b = 841.54 \pm 0.02$, $c = 857.54 \pm 0.02$ pm, $\beta = 108.665 \pm 0.003°$, and a calculated density of 6.365 g·cm^{-3} (Meisser et al. 2016, 2017).

$TlPbAs_3S_6$ (mineral edenharterite) crystallizes in the orthorhombic structure with the lattice parameters $a = 1547.64 \pm 0.08$, $b = 4760.2 \pm 0.3$, and $c = 584.89 \pm 0.04$ pm (Berlepsch 1996; Jambor et al. 1997) [$a = 4745.3 \pm 0.2$, $b = 1547.6 \pm 0.7$, $c = 584.7 \pm 0.2$ pm, and a calculated density of 5.13 g·cm^{-3} (Balić-Žunić and Engel 1983); $a = 1546.5 \pm 0.3$, $b = 4750.7 \pm 0.8$, $c = 584.3 \pm 0.2$ pm, and a calculated density of 5.09 g·cm^{-3} (Graeser and Schwander 1992; Jambor and Burke 1993); $a = 584.8 \pm 0.1$, $b = 1547.8 \pm 0.2$, and $c = 4760.0 \pm 0.9$ pm (Berlepsch 1995)]. This compound can be prepared by hydrothermal synthesis in the PbS–Tl_2S–As_2S_3 system (Balić-Žunić and Engel 1983).

$(Pb,Tl)_2As_5S_9$ (mineral hutchinsonite) also crystallizes in the orthorhombic structure with the lattice parameters $a = 1078.6 \pm 0.3$, $b = 3538.9 \pm 0.8$, $c = 814.1 \pm 0.3$ pm, and a calculated density of 4.59 g·cm^{-3} (Matsushita and Takéuchi 1994) [$a = 1081 \pm 1$, $b = 3536 \pm 4$, and $c = 816 \pm 1$ pm (Takéuchi et al. 1965)].

$(Pb,Tl)_3As_5S_{10}$ (mineral rathite) crystallizes in the monoclinic structure with the lattice parameters $a = 2452 \pm 5$, $b = 791 \pm 1$, $c = 843 \pm 1$ pm, $\beta = 90.0 \pm 0.5°$, and the calculated and experimental densities of 5.35 ± 0.03 and 5.38 ± 0.8 g·cm^{-3}, respectively (Le Bihan 1960, 1962) [$a = 843$, $b = 2580$, $c = 791$ pm, and $\beta = 90.00 \pm 0.25°$ (Nowacki et al. 1960); crystallizes in the triclinic structure with the lattice parameters $a = 2516 \pm 2$, $b = 794 \pm 1$, $c = 847 \pm 1$ pm, $\alpha = 90°00' \pm 10'$, $\beta = 100°28' \pm 10'$, $\gamma = 90°00' \pm 10'$, and a calculated density of 5.37 g·cm^{-3} (Marumo and Nowacki 1965)].

$TlPb_{58}As_{97}S_{204}$ (mineral dekatriasartorite) also crystallizes in the monoclinic structure with the lattice parameters $a = 5457.6 \pm 0.5$, $b = 789.47 \pm 0.06$, $c = 2010.2 \pm 1.6$ pm, and $\beta = 78.153 \pm 0.001°$ (Topa et al. 2017b,c).

$Tl_{1.5}Pb_8As_{17.5}S_{35}$ (mineral sartorite containing up to 6.5 mass% Tl) is characterized by ninefold monoclinic superstructure with $a = 3771 \pm 2$, $b = 789.8 \pm 0.3$, $c = 2010.6 \pm 0.8$ pm, and $\beta = 101.993 \pm 0.007°$ and has the monoclinic subcell ($a = 2011$, $b = 790$, $c = 419$ pm, and $\beta = 102.0°$) (Berlepsch et al. 2003).

$Tl_2Pb_{48}As_{82}S_{172}$ (mineral hendekasartorite) also crystallizes in the monoclinic structure with the lattice parameters $a = 3180.6 \pm 0.5$, $b = 788.90 \pm 0.12$, $c = 2855.6 \pm 0.4$ pm, and $\beta = 99.034 \pm 0.002°$ (Topa et al. 2015b, 2017a).

$Tl_6Pb_{32}As_{70}S_{140}$ (mineral enneasartorite) also crystallizes in the monoclinic structure with the lattice parameters $a = 3761.2 \pm 0.6$, $b = 787.77 \pm 0.01$, $c = 2007.1 \pm 0.3$ pm, $\beta = 101.930 \pm 0.002°$, and a calculated density of 5.1 g·cm^{-3} (Topa et al. 2015a, 2017a; Makovicky et al. 2018).

$Tl_6Pb_{144}As_{246}S_{516}$ (mineral incomsartorite) also crystallizes in the monoclinic structure with the lattice parameters $a = 4599.44 \pm 0.02$, $b = 787.93 \pm 0.01$, $c = 5867.16 \pm 0.08$ pm, and $\beta = 90.153 \pm 0.001°$ (Topa et al. 2016c).

$Tl_7Pb_{22}As_{55}S_{108}$ (mineral heptasartorite) also crystallizes in the monoclinic structure with the lattice parameters $a = 2926.9 \pm 0.2$, $b = 787.68 \pm 0.05$, $c = 2012.8 \pm 0.2$ pm, $\beta = 102.065 \pm 0.002°$, and a calculated density of 4.9 g·cm^{-3} (Topa et al. 2015c, 2017a; Makovicky et al. 2018).

10.79 Lead–Thallium–Antimony–Sulfur

The $TlPbSbS_3$ quaternary compound, which melts incongruently at 480°C (Balić-Žunić et al. 1992) and has a polymorphic transformation, is formed in the Pb–Tl–Sb–S system. α-$TlPbSbS_3$ crystallizes in the monoclinic structure with the lattice parameters $a = 417.07 \pm 0.04$, $b = 428.56 \pm 0.04$, $c = 1215.7 \pm 0.1$ pm, and $\beta = 105.49 \pm 0.01°$ (Balić-Žunić and Makovicky 1994; Balić-Žunić and Bente 1995) [$a = 416.9 \pm 0.2$, $b = 428.6 \pm 0.4$, $c = 1215.4 \pm 0.4$ pm, and $\beta = 105.5°$ (Bente and Anton 1995)]. β-$TlPbSbS_3$ crystallizes in the orthorhombic structure with the lattice parameters $a = 411.5 \pm 0.2$, $b = 1170.4 \pm 0.6$, and $c = 433.8 \pm 0.1$ pm (Bente and Anton 1995) [$a = 411.2 \pm 0.3$, $b = 1169 \pm 1$, and $c = 433.8 \pm 0.2$ pm (Balić-Žunić et al. 1992)]. At the cooling, β-$TlPbSbS_3$ inverts at about 347°C to the low-temperature α-$TlPbSbS_3$ (Balić-Žunić et al. 1992; Balić-Žunić and Bente 1995).

β-$TlPbSbS_3$ was prepared from mixtures of Pb, Tl, Sb_2S_3, and S in sealed silica tubes under dry condensed conditions (Balić-Žunić et al. 1992; Balić-Žunić and Bente 1995). After an initial reaction at 200°C for one week, the specimen was homogenized and subsequently heated to 480°C. After 220 days at this temperature, the specimen was quenched to room temperature.

α-$TlPbSbS_3$ was obtained by annealing high-temperature β-$TlPbSbS_3$ in evacuated and sealed silica tubes at 250°C for one month or by the hydrothermal synthesis at 280°C and 10

MPa for 21 days, using the powder of β-TlPbSbS₃ as the starting material (Balić-Žunić and Makovicky 1994; Balić-Žunić and Bente 1995).

10.80 Lead–Thallium–Bismuth–Sulfur

PbS–TlBiS₂. The phase diagram of this system, constructed through DTA, XRD, metallography, and measuring of the microhardness, belongs to the type I according to the Roseboom's classification (Malevskiy 1966; Gitsu et al. 1975). At temperatures above 600°C, a continuous series of solid solutions with a cubic structure is formed in the system. At temperatures below 600°C, in the interval of 12–40 mol% 2PbS, solid solution decomposition is observed. The **TlPbBiS₃** quaternary compound, which melts at 520°C and crystallizes in the cubic structure with the lattice parameter $a = 608.0$ pm and the calculated and experimental densities of 5.29 and 5.24 g·cm⁻³, respectively (Guseynov et al. 1972), was not found in the system. The ingots for the investigations were annealed at 650°C for 300 h (Gitsu et al. 1975) or at 700°C for 60 h (Malevskiy 1966).

10.81 Lead–Thallium–Oxygen–Sulfur

The **Tl₂Pb(SO₄)₂** quaternary compound, which crystallizes in the trigonal structure with the lattice parameters $a = 559 \pm 1$ and $c = 2212 \pm 5$ pm, is formed in the Pb–Tl–O–S system (Møller 1954). It was synthesized in the microcrystalline state by stirring a precipitate of PbSO₄ with a sufficiently concentration solution of Tl₂SO₄, filtering, and washing with dilute alcohol

10.82 Lead–Thallium–Selenium–Sulfur

PbS–Tl₄PbSe₃. The phase diagram of this system, constructed through DTA, XRD, and metallography using the alloys annealing at 300°C for 168 h, is a eutectic type (Figure 10.35) (Filep et al. 2012a). The eutectic contains 5 mol% PbS and crystallizes at 520°C. The extent of solid solutions based on Tl₄PbSe₃ does not exceed 3 mol% PbS, and the solubility of Tl₄PbSe₃ in PbS is not higher than 10 mol%.

Tl₄PbSe₃–Tl₂S. The phase diagram of this system, constructed through DTA, XRD, and metallography using the alloys annealing at 300°C for 168 h, is a eutectic type (Filep et al. 2012a). The eutectic contains 12 mol% Tl₄PbSe₃ and crystallizes at 394°C. The extent of solid solutions based on Tl₄PbSe₃ is 68 mol% PbS, and the solubility of Tl₄PbSe₃ in Tl₂S does not exceed 10 mol%.

PbS–PbSe–Tl₄PbSe₃. The isothermal section at 300°C and the liquidus surface of this quasiternary system are given in Figures 10.36 and 10.37 (Filep and Sabov 2018). The liquidus surface consists of two fields of primary crystallization: the field of primary crystallization of solid solution based on Tl₄PbSe₃ as well as PbS$_x$Se$_{1-x}$ solid solution, which intersect

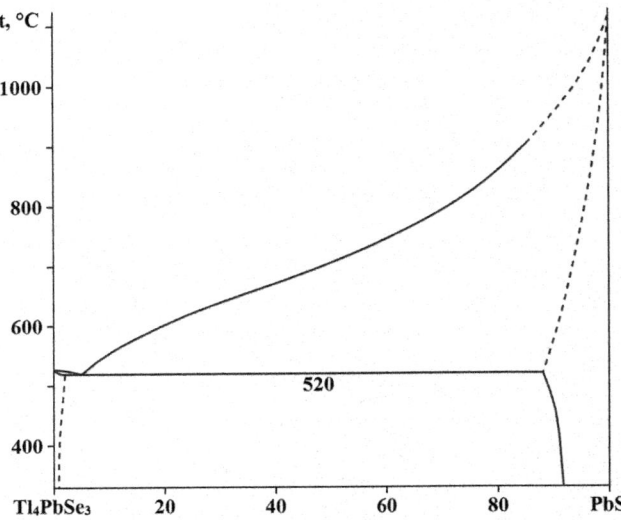

FIGURE 10.35 Phase diagram of the PbS–Tl₄PbSe₃ system. (From Filep, M.Y., et al., *Nauk. visn. Uzhgorod. Univ., Ser. Khim.*, [1(27)], 22, 2012.)

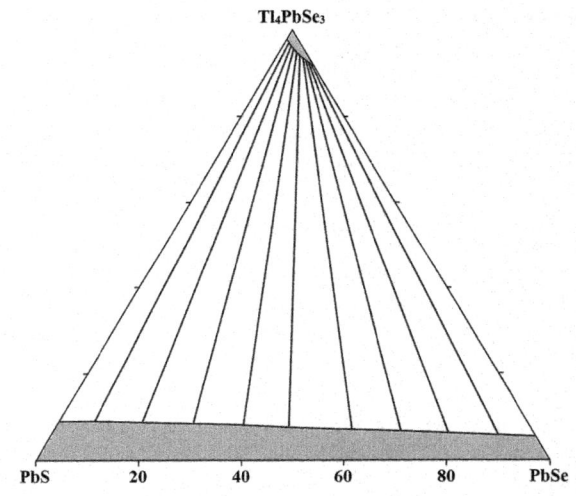

FIGURE 10.36 Isothermal section of the PbS–PbSe–Tl₄PbSe₃ quasiternary system at 300°C. (From Filep, M.Y., and Sabov, M.Yu., *Nauk. visn. Uzhgorod. Univ., Ser. Khim.*, [1(39)], 17, 2018.)

along a monovariant line. The ingots for the investigations were annealed at 300°C for 24 h followed by quenching in air.

Tl₄PbSe₃–Tl₂S–Tl₂Se. The isothermal section at 300°C of this quasiternary system is presented in Figure 10.38 (Filep and Sabov 2017b). It can be seen that the Tl₄₋₂ₓPb₁₋ₓSe₃₋₂ₓ ($x = 0–1$) solid solution is characterized by the largest region of homogeneity and dissolves 19–42 mol% Tl₂S. The liquidus surface of this system (Figures 10.39) consists of two fields of primary crystallization: one belongs to the solid solution based on Tl₂S and another to the Tl₄₋₂ₓPb₁₋ₓSe₃₋₂ₓ solid solution, which intersect along a monovariant line of the eutectic equilibrium.

The **Tl₄PbS₃–Tl₄PbSe₃** system is a nonquasibinary section of the PbS + Tl₂Se ↔ PbSe + Tl₂S ternary mutual system (Filep et al. 2016).

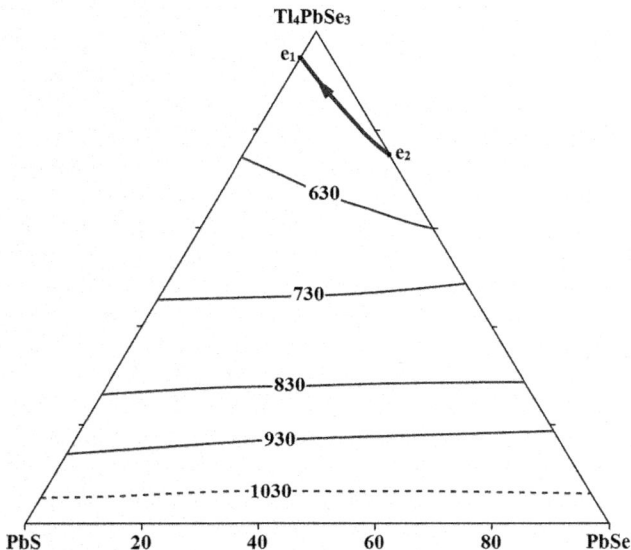

FIGURE 10.37 Liquidus surface of the PbS–PbSe–Tl₄PbSe₃ quasiternary system. (From Filep, M.Y., and Sabov, M.Yu., *Nauk. visn. Uzhgorod. Univ., Ser. Khim.*, [1(39)], 17, 2018.)

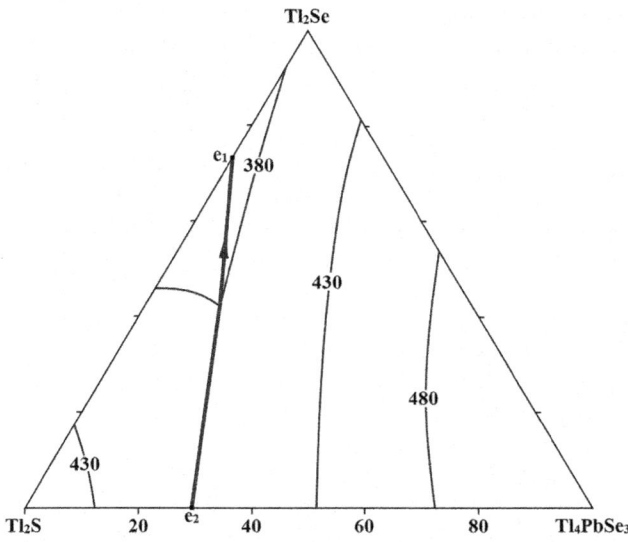

FIGURE 10.39 Liquidus surface of the Tl₄PbSe₃–Tl₂S–Tl₂Se quasiternary system. (From Filep, M.Y., and Sabov, M.Yu., *Nauk. visn. Uzhgorod. Univ., Ser. Khim.*, [1(37)], 14, 2017.)

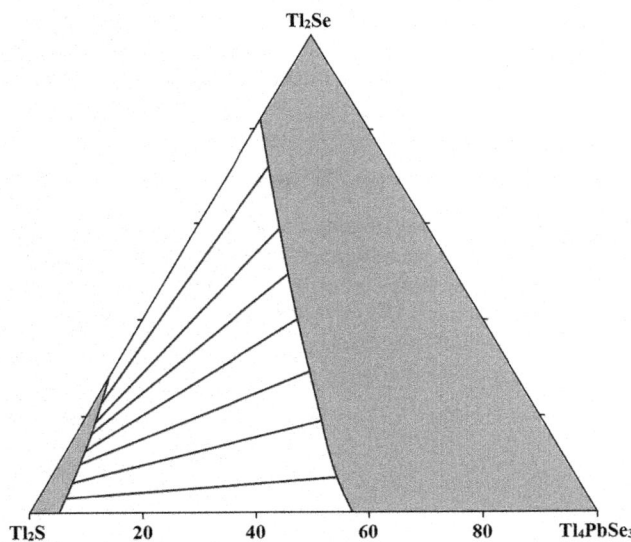

FIGURE 10.38 Isothermal section of the Tl₄PbSe₃–Tl₂S–Tl₂Se quasiternary system at 300°C. (From Filep, M.Y., and Sabov, M.Yu., *Nauk. visn. Uzhgorod. Univ., Ser. Khim.*, [1(37)], 14, 2017.)

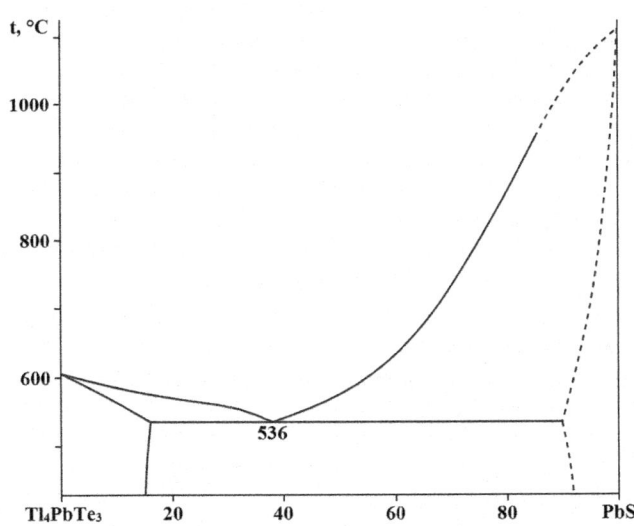

FIGURE 10.40 Phase diagram of the PbS–Tl₄PbTe₃ system. (From Filep, M.Y., et al., *Nauk. visn. Uzhgorod. Univ., Ser. Khim.*, [1(27)], 22, 2012.)

10.83 Lead–Thallium–Tellurium–Sulfur

PbS–Tl₄PbTe₃. The phase diagram of this system, constructed through DTA, XRD, and metallography using the alloys annealing at 200°C for 168 h, is a eutectic type (Figure 10.40) (Filep et al. 2012a). The eutectic contains 38 mol% PbS and crystallizes at 536°C. The extent of solid solutions based on Tl₄PbTe₃ does not exceed 16 mol% PbS, and the solubility of Tl₄PbTe₃ in PbS is not higher than 10 mol%.

Tl₄PbTe₃–Tl₂S. The phase diagram of this system, constructed through DTA, XRD, and metallography using the alloys annealing at 200°C for 168 h, is also a eutectic type (Filep et al. 2012a). The eutectic contains 5 mol% Tl₄PbTe₃ and crystallizes at 356°C. The extent of solid solutions based on Tl₄PbTe₃ is 68 mol% PbS, and the solubility of Tl₄PbTe₃ in Tl₂S does not exceed 5 mol%.

The **Tl₄PbS₃–Tl₄PbTe₃** system is a nonquasibinary section of the PbS + Tl₂Te ↔ PbTe + Tl₂S ternary mutual system (Filep et al. 2016).

10.84 Lead–Lanthanum–Bismuth–Sulfur

La$_2$PbS$_4$–PbBi$_2$S$_4$. This system is a nonquasibinary section of the Pb–La–Bi–S system as PbBi$_2$S$_4$ melts incongruently (Aliev et al. 2017). The eutectic contains 30 mol% PbBi$_2$S$_4$ and crystallizes at 577°C. The solubility of PbBi$_2$S$_4$ in La$_2$PbS$_4$ at room temperature is 3 mol%. In this system, the **LaPbBiS$_4$** quaternary compound is formed, which has a region of homogeneity in the range of 45–57 mol% PbBi$_2$S$_4$, melts congruently at 752°C, and crystallizes in the orthorhombic structure with the lattice parameters $a = 1170$, $b = 1454$, $c = 412$ pm. The system was studied by means of DTA, XRD, metallography, and by measuring the microhardness using the ingots annealed at 330°C for 120 h and at 130°C for 200 h.

The standard thermodynamic functions of the LaPbBiS$_4$ compound were calculated by Aliyev et al. (2014): $S^0_{298} = 285 \pm 10$ J·M^{-1}K^{-1}, $\Delta S^0_{298} = -21.5 \pm 4.0$ J·M^{-1}K^{-1}, $\Delta H^0_{298} = -876 \pm 25$ kJ·M^{-1}, and $\Delta G^0_{298} = -870 \pm 25$ kJ·M^{-1}. LaPbBiS$_4$ was prepared by interaction of LaPb$_2$S$_4$ and PbBi$_2$S$_4$ at 540–580°C for 2 weeks.

The **La$_x$Pb$_2$Bi$_{8-x}$S$_{14}$** ($x \approx 2.1$) quaternary compound, which melts congruently at ~800°C and crystallizes in the orthorhombic structure with the lattice parameters $a = 2125.92 \pm 0.04$, $b = 404.18 \pm 0.01$, $c = 2817.18 \pm 0.03$ pm, a calculated density of 6.553 g·cm^{-3}, and an energy gap of 0.95 eV, is formed in the Pb–La–Bi–S system (Iordanidis and Kanatzidis 2001). To synthesize this compound, a mixture of Pb (0.376 mM), Bi (1.120 mM), S (2.058 mM), and La$_2$S$_3$ (0.374 mM) was transferred to a carbon-coated silica tube that was flame-sealed under vacuum. The tube was heated at 950°C for 99 h and cooled to 550°C at a rate of 10°C·h^{-1} and then to 50°C in 10 h. The product consisted of long silver-gray needles. All manipulations were carried out under a dry N$_2$ atmosphere in a glove box.

10.85 Lead–Praseodymium–Bismuth–Sulfur

The **PrPbBiS$_4$** quaternary compound is formed in the Pb–Pr–Bi–S system. The standard thermodynamic functions of this compound were calculated by Aliyev et al. (2014): $S^0_{298} = 301 \pm 10$ J·M^{-1}K^{-1}, $\Delta S^0_{298} = -22.7 \pm 4.0$ J·M^{-1}K^{-1}, $\Delta H^0_{298} = -857 \pm 25$ kJ·M^{-1}, and $\Delta G^0_{298} = -850 \pm 25$ kJ·M^{-1}. PrPbBiS$_4$ was prepared by interaction of Pr$_2$PbS$_4$ and PbBi$_2$S$_4$ at 540–580°C for 2 weeks.

10.86 Lead–Neodymium–Bismuth–Sulfur

Nd$_2$PbS$_4$–PbBi$_2$S$_4$. This system is nonquasibinary section of the Pb–Nd–Bi–S system as PbBi$_2$S$_4$ melts incongruently (Aliyev et al. 2014). The eutectic contains 30 mol% PbBi$_2$S$_4$ and crystallizes at 707°C. In this system, the **NdPbBiS$_4$** quaternary compound is formed, which has a region of homogeneity in the range of 45–56 mol% PbBi$_2$S$_4$ and melts congruently at 842°C. The system was studies through DTA, XRD, and metallography. NdPbBiS$_4$ was prepared by interaction of Nd$_2$PbS$_4$ and PbBi$_2$S$_4$ at 540–580°C for 2 weeks. The standard thermodynamic functions of the NdPbBiS$_4$ compound were calculated:

$S^0_{298} = 296 \pm 10$ J·M^{-1}K^{-1}, $\Delta S^0_{298} = -23.7 \pm 5.0$ J·M^{-1}K^{-1}, $\Delta H^0_{298} = -848 \pm 25$ kJ·M^{-1}, and $\Delta G^0_{298} = -841 \pm 25$ kJ·M^{-1}.

According to the data of Khamiyev et al. (2011), one more quaternary compound, **NdPb$_2$BiS$_5$**, which crystallizes in the orthorhombic structure with the lattice parameters $a = 1916$, $b = 2391$, $c = 411$ pm, and the calculated and experimental densities of 6.57 and 6.65 g·cm^{-3}, respectively, is formed in the Pb–Nd–Bi–S system. This compound was obtained by direct synthesis from the elements, and its single crystals were grown by the method of directed crystallization.

10.87 Lead–Samarium–Bismuth–Sulfur

Two quaternary compounds, **SmPbBiS$_4$** and **SmPb$_2$BiS$_5$**, are formed in the Pb–Sm–Bi–S system. The standard thermodynamic functions of SmPbBiS$_4$ were calculated by Aliyev et al. (2014): $S^0_{298} = 302 \pm 10$ J·M^{-1}K^{-1}, $\Delta S^0_{298} = -17.8 \pm 3.0$ J·M^{-1}K^{-1}, $\Delta H^0_{298} = -882 \pm 25$ kJ·M^{-1}, and $\Delta G^0_{298} = -877 \pm 25$ kJ·M^{-1}. It was prepared by interaction of Sm$_2$PbS$_4$ and PbBi$_2$S$_4$ at 540–580°C for 2 weeks.

SmPb$_2$BiS$_5$ crystallizes in the orthorhombic structure with the lattice parameters $a = 1914$, $b = 2389$, $c = 409$ pm, and the calculated and experimental densities of 6.66 and 6.72 g·cm^{-3}, respectively (Khamiyev et al. 2011). This compound was obtained by direct synthesis from the elements, and its single crystals were grown by the method of directed crystallization.

10.88 Lead–Europium–Tellurium–Sulfur

PbTe–EuS. The phase diagram of this system, constructed through DTA, XRD, metallography, and measuring the microhardness using the alloys annealing at 600°C for 300 h, is a eutectic type (Figure 10.41) (Abilov et al. 1993). The eutectic contains 12–20 mol% EuS and crystallizes at 647°C. The solubility of EuS in PbTe at room temperature reaches 1 mol%.

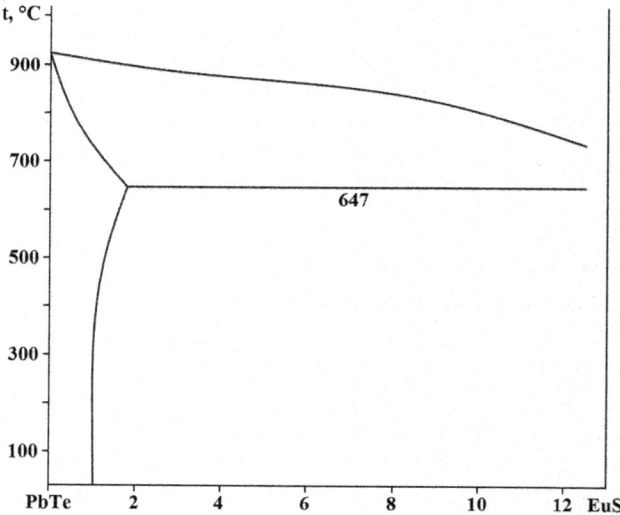

FIGURE 10.41 Part of the phase diagram of the PbTe–EuS system. (From Abilov, Ch.I., et al., *Neorgan. mater.*, **29**(2), 285, 1993.)

10.89 Lead–Europium–Bismuth–Sulfur

The **EuPbBi$_2$S$_5$** quaternary compound, which crystallizes in the orthorhombic structure with the lattice parameters a = 1909.2, b = 2379.3, c = 404.8 pm, and the calculated and experimental densities of 6.53 and 6.50 g·cm^{-3}, respectively, is formed in the Pb–Eu–Bi–S system (Gasymov et al. 2007). It was synthesized stepwise from a mixture of the elements. The mixture was placed in an evacuated and sealed ampoule in a rotary furnace. First, the temperature was raised to 130–180°C and kept for 2–3 h. Then heating was continued up to 775°C and kept at this temperature for 2 h, after which the furnace was slowly cooled. To bring it to equilibrium, the sample was annealed at 380–430°C for 240–250 h.

10.90 Lead–Gadolinium–Bismuth–Sulfur

Two quaternary compounds, **GdPbBiS$_4$** and **GdPb$_2$BiS$_5$**, are formed in the Pb–Gd–Bi–S system. The standard thermodynamic functions of GdPbBiS$_4$ were calculated by Aliyev et al. (2014): S^0_{298} = 294 ± 10 J·M^{-1}K^{-1}, ΔS^0_{298} = −23.6 ± 5.0 J·M^{-1}K^{-1}, ΔH^0_{298} = −888 ± 25 kJ·M^{-1}, and ΔG^0_{298} = −881 ± 25 kJ·M^{-1}. It was prepared by interaction of Gd$_2$PbS$_4$ and PbBi$_2$S$_4$ at 540–580°C for 2 weeks.

GdPb$_2$BiS$_5$ crystallizes in the orthorhombic structure with the lattice parameters a = 1910, b = 2387, c = 407 pm, and the calculated and experimental densities of 6.76 and 6.81 g·cm^{-3}, respectively (Khamiyev et al. 2011). This compound was obtained by direct synthesis from the elements, and its single crystals were grown by the method of directed crystallization.

10.91 Lead–Terbium–Bismuth–Sulfur

The **TbPbBiS$_4$** quaternary compound is formed in the Pb–Tb–Bi–S system. The standard thermodynamic functions of this compound were calculated by Aliyev et al. (2014): S^0_{298} = 302 ± 15 J·M^{-1}K^{-1}, ΔS^0_{298} = −25.6 ± 5.0 J·M^{-1}K^{-1}, ΔH^0_{298} = −892 ± 25 kJ·M^{-1}, and ΔG^0_{298} = −884 ± 25 kJ·M^{-1}. TbPbBiS$_4$ was prepared by interaction of Tb$_2$PbS$_4$ and PbBi$_2$S$_4$ at 540–580°C for 2 weeks.

10.92 Lead–Dysprosium–Bismuth–Sulfur

The **DyPbBiS$_4$** quaternary compound is formed in the Pb–Dy–Bi–S system. The standard thermodynamic functions of this compound were calculated by Aliyev et al. (2014): S^0_{298} = 300 ± 10 J·M^{-1}K^{-1}, ΔS^0_{298} = −24.7 ± 5.0 J·M^{-1}K^{-1}, ΔH^0_{298} = −888 ± 25 kJ·M^{-1}, and ΔG^0_{298} = −881 ± 25 kJ·M^{-1}. DyPbBiS$_4$ was prepared by interaction of Dy$_2$PbS$_4$ and PbBi$_2$S$_4$ at 540–580°C for 2 weeks.

10.93 Lead–Holmium–Bismuth–Sulfur

The **HoPbBiS$_4$** quaternary compound is formed in the Pb–Ho–Bi–S system. The standard thermodynamic functions of this compound were calculated by Aliyev et al. (2014): S^0_{298} = 304

± 15 J·M^{-1}K^{-1}, ΔS^0_{298} = −21.6 ± 4.0 J·M^{-1}K^{-1}, ΔH^0_{298} = −892 ± 25 kJ·M^{-1}, and ΔG^0_{298} = −886 ± 25 kJ·M^{-1}. HoPbBiS$_4$ was prepared by interaction of Ho$_2$PbS$_4$ and PbBi$_2$S$_4$ at 540–580°C for 2 weeks.

10.94 Lead–Erbium–Bismuth–Sulfur

The **ErPbBiS$_4$** quaternary compound is formed in the Pb–Er–Bi–S system. The standard thermodynamic functions of this compound were calculated by Aliyev et al. (2014): S^0_{298} = 301 ± 15 J·M^{-1}K^{-1}, ΔS^0_{298} = −21.9 ± 4.0 J·M^{-1}K^{-1}, ΔH^0_{298} = −902 ± 28 kJ·M^{-1}, and ΔG^0_{298} = −895 ± 28 kJ·M^{-1}. ErPbBiS$_4$ was prepared by interaction of Er$_2$PbS$_4$ and PbBi$_2$S$_4$ at 540–580°C for 2 weeks.

10.95 Lead–Ytterbium–Bismuth–Sulfur

The **YbPbBi$_2$S$_5$** quaternary compound, which crystallizes in the orthorhombic structure with the lattice parameters a = 1907.5, b = 2370.4, c = 404.1 pm, and the calculated and experimental densities of 6.81 and 6.75 g·cm^{-3}, respectively, is formed in the Pb–Yb–Bi–S system (Gasymov et al. 2007). It was synthesized stepwise from a mixture of the elements. The mixture was placed in an evacuated and sealed ampoule in a rotary furnace. First, the temperature was raised to 130–180°C and kept for 2–3 h. Then heating was continued up to 775°C and kept at this temperature for 2 h, after which the furnace was slowly cooled. To bring it to equilibrium, the sample was annealed at 380–430°C for 240–250 h.

10.96 Lead–Carbon–Oxygen–Sulfur

The **Pb$_2$(CO$_3$)(S$_2$O$_3$)** quaternary compound (mineral fassinaite), which crystallizes in the orthorhombic structure with the lattice parameters a = 1632.0 ± 0.2, b = 876.16 ± 0.06, c = 458.09 ± 0.07 pm, and a calculated density of 5.947 g·cm^{-3}, is formed in the Pb–C–O–S system (Bindi et al. 2011a,b).

10.97 Lead–Arsenic–Antimony–Sulfur

PbS–As$_2$S$_3$–Sb$_2$S$_3$. Phase relations in this quasiternary system at 250°C and 500°C were determined by Roland (1968). It has been assumed that the PbS/(As$_2$S$_3$, Sb$_2$S$_3$) ratio in mineral geocronite is constant at 9:2. According to the data of Walia and Chang (1973), at 400°C in this system three ternary phases, **Pb(Sb,As)$_2$S$_4$, Pb$_2$(Sb,As)$_2$S$_5$**, and **Pb$_{18}$(Sb,As)$_{15}$S$_{41}$**, are stable. Pb$_2$(Sb,As)$_2$S$_5$ was also found to form a complete solid solution series with Pb$_2$As$_2$S$_5$. **Pb$_{14}$(As,Sb)$_6$S$_{23}$** extends its stability region into the ternary system, taking up to 58 mol% Sb$_2$S$_3$ in its structure, and Pb$_9$Sb$_{22}$S$_{42}$ has a solid solution range with a maximum of 30 mol% As$_2$S$_3$.

Some quaternary compounds are formed in the Pb–As–Sb–S system. Pb(Sb,As)$_2$S$_4$ (mineral guettardite) crystallizes in the monoclinic structure with the lattice parameters a = 852.7 ± 0.4, b = 797.1 ± 0.1, c = 2010.2 ± 1.0 pm, β = 101.814 ± 0.007°, and a calculated density of 5.29 g·cm^{-3} (Makovicky et al. 2012) [a = 2000 ± 40, b = 794 ± 3, c = 872 ± 6 pm, β = 101°35′ ± 30′, and a calculated density of 5.49 g·cm^{-3} (Jambor 1967b; Fleischer et al. 1968; Pierrot 1968); a = 2005 ± 5, b = 795 ± 2, c = 844 ± 2 pm, and β = 102°46′ ± 10′ (Bracci et al. 1980)].

Pb(Sb,As)$_2$S$_4$ (mineral twinnite, may be a polymorph of guettardite) also crystallizes in the monoclinic structure with the lattice parameters a = 799.7 ± 0.2, b = 1951.7 ± 0.5, c = 863.4 ± 0.2 pm, β = 91.061 ± 0.004°, and a calculated density of 5.16 g·cm^{-3} (Makovicky and Topa 2012) [crystallizes in the orthorhombic structure with the lattice parameters a = 1960 ± 20, b = 799 ± 5, c = 860 ± 5 pm, and a calculated density of 5.323 g·cm^{-3} (Jambor 1967b; Fleischer et al. 1968; Pierrot 1968); a = 1854 ± 1, b = 795.5 ± 0.3, and c = 851.2 ± 0.3 pm for PbAsSbS$_4$ composition (Johan and Mantienne 2000)].

Pb$_2$(Sb,As)$_2$S$_5$ (mineral veenite) also crystallizes in the monoclinic structure with the lattice parameters a = 842.9 ± 0.2, b = 2606.9 ± 0.5, c = 896.2 ± 0.2 pm, β = 117.447 ± 0.002°, and a calculated density of 5.91 g·cm^{-3} (Topa and Makovicky 2017) [crystallizes in the orthorhombic structure with the lattice parameters a = 422, b = 2620, c = 790 pm, and the calculated and experimental densities of 5.96 and 5.92 g·cm^{-3}, respectively (Jambor 1967a; Fleischer et al. 1968)].

Pb$_5$Sb$_3$AsS$_{11}$ (mineral lopatkaite) also crystallizes in the monoclinic structure with the lattice parameters a = 808.06 ± 0.06, b = 2336.0 ± 0.2, c = 2148.8 ± 0.2 pm, and β = 100.709 ± 0.001° (Topa et al. 2013a).

Pb$_6$(As$_5$Sb$_3$)S$_{18}$ (mineral bernarlottiite) crystallizes in the triclinic structure with the lattice parameters a = 2370.4 ± 0.8, b = 838.6 ± 0.2, c = 2350.1 ± 0.8 pm, α = 89.9 ± 0.1°, β = 102.93 ± 0.01°, γ = 89.88 ± 0.01°, and a calculated density of 5.601 g·cm^{-3} (Orlandi et al. 2014, 2017).

Pb$_8$Sb$_{10}$As$_6$S$_{32}$ (mineral hyršlite) crystallizes in the monoclinic structure with the lattice parameters a = 847.5 ± 0.3, b = 791.7 ± 0.3, c = 2003.9 ± 0.8 pm, β = 102.070 ± 0.006°, and a calculated density of 5.26 g·cm^{-3} (Keutsch et al. 2017, 2018).

Pb$_{12}$(As$_{3.2}$Sb$_{2.8}$)S$_{21}$ (mineral arsenmarcobaldiite) crystallizes in the triclinic structure with the lattice parameters a = 897.36 ± 0.09, b = 2933.4 ± 0.3, c = 849.25 ± 0.10 pm, α = 98.369 ± 0.006°, β = 118.705 ± 0.006°, and γ = 90.874 ± 0.006° (Biagioni et al. 2016b, 2019).

Pb$_{14}$(As,Sb)$_6$S$_{23}$ (mineral jordanite) crystallizes in the monoclinic structure with the lattice parameters a = 893.38 ± 0.08, b = 3189.1 ± 0.3, c = 847.20 ± 0.08 pm, β = 117.943 ± 0.001° for Pb$_{14}$As$_{3.78}$Sb$_{2.22}$S$_{23}$ (Sb-bearing jordanite); a = 895.54 ± 0.03, b = 3192.28 ± 0.11, c = 849.37 ± 0.03 pm, β = 117.981 ± 0.001° for Pb$_{14}$As$_{3.20}$Sb$_{2.80}$S$_{23}$ (Sb-reach jordanite); a = 897.20 ± 0.06, b = 3195.35 ± 0.22, c = 848.88 ± 0.06 pm, β = 117.964 ± 0.004° for Pb$_{14}$As$_{3.46}$Sb$_{2.54}$S$_{23}$ (As-bearing geocronite) (Biagioni et al. 2016a) [a = 892, b = 3188, c = 845.7 pm, β = 117°43′, and a calculated density of 6.38 g·cm^{-3} (Jambor 1968); a = 891.8 ± 0.1, b = 3189.9 ± 0.4, c = 846.2 ± 0.1 pm, β = 117.79 ± 0.01° (Ito and Nowacki 1974b)]. **Pb$_{14}$(Sb,As)$_6$S$_{23}$** (mineral geocronite) also crystallizes in the monoclinic structure with the lattice parameters a = 896.3 ± 0.2, b = 3193 ± 2, c = 850.0 ± 0.2 pm, β = 118.02 ± 0.01°, and a calculated density of 6.52 g·cm^{-3} (Birnie and Burnham 1976) [a = 896–900, b = 3185–3194, c = 848–852 pm, β = 118°00′ ± 0.10′, and an experimental density of 6.46

± 0.05 g·cm^{-3} (Douglass et al. 1954); a = 897, b = 3203, c = 847 pm, β = 117°45′, and a calculated density of 6.81 g·cm^{-3} (Jambor 1968)]. The crystal-chemical characterization of the members of the jordanite–geocronite isotypic series allowed its identification ranging from Pb$_{14}$As$_2$Sb$_4$S$_{23}$ (As-bearing geocronite) to Pb$_{14}$As$_4$Sb$_2$S$_{23}$ (Sb-bearing jordanite) (Biagioni et al. 2016a).

Pb$_{18}$(Sb,As)$_{15}$S$_{41}$ (mineral madocite) crystallizes in the orthorhombic structure with the lattice parameters a = 2720 ± 20, b = 3410 ± 20, c = 812 ± 5 pm, and a calculated density of 5.98 g·cm^{-3} (Jambor 1967a; Fleischer et al. 1968). According to the data of Wei et al. (2017), the composition of this mineral is Pb$_{17}$(Sb$_{0.75}$As$_{0.25}$)$_{16}$S$_{41}$. Its decomposition temperature is 395°C and an energy gap has a value 1.32 eV. This phase was prepared by reaction of the PbS, As$_2$S$_3$, and Sb$_2$S$_3$ (molar ratio 8.5:1:3) powders which were loaded into a silica ampoule inside a N$_2$-filled glove box. The ampoule was sealed in a quartz tube, heated to 400°C, and held at this temperature for 4 days. The furnace was subsequently turned off and the product slowly cooled to room temperature. The product was then ground into fine powders, cold pressed into a pellet and annealed at 350°C for 2 weeks to promote homogeneity. The resulting pellet was then ground into fine powders inside a glove box and loaded into a graphite die for hot pressing. Densification was accomplished by hot pressing at 350°C and 160 MPa for 3 h under N$_2$ flow.

Pb$_{19}$(Sb,As)$_{20}$S$_{49}$ (mineral sorbyite) crystallizes in the monoclinic structure with the lattice parameters a = 4490 ± 50, b = 828 ± 1, c = 2640 ± 50 pm, β = 113°25′ ± 30′, and a calculated density of 5.52 g·cm^{-3} (Jambor 1967b; Fleischer et al. 1968; Pierrot 1968).

10.98 Lead–Arsenic–Bismuth–Sulfur

Three quaternary compounds were found in the Pb–As–Bi–S system. **Pb$_{10}$(As,Bi)$_6$S$_{19}$** (mineral kirkiite) crystallizes in the monoclinic structure with the lattice parameters a = 870.0 ± 0.2, b = 2623.7 ± 0.6, c = 877.4 ± 0.3 pm, and β = 119.653 ± 0.004° (Pinto et al. 2006; Piilonen et al. 2007) [a = 862.1 ± 0.4, b = 2603 ± 1, c = 881.0 ± 0.4 pm, β = 119.21 ± 0.01°, and a calculated density of 6.580 g·cm^{-3} (Makovicky et al. 2006a; Locock et al. 2006a); in the pseudohexagonal structure with the lattice parameters a = 869 ± 5, c = 2606 ± 10, and a calculated density of 6.82 g·cm^{-3} but the true symmetry is likely orthorhombic or monoclinic (Moëlo et al. 1985; Hawthorne et al. 1986b)].

Two new compounds were found in the PbS–As$_2$S$_3$–Sb$_2$S$_3$ system by Walia and Chang (1973): **Pb$_{27}$As$_{10.5–9.1}$Bi$_{3.5–4.9}$S$_{48}$**, which is in equilibrium with PbS, Pb$_{14}$As$_6$S$_{23}$, Pb$_2$As$_2$S$_5$, Pb$_3$Bi$_2$S$_6$, Bi$_2$S$_3$, and **Pb$_2$As$_{1.2}$Bi$_{0.8}$S$_5$**. The last compound shows no solid solution range.

10.99 Lead–Arsenic–Selenium–Sulfur

PbS–As$_2$Se$_3$. The phase diagram of this system, constructed through DTA, XRD, metallography, and measuring of the microhardness, is given in Figure 10.42 (Gurshumov et al. 1986a). The eutectic contains 8 mol% PbS and crystallizes

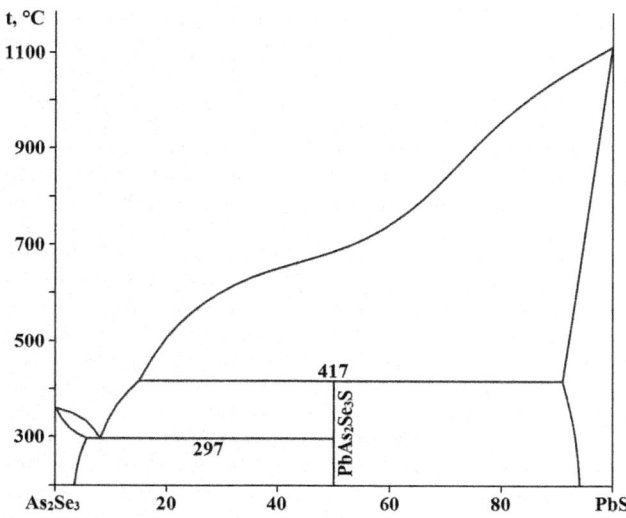

FIGURE 10.42 Phase diagram of the PbS–As$_2$Se$_3$ system. (From Gurshumov, A.P., et al., *Zhurn. neorgan. khimii*, **31**(1), 264, 1986.)

at 297°C. The solubility of PbS in As$_2$Se$_3$ and As$_2$Se$_3$ in PbS at room temperature is 1.5 and 2 mol%, respectively. The **PbAs$_2$Se$_3$S** quaternary compound, which melts incongruently at 417°C, is formed in this system. The ingots for the investigations were annealed at temperatures below the solidus temperature for 450 h.

10.100 Lead–Arsenic–Chlorine–Sulfur

The **Pb$_{28}$As$_{15}$S$_{50}$Cl** [mineral tsugaruite (Miyawaki et al. 2019a,b)] quaternary compound, which crystallizes in the orthorhombic structure with the lattice parameters $a = 807.74 \pm 0.10$, $b = 1517.72 \pm 0.16$, and $c = 3812.9 \pm 0.4$ pm (Biagioni et al. 2021) [$a = 1517.9 \pm 0.1$, $b = 3819.5 \pm 0.1$, $c = 807.45 \pm 0.01$ pm, and a calculated density of 6.83 g·cm^{-3} (Shimizu et al. 1998b)], is formed in the Pb–As–Cl–S system.

10.101 Lead–Arsenic–Iron–Sulfur

PbS–Fe$_2$As. The interaction in this system was investigated through DTA, XRD, metallography, and EPMA (Kopylov and Kaminskiy 1998). In the range of 12–89 mass% PbS, an immiscibility region was found. The initial components chemically interact with the formation of FeS, FeAs, and Pb, the compositions of which go beyond the cross-section into the concentration volume of the Pb–As–Fe–S quaternary system.

10.102 Lead–Antimony–Bismuth–Sulfur

The **PbSbBiS$_4$** quaternary compound, which melts congruently at 597°C, has no homogeneity range along the PbSb$_2$S$_4$–PbBi$_2$S$_4$ section, and crystallizes in the orthorhombic structure with the lattice parameters $a = 1572$, $b = 1136$, $c = 441$ pm, and an energy gap of 0.60 eV, is formed in the Pb–Sb–Bi–S system.

Single crystals of this compound have been prepared by chemical-vapor transport using I$_2$ as a transport agent (Gurbanov 2012b).

PbSbBiS$_4$–Sb$_2$S$_3$. The phase diagram of this system, constructed through DTA, XRD, metallography, and measuring of the microhardness and density, is a eutectic type (Gurbanov 2012c). The eutectic contains 60 mol% Sb$_2$S$_3$ and crystallizes at 352°C. At room temperature, the solubility of Sb$_2$S$_3$ in PbSbBiS$_4$ is 10 mol%, and PbSbBiS$_4$ dissolves up to 5 mol% Sb$_2$S$_3$. The ingots for the investigations were annealed at 300°C, followed by quenching in ice water.

PbSbBiS$_4$–Bi$_2$S$_3$. The phase diagram of this system, constructed through DTA, XRD, metallography, and measuring of the microhardness and density, is a eutectic type (Gurbanov 2012c). The eutectic contains 35 mol% Bi$_2$S$_3$ and crystallizes at 477°C. At room temperature, the solubility of Bi$_2$S$_3$ in PbSbBiS$_4$ is 8 mol% and PbSbBiS$_4$ dissolves up to 5 mol% Bi$_2$S$_3$. The ingots for the investigations were annealed at 370°C followed by quenching in ice water.

PbS–Sb$_2$S$_3$–Bi$_2$S$_3$. To study this system, the evacuated glass capsule technique was used (Chang et al. 1980). Starting compositions were prepared from Pb, Sb, Bi, and S. The samples were heated in a horizontal muffle furnace. Generally, 50 days were used for equilibration at 500°C, and 80 to 90 days at 450°C and below. At the end of the heat treatment, the samples were quenched in air. The system was studied through XRD, metallography, and EPMA. As a result, the isothermal sections at 400°C, 450°C, and 500°C were constructed. Seven phases were synthesized in this system: **2PbS·(Bi$_{0.80}$Sb$_{0.20}$)S$_3$**, **3PbS·Sb$_2$S$_3$·Bi$_2$S$_3$** (decomposes at ~430°C and melts incongruently at 685°C), **PbS·(Sb$_{0.50}$Bi$_{0.50}$)S$_3$** (is stable up to 442°C and crystallizes in the triclinic structure with the lattice parameters $a = 1647$, $b = 1763$, $c = 397$ pm, $\alpha = 96.11°$, $\beta = 96.42°$, and $\gamma = 90.76°$), **3PbS·(Sb$_{0.80}$Bi$_{0.20}$)S$_3$**, **9PbS·11(Bi$_{0.64}$Sb$_{0.36}$)$_2$S$_3$** (decomposes at ~430°C and melts incongruently at 645°C), and high- and low-temperature solid solutions between **2PbS·(Sb$_{0.50}$Bi$_{0.50}$)S$_3$** and **2PbS·(Sb$_{0.76}$Bi$_{0.24}$)S$_3$** (the transformation temperature is between 400°C and 415°C; melting temperature is 725°C for 2PbS·(Sb$_{0.70}$Bi$_{0.30}$)S$_3$ composition).

According to the data of Fleischer (1956b), the **Pb(Bi,Sb)$_6$S$_{10}$** quaternary compound (mineral ustarasite) is also formed in the Pb–Sb–Bi–S system.

10.103 Lead–Antimony–Niobium–Sulfur

The **[Pb,Sb)S]$_{2.28}$NbS$_2$** quaternary compound, which belongs to the layer misfit family and is incommensurable, is formed in the Pb–Sb–Nb–S system. Its structure is built from two [(Pb,Sb)S] layers alternating with one [NbS$_2$] layer along the c-axis. The [(Pb,Sb)S] sublattice crystallizes in the triclinic structure with the lattice parameters $a = 596.4 \pm 0.2$, $b = 582.85 \pm 0.09$, $c = 1764.9 \pm 0.6$ pm, $\alpha = 86.41 \pm 0.02°$, $\beta = 86.55 \pm 0.03°$, and $\gamma = 89.97 \pm 0.02°$ and the [NbS$_2$] sublattice crystallizes in the hexagonal structure with the lattice parameters $a = 333$ and $c = 1764.9 \pm 0.6$ pm (Lafond et al. 1997; Bengel et al. 2000) [$a = 584.9 \pm 0.4$, $b = 582.6 \pm 0.5$, $c = 1761 \pm 3$ pm, $\alpha = 84.9 \pm 0.1°$, $\beta = 96.1 \pm 0.1°$, and $\gamma = 90.1 \pm 0.1°$ for the [(Pb,Sb)S] sublattice and $a = 332.0 \pm 0.3$ and $c = 1761 \pm 3$ pm the [NbS$_2$} sublattice (Lafond et al. 1996)].

TABLE 10.2

Crystallographic Data for $Pb_{23}Sb_{23}S_{60}Cl$ (Mineral Dadsonite)

a, pm	b, pm	c, pm	β, °	$d_{calc.}$ g·cm⁻³	$d_{meas.}$ g·cm⁻³	References
1906	411	1726	95.83	—	—	Coleman (1953)
1905	411	1733	96.33	5.76	—	Jambor (1969); Fleischer (1970b)
1904.1 ± 0.7	822.6 ± 0.3	1372.7 ± 0.7	96.3	5.51	5.68	Cervelle et al. (1979)
1942	406	1519	94.67	6.01	—	Moëlo (1979)
1905.8 ± 0.3	823.4 ± 0.2	1735.2 ± 0.3	96.33 ± 0.02°	—	—	Orlandi et al. (2010)

This phase has been synthesized from mixtures of PbS, Sb_2S_3, NbS_2, and Nb around the hypothetic composition $(Pb_{15}SbS_{16})\cdot7NbS_2$ (Lafond et al. 1996). These mixtures were slowly heated (5°C·h⁻¹) to 800°C in a silica tube sealed under vacuum, held at this temperature for 10 days, and then quenched in cool water. The final products were finely crystallized. In the second step, iodine was used as a transport agent to favor crystallization.

10.104 Lead–Antimony–Oxygen–Sulfur

Two quaternary compounds, $Pb_{14}Sb_{30}O_5S_{54}$ (mineral scainiite) and $Pb_{15-2x}Sb_{14+2x}O_xS_{36}$ ($x \sim 0.2$) (mineral chovanite), are formed in the Pb–Sb–O–S system. First of them crystallizes in the monoclinic structure with the lattice parameters $a = 5199.6 \pm 0.8$, $b = 814.8 \pm 0.1$, $c = 2431.1 \pm 0.4$ pm, and $\beta = 104.09 \pm 0.01°$ (Moëlo et al. 2000) [$a = 5201 \pm 3$, $b = 813 \pm 1$, $c = 2434 \pm 2$ pm, $\beta = 104.03 \pm 0.07°$, and a calculated density of 5.56 g·cm⁻³ (Orlandi et al. 1999; Jambor et al. 2000b)].

$Pb_{15-2x}Sb_{14+2x}O_xS_{36}$ ($x \sim 0.2$) also crystallizes in the monoclinic structure with the lattice parameters $a = 4829.3 \pm 1.5$, $b = 411.07 \pm 0.13$, $c = 3422.3 \pm 1.1$ pm, $\beta = 106.168 \pm 0.005°$, and a calculated density of 5.91 g·cm⁻³ (Makovicky and Topa 2009) [$a = 4818.9 \pm 4.8$, $b = 411.04 \pm 0.40$, $c = 3423.5 \pm 3.5$ pm, $\beta = 106.059 \pm 0.015°$, and a calculated density of 7.14 g·cm⁻³ (Topa et al. 2012; Cámara et al. 2015); $a = 4838 \pm 5$, $b = 411 \pm 4$, $c = 3418 \pm 4$ pm, and $\beta = 106.26 \pm 0.02°$ (the mineral may contain minor thallium) (Biagioni and Moëlo 2017)].

10.105 Lead–Antimony–Chlorine–Sulfur

Some quaternary compounds are formed in the Pb–Sb–Cl–S. $Pb_{3+x}Sb_{3-x}S_{7-x}Cl_{1+x}$ ($x \sim 0.45$) crystallizes in the orthorhombic structure with the lattice parameters $a = 1519.4 \pm 0.3$, $b = 2303.5 \pm 0.5$, $c = 405.91 \pm 0.08$ pm, and a calculated density of 6.016 g·cm⁻³ (Doussier et al. 2008). This compound was synthesized by solid-state reaction of stoichiometric amounts of PbS, $PbCl_2$, and Sb_2S_3 (molar ratio 5:1:3). The starting materials were homogenized by grinding, put into a sealed evacuated silica tube, and heated at 500°C for 10 days. Some individual needles could be extracted from the massive powder.

$Pb_{4.32}Sb_{3.68}S_{8.68}Cl_{2.32}$ also crystallizes in the orthorhombic structure with the lattice parameters $a = 1504 \pm 4$, $b = 1551$

± 3, $c = 409 \pm 1$ pm, and a calculated density of 5.83 g·cm⁻³ (Kostov-Kytin et al. 1997). It was prepared at 500°C by gas-transport reaction.

$Pb_{12.65}Sb_{11.35}S_{28.35}Cl_{2.65}$ crystallizes in the monoclinic structure with the lattice parameters $a = 1951.2 \pm 1.2$, $b = 405.3 \pm 0.3$, $c = 3513.4 \pm 1.7$ pm, $\beta = 96.34 \pm 0.05°$, and a calculated density of 6.032 g·cm⁻³ (Kostov and Macíček 1995). The crystals of this compound were prepared at 420°C by dry synthesis.

$Pb_{19}Sb_{13}S_{35}Cl_7$ (mineral ardaite) also crystallizes in the monoclinic structure with the lattice parameters $a = 2209$, $b = 2111$, $c = 805$ pm, and $\beta = 103°01'$ (Breskovska et al. 1981, 1982; Dunn and Fleischer 1983).

$Pb_{23}Sb_{23}S_{60}Cl$ (mineral dadsonite) also crystallizes in the monoclinic structure with the lattice parameters given in Table 10.2. There is evidence in the literature that the mineral dadsonite crystallizes in the triclinic structure with the lattice parameters $a = 827.6 \pm 0.2$, $b = 1739.2 \pm 0.4$, $c = 1950.5 \pm 0.4$ pm, $\alpha = 83.527 \pm 0.007°$, $\beta = 77.882 \pm 0.008°$, $\gamma = 89.125 \pm 0.008°$, and a calculated density of 5.88 g·cm⁻³ (Makovicky et al. 2006b) [$a = 1733$, $b = 411$, $c = 1905$ pm, $\alpha = 90.0°$, $\beta = 96.3°$, $\gamma = 90.4°$ for $Pb_{10+x}Sb_{14-x}S_{31-x}Cl_x$ composition Makovicky et al. 1984]]. $Pb_{23}Sb_{23}S_{60}Cl$ can be synthesized hydrothermally at 200°C (Moëlo 1979) or grown by chemical-vapor transport using NH_4Cl as a transport agent (Kaden et al. 2012).

During the crystallization of mixtures PbS and Sb_2S_3 in $LiCl+NH_4Cl$ melt or in aqueous solutions of chlorides in autoclaves at 250–400°C and different concentrations of chlorine in the solutions, three more sulfosalts were synthesized: $Pb_2Sb_2S_{4.76}Cl_{0.31}$, $Pb_7Sb_8S_{16}Cl_{3.4}$, and $Pb_9Sb_8S_{19}Cl_{4.5}$ (Bortnikov et al. 1979).

The existence of $Pb_{17}Sb_{18}S_{43}Cl_2$ was determined by Moëlo et al. (1989) as mineral species, and two more quaternary compounds, $Pb_{16}Sb_{15}S_{35}Cl_7$ and $Pb_{19}Sb_{18}S_{42}Cl_8$, can be synthesized hydrothermally at 200°C (Moëlo 1979).

10.106 Lead–Antimony–Iodine–Sulfur

Three quaternary compounds, $Pb_2SbS_2I_3$, $Pb_{9.6}Sb_{7.9}S_{23}I_{1.55}$, and $Pb_{11}Sb_4S_{11}I_{12}$, are formed in the Pb–Sb–I–S system. The first melts congruently at 410°C, has a noticeable area of homogeneity along the PbI_2–$PbSb_2S_4$ section, and crystallizes in the orthorhombic structure with the lattice parameters $a = 433.4$, $b = 1419.5$, $c = 1655.2$ pm, and an energy gap of 2.0 eV (Dolgikh

1985a,b; Gorak and Starosta 1991). The melting enthalpy of this compound is 55.2 kJ·M⁻¹ (Gorak and Starosta 1991). $Pb_2SbS_2I_3$ was synthesized by heating in an evacuated quartz ampoule, a stoichiometric mixture of the elements, using a rocking furnace (Gorak and Starosta 1991). The mixture was heated to 650°C, kept at this temperature for 24–36 h, and cooled by switching off the furnace. Its single crystals were grown by recrystallization through the vapor phase (Dolgikh 1985b). This compound is stable in air.

The crystals of $Pb_{9.6}Sb_{7.9}S_{23}I_{1.55}$ were grown by chemical-vapor transport using I_2 as a transport agent (Kaden et al. 2012), and $Pb_{11}Sb_4S_{11}I_{12}$ was found in the system by Dolgikh (1985a).

10.107 Lead–Antimony–Manganese–Sulfur

$PbS–Sb_2S_3–MnS$. The phase relations in this quasiternary system were studied by Chang et al. (1987). The isothermal section at 500°C was constructed and it was shown that $Pb_4MnSb_6S_{14}$ has a range of solid solution elongated toward Sb_2S_3, and forms equilibrium assemblages with all phases on the related joins except PbS. To study this system, starting compositions were prepared from Pb, Sb, Mn, and S. Synthesis and heat treatment were made in muffle furnace using the conventional technique of sealed, evacuated glass ampoules. Generally, the duration of treatment ranged from 120 days at 400°C to 45 days at 500°C and to 5 days at and above 600°C.

The $Pb_4MnSb_6S_{14}$ quaternary compound, which forms in this system, melts congruently at 592°C (Chang et al. 1987) and crystallizes in the monoclinic structure with the lattice parameters $a = 402.16 \pm 0.08$, $b = 1917.8 \pm 0.4$, $c = 1583.7 \pm 0.3$ pm, $\beta = 91.89 \pm 0.03°$, and a calculated density of 5.612 g·cm⁻³ (Léone et al. 2003) [$a = 1574$, $b = 1914$, $c = 404$ pm, and $\beta = 91°20'$ (Chang et al. 1987)]. This compound was synthetized as follows (Léone et al. 2003). A mixture of PbS, Sb_2S_3, and MnS (molar ratio 4:3:1) was first loaded in a quartz tube sealed under

vacuum and heated at 600°C for 12 days. $Pb_4MnSb_6S_{14}$ was relatively homogeneous with small inclusions of $Pb_9Sb_{22}S_{44}$ and Sb_2S_3.

10.108 Lead–Antimony–Iron–Sulfur

$PbS–Sb_2S_3–FeS$. The phase relations in this quasiternary system were studied by Chang and Knowles (1977) and Chang et al. (1987). The isothermal section at 500°C was constructed, and it was shown that $Pb_4FeSb_6S_{14}$ has a range of solid solution elongated toward Sb_2S_3, and forms equilibrium assemblages with all phases on the related joins, except PbS. The extent of solid solution is very sensitive to temperature change. At 400°C, it is limited to within 1–2 mol% around the ideal composition. To study this system, starting compositions were prepared from Pb, Sb, Fe, and S. Synthesis and heat treatment were made in a muffle furnace using the conventional technique of sealed, evacuated glass ampoules. Generally, the duration of treatment ranged from 120 days at 400°C to 45 days at 500°C and to 5 days at and above 600°C.

The $Pb_4MnSb_6S_{14}$ quaternary compound (mineral jamesonite), which forms in this system, melts congruently at 592°C (Matsushita and Ueda 2003) [at 583°C (Chang et al. 1987); at 545°C (Chang and Knowles 1977)] and crystallizes in the monoclinic structure with the lattice parameters given in Table 10.3. This compound was synthesized under dry conditions (Matsushita and Ueda 2003; Derakhshan et al. 2005). A stoichiometric amount of Fe, Pb, Sb, and S was roughly mixed without any further treatments before use. The mixture was transferred to a quartz tube and was subsequently flame-sealed under vacuum (~10⁻³ Pa). The reaction was performed by the following two different heating methods (Matsushita and Ueda 2003).

Method *A* is a conventional heating method using a furnace. The tube was slowly heated to 500°C over a 12-h period and

TABLE 10.3

Crystallographic Data and an Energy Gap for $Pb_4MnSb_6S_{14}$ (Mineral Jamesonite)

a, pm	b, pm	c, pm	β, °	E_g, eV	$d_{calc.}$	$d_{meas.}$	References
					g·cm⁻³		
1568 ± 5	1901 ± 5	403 ± 1	91.8 ± 0.5	—	5.67	5.63	Berry (1940)
1557	1898	403	91.8	—	—	—	Niizeki and Buerger (1957)
1575	1917	398	91.37[a]	—	—	—	Chang and Knowles (1977); Chang et al. (1987)
1578	1892	402	92.73[b]				
1627	1873	399	92.77[c]				
402.35 ± 0.08	1907.4 ± 0.4	1573.7 ± 0.3	91.89 ± 0.03	—	5.679	—	Léone et al. (2003)
1575.0 ± 0.6	1912.5 ± 0.3	403.0 ± 0.4	91.68 ± 0.08	0.48	5.651	—	Matsushita and Ueda (2003)
403.4	1908.1	1571.0	91.80[d]	—	—	—	Léone et al. (2004)
401.1	1908.1	1570.4	91.70[e]				
—	—	—	—	0.12	—	—	Derakhshan et al. (2005)

Note:

[a] for the $Pb_{8.04}Fe_{1.96}Sb_{12.13}S_{28}$ composition

[b] for the $Pb_{7.00}Fe_{2.33}Sb_{12.44}S_{28}$ composition

[c] for the $Pb5_{8.74}Fe_{2.15}Sb_{12.92}S_{28}$ composition

[d] at room temperature

[e] at 44 K

then annealed at this temperature for 3 days, followed by cooling to room temperature over 12 h. This primary product was reground up and again heated in the vacuum-sealed tube using a 550°C furnace with a slight temperature gradient for an extra 3 days to make a pure compound. In this process, the temperature at the sample position in the furnace was 550°C, and the end part was ~500°C. The target compound of shiny metallic gray bulk together with needle-shaped crystals was obtained at the initial sample position.

Method *B* is direct heating by flame. The slight Pb–Sb–S-rich starting material put in an evacuated quartz tube was moderately heated by a direct flame and melted completely. The melt was stirred by shaking gently for good mixing and then quenched to room temperature. This is a good method to synthesize the metallic gray bulk of this compound very quickly.

10.109 Lead–Antimony–Nickel–Sulfur

PbS–NiSb. The phase diagram of this system, constructed through DTA, XRD, metallography, and measuring of the microhardness, is a eutectic type (Figure 10.43) (Zargarova et al. 1975b). The eutectic contains 46 mol% PbS and crystallizes at 910°C. The solubility of NiSb in PbS reaches 2–3 mol%, and PbS dissolves 1 mol% NiSb.

The **Ni$_9$PbSbS$_8$** quaternary compound (mineral hrabákite), which crystallizes in the tetragonal structure with the lattice parameters $a = 730.85 \pm 0.04$ and $c = 539.69 \pm 0.03$, is formed in the Pb–Sb–Ni–S system (Sejkora et al. 2020a, b).

10.110 Lead–Bismuth–Selenium–Sulfur

PbS + Bi$_2$Se$_3$ ↔ PbSe + Bi$_2$S$_3$. Phase relations in this system were examined in the temperature range of 500–900°C by Liu and Chang (1994a), and the isothermal section of this system at 500°C is given in Figure 10.44. A complete series of solid solutions exists between Pb$_9$Bi$_4$S$_{15}$ and Pb$_9$Bi$_4$Se$_{15}$, whereas along

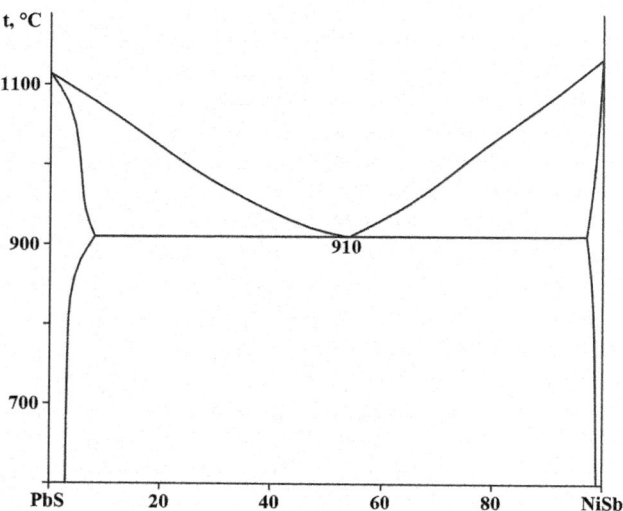

FIGURE 10.43 Phase diagram of the PbS–NiSb system. (From Zargarova, M.I., et al., *Izv. AN SSSR. Neorgan. mater.*, **11**(1), 165 (1975.)

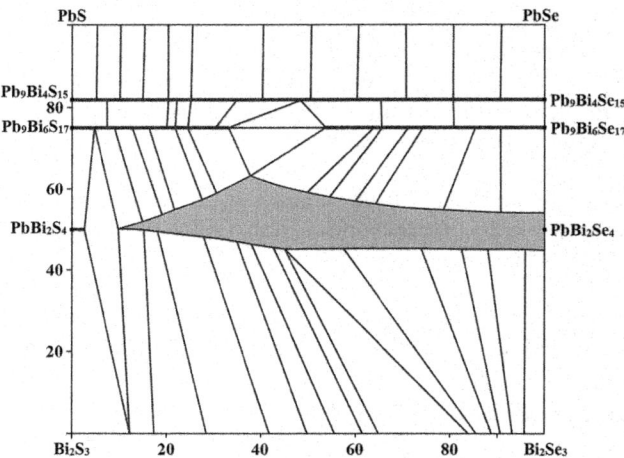

FIGURE 10.44 Isothermal section of the PbS + Bi$_2$Se$_3$ ↔ PbSe + Bi$_2$S$_3$ ternary mutual system at 500°C. (From Liu, H., and Chang, L.L.Y., *Amer. Mineralog.*, **79**(11–12), 1159, 1994.)

the join from Pb$_8$Bi$_6$S$_{17}$ to Pb$_8$Bi$_6$Se$_{17}$ there are two terminally solid solutions. The join PbBi$_2$S$_4$–PbBi$_2$Se$_4$ is characterized by an extensive solid solution based on PbBi$_2$S$_4$. The duration of the alloy treatment ranged from 60 days at 500°C to 1 day at 900°C. At the end of the heat treatment, the samples were quenched to room temperature by compressed air to preserve phase assemblages synthesized at high temperatures.

Some quaternary compounds are formed in the Pb–Bi–Se–S system. **PbBi$_4$(S,Se)$_7$** (mineral mozgovaite, new formula PbBi$_4$S$_7$) crystallizes in the orthorhombic structure with the lattice parameters $a = 1318 \pm 6$, $b = 3740 \pm 20$, $c = 405 \pm 3$ pm, and a calculated density of 6.26 ± 0.06 g·cm^{-3} (Vurro et al. 1999; Jambor et al. 2000a).

Pb$_2$Bi$_2$(S,Se)$_3$ (mineral babkinite) crystallizes in the trigonal structure with the lattice parameters $a = 419.1 \pm 0.2$, $c = 3960 \pm 3$ pm, and a calculated density of 8.096 g·cm^{-3} (Bryzgalov et al. 1996; Jambor et al. 1996a).

The structure of **Pb$_3$Bi$_4$(S,Se)$_9$** (mineral wittite) consists of pseudotetragonal (*Q* type) and pseudohexagonal (*H* type) layers: *H*-subcell is monoclinic with the lattice parameters $a = 1553 \pm 3$, $b = 407 \pm 1$, $c = 718 \pm 1$ pm, $\beta = 97.83 \pm 0.01°$, and *Q*-subcell is also monoclinic with the lattice parameters $a = 1570 \pm 4$, $b = 408 \pm 0$, $c = 414 \pm 1$ pm, $\beta = 101.34 \pm 0.01°$ (Borodaev et al. 2000) [$a = 3196 \pm 1$, $b = 412 \pm 1$, $c = 3669 \pm 3$ pm, $\beta = 109.52 \pm 0.03°$, and a calculated density of 7.08 g·cm^{-3} (Large and Mumme 1975); $a = 1803.4 \pm 0.3$, $b = 404.0 \pm 0.2$, $c = 1753.0 \pm 0.3$ pm, $\beta = 94°31' \pm 5'$, and a calculated density of 7.21 g·cm^{-3} (Johan and Picot 1976; Fleischer and Jambor 1977); $a = 721 \pm 2$, $b = 408 \pm 1$, $c = 1550 \pm 5$ pm, $\beta = 98.75 \pm 0.15°$ for *H*-subcell and $a = 419 \pm 1$, $b = 408 \pm 1$, $c = 1556 \pm 4$ pm, $\beta = 101.35 \pm 0.16°$ for *Q*-subcell (Mumme 1980a)]

Pb$_6$Bi$_8$(S,Se)$_{18}$ (mineral weibullite, may be a polymorph of wittite) crystallizes in the orthorhombic structure with the lattice parameters $a = 5368 \pm 9$, $b = 411 \pm 1$, and $c = 1540 \pm 3$ pm (Mumme 1980a,b) [$a = 1539 \pm 2$, $b = 406.8 \pm 0.4$, $c = 5380 \pm 50$ pm, and a calculated density of 7.08 g·cm^{-3} (Johan and Picot 1976; Fleischer and Jambor 1977); a calculated density of 6.96 g·cm^{-3} (Fleischer 1971)].

$Pb_4Bi_{12}Se_{6.2}S_{15.8}$ crystallizes in the orthorhombic structure with the lattice parameters $a = 408.424 \pm 0.010$, $b = 1342.73 \pm 0.03$, $c = 3747.61 \pm 0.08$ pm, and a calculated density of 7.02 g·cm^{-3} (Mumme et al. 2017). This compound was prepared through the hydrothermal synthesis in gold tubes at 535°C and 0.15 GPa using a started composition near to $Ag_{0.32}Pb_{5.09}Bi_{8.55}Se_{6.08}S_{11.92}$.

$Pb_6Bi_2(S_{8.53}Se_{0.44})$ (selenium heyrovskýite) also crystallizes in the orthorhombic structure with the lattice parameters $a = 1373.4 \pm 0.7$, $b = 3128 \pm 3$, and $c = 413.5 \pm 0.3$ pm (Borodaev et al. 2003).

10.111 Lead–Bismuth–Tellurium–Sulfur

$PbS + Bi_2Te_3 \leftrightarrow PbTe + Bi_2S_3$. Phase relations in this system were examined in the temperature range of 500–900°C by Liu and Chang (1994a) and the isothermal section of this system at 500°C is presented in Figure 10.45. Four phases are formed along the $PbS–Bi_2Te_2S$ section which crystallizes in the trigonal structure with the lattice parameters $a = 423$, $c = 1671$ pm for $Pb_2Bi_2Te_2S_3$, $a = 423$, $c = 6000$ pm for $Pb_3Bi_4Te_4S_5$, $a = 423$, $c = 3983$ pm for $PbBi_2Te_2S_2$, and $a = 424$, $c = 2312$ pm for $PbBi_4Te_4S_3$. Both phase A ($PbBi_2Te_4$) and phase M ($PbBi_4Te_7$) of the binary join $PbTe–Bi_2Te_3$ have ranges of solid solution by sulfur substitution. The A series extends from $PbBi_2Te_4$ to $PbBi_2Te_{3.65}S_{0.35}$ and the M series from $PbBi_4Te_7$ to $PbBi_4Te_{6.475}S_{0.525}$. In addition to tetradymite along the $Bi_2S_3–Bi_2Te_3$ join, five lead bismuth sulfide telluride phases are stable in the system. They are designated as phase K with compositions between $Pb_2Bi_2Te_{4.5}S_{0.5}$ and $Pb_2Bi_2Te_{4.7}S_{0.3}$, phase D with compositions between $PbBi_4Te_{3.94}S_{3.06}$ and $PbBi_4Te_{3.68}S_{3.32}$, phase E with compositions between $PbBi_2Te_{1.8}S_{2.2}$ and $PbBi_2Te_2S_2$, phase F with compositions between $Pb_3Bi_4Te_4S_5$ and $Pb_3Bi_4Te_{3.6}S_{5.4}$, and phase J with compositions between $Pb_2Bi_2Te_{1.8}S_{3.2}$ and $Pb_2Bi_2Te_2S_3$. The duration of the treatment ranged from 60 days at 500°C to one day at 900°C. At the end of the heat treatment, the samples were quenched to room temperature by compressed air to preserve phase assemblages synthesized at high temperatures.

According to the data of Datsenko et al. (1981), the phase diagram of the $PbS–Bi_2Te_3$ system, constructed through DTA and XRD, is a eutectic type with peritectic transformation. The eutectic crystallizes at 562°C. The $PbBi_2Te_3S$ quaternary compound, which forms in the system, melts incongruently at 575°C.

Some quaternary compounds of the Pb–Bi–Te–S system are found in the form of minerals. $PbBi_2Te_2S_2$ (mineral aleksite) crystallizes in the trigonal structure with the lattice parameters $a = 424$ and $c = 7964$ pm (Cook et al. 2007) [$a = 423.8 \pm 0.1$, $c = 7976 \pm 2$ pm and a calculated density of 7.80 g·cm^{-3} (Lipovetskiy et al. 1978, 1979; Fleischer et al. 1979b); $a = 424.23 \pm 0.25$ and $c = 7973 \pm 5$ pm (Bayliss 1991)].

$PbBi_4Te_4S_3$ (unnamed mineral species) also crystallizes in the trigonal structure with the lattice parameters $a = 425$ and $c = 6971$ pm (Cook et al. 2007).

$PbBi_5Te_2S_3$ (mineral plumbian baksanite) also crystallizes in the trigonal structure with the lattice parameters $a = 425.1 \pm 0.1$, $c = 6419 \pm 3$ pm, and a calculated density of 7.51 g·cm^{-3} (Bindi and Cipriani 2003).

$Pb_5Bi_2Te_2S_6$ (mineral hitachiite) also crystallizes in the trigonal structure with the lattice parameters $a = 422.00 \pm 0.13$, $c = 2702 \pm 4$ pm, and a calculated density of 7.54 g·cm^{-3} (Kuribayashi 2018a,b, 2019).

$Pb_2Bi_2Te_2S_3$ (mineral saddlebackite) crystallizes in the hexagonal structure with the lattice parameters $a = 423.0 \pm 0.4$, $c = 3343 \pm 2$ pm, and a calculated density of 7.61 g·cm^{-3} (Clark 1997; Jambor et al. 1998a).

10.112 Lead–Bismuth–Iron–Sulfur

$PbS–Bi_2S_3–FeS$. The phase relations in this quasiternary system were studied by Chang and Knowles (1977), and the isothermal section at 500°C was constructed. It was shown that this system is divided by the $FeS–Pb_6Bi_2S_9$, $FeS–Pb_3Bi_2S_6$, and $FeS–PbBi_2S_4$ into four subsystems. The experiments made at 450°C and 400°C produced phase relations no different from those obtained at 500°C. To study this system, starting compositions were prepared from Pb, Sb, Fe, and S. Synthesis and heat treatment were made in a muffle furnace using the conventional technique of sealed, evacuated glass ampoules. Generally, the duration of treatment ranged from 120 days at 400°C to 45 days at 500°C and to 5 days at and above 600°C.

10.113 Lead–Bismuth–Nickel–Sulfur

The $(Bi,Pb)_2Ni_3S_2$ quaternary compound, which apparently has two polymorphic modifications, is formed in the Pb–Bi–Ni–S system. The first crystallizes in the monoclinic structure with the lattice parameters $a = 1106.6 \pm 0.1$, $b = 808.5 \pm 0.1$, $c = 796.5 \pm 0.1$ pm, $\beta = 134.0°$, and the calculated and experimental densities of 8.53 and 8.50 g·cm^{-3}, respectively (Brower et al. 1974). It was prepared by heating the mixture of the elements in an evacuated silica tube. Heat treatment was varied

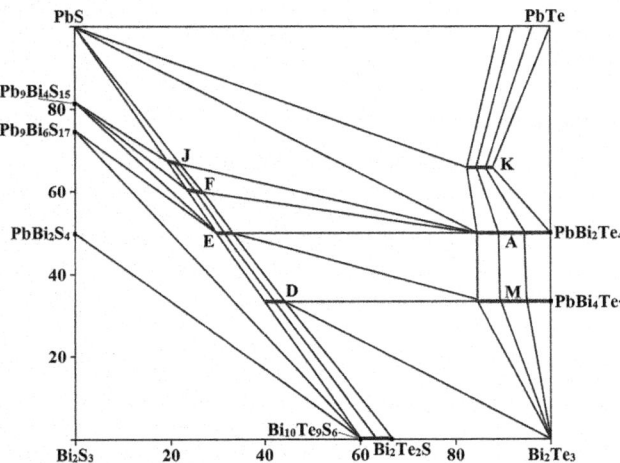

FIGURE 10.45 Isothermal section of the $PbS + Bi_2Te_3 \leftrightarrow PbTe + Bi_2S_3$ ternary mutual system at 500°C. (From Liu, H., and Chang, L.L.Y., *Amer. Mineralog.*, **79**(11–12), 1159, 1994.)

from 500–600°C to 1100–1100°. In general, an attempt was made to obtain complete melting of the mixture, which was then heat treated in a rocking furnace in order to ensure chemical homogeneity.

The second modification of this compound (mineral parkerite) crystallizes in the orthorhombic structure with the lattice parameters $a = 554.5 \pm 0.4$, $b = 573.1 \pm 0.3$, and $c = 405.2 \pm 0.3$ pm (Fleet 1973) [$a = 402$, $b = 552$, $c = 572$ pm, and the calculated and experimental densities of 8.50 and 8.40–8.44 g·cm⁻³, respectively (Michener and Peacock 1943)].

10.114 Lead–Niobium–Tantalum–Sulfur

The **Pb(Nb₁₋ₓTaₓ)S₃** solid solution ($0 \leq x \leq 1$) is formed in the Pb–Nb–Ta–S system (Schmidt et al. 1970). Little changes in the tetragonal lattice dimensions have been observed; a remains almost constant (407 pm) while c increases from 1191 to 1197 pm with increasing x. No indication of superstructures was found throughout the system at any preparation temperature.

10.115 Lead–Niobium–Iron–Sulfur

The **(Pb₂FeS₃)₀.₅₈NbS₂** quaternary phase is formed in the Pb–Nb–Fe–S system. It crystallizes in the orthorhombic structure with the lattice parameters $a = 576.32 \pm 0.07$, $b = 579.50 \pm 0.09$, and $c = 1408.05 \pm 0.12$ pm for Pb₂FeS₃ sublattice and $a = 332.79 \pm 0.01$, $b = 579.50 \pm 0.09$, and $c = 1408.05 \pm 0.12$ pm for NbS₂ sublattice (Lafond et al. 1999) [$a = 577.3 \pm 0.1$, $b = 580.3 \pm 0.1$, and $c = 1410.6 \pm 0.2$ pm for Pb₂FeS₃ sublattice and $a = 334.61 \pm 0.07$, $b = 580.3 \pm 0.1$, and $c = 1410.6 \pm 0.2$ pm for NbS₂ sublattice (Moëlo et al. 1997).

The structure of this phase can be described on the basis of a regular alternation of [Pb₂FeS₃] and [NbS₂] slabs stacked along the c direction (Lafond et al. 1999). The [Pb₂FeS₃] part has a thickness of three layers, the central |FeS| layer being sandwiched between the two external |PbS| layers. It was prepared from a mixture of Pb, S, FeS, and NbS₂ at 800°C over 10 days.

10.116 Lead–Oxygen–Selenium–Sulfur

The **Pb₂(SeO₄)(SO₄)** quaternary compound (mineral olsacherite), which crystallizes in the orthorhombic structure with the lattice parameters $a = 842 \pm 1$, $b = 1096 \pm 1$, $c = 700 \pm 1$ pm, and a calculated density of 6.55 g·cm⁻³, is formed in the Pb–O–Se–S system (Hurlbut and Aristarain 1969).

10.117 Lead–Oxygen–Tellurium–Sulfur

Three quaternary compounds, **Pb₂(TeO₃)(SO₄)**, **Pb₆(TeO₃)₅(SO₃S)**, and **Pb₁₂(TeO₃)₁₁(SO₄)**, are formed in the Pb–O–Te–S system. The first of them (mineral adanite) crystallizes in the monoclinic structure with the lattice parameters $a = 736.480 \pm 0.010$, $b = 1072.96 \pm 0.02$, $c = 932.68 \pm 0.02$ pm, $\beta = 111.463 \pm 0.001°$, and a calculated density of 6.643 g·cm⁻³

(Weil and Shirkhanlou 2017) [$a = 738.30 \pm 0.03$, $b = 1075.45 \pm 0.05$, $c = 935.17 \pm 0.07$ pm, $\beta = 111.500 \pm 0.008°$, and a calculated density of 6.596 g·cm⁻³ (Kampf 2020b,c,d)]. It was prepared through hydrothermal synthesis, which was conducted in sealed Teflon containers (capacity ca. 10 mL) in a steel autoclave at autogenous pressure (Weil and Shirkhanlou 2017). Pb(CH₃COO)₂ (0.319 g) was mixed in stoichiometric amounts with 94 mass% H₂SO₄ (0.15 mL), and TeO₂ (0.078 g) (molar ratio 2:2:1). The mixture was placed in the Teflon container, filled up with water to about two-thirds of the container volume and sealed with a Teflon lid. The sealed container was placed in the autoclave and heated at 210°C for 1 week. Afterwards, the autoclave was cooled to room temperature overnight. The solid products were filtered off, and washed subsequently with mother liquor, water, and ethanol. The reaction batch consisted of phase mixtures when inspected optically under a polarizing microscope.

Pb₆(TeO₃)₅(SO₃S) crystallizes in the hexagonal structure with the lattice parameters $a = 1024.95 \pm 0.05$, $c = 1166.77 \pm 0.08$ pm, and a calculated density of 6.987 g·cm⁻³ (Kampf et al. 2020a) [$a = 1025.3 \pm 0.1$ and $c = 1167.47 \pm 0.08$ pm (Kampf et al. 2019a, b)]

Pb₁₂(TeO₃)₁₁(SO₄) [mineral fairbankite (Miyawaki et al. 2020a, b)], crystallizes in the triclinic structure with the lattice parameters $a = 781$, $b = 711$, $c = 696$ pm, $\alpha = 117°12'$, $\beta = 93°47'$, $\gamma = 93°24'$, and a calculated density of 7.45 g·cm⁻³ (Williams 1979; Fleischer et al. 1980).

10.118 Lead–Oxygen–Chromium–Sulfur

PbSO₄–PbCrO₄. XRD studies have shown that solid solution in this system is incomplete at 25°C with an immiscibility gap bounded by Pb(CrO₄)₀.₈(SO₄)₀.₂ and Pb(CrO₄)₀.₁(SO₄)₀.₉ (Jaeger and Germs 1921; Crane et al. 2001). All PbCrO₄-rich compositions are metastable with respect to the monoclinic polymorph. The extent of solid solution of PbSO₄ in monoclinic PbCrO₄ at 25°C is Pb(CrO₄)₀.₆(SO₄)₀.₄. The transformation of orthorhombic and Pb(CrO₄)ₓ(SO₄)₁₋ₓ to the monoclinic phase is slow in the solid state, as compared to when solid samples are allowed to remain in contact with the solution from which they crystallized.

10.119 Lead–Oxygen–Molybdenum–Sulfur

PbSO₄–PbMoO₄. The phase diagram of this system, constructed through DTA, is a eutectic type (Jaeger and Germs 1921). The eutectic contains 43 mol% PbSO₄ and crystallizes at 962°C.

10.120 Lead–Oxygen–Tungsten–Sulfur

PbSO₄–PbWO₄. The phase diagram of this system, constructed through DTA, is a eutectic type (Jaeger and Germs 1921). The eutectic contains 49 mol% PbSO₄ and crystallizes at 995°C.

10.121 Lead–Oxygen–Fluorine–Sulfur

Two quaternary compounds, **Pb₂(SO₄)F₂** and **Pb₆(SO₄)F₁₀**, are formed in the Pb–O–F–S system. First of them (mineral grandreefite) crystallizes in the monoclinic structure with the lattice parameters $a = 866.7 \pm 0.1$, $b = 444.19 \pm 0.06$, $c = 1424.2 \pm 0.2$ pm, $\beta = 107.418 \pm 0.002°$, and an experimental density of 7.0 ± 0.1 g·cm⁻³ (Kampf 1991; Kampf and Foord 1996) [in the orthorhombic structure with the lattice parameters $a = 443.9 \pm 0.4$, $b = 1357.5 \pm 1.3$, and $c = 433.3 \pm 0.4$ pm, and the calculated and experimental densities of 7.15 and 7.0 ± 0.1 g·cm⁻³, respectively (Kampf et al. 1989)]. This compound decomposes in cold water.

Pb₆(SO₄)F₁₀ (mineral pseudograndreefite) crystallizes in the orthorhombic structure with the lattice parameters $a = 851.82 \pm 0.05$, $b = 1957.36 \pm 0.11$, and $c = 849.26 \pm 0.05$ pm, and the calculated and experimental densities of 7.08 and 7.0 ± 0.1 g·cm⁻³, respectively (Kampf et al. 1989; Kampf and Foord 1996)]. This compound also decomposes in cold water.

10.122 Lead–Oxygen–Chlorine–Sulfur

Two quaternary compounds, **Pb₄O₃(Cl,SO₄)₂** and **Pb₁₀(SO₄)O₈Cl₂**, are formed in the Pb–O–Cl–S system. The first is a mineral species and crystallizes in the monoclinic structure with the lattice parameters $a = 394.5 \pm 0.3$, $b = 393.3 \pm 0.5$, $c = 1311.9 \pm 0.5$ pm, and $\beta = 91.11 \pm 0.04°$ (Jambor and Puziewicz 1991).

Pb₁₀(SO₄)O₈Cl₂ (mineral sundiusite) also crystallizes in the monoclinic structure with the lattice parameters $a = 2467 \pm 1$, $b = 378.1 \pm 0.1$, $c = 1188.1 \pm 0.5$ pm, $\beta = 100.07 \pm 0.04°$, and the calculated and experimental densities of 7.20 and 7.0 ± 0.2 g·cm⁻³, respectively (Dunn and Rouse 1980).

10.123 Lead–Oxygen–Iron–Sulfur

The **(Fe,Pb)₄S₈O** quaternary compound (mineral viaeneite), which crystallizes in the monoclinic structure with the lattice parameters $a = 971.7 \pm 0.8$, $b = 728.0 \pm 0.6$, $c = 655.9 \pm 0.7$ pm, $\beta = 95.00 \pm 0.03°$, and the calculated and experimental densities of 3.65 and 3.8 ± 0.1 g·cm⁻³, respectively, is formed in the Pb–O–Fe–S system (Jambor et al. 1996b,c; Kucha et al. 1996).

10.124 Lead–Selenium–Tellurium–Sulfur

PbS–PbSe–PbTe. Phase relations in this quasiternary system were investigated in the temperature range between 300°C and 1150°C (Liu and Chang 1994b). Miscibility gaps from PbS–PbSe and PbSe–PbTe systems extend into the ternary system, and the range of the gap between PbS and PbTe reduces in size with temperature as well as Se content. At 300°C, the limits of two terminal solid solutions are at 18 and 68 mol% PbTe, and at 500°C, the gap disappears at the composition of 65 mol% PbSe. When the temperature increases from 500°C to 600°C, the amount of PbSe required eliminating the gap is reduced to 26 mol%. At 700°C, the miscibility gap becomes distinctly asymmetric. With 10 mol% PbSe, the ranges of the solid solution in the PbS-rich end are 17 mol% and in the PbTe-rich end 35 mol%. Beyond 13 mol% PbSe the gap disappears, and at 800°C, the gap is limited to a narrow region between 73 and 52 mol% PbS along the PbS–PbTe join and within a maximum of 3 mol% PbSe. A complete series of solid solution occupies the entire system between 815°C and 870°C. Starting compositions for the investigations were prepared from Pb, S, Se, and Te. The

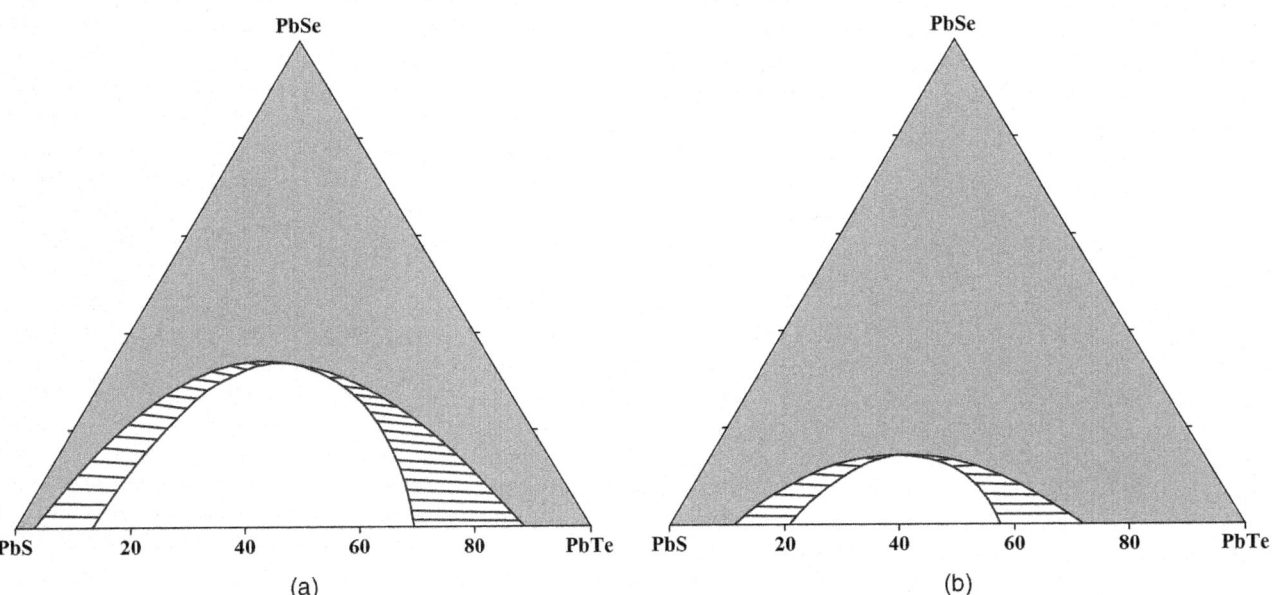

FIGURE 10.46 Isothermal section of the PbS–PbSe–PbTe quasiternary system at (a) 550°C and (b) 700°C with equilibrium curve for spinodal demixing. (From Leute, V., *Z. Naturforsch.*, **50A**(4–5), 357, 1995.) Open access

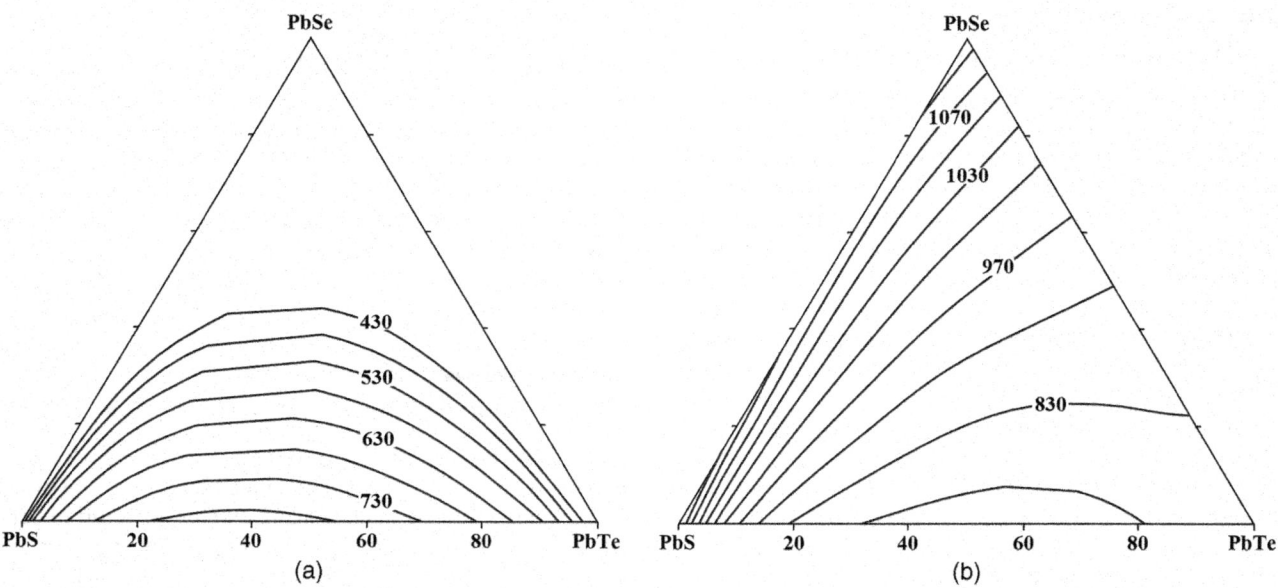

FIGURE 10.47 Calculated (a) solidus and (b) liquidus surface of the PbS–PbSe–PbTe quasiternary system. (From Volykhov, A.A., et al., *Inorg. Mater.*, **46**(5), 464, 2010.)

duration of heat treatment ranged from 60 days at 500°C, 30 days at 600°C, 7 days at 700°C, and 3 days at 800°C, one day at 900°C to 0.25 of a day at and above 1000°C. Experimental runs at 300°C were made with chloride flux for a period of 150 days. At the end of heat treatment, the samples were quenched by compressed air.

The isothermal sections of the PbS–PbSe–PbTe ternary system at 550°C, 630°C, and 700°C were calculated by Leute (1995), and two of them are shown in Figure 10.46.

At temperatures above 801°C, there is unlimited solid solubility in this system (Volykhov et al. 2010). The solidus and liquidus surfaces of the system, constructed using the interaction parameters of the quasibinary systems and taking into account ternary interactions, are presented in Figure 10.47.

The Pb–S–Se–Te quaternary system is divided by the PbS–PbSe–PbTe, PbS–PbSe–Te, and PbS–Se–Te quasiternary system into four secondary subsystems: Pb–PbS–PbSe–PbTe, Te–PbS–PbSe–PbTe, PbS–PbSe–Se–Te, and PbS–S–Se–Te (Filep and Sabov 2017a).

10.125 Lead–Tellurium–Chlorine–Sulfur

The **PbTe₃(Cl,S)₂** quaternary compound (mineral radhakrishnaite), which crystallizes in the tetragonal structure with the lattice parameters $a = 571 \pm 5$, $c = 377 \pm 5$, and a calculated density of 8.89 g·cm⁻³, is formed in the Pb–Te–Cl–S system (Genkin et al. 1985; Hawthorne et al. 1986a).

11

Systems Based on Lead Selenide

11.1 Lead–Hydrogen–Oxygen–Selenium

The $Pb(HSeO_3)_2$ quaternary compound, which is thermally stable up to 110–120°C (Markovskiy and Sapozhnikov 1960) and crystallizes in the orthorhombic structure with the lattice parameters $a = 1638.1 \pm 0.5$, $b = 611.1 \pm 0.1$, $c = 589.6 \pm 0.1$ pm, and a calculated density of 5.21 g·cm^{-3} (Koskenlinna and Valkonen 1995) [an experimental density of 5.19 ± 0.01 g·cm^{-3} (Markovskiy and Sapozhnikov 1960)], is formed in the Pb–H–O–Se system (Oykova and Gospodinov 1981).

The title compound can be crystallized by allowing crystals of either $PbSeO_3$ or $PbSe_2O_5$ to stand for several weeks in an aqueous solution of H_2SeO_3, partly neutralized with KOH or $MgCO_3$, at a temperature between 80°C and 100°C (Koskenlinna and Valkonen 1995). The fragile colorless needles grow on the crystals of the starting material, making the solids mixed phase. It can also be obtained by reacting $Pb(CH_3COO)_2$ or $Pb(NO_3)_2$ or $PbCO_3$ with at least a 10-fold excess of H_2SeO_3 (Markovskiy and Sapozhnikov 1960).

11.2 Lead–Lithium–Niobium–Selenium

Lithium intercalation of $PbNb_2Se_5$ was performed using n-buthyllithium technique in a hexane solution (1.6 M·L^{-1}) at room temperature under an inert atmosphere (Hernán et al. 1992). Unreacted n-buthyllithium was removed by washing off the product with dry hexane with suction filtering. The sample obtained after 3 h of lithiation was a mixture of phases that included traces of a partially lithiated misfit layered phase $Li_xPbNb_2Se_5$, PbSe, and $NbSe_2$. The obtained results indicate that $PbNb_2Se_5$ is unstable to lithiation.

11.3 Lead–Sodium–Phosphorus–Selenium

The $Na_{0.5}Pb_{0.75}PSe_4$ quaternary compound, which crystallizes in the cubic structure with the lattice parameter $a = 1434.79 \pm 0.02$, a calculated density of 4.828 g·cm^{-3} at 173 K, and an energy gap of 2.09 eV (Aitken et al. 2000), is formed in the Pb–Na–P–Se system. To prepare this compound, in a N_2-filled glove box Pb (0.9 mM), P_2Se_5 (0.6 mM), Na_2Se (0.9 mM), and Se (0.9 mM) were loaded into a Pyrex tube (Aitken et al. 2000). This tube was flame-sealed under vacuum (approximately 0.02 Pa) and inserted into a furnace. The temperature was raised from 50°C to 540°C in 24 h. It was kept at 540°C for 30 h, and then cooled at 3°C·h^{-1} to 300°C followed by cooling at 50°C·h^{-1} to 50°C. The product was isolated by washing away the excess flux with DMF and washed with ether.

11.4 Lead–Potassium–Antimony–Selenium

Two quaternary phases, $K_{1.45}Pb_{3.10}Sb_{7.45}Se_{15}$ and $K_{2.15}Pb_{1.70}Sb_{8.15}Se_{15}$, are formed in the Pb–K–Sb–Se system (Chung et al. 1998; Choi et al. 2001). Both phases are narrow band gap semiconductors, melt congruently at 576°C and crystallize in the monoclinic structure with the lattice parameters $a = 1712.04 \pm 0.06$, $b = 415.68 \pm 0.02$, $c = 2163.62 \pm 0.08$ pm, $\beta = 98.706 \pm 0.001°$, a calculated density of 6.089 g·cm^{-3} at 171 K, and an energy gap of 0.45 eV for the first phase and $a = 1716.4 \pm 0.4$, $b = 414.94 \pm 0.09$, $c = 2168.4 \pm 0.5$ pm, $\beta = 98.664 \pm 0.003°$, a calculated density of 5.683 g·cm^{-3} at room temperature, and an energy gap of 0.60 eV for the second one. Both compounds were synthesized from a mixture of K_2Se (2 mM), Pb (2 mM), Sb_2Se_3 (2 mM), and Se (10 mM) for the first phase and K_2Se (4 mM), Pb (1 mM), Sb_2Se_3 (3 mM), and Se (10 mM) for $K_{2.15}Pb_{1.70}Sb_{8.15}Se_{15}$. The reagents were thoroughly mixed, sealed in an evacuated Pyrex tube, and heated at 540°C for 5 days (cooling rate of 2°C·h^{-1}). Isolation in degassed DMF and water gave pure silver-colored needles of each compound. The obtained crystals are air- and water-stable.

11.5 Lead–Potassium–Bismuth–Selenium

Some quaternary compounds are formed in the Pb–K–Bi–Se system. $K_{0.7}Pb_{2.8}Bi_{11.1}Se_{20}$ melts congruently at 694°C and has an energy gap of 0.55 eV (Mrotzek et al. 2001c). To prepare this compound, a mixture of K_2Se (1 mM), Pb (3 mM), Se (3 mM), and Bi_2Se_3 (5.5 mM) was loaded in a carbon-coated quartz tube and sealed at a residual pressure of <0.01 Pa. The starting materials were heated within 12 h to 800°C and kept there for 24 h, followed by slow cooling to 400°C at a rate of 0.1°C·min^{-1} and then to 50°C in 10 h. A silver, shiny, polycrystalline ingot of the title compound was obtained after washing any impurities with DMF, methanol, and diethyl ether.

$KPbBiSe_3$ melts at 923°C, crystallizes in the cubic structure, and has an energy gap of 0.58 eV (Kanatzidis et al. 1998).

$KPbBi_3Se_6$ melts at 705°C and crystallizes in the hexagonal structure with the lattice parameters $a = 419 \pm 1$, $c = 2510 \pm 2$ pm, and an energy gap of 0.30 eV (Chung et al. 1999). This compound is an n-type semiconductor and did not show any sign of decomposition or phase transformation in differential thermal analysis (DTA). To synthesize this compound, a mixture of K_2Se (0.484 mM), Pb (0.241 mM), Bi (0.718 mM), and Se (1.912 mM) was thoroughly mixed, loaded into an alumina thimble, and subsequently sealed inside a carbon-coated quartz tube at a residual pressure of <0.01 Pa. The mixture was heated to 720°C in 24 h and kept there for 6 days, followed by slow cooling to 200°C at 4°C·h^{-1} and then to 50°C in 6 h. Thin metallic black plates of the title compound were isolated by washing with DMF. All manipulations were carried out under a dry N$_2$ atmosphere in a glove box.

$K_{1.25}Pb_{3.50}Bi_{7.25}Se_{15}$ melts congruently at 685°C and crystallizes in the monoclinic structure with the lattice parameters a = 1744.81 ± 0.08, b = 419.64 ± 0.02, c = 2169.45 ± 0.10 pm, β = 98.850 ± 0.001°, a calculated density of 7.350 g·cm^{-3}, and an energy gap of 0.53 eV (Chung et al. 1998; Kanatzidis et al. 1998; Choi et al. 2001). This compound was synthesized from a mixture of K$_2$Se (2 mM), Pb (2 mM), Bi$_2$Se$_3$ (3 mM), and Se (10 mM). The reagents were thoroughly mixed, sealed in an evacuated Pyrex tube, and heated at 540°C for 5 days (cooling rate of 2°C·h^{-1}). Isolation in degassed DMF and water gave pure silver-colored needles of this compound. The obtained crystals are air- and water-stable.

$K_{1.46}Pb_{3.08}Bi_{11.46}Se_{21}$ melts congruently at 674°C and crystallizes in the monoclinic structure with the lattice parameters a = 3188.3 ± 0.5, b = 418.63 ± 0.07, c = 1738.7 ± 0.3 pm, and β = 112.098 ± 0.003°, and a calculated density of 7.485 g·cm^{-3} (Mrotzek et al. 2001a). To prepare this compound, a mixture of K$_2$Se (0.5 mM), Pb (4 mM), Se (4 mM), and Bi$_2$Se$_3$ (5.5 mM) was loaded in a carbon-coated quartz tube and sealed at a residual pressure of <0.01 Pa. The starting materials were heated over 24 h to 800°C and kept at this temperature for a further 24 h, followed by slow cooling to 400°C at a rate of 0.1°C·min^{-1} and then to 50°C over a period of 10 h. A silver, shiny polycrystalline ingot of pure title compound was obtained after washing away any impurities with DMF, methanol, and diethyl ether. All manipulations were carried out under a dry nitrogen atmosphere in a glove box.

$K_{1.6}Pb_{2.3}Bi_{8.0}Se_{14}$ melts congruently at 692°C and has an energy gap of 0.56 eV (Mrotzek et al. 2001c). This compound was obtained as follows. A mixture of K$_2$Se (1 mM), Pb (7 mM), Se (7 mM), and Bi$_2$Se$_3$ (7 mM) was loaded in a carbon-coated quartz tube and sealed at a residual pressure of <0.01 Pa. The starting materials were heated within 24 h to 800°C and kept there for 24 h, followed by slow cooling to 50°C at a rate of 0.5°C·min^{-1}. A silver, shiny, polycrystalline ingot of pure $K_{1.6}Pb_{2.3}Bi_{8.0}Se_{14}$ was obtained after washing any impurities with DMF, methanol, and diethyl ether.

11.6 Lead–Potassium–Oxygen–Selenium

The $K_2Pb(SeO_3)_3$ quaternary compound is formed in the Pb–K–O–Se system (Frydrych 1976). To prepare this compound, a 150 mL beaker equipped with a thermometer and glass electrode for pH measurement was placed on a heatable magnetic stirrer and filled with 50 mL of H$_2$O. SeO$_2$ (13 g) was dissolved (pH ~ 0.8) and mixed with KCH$_3$COO (0.13 M). The pH value rises to around 3.6. The solution was then reacted with KOH or K$_2$CO$_3$ with heating to 60°C until the pH rose to 6.5. Freshly prepared Pb(SeO$_3$)$_2$ (6 g) was gradually added to this solution, and the temperature was increased to 70°C. The Pb(SeO$_3$)$_2$ dissolved within 2–3 min. If this is method not completely successful, the remaining was sucked off in devices

preheated to around 80°C. K$_2$Pb(SeO$_3$)$_3$ separated in crystalline form from a clear yellow solution on cooling to the low temperature. The product separated out after 3 to 24 h was filtered off with suction, dried with a little of 12 mass% CH$_3$COOH, then washed with CH$_3$OH and ether at 40°C.

11.7 Lead–Rubidium–Phosphorus–Selenium

Two quaternary compounds, $RbPbPSe_4$ and $Rb_4Pb(PSe_4)_2$, are formed in the Pb–Rb–P–Se system (Chondroudis et al. 1996). The first of them melts at 597°C and has an energy gap of 2.07 eV, and the second one melts at 615°C and crystallizes in the orthorhombic structure with the lattice parameters a = 1913.4 ± 0.9, b = 936.9 ± 0.3, c = 1048.8 ± 0.3 pm, a calculated density of 4.390 g·cm^{-3}, and an energy gap of 2.22 eV.

$RbPbPSe_4$ was synthesized from a mixture of Pb (0.5 mM), P (0.5 mM), Rb$_2$Se (0.25 mM), and Se (1.75 mM) that was sealed under vacuum in a Pyrex tube and heated to 500°C for 4 days followed by cooling to 150°C at a rate 4°C·h^{-1}. The product was washed with H$_2$O to reveal analytically pure dark red microcrystalline RbPbPSe$_4$. The crystals are air- and water-stable. Single crystals were synthesized from a mixture of Pb (0.15 mM), P$_2$Se$_5$ (0.30 mM), Rb$_2$Se (0.30 mM), and Se (1.5 mM) that was sealed under vacuum in a Pyrex tube and heated to 450°C for 4 days followed by cooling to 150°C at a rate 4°C·h^{-1}. The excess Rb$_x$P$_y$Se$_z$ flux was removed with DMF.

$Rb_4Pb(PSe_4)_2$ was prepared similarly from a mixture of Pb (0.15 mM), P$_2$Se$_5$ (0.225 mM), Rb$_2$Se (0.60 mM), and Se (1.5 mM) with the same heating conditions. The crystals of this compound are orange and disintegrate in water and over long exposure to air.

11.8 Lead–Rubidium–Antimony–Selenium

The $Rb_{1.45}Pb_{3.1}Sb_{7.45}Se_{15}$ quaternary compound, which melts congruently at 578°C and crystallizes in the monoclinic structure with the lattice parameters a = 1731.60 ± 0.07, b = 414.06 ± 0.02, c = 2164.01 ± 0.08 pm, β = 99.139 ± 0.001°, a calculated density of 6.195 g·cm^{-3} at 171 K, and an energy gap of 0.36 eV, is formed in the Pb–Rb–As–Se system (Chung et al. 1998; Choi et al. 2001). This compound was synthesized from a mixture of Rb$_2$Se (2 mM), Pb (2 mM), Bi$_2$Se$_3$ (3 mM), and Se (10 mM). The reagents were thoroughly mixed, sealed in an evacuated Pyrex tube, and heated at 540°C for 5 days (cooling rate of 2°C·h^{-1}). Isolation in degassed DMF and water gave pure silver-colored needles of this compound. The obtained crystals are air- and water-stable.

11.9 Lead–Rubidium–Bismuth–Selenium

Three quaternary compounds, $Rb_{0.69}Pb_{3.68}Bi_{11.31}Se_{21}$, $RbPbBi_3Se_6$, and $RbPb_{3.5}Bi_{11.4}Se_{20}$, are formed in the Pb–Rb–Bi–Se system. The first of them melts congruently at 680°C and crystallizes in the monoclinic structure with the lattice parameters a = 3199.1 ± 0.5, b = 419.37 ± 0.07, c = 1748.1

± 0.3 pm, and β = 112.519 ± 0.003°, a calculated density of 7.384 g·cm^{-3}, and an energy gap of 0.62 eV (Mrotzek et al. 2001a). This compound was synthesized in the same way as $K_{1.6}Pb_{2.3}Bi_{8.0}Se_{14}$ was obtained using Rb_2Se instead of K_2Se.

$RbPbBi_3Se_6$ melts at 695°C and crystallizes in the hexagonal structure with the lattice parameters a = 418.5 ± 0.2, c = 2498.5 ± 0.5 pm at 173 K, and a = 417.9 ± 0.2, c = 2538.3 ± 0.9 pm at room temperature, and an energy gap of 0.71 eV (Chung et al. 1999). This compound is an *n*-type semiconductor and did not show any sign of decomposition or phase transformation in DTA. It was synthesized through two various methods. According to the first method, a mixture of Rb_2Se (1.437 mM), Pb (0.241 mM), Bi (0.718 mM), and Se (3.850 mM) was loaded into a Pyrex tube and subsequently flame-sealed at a residual pressure of <0.1 Pa. The mixture was heated from room temperature to 400°C over 12 h and kept there for 6 days, followed by slowly cooling to 200°C at a rate of 2°C·h^{-1} and then to 50°C in 6 h. According to the second method, a mixture of Rb_2Se (0.480 mM), Pb (0.241 mM), Bi (0.718 mM), and Se (1.912 mM) was heated to 720°C over 24 h and kept there for 6 days, followed by slow cooling to 200°C at a rate of 4°C·h^{-1} and then to 50 °C in 6 h. Thin black hexagonal plates of $RbPbBi_3Se_6$ were obtained by isolation in DMF.

$RbPb_{3.5}Bi_{11.4}Se_{20}$ melts congruently at 667°C and has an energy gap of 0.54 eV (Mrotzek et al. 2001c). To obtain this compound, a mixture of Rb_2Se (1 mM), Pb (3 mM), Se (3 mM), and Bi_2Se_3 (5.5 mM) was loaded in a carbon-coated quartz tube and sealed at a residual pressure of <0.01 Pa. The starting materials were heated within 12 h to 800°C and kept there for 24 h, followed by slow cooling to 400°C at a rate of 0.1°C·min^{-1} and then to 50°C in 10 h. A silver, shiny, polycrystalline ingot of the title compound was obtained after washing any impurities with DMF, methanol, and diethyl ether.

11.10 Lead–Rubidium–Oxygen–Selenium

The $Rb_2Pb(SeO_3)_3$ quaternary compound is formed in the Pb–Rb–O–Se system (Frydrych 1976). It was synthesized in the same way as $K_2Pb(SeO_3)_3$ was prepared using $RbCH_3COO$ and RbOH or Rb_2CO_3 instead of KCH_3COO and KOH or K_2CO_3, respectively.

11.11 Lead–Cesium–Phosphorus–Selenium

Two quaternary compounds, $CsPbPSe_4$ and $Cs_4Pb(PSe_4)_2$, are formed in the Pb–Cs–P–Se system (Chondroudis et al. 1996). The first melts at 612°C, and crystallizes in the orthorhombic structure with the lattice parameters a = 1860.7 ± 0.4, b = 709.6, c = 661.2 ± 0.4 pm, a calculated density of 5.226 g·cm^{-3}, and an energy gap of 2.08 eV, and the second one melts at 616°C and has an energy gap of 2.26 eV. Both compounds were synthesized in the same way as $RbPbPSe_4$ and $Rb_4Pb(PSe_4)_2$ were prepared using Cs_2Se instead of Rb_2Se. The excess $Cs_xP_ySe_z$ flux was also removed with DMF and $Cs_4Pb(PSe_4)_2$ also disintegrates in water and over long exposure to air.

11.12 Lead–Cesium–Bismuth–Selenium

Four quaternary compounds, $Cs_{0.65}Pb_{3.65}Bi_{11.35}Se_{21}$, $Cs_{0.7}Pb_{3.5}Bi_{10.6}Se_{20}$, $CsPbBi_3Se_6$, and $CsPb_{2.7}Bi_{7.6}Se_{14}$, are formed in the Pb–Cs–Bi–Se system. First of them melts congruently at 687°C and crystallizes in the monoclinic structure with the lattice parameters a = 3217.2 ± 0.9, b = 415.9 ± 0.1, c = 1749.9 ± 0.5 pm, and β = 112.672 ± 0.006°, a calculated density of 7.394 g·cm^{-3}, and an energy gap of 0.62 eV (Mrotzek et al. 2001a). This compound was synthesized in the same way as $K_{1.6}Pb_{2.3}Bi_{8.0}Se_{14}$ was obtained using Cs_2Se instead of K_2Se.

$Cs_{0.7}Pb_{3.5}Bi_{10.6}Se_{20}$ melts congruently at 714°C (Mrotzek et al. 2001c). This compound was prepared as follows. A mixture of Cs_2Se (1 mM), Pb (3 mM), Se (3 mM), and Bi_2Se_3 (5.5 mM) was loaded in a carbon-coated quartz tube and sealed at a residual pressure of <0.01 Pa. The starting materials were heated within 12 h to 800°C and kept there for 24 h, followed by slow cooling to 400°C at a rate of 0.1°C·min^{-1} and then to 50°C in 10 h. A silver, shiny, polycrystalline ingot of $Cs_{0.7}Pb_{3.5}Bi_{10.6}Se_{20}$ was obtained after washing any impurities with DMF, methanol, and diethyl ether.

$CsPbBi_3Se_6$ has two polymorphic modifications (Chung et al. 1999). α-$CsPbBi_3Se_6$ melts congruently at 711°C and crystallizes in the orthorhombic structure with the lattice parameters a = 2356.4 ± 0.6, b = 421.0 ± 0.2, c = 1379.8 ± 0.3 pm, a calculated density of 6.991 g·cm^{-3}, and an energy gap of 0.55 eV. β-$CsPbBi_3Se_6$ melts at 691°C and crystallizes in the hexagonal structure with the lattice parameters a = 421.3 ± 0.3, c = 2522 ± 1 pm, a calculated density of 6.169 g·cm^{-3} at 173 K, and an energy gap of 0.71 eV. This compound is *n*-type semiconductor and did not show any sign of decomposition or phase transformation in DTA.

To synthesize α-$CsPbBi_3Se_6$, a mixture of Cs_2Se (0.719 mM), Pb (0.241 mM), Bi (0.718 mM), and Se (3.597 mM) was loaded into an alumina thimble and subsequently sealed inside a carbon-coated quartz tube at a residual pressure of <0.01 Pa. The mixture was heated to 720°C over 24 h and kept there for 6 days, followed by slow cooling to 200°C at a rate of 4°C·h^{-1} and then to 50°C in 6 h. Black needles of α-$CsPbBi_3Se_6$ were obtained after removing the Cs_2Se_x flux with DMF and washing with diethyl ether.

β-$CsPbBi_3Se_6$ was prepared through two various methods. According to the first method, a mixture of Cs_2Se (1.436 mM), Pb (0.241 mM), Bi (0.718 mM), and Se (3.850 mM) was loaded into a Pyrex tube and subsequently flame-sealed at a residual pressure of <0.1 Pa. The mixture was heated from room temperature to 400°C for 12 h and kept there for 6 days, followed by slow cooling to 200°C at a rate of 2°C·h^{-1} and then to 50°C in 6 h. According to the second method, a mixture of Cs_2Se (0.479 mM), Pb (0.241 mM), Bi (0.718 mM), and Se (1.912 mM) was heated to 720°C in 24 h and kept there for 6 days, followed by slow cooling to 200°C at a rate of 4°C·h^{-1} and then to 50°C in 6 h. Thin black hexagonal plates of $CsPbBi_3Se_6$ were obtained by isolation in DMF.

All manipulations for the preparation of both modifications of $CsPbBi_3Se_6$ were carried out under a dry N_2 atmosphere in a glove box.

CsPb$_{2.7}$Bi$_{7.6}$Se$_{14}$ melts congruently at 716°C and has an energy gap of 0.62 eV (Mrotzek et al. 2001c). This compound was obtained as follows. A mixture of Cs$_2$Se (1 mM), Pb (7 mM), Se (7 mM), and Bi$_2$Se$_3$ (7 mM) was loaded in a carbon-coated quartz tube and sealed at a residual pressure of <0.01 Pa. The starting materials were heated within 24 h to 800°C and kept there for 24 h, followed by slow cooling to 50°C at a rate of 0.5°C·min^{-1}. A silver, shiny, polycrystalline ingot of pure CsPb$_{2.7}$Bi$_{7.6}$Se$_{14}$ was obtained after washing any impurities with DMF, methanol, and diethyl ether.

11.13 Lead–Cesium–Oxygen–Selenium

The **Cs$_2$Pb(SeO$_3$)$_3$** quaternary compound is formed in the Pb–Cs–O–Se system (Frydrych 1976). It was synthesized in the same way as K$_2$Pb(SeO$_3$)$_3$ was prepared using CsCH$_3$COO and CsOH or Cs$_2$CO$_3$ instead of KCH$_3$COO and KOH or K$_2$CO$_3$, respectively.

11.14 Lead–Copper–Indium–Selenium

Pb–CuInSe$_2$–Se. This quasiternary system was studied through DTA, X-ray diffraction (XRD), metallography, and measuring of the microhardness by Allazova and Ilyasly (2015). The liquidus surface of the system was constructed, where the boundaries of the primary crystallization field of the low-temperature chalcopyrite phase were determined. The existence of the **CuInPbSe$_2$** quaternary compound, which melts incongruently at 650°C and crystallizes in the tetragonal structure, was established.

11.15 Lead–Copper–Scandium–Selenium

PbSe–Cu$_2$Se–Sc$_2$Se$_3$. The isothermal section of this quasiternary system at 600°C was constructed using the alloys annealed at this temperature for 240 h (Gulay et al. 2008a). No quaternary compounds exist in the system. The system is divided by the PbSe–ScCuSe$_2$ and ScCuSe$_2$–Sc$_2$PbSe$_4$ quasibinary sections into three subsystems: PbSe–Cu$_2$Se–ScCuSe$_2$, PbSe–ScCuSe$_2$–Sc$_2$PbSe$_4$, and ScCuSe$_2$–Sc$_2$PbSe$_4$–Sc$_2$Se$_3$.

11.16 Lead–Copper–Yttrium–Selenium

PbSe–Cu$_2$Se–Y$_2$Se$_3$. The isothermal section of this quasiternary system at 600°C was constructed using the alloys annealed at this temperature for 240 h (Figure 11.1) (Shemet et al. 2005). Two quaternary compounds, **CuYPbSe$_3$** and **CuY$_{3.33}$Pb$_{1.5}$Se$_7$**, are formed in this system. First of them crystallizes in the orthorhombic structure with the lattice parameters $a = 1054.38 \pm 0.05$, $b = 405.24 \pm 0.02$, $c = 1338.40 \pm 0.05$ pm, and a calculated density of 6.9279 g·cm^{-3} (Gulay et al. 2004c; Shemet et al. 2005) and the second one crystallizes in the monoclinic structure with the lattice parameters $a = 1356.75 \pm 0.07$, $b = 409.59 \pm 0.02$, $c = 1260.26 \pm 0.06$ pm, β

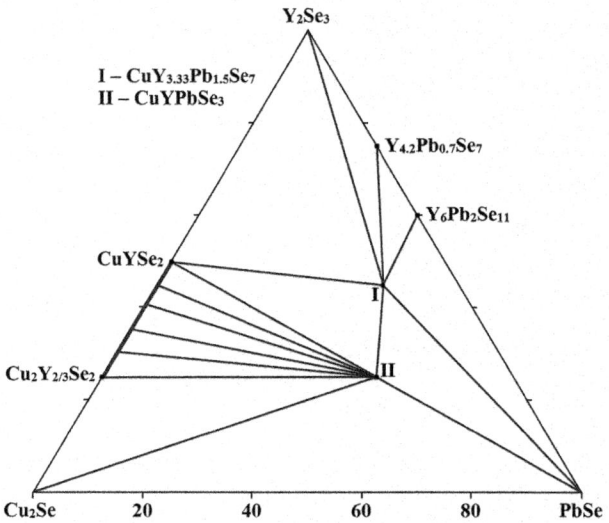

FIGURE 11.1 Isothermal section of the PbSe–Cu$_2$Se–Y$_2$Se$_3$ quasiternary system at 600°C. (From Shemet, V.Ya., et al., *Pol. J. Chem.*, **79**(8), 1315, 2005.)

$= 104.661 \pm 0.002°$, and a calculated density of 5.9963 g·cm^{-3} (Gulay et al. 2005h; Shemet et al. 2005).

CuYPbSe$_3$ was prepared by fusion of the high-purity elements (Gulay et al. 2004c). The calculated amounts of the components were sealed in an evacuated quartz ampoule. The synthesis was realized in a shaft furnace with a heating rate of 30°C·h^{-1}. The maximal temperature of the synthesis was about 1150°C. The sample was kept at this temperature for 4 h. After that, it was cooled slowly to 600°C at a rate of 10°C·h^{-1}. The alloys were annealed at this temperature for 240 h. After annealing, the ampoule with the sample was quenched in cold water.

CuY$_{3.33}$Pb$_{1.5}$Se$_7$ was prepared by fusion of the high-purity constituent elements in evacuated quartz ampoules (Gulay et al. 2005h). The synthesis was realized in a shaft furnace with a heating rate of 20°C·h^{-3}. The ampoule with the sample was heated to a maximal temperature of 1150°C. The sample was kept at the maximal temperature for 4 h. After that, it was cooled slowly to 600°C with a rate of 10°C·h^{-3} and annealed at this temperature for 240 h. After annealing, the ampoule with the sample was quenched in cold water.

11.17 Lead–Copper–Praseodymium–Selenium

PbSe–Cu$_2$Se–Pr$_2$Se$_3$. The isothermal section of this quasiternary system at 600°C was constructed using the alloys annealed at this temperature for 240 h (Figure 11.2) (Marchuk et al. 2006b). No quaternary compounds were found in the system.

11.18 Lead–Copper–Gadolinium–Selenium

The **CuGdPbSe$_3$** quaternary compound, which crystallizes in the orthorhombic structure with the lattice parameters $a = 1068.7 \pm 0.2$, $b = 408.58 \pm 0.05$, $c = 1344.3 \pm 0.2$ pm, and a

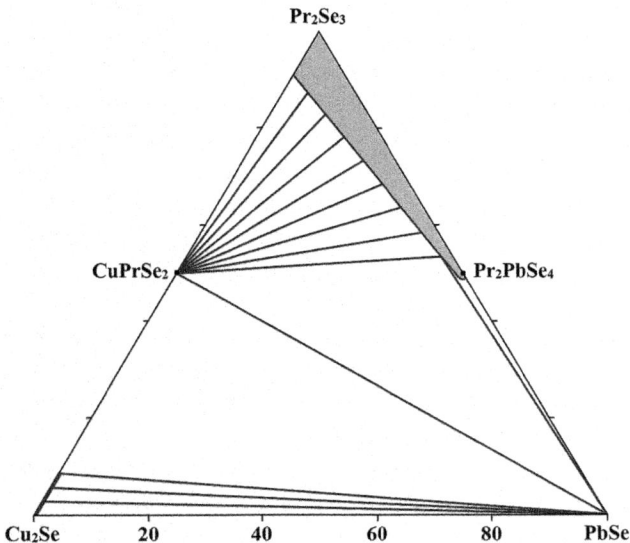

FIGURE 11.2 Isothermal section of the PbSe–Cu$_2$Se–Pr$_2$Se$_3$ quasiternary system at 600°C. (From Marchuk, O.V., et al., *J. Alloys Compd.*, **416**(1–2), 106, 2006.)

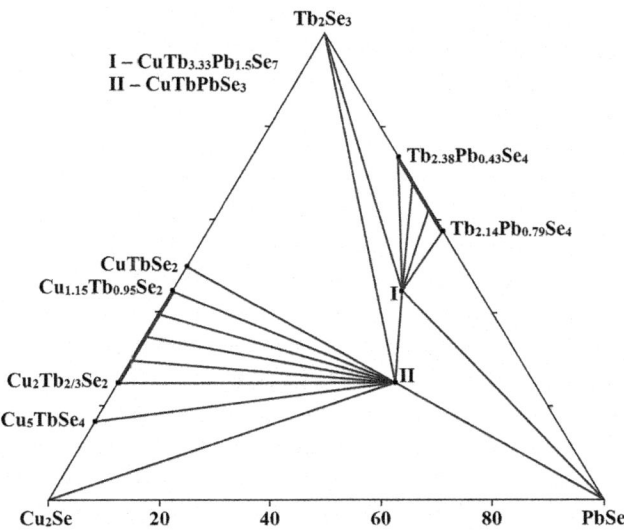

FIGURE 11.3 Isothermal section of the PbSe–Cu$_2$Se–Tb$_2$Se$_3$ quasiternary system at 600°C. (From Gulay, L.D., et al., *Pol. J. Chem.*, **80**(5), 805, 2006.)

calculated density of 7.523 g·cm^{-3}, is formed in the Pb–Cu–Gd–Se system (Gulay and Olekseyuk 2005a). It was prepared by fusion of the high-purity elements in an evacuated quartz ampoule. The synthesis was realized in a shaft furnace. The ampoule was heated with a heating rate of 30°C·h^{-1} to a maximal temperature of 1150°C. The sample was kept at the maximal temperature for 4 h. After that, it was cooled slowly (10°C·h^{-1}) to 600°C and annealed at this temperature for 240 h. After annealing, the ampoule with the sample was quenched in cold water.

11.19 Lead–Copper–Terbium–Selenium

PbSe–Cu$_2$Se–Tb$_2$Se$_3$. The isothermal section of this quasiternary system at 600°C was constructed using the alloys annealed at this temperature for 240 h (Figure 11.3) (Gulay et al. 2006g). Two quaternary compounds, **CuTbPbSe$_3$** and **CuTb$_{3.33}$Pb$_{1.5}$Se$_7$**, are formed in this system. First of them crystallizes in the orthorhombic structure with the lattice parameters $a = 1060.3 \pm 0.1$, $b = 406.80 \pm 0.03$, $c = 1341.4 \pm 0.1$ pm, and a calculated density of 7.651 g·cm^{-3} (Gulay and Olekseyuk 2005a; Gulay et al. 2005g) and the second one crystallizes in the monoclinic structure with the lattice parameters $a = 1362.4 \pm 0.3$, $b = 411.44 \pm 0.08$, $c = 1264.5 \pm 0.3$ pm, and $\beta = 104.68 \pm 0.01°$ (Gulay and Olekseyuk 2005b; Gulay et al. 2005g). Both compounds were synthesized in the same way as CuGdPbSe$_3$ was obtained.

11.20 Lead–Copper–Dysprosium–Selenium

PbSe–Cu$_2$Se–Dy$_2$Se$_3$. The isothermal section of this quasiternary system at 600°C was constructed using the alloys annealed at this temperature for 240 h (Figure 11.4) (Gulay

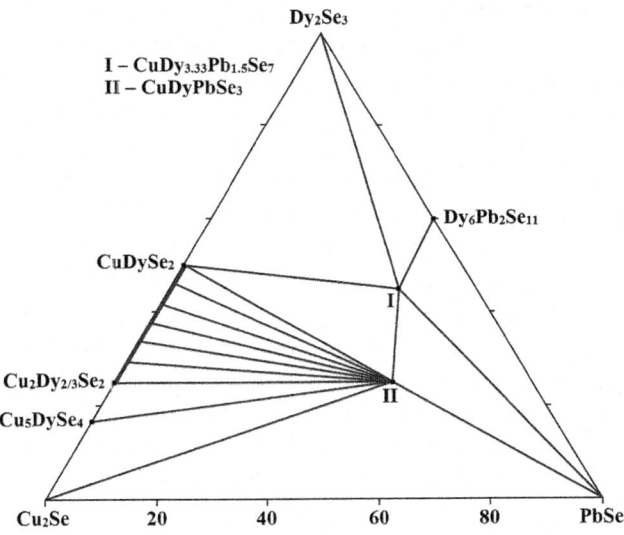

FIGURE 11.4 Isothermal section of the PbSe–Cu$_2$S–Dy$_2$Se$_3$ quasiternary system at 600°C. (From Gulay, L.D., et al., *Pol. J. Chem.*, **80**(5), 805, 2006.)

et al. 2006g). Two quaternary compounds, **CuDyPbSe$_3$** and **CuDy$_{3.33}$Pb$_{1.5}$Se$_7$**, are formed in this system. First of them crystallizes in the orthorhombic structure with the lattice parameters $a = 1056.06 \pm 0.09$, $b = 405.94 \pm 0.03$, $c = 1342.36 \pm 0.09$ pm, and a calculated density of 7.734 g·cm^{-3} (Gulay and Olekseyuk 2005a; Gulay et al. 2005g) and the second one crystallizes in the monoclinic structure with the lattice parameters $a = 1355.7 \pm 0.5$, $b = 409.1 \pm 0.1$, $c = 1257.4 \pm 0.4$ pm, and $\beta = 104.59 \pm 0.02°$ (Gulay and Olekseyuk 2005b; Gulay et al. 2005g). Both compounds were synthesized in the same way as CuGdPbSe$_3$ was obtained.

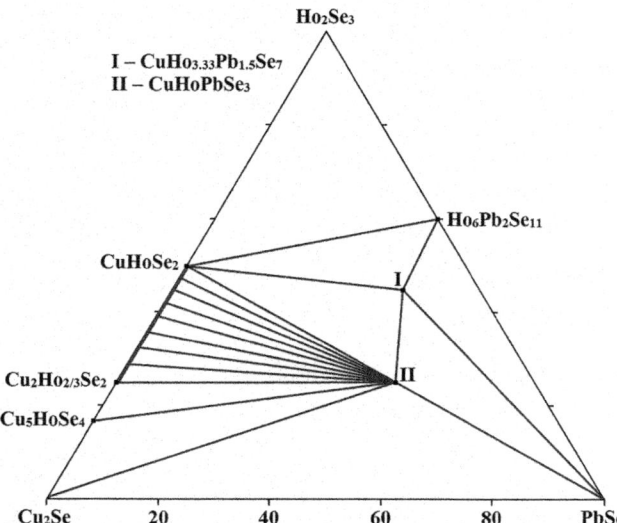

FIGURE 11.5 Isothermal section of the PbSe–Cu₂S–Ho₂Se₃ quasiternary system at 600°C. (From Gulay, L.D., et al., *J. Alloys Compd.*, **416**(1–2), 173, 2006.)

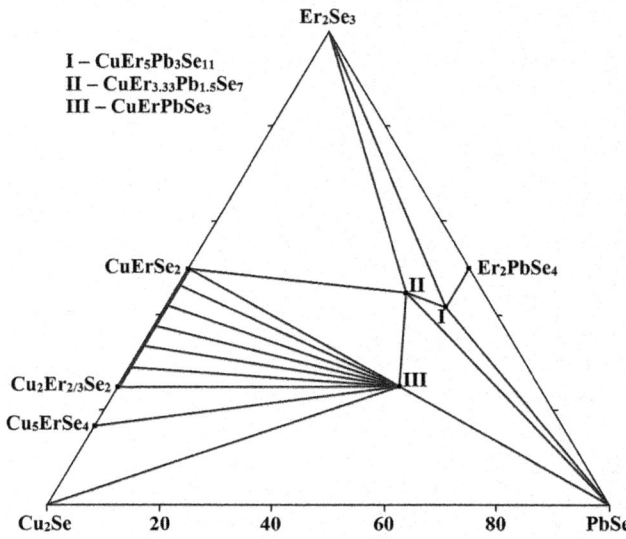

FIGURE 11.6 Isothermal section of the PbSe–Cu₂S–Er₂Se₃ quasiternary system at 600°C. (From Gulay, L.D., et al., *J. Alloys Compd.*, **416**(1–2), 173, 2006.)

11.21 Lead–Copper–Holmium–Selenium

PbSe–Cu₂Se–Ho₂Se₃. The isothermal section of this quasiternary system at 600°C was constructed using the alloys annealed at this temperature for 240 h (Figure 11.5) (Gulay et al. 2006e). Two quaternary compounds, **CuHoPbSe₃** and **CuHo₃.₃₃Pb₁.₅Se₇**, are formed in this system. The first crystallizes in the orthorhombic structure with the lattice parameters $a = 1051.6 \pm 0.2$, $b = 404.70 \pm 0.08$, $c = 1341.03 \pm 0.03$ pm, and a calculated density of 7.827 g·cm⁻³ (Gulay et al. 2006e) [$a = 1052.8 \pm 0.1$, $b = 405.16 \pm 0.05$, $c = 1342.0 \pm 0.2$ pm, and a calculated density of 7.806 g·cm⁻³ (Gulay and Olekseyuk 2005a)] and the second one crystallizes in the monoclinic structure with the lattice parameters $a = 1353.14 \pm 0.08$, $b = 408.19 \pm 0.02$, $c = 1256.09 \pm 0.07$ pm, $\beta = 104.577 \pm 0.003°$, and a calculated density of 7.305 g·cm⁻³ (Gulay and Olekseyuk 2005b; Gulay et al. 2006e). Both compounds were synthesized in the same way as CuGdPbSe₃ was obtained.

11.22 Lead–Copper–Erbium–Selenium

PbSe–Cu₂Se–Er₂Se₃. The isothermal section of this quasiternary system at 600°C was constructed using the alloys annealed at this temperature for 240 h (Figure 11.6) (Gulay et al. 2006e). Three quaternary compounds, **CuErPbSe₃**, **CuEr₃.₃₃Pb₁.₅Se₇**, and **CuEr₅Pb₃Se₁₁**, are formed in this system. The first crystallizes in the orthorhombic structure with the lattice parameters $a = 1048.46 \pm 0.08$, $b = 404.24 \pm 0.03$, $c = 1341.43 \pm 0.07$ pm, and a calculated density of 7.884 g·cm⁻³ (Gulay et al. 2006e; Gulay and Olekseyuk 2005a). CuEr₃.₃₃Pb₁.₅Se₇ crystallizes in the monoclinic structure with the lattice parameters $a = 1350.18 \pm 0.07$, $b = 406.93 \pm 0.02$, $c = 1254.33 \pm 0.06$ pm, $\beta = 104.492 \pm 0.002°$, and a calculated density of 7.390 g·cm⁻³ (Gulay and Olekseyuk 2005b; Gulay et al. 2006e). CuEr₅Pb₃Se₁₁ crystallizes

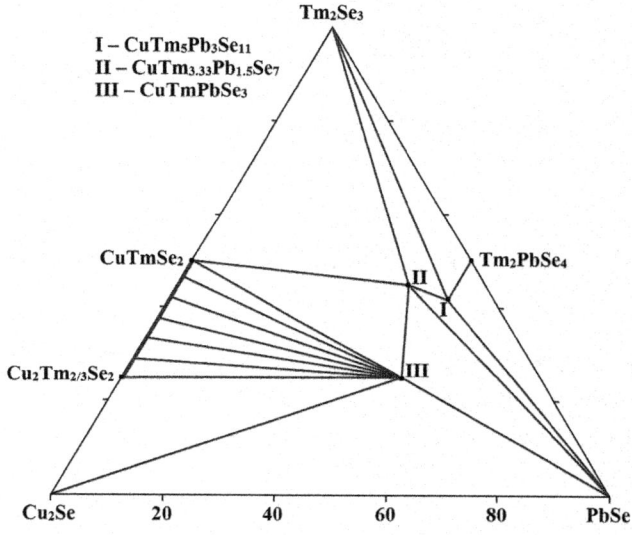

FIGURE 11.7 Isothermal section of the PbSe–Cu₂S–Tm₂Se₃ quasiternary system at 600°C. (From Gulay, L.D., *Pol. J. Chem.*, **80**(10), 1703, 2006.)

in the orthorhombic structure with the lattice parameters $a = 406.88 \pm 0.02$, $b = 1346.07 \pm 0.07$, and $c = 380.40 \pm 0.02$ pm (Gulay et al. 2006e) [$a = 407.10 \pm 0.08$, $b = 1348.0 \pm 0.2$, $c = 3809.2 \pm 0.7$ pm, and a calculated density of 7.594 g·cm⁻³ (Gulay et al. 2006f)]. All these compounds were synthesized in the same way as CuGdPbSe₃ was obtained.

11.23 Lead–Copper–Thulium–Selenium

PbSe–Cu₂Se–Tm₂Se₃. The isothermal section of this quasiternary system at 600°C was constructed using the alloys annealed at this temperature for 240 h (Figure 11.7) (Gulay et al. 2006h).

Three quaternary compounds, **CuTmPbSe₃**, **CuTm₃.₃₃Pb₁.₅Se₇**, and **CuTm₅Pb₃Se₁₁**, are formed in this system. First of them crystallizes in the orthorhombic structure with the lattice parameters $a = 1044.7 \pm 0.1$, $b = 404.00 \pm 0.04$, $c = 1342.3 \pm 0.1$ pm, and a calculated density of 7.931 g·cm⁻³ (Gulay and Olekseyuk 2005a). CuTm₃.₃₃Pb₁.₅Se₇ crystallizes in the monoclinic structure with the lattice parameters $a = 1345.84 \pm 0.07$, $b = 405.60 \pm 0.02$, $c = 1250.83 \pm 0.06$ pm, $\beta = 104.342 \pm 0.003°$, and a calculated density of 7.482 g·cm⁻³ (Gulay and Olekseyuk 2005b). CuTm₅Pb₃Se₁₁ crystallizes in the orthorhombic structure with the lattice parameters $a = 405.82 \pm 0.03$, $b = 1342.9 \pm 0.1$, $c = 379.79 \pm 0.02$ pm, and a calculated density of 7.696 g·cm⁻³ (Gulay et al. 2006f). All these compounds were synthesized in the same way as CuGdPbSe₃ was obtained.

11.24 Lead–Copper–Ytterbium–Selenium

Three quaternary compounds, **CuYbPbSe₃**, **CuYb₃.₃₃Pb₁.₅Se₇**, and **CuYb₅Pb₃Se₁₁**, are formed in the Pb–Cu–Yb–Se system. First of them crystallizes in the orthorhombic structure with the lattice parameters $a = 1040.7 \pm 0.2$, $b = 403.50 \pm 0.08$, $c = 1340.1 \pm 0.3$ pm, and a calculated density of 8.034 g·cm⁻³ (Gulay et al. 2006a) [$a = 1039.88 \pm 0.05$, $b = 402.85 \pm 0.02$, $c = 1340.89 \pm 0.05$ pm, and a calculated density of 8.048 g·cm⁻³ (Gulay and Olekseyuk 2005a)]. CuYb₃.₃₃Pb₁.₅Se₇ crystallizes in the monoclinic structure with the lattice parameters $a = 1342.5 \pm 0.2$, $b = 404.37 \pm 0.04$, $c = 1248.4 \pm 0.1$ pm, and $\beta = 104.381 \pm 0.008°$ (Gulay and Olekseyuk 2005b). CuYb₅Pb₃Se₁₁ crystallizes in the orthorhombic structure with the lattice parameters $a = 404.87 \pm 0.03$, $b = 1339.93 \pm 0.09$, $c = 378.76 \pm 0.02$ pm, and a calculated density of 7.818 g·cm⁻³ (Gulay et al. 2006f). All these compounds were synthesized in the same way as CuGdPbSe₃ was obtained.

11.25 Lead–Copper–Lutetium–Selenium

PbSe–Cu₂Se–Lu₂Se₃. The isothermal section of this quasiternary system at 600°C was constructed using the alloys annealed at this temperature for 240 h (Figure 11.8) (Gulay et al. 2006h). Two quaternary compounds, **CuLuPbSe₃** and **CuLu₃.₃₃Pb₁.₅Se₇**, are formed in this system. First of them crystallizes in the orthorhombic structure with the lattice parameters $a = 1038.57 \pm 0.06$, $b = 402.48 \pm 0.02$, $c = 1340.56 \pm 0.06$ pm, and a calculated density of 8.090 g·cm⁻³ (Gulay and Olekseyuk 2005a) and the second one crystallizes in the monoclinic structure with the lattice parameters $a = 1340.4 \pm 0.2$, $b = 403.07 \pm 0.06$, $c = 1247.5 \pm 0.2$ pm, $\beta = 104.36 \pm 0.01°$, and a calculated density of 7.683 g·cm⁻³ (Gulay and Olekseyuk 2005b). Both compounds were synthesized in the same way as CuGdPbSe₃ was obtained.

11.26 Lead–Copper–Arsenic–Selenium

PbSe–CuAsSe₂. This system is nonquasibinary section of the PbSe–Cu₂Se–As₂Se₃ quasiternary system as CuAsSe₂ melts incongruently (Figure 11.9) (Ostapyuk et al. 2014). The

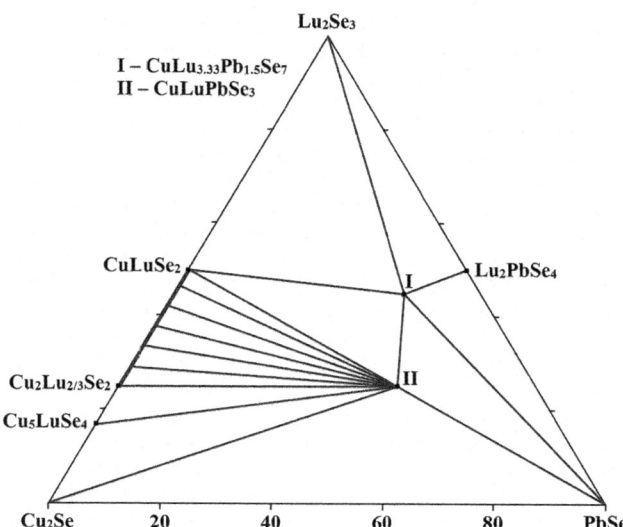

FIGURE 11.8 Isothermal section of the PbSe–Cu₂Se–Lu₂Se₃ quasiternary system at 600°C. (From Gulay, L.D., *Pol. J. Chem.*, **80**(10), 1703, 2006.)

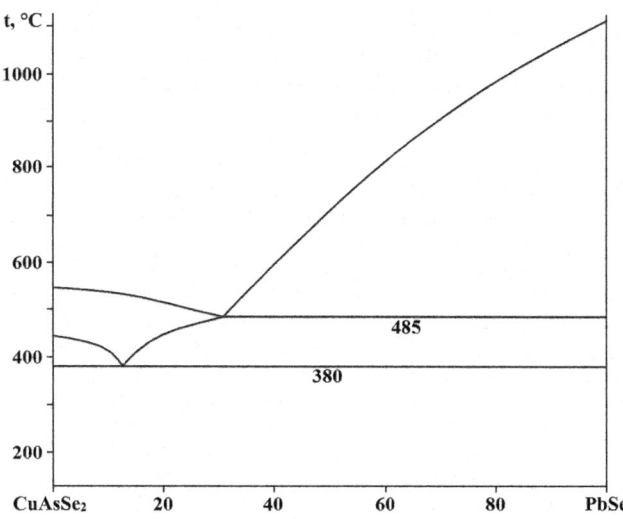

FIGURE 11.9 Phase relations in the PbSe–CuAsSe₂ system. (From Ostapyuk, T.A., et al., *Chem. Met. Alloys*, **7**(1–2), 164, 2014.)

liquidus of this section consists of the lines of primary crystallization of Cu₂Se and PbSe. The two fields of secondary crystallization are L + CuAsS₂ + Cu₂Se and L + Cu₂Se + PbSe. The horizontal line at 380°C results from the intersection with the plane of the transition process L (U₁) +Cu₂Se ↔ CuAsSe₂ + PbSe. This section ends with the disappearance of the liquid and Cu₂Se crystals, so the alloys are two-phase below 380°C. This horizontal line is the section solidus.

PbSe–Cu₂Se–As₂Se₃. The liquidus surface of this quasiternary system (Figure 11.10) consists of the five fields of primary crystallization of Cu₂Se, PbSe, As₂Se₃, and CuAsSe₂ (two of them belong to PbSe) (Ostapyuk et al. 2014). A region of immiscibility exists in the system. The major part of the triangle is occupied by the fields of primary crystallization of PbSe and Cu₂Se, which have the highest melting points. The fields

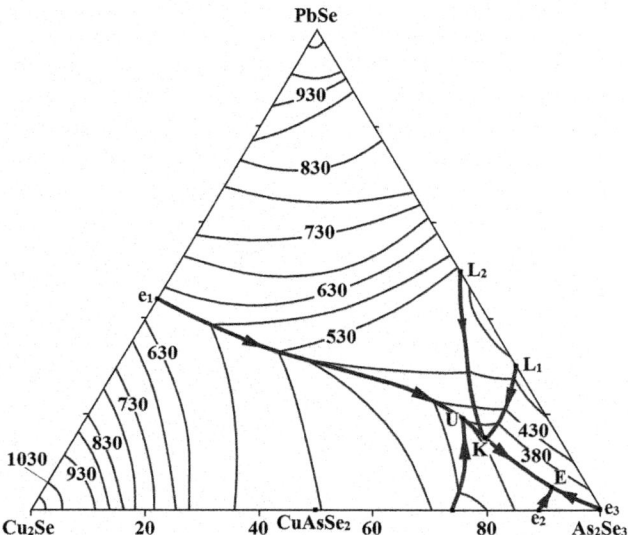

FIGURE 11.10 Liquidus surface of the PbSe–Cu$_2$Se–As$_2$Se$_3$ quasiternary system. (From Ostapyuk, T.A., et al., *Chem. Met. Alloys*, 7(1–2), 164, 2014.)

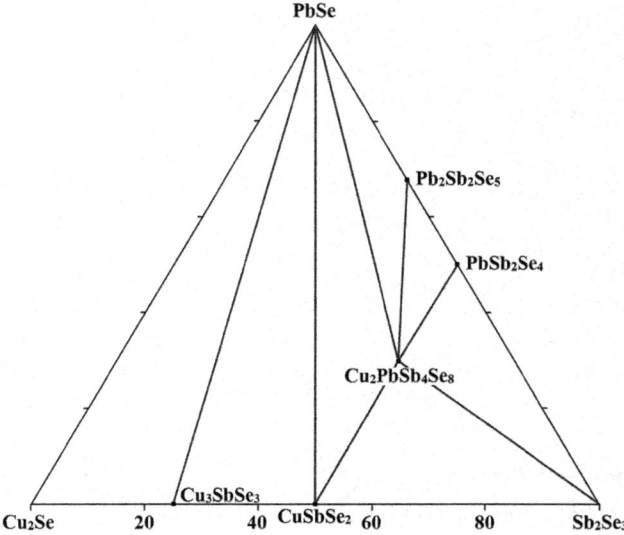

FIGURE 11.11 Isothermal section of the PbSe–Cu$_2$Se–Sb$_2$Se$_3$ quasiternary system at 600°C. (From Olekseyuk, I.D., et al., *Nauk. Visnyk Volyns'k. Nats. Univ. im. Lesi Ukrainky. Ser. Khim. nauky*, (16), 38, 2010.)

of the primary crystallization are separated by monovariant lines that cross in the invariant points *U* and *E*. Some vertical sections were also constructed, and the reaction scheme was given.

11.27 Lead–Copper–Antimony–Selenium

PbSe–Cu$_2$Se–Sb$_2$Se$_3$. The isothermal section of this quasiternary system at 350°C was constructed using the alloys annealed at this temperature for 600 h (Figure 11.11) (Olekseyuk et al. 2010b). The **Cu$_2$PbSb$_4$Se$_8$** quaternary compound exists in this system.

11.28 Lead–Copper–Bismuth–Selenium

Two quaternary compounds, **CuPbBiSe$_3$** (mineral cerromojonite) and **(Cu,□)$_6$(Pb,Bi)Se$_4$** (mineral schlemaite), are formed in the Pb–Cu–Bi–Se system. CuPbBiSe$_3$ melts at 505°C (Guseynov et al. 1972), and has apparently two polymorphic modifications. The first, the mineral cerromojonite, crystallizes in the orthorhombic structure with the lattice parameters *a* = 820.2 ± 0.1, *b* = 874.1 ± 0.1, *c* = 802.9 ± 0.1 pm, and a calculated density of 7.035 g·cm^{-3} (Förster et al. 2018a, b, c). The second modification of this compound crystallizes in the cubic structure with the lattice parameter *a* = 604.6 pm, the calculated and experimental density of 5.38 and 5.34 g·cm^{-3}, respectively, and an energy gap of 0.07 eV (Guseynov et al. 1972).

(Cu,□)$_6$(Pb,Bi)Se$_4$ crystallizes in the monoclinic structure with the lattice parameters *a* = 952.9 ± 0.4, *b* = 411.5 ± 0.2, *c* = 1023.7 ± 0.5 pm, *β* = 100.29 ± 0.05°, and a calculated density of 7.54 g·cm^{-3} from powder XRD and *a* = 953.41 ± 0.08, *b* = 410.04 ± 0.03, *c* = 1025.46 ± 0.08 pm, *β* = 100.066 ± 0.002° from single crystal data (Förster et al. 2003; Jambor et al. 2004).

11.29 Lead–Copper–Oxygen–Selenium

The **PbCu$_2$(SeO$_3$)$_3$** quaternary compound, which crystallizes in the triclinic structure with the lattice parameters *a* = 781.3 ± 0.1, *b* = 911.6 ± 0.1, *c* = 1257.0 ± 0.1 pm, *α* = 82.27 ± 0.01°, *β* = 72.90 ± 0.01°, *γ* = 89.69 ± 0.01° and a calculated density of 5.61 g·cm^{-3}, is formed in the Pb–Cu–O–Se system (Effenberger 1988). The crystals of this compound were synthesized under hydrothermal conditions in a steel vessel lined with Teflon. 2 g of an equimolar mixture of PbO, CuO, and SeO$_2$ were put into the vessel of ~6 mL capacity; 1 mL H$_2$O$_2$ was added, and the vessel was filled with H$_2$O to about 80 vol%. After heating for 2 days at 220°C and after cooling to room temperature (12 h), the green crystals of the title compound were obtained.

11.30 Lead–Copper–Tellurium–Selenium

PbTe–Cu$_2$Se. This system is nonquasibinary section of the PbSe + Cu$_2$Te ↔ PbTe + Cu$_2$Se ternary mutual system (Grytsiv et al. 1996). The PbSe$_x$Te$_{1-x}$ and Cu$_2$Se$_x$Te$_{1-x}$ solid solutions primarily crystallize, and the latter, upon cooling, decompose into solid solutions enriched, respectively, in Cu$_2$Se and Cu$_2$Te. The liquidus curve of this system intersects the line of secondary precipitation of the eutectic alloy at 550°C and 64 mol% Cu$_2$Se. The samples for the investigations were annealed at temperatures 50°C lower than solidus temperatures for 500–600 h.

11.31 Lead–Copper–Iron–Selenium

The **Cu$_2$FePbSe$_4$** quaternary compound, which melts congruently at 557°C and crystallizes in the orthorhombic structure with the lattice parameters *a* = 767.6, *b* = 669.1, and *c* = 635.5

pm (Quintero et al. 1999), is formed in the Pb–Cu–Fe–Se system. The samples of this compound were prepared by the melt and anneals technique. The components of 1-g sample were made from the appropriate amounts of the elements and were sealed under vacuum in a small quartz ampoule, which had previously been carbonized to prevent interaction of the components with the quartz. The components were melted together at 1150°C for about an hour, annealed to equilibrium at 500°C, and then cooled to room temperature by leaving the ampoule in the switched-off furnace.

11.32 Lead–Silver–Gallium–Selenium

PbSe–AgGaSe₂. The phase diagram of this system, constructed through DTA and XRD, is a eutectic type (Figure 11.12) (Vlayet and Roubin 1978). The eutectic contains 45 mol% AgGaSe₂ and crystallizes at 655°C. The mutual solubility of PbSe and AgGaSe₂ is negligible.

PbSe–Ag₉GaSe₆. The phase diagram of this system, constructed through DTA and XRD, is a eutectic type (Figure 11.13) (Vlayet and Roubin 1978). The eutectic contains 65 mol% PbSe and crystallizes at 607°C. At high temperatures, the limited solid solution is formed on the base of Ag₉GaSe₆. The solubility of PbSe in Ag₉GaSe₆ at room temperature is negligible.

11.33 Lead–Silver–Praseodymium–Selenium

PbSe–Ag₂Se–Pr₂Se₃. The isothermal section of this quasiternary system at 600°C was constructed using the alloys annealed at this temperature for 240 h (Figure 11.14) (Marchuk et al. 2006b). No quaternary compounds were found in the system. A solid solution range for PbSe was established in the system.

11.34 Lead–Silver–Antimony–Selenium

PbSe–AgSbSe₂. The phase diagram of this system, constructed through DTA, XRD, and metallography, is given in Figure 11.15 (Mansimova et al. 2018). This system is characterized by peritectic equilibrium at 18 mol% AgSbSe₂ and 947°C. The solubility of PbSe in AgSbSe₂ is 87 mol% at 997°C and decreases to 80 mol% at room temperature. PbSe dissolves 5 mol% AgSbSe₂ at 997°C and ~2 mol% at room temperature. According to the data of Mansimova et al. (2019b), in this system at 530°C there is a two-phase region (PbSe) + (AgSbSe₂) within the interval 3–39 mol% AgSbS₂.

The results of the study of this system (of measuring the EMF of concentration chains with respect to PbSe in a temperature

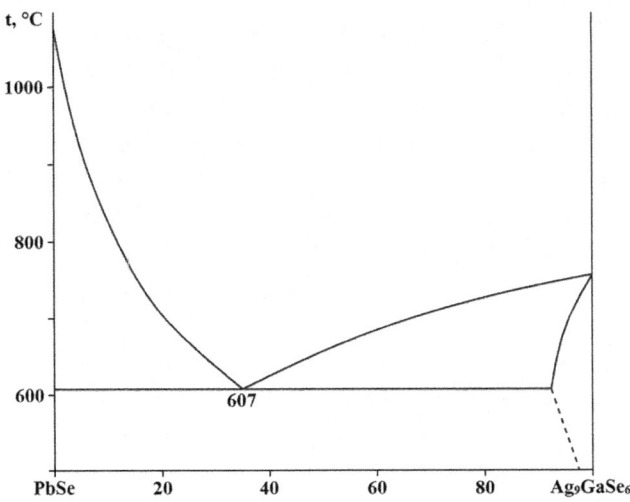

FIGURE 11.13 Phase diagram of the PbSe–Ag₉GaSe₆ system. (From Vlayet, F., and Roubin, M., *C. r. Acad. Sci Ser. C.*, **286**(9), 285, 1978.)

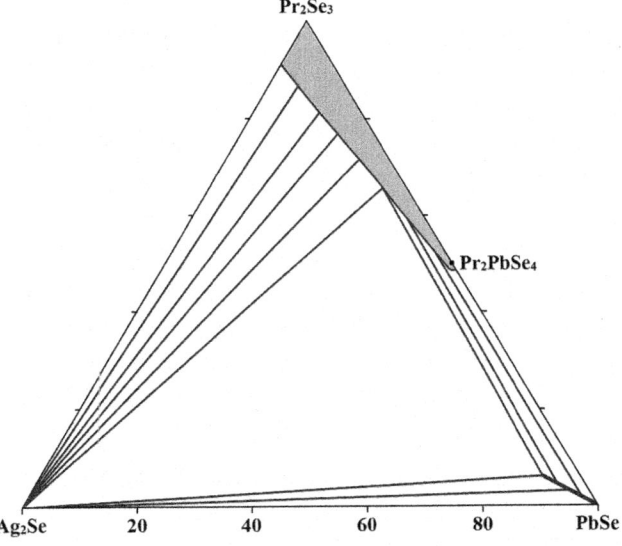

FIGURE 11.14 Isothermal section of the PbSe–Ag₂Se–Pr₂Se₃ quasiternary system at 600°C. (From Marchuk, O.V., et al., *J. Alloys Compd.*, **416**(1–2), 106, 2006.)

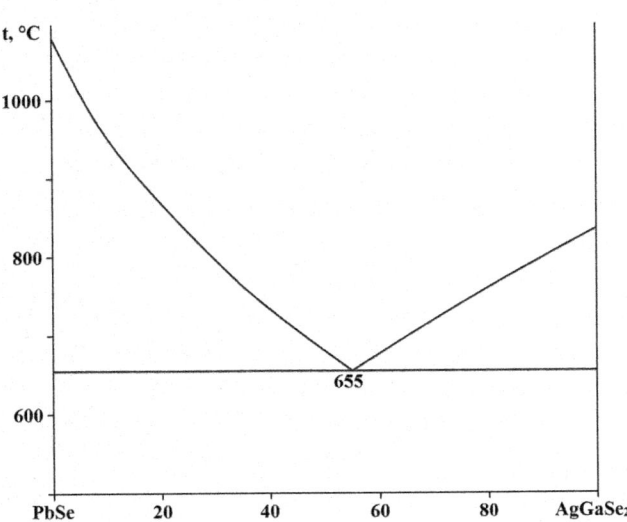

FIGURE 11.12 Phase diagram of the PbSe–AgGaSe₂ system. (From Vlayet, F., and Roubin, M., *C. r. Acad. Sci Ser. C.*, **286**(9), 285, 1978.)

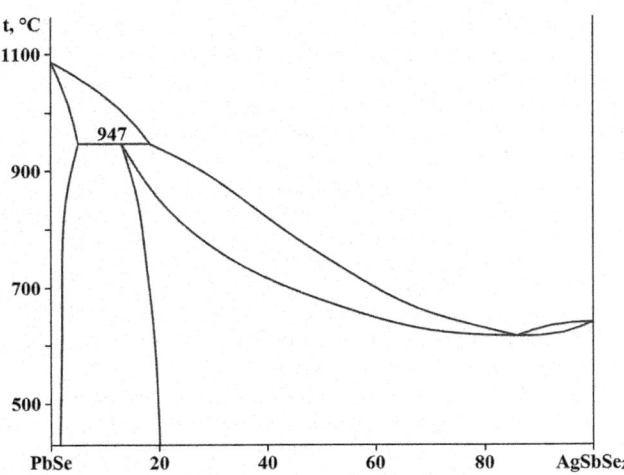

FIGURE 11.15 Phase diagram of the PbSe–AgSbSe₂ system. (From Mansimova, Sh.H., et al., *Chem. Probl.*, [4(16)], 530, 2018.)

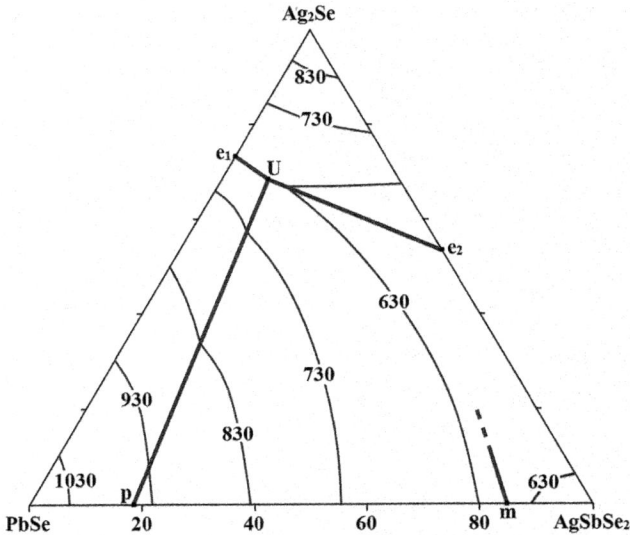

FIGURE 11.17 Liquidus surface of the PbSe–Ag₂Se–AgSbSe₂ quasiternary system. (From Mansimova, et al., *Chem. Probl.*, (1), 41, 2019.)

three-phase areas. The liquidus surface of this system (Figure 11.17) consists of three fields of primary crystallization, and the field of primary crystallization of solid solution based on AgSbSe₂ has the largest area. There is one transition point in the system: U (637°C) – L + (PbSe) ↔ Ag₂Se + (AgSbSe₂). Two vertical sections of this quasiternary system were also constructed.

Electron diffraction and high-resolution transmission electron microscopy studies of the **AgPb₁₈SbSe₂₀** crystals reveal that it is nanostructured rather than a solid solution (Lioutas et al. 2010). Nanocrystals of varying sizes were found, endotaxially grown in the matrix of PbSe, which consist of two phases, a cubic one and a tetragonal one. Well-defined coherent interfaces between the phases in the same nanocrystals were observed. Ingots with a nominal composition AgPb₁₈SbSe₂₀ (18PbSe + AgSbSe₂) were prepared by annealing, in quartz tubes under vacuum, mixtures of appropriate stoichiometric quantities of Ag, Pb, Sb, and Se at 950°C for 4 h and cooling to 450°C in 40 h or to room temperature over 48 h.

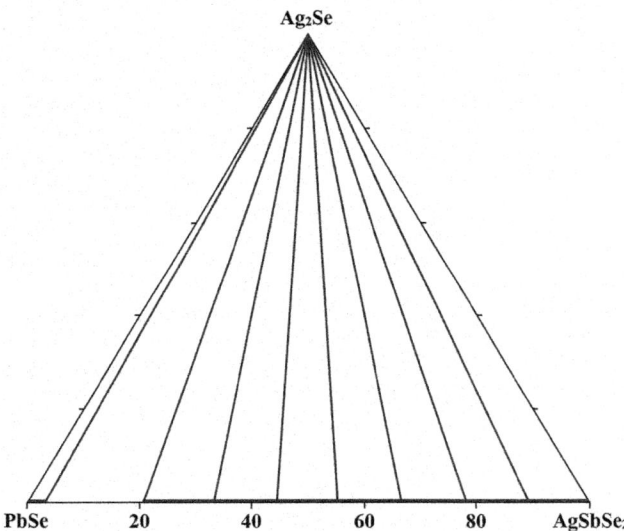

FIGURE 11.16 Isothermal section of the PbSe–Ag₂Se–AgSbSe₂ quasiternary system at room temperature. (From Mansimova, et al., *Chem. Probl.*, (1), 41, 2019.)

11.35 Lead–Silver–Bismuth–Selenium

PbSe–AgBiSe₂. The phase diagram of this system, constructed through DTA, XRD, and measuring of the microhardness using the alloys annealed at 630°C for 400 h, is given in Figure 11.18 (Aliev et al. 2008). This system contains a continuous series of cubic solid solutions with the NaCl-type structure. Their lattice parameter is an almost linear function of composition. The formation of the solid solutions stabilizes the high-temperature phase of AgBiSe₂: PbSe dissolution in this compound markedly reduces its polymorphic transformation temperature (317°C) down to room temperature at ≈10 mol% PbSe.

This system was also studied by the EMF method with solid electrolyte Ag₄RbI₅ (Babanly et al. 2012). On the basis of experimental data, the formation of a wide range of solid

range of 30–180°C) were presented by Mashadiyeva et al. (2020a). The formation in the system of a wide (37–100 mol% AgSbSe₂) region of solid solutions based on AgSbSe₂ was shown. The partial thermodynamic functions of PbSe and lead in the alloys were calculated from the equations of the temperature dependences of the EMF. The standard thermodynamic functions of formation and standard entropies of solid solutions (2PbSe)ₓ(AgSbSe₂)₁₋ₓ were calculated by the integration of the Gibbs–Duhem equation.

PbSe–Ag₂Se–AgSbSe₂. The isothermal section of this quasiternary system at room temperature was constructed using the alloys annealed at 500°C for 500 h (Figure 11.16) (Mansimova et al. 2019a). The section is divided into 2 two-phase and 1

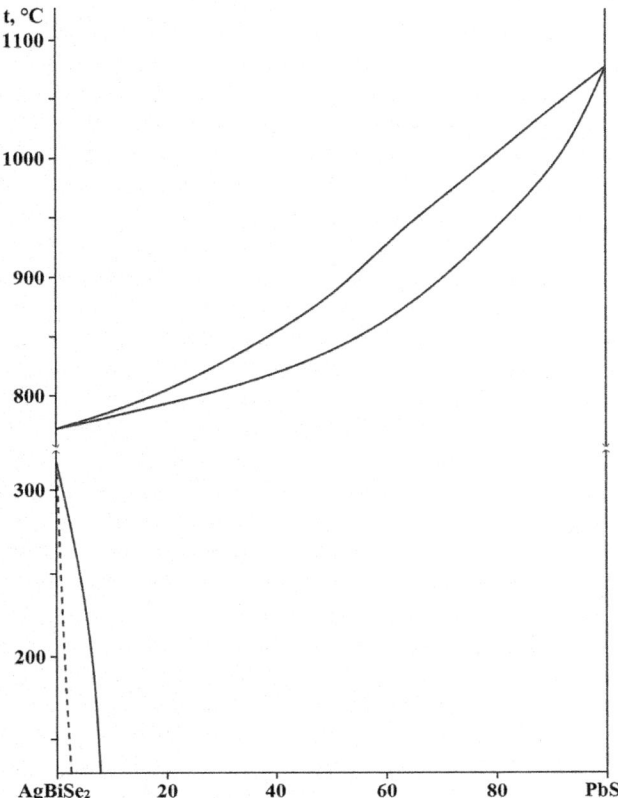

FIGURE 11.18 Phase diagram of the PbSe–AgBiSe₂ system. (From Aliev I.I., et al., *Inorg. Mater.*, **44**(11), 1179, 2008.)

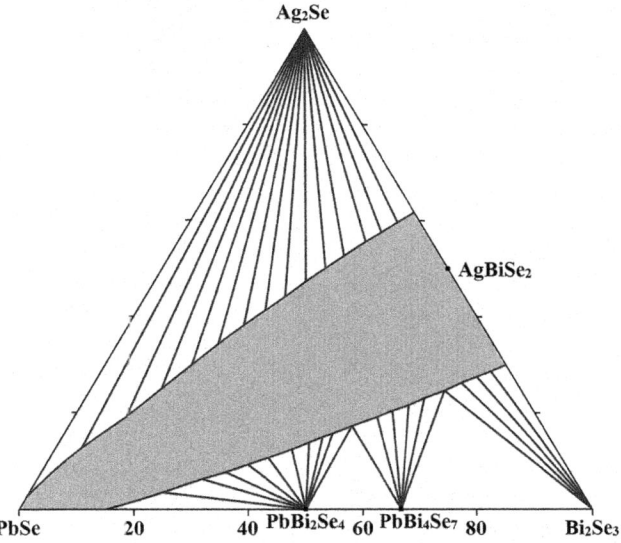

FIGURE 11.19 Isothermal section of the PbSe–Ag₂Se–Bi₂Se₃ quasiternary system at 630°C. (From Aliev I.I., et al., *Inorg. Mater.*, **44**(11), 1179, 2008.)

solutions (up to 90 mol% AgBiSe₂) based on PbSe was established. The partial thermodynamic functions of Ag in alloys were calculated, and the thermodynamic functions of formation and standard entropies of solid solutions were determined. The ingots for the investigations were annealed at 430°C for 1000 h.

Pb₅Bi₆Se₁₄–Ag₂Se. This system is nonquasibinary, as Pb₅Bi₆Se₁₄ melts incongruently (Babanly 2011). The solubility of the initial components in each other is negligible. An intermediate phase of variable composition is formed in the system, which crystallizes in a cubic structure with a region of homogeneity over this section of 25 mol% at 630°C (13–38 mol% Ag₂Se). As the temperature decreases, the homogeneity region of the phase sharply narrows, and at 130°C, it degenerates at a composition of ~28 mol% Ag₂Se.

PbSe–Ag₂Se–Bi₂Se₃. The isothermal section of this quasiternary system at 630°C was constructed using the alloys annealed at this temperature for 400 h (Figure 11.19) (Aliev et al. 2008). In the system, the solid solution exists in a broad region around the PbSe–AgBiSe₂ join.

Two quaternary compounds, **AgPbBiSe₃** and **Ag₂PbBi₄Se₈**, were found in this system. The first does not melt below 900°C [melts at 741°C (Guseynov et al. 1972)], has a phase transformation at around 820°C, and crystallizes in the cubic structure with the lattice parameter *a* = 595.4 ± 0.1 pm and an energy gap of 0.48 eV (Sportouch et al. 1998) [0.13 eV (Guseynov et al. 1972)]. It was synthesized by combining the chemical elements in stoichiometric ratios under a static vacuum at 900°C

for 3 days (Sportouch et al. 1998). According to the data of Aliev et al. (2008), this compound is one of the compositions of a continuous solid solution.

Ag₂PbBi₄Se₈ (mineral litochlebite) crystallizes in the monoclinic structure with the lattice parameters *a* = 1318.2 ± 0.2, *b* = 418.40 ± 0.08, *c* = 1529.9 ± 0.2 pm, *β* = 109.11 ± 0.01°, and a calculated density of 7.90 g·cm⁻³ (Sejkora et al. 2011a; Gatta et al. 2012).

11.36 Lead–Silver–Iodine–Selenium

PbI₂–Ag₂Se. The phase diagram of this system, constructed through DTA, XRD, and metallography, is a eutectic type (Figure 11.20) (Odin et al. 2005). The eutectic contains 13 mol% Ag₂Se and crystallizes at 386°C. The thermal effects at 128°C correspond to the polymorphic transformation of Ag₂Se. Ag₂Se dissolves less than 0.5 mol% PbI₂ and PbI₂ dissolves less than 1 mol% Ag₂Se. A metastable phase μ, which crystallizes in the rhombohedral structure, is formed in this system within the interval of 13–25 mol% Ag₂Se. This phase was obtained by the quenching method.

PbSe–Ag₂Se–PbI₂. On the isothermal section of this quasiternary system at 360°C, solid solutions based on the starting compounds are in equilibrium. On the liquidus surface of this system (Figure 11.21), there are three regions of primary crystallization of solid solutions based on PbSe, Ag₂Se, and PbI₂ (Odin et al. 2005). The ternary eutectic *E* crystallizes at 372°C.

11.37 Lead–Silver–Iron–Selenium

The **Ag₂FePbSe₄** quaternary compound, which melts congruently at 562°C and crystallizes in the orthorhombic structure with the lattice parameters *a* = 711.6, *b* = 693.0, and *c* = 684.0

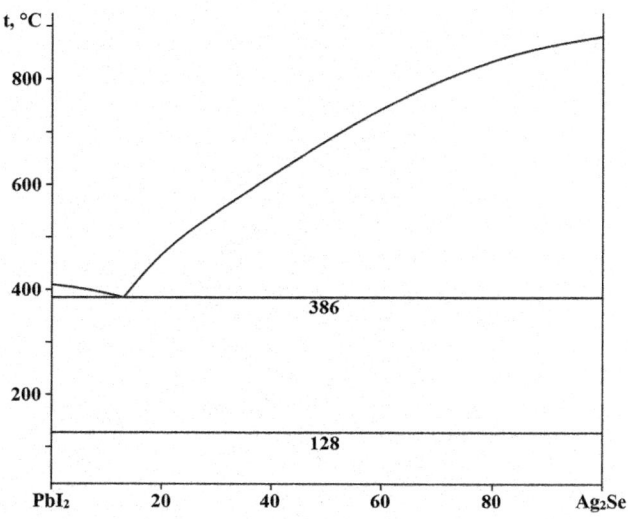

FIGURE 11.20 Phase diagram of the PbI₂–Ag₂Se system. (From Odin, I.N., et al., *Zhurn. neorgan. khimii*, **50**(5), 843, 2005.)

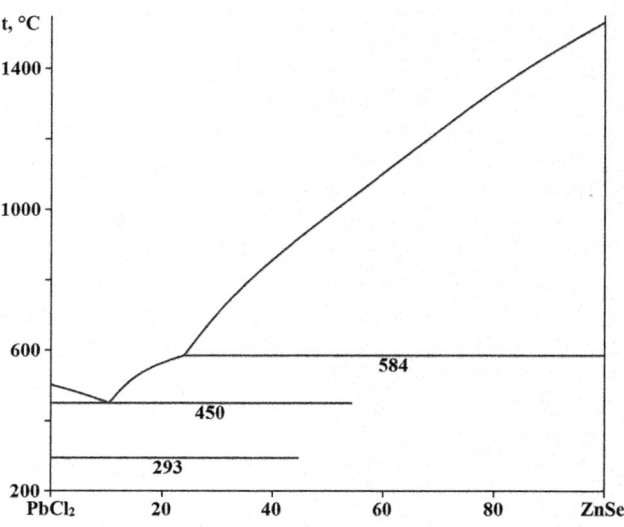

FIGURE 11.22 Phase diagram of the PbCl₂–ZnSe system. (From Triboulet, R., et al., *J. Cryst. Growth*, **59**(1–2), 172, 1982.)

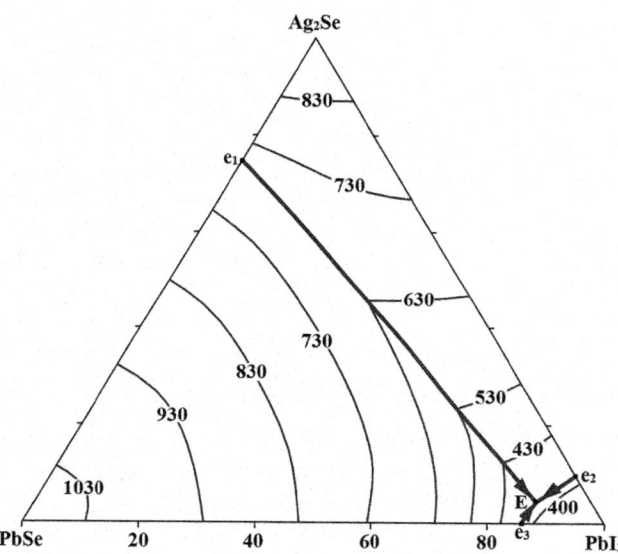

FIGURE 11.21 Liquidus surface of the PbSe–Ag₂Se–PbI₂ quasiternary system. (From Odin, I.N., et al., *Zhurn. neorgan. khimii*, **50**(5), 843, 2005.)

pm (Quintero et al. 1999), is formed in the Pb–Ag–Fe–Se system. The samples of this compound were prepared in the same way as Cu₂FePbSe₄ was synthesized.

The calculated standard thermodynamic values of the Gibbs energy, enthalpy of formation, and entropy of Ag₂FePbSe₄ are the next: $\Delta G^0_{f,298}$ = −248.9 ± 15.0 and −250.5 ± 15.0 kJ·M⁻¹, $\Delta H^0_{f,298}$ = −246.3 ± 15.0 and −248.5 ± 15.0 kJ·M⁻¹, and $S^0_{f,298}$ = 355.0 ± 8.1 and 353.1 ± 8.0 J·M⁻¹K⁻¹ (Moroz et al. 2020b). The calculation was performed in two fundamentally different potential-forming processes, the decomposition of Ag₂FePbSe₄ into Ag₂Se, PbSe, and AgFeSe₂; and the synthesis of Ag₂FePbSe₄ from PbSe, FeSe₂, Ag, and Se. This compound obtained from the melt decomposes during annealing.

11.38 Lead–Strontium–Bismuth–Selenium

The **Sr₂Pb₂Bi₆Se₁₃** quaternary compound, which crystallizes in the monoclinic structure, is formed in the Pb–Sr–Bi–Se system (Kanatzidis et al. 1998).

11.39 Lead–Barium–Bismuth–Selenium

Two quaternary compounds, **Ba₃PbBi₆Se₁₃** and **Ba₃₊ₓPb₃₊ₓBi₆Se₁₅**, are formed in the Pb–Ba–Bi–S system. First of them is a no degenerate semiconductor and crystallizes in the monoclinic structure with the lattice parameters a = 1724.27 ± 0.04, b = 427.36 ± 0.01, c = 1845.60 ± 0.04 pm, β = 90.861 ± 0.001°, and a calculated density of 7.081 g·cm⁻³ at 168 K (Wang and DiSalvo 2000). This compound was synthesized from a stoichiometric mixture of the elements, which was put in a vitreous carbon crucible and sealed in an evacuated quartz tube. Then it was gradually heated to 710°C in 48 h, kept at that temperature for 100 h, and slowly cooled to 510°C at 2°C·h⁻¹. The black ingot of Ba₃PbBi₆Se₁₃ was obtained. Due to the air sensitivity of barium, all work was carried out under an argon atmosphere in a vacuum-dry box.

Ba₃₊ₓPb₃₊ₓBi₆Se₁₅ also crystallizes in the monoclinic structure with the lattice parameters a = 1734.4 ± 0.2, b = 429.25 ± 0.06, c = 2183.5 ± 0.3 pm, β = 98.629 ± 0.002°, and an energy gap of 0.57 eV (Iordanidis et al. 1998). It was synthesized by reacting BaSe, Pb, Se, and Bi₂Se₃ (molar ratio 1:1:1:1.5) at 900°C for 6 days.

11.40 Lead–Zinc–Chlorine–Selenium

PbCl₂–ZnSe. The phase diagram of this system, constructed through DTA, is shown in Figure 11.22 (Triboulet et al. 1982). The eutectic composition and temperature are 10.5 mol% ZnSe and 450°C, respectively. At 584°C, a peritectic transformation takes place in this system. The composition of peritectic point corresponds to 24 mol% ZnSe (eutectic and peritectic compositions and temperatures are taken from Figure 11.22).

11.41 Lead–Zinc–Bromine–Selenium

PbBr₂–ZnSe. According to the data of (Triboulet et al. 1982), the phase diagram of this system, constructed through DTA, is a eutectic type with a eutectic that crystallizes at approximately 340°C and is situated near PbBr₂ composition.

11.42 Lead–Cadmium–Gallium–Selenium

PbSe–CdGa₂Se₄. The phase diagram of this system, constructed through DTA, XRD, and metallography, is a eutectic type (Figure 11.23) (Sosovska et al. 2006). The eutectic contains ~63 mol% PbSe and crystallizes at 700°C. The solid solubility of both components does not exceed 2 mol%. The invariant line at 817°C corresponds to the phase transformation of CdGa₂Se₄. The alloys for the investigations were annealed at 400°C for 250 h followed by quenching in cold water.

PbSe–CdSe–Ga₂Se₃. The isothermal section of this system at 600°C is shown in Figure 11.24 (Sosovska et al. 2008). The alloys for the investigations were annealed at this temperature for over 250 h, followed by rapid quenching in cold water.

The liquidus surface of this system is shown in Figure 11.25 (Sosovska et al. 2008). It includes six fields of primary crystallization. Three of them belong to the system components (namely, to their solid solutions). The other three fields correspond to the primary crystallization of the ternary compounds α- and β-CdGa₂Se₄ and PbGa₂Se₄. The aforementioned fields are separated with 11 monovariant lines. There are 7 binary and 5 ternary invariant points. Temperatures and the corresponding

reaction of the ternary invariant points are the next: E_1 (660°C) – L ↔ β-CdGa₂Se₄+ PbGa₂Se₄+ PbSe; E_2 (702°C) – L ↔ CdSe + β-CdGa₂Se₄+ PbSe; U_1 (712°C) – L + Ga₂Se₃ ↔ β-CdGa₂Se₄+ PbGa₂Se₄; U_2 (816°C) – L + α-CdGa₂Se₄ ↔ Ga₂Se₃ + β-CdGa₂Se₄; U_3 (816°C) – L + α-CdGa₂Se₄ ↔ CdSe + β-CdGa₂Se₄.

11.43 Lead–Cadmium–Tellurium–Selenium

PbSe–CdTe. This section is nonquasibinary section of the PbSe + CdTe ↔ PbTe + CdSe ternary mutual system (Raevskiy et al. 1982). The CdSe$_x$Te$_{1-x}$ solid solutions undergo polymorphous transformation within the interval of 60–70 mol% CdTe;

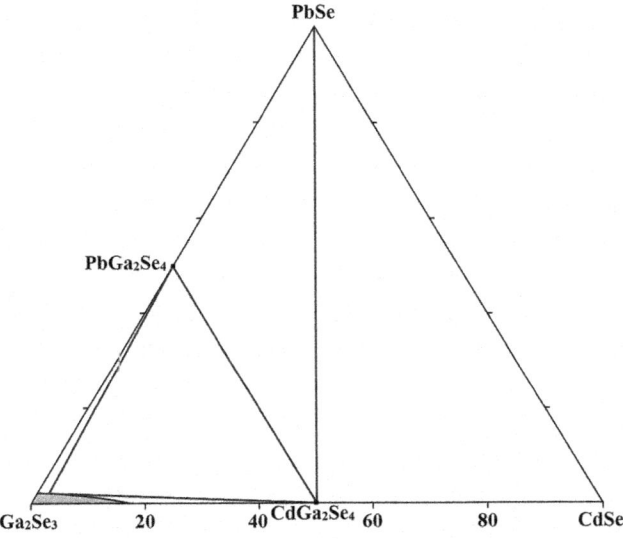

FIGURE 11.24 Isothermal section of the PbSe–CdSe–Ga₂Se₃ quasiternary system at 600°C. (From Sosovska, S.M. et al., *J. Alloys Compd.*, **453**(1–2), 115, 2008.)

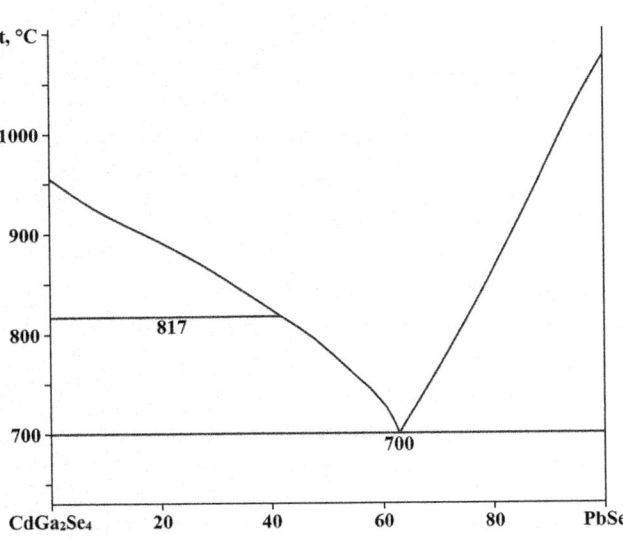

FIGURE 11.23 Phase diagram of the PbSe–CdGa₂Se₄ system. (From Sosovska, S.M., et al., *J. Alloys Compd.*, **425**, 206, 2006.)

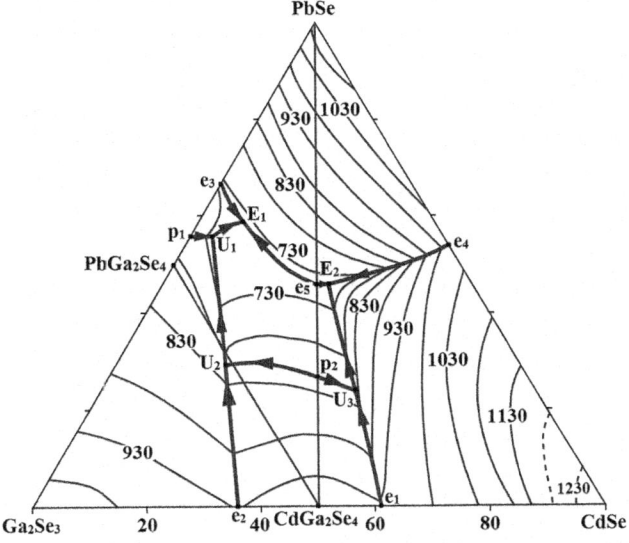

FIGURE 11.25 Liquidus surface of the PbSe–CdSe–Ga₂Se₃ quasiternary system. (From Sosovska, S.M. et al., *J. Alloys Compd.*, **453**(1–2), 115, 2008.)

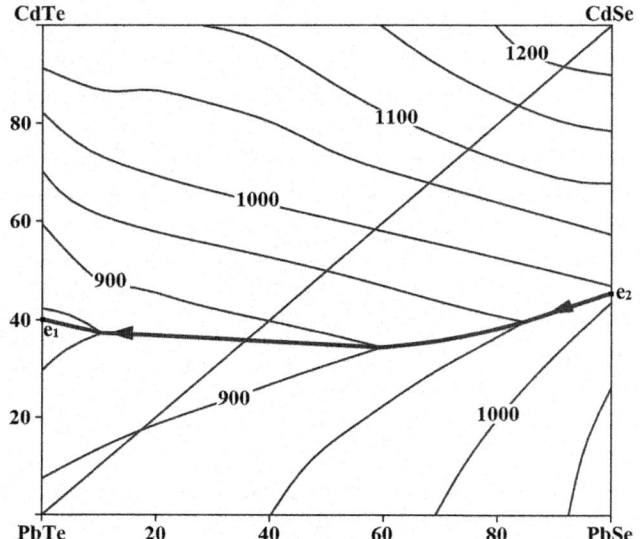

FIGURE 11.26 Liquidus surface of the PbSe + CdTe ↔ PbTe + CdSe ternary mutual system. (From Tomashik, Z.F., *Izv. AN SSSR. Neorgan. mater.*, **17**(9), 1575, 1981.)

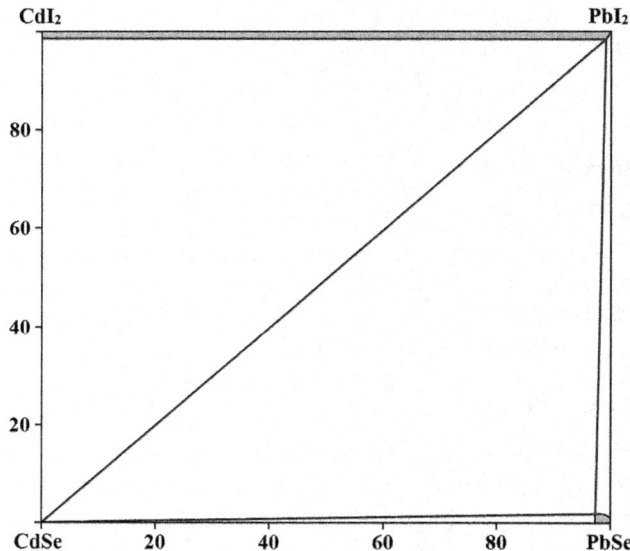

FIGURE 11.27 Isothermal section of the PbSe + CdI₂ ↔ PbI₂ + CdSe ternary mutual system at 350°C. (From Odin, I.N., *Zhurn. neorgan. khimii*, **46**(10), 1733, 2001.)

therefore, three phases (solid solutions based on PbSe and CdSe$_x$Te$_{1-x}$ solid solutions with the sphalerite and wurtzite structures) are in equilibrium in the narrow region of compositions (not higher than 3 mol%). Three-phase equilibrium takes place at 885°C and 38 mol% CdTe. The maximal solubility of CdTe in PbSe is equal to 27.5 mol% at 880°C, and the solubility of PbSe in CdTe at 850°C is not higher than 2.5 mol%. Lattice parameters of the solid solutions based on PbSe change linearly with composition. This system was investigated through DTA, metallography, XRD, and measuring of the microhardness. The ingots for the investigations were annealed at 590°C, 810°C, and 830°C.

PbTe–CdSe. This section is also nonquasibinary section of the PbSe + CdTe ↔ PbTe + CdSe ternary mutual system (Raevskiy et al. 1983a, b). Three phases are in equilibrium in the solid state: PbSe$_x$Te$_{1-x}$ solid solutions with the cubic structure of NaCl type, and CdSe$_x$Te$_{1-x}$ solid solutions with the sphalerite and wurtzite structure. The solubility of CdSe in PbTe at 870°C is equal to 16 mol% [30 mol% at 600°C (Nikolić 1966)], and the solubility of PbTe in CdSe is not higher than 2 mol%. Lattice parameters of the solid solutions based on PbTe change linearly with composition (Raevskiy et al. 1983a,b). This system was investigated through DTA, metallography, XRD, and measuring of the microhardness. The ingots were annealed at 650°C and 830°C for 340 and 120 h, respectively.

PbSe + CdTe ↔ PbTe + CdSe. The fields of primary crystallization of the CdSe$_x$Te$_{1-x}$ and PbSe$_x$Te$_{1-x}$ solid solutions exist on the liquidus surface of this system (Figure 11.26) (Tomashik 1981). The system was studied through DTA, metallography, and using mathematical simulation of the experiment.

11.44 Lead–Cadmium–Iodine–Selenium

PbI₂–CdSe. The phase diagram of this system, constructed through DTA, XRD, and metallography using the ingots annealed for 1000 h, is a eutectic type (Odin 2001). The eutectic

composition and temperature are 19 mol% CdSe and 372°C, respectively. The mutual solubility of CdSe and PbI₂ is negligible: the solubility of PbI₂ in CdSe is equal to 0.15 mol%. The crystallization of the melts from the PbI₂ side leads to the formation of two polytypic forms of this compound: 6*R*-PbI₂ and 2*H*-PbI₂ (Odin 2001; Odin and Chukichev 2001). In this system, at 20 mol% CdSe, a metastable phase μ was obtained by quenching, which crystallizes in a rhombohedral structure with lattice parameters $a = 745.3 \pm 0.6$ and $\alpha = 35.24 \pm 0.08'$ or $a = 454.6 \pm 0.5$ and $c = 2099 \pm 2$ in hexagonal setting (Odin et al. 2005).

PbSe + CdI₂ ↔ PbI₂ + CdSe. The isothermal section of this ternary mutual system at 350°C is shown in Figure 11.27 (Odin 2001). The fields of primary crystallization of solid solutions based on PbSe and CdSe and Cd$_x$Pb$_{1-x}$I₂ solid solutions exist on the liquidus surface of this ternary mutual system (Figure 11.28). The ternary eutectic *E* at 361°C is formed by these three phases. A minimum *m* exists on the e_4e_2 line of the secondary crystallization. The metastable phase can form at the crystallization of the melts in the PbSe–CdSe–PbI₂ system in the region enriched in PbI₂ (Odin 2001; Odin et al. 2005). This phase exists at room temperature for some months, but heating leads to its decomposition with the formation of the phases according to the equilibrium phase diagram.

11.45 Lead–Mercury–Tellurium–Selenium

PbSe + HgTe ↔ PbTe + HgSe. The phase diagram is not constructed. Solid solutions with sphalerite structure based on mercury chalcogenides and with rock salt structure based on lead chalcogenides exist in this ternary mutual system (Leute and Köller 1986). The solubility of mercury chalcogenides in lead chalcogenides at 630°C reaches 5 mol%, and the solubility of lead chalcogenides in mercury chalcogenides is insignificant. Between the temperatures 430°C and 630°C, equilibrium is shifted to the formation of PbSe and HgTe.

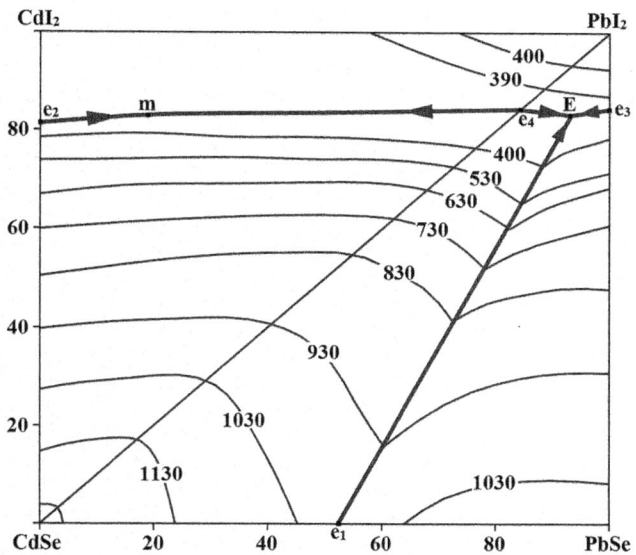

FIGURE 11.28 Liquidus surface of the PbSe + CdI₂ ↔ PbI₂ + CdSe ternary mutual system. (From Odin, I.N., *Zhurn. neorgan. khimii*, **46**(10), 1733, 2001.)

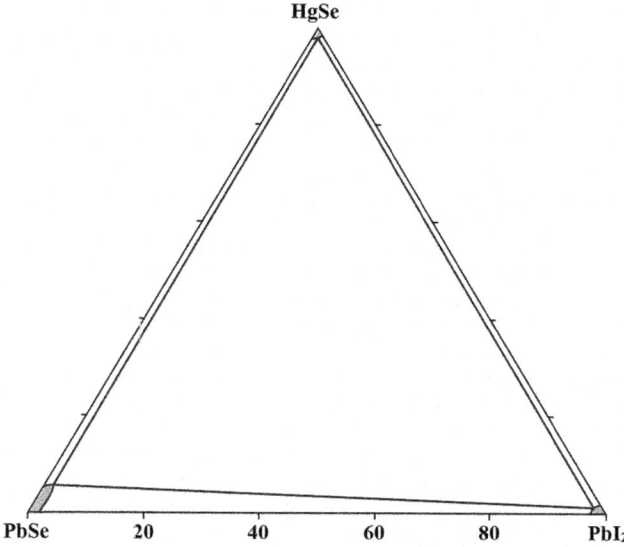

FIGURE 11.29 Isothermal section of the PbSe–HgSe–PbI₂ quasiternary system at 360°C. (From Odin, I.N. et al., *Zhurn. neorgan. khimii*, **49**(9), 1562, 2004.)

11.46 Lead–Mercury–Iodine–Selenium

PbI₂–HgSe. The phase diagram of this system, constructed through DTA and metallography is a eutectic type (Odin et al. 2004). The eutectic crystallizes at 386°C and contains 14 mol% HgSe. The solubility of PbI₂ in HgSe is less than 0.5 mol%, and the solubility of HgSe in PbI₂ is not higher than 1 mol%. Solid solution based on PbI₂ has a "mixed" structure in which the layers of 2*H*-PbI₂ and 6*R*-PbI₂ polytypes alternate. A new metastable phase μ, which contains 20 mol% HgSe and 80 mol% PbI₂, is formed in this system. It crystallizes in the rhombohedral structure with the lattice parameters $a = 746.3 \pm 0.6$ pm and $\alpha = 35.31° \pm 0.08°$ or in the hexagonal structure with the lattice parameters $a = 455.9 \pm 0.4$ and $c = 2103 \pm 2$ pm. Metastable phase μ was obtained by the quenching of the melts from temperatures of 810–947°C.

PbSe–HgSe–PbI₂. The isothermal section of this system at 360°C is given in Figure 11.29, and the liquidus surface (Figure 11.30) consists of the fields of the primary crystallization of HgSe, PbSe, and PbI₂ (Odin et al. 2004). Ternary eutectic *E* crystallizes at 371°C.

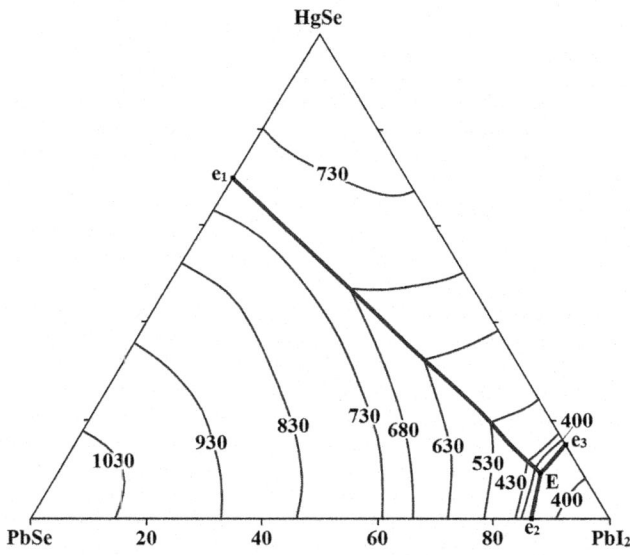

FIGURE 11.30 Liquidus surface of the PbSe–HgSe–PbI₂ quasiternary system. (From Odin, I.N. et al., *Zhurn. neorgan. khimii*, **49**(9), 1562, 2004.)

11.47 Lead–Gallium–Indium–Selenium

PbGa₂Se₄–PbIn₂Se₄. This system is nonquasibinary section of the PbSe–Ga₂Se₃–In₂Se₃ quasiternary system as PbGa₂Se₄ melts incongruently (Alidzhanov et al. 1987). The solubility of PbGa₂Se₄ in PbIn₂Se₄ at 480°C reaches 6 mol% and decreases to 2 mol% at room temperature. The system was studied through DTA, metallography, and measuring the microhardness.

11.48 Lead–Gallium–Yttrium–Selenium

PbSe–Ga₂Se₃–Y₂Se₃. The isothermal section of this quasiternary system at 500°C was constructed using the alloys annealed at this temperature for 250 h (Figure 11.31) (Filyuk et al. 2008). There are three three-phase regions limited by two-phase areas in the system. No quaternary compounds were found.

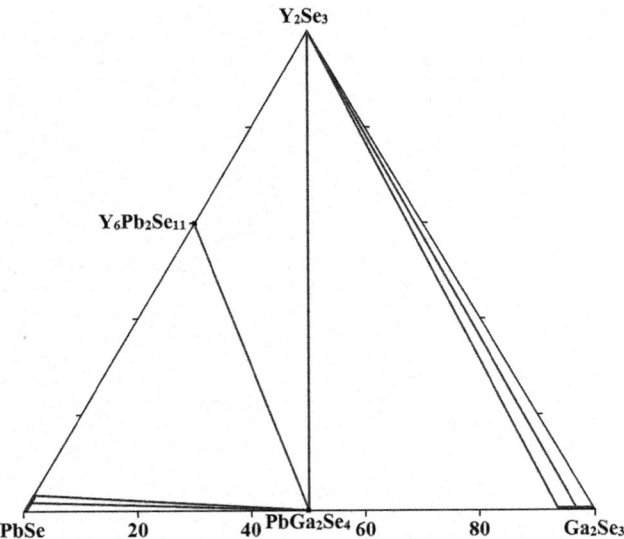

FIGURE 11.31 Isothermal section of the PbSe–Ga$_2$Se$_3$–Y$_2$Se$_3$ quasiternary system at 500°C. (From Filyuk, T.O., et al., *Nauk. Visnyk Volyns'k. Nats. Univ. im. Lesi Ukrainky. Ser. Khim. nauky*, (16), 51, 2008.)

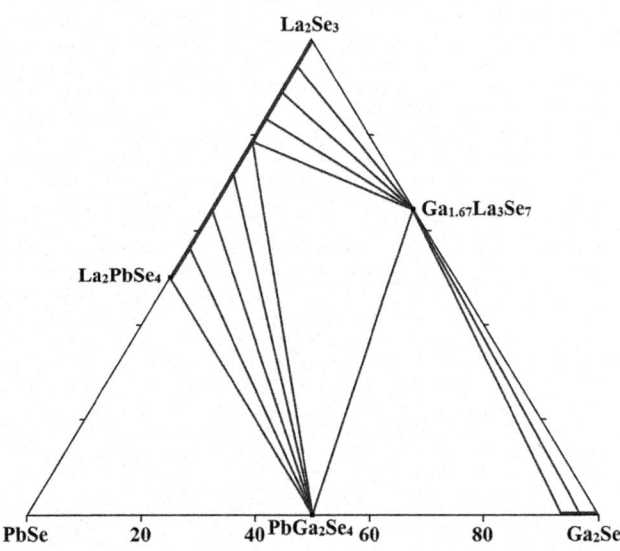

FIGURE 11.32 Isothermal section of the PbSe–Ga$_2$Se$_3$–La$_2$Se$_3$ quasiternary system at 500°C. (From Filyuk, T.O., et al., *Nauk. Visnyk Volyns'k. Nats. Univ. im. Lesi Ukrainky. Ser. Khim. nauky*, (16), 51, 2008.)

11.49 Lead–Gallium–Lanthanum–Selenium

PbSe–Ga$_2$Se$_3$–La$_2$Se$_3$. The isothermal section of this quasiternary system at 500°C was constructed using the alloys annealed at this temperature for 250 h (Figure 11.32) (Filyuk et al. 2008). There are three three-phase regions limited by two-phase areas in the system. No quaternary compounds were found.

11.50 Lead–Gallium–Tellurium–Selenium

PbTe–Ga$_2$Se$_3$. The phase diagram of this system, constructed through DTA and XRD, is a eutectic type (Figure 11.33) (Mayet and Roubin 1979). The eutectic contains 40 mol%

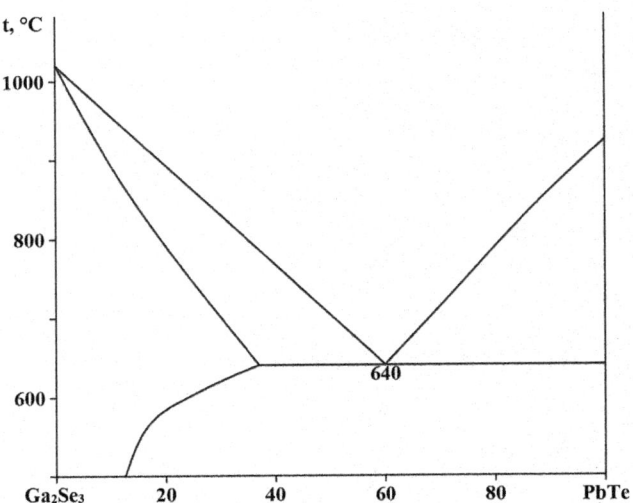

FIGURE 11.33 Phase diagram of the PbTe–Ga$_2$Se$_3$ system. (From Mayet, F., and Roubin, M., *C. r. Acad. sci. Sér. C*, **288**(17), 433, 1979.)

Ga$_2$Se$_3$ and crystallizes at 640°C. At the eutectic temperature, the solubility of PbTe in Ga$_2$Se$_3$ reaches 38 mol%.

11.51 Lead–Gallium–Manganese–Selenium

The **Pb$_{0.72}$Mn$_{2.84}$Ga$_{2.95}$Se$_8$** quaternary compound, which crystallizes in the hexagonal structure with the lattice parameters a = 1755.0 ± 0.3, c = 389.16 ± 0.08 pm, a calculated density of 5.486 g·cm^{-3}, and an energy gap of 1.65 eV, is formed in the Pb–Ga–Mn–Se system (Zhou et al. 2017a). It was obtained from a reacting mixture of PbSe (0.72 mM), MnSe (2.84 mM), and Ga$_2$Se$_3$ (1.475 mM). The raw materials were carefully ground and loaded into a fused silica tube, which was evacuated to a high vacuum of 10^{-3} Pa and flame-sealed. Afterward, the tube was heated from room temperature to 800°C within 15 h in a furnace, kept at that temperature for 100 h, and then the furnace was turned off. Single crystals were obtained by the spontaneous crystallization method from the solid-state reactions. The mixture of PbSe, MnSe, and Ga$_2$Se$_3$ (molar ratio 1:2:1) was blended and loaded into a fused silica tube in a glove box. Then, the tube was evacuated to a high vacuum of 10^{-3} Pa and flame-sealed. Subsequently, the tube was gradually heated to 950°C, kept for 72 h, then slowly cooled at a rate of 3°C·h^{-1}, and finally cooled to room temperature by turning off the furnace. All operations were conducted in an argon-filled glove box with controlled oxygen and moisture levels of <0.1 ppm.

11.52 Lead–Indium–Yttrium–Selenium

PbSe–In$_2$Se$_3$–Y$_2$Se$_3$. The isothermal section of this quasiternary system at 600°C was constructed using the alloys annealed at this temperature for 240 h, followed by quenching in cold water (Figure 11.34) (Pashyns'kyi et al. 2008). There are four three-phase, nine two-phase, and six single-phase fields in the system.

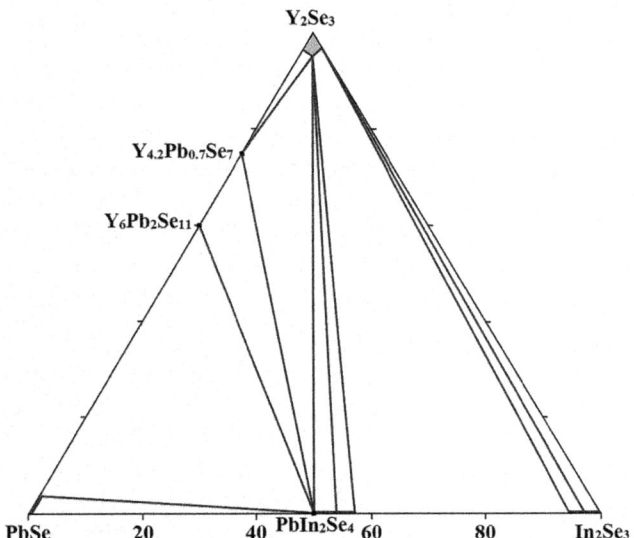

FIGURE 11.34 Isothermal section of the PbSe–In$_2$Se$_3$–Y$_2$Se$_3$ quasiternary system at 600°C. (From Pashyns'kyi, I.V., et al., *Nauk. Visnyk Volyns'k. Nats. Univ. im. Lesi Ukrainky. Ser. Khim. nauky*, (16), 43, 2008.)

FIGURE 11.35 Isothermal section of the PbSe–In$_2$Se$_3$–La$_2$Se$_3$ quasiternary system at 600°C. (From Pashyns'kyi, I.V., et al., *Nauk. Visnyk Volyns'k. Nats. Univ. im. Lesi Ukrainky. Ser. Khim. nauky*, (16), 43, 2008.)

11.53 Lead–Indium–Lanthanum–Selenium

PbSe–In$_2$Se$_3$–La$_2$Se$_3$. The isothermal section of this quasiternary system at 600°C was constructed using the alloys annealed at this temperature for 240 h followed by quenching in cold water (Figure 11.35) (Pashyns'kyi et al. 2008). There are four three-phase, nine two-phase, and six single-phase fields in the system.

11.54 Lead–Indium–Antimony–Selenium

The **In$_2$Pb$_4$Sb$_4$Se$_{13}$** quaternary compound, which crystallizes in the orthorhombic structure with the lattice parameters $a = 2172.2 \pm 0.4$, $b = 2714.0 \pm 0.5$, $c = 406.21 \pm 0.08$ pm, a calculated density of 7.136 g·cm^{-3}, and an energy gap less than 0.62 eV, is formed in the Pb–In–Sb–Se system (Wang et al. 2009b). This compound was obtained from a reaction with a molar ratio Pb/In/Sb/Se = 2:1:3:8. The mixture was heated from room temperature to 750°C over 8 h and maintained at this temperature for 16 h before natural cooling to room temperature. All operations on compound were performed in a glove box with a dry nitrogen atmosphere.

11.55 Lead–Indium–Bismuth–Selenium

The **In$_x$Pb$_4$Bi$_{6-x}$Se$_{13}$** ($x = 2.1–2.8$) quaternary compound, which crystallizes in the orthorhombic structure with the lattice parameters $a = 2227.3 \pm 0.5$, $b = 2748.8 \pm 0.6$, $c = 414.18 \pm 0.08$ pm, and a calculated density of 7.626 g·cm^{-3} for $x = 2.1$ and $a = 2206.5 \pm 0.4$, $b = 2742.0 \pm 0.6$, $c = 412.72 \pm 0.08$ pm, and a calculated density of 7.569 g·cm^{-3} for $x = 2.8$, is formed in the Pb–In–Bi–Se system (Wang et al. 2009b). It is

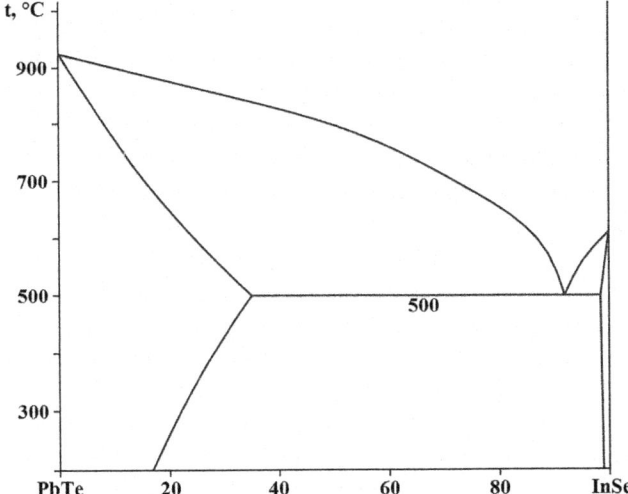

FIGURE 11.36 Phase diagram of the PbTe–InSe system. (From Gurshumov, A.P., et al., *Zhurn. neorgan. khimii*, **29**(9), 2433, 1984.)

a semiconductor with a narrow band gap of less than 0.62 eV. This compound was obtained in the same way as In$_2$Pb$_4$Sb$_4$Se$_{13}$ was prepared.

11.56 Lead–Indium–Tellurium–Selenium

PbTe–InSe. The phase diagram of this system, constructed through DTA, XRD, metallography, and measuring of the microhardness using the alloys annealed at 400°C for 220 h, is a eutectic type (Figure 11.36) (Gurshumov et al. 1984b). The eutectic contains 8 mol% PbTe and crystallizes at 500°C. The solubility of InSe in PbTe at the eutectic temperature reaches 35 mol%, and InSe dissolves less than 2 mol% PbTe.

11.57 Lead–Indium–Iron–Selenium

The $In_{17.37}Pb_{8.04}Fe_{0.47}Se_{34}$ quaternary compound, which melts congruently at 714°C and crystallizes in the monoclinic structure with the lattice parameters $a = 1320.98 \pm 0.02$, $b = 406.725 \pm 0.004$, $c = 2839.09 \pm 0.03$ pm, $\beta = 96.6427 \pm 0.0003°$, and a calculated density of 6.959 g·cm⁻³ at room temperature and an energy gap of 0.078 eV at 120 K, is formed in the Pb–In–Fe–Se system (Matsushita et al. 2006). The crystals of this compound have been obtained by a solid-state reaction using Fe, PbSe, In₂Se₃, and Se (molar ratio 1.5:5.5:5:1.5). The mixture was sealed into an evacuated silica tube (10⁻³ Pa), slowly heated to 900°C over 12 h, annealed at this temperature for 72 h, cooled to 400°C for 100 h, and then cooled to room temperature within 10 h.

11.58 Lead–Thallium–Antimony–Selenium

PbSe–TlSbSe₂. This system is a nonquasibinary section of the PbSe–Ti₂Se–Sb₂Se₃ quaternary system as TlSbSe₂ melts incongruently (Gitsu et al. 1981). From side PbSe, up to a content of 40 mol% 2PbSe, there is a region of solid solutions. The system was studied through DTA, XRD, and metallography. The initial samples were annealed in a stepwise mode at 420°C and 300°C for 240 h at each temperature, since TlSbSe₂ undergoes a polymorphic transformation at 380°C.

11.59 Lead–Thallium–Bismuth–Selenium

PbSe–TlBiSe₂. The phase diagram of this system (Figure 11.37) belongs to type I according to Roseboom's classification (Gitsu et al. 1980a). In this system, a continuous series of solid solutions is formed, which decompose at temperatures below 500°C. The **TlPbBiSe₃** quaternary compound [melting temperature 600°C, cubic structure, $a = 627.4$ pm, the calculated and experimental densities of 5.76 and 5.73 g·cm⁻³, respectively (Guseynov et al. 1972)] was not found in this system.

The PbSe–TlBiSe₂, **PbSe–Tl₉BiSe₆**, and **Tl₄PbSe₃–Tl₉BiSe₆** quasibinary systems divide the **PbSe–Tl₂Se–Bi₂Se₃** quasiternary system into four quasiternary subsystems (Masalovych et al. 2011a). The continuous series of solid solutions is formed in the Tl₄PbSe₃–Tl₉BiSe₆ system (Figure 11.38) (Masalovych et al. 2014a). As such, solid solutions are also formed in the Tl₄PbSe₃–Tl₂Se and Tl₂Se–Tl₉BiSe₆ systems, they are also formed in the **Tl₄PbSe₃–Tl₂Se–Tl₉BiSe₆** quasiternary subsystem, the liquidus surface of which is given in Figure 11.39.

PbSe–Tl₉BiSe₆. The phase diagram of this system, constructed through DTA, XRD, and metallography using the alloys homogenized at 330°C for 336 h with the next quenching in water, is a eutectic type (Figure 11.40) (Masalovich et al. 2016). The eutectic degenerates from the Tl₉BiSe₆ side at 512°C. The monotectic reaction exists in this system at 525°C within the interval of 4–48 mol% PbSe. PbSe dissolves up to 5 mol% Tl₉BiSe₆. The liquidus surface of the **PbSe–Tl₄PbSe₃–Tl₉BiSe₆** quasiternary subsystem is given in Figure 11.41.

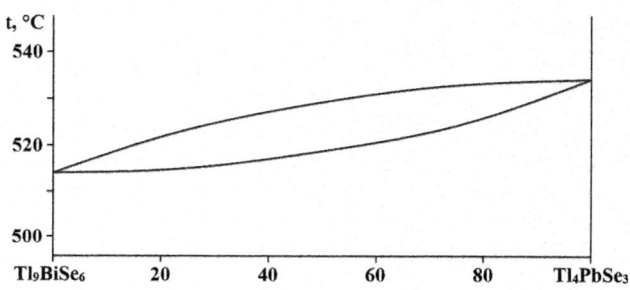

FIGURE 11.38 Phase diagram of the Tl₄PbSe₃–Tl₉BiSe₆ system. (From Masalovych, O.O., et al., *Nauk. visn. Uzhgorod. Univ., Ser. Khim.*, [1(31)], 25, 2014.)

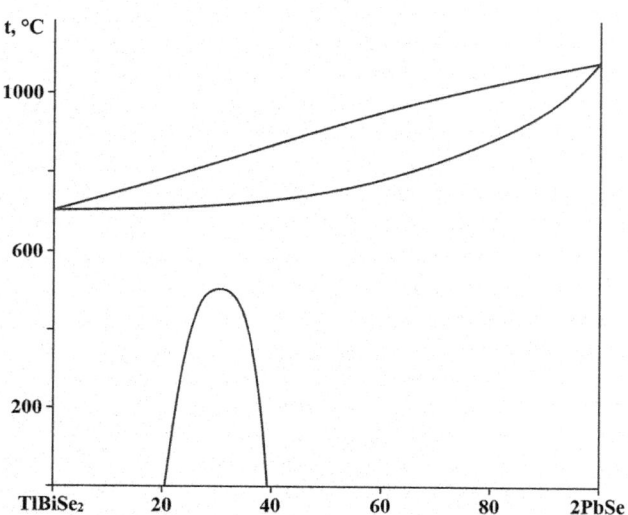

FIGURE 11.37 Phase diagram of the 2PbSe–TlBiSe₂ system. (From Gitsu, D.V., et al., *Izv. AN SSSR. Neorgan. mater.*, **16**(6), 988, 1980.)

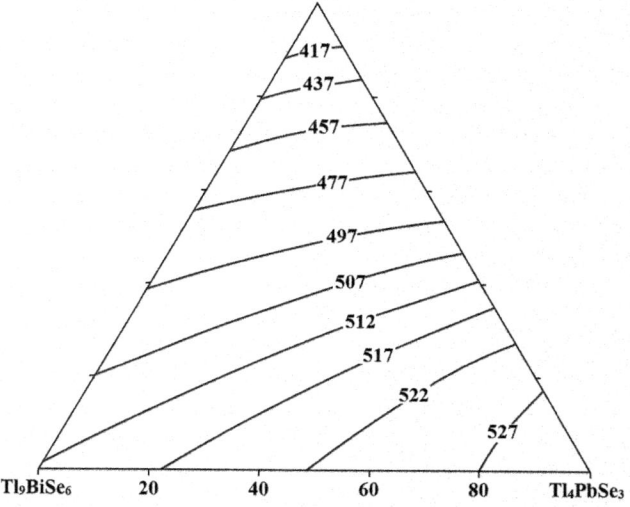

FIGURE 11.39 Liquidus surface of the Tl₄PbSe₃–Tl₂Se–Tl₉BiSe₆ quasiternary system. (From Masalovych, O.O., et al., *Nauk. visn. Uzhgorod. Univ., Ser. Khim.*, [1(31)], 25, 2014.)

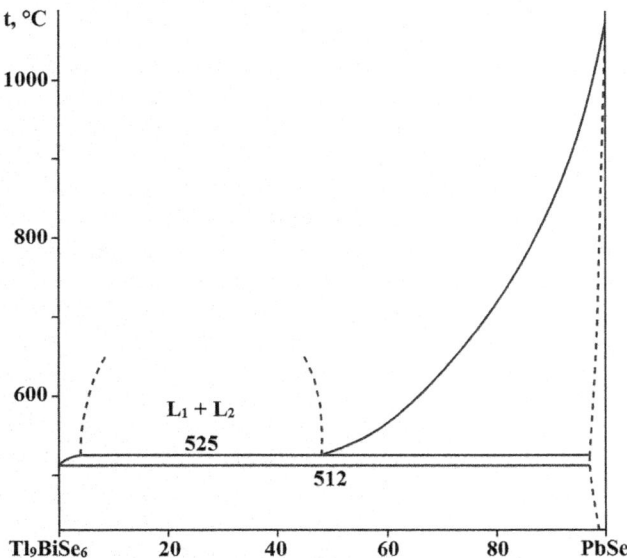

FIGURE 11.40 Phase diagram of the PbSe–Tl₉BiSe₆ system. (From Masalovich, E.E., et al., *Russ. J. Inorg. Chem.*, **61**(4), 507, 2016.)

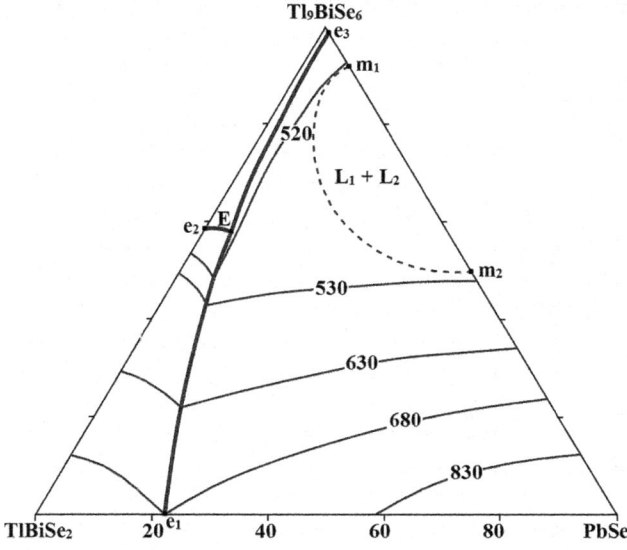

FIGURE 11.42 Liquidus surface of the PbSe–TlBiSe₂–Tl₉BiSe₆ quasiternary system. (From Masalovich, E.E., et al., *Russ. J. Inorg. Chem.*, **61**(4), 507, 2016.)

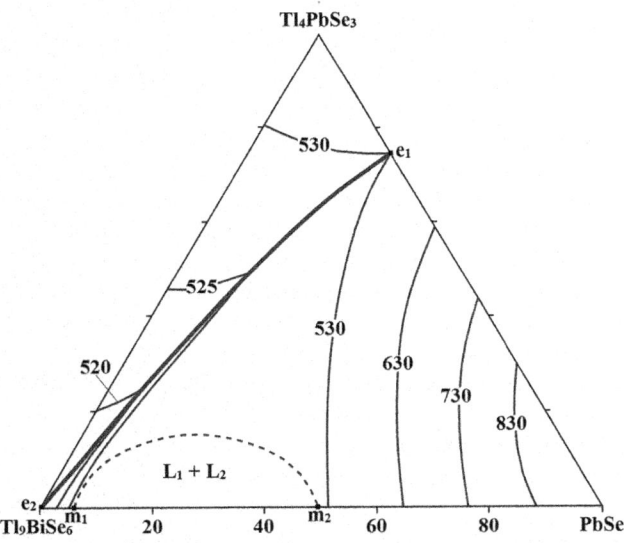

FIGURE 11.41 Liquidus surface of the PbSe–Tl₄PbSe₃–Tl₉BiSe₆ quasiternary system. (From Masalovich, E.E., et al., *Russ. J. Inorg. Chem.*, **61**(4), 507, 2016.)

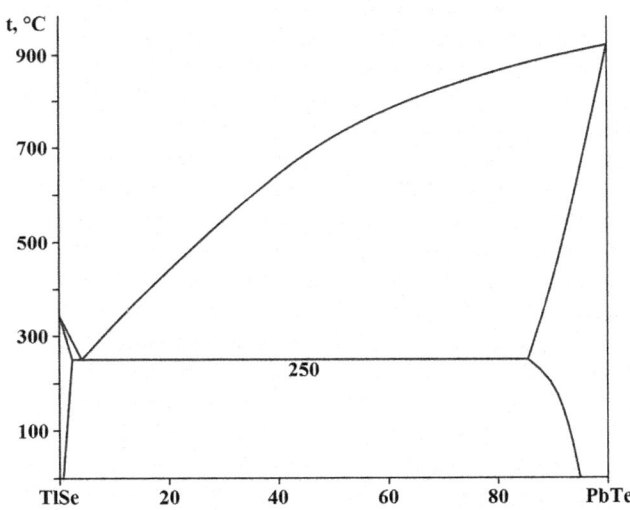

FIGURE 11.43 Phase diagram of the PbTe–TlSe system. (From Gurshumov, A.P., et al., *Zhurn. neorgan. khimii*, **29**(9), 2432, 1984.)

PbSe–TlBiSe₂. The most reliable phase diagram of this system, constructed through DTA, XRD, and metallography using the alloys homogenized at 330°C for 336 h with the next quenching in water, is a eutectic type (Masalovich et al. 2016). The eutectic contains 23 mol% PbSe and crystallizes at 677°C. TlBiSe₂ dissolves up to 5 mol% PbSe, and PbSe dissolves 55 mol% TlBiSe₂. The liquidus surface of the **PbSe–TlBiSe₂–Tl₉BiSe₆** quasiternary subsystem is given in Figure 11.42. Isothermal sections of the PbSe–Tl₄PbSe₃–Tl₉BiSe₆ and PbSe–TlBiSe₂–Tl₉BiSe₆ at 330°C were also constructed by Masalovich et al. (2016).

11.60 Lead–Thallium–Tellurium–Selenium

PbTe–TlSe. The phase diagram of this system, constructed through DTA, XRD, metallography, and measuring of the microhardness, is a eutectic type (Figure 11.43) (Gurshumov et al. 1984a). The eutectic contains 4 mol% PbTe and crystallizes at 250°C. The solubility of TlSe in PbTe at the room temperature is 6 mol% and TlSe dissolves less than 1 mol% PbTe. The ingots for research were annealed at temperatures 50–60°C below the solidus temperature for one month.

Tl₄PbSe₃–Tl₄PbTe₃. A continuous solid solution series are formed in this system (Filep and Sabov 2008).

11.61 Lead–Europium–Bismuth–Selenium

Two quaternary compounds, **$Eu_2Pb_2Bi_4Se_{10}$** and **$Eu_2Pb_2Bi_6Se_{13}$**, are formed in the Pb–Eu–Bi–Se system (Kanatzidis et al. 1998). First of them crystallizes in the orthorhombic structure and has an energy gap of less than 0.2 eV, and the second one crystallizes in the monoclinic structure and also has an energy gap of less than 0.2 eV.

11.62 Lead–Nitrogen–Oxygen–Selenium

Two quaternary compounds, **$Pb_2(NO_3)_2(SeO_3)$** and **$Pb_2(NO_2)$ $(NO_3)(SeO_3)$**, are formed in the Pb–N–O–Se system. The first is thermally stable up to 410°C (Markovskiy and Sapozhnikov 1960) and crystallizes in the orthorhombic structure with the lattice parameters $a = 546.69 \pm 0.03$, $b = 1032.77 \pm 0.06$, $c = 726.10 \pm 0.04$ pm, a calculated density of 5.390 g·cm^{-3}, and an energy gap of 3.76 eV (Meng et al. 2015) [an experimental density of 5.37 ± 0.03 g·cm^{-3} (Markovskiy and Sapozhnikov 1960)]. This compound was synthesized by conventional facile hydrothermal method at a middle temperature (Meng et al. 2015). A mixture of Pb(NO₃)₂ (4 mM), SeO₂ (2 mM), and H₂O (3 mL) was loaded into a 15 mL Teflon-lined autoclave and subsequently sealed. The autoclave was gradually heated to 200°C in an oven, held for 4 days, and followed by slow cooling to room temperature at a rate of 6°C·h^{-1}. The products were filtered and washed with deionized water, and many colorless block-shaped crystals with single phases were obtained. It can also be obtained by the interaction of Pb(CH₃COO)₂ and Pb(NO₃)₂ with K₂SeO₃ at an excess of Pb(CH₃COO)₂, by the interaction of an excess of Pb(NO₃)₂ with an H₂SeO₃ solution at the co-precipitation from hot concentration solutions, or by evaporating of the Pb(NO₃)₂ solution with the stoichiometric quantity of PbSeO₃ (Markovskiy and Sapozhnikov 1960). This compound is stable under air and moisture.

$Pb_2(NO_2)(NO_3)(SeO_3)$ also crystallizes in the orthorhombic structure with the lattice parameters $a = 552.9 \pm 0.2$, $b = 1035.7 \pm 0.3$, $c = 681.1 \pm 0.2$ pm, and a calculated density of 3.77 g·cm^{-3} (Effenberger 1987). The crystals of this compound were obtained under hydrothermal conditions as follows. 2 g of an equimolar mixture consisting of Pb(NO₃)₂, SeO₂, and CuO were placed in a Teflon-lined autoclave (~6 mL). About 1 g of thin Cu sheet was added; the reaction vessel was filled with H₂O to 80 vol% and heated to 230° C for 48 h. After cooling, besides unreacted Cu and CuO, pale-yellow crystals of the title compound were formed.

11.63 Lead–Antimony–Bismuth–Selenium

$PbSb_2Se_4$–$Pb_5Bi_6Se_{14}$. This system is a nonquasibinary section of the PbSe–Sb₂Se₃–Bi₂Se₃ quasiternary system as Pb₅Bi₆Se₁₄ melts incongruently (Gurbanov et al. 2017). The eutectic contains 30 mol% Pb₅Bi₆Se₁₄ and crystallizes at 527°C. The solubility of Pb₅Bi₆Se₁₄ in PbSb₂Se₄ at room temperature is 10 mol%. The **$Pb_6Sb_2Bi_6Se_{18}$** quaternary compound, which melts congruently at 677°C and crystallizes in the orthorhombic

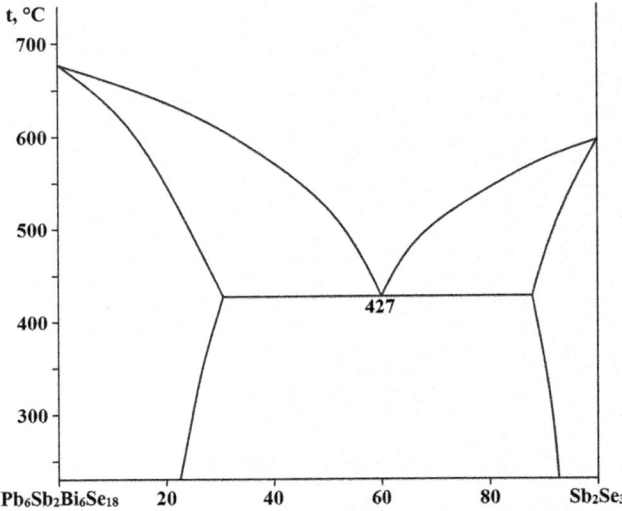

FIGURE 11.44 Phase diagram of the Pb₆Sb₂Bi₆Se₁₈–Sb₂Se₃ system. (From Gurbanov, G.R. and Mamedov, Sh.G., *Russ. J. Inorg. Chem.*, **64**(3), 383, 2019.)

structure with the lattice parameters $a = 1443$, $b = 2142$, $c = 390$ pm, and an experimental density of 7.41 g·cm^{-3}. Single crystals of this compound were grown through chemical transport reactions using iodine as a transport agent. The thermodynamic functions of Pb₆Sb₂Bi₆Se₁₈ were calculated: $S^0_{298} = 1472.8$ J·M^{-1}K^{-1}, $\Delta H^0_{298} = -1215.8$ kJ·M^{-1}, $\Delta G^0_{298} = -1193.6$ kJ·M^{-1}, and $\Delta S^0_{298} = -74.24$ J·M^{-1}K^{-1}.

$Pb_6Sb_2Bi_6Se_{18}$–Sb_2Se_3. The phase diagram of this system, constructed through DTA, XRD, metallography, and measuring of the microhardness and density using the alloys annealing at 280°C for 160–300 h, is a eutectic type (Figure 11.44) (Gurbanov and Mamedov 2019). The eutectic contains 60 mol% Sb₂Se₃ and crystallizes at 427°C. The solubility of Sb₂Se₃ in Pb₆Sb₂Bi₆Se₁₈ is 30 mol% at the eutectic temperature and 20 mol% at room temperature. Sb₂Se₃ dissolves 12 mol% Pb₆Sb₂Bi₆Se₁₈ at the eutectic temperature and 5 mol% at room temperature.

$Pb_6Sb_2Bi_6Se_{18}$–Bi_2Se_3. The phase diagram of this system, constructed through DTA, XRD, metallography, and measuring of the microhardness and density using the alloys annealing at 280°C for 160–300 h, is a eutectic type (Figure 11.45) (Gurbanov and Mamedov 2019). The eutectic contains 60 mol% Bi₂Se₃ and crystallizes at 452°C. The solubility of Bi₂Se₃ in Pb₆Sb₂Bi₆Se₁₈ at room temperature is 12 mol%, and Bi₂Se₃ dissolves 10 mol% Pb₆Sb₂Bi₆Se₁₈ at the same temperature.

11.64 Lead–Antimony–Iron–Selenium

Three quaternary compounds, **$Fe_{0.75}Pb_{3.25}Sb_4Se_{10}$, $FePb_3Sb_4Se_{10}$,** and **$FePb_4Sb_6Se_{14}$**, are formed in the Pb–Sb–Fe–Se system. The first crystallizes in the orthorhombic structure with the lattice parameters $a = 2433.3 \pm 0.5$, $b = 411.10 \pm 0.08$, $c = 1967.3 \pm 0.4$ pm, and a calculated density of 6.72 g·cm^{-3} and the second one crystallizes in the monoclinic structure with the lattice parameters $a = 1969.18 \pm 0.09$, $b = 411.53 \pm 0.02$, $c = 2436.07 \pm 0.11$ pm, $\beta = 90.02 \pm 0.01°$, and a calculated density

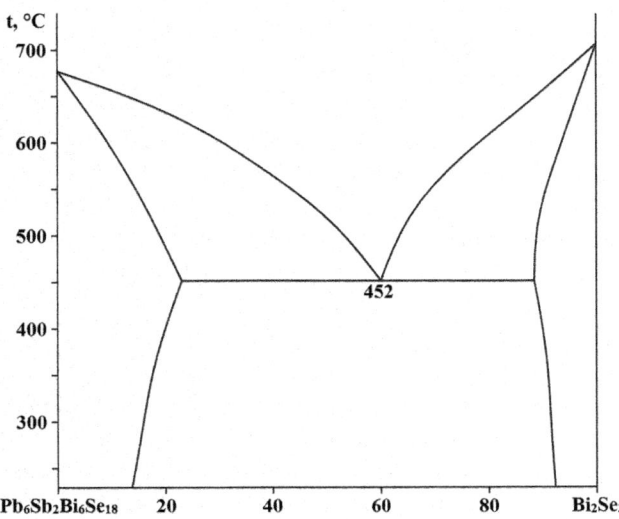

FIGURE 11.45 Phase diagram of the $Pb_6Sb_2Bi_6Se_{18}$–Bi_2Se_3 system. (From Gurbanov, G.R. and Mamedov, Sh.G., *Russ. J. Inorg. Chem.*, **64**(3), 383, 2019.)

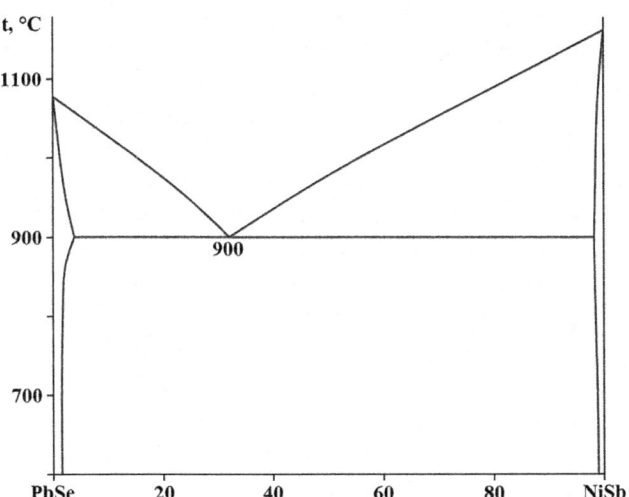

FIGURE 11.46 Phase diagram of the PbSe–NiSb system. (From Zargarova, M.I., et al., *Azerb. khim. zhurn.*, (6), 138, 1975.)

of 6.59 g·cm^{-3} (Poudeu et al. 2010). Single phases of these two compounds were obtained from solid-state reactions involving Fe, Pb, Sb, and Se. All four components in their powder form, weighed in the desired ratio under an ambient atmosphere, were ground with an agate mortar and pestle and transferred into a fused silica tube, which was flame-sealed under a residual pressure of ~ 0.01 Pa. The sealed tube was placed into a tube furnace and heated stepwise to the target temperature 550°C, at which a dwelling time of 72 h was imposed. The furnace was then slowly cooled to room temperature over 24 h. The resulting products were polycrystalline dark gray powders containing tiny needle-like single crystals. Alternatively, a single phase of $FePb_3Sb_4Se_{10}$ was obtained using the same reaction procedure, by combining binary FeSe and elemental Pb, Sb, and Se. Single crystals of both compounds were grown by annealing in a slight temperature gradient the synthesized polycrystalline powders, sealed in a quartz tube under a residual pressure of 0.01 Pa. The temperature at the sample position was 550°C while the temperature at the other end was maintained at 600°C. Single crystals were obtained at the initial sample position.

$FePb_4Sb_6Se_{14}$ also crystallizes in the monoclinic structure with the lattice parameters $a = 412.44 \pm 0.02$, $b = 1978.1 \pm 0.2$, $c = 1659.6 \pm 0.3$ pm, $\beta = 91.6 \pm 0.2°$, and a calculated density of 6.68 g·cm^{-3} (Poudeu et al. 2012). Single crystals of this compound were obtained from the combination of high purity Fe, Pb, Sb, and Se powders by solid-state reaction at moderate temperatures. All four components, weighed in the desired ratio under Ar atmosphere in a dry glove box, were well mixed in an agate mortar with pestle and transferred into a fused silica tube, which was flame-sealed under a residual pressure of ~0.1 Pa. The sample was heated over 18 h from room temperature to 600°C, annealed at this temperature for 72 h, and cooled down to room temperature in 1 h. The resulting polycrystalline powder was ground again for further annealing. The powder sample, sealed under vacuum in a fused silica tube, was heated from room temperature to 550°C in 6 h, kept

at this temperature for about 72 h and cooled down to room temperature over 72 h. Products from this annealing step contained several needle-like single crystals of the title compound coexisting with several crystals of $FePb_3Sb_4Se_{10}$.

11.65 Lead–Antimony–Nickel–Selenium

PbSe–NiSb. The phase diagram of this system, constructed through DTA, metallography, and measuring of the microhardness using the alloys annealed at 600°C for 3 days, is a eutectic type (Figure 11.46) (Zargarova et al. 1975d). The eutectic contains 32 mol% NiSb and crystallizes at 900°C. The solubility of NiSb in PbSe and PbSe in NiSb reaches 2 and 1 mol%, respectively

11.66 Lead–Bismuth–Tellurium–Selenium

PbSe–Bi$_2$Te$_3$. The phase diagram of this system, constructed through DTA and XRD, is given in (Figure 11.47) (Datsenko et al. 1981). The eutectic contains 22 mol% PbSe and crystallizes at 578°C, and **2PbSe.3Bi$_2$Te$_3$** quaternary compound, which melts incongruently at 610°C, is formed in this system.

PbTe–BiSe. At 450–500°C, BiSe dissolves 25 mol% PbTe (Abrikosov et al. 1972). The ingots for the investigations were annealed at these temperatures for 10–30 days.

PbBi$_2$Te$_4$–PbBi$_2$Se$_4$. This system is a nonquasibinary section of the PbSe–Bi$_2$Se$_3$–Bi$_2$Te$_3$ quasiternary system as PbBi$_2$Se$_4$ melts incongruently (Aliev 2019a). The PbSe$_{1-x}$Te$_x$ solid solution primarily crystallizes from the liquid phase. The system was studied through DTA and XRD and the ingots for the investigations were annealed at 550°C for 200 h.

PbSe + Bi$_2$Te$_3$ ↔ PbTe + Bi$_2$Se$_3$. Phase relations in this system were examined in the temperature range of 500–900°C by Liu and Chang (1994a) and the isothermal section of this system at 500°C is given in Figure 11.48. Four extensive ranges

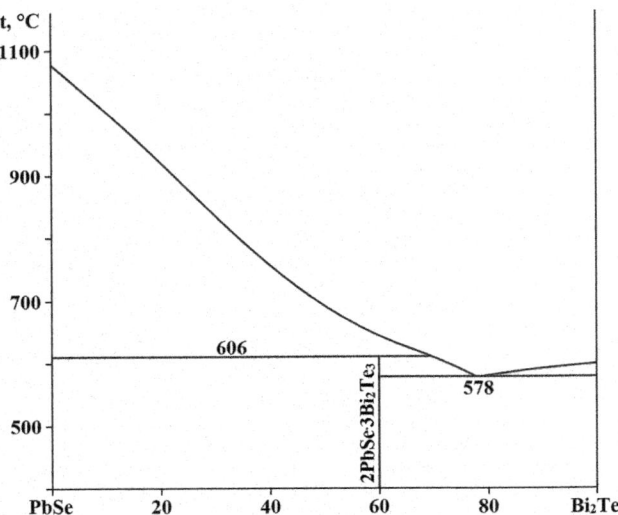

FIGURE 11.47 Phase diagram of the PbSe–Bi₂Te₃ system. (From Datsenko, A.M., et al., *Tr. Mosk. khim.-tehnol. in-t*, (120), 32, 1981.)

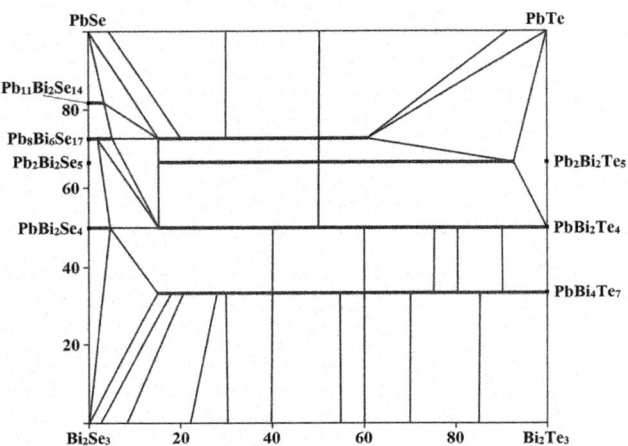

FIGURE 11.48 Isothermal section of the PbSe + Bi₂Te₃ ↔ PbTe + Bi₂Se₃ ternary mutual system at 500°C. (From Liu, H., and Chang, L.L.Y., *Amer. Mineralog.*, 79(11–12), 1159, 1994.)

of solid solutions exist in the system. Solid solutions based on PbBi₂Te₄ extend to PbBi₂(Se₀.₇₈Te₀.₂₂)₄ and solid solutions based on PbBi₄Te₇ extend to PbBi₄(Se₀.₇₅Te₀.₂₅)₇. The PbBi₄(SeₓTe₁₋ₓ)₇ solid solutions crystallize in the trigonal structure with the lattice parameters $a = 439.2 \pm 0.1$, $c = 2389 \pm 1$ pm for $x = 0.05$, $a = 438.0 \pm 0.1$, $c = 2390 \pm 1$ pm for $x = 0.10$, $a = 437.3 \pm 0.1$, $c = 2384 \pm 1$ pm for $x = 0.15$, and $a = 436.5 \pm 0.2$, $c = 2379 \pm 2$ pm for $x = 0.20$ (Shelimova et al. 2004). The band gap of these solid solutions tends to increase upon Se substitution for Te. They were annealed at 500°C for 1000 h and then quenched in ice water.

The solid solutions along the section Pb₂Bi₂Se₅–Pb₂Bi₂Te₅ exist between Pb₂Bi₂(Se₀.₀₆Te₀.₉₄)₅ and Pb₂Bi₂(Se₀.₇₅Te₀.₂₅)₅ and solid solutions with compositions Pb₈Bi₆(SeₓTe₁₋ₓ)₁₇ is stable in the system from Pb₈Bi₆(Se₀.₂₆Te₀.₇₄)₁₇ to Pb₈Bi₆(Se₀.₇₂Te₀.₂₈)₁₇ (Liu and Chang 1994a). The duration of the samples' treatment ranged from 60 days at 500°C to one day at 900°C. At the end of the heat treatment, the samples were quenched to room

temperature by compressed air to preserve phase assemblages synthesized at high temperatures.

The isothermal section of the PbSe + Bi₂Te₃ ↔ PbTe + Bi₂Se₃ ternary mutual system at 530°C was also constructed by Aliev (2019b), but some discrepancies exist between obtained data and the results of Liu and Chang (1994a).

11.67 Lead–Vanadium–Oxygen–Selenium

Three quaternary compounds, **Pb₂V₂Se₂O₁₁**, **Pb₂V₃Se₅O₁₈**, and **Pb₄(VO₂)₂(SeO₃)(Se₂O₅)** or **Pb₄V₂Se₆O₂₁**, are formed in the Pb–V–O–Se system. The first is thermally stable up to 260°C and crystallizes in the triclinic structure with the lattice parameters $a = 683.2 \pm 0.3$, $b = 718.1 \pm 0.3$, $c = 985.7 \pm 0.5$ pm, $\alpha = 96.031 \pm 0.004°$, $\beta = 93.657 \pm 0.006°$, $\gamma = 95.455 \pm 0.007°$, a calculated density of 5.914 g·cm⁻³, and an energy gap of 2.39 eV (Li et al. 2010). It was hydrothermally synthesized by reactions of a mixture of PbO (0.30 mM), V₂O₃ (0.15 mM), and SeO₂ (1.02 mM) in 5 mL of distilled water, at 200°C for 4–5 days. The initial and final pH values of the reaction media were close to 2.0. The excess SeO₂ aided in the crystallization because of its high solubility in water. V₂O₃ has been oxidized for the higher value states (5+) at 200°C, probably by oxidization of SeO₂. When V₂O₅ was used instead of V₂O₃, Pb₂V₂Se₂O₁₁ could not be obtained.

Pb₂V₃Se₅O₁₈ is thermally stable up to 290°C and crystallizes in the orthorhombic structure with the lattice parameters $a = 1701.3 \pm 0.3$, $b = 1071.0 \pm 0.2$, $c = 905.9 \pm 0.1$ pm, a calculated density of 5.030 g·cm⁻³, and an energy gap of 3.14 eV (Li et al. 2010). It was hydrothermally synthesized by reactions of a mixture of PbO (0.30 mM), V₂O₃ (0.15 mM), and SeO₂ (1.10 mM) in 5 mL of distilled water, at 230°C for 4–5 days. The initial and final pH values of the reaction media were close to 2.0. The excess SeO₂ aided in the crystallization because of its high solubility in water. V₂O₃ has been oxidized for the higher value states (5+) at 230°C, probably by oxidization of SeO₂. When V₂O₅ was used instead of V₂O₃, the title compound could not be obtained.

Pb₄(VO₂)₂(SeO₃)₄(Se₂O₅) also crystallizes in the orthorhombic structure with the lattice parameters $a = 2502.9 \pm 0.2$, $b = 1221.47 \pm 0.10$, and $c = 1301.54 \pm 0.10$ pm (Yeon et al. 2012). The decomposition of this compound starts at approximately 305°C. The bulk compound was prepared by conventional solid-state methods. A stoichiometric mixture of PbO (8.00 mM), V₂O₅ (2.00 mM), and SeO₂ (12.0 mM) was thoroughly ground, pressed into a pellet, and introduced into a Pyrex tube that was subsequently evacuated under vacuum and sealed to prevent vaporization of SeO₂. The tube was heated to 270°C (48 h) and then cooled slowly to room temperature. Single crystals of the title compound were grown by using hydrothermal techniques. PbO (0.702 mM), V₂O₅ (0.175 mM), and SeO₂ (2.81 mM) were combined with 5 mL of H₂O. The solutions were placed in a 23 mL Teflon-lined autoclave. The autoclave was closed, gradually heated to 230°C, held for 4 days, and cooled slowly to room temperature at a rate of 6°C·h⁻¹. The mother liquor was decanted from the product, which was recovered by filtration and washed with distilled water and acetone. Pale yellow crystals were obtained.

11.68 Lead–Niobium–Oxygen–Selenium

The $Pb_2Nb_2Se_4O_{15}$ quaternary compound, which is thermally stable up to 390°C and crystallizes in the monoclinic structure with the lattice parameters $a = 2649 \pm 1$, $b = 700.1 \pm 0.3$, $c = 768.6 \pm 0.3$ pm, $\beta = 94.503 \pm 0.007°$, a calculated density of 5.403 g·cm^{-3}, and an energy gap of 3.15 eV, is formed in the Pb–Nb–O–Se system (Li et al. 2010). This compound was hydrothermally synthesized by reactions of a mixture of PbO (0.16 mM), Nb_2O_5 (0.16 mM), and SeO_2 (0.51 mM) in 8 mL of distilled water, at 200°C for 4–5 days. The initial and final pH values of the reaction media were close to 2.0. The excess SeO_2 aided in the crystallization because of its high solubility in water.

11.69 Lead–Oxygen–Tellurium–Selenium

Two quaternary compounds, $Pb_3(SeO_4)(TeO_3)_2$ and $Pb_7O_4(SeO_4)_2(TeO_3)$, are formed in the Pb–O–Te–Se system (Weil and Shirkhanlou 2017). First of them crystallizes in the monoclinic structure with the lattice parameters $a = 1608.53 \pm 0.04$, $b = 1173.65 \pm 0.03$, $c = 607.88 \pm 0.01$ pm, $\beta = 107.599 \pm 0.001°$, and a calculated density of 6.775 g·cm^{-3}. Second compound crystallizes in the triclinic structure with the lattice parameters $a = 731.91 \pm 0.05$, $b = 1036.96 \pm 0.06$, $c = 1163.17 \pm 0.08$ pm, $\alpha = 73.534 \pm 0.002°$, $\beta = 89.982 \pm 0.002°$, $\gamma = 87.431 \pm 0.002°$, and a calculated density of 7.759 g·cm^{-3}.

Both compounds were prepared through hydrothermal synthesis, which was conducted in sealed Teflon containers (capacity ca. 10 mL) in steel autoclave at autogenous pressure (Weil and Shirkhanlou 2017). PbO (0.180 g) was mixed in stoichiometric amounts with 80 mass% H_2SeO_4 (0.17 mL), TeO_2 (0.084 g), and KOH (0.09 g for the first compound or 0.18 g for the second one) (molar ratio 1.5:2:1:2 or 1.5:2:1:4 for the first and the second compound, respectively). The mixtures were placed in the Teflon containers, filled up with water to about two-thirds of the container volume, and sealed with a Teflon lid. The sealed containers were placed in the autoclave and heated at 210°C for 1 week. Afterward, the autoclave was cooled to room temperature overnight. The solid products were filtered off and washed subsequently with mother liquor, water, and ethanol. The reaction batch consisted of phase mixtures when inspected optically under a polarizing microscope.

Using the method of diagrams of coexisting phases, the process of oxidation of $PbTe_{1-x}Se_x$ solid solution was analyzed by Medvedev et al. (1987). A good agreement with the results of experimental studies of the natural aging process has been obtained.

11.70 Lead–Oxygen–Molybdenum–Selenium

Two quaternary compounds, $PbMoSeO_6$ and $PbMo_2O_5(SeO_3)_2$, are formed in the Pb–O–Mo–Se system. The first compound crystallizes in the triclinic structure with the lattice parameters $a = 689.44 \pm 0.06$, $b = 722.19 \pm 0.06$, $c = 1082.94 \pm 0.09$ pm, $\alpha = 99.751 \pm 0.002°$, $\beta = 99.996 \pm 0.002°$, $\gamma = 90.041 \pm 0.002°$, and a calculated density of 6.071 g·cm^{-3} (Oh et al. 2012b). A polycrystalline sample of $PbMoSeO_6$ was

synthesized through standard solid-state techniques. PbO (2.00 mM), MoO_3 (2.00 mM), and SeO_2 (3.00 mM) were thoroughly ground and pressed into a pellet. The pellet was introduced into a fused silica tube that was evacuated and sealed. The tube was gradually heated to 400°C, held for 24 h, and cooled to room temperature. Crystals of $PbMoSeO_6$ were prepared by placing a mixture of PbO (1.35 mM), MoO_3 (0.680 mM), and SeO_2 (2.71 mM) into a fused silica tube that was subsequently evacuated and sealed. The tube was gradually heated to 500°C, held for 24 h, and then cooled slowly to 300°C at a rate of 6°C·h^{-1} before being quenched to room temperature. Crystals of the title compound as pale yellow needles were recovered with some unknown amorphous materials. This compound is not stable at elevated temperatures and decomposes in a single step to evaporation of SeO_2 and formation of $PbMoO_4$.

$PbMo_2O_5(SeO_3)_2$ also crystallizes in the triclinic structure with the lattice parameters $a = 785.60 \pm 0.06$, $b = 823.07 \pm 0.06$, $c = 838.81 \pm 0.06$ pm, $\alpha = 82.881 \pm 0.004°$, $\beta = 64.843 \pm 0.004°$, $\gamma = 66.419 \pm 0.004°$, and a calculated density of 5.143 g·cm^{-3} at 200 K (Oh et al. 2012a). Single crystals of this compound were prepared by standard solid-state reactions. PbO (1.50 mM), MoO_3 (3.00 mM), and SeO_2 (3.30 mM) were thoroughly mixed with an agate mortar and pestle and pressed into a pellet. The pellet was introduced into a fused silica tube, which was subsequently evacuated and sealed. The tube was gradually heated to 400°C for 36 h and cooled at a rate of 12°C·h^{-1} to room temperature. The product contained colorless block-shaped crystals with white polycrystalline $PbMo_2O_5(SeO_3)_2$. This compound is stable up to 400°C. Above this temperature, decomposition occurs attributable to the sublimation of SeO_2.

11.71 Lead–Oxygen–Tungsten–Selenium

The $PbWSeO_6$ quaternary compound, which crystallizes in the triclinic structure with the lattice parameters $a = 686.89 \pm 0.02$, $b = 723.98 \pm 0.02$, $c = 1090.37 \pm 0.03$ pm, $\alpha = 99.699 \pm 0.004°$, $\beta = 100.348 \pm 0.003°$, and $\gamma = 90.139 \pm 0.004°$, is formed in the Pb–O–W–Se system (Oh et al. 2012b). A polycrystalline sample of $PbWSeO_6$ was synthesized in the same way as $PbMoSeO_6$ was prepared using WO_3 instead of MoO_3 and gradually heating the tube to 450°C. This compound is also not stable at elevated temperatures and decomposes in a single step to evaporation of SeO_2 and formation of $PbWO_4$.

11.72 Lead–Oxygen–Chlorine–Selenium

The $Pb_3(SeO_3)_2Cl_2$ quaternary compound, which crystallizes in the monoclinic structure with the lattice parameters $a = 1342.13 \pm 0.11$, $b = 558.09 \pm 0.05$, $c = 1300.00 \pm 0.11$ pm, and $\beta = 94.301 \pm 0.014°$, is formed in the Pb–O–Cl–Se system (Porter and Halasyamani 2001). This compound was synthesized by using a low-temperature aqueous method. Na_2SeO_3 (1.69 mM) and $PbCl_2$ (2.54 mM) were dissolved in 5 mL of distilled water. The respective solution was refluxed overnight at 160°C and cooled to room temperature over a period of 2 h. The colorless prismatic crystals of the title compound were obtained. They were thoroughly washed with H_2O.

11.73 Lead–Oxygen–Bromine–Selenium

Two quaternary compounds, **Pb₃(SeO₃)Br₄** and **Pb₃(SeO₃)₂Br₂**, are formed in the Pb–O–Br–Se system. First of them is thermally stable up to about 230°C and crystallizes in the orthorhombic structure with the lattice parameters $a = 772.90 \pm 0.15$, $b = 847.03 \pm 0.17$, $c = 1652.1 \pm 0.3$ pm, a calculated density of 6.560 g·cm⁻³, and an energy gap of 3.35 eV (Wang et al. 2018). To prepare this compound, PbBr₂ (2 mM), SeO₂ (1.5 mM), and 2 mL of deionized water were put into a 23 mL Teflon-lined autoclave and subsequently sealed. The autoclave was heated to 210°C and held for two days. After that, the autoclave was cooled to 100°C at a rate of 1.5°C·h⁻¹, followed by slow cooling to room temperature by turning off the muffle furnace. After filtration, the colorless transparent rod-like crystals of Pb₃(SeO₃)Br₄ were obtained and further washed with deionized water and dried.

Pb₃(SeO₃)₂Br₂ is stable up to 530°C and crystallizes in the monoclinic structure with the lattice parameters $a = 1343.69 \pm 0.17$, $b = 566.26 \pm 0.08$, $c = 1332.1 \pm 0.2$ pm, and $\beta = 92.995 \pm 0.009°$, a calculated density of 6.794 g·cm⁻³, and an energy gap of 3.73 eV (Zhang et al. 2012b) [$a = 1340.1 \pm 0.2$, $b = 565.48 \pm 0.09$, $c = 1330.6 \pm 0.4$ pm, and $\beta = 92.94 \pm 0.02°$, and a calculated density of 6.829 g·cm⁻³ (Berdonosov et al. 2012)]. To synthesize this compound, a mixture of Pb(NO₃)₂ (0.3 mM), KBr (0.4 mM), SeO₂ (0.6 mM), and H₂O (5 mL) was sealed in an autoclave equipped with a Teflon liner (23 mL) and heated at 200°C for 4 days, followed by slow cooling to room temperature at a rate of 6°C·h⁻¹. Colorless block-shaped crystals of the title compound were recovered along with tiny PbSeO₃ crystals as an impurity. After proper structural analysis, Pb₃(SeO₃)₂Br₂ was obtained as a single phase by the hydrothermal reaction of a mixture of Pb(NO₃)₂ (0.3 mM), KBr (2.0 mM) and SeO₂ (0.4 mM) in 5 mL H₂O at 200°C for 4 days. The large excess amount of KBr is needed to prevent the formation of PbSeO₃.

Pb₃(SeO₃)₂Br₂ can also be prepared as follows (Berdonosov et al. 2012). The mixtures of PbO, PbBr₂, and SeO₂ (molar ratio 2:1:2) were carefully ground in an agate mortar and placed in a quartz ampoule in a dry chamber filled with argon. The ampoule was then sealed under vacuum (~0.01 Pa). The sample was annealed at 300°C for 24 h and at 490°C for 12 days, after which it was cooled with the furnace switched off.

11.74 Lead–Oxygen–Iodine–Selenium

The **Pb₃(SeO₃)₂I₂** quaternary compound, which crystallizes in the monoclinic structure with the lattice parameters $a = 1349.3 \pm 0.3$, $b = 573.60 \pm 0.11$, $c = 1418.8 \pm 0.3$ pm, and $\beta = 92.98 \pm 0.03°$, and a calculated density of 6.840 g·cm⁻³, is formed in the Pb–O–I–Se system (Berdonosov et al. 2012). This compound was prepared in the same way as Pb₃(SeO₃)₂Br₂ was synthesized using PbI₂ instead of PbBr₂.

11.75 Lead–Oxygen–Iron–Selenium

The **PbFe₂(SeO₃)₄** quaternary compound, which crystallizes in the triclinic structure with the lattice parameters $a = 523.18 \pm 0.05$, $b = 679.25 \pm 0.06$, $c = 764.45 \pm 0.07$ pm, $\alpha = 94.300 \pm$ 0.002°, $\beta = 90.613 \pm 0.002°$, and $\gamma = 95.524 \pm 0.002°$, and a calculated density of 5.09 g·cm⁻³, is formed in the Pb–O–Fe–Se system (Johnston and Harrison 2004). A rational hydrothermal synthesis of this compound started from Pb(NO₃)₂ (2 mM), Fe(NO₃)₃·9H₂O (4 mM), SeO₂ (8 mM), and H₂O (15 mL), which were sealed in a 23 mL bomb and heated to 200°C for 2 days, followed by cooling to room temperature over a few hours. Product recovery by vacuum filtration and washing with water and acetone led to a mixture of tiny yellow cuboids of PbFe₂(SeO₃)₄ accompanied by brown whiskers and needles of Fe₂O(SeO₃)₄.

11.76 Lead–Oxygen–Cobalt–Selenium

The **PbCo(SeO₃)₂** quaternary compound, which apparently has two polymorphic modifications, is formed in the Pb–O–Co–Se system (Kovrugin et al. 2015). α-PbCo(SeO₃)₂ crystallizes in the orthorhombic structure with the lattice parameters $a = 1282.08 \pm 0.04$, $b = 549.02 \pm 0.02$, $c = 790.85 \pm 0.04$ pm, and a calculated density of 6.205 g·cm⁻³. This modification has been prepared by hydrothermal techniques. After weighing and grinding PbO, SeO₂, and CoO, the reagents were mixed in 6 mL of distilled water. When necessary, a solution of NaOH was used to adjust the pH to its desired value. It has been experimentally established that only acidic conditions favor crystallization. A large stoichiometric excess of SeO₂ was necessary to achieve reactions, where the amount of SeO₂ is multiplied by five. The pH values increase from ~1 to ~5.5–6.0 on decreasing the SeO₂ content. The chemical reactions were performed for 36 h in 23 mL Teflon-lined reaction vessels heated in an oven at 200°C. At the end of the experiment time, the vessels were cooled for 48 h. The precipitate was filtered through filter paper. Purple needles of α-PbCo(SeO₃)₂ were obtained.

11.77 Lead–Oxygen–Nickel–Selenium

The **PbNi(SeO₃)₂** quaternary compound, which has two polymorphic modifications, is formed in the Pb–O–Ni–Se system (Kovrugin et al. 2015). Both modifications crystallize in the orthorhombic structure with the lattice parameters $a = 1274.76 \pm 0.04$, $b = 545.62 \pm 0.02$, $c = 783.32 \pm 0.02$ pm, and a calculated density of 6.337 g·cm⁻³ for α-PbNi(SeO₃)₂ and $a = 547.15 \pm 0.04$, $b = 919.63 \pm 0.06$, $c = 1144.36 \pm 0.09$ pm, and a calculated density of 5.996 g·cm⁻³ for β-PbNi(SeO₃)₂. Both modifications were synthesize in the same way as PbCo(SeO₃)₂ was prepared using NiO instead of CoO. Yellow needles of α-PbNi(SeO₃)₂ and yellow prisms of β-PbNi(SeO₃)₂ were obtained.

11.78 Lead–Tellurium–Nickel–Selenium

PbTe–NiSe₂. The phase diagram of this system, constructed through DTA, XRD, metallography, and measuring of the microhardness using the alloys annealed at 500°C for 300 h, is a eutectic type (Figure 11.49) (Alidzhanov et al. 1995). The

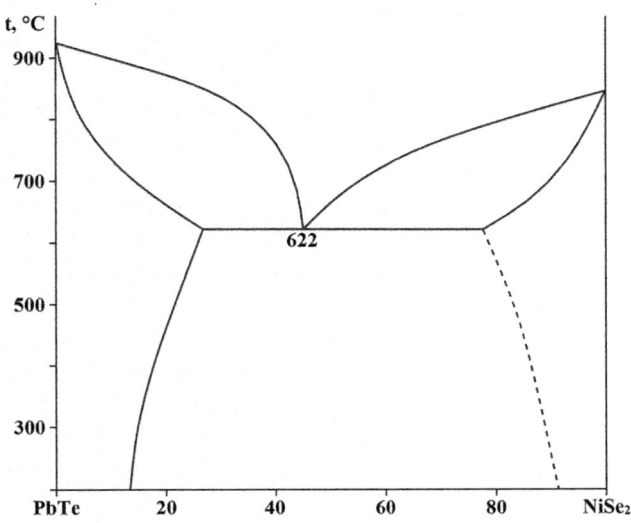

FIGURE 11.49 Phase diagram of the PbTe–NiSe₂ system. (From Alidzhanov, M.A., et al., *Neorgan. mater.*, **31**(9), 1174, 1995.)

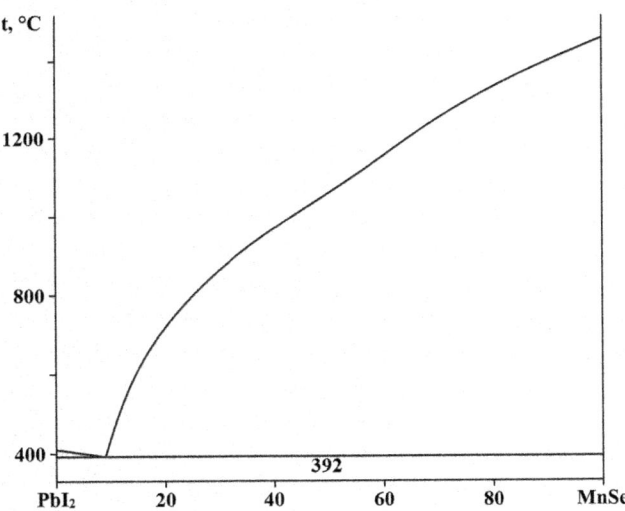

FIGURE 11.50 Phase diagram of the PbI₂–MnSe system. (From Odin, I.N., et al., *Zhurn. neorgan. khimii*, **48**(5), 839, 2003.)

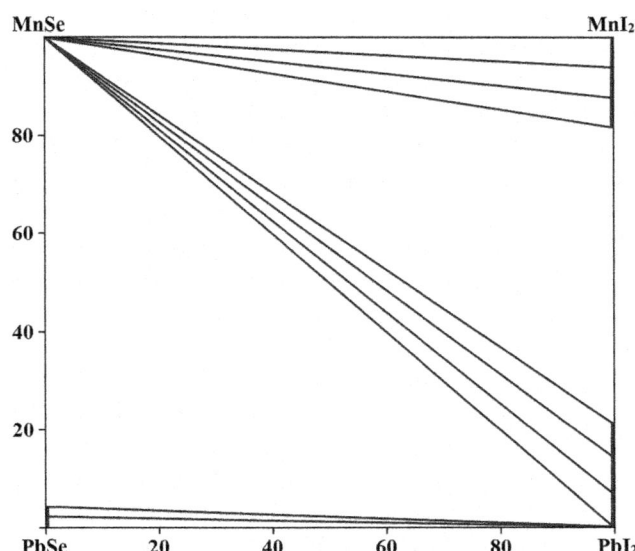

FIGURE 11.51 Isothermal section of the PbSe + MnI₂ ↔ PbI₂ + MnSe ternary mutual system at 363°C. (From Odin, I.N., et al., *Zhurn. neorgan. khimii*, **48**(5), 839, 2003.)

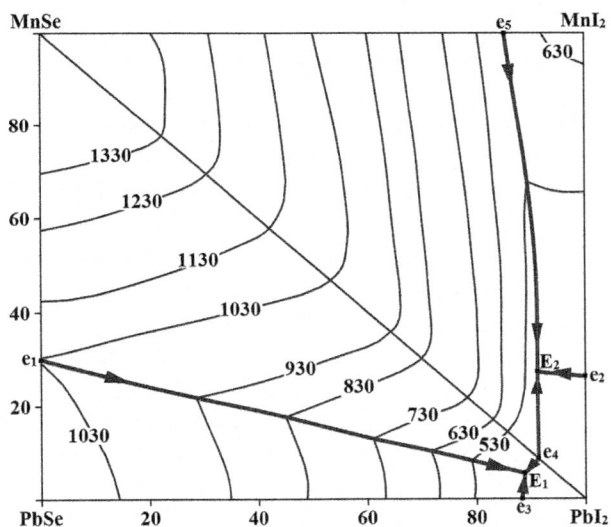

FIGURE 11.52 Liquidus surface of the PbSe + MnI₂ ↔ PbI₂ + MnSe ternary mutual system. (From Odin, I.N., et al., *Zhurn. neorgan. khimii*, **48**(5), 839, 2003.)

eutectic contains 45 mol% NiSe₂ and crystallizes at 622°C. The solubility of NiSe₂ in PbTe increases from 10 mol% at room temperature to 20 mol% at 500°C.

11.79 Lead–Iodine–Manganese–Selenium

PbI₂–MnSe. The phase diagram of this system, constructed through DTA, XRD, and metallography using the alloys annealed at 500°C for 300 h, is a eutectic type (Figure 11.50) (Odin et al. 2003b). The eutectic contains 9 mol% MnSe and crystallizes at 392°C. MnSe dissolves less than 0.4 mol% PbI₂, and PbI₂ dissolves less than 1 mol% MnSe. According to the data of Odin et al. (2004), a metastable phase μ, which

crystallizes in the rhombohedral structure, is formed in this system within the interval of 12–22 mol% MnSe. This phase is kept at room temperature for months.

PbSe + MnI₂ ↔ PbI₂ + MnSe. The isothermal section of this ternary mutual system is presented in (Figure 11.51) (Odin et al. 2003b). The solid solutions based on MnSe and PbSe and Pb₁₋ₓMnₓI₂ and Mn₁₋ᵧPbᵧI₂ solid solutions are in equilibrium in the system. The field of primary crystallization of MnSe occupies the largest part of the liquidus surface (Figure 11.52). The ternary eutectics crystallize at 380°C (E_1) and 369°C (E_2). The phases based on PbSe and MnSe are stable (Odin et al. 2004). 2*H*-PbI₂ and 6*R*-PbI₂ also crystallize from the liquid. It should be noted that these polytypes do not transform into each other.

12

Systems Based on Lead Telluride

12.1 Lead–Hydrogen–Oxygen–Tellurium

Two quaternary compounds, **PbH₄TeO₆** and **3PbTeO₃·2H₂O**, are formed in the Pb–H–O–Te system (Tananaeva and Novoselova 1967). The first was obtained by adding a 5% solution of $Pb(CH_3COO)_2$, previously acidified to pH = 4–5 with acetic acid, to a heated 20% solution of H_6TeO_6. 3PbTeO₃·2H₂O was obtained by slowly dropping a 5% solution of $Pb(NO_3)_2$ into a 10% heated solution of Na_2TeO_3 with continuous stirring. This compound is oxidized when heated in air starting from 400°C.

Mineral girdite, **H₂Pb₃(TeO₃)(TeO₆)**, was discredited as its original description was based on data obtained from at least two and probably more different phases (Kampf et al. 2017).

12.2 Lead–Sodium–Chlorine–Tellurium

The solubility of PbTe in the NaCl+PbCl₂ melts, containing up to 70 mol% PbCl₂, is less than 10 mass% at 860°C (Shurygin and Serebryakova 1975). After increasing of PbCl₂ content from 70 to 100 mol%, the solubility of PbTe at 820°C increases from 10 to 25 mol%.

12.3 Lead–Potassium–Bismuth–Tellurium

The **KPbBiTe₃** quaternary compound, which crystallizes in the cubic structure, is formed in the Pb–K–Bi–Te system (Kanatzidis et al. 1998).

12.4 Lead–Potassium–Chlorine–Tellurium

The solubility of PbTe in the KCl+PbCl₂ melts, containing up to 70 mol% PbCl₂ is less than 10 mass% at 860°C (Shurygin and Serebryakova 1975). After increasing PbCl₂ content from 70 to 100 mol%, the solubility of PbTe at 820°C increases from 10 to 25 mol%. At 600°C, the solubility of PbTe in the KCl+PbCl₂ melt, determined by the method of isothermal saturation followed by an analysis of the salt phase, is 0.06 ± 0.02 mol% (Rodionov et al. 1972). Dilution of lead chloride with potassium chloride sharply reduces the solubility of PbTe in the salt phase.

12.5 Lead–Rubidium–Bismuth–Tellurium

The **RbPb₂Bi₃Te₇** quaternary compound, which melts at 579°C, crystallizes in the orthorhombic structure with the lattice parameters $a = 443.51 \pm 0.09$, $b = 3271.0 \pm 0.7$, $c = 1267.3 \pm 0.3$ pm, and an energy gap of <0.05 eV, is formed in the Pb–Rb–Bi–Te system (Chung et al. 1998; Kanatzidis et al. 1998). Silver needles of this compound were prepared by a stoichiometric reaction of Pb, Bi, Te, and Rb₂Te at 720°C.

12.6 Lead–Cesium–Bismuth–Tellurium

Four quaternary compounds, **CsPbBi₃Te₆**, **CsPb₂Bi₃Te₇**, **CsPb₃Bi₃Te₈**, and **CsPb₄Bi₃Te₉**, are formed in the Pb–Cs–Bi–Te system. First of them crystallizes in the orthorhombic structure with the lattice parameters $a = 633.26 \pm 0.06$, $b = 2866.7 \pm 0.3$, and $c = 436.37 \pm 0.04$ pm (Hsu et al. 2001, 2002a, 2002b). This compound was prepared by reacting CsBi₄Te₆ (0.145 mM), Pb (0.145 mM), and Te (0.146 mM) at 600°C for 1 h and slow cooling to 50°C.

CsPb₂Bi₃Te₇ melts at 582°C (Chung et al. 1998; Kanatzidis et al. 1998) and also crystallizes in the orthorhombic structure with the lattice parameters $a = 434.56 \pm 0.06$, $b = 3247.6 \pm 0.5$, $c = 1250.8 \pm 0.2$ pm (Hsu et al. 2001, 2002a, 2002b), and an energy gap of <0.05 eV (Kanatzidis et al. 1998) [$a = 440.53 \pm 0.09$, $b = 3279.9 \pm 0.7$, and $c = 1266.7 \pm 0.3$ pm at 170 K (Chung et al. 1998)]. It was prepared by reacting CsBi₄Te₆ (0.117 mM), Pb (0.351 mM), and Te (0.351 mM) at 600°C for 1 h and slow cooling to 50°C. Both compounds can also be prepared by the stoichiometric reactions of Cs₂Te, Pb, Bi, and Te (Chung et al. 1998; Hsu et al. 2001, 2002a, 2002b).

CsPb₃Bi₃Te₈ and CsPb₄Bi₃Te₉ also crystallize in the orthorhombic structure with the lattice parameters $a = 637.36 \pm 0.08$, $b = 3773.1 \pm 0.5$, $c = 444.16 \pm 0.06$ pm, and a calculated density of 15.765 g·cm⁻³ for the first compound and 445.24 ± 0.06, $b = 4213.2 \pm 0.6$, $c = 1274.2 \pm 0.2$ pm, and a calculated density of 15.23 g·cm⁻³ for the second one (Hsu et al. 2001, 2002b). Crystals of both compounds were obtained by reacting CsBi₄Te₆ with PbTe in molar ratios 1:4 for CsPb₃Bi₃Te₈ and 1:6 for CsPb₄Bi₃Te₉. A single phase for CsPb₃Bi₃Te₈ was also prepared by reacting Cs and "Pb₂.₈₅Bi₃Te₈" (molar ratio 0.85:1), and a single phase for CsPb₄Bi₃Te₉ can also be prepared by reacting Cs and "Pb₃.₉Bi₃Te₉" (molar ratio 0.90:1). The two reactions were heated at 700°C for 2 h followed by cooling to 500°C in 2 h and finally quenching to room temperature.

12.7 Lead–Cesium–Chlorine–Tellurium

PbTe–CsCl. At 640°C, the solubility of PbTe in the CsCl melt, determined by the method of isothermal saturation followed by analysis of the salt phase, is 0.001 mol% (Rodionov et al. 1972).

12.8 Lead–Cesium–Iodine–Tellurium

PbTe–CsI. The phase diagram of this system, constructed through differential thermal analysis (DTA), X-ray diffraction (XRD), and metallography using the alloys annealing at 300°C for 300 h, is a eutectic type (Burmistrova et al. 1980). The eutectic degenerates from the CsI side and crystallizes at 640°C. There is a wide immiscibility region in the system at 910°C. The solubility of CsI in PbTe reaches 0.4 mol%.

12.9 Lead–Copper–Gold–Tellurium

Two quaternary compounds, **Au₃Cu₂PbTe₂** and **Au₄Cu(Te,Pb)**, are formed in the Pb–Cu–Au–Te system. The first, mineral bilibinskite, crystallizes in the cubic structure with the lattice parameter $a = 410$ pm (Spiridonov et al. 1978a; Fleischer et al. 1979b; Spiridonov 2011) [$a = 1003$ pm and a calculated density of 12.7 g·cm⁻³ (Spiridonov et al. 1978b)].

Au₄Cu(Te,Pb) (mineral bezsmertnovite) crystallizes in the orthorhombic structure with the lattice parameters $a = 403.6$, $b = 402.5$, $c = 406.1$ pm, and a calculated density of 16.3 g·cm⁻³ (Spiridonov and Chvileva 1979; Fleischer et al. 1981).

12.10 Lead–Copper–Yttrium–Tellurium

PbTe–Cu₂Te–Y₂Te₃. The isothermal section of this quasiternary system at 600°C is given in Figure 12.1 (Shemet et al. 2006c). The formation of quaternary compounds was not observed. The alloys for the investigations were annealed at 600°C for 240 h, with the next quenching in cold water.

12.11 Lead–Copper–Terbium–Tellurium

PbTe–Cu₂Te–Tb₂Te₃. The isothermal section of this quasiternary system at 500°C is shown in Figure 12.2 (Marchuk et al. 2008a). PbTe dissolves ~9 mol% Tb₂Te₃. No quaternary compounds were found. The alloys for the investigations were annealed at 500°C for 500 h and quenched in cold water.

12.12 Lead–Copper–Dysprosium–Tellurium

PbTe–Cu₂Te–Dy₂Te₃. The isothermal section of this quasiternary system at 500°C was constructed by Marchuk et al. (2008a) and is analogous to the isothermal section of the PbTe–Cu₂Te–Tb₂Te₃ system. PbTe dissolves 8 mol% Dy₂Te₃. The formation of quaternary compounds was not observed.

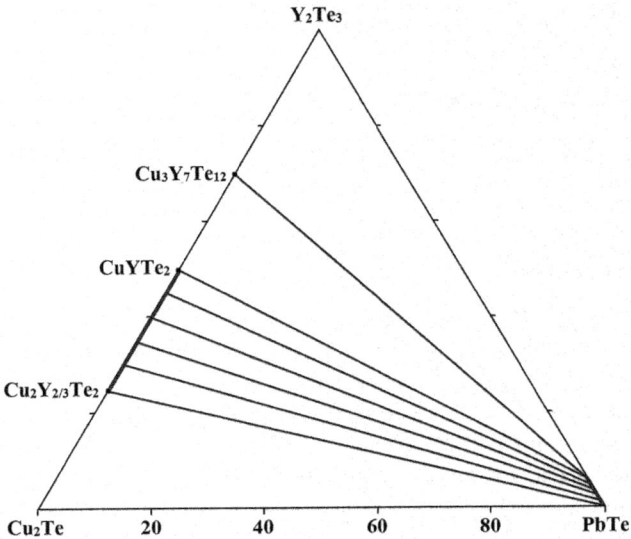

FIGURE 12.1 Isothermal section of the PbTe–Cu₂Te–Y₂Te₃ quasiternary system at 600°C. (From Shemet, V.Ya., et al., *Nauk. Visnyk Volyns'k. Derzh. Univ. im. Lesi Ukrainky. Ser. Khim. nauky*, (4), 124, 2006.)

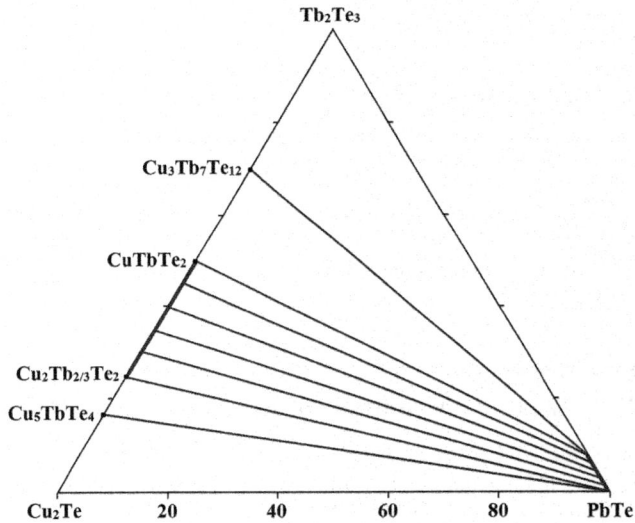

FIGURE 12.2 Isothermal section of the PbTe–Cu₂Te–Tb₂Te₃ quasiternary system at 500°C. (From Marchuk, O.V., et al., *J. Alloys Compd.*, **455**(1–2), 186, 2008.)

The alloys for the investigations were annealed at 500°C for 500 h and quenched in cold water.

12.13 Lead–Copper–Holmium–Tellurium

PbTe–Cu₂Te–Ho₂Te₃. The isothermal section of this quasiternary system at 600°C was constructed by Gulay and Olekseyuk (2006b) and is analogous to the isothermal section of the PbTe–Cu₂Te–Y₂Te₃ system. PbTe dissolves 7 mol% Ho₂Te₃. No quaternary compounds were found. The alloys for the investigations were annealed at 600°C for 240 h and quenched in cold water.

12.14 Lead–Copper–Erbium–Tellurium

PbTe–Cu₂Te–Er₂Te₃. The isothermal section of this quasiternary system at 600°C was constructed by Gulay and Olekseyuk (2006b). PbTe dissolves 4 mol% Er_2Te_3. The formation of quaternary compounds was not observed. This section could be divided by the PbTe–$Er_{2/3}Cu_2Te_2$, PbTe–$ErCuTe_2$, and PbTe–$Er_7Cu_3Te_{12}$ quasibinary sections into four subsystems. The alloys for the investigations were annealed at 600°C for 240 h and quenched in cold water.

12.15 Lead–Copper–Thulium–Tellurium

PbTe–Cu₂Te–Tm₂Te₃. The isothermal section of this quasiternary system at 600°C was constructed by Gulay and Olekseyuk (2006b). PbTe dissolves 6 mol% Tm_2Te_3. No quaternary compounds were found. This section could be divided by the PbTe–$Tm_{2/3}Cu_2Te_2$, PbTe–$TmCuTe_2$, and PbTe–$Tm_7Cu_3Te_{12}$ quasibinary sections into four subsystems. The alloys for the investigations were annealed at 600°C for 240 h and quenched in cold water.

12.16 Lead–Copper–Oxygen–Tellurium

Some quaternary compounds are formed in the Pb–Cu–O–Te system. **CuPbTeO₅** has two polymorphic modifications. First of them crystallizes in the monoclinic structure with the lattice parameters $a = 1226.80 \pm 0.03$, $b = 639.16 \pm 0.02$, $c = 1128.81 \pm 0.03$ pm, $\beta = 107.996 \pm 0.002°$, and a calculated density of 7.548 g·cm⁻³ and the second one crystallizes triclinic structure with the lattice parameters $a = 1226.86 \pm 0.08$, $b = 641.19 \pm 0.04$, $c = 1126.01 \pm 0.08$ pm, $\alpha = 90.452 \pm 0.005°$, $\beta = 107.708 \pm 0.005°$, $\gamma = 90.939 \pm 0.005°$, and a calculated density of 7.532 g·cm⁻³ (Weil et al. 2019). These modifications were obtained via solid-state reactions of Cu(NO₃)₂·2.5H₂O (0.232 g), Pb(NO₃)₂ (0.334 g), and H_6TeO_6 (0.229 g) (molar ratio 1:1:1). The starting materials were weighted, homogenously mixed by grinding, and transferred into ceramic crucibles. The open crucibles were placed in a furnace under the following temperature/time conditions under atmospheric conditions: $t_1 = 300$ min [25 → 800°C], $t_2 = 3000$ min [800°C], and $t_3 = 1000$ min [800°C → 25°C]. The solid product was evaluated under a polarizing microscope to choose turquoise crystals of $CuPbTeO_5$ with unspecific forms for the diffraction experiments. Two different modifications of this compound were identified by single-crystal XRD.

CuPb(TeO₃)₂ crystallizes in the cubic structure with the lattice parameter $a = 1260.10 \pm 0.08$ pm and a calculated density of 6.194 g·cm⁻³ (Weil et al. 2019) [$a = 1251.4$ pm and the calculated and experimental densities of 6.323 and 6.26 ± 0.08 g·cm⁻³, respectively (Powell et al. 1994; Jambor et al. 1995); $a = 1252.0 \pm 0.4$ pm for mineral choloalite of the same composition (Lam et al. 1999)].

The synthesis of this compound was carried out under hydrothermal conditions (Weil et al. 2019). The initial materials, CuO (0.063 g), PbO (0.179 g), TeO₂ (0.257 g), and KOH (0.09 g) (molar ratio 1:1:2:2) were placed in a Teflon container that was filled up to three-fourth of its volume with water. The container was sealed and loaded into a stainless-steel autoclave and heated in an oven at 210°C for 1 week. Dark-brown $CuPb(TeO_3)_2$ crystals with an octahedral form were present in minor amounts and were manually separated from the other solid products.

The title compound was also synthesized by fusion of stoichiometric amounts of CuO, PbO, and TeO₂ in a silica crucible (Powell et al. 1994; Jambor et al. 1995). The product was annealed at 500°C for several hours. A deep green, crystalline, but somewhat porous product was obtained.

CuPbTe₂O₇ crystallizes in the orthorhombic structure with the lattice parameters $a = 720.33 \pm 0.05$, $b = 1504.68 \pm 0.10$, and $c = 546.91 \pm 0.04$ pm (Yeon et al. 2011). The decomposition of this compound starts at approximately 680°C. Pure and polycrystalline $CuPbTe_2O_7$ was prepared by conventional solid-state methods. Stoichiometric amounts of PbO (3.00 mM), CuO (3.00 mM), TeO₂ (3.00 mM), and $H_2TeO_4·2H_2O$ (3.00 mM) were thoroughly ground and pressed into pellets. The pellet was placed in an alumina crucible and heated to 550°C in air, held for 3 days, and then cooled to room temperature. To obtain the single phase, several intermittent grindings and refirings were essential. Single crystals of $CuPbTe_2O_7$ were grown from a reaction mixture with an excess amount of TeO₂. PbO (1.00 mM), CuO (1.00 mM), and TeO₂ (5.00 mM) were placed in a platinum crucible. The crucible was gradually heated to 750°C in air, held for 24 h, and then cooled slowly to 450°C at a rate of 6°C·h⁻¹, followed by rapid cooling to room temperature. Green plate-shaped crystals were obtained.

Cu₃PbTeO₇ also crystallizes in the orthorhombic structure with the lattice parameters $a = 1048.8 \pm 0.1$, $b = 635.3 \pm 0.1$, and $c = 881.4 \pm 0.2$ pm (Wedel and Müller-Buschbaum 1996). To produce this compound, PbO, CuO, and TeO₂ (molar ratio 1:2:6) were intimately mixed in a corundum boat and quickly heated to 720°C in air. TeO₂ was added in excess because of its good flux properties. If the proportion of CuO in the reaction is adapted to that of the gross formula of Cu_3PbTeO_7, the homogeneity of the product improves. Since CuO single crystals were always present in the reaction product, it was not possible to obtain phase-pure product regardless of the TeO₂ amount used. Brown, air-resistant single crystals of the title compound were isolated from the reaction mixture.

12.17 Lead–Silver–Terbium–Tellurium

PbTe–Ag₂Te–Tb₂Te₃. The isothermal section of this quasiternary system at 500°C is shown in Figure 12.3 (Marchuk et al. 2008a). No quaternary compounds were found. The alloys for the investigations were annealed at 500°C for 500 h and quenched in cold water.

12.18 Lead–Silver–Dysprosium–Tellurium

PbTe–Ag₂Te–Dy₂Te₃. The isothermal section of this quasiternary system at 500°C was constructed by Marchuk et al. (2008a) and is analogous to the isothermal section of the

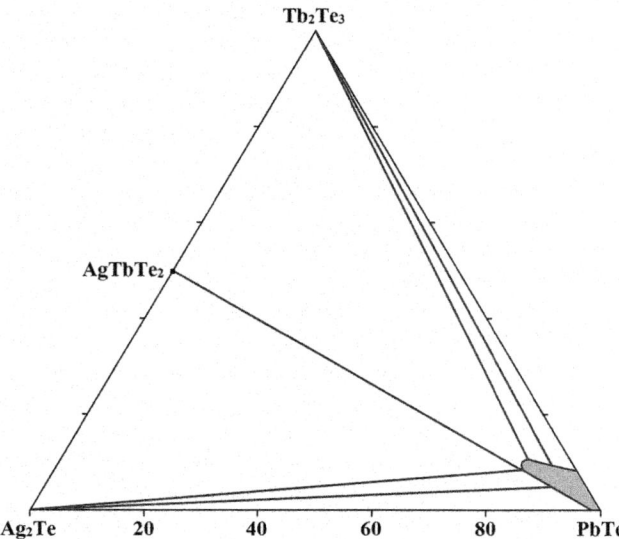

FIGURE 12.3 Isothermal section of the PbTe–Ag₂Te–Tb₂Te₃ quasiternary system at 500°C. (From Marchuk, O.V., et al., *J. Alloys Compd.*, **455**(1–2), 186, 2008.)

PbTe–Ag₂Te–Tb₂Te₃ system. The formation of quaternary compounds was not observed. The alloys for the investigations were annealed at 500°C for 500 h and quenched in cold water.

12.19 Lead–Silver–Antimony–Tellurium

PbTe–AgSbTe₂. This system is a nonquasibinary section of the Pb–Ag–Sb–Te quaternary system (Fleischmann et al. 1963; Maier 1963). At 500°C and 400°C, the solid solution region based on PbTe reaches respectively 74 and 79 mol% AgSbTe₂ and increases with the temperature decreasing [according to the data of Rodot (1959), the solubility of AgSbTe₂ in PbTe is 60 mol%]. The lattice parameter of this solid solution changes linearly with composition. The ingots for the investigations were annealed at 400°C, 500°C, and 550°C for 750, 500, and 300 h, respectively. The system was studied through DTA, metallography, and dilatometry.

These results were confirmed later by Mansimova (2019a, b), Mashadieva et al. (2018), and Mashadiyeva et al. (2020b) through DTA and measuring the microhardness and electromotive force (EMF) of the concentration chains. The partial thermodynamic functions of PbTe, Pb, and Ag in the alloys were calculated at room temperature using the temperature dependences of EMF. The standard thermodynamic functions of formation and standard entropies of $(2\text{PbTe})_x(\text{AgSbTe}_2)_{1-x}$ solid solutions were calculated. The alloys for the investigations were annealed at 480°C for 500 h and then at 210°C for another 500 h.

The ingots of AgPb$_m$SbTe$_{2+m}$ with nominal compositions of AgPb₆SbTe₈, AgPb₁₂SbTe₁₄, Ag₀.₉₅Pb₁₅SbTe₁₇, AgPb₁₈SbTe₂₀, and Ag₀.₈₆Pb₁₈SbTe₂₀ were prepared by annealing, in quartz tubes under vacuum, mixtures of Ag, Pb, Sb, and Te at 950°C for 4 h and cooling to 450°C in 40 h (Quarez et al. 2005). Ag₀.₅Pb₃Sb₀.₅Te₄, Ag₀.₃₈Pb₃.₂₃Sb₀.₃₈Te₄, and Ag₀.₂₅Pb₃.₅Sb₀.₂₅Te₄

single crystals were obtained by synthesizing $m = 1$, $m = 17$, and $m = 18$ compounds, respectively. Ag₁.₃₃Pb₁.₃₃Sb₁.₃₃Te₄ single crystals were grown by annealing at 1000°C a mixture of AgSbTe₂ and PbTe and cooling at a 20°C·h⁻¹ rate. Single crystals generally adopted a square bypyramidal shape or sometimes more complex polyhedral shapes. All synthesize compounds crystallize in the cubic or tetragonal structures with the lattice parameter $a = 621.73 \pm 0.08$ pm and a calculated density of 7.551 g·cm⁻³ or $a = 439.63 \pm 0.06$, $c = 621.73 \pm 0.12$ pm, and a calculated density of 7.551 g·cm⁻³ for Ag₁.₃₃Pb₁.₃₃Sb₁.₃₃Te₄; $a = 641.73 \pm 0.08$ pm and a calculated density of 7.834 g·cm⁻³ or $a = 453.77 \pm 0.06$, $c = 641.73 \pm 0.13$ pm, and a calculated density of 7.834 g·cm⁻³ for Ag₀.₅Pb₃Sb₀.₅Te₄; $a = 642.80 \pm 0.11$ pm and a calculated density of 7.910 g·cm⁻³ or $a = 454.53 \pm 0.06$, $c = 642.80 \pm 0.13$ pm, and a calculated density of 7.910 g·cm⁻³ for Ag₀.₃₈Pb₃.₂₃Sb₀.₃₈Te₄; $a = 644.04 \pm 0.12$ pm and a calculated density of 8.037 g·cm⁻³ or $a = 455.38 \pm 0.06$, $c = 644.04 \pm 0.14$ pm, and a calculated density of 8.038 g·cm⁻³ for Ag₀.₂₅Pb₃.₅Sb₀.₂₅Te₄. It is necessary to note that the AgPb$_m$SbTe$_{2+m}$ samples do not behave as classical solid solutions between AgSbTe₂ and PbTe but exhibit nanoscopic inhomogeneities with a large range of structural features from atomic ordering, revealed by single-crystal XRD, to nanostructures highlighted by electron diffraction and high-resolution imaging.

AgPb₁₈SbTe₂₀ can also be prepared as follows (Li et al. 2008). A certain amount of Pb(NO₃)₂, AgNO₃, and Sb(NO₃)₃ were dissolved in deionized water firstly under constant stirring. Then Na₂TeO₃ was added in the solution followed by adding KOH and KBH₄ in proportion. The mixture was kept under stirring with a magnetic bar for 5 min, and then was poured into a 100-mL autoclave to reach 80% of its volume. The autoclave was sealed and then placed in an oven to heat up to 180°C and held for 20 h, and finally naturally cooled to room temperature. Black precipitate at the bottom of the autoclave was washed with deionized water and absolute ethanol in sequence for several times and then filtered. The precipitate was dried in vacuum at 80°C for 10 h. The dried powder was cold-pressed into a pellet under 10 MPa, and then the pellet was sintered at 450°C or 520°C for 3 h in Ar.

The AgPb$_m$SbTe$_{2+m}$ nanocrystals with six different atomic compositions ($m = 1 - 18$) were also prepared by varying the molar ratio of Pb, Ag and Sb precursors (Arachchige et al. 2008). Nanocrystals with higher m values ($m = 4, 6, 8$, and 18) form stable colloidal suspensions in most of the common nonpolar solvents (hexane, chloroform, and toluene). However the stability of the nanocrystals suspensions decreases with decreasing m values. AgPbSbTe₃ and AgPb₂SbTe₄ nanoparticles are less stable in colloidal solutions and tend to precipitate within 1–2 days. The lattice parameters lie between AgSbTe₂ and PbTe and decrease systematically from higher to lower m values (Vegard's law is generally obeyed) suggesting a solid solution behavior. These nanocrystals, which are metastable solid solutions at room temperature, tend to phase-separate or segregate at moderately high temperature (150–200°C).

In a typical synthesis of AgPb$_m$SbTe$_{2+m}$ nanocrystals, lead acetate trihydrate, antimony acetate, octadecene (20 mL), and oleic acid (3.8 mL) were loaded into a Schlenk flask and heated under vacuum at 80°C for 3 h to generate a homogeneous colorless solution (Arachchige et al. 2008). The reaction

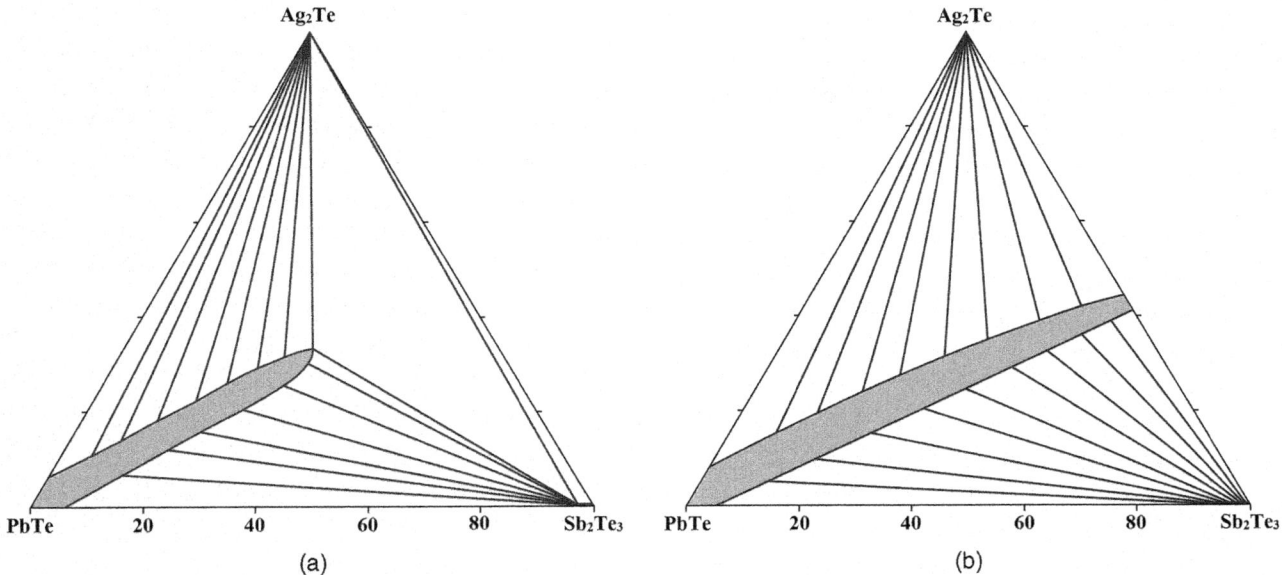

FIGURE 12.4 Isothermal sections of the PbTe–Ag$_2$Te–Sb$_2$Te$_3$ quasiternary system at (a) room temperature and (b) at 500°C. (From Mashadiyeva, L.F., et al., *Acta Chim. Slov.*, **67**(3), 799, 2020.)

mixture was cooled down to room temperature, and a stoichiometric amount of thiolate-stabilized silver complex solution was added to it. Then the resulting mixture was slowly heated to 50–60°C for 3 h under vacuum to remove the hexane solvent. The reaction mixture was flushed with nitrogen, and the temperature was raised to 150°C. At 150°C, a stock solution containing tellurium dissolved in trioctylphosphine (0.75 M) was quickly injected into the reaction mixture. After injection temperature of the reaction was maintained at 130–150°C for 2 min, and the reaction was terminated by rapid cooling back to room temperature using a cold-water bath. As-synthesized nanocrystals were precipitated from the crude mixture by adding excess acetone, and the nanocrystal precipitate was collected by centrifugation. The resultant nanocrystals were purified by dispersing in anhydrous hexane and precipitating with anhydrous ethanol twice. Finally, the purified nanocrystals were suspended in anhydrous hexane, chloroform, or carbon tetrachloride to form colloidal solutions. The nanocrystals with different atomic compositions were prepared by varying the molar ratio of Pb, Ag, and Sb precursors.

The Ag$_{1-x}$Pb$_{18}$SbTe$_{20}$ polycrystalline sintered samples ($x = 0, 0.1, 0.3$) were prepared as follows (Kosuga et al. 2005). Appropriate ratios of Ag, Pb, Sb, and Te were melted in a sealed evacuated quartz ampoule at 850°C for a few hours. The obtained sample was hot pressed at a pressure of 80 MPa for 2 h at 450°C under N$_2$ atmosphere. These phases crystallize in the cubic structure with the lattice parameter $a = 645.99$ pm and a calculated density of 8.02 g·cm^{-3} for $x = 0$; $a = 645.86$ pm and a calculated density of 8.01 g·cm^{-3} for $x = 0.1$; and $a = 645.46$ pm and a calculated density of 7.95 g·cm^{-3} for $x = 0.3$.

PbTe–Ag$_{19}$Sb$_{29}$Te$_{52}$. This system is a nonquasibinary section of the Pb–Ag–Sb–Te quaternary system (Kuchar and Heinrich 1967). Sb$_2$Te$_3$ is formed in the system from the Ag$_{19}$Sb$_{29}$Te$_{52}$ side. The continuous solid solutions are formed in the system at the temperature higher than 530°C. At 250°C, 340°C, and

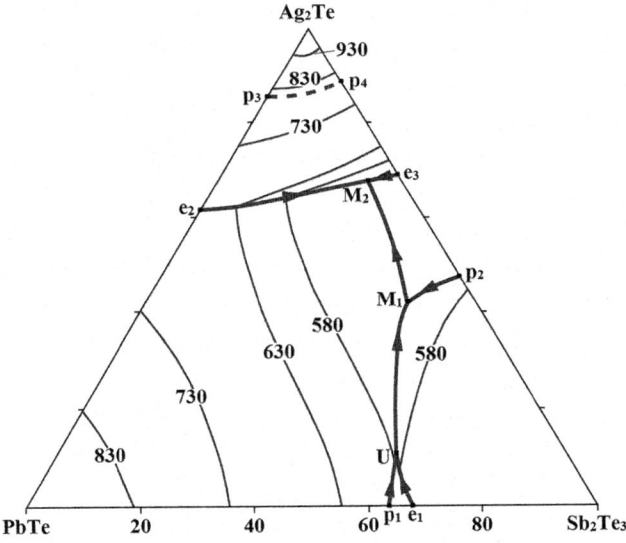

FIGURE 12.5 Liquidus surface of the PbTe–Ag$_2$Te–Sb$_2$Te$_3$ quasiternary system. (From Mashadiyeva, L.F., et al., *Acta Chim. Slov.*, **67**(3), 799, 2020.)

400°C, the area of solid solutions based on PbTe is 85, 77, and 64 mol% Ag$_{19}$Sb$_{29}$Te$_{52}$, respectively. The ingots for the investigations were annealed at 210–530°C for 1300–500 h. The system was studied through DTA and metallography.

PbTe–Ag$_2$Te–Sb$_2$Te$_3$. The isothermal sections of this quasiternary system at room temperature and at 480°C are shown in Figure 12.4 (Mashadiyeva et al. 2020b). The liquidus surface of the system consists of the fields of primary crystallization of the solid solutions based on the starting binary compounds, as well as Pb$_2$Sb$_6$Te$_{11}$ ternary compound (Figure 12.5). Largest crystallization field in the system belongs to the solid solution based on PbTe. This region is divided into two parts by

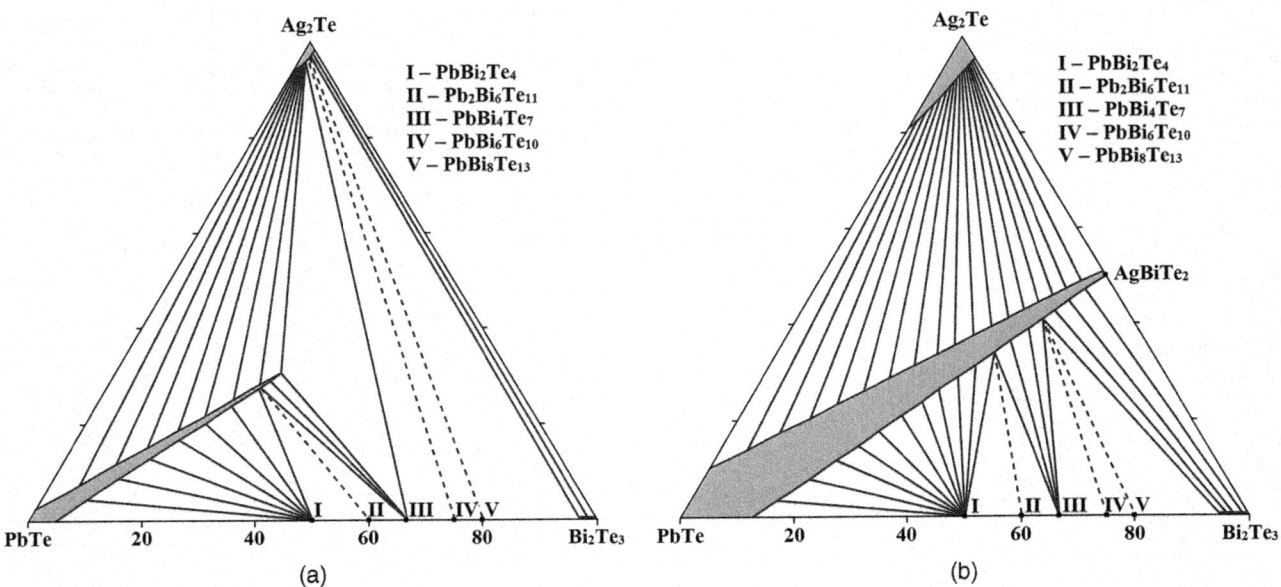

FIGURE 12.6 Isothermal sections of the PbTe–Ag$_2$Te–Bi$_2$Te$_3$ quasiternary system at (a) 330°C and (b) at 530°C. (From Babanly, D.M., et al., *Russ. J. Inorg. Chem.*, **56**(9), 1472, 2011.)

the curve M_1M_2 connecting the minimum points of M_1 and M_2. The primary crystallization area of the ternary Pb$_2$Sb$_6$Te$_{11}$ compound is very small. Transition equilibrium U [574°C; L + Pb$_2$Sb$_6$Te$_{11}$ ↔ (PbTe) + (Sb$_2$Te$_3$)] limits the extent of this area inside the concentration triangle. This is in good agreement with the literature data on a narrow temperature range for the existence of Pb$_2$Sb$_6$Te$_{11}$. Five vertical sections of this quasiternary system were also constructed.

The formation of the solid solution near the PbTe corner of the PbTe–Ag$_2$Te–Sb$_2$Te$_3$ system was also studied by Viskova et al. (1966).

12.20 Lead–Silver–Bismuth–Tellurium

PbTe–AgBiTe$_2$. This system is a nonquasibinary section of the Pb–Ag–Bi–Te quaternary system (Stegherr et al. 1963; Babanly et al. 2011a). This section is characterized by the formation of a continuous series of high-temperature solid solutions (Fleischmann et al. 1959; Stegherr et al. 1963; Babanly et al. 2011a). At 370–445°C, the decomposition of the solid solution based on AgBiTe$_2$ takes place. The lattice parameter of these solid solutions changes linearly with composition.

The **AgPbBiTe$_3$** quaternary compound is formed in this system. It does not melt below 900°C [melts at 705°C (Guseynov et al. 1972)] and crystallizes in the cubic structure with the lattice parameter $a = 629.0 \pm 0.2$ pm and an energy gap of 0.28 eV (Sportouch et al. 1998) [$a = 630.6$ pm and the calculated and experimental densities of 6.00 and 5.96 g·cm^{-3}, respectively (Guseynov et al. 1972)]. It was synthesized by combining the chemical elements in stoichiometric ratios under static vacuum at 900°C for 3 days (Sportouch et al. 1998).

The ingots for the investigations were annealed at 150°C, 470°C, and 520°C for 60, 8, and 9 days, respectively (Stegherr

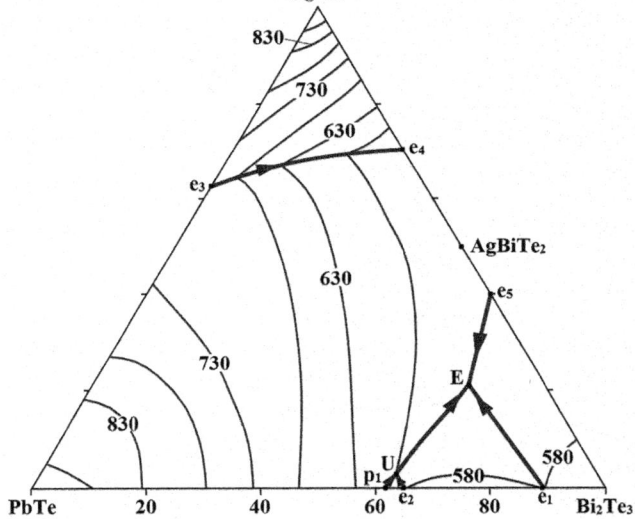

FIGURE 12.7 Liquidus surface of the PbTe–Ag$_2$Te–Bi$_2$Te$_3$ quasiternary system. (From Babanly, D.M., et al., *Russ. J. Inorg. Chem.*, **56**(9), 1472, 2011.)

et al. 1963) [at 530°C for 500 h and at 330°C for 1000 h with the next quenching in cold water (Babanly et al. 2011a)]. This system was studied through DTA, XRD, and metallography.

PbTe–Ag$_2$Te–Bi$_2$Te$_3$. The isothermal sections of this quasiternary system at 330°C and 530°C are shown in Figure 12.6 (Babanly et al. 2011a). Some vertical sections and the liquidus surface (Figure 12.7) of this system were also constructed. The liquidus surface consists of the fields of primary crystallization of the (PbTe)$_x$(AgBiTe$_2$)$_{1-x}$ solid solutions, solid solutions based on Ag$_2$Te and Bi$_2$Te$_3$, and PbBi$_2$Te$_4$ and PbBi$_4$Te$_7$ ternary compounds. The ternary eutectic E crystallizes at 542°C.

12.21 Lead–Silver–Iron–Tellurium

The $Ag_2FePbTe_4$ quaternary compound, which melts congruently at 627°C and crystallizes in the orthorhombic structure with the lattice parameters $a = 819.6$, $b = 699.2$, and $c = 617.2$ pm (Quintero et al. 1999), is formed in the Sn–Ag–Fe–Te system. The samples of this compound were prepared by the melt and anneals technique. The components of 1-g sample were made from the appropriate amounts of the elements and were sealed under vacuum in a small quartz ampoule, which had previously been carbonized to prevent interaction of the components with the quartz. The components were melted together at 1150°C for about an hour, annealed to equilibrium at 500°C, and then cooled to room temperature by leaving the ampoule in the switched-off furnace.

12.22 Lead–Magnesium–Oxygen–Tellurium

The $MgPb_2TeO_6$ quaternary compound is formed in the Pb–Mg–O–Te system. It undergoes two structural phase transitions at about 194 and 142 K (Baldinozzi et al. 1998). High-temperature modification crystallizes in the cubic structure with the lattice parameter $a = 798.38 \pm 0.05$ and 799.76 ± 0.05 pm at 220 and 350 K, respectively (Baldinozzi et al. 1998) [$a = 799 \pm 1$ (Bayer 1963; Politova and Venevtsev 1973); $a = 799.36 \pm 0.20$ pm at room temperature (Baldinozzi et al. 1994)]. Low-temperature modifications crystallizes in the rhombohedral structure with the lattice parameters $a = 564.47 \pm 0.05$ and $\alpha = 59.923 \pm 0.004°$ at 6 K (Baldinozzi et al. 1998) [a is not given and $\alpha = 89.933 \pm 0.005°$ at 80 K (Baldinozzi et al. 1994)].

Powder samples of this compound were prepared by solid-state reaction of the oxides (Bayer 1963; Baldinozzi et al. 1994). TeO_2, MgO, and PbO in stoichiometric proportions were put in a mortar to obtain an intimate mixture. The mixture was then heated in a gold crucible under ambient atmosphere at 650°C for 12 h to allow a complete oxidation of Te(IV) into Te(VI). The color of the mixture changes to pale yellow and the weight increases. A final annealing, performed under atmosphere at 870°C for 12 h, gives a well-crystallized powder of Pb_2MgTeO_6.

Synthesis of this compound can also be carried out according to ceramic technology by two-stage annealing (Politova and Venevtsev 1973). The initial mixture was prepared from stoichiometric amounts of $PbCO_3$, $MgCO_3$, and TeO_3. The mixture was annealed first at 700°C for 6 h, and then at 840°C for another 6 h.

12.23 Lead–Calcium–Oxygen–Tellurium

The $CaPb_2TeO_6$ quaternary compound, which exhibits reversible phase transition at 225.3°C and starts to decompose above 870°C, is formed in the Pb–Ca–O–Te system (Artner and Weil 2019). It crystallizes in the monoclinic structure with the lattice parameters $a = 1022.93 \pm 0.03$, $b = 579.26 \pm 0.02$, $c = 1016.98 \pm 0.03$ pm, $\beta = 108.6380 \pm 0.0010°$, and a calculated density

of 7.888 g·cm^{-3} (Artner and Weil 2019) [in the orthorhombic structure with the lattice parameters $a = c = 826 \pm 1$, $b = 824 \pm 1$, $\beta = 91°50'$ (Politova and Venevtsev 1973)].

To prepare this compound, $2PbCO_3·Pb(OH)_2$ (1.16 g), $CaCO_3$ (0.075 g), and TeO_2 (0.117 g) were fired within 18 h from room temperature to 960°C and held at that temperature for 6 h before cooling down to room temperature within 18 h (Artner and Weil 2019). Small colorless single crystals of Pb_2CaTeO_6 with an irregular form could be separated from a brick-red powder. Microcrystalline single-phase Pb_2CaTeO_6 was obtained by heating a well-ground and pressed mixture of $2PbCO_3·Pb(OH)_2$ (0.39 g), $CaCO_3$ (0.05 g) and TeO_2 (0.12 g) in a platinum crucible within 6 h from room temperature to 750°C where that was held for 99 h. Then the mixture was allowed to cool down in the turned-off furnace. After regrinding, a second identical heating protocol was carried out.

Synthesis of this compound can also be carried out according to ceramic technology by two-stage annealing (Politova and Venevtsev 1973). The initial mixture was prepared from stoichiometric amounts of $PbCO_3$, $CaCO_3$, and TeO_3. The mixture was annealed first at 700°C for 6 h, and then at 860°C for another 6 h.

12.24 Lead–Strontium–Oxygen–Tellurium

The $Sr_{3.53}Pb_{7.47}Te_4O_{23}$ quaternary compound, which crystallizes in the cubic structure with the lattice parameter $a = 1673.54 \pm 0.08$ pm and a calculated density of 7.752 g·cm^{-3}, is formed in the Pb–Sr–O–Te system (Artner and Weil 2019). To prepare this compound, PbO (2.23 g), $SrCO_3$ (0.74 g) and TeO_2 (1.0 g) were heated from room temperature to 800°C within 6 h and kept at that temperature for one day. The greyish-yellow material was reground and heated in two subsequent steps at 900°C for 3 days and with intermittent grinding at 1000°C for another 3 days. The color changed from greenish-orange for material treated at 900°C to dark orange for material treated at 1000°C, together with the formation of single crystals suitable for XRD at the higher temperature.

12.25 Lead–Barium–Oxygen–Tellurium

The $BaPbTe_2O_6$ quaternary compound, which melts incongruently at 561°C and crystallizes in the orthorhombic structure with the lattice parameters $a = 594.14 \pm 0.03$, $b = 859.79 \pm 0.04$, $c = 1385.56 \pm 0.06$ pm, a calculated density of 6.529 g·cm^{-3}, and an energy gap of 3.65 eV, is formed in the Pb–Ba–O–Te system (Yang et al. 2022b). Colorless bulk crystals of this compound were synthesized through a hydrothermal reaction by using a mixture of TeO_2 (2 mM), $Ba(OH)_2·8H_2O$ (2.2 mM), $PbCl_2$ (0.5 mM), and 4 mL of deionized water, which was placed in an autoclave equipped with a Teflon liner (25 mL). The mixture was heated at 200°C for 5 days and cooled to 30°C at a rate of 3°C·h^{-1}. The obtained crystals were washed with deionized water and then dried in air.

12.26 Lead–Zinc–Oxygen–Tellurium

The **ZnPbTe₂O₆** quaternary compound, which crystallizes in the cubic structure with the lattice parameter $a = 801 \pm 1$ pm, is formed in the Pb–Zn–O–Te system (Politova and Venevtsev 1973). Synthesis of this compound can be carried out according to ceramic technology by two-stage annealing. The initial mixture was prepared from stoichiometric amounts of PbCO₃, ZnCO₃, and TeO₃. The mixture was annealed first at 700°C for 6 h, and then at 820°C for another 6 h.

12.27 Lead–Cadmium–Oxygen–Tellurium

Two quaternary compounds, **CdPb₂TeO₆** and **CdPb₆TeO₁₀**, are formed in the Pb–Cd–O–Te system. First of them exhibits reversible phase transition at 270.9°C, starts to decompose above 950°C, and crystallizes in the monoclinic structure with the lattice parameters $a = 1007.16 \pm 0.03$, $b = 571.97 \pm 0.02$, $c = 1015.77 \pm 0.03$ pm, $\beta = 107.528 \pm 0.002°$, and a calculated density of 8.932 g·cm⁻³ (Artner and Weil 2019) [in the orthorhombic structure with the lattice parameters $a = c = 824 \pm 1$, $b = 828 \pm 1$ pm, $\beta = 92°35'$ (Politova and Venevtsev 1973)].

Single crystals of CdPb₂TeO₆ were obtained by heating a powder mixture in the molar ratio Pb:Cd:Te = 6:1:1 (Artner and Weil 2019). The mixture of PbO (2.01 g), CdO (0.194 g), and TeO₂ (0.238 g) was treated according to the following temperature program: room temperature → 800°C [6 h] → 800°C [6 h] → 600°C [12 h] → 600°C [2 h] → 400°C [12 h] → 400°C [2 h] → room temperature [24 h]. The obtained dark yellow polycrystalline material was reground, pressed again, and heat-treated in a second temperature program: room temperature → 900°C [6 h] → 900°C [6 h] → 700°C [12 h] → 700°C [2 h] → 500°C [12 h] → 500°C [2 h] → 300°C [12 h] → 300°C [2 h] → room temperature [12 h]. Besides crystals of CdPb₂TeO₆, translucent yellow crystals of Pb₅TeO₈ were also identified. The microcrystalline single phase of CdPb₂TeO₆ was prepared in a platinum crucible by heating a finely ground and pressed mixture of PbO (0.45 g), CdCO₃ (0.287 g), and H₆TeO₆ (0.229 g) within 6 h from room temperature to 750°C that was held for 99 h. Then the mixture was allowed to cool down in the turned-off furnace.

Synthesis of this compound can also be carried out according to ceramic technology by two-stage annealing (Politova and Venevtsev 1973). The initial mixture was prepared from stoichiometric amounts of PbCO₃, CdCO₃, and TeO₃. The mixture was annealed first at 700°C for 6 h, and then at 860°C for another 6 h.

CdPb₆TeO₁₀ melts incongruently at 845°C and no first-order phase transition was revealed for this compound in the temperature range 30–1000°C (Artner and Weil 2013). It crystallizes in the orthorhombic structure with the lattice parameters $a = 912.06 \pm 0.2$, $b = 1156.74 \pm 0.03$, and $c = 1131.13 \pm 0.03$ pm. Single-crystal growth experiments of CdPb₆TeO₁₀ were performed using the mixture of PbCO₃ (8.5 mM), CdCO₃ (1.4 mM), and TeO₃, (1.4 mM) which was heated in a corundum crucible under the following conditions: room temperature → 900°C [6 h] → 900°C [1 h] → 700°C [99 h] → 700°C [1 h]

→ 500°C [99 h] → 300°C [99 h] → room temperature [5 h]. An orange fragment was isolated from the recrystallized melt. A microcrystalline pure-phase sample of this compound was obtained by heating a well-ground and pressed mixture of PbO (9.0 mM), CdCO₃ (1.7 mM), and TeO₂ (1.5 mM) heated from room temperature to 750°C in 6 h and treated at this temperature for 99 h. Two further heat treatments were carried out, under the same conditions as the first one, with regrinding after every step.

12.28 Lead–Cadmium–Iodine–Tellurium

PbI₂–CdTe. The phase diagram of this system, constructed through DTA, XRD, and metallography using the ingots annealed for 1000 h, is a eutectic type (Odin 2001). The eutectic composition and temperature are 15 mol% CdTe and 384°C, respectively. The mutual solubility of CdSe and PbI₂ is negligible: the solubility of PbI₂ in CdTe is equal to 0.2 mol%. The crystallization of the melts from the PbI₂ side leads to the formation of 2H-PbI₂ metastable phase (Odin 2001; Odin and Chukichev 2001).

PbTe + CdI₂ ↔ CdTe + PbI₂. The isothermal section of this ternary mutual system at 350°C is analogous to one of the PbSe + CdI₂ ↔ CdSe + PbI₂ ternary mutual systems (Figure 11.27) (Odin 2001). The fields of primary crystallization of the solid solutions based on CdSe and PbTe and Cd$_x$Pb$_{1-x}$I₂ solid solutions exist on the liquidus surface of this system (Figure 12.8) (Odin 2001). The ternary eutectic E at 368°C is formed by the solid solutions based on CdTe and PbTe and Cd$_x$Pb$_{1-x}$I₂ solid solutions. A minimum m exists on the e_4e_2 line of the secondary crystallization.

Metastable phases of 2H-PbI₂ and 6R-PbI₂ can form at the crystallization of the melts in the CdTe–PbTe–PbI₂ quasiternary system in the region enriched in PbI₂ (Odin and Chukichev 2001).

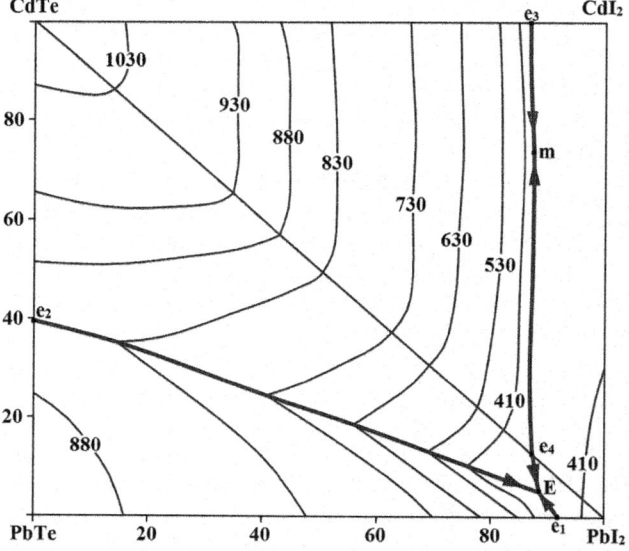

FIGURE 12.8 Liquidus surface of the PbTe + CdI₂ ↔ CdTe + PbI₂ ternary mutual system. (From Odin, I.N., *Zhurn. neorgan. khimii*, **46**(10), 1733, 2001.)

12.29 Lead–Gallium–Yttrium–Tellurium

PbTe–Ga$_2$Te$_3$–Y$_2$Te$_3$. The isothermal section of this quasiternary system at 500°C was constructed using the alloys annealed at this temperature for 250 h (Filyuk et al. 2008). This system is divided into two subsystems by the PbGa$_6$Te$_{10}$–Y$_2$Te$_3$ quasibinary system. No quaternary compounds were found. The solubility of Y$_2$Te$_3$ in PbTe at 500°C is 6 mol%.

12.30 Lead–Gallium–Lanthanum–Tellurium

PbTe–Ga$_2$Te$_3$–La$_2$Te$_3$. The isothermal section of this quasiternary system at 500°C was constructed using the alloys annealed at this temperature for 250 h (Filyuk et al. 2008). This system is divided into two subsystems by the PbGa$_6$Te$_{10}$–La$_2$Te$_3$ quasibinary system. No quaternary compounds were found.

12.31 Lead–Gallium–Arsenic–Tellurium

PbTe–GaAs. The phase diagram of this system, constructed through DTA, XRD, and metallography using the alloys annealed at 700°C for 100 h, is a eutectic type (Figure 12.9) (Raevskiy 1981). The eutectic contains 11.5 mol% GaAs and crystallizes at 875°C. The solubility of GaAs in PbTe is 0.7 mol%.

12.32 Lead–Indium–Antimony–Tellurium

PbSb$_2$Te$_4$–InSb. The phase equilibria in this system were studied by Vassilev et al. (2011). **(2±δ)InSb·PbSb$_2$Te$_4$** quaternary compound of variable composition, which melts incongruently at 530°C and crystallizes in the orthorhombic structure with the lattice parameters $a = 776.40 \pm 1.85$, $b = 641.44 \pm 0.30$, and $c = 360.10 \pm 0.03$ pm (for $\delta = 0$), is formed in this system. The title compound has an asymmetric area of homogeneity shifted to the side rich in InSb. The system also contains InSb- and PbSb$_2$Te$_4$-based boundary solid solutions extended from

0 to 10 mol% and from 95 to 100 mol% InSb, respectively, at room temperature. There are three nonvariant equilibria in this system: syntectic equilibrium (66.7 mol% InSb and 530°C) and two eutectic equilibria (90 mol% InSb and 460°C, and 40 mol% InSb and 500°C). This system was investigated through DTA, XRD, and measuring of the microhardness and density.

12.33 Lead–Thallium–Samarium–Tellurium

Tl$_4$PbTe$_3$–Tl$_9$SmTe$_6$. Using the multipurpose genetic algorithm approach, the analytical models of the phase diagrams of this system as temperature dependencies of compositions of the equilibrium phases were obtained and the boundaries of the heterogeneous equilibrium of liquid-solid have been determined in a wide range of temperatures (30–930°C) (Imamaliyeva et al. 2020).

12.34 Lead–Thallium–Terbium–Tellurium

Tl$_4$PbTe$_3$–Tl$_9$TbTe$_6$. This system is characterized by the formation of continuous solid solutions at low temperatures (Imamaliyeva et al. 2018). It is nonquasibinary section of the Pb–Tl–Tb–Te system, as Tl$_9$TbTe$_6$ melts incongruently (Imamaliyeva et al. 2018, 2021). This leads to the primary crystallization of TlTbTe$_2$ in a wide composition range. The system was studied through DTA, XRD, and metallography using the alloys annealed at 480°C for 1000 h.

12.35 Lead–Thallium–Antimony–Tellurium

PbTe–TlSbTe$_2$. This system is nonquasibinary section of the Pb–Tl–Sb–Te system, as TlTbTe$_2$ melts incongruently (Gitsu et al. 1980c). At temperatures above 330°C, there is a continuous series of solid solutions in the system. When the temperature is lowered in the range of 20–40 mol % 2PbTe, solid solutions decompose. On the PbTe side, solid solutions crystallize in a cubic structure, and on the TlTbTe$_2$ side, in a rhombohedral structure (Mazelsky and Lubell 1962; Gitsu et al. 1980c). This system was studied through DTA, XRD, and metallography (Gitsu et al. 1980c). The alloys for the investigations were annealed 430°C for 240 h, and some of them in the stepped mode at 600°C and 400°C for 300 and 400 h, respectively.

12.36 Lead–Thallium–Bismuth–Tellurium

PbTe–TlBiTe$_2$. This system is nonquasibinary section of the Pb–Tl–Bi–Te system, as TlBiTe$_2$ melts incongruently (Gitsu et al. 1980b). At temperatures above 380°C, there is a continuous series of solid solutions in the system. When the temperature is lowered in the range of 20–30 mol % 2PbTe, solid solutions decompose. On the PbTe side, solid solutions crystallize in a cubic structure, and on the TlBiTe$_2$ side, in a rhombohedral structure (Mazelsky and Lubell 1962; Gitsu et al. 1980c). The **TlPbBiTe$_3$** quaternary compound [melting temperature

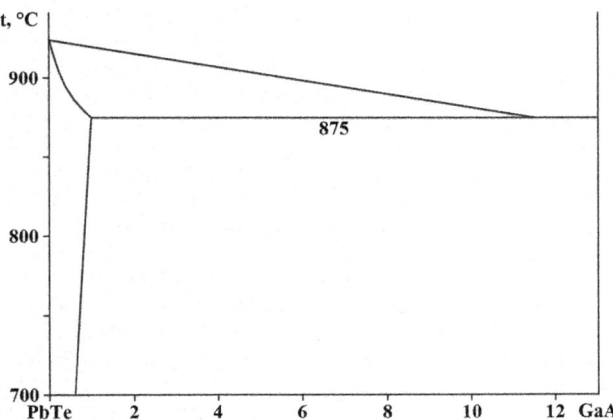

FIGURE 12.9 Part of the phase diagram of the PbTe–GaAs system. (From Raevskiy, S.D., *Izv. AN SSSR. Neorgan. mater.*, **17**(8), 1385, 1981.)

635°C, cubic structure, a = 648.8 pm and the calculated and experimental densities of 8.042 and 8.038 g·cm⁻³, respectively (Guseynov et al. 1972)] was not found in the system (Mazelsky and Lubell 1962; Gitsu et al. 1980c). This system was studied through DTA, XRD, and metallography (Gitsu et al. 1980c). The alloys for the investigations were annealed at 500°C for 560 h.

These results were confirmed later by Guseynov (2011) by measuring of EMF of the concentration chains within the temperature interval 30–160°C. The partial thermodynamic functions of Tl in the alloys were calculated at room temperature using the temperature dependencies of EMF. The standard thermodynamic functions of formation and standard entropies of $(2PbTe)_x(TlBiTe_2)_{1-x}$ solid solutions were calculated. The alloys for the investigations were annealed at 430°C for 1000 h.

According to the data of Malakhovska et al. (2007), the PbTe–TlBiTe₂ is a quasibinary section of the PbTe–Tl₂Te–Bi₂Te₃ quasiternary system.

Tl₄PbTe₃–TlBiTe₂. The phase diagram of this system, constructed through DTA and XRD, is a eutectic type (Figure 12.10) (Barchiy et al. 2010). The eutectic contains 36 mol% TlBiTe₂ and crystallizes at 483°C. The solubility of TlBiTe₂ in Tl₄PbTe₃ and Tl₄PbTe₃ in TlBiTe₂ is respectively 15 and 45 mol% at the eutectic temperature and 7 and 25 mol% at 200°C. According to the data of Guseynov et al. (2009), this system is a nonquasibinary section of the PbTe–Tl₂Te–Bi₂Te₃ quasiternary system.

Tl₄PbTe₃–Tl₉BiTe₆. The phase diagram of this system, constructed through DTA and XRD, belongs to type I according to Roseboom's classification and is given in Figure 12.11 (Barchiy et al. 2010).

PbTe–Tl₂Te–Bi₂Te₃. The isothermal sections of this quasiternary system at 330°C and 530°C are presented in Figure 12.12 (Guseinov et al. 2012). At 330°C, wide regions of the solid solutions exist along the PbTe–TlBiTe₂ and

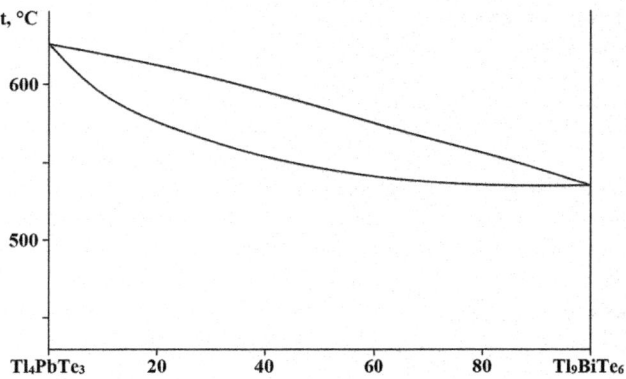

FIGURE 12.11 Phase diagram of the Tl₄PbTe₃–Tl₉BiTe₆ system. (From Barchiy, I.Ye., et al., *Nauk.Visnyk Volyns'k. Nats. Univ. im. Lesi Ukrainky. Ser. Khim. nauky.*, (16), 18, 2010.)

Tl₄PbTe₃–Tl₉BiTe₆ sections. Near Tl₂Te, two regions, α and X, were identified. Alloys in the α region consist of a monoclinic phase with the Tl₂Te structure. A two-phase region between α and solid solutions based on the Tl₄PbTe₃–Tl₉BiTe₆ section was not determined. Equilibrium alloys in the X region contain not only α and solid solutions based on the Tl₄PbTe₃–Tl₉BiTe₆ section, but also a metallic phase based on Tl. The alloys for the investigations were annealed at 330°C and 530°C for 1000 h. The alloys annealed at 330°C were cooled in a switched-off furnace, and the alloys annealed at 530°C were quenched in cold water.

Some vertical sections and the liquidus surface (Figure 12.13) of this system were also constructed (Guseinov et al. 2012). The liquidus surface consists of six fields of primary crystallization of solid solutions based on PbTe, Tl₂Te, Bi₂Te₃, and TlBiTe₂, PbBi₂Te₄ and PbBi₄Te₇ ternary compounds. The field of primary crystallization of solid solutions based on PbTe occupies the biggest part of the concentration triangle. The ternary eutectics E_1, E_2, and E_3 crystallize at 572°C, 532°C, and 484°C, respectively.

The isothermal section at 200°C, liquidus surface, and space phase diagram of the Tl₄PbTe₃–TlBiTe₂–Tl₉BiTe₆ subsystem as a part of the PbTe–Tl₂Te–Bi₂Te₃ quasiternary system was constructed by Malakhovska et al. (2019). Two vertical sections, the isothermal sections at 330°C and 530°C, and the liquidus surface of another subsystem (PbTe–Tl₂Te–TlBiTe₂) of this quasiternary system were constructed by Guseynov et al. (2009).

12.37 Lead–Scandium–Oxygen–Tellurium

The **Sc₂Pb₃TeO₉** quaternary compound, which crystallizes in the cubic structure with the lattice parameter a = 812.16 ± 0.01 pm, is formed in the Pb–Sc–O–Te system (Larrégola et al. 2010). It was obtained by solid-state reaction. As starting materials, PbO, Sc₂O₃, and TeO₂ were used. They were weighed out in the appropriate metal ratios and well mixed in an agate mortar. The mixture was annealed at 550°C for 24 h in order to oxidize Te(IV) to Te(VI). Subsequently, the product was

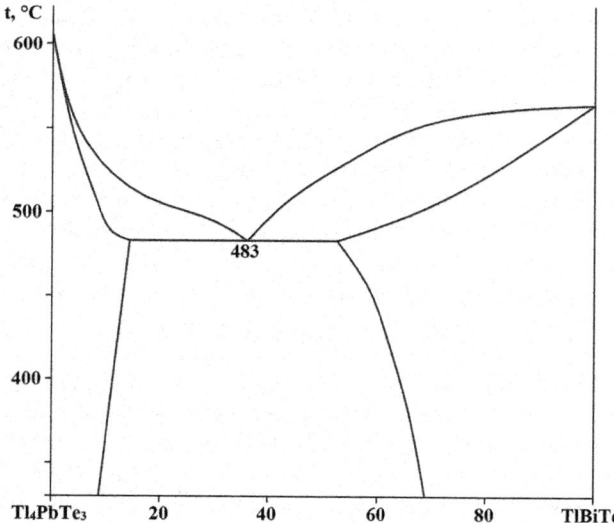

FIGURE 12.10 Phase diagram of the Tl₄PbTe₃–TlBiTe₂ system. (From Barchiy, I.Ye., et al., *Nauk.Visnyk Volyns'k. Nats. Univ. im. Lesi Ukrainky. Ser. Khim. nauky.*, (16), 18, 2010.)

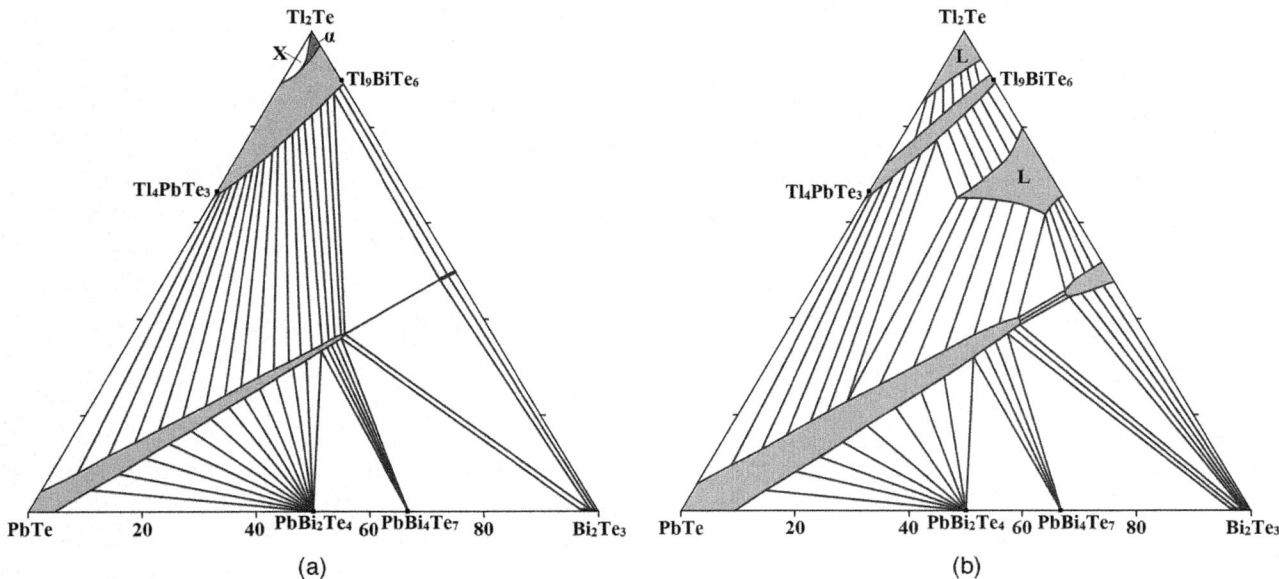

(a) (b)

FIGURE 12.12 Isothermal sections of the PbTe–Tl$_2$Te–Bi$_2$Te$_3$ quasiternary system at (a) 330°C and (b) at 530°C. (From Guseinov, F.N., et al., *Russ. J. Inorg. Chem.*, **57**(10), 1387, 2012.)

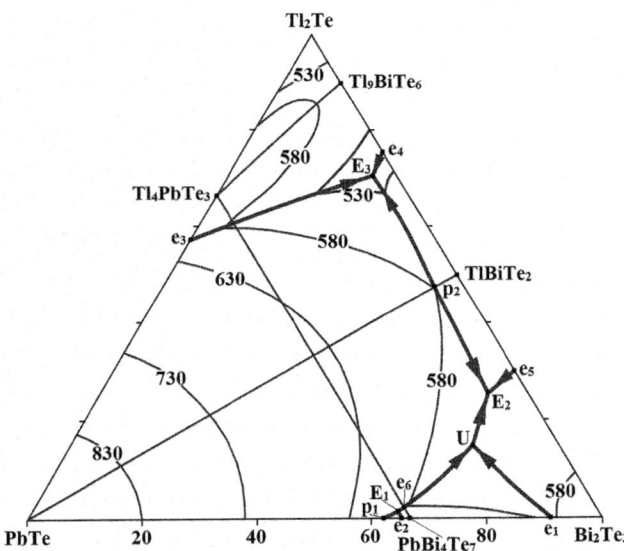

FIGURE 12.13 Liquidus surface of the PbTe–Tl$_2$Te–Bi$_2$Te$_3$ quasiternary system. (From Guseinov, F.N., et al., *Russ. J. Inorg. Chem.*, **57**(10), 1387, 2012.)

homogenized at 800°C for 12 h and finally at 950°C for 12 h with intermediate regrinding until a single phase of the title compound was obtained. All the calcination steps were carried out in sintered alumina crucibles and air atmosphere.

12.38 Lead–Ytterbium–Bismuth–Tellurium

PbBi$_2$Te$_4$–YbTe. This system is nonquasibinary section of the PbTe–YbTe–Bi$_2$Te$_3$ quasiternary system as PbBi$_2$Te$_4$ melts incongruently (Rasulova et al. 2013b).

PbBi$_4$Te$_7$–YbTe. The phase diagram of this system is a eutectic type (Rasulova et al. 2013b). The eutectic contains ~5 mol% YbTe and crystallizes at 582°C.

Both systems were studied through DTA and XRD using the alloys annealed at 530°C for 1000 h.

PbTe–YbTe–Bi$_2$Te$_3$. The isothermal sections of this quasiternary system at room temperature and at 530°C are shown in Figure 12.14 (Rasulova et al. 2013a). The existence of the YbPbTe$_2$, YbBi$_2$Te$_4$, and YbBi$_4$Te$_7$ ternary compounds was not confirmed. The wide regions of the solid solutions based on PbTe and Bi$_2$Te$_3$ are formed in the system. The alloys for the investigations were annealed at 530°C for 1000 h.

12.39 Lead–Uranium–Oxygen–Tellurium

Three quaternary compounds, **Pb(UO$_2$)(TeO$_3$)$_2$**, **Pb(UO$_2$) TeO$_6$**, and **Pb$_2$(UO$_2$)(TeO$_3$)$_3$**, are formed in the Pb–U–O–Te system. First of them (mineral moctezumite) crystallizes in the monoclinic structure with the lattice parameters $a = 781.3 \pm 0.5$, $b = 706.1 \pm 0.2$, $c = 1377.5 \pm 0.4$ pm, $\beta = 93.71 \pm 0.02°$, and a calculated density of 7.256 g·cm^{-3} (Swihart et al. 1993) [$a = 781.9 \pm 0.7$, $b = 707.0 \pm 0.3$, $c = 1383.6 \pm 1.3$ pm, $\beta = 93°37.5'$, and the calculated and experimental densities of 5.41 and 5.73 g·cm^{-3}, respectively (Gaines 1965)].

Pb$_2$(UO$_2$)TeO$_6$ (mineral markcooperite) also crystallizes in the monoclinic structure with the lattice parameters $a = 572.2 \pm 0.2$, $b = 774.8 \pm 0.2$, $c = 788.9 \pm 0.2$ pm, $\beta = 90.833 \pm 0.005°$, and a calculated density of 8.361 g·cm^{-3} (Kampf et al. 2010a).

Pb$_2$(UO$_2$)(TeO$_3$)$_3$ also crystallizes in the monoclinic structure with the lattice parameters $a = 1160.5 \pm 0.4$, $b = 1338.9 \pm 1.7$, $c = 698.1 \pm 0.1$ pm, $\beta = 91.23 \pm 0.03°$, and a calculated density of 7.42 ± 0.01 g·cm^{-3} (Brandstätter 1981). It was synthesized by hydrothermal treatment of a stoichiometric mixture

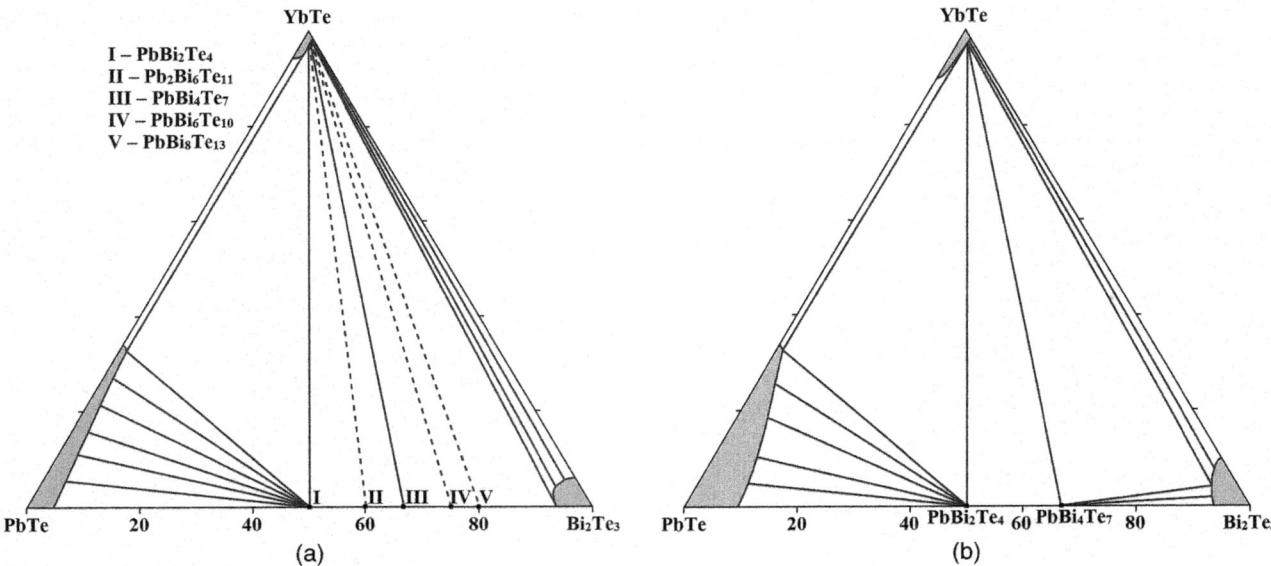

FIGURE 12.14   Isothermal sections of the PbTe–YbTe–Bi₂Te₃ quasiternary system at (a) room temperature and (b) at 530°C. (From Rasulova, K.D., et al., *Kimya Problem.*, (2), 190, 2013.)

of PbO, UO₂(CH₃COO)·2H₂O, and TeO₂ in a Teflon lined reactor. A temperature of about 230°C was maintained for 60 h. After slow cooling the reaction product contained rosettes and single-crystal needles, bright orange to brownish orange in color.

12.40 Lead–Titanium–Oxygen–Tellurium

The **Pb₂(Ti₀.₅Te₀.₅)₆.₅** quaternary compound, which crystallizes in the cubic structure with the lattice parameter $a = 1035.29 \pm 0.01$ pm and a calculated density of 7.83 g·cm⁻³, is formed in the Pb–Ti–O–Te system (Alonso et al. 1986). It was obtained from stoichiometric mixtures of PbO, TiO₂, and a light excess (15%) of TeO₂ to offset its partial sublimation. The mixture was ground and heated in air at 500°C for 3 days, at 650°C and 750°C for 24 h each, and 830°C also for 24 h. After each thermal treatment the materials were quenched, weighed, and ground. This compound was obtained as strong yellow powder.

12.41 Lead–Zirconium–Oxygen–Tellurium

The **Pb₂(Zr₀.₅Te₀.₅)₆.₅** quaternary compound, which crystallizes in the cubic structure with the lattice parameter $a = 1074.06 \pm 0.01$ pm and a calculated density of 7.71 g·cm⁻³, is formed in the Pb–Zr–O–Te system (Alonso et al. 1986). It was obtained from stoichiometric mixtures of PbO, ZrO₂, and a light excess (15%) of TeO₂ to offset its partial sublimation. The mixture was ground and heated in air at 500°C for 3 days, at 650°C and 750°C for 24 h each, and 890°C also for 24 h. After each thermal treatment the materials were quenched, weighed, and ground. This compound was obtained as pale-yellow powder.

12.42 Lead–Hafnium–Oxygen–Tellurium

The **Pb₂(Hf₀.₅Te₀.₅)₆.₅** quaternary compound, which crystallizes in the cubic structure with the lattice parameter $a = 1070.72 \pm 0.02$ pm and a calculated density of 9.20 g·cm⁻³, is formed in the Pb–Hf–O–Te system (Alonso et al. 1986). It was obtained from stoichiometric mixtures of PbO, HfO₂, and a light excess (15%) of TeO₂ to offset its partial sublimation. The mixture was ground and heated in air at 500°C for 3 days, at 650°C and 750°C for 24 h each, and 930°C also for 24 h. After each thermal treatment the materials were quenched, weighed, and ground. This compound was obtained as pale-yellow powder.

12.43 Lead–Antimony–Bismuth–Tellurium

PbTe–Sb₂Te₃–Bi₂Te₃. The isothermal section of this quasiternary system at room temperature is shown in Figure 12.15 (Aghazade 2020). A wide area of solid solutions is formed along the PbSb₂Te₄–PbBi₂Te₄ section. Solid solution based on PbTe primarily crystallizes from the liquid in the entire range of composition of this section.

12.44 Lead–Antimony–Nickel–Tellurium

PbTe–NiSb. The phase diagram of this system, constructed through DTA, metallography, and measuring of the microhardness using the alloys annealed at 700°C for 100 h, is a eutectic type (Zargarova et al. 1975a). The eutectic contains 8 mol% NiSb and crystallizes at 870°C. The solubility of NiSb in PbTe is less than 0.5 mol%, and the solubility of PbTe in NiSb is 1 mol%.

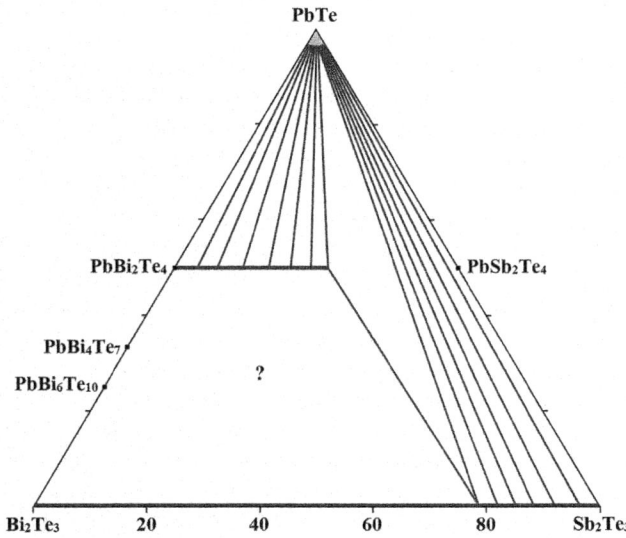

FIGURE 12.15 Isothermal sections of the PbTe–Sb$_2$Te$_3$–Bi$_2$Te$_3$ quasiternary system at room temperature. (From Aghazade, A.I., *Azerb. Chem. J.*, (4), 53, 2020.)

12.45 Lead–Niobium–Oxygen–Tellurium

The **Pb$_4$Te$_6$Nb$_{10}$O$_{41}$** quaternary compound, which crystallizes in the monoclinic structure with the lattice parameters a = 2334.0 ± 0.8, b = 2006.8 ± 0.5, c = 747.2 ± 0.2 pm, and β = 99.27 ± 0.03°, is formed in the Pb–Nb–O–Te system (Ok and Halasyamani 2004). This compound was synthesized through standard solid-state techniques. A stoichiometric mixture of PbO (4.00 mM), Nb$_2$O$_5$ (5.00 mM), and TeO$_2$ (6.00 mM) was thoroughly ground and pressed into a pellet. The pellet was introduced into a fused silica tube that was evacuated and sealed. The tube was gradually heated to 750°C, held for 24 h, and cooled to room temperature with an intermediate regrinding. Crystals of Pb$_4$Te$_6$Nb$_{10}$O$_{41}$ were prepared by placing an intimate mixture of PbO (1.00 mM), Nb$_2$O$_5$ (0.50 mM), and TeO$_2$ (3.00 mM) into a gold tube that was subsequently sealed. The gold tube was gradually heated to 800°C, held for 15 h, and then cooled slowly to 500°C at 6°C·h^{-1} before being quenched to room temperature. Crystals of Pb$_4$Te$_6$Nb$_{10}$O$_{41}$ (pale yellow blocks) were recovered with TeO$_2$ from the tube. This compound is not stable at higher temperatures: single step decomposition occurs indicating on volatilization above 790°C.

12.46 Lead–Tantalum–Oxygen–Tellurium

The **Pb$_4$Te$_6$Ta$_{10}$O$_{41}$** quaternary compound, which crystallizes in the monoclinic structure with the lattice parameters a = 2341.2 ± 0.3, b = 2011.4 ± 0.3, c = 750.08 ± 0.10 pm, β = 99.630 ± 0.004°, and a calculated density of 6.064 g·cm^{-3}, is formed in the Pb–Ta–O–Te system (Ok and Halasyamani 2004). This compound was prepared in the same way as Pb$_4$Te$_6$Nb$_{10}$O$_{41}$ using Ta$_2$O$_5$ instead of Nb$_2$O$_5$ and heated the mixture to 800°C. All efforts to grow single crystals of Pb$_4$Te$_6$Ta$_{10}$O$_{41}$ under similar reaction conditions as in the case of Pb$_4$Te$_6$Nb$_{10}$O$_{41}$

produced PbTeO$_3$ and a mixture of reagents. This compound is not stable at higher temperatures: single-step decomposition occurs, indicating volatilization above 810°C.

12.47 Lead–Oxygen–Chlorine–Tellurium

Three quaternary compounds, **Pb$_3$TeO$_4$Cl$_2$**, **Pb$_3$(TeO$_3$)Cl$_4$**, and **Pb$_3$(TeO$_3$)$_2$Cl$_2$**, are formed in the Pb–O–Cl–Te system. Pb$_3$TeO$_4$Cl$_2$ apparently has two polymorphic modifications. The first modification crystallizes in the tetragonal structure with the lattice parameters a = 393.82 ± 0.01, c = 1250.34 ± 0.02 pm, and a calculated density of 7.49 g·cm^{-3} (Charkin et al. 2006). Second modification (mineral telluroperite) crystallizes in the orthorhombic structure with the lattice parameters a = 556.49 ± 0.06, b = 555.65 ± 0.06, c = 1247.50 ± 0.14 pm, and a calculated density of 7.599 g·cm^{-3} (Kampf et al. 2010b) [a = 557.6 ± 0.01, b = 555.9 ± 0.1, and c = 1249.29 ± 0.06 pm (Porter and Halasyamani 2003)].

Tetragonal Pb$_3$TeO$_4$Cl$_2$ was prepared in two steps (Charkin et al. 2006). First, PbO and PbCl$_2$ were annealed at 500°C in alumina crucibles sealed into an evacuated silica ampoule for 48 h to yield Pb$_3$O$_2$Cl$_2$. This was annealed with TeO$_2$ at 600°C for 120 h. The sample was light yellow and well-crystallized. Twinned crystals of this modification could be obtained by heating a mixture of Pb$_3$O$_2$Cl$_2$ and TeO$_2$ (molar ratio 1:1) with a 10-fold amount of NaCl to 760°C within 12 h, soaked at this temperature for 10 min, and cooled to 510°C with a rate of 1°C·h^{-1}. The solidified melt was washed with distilled water to remove excess NaCl.

Orthorhombic Pb$_3$TeO$_4$Cl$_2$ was synthesized as a polycrystalline powder (Porter and Halasyamani 2003). All attempts to grow single crystals were unsuccessful. This modification was prepared by combining stoichiometric amounts of Pb$_3$O$_2$Cl$_2$ with TeO$_2$. The mixture was introduced into a fused silica tube that was subsequently evacuated and sealed. The tube was heated to 650°C for 1 day and cooled at a rate of 6°C·h^{-1} to room temperature. Yellow polycrystalline powder was recovered.

Pb$_3$(TeO$_3$)Cl$_4$ is stable up to 530°C and also crystallizes in the orthorhombic structure with the lattice parameters a = 742.7 ± 0.4, b = 1599.3 ± 0.8, c = 848.3 ± 0.4 pm, a calculated density of 6.190 g·cm^{-3} and an energy gap of 3.79 eV (Zhang et al. 2012b). Single crystals of this compound were initially obtained, along with some unidentified powder, by hydrothermal reactions of a mixture of Ga$_2$O$_3$, PbCl$_2$ and TeO$_2$ (molar ratio 1:1:10) in H$_2$O (4 mL), which was sealed in an autoclave equipped with a Teflon liner (23 mL) at 230°C for 4 days, followed by slow cooling to room temperature at a rate of 6°C·h^{-1}. The final pH value of the reaction media was close to 2.0. The single phase of Pb$_3$(TeO$_3$)Cl$_4$ was obtained after washing away the above powder in water by ultrasonic cleaning.

Pb$_3$(TeO$_3$)$_2$Cl$_2$ crystallizes in the monoclinic structure with the lattice parameters a = 1643.49 ± 0.05, b = 562.55 ± 0.01, c = 1088.02 ± 0.03 pm, β = 103.050 ± 0.001°, and a calculated density of 7.059 g·cm^{-3} (Porter and Halasyamani 2003). Single crystals of this compound were grown by combining Pb$_3$O$_2$Cl$_2$ (0.367 mM) and TeO$_2$ (1.47 mM) in a fused silica tube that was subsequently evacuated and sealed. The mixture was heated at

550°C for 1 day and cooled at a rate of 6°C·h⁻¹ to room temperature. A few clear colorless crystals were observed (roughly 2% of the bulk powder). Bulk, polycrystalline $Pb_3(TeO_3)_2Cl_2$ was synthesized by combining $Pb_3O_2Cl_2$ (0.479 mM) and TeO_2 (0.958 mM). The mixture was introduced into a fused silica tube that was subsequently evacuated and sealed. The tube was heated to 575°C for 1 day, and then furnace cooled to room temperature. An off-white powder of the title compound was recovered.

12.48 Lead–Oxygen–Bromine–Tellurium

Two quaternary compounds, $Pb_3TeO_4Br_2$ and $Pb_3(TeO_3)_2Br_2$, are formed in the Pb–O–Br–Te system. $Pb_3TeO_4Br_2$ has apparently two polymorphic modifications. First modification crystallizes in the tetragonal structure with the lattice parameters $a = 398.7 \pm 0.4$, $c = 1292 \pm 3$ pm, and a calculated density of 7.868 g·cm⁻³ (Charkin et al. 2006). Second modification crystallizes in the orthorhombic structure with the lattice parameters $a = 564.34 \pm 0.04$, $b = 564.34 \pm 0.05$, and $c = 1291.72 \pm 0.06$ pm (Porter and Halasyamani 2003)].

Tetragonal $Pb_3TeO_4Br_2$ was prepared in two steps (Charkin et al. 2006). First, PbO and $PbBr_2$ were annealed at 500°C in alumina crucibles sealed into an evacuated silica ampoule for 48 h to yield $Pb_3O_2Br_2$. This was annealed with TeO_2 at 600°C for 120 h. The sample has a greenish tint and well-crystallized. Single crystals of this modification were obtained as follows. 1.75 mM of yellow PbO and TeO_2, and 10.5 mM of NaBr were ground, placed in an evacuated silica ampoule, heated to 760°C within 12 h, soaked for 10 min, and cooled to 510°C with a rate of 1°C·h⁻¹. The solidified melt was washed with distilled water, to remove excess NaBr. Small shiny platelets of tetragonal $Pb_3TeO_4Br_2$ were found.

Orthorhombic $Pb_3TeO_4Br_2$ was synthesized as a polycrystalline powder (Porter and Halasyamani 2003). All attempts to grow single crystals were unsuccessful. This modification was prepared by combining stoichiometric amounts of $Pb_3O_2Br_2$ with TeO_2. The mixture was introduced into fused silica tube that was subsequently evacuated and sealed. The tube was heated to 650°C for 1 day and cooled at a rate of 6°C·h⁻¹ to room temperature. Yellow polycrystalline powder was recovered.

$Pb_3(TeO_3)_2Br_2$ crystallizes in the monoclinic structure with the lattice parameters $a = 1691.51 \pm 0.09$, $b = 568.13 \pm 0.03$, $c = 1106.23 \pm 0.06$ pm, $\beta = 104.046 \pm 0.001°$, and a calculated density of 7.295 g·cm⁻³ (Weil and Stöger 2010) [$a = 1689.11 \pm 0.08$, $b = 568.04 \pm 0.02$, $c = 1104.18 \pm 0.05$ pm, and $\beta = 104.253 \pm 0.002°$ (Porter and Halasyamani 2003)].

This compound was synthesized as polycrystalline powder (Porter and Halasyamani 2003). All attempts to grow single crystals were unsuccessful. It was prepared by combining stoichiometric amounts of $Pb_3O_2Br_2$ with TeO_2. The mixture was introduced into a fused silica tube that was subsequently evacuated and sealed. The tube was heated to 575°C for 1 day and cooled at a rate of 6°C·h⁻¹ to room temperature. An off-white polycrystalline powder was recovered.

Single crystals of the title compound have been grown under hydrothermal conditions (Weil and Stöger 2010). $PbBr_2$ (1 mM) and TeO_2 (1 mM) were placed in a Teflon inlay that was filled with a hydrous NH_4OH solution (10 mass%) to two-thirds of its volume. The inlay was placed in a steel autoclave and heated at 220°C for 4 weeks. The reaction product consisted of small colorless crystals of the title compound with rod-like habit.

12.49 Lead–Oxygen–Iodine–Tellurium

The $Pb_3TeO_4I_2$ quaternary compound, which crystallizes in the tetragonal structure with the lattice parameters $a = 409.19 \pm 0.01$, $c = 1369.83 \pm 0.02$ pm, and a calculated density of 7.73 g·cm⁻³, is formed in the Pb–O–I–Te system (Charkin et al. 2006). It was prepared in two steps. First, PbO and PbI_2 were annealed at 400°C in alumina crucibles sealed into an evacuated silica ampoule for 48 h to yield $Pb_3O_2I_2$. This was annealed with TeO_2 at 400°C for 240 h with one intermediate grinding. The sample was an intense yellow microcrystalline powder.

12.50 Lead–Oxygen–Manganese–Tellurium

Two quaternary compounds, $PbMnTeO_6$ and Pb_2MnTeO_6, are formed in the Pb–O–Mn–Te system. The first of them (mineral kuranakhite) crystallizes in the orthorhombic structure with the lattice parameters $a = 511 \pm 1$, $b = 891 \pm 2$, $c = 532 \pm 1$ pm, and the calculated and experimental densities of 6.66 ± 0.01 and 6.72 ± 0.02 g·cm⁻³, respectively (Xinchun et al. 1998) [$a = 510 \pm 10$, $b = 890 \pm 10$, and $c = 530 \pm 10$ pm (Yablokova et al. 1975; Fleischer et al. 1976)].

Pb_2MnTeO_6 apparently has three polymorphic modifications. The first modification crystallizes in the cubic structure with the lattice parameter $a = 810.54 \pm 0.01$ pm and a calculated density of 8.643 g·cm⁻³ (Artner and Weil 2019) [$a = 806 \pm 1$ pm (Politova and Venevtsev 1973); $a = 404.5 \pm 0.3$ pm (Wulff et al. 1998)]. Second modification of Pb_2MnTeO_6 undergoes a structural phase transition at ~120 K when the temperature is lowered (Retuerto et al. 2016). Both phases crystallize in the monoclinic structure with the lattice parameters $a = 576.346 \pm 0.007$, $b = 571.910 \pm 0.008$, $c = 808.374 \pm 0.010$ pm, and $\beta = 89.9310 \pm 0.0015°$ at room temperature and $a = 989.484 \pm 0.018$, $b = 568.523 \pm 0.010$, $c = 992.111 \pm 0.018$ pm, and $\beta = 108.4296 \pm 0.0013°$ at 14 K.

Preparation of the cubic $PbMnTeO_6$ was carried out in a corundum crucible using pressed pellets of finely ground powder mixture that comprised the elements in a molar Pb/Mn/Te ratio of 2:1:1 (Artner and Weil 2019). The mixture of PbO (1.286 g), $MnCO_3$ (0.333 g), and TeO_2 (0.463 g) were treated according to the following temperature program: room temperature → 800°C [6 h] → 800°C [6 h] → 600°C [12 h] → 600°C [2 h] → 400°C [12 h] → 400°C [2 h] → room temperature [24 h]. After regrinding the yellowish-greenish product, a second temperature program was applied: room temperature → 900°C [6 h] → 900°C [6 h] → 700°C [12 h] → 700°C [2 h] → 500°C [12 h] → 500°C [2 h] → 300°C [12 h] → 300°C [2 h] → room temperature [12 h]. A few dark-brown single crystals were obtained.

Synthesis of this modification can also be carried out according to ceramic technology by two-stage annealing (Politova and Venevtsev 1973). The initial mixture was prepared from stoichiometric amounts of $PbCO_3$, MnO, and TeO_3. The mixture was annealed first at 700°C for 6 h, and then at 830°C for another 6 h. It was also obtained by rapidly cooling a melt of PbO, PbF_2, TeO_2, and $MnCO_3$ (molar ratio 1:1:1:1) heated to 730°C. PbF_2 served as a flux (Wulff et al. 1998). Within 2 days, many dark-brown cube-shaped crystals formed, which were separated mechanically.

Monoclinic Pb_2MnTeO_6 was prepared by a standard solid-state technique (Retuerto et al. 2016). A stoichiometric mixture of PbO, $MnCO_3$, and TeO_2 was thoroughly ground and heated in an oxygen flow at 800°C for 12 h to obtain a pure sample.

12.51 Lead–Oxygen–Iron–Tellurium

The $Pb_3Fe_2Te_2O_{12}$ quaternary compound, which crystallizes in the monoclinic structure with the lattice parameters $a = 986.6 \pm 0.3$, $b = 1533.2 \pm 0.4$, $c = 717.2 \pm 0.2$ pm, and $\beta = 111.34 \pm 0.03°$, is formed in the Pb–O–Fe–Te system (Wedel and Müller-Buschbaum 1997). To prepare this compound, PbO, Fe_2O_3, and TeO_2 (molar ratio 3:1:6) were intimately mixed and heated to 730°C in a corundum boat in the air for 48 h. TeO_2 was added in excess because of its good flux properties and its volatility. The reaction product was inhomogeneous and contained excess TeO_2, α-$PbTeO_3$, and $Pb_2Te_3O_8$. Red rods of the title compound were isolated from the reaction.

12.52 Lead–Oxygen–Cobalt–Tellurium

Two quaternary compounds, Pb_2CoTeO_6 and $Pb_6Co_9(TeO_6)_5$, are formed in the Pb–O–Co–Te system. First of them undergoes a number of temperature-induced phase transitions and adopts four different structures in the temperature range 5–500 K (Ivanov et al. 2010): monoclinic structure (5 < T < 125 K), another monoclinic structure (125 < T < 210 K), rhombohedral structure (210 < T < 370 K), and finally cubic structure above 370 K. The first monoclinic structure has lattice parameters $a = 571.55 \pm 0.01$, $b = 564.20 \pm 0.01$, $c = 793.46 \pm 0.02$ pm, and $\beta = 90.12 \pm 0.01°$ at 5 K and second one $a = 570.34 \pm 0.01$, $b = 565.48 \pm 0.01$, $c = 796.96 \pm 0.02$ pm, and $\beta = 90.07 \pm 0.01°$ at 175 K. The tetragonal structure has lattice parameters $a = 567.83 \pm 0.01$ and $c = 1385.52 \pm 0.02$ pm at 295 K [$a = 566.1 \pm 0.5$ and $c = 800.4 \pm 0.7$ pm; may be one more tetragonal structure (Wedel and Müller-Buschbaum 1997)]. The cubic structure has a lattice parameter $a = 803.33 \pm 0.02$ at 450 K [$a = 801.380 \pm 0.010$ and a calculated density of 8.994 g·cm^{-3} (Artner and Weil 2019)] [$a = 800 \pm 1$ pm (Politova and Venevtsev 1973)].

A high-quality polycrystalline sample of Pb_2CoTeO_6 was prepared by a conventional solid-state sintering procedure (Ivanov et al. 2010). $PbCO_3$, $CoCO_3$, and TeO_3 were used as starting materials. The raw materials were weighed in appropriate proportions for this compound formula. After that, they were wet-ball milled for 24 h, dried, and then heated at 700°C for 10 h in an O_2 environment. The fired powders were ground again by wet-ball milling and pressed into a pellet, and then heated up at 780°C for 2 days and 820°C for another 2 days in a flow of oxygen with several intermediate grindings, and new pellets made after each grinding step. The samples were, finally, slowly cooled to room temperature for about 15 h.

To prepare tetragonal Pb_2CoTeO_6, PbO, CoO, and TeO_2 (molar ratio 2:1:4) were intimately mixed and heated to 730°C in a corundum boat in the air for 48 h (Wedel and Müller-Buschbaum 1997). TeO_2 was added in excess because of its good flux properties and its volatility. The reaction product was inhomogeneous and contained excess TeO_2, α-$PbTeO_3$, and $Pb_2Te_3O_8$. The turquoise octahedra of the title compound were isolated from the reaction.

Preparation of the cubic Pb_2CoTeO_6 was carried out in a corundum crucible using pressed pellets of finely ground powder mixture that comprised the elements in a molar Pb/Mn/Te ratio of 2:1:1 (Artner and Weil 2019). The mixture of PbO (1.282 g), CoO (0.216 g), and TeO_2 (0.914 g) was heated for 6 h to 750°C and held at that temperature for 48 h before cooling down to room temperature in the turned-off furnace. Blue block-shaped crystals of cubic Pb_2CoTeO_6 were obtained as the main phase. Synthesis of this modification can also be carried out according to ceramic technology by two-stage annealing (Politova and Venevtsev 1973). The initial mixture was prepared from stoichiometric amounts of $PbCO_3$, $CoCO_3$, and TeO_3. The mixture was annealed first at 700°C for 6 h, and then at 820°C for another 6 h.

The pressure effect on the crystal structure of Pb_2CoTeO_6 was systematically studied by employing *in situ* synchrotron XRD and Raman scattering techniques up to 60 GPa (Liu et al. 2019b). A structural phase transition from trigonal to monoclinic structure was observed at around 20 GPa, indicating that increasing the pressure has a similar effect on Pb_2CoTeO_6 as decreasing the temperature, i.e., promoting the distortion of the structure.

$Pb_6Co_9(TeO_6)_5$ crystallizes in the hexagonal structure with the lattice parameters $a = 1039.15 \pm 0.01$, $c = 1362.73 \pm 0.02$ pm, and a calculated density of 7.535 g·cm^{-3} (Artner and Weil 2012). Single crystals of the title compound were serendipitously obtained as a minority phase during phase formation studies intended on crystal growth of cubic Pb_2CoTeO_6. PbO (5.7 mM), CoO (2.9 mM), and TeO_2 (5.7 mM) were mixed and thoroughly ground and heated in an alumina crucible under atmospheric conditions from 6 h to 750°C and held at that temperature for 48 h. Then the furnace was shut off. Several crystal phases could be identified from the cooled reaction mixture by single-crystal diffraction: dark blue isometric crystals of Pb_2CoTeO_6, dark-red (nearly black) block-like crystals of Pb_5TeO_8, colorless crystals of α-Al_2O_3, and dark-red crystals of $Pb_6Co_9(TeO_6)_5$ with a block-like shape.

12.53 Lead–Oxygen–Nickel–Tellurium

Two quaternary compounds, Pb_2NiTeO_6 and $Pb_6Ni_9(TeO_6)_5$, are formed in the Pb–O–Ni–Te system. The first crystallizes in the cubic structure with the lattice parameter $a = 797 \pm 1$ pm

(Politova and Venevtsev 1973). Synthesis of this compound can be carried out according to ceramic technology by two-stage annealing. The initial mixture was prepared from stoichiometric amounts of $PbCO_3$, $NiCO_3$, and TeO_3. The mixture was annealed first at 700°C for 6 h, and then at 840°C for another 6 h.

$Pb_6Ni_9(TeO_6)_5$ crystallizes in the hexagonal structure with the lattice parameters $a = 1025.89 \pm 0.01$ and $c = 1355.4$ $\pm$ 0.5 pm (Wedel et al. 1998). For its preparation, PbO, $Ni_3(OH)_4CO_3 \cdot 4H_2O$, and TeO_2 (molar ratio 2:1:1) were intimately mixed, pressed to tablets, and heated to 730°C in a corundum boat in the air for 48 h. After the reaction time, the preparation was rapidly cooled. In an inhomogeneous product, numerous light green triangular platelets were present, which were separated under the microscope.

References

Abbasova V.A. Phase equilibria along the Cu_2GeSe_3–Ag_2GeSe_3 isopleth section of the Cu_2Se–Ag_2Se–$GeSe_2$ system, *Azerb. Chem. J.*, 3, 52–56 (2019).

Abbasova V.A., Alverdiyev I.J., Mashadiyeva L.F., Yusibov Y.A., Babanly M.B. Phase equilibria in the Cu_8GeSe_6–Ag_8GeSe_6 system, *Azerb. Khim. Zhurn.*, 1, 30–33 (2017a).

Abbasova V.A., Alverdiyev I.J., Rahimoglu E., Mirzoyeva R.J., Babanly M.B. Phase relations in the Cu_8GeS_6–Ag_8GeS_6 system and some properties of solid solutions, *Azerb. Khim. Zhurn.*, 2, 25–29 (2017b).

Abilov Ch.I., Iskender-Zade Z.A., Dao N.L. Properties of the $(PbTe)_{1-x}(EuS)_x$ alloys [in Russian], *Neorgan. mater.*, 29(2), 285–286 (1993).

Abrikosov N.H., Avilov E.S., Karpinskiy O.G., Radkevich O.V., Shelimova L.E. Investigation of the GeTe–PbSe system [in Russian], *Izv. AN SSSR. Neorgan. mater.*, 21(10), 1664–1669 (1985a).

Abrikosov N.H., Avilov E.S., Shelimova L.E. Physico-chemical investigation of the GeTe–MnTe–SnTe alloys in the region based on GeTe solid solutions [in Russian], *Izv. AN SSSR. Neorgan. mater.*, 14(2), 239–249 (1978).

Abrikosov N.H., Avilov E.S., Shelimova L.E. Physico-chemical investigation of the GeTe–MnTe–PbTe alloys in the region based on GeTe solid solutions [in Russian], *Izv. AN SSSR. Neorgan. mater.*, 15(10), 1757–1761 (1979).

Abrikosov N.H., Bankina V.F., Kolomoets L.A., Shubina G. Y., Tshadaya R.A. Investigation of the solid solutions based on bismuth monoselenide [in Russian], *Izv. AN SSSR. Neorgan. mater.*, 8(12), 2120–2124 (1972).

Abrikosov N.H., Dimitrova St K., Karpinskiy O.G., Plachkova S.K., Hristakudis G.H., Shelimova L.E. Phase transitions in the GeTe–Ag_2Te–Sb_2Te_3 system [in Russian], *Zhurn. neorgan. khimii*, 30(11), 2931–2935 (1985b).

Abrikosov N.H., Dimitrova St K., Karpinskiy O.G., Plachkova S.K., Shelimova L.E. Phase transitions and electrophysical properties of the solid solutions based on GeTe along the GeTe–$AgSbTe_2$ section [in Russian], *Izv. AN SSSR. Neorgan. mater.*, 20(1), 55–59 (1984).

Abrikosov N.H., Goncharova L.S., Gurova I.I. The PbTe–SnSe phase diagram [in Russian], *Izv. AN SSSR. Neorgan. mater.*, 9(7), 1146–1148 (1973).

Abrikosov N.H., Goncharova L.S. Investigation of interactions PbTe + SnSe ↔ PbSe + SnTe [in Russian], In: *Polumetally i poluprovodniki s uzkimi zapreshchonnymi zonami*. L'viv: L'viv Univ. Publish., 130–132 (1973).

Abrikosov N.H., Goncharova L.S. The PbSe–SnTe phase diagram [in Russian], *Izv. AN SSSR. Neorgan. mater.*, 10(8), 1533–1534 (1974).

Abrikosov N.H., Karpinskiy O.G., Makalatiya T.Sh., Shelimova L.E. Doping of germanium telluride by bismuth and copper [in Russian], *Izv. AN SSSR. Neorgan. mater.*, 18(9), 1504–1509 (1982).

Abrikosov N.H., Karpinskiy O.G., Radkevich O.V., Shelimova L.E., Baranov V.M. The $Ge_{0.98}Te$–GeSe system and $GeSe_{0.75}Te_{0.25}$ ternary compound [in Russian], *Izv. AN SSSR. Neorgan. mater.*, 24(1), 46–51 (1988).

Abrikosov N.H., Karpinskiy O.G., Shelimova L.E., Avilov E.S. Some peculiarities of solid solutions based on germanium telluride [in Russian], *Izv. AN SSSR. Neorgan. mater.*, 16(2), 237–240 (1980a).

Abrikosov N.H., Makalatiya T.Sh., Shelimova L.E., Avilov E.S. Electrophysical properties of the solid solutions based on GeTe in the GeTe–Bi_2Te_3–Cu_2Te quasiternary system [in Russian], *Izv. AN SSSR. Neorgan. mater.*, 16(8), 1398–1402 (1980b).

Abrikosov N.H., Makalatiya T.Sh., Shelimova L.E. Investigation of the GeTe–Bi_2Te_3–Cu_2Te system [in Russian], *Izv. AN SSSR. Neorgan. mater.*, 16(4), 648–651 (1980c).

Abrikosov N.H., Sokolova I.F. Investigation of the germanium telluride–bismuth telluride–antimony telluride system [in Russian], *Zhurn. neorgan. khimii*, 22(6), 1651–1655 (1977).

Abudurusuli A., Ding H., Wu K. Synthesis and characterization of two lead-containing metal chalcogenides: $Ba_5Pb_2Sn_3S_{13}$ and $Ba_5PbSn_3S_{13}$, *J. Solid State Chem.*, 255, 133–138 (2017).

Abudurusuli A., Huang J., Wang P., Yang Z., Pan S., Li J. $Li_4MgGe_2S_7$: the first alkali and alkaline-earth diamond-like infrared nonlinear optical material with exceptional large band gap, *Angew. Chem. Int. Ed.*, 60(45), 24131–24136 (2021).

Abudurusuli A., Wu K., Pan S. Four new quaternary chalcogenides $A_2B_7Sn_4Q_{16}$ (A = Li, Na; Q = S, Se): synthesis, crystal structures, noinlinear optical performances, *New J. Chem.*, 42(5), 3350–3355 (2018).

Ackermann J., Soled S., Wold A., Kostiner E. The preparation and characterization of the solid solution series $CuFe_xGe_{1-x}S_2$ (0.5 < x < 1.0). *J. Solid State Chem.*, 19(1), 75–80 (1976).

Agaguseinova M.M., Gurbanov K.R., Adygezalova M.B. Physicochemical interactions in the $GeSb_2Te_4$–$GeBi_2Te_4$ system, *Russ. J. Inorg. Chem.*, 57(3), 449–451 (2012).

Agaguseinova M.M., Gurbanov K.R., Adygezalova M.B. SnSb-BiS_4–SnS and $SnSbBiS_4$–$Sn_2Sb_6S_{11}$ sections of the SnS–Sb_2S_3–Bi_2S_3 ternary system, *Russ. J. Inorg. Chem.*, 56(8), 1331–1334 (2011).

Aghazade A.I. Phase equilibria of the $PbBi_2Te_4$–"$PbSb_2Te_4$" section of the PbTe–Bi_2Te_3–Sb_2Te_3 system and some properties of the solid solutions, *Azerb. Chem. J.*, (4), 53–59 (2020).

Aghazade A.I. Phase relations and characterization of solid solutions in the $SnBi_2Te_4$–$PbBi_2Te_4$ and $SnBi_4Te_7$–$PbBi_4Te_7$ systems, *Azerb. Chem. J.*, (3), 75–80 (2022).

Aitken J.A., Kanatzidis M.G. α-$Na_6Pb_3(PS_4)_4$, a noncentrosymmetric thiophosphate with the novel, saucer-shaped $[Pb_3(PS_4)_4]^{6-}$ cluster, and its metastable, 3-dimensionally polymerized allotrope β-$Na_6Pb_3(PS_4)_4$, *Inorg. Chem.*, 40(13), 2938–2939 (2001).

Aitken J.A., Larson P., Mahanti S.D., Kanatzidis M.G. Li_2PbGeS_4 and Li_2EuGeS_4: polar chalcopyrites with a severe tetragonal compression, *Chem. Mater.*, 13(12), 4714–4721 (2001).

Aitken J.A., Lekse J.W., Yao J.-L., Quinones R. Synthesis, structure and physicochemical characterization of a noncentrosymmetric, quaternary thiostannate: $EuCu_2SnS_4$, *J. Solid State Chem.*, 182(1), 141–146 (2009).

Aitken J.A., Marking G.A., Evain M., Iordanidis L., Kanatzidis M.G. Flux synthesis and isostructural relationship of cubic $Na_{1.5}Pb_{0.75}PSe_4$, $Na_{0.5}Pb_{1.75}GeS_4$, and $Li_{0.5}Pb_{1.75}GeS_4$, *J. Solid State Chem.*, **153**(1), 158–169 (2000).

Akopov G., Hewage N.W., Viswanathan G., Yox P., Wu K., Kovnir K. Non-linear optical properties of the $(RE)_3CuGeS_7$ family of compounds, *Z. anorg. und allg. Chem.*, **648**(15), e202200096 (2022a).

Akopov G., Viswanathan G., Hewage N.W., Yox P., Wu K., Kovnir K. Pd and octahedra do not get along: square planar $[PdS_4]$ units in non-centrosymmetric $La_6PdSi_2S_{14}$, *J. Alloys Compd.*, **902**, 163756 (2022b).

Alahmari F., Davaasuren B., Emwas A.-H., Rothenberger A. Thioaluminogermanate $M(AlS_2)(GeS_2)_4$ (M = Na, Ag, Cu): synthesis, crystal structures, characterization, ion-exchange and solid-state ^{27}Al and ^{23}Na NMR spectroscopy, *Inorg. Chem.*, **57**(7), 3713–3719 (2018).

Alakbarova T.M., Jafarov Y.I., Mustafayeva A.L., Babanly M.B. $Tl_2Te–Tl_9BiTe_6–Tl_8GeTe_5$ system, *Kimya problem.*, (4), 355–363 (2017).

Alakbarova T.M. Physicochemical interactions in $TlBiTe_2–Tl_8GeTe_5(Tl_2GeTe_2)$ systems, *New Mater., Compd. Appl.*, **6**(2), 162–168 (2022).

Albert S., Haines J., Granier D., Pradel A., Ribes M. Effect of pressure on the superionic argyrodite Ag_7GeSe_5I, *J Appl. Crystallogr.*, **42**(1), 93–100 (2009).

Albert S., Pillet S., Lecomte C., Pradel A., Ribes M. Disorder in Ag_7GeSe_5I, a superionic conductor: temperature-dependent anharmonic structural study, *Acta Crystallogr.*, **B64**(1), 1–11 (2008).

Al-Bloushi M., Davaasuren B., Emwas A.-H., Rothenberger A. Synthesis and characterization of the quaternary thioaluminogermanates $A(AlS_2)(GeS_2)$ (A = Na, K), *Z. anorg. und allg. Chem.*, **641**(7), 1352–1356 (2015).

Aldon L., Belin R., Pontillon Y. Refinement of the crystal structure of heptasilver(I) tetraseleniogermanate(IV) selenide iodide, Ag_7GeISe_5, *Z. Kristallogr., New Cryst. Str.*, **216**(2), 181–182 (2001).

Alekperova T.M., Amiraslanov I.R., Babanly M.B. Phase equilibriums in the $Tl_8GeTe_5–Tl_9BiTe_6$ system and some properties of the solid solutions [in Russian], *Kimiya Probl.*, (4), 376–381 (2015).

Alekseeva G.T., Efimova B.A., Lainer D.I., Ostrovskaya L.M. Physico-chemical properties of the PbTe–SnTe–PbSe and PbTe–GeTe–PbSe alloys [in Russian], *Izv. AN SSSR. Neorgan. mater.*, **5**(12), 2105–2109 (1969).

Alidzhanov M.A., Asadov M.M., Alieva F.M., Mamedov F.M. Electrophysical properties of the $(SnTe)_{1-x}(NiSe)_x$ solid solutions [in Russian], *Neorgan. mater.*, **33**(9), 1067–1069 (1997).

Alidzhanov M.A., Kurbanova R.D., Agdamskaya S.G. Interaction in the $PbTe–NiSe_2$ system [in Russian], *Neorgan. mater.*, **31**(9), 1174–1176 (1995).

Alidzhanov M.A., Rustamov P.G., Melikova Z.D. The $PbIn_2Se_4–PbGa_2Se_4$ section of the $PbSe–In_2Se_3–Ga_2Se_3$ quasiternary system [in Russian], *Izv. AN SSSR. Neorgan. mater.*, **23**(6), 900–901 (1987).

Alieva R.A., Bairamova S.T., Ragimova V.M., Aliev O.M., Bagieva M.R. Phase diagrams of the $CuSbS_2–MS$ (M = Pb, Eu, Yb) systems, *Inorg. Mater.*, **46**(7), 703–706 (2010).

Alieva Z.M., Bagkheri S.M., Alverdiev I.J., Yusibov Y.A., Babanly M.B. Phase equilibria in the pseudoternary system $Ag_2Se–Ag_8GeSe_6–Ag_8SnSe_6$, *Inorg. Mater.*, **50**(10), 981–986 (2014).

Aliev I.I., Babanly K.N., Babanly N.B. Solid solutions in the $Ag_2Se–PbSe–Bi_2Se_3$ system, *Inorg. Mater.*, **44**(11), 1179–1182 (2008).

Aliev M.I., Suleymanov Z.I., Arasly D.G., Ragimov R.N. Investigation of the $InSb–In_2GeTe$ system [in Russian], *Izv. AN SSSR. Neorgan. mater.*, **17**(10), 1763–1766 (1981).

Aliev O.M., Azhdarova D.S., Ragimova V.M., Agayeva R.M. Phase equilibria in the $PbLa_2S_4–PbBi_2S_4$ system, *Azerb. khim. zhurn.*, (1), 85–88 (2017).

Aliev O.M., Bayramova S.T., Azhdarova D.S., Mammadov Sh.H., Ragimova V.M., Maksudova T.F. Synthesis and properties of synthetic aikinite $PbCuBiS_3$ [in Russian], *Kondens. sredy i mezhfaz. granitsy*, **22**(2), 182–188 (2020).

Aliev V.O., Akhmedova N.R., Agapasheva S.M., Aliev O.M. Crystal growth and physicochemical properties of structural analogs of krupkaite, *Inorg. Mater.*, **45**(7), 717–722 (2009).

Aliev Z.S., Amiraslanov I.R., Record M.-C., Tedenac J.-C., Babanly M.B. The $YbTe–SnTe–Bi_2Te_3$ system, *J. Alloys Compd.*, **750**, 887–894 (2018).

Aliev Z.S., Ibadova G.I., Tedenac J.-C., Babanly M.B. Study of the $YbTe–SnTe–Sb_2Te_3$ quasi-ternary system, *J. Alloys Compd.*, **602**, 248–254 (2014).

Aliev Z.S. Novel variable phases in the quaternary Pb–Bi–Te–Se system along the $PbBi_2Te_4$–"$PbBi_2Se_4$" isopleth section, *Azerb. Chem. J.*, (4), 54–58 (2019a).

Aliev Z.S. The solid-state phase diagram of the $PbTe–PbSe–Bi_2Se_3–Bi_2Te_3$ reciprocal system, *New Mater., Compd. Appl.*, **3**(3), 180–186 (2019b).

Ali M.A., Hossain M.A., Rayhan M.A., Hossain M.M., Uddin M.M., Roknuzzaman M., Ostrikov K., Islam A.K.M.A., Naqib S.H. First-principles study of elastic, electronic, optical and thermoelectric properties of newly synthesized $K_2Cu_2GeS_4$ chalcogenide, *J. Alloys Compd.*, **781**, 37–46 (2019).

Alimova L.L., Belov N.V., Badikov V.V. Crystal structures of the Ag-germanogallium sulfide and GeS_2 [in Russian]. *Dokl. AN SSSR*, **257**(3), 611–614 (1981).

Aliyeva Z.M., Bagheri S.M., Aliev Z.S., Alverdiyev I.J., Yusibov Y.A., Babanly M.B. The phase equilibria in the $Ag_2S–Ag_8GeS_6–Ag_8SnS_6$ system, *J. Alloys Compd.*, **611**, 395–400 (2014).

Aliyev I.I., Babanly N.B., Osman C.K., Yusibov Yu.A., Suleymanova M.H. Nature of interaction in the $SnTe–AgBiTe_2$ system [in Azerbaijanian], *Chem. Probl.*, (3), 344–348 (2013).

Aliyev O.M., Ajdarova D.S., Agayeva R.M., Maksudova T.F., Mamedov Sh.H. Phase relations along the $Cu_2S(Sb_2S_3, PbSb_2S_4, Pb_5Sb_4S_{11})–PbCuSbS_3$ joins in the pseudoternary system $Cu_2S–PbS–Sb_2S_3$ and physical properties of $(Sb_2S_3)_{1-x}(PbCuSbS_3)_x$ solid solutions, *Inorg. Mater.*, **54**(12), 1199–1204 (2018).

Aliyev O.M., Ajdarova D.S., Bayramova S.T., Aliyeva S.I., Ragimova V.M. Nonstoichiometry in $PbCuSbS_3$ compound, *Azerb. khim. zhurn.*, (2), 51–54 (2016).

Aliyev O.M., Azhdarova D.S., Ragimova V.M., Guliyeva S.A., Bayramova S.T. Synthesis and physico-chemical properties of $PbLnBiS_4$ (Ln = La÷Er) compounds [in Russian], *Azerb. khim. zhurn.*, (1), 39–43 (2014).

Allazova N.M., Abbasova R.F., Ilyasli T.M., Aliyev I.I., Allazov M.R. Phase equilibria in the $CuInSe_2–Ge–Se$ quasiternary system, *Chem. Probl.*, **18**(2), 244–249 (2020).

Allazova N.M., Abbasova R.F., Ilyasly T.M. Primary crystallization region of the chalcopyrite phase in the $CuInSe_2–Sn–Se$ system, *Russ. J. Inorg. Chem.*, **56**(10), 1634–1639 (2011).

Allazova N.M., Ilyasly T.M. Determination of primary crystallization region of the chalcopyrite phase in the $CuInSe_2$–Pb–Se system, *Azerb. Chem. J.*, (*1*), 60–66 (2015).

Allemand J., Winterberger M. Propriétés structurales et magnétiques de quelques composés du type stannite, *Bull. Soc. fr. Minéral. Cristallogr.*, **93**(1), 14–17 (1970).

van Almsick T., Sheldrick W.S. Methanolothermal synthesis and structures of the quaternary group 14 – group 15 cesium selenidometalates $Cs_3AsGeSe_5$ and $Cs_4Ge_2Se_6$, *Z. anorg. und allg. Chem.*, **631**(10), 1746–1748 (2005).

Alnujaim S., Bouhemadou A., Bedjaoui A., Bin-Omran S., Al-Douri Y., Khenata R., Maabed S. *Ab initio* prediction of the elastic, electronic and optical properties of a new family of diamond-like semiconductors, Li_2HgMS_4 (M = Si, Ge and Sn), *J. Alloys Compd.*, **843**, 155991 (2020).

Alonso J.A., Cascales C., Rasines I. Preparation and crystal data of the new pyrochlores $Pb_2[M_{1.5}Te_{0.5}]O_{6.5}$ (M = Ti, Zr, Sn, Hf), *Z. anorg. und allg. Chem.*, **537**(6), 213–218 (1986).

Alverdiev I., Alverdiev J., Bagheri S.M., Aliyeva Z.M., Yusibov Yu. A., Babanly M.B. Phase equilibria in the Ag_2Se–$GeSe_2$–$SnSe_2$ system and thermodynamic properties of $Ag_8Ge_{1-x}Sn_xSe_6$ solid solutions, *Inorg. Mater.*, **53**(8), 786–796 (2017a).

Alverdiev, I. Dzh., Abbasova V.A., Yusibov Yu.A., Babanly M.B. Thermodynamic properties of the solid solutions in the Cu_8GeS_6–Ag_8GeS_6 [in Russian], *Kondens. sredy i mezhfaz. granitsy*, **19**(1), 22–26 (2017b).

Alverdiev I.J., Abbasova V.A., Yusibov Yu.A., Tagiev D.B., Babanly M.B. Thermodynamic study of Cu_2GeS_3 and $Cu_{2-x}Ag_xGeS_3$ solid solutions by the EMF method with a $Cu_4RbCl_3I_2$ solid electrolyte, *Russ. J. Electrochem.*, **54**(2), 195–200 (2018).

Alverdiev I.J., Aliev Z.S., Bagheri S.M., Mashadiyeva L.F., Yusibov Y.A., Babanly M.B. Study of the $2Cu_2S + GeSe_2 \leftrightarrow 2Cu_2Se + GeS_2$ reciprocal system and thermodynamic properties of the $Cu_8GeS_{6-x}Se_x$ solid solutions, *J. Alloys Compd.*, **691**, 255–262 (2017c).

Alverdiyev I.J. Solid-phase equilibria in the reciprocal system $6Ag_2S + Ag_8SnSe_6 \leftrightarrow 6Ag_2Se + Ag_8SnS_6$, *Azerb. Chem. J.*, (*4*), 70–75 (2019).

Amcoff Ö. Kinetics in two sulfide systems below 500°C, *Acta Univ. Upsal. Abstrs Uppsala Diss. Fac. Sci.*, (*494*), 10 (1978).

Aminov T.G., Shabunina G.G., Busheva E.V. Magnetic properties of the quaternary compounds of the Cu_2GeSe_3–Cr_2Se_3 join of the Cu_2Se–$GeSe_2$–Cr_2Se_3 system, *Inorg. Mater.*, **45**(3), 242–245 (2009).

Anglin C., Takas N., Callejas J., Poudeu P.F.P. Crystal structure and physical properties of the quaternary manganese-bearing pavonite homologue $Mn_{1.34}Sn_{6.66}Bi_8Se_{20}$, *J. Solid State Chem.*, **183**(7), 1529–1539 (2010).

An Y., Baiyin M., Liu X., Ji M., Jia C., Ning G. A solvothermal synthesis and structure of $K_2Ag_2GeS_4$ with the simplest helical chains, *Inorg. Chem. Comm.*, **7**(1), 114–116 (2004).

Anzai S., Fukazawa T. Lattice parameters and the first order structural phase transition in $Cu_4Sn_{1-x}Ge_xS_2$ ($x \leq 0.3$), *J. Phys. Soc. Jpn.*, **55**(2), 701–702 (1986).

Anzai S., Matoba M., Inoue I., Fukazawa T., Kamimura T., Kitamura K. Conductivity phase-transition in $Cu_4Sn(S_{1-y}Se_y)_4$, *J. Phys. Soc. Jpn.*, **59**(10), 3799–3800 (1990).

Arachchige I.U., Wu J., Dravid V.P., Kanatzidis M.G. Nanocrystals of the quaternary thermoelectric materials: $AgPb_mSbTe_{m+2}$ (m = 1–18): phase-segregated or solid solutions?, *Adv. Mater.*, **20**(19), 3638–3642 (2008).

Armbruster T., Makovicky E., Berlepsch P., Sejkora J. Crystal structure, cation ordering, and polytypic character of diaphorite, $Pb_2Ag_3Sb_3S_8$, a PbS-based structure, *Eur. J. Mineralog.*, **15**(1), 137–146 (2003).

Artner C., Weil M. Lead(II) oxidotellurates(VI) with double perovskite structures, *J. Solid State Chem.*, **276**, 75–86 (2019).

Artner C., Weil M. $Pb_6Co_9(TeO_6)_5$, *Acta Crystallogr.*, **E68**(9), i71 (2012).

Artner C., Weil M. Re-examination of "Pb_3TeO_6": determination of its correct composition as Pb_5TeO_8, *J. Solid State Chem.*, **199**, 240–247 (2013).

Asadov M.M. Dependences of physico-chemical properties of solid solutions in the systems $GeSe_2$–A^2B^6 (A^2 = Hg, Cd; B^6 = S, Te) versus composition [in Russian], *Azerb. khim. zhurn.*, (2), 77–81 (2006).

Asadov M.M., Mamedov F.M., Mirzoev A.Ch., Aliev O.M. *T-x* phase diagrams of the FeSe–GeTe, $GeSe_2$–CdTe systems [in Russian], *Konf. stran Inst. Singl. Cryst. Publish., SNG po rostu kristallov. Tez. dokl. RK SNG-2012*, Kharkov, 99 (2012).

Asadov S.M., Mamedov A.N., Kulieva S.A. Composition- and temperature-dependent thermodynamic properties of the $Cd,Ge||Se,Te$ system, containing $CdSe_{1-x}Te_x$ solid solutions, *Inorg. Mater.*, **52**(9), 876–885 (2016).

Ashirov A., Gurshumov A.P., Dovletov K., Mamedov N.A. The SnTe–InSe and SnTe–Tl_2Se systems [in Russian], *Zhurn. neorgan. khimii*, **31**(5), 1282–1284 (1986).

Ashirov G.M. Phase equilibria in the Ag_8SiTe_6–Ag_8GeTe_6 system, *Azerb. Chem. J.*, (*1*), 89–93 (2022).

Assoud A., Kleinke H. Unique barium selenostannate-selenide: $Ba_7Sn_3Se_{13}$ (and its variants $Ba_7Sn_3Se_{13-\delta}Te_\delta$) with $SnSe_4$ tetrahedra and isolated Se anions, *Chem. Mater.*, **17**(17), 4509–4513 (2005).

Assoud A., Sankar C.R., Kleinke H. Synthesis, crystal structure, electronic structure and electrical conductivity of $La_3GeSb_{0.31}Se_7$ and $La_3SnFe_{0.61}Se_7$, *Solid State Sci.*, **38**, 124–128 (2014).

Assoud A., Soheilnia N., Kleinke H. Band gap tuning in new strontium seleno-stannates, *Chem. Mater.*, **16**(11), 2215–2221 (2004a).

Assoud A., Soheilnia N., Kleinke H. From yellow to black: new semiconducting Ba chalcogeno-germanates, *Z. Naturforsch.*, **59B**(9), 975–979 (2004b).

Assoud A., Soheilnia N., Kleinke H. New quaternary barium copper/silver selenostannates: different coordination spheres, metal-metal interactions, and physical properties, *Chem. Mater.*, **17**(9), 2255–2261 (2005).

Avetisian H.K., Gnatyshenko G.I. Thermal and metallographic investigation of PbS–ZnS–FeS system [in Russian], *Izv. AN KazSSR. Ser. gorn. dela, stroimaterialov i metallurgii*, (6), 11–25 (1956).

Azam S., Khan S.A., Goumri-Said S. Exploring the electronic structure and optical properties of the quaternary selenide compound, $Ba_4Ga_4SnSe_{12}$: for photovoltaic applications, *J. Solid State Chem.*, **229**, 260–265 (2015).

Azhdarova D.S., Mehdiev I.G., Mamedov A.N., Zargarova M.I. Liquidus of the $CuInSe_2$–InSe–$SnSe_2$ ternary system [in Russian], *Neorgan. mater.*, **35**(8), 923–926 (1999).

Babanly D.M., Aliev I.I., Babanly K.N., Yusibov Yu.A. Phase equilibria in the Ag_2Te–PbTe–Bi_2Te_3 system, *Russ. J. Inorg. Chem.*, **56**(9), 1472–1477 (2011a).

Babanly D.M., Zlomanov V.P., Guseinov F.N., Dashdyeva G.B. Phase equilibria in the Tl_2Te–SnTe–Bi_2Te_3 system, *Russ. J. Inorg. Chem.*, **56**(12), 1981–1987 (2011b).

Babanly K.N., Aliyev I.I., Babanly N.B. Thermodynamic investigation of solid solutions $(2PbSe)_x(AgBiSe_2)_{1-x}$ by e.m.f method with solid electrolyte [in Russian], *Azerb. khim. zhurn.*, (*1*), 49–53 (2012).

Babanly K.N. Phase equilibria in the $Ag_2Se–Pb_5Bi_6Se_{14}$ system [in Azerbaijanian], *Kimya Problem.*, (*3*), 506–509 (2011).

Babanly M.B., Dashdieva G.B., Guseinov F.N. Phase equilibria in the system $Tl_2Te–SnTe–TlBiTe_2$, *Inorg. Mater.*, 44(10), 1060–1065 (2008).

Babanly M.B., Gotuk A.A., Kuliev A.A. The $Tl_5Te_3–Tl_4SnTe_3–Tl_4PbTe_3$ system [in Russian], *Izv. AN SSSR. Neorgan. mater.*, 15(7), 1292–1293 (1979).

Babitsyna A.A., Novotortsev V.M. On the interaction of chromium selenide Cr_2Se_3 with some selenides of copper and tin [in Russian], *Zhurn. neorgan. khimii*, 31(7), 1825–1828 (1986).

Babizhetskyy V., Levytskyy V., Smetana V., Wilk-Kozubek M., Tsisar O., Piskach L., Parasyuk O., Mudring A.-V. New cation-disordered quaternary selenides Ga_2TtSe_6 (Tt = Ge, Sn), *Z. Naturforsch.*, 75B(1–2), 135–142 (2020).

Badikov V.V., Badikov D.V., Laptev V.B., Mitin K.V., Shevyrdyaeva G.S., Shchebetova N.I., Petrov V. Crystal growth and characterization of new quaternary chalcogenide nonlinear crystals for the mid-IR: $BaGa_2GeS_6$ and $BaGa_2GeSe_6$, *Opt. Mater. Expr.*, 6(9), 2933–2938 (2016).

Badikov V.V., Badikov D.V., Shevyrdyaeva G.S., Kato K., Umemura N., Miyata K., Panyutin V.L., Petrov V. Crystal growth and characterization of a new quaternary hexagonal nonlinear crystal for the mid-IR: $Ba_2Ga_8GeS_{16}$, *J. Alloys Compd.*, 907, 164378 (2022a).

Badikov V.V., Badikov D.V., Shevyrdyaeva G.S., Laptev V.B., Melnikov A.A., Chekalin S.V. Optical and generation characteristics of new nonlinear $Ba_2Ga_8GeS_{16}$ and $Ba_2Ga_8(GeSe_2)S_{14}$ crystals for the mid-IR range, *Quant. Electron.*, 52(3), 296–300 (2022b).

Badikov V.V., Badikov D.V., Wang L., Shevyrdyaeva G.S., Panyutin V.L., Fintisova A.A., Sheina S.G., Petrov V. Crystal growth and characterization of a new quaternary chalcogenide nonlinear crystal for the mid-infrared: $PbGa_2GeSe_6$, *Cryst. Growth Des.*, 19(8), 4224–4228 (2019).

Badikov V.V., Tyulyupa A.G., Shevyrdyaeva G.S., Sheina S.G. Solid solutions in the $AgGaS_2–GeS_2$, $AgGaSe_2–GeSe_2$ systems [in Russian], *Izv. AN SSSR. Neorgan. mater.*, 27(2), 248–252 (1991).

Bagheri S.M., Alverdiyev I.J., Aliev Z.S., Yusibov Y.A., Babanly M.B. Phase relationships in the $1.5GeS_2 + Cu_2GeSe_3 \leftrightarrow 1.5GeSe_2 + Cu_2GeS_3$ reciprocal system, *J. Alloys Compd.*, 625, 131–137 (2015).

Bagheri S.M., Alverdiyev I.J., Babanly M.B. Phase equilibria in the $Ag_8GeS_6–Ag_8GeSe_6$ system and some properties of solid solutions [in Azerbaijanian], *Azerb. khim. zhurn.*, (*3*), 15–21 (2014a).

Bagheri S.M., Imamaliyeva S.Z., Mashadiyeva L.F., Babanly M.B. Phase equilibria in the $Ag_8SnS_6–Ag_8SnSe_6$ system, *Int. J. Adv. Sci. Tech. Res.*, 2(4), 291–296 (2014b).

Bairamova S.T., Bagieva M.R., Agapashaeva S.M., Aliev O.M. Phase relations in the $CuAsS_2–MS$ (M – Pb, Eu, Yb) systems, *Inorg. Mater.*, 47(3), 231–234 (2011a).

Bairamova S.T., Bagieva M.R., Agapashaeva S.M., Aliev O.M. Synthesis and properties of structural analogs of the mineral bournonite, *Inorg. Mater.*, 47(4), 345–348 (2011b).

Baiyin M., An Y., Liu X., Ji M., Jia C., Ning G. $K_2Ag_6Sn_3S_{10}$: a quaternary sulfide composed of silver sulfide layers pillared by zigzag chains $^1_\infty[SnS_3]^{2-}$. *Inorg. Chem.*, 43(13), 3764–3765 (2004).

Baiyin M.-H., Gang G., Naren J.-R.-G. Solvothermal synthesis and crystal structure of K_2CdSnS_4 [in Chinese], *Chin. J. Inorg. Chem.*, 30(2), 405–410 (2014).

Bakakin V.V., Godovikov A.A. On the crystal structure and twinning of seligmanite and bournonite [in Russian], *Dokl. AN SSSR*, 251(2), 345–347 (1980).

Bakhtiyarly I.B., Azhdarova D.S., Mamedov Sh.G., Kurbanov G.R. The $SnPbSb_4S_8–4SnS$ system [in Russian], *ChemChemTech [Izv. Vyssh. Uchebn. Zaved. Khim. Khim. Tekhnol.]*, 52(4), 120–122 (2009).

Bakhtiyarov A.Sh., Vasil'ev L.N. NGR spectrum of glasses in the $Tl_2Se–As_2Se_3–SnSe_2$ system [in Russian], *Izv. AN SSSR. Neorgan. mater.*, 10(11), 2079–2081 (1974).

Bakhtiyarov A.P., Vasil'ev L.N. The gamma-resonance investigation of the $As–Se–Ge–Sn$ system [in Russian], *Izv. AN SSSR. Neorgan. mater.*, 11(4), 741–742 (1975).

Baldinozzi G., Grebille D., Sciau Ph., Kiat J.-M., Moret J., Bérar J.-F. Rietveld refinement of the incommensurate structure of the elpasolite (ordered perovskite), *J. Phys.: Condens. Matter*, 10(29), 6461–6472 (1998).

Baldinozzi G., Sciau Ph., Moret J., Buffat P.A. A new incommensurate phase in a lead ordered perovskite: Pb_2MgTeO_6, *Solid State Commun.*, 89(5), 441–445 (1994).

Balić-Žunić T., Bente K., Edenharter A. Crystal structure of high temperature thallium lead antimony sulfide, $Tl_{0.333}Pb_{0.333}Sb_{0.333}S$, *Z. Kristallogr.*, 202(1–4), 145–146 (1992).

Balić-Žunić T., Bente K. The two polymorphs of $TlPbSbS_3$ and the structural relations of phases in the system $TlSbS_2–PbS$, *Mineral. Petrol.*, 53(4), 265–276 (1995).

Balić-Žunić T., Engel P. Crystal structure of synthetic $PbTlAs_3S_6$, *Z. Kristallogr.*, 165(1–4), 261–269 (1983).

Balić-Žunić T., Makovicky E. Determination of the crystal structure of $TlPbSbS_3$, $(Tl,K)Fe_3(SO_4)_2(OH)_6$ and Cu_3SbS_3 from X-ray powder diffraction data, *Mater. Sci. Forum*, 166–169, 659–664 (1994).

Balić-Žunić T., Topa D., Makovicky E. The crystal structure of emilite, $Cu_{10.7}Pb_{10.7}Bi_{21.3}S_{48}$, the second 45 Å derivative of the bismuthinite–aikinite solid-solution series. *Canad. Mineralog.*, 40(1), 239–245 (2002).

Balić-Žunić T., Topa D., Makovicky E. The crystal structure of kudriavite, $(Cd,Pb)Bi_2S_4$. *Canad. Mineralog.*, 45(3), 437–443 (2007).

Balijapelli S., Craig A.J., Cho J.B., Jang J.I., Ghosh K., Aitken J.A. Chernatynskiy A.V., Choudhury A. Building-block approach to the discovery of $Na_8Mn_2(Ge_2Se_6)_2$: a polar chalcogenide exhibiting promising harmonic generation signals with a high laser-induced damage threshold, *J. Alloys Compd.*, 900, 163392 (2022).

Banerjee S., Malliakas C.D., Kanatzidis M.G. New layered tin(II) thiophosphates $ASnPS_4$ (A = K, Rb, Cs): synthesis, structure, glass formation, and the modulated $CsSnPS_4$, *Inorg. Chem.*, 51(21), 11562–11573 (2012).

Barchiy I.Ye., Hlukh O.S., Peresh E.Yu., Tsigika V.V., Sabov M.Yu. The $Tl_2GeSe_3–Tl_4Ge_xSn_{1-x}Se_4–Tl_2SnSe_3$ system [in Ukrainian], *Ukr. Khim. Zhurn.*, 72(7), 6–10 (2006).

Barchiy I.Ye., Hlukh O.S., Peresh E.Yu., Tsigika V.V. The $Tl_4GeSe_4–Tl_2Se–Tl_4SnSe_4$ system [in Russian]. *Zhurn. neorgan. khimii*, 50(5), 835–837 (2005).

Barchiy I.Ye., Sabov M.Yu., Glukh O.S., Malakhovs'ka-Rosokha T.O. New thermoelectric materials based on the solid solutions of complex compounds of the $Tl_2Te–PbTe–Bi_2Te_3$ system [in Ukrainian], *Nauk.Visnyk Volyns'k. Nats. Univ. im. Lesi Ukrainky. Ser. Khim. nauky.*, (*16*), 18–24 (2010).

Barnier S., Guittard M. Flahaut J. Étude de verres de chalcogenures contenant de l'europium divalent: systéme EuS–Ga$_2$S$_3$–GeS$_2$, *Mater. Res. Bull.*, **15**(6), 689–705 (1980).

Barnier S., Guittard M., Flahaut J. Etude des verres de chalcogenures contenant du manganese divalent systeme MnS–Ga$_2$S$_3$–GeS$_2$. Preparations, proprietes thermiques et optiques, *Mater. Res. Bull.*, **19**(7), 837–848 (1984).

Barnier S., Guittard M., Julien C. Glass formation and structural studies of chalcogenide glasses in the CdS–Ga$_2$S$_3$–GeS$_2$ system, *Mater. Sci. Eng.*, **B7**(3), 209–214 (1990).

Barton A.T., Liang M., Craig A.J., Zhang W.,Stoyko S.S., Radzanowski A.N., Fingerlow D., Halasyamani P.S., MacNeil J.H., Aitken J.A. Li$_2$Mg$_2$Si$_2$S$_6$ and Li$_2$Mg$_2$Ge$_2$S$_6$: two nonlinear optical sulfides featuring a unique, polar trigonal structure incorporating ethane-like anions, *Z. anorg. und allg. Chem.*, **648**(15), e202200071 (2022).

Bayer G. New perovskite-type compounds A$_2$BTeO$_6$, *J. Am. Ceram. Soc.*, **46**(12), 604–605 (1963).

Bayliss P. Crystal chemistry and crystallography of some minerals in the tetradymite group, *Am. Mineralog.*, **76**(1–2), 257–265 (1991).

Bayramova S.T., Alieva S.I., Azhdarova L.S., Aliev O.M. Phase equilibria in the Cu$_2$S–PbS–Gd$_2$S$_3$ system along the CuGdS$_2$–PbS and Cu$_2$S–PbCuGdS$_3$ sections, *Kimya Problem.*, (4), 424–427 (2015).

Bayramova S.T., Guliyeva S.A., Aliyev O.M. Synthesis and physico-chemical properties of the quaternary sulfides of the PbLnCuS$_3$ type [in Russian], *Kimya Problem.*, (2), 249–252 (2010).

Becker R., Brockner W., Eisenmann B. Kristallstruktur und Schwingungsspektrum des Thallium(I)-Zinn(II)-ortho-Thiophosphates TlSnPS$_4$, *Z. Naturforsch.*, **42A**(11), 1309–1312 (1987).

Bedjaoui A., Bouhemadou A., Aloum, S., Khenata R., Bin-Omran S., Al-Douri Y., Saoud F.S., Bensalem S. Structural, elastic, electronic and optical properties of the novel quaternary diamond-like semiconductors Cu$_2$MgSiS$_4$ and Cu$_2$MgGeS$_4$, *Solid State Sci.*, **70**, 21–35 (2017).

Behrens H., Ostermeier J. Zur Kenntnis des Verhaltens von Nichtmetallchalkogeniden gegenüber flüssigem Ammoniak, VIII. Über Reaktionen der Sulfide des Siliciums, Germaniums und Zinns im Ammonosystem, *Chem. Ber.*, **95**(2), 487–499 (1962).

Belakovskiy D., Cámara F., Uvarova Y., Gagne O.C. New mineral names, *Am. Mineralog.*, **99**(8–9), 1806–1813 (2014).

Belin R., Aldon L., Zerouale A., Belin C., Ribes M. Crystal structure of the non-stoichiometric argyrodite compound Ag$_{7-x}$GeSe$_5$I$_{1-x}$ (x = 0.31). A highly disordered silver superionic conducting material, *Solid State Sci.*, **3**(3), 251–265 (2001).

Belkyal I., El Azhari M., Bensch W., Depmeier W. TlPbPS4, *Acta Crystallogr.*, **E62**(10), i210–i212 (2006a).

Belkyal I., El Azhari M., Wu Y., Bensch W., Depmeier W. Phase transition of rubidium lead tetrathiophosphate RbPbPS$_4$, *Z. anorg. und allg. Chem.*, **635**(11), 1600–1603 (2009).

Belkyal I., El Azhari M., Wu Y.-D., Bensch W., Hesse K.-F., Depmeier W. Crystal structure of cesium lead tetrathiophosphate, CsPbPS$_4$, *Z. Kristallogr., New Cryst. Str.*, **220**(2), 127–128 (2005).

Belkyal I., El Azhari M., Wu Y.-D., Bensch W., Hesse K.-F., Depmeier W. Synthesis and crystal structure of the new quaternary thiophosphate KPbPS$_4$, *Solid State Sci.*, **8**(1), 59–63 (2006b).

Bell M.C., Flengas S.N. The electrical conductivities and the structural properties of molten PbCl$_2$–PbS mixtures. I. Structural properties, *J. Electrochem. Soc.*, **113**(1), 27–31 (1966).

Bénazeth S., Guittard M., Laruelle P. Structure de l'oxysulfure de lanthane et d'étain (LaO)$_4$Sn$_2$S$_6$, *Acta Crystallogr.*, **C41**(5), 649–651 (1985).

Bengel H., Jobic S., Moölo Y., Lafond A., Rouxel J., Seo D.-K., Whangbo M.-H. Distribution of the Pb and Sb atoms in the (Pb,Sb)S layers of the franckeite-type misfit compound [(Pb,Sb)S]$_{2.28}$NbS$_2$ examined by scanning tunneling and atomic force microscopy, *J. Solid State Chem.*, **149**(2), 370–377 (2000).

Bensalem S., Chegaar M., Maouche D., Bouhemadou A. Theoretical study of structural, elastic and thermodynamic properties of CZTX (X = S and Se) alloys, *J. Alloys Compd.*, **589**, 137–142 (2014).

Bente K., Anton R. Crystal chemistry and thin film epitaxy of sulfides and Bi–Sb-sulfosalts and their geological application, *Mineral. Petrol.*, **53**(4), 209–228 (1995).

Bente K., Engel M., Steins M. Crystal structure of lead bismuth silver sulfide, PbAgBi$_3$S$_6$, *Z. Kristallogr.*, **205**(2), 327–328 (1993).

Beraich M., Shaili H., Hafidi Z., Benhsina E., Majdoubi H., Taibi M., Guenbour A., Bellaouchou A., Mzerd A., Bentiss F., Zarrouk A., Fahoume M. Facile synthesis of the wurtz stannite (orthorhombic) Cu$_2$MnGeS$_4$ thin film via spray ultrasonic method: structural, Raman, optical and electronic study, *J. Alloys Compd.*, **845**, 156216 (2020).

Berdonosov P.S., Olenev A.V., Dolgikh V.A. Lead (II) selenite halides Pb$_3$(SeO$_3$)$_2$X$_2$ (X = Br, I): synthesis and crystal structure, *Crystallogr. Rep.*, **57**(2), 200–204 (2012).

Bereznyuk O.P., Olekseyuk I.D., Petrus I.I., Smitiukh O.V. The Ag$_2$S–SnS$_2$–P$_2$S$_5$ system [in Ukrainian], *Visnyk Odes'k. Nats. Univ.*, **25**[4(76)], 32–44 (2020).

Bereznyuk O.P., Petrus I.I. Glass formation in the A$^I_2$S–B^{IV}S$_2$–P$_2$S$_5$ (A^I – Cu, Ag; B^{IV} – Ge, Sn) quasiternary system [in Ukrainian], *Nauk. Visn. Uzhgorod. Univ., Ser. Khim.*, [2(44)], 41–44 (2020).

Berlepsch P., Armbruster T., Makovicky E., Topa D. Another step toward understanding the true nature of sartorite: determination and refinement of a ninefold superstructure, *Am. Mineralog.*, **88**(2–3), 450–461 (2003).

Berlepsch P. Chemical and crystallographical investigations on edenharterite (TlPbAs$_3$S$_6$), *Schweiz. Mineral. Petrogr. Mitt.*, **75**(2), 277–281 (1995).

Berlepsch P. Crystal structure and crystal chemistry of the homeotypes edenharterite (TlPbAs$_3$S$_6$) and jentschite (TlPbAs$_2$SbS$_6$) from Lengenbach, Binntal (Switzerland), *Schweiz. Mineral. Petrogr. Mitt.*, **76**(2), 147–157 (1996).

Bernert Th., Pfitzner A. Characterization of mixed crystals in the system Cu$_2$Mn$_x$Co$_{1-x}$GeS$_4$ and investigations of the tetrahedra volumes, *Z. anorg. und allg. Chem.*, **632**(7), 1213–1218 (2006).

Bernert Th., Pfitzner A. Cu$_2$MnM^{IV}S$_4$ (M^{IV} = Si, Ge, Sn) – analysis of crystal structures and tetrahedra volumes of normal tetrahedral compounds, *Z. Kristallogr.*, **220**(11), 968–972 (2005).

Berry L.G. Studies of mineral sulpho-salts: II. Jamesonite from Cornwall and Bolivia, *Mineralog. Mag.*, **25**(170), 597–608 (1940).

Biagioni C., Bindi L., Moëlo Y. Another step toward the solution of the real structure of zinkenite, *Z. Kristallogr.*, **233**(3–4), 269–277 (2018).

Biagioni C., Bindi L., Momma K., Miyawaki R., Matsushita Y., Moëlo Y. Determination of the crystal structure and redefinition of tsugaruite, Pb$_{28}$As$_{15}$S$_{50}$Cl, the first lead-arsenic chloro-sulfosalt, *Canad. Mineralog.*, **59**(1), 125–137 (2021).

Biagioni C., Dini A., Orlandi P., Moëlo Y., Pasero M., Zaccarini F. Lead-antimony sulfosalts from Tuscany (Italy). XX. Members of the jordanite–geocronite series from the Pollone mine, Valdicastello Carducci: occurrence and crystal structures, *Minerals*, **6**(1), 15 (2016a).

Biagioni C., Merlino S., Moëlo Y., Pasero M., Paar W.H., Vezzoni S., Zaccarini F. Arsenmarcobaldiite, IMA 2016-045. CNMNC Newsletter No. 33, October 2016, page 1138, *Mineralog. Mag.*, **80**(6), 1135–1144 (2016b).

Biagioni C., Moëlo Y. Lead-antimony sulfosalts from Tuscany (Italy). XIX. Crystal chemistry of chovanite from two new occurrences in the Apuan Alps and its 8 Å crystal structure, *Mineralog. Mag.*, **81**(4), 811–831 (2017).

Biagioni C., Moëlo Y., Merlino S., Pasero M., Paar W.H., Vezzoni S., Zaccarini F. Arsenmarcobaldiite, $Pb_{12}(As_{3.2}Sb_{2.8})_{\Sigma=6}S_{21}$, a new N = 3.5 jordanite homologue from the Sant'Anna tectonic window, Apuan Alps (Tuscany, Italy), *Eur. J. Mineral.*, **31**(5–6), 1067–1077 (2019).

Bibenina G.A., Smirnov M.P. Guselnikova N Yu. Investigation of phase diagram of the $PbS–Na_2S–Cu_2S$ system [in Russian], In: *Fiz.-khim. osnovy metallurg. processov*. Moscow, Metallurgiya Publish., 35–48 (1967).

Bibenina G.A., Smirnov M.P. The $PbS–Na_2S–Cu_2S$ system [in Russian], *Zhurn. neorgan. khimii*, **11**(9), 2133–2138 (1966).

Bilousov O.V., Gurs'kyi A.V., Kogut Yu.M., Piskach L.V. Phase diagrams of the $Ag_8GeS(Se)_6–PbS(Se)$ systems [in Ukrainian], *Nauk. Visnyk Volyns'k. Derzh. Univ. im. Lesi Ukrainky. Ser. Khim. nauky*, **4**, 128–132 (2006).

Bindi L., Cipriani C. Plumbian baksanite from the Tyrnyauz W–Mo deposit, Baksan River valley, Northern Caucasus, Russian Federation, *Canad. Mineralog.*, **41**(6), 1475–1479 (2003).

Bindi L., Nestola F., Guastoni A., Zorzi F., Nasdala L. Fassinaite, IMA 2011-048. CNMNC Newsletter No. 10, October 2011, page 2559, *Mineralog. Mag.*, **75**(5), 2549–2561 (2011a).

Bindi L., Nestola F., Guastoni A., Zorzi F., Peruzzo L., Raber T. Te-rich canfieldite, $Ag_8Sn(S,Te)_6$, from the Lengenbach quarry, Binntal, canton Valais, Switzerland: occurrence, description and crystal structure, *Canad. Mineralog.*, **50**(1), 111–118 (2012).

Bindi L., Nestola F., Kolitsch U., Guastoni A., Zorzi F. Fassinaite, $Pb_2^{2+}(S_2O_3)(CO_3)$, the first mineral with coexisting thiosulphate and carbonate groups: description and crystal structure, *Mineralog. Mag.*, **75**(6), 2721–2732 (2011b).

Bindi L., Paar W.H. Jaszczakite, $[(Bi,Pb)_3S_3][AuS_2]$, a new mineral species from Nagybörzsöny, Hungary, *Eur. J. Mineral.*, **29**(4), 373–677 (2017).

Bindi L., Paar W.H. Jaszczakite, IMA 2016-077. CNMNC Newsletter No. 34, December 2016, page 1319, *Mineralog. Mag.*, **80**(7), 1315–1321 (2016).

Bindi L., Petříček V., Biagioni C., Plášil J., Moëlo Y. Could incommensurability in sulfosalts be more common than thought? The case of meneghinite, $CuPb_{13}Sb_7S_{24}$, *Acta Crystallogr.*, **B73**(3), 369–376 (2017).

Birnie R.W., Burnham C.W. The crystal structure and extent of solid solution of geocronite, *Am. Mineralog.*, **61**(9–10), 963–970 (1976).

Blachnik R., Buchmeier W., Dreisbach H.A. Structure of lead(II) dimercury(II) diiodide disulfide, *Acta Crystallogr.*, **C42**(5), 515–517 (1986).

Blachnik R., Lytze K., Reuter H. A new quaternary chalcogenide halide: synthesis and structure of $Hg_2SnS_2Br_2$, *J. Solid State Chem.*, **126**(1), 95–98 (1996).

Blashko N.M., Gulay L.D., Marchuk O.V., Olekseyuk I.D. The $Pr_3Ga_{1.67}Se_7–Pr_3Ge_{1.25}Se_7$ system [in Ukrainian], *Visnyk Odes'k. Nats. Univ.*, **25**[1(73)], 24–30 (2020).

Bode H., Voss E. Basische Bleisulfate und deren Bildung bei der Herstellung von Elektroden für Bleiakkumulatoren, *Electrochim. Acta*, **1**(4), 318–325 (1959).

Bogdashevskaya N.N., Mel'Nichenko T.N., Migolinets I.M., Turyanitsa I.D. Liquation phenomena in the $Sb_2S_3–SbI_3–GeSe_2$ systems [in Russian], *Ukr. khim. zhurn.*, **47**(7), 702–705 (1981).

Bordas S., Casas-Vazquez J., Clavaguera N., Clavaguera-Mora M.T. Glass-forming ability in the Ge–Sb–Te–Se quaternary system, *Thermochim. Acta*, **28**(2), 387–393 (1979).

Borisova L., Decheva St., Dimitrova St., Kristev V., Moraliyski P. Solid solutions in the $GeTe–AgSbTe_2$ system [in Bulgarian], *Godishn. Sofiysk. Univ. Phys. Fak.*, 1968/1969, **63**, 137–144 (1971).

Borodaev Y.S., Garavelli A., Garbarino C., Grillo S.M., Mozgova N.N., Organova N.I., Trubkin N.V., Vurro F. Rare sulfosalts from Vulcano, Aeolian Islands, Italy. III. Wittite and cannizzarite, *Canad. Mineralog.*, **38**(1), 23–34 (2000).

Borodaev Y.S., Garavelli A., Garbarino C., Grillo S.M., Mozgova N.N., Paar W.H., Topa D. Rare sulfosalts from Vulcano, Aeolian Islands, Italy. V. Selenian heyrovskýite, *Canad. Mineralog.*, **41**(2), 429–440 (2003).

Borovikova R.P., Dudkin L.D., Kazanskaya A.O., Kosolapova E.F. Investigation of the SnTe–PbSe system [in Russian], *Izv. AN SSSR. Neorgan. mater.*, **8**(10), 1762–1764 (1972).

Bortnikov N.S., Mozgova N.N., Tsepin A.I., Breskovska V.V. The first experience in the synthesis of lead chlorosulfoantimonites [in Russian], *Dokl. AN SSSR*, **244**(4), 955–958 (1979).

Bourgès C., Al Rahal Al Orabi R., Miyazaki Y. Off-stoichiometry effect on thermoelectric properties of the new *p*-type sulfides compounds Cu_2CoGeS_4, *J. Alloys Compnd.*, **826**, 154240 (2020).

Bouyrie Y., Ohta M., Suekuni K., Kikuchi Y., Jood P., Yamamotoa A., Takabatake T. Enhancement in the thermoelectric performance of colusites $Cu_{26}A_2E_6S_{32}$ (A = Nb, Ta; E = Sn, Ge) using E-site non-stoichiometry, *J. Mater. Chem. C*, **5**(17), 4174–4184 (2017).

Boycheva S.V., Vassilev V.S., Ivanova Z.G. A Zn(II) ion-selective electrode based on chalcogenide $As_2Se_3–Sb_2Se_3–ZnSe$ and $GeSe_2–ZnSe–ZnTe$ glasses, *J. Appl. Electrochem.*, **32**(3) 281–285 (2002).

Boycheva S.V., Vassilev V.S., Ivanova Z.G. On the glass formation in the semiconducting $GeSe_2–Sb_2Se_3–ZnTe$, $As_2Se_3–Sb_2Se_3–ZnSe$ and $GeSe_2–ZnTe–ZnSe$ systems, *J. Phys. D: Appl. Phys.*, **32**(4), 529–532 (1999).

Bracci G., Dalena D., Orlandi P., Duchi G., Vezzalini G. Guettardite from Tuscany, Italy: a second occurrence, *Canad. Mineralog.*, **18**(1), 13–15 (1980).

Brandmayer M.K., Clérac R., Weigend F., Dehnen S. Ortho-chalcogenostannates as ligands: syntheses, crystal structures, electronic properties, and magnetism of novel compounds containing ternary anionic substructures $[M_4(\mu_4-Se)(SnSe_4)_4]^{10-}$ (M = Mn, Zn, Cd, Hg), $^3_\infty\{[Hg_4(\mu_4-Se)(SnSe_4)_3]^{6-}\}$, or $^1_\infty\{[HgSnSe_4]^{2-}\}$, *Chem. Eur. J.*, **10**(20), 5147–5157 (2004).

Brandstätter F. Synthesis and crystal structure determination of $Pb_2[UO_2][TeO_3]_3$, *Z. Kristallogr.*, **155**(1–4), 193–200 (1981).

Brant J.A., Clark D.J., Kim Y.S., Jang J.I., Weiland A., Aitken J.A. Outstanding laser damage threshold in Li_2MnGeS_4 and tunable optical nonlinearity in diamond-like semiconductors, *Inorg. Chem.*, **54**(6), 2809–2819 (2015a).

Brant J.A., Clark D.J., Kim Y.S., Jang J.I., Zhang J.-H., Aitken J.A. Li$_2$CdGeS$_4$, a diamond-like semiconductor with strong second-order optical nonlinearity in the infrared and exceptional laser damage threshold, *Chem. Mater.*, **26**(10), 3045–3048 (2014a).

Brant J.A., Cruz Dela C., Yao J., Douvalis A.P., Bakas T., Sorescu M., Aitken J.A. Field-induced spin-flop in antiferromagnetic semiconductors with commensurate and incommensurate magnetic structures: Li$_2$FeGeS$_4$ (LIGS) and Li$_2$FeSnS$_4$ (LITS), *Inorg. Chem.*, **53**(23), 12265–12274 (2014b).

Brant J.A., Devlin K.P., Bischoff C., Watson D., Martin S.W., Gross M.D., Aitken J.A. A new class of lithium ion conductors with tunable structures and compositions: quaternary diamond-like thiogermanates, *Solid State Ionics*, **278**, 268–274 (2015b).

Brennan T.H., Ibers J.A. Lanthanum orthosilicate selenide, La$_2$SeSiO$_4$, *Acta Crystallogr.*, **C47**(5), 1062–1064 (1991).

Brennan T.D., Ibers J.A. LaPbCuS$_3$: Cu(I) insertion into the α-La$_2$S$_3$ framework, *J. Solid State Chem.*, **97**(2), 377–382 (1992).

Breskovska V.V., Mozgova N.N., Bortnikov N.S., Gorshkov A.I., Tsepin A.I. Ardaite – a new lead-antimony chlorosulphosalt, *Mineralog. Mag.*, **46**(340), 357–361 (1982).

Breskovska V.V., Mozgova N.N., Bortnikov N.S., Gorshkov A.I., Tzepin A.I. New data on chlorine sulfosalts, *Bull. Minéralog.*, **104**(6), 757–762 (1981).

Brockwey L.O. The crystal structure of stannite, Cu$_2$FeSnS$_4$, *Z. Kristallogr.*, **89**(1–6), 434–441 (1934).

Bron P., Johansson S., Zick K., Schmedt auf der Günne J., Dehnen S., Roling B. Li$_{10}$SnP$_2$S$_{12}$: an affordable lithium superionic conductor, *J. Am. Chem. Soc.*, **135**(42), 15694–15697 (2013).

Brotherton P.D., Epstein J.M., Pryce M.W., White A.H. Crystal structure of 'calcium sulphosilicate', Ca$_5$(SiO$_4$)$_2$SO$_4$, *Aust. J. Chem.*, **27**(3), 657–660 (1974).

Brower W.S., Parker H.S., Roth R.S. Reexamination of synthetic parkerite and shandite, *Am. Mineralog.*, **59**(3–4), 296–301 (1974).

Brunetta C.D., Brant J.A., Rosmus K.A., Henline K.M., Karey E., MacNeil J.H., Aitken J.A. The impact of three new quaternary sulfides on the current predictive tools for structure and composition of diamond-like materials, *J. Alloys Compd.*, **574**, 495–503 (2013).

Brunetta C.D., Karuppannan B., Rosmus K.A., Aitken J.A. The crystal and electronic band structure of the diamond-like semiconductor Ag$_2$ZnSiS$_4$, *J. Alloys Compd.*, **516**, 65–72 (2012a).

Brunetta C.D., Minsterman III W.C., Lake C.H., Aitken J.A. Cation ordering and physicochemical characterization of the quaternary diamond-likesemiconductor Ag$_2$CdGeS$_4$, *J. Solid State Chem.*, **187**, 177–185 (2012b).

Bryzgalov I.A., Spiridonov E.M., Petrova I.V., Sakharova M.S. Babkinite Pb$_2$Bi$_2$(S,Se)$_3$ – a new mineral [in Russian], *Dokl. AN SSSR*, **346**(5), 656–659 (1996).

Bucher C.K., Hwu S.-J. CsSmGeS$_4$: A novel layered mixed-metal sulfide crystallizing in the noncentrosymmetric space group *P*2$_1$2$_1$2$_1$, *Inorg. Chem.*, **33**(25), 5831–5835 (1994).

Burbank J. Anodic oxidation of the basic sulfates of lead, *J. Electrochem. Soc.*, **113**(1), 10–14 (1966).

Burmistrova N.P., Fitseva R.G., Gol'mgreyn L.A., Davletshin R.Yu. Investigation of the lead telluride–cesium iodine system [in Russian], *Izv. AN SSSR. Neorgan. mater.*, **16**(10), 1768–1770 (1980).

Bushmarina G.S., Dedegkaev T.T., Drabkin I.A., Zhukova T.B., Konstantinov P.P., Lev E.Ya., Sysoeva L.M. Investigation of solid solutions in the GeTe–AgBiTe$_2$ and GeTe–AgSbTe$_2$ systems [in Russian], *Izv. AN SSSR. Neorgan. mater.*, **17**(8), 1392–1397 (1981).

Bushmarina G.S., Gruzinov B.F., Dedegkaev T.T., Drabkin I.A., Zhukova T.B., Lev E.Ya. Peculiarities of gallium doping of PbTe and PbTe–SnTe solid solution [in Russian], *Izv. AN SSSR. Neorgan. mater.*, **16**(12), 2136–2140 (1980).

Caldera D., Quintero M., Morocoima M., Moreno E., Quintero E., Grima-Gallardo P., Bocaranda P., Henao J.A., Macías M.A., Briceño J.M., Mora A.E. Lattice parameters values and phase diagram for the Cu$_2$Zn$_{1-z}$Mn$_z$GeSe$_4$ alloy system, *J. Alloys Compd.*, **614**, 253–257 (2014).

Caldera D., Quintero M., Morocoima M., Quintero E., Grima P., Marchan N., Morenoa E., Bocaranda P., Delgado G.E., Mora A.E., Briceño J.M., Fernandez J.L. Lattice parameters values and phase diagram for the Cu$_2$Zn$_{1-z}$Fe$_z$GeSe$_4$ alloy system, *J. Alloys Compd.*, **457**(1–2), 221–224 (2008).

Cámara F., Gagné O.C., Belakovskiy D.I., Uvarova Y. New mineral names, *Am. Mineralog.*, **100**(5–6), 1319–1332 (2015).

Cervelle B., Cesbron F., Sichère M.C., Dietrich J. La chalcostibite et la dadsonite de Saint-Pons, Alpes de Haute Provence, France, *Canad. Mineralog.*, **17**(3), 601–605 (1979).

Chan B.C., Dorhout P.K. Crystal structures of potassium terbium(III) tetrasulfidogermanate, KTbGeS$_4$, and potassium praseodymium(III) tetraselenidogermanate, KPrGeSe$_4$, *Z. Kristallogr., New Cryst. Str.*, **220**(1), 7–8 (2005).

Chang L.L.Y. Ag$_{1.2}$Sn$_{0.9}$Sb$_3$S$_6$, a tin-bearing andorite phase, *Mineralog. Mag.*, **51**(363), 741–743 (1987).

Chang L.L.Y., Hoda S.H. Phase relations in the systems PbS–Cu$_2$S–Bi$_2$S and the stability of galenobismutite, *Am. Mineralog.*, **62**(3–4), 346–350 (1977).

Chang L.L.Y., Knowles C.R. Phase relations in the systems PbS–Fe$_{1-x}$S–Sb$_2$S$_3$ and PbS–Fe$_{1-x}$S–Bi$_2$S$_3$, *Canad. Mineralog.*, **15**(3), 374–379 (1977).

Chang L.L.Y., Li X., Zheng C. The jamesonite – benavidesite series, *Canad. Mineralog.*, **25**(4), 667–672 (1987).

Chang L.L.Y., Walia D.S., Knowles C.R. Phase relations in the systems PbS–Sb$_2$S$_3$–Bi$_2$S$_3$ and PbS–FeS–Sb$_2$S$_3$–Bi$_2$S$_3$, *Econ. Geol.*, **75**(2), 317–328 (1980).

Chang L.L.Y., Wu D., Knowles Ch.R. Phase relations in the system Ag$_2$S–Cu$_2$S–PbS–Bi$_2$S$_3$, *Econ. Geol.*, **83**(2), 405–418 (1988).

Chaplygin I.V., Mozgova N.N., Bryzgalov I.A. Belakovsky D.I., Pervukhina N.V., Borisov S.V., Magarill S.A. Znamenskyite, IMA 2014-026. CNMNC Newsletter No. 21, August 2014, page 801, *Mineralog. Mag.*, **78**(4), 797–804 (2014).

Chaplygin I.V., Mozgova N.N., Magazina L.O., Kuznetsova O.Yu, Safonov Y.G., Bryzgalov I.A., Makovicky E., Balić-Žunić T. Kudriavite, (Cd,Pb)Bi$_2$S$_4$, a new mineral species from Kudriavy volcano, Iturup Island, Kurile arc, Russia, *Canad. Mineralog.*, **43**(2), 695–701 (2005).

Chapuis G., Niggli A. Die idealisierte Kristallstruktur von Cu$_2$CdSiS$_4$, *Naturwissenschaften*, **55**(9), 441–442 (1968).

Chapuis G., Niggli A. The crystal structure of the 'normal tetrahedral' compound Cu$_2$CdSiS$_4$, *Acta Crystallogr.*, **B28**(5), 1626–1628 (1972).

Charkin D.O., Morozov O.S., Ul'yanova E.A., Berdonosov P.S., Dolgikh V.A., Dickinson C., Zhou W., Lightfoot P. A reinvestigation of Sillén X$_1$-type lead tellurium oxyhalides, Pb$_3$TeO$_4$X$_2$ (X = Cl, Br, I), *Solid State Sci.*, **8**(9), 1029–1034 (2006).

Chbani N., Loireau-Lo'zach A.-M., Rivet J., Dugué J. Système pseudo-ternaire $Ag_2S–Ga_2S_3–GeS_2$: diagramme de phases – Domaine vitreux, *J. Solid State Chem.*, **117**(1), 189–200 (1995).

Chen D., Ravindra N.M. Electronic and optical properties of Cu_2ZnGeX_4 (X = S, Se and Te) quaternary semiconductors, *J. Alloys Compd.*, **579**, 468–472 (2013).

Chen G.-R., Wang M.-F., Lee C.-S. Synthesis and characterization of new multinary selenides $Sn_4In_5Sb_9Se_{25}$ and $Sn_{6.13}Pb_{1.87}In_{5.00}Sb_{10.12}Bi_{2.88}Se_{35}$, *J. Solid.State Chem.*, **307**, 122855 (2022a).

Chen H., Chen Y.-K., Lin H., Shen J.-N., Wu L.-M., Wu X.-T. Quaternary layered semiconductor $Ba_2Cr_4GeSe_{10}$: synthesis, crystal structure, and thermoelectric properties, *Inorg. Chem.*, **57**(3), 916–920 (2018a).

Chen H., Liu P.F., Li B.-X., Lin H., Wu L.-M., Wu X.-T. Two quaternary non-centrosymmetric chalcogenides $BaAg_2GeS_4$ and $BaAg_2SnS_4$: experimental and theoretical studies on the NLO properties, *Dalton Trans.*, **47**(2), 429–437 (2018b).

Chen K.-B., Lee C.-S. Experimental and theoretical studies of $Sn_{3-8}Pb_8Bi_2Se_6$ (δ = 0.0–0.7), *J. Solid State Chem.*, **183**(4), 807–813 (2010a).

Chen K.-B., Lee C.-S. Synthesis and characterization of quaternary selenides $Sn_2Pb_5Bi_4Se_{13}$ and $Sn_{8.65}Pb_{0.35}Bi_4Se_{15}$, *Solid State Sci.*, **11**(9), 1666–1672 (2009).

Chen K.-B., Lee C.-S. Synthesis and phase width of new quaternary selenides $Pb_xSn_{6-x}Bi_2Se_9$ (x = 0–4.36), *J. Solid State Chem.*, **183**(11), 2616–2622 (2010b).

Chen M.-C., Li P., Zhou L.-J., Li L.-H., Chen L. Structure change induced by terminal sulfur in noncentrosymmetric $La_2Ga_2GeS_8$, $Eu_2Ga_2GeS_7$ and nonlinear optical responses in middle infrared, *Inorg. Chem.*, **50**(24), 12402–12404 (2011).

Chen S., Gong X.G., Walsh A., Wei S.-H. Crystal and electronic band structure of Cu_2ZnSnX_4 (X = S and Se) photovoltaic absorbers: first-principles insights, *Appl. Phys. Let.*, **94**(4), 041903_1–041903_3 (2009).

Chen S., Walsh A., Luo Y., Yang J.-H., Gong X.G., Wei S.-H. Wurtzite-derived polytypes of kesterite and stannite quaternary chalcogenide semiconductors, *Phys. Rev. B*, **82**(19), 195203_1–195203_8 (2010).

Chen T.T., Kirchner E., Paar W. Friedrichite, $Cu_5Pb_5Bi_7S_{18}$, a new member of the aikinite-bismuthinite series, *Canad. Mineralog.*, **16**(2), 127–130 (1978).

Chen X., Huang X., Fu. A., Li J., Zhang L.-D., Guo H.-Y. From 1D chain to 3D network: syntheses, structures, and properties of $K_2MnSn_2Se_6$, $K_2MnSnSe_4$, and $K_2Ag_2SnSe_4$, *Chem. Mater.*, **12**(8), 2385–2391 (2000a).

Chen X., Huang X., Li J. Potassium sodium tin selenide, $K_3NaSn_3Se_8$, *Acta Crystallogr.*, **C56**(10), 1181–1182 (2000b).

Chen Y.-K., Chen M.-C., Zhou L.-J., Chen L., Wu L.-M. Syntheses, structures, and nonlinear optical properties of quaternary chalcogenides: $Pb_4Ga_4GeQ_{12}$ (Q = S, Se), *Inorg. Chem.*, **52**(15), 8334–8341 (2013).

Chen Z., Mei D., Jiang X., Zhao J., Wu Y., Wang J., Wen S. New quaternary sulfide $LiGaSiS_4$: synthesis, structure and optical properties, *J. Solid State Chem.*, **312**, 123230 (2022b).

Chi Y., Guo S.-P. Syntheses, crystal and electronic structure of a series of quaternary rare-earth sulfides $MgRE_6Si_2S_{14}$ (RE = Y, Ce, Pr, Nd and Sm), *J. Mol. Struct.*, **1127**, 53–58 (2017).

Chi Y., Guo S.-P., Xue H.-G. Band gap tuning from an indirect $EuGa_2S_4$ to a direct $EuZnGeS_4$ semiconductor: syntheses, crystal and electronic structures, and optical properties, *RSC Adv.*, **7**(9), 5039–5045 (2017).

Choi K.-S., Chung D.-Y., Mrotzek A., Brazis P., Kannewurf C.R., Uher C., Chen W., Hogan T., Kanatzidis M.G. Modular construction of $A_{1+x}M_{4-2x}M'_{7+x}Se_{15}$ (A = K, Rb; M = Pb, Sn; M' = Bi, Sb): a new class of solid state quaternary thermoelectric compounds, *Chem. Mater.*, **13**(3), 756–764 (2001).

Choi K.-S., Kanatzidis M.G. Si extraction from silica in a basic polychalcogenide flux. Stabilization of $Ba_4SiSb_2Se_{11}$, a novel mixed selenosilicate/selenoantimonate with a polar structure, *Inorg. Chem.*, **40**(1), 101–104 (2001).

Choi K.-S., Kanatzidis M.G. Sulfosalts with alkaline earth metals. Centrosymmetric vs acentric interplay in $Ba_3Sb_{4.66}S_{10}$ and $Ba_{2.62}Pb_{1.38}Sb_4S_{10}$ based on the Ba/Pb/Sb ratio. Phases related to arsenosulfide minerals of the rathite group and the novel polysulfide $Sr_6Sb_6S_{17}$, *Inorg. Chem.*, **39**(25), 5655–5662 (2000).

Chondroudis K., Kanatzidis M.G. Isolation of $[Sn(PSe_5)_3]^{5-}$ and $[Sn_2Se_4(PSe_5)]^{6-}$; the first discrete complexes from molten alkali-metal polyselenophosphate fluxes, *Chem. Commun.*, (*11*), 1371–1372 (1996).

Chondroudis K., Kanatzidis M.G. $K_4In_2(PSe_5)_2(P_2Se_6)$ and $Rb_3Sn(PSe_5)(P_2Se_6)$: one-dimensional compounds with mixed selenophosphate anions, *J. Solid State Chem.*, **136**(1), 79–86 (1998).

Chondroudis K., McCarthy T.J., Kanatzidis M.G. Chemistry in molten alkali metal polyselenophosphates fluxes. Influence of flux composition on dimensionality. Layers and chains in $APbPSe_4$, $A_4Pb(PSe_4)_2$ (A = Rb, Cs), and $K_4Eu(PSe_4)_2$, *Inorg. Chem.*, **35**(4), 840–844 (1996).

Choubrac L., Lafond A., Guillot-Deudon C., Moëlo Y., Jobic S. Structure flexibility of the Cu_2ZnSnS_4 absorber in low-cost photovoltaic cells: from the stoichiometric to the copper-poor compounds, *Inorg. Chem.*, **51**(6), 3346–3348 (2012).

Choudhury A., Dorhout P.K. Alkali-metal thiogermanates: sodium channels and variations on the La_3CuSiS_7 structure type, *Inorg. Chem.*, **54**(3), 1055–1065 (2015).

Choudhury A., Dorhout P.K. An ordered assembly of filled nanoscale tubules of europium seleno-silicate in the crystal structure of a quaternary compound, *J. Am. Chem. Soc.*, **129**(30), 9270–9271 (2007).

Choudhury A., Dorhout P.K. Destruction of noncentrosymmetry through chalcogenide salt inclusion, *Inorg. Chem.*, **45**(14), 5245–5247 (2006).

Choudhury A., Ghosh K., Grandjean F., Long G.J., Dorhout P.K. Structural, optical, and magnetic properties of $Na_8Eu_2(Si_2S_6)_2$ and $Na_8Eu_2(Ge_2S_6)_2$: Europium(II) quaternary chalcogenides that contain an ethane-like $(Si_2S_6)^{6-}$ or $(Ge_2S_6)^{6-}$ moiety, *J. Solid State Chem.*, **226**, 74–80 (2015).

Choudhury A., Gradjean F., Long G.J., Dorhout P.K. $Na_{1.515}EuGeS_4$, a three-dimensional crystalline assembly of empty nanotubes constructed with europium(II/III) mixed valence ions, *Inorg. Chem.*, **51**(21), 11779–11786 (2012).

Choudhury A., Polyakova L.A., Hartenbach I., Schleid T., Dorhout P.K. Synthesis, structures, and properties of layered quaternary chalcogenides of the general formula $ALnEQ_4$ (A = K, Rb; Ln = Ce, Pr, Eu; E = Si, Ge; Q = S, Se), *Z. anorg. und. allg Chem.*, **632**(15), 2395–2401 (2006).

Choudhury A., Polyakova L.A., Strobel S., Dorhout P.K. Two noncentrosymmetric cubic seleno-germanates related to CsCl-type structure: synthesis, structure, magnetic and optical properties, *J. Solid State Chem.*, **180**(4), 1381–1389 (2007a).

Choudhury A., Strobel S., Martin B.R., Karst A.L., Dorhout P.K. Synthesis of a family of solids through the building-block approach: a case study with Ag+ substitution in the ternary Na–Ge–Se system, *Inorg. Chem.*, **46**(6), 2017–2027 (2007b).

Chung D.-Y., Choi K.-S., Brazis P.W., Kannewurf C.R., Kanatzidis M.G. Flux synthesis of new multinary bismuth chalcogenides and their thermoelectric properties, *Mater. Res. Soc. Symp. Proc.*, **545**, 65–74 (1998).

Chung D.-Y., Iordanidis L., Rangan K.K., Brazis P.W., Kannewurf C.R., Kanatzidis M.G. First quaternary A–Pb–Bi–Q (A = K, Rb, Cs; Q = S, Se) compounds: synthesis, structure and properties of α- and β-CsPbBi$_3$Se$_6$, APbBi$_3$Se$_6$ (A = K, Rb) and APbBi$_3$S$_6$ (A = K, Rb), *Chem. Mater.*, **11**(5), 1352–1362 (1999).

Chung I., Biswas K., Song J.-H., Androulakis J., Chondroudis K., Paraskevopoulos K.M., Freeman A.J., Kanatzidis M.G. Rb$_4$Sn$_5$P$_4$Se$_{20}$: a semimetallic selenophosphate, *Angew. Chem. Int. Ed.*, **50**(38), 8834–8838 (2011).

Chung I., Kanatzidis M.G. Stabilization of Sn^{2+} in K$_{10}$Sn$_3$(P$_2$Se$_6$)$_4$ and Cs$_2$SnP$_2$Se$_6$ derived from a basic flux, *Inorg. Chem.*, **50**(2), 412–414 (2011).

Chung M.-Y., Lee C.-S. Multinary selenides with unusual coordination environment of bismuth, *Inorg. Chem.*, **51**(24), 13328–13333 (2012).

Chutas N.I., Kress V.C., Ghiorso M.S., Sack R.O. A solution model for high-temperature PbS–AgSbS$_2$–AgBiS$_2$ galena, *Am. Mineralog.*, **93**(10), 1630–1640 (2008).

Chykhriy S.I., Sysa L.V., Parasyuk O.V., Piskach L.V. Crystal structure of the Cu$_2$CdSn$_3$S$_8$ compound, *J. Alloys Compd.*, **307**(1–2), 124–126 (2000).

Clark R.M. Saddlebackite, Pb$_2$Bi$_2$Te$_2$S$_3$, a new mineral species from the Boddinton gold deposit, Western Australia, *Aust. J. Mineral.*, **3**(2), 119–124 (1997).

Coleman L.C. Mineralogy of the Yellowknife Bay area, N.W.T., *Am. Mineralog.*, **38**(5–6), 506–527 (1953).

Collin G., Etienne J., Laruelle P. Composés hexagonaux L$_6$B$_2$C$_2$X$_{14}$ lacunaires ordonnés, *Bull. Soc. fr. Minéral. Cristallogr.*, **96**(1), 12–17 (1973).

Collin G., Flahaut J. Sur plusiers séries des composés non lacunaires de formule L$_2$B$_2$C$_2$X$_{14}$, *Bull. Soc. chim. France*, (6), 2207–2209 (1972).

Collin G., Laruelle P. Structure crystalline de La$_6$MnSi$_2$S$_{14}$, *C. r. Acad. Sci., Ser. C*, **270**(4), 410–412 (1970).

Collin G., Laruelle P. Structure de La$_6$Cu$_2$Si$_2$S$_{14}$, *Bull. Soc. fr. Minéral. Cristallogr.*, **94**(2), 175–176 (1971).

Colombara D., Delsante S., Borzone G., Mitchels J.M., Thomas L.H., Mendis B.G., Cummings C.Y., Marken F., Peter L.M. Crystal growth of Cu$_2$ZnSnS$_4$ solar cell absorber by chemically vapour transport with I$_2$, *J. Cryst. Growth*, **364**, 101–110 (2013).

Cook N.J., Ciobanu C.L., Stanley C.J., Paar W.H., Sundblad K. Compositional data for Bi-Pb tellurosulfides, *Canad. Mineralog.*, **45**(3), 417–435 (2007).

Cooper M.A., Hawthorne F.C. The structure topology of sidpietersite, Pb$_2$$^{+4}$(S^{6+}O$_3$S^{2-})O$_2(OH)_2$, a novel thiosulfate structure, *Canad. Mineralog.*, **37**(5), 1275–1282 (1999).

Corps J., Vaqueiro P., Aziz A., Grau-Crespo R., Kockelmann W., Jumas J.-C., Powell A.V. Interplay of metal-atom ordering, Fermi level tuning, and thermoelectric properties in cobalt shandites Co$_3$M$_2$S$_2$ (M = Sn, In), *Chem. Mater.*, **27**(11), 3946–3956 (2015).

Corps J., Vaqueiro P., Powell A.W. Co$_3$M$_2$S$_2$ (M = Sn, In) shandites as tellurium-free thermoelectrics, *J. Mater. Chem. A*, **1**(22), 6553–6557 (2013).

Craig A.J., Bin C.J., Shin S.H., Ha S.H., Stoyko S.S., Jang J.I., Aitken J.A. Homovalent cation substitutions leading to new lithium-containing La$_3$CuSiS$_7$ family members having chiral, uniaxial structures and exhibiting second- and third-harmonic generation, *J. Alloys Compd.*, **910**, 164855 (2022).

Craig A.J., Stoyko S.S., Bonnoni A., Aitken J.A. Syntheses and crystal structures of the quaternary thiogermanates Cu4FeGe2S7 and Cu4CoGe2S7, *Acta Crystallogr.*, **E76**(7), 1117–1121 (2020).

Craig J.R. Phase relation and mineral assemblages in the Ag–Bi–Pb–S, *Mineral. Deposita*, **1**(4), 278–306 (1967).

Crane M., Frost R.L., Kloprogge J.Th., Leverett P., Shaddick L., Williams P. The PbCrO$_4$–PbSO$_4$ system and its mineralogical significance, *N. Jb. Miner. Mh.*, **11**(11), 505–519 (2001).

Cros B.., Laqibi M., Peytavin S., Ribes M. Nouveaux conducteurs superioniques a l'argent Ag$_7$XS$_5$Z (X = Si, Ge, Sn; Z = Cl, Br, I). Étude structurale, *Rev. chim. minér.*, **23**(6), 796–809 (1986).

Cui Y., Assoud A., Kleinke H. Synthesis and structural and physical properties of new semiconducting quaternary tellurides: Ba$_4$Ag$_{3.95}$Ge$_2$Te$_9$ and Ba$_4$Cu$_{3.71}$Ge$_2$Te$_9$, *Inorg. Chem.*, **48**(12), 5313–5319 (2009).

Cui Y., Mayasree A., Assoud A., Kleinke H. Different clusters within the Ba$_4$M$_{4-x}$A$_2$Te$_9$ (M = Cu, Ag, Au; A= Si, Ge) series: crystal structures and transport properties, *J. Alloys Compd.*, **493**(1–2), 70–76 (2010).

Danot M., Colombet P., Tremblet M., Soubeyroux J.-L. Crystal structure of a metal excess spinel: Cu$_{1.10}$Cr$_{1.30}$Sn$_{0.70}$S$_{3.90}$, *Mater. Res. Bull.*, **20**(4), 463–468 (1985).

Das S., Krishna R.M., Ma S., Mandal K.C. Single phase polycrystalline Cu$_2$ZnSnS$_4$ grown by vertical gradient freeze technique, *J. Cryst. Growth*, **381**, 148–152 (2013).

Daszkiewicz M., Gulay L.D. Accidental formation of Gd$_4$(SiO$_4$)$_2$OTe: crystal structure and spectroscopic properties, *Acta Crystallogr.*, **C71**(7), 598–601 (2015).

Daszkiewicz M., Gulay L.D., Lychmanyuk O.S. Ln$_3$M$_{1-\delta}$TX$_7$ – quasi-isostructural compounds: stereochemistry and silver-ion motion in the Ln$_3$Ag$_{1-\delta}$GeS$_7$ (Ln = La–Nd, Sm, Gd–Er and Y; δ = 0.11–0.50) compounds, *Acta Crystallogr.*, **B65**(2), 126–133 (2009a).

Daszkiewicz M., Gulay L.D., Lychmanyuk O.S., Pietraszko A. Crystal structure of the R$_3$Ag$_{1-\delta}$SiS$_7$ (R = La, Ce, Pr, Nd, Sm, δ = 0.10–0.23) compounds, *J. Alloys Compd.*, **460**, 201–205 (2008a).

Daszkiewicz M., Gulay L.D., Lychmanyuk O.S., Pietraszko A. Crystal structures of the R$_3$Ag$_{1-\delta}$TSe$_7$ (R = La–Nd, Sm, Gd–Dy, δ = 0–0.30; T = Ge, Si) compounds, *J. Alloys Compd.*, **467**, 168–172 (2009b).

Daszkiewicz M., Gulay L.D., Pietraszko A., Shemet V.Ya. Crystal structures of the La$_3$AgSnSe$_7$ and R$_3$Ag$_{1-\delta}$SnS$_7$ (R = La, Ce; δ = 0.18–0.19) compounds, *J. Solid State Chem.*, **180**(7), 2053–2060 (2007).

Daszkiewicz M., Gulay L.D. Pressure induced silver ion displacement in La$_3$Ag$_{0.82}$SnS$_7$, *Mater. Res. Bull.*, **47**(2), 497–499 (2012).

Daszkiewicz M., Gulay L.D., Shemet V.Ya., Pietraszko A. Dy$_8$SnS$_{13.61}$O$_{0.39}$ from single-crystal data, *Acta Crystallogr.*, **E64**(1), i2 (2008b).

Daszkiewicz M., Marchuk O.V., Gulay L.D., Kaczorowski D. Crystal structure and magnetic properties of R$_3$Mn$_{0.5}$GeS$_7$ (R = Y, Ce, Pr, Nd, Sm, Gd, Tb, Dy, Ho and Er), *J. Alloys Compd.*, **610**, 258–263 (2014a).

Daszkiewicz M., Marchuk O.V., Gulay L.D., Kaczorowski D. Crystal structures and magnetic properties of R$_2$PbSi$_2$S$_8$ (R = Y, Ce, Pr, Nd, Sm, Gd, Tb, Dy, Ho), R$_2$PbSi$_2$Se$_8$ (R = La, Ce, Pr, Nd, Sm, Gd) and R$_2$PbGe$_2$S$_8$ (R = Ce, Pr) compounds, *J. Alloys Compd.*, **519**, 85–91 (2012).

Daszkiewicz M., Pashynska Yu.O., Marchuk O.V., Gulay L.D., Kaczorowski D. Crystal structure and magnetic properties of R$_3$Co$_{0.5}$GeS$_7$ (R = Y, La, Ce, Pr, Nd, Sm, Gd, Tb, Dy, Ho, Er and Tm) and R$_3$Ni$_{0.5}$GeS$_7$ (R = Y, Ce, Sm, Gd, Tb, Dy, Ho, Er and Tm), *J. Alloys Compd.*, **647**, 445–455 (2015).

Daszkiewicz M., Pashynska Yu.O., Marchuk O.V., Gulay L.D., Kaczorowski D. Crystal structure and magnetic properties of $R_3Fe_{0.5}GeS_7$ (R = Y, La, Ce, Pr, Sm, Gd, Tb, Dy, Ho, Er and Tm), *J. Alloys Compd.*, **616**, 243–249 (2014b).

Daszkiewicz M., Smitiukh O.V., Marchuk O.V., Gulay L.D. The crystal structure of $Er_{2.34}La_{0.66}Ge_{1.28}S_7$ and the $La_xR_yGe_3S_{12}$ phases (R – Tb, Dy, Ho and Er), *J. Alloys Compd.*, **738**, 263–269 (2018).

Datsenko A.M., Razvazhnoy E.M., Chashchin V.A. Investigation of interaction in the Bi_2Te_3–(Cd,Pb)(S,Se,Te) systems [in Russian], *Tr. Mosk. khim.-tehnol. in-t*, (120), 32–34 (1981).

Davaasuren B., Emwas A.-H., Rothenberger A. MAu_2GeS_4-chalcogel (M = Co, Ni): heterogeneous intra- and intermolecular hydroamination catalysts, *Inorg. Chem.*, **56**(16), 9609–9616 (2017).

Davarashvili O.I., Kuznetsov V.V., Selin A.A., Shotov A.P. Calculation of equilibrium compositions of liquid and solid phases in the Pb–Sn–Se–Te system [in Russian], *Zhurn. neorgan. khimii*, **37**(6), 1362–1366 (1992).

Davies C.G., Donaldson J.D. Basic tin(II) sulphates, *J. Chem. Soc. A: Inorg., Phys., Theor.*, 1790–1793 (1967).

Davies C.G., Donaldson J.D., Laughlin D.R., Howie R.A., Beddoes R. Crystal structure of tritin(II) dihydroxide oxide sulphate, *J. Chem. Soc., Dalton Trans.*, (21), 2241–2244 (1975).

Davydyuk G.Ye., Olekseyuk I.D., Parasyuk O.V., Piskach L.V., Semenyuk S.A., Kevshyn A.G., Pekhnyo V.I. Obtaining and investigation of the physical properties of the single crystals of the Cu_2CdGeS_4 and Cu_2CdSnS_4 compounds [in Ukrainian], *Nauk. Visnyk Volyns'k. Derzh. Univ. im. Lesi Ukrainky, Ser. Khim. nauky*, **1**, 25–29 (2005).

Davydyuk G.Ye., Parasyuk O.V., Romanyuk Ya E., Semenyuk S.A., Zaremba V.I., Piskach L.V., Kozioł J.J., Halka V.O. Single crystal growth and physical properties of the Cu_2CdGeS_4 compound, *J. Alloys Compd.*, **339**(1–2), 40–45 (2002).

Delgado G.E., Sierralta N., Quintero M., Quintero E., Moreno E., Flores-Cruz J.A., Rincón C. Synthesis, structural characterization and differential thermal analysis of the quaternary compound Ag_2MnSnS_4, *Rev. Mex. Fís.*, **64**(3), 216–221 (2018).

Deng B., Yao J., Ibers J.A. Dicerium orthosilicate selenide and dicerium orthosilicate telluride, $Ce_2(SiO_4)Q$ (Q = Se or Te), *Acta Crystallogr.*, **C60**(11), i110–i112 (2004).

Derakhshan S., Assoud A., Soheilnia N., Kleinke H. Electronic structure and thermoelectric properties of the thioantimonate $FePb_4Sb_6S_{14}$, *J. Alloys Compd.*, **390**(1–2) 51–54 (2005).

Deudon C., Meerschaut A., Rouxel J. Structure determination of lanthanum seleno-silicate, $La_4Se_3Si_2O_7$, *J. Solid State Chem.*, **104**(2), 282–288 (1993).

Devi M.S., Vidyasagar K. First examples of sulfides in the quaternary A/Cd/Sn/S (A = Li, Na) systems: molten flux synthesis and single crystal X-ray structures of Li_2CdSnS_4, Na_2CdSnS_4 and $Na_6CdSn_4S_{12}$, *J. Chem. Soc., Dalton Trans.*, (9), 2092–2096 (2002).

Devlin K.P., Glaid A.J., Brant J.A., Zhang J.-H., Srnec M.N., Clark D.J., Kim Y.S., Jang J.I., Daley K.R., Moreau M.A., Madura J.D., Aitken J.A. Polymorphism and second harmonic generation in a novel diamond-like semiconductor: Li_2MnSnS_4, *J. Solid State Chem.*, **231**, 256–266 (2015).

Dhingra S.S., Haushalter R.C. One-dimensional inorganic polymers: synthesis and structural characterization of the main-group metal polymers $K_2HgSnTe_4$, $(Et_4N)_2HgSnTe_4$, $(Ph_4P)GeInTe_4$, and $RbInTe_2$, *Chem. Mater.*, **6**(12), 2376–2381 (1994).

Ding N., Chung D.-Y., Kanatzidis M.G. $K_6Cd_4Sn_3Se_{13}$: a polar open-framework compound based on the partially destroyed supertetrahedral $[Cd_4Sn_4Se_{17}]^{10-}$ cluster, *Chem. Comm.*, (10), 1170–1171 (2004).

Ding N., Kanatzidis M.G. Acid-induced conversions in open-framework semiconductors: from $[Cd_4Sn_3Se_{13}]^{6-}$ to $[Cd_{15}Sn_{12}Se_{46}]^{14-}$, a remarkable disassembly/reassembly process, *Angew. Chem.*, **118**(9), 1425–1429 (2006a).

Ding N., Kanatzidis M.G. Acid-induced conversions in open-framework semiconductors: from $[Cd_4Sn_3Se_{13}]^{6-}$ to $[Cd_{15}Sn_{12}Se_{46}]^{14-}$, a remarkable disassembly/reassembly process, *Angew. Chem. Int. Ed.*, **45**(9), 1397–1401 (2006b).

Dolgikh V.A. New chalcogengalogenides of the $M_2SbS_2I_3$ type [in Russian], *Izv. AN SSSR. Neorgan. mater.*, **21**(7), 1211–1214 (1985a).

Dolgikh V.A. Obtaining of single crystals and dielectric properties of the $Sn_2SbS_2I_3$ and $Pb_2SbS_2I_3$ [in Russian], *Izv. AN SSSR. Neorgan. mater.*, **21**(7), 1215–1218 (1985b).

Dong Y., Do Y., Yun H. Synthesis and crystal structures of the first quaternary tantalum thiogermanates, $ATaGeS_5$ (A = K, Rb, Cs), *Z. anorg. und allg. Chem.*, **635**(15), 2676–2681 (2009).

Dong Y., Khabibullin A.R., Wei K., Salvador J.R., Nolas G.S., Woods L.M. Bournonite $PbCuSbS_3$: stereochemically active lone-pair electrons that induce low thermal conductivity, *ChemPhysChem.*, **16**(15), 3264–3270 (2015a).

Dong Y., Wang H., Nolas G.S. Synthesis and thermoelectric properties of Cu excess $Cu_2ZnSnSe_4$, *Phys. Stat. Sol. (RRL)*, **8**(1), 61–64 (2014).

Dong Y., Wojtas L., Martin J., Nolas G.S. Synthesis, crystal structure, and transport properties of quaternary tetrahedral chalcogenides, *J. Mater. Chem. C*, **3**(40), 10436–10441 (2015b).

Dornberger-Schiff K., Höhne E. Die Kristallstruktur des Betechtinit $Pb_2(Cu,Fe)_{21}S_{15}$, *Acta Crystallogr.*, **12**(9), 646–651 (1959).

Douglass R.M., Murphy M.J., Pabst A. Geocronite, *Am. Mineralog.*, **39**(11–12), 908–928 (1954).

Doussier C., Moëlo Y., Meerschaut A., Léone P., Guillot-Deudon C. Crystals structure of the new compound $Pb_{3+x}Sb_{3-x}S_{7-x}Cl_{1+x}$ (x ~ 0.45). The homologous series $Pb_{(2+2N)}(Sb,Pb)_{(2+2N)}S_{(2+2N)}(S,Cl)_{(4+2N)}Cl_N$ and its polychalcogenide derivatives (N = 1–3), *J. Solid State Chem.*, **181**(4), 920–934 (2008).

Dou Y., Chen Y., Li Z., Iyer A.K., Kang B., Yin W., Yao J., Mar A. $SrCdGeS_4$ and $SrCdGeSe_4$: promising infrared nonlinear optical materials with congruent-melting behavior, *Cryst. Growth Des.*, **19**(2), 1206–1214 (2019).

Doverspike K., Dwight K., Wold A. Preparation and characterization of $Cu_2ZnGeS_{4-y}Se_y$, *Chem. Mater.*, **2**(2), 194–197 (1990).

Dovletmuradov Ch., Dovletov K., Krzhivitskaya S.N., Mamaev S., Allanazarov A., Ashirov A. Obtaining and investigation of physico-chemical and electrical properties of $CdSnAs_2$–$A^{II}B^{VI}$ (2CdSe, 2CdTe) solid solutions [in Russian], In: *Troin. poluprovodniki $A^{II}B^{IV}C_2^V$ i $A^{II}B_2^{III}C_4^{VI}$*, Kishinev: Shtiintsa Publish., 96–98 (1972).

Dovletmuradov Ch., Krzhivitskaya S.N., Dovletov K., Mamaev S., Allanazarov A., Ashirov A. Some properties of $CdSnAs_2$–$A^{II}B^{VI}$ (CdSe, CdTe) solid solutions [in Russian], *Izv. AN TurkmSSR. Ser. fiz.-tehn., khim. i geol. nauk*, (5), 111–114 (1971).

Dovletov K., Mkrtchian S.A., Zhukov E.G., Melikdzhanian A.G. Electrophysical properties of $(CdSe)_x[Cu_2Ge(Sn)Se_3]_{1-x}$ solid solutions [in Russian], *Izv. AN SSSR. Neorgan. materialy*, **23**(5), 857–860 (1987).

Duan R.-H., Li R.-A., Liu P.-F., Lin H., Wang Y., Wu L.-M. Modifying disordered sites with rational cations to regulate bandgaps and second-harmonic-generation responses markedly: $Ba_6Li_2ZnSn_4S_{16}$ vs. $Ba_6Ag_2ZnSn_4S_{16}$ vs. $Ba_6Li_{2.67}Sn_{4.33}S_{16}$, *Cryst. Growth Des.*, **18**(9), 5609–5616 (2018).

Duan R.-H., Liu P.-F., Lin H., Huangfu S.-X., Wu L.-M. Syntheses and characterization of three new sulfides with large band gaps: acentric $Ba_4Ga_4SnS_{12}$, centric $Ba_{12}Sn_4S_{23}$ and $Ba_7Sn_3S_{13}$, *Dalton Trans.*, **46**(43), 14771–14778 (2017).

Duan R.-H., Yu J.-S., Lin H., Zheng Y.-J., Zhao H.-J., Huang-Fu S.-X., Khan M.A., Chen L., Wu L.-M. $Pb_5Ga_6ZnS_{15}$: a noncentrosymmetric framework with chains of T2-supertetrahedra, *Dalton Trans.*, **45**(31), 12288–12291 (2016).

Dubrovin I.V., Budionnaya L.D., Mizetskaya I.B., Sharkina E.V. Interaction along the SnTe–CdS section in the SnTe+CdS ↔ SnS+CdTe ternary mutual system [in Russian], *Izv. AN SSSR. Neorgan. mater.*, **21**(11), 1873–1878 (1985).

Dubrovin I.V., Budionnaya L.D., Mizetskaya I.B., Sharkina E.V. Phase diagram of the SnTe–CdSe section in the SnTe+CdSe ↔ SnSe+CdTe ternary mutual system [in Russian], *Izv. AN SSSR. Neorgan. mater.*, **22**(4), 590–595 (1986).

Dubrovin I.V., Budionnaya L.D., Mizetskaya I.B., Sharkina E.V. Phase equilibria in the SnTe–ZnS system [in Russian], *Izv. AN SSSR. Neorgan. mater.*, **19**(11), 1816–1819 (1983).

Dubrovin I.V., Budionnaya L.D., Mizetskaya I.B., Sharkina E.V. The SnTe–ZnSe system [in Russian], *Izv. AN SSSR. Neorgan. mater.*, **20**(4), 571–573 (1984).

Dubrovin I.V., Budionnaya L.D., Oleynik N.D., Sharkina E.V. Interaction of Cd_4SiS_6 with ZnS and SiS_2 along the section Cd_4SiS_6–$(ZnS)_{0.8}(SiS_2)_{0.2}$ [in Russian], *Izv. AN SSSR. Neorgan. mater.*, **25**(5), 722–725 (1989).

Dubrovin I.V., Budennaya L.D., Sharkina E.V. Interaction of the Cd_4GeS_6 with ZnS and GeS_2 along the Cd_4GeS_6–$(ZnS)_{0.8}(GeS)_{0.2}$ section in the CdS–ZnS–GeS_2 system [in Russian], *Izv. AN SSSR. Neorgan. mater.*, **27**(2), 244–247 (1991).

Dudchak I.V., Piskach L.V. Phase equilibria in the Ag_8SnS_6–ZnS section [in Ukrainian], *Nauk. Visnyk Volyns'k. Derzh. Univ. im. Lesi Ukrainky. Ser. Khim. nauky*, (6), 59–60 (2001a).

Dudchak I.V., Piskach L.V. Phase equilibria in the Cu_2SnSe_3–$SnSe_2$–ZnSe system, *J. Alloys Compd.*, **351**, 145–150 (2003).

Dudchak I., Piskach L. Phase equilibria in the Cu_2SnSe_3–ZnSe system [in Ukrainian], *Visnyk L'viv. Univ. Ser. khim.*, (40), 73–76 (2001b).

Dumail R., Ribes M., Philippot E. Sur l'étude d'une solution solide entre orthothiogermanate et orthothiosilicate de baryum $Ba_2(Si_{1-x}Ge_x)S_4$ ($0 \leq x \leq 1$). Structure cristalline Ba_2SiS_4, *C.r. Acad. sci., Sér. C*, **272**(3), 303–306 (1971).

Dunn P.J., Cabri L.J., Chao G.Y., Fleischer M., Francis C.A., Grice J.D., Jambor J.L., Pabst A. New mineral names, *Am. Mineralog.*, **69**(3–4), 406–412 (1984a).

Dunn P.J., Chao G.Y., Fleischer M., Ferraiolo J.A., Langley R.H., Pabst A., Zilczer J.A. New mineral names, *Am. Mineralog.*, **70**(1–2), 214–221 (1985a).

Dunn P.J., Chao G.Y., Grice J.D., Ferraiolo J.A., Fleischer M., Pabst A., Zilczer J.A. New mineral names, *Am. Mineralog.*, **69**(5–6), 565–569 (1984b).

Dunn P.J., Fleischer M., Chao G.Y., Cabri L.J., Mandarino J.A. New mineral names, *Am. Mineralog.*, **68**(11–12), 1248–1252 (1983).

Dunn P.J., Fleischer M. New mineral names, *Am. Mineralog.*, **68**(5–6), 642–645 (1983).

Dunn P.J., Gobel V., Grice J.D., Puziewicz J., Shigley J.E., Vanko D.A., Zilczer J. New mineral names, *Am. Mineralog.*, **70**(3–4), 436–441 (1985b).

Dunn P.J., Rouse R.S. Sundiusite, a new lead sulfate oxychloride from Lengban, Sweden, *Am. Mineralog.*, **65**(5–6), Pt. 1, 506–508 (1980).

Dutrizac J.E. The Fe_{1-x}S–PbS–ZnS phase system, *Can. J. Chem.*, **58**(7), 739–743 (1980).

Edenharter A., Nowacki W. Verfeinerung der Kristallstruktur von Bournonit $[(SbS_3)_2|Cu^{IV}_2Pb^{VII}Pb^{VIII}]$ und von Seligmannit $[(AsS_3)_2|Cu^{IV}_2Pb^{VII}Pb^{VIII}]$, *Z. Kristallog.*, **131**(1–6), 397–417 (1970).

Edwards R., Gillard R.D., Williams P.A. The stabilities of secondary tin minerals. Part 2: the hydrolysis of tin(II) sulphate and the stability of $Sn_3O(OH)_2SO_4$, *Mineralog. Mag.*, **60**(3), 427–432 (1996).

Effenberger H. Contribution to the stereochemistry of copper. The transition from a tetragonal pyramidal to a trigonal bipyramidal $Cu(II)O_5$ coordination figure with a structure determination of $PbCu_2(SeO_3)_3$, *J. Solid State Chem.*, **73**(1), 118–126 (1988).

Effenberger H. Darstellung und Kristallstruktur von $Pb_2(NO_2)(NO_3)(SeO_3)$, *Monatsh. Chem.*, **118**(2), 211–216 (1987).

Eibschütz M., Herman E., Shtrikman S. Determination of cation valencies in $Cu_2^{57}Fe^{119}SnS_4$ by Mössbauer effect and magnetic susceptibility measurements, *J. Phys. Chem. Solids*, **28**(9), 1633–1636 (1967).

Eriç H., Timuçin M. Activities in Cu_2S–FeS–PbS melts at 1200°C, *Metall. Trans. B*, **12**(3), 493–500 (1981).

Eulenberger G. Darstellung und Kristallstruktur des Dithallium(I) blei(II)tetrathiogermanats(IV) Tl_2PbGeS_4, *Z. Naturforsch.*, **35B**(3), 335–339 (1980).

Euler R., Hellner E. Über komplex zusammengesetzte sulfidische Erze. VI. Zur Kristallstruktur des Meneghinits, $CuPb_{13}Sb_7S_{24}$, *Z. Kristallogr.*, **113**(1–6), 345–372 (1960).

Evenson IV C.R., Dorhout P.K. Synthesis and characterization of four new europium group XIV chalcogenides: K_2EuTSe_5 and $KEuTS_4$ (T = Si, Ge), *Inorg. Chem.*, **40**(10), 2409–2414 (2001).

Evstigneeva T.L., Kabalov Yu.K. Crystal structure of the Cu_2FeSnS_4 cubic modification [in Russian], *Kristallografiya*, **46**(3), 418–422 (2001).

Fahrnbauer F., Urban P., Welzmiller S., Schröder T., Rosenthal T., Oeckler O. $(GeTe)_nSbInTe_3$ ($n \leq 3$) – Element distribution and thermal behavior, *J. Solid State Chem.*, **208**, 20–26 (2013).

Falmbigl M., Hay Z., Ditto J., Mitchson G., Johnson D.C. Modifying a charge density wave transition by modulation doping: ferecrystalline compounds $([Sn_{1-x}Bi_xSe]_{1.15})_1(VSe_2)_1$ with $0 \leq x \leq 0.66$, *J. Mater. Chem. C*, **3**(47), 12308–12315 (2015).

Fard Z.H., Kanatzidis M.G. Phase-change materials exhibiting tristability: interconverting forms of crystalline α-, β-, and glassy $K_2ZnSn_3S_8$, *Inorg. Chem.*, **51**(15), 7963–7965 (2012).

Fedorchuk A.O., Parasyuk O.V., Cherniushok O., Andriyevsky B., Myronchuk G.L., Khyzhun O.Y., Lakshminarayana G., Jedryka J., Kityk I.V., ElNaggar A.M., Albassam A.A., Piasecki M., $PbGa_2GeS_6$ crystal as a novel nonlinear optical material: band structure aspects, *J. Alloys Compd.*, **740**, 294–304 (2018).

Feltz A., Burckhardt W., Senf L. New vitreous semiconductors, In: *Tr. 6th Mezhdunar. konf. po amorf. i zhidk. poluprovodn: Struktura i sv-va nekristal. poluprovodn.* Leningrad, Nov. 18-24, 1975, L. : Nauka Publish. House, 24–31 (1976).

Feltz A., Pfaff G. Über Glasbildung und Eigenschaften in Chalkogenidsystemen; 1 Die Verbindungen $Na_6Si_2S_6$ und $Na_6Si_2Se_6$ und deren Bromspaltungsprodukte Na_3SiS_3Br und Na_3SiSe_3Br, *Z. Chem.*, **23**(2), 68 (1983).

Feltz A., Pfaff G. Über Glasbildung und Eigenschaften von Chalkogenidsystemen. XXI. Na_3GeS_3Br und Na_3GeSe_3Br und die amorphen Verbindungen Ge_2S_3 und Ge_2Se_3, *Z. anorg. und allg. Chem.*, **467**(1), 211–217 (1980).

Feng K., Wang W., He R., Kang L., Yin W., Lin Z., Yao J., Shi Y., Wu Y. $K_2FeGe_3Se_8$: a new antiferromagnetic iron selenide, *Inorg. Chem.*, **52**(4), 2022–2028 (2013).

Feng K., Zhang X., Yin W., Shi Y., Yao J., Wu Y. New quaternary rare-earth chalcogenides $BaLnSn_2Q_6$ (Ln = Ce, Pr, Nd, Q = S; Ln = Ce, Q = Se): synthesis, structure, and magnetic properties, *Inorg. Chem.*, **53**(4), 2248–2253 (2014).

Filep M., Sabov M. Polyhedration of the Pb–S–Se–Te system, *Chem. Met. Alloys*, **10**(3–4), 120–125 (2017a).

Filep M.Y., Barchiy I.E., Sabov M.Yu. Interaction of the components in the $Tl_2S+PbSe \leftrightarrow Tl_2Se+PbS$ and $Tl_2S+PbTe \leftrightarrow Tl_2Te+PbS$ ternary mutual systems [in Ukrainian], *Nauk. Visn. Uzhgorod. Univ., Ser. Khim.*, **1**(27), 22–24 (2012a).

Filep M.Y., Makakhovs'ka T.O., Pogodin A.I., Sabov M.Yu. Determination of the quasibinary sections in the $Tl_2Se+SnTe \leftrightarrow Tl_2Te+SnSe$ ternary mutual system [in Ukrainian], *Nauk. Visn. Uzhgorod. Univ., Ser. Khim.*, [*1*(41)], 43–48 (2019).

Filep M.Y., Sabov M.Yu., Barchii I.E., Solomon A.M. Interaction in the $Tl_2S–SnS–PbS$ quasi-ternary system, *Russ. J. Inorg. Chem.*, **59**(9), 1026–1029 (2014).

Filep M.Y., Sabov M.Yu., Barchiy I.E. Physico-chemical interaction in the $Tl_2Se–SnSe–PbSe$ quasiternary system, *Chem. Met. Alloys*, **5**(3–4), 118–122 (2012b).

Filep M.J., Sabov M.Yu., Barchiy I.E., Plucinski K.J., Solomon A.M. Interactions in the ternary reciprocal system $Tl_2S + SnTe \leftrightarrow Tl_2Te + SnS$, *Chem. Met. Alloys*, **6**(3–4), 125–129 (2013).

Filep M.Y., Sabov M.Yu., Barchiy I.E., Solomon A.M. Establishment of the quasi-binary sections of the concentration plane $Tl_2S–Tl_2Se–SnS–SnSe$ [in Ukrainian], *Nauk. Visn. Uzhgorod. Univ., Ser. Khim.*, [*1*(25)], 7–10 (2011).

Filep M.Y., Sabov M.Yu., Makakhovs'ka T.O., Solomon A.M. Physico-chemical interaction in the $Tl_4PbS_3–Tl_4PbSe_3$ and $Tl_4PbS_3–Tl_4PbTe_3$ systems [in Ukrainian], *Nauk. Visn. Uzhgorod. Univ., Ser. Khim.*, [*1*(35)], 34–36 (2016).

Filep M.Y., Sabov M.Yu. On the possibility of solid solution formation in the systems based on Tl_4XY_3 (X – Sn, Pb; Y – S, Se, Te) type mixed chalcogenides [in Ukrainian], *Nauk. Visn. Uzhgorod. Univ., Ser. Khim.*, (20), 102–104 (2008).

Filep M.Y., Sabov M.Yu. The $PbS–PbSe–Tl_4PbSe_3$ quasiternary system [in Ukrainian], *Nauk. Visn. Uzhgorod. Univ., Ser. Khim.*, **1**(39), 17–20 (2018).

Filep M.Y., Sabov M.Yu. The $Tl_2S–Tl_2Se–Tl_4PbSe_3$ quasiternary system [in Ukrainian], *Nauk. Visn. Uzhgorod. Univ., Ser. Khim.*, **1**(37), 14–16 (2017b).

Filimonov D.S., Kesler Ya A., Pokholok K.V. Spinel phases in the $FeS–Cr_2S_3–SnS_2$ system [in Russian], *Neorgan. mater.*, **32**(8), 930–936 (1996).

Filonenko V.V., Nechiporuk B.D., Novoseletski N.E., Yuhimchuk V.A., Lavorik Yu.F. Obtaining and some properties of Cu_2CdGeS_4 crystals [in Russian], *Izv. AN SSSR. Neorgan. mater.*, **27**(6), 1166–1168 (1991).

Filyuk T.O., Olekseyuk I.D., Gulay L.D., Mazurets I.I. Phase equilibria in the $R_2X_3–Ga_2X_3–PbX$ (R = Y, La; X = S, Se, Te) at 770 K [in Ukrainian], *Nauk. Visnyk Volyns'k. Nats. Univ. im. Lesi Ukrainky. Ser. Khim. nauky*, (16), 51–56 (2008).

Filyuk T.O., Olekseyuk I.D., Mazurets I.I. Isothermal section of the $HgS–Ga_2S_3–PbS$ system at 670 K [in Ukrainian], *Nauk. Visnyk Volyns'k. Derzh. Univ. im. Lesi Ukrainky. Ser. Khim. nauky*, (13), 12–24 (2007).

Fleet M.E. The crystal structure of parkerite ($Ni_3Bi_2S_2$), *Am. Mineralog.*, **58**(5–6), 435–439 (1973).

Fleischer M., Cabri L.J., Chao G.Y., Mandarino J.A., Pabst A. New mineral names, *Am. Mineralog.*, **67**(5–6), 621–624 (1982a).

Fleischer M., Cabri L.J., Chao G.Y., Mandarino J.A., Pabst A. New mineral names, *Am. Mineralog.*, **67**(9–10), 1074–1082 (1982b).

Fleischer M., Cabri L.J., Chao G.Y., Pabst A. New mineral names, *Am. Mineralog.*, **65**(7–8), 808–814 (1980).

Fleischer M., Chao G.Y., Cabri L.J. New mineral names, *Am. Mineralog.*, **60**(7–8), 736–739 (1975).

Fleischer M., Chao G.Y., Francis C.A. New mineral names, *Am. Mineralog.*, **66**(7–8), 878–879 (1981).

Fleischer M., Chao G.Y., Mandarino J.A. New mineral names, *Am. Mineralog.*, **61**(3–4), 338–341 (1976).

Fleischer M., Chao G.Y., Pabst A. New mineral names, *Am. Mineralog.*, **64**(1–2), 241–245 (1979a).

Fleischer M., Jambor J. New mineral names, *Am. Mineralog.*, **62**(3–4), 395–397 (1977).

Fleischer M., Mandarino J.A., Chao G.Y. New mineral names, *Am. Mineralog.*, **64**(5–6), Pt. 1, 652–659 (1979b).

Fleischer M., Mandarino J.A., Kato A. New mineral names, *Am. Mineralog.*, **53**(7–8), 1421–1427 (1968).

Fleischer M., Mandarino J.A. New mineral names, *Am. Mineralog.*, **59**(3–4), 381–384 (1974).

Fleischer M. New mineral names, *Am. Mineralog.*, **41**(3–4), 370–372 (1956a).

Fleischer M. New mineral names, *Am. Mineralog.*, **41**(9–10), 814–816 (1956b).

Fleischer M. New mineral names, *Am. Mineralog.*, **50**(5–6), 805–812 (1965).

Fleischer M. New mineral names, *Am. Mineralog.*, **54**(7–8), 1218–1223 (1969).

Fleischer M. New mineral names, *Am. Mineralog.*, **55**(3–4), Pt. 1, 533–535 (1970a).

Fleischer M. New mineral names, *Am. Mineralog.*, **55**(7–8), 1444–1449 (1970b).

Fleischer M. New mineral names, *Am. Mineralog.*, **56**(3–4), 631–640 (1971).

Fleischer M., Pabst A. New mineral names, *Am. Mineralog.*, **66**(3–4), 436–439 (1981).

Fleischmann H., Folberth O.G., Pfister H. Halbleitende Mischkristalle vom Type ($A^I_{x/2}B^{VI}_{(1-x)}C^V_{x/2}$)$D^{VI}$, *Z. Naturforsch.*, **14A**(11), 999–1000 (1959).

Fleischmann H., Luy H., Rupprecht J. Neuere Untersuchungen an halbleitenden IV VI–I VI_2-Mischkristallen, *Z. Naturforsch.*, **18A**(5), 646–649 (1963).

Förster H.-J., Bindi L., Grundmann G., Stanley C.J. Cerromojonite, $CuPbBiSe_3$, from El Dragón (Bolivia): a new member of the bournonite group, *Minerals*, **8**(10), 420 (2018a).

Förster H.-J., Bindi L., Grundmann G., Stanley C.J. Cerromojonite, IMA 2018-040. CNMNC Newsletter No. 44, August 2018, page 882, *Eur. J. Mineral.*, **30**(4), 877–882 (2018b).

Förster H.-J., Bindi L., Grundmann G., Stanley C.J. Cerromojonite, IMA 2018-040. CNMNC Newsletter No. 44, August 2018, page 1021, *Mineralog. Mag.*, **82**(4), 1015–1021 (2018c).

Förster H.-J., Cooper M.A., Roberts A.C., Stanley C.J., Criddle A.J., Hawthorne F.C., Laflamme J.H.G. Schlemaite, $(Cu,\square)_6(Pb,Bi)Se_4$, a new mineral species from Niederschlema–Alberoda, Erzgebirge, Germany: description and crystal structure, *Canad. Mineralog.*, **41**(6), 1433–1444 (2003).

Frank D., Pfitzner A. $Ag_8SiS_4Te_2$, a new thiosilicate telluride, *Z. anorg. und allg. Chem.*, **638**(10), 1569 (2012).

Friedrich D., Byun H.R., Hao S., Patel S., Wolverton C., Jang J.I., Kanatzidis M.G. Layered and cubic semiconductors $AGaM'Q_4$ (A^+ = K^+, Rb^+, Cs^+, Tl^+; M'^{4+} = Ge^{4+}, Sn^{4+}; Q^{2-} = S^{2-}, Se^{2-}) and high third-harmonic generation, *J. Am. Chem. Soc.*, **142**(41), 17730–17742 (2020).

Friedrich D., Greil S., Block T., Heletta L., Pöttgen R., Pfitzner A. Synthesis and characterization of Ag_2MnSnS_4, a new diamond-like semiconductor, *Z. anorg. und allg. Chem.*, **644**(24), 1707–1714 (2018).

Friedrich D., Hao S., Patel S., Wolverton C., Kanatzidis M.G. Vast structural and polymorphic varieties of semiconductors $AMM'Q_4$ (A = K, Rb, Cs, Tl; M = Ga, In; M' = Ge, Sn; Q = S, Se), *Chem. Mater.*, **33**(16), 6572–6583 (2021).

Fröhlich R. $Pb_8[O_2(SO_4)(Si_4O_{13})]$, a new tetrasilicate, *Acta Crystallogr.*, **A40**, Suppl., C-224 (1984).

Frondel C. Unit cell and space group of vrbaite ($Tl(As,Sb)_3S_5$), seligmannite ($CuPbAsS_3$) and samsonite ($Ag_4MnSb_2S_6$), *Mineralog. Mag.*, **26**(1), 25–28 (1941).

Frydrych R. Darstellung und Eigenschaften von Blei(IV)-selenit, $Pb(SeO_3)_2$, und der Alkali-triselenitoplumbate(IV), $M_2[Pb(SeO_3)_3]$, *Z. anorg. und allg. Chem.*, **427**(3), 260–264 (1976).

Fuhrmann J., Pickardt J. Synthesis and structure of magnesium iron thiosilicates ($Mg_{1-x}Fe_x)_2SiS_4$, *Acta Crystallogr.*, **C46**(11), 1996–1998 (1990a).

Fuhrmann J., Pickardt J. Züchtung und röntgenographische Untersuchung von Einkristallen der Thiosilicate $(Mn,Mg)_2SiS_4$, $(Mn,Fe)_2SiS_4$ und $(Mg,Fe)_2SiS_4$, *Z. anorg. und allg. Chem.*, **586**(1), 73–78 (1990b).

Gaines R.V. Moctezumite, a new lead uranyl tellurite, *Am. Mineralog.*, **50**(9), 1158–1163 (1965).

Garbato L., Geddo-Lehmann A., Ledda F. Growth and structural properties of quaternary copper thiostannates, *J. Cryst. Growth*, **114**(3), 299–306 (1991).

Garg G., Bobev S., Ganguli A.K. Single crystal structure and electrical properties of $Cu_8Ni_4Sn_{12}S_{32}$, *J. Alloys Compd.*, **327**(1–2), 113–115 (2001a).

Garg G., Bobev S., Ganguli A.K. Single crystal structures of two new cation-deficient thiospinels: $Cu_{7.38(11)}Mn_4Sn_{12}S_{32}$ and $Cu_{7.07(6)}Ni_4Sn_{12}S_{32}$, *Solid State Ionics*, **146**(1–2), 195–198 (2002).

Garg G., Bobev S., Roy A., Ghose J., Das D., Ganguli A.K. Single crystal structure and Mössbauer studies of a new cation-deficient thiospinel: $Cu_{5.47}Fe_{2.9}Sn_{13.1}S_{32}$, *Mater. Res. Bull.*, **36**(13–14), 2429–2435 (2001b).

Garg G., Gupta S., Maddanimath T., Gascoin F., Ganguli A.K. Single crystal structure, electrical and electrochemical properties of the quaternary thiospinel: $Ag_2FeSn_3S_8$, *Solid State Ionics*, **164**(3–4), 205–209 (2003a).

Garg G., Gupta S., Ramanujachary K.V., Lofland S.E., Ganguli A.K. Investigation of cation-deficient quaternary thiospinels: single crystal study of $Ag_{1.4}Cr_{1.47}Sn_{2.53}S_8$, *J. Alloys Compd.*, **390**(1–2), 46–50 (2005).

Garg G., Ramanujachary K.V., Lofland S.E., Lobanov M.V., Greenblatt M., Maddanimath T., Vijayamohanan K., Ganguli A.K. Crystal structure, magnetic and electrochemical properties of a quaternary thiospinel: $Ag_2MnSn_3S_8$, *J. Solid State Chem.*, **174**(1), 229–232 (2003b).

Gasymov V.A., Aliev O.M., Ismailova G.N., Dzhalaliddinov F.F. Synthesis and X-ray diffraction ibvestigation of $PbEuBi_2S_5$ and $PbYbBi_2S_5$ – analogues of cosalite mineral [in Russian], *Azerb. khim. zhurn.*, (*1*), 116–117 (2007).

Gatta G.D., Cámara F., Tait K.T., Belakovskiy D. New mineral names, *Am. Mineralog.*, **97**(11–12), 2064–2072 (2012).

Gauthier G., Guillen F., Jobic S., Deniard P., Macaudière P., Fouassier C., Brec R. Synthesis, structure and electronic properties of the cerium and potassium thiosilicate: $KCeSiS_4$, *C.r. Acad. sci. Ser. 2. Fasc. c*, **2**(11–13), 611–616 (1999).

Gauthier G., Kawasaki S., Jobic S., Macaudiere P., Brec R., Rouxel J. Characterization of $Ce_3(SiS_4)_2I$, a compound with a new structure type, *J. Mater. Chem.*, **8**(1), 179–186 (1998a).

Gauthier G., Kawasaki S., Jobic S., Macaudiere P., Brec R., Rouxel J. Synthesis and structure of the first cerium iodothiosilicate: $Ce_3(SiS_4)_2I$, *J. Alloys Compd.*, **275–277**, 46–49 (1998b).

Geng L. $Ba_2Sb_4GeS_{10}$. *Acta Crystallogr.*, **E69**(5), i24 (2012).

Genkin A.D., Safonov Y.G., Vasudev V.N., Rao B.K., Boronikhin V.A., Vyalsov L.N., Gorshkov A.I., Mokhov A.V. Kolarite $PbTeCl_2$ and radhakrishnaite $PbTe_3(Cl,S)_2$, new mineral species from the Kolar Gold Deposit, India, *Canad. Mineralog.*, **23**(3), 501–506 (1985).

Gilde E.R., Gymez D.C., Rivera A.V., Lopez-Rivera A. X-ray crystal structure of the quaternary semiconductor compound: $AgGaSn()Se_4$, *Progr. Cryst. Growth Charact.*, **10**(1–4), 217–223 (1985).

Gitzendanner R.L., Spencer C.M., DiSalvo F.J., Pell M.A., Ibers J.A. Synthesis and structure of a new quaternary rare-earth sulfide, $La_6MgGe_2S_{14}$, and the related compound $La_6MgSi_2S_{14}$, *J. Solid State Chem.*, **131**(2), 399–404 (1997).

Gitsu D.V., Popovich N.S., Chebanovski A.V. Physico-chemical and electrophysical properties of alloys in the $TlSbSe_2$–PbSe section [in Russian], *Izv. AN SSSR. Neorgan. mater.*, **17**(7), 1183–1185 (1981).

Gitsu D.V., Popovich N.S., Chebanovski A.V. The phase diagram and electrophysical properties of alloys in the $TlBiSe_2$–PbSe section [in Russian], *Izv. AN SSSR. Neorgan. mater.*, **16**(6), 988–990 (1980a).

Gitsu D.V., Popovich N.S., Shura V.K. Physico-chemical and electrophysical properties of alloys in the $TlBiTe_2$–2PbTe section [in Russian], *Izv. AN SSSR. Neorgan. mater.*, **16**(12), 2130–2132 (1980b).

Gitsu D.V., Popovich N.S., Shura V.K. The $TlSbTe_2$–2PbTe phase diagram [in Russian], *Izv. AN SSSR. Neorgan. mater.*, **16**(8), 1387–1389 (1980c).

Gitsu D.V., Popovich N.S., Zbigli K.R., Chumak G.D. The $(TlBiS_2)_{1-x}$–$(2PbS)_x$ system [in Russian], *Izv. AN SSSR. Neorgan. mater.*, **11**(11), 2068–2069 (1975).

Glasser L.S.D., Lee C.K. The structure of jasmundite, $Ca_{22}(SiO_4)_8O_4S_2$, *Acta Crystallogr.*, **B37**(4), 803–806 (1981).

Glenn J.R., Cho J.B., Wang Y., Craig A.J., Zhang J.-H., Cribbs M., Stoyko S.S., Rosello K.E., Barton C., Bonnoni A., Grima-Gallardo P., MacNeil J.H., Rondinelli J.M., Jang J.I., Aitken J.A. $Cu_4MnGe_2S_7$ and Cu_2MnGeS_4: two polar thiogermanates exhibiting second harmonic generation in the infrared and structures derived from hexagonal diamond, *Dalton Trans.*, **50**(47), 17524–17537 (2021).

Glukh O.S., Barchiy I.Ye., Sabov M.Yu, Tsygyka V.V. Physico-chemical interaction in the $GeSe_2$–Tl_2SnSe_3 system [in Ukrainian], *Nauk. Visn. Uzhgorod. Univ., Ser. Khim.*, (*19*), 37–39 (2008).

Godovikov A.A., Fedorova Zh.N. The Cu_2S–PbS–Bi_2S_3 system [in Russian], In: *Eksper. issled. po mineralogii (1968-1969)*. Novosibirsk, Nauka Publish., 42–49 (1969).

Godovikov A.A., Il'yasheva N.A., Kuryaeva R.G., Nenasheva S.N., Fedorova Zh.N. Study of dry sulfide systems [in Russian], In: *Fiz.-khim. usloviya protsessov mineraloobrazovaniya po teor. i eksp. dannym*, Novosibirsk, Nauka Publish., 5–27 (1976).

Godovikov A.A., Nenasheva S.N. The $AgSbS_2$–PbS system at temperature higher than 480°C [in Russian], *Dokl. AN SSSR*, **185**(1), 159–162 (1969a).

Godovikov A.A., Nenasheva S.N. The $AgSbS_2$–PbS system [in Russian], In: *Eksper. issled. po mineralogii (1968–1969)*. Novosibirsk, 61–64 (1969b).

Goehring M., Weiss J., Zirker G. Über Metall-Thionitrosylverbindungen. II. Die Thionitrosylate von Blei, Thallium, Kupfer und Silber, *Z. anorg. und allg. Chem.*, **278**(1–2), 1–11 (1955).

Goodchild R.G., Hughes O.H., Lopez-Rivera S.A., Woolley J.C. Energy gap values by optical absorption in I III IV Se_4 compounds, *Can. J. Phys.*, **60**(8), 1096–1100 (1982).

Goodchild R.G., Hughes O.H., Woolley J.C. Crystal structure of I III IV Se_4 compounds, *Phys. stat. sol. (a)*, **68**(1), 239–244 (1981).

Gorak, Ya., Starosta V.I. Obtaining and some properties of the $Pb_2SbS_2I_3$ and $Sn_2SbS_2I_3$ [in Russian], In: *Poluch. i svoystva slozhn. poluprovod*. Uzhgor. Gos. Univ. Publish., 76–77 (1991).

Gorgut G.P., Fedorchuk A.O., Kityk I.V., Sachanyuk V.P., Olekseyuk I.D., Parasyuk O.V. Synthesis and structural properties of $CuInGeS_4$, *J. Cryst. Growth*, **324**(1), 212–216 (2011).

Gospodinov G., Barkov D. Die Phasengleichgewichte im Dreistoffsystem GeO_2–SeO_2–H_2O, *Z. Chem.*, **22**(3), 114–115 (1982).

Gospodinov G.G., Odin I.N., Novoselova A.V. Physico-chemical investigation of the SnSe–Sb_2S_3 system [in Russian], *Dokl. Bolg. AN*, **27**(8), 1061–1064 (1974a).

Gospodinov G.G., Odin I.N., Novoselova A.V. The Sn_3S_3+Sb_2Se_3 ↔ Sn_3Se_3+Sb_2S_3 system [in Russian], *Zhurn. neorgan. khimii*, **19**(6), 1644–1647 (1974b).

Gospodinov G.G., Odin I.N., Popovkin B.A., Novoselova A.V. Investigation of SnS–Sb_2Se_3 section in the Sn_3S_3 + Sb_2Se_3 ↔ Sn_3Se_3 + Sb_2S_3 ternary mutual system [in Russian], *Izv. AN SSSR. Neorgan. mater.*, **8**(1), 175–176 (1972).

Graeser S., Schwander H. Edenharterite ($TlPbAs_3S_6$): a new mineral from Lengenbach, Binntal (Switzerland), *Eur. J. Mineral.*, **4**(6), 1265–1270 (1992).

Graeser S., Schwander H., Wulf R., Edenharter A. Erniggliite ($Tl_2SnAs_2S_6$), a new mineral from Lengenbach, Binntal (Switzerland): description and crystal structure determination based on data from synchrotron radiation, *Schweiz. Mineral. Petrogr. Mitt.*, **72**(3), 293–305 (1992).

Gray A.K., Knaust J.M., Chan B.C., Polyakova L.A., Dorhout P.K. Crystal structure of potassium ytterbium(III) tetrathiosilicate, $KYbSiS_4$, *Z. Kristollagr. New Cryst. Str.*, **220**(3), 293 (2005).

Greil S., Pfitzner A. Synthesis and crystal structure of Ag_2MnSnS_4, *Z. anorg. und allg. Chem.*, **638**(10), 1570 (2012).

Grimvall S. On the crystal structure of $Sn_3O(OH)_2SO_4$, *Acta Chem. Scand.*, **27**(4), 1447 (1973).

Grimvall S. The crystal structure of $Sn_7(OH)_{12}(SO_4)_2$, *Acta Chem. Scand. A*, **36**(4), 361–364 (1982).

Grimvall S. The crystal structure of $Sn_3O(OH)_2SO_4$, *Acta Chem. Scand. A*, **29**(6), 590–598 (1975).

Grupe M., Lissner F., Schleid T., Urland, W. Chalkogenid-Disilicate der Lanthanoide vom Typ $M_4X_3[Si_2O_7]$ (M = Ce-Er; X = S, Se), *Z. anorg. und allg. Chem.*, **616**(10), 53–60 (1992).

Grupe M., Urland W. Darstellung und Kristallstruktur von Nd_2SeSiO_4, *Z. Naturforsch.*, **45B**(4), 465–468 (1990).

Grupe M., Urland W. Die ersten Oxoselenosilicate der Seltenen Erden (SE = Ce, Nd), *Natutwisseschaften*, **76**(7), 327–329 (1989).

Gruzdev V.S., Volgin V.Yu., Spiridonov E.M., Evstigneeva T.L., Kabalov Yu.K., Sorokin V.I., Osadchyi E.G., Chvileva T.N., Chernitsova N.M. Velikite Cu_2HgSnS_4 (mercurian member of the stannite group) – a new mineral [in Russian], *Zap. Ros. mineralog. obshch.*, **126**(4), 71–75 (1997).

Gruzdev V.S., Volgin V.Y., Spiridonov E.M., Kaplunnik L.N., Pobedimskaya Y.A., Chvileva T.N., Chernitsova N.M. Velikite Cu_2HgSnS_4 – the mercury member of the stannite group [in Russian], *Dokl. AN SSSR*, **300**(2), 432–435 (1988).

Grytsiv V.I., Vengel' P.F., Tomashyk V.N. The Cu,Pb‖Se,Te ternary mutual system, *Neorgan. mater.*, **32**(2), 142–143 (1996).

Grytsiv V.I., Vengel' P.F., Tomashik Z.F., Tomashik V.N. The PbTe–Cu_2S diagonal section of the PbTe+Cu_2S ↔ PbS–Cu_2Te ternary mutual system [in Russian], *Neorgan. mater.*, **30**(3), 346–349 (1994).

Guen L., Glaunsinger W.S. Electrical, magnetic and EPR studies of the quaternary chalcogenides $Cu_2A^{II}B^{VI}X_4$ prepared by iodin transport, *J. Solid State Chem.*, **35**(1), 10–21 (1980).

Guen L., Glaunsinger W.S., Wold A. Physical properties of the quarternary chalcogenides $Cu^I_2B^{II}C^{IV}X_4$ (B^{II} = Zn, Mn, Fe, Co; C^{IV} = Si, Ge, Sn; X = S, Se), *Mater. Res. Bull.*, **14**(4), 463–467 (1979).

Guittard M., Julien-Pouzol M., Laruelle P., Flahaut J. Sur de nouvelles familles de combinaisons soufrées et séléniées formées par les terres rares, *C. r. Acad. sci. Sér. C*, **267**(13), 767–769 (1968).

Guittard M., Julien-Pouzol M. Les composés haxagonaux de type La_3CuSiS_7, *Bull. Soc. chim. France*, (7), 2467–2469 (1970).

Gulay L.D., Daszkiewicz M., Huch M.R., Pietraszko A. $Ce_3Mg_{0.5}GeS_7$ from single-crystal data, *Acta Crystallogr.*, **E63**(11), i187 (2007a).

Gulay L.D., Daszkiewicz M., Ruda I.P., Marchuk O.V. $La_2Pb(SiS_4)_2$, *Acta Crystallogr.*, **C66**(3), i19–i21 (2010).

Gulay L.D., Daszkewicz A., Shemet V.Ya., Pietraszko A. Investigation of the Sc_2Se_3–Cu_2Se–$SnSe_2$ and Sc_2Se_3–Cu_2Se–PbSe systems at 870 K, *Pol. J. Chem.*, **82**(5), 1001–1014 (2008a).

Gulay L.D., Kaczorowski D., Pietraszko A. Crystal structure and magnetic properties of $Ce_3CuSnSe_7$, *J. Alloys Compd.*, **403**(1–2), 49–52 (2005a).

Gulay L.D., Kaczorowski D., Pietraszko A. Crystal structure and magnetic properties of $YbCuPbSe_3$, *J. Alloys Compd.*, **413**, 26–28 (2006a).

Gulay L.D., Lychmanyuk O.S. Investigation of the Sm_2S_3–Cu_2S–SiS_2 and Er_2S_3–Cu_2S–GeS_2 system at 870 K [in Ukrainian], *Nauk. Visnyk Volyns'k. Nats. Univ. im. Lesi Ukrainky. Ser. Khim. nauky*, (29), 29–35 (2009).

Gulay L.D., Lychmanyuk O.S., Olekseyuk I.D., Daszkiewicz M., Stępień-Damm J., Pietraszko A. Crystal structures of the compounds R_3CuSiS_7 (R = Ce, Pr, Nd, Sm, Tb, Dy and Er) and $R_3CuSiSe_7$ (R = La, Ce, Pr, Nd, Sm, Gd, Tb and Dy), *J. Alloys Compd.*, **431**(1–2), 185–190 (2007b).

Gulay L.D., Lychmanyuk O.S., Olekseyuk I.D., Pietraszko A. Crystal structures of the $R_3CuGeSe_7$ (R = Ce, Pr, Nd, Sm, Gd, Tb and Ho) compounds, *J. Alloys Compd.*, **422**(1–2), 203–207 (2006b).

Gulay L.D., Lychmanyuk O.S., Stępień-Damm J., Pietraszko A., Olekseyuk I.D. Crystal structures of the Y_3CuSiS_7 and $Y_3CuSiSe_7$ compounds, *J. Alloys Compd.*, **402**, 201–203 (2005b).

Gulay L.D., Lychmanyuk O.S., Stępień-Damm J., Pietraszko A., Olekseyuk I.D. Isothermal section of the Y_2S_3–Cu_2S–GeS_2 at 870 K and crystal structures of the $Y_3Ge_{1.25}S_7$ and Y_3CuGeS_7 compounds, *J. Alloys Compd.*, **414**(1–2), 113–117 (2006c).

Gulay L.D., Lychmanyuk O.S., Wołcyrz M., Pietraszko A., Olekseyuk I.D. The crystal structures of R$_3$CuGeS$_7$ (R = Ce–Nd, Sm, Gd–Dy and Er), *J. Alloys Compd.*, **425**, 159–163 (2006d).

Gulay L.D., Marchuk O.V. Phase equilibria in the Sm(Ho)$_2$S$_3$–PbS–SnS$_2$ systems at 770 K [in Ukrainian], *Nauk. Visnyk Volyns'k. Nats. Univ. im. Lesi Ukrainky. Ser. Khim. nauky*, (*16*), 50–54 (2010).

Gulay L.D., Nazarchuk O.P., Olekseyuk I.D. Crystal structures of the compounds Cu$_2$CoSi(Ge,Sn)S$_4$ and Cu$_2$CoGe(Sn)Se$_4$, *J. Alloys Compd.*, **377**, 306–311 (2004a).

Gulay L.D., Olekseyuk I.D. Crystal structures of the RCuPbSe$_3$ (R = Gd, Tb, Dy, Ho, Er, Tm, Yb and Lu) compounds, *J. Alloys Compd.*, **387**, 160–164 (2005a).

Gulay L.D., Olekseyuk I.D. Crystal structures of the R$_{3.33}$CuPb$_{1.5}$Se$_7$ (R = Tb, Dy, Ho, Er, Tm, Yb and Lu) compounds, *J. Alloys Compd.*, **396**, 233–239 (2005b).

Gulay L.D., Olekseyuk I.D. Crystal structures of the R$_{3.33}$CuPb$_{1.5}$S$_7$ (R = Tb, Dy, Ho, Er and Lu) compounds, *J. Alloys Compd.*, **413**(1–2), 122–126 (2006a).

Gulay L.D., Olekseyuk I.D. Crystal structures of the R$_3$CuSnSe$_7$ (R = La, Ce, Pr, Nd, Sm, Gd, Tb and Dy) compounds, *J. Alloys Compd.*, **388**, 274–278 (2005c).

Gulay L.D., Olekseyuk I.D., Huch M.R., Wołcyrz M. Crystal structure of selenosilicates Pb$_{1.75}$M$_{0.5}$SiSe$_4$ (M = Cu and Ag), *J. Alloys Compd.*, **402**, 115–117 (2005c).

Gulay L.D., Olekseyuk I.D. Investigation of the R$_2$Te$_3$–Cu$_2$Te–PbTe (R = Ho, Er, Tm) system at 870 K [in Ukrainian], *Nauk. Visnyk Volyns'k. Derzh. Univ. im. Lesi Ukrainky. Ser. Khim. nauky*, (*4*), 101–108 (2006b).

Gulay L.D., Olekseyuk I.D., Parasyuk O.V. Crystal structures of the Ag$_4$HgGe$_2$S$_7$ and Ag$_4$CdGe$_2$S$_7$ compounds, *J. Alloys Compd.*, **340**(1–2), 157–166 (2002a).

Gulay L.D., Olekseyuk I.D., Parasyuk O.V. Crystal structures of the Ag$_6$HgGeSe$_6$ and Ag$_6$HgSiSe$_6$ compounds, *J. Alloys Compd.*, **343**(1–2), 116–121 (2002b).

Gulay L.D., Olekseyuk I.D., Wołcyrz M., Stępień-Damm J. Crystal structures of the RCuPbS$_3$ (R = Tb, Dy, Ho, Er, Tm, Yb and Lu) compounds, *J. Alloys Compd.*, **399**, 189–195 (2005d).

Gulay L.D., Olekseyuk I.D., Wołcyrz M., Stępień-Damm J., Pietraszko A. Investigation of the Ho$_2$Se$_3$–Cu$_2$Se–PbSe and Er$_2$Se$_3$–Cu$_2$Se–PbSe systems at 870 K, *J Alloys Compd.*, **416**(1–2), 173–178 (2006e).

Gulay L.D., Olekseyuk I.D., Wołcyrz M., Stępień-Damm J. The crystal structures of R$_3$CuSnS$_7$ (R = La-Nd, Sm, Gd-Ho), *Z. anorg. und. allg. Chem.*, **631**(10), 1919–1923 (2005e).

Gulay L.D., Parasyuk O.V. Crystal structure of the Ag$_6$Hg$_{0.82}$GeS$_{5.82}$ compound, *J. Alloys Compd.*, **327**(1–2), 100–103 (2001).

Gulay L.D., Parasyuk O.V. Crystal structure of the Ag$_{2.66}$Hg$_2$Sn$_{1.34}$Se$_6$ and Hg$_2$SnSe$_4$ compounds, *J. Alloys Compd.*, **337**(1–2), 94–98 (2002).

Gulay L.D., Parasyuk O.V., Olekseyuk I.D. Crystal structures of the Cu$_6$Hg$_{0.973}$SiS$_{5.973}$ and Ag$_6$Hg$_{0.897}$SiS$_{5.897}$ compounds, *J. Alloys Compd.*, **335**(1–2), 111–114 (2002c).

Gulay L.D., Parasyuk O.V., Romanyuk Ya E., Olekseyuk I. D. Crystal structure of the Cu$_{5.976}$Hg$_{0.972}$SiSe$_6$ compound, *J. Alloys Compd.*, **367**(1–2), 121–125 (2004b).

Gulay L.D., Romanyuk Ya E., Parasyuk O.V. Crystal structures of low- and high-temperature modifications of Cu$_2$CdGeSe$_4$, *J. Alloys Compd.*, **347**(1–2), 193–197 (2002d).

Gulay L.D., Ruda I.P., Marchuk O.V., Olekseyuk I.D. Crystal structures of the R$_2$Pb$_3$Sn$_3$S$_{12}$ (R = La, Ce, Pr, Nd, Sm, Gd, Tb, Dy, Ho, Er and Tm) compounds, *J. Alloys Compd.*, **457**, 204–208 (2008b).

Gulay L.D., Shemet V.Ya. Investigations the Dy$_2$S$_3$–Cu$_2$S–SnS$_2$ and Dy$_2$Se$_3$–Cu$_2$Se–SnSe$_2$ systems at 770 K [in Ukrainian], *Nauk. Visnyk Volyns'k. Nats. Univ. im. Lesi Ukrainky. Ser. Khim. nauky*, **17**, 69–75 (2012).

Gulay L.D., Shemet V.Ya., Olekseyuk I.D. Crystal structures of the compounds YCuPbSe$_3$, Y$_3$CuSnSe$_7$ and Y$_3$Cu$_{0.685}$Se$_6$, *J. Alloys Compd.*, **385**, 160–168 (2004c).

Gulay L.D., Shemet V.Ya., Olekseyuk I.D. Crystal structures of the compounds YCuS$_2$, Y$_3$CuSnS$_7$ and YCuPbS$_3$, *J. Alloys Compd.*, **388**, 59–64 (2005f).

Gulay L.D., Shemet V.Ya., Olekseyuk I.D. Crystal structures of the ScCuSe$_2$ and Sc$_3$CuSn$_3$Se$_{11}$ compounds, *J. Alloys Compd.*, **393**, 174–179 (2005g).

Gulay L.D., Shemet V.Ya. Olekseyuk I.D. Crystal structures of the Y$_{3.33}$CuPb$_{1.5}$X$_7$ (X = S, Se) compounds, *J. Alloys Compd.*, **394**, 250–254 (2005h).

Gulay L.D., Shemet V Ya., Olekseyuk I. D., Stępień-Damm J., Pietraszko A., Koldun L.V., Filimonyuk J.O. Investigation of the R$_2$S$_3$–Cu$_2$S–PbS (R = Y, Dy, Ho and Er) systems, *J. Alloys Compd.*, **431**(1–2), 77–84 (2007c).

Gulay L.D., Stępień-Damm J., Pietraszko A., Olekseyuk I.D. Crystal structure of the R$_5$CuPb$_3$Se$_{11}$ (R = Er, Tm and Yb) compounds, *J. Alloys Compd.*, **413**, 90–95 (2006f).

Gulay L.D., Wołcyrz M., Olekseyuk I.D. Investigation of the Tb$_2$Se$_3$–Cu$_2$Se–PbSe and Dy$_2$Se$_3$–Cu$_2$Se–PbSe systems at 870 K, *Pol. J. Chem.*, **80**(5), 805–815 (2006g).

Gulay L.D., Wołcyrz M., Pietraszko A., Olekseyuk I.D. Investigation of the Tm$_2$Se$_3$–Cu$_2$Se–PbSe and Lu$_2$Se$_3$–Cu$_2$Se–PbSe systems at 870 K, *Pol. J. Chem.*, **80**(10), 1703–1714 (2006h).

Guo H., Li Z., Yang L., Wang P., Huang X., Li J. Potassium silver tin selenide, K$_2$Ag$_2$Sn$_2$Se$_6$, *Acta Crystallogr.*, **C57**(11), 1237–1238 (2001).

Guo S.-P., Guo G.-C., Huang J.-S. Syntheses, structures and properties of five chiral quaternary sulfides, Al$_x$Ln$_3$(Si$_y$Al$_{1-y}$)S$_7$ (Ln = Y, Gd, Dy) and In$_{0.33}$Sm$_3$SiS$_7$, *Sci. China. Ser. B. Chem.*, **52**(10), 1609–1615 (2009a).

Guo S.-P., Guo G.-C., Wang M.-S., Zou J.-P., Xu G., Wang G.-J., Long X.-F., Huang J.-S. A series of new infrared NLO semiconductors, ZnY$_6$Si$_2$S$_{14}$, Al$_x$Dy$_3$(Si$_y$Al$_{1-y}$)S$_7$, and Al$_{0.33}$Sm$_3$SiS$_7$, *Inorg. Chem.*, **48**(15), 7059–7065 (2009b).

Guo S.-P., Zeng H.-Y., Guo G.-C., Zou J.-P., Xu G., Huang J.-S. Syntheses, structures and band gaps of KLnSiS$_4$ (Ln = Sm, Yb), *Chin. J. Struct. Chem.*, **27**(12), 1543–1548 (2008).

Guo S.-P., Zeng H.-Y., Jiang X.-M., Guo G.-C. Crystal structure and magnetic property of a quaternary sulfide, Al$_{0.36}$Sm$_3$Ge$_{0.98}$S$_7$, *Chin. J. Struct. Chem.*, **28**(11), 1448–1452 (2009c).

Guo Y., Liang F., Li Z., Xing W., Lin Z.-S., Yao J., Mar A., Wu Y. AHgSnQ$_4$ (A = Sr, Ba; Q = S, Se): a series of Hg-based infrared nonlinear-optical materials with strong second-harmonic-generation response and good phase matchability, *Inorg. Chem.*, **58**(15), 10390–10398 (2019a).

Guo Y., Liang F., Li Z., Xing W., Lin Z., Yao J., Wu Y. Li$_4$HgSn$_2$Se$_7$: the first second-order nonlinear optical-active selenide in the I$_4$–II–IV$_2$–VI$_7$ diamond-like family, *Cryst. Growth Des.*, **19**(10), 5494–5497 (2019b).

Guo Y., Liang F., Yin W., Li Z., Luo X., Lin Z.-S., Yao Y., Mar A., Wu Y. BaHgGeSe$_4$ and SrHgGeSe$_4$: two new Hg-based infrared nonlinear optical materials, *Chem. Mater.*, **31**(8), 3034–3040 (2019c).

Guo Y., Li X., Feng K., Li C., Zhou M., Wu Y., Yao J. Li$_7$Cd$_{4.5}$Ge$_4$Se$_{16}$ and Li$_{6.4}$Cd$_{4.8}$Sn$_4$Se$_{16}$: strong nonlinear optical response in quaternary diamond-like selenide networks, *Chem. Asian J.*, **13**(7), 871–876 (2018).

Gurbanov G.R., Adygezalova M.B. The sections GeSnSb$_4$Te$_8$–GeTe and GeSnSb$_4$Te$_8$–SnTe of the quasi-ternary system GeTe–Sb$_2$Te$_3$–SnTe, *Russ. J. Inorg. Chem.*, **63**(1), 111–115 (2018).

Gurbanov G.R., Agaguseynova M.M., Adygezalova N.B. The SnSbBiTe$_4$–SnBi$_4$Te$_7$ system [in Russian], *Azerb. khim. zhurn.*, (*4*), 183–185 (2010a). Гурбанов Г.Р., Агагусейнова М.М., Адыгезалова Н.Б. Система SnSbBiTe$_4$–SnBi$_4$Te$_7$. *Азерб. хим. журн.*, (4), 183–185 (2010).

Gurbanov G.R., Bakhtiyarly I.B., Kurbanova R.D. The SnSbBiTe$_4$–2Bi$_2$Te$_3$ section of the SnTe–Bi$_2$Te$_3$–Sb$_2$Te$_3$ quasiternary system, *Russ. J. Inorg. Chem.*, **55**(7), 1149–1152 (2010b).

Gurbanov G.R. Investigation of the PbS–Bi$_2$S$_3$–SnS quasiternary system [in Russian], *Kimiya Probl.*, (*4*), 461–471 (2012a).

Gurbanov G.R. Mamedov Sh.G., Adygezalova M.B., Mamedov A.N. The PbSb$_2$Se$_4$–Pb$_5$Bi$_6$Se$_{14}$ section of the Sb$_2$Se$_3$–PbSe–Bi$_2$Se$_3$ quasi-ternary system, *Russ. J. Inorg. Chem.*, **62**(12), 1659–1664 (2017).

Gurbanov G.R., Mamedov Sh.G. Phase equilibria in the PbBi$_2$S$_4$–PbSnS$_2$ system, *Russ. J. Inorg. Chem.*, **61**(5), 689–691 (2016).

Gurbanov G.R., Mamedov Sh.G. Sections Pb$_6$Sb$_2$Bi$_6$Se$_{18}$–Sb$_2$Se$_3$ and Pb$_6$Sb$_2$Bi$_6$Se$_{18}$–Bi$_2$Se$_3$ of the quasi-ternary system Sb$_2$Se$_3$–PbSe–Bi$_2$Se$_3$, *Russ. J. Inorg. Chem.*, **64**(3), 383–388 (2019).

Gurbanov G.R. New quaternary compound PbSbBiS$_4$ [in Russian], *Neorgan. mater.*, **48**(7), 662–664 (2012b).

Gurbanov G.R. PbSbBiSb$_4$–Sb$_2$S$_3$ and PbSbBiS$_4$–Bi$_2$S$_3$ joins in the PbS–Sb$_2$S$_3$–Bi$_2$S$_3$ system [in Russian], *Neorgan. mater.*, **48**(9), 888–890 (2012c).

Gurbanov G.R. Phase equilibria in the GeTe–Sb$_2$Te$_3$–Bi$_2$Te$_3$ quasiternary system, *Russ. J. Inorg. Chem.*, **58**(1), 91–95 (2013).

Gurbanov G.R. Physico-chemical investigation of the SnSbBiS$_4$–Sn$_2$Sb$_2$S$_5$ system [in Azerbaijanian], *Chem. Probl.*, (*1*), 101–103 (2010).

Gurbanov G.R. The nature of interaction in the SnSbBiS$_4$–Bi$_2$S$_3$ system [in Russian], *ChemChemTech* [*Izv. Vyssh. Uchebn. Zaved. Khim. Khim. Tekhnol.*], **55**(8), 47–49 (2012d).

Gurbanov G.R. Thermoelectric properties of tetradymite-like layered materials in the pseudoternary system Sb$_2$Te$_3$–GeTe–Bi$_2$Te$_3$, *Inorg. mater.*, **53**(7), 691–697 (2017).

Gurshumov A.P., Aliev A.S., Mamedov N.A., Alidzhanov M.A., Gadzhiev T.G. The SnTe–InSe system [in Russian], *Izv. AN SSSR. Neorgan. mater.*, **21**(10), 1670–1672 (1985).

Gurshumov A.P., Asadov Yu., Murguzov M.I., Ahmedov A.M., Sadyhova L.N. Interaction in the PbTe–TlSe [in Russian], *Zhurn. neorgan. khimii*, **29**(9), 2432–2433 (1984a).

Gurshumov A.P., Murguzov M.I., Asadov Yu., Sadyhova L.N., Ahmedov A.M. The PbTe–InSe system [in Russian], *Zhurn. neorgan. khimii*, **29**(9), 2433–2435 (1984b).

Gurshumov A.P., Murguzov M.I., Nadzhafova Z.Z., Ahmedov A.M., Eyvazov S.Sh. The As$_2$Se$_3$–PbS system [in Russian], *Zhurn. neorgan. khimii*, **31**(1), 264–266 (1986a).

Gurshumov A.P., Murguzov M.I., Nadzhafova Z.Z., Ahmedov A.M., Eyvazov S.Sh. The As$_2$S$_3$–SnSe system [in Russian], *Zhurn. neorgan. khimii*, **31**(1), 266–268 (1986b).

Gurukrishna K., Rao A., Chung Y.-C., Kao Y.-K. Manipulating the phonon transport towards reducing thermal conductivity via replacement of Cu by Mn in Cu$_2$SnSe$_3$ thermoelectric system, *J. Solid State Chem.*, **307**, 122755 (2022).

Guseinov F.N., Babanly M.B., Zlomanov V.P., Yusibov Yu.A. Phase equilibria in the Tl$_2$Te–PbTe–Bi$_2$Te$_3$ system, *Russ. J. Inorg. Chem.*, **57**(10), 1387–1392 (2012).

Guseynov F.N., Dashdyeva G.B., Babanly M.B. Phase equilibria in the Tl$_2$Te–PbTe–TlBiTe$_2$ system [in Russian], *Azerb. khim. zhurn.*, (*3*), 61–65 (2009).

Guseynov F.N. Thermodynamic properties of the solid solutions in the PbTe–TlBiTe$_2$ system [in Russian], *Azerb. khim. zhurn.*, (*2*), 155–158 (2011).

Guseynov G.D., Godzhaev E.M., Khalilov H.Ya., Seidov F.M., Pashaev A.M. Complex semiconducting chalcogenides [in Russian], *Izv. AN SSSR. Neorgan. mater.*, **8**(9), 1569–1572 (1972).

Guseynov G.G., Amiraslanov I.R., Yusifov A.G., Mamedov H.N. Crystal structure of the SnBi$_{12}$S$_{18}$I$_2$ and about some stable structure fragments with trigonal symmetry [in Russian], *Dokl. AN AzSSR*, **45**(9), 36–40 (1989).

Guseynov Z.A., Guseynov F.N., Mashadieva L.F., Il'yasov T.M., Babanly M.B. Phase equilibria in the Tl$_2$X–SnX–PbS (X = Se, Te) systems [in Russian], *Zhurn. neorgan. khimii*, **46**(11), 1927–1930 (2001).

Gu S., Hu H., Tao H., Zhao X. Properties and microstructure of GeS$_2$–Ga$_2$S$_3$–CdS glasses, *J. Wuhan Univ. Technol. Mater. Sci. Ed.*, **20**(4), 120–122 (2005).

Gutt W., Smith M.A. A new calcium silicosulphate, *Nature*, **210**(5034), 408–409 (1966).

Haeuseler H., Himmrich M. Neue Verbindungen Ag$_2$HgMX$_4$ mit Wurtzstannitstruktur *Z. Naturforsch.*, **44B**(9), 1035–1036 (1989).

Haeuseler H., Ohrendori F., Himmrich M. Zur Kenntnis quaternärer Telluride Cu$_2$MM'Te$_4$ mit Tetrahederstrukturen, *Z. Naturforsch.*, **46B**(8), 1049–1052 (1991).

Haeuseler H., Schmidt Ch. Glass formation in the system BaS–Ga$_2$S$_3$–GeS$_2$ and the structure of Ba$_{2,7}$Ga$_{5,4}$Ge$_{3,6}$S$_{1,8}$, *J. Alloys Compd.*, **204**(1–2), 209–213 (1994).

Hahn H., Schulze H. Über quaternäre Chalcogenide des Germaniums und Zinns, *Naturwissenschaften*, **52**(14), 426 (1965).

Hahn H., Strick G. Über einige quaternäre Chalkogenide mit Spinellstruktur, *Naturwissenschaften*, **54**(2), 42 (1967a).

Hahn H., Strick G. Über quaternäre Chalkogenide zinkblendeähnlicher Struktur, *Naturwissenschaften*, **54**(9), 225–226 (1967b).

Hålenius U., Hatert F., Pasero M., Mills S.J. IMA Commission on New Minerals, Nomenclature and Classification (CNMNC). Newsletter No. 25. New minerals and nomenclature modifications approved in 2015, *Mineralog. Mag.*, **79**(3), 529–535 (2015).

Hall S.R., Szymański J.T., Stewart J.M. Kesterite, Cu$_2$(Zn,Fe)SnS$_4$, and stannite, Cu$_2$(Fe,Zn)SnS$_4$, structurally similar but distinct minerals, *Canad. Mineralog.*, **16**(2), 131–137 (1978).

Harada S. Some new sulfo-spinels containing iron-group transition metals, *Mater. Res. Bull.*, **8**(12), 1361–1369 (1973).

Harris D.C., Chen T.T. Gustavite: two Canadian occurrences, *Canad. Mineralog.*, **13**(4), 411–414 (1975).

Harris D.C., Owens D.R. A tellurium-bearing canfieldite, from Revelstoke, B.C., *Canad. Mineralog.*, **10**(5), 895–898 (1971).

Harris D.C., Roberts A.C., Criddle A.J., Stanley C.J. Kiddcreekite, a new mineral from the Kidd Creek mine, Timmins, Ontario, and from the Campbell orebody, Bisbee, Arizona, *Canad. Mineralog.*, **22**(2), 227–232 (1984).

Hartenbach I., Lauxmann P., Schleid T. Er$_2$S[SiO$_4$]: Ein Erbiumsulfid-*ortho*-Oxosilicat mit ungewöhnlicher Sulfidionen-Koordination, *Z. anorg. und allg. Chem.*, **630**(10), 1408–1412 (2004).

Hartenbach I., Meier S.F., Wontcheu J., Schleid T. Ho$_2$O[SiO$_4$] und Ho$_2$S[SiO$_4$]: Zwei Chalkogenid-Derivate von Holmium(III)-*ortho*-Oxosilicat, *Z. anorg. und allg. Chem.*, **628**(13), 2907–2913 (2002).

Hartenbach I., Müller A.C., Schleid T. CuGd$_3$SiS$_7$: Ein Gadoliniumsulfid mit zwei isolierten komplexen Thioanionen gemäß Gd$_3$[CuS$_3$][SiS$_4$], *Z. anorg. und. allg. Chem.*, **632**(12–13), 2147 (2006).

Hartenbach I., Nilges T., Schleid T. Thiosilicate der Selten-Erd-Elemente: IV. Die *quasi*-isostrukturellen Verbindungen NaSm$_3$S$_3$[SiS$_4$], CuCe$_3$S$_3$[SiS$_4$], Ag$_{0,63}$Ce$_3$S$_{2,63}$Cl$_{0,37}$[SiS$_4$] und Sm$_3$S$_2$Cl[SiS$_4$] – Synthese, Kristallstruktur und Untersuchungen zur Silberionendynamik, *Z. anorg. und allg. Chem.*, **633**(13–14), 2445–2452 (2007).

Hartenbach I., Schleid T. NaY$_3$S$_3$[SiS$_4$]: a sodium-containing yttrium sulfide thiosilicate with channel structure, *J. Solid State Chem.*, **171**(1–2), 382–386 (2003a).

Hartenbach I., Schleid T. Crystal structure of tetracerium(III) trisulfide heptaoxodisilicate(IV), Ce$_4$S$_3$[Si$_2$O$_7$], *Z. Kristallogr., New Cryst. Str.*, **217**(2), 175–176 (2002a).

Hartenbach I., Schleid T. M$_{4,667}$Ch[SiO$_4$]$_3$ (M = Nd, Sm; Ch = O, S): structural comparison of two *apatite*-type lanthanoid chalcogenide *ortho*-oxosilicates, *Z. Kristallogr.*, **220**(2–3), 206–210 (2005a).

Hartenbach I., Schleid T. The non-isotypic pair CsCe[SiS$_4$] and CsCe[SiSe$_4$]: a structural comparison, *J. Alloys Compd.*, **418**, 95–100 (2006).

Hartenbach I., Schleid T. Thiosilicate der Selten-Erd-Elemente: I. Die isotypen Verbindungen KCe[SiS$_4$] und Eu$_2$[SiS$_4$], *Z. anorg. und allg. Chem.*, **628**(6), 1327–1331 (2002b).

Hartenbach I., Schleid T. Thiosilicate der Selten-Erd-Elemente: II. Die nichtzentrosymmetrischen Caesium-Derivate CsM[SiS$_4$] (M = Sm-Tm), *Z. anorg. und allg. Chem.*, **629**(3), 394–398 (2003b).

Hartenbach I., Schleid T. Thiosilicate der Selten-Erd-Elemente: III. KLa[SiS$_4$] und RbLa[SiS$_4$] – Ein struktureller Vergleich, *Z. anorg. und allg. Chem.*, **631**(8), 1365–1370 (2005b).

Hasanova U.A., Mammadov Sh.H., Aliyev O.M., Bakhtiyarly I.B. Quasi-binary sections and triangulation of FeS–Ga$_2$S$_3$–PbS quasi-ternary system, *Azerb. Chem. J.*, (2), 58–61 (2019).

Hatscher S.T., Urland W. Lanthanum iodide thiosilicate, La$_3$I[SiS$_4$]$_2$, *Acta Crystallogr.*, **E58**(11), i100–i102 (2002a).

Hatscher S.T., Urland W. Samarium chloride sulfide thiosilicate, Sm$_3$ClS$_2$[SiS$_4$], *Acta Crystallogr.*, **E58**(12), i124–i126 (2002b).

Hatscher S.T., Urland W. Synthese und Kristallstrukturen von Bromid-Thiosilicaten Ln$_3$Br[SiS$_4$]$_2$ der Lanthanide (Ln = La, Ce, Pr, Nd, Sm, Gd), *Z. anorg. und allg. Chem.*, **628**(3), 608–611 (2002c).

Hatscher S.T., Urland W. Synthese und Kristallstrukturen von Ln$_3$I[SiS$_4$]$_2$ (Ln = Pr, Nd, Sm, Tb), *Z. anorg. und allg. Chem.*, **627**(9), 2198–2200 (2001).

Hatscher S.T., Urland W. Synthesis and structures of chloride thiosilicates with lanthanides Ln$_3$Cl[SiS$_4$]$_2$ (Ln = La, Ce, Pr), *Mater. Res. Bull.*, **37**(7), 1239–1247 (2002d).

Hatscher S.T., Urland W. Synthesis, structure, and magnetic behavior of a new chloride thiosilicate with neodymium Nd$_3$ClS$_2$[SiS$_4$], *Mater. Res. Bull.*, **38**(1), 99–112 (2003).

Hawes L.L. Sulphur-selenium and sulphur-tellurium cyclic interchalcogen compounds, *Nature*, **198**(4887), 1267–1270 (1963).

Hawthorne F.C., Bladh K.W., Burke E.A.J., Ercit T.S., Grew E.S., Grice J.D., Puziewicz J., Roberts A.C., Schedler R.A., Shigley J.E. New mineral names, *Am. Mineralog.*, **71**(11–12), 1545–1548 (1986a).

Hawthorne F.C., Bladh K.W., Burke E.A.J., Grew E.S., Langley R.H., Puziewicz J., Roberts A.C., Schedler R.A., Shigley J.E., Vanko D.A. New mineral names, *Am. Mimeralog.*, **72**(1–2), 222–230 (1987).

Hawthorne F.C., Fleischer M., Grew E.S., Grice J.D., Jambor J.L., Puziewicz J., Roberts A.C., Vanko D.A., Zilczer J.A. New mineral names, *Am. Mimeralog.*, **71**(9–10), 1277–1282 (1986b).

Haynes A.S., Liu T.-K., Frazer L., Lin J.-F., Wang S.-Y., Ketterson J.B., Kanatzidis M.G., Hsu K.-F. Second harmonic generation response of the cubic chalcogenide Ba$_{(6-x)}$Sr$_x$[Ag$_{(4-y)}$Sn$_{(y/4)}$](SnS$_4$)$_4$, *J. Solid State Chem.*, **248**, 119–125 (2017).

He J., Guo Y., Huang W., Zhang X., Yao J., Zhai T., Huang F. Synthesis, crystal structure, and optical properties of noncentrosymmetric Na$_2$ZnSnS$_4$, *Inorg. Chem.*, **57**(16), 9918–9924 (2018a).

He J., Wang Z., Zhang X., Cheng Y., Gong Y., Lai X., Zheng C., Lin J., Huang F. Synthesis, structure, magnetic and photoelectric properties of Ln$_3$M$_{0.5}$M'Se$_7$ (Ln = La, Ce, Sm; M = Fe, Mn; M' = Si, Ge) and La$_3$MnGaSe$_7$, *RSC Adv.*, **5**(65), 52629–52635 (2015).

He J., Zhang X., Zheng C., Huang F. Synthesis, structure, magnetic and optoelectric properties of layered NaM$_{0.5}$Sn$_{0.5}$S$_2$ (M = Mn, Fe), *J. Alloys Compd.*, **726**, 328–334 (2018b).

Hellner E., Leineweber G. Über komplex zusammengesetzte sulfidische Erze. I. Zur Struktur des Bournonits, CuPbSbS$_3$, und Seligmannits, CuPbAsS$_3$, *Z. Kristallogr.*, **107**(1–2), 150–154 (1956).

Hellner E. Über komplex zusammengesetzte Spießglanze III. Zur Struktur des Diaphorits, Ag$_3$Pb$_2$Sb$_3$S$_8$, *Z. Kristallogr.*, **110**(1–6), 169–174 (1958).

Hentschel G., Glasser L.S.D., Lee C.K. Jasmundite, Ca$_{22}$(SiO$_4$)$_8$O$_4$S$_2$, a new mineral, *N. Jb. Miner., Mh.*, (8), 337–342 (1983).

Heppke E.M., Klenner S., Janka O., Pöttgen R., Lerch M. Mechanochemical synthesis of Cu$_2$MgSn$_3$S$_8$ and Ag$_2$MgSn$_3$S$_8$, *Z. anorg. und. allg. Chem.*, **646**(1), 5–9 (2020).

Hernán L., Lavela P., Morales J., Pattanayak J., Tirado J.L. Structural aspects of lithium intercalated PbVS$_3$, PbTiS$_3$, PbTi$_2$S$_5$ and SnNbS$_3$ misfit layer compounds, *Mater. Res. Bull.*, **26**(11), 1211–1218 (1991).

Hernán L., Morales J., Pattanayak J., Tirado J.L. Lithium intercalation into PbNb$_2$S$_5$, PbNbS$_3$, SnNb$_2$Se$_5$, BiVS$_3$, SnVSe$_3$, and PbNb$_2$Se$_5$ misfit layer chalcogenides, *J. Solid State Chem.*, **100**(2), 262–271 (1992).

Hirai K., Tatsumisago M., Minami T. Thermal and electrical properties of rapidly quenched glasses in the systems Li$_2$S–SiS$_2$–Li$_x$MO$_y$ (Li$_x$MO$_y$ = Li$_4$SiO$_4$, Li$_2$SO$_4$), *Solid State Ionics*, **78**(3–4), 269–273 (1995).

Hirai T., Kurata K., Takeda Y. Derivation of new semiconducting compounds by cross substitution for group IV semiconductors and their semiconducting and thermal properties, *Solid-State Electronics*, **10**(10), 975–981 (1967).

Hoda S.N., Chang L.L.Y. Phase relations in the pseudo-ternary system PbS–Cu$_2$S–Sb$_2$S$_3$ and the synthesis of meneghinite, *Canad. Mineralog.*, **13**(4), 388–393 (1975a).

Hoda S.N., Chang L.L.Y. Phase relations in the systems PbS–Ag$_2$S–Sb$_2$S$_3$ and PbS–Ag$_2$S–Bi$_2$S$_3$, *Am. Mineralog.*, **60**(7–8), 621–633 (1975b).

Honig E., Shen H.-S., Yao G.-Q., Doverspike K., Kershaw R., Dwight K., Wold A. Preparation and characterization of Cu$_2$Zn$_{1-x}$Mn$_x$GeS$_4$, *Mater. Res. Bull.*, **23**(3), 307–312 (1988).

Hori S., Kato M., Suzuki K., Hirayama M., Kato Y., Kanno R. Phase diagram of the Li_4GeS_4–Li_3PS_4 quasi-binary system containing the superionic conductor $Li_{10}GeP_2S_{12}$, *J. Am. Ceram. Soc.*, **98**(10), 3352–3360 (2015a).

Hori S., Suzuki K., Hirayama M., Kato Y., Saito T., Yonemura M., Kanno R. Synthesis, structure, and ionic conductivity of solid solution, $Li_{10+\delta}M_{1+\delta}P_{2-\delta}S_{12}$ (M = Si, Sn), *Faraday Discuss.*, **176**, 83–94 (2014).

Hori S., Taminato S., Suzuki K., Hirayama M., Kato Y., Kanno R. Structure–property relationships in lithium superionic conductors having a $Li_{10}GeP_2S_{12}$-type structure, *Acta Crystallogr.*, **B71**(6), 727–736 (2015b).

Horiuchi H., Wuensch B.J. Lindströmite, $Cu_3Pb_3Bi_7S_{15}$: its space group and ordering scheme for metal atoms in the crystal structure, *Canad. Mineralog.*, **15**(4), 527–535 (1977).

Horiuchi H., Wuensch B.J. The ordering scheme for metal atoms in the crystal structure of hammarite, $Cu_2Pb_2Bi_4S_9$, *Canad. Mineralog.*, **14**(4), 536–539 (1976).

Hsu K.-F., Chung D.-Y., Lal S., Hogan T., Kanatzidis M.G. Structure and thermoelectric properties of new layered compounds in the quaternary system Cs–Pb–Bi–Te, *Mat. Res. Soc. Symp. Proc.*, **691**, G8.25.2–G8.25.6 (2001).

Hsu K.-F., Chung D.-Y., Lal S., Mrotzek A., Kyratsi T., Hogan T., Kanatzidis M.G. $CsMBi_3Te_6$ and $CsM_2Bi_3Te_7$ (M = Pb, Sn): new thermoelectric compounds with low-dimensional structures, *J. Am. Chem. Soc.*, **124**(11), 2410–2411 (2002a).

Hsu K.-F., Lal S., Hogan T., Kanatzidis M.G. $CsPb_3Bi_3Te_8$ and $CsPb_4Bi_3Te_9$: low-dimensional compounds and the homologous series $CsPb_mBi_3Te_{5+m}$, *Chem. Comm.*, (13), 1380–1381 (2002b).

Huang F.Q., Ibers J.A. $Dy_3CuGeSe_7$, *Acta Crystallogr.*, **C55**(8), 1210–1212 (1999).

Huang W., He Z., Zhao B., Zhu S., Chen B. Crystal growth, structure, and optical properties of new quaternary chalcogenide nonlinear optical crystal $AgGaGeS_4$, *J. Alloys Compd.*, **796**, 138–145 (2019a).

Huang W., He Z., Zhu S., Zhao B., Chen B., Zhu S. Polycrystal synthesis, crystal growth, structure, and optical properties of $AgGaGe_nS_{2(n+1)}$ (n = 2, 3, 4, and 5) single crystals for mid-IR laser applications, *Inorg. Chem.*, **58**(9), 5865–5874 (2019b).

Huang W., Wu J., Chen B., Li J., He Z. Crystal growth and thermal annealing of $AgGaGe_5Se_{12}$ crystal, *J. Alloys Compd.*, **862**, 158002 (2021).

Huang W., Yoshino K., Hori S., Suzuki K., Yonemura M., Hirayama M., Kanno R. Superionic lithium conductor with a cubic argyrodite-type structure in the Li–Al–Si–S system, *J. Solid State Chem.*, **270**, 487–492 (2019c).

Huang W., Zhao B., Zhu S., He Z., Chen B., Pu Y., Lin L., Zhao Z., Zhong Y. Synthesis of $AgGaGeS_4$ polycrystalline materials by vapor transporting and mechanical oscillation method, *J. Cryst. Growth*, **468**, 469–472 (2017).

Huang Y., Wu K., Cheng J., Chu Y., Yang Z., Pan S. Li_2ZnGeS_4: a promising diamond-like infrared nonlinear optical material with high laser damage threshold and outstanding second-harmonic generation response, *Dalton Trans.*, **48**(14), 4484–4488 (2019d).

Huang Y.-Z., Zhang H., Lin C.-S., Cheng W.D., Guo Z., Chai G.-L. $PbGa_2GeS_6$: an infrared nonlinear optical material synthesized by an intermediate-temperature self-fluxing method, *Cryst. Growth Des.*, **18**(2), 1162–1167 (2018).

Huch M.R., Gulay L.D., Olekseyuk I.D. Crystal structures of the $R_3Mg_{0.5}GeS_7$ (R = Y, Ce, Pr, Nd, Sm, Gd, Tb, Dy, Ho and Er) compounds, *J. Alloys Compd.*, **424**(1–2), 114–118 (2006).

Hughes O.H., Woolley J.C., Lopez-Rivera S.A., Pamplin B.R. Quaternary adamantine selenides and tellurides of the form I III IV VI_4, *Solid State Commun.*, **35**(8), 573–575 (1980).

Hurlbut C.S., Jr., Aristarain L.F. Olsacherite, $Pb_2(SO_4)(SeO_4)$, a new mineral from Bolivia, *Am. Mineralog.*, **54**(11–12), 1519–1527 (1969).

Hu X.-N., Xiong L., Wu L.-M. Six new members of $A_2M^{II}M^{IV}_3Q_8$ family and their structural relationship, *Cryst. Growth Des.*, **18**(5), 3124–3131 (2018).

Hwang S.-J., Iyer R.G., Kanatzidis M.G. Quaternary selenostannates $Na_{2-x}Ga_{2-x}Sn_{1+x}Se_6$ and $AGaSnSe_4$ (A = K, Rb, and Cs) through rapid cooling of melts. Kinetics versus thermodynamics in the polymorphism of $AGaSnSe_4$, *J. Solid State Chem.*, **177**(10), 3640–3649 (2004a).

Hwang S.-J., Iyer R.G., Trikalitis P.N., Ogden A.G., Kanatzidis M.G. Cooling of melts: kinetic stabilization and polymorphic transitions in the $KInSnSe_4$ system, *Inorg. Chem.*, **43**(7), 2237–2239 (2004b).

Hwu S.-J., Bucher C.K., Carpenter J.D., Taylor S.P. A solid-state diastereomer, $AgLa_3GeS_7$, *Inorg. Chem.*, **34**(8), 1979–1980 (1995).

Ibadova G.I., Aliyev Z.S., Kuliyeva U.A., Shukurova G.M., Hasanova Z.T., Babanly M.B. Physico-chemical study of the $YbTe$–$SnSb_2Te_4$ system [in Azerbaijanian], *Chem. Probl.*, (4), 418–422 (2013).

Ibañez A., Jumas J.C., Olivier-Fourcade J., Philippot E., Maurin M. Propriétès electriques et optiques dans les systèmes tenaires V–VI–VII ou quarternaires IV–V–VI–VII, *Mater. Res. Bull.*, **19**(8), 1005–1013 (1984).

Ichikawa T., Maeda T., Matsushita H., Katsui A. Crystal growth and characterization of Cu_2-II-IV-S_4 compound semiconductors [in Japanese], *J. Adv. Sci.*, **12**(1–2), 99–100 (2000).

Ijjaali I., Ibers J.A. Crystal structure of diterbium orthosilicate selenide, $Tb_2(SiO_4)Se$, *Z. Kristallogr., New Cryst. Str.*, **217**(2), 157–158 (2002).

Ijjaali I., Mitchell K., Ibers J.A. Crystal structure of digadolinium orthosilicate telluride, $Gd_2(SiO_4)Te$, *Z. Kristallogr., New Cryst. Str.*, **216**(4), 487–488 (2001).

Ilinca G., Makovicky E., Topa D., Zagler G. Cuproneyite, $Cu_7Pb_{27}Bi_{25}S_{68}$, a new mineral species from Băiţa Bihor, Romania, *Canad. Mineralog.*, **50**(2), 353–370 (2012).

Imamaliyeva S.Z., Alakbarzade G.I., Gasymov V.A., Babanly M.B. Experimental study of the Tl_4PbTe_3–Tl_9TbTe_6–Tl_9BiTe_6 section of the Tl–Pb–Bi–Tb–Te system, *Mater. Res.*, **21**(4), e20180189 (2018).

Imamaliyeva S.Z., Alakbarzade G.I., Mamedov A.N., Babanly M.B. Modeling the phase diagrams of the Tl_9SmTe_6–Tl_4PbTe_3 and Tl_9SmTe_6–Tl_9BiTe_6 systems. *Azerb. Chem. J.*, **4**, 12–16 (2020).

Imamaliyeva S.Z., Alakbarzade G.I., Mamedov A.N., Babanly M.B. Modeling the phase diagram of the Tl_9TbTe_6–Tl_4PbTe_3–Tl_9BiTe_6 system. *Azerb. Chem. J.*, **2**, 6–12 (2021).

Imaoka M., Jamadzaki T. Investigation of chalcogenide glass system, containing GeS_2. The range of glass formation [in Japanese], *Monthly J. Inst. Industr. Sci., Univ. Tokyo*, **19**(9), 261–262 (1967).

Infante E.R., Delgado J.M., López Rivera S.A. Synthesis and crystal structure of $Cu_2FeSnSe_4$, a I_2 II IV VI_4 semiconductor, *Mater. Lett.*, **33**(1–2), 67–70 (1997).

Inoue Y., Suzuki K., Matsuia N., Hirayama M., Kanno R. Synthesis and structure of novel lithium-ion conductor $Li_7Ge_3PS_{12}$, *J. Solid State Chem.*, **246**, 334–340 (2017).

Iordanidis L., Brazis P.W., Kannewurf C.K., Kanatzidis M.G. Synthesis and thermoelectric properties of $Cs_2Bi_{7.33}Se_{12}$, $A_2Bi_8Se_{13}$ (A = Rb, Cs), $Ba_{4-x}Bi_{6+2/3x}Se_{13}$, and $Ba_{3\pm x}Pb_{3\pm x}Bi_8Se_{15}$, *Mat. Res. Soc. Symp. Proc.*, **545**, 189–196 (1998).

Iordanidis L., Kanatzidis M.G. Novel quaternary lanthanum bismuth sulfides $Pb_2La_xBi_{8-x}S_{14}$, $Sr_2La_xBi_{8-x}S_{14}$, and $Cs_2La_xBi_{10-x}S_{16}$ with complex structures, *Inorg. Chem.*, **40**(8), 1878–1887 (2001).

Irran E., Tillmanns E., Hentschel G. Ternesite, $Ca_5(SiO_4)_2SO_4$, a new mineral from the Ettringer Bellerberg/Eifel, Germany, *Mineral. Petrol.*, **60**(1–2), 121–132 (1997).

Isaenko L.I., Yelisseyev A.P., Lobanov S.I., Krinitsin P.G., Molokeev M.S. Structure and optical properties of $Li_2Ga_2GeS_6$ nonlinear crystal, *Opt. Mater.*, **47**, 413–419 (2015).

Ismayilova E.N., Baladzhayeva A.N., Mashadiyeva L.F. Phase equilibria along the Cu_3SbSe_4–$GeSe_2$ section of the Cu–Ge–Sb–Se system, *New Mater., Compd. Appl.*, **5**(1), 52–58 (2021).

Ismayilova E.N., Mashadieva L.F., Bakhtiyarly I.B., Babanly M.B. Phase equilibria in the Cu_2Se–$SnSe$–$CuSbSe_2$ system, *Russ. J. Inorg. Chem.*, **64**(6), 801–809 (2019).

Ismayilova E.N., Mashadieva L.F., Bakhtiyarly I.B., Babanly M.B. Phase equilibria in the Cu_2Se–$SnSe$–Sb_2Se_3 system, *Azerb. Chem. J.*, **1**, 73–82 (2022).

Ismayilova E.N., Mashadieva L.F. Phase equilibria in the Cu_2Se–$SnSe$–Sb_2Se_3 system along the $SnSe$–Cu_3SbSe_3 section, *Kondens. sredy i mezhfaz. granitsy*, **20**(2), 218–221 (2018).

Ismayilova E.N., Shukurova G.M., Aliyeva E.R., Mashadieva L.F. Polythermal section $SnSe$–$CuSbSe_2$ of phase diagram of the Cu_2Se–$SnSe$–Sb_2Se_3 system, *Azerb. khim. zhurn.*, **4**, 29–32 (2018).

Ito T., Nowacki W. The crystal structure of freieslebenite, $PbAgSbS_3$, *Z. Kristallogr.*, **139**(1–2), 85–102 (1974a).

Ito T., Nowacki W. The crystal structure of jordanite, $Pb_{28}As_{12}S_{46}$, *Z. Kristallogr.*, **139**(3–5), 161–185 (1974b).

Ivanova Z.G., Vassilev V.S., Vassileva, K.G. New IR-transmitting chalcohalide glasses, *J. Non-Cryst. Solids*, **162**(1–2), 123–127 (1993).

Ivanov S.A., Nordblad P., Mathieu R. Tellgren R., Ritter C. Structural and magnetic properties of the ordered perovskite Pb_2CoTeO_6, *Dalton Trans.*, **39**(36), 11136–11148 (2010).

Ivashchenko I.A., Klymovych O.S., Olekseyuk I.D., Gulay L.D., Halyan V.V., Strok O.M. Quasi-ternary system Ag_2Se–$GeSe_2$–As_2Se_3, *J. Phase Equilib. Diffus.*, **43**(4), 483–494 (2022).

Ivashchenko I.A., Ostapyuk T.A., Olekseyuk I.D., Zmiy O.F., Strok O.M. Phase equilibria in the quasi-ternary system Cu_2Se–$GeSe_2$–Sb_2Se_3, *J. Phase Equilib. Diffus.*, **41**(6), 827–834 (2020).

Iyer R.G., Aitken J.A., Kanatzidis M.G. Noncentrosymmetric cubic thio- and selenogermanates: $A_{0.5}M_{1.75}GeQ_4$ (A = Ag, Cu, Na; M= Pb, Eu; Q = S, Se), *Solid State Sci.*, **6**(5), 451–459 (2004).

Iyer R.G., Do J., Kanatzidis M.G. Flux synthesis of the noncentrosymmetric cluster compounds $Cs_2SnAs_2Q_9$ (Q = S, Se) containing two different polychalcoarsenite β-$[AsQ_4]^{3-}$ and $[AsQ_5]^{3-}$ ligands, *Inorg. Chem.*, **42**(5), 1475–1482 (2003).

Iyer R.G., Kanatzidis M.G. Controlling Lewis basicity in polythioarsenate fluxes: stabilization of $KSnAsS_5$ and $K_2SnAs_2S_6$. Extended chains and slabs based on pyramidal β-$[AsS_4]^{3-}$ and $[AsS_3]^{3-}$ units, *Inorg. Chem.*, **41**(14), 3605–3607 (2002).

Iyer A.K., Yin W., Lee E.J., Bernard G.M., Michaelis V.K., Mar A. Quaternary chalcogenides $La_3Sn_{0.5}InS_7$ and $La_3Sn_{0.5}InSe_7$, *Z. anorg. und allg. Chem.*, **643**(23), 1867–1873 (2017a).

Iyer A.K., Yin W., Lee E.J., Lin X., Mar A. Quaternary rare-earth sulfides $RE_3M_{0.5}GeS_7$ (RE = La–Nd, Sm; M = Co, Ni) and $Y_3Pd_{0.5}SiS_7$, *J. Solid State Chem.*, **250**, 14–23 (2017b).

Jaeger F.M., Germs H.C. III. Über die binären Systeme der Sulfate, Chromate, Molybdate und Wolframate des Bleies, *Z. anorg. und allg. Chem.*, **119**(1), 145–173 (1921).

Jahangirova S.K., Mammadov Sh.H., Gurbanov G.R., Aliyev O.M. Interaction in the system $CuGaS_2$–$PbGa_2S_4$, *Azerb. Chem. J.*, (*1*), 46–49 (2019).

Jakubcová P., Johrendt D., Sebastian C.P., Rayaprol S., Pöttgen R. Structure, magnetic properties and ^{151}Eu, ^{119}Sn Mössbauer spectroscopy of $Eu_5Sn_3S_{12}$ and $Eu_4LuSn_3S_{12}$, *Z. Naturforsch.*, **62B**(1), 5–14 (2007).

Jambor J.L., Burke E.A.J. New mineral names, *Am. Mineralog.*, **78**(7–8), 845–849 (1993).

Jambor J.L. Dadsonite (minerals Q and QM), a new lead sulphantimonide, *Mineralog. Mag.*, **37**(288), 437–441 (1969).

Jambor J.L., Grew E.S. New mineral names, *Am. Mineralog.*, **75**(7–8), 931–937 (1990).

Jambor J.L., Grew E.S., Roberts A.C. New mineral names, *Am. Mineralog.*, **81**(11–12), 1513–1518 (1996a).

Jambor J.L., Grew E.S., Roberts A.C. New mineral names, *Am. Mineralog.*, **85**(10), 1561–1565 (2000a).

Jambor J.L., Grew E.S., Roberts A.C. New mineral names, *Am. Mineralog.*, **89**(9–10), 1574–1578 (2004).

Jambor J.L., Kovalenker V.A., Roberts A.C. New mineral names. *Amer. Mineralog*, **80**(5–6), 630–635 (1995).

Jambor J.L., Kovalenker V.A., Puziewicz J., Roberts A.C. New mineral names, *Am. Mineralog.*, **81**(9–10), 1282–1286 (1996b).

Jambor J.L., Kovalenker V.A., Roberts A.C. New mineral names, *Am. Mineralog.*, **81**(7–8), 1013–1017 (1996c).

Jambor J.L., Kovalenker V.A., Roberts A.C. New mineral names, *Am. Mineralog.*, **83**(9–10), 1117–1121 (1998a).

Jambor J.L. New lead sulfantimonides from Madoc, Ontario – Part 1, *Canad. Mineralog.*, **9**(1) 7–24 (1967a).

Jambor J.L. New lead sulfantimonides from Madoc, Ontario. Part 2 – mineral descriptions, *Canad. Mineralog.*, **9**(2), 191–213 (1967b).

Jambor J.L. New lead sulfantimonides from Madoc, Ontario. Part 3. Syntheses, paragenesis, origin, *Canad. Mineralog.*, **9**(4), 505–521 (1968).

Jambor J.L., Pertsev N.N., Roberts A.C. New mineral names, *Am. Mineralog.*, **82**(3–4), 430–433 (1997).

Jambor J.L., Pertsev N.N., Roberts A.C. New mineral names, *Am. Mineralog.*, **85**(9), 1321–1325 (2000b).

Jambor J.L., Puziewicz J. New mineral names, *Am. Mineralog.*, **76**(9–10), 1728–1735 (1991).

Jambor J.L., Puziewicz J. New mineral names, *Am. Mineralog.*, **77**(9–10), 1116–1121 (1992).

Jambor J.L., Puziewicz J., Roberts A.C. New mineral names, *Am. Mineralog.*, **83**(5–6), 652–656 (1998b).

Jambor J.L., Roberts A.C. New mineral names, *Am. Mineralog.*, **86**(1–2), 197–200 (2001).

Jambor J.L., Roberts A.C. New mineral names, *Am. Mineralog.*, **89**(11–12), 1826–1834 (2004).

Jambor J.L., Vanko D.A. New mineral names, *Am. Mineralog.*, **77**(3–4), 446–452 (1992).

Jana S., Panigrahi G., Ishtiyak M., Narayanswamy S., Bhattacharjee P.P., Niranjan M.K., Prakash J. Germanium antimony bonding in $Ba_4Ge_2Sb_2Te_{10}$ with low thermal conductivity, *Inorg. Chem.*, **61**(2) 968–981 (2022a).

Jana S., Panigrahi G., Tripathy B., Malladi S.K., Niranjan M.K. Prakash J. A new non-stoichiometric quaternary sulfide $Ba_{3.14(4)}Sn_{0.61(1)}Bi_{2.39(1)}S_8$: synthesis, crystal structure, physical properties, and electronic structure, *J. Solid State, Chem.*, **308**, 122914 (2022b).

Jiang H., Dai P., Feng Z., Fan W., Zhan J. Phase selective synthesis of metastable orthorhombic Cu_2ZnSnS_4, *J. Mater. Chem.*, **22**(15), 7502–7506 (2012).

Ji B., Pandey K., Harmer C.P., Wang F., Wu K., Hu J., Wang J. Centrosymmetric or noncentrosymmetric? Transition metals talking in $K_2TGe_3S_8$ (T = Co, Fe), *Inorg. Chem.*, **60**(14), 10603–10613 (2021a).

Ji M., Baiyin M., Ji S., An Y. Solvothermal syntheses and structures of $A_3AgSn_3Se_8$ (A = Rb, K), *Inorg. Chen. Commun.*, **10**(5), 555–557 (2007).

Jin X., Zhang L., Shu G., Wang R., Guo H. Synthesis and characterization of a novel quaternary metal selenide, $K_2Hg_3Ge_2Se_8$, *J. Alloys Compd.*, **347**(1–2), 67–71 (2002).

Jin Z., Li Z., Du Y. Synthesis and thecrystal structure of $La_6NiSi_2S_{14}$ and $La_6CoSi_2S_{14}$ [in Chinese], *Chin. J. Appl. Chem.*, **2**(4), 42–46 (1985).

Ji X., Wu H., Zhang B., Yu H., Hu Z., Wang J., Wu Y. Intriguing dimensional transition inducing variable birefringence in $K_2Na_2Sn_3S_8$ and $Rb_3NaSn_3Se_8$, *Inorg. Chem.*, **60**(2), 1055–1061 (2021b).

Johan Z., Mantienne J. Thallium-rich mineralization at Jas Roux, Hautes-Alpes, France: a complex epithermal, sediment-hosted, ore-forming system, *J. Czech Geol. Soc.*, **45**(1–2), 63–77 (2000).

Johan Z., Picot P. Définition nouvelle de la weibullite et de la wittite, *C. r. Acad. sci., Sér. D.*, **282**(1), 137–139 (1976).

Johan Z., Picot P. La pirquitasite, Ag_2ZnSnS_4, un nouveau membre du groupe de la stannite, *Bull. Minéral.*, **105**(3), 229–235 (1982).

Johan Z., Picot P., Ruhlmann F. The ore mineralogy of the Otish Mountains uranium deposit, Quebec: skippenite, Bi_2Se_2Te, and watkinsonite, $Cu_2PbBi_4(Se,S)_8$, two new mineral species, *Canad. Mineralog.*, **25**(4), 625–638 (1987).

Johnston M.G., Harrison W.T.A. Two new octahedral/pyramidal frameworks containing both cation channels and lone-pair channels: syntheses and structures of $Ba_2Mn^{II}Mn_2^{III}(SeO_3)_6$ and $PbFe_2(SeO_3)_4$, *J. Solid State Chem.*, **177**(12), 4680–4686 (2004).

Johrendt D. Barium palladium selenostannate perselenostannate diselenide, $Ba_8Pd(SnSe_4)_{3.75}(SnSe_5)_{0.25}(Se_2)$, *Acta Crystallogr.*, **E58**(6), i52–i55 (2002).

de Jong W.F. Die Kristallstruktur von Germanit, *Z. Kristallogr.*, **73**(1–6), 176–180 (1930).

Julien C., Barnier S., Massot M., Chbani N., Cai X., Loireau-Lozac'h A.M., Guittard M. Raman and infrared spectroscopic studies of Ge–Ga–Ag sulphide glasses, *Mater. Sci. Eng.*, **B22**(2–3), 191–200 (1994).

Jumas J.C., Philippot E., Maurin M. Structure du rhodostannite synthétique, *Acta Crystallogr.*, **B35**(9), 2195–2197 (1979).

Jumas J.C., Ribes M., Philippot E., Maurin M. Sur le système SnS_2–PbS. Structure cristalline de $PbSnS_3$, *C. r. Acad. sci. Sér. C*, **275**(4), 269–272 (1972).

Kabalov Yu.K., Evstigneeva T.L., Spiridonov E.M. Crystal structure of Cu_2HgSnS_4, a synthetic analogue of mineral velikite, *Crystallogr. Rep.*, **43**(1), 16–20 (1998).

Kabbour K., Cario L., Danot M., Meerschaut A. Design of a new family of inorganic compounds $Ae_2F_2SnX_3$ (Ae = Sr, Ba; X = S, Se) using rock salt and fluorite 2D building blocks, *Inorg. Chem.*, **45**(2), 917–922 (2006).

Kaden R., Wagner G., Bente K. Crystal chemistry and electrical conductivity of boulangerite, dadsonite and iodine-substituted pillaite grown by chemical vapor transport, *Canad. Mineralog.*, **50**(2), 219–233 (2012).

Kaib T., Haddadpour S., Andersen H.F., Mayrhofer L., Järvi T.T., Moseler M., Möller K.C., Dehnen S. Quaternary diamond-like chalcogenidometalate networks as efficient anode material in lithium-ion batteries, *Adv. Funct. Mater.*, **23**(46), 5693–5699 (2013).

Kalashnikov A.A., Drovart Zh., Burdeynyy A.N., Pashinkin A.S. Investigation of the PbSe–SnTe solid solutions sublimation by mass-spectrometric method [in Russian], *Dokl. AN SSSR*, **261**(4), 905–907 (1981).

Kala T., Frumar M., Klikorka J. Herstellung und einige physikalische Eigenschaften von Monokristallen der Halbleiterverbindung $CuPbAsS_3$, *Collect. Czech. Chem. Comm.*, **36**(11), 3824–3833 (1971).

Kamaya N., Homma K., Yamakawa Y., Hirayama M., Kanno R., Yonemura M., Kamiyama T., Kato Y., Hama S., Kawamoto K., Mitsui A. A lithium superionic conductor, *Nature Mater.*, **10**(9), 682–686 (2011).

Kamimura T., Matoba M., Anzai S. Conductivity phase transition with shift of valence-band edge and with structural distortion in $Cu_4Sn_{1-x}Ge_xS_4$, *J. Phys. Soc. Jpn.*, **59**(9), 3045–3048 (1990).

Kampf A.R., Dunn P.J., Foord E.E. Grandreefite, pseudograndreefite, laurelite, and aravaipaite: four new minerals from the Grand Reef mine, Graham County, Arizona, *Am. Mineralog.*, **74**(7–8), 927–933 (1989).

Kampf A.R., Foord E.E. Calcioaravaipaite, a new mineral, and associated lead fluoride minerals from the Grand Reef mine, Graham County, Arizona, *Mineralog. Rec.*, **27**(4), 293–300 (1996).

Kampf A.R. Grandreefite, $Pb_2F_2SO_4$: crystal structure and relationship to the lanthanide oxide sulfates, $Ln_2O_2SO_4$, *Am. Mineralog.*, **76**(1–2), 278–282 (1991).

Kampf A.R., Housley R.M., Rossman G.R. Northstarite, a new lead-tellurite-thiosulfate mineral from the North Star mine, Tintic, Utah, USA, *Canad. Mineralog.*, **58**(4), 533–542 (2020a).

Kampf A.R., Housley R.M., Rossman G.R. Northstarite, IMA 2019-031. CNMNC Newsletter No. 51, *Eur. J. Mineral.*, **31**(5–6), 1100 (2019a).

Kampf A.R., Housley R.M., Rossman G.R. Northstarite, IMA 2019-031. CNMNC Newsletter No. 51, *Mineralog. Mag.*, **83**(5), 758 (2019b).

Kampf A.R., Housley R.M., Rossman G.R., Yang H., Downs R.T. Adanite, a new lead-tellurite-sulfate mineral from the North Star mine, Tintic, Utah, and Tombstone, Arizona, USA, *Canad. Mineralog.*, **58**(3), 403–410 (2020b).

Kampf A.R., Housley R.M., Rossman G.R., Yang H., Downs R.T. Adanite, IMA 2019-088. CNMNC Newsletter No. 53, *Eur. J. Mineral.*, **32**(1), 210–211 (2020c).

Kampf A.R., Housley R.M., Rossman G.R., Yang H., Downs R.T. Adanite, IMA 2019-088. CNMNC Newsletter No. 53, *Mineralog. Mag.*, **84**(1), 160 (2020d).

Kampf A.R., Mills S.J., Housley R.M., Marty J., Thorne B. Lead-tellurium oxysalts from Otto Mountain near Baker, California: IV. Markcooperite, $Pb(UO_2)Te^{6+}O_6$, the first natural uranyl tellurate. *Am. Mineralog.*, **95**(10), 1554–1559 (2010a).

Kampf A.R., Mills S.J., Housley R.M., Marty J., Thorne B. Lead-tellurium oxysalts from Otto Mountain near Baker, California: VI. Telluroperite, $Pb_3Te^{4+}O_4Cl_2$, the Te analog of perite and nadorite, *Am. Mineralog.*, **95**(10), 1569–1573 (2010b).

Kampf A.R., Mills S.J., Rumsey M.S. The discreditation of girdite, *Mineralog. Mag.*, **81**(5), 1125–1128 (2017).

Kampf A.R., Smith J.B., Hughes J.M., Ma C., Emproto C. Boojumite, IMA 2022-028. CNMNC Newsletter No 68, *Eur. J. Mineral.*, **34**(5), 389 (2022a).

Kampf A.R., Smith J.B., Hughes J.M., Ma C., Emproto C. Boojumite, IMA 2022-028. CNMNC Newsletter No 68, *Mineralog. Mag.*, **86**(5), 857–858 (2022b).

Kampf A.R., Smith J.B., Hughes J.M., Ma C., Emproto C. Cubothioplumbite, IMA 2021-091. CNMNC Newsletter No 65, *Eur. J. Mineral.*, **34**(1), 146 (2022c).

Kampf A.R., Smith J.B., Hughes J.M., Ma C., Emproto C. Cubothioplumbite, IMA 2021-091. CNMNC Newsletter No 65, *Mineralog. Mag*, **86**(2), 356 (2022d).

Kampf A.R., Smith J.B., Hughes J.M., Ma C., Emproto C. Finescreekite, IMA 2022-030. CNMNC Newsletter No 68, *Eur. J. Mineral.*, **34**(5), 389–390 (2022e).

Kampf A.R., Smith J.B., Hughes J.M., Ma C., Emproto C. Finescreekite, IMA 2022-030. CNMNC Newsletter No 68, *Mineralog. Mag.*, **86**(5), 858 (2022f).

Kampf A.R., Smith J.B., Hughes J.M., Ma C., Emproto C. Hexathioplumbite, IMA 2021-092. CNMNC Newsletter No 65, *Eur. J. Mineral.*, **34**(1), 146 (2022g).

Kampf A.R., Smith J.B., Hughes J.M., Ma C., Emproto C. Hexathioplumbite, IMA 2021-092. CNMNC Newsletter No 65, *Mineralog. Mag*, **86**(2), 356–357 (2022h).

Kampf A.R., Smith J.B., Hughes J.M., Ma C., Emproto C. Kennygayite, IMA 2022-032. CNMNC Newsletter No 68, *Eur. J. Mineral.*, **34**(5), 390 (2022i).

Kampf A.R., Smith J.B., Hughes J.M., Ma C., Emproto C. Kennygayite, IMA 2022-032. CNMNC Newsletter No 68, *Mineralog. Mag.*, **86**(5), 858 (2022k).

Kanatzidis M.G., Chung D.-Y., Iordanidis L., Choi K.-S., Brazis P., Rocci M., Hogan T., Kannewurf C. Solid state chemistry approach to advanced thermoelectrics. Ternary and quaternary alkali metal bismuth chalcogenides as thermoelectric materials, *Mater. Res. Soc. Symp. Proc.*, **545**, 233–246 (1998).

Kanatzidis M.G., Liao J.H., Marking G.A. Alkali metal quaternary chalcogenides and process for the preparation of thereof, USA Patent 5 614 128. Appl. № 606565. Filed 26.02.96; Data of Patent 25.03.1997 (1997a).

Kanatzidis M.G., Liao J.H., Marking G.A. Alkali metal quaternary chalcogenides and process for the preparation of thereof, USA Patent 5 618 471. Appl. № 606886. Filed 26.02.96; Data of Patent 08.04.1997 (1997b).

Kanno R., Murayama M. Lithium ionic conductor thio-LISICON: the Li_2S–GeS_2–P_2S_5 system, *J. Electrochem. Sod.*, **148**(7), A742–A746 (2001).

Kanno R., Hata T., Kawamoto Y., Irie M. Synthesis of a new lithium ionic conductor, thio-LISICON – lithium germanium sulfide system, *Solid State Ionics*, **130**(1–2), 97–104 (2000).

Kaplunnik L.N., Pobedimskaya E.A., Belov N.V. Crystal structure of the lindstremite, $CuPbBi_3S_6$ [in Russian], *Kristallografiya*, **20**(5), 1037–1039 (1975).

Kaplunnik L.N., Pobedimskaya E.A., Belov N.V. Crystal structure of velikite $Cu_{3.75}Hg_{1.75}Sn_2S_8$ [in Russian], *Kristallografiya*, **22**(1), 175–177 (1977).

Karup-Møller S. Gustavite, a new sulphosalt mineral from Greenland, *Canad. Mineralog.*, **10**(2), 173–190 (1970).

Karup-Møller S. Makovicky E. On pavonite, cupropavonite, benjaminite, and "oversubstituted" gustavite, *Bull. Minéralog.*, **102**(4), 351–367 (1979).

Karup-Møller S. Mineralogy of some Ag–(Cu)–Pb–Bi sulfide associations, *Bull. Geol. Soc. Denmark*, **26**(1–2), 41–68 (1977).

Karup-Møller S. New data on schirmerite, *Canad. Mineralog.*, **11**(5), 952–957 (1973).

Kasatkin A.V., Makovicky E., Plášil J., Škoda R., Agakhanov A.A., Chaikovskiy I.I., Vlasov E.A., Pekov I.V. Chukotkaite, $AgPb_7Sb_5S_{15}$, a new sulfosalt mineral from Eastern Chukotka, Russia, *Canad. Mineralog.*, **58**(5), 587–596 (2020a).

Kasatkin A.V., Plášil J., Makovicky E., Škoda R., Agakhanov A.A., Chaikovskiy I.I., Vlasov E.A., Pekov I.V. Chukotkaite, IMA 2019-124. CNMNC Newsletter No. 54, *Eur. J. Mineral.*, **32**(2), 281 (2020b).

Kasatkin A.V., Plášil J., Makovicky E., Škoda R., Agakhanov A.A., Chaikovskiy I.I., Vlasov E.A., Pekov I.V. Chukotkaite, IMA 2019-124. CNMNC Newsletter No. 54, *Mineralog. Mag.*, **84**(2), 364 (2020c).

Kassem M.A., Tabata Y., Waki T., Nakamura H. Single crystal growth and characterization of kagomé-lattice shandites $Co_3Sn_{2-x}In_xS_2$, *J. Crys. Growth*, **426**, 208–213 (2015).

Kassem M.A., Tabata Y., Waki T., Nakamura H. Structure and magnetic properties of flux grown single crystals of $Co_{3-x}Fe_xSn_2S_2$ shandites, *J. Solid State Chem.*, **233**, 8–13 (2016).

Kato Y., Kawamoto K., Kanno R., Hirayama M. Discharge performance of all-solid-state battery using a lithium superionic conductor $Li_{10}GeP_2S_{12}$, *Electrochemistry*, **80**(10), 749–751 (2012).

Kennedy J.H., Yang Y. Glass-forming region and structure in SiS_2–Li_2S–LiX (X = Br, I), *J. Solid State Chem.*, **69**(2), 252–257 (1987).

Keutsch F.N., Topa D., Fredrickson R.T., Makovicky E., Paar W. Agmantinite, Ag_2MnSnS_4, a new mineral with a wurtzite derivative structure from the Uchucchacua polymetallic deposit, Lima Department, Peru, *Mineralog. Mag.*, **83**(2), 233–238 (2019).

Keutsch F.N., Topa D., Fredrickson R.T., Makovicky E., Paar W. Agmantinite, IMA 2014-083. CNMNC Newsletter No. 23, February 2015, page 57, *Mineralog. Mag.*, **79**(1), 51–58 (2015).

Keutsch F.N., Topa D., Makovicky E. Hyršlite, IMA 2016-097. CNMNC Newsletter No. 36, April 2017, page 404, *Mineralog. Mag.*, **81**(2), 403–409 (2017).

Keutsch F.N., Topa D., Makovicky E. Hyršlite, $Pb_8As_{10}Sb_6S_{32}$, a new N = 3;3 member of the sartorite homologous series from the Uchucchacua polymetallic deposit, Peru, *Eur. J. Mineral.*, **30**(6), 1155–1162 (2018).

Kevshyn A.H., Halyan V.V., Davydyuk H.Ye., Parasyuk O.V., Mazurets I.I. Concentration dependence of the optical properties of glassy alloys in the HgS–Ga_2S_3–GeS_2 system, *Glass Phys. Chem.*, **36**(1), 27–32 (2010).

Khamiyev M.S., Aliyev O.M., Agayeva R.M. Synthesis, single crystals growth and some physic.co-chemical properties of the quaternary compounds of the Pb_2LnBiS_5 type (Ln = Nd, Sm, Gd) [in Azerbaijanian], *Kimya Problem.*, (4), 614–616 (2011).

Kheraj V., Patel K.K., Patel S.J., Shah D.V. Synthesis and characterization of copper zinc tin sulphide (CZTS) compound for absorber material in solar-cells, *J. Cryst. Growth*, **362**, 174–177 (2013).

Khiminets V.V. Investigation of interaction and glass formation in the $AsSI$–GeS_2 system [in Russian], *Zhurn. neorgan. khimii*, **27**(8), 2094–2097 (1982).

Khiminets V.V., Tsigika V.V., Khiminets O.V., Dobosh M.V. Interaction and glassforming in the Ge–As–S–Br system [in Russian], *Zhurn. neorgan. khimii*, **35**(11), 2945–2947 (1990).

Khvaleba N.V., Gulay L.D., Zmiy O.F., Olekseyuk I.D. Investigation of the Tb$_2$S$_3$–Cu$_2$S–PbS system at 870 K [in Ukrainian], *Nauk. Visnyk Volyns'k. Derzh. Univ. im. Lesi Ukrainky. Ser. Khim. nauky*, **4**, 112–118 (2006).

Khyzhun O.Y., Babizhetskyy V.S., Kityk I.V., Myronchuk G.L., Jędryka J., Lakshinarayana G., Levytskyy V.O., Tsisar O.V., Piskach L.V., Parasyuk O.V., ElNaggar A.M., Albasssam A.A., Piasecki M. Thallium indium germanium sulphide (TlInGe$_2$S$_6$) as efficient material for nonlinear optical application, *J. Alloys Compd.*, **735**, 1694–1702 (2018a).

Khyzhun O.Y., Bekenev V.L., Ocheretova V.A., Fedorchuk A.O., Parasyuk O.V. Electronic structure of Cu$_2$ZnGeSe$_4$ single crystal: ab initio FP-LAPW calculations and X-ray spectroscopy measurements, *Physica B: Condens. Matter*, **461**, 75–84 (2015).

Khyzhun O.Y., Fedorchuk A.O., Kityk I.V., Piasecki M., Mozolyuk M.Y., Piskach L.V., Parasyuk O.V., ElNaggar A.M., Albasssam A.A., Karasinski P. Electronic structure and laser induced piezoelectricity of a new quaternary compound TlInGe$_3$S$_8$, *Mater. Chem. Phys.*, **204**, 336–344 (2018b).

Khyzhun O.Y., Parasyuk O.V., Fedorchuk A.O. Single crystal growth and electronic structure of thiogermanate AgGaGeS$_4$, a novel nonlinear optical material, *Adv. Alloys Compd.*, **1**(1), 15–29 (2014).

Khyzhun O.Y., Parasyuk O.V., Tsisar O.V., Piskach L.V., Myronchuk G.L., Levytskyy V.O., Babizhetskyy V.S. New quaternary thallium indium germanium selenide TlInGe$_2$Se$_6$: crystal and electronic structure, *J. Solid State Chem.*, **254**, 103–108 (2017).

Kikuchi Y., Bouyrie Y., Ohta M., Suekuni K., Aihara M., Takabatake T. Vanadium-free colusites Cu$_{26}$A$_2$Sn$_6$S$_{32}$ (A = Nb, Ta) for environmentally-friendly thermoelectrics, *J. Mater. Chem. A*, **4**(39), 15207–15214 (2016).

Kim Y., Seo I.-S., Martin S.W., Baek J., Halasyamani P.S., Arumugam N., Steinfink H. Characterization of new infrared nonlinear optical material with high laser damage threshold, Li$_2$Ga$_2$GeS$_6$, *Chem. Mater.*, **20**(19), 6048–6052 (2008).

Kiryakov A.S., Piryazev D.A., Tarasenko N.S., Naumov N.G. Crystal structures of new chalcogenidecontaining yttrium orthosilicates Y$_2$SiO$_4$Q (Q = S, Se), *J. Struct. Chem.*, **59**(3), 635–640 (2018).

Kislinskaya G.E. Investigation of coprecipitation of cadmium, indium and gallium with sulfides of some metals [in Russian], *Avtoref. dis.... kand. khim. nauk*, Kiev, 27 p. (1974).

Kissin S.A., Owens de A.R. The relatives of stannite in the light of new data, *Canad. Mineralog.*, **27**(4), 673–688 (1989).

Kissin S.A., Owens D.R. New data on stannite and related tin sulfide minerals, *Canad. Mineralog.*, **17**(1), 125–135 (1979).

Kissin S.A., Owens D.R., Roberts W.L. Černýite, a copper cadmium tin sulfide with the stannite structure, *Canad. Mineralog.*, **16**(2), 139–146 (1978)

Klymovych O., Ivashchenko I., Olekseyuk I., Zmiy O., Lavrynyuk Z. Quasi-ternary system Cu$_2$Se–GeSe$_2$–As$_2$Se$_3$, *J. Phase Equilib. Diffus.*, **41**(2), 157–163 (2020a).

Klymovych O., Ivashchenko I., Olekseyuk I., Zmiy O. Phase equilibria in the Cu$_2$Se–SnSe$_2$–As$_2$Se$_3$ quasiternary system, *Visnyk Odes'k. Nats. Univ.*, **25**[1(73)], 31–42 (2020b).

Klymovych O., Zmiy O., Olekseyuk I. Phase equlibria and the glass formation in the Cu$_2$Se–SnSe$_2$–As$_2$Se$_3$ system [in Ukrainian], *Nauk. Visnyk Skhidnoyevrop. Nats. Univ. im. Lesi Ukrainky. Ser. Khim. nauky*, **23**(272), 89–94 (2013).

Klymovych O.S., Zmiy O.F., Olekseyuk I.D. The glass formation in the Cu$_2$Se–GeSe$_2$–As$_2$Se$_3$ system [in Ukrainian], *Nauk. Visnyk Volyns'k. Derzh. Univ. im. Lesi Ukrainky. Ser. Khim. nauky*, **15**, 14–18 (2007).

Knaust J.M., Polyakova L.A., Dorhout P.K. Crystal structures of octasodium dieuropium(II) bis[hexatellurodisilicate], NasEu$_2$[Si$_2$Te$_6$]$_2$, octasodium dieuropium(II) bis[hexaselenodigermanate], Na$_8$Eu$_2$[Ge$_2$Se$_6$]$_2$, and nonasodium samarium(III) bis[hexaselenodisilicate], Na$_9$Sm[Si$_2$Se$_6$]$_2$, *Z. Kristallogr., New. Cryst. Str.*, **220**(3), 295–297 (2005).

Kogut Yu., Fedorchuk A., Zhbankov O., Romanyuk Ya.., Kityk I., Piskach L., Parasyuk O. Isothermal section of the Ag$_2$S–PbS–GeS$_2$ system at 300 K and the crystal structure of Ag$_2$PbGeS$_4$, *J. Alloys Compd.*, **509**(11), 4264–4267 (2011).

Kogut Yu.M., Piskach L.V., Olekseyuk I.D., Parasyuk O.V. Isothermal sections of the Ag(Cu)$_2$X–PbX–SnX$_2$ (X = S, Se) [in Ukrainian], *Nauk. Visnyk Volyns'k. Derzh. Univ. im. Lesi Ukrainky. Ser. Khim. nauky*, **(4)**, 63–66 (2006).

Kohatsu I., Wuensch B.J. The crystal structure of aikinite, PbCuBiS$_3$, *Acta Crystallogr.*, **B27**(6), 1245–1252 (1971).

Kohatsu I., Wuensch B.J. The crystal structure of gladite, PbCuBi$_5$S$_9$, *Am. Mineralog.*, **58**(11–12), 1098 (1973).

Kohatsu I., Wuensch B.J. The crystal structure of gladite, PbCuBi$_5$S$_9$, a superstructure intermediate in the series Bi$_2$S$_3$–PbCuBiS$_3$ (bismuthinite–aikinite), *Acta. Crystallogr.*, **B32**(8), 2401–2409 (1976).

Koike K., Yazawa A. Thermodynamic studies of the molten Cu$_2$S–FeS–SnS systems [in Japanese], *Shigen-to-Sozai*, **110**(1), 43–47 (1994).

Kong F., Jiang H.-L., Mao J.-G. La$_4$(Si$_{5.2}$Ge$_{2.8}$O$_{18}$)(TeO$_3$)$_4$ and La$_2$(Si$_6$O$_{13}$)(TeO$_3$)$_2$: intergrowth of the lanthanum(III) tellurite layer with the XO$_4$ (X = Si/Ge) tetrahedral layer, *J. Solid State Chem.*, **181**(2), 263–268 (2008).

Konstantinova N.N., Medvedkin G.A., Polushina I.K., Smirnova A.D., Sokolova V.I., Tairov M.A. Optical and electrical properties of the Cu$_2$CdSnSe$_4$ and Cu$_2$CdGeSe$_4$ crystals [in Russian], *Izv. AN SSSR. Neorgan. mater.*, **25**(9), 1445–1448 (1989).

Kopinets I.F., Migolinets I.M., Popovich I.L., Romanovich M.E. The glasses of the SbSI–GeS$_2$ system [in Russian], *Izv. AN SSSR. Neorgan. mater.*, **11**(3), 952–953 (1975).

Kopylov N.I., Kaminskiy Yu.D. The PbS–Fe$_2$As system [in Russian], *Zhurn. neorgan. khimii*, **43**(9), 1552–1554 (1998).

Kopylov N.I., Kostenetskiy V.P., Toguzov M.Z., Rozhin A.V., Letyagin Yu.I. Interfacial equilibria in complex systems determining the modes of electric melting of semi-products of lead production [in Russian], In: *Sulfid. rasplavy tyazh. met.* Moscow, Metallurgiya Publish., 128–135 (1982).

Kopylov N.I., Smailov B.D., Toguzov M.Z., Minkevich S.M., Kostenetskiy V.P. The Cu$_3$As–Cu$_2$S–PbS system [in Russian], *Zhurn. neorgan. khimii*, **21**(11), 3068–3072 (1976a).

Kopylov N.I. The PbS–Cu$_2$S–FeS–Na$_2$S system [in Russian], *Zhurn. neorgan. khimii*, **14**(6), 1702–1704 (1969).

Kopylov N.I. The PbS–FeS–Na$_2$S system [in Russian], *Zhurn. neorgan. khimii*, **12**(10), 2832–2837 (1967).

Kopylov N.I., Toguzov M.Z., Yarygin V.I. Investigation of the PbS–Cu$_{1.8}$S–ZnS system [in Russian], *Izv. AN SSSR. Metally*, **(6)**, 80–83 (1976b).

Kopylov N.I., Toguzov M.Z., Yarygin V.I., Smailov B.D., Minkevich S.M., Kalganov I.M. The Cu$_3$As–PbS system [in Russian], *Zhurn. neorgan. khimii*, **21**(6), 1698–1701 (1976c).

Koscielski L.A., Ibers J.A. Tetrayttrium(III) trisulfide disilicate, *Acta Crystallogr.*, **E67**(2), i16 (2011).

Koskenlinna M., Valkonen J. Lead(II) hydrogenselenite, *Acta Crystallogr.*, **C51**(9), 1737–1739 (1995).

Kostov-Kytin V.V., Petrova R., Macíček J. Crystal structure of synthetic $Pb_{4.32}Sb_{3.68}S_{8.68}Cl_{2.32}$. A chlorine-bearing alternative to $Pb_4Sb_4S_{11}$, *Eur. J. Mineral.*, **9**(6), 1191–1197 (1997).

Kostov V.V., Macíček J. Crystal structure of synthetic $Pb_{12.65}Sb_{11.35}$ $S_{28.35}Cl_{2.65}$ – A new view of the crystal chemistry of chlorine-bearing lead-antimony sulphosalts, *Eur. J. Mineral.*, **7**(4), 1007–1018 (1995).

Kosuga A., Uno M., Kurosaki K., Yamanaka S. Thermoelectric properties of $Ag_{1-x}Pb_{18}SbTe_{20}$ (x = 0, 0.1, 0.3), *J. Alloys Compd.*, **387**(1–2), 52–53 (2005).

Kovalenker V.A., Evstigneeva T.L., Malov V.S., Vyal'sov L.N. Chatkalite, $Cu_6FeSn_2S_8$, a new mineral [in Russian], *Mineralog. zhurn.*, **3**(5), 79–86 (1981).

Kovrugin V.M., Colmont M., Terryn C., Colis S., Siidra O.I., Krivovichev S.V., Mentré O. pH Controlled pathway and systematic hydrothermal phase diagram for elaboration of synthetic lead nickel selenites, *Inorg. Chem.*, **54**(5), 2425–2434 (2015).

Koz'ma A.A., Barchiy I.Ye., Peresh Ye. Yu, Tsygyka V.V. Physico-chemical interaction in the $SnSe_2–TlBiSe_2–Bi_2Se_3$ quasiternary system [in Ukraininan], *Nauk. Visn. Uzhgorod. Univ. Ser. Khimiya*, (*21*), 6–12 (2009).

Koz'ma A.A. Interaction of the components in the $Tl_4SnSe_4–Tl_2Se–Tl_9BiSe_6$ quasiternary system [in Ukraininan], *Nauk. Visn. Uzhgorod. Univ. Ser. Khimiya*, **2**(30), 15–22 (2013).

Koz'ma A.A., Peresh Ye Yu., Barchiy I. Ye., Tsygyka V.V., Barchiy O.I. The $SnSe_2–TlBiSe_2$ system [in Ukraininan], *Nauk. Visn. Uzhgorod. Univ. Ser. Khimiya*, **19–20**, 89–92 (2008).

Krämer V. Lead indium bismuth chalcogenides. I. Structure of $Pb_{1.6}In_8Bi_4S_{19}$, *Acta Crystallogr.*, **C39**(10), 1328–1329 (1983).

Krämer V. Lead indium bismuth chalcogenides. II. Structure of $Pb_4In_3Bi_7S_{18}$, *Acta Crystallogr.*, **C42**(2), 249–251 (1986a).

Krämer V. Lead indium bismuth chalcogenides. III. Structure of $Pb_4In_2Bi_4S_{13}$, *Acta Crystallogr.*, **C42**(8), 1089–1091 (1986b).

Krebs H., Grün K., Kallen D. Über Structur und Eigenschaften der Halbmetalle. XIV. Mischkristallsysteme zwischen halbleitenden Chalkogeniden der vierten Hauptgruppe, *Z. anorg. und allg. Chem.*, **312**(5–6), 307–313 (1961).

Krebs H., Langner D. Über Structur und Eigenschaften der Halbmetalle. XVI. Mischkristallsysteme zwischen halbleitenden Chalkogeniden der vierten Hauptgruppe. II, *Z. anorg. und allg. Chem.*, **334**(1–2), 37–49 (1964).

Krykhovets O.V., Gulay L.D., Olekseyuk I.D. Crystal structure of the $Ag_{0.735}InGeSe_4$ compound, *J. Alloys Compd.*, **337**(1–2), 182–188 (2002).

Krykhovets A.V., Olekseyuk I.D., Sysa L.V. $Ag_2Se–In_2Se_3–GeSe_2$ system [in Russian], *Zhurn. neorgan. khimii*, **46**(7), 1180–1188 (2001).

Krykhovets O.V., Sysa L.V., Olekseyuk I.D., Glowyak T. Crystal structure of $Ag_2In_2GeSe_6$, *J. Alloys Compd.*, **287**(1–2), 181–184 (1999).

Kucha H., Osuch W., Elsen J. Viaeneite, $(Fe,Pb)_4S_8O$, a new mineral with mixed sulfur valencies from Engis, Belgium, *Eur. J. Mineral.*, **8**(1), 93–102 (1996).

Kuchar F., Heinrich H. Phasengrenzen in System $Ag_{19}Sb_{29}Te_{52}–PbTe$, *Z. Naturforsch.*, **22B**(9), 982–983 (1967).

Kuhn A., Duppel V., Lotsch B.V. Tetragonal $Li_{10}GeP_2S_{12}$ and Li_7GePS_8 – exploring the Li ion dynamics in LGPS Li electrolytes, *Energy Environ. Sci.*, **6**(12), 3548–3552 (2013a).

Kuhn A., Gerbig O., Zhu C., Falkenberg F., Maier J., Lotsch B.V. A new ultrafast superionic Li-conductor: ion dynamics in $Li_{11}Si_2PS_{12}$ and comparison with other tetragonal LGPS-type electrolytes, *Phys. Chem. Chem. Phys.*, **16**(28), 14669–14674 (2014).

Kuhn A., Köhler J., Lotsch B.V. Single-crystal X-ray structure analysis of the superionic conductor $Li_{10}GeP_2S_{12}$, *Phys. Chem. Chem. Phys.*, **15**(28), 11620–11622 (2013b).

Kuhs W.F., Nitsche R., Scheunemann K. The argyrodites – a new family of tetrahedrally close-packed structures, *Mater. Res. Bull.*, **14**(2), 241–248 (1979).

Kuliev A.Z., Kahramanov K.Sh., Babaev G.M. Investigation of physico-chemical compatibility of $SnTe–CuSbTe_2$ with iron, cobalt, nickel and nickel monoantimonide [in Russian], *Izv. AN AzSSR. Ser. phys.-techn. i mat. nauk*, **4**, 82–84 (1974).

Kuliev B.B., Gurshumov A.P., Kuliev E.M., Khalilov Kh.Ya. The $As_2S_3–SnTe$ system [in Russian], *Izv. AN SSSR. Neorgan. mater.*, **18**(5), 738–741 (1982).

Kumari A., Vidyasagar K. Solid-state synthesis, structural variants and transformation of three-dimensional sulfides, $AGaSnS_4$ (A = Na, K, Rb, Cs, Tl) and $Na_{1.263}Ga_{1.263}Sn_{0.737}S_4$, *J. Solid State Chem.*, **180**(7), 2013–2019 (2007).

Kumar S., Sharma V. Structural transition on doping rare earth *Sm* to $Ge_2Sb_2Te_5$ phase change material, *J. Alloys Compd.*, **877**, 160246 (2021).

Kumar S., Singh K. Glass transition, thermal stability and glass-forming tendency of $Se_{90-x}Te_5Sn_5In_x$ multi-component chalcogenide glasses, *Thermochim. Acta*, **528**, 32–37 (2012).

Kumar S., Singh K. The effect of indium additive on crystallization kinetics and thermal stability of Se–Te–Sn chalcogenide glasses, *Physica B: Condens. Matter*, **406**(8), 1519–1524 (2011).

Kumar V.P., Guilmeau E., Raveau B., Caignaert V., Varadaraju U.V. A new wide band gap thermoelectric quaternary selenide $Cu_2MgSnSe_4$, *J. Appl. Phys.*, **118**(15), 155101_1–155101_8 (2015).

Kumar V.P., Paradis-Fortin L., Lemoine P., Caignaert V., Raveau B., Malaman B., Le Caër G., Cordier S., Guilmeau E. Designing a thermoelectric copper-rich sulfide from a natural mineral: synthetic germanite $Cu_{22}Fe_8Ge_4S_{32}$, *Inorg. Chem.*, **56**(21), 13376–13381 (2017).

Kupčik V. $CuPb_9Bi_{12}S_{28}$ – Ein Naturpuzzle, *Z. Kristallogr.*, **162**(1–4) 155–156 (1983).

Kupčik V., Mariolacos K., Bente K. Feste Lösungen im pseudoternären System $Cu_2S–Bi_2S_3–PbS$, *Fortschr. Mineral.*, **56**(1), 73–74 (1978).

Kupčik V., Wendschuh M. The structure of antimony bismuth tin sulphide $Bi_xSb_{2-x}Sn_2S_5$, *Acta Crystallogr.*, **B38**(12), 3070–3071 (1982).

Kurbanov G.R. Synthesis and crystal structure of $SnSbBiS_4$, *Russ. J. Inorg. Chem.*, **55**(3), 495–496 (2010).

Kuribayashi T., Nagase T., Nozaki T., Ishibashi J., Shimada K., Shimizu M. Momma K. Hitachiite, IMA 2018-027. CNMNC Newsletter No. 44, August 2018, page 879, *Eur. J. Mineral.*, **31**(4), 877–882 (2018a).

Kuribayashi T., Nagase T., Nozaki T., Ishibashi J., Shimada K., Shimizu M. Momma K. Hitachiite, IMA 2018-027. CNMNC Newsletter No. 44, August 2018, page 1018, *Mineralog. Mag.*, **82**(4), 1015–1021 (2018b).

Kuribayashi T., Nagase T., Nozaki T., Ishibashi J., Shimada K., Shimizu M. Momma K. Hitachiite, $Pb_5Bi_2Te_2S_6$, a new mineral from the Hitachi mine, Ibaraki Prefecture, Japan, *Mineralog. Mag.*, **83**(5), 733–739 (2019).

Kuznetsov V.L., Zlomanov V.P., Chemleva T.A., Gas'kov A.M. A part of liquidus surface of the Pb–Sn–Te–Se system [in Russian], *Izv. AN SSSR. Neorgan. mater.*, **19**(1), 47 50 (1983).

Kwon O., Hirayama M., Suzuki K., Kato Y., Saito T., Yonemura M., Kamiyama T., Kanno R. Synthesis, structure, and conduction mechanism ofthe lithium superionic conductor $Li_{10+\delta}Ge_{1+\delta}P_{2-\delta}S_{12}$, *J. Mater. Chem. A*, **3**(1), 438–446 (2015).

Lafond A., Choubrac L., Guillot-Deudon C., Deniard P., Jobic S. Crystal structures of photovoltaic chalcogenides, an intricate puzzle to solve: the cases of CIGSe and CZTS materials, *Z. anorg. und allg. Chem.*, **638**(15), 2571–2577 (2012).

Lafond A., Choubrac L., Guillot-Deudon C., Fertey P., Evain M., Jobic S. X-ray resonant single-crystal diffraction technique, a powerful tool to investigate the kesterite structure of the photovoltaic Cu_2ZnSnS_4 compound, *Acta Crystallogr.*, **C70**(2), 390–394 (2014).

Lafond A., Deudon C., Meerschaut A., Pavladeau P., Moëlo Y., Briggs A. Structure determination and physical properties of misfit layered compound $(Pb_2FeS_3)_{0.58}NbS_2$, *J. Solid State Chem.*, **142**(2), 461–469 (1999).

Lafond A., Meerschaut A., Moëlo Y., Rouxel J. Premier composite bicouche incommensurable de type franckéite dans le système Pb–Sb–Nb–S, *C. r. Acad. sci., Sér. II, Fasc. b*, **322**(2), 165–173 (1996).

Lafond A., Nader A., Moëlo Y., Meerschaut A., Briggs A., Perrin S., Monceau P., Rouxel J. X-Ray structure determination and superconductivity of a new layered misfit compound with a franckeite-like stacking, $[(Pb,Sb)S]_{2.28}NbS_2$, *J. Alloys Compd.*, **261**(1–2), 114–122 (1997).

Laitinen R., Steidel J., Steudel R. The crystal structures and Raman spectra of $2S_8 \cdot SnI_4$ and $2S_nSe_{8-n} \cdot SnI_4$, *Acta Chem. Scand. A*, **34**(9), 687–693 (1980).

Lai W.-H., Haynes A.S., Frazer L., Chang Y.-M., Liu T.-K., Lin J.-F., Liang I.-C., Sheu H.-S., Ketterson J.B., Kanatzidis M.G., Hsu K.-F. Second harmonic generation response optimized at various optical wavelength ranges through a series of cubic chalcogenides $Ba_6Ag_{2.67+4\delta}Sn_{4.33-\delta}S_{16-x}Se_x$, *Chem. Mater.*, **27**(4), 1316–1326 (2015).

Lam A.E., Groat L.A., Grice J.D., Ercit T.S. The crystal structure of choloalite, *Canad. Mineralog.*, **37**(3), 721–729 (1999).

Lamarche A.-M., Willsher A., Chen L., Lamarche G., Woolley J.C. Crystal structures of I_2 Mn IV VI_4 compounds, *J. Solid State Chem.*, **94**(2), 313–318 (1991).

Lander J.J. The basic sulfates of lead, *J. Electrochem. Soc.*, **95**(4), 174–186 (1949).

Lanza A.E., Gemmi M., Bindi L., Mugnaioli E., Paar W.H. Daliranite, $PbHgAs_2S_5$: determination of the incommensurately modulated structure and revision of the chemical formula, *Acta Crystallogr.*, **B75**(4), 711–716 (2019).

Laqibi M., Cros B., Peytavin S., Ribes M. New silver superionic conductors Ag_7XY_5Z (X = Si, Ge, Sn; Y = S, Se; Z = Cl, Br, I) – synthesis and electrical studies, *Solid State Ionics*, **23**(1–2), 21–26 (1987).

Large R.R., Mumme W.G. Junoite, "wittite", and related seleniferous bismuth sulfosalts from Juno Mine, Northern Territory, Australia, *Econ. Geol.*, **70**(2), 369–383 (1975).

Larrégola S.A., Alonso J.A., Algueró M., Jiménez R., Suard E., Porcher F., Pedregosa J.C. Effect of the Pb^{2+} lone electron pair in the structure and properties of the double perovskites $Pb_2Sc(Ti_{0.5}Te_{0.5})O_6$ and $Pb_2Sc(Sc_{0.33}Te_{0.66})O_6$: relaxor state due to intrinsic partial disorder, *Dalton Trans.*, **39**(21), 5159–5165 (2010).

Laufek F., Pažout R., Makovicky E. Crystal structure of owyheeite, $Ag_{1.5}Pb_{4.43}Sb_{6.07}S_{14}$: refinement from powder synchrotron X-ray diffraction, *Eur. J. Mineral.*, **19**(4), 557–566 (2007a).

Laufek F., Sejkora J., Fejfarová R., Dušek M., Ozdín D. The mineral marrucciite: monoclinic $Hg_3Pb_{16}Sb_{18}S_{46}$, *Acta Crystallogr.*, **E63**(11), i190 (2007b).

Lavela P., Tirado J.L., Morales J., Olivier-Fourcade J., Jumas J.-C. Lithium intercalation and copper extraction in spinel sulfides of general formula $Cu_2MSn_3S_8$ (M = Mn, Fe, Co, Ni), *J. Mater. Chem.*, **6**(1), 41–47 (1996).

Layner D.I., Samedov G.K. On correlation of thermal and structural properties of the PbTe–SnTe–PbS alloys [in Russian], In: *Teplofiz. svoistva tverd. veshchestva*, M.: Nauka Publish., 32–34 (1971).

Lebedev A.I., Sluchinskaya I.A. Low-temperature phase transitions in some quaternary solid solution of IV-VI semiconductors, *J. Alloys Compd.*, **203**(1–2), 51–54 (1994).

Le Bihan M.-Th. Contribution à l'étude structurale des sulfo-arséniures de plomb du gisement de Binn. Structure crystalline de la rathite III, *C. r. Acad. sci.*, **251**(20), 2196–2198 (1960).

Le Bihan M.-Th. Étude structurale de quelques sulfures de plomb et d'arsenic naturels du gisement de Binn, *Bull. Soc. fr. Minéral. Cristallogr.*, **85**(1), 15–47 (1962).

Lei H., Yamaura J.-I., Guo J., Qi Y., Toda Y., Hosono H. Layered compounds $BaM_2Ge_4Ch_6$ (M = Rh, Ir and Ch = S, Se) with pyrite-type building blocks and Ge−Ch heteromolecule-like anions, *Inorg. Chem.*, **53**(11), 5684–5691 (2014).

Lekse J.W., Leverett B.M., Lake CH., Aitken J.A. Synthesis, physicochemical characterization and crystallographic twinning of Li_2ZnSnS_4, *J. Solid State Chem.*, **181**(12), 3217–3222 (2008).

Lekse J.W., Moreau M.A., McNerny K.L., Yeon J., Halasyamani P.S., Aitken J.A. Second-harmonic generation and crystal structure of the diamond-like semiconductors Li_2CdGeS_4 and Li_2CdSnS_4, *Inorg. Chem.*, **48**(16), 7516–7518 (2009).

Lemoine P., Kumar P.V., Guélou G., Nassif V., Raveau B., Guilmeau E. Thermal stability of the crystal structure and electronic properties of the high power factor thermoelectric colusite $Cu_{26}Cr_2Ge_6S_{32}$, *Chem. Mater.*, **32**(2), 830–840 (2020).

Léone P., André G., Doussier C., Moëlo Y. Neutron diffraction study of the magnetic ordering of jamesonite ($FePb_4Sb_6S_{14}$), *J. Magn. Magn. Mater.*, **284**, 92–96 (2004).

Léone P., Le Leuch L.-M., Palvadeau P., Moliniém Ph., Moëlo Y. Single crystal structures and magnetic properties of two iron or manganese-lead-antimony sulfides: $MPb_4Sb_6S_{14}$ (M: Fe, Mn), *Solid State Sci.*, **5**(5), 771–776 (2003).

Leute V., Böttner H. Phase boundary processes in the system (Cd, Pb) (S, Se), *Ber. Bunsenges. phys. Chem.*, **82**(3), 302–306 (1978).

Leute V., Brinkmann S., Linnenbrink J., Schmidtke H.M. The phase diagram of the quasiternary system (Sn,Pb)(S,Te), *Z. Naturforsch.*, **50A**(4–5), 459–467 (1995).

Leute V., Köller H.-J. Thermodynamic properties of the quasiternary system $Hg_kPb_{(1-k)}Se_lTe_{(1-l)}$, *Z. phys. Chem. (BRD)*, **150**(2), 227–243 (1986).

Leute V., Menge D. The quasiternary system $(Cd_kSn_{1-k})(Se_lTe_{1-l})$, *Z. Phys. Chem. (Munchen)*, **176**(1), 65–76 (1992).

Leute V., Schmidtke H. M., Brokamp H., Kebsch K. A contribution to the phase diagram of the quasiternary system $(Sn_kPb_{(1-k)})(S_lSe_{(1-l)})$, *Mater. Sci. Forum*, **133–136**, 473–478 (1993).

Leute V. Thermodynamic excess functions for quasiternary solid solutions of type M(X,Y,Z). I. The system Pb(S,Te,Se), *Z. Naturforsch.*, **50A**(4–5), 357–367 (1995).

Liao J.-H., Kanatzidis M.G. Quaternary $Rb_2Cu_2SnS_4$, $A_2Cu_2Sn_2S_6$ (A = Na, K, Rb, Cs), $A_2Cu_2Sn_2Se_6$ (A = K, Rb), $K_2Au_2SnS_4$, and $K_2Au_2Sn_2S_6$. Syntheses, structures, and properties of new solid-state chalcogenides based on tetrahedral $[SnS_4]^{4-}$ units, *Chem. Mater.*, **5**(10), 1561–1569 (1993).

Liao J.H., Marking G.M., Hsu K.F., Matsushita Y., Ewbank M.D., Borwick R., Cunningham P., Rosker M.J., Kanatzidis M.G. α- and β-$A_2Hg_3M_2S_8$ (A = K, Rb; M = Ge, Sn): polar quaternary chalcogenides with strong nonlinear optical response, *J. Am. Chem. Soc.*, **125**(31), 9484–9496 (2003).

Li G., Chu Y., Zhou Z. From $AgGaS_2$ to Li_2ZnSiS_4: realizing impressive high laser damage threshold together with large second-harmonic generation response, *Chem. Mater.*, **30**(3), 602–606 (2018a).

Li G.-M., Chu Y., Li J., Zhou Z.-X. Li_2CdSiS_4, a promising IR NLO material with balanced E_g and SHG response originated from the effect of Cd with d^{10} configuration, *Dalton Trans.*, **49**(6), 1975–1980 (2020a).

Li G.-M., Liu Q., Wu K., Yang Z.-H., Pan S.-L. $Na_2CdGe_2Q_6$ (Q = S, Se): two metal-mixed chalcogenides with phase-matching abilities and large second-harmonic generation responses, *Dalton Trans.*, **46**(9), 2778–2784 (2017a).

Li G.-M., Wu K., Liu Q., Yang Z.-H., Pan S.-L. $Na_2ZnSn_2S_6$: a mixed-metal thiostannate with large second-harmonic generation response activated by penta-tetrahedral $[ZnSn_4S_{14}]^{10-}$ clusters, *Sci. China. Tech. Sci.*, **60**(10), 1465–1472 (2017b).

Li G., Wu K., Liu Q., Yang Z., Pan S. $Na_2ZnGe_2S_6$: a new infrared nonlinear optical material with good balance between large second-harmonic generation response and high laser damage threshold, *J. Am. Chem. Soc.*, **138**(23), 7422–7428 (2016a).

Li H., Caib K.F., Wang H.F., Wang L., Li X.L. Preparation and thermoelectric properties of $AgPb_{18}SbTe_{20}$ materials via hydrothermal synthesis method, *Key Eng. Mater.*, **368–372**, 550–552 (2008).

Li H., Rankin W.J. Thermodynamics and phase relations of the $Fe–O–S–SiO_2$(sat) system at 1200°C and the effect of copper, *Metall. Mater. Trans. B*, **25**(1), 79–89 (1994).

Li J., Guo H.-Y., Proseprio D.M., Sironi A. Exploring tellurides: synthesis and characterization of new binary, ternary, and quaternary compounds, *J. Solid State Chem.*, **117**(2), 247–255 (1995).

Lin H., Chen H., Zheng Y.-J., Yu J.-S., Wu X.-T., Wu L.-M. Two excellent phase-matchable infrared nonlinear optical materials based on 3D diamond-like frameworks: $RbGaSn_2Se_6$ and $RbInSn_2Se_6$, *Dalton Trans.*, **46**(24), 7714–7721 (2017a).

Lin H., Chen L., Yu J.-S., Chen H., Wu L.-M. Infrared SHG materials CsM_3Se_6 (M = Ga/Sn, In/Sn): phase matchability controlled by dipole moment of the asymmetric building unit, *Chem. Mater.*, **29**(2), 499–503 (2017b).

Lin Q., Smalley A.L.E., Johnson D.C., Martin J., Nolas G.S. Synthesis and properties of $Ce_xCo_4Ge_6Se_6$, *Chem. Mater.*, **19**(26), 6615–6620 (2007).

Lin S.-H., Mao J.-G., Guo G.-C., Huang J.-S. Synthesis and crystal structure of a new quaternary compound: La_3AgSe_7Si, *J. Alloys Compd.*, **252**(1–2), L8–L11 (1997).

Lin X., Guo Y., Ye N. $BaGa_2GeX_6$ (X = S, Se): new mid-IR nonlinear optical crystals with large band gaps, *J. Solid State Chem.*, **195**, 172–177 (2012).

Lin Y.-J., Liu B.-W., Ye R., Jiang X.-M., Yang L.-Q., Zeng H.-Y., Guo G.-C. $SrCdSnQ_4$ (Q = S, Se): infrared nonlinear optical chalcogenides with mixed NLO-active and synergetic distorted motifs, *J. Mater. Chem. C*, **7**(15), 4459–4465 (2019a).

Lin Y.-J., Ye R., Yang L.-Q., Jiang X.-M., Liu B.-W., Zeng H.-Y., Guo G.-C. $BaMnSnS_4$ and $BaCdGeS_4$: infrared nonlinear optical sulfides containing highly distorted motifs with centers of moderate electronegativity, *Inorg. Chem. Front.*, **6**(9), 2365–2368 (2019b).

Lin Z., Li C., Kang L., Lin Z., Yao J., Wu Y. $SnGa_2GeS_6$: synthesis, structure, linear and nonlinear optical properties, *Dalton Trans.*, **44**(16), 7404–7410 (2015).

Lin Z.-S., Chen L., Wang L.-M., Zhao J.-T., Wu L.-M. A promising mid-temperature thermoelectric material candidate: Pb/Sn-codoped $In_4Pb_xSn_ySe_3$, *Adv. Mater.*, **25**(34), 4800–4806 (2013).

Lioutas C.B., Frangis N., Todorov I., Chung D.Y., Kanatzidis M.G. Understanding nanostructures in thermoelectric materials: an electron microscopy study of $AgPb_{18}SbSe_{20}$ crystals, *Chem. Mater.*, **22**(19), 5630–5635 (2010).

Lipovetskiy A.G., Borodayev Y.S., Zav'yalov Ye N. Aleksite, $PbBi_2Te_2S_2$, a new mineral [in Russian], *Zap. Vses. Mineralog. Obshch.*, **107**(3), 315–321 (1978).

Lipovetskiy A.G., Borodayev Y.S., Zav'yalov Ye N. Aleksite, $PbBi_2Te_2S_2$, a new mineral, *Int. Geol. Rev.*, **21**(10), 1223–1228 (1979).

Li P.-X., Kong F., Hu C.-L., Zhao N., Mao J.-G. A series of new phases containing three different asymmetric building units, *Inorg. Chem.*, **49**(13), 5943–5952 (2010).

Li S.-F., Jiang X.-M., Fan Y.-H., Liu B.-W., Zeng H.-Y., Guo G.-C. New strategy for designing promising mid-infrared nonlinear optical materials: narrowing band gap for large nonlinear optical efficiency and reducing thermal effect for high laser-induced damage threshold, *Chem. Sci.*, **9**(26), 5700–5708 (2018b).

Li S.-F., Jiang X.-M., Liu B.-W., Yan D., Zeng H-Y., Guo G.-C. Strong infrared nonlinear optical efficiency and high laser damage threshold realized in quaternary alkali metal sulfides $Na_2Ga_2MS_6$ (M = Ge, Sn) containing mixed nonlinear optically active motifs, *Inorg. Chem.*, **57**(12), 6783–6786 (2018c).

Li S.-F., Liu B.-W., Zhang M.-J., Fan Y.-H., Zeng H.-Y., Guo G.-C. Syntheses, structures, and nonlinear optical properties of two sulfides $Na_2In_2MS_6$ (M = Si, Ge), *Inorg. Chem.*, **55**(4), 1480–1485 (2016b).

Li S.-F., Yan D. Synthesis, crystal structure and physical properties of a new chalcogenides $Rb_3Ga_3Ge_7S_{20}$, *J. Solid State Chem.*, **296**, 121945 (2021).

Lissner F., Schleid T. $Pr_3Se_2SiN_3$: das erste Nitrido-*cyclo*-Trisilicat mit diskreten $[Si_3N_9]^{15-}$-Anionen gemäß $Pr_9Se_6[Si_3N_9]$, *Z. anorg. und allg. Chem.*, **630**(13–14), 2226–2230 (2004).

Liu B.-W., Jiang X.-M., Wang G.-E., Zeng H.-Y., Zhang M.-J., Li S.-F., Guo W.-H., Guo G.-C. Oxychalcogenide $BaGeOSe_2$: highly distorted mixed-anion building units leading to a large second-harmonic generation response, *Chem. Mater.*, **27**(24), 8189–8192 (2015a).

Liu B.-W., Zeng H.-Y., Zhang M.-J., Fan Y.-H., Guo G.-C., Huang J.S., Dong Z.-C. Syntheses, structures, and nonlinear-optical properties of metal sulfides $Ba_2Ga_8MS_{16}$ (M = Si, Ge), *Inorg. Chem.*, **54**(3), 976–981 (2015b).

Liu B.-W., Zhang M.-J., Zhao Z.-Y., Zeng H.-Y., Zheng F.-K., Guo G.-C., Huang J.-S. Synthesis, structure, and optical properties of the quaternary diamond-like compounds $I_2–II–IV–VI_4$ (I = Cu; II = Mg; IV = Si, Ge; VI = S, Se), *J. Solid State. Chem.*, **204**, 251–256 (2013).

Liu C., Mei D., Cao W., Yang Y., Wu Y., Li G., Lin Z. Mn-based tin sulfide $Sr_3MnSn_2S_8$ with wide band gap and strong nonlinear optical response. *J. Mater. Chem. C*, **7**(5), 1146–1150 (2019a).

Liu H., Chang L.L.Y. Lead and bismuth chalcogenide systems, *Am. Mineralog.*, **79**(11–12), 1159–1166 (1994a).

Liu H., Chang L.L.Y. Phase relations in the system $PbS–PbSe–PbTe$, *Mineralog. Mag.*, **58**(393), 567–578 (1994b).

Liu H., Chang L.L.Y. Phase relations in systems of tin chalcogenides, *J. Alloys Compd.*, **185**(1), 183–190 (1992).

Liu L., Ivanov S., Mathieu R., Weil M., Li X., Lazor P. Pressure tuning of octahedral tilt in the ordered double perovskite Pb_2CoTeO_6, *J. Alloys Compd.*, **801**, 310–317 (2019b).

Liu P.-F., Li Y.-Y., Zheng Y.-J., Yu J.-S., Duan R.-H., Chen H., Lin H., Chen L., Wu L.-M. Tailored synthesis of nonlinear optical quaternary chalcohalides: $Ba_4Ge_3S_9Cl_2$, $Ba_4Si_3Se_9Cl_2$ and $Ba_4Ge_3Se_9Cl_2$, *Dalton Trans.*, **46**(8), 2715–2721 (2017).

Liu Q.-Q., Liu X., Chen L., Wu L.-M. $AGaSnS_4$ (A = Rb, Cs): three sulfides and their structure diversity, *J. Solid State Chem.*, **285**, 121233 (2020a).

Liu Q., Zhang P. IR $Li_2Ga_2GeS_6$ nanocrystallized GeS_2–Ga_2S_3–Li_2S electroconductive chalcogenide glass with good nonlinearity, *Sci. Rep.*, **4**, 5719 (2014).

Liu Y., Li Y., Zhao J., Zhang R., Ji M., You Z., An Y. Solvothermal syntheses, characterizations and semiconducting properties of four quaternary thioargentates Ba_2AgInS_4, $Ba_3Ag_2Sn_2S_8$, $BaAg_2MS_4$ (M = Sn, Ge), *J. Alloys Compd.*, **815**, 152413 (2020b).

Li X., Li C., Gong P., Lin Z., Yao J., Wu Y. $BaGa_2SnSe_6$: a new phase-matchable IR nonlinear optical material with strong second harmonic generation response, *J. Mater. Chem. C*, **3**(42), 10998–11004 (2015).

Li X., Li C., Gong P., Lin Z., Yao J., Wu Y. Syntheses, crystal structures and physical properties of three new chalcogenides: $NaGaGe_3Se_8$, $K_3Ga_3Ge_7S_{20}$, and $K_3Ga_3Ge_7Se_{20}$, *Dalton Trans.*, **45**(2), 532–538 (2016c).

Li X., Li C., Zhou M., Wu Y., Yao J. $Li_2MnSnSe_4$: a new quaternary diamond-like semiconductor with nonlinear optical response and antiferromagnetic property, *Chem. Asian J.*, **12**(24), 3172–3177 (2017c).

Li X., Peng W., Fu H. Theoretical investigations on the elastic, electronic and thermal properties of orthorhombic Li_2CdGeS_4 under pressure, *J. Alloys Compd.*, **581**, 867–872 (2013).

Li Y., Fan W., Sun H., Cheng X., Li P., Zhao X. Electronic, optical and lattice dynamic properties of the novel diamond-like semiconductors Li_2CdGeS_4 and Li_2CdSnS_4, *J. Phys.: Condens. Matter*, **23**(22), 225404_1–225404_11 (2011).

Li Z., Liu Y., Zhang S., Xing W., Yin W., Lin Z., Yao J., Wu Y. Functional chalcogenide $Na_2HgSn_2Se_6$ and $K_2MnGe_2Se_6$ exhibiting flexible chain structure and intriguing birefringence tenability, *Inorg. Chem.*, **59**(11), 7614–7621 (2020b).

Li Z., Jiang X., Yi C., Zhou M., Guo Y., Luo X., Lin Z., Wu Y., Shi Y., Yao J. $K_2MnGe_3S_8$: a new multifunctional semiconductor featuring $[MnGe_3S_8]^{2-}$ layer and demonstrating interesting nonlinear optical response and antiferromagnetic property, *J. Mater. Chem.C*, **6**(37), 10042–10049 (2018d).

Li Z., Yang Y., Guo Y., Xing W., Luo X., Lin Z., Yao J., Wu Y. Broadening frontiers of infrared nonlinear optical materials with π-conjugated trigonal-planar groups, *Chem. Mater.*, **31**(3), 1110–1117 (2019a).

Li Z., Zhang S., Guo Y., Lin Z., Yao J., Wu Y. $SnGa_2GeSe_6$, a benign addition to the $AM^{III}_2M^{IV}Q_6$ family: synthesis, crystal structure and nonlinear optical performance, *Dalton Trans.*, **48**(19), 6638–6644 (2019b).

Llanos J., Mujica C., Sánchez V., Peña O. Physical and optical properties of the quaternary sulfides $SrCu_2MS_4$ and $EuCu_2MS_4$ (M = Ge and Sn), *J. Solid State Chem.*, **173**(1), 78–82 (2003).

Llanos J., Tapia M., Mujica C., Oro-Sole J., Gomez-Romero P. A new structural modification of stannite, *Bol. Soc. Chil. Quim.*, **45**(4), 605–609 (2000).

Locock A.J., Ercit T.S., Kjellman J., Piilonen P.C. New mineral names, *Am. Mineralog.*, **91**(11–12), 1945–1954 (2006a).

Locock A.J., Piilonen P.C., Ercit T.S., Rowe R. New mineral names, *Am. Mineralog.*, **91**(1), 216–224 (2006b).

Loireau-Lozac'h A.-M., Guittard M. Système ternaire La_2Se_3–Ga_2Se_3–$GeSe_2$. Diagramme de phase – Etude des verres, *Mater. Res. Bull.*, **12**(9), 887–893 (1977).

Löken S., Tremel W. Synthesis, structures and properties of new quaternary gold-chalcogenides: $K_2Au_2Ge_2S_6$, $K_2Au_2Sn_2Se_6$, and $Cs_2Au_2SnS_4$, *Gold Bull. (Gr. Brit.)*, **32**(1), 31 (1999).

Löken S., Tremel W. Synthesis, structures and properties of new quaternary gold-chalcogenides: $K_2Au_2Ge_2S_6$, $K_2Au_2Sn_2Se_6$, and $Cs_2Au_2SnS_4$, *Z. anorg. und allg. Chem.*, **624**(10), 1588–1594 (1998).

López-Rivera S.A., Pamplin B.R., Woolley J.C. High-temperature lattice parameters and DTA of the quaternary compound $CuGaSn\square Se_4$, *Nuovo cim.*, **D2**(6), 1728–1735 (1983).

López-Vergara F., Galdámez A., Manríquez V., Barahona P., Peña O. $Cu_2Mn_{1-x}Co_xSnS_4$: Novel kësterite type solid solutions, *J. Solid State. Chem.*, **198**, 386–391 (2013).

López-Vergara F., Galdámez A., Manríquez V., Barahona P., Peña O. Magnetic properties and crystal structure of solid-solution $Cu_2Mn_xFe_{1-x}SnS_4$ chalcogenides with stannite-type structure, *Phys. stat. sol. (b)*, **251**(5), 958–964 (2014).

Luo X., Liang F., Zhou M., Guo Y., Li Z., Lin Z., Yao J., Wu Y. $K_2ZnGe_3S_8$: a congruent-melting infrared nonlinear-optical material with a large band gap, *Inorg. Chem.*, **57**(15), 9446–9452 (2018).

Luo X., Li Z., Liang F., Guo Y., Wu Y., Lin Z., Yao J. Synthesis, structure, and characterization of two mixed-cation quaternary chalcogenides K_2BaSnQ_4 (Q = S, Se), *Inorg. Chem.*, **58**(10), 7118–7125 (2019).

Luo Z.-Z., Lin C.-S., Cui H.-H., Zhang W.-L., Zhang H., Chen H., He Z.Z., Chen W.-D. $PbGa_2MSe_6$ (M = Si, Ge): two exceptional infrared nonlinear optical crystals, *Chem. Mater.*, **27**(3), 914–922 (2015).

Lu X., Zhuang Z., Peng Q., Li Y. Wurtzite Cu_2ZnSnS_4 nanocrystals: a novel quaternary semiconductor, *Chem. Commun.*, **47**(11) 3141–3143 (2011).

Lychmanyuk O.S., Gulay L.D., Olekseyuk I.D. Investigation of the Pr_2X_3–Cu_2X–ZX_2 (Z = Si, Ge; X = S, Se) systems at 870 K [in Ukrainian], *Nauk. Visnyk Volyns'k. Derzh. Univ. im. Lesi Ukrainky. Ser. Khim. nauky*, (*15*), 10–13 (2007a).

Lychmanyuk O.S., Gulay L.D., Olekseyuk I.D. Investigation of the Y_2S_3–Cu_2S–SiS_2 and Y_2Se_3–Cu_2Se–$SiSe_2$ systems at 870 K [in Ukrainian], *Nauk. Visnyk Volyns'k. Derzh. Univ. im. Lesi Ukrainky. Ser. Khim. nauky*, (*4*), 118–124 (2006a).

Lychmanyuk O.S., Gulay L.D., Olekseyuk I.D. Isothermal section of the Y_2Se_3–Cu_2Se–$GeSe_2$ system at 870 K and crystal structure of the $Y_3CuGeSe_7$ compound, *Pol. J. Chem.*, **80**(3), 463–469 (2006b).

Lychmanyuk O.S., Gulay L.D., Olekseyuk I.D., Stępień-Damm J., Daszkiewicz M., Pietraszko A. Investigation of the Ho_2X_3–Cu_2X–ZX_2 (X = S, Se; Z = Si, Ge) systems, *Polish J. Chem.*, **81**(3), 353–367 (2007b).

MacLean W.H. Liquidus phase relations in the FeS–FeO–Fe_3O_4–SiO_2 system, and their application in geology. *Econ. Geol.*, **64**(8), 865–884. (1969).

Magunov R.L., Zakolodyazhnaya O.V., Kovalevskaya N.I., Sherstyuk L.G. The $GaSe$–$GeTe$ system [in Russian], *Zhurn. neorgan. khimii*, **24**(11), 3133–3134 (1979).

Ma H., Guimond Y., Zhang X., Lucas J. Ga–Ge–Sb–Se based glasses and influence of alkaline halide addition, *J. Non-Cryst. Solids*, **256–257**, 165–169 (1999).

Mähl D., Pickard J., Reuter B. Züchtung und Untersuchung von Einkristallen der Verbindungen CuCrZrSe$_4$ und CuCrSnSe$_4$, *Z. anorg. allg. Chem.*, **508**(1), 197–200 (1984a).

Mähl D., Pickardt J., Reuter B. Züchtung und Untersuchung von Einkristallen einiger Ternärer und quaternärer Kupferthiospinelle, *Z. anorg. und allg. Chem.*, **491**(1), 203–207 (1982).

Mähl D., Pickardt J., Reuter B. Züchtung und Unterzuchung von Einkristallen in den pseudobinären Systemen ABS$_2$–SnS$_2$ (A = Cu, Ag; B = Al, Cr), *Z. anorg. und allg. Chem.*, **516**(9), 102–106 (1984b).

Maier R.G. Zur Kenntnis des Systems PbTe–AgSbTe$_2$, *Z. Metallkd.*, **54**(5), 311–312 (1963).

Makovicky E., Balić-Žunić T., Karanović L., Poleti D. The crystal structure of kirkiite, Pb$_{10}$Bi$_3$As$_3$S$_{19}$, *Canad. Mineralog.*, **44**(1), 177–188 (2006a).

Makovicky E., Karup-Møller S. Ourayite from Ivigtut, Greenland, *Canad. Mineralog.*, **22**(4), 565–575 (1984).

Makovicky E., Mumme W.G., Gable R.W. The crystal structure of ramdohrite, Pb$_{5.9}$Te$_{0.1}$Mn$_{0.1}$In$_{0.1}$Cd$_{0.2}$Ag$_{2.8}$Sb$_{10.8}$S$_{24}$: a new refinement, *Am. Mineralog.*, **98**(4), 773–779 (2013).

Makovicky E., Mumme W.G., Norrestam R. The crystal structures of izoklakeite, dadsonite and jaskolskiite, *Acta Crystallogr.*, **A40**(1), Suppl., C-246 (1984).

Makovicky E., Olsen P.N. The order–disorder character of owyheeite, *Canad. Mineralog.*, **53**(5), 879–884 (2015).

Makovicky E., Petříček V., Dušek M., Topa D. Crystal structure of a synthetic tin-selenium representative of the cylindrite structure type, *Am. Mineralog.*, **93**(11–12), 1787–1798 (2008).

Makovicky E., Topa D., Balić-Žunić T. The crystal structure of paarite, the newly discovered 56 Å derivative of the bismuthinite–aikinite solid-solution series, *Canad. Mineralog.*, **39**(5), 1377–1382 (2001).

Makovicky E., Topa D., Mumme W.G. The crystal structure of dadsonite, *Canad. Mineralog.*, **44**(6), 1499–1512 (2006b).

Makovicky E., Topa D., Stoeger B. The crystal structures of heptasartorite, Tl$_7$Pb$_{22}$As$_{55}$S$_{108}$, and enneasartorite, Tl$_6$Pb$_{32}$As$_{70}$S$_{140}$, two members of an anion-omission series of complex sulfosalts from Lengenbach, the Swiss Alps, and comparison with the structures of As–Sb sartorite homologues, *Eur. J. Mineral.*, **30**(1), 149–164 (2018).

Makovicky E., Topa D. Tajjedin H., Rastad E., Yaghubpur A. The crystal structure of guettardite, PbAsSbS$_4$, and the twinnite–guettardite problem, *Canad. Mineralog.*, **50**(2), 253–265 (2012).

Makovicky E., Topa D. The crystal structure of gustavite, PbAgBi$_3$S$_6$. Analysis of twinning and polytypism using the OD approach, *Eur. J. Mineral.*, **23**(4), 537–550 (2011).

Makovicky E., Topa D. The crystal structure of sulfosalts with the boxwork architecture and their new representative, Pb$_{15-2x}$Sb$_{14+2x}$S$_{36}$O$_x$, *Canad. Mineralog.*, **47**(1), 3–24 (2009).

Makovicky E., Topa D. Twinnite, Pb$_{0.8}$Tl$_{0.1}$Sb$_{1.3}$As$_{0.8}$S$_4$, the OD character and the question of its polytypism, *Z. Kristallogr.*, **227**(7), 468–475 (2012).

Makovskaya Z.G., Zhukov E.G. The GeSe$_2$–Sb$_2$Te$_3$ system [in Russian], *Zhurn. neorgan. khimii*, **28**(3), 805–807 (1983a).

Makovskaya Z.G., Zhukov E.G. The TlSbSe$_2$–GeSe$_2$ system [in Russian], *Zhurn. neorgan. khimii*, **28**(4), 1075–1076 (1983b).

Malakhovska T.O., Barchiy O.I., Peresh E.Yu., Barchiy I.Ye., Sabov M.Yu. Triangulation of the Tl$_2$Te–PbTe–Bi$_2$Te$_3$ [in Ukrainian], *Nauk. vìsn. Uzhgorod. Univ., Ser. Khim.*, (17–18), 15–19 (2007).

Malakhovska T.O., Glukh O.S., Pogodin A.I., Filep M.Yo., Sabov M.Yu., Stasyuk Yu. M., Barchiy I.Ye. Physico-chemical

interaction in the Tl$_4$PbTe$_3$–Tl$_9$BiTe$_6$–TlBiTe$_2$ system [in Ukrainian], *Nauk. Visn. Uzhgorod. Univ., Ser. Khim.*, [*1*(41)], 32–37 (2019).

Malevskiy A.Yu. On the isomorphic penetration of tallium in the galenite [in Russian], *Dokl. AN SSSR*, **169**(6), 1324–1327 (1966).

Malinowski C., Malinowska K., Małecki S. Analysis of the chemical processes occuring in the system PbSO$_4$–ZnS, *Thermochim. Acta*, **275**(1), 117–130 (1996).

Malinowski C. Phenomenologic analysis of chemical process in the systems ZnSO$_4$–ZnS and PbSO$_4$–ZnS [in Polish], *Zesz. nauk. AGH im. Stanisława Staszika. Met. i odlew.*: [*Monogr.*], (141), 1–120 (1992).

Mamedov A.N., Tagiev E.R., Aliev Z.S., Babanly M.B. Phase boundaries of the (YbTe)$_x$ (PbTe)$_{1-x}$ and (YbTe)$_x$ (SnTe)$_{1-x}$ solid solution series, *Inorg. Mater.*, **52**(6), 543–545 (2016).

Mamedov Sh.G., Baktiyarly I.B., Kurbanov G.R. Phase equilibrium in the PbSnS$_2$–PbSb$_2$S$_4$ system, *Russ. J. Inorg. Chem.*, **55**(4), 626–628 (2010).

Mamedov Sh.G. Phase equilibrium in the Pb$_2$SnSb$_2$S$_6$–SnS system, *Russ. J. Inorg. Chem.*, **55**(8), 1292–1294 (2010).

Mammadov Sh.G., Bakhtiyarli I.B., Mammadov A.N., Mammadov V.S. The thermodynamic functions of the compounds, formed in the PbS–Sb$_2$S$_3$–SnS pseudoternary system [in Azerbaijanian], *Azerb. khim. zhurn.*, (2), 46–48 (2014).

Mammadov Sh.G., Mammadov A.N., Kurbanova R.C. Quasibinary section Ag$_2$SnS$_3$–AgSbS$_2$, *Russ. J. Inorg. Chem.*, **65**(2), 217–221 (2020).

Mandarino J.A. Abstracts of new mineral descriptions (department), *Mineral. Rec.*, **29**(5), 476–479 (1998).

Manos M.J., Ding N., Kanatzidis M.G. Layered metal sulfides: exceptionally selective agents for radioactive strontium removal, *Proc. Nat. Acad. Sci. USA*, **105**(10), 3696–3699 (2008).

Manos M.J., Iyer R.G., Quarez E., Liao J.H., Kanatzidis M.G. {Sn[Zn$_4$Sn$_4$S$_{17}$]}$^{6-}$: A robust open framework based on metal-linked penta-supertetrahedral [Zn$_4$Sn$_4$S$_{17}$]$^{10-}$ clusters with ion-exchange properties, *Angew. Chem. Int. Ed.*, **44**(23), 3552–3555 (2005).

Manos M.J., Kanatzidis M.G. Highly efficient and rapid Cs$^+$ uptake by the layered metal sulfide K$_{2x}$Mn$_x$Sn$_{3-x}$S$_6$ (KMS-1), *J. Am. Chem. Soc.*, **131**(18), 6599–6607 (2009).

Manos M.J., Petkov V.G., Kanatzidis M.G. H$_{2x}$Mn$_x$Sn$_{3-x}$S$_6$ (x = 0.11–0.25): a novel reusable sorbent for highly specific mercury capture under extreme pH conditions, *Adv. Funct. Mater.*, **19**(7), 1087–1092 (2009).

Mansimova Sh.H., Babanly K.N., Mashadiyeva L.F., Mirzoyeva R.J., Babanly M.B. Phase equilibria in the Ag$_2$Se–PbSe–AgSbSe$_2$ system, *Chem. Probl.*, (*1*), 41–49 (2019a).

Mansimova Sh.H., Babanly K.N., Mashadiyeva L.F. Phase equilibria in the PbSe–AgSbSe$_2$ system, *Chem. Probl.*, [*4*(16)], 530–536 (2018).

Mansimova Sh.H., Orujlu E.N., Babanly M.B. Roentgenographic investigation of solid-phase equilibria in the PbSe–AgSbSe$_2$ system, *Appl. Chem. Eng.*, **2**(2), 1–4 (2019b).

Mansimova Sh.H. Phase relations in the PbTe–AgSbTe$_2$ system, *Chem. Probl.*, (*3*), 366–372 (2019a).

Mansimova Sh.H. Thermodynamic investigation of the PbTe–AgSbTe$_2$ system by means of EMF method, *Azerb. Chem. J.*, (*3*), 41–47 (2019b).

Ma N., Xiong L., Chen L., Wu L.-M. Vibration uncoupling of germanium with different valence states lowers thermal conductivity of Cs$_2$Ge$_3$Ga$_6$Se$_{14}$, *Sci. China Mater.*, **62**(12), 1788–1797 (2019).

Marchuk O.V., Daszkiewicz M., Gulay L.D., Olekseyuk I.D., Pietraszko A. Investigation of the R_2Te_3–M_2Te–PbTe (R = Tb, Dy; M = Cu, Ag) systems at 770 K, *J. Alloys Compd.*, **455**(1–2), 186–190 (2008a).

Marchuk O.V., Gulay L.D., Olekseyuk I.D. Phase equilibria in the $PrCuS_2$–PbS–Pr_2S_3 system [in Ukrainian], *Nauk. Visnyk Volyns'k. Derzh. Univ. im. Lesi Ukrainky. Ser. Khim. nauky*, (4), 96–101 (2006a).

Marchuk O., Gulay L., Olekseyuk I., Shemet V. Phase equilibria in $PbSe$–$Gd(Ho)_2Se_3$–$GeSe_2$ systems at 770 K [in Ukrainian], *Nauk. Visnyk Skhidnoyevrop. Nats. Univ. im. Lesi Ukrainky. Ser. Khim. nauky*, [20(297)], 29–33 (2014).

Marchuk O.V., Gulay L.D., Parasyuk O.V. The Cu_2S–HgS–GeS_2 system at 670 K and the crystal structure of the $Cu_6Hg_{0.92}GeS_{5.92}$ compound, *J. Alloys Compd.*, **333**(1–2), 143–146 (2002).

Marchuk O.V., Gulay L.D. Phase equilibria in the La_2X_3–PbX–SiX_2 (X – S, Se) at the temperature of 770 K [in Ukrainian], *Nauk. Visnyk Volyns'k. Nats. Univ. im. Lesi Ukrainky. Ser. Khim. nauky*, (17), 93–97 (2012).

Marchuk O.V., Gulay L.D., Shemet V.Y., Olekseyuk I.D. Investigation of the Pr_2Se_3–Cu_2Se–PbSe and Pr_2Se_3–Ag_2Se–PbSe systems, *J. Alloys Compd.*, **416**(1–2), 106–109 (2006b).

Marchuk O.V., Melnychuk Kh.O., Gulay L.D., Shemet V.Ya. Phase equilibria in the NiS–La_2S_3–GeS_2 systems at the temperature of 770 K [in Ukrainian], *Nauk. notatky*, (50), 176–179 (2015).

Marchuk O.V., Olekseyuk I.D., Grebenyuk A.G. Phase equilibrium in the system Cu_2Se–HgSe–$GeSe_2$, *J. Alloys Compd.*, **457**(1–2), 337–343 (2008b).

Marchuk O.V., Ruda I.P., Gulay L.D., Olekseyuk I.D. Investigation of the Y_2S_3–PbS–SnS_2 system at 770 K, *Pol. J. Chem.*, **81**(3), 425–432 (2007).

Marchuk O.V., Ruda I.P., Gulay L.D., Olekseyuk I.D. Phase equilibria in the $Y_2S(Se)_3$–$PbSe(Se)$–$SiS(Se)_2$ systems at 770 K [in Ukrainian], *Nauk. Visnyk Volyns'k. Nats. Univ. im. Lesi Ukrainky. Ser. Khim. nauky*, (13), 24–27 (2008c).

Marchuk O.V., Shemet V.Ya., Smitukh O.V., Gulay L.D. The CoS–Er_2S_3–SiS_2 system at the temperature of 770 K and the crystal structure of the $Er_3Co_{0.5}SiS_7$ compound [in Ukrainian], *Nauk. notatky*, (41), Pt. 2, 78–82 (2013).

Marchuk O.V., Smitiukh O.V., Kogut Y.M. Quasi-ternary system Cu_2S–HgS–SnS_2, *J. Phase Equilib. Diffus.*, **42**(2), 245–253 (2021).

Marchuk O.V. The $La_2PbSi_2S_8$ structure type [in Ukrainian], *Nauk. Visn. Uzhgorod. Univ., Ser. Khim.*, [1(41)], 20–24 (2019).

Marchuk O.V. The $Y_{1.32}Pb_{1.68}Ge_{1.67}Se_7$ structure type [in Ukrainian], *Nauk. Visn. Uzhgorod. Univ., Ser. Khim.*, [1(39)], 26–29 (2018).

Markham N.L., Lawrence L.J. Mawsonite, a new copper–iron–tin sulfide form Mt. Lyell, Tasmania and Tingha, New South Wales, *Am. Mineralog.*, **50**(7–8), 900–908 (1965).

Marking G.A., Hanko J.A., Kanatzidis M.G. New quaternary thiostannates and thiogermanates $A_2Hg_3M_2S_8$ (A = Cs, Rb, M = Sn, Ge) through molten A_2S_x. Reversibile glass formation in $Cs_2Hg_3M_2S_8$, *Chem. Mater.*, **10**(4), 1191–1199 (1998).

Marking G.A., Kanatzidis M.G. The ethane-like $[Ge_2S_6]^{6-}$ and $[Si_2S_6]^{6-}$ ligands bound to main-group metals in $Na_8Pb_2[Ge_2S_6]_2$, $Na_8Sn_2[Ge_2S_6]_2$, $Na_8Pb_2[Si_2Se_6]_2$, *J. Alloys Compd.*, **259**(1–2), 122–128 (1997).

Markovskiy L.Ya., Sapozhnikov Yu.P. On some properties of lead selenite [in Russian], *Zhurn. neorgan. khimii*, **5**(12), 2655–2661 (1960).

Márquez J.A., Sun J.-P., Stange H., Ali H., Choubrac L., Schäfer S., Hages C.J., Leifer K., Unold T., Mitzi D.B., Mainz R. High-temperature decomposition of Cu_2BaSnS_4 with Sn loss reveals newly identified compound $Cu_2Ba_3Sn_2S_8$, *J. Mater. Chem., A*, **8**(22), 11346–11353 (2020).

Martan H., Weiss J. Metall-Schwefelstickstoff-Verbindungen. 16. Reaktionsprodukte von Blei- und Zinnsalzen mit S_4N_4. Die Strukturen von $PbN_2S_2\cdot NH_3$, PbN_2S_2 und $SnCl_4\cdot2S_4N_4$, *Z. anorg. und allg. Chem.*, **514**(7), 107–114 (1984).

Martin B.J., Dorhout P.K. Molten flux synthesis of an analogous series of layered alkali samarium selenogermanate compounds, *Inorg. Chem.*, **43**(1), 385–391 (2004).

Martin B.R., Knaust J.M., Dorhout P.K. Crystal structure of nonasodium lanthanum(III) bis[hexaselenodigermanate], $Na_9La[Ge_2Se_6]_2$, *Z. Kryst. Str.*, **220**(3), 294 (2005).

Martin B.R., Polyakova L.A., Dorhout P.K. Synthesis and characterization of a family of two related quaternary selenides: $Na_8Eu_2(Si_2Se_6)_2$ and $Na_9Sm(Ge_2Se_6)_2$, *J. Alloys Compd.*, **408–412**, 490–495 (2006).

Marumo F., Nowacki W. The crystal structure of rathite-I, *Z. Kristallogr.*, **122**(5–6), 433–456 (1965).

Marushko D.P., Piskach L.V. The Cu_2GeS_3–Cu_2SnS_3 system [in Uraininan], *Nauk. Visnyk Volyns'k. Derzh. Univ. im. Lesi Ukrainky. Ser. Khim. nauky*, (4), 133–137 (2006).

Masalovich E.E., Sabov M.Yu., Barchii I.E., Solomon A.M. Interaction in the systems $TlBiSe_2$–Tl_9BiSe_6–PbSe and Tl_9BiSe_6–Tl_4PbSe_3–PbSe, *Russ. J. Inorg. Chem.*, **61**(4), 507–510 (2016).

Masalovych O.O., Rak D.M., Sabov M.Yu. Pecularities of the interaction in Tl_2Se–SnSe–Bi_2Se_3 system [in Uraininan], *Nauk. Visn. Uzhgorod. Univ., Ser. Khim.*, [2(26)], 13–15 (2011a).

Masalovych O.O., Rak D.M., Sabov M.Yu. Triangulation of the Tl_2Se–PbSe–Bi_2Se_3 system [in Uraininan], *Nauk. Visn. Uzhgorod. Univ., Ser. Khim.*, [1(25)], 14–16 (2011b).

Masalovych O.O., Sabov M.Yu., Barchiy I.E., Malakhovska T.O., Solomon A.M. Phase equilibria in the Tl_2Se–Tl_9BiSe_6–Tl_4PbSe_3 system [in Uraininan], *Nauk. Visn. Uzhgorod. Univ., Ser. Khim.*, [1(31)], 25–27 (2014).

Mashadieva L.F., Alieva Z.M., Mirzoeva R. Dzh., Yusibov Yu. A., Shevel'kov A.V., Babanly M.B. Phase equilibria in the Cu_2Se–$GeSe_2$–$SnSe_2$ system, *Russ. J. Inorg. Chem.*, **67**(5), 670–682 (2022).

Mashadieva L.F., Kevser J.O., Aliev I.I., Yusibov Y.A., Tagiyev D.B., Aliev Z.S., Babanly M.B. Phase equilibria in the Ag_2Te–$SnTe$–Sb_2Te_3 system and thermodynamic properties of the $(2SnTe)_{1-x}(AgSbTe_2)_x$ solid solution, *J. Phase Equilib. Diffus.*, **38**(5), 603–614 (2017a).

Mashadieva L.F., Kevser J.O., Aliev I.I., Yusibov Y.A., Tagiyev D.B., Aliev Z.S., Babanly M.B. The Ag_2Te–$SnTe$–Bi_2Te_3 system and thermodynamic properties of the $(2SnTe)_{1-x}(AgBiTe_2)_x$ solid solutions series, *J. Alloys Compd.*, **724**, 641–648 (2017b).

Mashadieva L.F., Mansimova Sh.G., Yusibov Yu.A., Babanly M.B. Thermodynamic study of the $2PbTe$–$AgSbTe_2$ system using EMF technique with the Ag_4RbI_5 solid electrolyte, *Russ. J. Electrochem.*, **54**(1), 106–111 (2018).

Mashadieva L.F., Yusibov Yu.A., Kevser Dzh., Babanly M.B. Thermodynamic study of solid solutions in the $SnTe$–$AgSbTe_2$ system by means of EMF with solid electrolyte Ag_4RbI_5, *Russ. J. Phys. Chem. A*, **91**(9), 1642–1646 (2017c).

Mashadiyeva L.F., Mansimova Sh.G., Babanly K.N., Yusibov Yu.A., Babanly M.B. Thermodynamic properties of solid solutions in the $PbSe$–$AgSbSe_2$ system, *Russ. Chem. Bull., Int. Ed.*, **69**(4), 660–664 (2020a).

Mashadiyeva L.F., Mansimova Sh.H., Babanly D.M., Yusibov Yu.A., Tagiyev D.B., Babanly M.B. Phase equilibria in the $Ag_2Te–PbTe–Sb_2Te_3$ system and thermodynamic properties of the $(2PbTe)_{1-x}(AgSbTe_2)_x$ solid solutions, *Acta Chim. Slov.*, **67**(3), 799–811 (2020b).

de Matos Gomes E. Crystal structures of strontium and lead dithionate tetrahydrate, *Acta Crystallogr.*, **B47**(1), 12–17 (1991).

Matsunaga T., Yamada N. A study of highly symmetrical crystal structures, commonly seen in high-speed phase-change materials, using synchrotron radiation, *Jpn. J. Appl. Phys.*, **41**(3B), 1674–1678 (2002).

Matsushita H., Ichikawa T., Katsui A. Structural, thermodynamic and optical properties of $Cu_2–II–IV–VI_4$ quaternary compounds, *J. Mater. Sci.*, **40**(8), 2003–2005 (2005).

Matsushita H., Katsui A. Materials design for Cu-based quaternary compounds derived from chalcopyrite-rule, *J. Phys. Chem. Solids*, **66**(11), 1933–1936 (2005).

Matsushita H., Maeda T., Katsui A., Takizawa T. Thermal analysis and synthesis from the melts of Cu-based quaternary compounds $Cu–III–IV–VI_4$ and $Cu_2–II–IV–VI_4$ (II – Zn, Cd; III – Ga, In; IV – Ge, Sn; VI – Se), *J. Cryst. Growth*, **208**(1–4), 416–422 (2000a).

Matsushita H., Maeda T., Katsui A., Takizawa T. Thermal analysis of $CuInGeSe_4$ quaternary compound for crystal growth by solution method, *Jpn. J. Appl. Phys.*, **39**, Suppl. 39-1, 62–64 (2000b).

Matsushita H., Sugiyama K., Ueda Y. Stabilization by a divalent transition metal in lead indium quaternary selenide, $Fe_{0.47}Pb_{8.04}In_{17.37}Se_{34}$, and specific indium coordination, *Inorg. Chem.*, **45**(17), 6598–6600 (2006).

Matsushita H., Ueda Y. Crystal structure and physical properties of $Fe_{1.5}Pb_{5.5}In_{10}S_{22}$, *Inorg. Chem.*, **45**(5), 2022–2026 (2006).

Matsushita Y., Takéuchi Y. Refinement of the crystal structure of hutchinsonite, $TlPbAs_5S_9$, *Z. Kristallogr.*, **209**(6), 475–478 (1994).

Matsushita Y., Ueda Y. Structure and physical properties of 1D magnetic chalcogenide, jamesonite ($FePb_4Sb_6S_{14}$), *Inorg. Chem.*, **42**(24), 7830–7838 (2003).

Matyas E.E., Borisenko T.E. The PbS–SnTe system [in Russian], *Izv. AN SSSR. Neorgan. mater.*, **26**(3), 643–644 (1990).

Matyas E.E. The PbTe–SnS system [in Russian], *Izv. AN SSSR. Neorgan. mater.*, **21**(1), 144–145 (1985).

Mayet F., Roubin M. Constribution à l'étude du système $Ag_2Te–Ga_2Se_3–PbTe$, *C. r. Acad. sci. Sér. C*, **288**(17), 433–436 (1979).

Mazelsky R., Lubell M.S. Solid solution study of some post-transition metal tellurides of the rock salt structural type, *J. Phys. Chem.*, **66**(8), 1408–1411 (1962).

McGuire M.A., Reynolds T.K., DiSalvo F.J. Exploring thallium compounds as thermoelectric materials: seventeen new thallium chalcogenides, *Chem. Mater.*, **17**(11), 2875–2884 (2005a).

McGuire M.A., Scheidemantel T.J., Badding J.V., DiSalvo F.J. Tl_2AXTe_4 (A = Cd, Hg, Mn; X = Ge, Sn): crystal structure, electronic structure, and thermoelectric properties, *Chem. Mater.*, **17**(24), 6186–6191 (2005b).

McKeown Wessler G.C., Wang T., Blum V., Mitzi D.B. Cubic crystal structure formation and optical properties within the Ag-$B^{II}–M^{IV}–X$ (B^{II} = Sr, Pb; M^{IV} = Si, Ge, Sn; X = S, Se) family of semiconductors, *Inorg. Chem.*, **61**(6), 2929–2944 (2022).

McKeown Wessler G.C., Wang T., Sun J.-P., Liao Y., Fischer M.C., Blum V., Mitzi D.B. Structural, optical, and electronic properties of two quaternary chalcogenide semiconductors: Ag_2SrSiS_4 and Ag_2SrGeS_4, *Inorg. Chem.*, **60**(16), 12206–12217 (2021).

Mecholsky J.J., Srinivasan G.R., Meynihan C.T. Macedo P.B. Immiscibility and liquidus temperatures in the pseudobinary chalcogenide system $PbSe–Ge_{1.5}Ag_{0.5}Se_3$, *J. Non-Cryst. Solids*, **11**(4), 331–340 (1973).

Medvedev Yu.V., Berchenko N.N., Kostikov Yu.P., Matveenko A.V., Olesk A.O. Phase equilibria in the Pb–Sn–Te–O, Pb–Sn–Se–O, Pb–Te–Se–O systems [in Russian], *Izv. AN SSSR. Neorgan. mater.*, **23**(1), 108–111 (1987).

Medzhidov G.A., Grincheshen I.N., Zbigli K.R., Popovich N.S. The $TlSbSe_2–SnSe$ system [in Russian], *Neorgan. mater.*, **29**(9), 1304–1306 (1993).

Meetsma A., Wiegers G.A., Haange R.J., de Boer J.L., Boom G. Structure of two modifications of dysprosium sesquisulfide, Dy_2S_3, *Acta Crystallogr.*, **C47**(11), 2287–2291 (1991).

Mei D., Gong P., Lin Z., Feng K., Yao J., Huang F., Wu Y. $Ag_3Ga_3SiSe_8$: a new infrared nonlinear optical material with a chalcopyrite structure. *CrystEngComm*, **16**(30), 6836–6840 (2014).

Mei D., Lin Z., Bai L., Yao J., Fu P., Wu Y. $KBiMS_4$ (*M* = Si, Ge): synthesis, structure, and electronic structure, *J. Solid State Chem.*, **183**(7), 1640–1644 (2010).

Mei D., Yin W., Feng K., Lin Z., Bai L., Yao J., Wu Y. $LiGaGe_2Se_6$: a new IR nonlinear optical material with low melting point, *Inorg. Chem.*, **51**(2), 1035–1040 (2012).

Mei D., Zhang S., Liang F., Zhao S., Jiang J., Zhong J., Lin Z., Wu Y. $LiGaGe_2S_6$: a chalcogenide with good infrared nonlinear optical performance and low melting point, *Inorg. Chem.*, **56**(21), 13267–13273 (2017).

Meisser N., Roth P., Nestola F., Biagioni C., Bindi L., Robyr M. Richardsollyite, IMA 2016-043. CNMNC Newsletter No. 33, October 2016, page 1138, *Mineralog. Mag.*, **80**(6), 1135–1144 (2016).

Meisser N., Roth P., Nestola F., Biagioni C., Bindi L., Robyr M. Richardsollyite, $TlPbAsS_3$, a new sulfosalt from the Lengenbach quarry, Binn Valley, Switzerland, *Eur. J. Mineral.*, **29**(4), 679–688 (2017).

Mel'nikova N.V., Kobelev L.Ya., Zlokazov V.B. Crystal structure of the $(GeSe)_{1-x}(CuAsSe_2)_x$ compounds [in Russian], *Pis'ma v ZhTF*, **21**(1), 9–13 (1995).

Melnychuk Kh.O., Marchuk O.V., Gulay L.D., Olekseyuk I.D. The crystal structure of the $Sm_3Co_{0.5}SiS_7$ and $Tb_3Co_{0.5}SiS_7$ compounds [in Ukrainian], *Nauk. Visn. Uzhgorod. Univ., Ser. Khim.*, [*1*(37)], 34–37 (2017).

Melnychuk Kh.O., Marchuk O.V., Gulay L.D. The crystal structure of the $Dy_3Co(Ni)_{0.5}SiS_7$ compounds [in Ukrainian], *Nauk. Visn. Cherniv. Univ. Khimiya*, **781**, 80–83 (2016a).

Melnychuk Kh.O., Marchuk O.V., Gulay L.D. The crystal structure of the $Ho_3Ni_{0.5}SiS_7$ compound [in Ukrainian], *Nauk. Visn. Uzhgorod. Univ., Ser. Khim.*, [*2*(36)], 10–13 (2016b).

Melnychuk K., Marchuk O., Daszkiewicz M., Gulay L. Crystal structure of novel $R_3Fe(Co,Ni)_{0.5}SnS_7$ (R = Y, La, Ce, Pr, Nd, Sm, Gd, Tb, Dy and Ho) compounds, *Struct. Chem.*, **31**(5), 1945–1957 (2020).

Meng C.-Y., Geng L., Chen W.-T., Wei M.-F., Dai K., Lu H.-Y., Cheng W.-D. Syntheses, structures, and characterizations of a new second-order nonlinear optical material: $Pb_2(SeO_3)(NO_3)_2$, *J. Alloys Compd.*, **640**, 39–44 (2015).

Meng Y., Fan Y., Lu R., Cui Q., Zou G. Structural studies of single crystals $La_6NiSi_2S_{14}$ and $La_6CoSi_2S_{14}$ under high pressure, *Physica B + C*, **139–140**, 337–340 (1986).

Meng Y., Gu X. Ruizhongite, IMA 2022-066. CNMNC Newsletter No 70, *Eur. J. Mineral.*, **34**(6), 594–595 (2022).

Meng Y., Gu X. Ruizhongite, IMA 2022-066. CNMNC Newsletter No 70, *Mineralog. Mag.*, **87**(61), 160 (2023).

Mertz J.L., Fard Z.H., Malliakas C.D., Manos M.J., Kanatzidis M.G. Selective removal of Cs^+, Sr^{2+}, and Ni^{2+} by $K_{2x}Mg_xSn_{3-x}S_6$ ($x = 0.5$-1) (KMS-2) relevant to nuclear waste remediation, Chem. Mater., 25(10), 2116–2127 (2013).

Michelet A., Flahaut J. Sur les composés du type $La_6MnSi_2S_4$, C. r. Acad. sci., Ser. C, 269(20), 1203–1205 (1969).

Michener C.E., Peacock M.A. Parkerite ($Ni_3Bi_2S_2$) from Sudbury, Ontario: redefinition of the species, Am. Mineralog., 28(6), 343–355 (1943).

Mikolaichuk O.G., Moroz N.V., Demchenko P.Yu., Akselrud L.G., Gladyshevskii R.E. Phase relations in the Ag_8SnS_6–Ag_2SnS_3–AgBr system and crystal structure of $Ag_6SnS_4Br_2$, Inorg. Mater., 46(6), 590–597 (2010).

Mikolaychuk A.G., Moroz V.N. The Ag_8GeS_6–Ag_8GeSe_6 section of the Ag–Ge–S–Se system [in Russian], Izv. AN SSSR. Neorgan. mater., 21(5), 766–769 (1985a).

Mikolaychuk A.G., Moroz V.N. The Ag_8GeSe_6–Ag_8GeTe_6 section of the Ag–Ge–Se–Te system [in Russian], Izv. AN SSSR. Neorgan. mater., 21(5), 770–773 (1985b).

Mirzoyev A.J., Asadov M.M., Aliyev O.M. Phase equilibria in the CdS–$GeSe_2$ system and formed phases properties [in Azerbaijanian], Zhurn. khim. probl., (2), 322–325 (2006).

Mishra B., Pruseth K.L. Phase equilibrium study in the system Cu_2S–PbS–Sb_2S_3: non-stoichiometry in sulfosalts and isothermal variation in sulfur fugacity, Contrib. Mineral. Petrol., 118(1), 92–98 (1994).

Miyawaki R., Hatert F., Pasero M., Mills S.J. IMA Commission on New Minerals, Nomenclature and Classification (CNMNC). Newsletter No. 49, Eur. J. Mineral., 31(3), 653–658 (2019a).

Miyawaki R., Hatert F., Pasero M., Mills S.J. IMA Commission on New Minerals, Nomenclature and Classification (CNMNC). Newsletter No. 49, Mineralog. Mag., 83(3), 479–483 (2019b).

Miyawaki R., Hatert F., Pasero M., Mills S.J. IMA Commission on New Minerals, Nomenclature and Classification (CNMNC). Newsletter No. 50, Eur. J. Mineral., 31(4), 847–853 (2019c).

Miyawaki R., Hatert F., Pasero M., Mills S.J. IMA Commission on New Minerals, Nomenclature and Classification (CNMNC). Newsletter No. 50, Mineralog. Mag., 83(4), 615–620 (2019d).

Miyawaki R., Hatert F., Pasero M., Mills S.J. IMA Commission on New Minerals, Nomenclature and Classification (CNMNC). Newsletter No. 54, Eur. J. Mineral., 32(2), 275–283 (2020a).

Miyawaki R., Hatert F., Pasero M., Mills S.J. IMA Commission on New Minerals, Nomenclature and Classification (CNMNC). Newsletter No. 54, Mineralog. Mag., 84(2), 359–365 (2020b).

Mkrtchian S.A., Dovletov K., Zhukov E.G., Melikdzhanian A.G., Nuryiev S. Electrophysical properties of the $Cu_2A^{II}B^{IV}Se_4$ (A^{II} – Cd, Hg; B^{IV} – Ge, Sn) compounds [in Russian], Izv. AN SSSR. Neorgan. mater., 24(7), 1094–1096 (1988a).

Mkrtchian S.A., Dovletov K., Zhukov E.G., Melikdzhanian A.G., Nuryiev S. Interaction in the $Cu_2Ge(Sn)Se_3$–HgSe systems, Zhurn. neorgan. khimii, 33(9), 2379–2384 (1988b).

Moëlo Y., Balitskaya O., Mozgova N., Sivtsov A. Chloro-sulfosels de l'indice plombo-antimonifère des Cougnasses (Hautes-Alpes), Eur. J. Mineral., 1(3), 381–390 (1989).

Moëlo Y., Lafond A., Deudon C., Coulon N., Lancin M., Meerschaut A. Un nouveau chalcogenure composite à feuillets désaccordès dans le système Pb–Fe–Nb–S, C.r. Acad. sci., Ser. 2b, 325(5), 287–296 (1997).

Moëlo Y., Meerschaut A., Orlandi P., Palvadeau P. Lead-antimony sulfosalts from Tuscany (Italy). II. Crystal structure of scainiite, $Pb_{14}Sb_{30}S_{54}O_5$, an expanded monoclinic derivative of $Ba_{12}Bi_{24}S_{48}$ hexagonal sub-type (zinkenite group), Eur. J. Mineral., 12(4), 835–846 (2000).

Moëlo Y., Mozgova N., Picot P., Bortnikov N., Vrublevskaya Z. Cristallochimie de l'owyheeite: nouvelles données, Tschermaks Miner. Petr. Mitt., 32(4), 271–284 (1984).

Moëlo Y., Oudin E., Makovicky E., Karup-Møller S., Pillard F., Bornuat M., Evanghelou E. La kirkiite, $Pb_{10}Bi_3As_3S_{19}$, une nouvelle espèce minérale homologue de la jordanite, Bull. Minéralog., 108, 667–677 (1985).

Moëlo Y., Pecorini R., Ciriotti M.E., Meisser N., Caldes M.T., Orlandi P., Petit P.-E., Martini B., Salvetti A. Tubulite, $\sim Ag_2Pb_{22}Sb_{20}S_{53}$, a new Pb-Ag-Sb sulfosalt from Le Rivet quarry, Peyrebrune ore field (Tarne, France) and Biò, Borgofranco mines, Borgofranco d'Ivrea (Piedmont, Italy), Eur. J. Mineral., 25(6), 1017–1030 (2013).

Moëlo Y., Pecorini R., Ciriotti M.E., Meisser N., Caldes-Rouillon M., Orlandi P., Petit P.E., Martini B., Salvetti A. (2012) Tubulite, IMA 2011-109. CNMNC Newsletter No. 13, June 2012, page 811, Mineralog. Mag., 76(3), 807–817 (2012).

Moëlo Y. Quaternary compounds in the system Pb-Sb-S-Cl: dadsonite and synthetic phases, Canad. Mineralog., 17(3), 595–600 (1979).

Moh G.H. Tin-containing mineral systems. Part II: phase relations and mineral assemblages in Cu–Fe–Zn–Sn–S system, Chem. Erde, 34(1), 1–61 (1975).

Møller C.K. The structure of $Pb(NH_4)_2(SO_4)_2$ and related compounds, Acta Chem. Scand., 8(1), 81–87 (1954).

Moodie A.F., Whitefield H.J. Determination of the structure of Cu_2ZnGeS_4 polymorphs by lattice imaging and convergent-beam electron diffraction, Acta Crystallogr., B42(3), 236–247 (1986).

Morihama M., Maeda T., Yamauchi I., Wada T. Crystallographic and optical properties of narrow band gap Cu_2GeSe_3 and $Cu_2(Sn_{1-x}Ge_x)Se_3$ solid solution, Jpn. J. Appl. Phys., 53(5S1), 05FW06_1–05FW06_5 (2014).

Moroz M., Demchenko P., Romaka V., Serkiz R., Akselrud L., Gladyshevskii R. Mykolaychuk O. $Ag_3Ge_2S_5Br$: synthesis, structure and ionic conductivity, Chem. Met. Alloys, 7(3–4), 139–148 (2014a).

Moroz M., Tesfaye F., Demchenko P., Prokhorenko M., Kogut Y., Pereviznyk O., Prokhorenko S., Reshetnyak O. Solid-state electrochemical synthesis and thermodynamic properties of selected compounds in the Ag–Fe–Pb–Se system, Solid State Sci., 107, 106344 (2020a).

Moroz M., Tesfaye F., Demchenko P., Prokhorenko M., Lindberg D., Reshetnyak O., Hupa L. Determination of the thermodynamic properties of the $Ag_2CdSn_3S_8$ and Ag_2CdSnS_4 phases in the Ag–Cd–Sn–S system by the solid-state electrochemical cell method, J. Chem. Thermodyn., 118, 255–262 (2018a).

Moroz M., Tesfaye F., Demchenko P., Prokhorenko M., Lindberg D., Reshetnyak O., Hupa L. Phase equilibria and thermodynamics of selected compounds in the Ag–Fe–Sn–S system, J. Electron. Mater., 47(9), 5433–5442 (2018b).

Moroz M., Tesfaye F., Demchenko P., Prokhorenko M., Lindberg D., Reshetnyak O., Hupa L. Thermal stability and thermodynamics of the Ag_2ZnGeS_4 compound, Mater. Proc. Fund., Miner. Met. Mater. Ser., 215–226 (2019).

Moroz M., Tesfaye F., Demchenko P., Prokhorenko M., Lindberg D., Reshetnyak O., Hupa L. Thermodynamic properties of magnetic semiconductors $Ag_2FeSn_3S_8$ and Ag_2FeSnS_4 determined by the EMF method, Mater. Proc. Fund., Miner. Met. Mater. Ser., 87–98 (2018c).

Moroz M., Tesfaye F., Demchenko P., Prokhorenko M., Rudyk B., Soliak L., Lindberg D., Reshetnyak O., Hupa L. Thermodynamic examination of quaternary compounds in the Ag–Fe–

(Ge, Sn)–Se systems by the solid-state EMF method, Lee J. et al. (Eds.), *The Minerals, Metals and Materials Series, Mater. Proc. Fund.*, 271–283 (2021a).

Moroz M., Tesfaye F., Demchenko P., Prokhorenko M., Yarema N., Lindberg D., Reshetnyak O., Hupa L. The equilibrium phase formation and thermodynamic properties of functional tellurides in the Ag–Fe–Ge–Te system, *Energies*, 14(5), 1314 (2021b).

Moroz M.V., Demchenko P.Yu., Akselrud L.G., Mykolaychuk O.G., Gladyshevskii R.E. Phase relation along the Ag_8GeS_6–[$(AgBr)_4 GeS_2$] cross-section. Crystal structure and electric conductivity of $Ag_6GeS_4Br_2$ in bulk, *Chem. Met. Alloys*, 3(3–4), 161–168 (2010).

Moroz M.V., Demchenko P.Yu., Mykolaychuk O.G., Akselrud L.G., Gladyshevskii R.E.. Synthesis and electrical conductivity of crystalline and glassy alloys in the Ag_3GeS_3Br–GeS_2 system, *Inorg. Mater.*, 49(9), 867–871 (2013a).

Moroz M.V., Demchenko P.Yu., Prokhorenko S.V., Moroz V.M. Physical properties of glasses in the Ag_2GeS_3–$AgBr$ system, *Phys. Solid State*, 55(8), 1613–1618 (2013b).

Moroz M.V., Demchenko P.Yu., Prokhorenko M.V., Reshetnyak O.V. Thermodynamic properties of saturated solid solutions of the phases Ag_2PbGeS_4, $Ag_{0.5}Pb_{1.75}GeS_4$ and $Ag_{6.72}Pb_{0.16}Ge_{0.84}S_{5.20}$ of the Ag–Pb–Ge–S system determined by EMF method, *J. Phase Equilib. Diffus.*, 38(4), 426–433 (2017a).

Moroz M.V., Prokhorenko M.V., Demchenko P.Yu., Reshetnyak O.V. Thermodynamic properties of saturated solid solutions of Ag_7SnSe_5Br and Ag_8SnSe_6 compounds in the Ag–Sn–Se–Br system measured by the EMF method, *J. Chem. Thermodyn.*, 106, 228–231 (2017b).

Moroz M.V., Prokhorenko M.V., Moroz V.M. $Ag_{0.225}Ge_{0.260}S_{0.515}$–$AgBr$ glasses, *Inorg. Mater.*, 50(5), 532–536 (2014b).

Moroz M.V., Prokhorenko M.V. Phase equilibria and thermodynamic properties of phases in the Ag–Cd–Sn–Se system, *Inorg. Mater.*, 51(8), 799–805 (2015).

Moroz M.V., Prokhorenko M.V., Reshetnyak O.V., Demchenko P.Yu. Electrochemical determination of thermodynamic properties of saturated solid solutions of Hg_2GeSe_3, Hg_2GeSe_4, $Ag_2Hg_3GeSe_6$, and $Ag_{1.4}Hg_{1.3}GeSe_6$ compounds in the Ag–Hg–Ge–Se system, *J Solid State Electrochem.*, 21(3), 833–837 (2017c).

Moroz M.V., Reshetnyak O.V., Demchenko P.Yu., Prokhorenko M.V., Soliak L.V., Rudyk B.P., Pereviznyk O.B., Prokhorenko S.V. Thermodynamic properties of silver-containing compounds of the Ag–Fe–Sn–S system obtained by low-temperature solid-state synthesis [in Ukrainian], *Ukr. khim. zhurn.*, 86(11), 34–50 (2020b).

Morozov I.S., Li C.-F. The $SnCl_2$+PbS↔SnS+$PbCl_2$ mutual system [in Russian], *Zhurn. neorgan. khimii*, 9(7), 1688–1692 (1963).

Moroz V.N. The Ag_8GeSe_6–Cu_8GeSe_6 section of the Cu–Ag–Ge–Se system [in Russian], *Izv. AN SSSR. Neorgan. mater.*, 26(9), 1830–1833 (1990).

Morris C.D., Chung I., Park S., Harrison C.M., Clark D.J., Jang J.I., Kanatzidis M.G. Molecular germanium selenophosphate salts: phase-change properties and strong second harmonic generation, *J. Am. Chem. Soc.*, 134(51), 20733–20744 (2012).

Morris C.D., Li H., Jin H., Malliakas C.D., Peters J.A., Trikalitis P.N., Freeman A.J., Wessels B.W., Kanatzidis M.G. $Cs_2M^{II}M^{IV}_3Q_8$ (Q = S, Se, Te): an extensive family of layered semiconductors with diverse band gaps, *Chem. Mater.*, 25(16), 3344–3356 (2013).

Mozgova N.N., Efimov A.V., Nenasheva S.N., Golovanova T.I., Sivtsov A.V., Tsepin A.I., Dobretsova I.G. New data on diaphorite and brogniardite [in Russian], *Zap. Vses. Mineralog. Obsh.*, 118(5), 47–63 (1989).

Mozolyuk M.Yu., Olekseyuk I.D., Piskach L.V., Parasyuk O.V. X-ray diffraction study of the Tl_2GeS_3–{Zn,Cd,Hg}S sections [in Ukrainian], *Nauk. Visnyk Volyns'k. Nats. Univ. im. Lesi Ukrainky. Ser. Khim. nauky*, (17), 75–78 (2012a).

Mozolyuk M.Yu., Piskach L.V., Fedorchuk A.O., Kityk I.V., Olekseyuk I.D., Parasyuk O.V. Phase diagram of the quasibinary system $TlInSe_2$–$SnSe_2$, *J. Alloys Compd.*, 509(6), 2693–2696 (2011).

Mozolyuk M.Yu., Piskach L.V., Fedorchuk A.O., Olekseyuk I.D., Parasyuk O.V., Khyzhun O.Y. The Tl_2Se–$PbSe$–SiS_2 system and the crystal and electronic structure of quaternary chalcogenide Tl_2PbSiS_4, *Mater. Chem. Phys.*, 195, 132–142 (2017).

Mozolyuk M.Yu., Piskach L.V., Fedorchuk A.O., Olekseyuk I.D., Parasyuk O.V. Phase equilibria in the Tl_2S–PbS–GeS_2 system and crystal structure of $Tl_{0.5}Pb_{1.17}GeS_4$, *Chem. Met. Alloys*, 5(1–2), 37–41 (2012b).

Mozolyuk M.Yu., Piskach L.V., Fedorchuk A.O., Olekseyuk I.D., Parasyuk O.V. Physico-chemical interaction in the Tl_2Se–$HgSe$–$D^{IV}Se_2$ systems (D^{IV} – Si, Sn), *Mater. Res. Bull.*, 47(11), 3830–3834 (2012c).

Mozolyuk M.Yu., Piskach L.V., Fedorchuk A.O., Olekseyuk I.D., Parasyuk O.V. The Tl_2Se–$HgSe$–$GeSe_2$ system and the crystal structure of $Tl_2HgGeSe_4$, *Chem. Met. Alloys*, 6(1–2), 55–62 (2013).

Mrotzek A., Chung D.-Y., Ghelani N., Hogan T., Kanatzidis M.G. Structure and thermoelectric properties of the new quaternary bismuth selenides $A_{1-x}M_{4+x}Bi_{11+x}Se_{21}$ (A = K and Rb and Cs; M = Sn and Pb) – members of the grand homologous series $K_m(M_6Se_8)_m(M_{5+n}Se_{9+n})$, *Chem. Eur. J.*, 7(9), 1915–1926 (2001a).

Mrotzek A., Chung D.-Y., Hogan T., Kanatzidis M.G. Structure and thermoelectric properties of the new quaternary tin selenide $K_{1-x}Sn_{5-x}Bi_{11+x}Se_{22}$, *J. Mater. Chem.*, 10(7), 1667–1672 (2000).

Mrotzek A., Iordanidis L., Kanatzidis M.G. $Cs_{1-x}Sn_{1-x}Bi_{9+x}Se_{15}$ and $Cs_{1.5-3x}Bi_{9.5+x}Se_{15}$: members of the homologous superseries $A_m[M_{1+l}Se_{2+l}]_{2m}[M_{l+2l+n}Se_{3+3l+n}]$ (A = alkali metal, M = Sn and Bi) allowing structural evolution in three different dimensions, *Chem. Commun.*, (17), 1648–1649 (2001b).

Mrotzek A., Iordanidis L., Kanatzidis M.G. New members of the homologous series $A_m[M_6Se_8]_m[M_{5+n}Se_{9+n}]$: the quaternary phases $A_{1-x}M'_{3-x}Bi_{11+x}Se_{20}$ and $A_{1+x}M'_{3-2x}Bi_{7+x}Se_{14}$ (A = K, Rb, Cs; M' = Sn, Pb), *Inorg. Chem.*, 40(24), 6204–6211 (2001c).

Mrotzek A., Kanatzidis M.G. Design in solid state synthesis based on phase homologies: $A_{1-x}Sn_{9-x}Bi_{11+x}Se_{26}$ (A = K, Rb, Cs) –a new member of the grand homologous series $A_m[M_6Se_8]_m[M_{5+n}Se_{9+n}]$ with M = Sn and Bi, *J. Solid State Chem.*, 167(2), 299–301 (2002).

Mrotzek A., Kanatzidis M.G. Tropochemical cell-twinning in the new quaternary bismuth selenides $K_xSn_{6-2x}Bi_{2+x}Se_9$ and $KSn_5Bi_5Se_{13}$, *Inorg. Chem.* 42(22), 7200–7206 (2003).

Mumbaraddi D., Iyer A.K., Üzer E., Mishra V., Oliynyk A.O., Nilges T., Mar A. Synthesis, structure, and properties of rare-earth germanium sulfide iodides $RE_3Ge_2S_8I$ (RE = La, Ce, Pr), *J. Solid State Chem.*, 274, 162–167 (2019).

Mumme W.G., Gable R.W., Lindquist B. Crystal structure of synthetic $Pb_4Bi_{12}Se_{6.2}S_{15.8}$, *N. Jb. Miner. Abh. (J. Min. Geochem.)*, 194(1), 27–33 (2017).

Mumme W.G. Seleniferous lead-bismuth sulphosalts from Falun, Sweden: weibullite, wittite, and nordstriimite, *Am. Mineralog.*, **65**(7–8), 789–796 (1980a).

Mumme W.G. The crystal structure of krupkaite, $CuPbBi_3S_6$, from the Juno Mine at Tennant Creek, Northern Territory, Australia, *Am. Mineralog.*, **60**(3–4), 300–308 (1975).

Mumme W.G., Watts J.A. Additional physical, optical and X-ray data for pekoite, *Canad. Mineralog.*, **14**(4), 578 (1976a).

Mumme W.G. Watts J.A. Pekoite, $CuPbBi_{11}S_{18}$, a new member of the bismuthinite-aikinite mineral series: its crystal structure and relationship with naturally- and synthetically-formed members, *Canad. Mineralog.*, **14**(3), 322–333 (1976b).

Mumme W.G. Weibullite $Ag_{0.32}Pb_{5.09}Bi_{8.55}Se_{6.08}S_{11.92}$ from Falun, Sweden: a higher homologue of galenobismutite, *Canad. Mineralog.*, **18**(1), 1–12 (1980b).

Mumme W.G., Welin E., Wuensch B.J. Crystal chemistry and proposed nomenclature for sulfosalts intermediate in the system bismuthinite–aikinite (Bi_2S_3–$CuPbBiS_3$), *Am. Mineralog.*, **61**(1–2), 15–20 (1976).

Murayama M., Kanno R., Irie M., Ito S., Hata T., Sonoyama N., Kawamoto Y. Synthesis of new lithium ionic conductor thio-LISICON – lithium silicon sulfides system, *J. Solid State Chem.*, **168**(1), 140–148 (2002).

Murdoch J. X-ray investigation of colusite, germanite and renierite, *Am. Mineralog.*, **38**(9–10), 794–801 (1953).

Murguzov M.I., Gurshumov A.P., Guliev M.M. The SnTe–FeSe system [in Russian], *Zhurn. neorgan. khimii*, **29**(11), 2968–2970 (1984).

Murguzov M.I., Gurshumov A.P., Guliev M.M. The SnTe–NiSe and SnTe–CoSe systems [in Russian], *Zhurn. neorgan. khimii*, **30**(1), 186–189 (1985).

Nagaoka A., Yoshino K., Kakimoto K., Nishioka K. Phase diagram of the Ag_2SnS_3–ZnS pseudobinary system for Ag_2ZnSnS_4 crystal growth, *J. Cryst. Growth*, **555**, 125967 (2021).

Nagaoka A., Yoshino K., Taniguchi H., Taniyama T., Kakimoto K., Miyake H. Growth and characterization of Cu_2ZnSnS_4 single crystals, *Phys. stat. sol. (a)*, **210**(7), 1328–1331 (2013).

Nagaoka A., Yoshino K., Taniguchi H., Taniyama T., Miyake H. Growth of Cu_2ZnSnS_4 single crystal by travelling heater method, *Jpn. J. Appl. Phys.*, **50**(12), 128001_1–128001_2 (2011).

Nagaoka A., Yoshino K., Taniguchi H., Taniyama T., Miyake H. Preparation of Cu_2ZnSnS_4 single crystals from Sn solutions, *J. Cryst. Growth*, **341**(1), 38–41 (2012).

Nagel A., Range K.-J. Die Kristallstruktur von Ag_7GeS_5I, *Z. Naturforsch.*, **34B**(2), 360–362 (1979).

Nagel A., Range K.-J. Verbindungsbildung in system Ag_2S–GeS_2–AgI, *Z. Naturforsch.*, **33B**(12), 1461–1464 (1978).

Nakamura Y., Aruga A., Nakai I., Nagashima K. The crystal structure of a new thiosilicate of thallium, $TlInSiS_4$, *Bull. Chem. Soc. Jpn.*, **57**(7), 1718–1722 (1984a).

Nakamura Y., Nakai I., Nagashima K. Preparation and characterization of the new quaternary chalcogenides Tl–III–IV–S_4 (III = Al, Ga, In; IV = Si, Ge), *Mater. Res. Bull.*, **19**(5), 563–570 (1984b).

Naseri M., Hoat D.M., Ponce-Pérez R., Rivas-Silva J.F., Cocoletzi G.H. An assessment of the structural, electronic, optical and thermoelectric properties of the $BaAg_2GeS_4$ compound, *J. Solid State Chem.*, **285**, s (2020).

Nasonova D.I., Sobolev A.V., Presniakov I.A., Andreeva K.D., Shevelkov A.V. Position and oxidation state of tin in Sn-bearing tetrahedrites $Cu_{12-x}Sn_xSb_4S_{13}$, *J. Alloys Compd.*, **778**, 774–778 (2019).

Natarajan S., Rao G.V.S., Baskaran R., Radhakrishnan T.S. Synthesis and electrical properties of shandite–parkerite phases, $A_2M_3Ch_2$, *J. Less-Common Metals*, **138**(2), 215–224 (1988).

Nazarchuk O.P., Mazurets I.I., Olekseyuk I.D., Gulay L.D. The $Cu_2S(Se)$–$NiS(Se)$–$SiS_2(Se)$ systems and crystal structure of $Cu_4NiSi_2S_7$ [in Ukrainian], *Nauk. Visnyk Volyns'k. Nats. Univ. im. Lesi Ukrainky. Ser. Khim. nauky*, (16), 21–26 (2008).

Nenasheva S.N., Kalinina T.A. Investigation of melting peculiarities of the stannite Cu_2FeSnS_4 [in Russian], *Izv. AN SSSR. Neorgan. mater.*, **22**(1), 22–25 (1986).

Nespolo M., Ozawa T., Kawasaki Y., Sugiyama K. Structural relations and pseudosymmetries in the andorite homologous series, *J. Mineral. Petrol. Sci.*, **107**(6), 226–243 (2012).

Nesterov V.N., Isakova R.A., Shendyapin A.S. The lead activity in sulfide melts [in Russian], *Zhurn. phys. khimii*, **43**(12), 3181–3183 (1969).

Neves F., Correia J.B., Hanada K., Santos L.F., Gunder R., Schorr S. Structural characterization of Cu_2SnS_3 and $Cu_2(Sn,Ge)S_3$ compounds, *J. Alloys Compd.*, **682**, 489–494 (2016).

Nhalil H., Han D., Du M.-H., Chen S., Antonio D., Gofryk K., Saparov B. Optoelectronic properties of candidate photovoltaic Cu_2PbSiS_4, Ag_2PbGeS_4 and KAg_2SbS_4 semiconductors, *J. Alloys Compd.*, **726**, 405–412 (2018).

Nian L., Huang J., Wu K., Su Z., Yang Z., Pan S. $BaCu_2M^{IV}Q_4$ (M^{IV} = Si, Ge, and Sn; Q = S, Se): synthesis, crystal structures, optical performances and theoretical calculations, *RSC Adv.*, **7**(47), 29378–29385 (2017).

Nian L., Wu K., He G., Yang Z., Pan S. Effect of element substitution on structural transformation and optical performances in $I_2BaM^{IV}Q_4$ (I = Li, Na, Cu, and Ag; M^{IV} = Si, Ge, and Sn; Q = S and Se), *Inorg. Chem.*, **57**(6), 3434–3442 (2018).

Niizeki N., Buerger M.J. The crystal structure of jamesonite, $FePb_4Sb_6S_{14}$, *Z. Kristallogr.*, **109**(1–6), 161–183 (1957).

Nikolaev R.E., Vasilyeva I.G. A new way of phase identification, of $AgGaGeS_4 \cdot nGeS_2$ crystals, *J. Solid State. Chem.*, **203**, 340–344 (2013).

Nikolić P.M. Optical energy gaps, lattice parameters and solubility limits of solid solutions of SnSe and GeSe in PbTe, and GeSe in SnTe, *Brit. J. Appl. Phys.*, **16**(8), 1075–1079 (1965).

Nikolić P.M. Optical energy gaps of PbSe–SnTe, PbSe–SnSe, PbTe–SnTe and PbTe–SnSe, *Brit. J. Appl. Phys.*, **18**(7), 897–903 (1967).

Nikolić P.M. Solid solutions of CdSe and CdTe in PbTe and their optical properties, *Brit. J. Appl. Phys.*, **17**(3), 341–344 (1966).

Nikolić P.M. Solid solution of lead-germanium chalcogenide alloys and some of their optical properties, *J. Phys. D: Appl. Phys.*, **2**(3), 383–388 (1969).

Nilges T., Pfitzner A. A structural differentiation of quaternary copper argyrodites: structure–property relations of high temperature ion conductors, *Z. Kristallogr.*, **220**(2–3), 281–294 (2005).

Nitsche R., Sargent D.F., Wild P. Crystal growth of quaternary 1_2246_4 chalcogenides by iodine vapor transport, *J. Cryst. Growth*, **1**(1), 52–53 (1967).

Ni Y., Wu H., Wang Z., Mao M., Cheng G., Fei H. Synthesis and growth of nonlinear infrared crystal material $AgGeGaS_4$ via a new reaction route, *J. Cryst. Growth*, **311**(5), 1404–1406 (2009).

Nomura T., Maeda T., Takei K., Morihama M., Wada T. Crystal structures and band-gap energies of $Cu_2Sn(S,Se)_3$ ($0 \leq x \leq 1.0$) solid solution, *Phys. stat. sol. (c)*, **10**(7–8), 1093–1097 (2013a).

Nomura T., Maeda T., Wada T. Preparation of narrow band-gap $Cu_2Sn(S,Se)_3$ and fabrication of film by non-vacuum process, *Jpn. J. Appl. Phys.*, **52**(4S), 04CR08_1–04CR08_4 (2013b).

Nowacki W., Iitaka Y., Bürki H. Structural investigations on sulfo-salts from the Lengenbach, Binn Valley, Switzerland, *Acta Crystallogr.*, **13**(12), 1006–1007 (1960).

Odin I.N., Chukichev M.V. Metastable phases crystallizing from the melts in the $PbX + CdI_2 = CdX + PbI_2$ (X = S, Se, Te) mutual systems [in Russian], *Zhurn. neorgan. khimii*, **46**(12), 2083–2087 (2001).

Odin I.N., Chukichev M.V. Physico-chemical analysis of the Cd–Sb(Bi)–S systems and properties of photosensitivety solid solutions based on cadmium sulfide compounds [in Russian], *Zhurn. neorgan. khimii*, **45**(2), 255–260 (2000).

Odin I.N., Gapanovich M.V., Urkhanov O.Yu., Chukichev M.V., Novikov G.F. Crystallographic and luminescence characteristics of the $Cu_2MgSnSe_4$ quaternary compound and $Cu_{2-x}MgSnSe_4$ ($0 < x \leq 0.15$) copper-deficient solid solutions, *Inorg. Mater.*, **57**(1), 1–7 (2021).

Odin I.N., Grin'ko V.V., Kozlovskiy V.F., Demidova E.D. The stable and metastable equilibria in the $PbSe + SnI_2 \leftrightarrow SnSe + PbI_2$ mutual system [in Russian], *Zhurn. neorgan. khimii*, **48**(5), 842–845 (2003a).

Odin I.N., Grin'ko V.V., Kozlovskiy V.F., Safronov E.V., Gapanovich M.V. Formation of the stable and metastable phases in the $PbSe + MI_2 \leftrightarrow MSe + PbI_2$ (M = Hg, Mn, Sn) mutual systems [in Russian], *Zhurn. neorgan. khimii*, **49**(9), 1562–1567 (2004).

Odin I.N., Grin'ko V.V., Kozlovskiy V.F. Stable and metastable phases in the $PbSe + Ag_2I_2 \leftrightarrow Ag_2Se + PbI_2$ and $PbSe + CdI_2 \leftrightarrow CdSe + PbI_2$ mutual system [in Russian], *Zhurn. neorgan. khimii*, **50**(5), 843–847 (2005).

Odin I.N. Physico-chemical analysis of ternary and ternary mutual systems, containing cadmium, zinc, silicon, and bismuth chalcogenides and properties of ingots in these systems [in Russian], *Zhurn. neorgan. khimii*, **41**(6), 941–953 (1996).

Odin I.N., Safronov E.V., Kozlovskiy V.F., Galiulin E.A. *T-x-y*-phase diagrams for the $PbSe + MnI_2 \leftrightarrow MnSe + PbI_2$ mutual system, *Zhurn. neorgan. khimii*, **48**(5), 839–841 (2003b).

Odin I.N. *T-x-y* diagrams for mutual systems $PbX + CdI_2 = CdX + PbI_2$ (X = S, Se, Te) [in Russian], *Zhurn. neorgan. khimii*, **46**(10), 1733–1738 (2001).

Oftedal I. Die Raumgruppe des Bournonits ($CuPbSbS_3$), *Z. Kristallogr.*, **83**(1–6), 157–158 (1932).

Ohachi T., Pamplin B.R. Growth of new spinel compounds $CuInSnS_4$ and $CuIn_{11}S_{17}$, *J. Cryst. Growth*, **42**, 598–601 (1977a).

Ohachi T., Pamplin B.R. Quaternary compounds of Ag^I, In^{III}, Sn^{IV} and S^{VI} on the system $Ag_2S–In_2S_3–SnS_2$, In: *Ternary compounds. Inv. Contrib. Pap. 3rd Int. Conf.*, Edinburgh, 1977. *Inst. Phys. Conf. Ser.*, (35), 21–34 (1977b).

Ohmasa M., Nowacki W. A redetermination of the crystal structure of aikinite $[BiS_2|S|Cu^{IV}Pb^{VII}]$, *Z. Kristallogr.*, **132**(1–6), 71–86 (1970).

Oh S.-J., Lee D.W., Ok K.M. Influence of the cation size on the framework structures and space group centricities in $AMo_2O_5(SeO_3)_2$ (A = Sr, Pb, and Ba), *Inorg. Chem.*, **51**(9), 5393–5399 (2012a).

Oh S.-J., Lee D.W., Ok K.M. $PbMSeO_6$ (M = Mo and W): new quaternary mixed metal selenites with asymmetric cationic coordination environments, *Dalton Trans.*, **41**(10), 2995–3000 (2012b).

Ohtsuki T., Kitakaze A., Sugaki A. Synthetic minerals with quaternary components in the system Cu–Fe–Sn–S – Synthetic sulfide minerals (X), *Sci. Repts. Tohoku Univ. Ser. III*, **14**(3), 269–282 (1980).

Ohtsuki T., Sugaki A., Kitakaze A. Synthetic sulfide minerals. XI. Three new phases in the system Cu–Fe–Sn–S, *Sci. Repts. Tohoku Univ. Ser. I*, **15**(1), 79–87 (1981).

Ok K.M., Halasyamani P.S. Asymmetric cationic coordination environments in new oxide materials: synthesis and characterization of $Pb_4Te_6M_{10}O_{41}$ (M = Nb^{5+} or Ta^{5+}). *Inorg. Chem.*, **43**(14), 4248–4253 (2004).

Oleinik G.S., Mizetski P.A., Nizkova A.I. Interaction between lead and zinc chalcogenides [in Russian], *Izv. AN SSSR. Neorgan. mater.*, **18**(5), 873–874 (1982).

Olekseyuk I.D., Dudchak I.V., Piskach L.V. Phase equilibria in the $Cu_2Se–ZnSe–Cu_2SnSe_3$ quasiternary system [in Ukrainian], *Fiz. i khim. tv. tila*, **2**(2), 195–200 (2001a).

Olekseyuk I.D., Dudchak I.V., Piskach L.V. Phase equilibria in the $Cu_2S–ZnS–SnS_2$ system, *J. Alloys Compd.*, **368**(1–2), 135–143 (2004).

Olekseyuk I.D., Gorgut G.P., Parasyuk O.V. The phase equilibria in the quasi-ternary $Ag_2Se–Ga_2Se_3–GeSe_2$, *J. Alloys Compd.*, **260**(1–2), 111–120 (1997a).

Olekseyuk I.D., Gorgut G.P., Prodous Yu.Ya. Glass formation in the $Ag_2Se–Ga_2Se_3–GeSe_2$ quasiternary system [in Ukraininan], *Ukr. khim. zhurn.*, **67**(5–6), 68–69 (2001b).

Olekseyuk I.D., Gorgut G.P., Shevchuk M.V. Phase equilibria in the $AgGaS_2–GeS_2$ system, *Pol. J. Chem.*, **76**(7), 915–919 (2002a).

Olekseyuk I.D., Gulay L.D., Dudchak I.V., Piskach L.V., Parasyuk O.V., Marchuk O.V. Single crystal preparation and crystal structure of the $Cu_2Zn/Cd,Hg/SnSe_4$ compounds. *J. Alloys Compd.*, **340**(1–2), 141–145 (2002b).

Olekseyuk I.D., Gulay L. D., Marchuk O. V. Isothermal iections of the $Sm(Er)_2Se_3–PbSe–GeSe_2$ systems at 770 K and crystal structure of the $Sm_{1.32}Pb_{1.68}Ge_{1.67}Se_7$ compound [in Ukraininan], *Nauk. Visnyk Volyns'k. Nats. Univ. im. Lesi Ukrainky. Ser. Khim. nauky*, (24), 14–19 (2009a).

Olekseyuk I.D., Gulyak A.V., Sysa L.V., Gorgut G.P., Lomzin A.F. Crystal chemical properties and preparation of single crystals of $AgGaSe_2–GeSe_2$ γ-solid solutions, *J. Alloys Compd.*, **241**(1–2), 187–190 (1996a).

Olekseyuk I.D., Gulyak A.V., Sysa L.V. Investigation of the $AgIn_5Se_8–GeSe_2$ section [in Russian], *Zhurn. neorgan. khimii*, **42**(8), 1392–1393 (1997b).

Olekseyuk I.D., Gulyak A.V., Sysa L.V. Investigation of the $AgInSe_2–SnSe_2$ and $AgIn_5Se_8–SnSe_2$ sections [in Russian], *Zhurn. neorgan. khimii*, **43**(12), 2084–2085 (1998a).

Olekseyuk I.D., Kogut Yu.M., Parasyuk O.V., Piskach L.V., Gorgut G.P., Kus'ko O.P., Pekhnyo V.I., Volkov S.V. Glass-formation in the $Ag_2Se–Zn(Cd, Hg)Se–GeSe_2$ systems, *Chem. Met. Alloys*, **2**(3–4), 146–150 (2009b).

Olekseyuk I.D., Kogut Yu.M., Yurchenko O.M., Parasyuk O.V., Volkov S.V., Pekhnyo V.I. Glass formation and optical properties of the glasses in the $Ag_2S–HgS–GeS_2$ system, *Chem. Met. Alloys*, **2**(1–2), 49–54 (2009c).

Olekseyuk I.D., Krykhovets O.V. Interaction in the $AgInSe_2–GeSe_2$ and $AgInSe_2–SnSe_2$ systems [in Ukraininan], *Ukr. khim. zhurn.*, **67**(7–8), 81–84 (2001a).

Olekseyuk I.D., Krykhovets O.V., Sysa L.V. Phase equilibria in the $Ag_2Se–In_2Se_3–GeSe_2(SnSe_2)$ system, *Pol. J. Chem.*, **73**(3), 431–436 (1999a).

Olekseyuk I.D., Krykhovets O.V. The $Ag_2Se–In_2Se_3–SnSe_2$ system, *J. Alloys Compd.*, **316**(1–2), 193–202 (2001b).

Olekseyuk I.D., Marchuk O.V., Bozhko V.V., Trofymchuk L.V. Electrophysical properties of the $Cu_2HgC^{VI}Se_4$ (C – Sn, Ge) quaternary compounds [in Ukrainian], *Nauk. Visnyk Volyns'k. Derzh. Univ. im. Lesi Ukrainky. Ser. Khim. nauky*, **6**, 34–37 (2001c).

Olekseyuk I., Marchuk O., Dudchak I., Parasyuk O., Piskach L. Phase equilibria in the Cu_2SnS_3–$Zn(Hg)S$ systems [in Ukrainian], *Visnyk L'viv. Univ. Ser. khim.*, (*39*), 48–52 (2000a).

Olekseyuk I.D., Marchuk O.V., Gulay L.D., Zhbankov O.Ye. Isothermal section of the Cu_2Se–$HgSe$–$GeSe_2$ system at 670 K and crystal structure of the compounds $Cu_2HgGeSe_4$ and HT-modification of Cu_2HgGeS_4, *J. Alloys Compd.*, **398**(1–2), 80–84 (2005a).

Olekseyuk I.D., Marchuk O.V., Parasyuk O.V., Bozhko V.V., Galyan V.V. Physico-chemical and physical properties of glasses in the Cu_2Se–$HgSe$–$GeSe_2$ system [in Ukrainian], *Fiz. i khim. tv. tila*, **2**(1), 69–76 (2001d).

Olekseyuk I.D., Mazurets I.I., Parasyuk O.V. Phase equilibria in the HgS–Ga_2S_3–GeS_2 system, *J. Alloys Compd.*, **417**(1–2), 131–137 (2006a).

Olekseyuk I.D., Mazurets I.I., Parasyuk O.V. Phase relations in the $ZnSe$–Ga_2Se_3–$GeSe_2$ system, *J. Alloys Compd.*, **351**(1–2), 171–175 (2003).

Olekseyuk I.D., Mozolyuk M.Y., Piskach L.V., Litvinchuk M.Yu., Parasyuk O.V. Interaction of the components in the systems formed by the chalcogenides of Tl(I), Hg(II), Pb(II), Si(IV) [in Ukrainian], *Nauk. Visnyk Volyns'k. Nats. Univ. im. Lesi Ukrainky. Ser. Khim. nauky*, (*17*), 62–69 (2012a).

Olekseyuk I.D., Mozolyuk M.Yu., Piskach L.V., Parasyuk O.V. Phase equilidria in the $Tl_2S(Se)$–$HgS(Se)$–$SnS(Se)_2$ systems at 520 K [in Ukrainian], *Nauk. Visnyk Volyns'k. nats. un-tu im. Lesi Ukrainky. Ser. Khim. nauky*, (*30*), 19–21 (2010a).

Olekseyuk I.D., Mozolyuk M.Y., Piskach L.V., Parasyuk O.V. Physico-chemical interaction in the Tl_2X–PbX–SnX_2 (X – S, Se) systems at 520 K [in Ukrainian], *Nauk. Visnyk Volyns'k. Nats. Univ. im. Lesi Ukrainky. Ser. Khim. nauky*, (*14*), 40–45 (2011).

Olekseyuk I.D., Ostap'yuk T.A., Viskunets' L.M., Zmiy O.F. Interaction of the components in the Cu_2Se–$PbSe$–Sb_2Se_3 system [in Ukrainian], *Nauk. Visnyk Volyns'k. Nats. Univ. im. Lesi Ukrainky. Ser. Khim. nauky*, (*16*), 38–42 (2010b).

Olekseyuk I.D., Ostap'yuk T.A., Yukhymuk T.V., Zmiy O.F. Phase interactions on isothermal sections of the Ag_2Se–$Ge(Sn)Se_2$–Sb_2Se_3 systems 570 K [in Ukraininan], *Nauk. Visnyk Volyns'k. Nats. Univ. im. Lesi Ukrainky. Ser. Khim. nauky*, (*29*), 35–40 (2009d).

Olekseyuk I.D., Ostap'yuk T.A., Zmiy O.F., Vlasyuk A.M. Phase diagram of the $AgSbSe2$–$SnSe$ system [in Ukraininan], *Nauk. Visnyk Volyns'k. Nats. Univ. im. Lesi Ukrainky. Ser. Khim. nauky*, (*17*), 105–110 (2012b).

Olekseyuk I.D., Parasyuk O.V., Bozhko V.V., Galyan V.V., Petrus I.I. Glass-forming in the $Zn(Cd,Hg)Se$–Ga_2Se_3–$GeSe_2$ systems [in Ukrainian], *Fizyka kondens. vysokomolek. system. Nauk. zap. Rivnens'kogo pedinstytutu*, (*3*), 148–152 (1997c).

Olekseyuk I.D., Parasyuk O.V., Bozhko V.V., Petrus I.I., Galyan V.V. Formation and properties of the quasiternary $Zn(Cd,Hg)Se$–Ga_2Se_3–$SnSe_2$ system glasses, *Funct. Mater.*, **6**(3), 474–477 (1999b).

Olekseyuk I.D., Parasyuk O.V. Phase equilibria in the $HgSe$–Ga_2Se_3–$SnSe_2$ system [in Russian], *Zhurn. neorgan. khimii*, **42**(5), 838–844 (1997).

Olekseyuk I.D., Parasyuk O.V., Salamakha P.S., Prots' Yu.M. The phase equilibria in the quasi-ternary $HgSe$–Ga_2Se_3–$GeSe_2$ system, *J. Alloys Compd.*, **238**(1–2), 141–148 (1996b).

Olekseyuk I.D., Parasyuk O.V., Sysa L.V., Yurchenko Yu.V. The $CdSe$–Ga_2Se_3–$GeSe_2$ system at 870 K, *Pol. J. Chem.*, **71**(6), 701–704 (1997e).

Olekseyuk I.D., Parasyuk O.V. The $CdSe$–Ga_2Se_3–$GeSe_2$ system [in Russian], *Zhurn. neorgan. khimii*, **40**(2), 315–319 (1995).

Olekseyuk I.D., Petrus I.I., Parasyuk O.V. Isothermal section of the HgS–Ga_2S_3–GeS_2 system at 670 K [in Ukrainian], *Nauk. Visnyk Volyns'k. Derzh. Univ. im. Lesi Ukrainky. Ser. Khim. nauky*, (*6*), 38–40 (2001e).

Olekseyuk I.D., Piskach L.V., Parasyuk O.V., Gorgut G.P., Volkov S.V., Pekhnyo V.I. Solid-liquid equilibria in the quasi-ternary system CdS–Ga_2S_3–GeS_2, *J. Alloys Compd.*, **421**(1–2), 91–97 (2006b).

Olekseyuk I.D., Piskach L.V., Parasyuk O.V., Melnyk O.M., Lyskovetz T.A. Cu_2Se–$CdSe$–$GeSe_2$ system, *J. Alloys Compd.*, **298**(1–2), 203–212 (2000b).

Olekseyuk I.D., Piskach L.V., Parasyuk O.V. Phase equilibria of $Ag_{0.33}Sn_{16.7}Se_{50}$–$CdSe$ section of the quasiternary Ag_2Se–$CdSe$–$SnSe_2$ system, *Pol. J. Chem.*, **71**(6), 721–724 (1997f).

Olekseyuk I.D., Piskach L.V., Parasyuk O.V. Phase equilibria in the $Cu_2SiSe_3(Te_3)$–$CdSe(Te)$ system [in Russian], *Zhurn. neorgan. khimii*, **43**(3), 516–519 (1998b).

Olekseyuk I.D., Piskach L.V. Phase equilibria in the Cu_2SnX_3–CdX (X = S, Se, Te) systems [in Russian], *Zhurn. neorgan. khimii*, **42**(2), 331–333 (1997).

Olekseyuk I.D., Piskach L.V., Sysa L.V. The Cu_2GeTe_3–$CdTe$ system and structure of the $Cu_2CdGeTe_4$ compound [in Russian], *Zhurn. neorgan. khimii*, **41**(9), 1420–1422 (1996c).

Olekseyuk I.D., Piskach L.V., Zhbankov O.Ye., Parasyuk O.V., Kogut Yu.M. Phase diagrams of the quasi-binary systems Cu_2S–SiS_2 and Cu_2SiS_3–PbS and the crystal structure of the new quaternary compound Cu_2PbSiS_4, *J. Alloys Compd.*, **399**(1–2), 149–154 (2005c).

Olekseyuk I.D., Sachanyuk V.P., Parasyuk O.V. X-ray powder diffraction refinement of $Ag_2In_2SiSe_6$ structure and phase diagram of the $AgInSe_2$–$SiSe_2$ system, *J. Alloys Compd.*, **414**(1–2), 73–77 (2006c).

Olekseyuk I.D., Strok O.M., Gulay L.D., Yakymchuk I.V., Romanishyna O.V. Isothermal sections of the $Cu_2S(Se)$–$La_2S(Se)_3$–$GeS(Se)_2$ quasiternary systems at 870 K [in Ukrainian], *Nauk. Visnyk Volyns'k. Derzh. Univ. im. Lesi Ukrainky. Ser. Khim. nauky*, (*29*), 13–20 (2009e).

Olekseyuk I.D., Strok O.M., Gulay L.D., Yakymchuk I.V., Romanishyna O.V. Isothermal sections of the Sm_2X_3–Cu_2X–GeX_2 (X – S, Se) systems at 870 K [in Ukrainian], *Nauk. Visnyk Volyns'k. Nats. Univ. im. Lesi Ukrainky. Ser. Khim. nauky*, (*16*), 54–62 (2010c).

Olekseyuk I.D., Strok O.M., Zmiy O.F. The polythermal $CuGaSe_2$–Cu_8GeSe_6 section of the quasiternary Cu_2Se–Ga_2Se_3–$GeSe_2$ system, *Pol. J. Chem.*, **75**(10), 1413–1416 (2001f).

Olekseyuk I., Zhbankov O. Phase equilibria in the Cu_2S–{Ge,Sn}S_2–Bi_2S_3 and Cu_2S–{Zn,Cd,Ge,Sn}S–Bi_2S_3 systems at 673 K [in Ukrainian], *Visnyk L'viv. Univ. Ser. khim.*, (*46*), 53–60 (2005).

Olekseyuk I.D., Zmiy O.F., Strok O.M., Kadykalo E.M. Isothermal section of the Cu_2Se–Ga_2Se_3–$GeSe_2$ quasiternary system at 770 K [in Ukrainian], *Nauk. Visnyk Volyns'k. Nats. Univ. im. Lesi Ukrainky. Ser. Khim. nauky*, (*16*), 30–35 (2008).

Olivier-Fourcade J., Jumas J.C., Maurin M., Philippot E. Mise en evidence d'un nouveau sulfoiodure d'étain et d'antimoine $Sn_2SbS_2I_3$: étude structurale, *Z. anorg. und allg. Chem.*, **468**(1), 91–98 (1980).

Omar M.S. Crystal growth and investigation of the solid solutions of the system $CuGe_2P_3$–I_2–IV–VI_3, *Mater. Res. Bull.*, **25**(6), 691–698 (1990).

Omloo W.P.F.A.M., Bommerson J.C., Heikens H.H., Risselada H., Vellinga M.B., Bruggen Van C.F., Haas C, Jellinek F. Europium chromium sulfide, EuCr$_2$S$_4$, and some isotypic compounds, *Phys. stat. sol. (a)*, **5**(2), 349–357 (1971).

Organova N.I., Kuz'mina O.V., Bortnikov N.S., Mozgova N.N. Crystal structure of the subcell of synthetic andorite-24 [in Russian], *Dokl. AN SSSR*, **267**(4), 939–942 (1982).

Orlandi P., Biagioni C., Bonaccorsi E., Moëlo Y., Paar W.H. Bernarlottiite, IMA 2013-133. CNMNC Newsletter No. 20, June 2014, page 553, *Mineralog. Mag.*, **78**(3), 549–558 (2014).

Orlandi P., Biagioni C., Bonaccorsi E., Moëlo Y., Paar W.H. Lead-antimony sulfosalts from Tuscany (Italy). XXI. Bernarlottiite, Pb$_{12}$(As$_{10}$Sb$_6$)$_{\Sigma16}$S$_{36}$, a new N = 3.5 member of the sartorite homologous series from the Ceragiola marble quarry: occurrence and crystal structure, *Eur. J. Mineral.*, **29**(4), 713–726 (2017).

Orlandi P., Moëlo Y., Biagioni C. Lead-antimony sulfosalts from Tuscany (Italy). X. Dadsonite from the Buca della Vena mine and Bi-rich izoklakeite from the Seravezza marble quarries, *Per. Mineral.*, **79**(1), 113–121 (2010).

Orlandi P., Moëlo Y., Campostrini I., Meerschaut A. Lead-antimony sulfosalts from Tuscany (Italy). IX. Marrucciite, Hg$_3$Pb$_{16}$Sb$_{18}$S$_{46}$, a new sulfosalt from Buca della Vena mine, Apuan Alps: definition and crystal structure, *Eur. J. Mineral.*, **19**(2), 267–279 (2007).

Orlandi P., Moëlo Y., Meerschaut A., Palvadeau P. Lead-antimony sulfosalts from Tuscany (Italy). I. Scainiite, Pb$_{14}$Sb$_{30}$S$_{54}$O$_5$, the first Pb–Sb oxy-sulfosalt, from Buca della Vena Mine, *Eur. J. Mineral.*, **11**(6), 949–954 (1999).

Orujlu E.N., Mammadov A.N., Babanly M.B. 3D analytical modeling of crystallization surfaces of the MnTe–SnTe–Sb$_2$Te$_3$ system, *Azerb. Chem. J.*, (2), 94–100 (2021).

Orujlu E.N. Phase equilibria in the SnBi$_2$Te$_4$–MnBi$_2$Te$_4$ system and characterization of the Sn$_{1-x}$Mn$_x$Bi$_2$Te$_4$ solid solutions, *Phys. Chem. Solid State*, **21**(1), 113–116 (2020a).

Orujlu E.N. Phase relations and characterization of solid solutions in the SnSb$_2$Te$_4$–MnSb$_2$Te$_4$ system, *New Mater., Compd. Appl.*, **4**(1), 38–43 (2020b).

Ostapyuk T.A., Yermiychuk I.M., Zmiy O.F., Olekseyuk I.D. Phase equilibria in the quasiternary system Cu$_2$Se–SnSe$_2$–Sb$_2$Se$_3$, *Chem. Met. Alloys*, **2**(3–4), 164–169 (2009a).

Ostapyuk T.A., Zmiy O.F., Ivashchenko I.A., Olekseyuk I.D. The Cu$_2$Se–PbSe–As$_2$Se$_3$ system, *Chem. Met. Alloys*, **7**(1–2), 164–169 (2014).

Ostapyuk T.A., Zmiy O.F., Olekseyuk I.D. Phase equilibria in the Cu$_2$Se–GeSe$_2$–Sb$_2$Se$_3$ quasi-ternary system [in Ukrainian], *Nauk. Visnyk Volyns'k. Nats. Univ. im. Lesi Ukrainky. Ser. Khim. nauky*, (24), 23–28 (2009b).

Ottenburgs R., Goethals H. Synthèse et polymorphisme de la briartite, *Bull. Soc. fr. Minéral. Cristallogr.*, **95**(4), 458–463 (1972).

Oykova T., Gospodinov G. Phase interactions in the PbO–SeO$_2$–H$_2$O and PbO–SeO$_3$–H$_2$O systems at 100°C [in Russian], *Zhurn. neorgan. khimii*, **26**(2), 491–494 (1981).

Paar W.H., Pring A., Moëlo Y., Stanley C.J., Putz H., Topa D., Roberts A.C., Braithwaite R.S.W. Daliranite, PbHgAs$_2$S$_6$, a new sulphosalt from the Zarshouran Au-As deposit, Takab region, Iran, *Mineralog. Mag.*, **73**(5), 871–881 (2009).

Paar W.H., Putz H., Topa D., Roberts A.C., Stanley C.J., Culetto F.J. Jonassonite, Au(Bi,Pb)$_5$S$_4$, a new mineral species from Nagybörzsöny, Hungry, *Canad. Mineralog.*, **44**(5), 1127–1136 (2006).

Paar W.H., Roberts A.C., Berlepsch P., Armbruster T., Topa D., Zagler G. Putzite, (Cu$_{4.7}$Ag$_{3.3}$)$_{\Sigma8}$GeS$_6$, a new mineral species from Capillitas, Catamarca, Argentina: description and crystal structure, *Canad. Mineralog.*, **42**(6), 1757–1769 (2004).

Palache C., Richmond W.E., Winchell H. Crystallographic studies of sulphosalts: baumhauerite, meneghinite, jordanite, diaphorite, freieslebenite, *Am. Mineralog.*, **23**(11), 821–836 (1938).

Palchik O., Iyer R.G., Canlas C.G., Weliky D.P., Kanatzidis M.G. K$_{10}$M$_4$M'$_4$S$_{17}$ (M = Mn, Fe, Co, Zn; M' = Sn, Ge) and Cs$_{10}$Cd$_4$Sn$_4$S$_{17}$: compounds with a discrete supertetrahedral cluster, *Z. anorg. und allg. Chem.*, **630**(13–14), 2237–2247 (2004).

Palchik O., Iyer R.G., Liao J.H. K$_{10}$M$_4$Sn$_4$S$_{17}$ (M = Mn, Fe, Co, Zn): soluble quaternary sulfides with the discrete [M$_4$Sn$_4$S$_{17}$]$^{10-}$ supertetrahedral clusters, *Inorg. Chem.*, **42**(17), 5052–5054 (2003).

Palchik O., Marking G.M., Kanatzidis M.G. Exploratory synthesis in molten salts: role of flux basicity in the stabilization of the complex thiogermanates Cs$_4$Pb$_4$Ge$_5$S$_{16}$, K$_2$PbGe$_2$S$_6$, and K$_4$Sn$_3$Ge$_3$S$_{14}$, *Inorg. Chem.*, **44**(12), 4151–4153 (2005).

Pamplin B.R., Ohachi T., Maeda S., Negrete P., Elworthy T.P., Sanderson R., Whitlow H.J. Solubility of the group IV chalcogenides in I-III-VI$_2$ compounds, In: *Ternary compounds. Inv. Contrib. Pap. 3rd Int. Conf.*, Edinburgh, 1977. Bristol–London: 35–42 (1977).

Pang X., Wang R., Che X., Huang F. SrZnSnSe$_4$: synthesis, crystal structure and nonlinear optical properties, *J. Solid State Chem.*, **297**, 122092 (2021).

Panyutin V., Badikov V., Shevyrdyaeva G., Mitin K., Seryogin A., Petrov V., Noack F. Quaternary nonlinear crystals of Ag$_x$Ga$_x$Ge$_{1-x}$Se$_2$ with orthorhombic symmetry for the mid-infrared spectral range, *Proc. SPIE*, **6875**, 68750A_1–68750A_12 (2008).

Paradis-Fortin L., Guélou G., Kumar V.P., Lemoine P., Prestipino C., Merdrignac-Conanec O., Durand G.R., Cordier S., Lebedev O.I., Guilmeau E. Structure, microstructure and thermoelectric properties of germanite-type Cu$_{22}$Fe$_8$Ge$_4$S$_{32}$ compounds, *J. Alloys Compd.*, **831**, 154767 (2020).

Parasyuk O.V., Babizhetskyy V.S., Khyzhun O.Y., Levytskyy V.O., Kityk I.V., Myronchuk G.L., Tsisar O.V., Piskach L.V., Jedryka J., Maciag A., Piasecki M. Novel quaternary TlGaSn$_2$Se$_6$ single crystal as promising material for laser operated infrared nonlinear optical modulators crystals, *Crystals*, **7**(11), 341 (2017).

Parasyuk O.V., Chykhrij S.I., Bozhko V.V., Piskach L.V., Bogdanyuk M.S., Olekseyuk I.D., Bulatetska L.V., Pekhnyo V.I. Phase diagram of the Ag$_2$S–HgS–SnS$_2$ system and single crystal preparation, crystal structure and properties of Ag$_2$HgSnS$_4$, *J. Alloys Compd.*, **399**(1–2), 32–37 (2005a).

Parasyuk O.V., Fedorchuk A.O., Gorgut G.P., Khyzhun O.Y., Wojciechowski A., Kityk I.V. Crystal growth, electron structure and photo induced optical changes in novel Ag$_x$Ga$_x$Ge$_{1-x}$Se$_2$ (x = 0.333, 0.250, 0.200, 0.167) crystals, *Opt. Mater.*, **35**(1), 65–73 (2012).

Parasyuk O.V., Fedorchuk A.O., Kogut Yu.M., Piskach L.V., Olekseyuk I.D. The Ag$_2$S–ZnS–GeS$_2$ system: phase di agram, glass-formation region and crystal structure of Ag$_2$ZnGeS$_4$, *J. Alloys Compd.*, <u>**500**</u>(1), 26–29 (2010).

Parasyuk O.V., Gulay L.D., Piskach L.V., Galagowska O.P. The Ag$_2$S–HgS–GeS$_2$ system at 670 K and the crystal structure of the Ag$_2$HgGeS$_4$ compound, *J. Alloys Compd.*, **336**(1–2), 213–217 (2002a).

Parasyuk O.V., Gulay L.D., Piskach L.V., Kumanska Yu.O. The $Ag_2Se-HgSe-SnSe_2$ system and the crystal structure of the $Ag_2HgSnSe_4$ compound, *J. Alloys Compd.*, 339(1–2), 140–143 (2002b).

Parasyuk O.V., Gulay L.D., Piskach L.V., Olekseyuk I.D. The $Ag_2Se-CdSe-SnSe_2$ system at 670 K and the crystal structure of the $Ag_2CdSnSe_4$ compound, *J. Alloys Compd.*, 335(1–2), 176–180 (2002c).

Parasyuk O.V., Gulay L.D., Romanyuk Ya. E., Olekseyuk I.D. Phase diagram of the quasi-binary Cu_2GeS_3-HgS system and crystals structure of the LT-modification of the Cu_2HgGeS_4 compound, *J. Alloys Compd.*, 334(1–2), 143–146 (2002d).

Parasyuk O.V., Gulay L.D., Romanyuk Ya.E, Olekseyuk I.D., Piskach L.V. The $Ag_2Se-HgSe-GeSe_2$ system and crystal structures of the compounds, *J. Alloys Compd.*, 351(1–2), 135–144 (2003a).

Parasyuk O.V., Gulay L.D., Romanyuk Ya.E, Olekseyuk I.D. The $Ag_2Se-HgSe-SiSe_2$ system in the 0–60 mol. % $SiSe_2$ region, *J. Alloys Compd.*, 348(1–2), 157–166 (2003b).

Parasyuk O.V., Gulay L.D., Romanyuk Ya.E., Piskach L.V. Phase diagram of the Cu_2GeSe_3-ZnSe system and crystal structure of the $Cu_2ZnGeSe_4$ compound, *J. Alloys Compd.*, 329(1–2), 202–207 (2001).

Parasyuk O.V., Olekseyuk I.D., Galka V.O., Marchuk O.V. Phase equilibria in the $A^{I}B^{III}Se_2-HgSe$ and $A^{I}_2C^{IV}Se_3-HgSe$ (A^{I} = Ag, Cu; B^{III} = Ga, In; C^{IV} = Si, Ge, Sn) systems [in Ukrainian], *Fizyka kondens. vysokomolek. system. Nauk. zap. Rivnens'kogo pedinstytutu*, (3), 158–162 (1997).

Parasyuk O.V., Olekseyuk I.D., Gulay L.D., Piskach L.V. Phase diagram of the $Ag_2Se-Zn(Cd)Se-SiSe_2$ systems and crystal structure of the Cd_4SiSe_6 compound, *J. Alloys Compd.*, 354(1–2), 138–142 (2003c).

Parasyuk O.V., Olekseyuk I.D., Marchuk O.V. Phase equilibria in the $Cu_2Si(Ge,Sn)Se_3-HgSe$ systems [in Ukrainian], *Ukr. khim. zhurn.*, 64(9), 20–23 (1998).

Parasyuk O.V., Olekseyuk I.D., Marchuk O.V. The $Cu_2Se-HgSe-SnSe_2$ system, *J. Alloys Compd.*, 287(1–2), 197–205 (1999a).

Parasyuk O.V., Olekseyuk I.D., Mazurets I.I., Piskach L.V. Phase equilibria in the quasi-ternary $ZnSe-Ga_2Se_3-SnSe_2$ system, *J. Alloys Compd.*, 379(1–2), 143–147 (2004).

Parasyuk O.V., Olekseyuk I.D., Piskach L.V. Crystal structure of the $Cu_2HgSi(Ge,Sn)Te_4$ compounds [in Ukrainian], *Nauk. Visnyk Volyns'k. Derzh. Univ. im. Lesi Ukrainky. Ser. Khim. nauky*, 4, 40–44 (2006a).

Parasyuk O.V., Olekseyuk I.D., Piskach L.V., Volkov S.V., Pekhnyo V.I. Phase relations in the $Ag_2S-CdS-SnS_2$ system and the crystal structure of the compounds, *J. Alloys Compd.*, 399(1–2), 173–177 (2005b).

Parasyuk O.V., Olekseyuk I.D., Piskach L.V. X-ray powder diffraction refinement of $Cu_2ZnGeTe_4$ structure and phase diagram of the Cu_2GeTe_3-ZnTe system, *J. Alloys Compd.*, 397(1–2), 169–172 (2005c).

Parasyuk O.V., Pavlyuk V.V., Khyzhun O.Y., Kozer V.R., Myronchuk G.L., Sachanyuk V.P., Dmytriv G.S., Krymus A., Kityk I.V., El-Naggar A.M., Albassamh A.A., Piasecki M. Synthesis and structure of novel $Ag_2Ga_2SiSe_6$ crystals: promising materials for dynamic holographic image recording, *RSC Adv.*, 6(93), 90958–90966 (2016).

Parasyuk O.V. Phase relations of the Ag_2SnS_3-HgS and $Ag_{33.3}Sn_{16.7}Se(Te)_{50}-HgSe(Te)$ sections in Ag–Hg–Sn–S(Se, Te) systems, *J. Alloys Compd.*, 291(1–2), 215–219 (1999).

Parasyuk O.V. Phase relations in the $Cu_2Si(Ge,Sn)Te_3-HgTe$ systems, *Pol. J. Chem.*, 72(11), 2440–2449 (1998).

Parasyuk O.V., Piskach L.V., Olekseyuk I.D. Crystal structure of the $Cu_2CdSi(Sn)Te_4$ compounds [in Ukrainian], *Nauk. Visnyk Volyns'k. Derzh. Univ. im. Lesi Ukrainky. Ser. Khim. nauky*, 4, 35–40 (2006b).

Parasyuk O.V., Piskach L.V., Olekseyuk I.D., Pekhnyo V.I. The quasiternary system $Ag_2S-CdS-GeS_2$ and the crystal structure of Ag_2CdGeS_4, *J. Alloys Compd.*, 397(1–2), 95–98 (2005d).

Parasyuk O.V., Piskach L.V., Olekseyuk I.D. The $Cu_2Se-CdSe-SnSe_2$ system [in Russian], *Zhurn. neorgan. khimii*, 44(8), 1363–1367 (1999b).

Parasyuk O.V., Piskach L.V. Phase equilibria in the Ag_2SiS_3-CdS system [in Russian], *Zhurn. neorgan. khimii*, 44(6), 1032–1033 (1999).

Parasyuk O.V., Piskach L.V., Romanyuk Y.E., Olekseyuk I.D., Zaremba V.I., Pekhnyo V.I. Phase relations in the quasi-binary Cu_2GeS_3-ZnS and quasi-ternary $Cu_2S-Zn(Cd)S-GeS_2$ systems and crystal structure of Cu_2ZnGeS_4, *J. Alloys Compd.*, 397(1–2), 85–94 (2005e).

Parasyuk O.V., Piskach L.V. The Ag_2SnS_3-CdS system, *Pol. J. Chem.*, 72(5), 966–968 (1998).

Parasyuk O.V., Romanyuk Ya. E., Olekseyuk I.D. Single-crystal growth of Cu_2CdGeS_4, *J. Cryst. Growth*, 275(1–2), e159–e162 (2005).

Parthé E., Yvon K., Deitch R.H. The crystal structure of Cu_2CdGeS_4 and other quaternary normal tetrahedral structure compounds, *Acta Crystallogr.*, B25(6), 1164–1174 (1969).

Pashyns'kyi I.V., Olekseyuk I.D., Gulay L.D. Isothermal sections of the $Y(La)_2Se_3-In_2Se_3-PbSe$ systems at 870 K [in Ukrainian], *Nauk. Visnyk Volyns'k. Nats. Univ. im. Lesi Ukrainky. Ser. Khim. nauky*, (16), 43–46 (2008).

Pazin A.V., Glushkov E.D., Borisova Z.U. Glass formation and electrical conductivity of glasses in the $GeSe_2-As_2Se_3-Sb_2Se_3$ system [in Russian], *Izv. AN SSSR. Neorgan. mater.*, 14(3), 417–421 (1978a).

Pazin A.V., Morozov V.A., Borisova Z.U. Glass formation and electrical conductivity of glasses in the $SnSe_2-GeSe_2-As_2Se_3$ system [in Russian], *Izv. AN SSSR. Neorgan. mater.*, 14(3), 422–426 (1978b).

Pazin A.V., Morozov V.A. The glass region and electrical conductivity of glasses in the $SnSe-GeSe_2-As_2Se_3$ system [in Russian], *Izv. AN SSSR. Neorgan. mater.*, 14(8), 1401–1405 (1978).

Pelton A.D., Flengas S.N. Phase relationships in lead sulfide-alkali chloride quasibinary systems, *Can. J. Chem.*, 48(13), 2016–2020 (1970).

Perez G., Darriet-Duale M., Hagenmuller P. Les systèmes ternaires $MS_2-CdS-Ln_2S_3$ à 1050°C (M = Si, Ge), (Ln = La… Gd), *J. Solid State Chem.*, 2(1), 42–48 (1970a).

Perez G., Darriet M. Étude structural des solutions solides formées dans les systèms $MS_2-CdS-Ln_2S_3$ (M = Si, Ge) (Ln = La … Gd), *C. r. Acad. sci.*, C270(4), 420–422 (1970b).

Person H., Grupe M., Urland W. Synthese und Kristallstruktur von Ln_2SeSiO_4 (Ln = Sm, Dy, Ho) und Sm_2TeSiO_4, *Z. anorg. und allg. Chem.*, 626(1), 280 283 (2000).

Persson C. Electronic and optical properties of Cu_2ZnSnS_4 and $Cu_2ZnSnSe_4$, *J. Appl. Phys.*, 107(5), 053710_1-053710_8 (2010).

Piasecki M., Myronchuk G.L., Parasyuk O.V., Khyzhun O.Y., Fedorchuk A.O., Pavlyuk V.V., Kozer V.R., Sachanyuk V.P., El-Naggar A.M., Albassam A.A., Jedryka J., Kityk I.V. Synthesis, structural, electronic and linear electro-optical features of new quaternary $Ag_2Ga_2SiS_6$ compound, *J. Solid State Chem.*, 246, 363–371 (2017).

Pierrot R. Revue bibliographique des modifications apportées à la nomenclature minéralogique, *Bull. Soc. fr. Minéral. Cristallogr.*, **91**(3), 300–307 (1968).

Pietak K., Jastrzebski C., Zberecki K., Jastrzebski D.J., Paszkowicz W., Podsiadlo S. Synthesis and structural characterization of Ag₂ZnSnS₄ crystals, *J. Solid State Chem.*, **290**, 121467 (2020).

Piilonen P.C., Grew E.S., Ercit T.S., Roberts A.C. New mineral names, *Am. Mineralog.*, **90**(7), 1227–1233 (2005a).

Piilonen P.C., Locock A., Grew E.S. New mineral names, *Am. Mineralog.*, **90**(11–12), 1945–1952 (2005b).

Piilonen P.C., Locock A.J., Rowe R., Ercit T.S. New mineral names, *Am. Mineralog.*, **92**(4), 703–707 (2007).

Piilonen P.C., Poirier G. New mineral names, *Am. Mineralog.*, **95**(8–9), 1357–1361 (2010).

Pinto D., Balić-Žunić T., Garavelli A., Garbarino C., Makovicky E., Vurro F. First occurrence of close-to-ideal kirkiite at Vulcano (Aeolian Islands, Italy): chemical data and single-crystal X-ray study, *Eur. J. Mineral.*, **18**(3), 393–401 (2006).

Piskach L., Mozolyuk M., Fedorchuk A., Olekseyuk I., Parasyuk O. Phase equilibria in the Tl₂S–HgS–SnS₂ systemat 520 K and crystal structure of Tl₂HgSnS₄, *Chem. Met. Alloys*, **10**(3–4), 136–141 (2017).

Piskach L.V., Olekseyuk I.D., Parasyuk O.V. Physico-chemical peculiarities of the Cu₂CdC^IV X₄ (C^IV - Si, Ge, Sn; X - S, Se, Te) quaternary phase formations [in Ukrainian], *Fiz. kondens. vysokomolek. system. Nauk. zap. Rivnens'kogo pedinstytutu*, (3), 153–157 (1997).

Piskach L.V., Parasyuk O.V., Olekseyuk I.D., Halahan V.Ya. CdSe–Ga₂Se₃–SnSe₂ system [in Ukrainian], *Fiz. i khim. tv. tila*, **3**(1), 25–32 (2002).

Piskach L.V., Parasyuk O.V., Olekseyuk I.D., Romanyuk Y.E., Volkov S.V., Pekhnyo V.I. Interaction of argyrodite family compounds with the chalcogenides of II-b elements, *J. Alloys Compd.*, **421**(1–2), 98–104 (2006).

Piskach L.V., Parasyuk O.V., Olekseyuk I.D. The Cu₂SiS₃–CdS system [in Russian], *Zhurn. neorgan. khimii*, **44**(5), 823–824 (1999).

Piskach L.V., Parasyuk O.V., Olekseyuk I.D. The phase equilibria in the quasi-ternary Cu₂S–CdS–SnS₂ system, *J. Alloys Compd.*, **279**(2), 142–152 (1998).

Piskach L.V., Parasyuk O.V., Romanyuk Ya.E. The phase equilibria in the quasi-binary Cu₂GeS₃(Se₃)–CdS(Se) systems, *J. Alloys Compd.*, **299**(1–2), 227–231 (2000).

Piskach L.V., Parasyuk O.V. The Ag₂GeS₃–CdS system, *Pol. J. Chem.*, **72**(6), 1112–1115 (1998).

Piskach L., Tsisar O.V., Marushko L.P. Phase equilibria in the PbGa₂X₄–SiX₂ (X – S, Se) systems [in Ukrainian], *Persp. tekhnol. ta prylady*, (14), 109–112 (2019).

Piskach L.V., Zmiy O.F., Olekseyuk I.D., Mokraya I.R. Phase equilibria in the Cu₂GeTe₃–CdTe system [in Russian], *Vestn. L'vov. un-ta. Ser. khim.*, (29), 40–42 (1988).

Plachkova S.K., Georgieva O.Zh., Odin I.N. Synthesis, physico-chemical investigation and electrophysical properties of the solid solutions based on GeTe along the (GeTe)₁₋ₓ–[(Ag₂Te)₀,₄₆₆₇(Bi₂Te₃)₀,₅₃₃₃]ₓ section [in Russian], *Zhurn. neorgan. khimii*, **37**(5), 1189–1193 (1992).

Plachkova S.K., Odin I.N. Electrophysical properties of the solid solutions based on germanium telluride in the GeTe–AgSbTe₂ system [in Russian], *Izv. AN SSSR. Neorgan. mater.*, **19**(4), 588–592 (1983).

Plachkova S.K., Odin I.N., Novoselova A.V. Physico-chemical investigation of the GeTe–AgSbTe₂ alloys [in Russian], *Zhurn. neorgan. khimii*, **28**(8), 2091–2096 (1983).

Plachkova S.K., Odin I.N., Novoselova A.V. Solid-phase transformation and electrophysical properties of the GeTe–AgBiTe₂ alloys [in Russian], *Izv. AN SSSR. Neorgan. mater.*, **20**(3), 403–408 (1984).

Plachkova S.K., Odin I.N., Sher A.A., Novoselova A.V. Physico-chemical investigation of the GeTe–AgBiTe₂ system [in Russian], *Izv. AN SSSR. Neorgan. mater.*, **16**(7), 1199–1202 (1980).

Pobedimskaya E.A., Alimova L.L., Belov N.V., Badikov V.V. Crystal structures of Ag-germanogallium sulfide and GeS₂ [in Russian], *Dokl. AN SSSR*, **257**(3), 611–614 (1981).

Poduska K.M., Cario L., DiSalvo F.J., Min K., Halasyamani P.S. Structural studies of a cubic, high-temperature (α) polymorph of Pb₂GeS₄ and the isostructural Pb₂₋ₓSnₓGeS₄₋ySe_y solid solution, *J. Alloys Compd.*, **335**(1–2), 105–110 (2002a).

Poduska K.M., DiSalvo F.J., Min K., Halasyamani P.S. Structure determination of La₃CuGeS₇ and La₃CuGeSe₇, *J. Alloys Compd.*, **335**(1–2), L5–L9 (2002b).

Pogu A., Jaschin P.W., Varma K.B.R., Vidyasagar K. Syntheses, structural variants and characterization of A₂CdSn₂S₆ (A = Cs, Rb and K) compounds. J. Solid State Chem., **277**, 713–720 (2019).

Pogu A., Vidyasagar K. Syntheses, structural variants and characterization of A₂ZnSn₃S₈ (A = Cs, Rb) and A₂CdSn₃S₈ (A = Cs, Rb, K, Na) compounds, *J. Solid State Chem.*, **291**, 121647 (2020).

Politova E.D., Venevtsev Yu.N. New tellurium containing ferro- and antiferroelectrics with the perovskite structure [in Russian], *Dokl. AN SSSR*, **209**(4), 838–847 (1973).

Porter Y., Halasyamani P.S. A low temperature method for the synthesis of new lead selenite chlorides: Pb₃(SeO₃)(SeO₂OH)Cl₃ and Pb₃(SeO₃)₂Cl₂, *Inorg. Chem.*, **40**(12), 2640–2641 (2001).

Porter Y., Halasyamani P.S. Syntheses, structures, and characterization of new lead(II)-tellurium(IV)-oxide halides: Pb₃Te₂O₆X₂ and Pb₃TeO₄X₂ (X = Cl or Br), *Inorg. Chem.*, **42**(1), 205–209 (2003).

Potoriy M.V., Seykovskaya L.A., Prits I.P., Voroshilov Yu.V. The Sn,Pb‖P₂S₆,P₂Se₆ ternary mutual system [in Russian], *ChemChemTech [Izv. Vyssh. Uchebn. Zaved. Khim. Khim. Tekhnol.]*, **31**(8), 21–24 (1988).

Poudeu P.F.P., Djieutedjeu H., Sahoo P. Crystal structure of FePb₄Sb₆Se₁₄ and its structural relationship with FePb₃Sb₄Se₁₀, *Z. anorg. und allg. Chem.*, **638**(15), 2549–2554 (2012).

Poudeu P.F.P., Takas N., Anglin C., Eastwood J., Rivera A. Fe_xPb₄₋_xSb₄Se₁₀: a new class of ferromagnetic semiconductors with quasi 1D {Fe₂Se₁₀} ladders, *J. Am. Chem. Soc.*, **132**(16), 5751–5760 (2010).

Povilaytis M.M., Mozgova N.N., Borodaev Yu.S., Senderova V.M., Ronami G.N. The first occurrence of hammarite in the USSR [in Russian], *Dokl. AN SSSR*, **187**(4), 886–889 (1969).

Powell D.W., Thomas R.G., Williams P.A., Birch W.D., Plimer I.R. Choloalite: synthesis and revised chemical formula, *Mineralog. Mag.*, **58**(392), 505–508 (1994).

Pring A. Annealing of synthetic hammarite, Cu₂Pb₂Bi₄S₉, and the nature of cation-ordering processes in the bismuthinite-aikinite series, *Am. Mineralog.*, **80**(11–12), 1166–1173 (1995).

Pring A., Graeser S. Polytypism in baumhauerite, *Am. Mineralog.*, **79**(3–4), 302–307 (1994).

Pring A., Grguric B.A., Criddle A.J. Lindströmite from Cobalt, Ontario, *Canad. Mineralog.*, **36**(4), 1139–1148 (1998).

Pruseth K.L., Mishra B., Bernhardt H.J. Phase relations in the $Cu_2S–PbS–Sb_2S_3$ system: an experimental appraisal and application to natural polymetallic sulfide ores, *Econ. Geol.*, **92**(6), 720–732 (1997).

Pruseth K.L., Mishra B., Bernhardt H.J. Solid solution in the synthetic zinkenite, robinsonite and meneghinite in the system $Cu_2S–PbS–Sb_2S_3$, *Canad. Mineralog.*, **36**(1), 207–213 (1998).

Pruseth K.L., Mishra B., Bernhardt H.-J. The minerals boulangerite, falkmanite and Cu-free meneghinite: synthesis, new powder diffraction data and stability relations, *Eur. J. Mineral.*, **13**(2), 411–419 (2001).

Pryce M.W. Calcium sulphosilicate in lime-kiln wall coating, *Mineralog. Mag.*, **38**(300), 968–971 (1972).

Puff H., Lorbacher G., Heine D. Quecksilberchalkogen-Fluorosilicate, *Naturwissenschaften*, **56**(9), 461 (1969).

Putz H., Paar W.H., Topa D., Makovicky E., Roberts A.C. Catamarcaite, Cu_6GeWS_8, a new germanium sulfide mineral species from Capillitas, Catamarca, Argentina: description, paragenesis and crystal structure, *Canad. Mineralog.*, **44**(6), 1481–1497 (2006).

Quarez E., Hsu K.-F., Pcionek R., Frangis N., Polychroniadis E.K., Kanatzidis M.G. Nanostructuring, compositional fluctuations, and atomic ordering in the thermoelectric materials $AgPb_mSbTe_{2+m}$. The myth of solid solutions, *J. Am. Chem. Soc.*, **127**(25), 9177–9190 (2005).

Quintero E., Quintero M., Moreno E., Lara L., Morocoima M., Pineda P., Grima P., Tovar R., Bocaranda P., Henao J.A., Macías M.A. Magnetic properties for the $Cu_2MnSnSe_4$ and $Cu_2FeSnSe_4$ compounds, *J. Phys. Chem. Solids*, **71**(7), 993–998 (2010).

Quintero E., Tovar R., Quintero M., Sánchez-Porras G., Bocaranda P., Broto J.M., Rakoto H., Barbaste R., Woolley J.C., Lamarche G., Lamarche A.-M. Crystallographic, electrical and magnetic properties of the $Cu_2FeGeSe_4$ compound, *Phys. stat. sol. (b)*, **220**(1), 417–424 (2000).

Quintero M., Barreto A., Grima P., Tovar R., Quintero E., Porras G.S., Ruiz J., Wooley J.C., Lamarche G., Lamarche A.-M. Crystallographic properties of $I_2–Fe–IV–VI_4$ magnetic semiconductor compounds, *Mater. Res. Bull.*, **34**(14–15), 2263–2270 (1999).

Quintero M., Marquina J., Quintero E., Moreno E., Alvarez S., Rincón C., Grima P., Bocaranda P., Rivero D., Henao J.A., Macías M.A. X-ray diffraction analysis of stannite, wurtz-stannite and pseudo-cubic quaternary compounds by Rietveld method, *Rev. Mex. Fís.*, **60**(2), 168–175 (2014a).

Quintero M., Moreno E., Alvarez S., Marquina J., Rincón C., Quintero E., Grima P., Heano J.-A., Macías M.A. Lattice parameter values and phase transitions for the $Cu_2–II–IV–S_4(Se_4)$ (II = Mn, Fe, Co; IV = Si, Ge, Sn) magnetic semiconductor compounds, *Rev. LatinAm. Metal. Mat.*, **34**(1), 28–38 (2014b).

Quintero M., Tovar R., Barreto A., Quintero E., Rivero A., Gonzalez J., Sánchez-Porras G., Ruiz J., Bocaranda P., Broto J.M., Rakoto H., Barbaste R. Crystallographic and magnetic properties of $Cu_2FeGeSe_4$ and $Cu_2FeGeTe_4$ compounds, *Phys. stat. sol. (b)*, **209**(1), 135–143 (1998).

Quintero M., Quintero E., Moreno E., Grima-Gallardo P., Marquina J., Alvarez S., Rincón C., Rivero D., Morocoima M., Henao J.A., Macías M.A., Briceño J.M., Rogríguez N. Magnetic susceptibility for the $Cu_2–II–IV–S_4$ (II = Mn, Fe; IV=Si, Ge or Sn) compounds: exchange interaction parameters, *Rev. LatinAm. Metal. Mat.*, **37**(1), 27–34 (2017).

Raevskiy S.D. The PbTe–GaAs phase diagram nearby lead telluride [in Russian], *Izv. AN SSSR. Neorgan. mater.*, **17**(8), 1385–1387 (1981).

Raevskiy S.D., Zbigli K.R., Kazak G.F., Prunich M.D. Phase diagram of the PbSe–CdTe system [in Russian], *Izv. AN SSSR. Neorgan. mater.*, **18**(8), 1267–1270 (1982).

Raevskiy S.D., Zbigli K.R., Kazak G.F., Prunich M.D. Phase diagram of the PbTe–CdSe system [in Russian], *Izv. AN SSSR. Neorgan. mater.*, **19**(6), 889–892 (1983a).

Raevskiy S.D., Zbigli K.R., Kazak G.F., Prunich M.D. Solid solutions of cadmium selenide in the lead telluride [in Russian], *Izv. AN MSSR. Ser. phys.-techn. i mat. nauk*, (2), 42–44 (1983b).

Raghavan V. Ag–Fe–Pb–S (Silver–Iron–Lead–Sulfur), *J. Phase Equilib. Diffus.*, **30**(1), 110 (2009a).

Raghavan V. Au–Fe–Pb–S (Gold–Iron–Lead–Sulfur), *J. Phase Equilib. Diffus.*, **30**(1), 111 (2009b).

Range K.-J., Andratschke M., Gietl A. Crystal structure of ytterbium(III) trisulfide disilicate, $Yb_4S_3(Si_2O_7)$, *Z. Kristallogr.*, **211**(11), 816 (1996).

Rasulova K.D., Aliyev Z.S., Babanly M.B. Diagrams of solid-phase equilibria in the $YbTe–PbTe–Bi_2Te_3$ system at 300 and 800 K [in Russian], *Kimya Problem.*, (3), 288–294 (2013a).

Rasulova K.D., Aliyev Z.S., Babanly M.B. Phase equilibria in the $YbTe–PbBi_4Te_7$ and $YbTe–PbBi_2Te_4$ [in Russian], *Kimya Problem.*, (2), 190–193 (2013b).

Reshak A.H., Khyzhun O.Y., Kityk I.V., Fedorchuk A.O., Kamarudin H., Auluck S., Parasyuk O.V. Electronic structure of quaternary chalcogenide $Ag_2In_2Ge(Si)S_6$ single crystals and the influence of replacing Ge by Si: experimental X-ray photoelectron spectroscopy and X-ray diffraction studies and theoretical calculations, *Sci. Adv. Mater.*, **5**(4), 316–327 (2013a).

Reshak A.H., Parasyuk O.V., Fedorchuk A.O., Kamarudin H., Auluck S., Chyský J. Optical spectra and band structure of $Ag_xGa_xGe_{1-x}Se_2$ (x = 0.333, 0.250, 0.200, 0.167) single crystals: experiment and theory, *J. Phys. Chem. B*, **117**(48), 15220–15231 (2013b).

Retuerto M., Skiadopoulou S., Li M.-R., Abakumov A.M., Croft M., Ignatov A., Sarkar T., Abbett B.M., Pokorný J., Savinov M., Nuzhnyy D., Prokleška J., Abeykoon M., Stephens P.W., Hodges J.P., Vaněk P., Fennie C.J., Rabe K.M., Kamba S., Greenblatt M. Pb_2MnTeO_6 double perovskite: an antipolar anti-ferromagnet, *Inorg. Chem.*, **55**(9), 4320–4329 (2016).

Riccardi R., Gout D., Gauthier G., Guillen F., Jobic S., Garcia A., Huguenin D., Macaudière P., Fouassier C., Brec R. Structural investigations and luminescence properties of the $Ce_3(SiS_4)_2X$ (X = Cl, Br, I) family and the $La_{3-x}Ce_x(SiS_4)_2I$ ($0 \leq x \leq 1$) solid solution, *J. Solid State Chem.*, **147**(1), 259–268 (1999).

Riedel E., Morlock W. Spinelle mit substituierten Nichtmetallteilgittern. VI. Röntgenographische und elektronische Eigenschaften, Mößbauer- und IR-Spektren des Spinellsystems $CuCrSn(S_{1-x}Se_x)_4$, *Z. anorg. und allg. Chem.*, **438**(1), 233–241 (1978).

Rincón C., Quintero M., Moreno E., Power Ch., Quintero E., Henao J.A., Macías M.A., Delgado G.E., Tovar R., Morocoima M. X-ray diffraction, Raman spectrum and magnetic susceptibility of the magnetic semiconductor Cu_2FeSnS_4, *Solid State Commun.*, **151**(13), 947–951 (2011).

Roberts A.C., Cooper M.A., Hawthorne F.C., Criddle A.J., Stanley C.J., Key C.L., Jambor J.L. Sidpietersite, $Pb^{2+}_4(S^{6+}O_3S^{2-})O_2(OH)_2$, a new thiosulfate-bearing mineral species from Tsumeb, Namibia, *Canad. Mineralog.*, **37**(5), 1269–1273 (1999).

Robinson S.C. Owyheeite, *Am. Mineralog.*, **34**(5–6), 398–402 (1949).

Rodionov Yu.I., Klokman V.R., Myakishev K.G. Solubility of the $A^{II}B^{VI}$, $A^{IV}B^{VI}$ and $A^{V}B^{VI}$ semiconductor compounds in galogenide melts [in Russian], *Zhurn. neorgan. khimii*, **17**(3), 846–851 (1972).

Rodot H. Étude et propriétés du système $AgSbTe_2$–$PbTe$, *C. r. Acad. sci.*, **249**(19), 1872–1874 (1959).

Roland G.W. The system Pb–As–S. Composition and stability of jordanite, *Mineral. Deposita*, **3**(3), 249–260 (1968).

Romanyuk Ya.E., Parasyuk O.V. Phase equilibria in the quasi-ternary Cu_2Se–$ZnSe$–$GeSe_2$ system, *J. Alloys Compd.*, **348**(1–2), 195–202 (2003).

Rosenthal T., Neudert L., Ganter P., de Boor J., Stiewe C., Oeckler O. Nanostructured rocksalt-type solid solution series $(Ge_{1-x}Sn_xTe)_nSb_2Te_3$ ($n = 4, 7, 12$; $0 \leq x \leq 1$): thermal behavior and thermoelectric properties, *J. Solid State Chem.*, **215**, 231–240 (2014a).

Rosenthal T., Welzmiller S., Neudert L., Urban P., Fitch A., Oeckler O. Novel superstructure of the rocksalt type and element distribution in germanium tin antimony tellurides, *J. Solid State Chem.*, **219**, 108–117 (2014b).

Rosenthal T., Welzmiller S., Oeckler O. The solid solution series $Ge_{12}M_2Te_3$ (M = Sb, In): nanostructures and thermoelectric properties, *Solid State Sci.*, **25**, 118–123 (2013).

Rosmus K.A., Aitken J.A. Cu_2ZnSiS_4, *Acta. Crystallogr.*, **E67**(4), i28 (2011).

Rosmus K.A., Brant J.A., Wisneski S.D., Clark D.J., Kim Y.S., Jang J.I., Brunetta C.D., Zhang J.-H., Srnec M.N., Aitken J.A. Optical nonlinearity in Cu_2CdSnS_4 and α/β-Cu_2ZnSiS_4: diamond-like semiconductors with high laser-damage thresholds, *Inorg. Chem.*, **53**(15), 7809–7811 (2014).

Rosmus K.A., Brunetta C.D., Srnec M.N., Karuppannan B., Aitken J.A. Synchrotron X-ray powder diffraction and electronic band structure of α- and β-Cu_2ZnSiS_4, *Z. anorg. und allg. Chem.*, **638**(15), 2578–2584 (2012).

Rothballer J., Bachhuber F., Pielnhofer F., Schappacher F.M., Pöttgen R., Weihrich R. Effect of In-Sn ordering on semiconducting properties in $InSnCo_3S_2$ – X-ray, ^{119}Sn Mößbauer spectroscopy, and DFT sudies, *Eur. J. Inorg. Chem.*, **2013**(2), 248–255 (*2013*).

Rothballer J., Bachhuber F., Rommel S.M., Söhnel T., Weihrich R. Origin and effect of In–Sn ordering in $InSnCo_3S_2$: a neutron diffraction and DFT study, *RSC Adv.*, **4**(79), 42183–42189 (2014).

Rothenberger A., Shafaei-Fallah M., Kanatzidis M.G., Aluminosilicate relatives: chalcogenoaluminogermanates $Rb_3(AlQ_2)_3(GeQ_2)_7$ (Q = S, Se). *Inorg. Chem.*, **49**(21), 9749–9751 (2010).

Ruan T.-T., Wang W.-W., Hu C.-L., Xu X., Mao J.-G. $Pb_4(BO_3)_2(SO_4)$ and $Pb_2[(BO_2)(OH)](SO_4)$: new lead(II) borate-sulfate mixed-anion compounds with two types of 3D network structures, *J. Solid State Chem.*, **260**, 39–45 (2018).

Ruda I.P., Marchuk O.V., Gulay L.D., Olekseyuk I.D. Crystal structure of the $R_{1.32}Pb_{1.68}Ge_{1.67}S_7$ (R = Y, La, Ce, Pr, Nd, Sm, Gd, Tb, Dy and Ho) [in Ukrainian], *Nauk. Visnyk Volyns'k. Derzh. Univ. im. Lesi Ukrainky. Ser. Khim. nauky*, **13**, 7–11 (2007).

Ruff O., Geisel E. Zur Constitution des Schwefelstickstoffs, *Ber. Deutsch. Chem. Ges.*, **37**(2), 1573–1595 (1904).

Ruiz E., Payne M.C. One-dimensional intercalation compound $2HgS\cdot SnBr_2$: ab initio electronic structure calculations and molecular dynamics simulation, *Chem. Eur. J.*, **4**(12), 2485–2492 (1998).

Rustamov P.G., Abilov Ch.I., Il'yasov T.M. The $PbTe$–$GeSe_2$ system [in Russian], *Izv. AN SSSR. Neorgan. mater.*, **24**(1), 52–54 (1988).

Rustamov P.G., Kahramanov K.Sh., Cherstvova V.B., Ahmedli G.T., Mamedova S.Yu. Investigation of the $SnTe$–$NiSb$ system [in Russian], *Azerb. khim. zhurn.*, (*1*), 116–119 (1975).

Rybkin S.G., Nikolaev Yu.L., Barankevich V.G. Isothermal sections at 1473 K through the Pb–Au–FeS and Pb–Ag–FeS phase diagrams, *Russ. J. Inorg. Chem.*, **51**(3), 470–473 (2006).

Rzaguliev V.A., Mamedov A.N., Kerimli O.S., Mamedov Sh.G. Phase equilibria in the Ag_2Se–Cu_2SnSe_3 and Ag_8SnSe_6–Cu_2SnSe_3 systems, *Russ. J. Inorg. Chem.*, **65**(12), 1899–1904 (2020).

Rzaguluyev V.A., Kerimli O.Sh., Azhdarova D.S., Mamedov Sh.H., Aliyev O.M. Phase equilibria in the Ag_8SnS_6–Cu_2SnS and Ag_2SnS_3–$Cu_2Sn_4S_9$ systems, *Kondens. sredy i mezhfaz. granitsy*, **21**(4), 544–551 (2019).

Sachanyuk V.P., Gorgut G.P., Atuchin V.V., Olekseyuk I.D., Parasyuk O.V. The Ag_2S–In_2S_3–$Si(Ge)S_2$ systems and crystal structure of quaternary sulfides $Ag_2In_2Si(Ge)S_6$, *J. Alloys Compd.*, **452**(2), 348–358 (2008).

Sachanyuk V.P., Olekseyuk I.D., Parasyuk O.V. Formation of an endothermal $Cu_2In_2SiS_6$ compound in the $CuInS_2$–SiS_2 system, *J. Alloys Compd.*, **443**, 61–67 (2007).

Sachanyuk V.P., Olekseyuk I.D., Parasyuk O.V. Phase diagram of the $CuGaSe_2$–$SiSe_2$ and $CuInSe_2$–$SiSe_2$ systems, *J. Alloys Compd.*, **420**, 54–57 (2006a).

Sachanyuk V.P., Olekseyuk I.D., Piskach L.V., Parasyuk O.V. Crystal structure of the $Cu_2Cd/Hg/SiS_4$ and $Cu_2MnSi/Ge,Sn/S_4$ compounds [in Ukrainian], *Nauk. Visnyk Volyns'k. Derzh. Univ. im. Lesi Ukrainky. Ser. Khim. nauky*, **4**, 45–54 (2006b).

Sachdev S.C., Chang L.L.Y. Phase relations in the system tin–antimony–lead sulfides and the synthesis of cylindrite and franckeite, *Econ. Geol.*, **70**(6), 1111–1122 (1975).

Sadykhova S.A., Gabib-Zade S.A., Ragimova V.M., Suleymanova A.U. Interaction of $CuInSe_2$ with $GeSe_2$ [in Russian], *Zhurn. neorgan. khimii*, **33**(11), 2964–2966 (1988).

Sakai Y., Tanakadate R., Matoba M., Yamada I., Nishiyama N., Irifune T., Funakoshi K., Kunimoto T., Higo Y., Kamihara Y. Magnetic properties of shandite-phase $Co_{3-x}Fe_xSn_2S_2$ ($x = 0$–1.0) obtained with high pressure synthesis, *J. Phys. Soc. Jpn.*, **84**(4), 044705_1–044705_6 (2015).

Sakamoto R., Tatsumisago M., Minami T. Preparation of fast lithium ion conducting glasses in the system Li_2S–SiS_2–Li_3N, *J. Phys. Chem.*, **B103**(20), 4029–4031 (1999).

Sashital S.R. Gentile A.L. Liquid phase epitaxial growth of $AgGaS_2$ using chalcogenide (sulphide) fluxes, *J. Cryst. Growth*, **69**(2–3), 379–387 (1984).

Sawada H., Kawada I., Hellner E., Tokonami M. The crystal structure of senandorite (andorite VI): $PbAgSb_3S_6$, *Z. Kristallogr.*, **180**(1–4), 141–150 (1987).

Schäfer W., Nitsche R. Tetrahedral quaternary chalcogenides of the type Cu_2-II–IV–$S_4(Se_4)$, *Mater. Res. Bull.*, **9**(5), 645–654 (1974).

Schäfer W., Nitsche R. Zur Systematik tetraedrischer Verbindungen vom Typ $Cu_2Me^{II}Me^{IV}Me^{VI}_4$ (Stannite und Wurtzstannite), *Z. Kristallogr.*, **145**(5–6), 356–370 (1977).

Schäfer W., Scheunemann K., Nitsche R. Crystal structure and magnetic properties of $Cu_4NiSi_2S_7$, *Mater. Res. Bull.*, **15**(7), 933–937 (1980).

Schleich D.M., Wold A. Optical and electrical properties of quaternary chalcogenides, *Mater. Res. Bull.*, **12**(2), 111–114 (1977).

Schmidt L., McCarthy S.L., Maita J.P. Superconducting behavior of the system Pb($Nb_{1-x}Ta_x$)S_3, *Solid State Commun.*, **8**(19), 1513–1515 (1970).

Schönegger S., Bruns J., Gartner B, Wurst K., Huppertz H. Synthesis and characterization of the first lead(II) borosulfate Pb[$B_2(SO_4)_4$], *Z. anorg. und allg. Chem.*, **644**(24), 1702–1706 (2018).

Schorr S., Gonzalez-Aviles G. In-situ investigation of the structural phase transition in kesterite, *Phys. stat. sol. (a)*, **206**(5), 1054–1058 (2009).

Schorr S. The crystal structure of kesterite type compounds: a neutron and X-ray diffraction study, *Sol. Energy Mater Sol. Cells*, **95**(6), 1482–1488 (2011).

Schröder T., Schwarzmüller S., Stiewe C., de Boor J., Hölzel M., Oeckler O. The solid solution series (GeTe)$_x$(LiSbTe$_2$)$_2$ ($1 \leq x \leq 11$) and the thermoelectric properties of (GeTe)$_{11}$(LiSbTe$_2$)$_2$, *Inorg. Chem.*, **52**(19), 11288–11294 (2013).

Schultze-Rhonhof E. Die Kristallstruktur des Silbersulfidorthosilicates, $Ag_8S_2SiO_4$, *Acta Crystallogr.*, **B30**(11), 2553–2558 (1974).

Schumer B.N., Downs R.T., Domanik K.J., Andrade M.B., Origlieri M.J. Pirquitasite, Ag_2ZnSnS_4, *Acta Crystallogr.*, **E69**(2), i8–i9 (2013).

Schunemann P.G., Zawilski K.T., Pollak T.M. Horizontal gradient freeze growth of $AgGaGeS_4$ and $AgGaGe_5Se_{12}$, *J. Cryst. Growth*, **287**(2), 248–251 (2006).

Schwarz H. Doppelverbindungen vom Typ Me$^I_2$Me^{II}(X^{VI}O$_4$)$_2$ mit der Struktur von Sr$_3$(PO$_4$)$_2$. I. Sulfate, *Z. anorg. allg. Chem.*, **344**(1–2), 41–55 (1966).

Schwarzmüller S., Yang F., Oeckler O. Structures and transport properties of metastable solid solutions (NaSbTe$_2$)$_{1-x}$(GeTe)$_x$, *J. Alloys Compd.*, **806**, 774–779 (2019).

Seidzade A.E., Babanly M.B. Phase diagram of the SnSb$_2$Te$_4$–SnBi$_2$Te$_4$ system and some properties of the SnSb$_{2-x}$Bi$_x$Te$_4$ solid solutions, *Azerb. Chem. J.*, (4), 6–10 (2019).

Seidzade A.E. Phase diagram of the SnSb$_4$Te$_7$–SnBi$_4$Te$_7$ system, *New Mater., Compd. Appl.*, **3**(3), 193–197 (2019).

Sejkora J., Makovicky E., Topa D., Putz H., Zagler G., Plášil J. Litochlebite, $Ag_2PbBi_4Se_8$, a new selenide mineral species from Zálesí, Czech Republic: description and crystal structure, *Canad. Mineralog.*, **49**(2), 639–650 (2011a).

Sejkora J., Ozdín D., Laufek F., Plášil J., Litochleb J. Marrucciite, a rare Hg-sulfosalt from the Gelnica ore deposit (Slovak Republic), and its comparison with the type occurrence at Buca della Vena mine (Italy), *J. Geosci.*, **56**(4), 399–408 (2011b).

Sejkora J., Škácha P., Plášil J., Dolníček Z., Ulmanová J. Hrabákite, IMA 2020-034. CNMNC Newsletter No. 57, *Eur. J. Mineral.*, **32**(5), 496 (2020a).

Sejkora J., Škácha P., Plášil J., Dolníček Z., Ulmanová J. Hrabákite, IMA 2020-034. CNMNC Newsletter No. 57, *Mineralog. Mag.*, **84**(5), 792 (2020b).

Selezen A.O., Olekseyuk I.D., Myronchuk G.L., Smitiukh O.V., Piskach L.V. Synthesis and structure of the new semiconductor compounds Tl$_2$B^{II}D^{IV}X$_4$ (B^{II} – Cd, Hg; D^{IV} – Si, Ge; X – Se, Te) and isothermal sections of the Tl$_2$Se–CdSe–Ge(Sn)Se$_2$ systems at 570 K, *J. Solid State Chem.*, **289**, 121422 (2020).

Selezen A.O., Piskach L.V., Parasyuk O.V., Olekseyuk I.D. The Tl$_2$SnSe$_3$–CdSe system and the crystal structure of the Tl$_2$CdSnSe$_4$ compound, *J. Phase Equilib. Diffus.*, **40**(6), 797–801 (2019).

Shabunina G.G., Aminov T.G. Investigation of CuInS$_2$ with SnS$_2$ interaction [in Russian], *Zhurn. neorgan. khimii*, **44**(5), 859–861 (1999).

Shabunina G.G., Busheva E.V., Aminov T.G. Interaction between Cr$_2$Se$_3$ and Cu$_2$GeSe$_3$, *Russ. J. Inorg. Chem.*, **51**(1), 126–130 (2006).

Sharda S., Sharma N., Sharma P., Sharma V. Thermal stability and crystallization kinetics of quaternary Sb–Se–Ge–In chalcogenide glasses, *J. Alloys Compd.*, **611**, 96–99 (2014).

Shelimova L.E., Abrikosov N.H. Investigation of the GeTe + PbSe ↔ GeSe + PbTe ternary mutual system [in Russian], *Izv. AN SSSR. Neorgan. mater.*, **4**(11), 1885–1889 (1968).

Shelimova L.E., Avilov E.S., Kretova M.A. Physico-chemical and electrophysical properties of the solid solutions based on GeTe in the GeTe–MnTe–PbTe system [in Russian], *Neorgan. mater.*, **29**(8), 1089–1092 (1993a).

Shelimova L.E., Karpinskiy O.G., Avilov E.S., Kretova M.A. Crystal structure, phase transitions and mechanical properties of solid solutions based on GeTe in the GeTe–PbTe–MTe (M = Mn, Sc, La) systems [in Russian], *Neorgan. mater.*, **29**(11), 1449–1457 (1993b).

Shelimova L.E., Karpinskiy O.G. Phase transitions in the solid solutions based on γ-germanium telluride with cation-anion substitution [in Russian], *Neorgan. mater.*, **27**(11), 2290–2294 (1991).

Shelimova L.E., Konstantinov P.P., Karpinskii O.G., Avilov E.S., Kretova M.A., Zemskov V.S. Thermoelectric properties of PbBi$_4$Te$_7$-based anion-substituted layered solid solutions, *Inorg. Mater.*, **40**(11), 1146–1152 (2004).

Shemet V.Ya., Gulay L.D., Olekseyuk I.D. Crystal structures of the ScAgSe and Sc$_{1.02}$Cu$_{0.54}$Sn$_{1.1}$S$_4$ compounds, *J. Alloys Compd.*, **426**, 186–189 (2006a).

Shemet V.Ya., Gulay L.D., Olekseyuk I.D. Isothermal sections of the Y$_2$Se$_3$–Cu$_2$Se–Sn(Pb)Se systems at 870 K and crystal structure of the Y$_{4.2}$Pb$_{0.7}$Se$_7$ compound, *Pol. J. Chem.*, **79**(8), 1315–1326 (2005).

Shemet V.Ya., Gulay L.D., Olekseyuk I.D. The isothermal sections of Y$_2$S$_3$–Cu$_2$S–SnS at 720 K and Y$_2$Te$_3$–Cu$_2$Te–SnTe at 870 K [in Ukrainian], *Nauk. Visnyk Volyns'k. Derzh. Univ. im. Lesi Ukrainky. Ser. Khim. nauky*, (4), 124–128 (2006b).

Shemet V.Ya., Gulay L.D., Stępień-Damm J., Pietraszko A., Olekseyuk I.D. Investigation of the Y$_2$Te$_3$–Cu$_2$Te–PbTe system at 870 K and crystal structures of the Y$_7$Cu$_3$Te$_{12}$ and YCu$_{0.264}$Te$_2$ compounds. *J. Alloys Compd.*, **420**(1–2), 58–62 (2006c).

Shemet V.Ya., Gulay L.D., Stępień-Damm J., Pietraszko A., Olekseyuk I.D. Investigation of the Y$_2$X$_3$–Cu$_2$X–SnX$_2$ (X = S, Se) systems, *Pol. J. Chem.*, **80**(6), 943–955 (2006d).

Shendyapin A.S., Isakova R.A., Nesterov V.N. Vapor pressure of lead sulfide in the PbS–FeS–Cu$_2$S system [in Russian], *Tr. In-ta metallurgii i obogaschen. AN KazSSR*, **13**, 25–31 (1965).

Shevchuk M.V., Olekseyuk I.D. Phase equilibria in the systems AgGaS$_2$–SnS$_2$, AgGaSe$_2$–SnSe$_2$, *J. Alloys Compd.*, **433**, 171–174 (2007).

Shevchuk M.V., Olekseyuk I.D. The AgGaSe$_2$–SnSe$_2$ system [in Ukrainian], *Nauk. Visnyk Volyns'k. Derzh. Univ. im. Lesi Ukrainky. Ser. Khim. nauky*, (6), 28–30 (2001).

Shimizu M., Kato A., Cioflica G., Lupulescu M., Shimizu M. Friedrichite from Bica Bihor, Romania, *Canad. Mineralog.*, **36**(3), 861–868 (1998a).

Shimizu M., Kato A., Matsubara S. Hemusite and paraguanajuatite from the Kawazu mine, Shizuoka Prefecture, Japan, *Mineralog. J.*, **14**(3), 92–100 (1988).

Shimizu M., Miyawaki R., Kato A., Matsubara S., Matsuyama F., Kiyota K. Tsugaruite, $Pb_4As_2S_7$, a new mineral species from the Yunosawa mine, Aomori Prefecture, Japan, *Mineralog. Mag.*, **62**(6), 793–799 (1998b).

Shiryaev V.S., Adam J.-L., Zhang X.-H. Calorimetric study of characteristic temperatures and crystallization behavior in Ge–As–Se–Te glass system, *J. Phys. Chem. Solids*, **65**(10), 1737–1744 (2004).

Shi Y.-F., Chen Y.-K., Chen M.-C., Wu L.-M., Lin H., Zhou L.-J., Chen L. Strongest second harmonic generation in the polar R_3MTQ_7 family: atomic distribution induced nonlinear optical cooperation, *Chem. Mater.*, **27**(5), 1876–1884 (2015).

Shi Z.-H., Yang M., Yao W.-D., Liu W., Guo S.-P. $SnPQ_3$ (Q = S, Se, S/Se): a series of lone-pair cationic chalcogenophosphates exhibiting balanced NLO activity originating from SnQ_8 units, *Inorg. Chem.*, **60**(18), 14390–14398 (2021).

Shokol A.A., Vereshchinskaya G.A., Stel'makh S.I. On the interaction of germanium dioxide with selenic acid [in Russian], *Zhurn. neorgan. khimii*, **12**(9), 2331–2336 (1967).

Shpak O., Kogut Yu., Fedorchuk A., Piskach L., Parasyuk O. The $Cu_2Se–PbSe–SiSe_2$ system and the crystal structure of $CuPb_{1.5}SiSe_4$ [in Ukrainian], *Nauk. Visnyk Skhidnoyevrop. Nats. Univ. im. Lesi Ukrainky. Ser. Khim. nauky*, [*21*(298)], 39–47 (2014).

Shurygin P.M., Serebryakova L.I. Kinetics of the PbSe and PbTe interaction with $PbCl_2$, KCl and NaCl melts [in Russian], *Izv. AN SSSR. Neorgan. mater.*, **11**(11), 1957–1960 (1975).

Siegrist T., Petter W., Hulliger F. Samarium pyrosilicate sulfide, $Sm_4S_3Si_2O_7$, *Acta Crystallogr.*, **B38**(11), 2872–2874 (1982).

Sieke C., Hartenbach I., Schleid T. Sulfidisch derivatisierte Oxodisilicate der schweren Lanthanide vom Formeltyp $M_4S_3[Si_2O_7]$ (M = Gd-Tm), *Z. Naturforsch.*, **57B**, 1427–1432 (2002).

Sieke C., Schleid T. $Pr_4S_3[Si_2O_7]$ und $Pr_3Cl_3[Si_2O_7]$: durch weiche Fremdanionen modifizierte Derivate von Praseodymdisilicat, *Z. anorg. und allg. Chem.*, **626**(1), 196–201 (2000).

Sieke C., Schleid T. $Sm_4S_3[Si_2O_7]$ und $NaSm_9S_2[SiO_4]_6$: Zwei Sulfidsilicate mit dreiwertigem Samarium, *Z. anorg. und allg. Chem.*, **625**(1), 131–136 (1999).

Sinagra III C.W., Saouma F.O., Otieno C.O., Lapidus S.H., Zhang J.-H., Craig A.J., Grima-Gallardo P., Brant J.A., Rosmus K.A., Rosello K.E., Jang J.I., Aitken J.A. Synthesis, structure, linear and nonlinear optical properties of noncentrosymmetric quaternary diamond-like semiconductors, $Cu_2ZnGeSe_4$ (CZGSe) and the novel $Cu_4ZnGe_2Se_7$, *J. Alloys Compd.*, **888**, 161499 (2021).

Smirnov M.P., Kudryashova L.N. Investigation of the $PbS–Na_2S$ and $PbS–Na_2SO_4$ phase diagrams [in Russian], *Tsvet. Met.*, (*12*), 36–42 (1956).

Smitiukh O.V., Gulay L.D., Marchuk O.V. Crystal structure of the $Er_{2.34}Ce(Pr)_{0.66}Ge_{1.28}S_7$ compounds [in Ukrainian], *Visnyk Odes'k. Nats. Univ., Хімія* **23**[2(66)], 86–94 (2018a).

Smitiukh O.V., Marchuk O.V., Fedorchuk A.O., Grebenyuk A.G. Crystal structure of $R_3Si_{1.75}Se_7$ (R – 1.5Y + 1.5La), *J. Alloys Compd.*, **765**, 731–735 (2018b).

Smitiukh O.V., Marchuk O.V., Olekseyuk I.D., Fedorchuk A.O. Crystal structure of the $Er_{1.5}La(Pr)_{1.5}Si_{1.67}Se_7$ compounds [in Ukrainian], *Nauk. Visn. Uzhgorod. Univ. Ser. Khimiya*, [*1*(37)], 44–47 (2017a).

Smitiukh O.V., Marchuk O.V., Olekseyuk I.D., Fedorchuk A.O. Crystal structure of the $Y_{1.5}Pr_{1.5}Si_{1.75}Se_7$ and $Dy_{1.5}La_{1.5}Si_{1.66}Se_7$ compounds [in Ukrainian], *Nauk. Visn. Uzhgorod. Univ. Ser. Khimiya*, [*2*(36)], 18–21 (2016).

Smitiukh O., Marchuk O., Olekseyuk I., Fedorchuk A. Crystal structure of $Y_{1.5}La_{1.5}Si_{1.75}Se_7$, *Chem. Met. Alloys*, **10**(1), 7–11 (2017b).

Smitiukh O.V., Marchuk O.V., Olekseyuk I.D., Gulay L.D. The $Y_2S_3–La_2S_3–GeS_2$ system at 770 K, *J. Alloys Compd.*, **698**, 739–742 (2017c).

Song Q., Qiu P., Chen H., Zhao K., Ren D., Shi X., Chen L. Improved thermoelectric performance in nonstoichiometric $Cu_{2+δ}Mn_{1-δ}SnSe_4$ quaternary diamond-like compounds, *ACS Appl. Mater. Interfaces*, **10**(12), 10123–10131 (2018).

Song Q., Qiu P., Hao F., Zhao K., Zhang T., Ren D., Sh, X., Chen L. Quaternary pseudocubic $Cu_2TMSnSe_4$ (TM = Mn, Fe, Co) chalcopyrite thermoelectric materials, *Adv. Electron. Mater.*, **2**(12), 1600312 (2016).

Sosovska S.M., Olekseyuk I.D., Parasyuk O.V. The $CdSe–Ga_2Se_3–PbSe$ system, *J. Alloys Compd.*, **453**(1–2), 115–120 (2008).

Sosovska S.M., Romanyuk Y.E., Olekseyuk I.D., Parasyuk O.V. Phase equilibria in the $CdGa_2Se_4–PbSe$ and $CdGa_2Se_4–As_2Se_3$ systems, *J. Alloys Compd.*, **425**, 206–209 (2006).

Spiridonov E.M., Bezsmertnaya M.S., Chileva T.N., Bezsmertnyi V.V. Bilibinskite, $Au_3Cu_2PbTe_2$, a new mineral of gold-telluride deposits [in Russian], *Zap. Vses. Mineral. Obshch.*, **107**(3), 310 315 (1978a).

Spiridonov E.M., Bezsmertnaya M.S., Chileva T.N., Bezsmertnyi V.V. Letter to the editors [in Russian], *Zap. Vses. Mineral. Obshch.*, **107**(4), 501 (1978b).

Spiridonov E.M. Bilibinskite, $(Au_{5-6}Cu_{3-2})_8(Te,Pb,Sb)_5$, from the cementation zone of the Aginskoe, Kamchatka and Pionerskoe, Sayan Mountains gold-telluride deposits, *New Data on Minerals*, Moscow, **46**, 162–164 (2011).

Spiridonov E.M., Chvileva T.N. Bezsmertnovite, $Au_4Cu(Te,Pb)$, a new mineral from the oxidation zone of the Far East deposits [in Russian], *Dokl. AN SSSR*, **249**(1), 185–189 (1979).

Spiridonov E.M., Petrova I.V., Dashevskaya D.M., Balashov E.P., Klimova L.M. Roshchinite $Ag_{19}Pb_{10}Sb_{51}S_{96}$, a new mineral [in Russian], *Dokl. AN SSSR*, **312**(1), 197–200 (1990a).

Spiridonov E.M., Petrova I.V., Dashevskaya D.M., Balashov E.P., Klimova L.M. Roshchinite $Ag_{19}Pb_{10}Sb_{51}S_{96}$, a new mineral of the andorite group [in Russian], *Zap. Vses. Mineralog. Obshch.*, **119**(5), 32–43 (1990b).

Sportouch S., Bastea M., Brazis P., Ireland J., Kannewurf C.R., Uher C., Kanatzidis M.G. Thermoelectric properties of the cubic family of compounds $AgPbBiQ_3$ (Q = S, Se, Te). Very low thermal conductivity materials, *Mat. Res. Soc. Symp. Proc.*, **545**, 123–130 (1998).

Springer G. Electronprobe analyses of stannite and related tin minerals, *Mineral. Mag.*, **36**(284), 1045–1051 (1968).

Staude S., Dorn A., Pfaff K., Markl G. Assemblages of Ag–Bi sulfosalts and conditions of their formation: the type locality of schapbachite ($Ag_{0.4}Pb_{0.2}Bi_{0.4}S$) and neighboring mines in the Schwarzwald ore district, southern Germany, *Canad. Mineralog.*, **48**(3), 441–466 (2010).

Starik P.M., Zayachuk D.M., Luchitski R.M., Lototski V.B. The quaternary solid solutions of the $A^{IV}B^{VI}$ compounds with constant lattice parameter [in Russian], *Izv. AN SSSR. Neorgan. mater.*, **15**(8), 1474–1475 (1979).

Steele I.M., Pluth J.J., Richardson, Jr. J.W. Crystal structure of tribasic lead sulfate ($3PbO\cdot PbSO_4\cdot H_2O$) by X-rays and neutrons: an intermediate phase in the production of lead acid batteries, *J. Solid State Chem.*, **132**(1), 173–181 (1997).

Stegherr A., Eckerlin P., Wald F. Untersuchung der Schnitte $Ag_2Te–Bi_2Te_3$ und $AgBiTe_2–PbTe$, *Z. Metallkd.*, **54**(10), 598–600 (1963).

Stolyarova T.A., Brichkina E.A., Baranov A.V., Osadchii E.G. Enthalpy of formation of $Cu_2ZnSnSe_4$ from its constituent elements, *Inorg. Mater.*, **55**(8). 755–757 (2019).

Stöwe K. Zur Struktur und Dotierung von Selenosilikaten: die Kristallstruktur von Er_2SeSiO_4 und $Er_{3.75}Ca_{0.25}Se_{2.75}Cl_{0.25}Si_2O_7$, *Z. Naturforsch.*, **49B**(6), 733–740 (1994).

Stoyko S.S., Craig A.J., Kotchey J.W., Aitken J.A. Synthesis, crystal structure, and electronic structure of Li_2PbSiS_4: a quaternary thiosilicate with a compressed chalcopyrite-like structure, *Acta crystallogr.*, **C77**(1), 1–10 (2021).

Strick G., Eulenberger G., Hahn H. Über einige quaternäre Chalcogenide mit Spinellstruktur, *Z. anorg. und allg. Chem.*, **357**(4–6), 338–344 (1968).

Strohfeldt E. Beiträge zu den Systemen Kupfer–Zink–Schwefel und Bleisulfid–Kupfersülfur–Zinksulfid, *Metall und Erz*, **33**(21), 561–572 (1936).

Strok O., Daszkiewicz M., Gulay L. Crystal structure of $R_3Mg_{0.5}DSe_7$ (R = Ce, Pr; D = Si, Ge), *Chem. Met. Alloys*, **8**(1–2), 16–21 (2015).

Strok O., Gulay L., Daszkiewicz M. Crystal structure of the $Ce_3Mg_{0.5}SiS_7$ compound [in Ukrainian], *Visnyk Skhidnoyevrop. Nats. Univ. im. Lesi Ukrainky. Ser. Khim. nauky*, [*24*(273)], 12–15 (2013a).

Strok O.M., Daszkiewicz M., Gulay L.D., Kaczorowski D. Crystal structure and magnetic properties of Sm_3CuGeS_7 and $Sm_3CuGeSe_7$, *J. Alloys Compd.*, **493**(1–2), 47–49 (2010).

Strok O.M. Isothermal section of the Pr_2Se_3–Cu_2Se–$SnSe_2$ system at the temperature of 870 K [in Ukrainian], *Nauk. Visnyk Volyns'k. Nats. Univ. im. Lesi Ukrainky. Ser. Khim. nauky*, (*17*), 100–104 (2012).

Strok O.M., Olekseyuk I.D., Zmiy O.F., Ivashchenko I.A., Gulay L.D. The quasi-ternary system Cu_2Se–Ga_2Se_3–$GeSe_2$, *J. Phase Equilib. Diffus.*, **34**(2), 94–103 (2013b).

Strok O., Zmiy O., Olekseyuk I. The interaction of the components along the $CuGaSe_2$–$GeSe_2$ section [in Ukrainian], *Visnyk L'viv. Univ. Ser. khim.*, **39**, 72–76 (2000).

Studenyak I.P., Kokhan O.P., Kranjčec M., Bilanchuk V.V., Panko V.V. Influence of S-Se substitution on chemical and physical properties of $Cu_7Ge(S_{1-x}Se_x)_5I$ superionic solid solutions, *J. Phys. Chem. Solids*, **68**(10), 1881–1884 (2007a).

Studenyak I.P., Kokhan O.P., Kranjčec M., Hrechyn M.I., Panko V.V. Crystal growth and phase interaction studies in the Cu_7GeS_5I–Cu_7SiS_5I superionic system, *J. Cryst. Growth*, **306**(2), 326–329 (2007b).

Suekuni K., Kim F.S., Nishiate H., Ohta M., Tanaka H.I., Takabatake T. High performance thermoelectric minerals: colusites $Cu_{26}V_2M_6S_{32}$ (M = Ge, Sn), *Appl. Phys. Lett.*, **105**(13), 132107 (2014).

Suekuni K., Tsuruta K., Fukuoka H., Koyano M. Structural and thermoelectric properties of $Cu_6Fe_4Sn_{12}Se_{32}$ single crystal, *J. Alloys Compd.*, **564**, 91–94 (2013).

Sugaki A., Kitakaze A., Odashima Y. Furutobeite, a new copper–silver–lead sulfide mineral, *Bull. Minéral.*, **104**(6), 737–741 (1981).

Suleimanov T.Zh., Onaev I.A., Kozhahmetov S.M. Interaction rate of lead oxide and zinc sulfide [in Russian], *Metallurgia i obogashchenie*, (9), 77–80 (1974a).

Suleimanov T.Zh., Onaev I.A., Kozhahmetov S.M., Kairbayeva Z.K. Thermographic investigation of the PbO–ZnS system [in Russian], *Metallurgia i obogashchenie*, (9), 74–76 (1974b).

Sun B., He J., Zhang X., Bu K., Zheng C., Huang F. Synthesis, crystal structure and optical properties of $K_2Cu_2GeS_4$, *J. Alloys Compd.*, **725**, 557–562 (2017).

Sun J.-P., McKeown Wessler G.C., Wang T., Zhu T., Blum V., Mitzi D.B. Correction to Structural tolerance factor approach to defect-resistant I_2–II–IV–X_4 semiconductor design, *Chem. Mater.*, **32**(13), 5925–5926 (2020a).

Sun J.-P., McKeown Wessler G.C., Wang T., Zhu T., Blum V., Mitzi D.B. Structural tolerance factor approach to defect-resistant I_2–II–IV–X_4 semiconductor design, *Chem. Mater.*, **32**(4), 1636–1649 (2020b).

Sun Y-L., Chi Y., Guo S.-P. Synthesis and crystal structure of a new quaternary sulfide $FeSm_6Si_2S_{14}$, *Chin. J. Struct. Chem.*, **35**(9), 1369–1375 (2016).

Sun Z.-D., Chi Y., Guo S.-P. Cu_2EuMQ_4 (M = Si, Ge; Q = S, Se): syntheses, structure study and physical properties determination, *J. Solid State Chem.*, **269**, 225–232 (2019).

Surapunt S., Nyamai C.M., Hino M., Itagaki K. Phase relations and distribution of minor elements in the Cu–Zn–S, Cu–Zn–Fe–S and Cu–Zn–Pb–S systems at 1473 K, *Metal Rev. MMIJ*, **12**(2), 84–97 (1995).

Surapunt S., Nyamai C.M., Hino M., Itagaki K. Phase relations in the Cu–Zn–S, Cu–Zn–Fe–S and Cu–Zn–Pb–S systems at 1473 K [in Japanese], *Shigen-to-Sozai*, **112**(1), 56–60 (1996).

Suriñach S., Baro M.D., Clavaguera-Mora M.T., Clavaguera N. Glass formation and crystallization in the $GeSe_2$–Sb_2Te_3 system, *J. Mater. Sci.*, **19**(9), 3005–3012 (1984).

Suriñach S., Baro M.D., Clavaguera-Mora M.T., Clavaguera N. "Kinetic study of isothermal and continuos heating crystallization in $GeSe_2$–$GeTe$–Sb_2Te_3 alloy glasses" *J. Non-Cryst. Solids*, **58**(2–3), 209–217 (1983).

Suriñach S., Baro M.D., Clavaguera-Mora M.T., Clavaguera N. Thermodynamic aspects of glassformation and crystallization in the $GeSe_2$–Sb_2Te_3 system, *Fluid Phase Equil.*, **20**, 341–346 (1985).

Swihart G.H., Sen Gupta P.K., Schlemper E.O., Back M.E., Gaines R.V. The crystal structure of moctezumite $[PbUO_2](TeO_3)_2$, *Am. Mineralog.*, **78**(7–8), 835–839 (1993).

Szymański J.T. The crystal structure of černýite, Cu_2CdSnS_4, a cadmium analogue of stannite, *Canad. Mineralog.*, **16**(2), 147–151 (1978).

Szymański J.T. The crystal structure of mawsonite, $Cu_6Fe_2SnS_8$, *Canad. Mineralog.*, **14**(4), 529–535 (1976).

Tait K., Poirier G., Rowe R., Piilonen P.C. New mineral names, *Am. Mineralog.*, **93**(1), 252–256 (2008).

Takéuchi Y., Ghose S., Nowacki W. The crystal structure of hutchinsonite, $(Tl,Pb)_2As_5S_9$, *Z. Kristallogr.*, **121**(5), 321–348 (1965).

Tămaş C.G., Grobety B., Bailly L., Bernhardt H.-J., Minut A. Alburnite, $Ag_8GeTe_2S_4$, a new mineral species from the Roşia Montana Au-Ag epithermal deposit, Apuseni Mountains, Romania, *Am. Mineralog.*, **99**(1), 57–64 (2014).

Tămas C.G., Grobety B., Bailly L., Bernhardt H.-J., Minuţ A. Alburnite, IMA 2012-073. CNMNC Newsletter No. 15, February 2013, page 10, *Mineralog. Mag.*, **77**(1), 1–12 (2013).

Tampier M., Johrendt D. $BaCu_6Ge_2S_8$ – Ein Thiogermanat als Variante der Li_3Bi-Struktur, *Z. Naturforsch.*, **53B**(12), 1483–1488 (1998).

Tampier M., Johrendt D. Neue azentrische Selenogermanate. I. Kristallstrukturen und chemische Bindung von AM_2GeSe_4 (A = Sr, Ba; M = Cu, Ag), *Z. anorg. und allg. Chem.*, **627**(3), 312–320 (2001).

Tananaeva O.I., Novoselova A.V. Investigation of the thermal stability of lead tellurite and tellurate [in Russian], *Izv. AN SSSR.. Neorgan. mater.*, **3**(1), 114–118 (1967).

Tang J., Xing W., Kang K., Zeng T., Yin W., Kang B. Quaternary rareearth sulfide LaSnGa$_3$S$_7$: synthesis, structure, thermal and optical properties, *J. Alloys Compd.*, **828**, 154380 (2020).

Tarasenko M.S., Berezin A.S., Kiryakov A.S., Piryazev D.A., Filatova I.Yu., Naumov N.G. Synthesis, crystal structure and photoluminescence of Eu^{3+} or Tb^{3+} doped solid solutions (Y$_{1-x}$RE$_x$)$_4$S$_3$(Si$_2$O$_7$), *J. Solid State Chem.*, **265**, 36–41 (2018).

Tats'kar A.R., Koz'ma A.A., Barchiy I.E., Solomon A.M., Rekita V.V. Triangulation of the Tl$_2$Se–SnSe$_2$–Sb$_2$Se$_3$ quasiternary system [in Ukrainian], *Nauk. Visn. Uzhgorod. Univ. Ser. Khimiya*, [*1*(29)], 14–19 (2013).

Tavera F.J., Davenport W.G. Equilibrations of copper matte and fayalite slag under controlled partial pressures of SO$_2$, *Metall. Trans. B*, **10**(2), 237–241 (1979).

Terziev G.I. Hemusite – a complex copper–tin–molybdenum sulfide fron the Chelopech ore deposite, Bulgaria, *Am. Mineralog.*, **56**(11–12), 1847–1854 (1971).

Teske C.L. Darstellung, Kristallstrukturdaten und Eigenschaften der quaternären Thiostannate(IV) BaZnSnS$_4$ und BaMnSnS$_4$, *Z. Naturforsch.*, **35B**(4), 509–510 (1980a).

Teske C.L. Darstellung und Kristallstruktur von Ba$_3$CdSn$_2$S$_8$ mit einer Anmerkung über Ba$_6$CdAg$_2$Sn$_4$S$_{16}$, *Z. anorg. und allg. Chem.*, **522**(3), 122–130 (1985).

Teske C.L. Darstellung und Kristallstruktur von Barium-Cadmium-Thiostannat(IV) BaCdSnS$_4$, *Z. anorg. und allg. Chem.*, **460**(1), 163–168 (1980b).

Teske C.L. Darstellung und Kristallstruktur von Barium–Quecksilber–Thiostannat(IV), BaHgSnS$_4$, *Z. Naturforsch.*, **35B**(1), 7–11 (1980c).

Teske C.L. Darstellung und Kristallstruktur von Cu$_2$SrSnS$_4$, *Z. anorg. und allg. Chem.*, **419**(1), 67–76 (1976).

Teske C.L. Darstellung und Kristallstruktur von Gold–Barium–Thiostannat(IV), Au$_2$BaSnS$_4$, *Z. anorg. und allg. Chem.*, **445**(1), 193–201 (1978).

Teske C.L. Darstellung und Kristallstruktur von Zilber Barium Thiogermanat(IV), Ag$_2$BaGeS$_4$, *Z. Naturforsch.*, **34B**(4), 544–547 (1979a).

Teske C.L. Über die Darstellung und röntgenographische Untersuchung von Cu$_2$SrGeS$_4$ und Cu$_2$BaGeS$_4$, *Z. Naturforsch.*, **34B**(3), 386–389 (1979b).

Teske C.L., Vetter O. Ergebnisse einer Röntgenstrukturanalyse von Silber–Barium–Thiostannat(IV), Ag$_2$BaSnS$_4$, *Z. anorg. und allg. Chem.*, **427**(3), 200–204 (1976a).

Teske C.L., Vetter O. Präparative und röntgenographische Untersuchung am System Cu$_{2-x}$Ag$_x$BaSnS$_4$, *Z. anorg. und allg. Chem.*, **426**(3), 281–287 (1976b).

Teske C.L. Zur Kenntnis von BaCdGeS$_4$ mit einem Beitrag zur Kristallchemie von Verbindungen des Typs BaABS$_4$, *Z. anorg. und allg. Chem.*, **468**(9), 27–34 (1980d).

Tettenhorst R.T., Corbato C.E. Crystal structure of germanite, Cu$_{26}$Ge$_4$Fe$_4$S$_{32}$, determined by powder X-ray diffraction, *Am. Mineralog.*, **69**(9–10), 943–947 (1984).

Tian T., Li Z., Wang N., Zhao S., Xu J., Lin Z., Mei D. Cs$_2$ZnSn$_3$S$_8$: a sulfide compound realizes a large birefringence by modulating the dimensional structure, *Inorg. Chem.*, **60**(13), 9248–9253 (2021).

Timofeevskiy D.A. On the lead and silver sulphoantimonates of the owyheeite group from the East Transbaikal deposits [in Russian], *Zap. Vses. Mineralog. Obsh.*, **96**(1), 30–44 (1967).

Tissot R.G., Rodriguez M.A., Sipola D.L., Voight J.A. X-ray powder diffraction study of synthetic palmierite, K$_2$Pb(SO$_4$)$_2$, *Powder Diffr.*, **16**(2), 92–97 (2001).

Tomashik Z.F. Liquidus surface of the CdSe + PbTe ↔ CdTe + PbSe ternary mutual system [in Russian], *Izv. AN SSSR. Neorgan. mater.*, **17**(9), 1575–1577 (1981).

Tomashik Z.F., Oleinik G.S., Tomashik V.N., Nizkova A.I. Phase equilibria in the PbSe–CdS system [in Russian], *Izv. AN SSSR. Neorgan. mater.*, **17**(12), 2155–2158 (1981).

Tomashik Z.F., Tomashik V.N. Liquidus surface of the PbTe+CdS ↔ PbS+CdTe ternary mutual system, *Izv. AN SSSR. Neorgan. mater.*, **23**(12), 1981–1984 (1987).

Tomashik Z.F., Tomashik V.N. The CdS+PbSe ↔ CdSe+PbS ternary mutual system [in Russian], *Izv. AN SSSR. Neorgan. mater.*, **20**(4), 568–570 (1984).

Tomm Y., Schorr S., Fiechter S. Crystal growth of argyrodite-type phases Cu$_{8-x}$GeS$_{6-x}$I$_x$ and Cu$_{8-x}$GeSe$_{6-x}$I$_x$ (0 ≤ x ≤ 0.8), *J. Crys. Growth*, **310**(7–9), 2215–2221 (2008).

Topa D., Balić-Žunić T., Makovicky E. The crystal structure of Cu$_{1.6}$Pb$_{1.6}$Bi$_{6.4}$S$_{12}$, a new 44.8 Å derivative of the bismuthinite–aikinite solid-solution series, *Canad. Mineralog.*, **38**(3), 611–616 (2000a).

Topa D., Berlepsch P., Makovicky E., Stroeger B., Stanley C. Enneasartorite, IMA 2015-074. CNMNC Newsletter No. 28, December 2015, page 1861, *Mineralog. Mag.*, **79**(7), 1859–1864 (2015a).

Topa D., Graeser S., Makovicky E., Stanley C. Argentoliveingite, IMA 2016-029. CNMNC Newsletter No. 32, August 2016, page 920, *Mineralog. Mag.*, **80**(5), 915–922 (2016a).

Topa D., Makovicky E. Argentobaumhauerite: name, chemistry, crystal structure, comparison with baumhauerite, and position in the Lengenbach mineralization sequence, *Mineralog. Mag.*, **80**(5), 819–840 (2016).

Topa D., Makovicky E., Balić-Žunić T., Berlepsch P. The crystal structure of Cu$_2$Pb$_6$Bi$_8$S$_{19}$, *Eur. J. Mineral.*, **12**(4), 825–833 (2000b).

Topa D., Makovicky E., Balić-Žunić T. Mineralogical data on salzburgite and paarite, two new members of the bismuthinite–aikinite series, *Canad. Mineralog.*, **43**(3), 909–917 (2005).

Topa D., Makovicky E., Balić-Žunić T. The structural role of excess Cu and Pb in gladite and krupkaite based on new refinements of their structure, *Canad. Mineralog.*, **40**(4), 1147–1159 (2002).

Topa D., Makovicky E., Berlepsch P., Stroeger B., Stanley C. Hendekasartorite, IMA 2015-075. CNMNC Newsletter No. 28, December 2015, page 1861, *Mineralog. Mag*, **79**(7), 1859–1864 (2015b).

Topa D., Makovicky E., Criddle A.J., Paar W.H., Balić-Žunić T. Felbertalite, Cu$_2$Pb$_6$Bi$_8$S$_{19}$, a new mineral species from Felbertal, Salzburg Province, Austria, *Eur. J. Mineral.*, **13**(5), 961–972 (2001).

Topa D., Makovicky E., Putz H., Zagler G. Lopatkaite, IMA 2012-083. CNMNC Newsletter No. 15, February 2013, page 11, *Mineralog. Mag.*, **77**(1), 1–12 (2013a).

Topa D., Makovicky E., Stanley C., Cannon R. Argentodufrénoysite, IMA 2016-046. CNMNC Newsletter No. 33, October 2016, page 1138, *Mineralog. Mag.*, **80**(6), 1135–1144 (2016b).

Topa D., Makovicky E., Stoeger B., Stanley C. Heptasartorite, Tl$_7$Pb$_{22}$As$_{55}$S$_{108}$, enneasartorite, Tl$_6$Pb$_{32}$As$_{70}$S$_{140}$ and hendekasartorite, Tl$_2$Pb$_{48}$As$_{82}$S$_{172}$, three members of the anion-omission series of 'sartorites' from the Lengenbach quarry at Binntal, Wallis, Switzerland, *Eur. J. Mineral.*, **29**(4), 701–712 (2017a).

Topa D., Makovicky E. Structures and classification of indium sulfosalts, and the crystal structure of Sn$_{12}$In$_{19}$(Se,S)$_{41}$, *Canad. Mineralog.*, **50**(2), 397–422 (2012).

Topa D., Makovicky E. The crystal structure of veenite, *Mineralog. Mag.*, **81**(2), 355–368 (2017).

Topa D., Makovicky E., Zagler G., Putz H., Paar W.H. Erzwiesite, IMA 2012-082. CNMNC Newsletter No. 15, February 2013, page 11, *Mineralog. Mag.*, **77**(1), 1–12 (2013b).

Topa D., Paar W.H., Balić-Žunić T. Emilite, $Cu_{10.72}Pb_{10.72}Bi_{21.28}S_{48}$, the last missing link of the bismuthinite–aikinite series? *Canad Mineralog.*, **44**(2), 459–464 (2006).

Topa D., Sejkora J., Makovicky E., Pršek J., Ozdín D., Putz H., Dittrich H., Karup-Møller S. Chovanite, $Pb_{15-2x}Sb_{14+2x}S_{36}O_x$ (x ~ 0.2), a new sulphosalt species from the Low Tatra Mountains, Western Carpathians, Slovakia, *Eur. J. Mineral.*, **24**(4), 727–740 (2012).

Topa D., Stroeger B., Makovicky E., Berlepsch P., Stanley C. Heptasartorite, IMA 2015-073. CNMNC Newsletter No. 28, December 2015, page 1861, *Mineralog. Mag.*, **79**(7), 1859–1864 (2015c).

Topa D., Stoeger B., Makovicky E., Stanley C. Dekatriasartorite, IMA 2017-071. CNMNC Newsletter No. 40, December 2017, page 1084, *Eur. J. Mineral.*, **29**(6), 1083–1087 (2017b).

Topa D., Stoeger B., Makovicky E., Stanley C. Dekatriasartorite, IMA 2017-071. CNMNC Newsletter No. 40, December 2017, page 1579, *Mineralog. Mag.*, **81**(6), 1577–1581 (2017c).

Topa D., Stoeger B., Makovicky E., Stanley C. Incomsartorite, IMA 2016-035. CNMNC Newsletter No. 33, October 2016, page 1136, *Mineralog. Mag.*, **80**(6), 1135–1144 (2016c).

Triboulet R., Rabago F., Legros R., Lozykowski H., Didier G. Low-temperature growth of ZnSe crystals, *J. Cryst. Growth*, **59**(1–2), 172–177 (1982).

Tsisar O.V., Babizhetskyy V.S., Levytskyy V.O., Piskach L.V., Olekseyuk I.D., Parasyuk O.V. Crystal structure of new quarternary sulfide $TlGaGe_3S_8$ [in Ukrainian], *Nauk. Visn. Uzhgorod. Univ. Ser. Khimiya*, [2(38)], 32–38 (2017a).

Tsisar O.V., Babizhetskyy V.S., Levytskyy V.O., Piskach L.V., Parasyuk O.V., Olekseyuk I.D., Mozolyuk M.Yu. Phase equilibria in the $Tl_2S–Ga_2S_3–GeS_2$ system [in Ukrainian], *Nauk. Visn. Uzhgorod. Univ. Ser. Khimiya*, [2(38)], 26–31 (2017b).

Tsisar O., Piskach L., Babizhetskyy V., Levytskyy V., Kotur B., Marushko L.,. Olekseyuk I., Parasyuk O. Phase equilibria in the $Tl_2Se–In_2Se_3–GeSe_2$ system at 520 K [in Ukrainian], *Visnyk L'viv. Univ. Ser. khim.*, (59), Pt. 1, 46–52 (2018).

Tsisar O.V., Piskach L.V., Marushko L.P., Olekseyuk I.D., Zamuruyeva O.V., Parasyuk O.V. Glass formation in the $Tl_2Se–Ga_2Se_3–GeSe_2$ system [in Ukrainian], *Nauk. Visn. Uzhgorod. Univ. Ser. Khimiya*, [1(37)], 63–67 (2017c).

Tsisar O.V., Piskach L.V., Parasyuk O.V., Marushko L.P., Olekseyuk I.D., Zamuruyeva O.V., Czaja P., Karasiński P., El-Naggar A.M., Albassam A.A., Lakshminarayana G. $Tl_2S–Ga_2S_3–GeS_2$ glasses for optically operated laser third harmonic generation, *J. Mater. Sci: Mater Electron.*, **28**(24), 19003–19009 (2017d).

Tsuji I., Shimodaira Y., Kato H., Kobayashi H., Kudo A. Novel stannite-type complex sulfide photocatalysts $A^I_2–Zn–A^{IV}–S_4$ (A^I = Cu and Ag; A^{IV} = Sn and Ge) for hydrogen evolution under visible-light irradiation, *Chem. Mater.*, **22**(4), 1402–1409 (2010).

Urazov G.G., Sokolova M.A. Investigation of the $Ag_2Cl_2+PbS \leftrightarrow PbCl_2+Ag_2S$ mutual system by thermal analysis and microstructure method [in Russian], *Izv. sektora phys.-khim. analiza*, **14**, 317–337 (1941).

Vakulovich A.P., Olekseyuk I.D. Phase equilibria in the $CuInSe_2–GeSe_2$ and $CuInSe_2–Cu_2GeSe_3$ sections of the quaternary $Cu_2Se–In_2Se_3–GeSe_2$ system, *J. Alloys Compd.*, **367**(1–2), 47–48 (2004).

Van Hook H.J. The ternary system $Ag_2S–Bi_2S_3–PbS$, *Econ. Geol.*, **55**(4), 759–788 (1960).

Vaqueiro P., Sobany G.G., Stindl M. Structure and electrical transport properties of the ordered skutterudites $MGe_{1.5}S_{1.5}$ (M = Co, Rh, Ir), *J. Solid State Chem.*, **181**(4), 768–776 (2008).

Vasil'eva I.G.., Nikolaev R.E. Saturated vapor pressure over $AgGaGeS_4$ crystals, *Inorg. Mater.*, **42**(12), 1299–1301 (2006),

Vasil'ev L.N., Bakhtiyarov A.Sh., Makeeva K.V., Sivkov V.P., Seregina L.N. The valent state of tin and some properties of arsenic–selenium–germanium–tin glasses [in Russian], *Fiz. i khim. stekla*, **3**(1), 77–81 (1977).

Vasil'ev L.N., Bakhtiyarov A.Sh. NGR spectra of the $Tl_2Se–As_2Se_3–SnSe$ alloys [in Russian], *Izv. AN SSSR. Neorgan. mater.*, **11**(11), 2074–2076 (1975).

Vasiliev V.I., Borisenko A.S., Mortsev N.K., Khoa C.C., Fyong N.T. Jonassonite ($AuBi_5S_4$) from the Dakripen gold deposit in Vietnam. *Geol. Ore Deposits*, **53**(7), 614–619 (2011).

Vassilev V., Aljihmani L., Hristova-Vasileva T. Phase equilibria in the $GeSe_2–Ag_4SSe$ system, *J. Phase Equilib. Diffus.*, **33**(2), 106–109 (2012).

Vassilev V., Aljihmani L., Parvanova V. Phase equilibria in the $GeSe_2–SnTe$ system, *J. Therm. Anal. Calorim.*, **76**(3), 727–735 (2004).

Vassilev V., Atanassova D., Mihaylova I., Parvanova V., Aljihmani L. Phase equilibria in the $PbSb_2Te_4–InSb$ system, *Thermochim. Acta*, **520**(1–2), 80–83 (2011).

Vassilev V.S., Boycheva S.V., Ivanova Z.G. Glass formation and physicochemical properties of the $GeSe_2–Sb_2Se_3–Ag_2Se(ZnSe)$ systems, *J. Mater. Sci. Lett.*, **17**(23), 2007–2008 (1998).

Vassilev V.S., Boycheva S.V., Petkov P. Glass formation in the $GeSe_2(As_2Se_3)–Sb_2Se_3–CdTe$, *Mater. Lett.*, <u>52</u>(1–2), 126–129 (2002a).

Vassilev V.S., Ivanova Z.G., Dospeiska E.S., Boycheva S.V. Multicomponent $GeSe_2–CdI_2–TeO_2$ (Bi_2O_3) systems: glass formation and properties, *J. Phys. Chem. Solids*, **63**(5), 815–819 (2002b).

Vassilev V.S., Ivanova Z.G. On the glass formation in the Ge–S–Cd and $GeS_2–Cd–I$ systems, *J. Phys. Chem. Solids*, **58**(4), 573–576 (1997).

Vassilev V., Parvanov S., Hristova-Vasileva T., Aljihmani L., Vachkov V., Vassileva-Evtimova T. Glass formation in the $As_2Te_3–As_2Se_3–SnTe$ system, *Mater. Let.*, **61**(17), 3676–3678 (2007a).

Vassilev V., Stephanova S., Markova I., Andreev R., Ivanova Y., Alexandrova N., Ivanova Z., Hadjinikolova S. Glass formation in the Se–Ge–Zn and $GeSe_3–ZnSe–Ag_2Se$ systems, *J. Mater. Sci.*, **32**(16), 4443–4445 (1997).

Vassilev V., Tomova K., Parvanova V., Boycheva S. Glass-formation in the $GeSe_2–Sb_2Se_3–SnSe$ system, *J. Alloys Compd.*, **485**(1–2), 569–572 (2009).

Vassilev V., Tomova K., Parvanov S. Phase equilibria in the $GeSe_2–PbTe$ system, *Thermochim. Acta*, **459**(1–2), 12–16 (2007b).

Viskova N.V., Kolomoets N.V., Rudnik I.M. Investigation of a section of the $PbTe–Ag_2Te–Sb_2Te_3$ system [in Russian], *Izv. AN SSSR. Neorgan. mater.*, **2**(9), 1560–1562 (1966).

Vlayet F., Roubin M. Étude de deux sections du systéme ternaire $Ag_2Se–PbSe–Ga_2Se_3$, *C. r. Acad. Sci Ser. C.*, **286**(9), 285–288 (1978).

Volykhov A.A., Yashina L.V., Shtanov V.I. Phase equilibria in pseudoternary systems of IV-VI compounds, *Inorg. Mater.*, **46**(5), 464–471 (2010).

Volykhov A.A., Yashina L.V., Tamm M.E., Ryzhenkov A.V. Phase equilibria in ternary reciprocal systems based on IV-VI compounds, *Inorg. Mater.*, **45**(9), 968–974 (2009).

Vurro F., Garavelli A., Garbarino C., Moëlo Y., Borodaev Y.S. Rare sulfosalts from Vulcano, Aeolian Islands, Italy. II. Mozgovaite, $PbBi_4(S,Se)_7$, a new minerals species, *Canad. Mineralog.*, **37**(6), 1499–1506 (1999).

Vu T.V., Lavrentyev A.A., Gabrelian B.V., Selezen A.O., Piskach L.V., Olekseyuk I.D., Myronchuk G.L., Denysyuk M., Tkach V.A., Hieu N.N., Pham K.D., Khyzhun O.Y. Quaternary $Tl_2CdGeSe_4$ selenide: electronic structure and optical properties of a novel semiconductor for potential application in optoelectronics, *J. Solid State Chem.*, **302**, 122453 (2021).

Vu T.V., Marchuk O.V., Smitiukh O.V., Tkach V.A., Myronchuk D., Myronchuk G.L., Khyzhun O.Y. High-temperature orthorhombic phase of Cu_2HgGeS_4: electronic structure and principal optical constants as evidenced from the experiment and theory, *J. Solid State Chem.*, **313**, 123313 (2022).

Vymazalová A., Subbotin V.V., Laufek F., Savchenko Y.E., Stanley C.J., Gabov D.A., Plášil J. Panskyite, IMA 2020-039. CNMNC Newsletter No. 57, *Eur. J. Mineral.*, **32**(5), 497 (2020a).

Vymazalová A., Subbotin V.V., Laufek F., Savchenko Y.E., Stanley C.J., Gabov D.A., Plášil J. Panskyite, IMA 2020-039. CNMNC Newsletter No. 57, *Mineralog. Mag.*, **84**(5), 792–793 (2020b).

Vymazalová A., Subbotin V.V., Laufek F., Savchenko Y.E., Stanley C.J., Gabov D.A., Plášil J. Panskyite, $Pd_9Ag_2Pb_2S_4$, a new platinum group mineral from the Southern Kievey ore occurrence of the Fedorova–Pana layered intrusion, Kola Peninsula, Russia, *Mineralog. Mag.*, **85**(2), 161–171 (2021).

Wagener M., Deiseroth H.-J., Reiner C. $Ag_6GeS_4X_2$ (X: Cl, Br): surprisingly no filled Laves phases but the first representatives of a new structure type, *Z. Kristallogr.*, **221**(5–7), 533–538 (2006).

Walia D.S., Chang L.L.Y. Investigation in the systems $PbS–Sb_2S_3–As_2S_3$ and $PbS–Bi_2S_3–As_2S_3$, *Canad. Mineralog.*, **12**(2), 113–119 (1973).

Wang Ch., Hughbanks T. Main group element size and substitution effects on the structural dimensionality of zirconium tellurides of the ZrSiS type, *Inorg. Chem.*, **34**(22), 5524–5529 (1995).

Wang K.-C., Lee C.-S. Effect of the transition metal on the synthesis of quaternary sulfides $MPb_8In_{17}S_{34}$ (M = Cu, Ag, Au), *Inorg. Chem.*, **45**(4), 1415–1417 (2006).

Wang M.-F., Huang W.-H., Lee C.-S. Synthesis and phase width of quaternary selenides $Pb_4In_xM_{6-x}Se_{13}$ (M = Bi, x = 2.1–2.8; Sb, x = 2), *Inorg. Chem.*, **48**(14), 6402–6408 (2009a).

Wang M.-F., Jang S.-M., Huang J.-C., Lee C.-S. Synthesis and characterization of quaternary chalcogenides $InSn_2Bi_3Se_8$ and $In_{0.2}Sn_6Bi_{1.8}Se_9$, *J. Solid State Chem.*, **182**(6), 1450–1456 (2009b).

Wang N. Investigation of synthetic rhodostannite $Cu_2FeSn_3S_8$, *N. Jb. Miner., Mh.*, (4), 166–171 (1975).

Wang W., Cao W., Zhang L., Li G., Wu Y, Wen S., Mei D. $Rb_2FeGe_3S_8$ and $Cs_2FeGe_3S_8$: new layered chalcogenides in $A_2M^{II}M^{IV}_3Q_8$ family with antiferromagnetic property, *J. Solid State Chem.*, **313**, 123276 (2022).

Wang X., Gua S., Yu J., Zhaoa X., Tao H. Formation and properties of chalcogenide glasses in the $GeS_2–Ga_2S_3–CdS$ system, *Mater. Chem. Phys.*, **83**(2–3), 284–288 (2004).

Wang X., Jiang X., Liu H., Yang L., Lin Z., Hu Z., Meng X., Chen X., Qin J. $Pb_3(SeO_3)Br_4$: a new nonlinear optical material with enhanced SHG response designed via ion-substitution strategy, *Dalton Trans.*, **47**(6), 1911–1917 (2018).

Wang Y.C., DiSalvo F.J. Synthesis and characterization of $Ba_3Bi_{6.67}Se_{13}$ and its filled variants $Ba_3Bi_6PbSe_{13}$ and $Ba_3Bi_6SnSe_{13}$, *Chem. Mater.*, **12**(4), 1011–1017 (2000).

Ward M.D., Pozzi E.A., Van Duyne R.P.,, Ibers J.A. Syntheses, structures, and optical properties of the indium/germanium selenides $Cs_4In_8GeSe_{16}$, $CsInSe_2$, and $CsInGeSe_4$, *J. Solid State Chem.*, **212**, 191–196 (2014).

Weber F.A., Schleid T. Zwei Formen (A- und B-Typ) von $Pr_2Te[SiO_4]$, *Z. anorg. und allg. Chem.*, **625**(12), 2071–2076 (1999).

Wedel B., Müller-Buschbaum HK. Über die Kristallstrukturen der Tellurate $Pb_3Fe_2Te_2O_{12}$ und Pb_2CoTeO_6, *Z. Naturforsch.*, **52B**(1), 35–39 (1997).

Wedel B., Müller-Buschbaum HK. Über ein Blei-Kupfer-Tellurat $PbCu_3TeO_7$ mit Cu^{2+} in deformiert tetragonal pyramidaler und tetraedrischer Sauerstoffkoordination, *Z. Naturforsch.*, **51B**(11), 1587–1590 (1996).

Wedel B., Sugiyama K., Müller-Buschbaum HK. Verknüpfung von $(TeO_6)^{6-}$ und $(TeO_6)(NiO)_3$-Sechsringen durch $TeNiO_9$-Oktaederdoppel in $Pb_3Ni_{4.5}Te_{2.5}O_{15}$, *Z. Naturforsch.*, **53B**(5–6), 527–531 (1998).

Weihrich R., Anusca I. Half antiperovskites. III. Crystallographic and electronic structure effects in $Sn_{2-x}In_xCo_3S_2$. , *Z. anorg. und allg Chem.*, **632**(2), 1531–1537 (2006).

Weihrich R., Yan W., Rothballer J., Peter P., Rommel S.M., Haumann S., Winter F., Schwickert C., Pöttgen R. Tuneable anisotropy and magnetism in $Sn_2Co_3S_{2-x}Se_x$ – probed by ^{119}Sn Mößbauer spectroscopy and DFT studies, *Dalton Trans.*, **44**(36), 15855–15864 (2015).

Wei K., Martin J., Maruyama S., Mori T., Nolas G.S. Physical properties of madocite: a quaternary chalcogenide with very low thermal conductivity, *J. Solid State Chem.*, **251**, 113–115 (2017).

Weiland A., Zhang J.-H., Clark D.J., Brant J.A., Sinagra III C.W., Kim Y.S., Jang J.I., Aitken J.A. Correction: infrared nonlinear optical properties of lithium-containing diamondlike semiconductors $Li_2ZnGeSe_4$ and $Li_2ZnSnSe_4$, *Dalton Trans.*, **46**(30), 10102–10104 (2017).

Weil M., Shirkhanlou M. Incorporation of sulfate or selenate groups into oxotellurates(IV): II. Compounds with divalent lead, *Z. anorg. und allg. Chem.*, **643**(12), 757–765 (2017).

Weil M., Shirkhanlou M., Stürzer T. Phase formation studies of lead(II) copper(II) oxotellurates: the crystal structures of dimorphic $PbCuTeO_5$, $PbCuTe_2O_6$, and $[Pb_2Cu_2(Te_4O_{11})]$ $(NO_3)_2$, *Z. anorg. und allg. Chem.*, **645**(3), 347–353 (2019).

Weil M., Stöger B. $Pb_3Te_2O_6Br_2$, *Acta Crystallogr.*, **E66**(2), i7 (2010).

Weiss J., Neubauer D. Die Kristallstruktur des $Pb(NS)_2NH_3$, Über Metall-thionitrosylverbindungen, IX, *Z. Naturforsch.*, **13B**(7), 459–460 (1958).

Weiss J., Schäfer H. $Na_3(AlSi)Te_4$ – das erste Hypotelluroalumosilikat, *Z. Naturforsch.*, **34B**(2), 176–178 (1979).

Weiss J. Zur Kenntnis der Metall-Schwefelstickstoff-Verbindungen. II. Zur Struktur von $PbN_2S_2NH_3$, *Z. anorg. allg. Chem.*, **343**(5–6), 315–322 (1966).

Welzmiller S., Rosenthal T., Ganter P., Neudert L., Fahrnbauer F., Urban P., Stiewe C., de Boor J., Oeckler O. Layered germanium tin antimony tellurides: element distribution, nanostructures and thermoelectric properties, *Dalton Trans.*, **43**(27), 10529–10540 (2014).

Wenyuan L., Cheng D., Xiangping G., Yu L., Xiaoping Q., Yuchuan C. The crystal structure of kiddcreekite solved using micro X-ray diffraction and the *EPCryst* program, *Mineralog. Mag.*, **78**(7), 1517–1525 (2014).

Wernick J.H. Constitution of the $AgSbS_2$–PbS, $AgBiS_2$–PbS and $AgBiS_2$–$AgBiSe_2$ systems, *Am. Mineralog.*, **45**(5–6), 591–611 (1960).

Westover R.D., Atkins R.A., Ditto J.J., Johnson D.C. Synthesis of $[(SnSe)_{1.16-1.09}]_1[(Nb_xMo_{1-x})Se_2]_1$ ferecrystal alloys, *Chem. Mater.*, **26**(11), 3443–3449 (2014).

Wickleder C., Hartenbach I., Lauxmann P., Schleid T. $Eu_5F[SiO_4]_3$ und $Yb_5S[SiO_4]_3$: Gemischtvalente Lanthanoid-Silicate mit Apatit-Struktur, *Z. anorg. und allg. Chem.*, **628**(7), 1602–1606 (2002).

Williams S.A. Girdite, oboyerite, fairbankite, and winstanleyite, four new tellurium minerals from Tombstone, Arizona, *Mineralog. Mag.*, **43**(328), 453–457 (1979).

Wintenberger M. Étude de la structure cristallographique et magnétique de Cu_2FeGeS_4 et remarque sur la structure magnétique de Cu_2MnSnS_4, *Mater. Res. Bull.*, **14**(9), 1195–1202 (1979).

Wong K.M., Khan W., Shoaib M., Shah U., Khan S.H., Murtaza G. *Ab initio* investigation of the structural, electronic and optical properties of the $Li_2In_2XY_6$ (X = Si, Ge; Y = S, Se) compounds, *J. Electron. Mater.*, **47**(1), 566–576 (2018).

Woolley J.C., Goodchild R.G., Hughes O.H., Lopez-Rivera S.A., Pamplin B.R. Quaternary defect chalcopyrite compounds I III IV VI_4, *Jpn. J. Appl. Phys.*, **19**, Suppl. 19-3, 145–148 (1980).

Woolley J.C., Lamarche A.-M., Lamarche G., Church C. Crystal symmetry of $Ag_2MnGeTe_4$ phases, *J. Solid State Chem.*, **115**(1), 192–196 (1995).

Woolley J.C., Williams E.W. Cross-substitutional alloys of InSb, *J. Electrochem. Soc.*, **111**(2), 210–215 (1964).

Wu D. Phase relations in the systems Ag_2S–Cu_2S–PbS and Ag_2S–Cu_2S–Bi_2S_3 and their mineral assemblages, *Chin. J. Geochem.*, **6**(3), 216–224 (1987a).

Wu D. Phase relations in the systems Cu_2S–PbS–Bi_2S_3 and Ag_2S–PbS–Bi_2S_3 and their mineral assemblages, *Chin. J. Geochem.*, **6**(3), 225–233 (1987b).

Wuensch B.J., Nowacki W. The crystal structure of marrite, $PbAgAsS_3$, *Z. Kristallogr.*, **125**(1–6), 459–488 (1967).

Wu K., Chu Y., Yang Z., Pan S. $A_2SrM^{IV}S_4$ (A = Li, Na; M^{IV} = Ge, Sn) concurrently exhibiting wide bandgaps and good nonlinear optical responses as new potential infrared nonlinear optical materials, *Chem. Sci.*, **10**(14), 3963–3968 (2019).

Wu K., Pan S. Li_2HgMS_4 (M = Si, Ge, Sn): new quaternary diamond-like semiconductors for infrared laser frequency conversion, *Crystals*, **7**(4), 107 (2017).

Wu K., Su X., Yang Z., Pan S. An investigation of new infrared nonlinear optical material: $BaCdSnSe_4$, and three new related centrosymmetric compounds: Ba_2SnSe_4, Mg_2GeSe_4, and $Ba_2Ge_2S_6$, *Dalton Trans.*, **44**(46), 19856–19864 (2015a).

Wu K., Yang Z., Pan S. A first quaternary diamond-like semiconductor with 10-membered LiS_4 rings exhibiting excellent nonlinear optical performances, *Chem. Commun.*, **53**(21), 3010–3013 (2017a).

Wu K., Yang Z., Pan S. Na_2BaMQ_4 (M = Ge, Sn; Q = S, Se): infrared nonlinear optical materials with excellent performances and that undergo structural transformations, *Angew. Chem.*, **128**(23), 6825–6827 (2016a).

Wu K., Yang Z., Pan S. Na_2BaMQ_4 (M = Ge, Sn; Q = S, Se): infrared nonlinear optical materials with excellent performances and that undergo structural transformations, *Angew. Chem. Int. Ed.*, **65**(23), 6713–6715 (2016b).

Wu K., Yang Z., Pan S. $Na_2Hg_3M_2S_8$ (M = Si, Ge, and Sn): new infrared nonlinear optical materials with strong second harmonic generation effects and high laser-damage thresholds, *Chem. Mater.*, **18**(8), 2795–2801 (2016c).

Wu K., Yang Z., Pan S. $Na_4MgM_2Se_6$ (M = Si, Ge): the first noncentrosymmetric compounds with special ethane-like $[M_2Se_6]^{6-}$ units exhibiting large laser-damage thresholds, *Inorg. Chem.*, **54**(21), 10108–10110 (2015b).

Wu K., Yang Z., Pan S. $Na_6Si_3S_8O$: first example of sulfide silicate exhibiting unusual tripolymerized $[Si_3S_8O]^{6-}$ units without the S−O bonds, *Dalton Trans.*, **46**(39), 13356–13359 (2017b).

Wu K., Zhang B., Yang Z., Pan S. New compressed chalcopyrite-like $Li_2BaM^{IV}Q_4$ (M^{IV} = Ge, Sn; Q = S, Se): promising infrared nonlinear optical materials, *J. Am. Chem. Soc.*, **139**(42), 14885–14888 (2017c).

Wu L.-B., Huang F.-Q. Crystal structure of trilanthanum monosilver monosilicon heptasulfide, La_3AgSiS_7, *Z. Kristallogr., New. Cryst. Str.*, **220**(3), 307–308 (2005).

Wulff L., Wedel B., Müller-Buschbaum HK. Zur Kristallchemie von Telluraten mit Mn^{2+} im kationischen und anionischen Teil der Kristallstruktur: $(Mn_{2,4}Cu_{0,6})TeO_6$, Ba_2MnTeO_6 und $Pb(Mn_{0,5}Te_{0,5})O_3$, *Z. Naturforsch.*, **53B**(1), 49–52 (1998).

Wu P., Ibers J.A. Synthesis and structures of the quaternary chalcogenides of the type $KLnMQ_4$ (Ln = La, Nd, Gd, Y; M = Si, Ge; Q = S, Se), *J. Solid State Chem.*, **107**(2), 347–355 (1993a).

Wu P., Ibers J.A. Synthesis and structures of the quaternary sulfides $K_{0.5}Zr_{0.5}In_{0.5}S_2$ and $K_{0.5}Ga_{0.5}Sn_{0.5}S$, *Acta Crystallogr.*, **C49**(1), 126–129 (1993b).

Wu P., Ibers J.A. Synthesis of the new quaternary sulfides $K_2Y_4Sn_2S_{11}$ and $BaLnAgS_3$ (Ln = Er, Y, Gd) and the structures of $K_2Y_4Sn_2S_{11}$ and $BaErAgS_3$, *J. Solid State Chem.*, **110**(1), 156–161 (1994).

Wu P., Lu Y.-J., Ibers J.A. Synthesis and structures of the quaternary sulfides $KGaSnS_4$, $KInGeS_4$, and $KGaGeS_4$, *J. Solid State Chem.*, **97**(2), 383–390 (1992).

Wu X., Hu Y., Pan H., Su Z. $Na_9Sb(Ge_2Q_6)_2$ (Q = S, Se): two new antimony(III) quaternary chalcogenides with ethane-like $[Ge_2Q_6]^{6-}$ ligands, *RSC Adv.*, **6**(101), 99475–99481 (2016d).

Wu Y.-Y., Xiong L., Jia F., Chen L. $Cs_2Ge_3In_6Se_{14}$: a structure transformation driven by the size preference and its properties, *Inorg. Chem.*, **57**(8), 4667–4672 (2018).

Xinchun Z., Liang L., Shizhong W., Yan W., Jiankun Y., Nenglin G., Guanghui L., Jianmin H. Kuranakhite discovered in China for the first time, *Chin. J. Geochem.*, **17**(1), 77–80 (1998).

Xing W., Tang C., Wang N., Li C., Li Z., Wu J., Lin Z., Yao J., Yin W., Kang B. $EuHgGeSe4$ and $EuHgSnS4$: two quaternary Eu-based infrared nonlinear optical materials with strong second-harmonic-generation responses, *Inorg. Chem.*, **59**(24), 18452–18460 (2020a).

Xing W., Wang N., Guo Y., Li Z., Tang J., Kang K., Yin W., Lin Z., Yao J., Kang B. Two rare-earth-based quaternary chalcogenides $EuCdGeQ_4$ (Q = S, Se) with strong second-harmonic generation, *Dalton Trans.*, **48**(47), 17620–17625 (2019).

Xing W., Wang N., Iyer A.K., Yin W., Lin Z., Yao J., Kang B., Mar A. Evaluation of nonlinear optical properties of quaternary chalcogenide halides $Ba_4Si_3Se_9Br_2$ and $Ba_4Ge_3Se_9Br_2$, *J. Alloys Compd.*, **846**, 156398 (2020b).

Yablokova S.V., Dubakina L.S., Dmitrik A.A., Sokolova G.V. Kuranakhite – a new hypergenic mineral of tellurium [in Russian], *Zap. Vses. Mineralog. Obsh.*, **104**(3), 310–313 (1975).

Yajima J., Ohta E., Kanazawa Y. Toyohaite, $Ag_2FeSn_3S_8$, a new mineral, *Mineralog. J.*, **15**(5), 222–232 (1991).

Yamanaka T., Kato A. Mössbauer effect study of ^{57}Fe and ^{119}Sn in stannite, stannoidite, and mawsonite, *Am. Mineralog.*, **61**(3–4), 260–265 (1976).

Yamashita M., Yamanaka H. Formation and ionic conductivity of Li_2S–GeS_2–Ga_2S_3 glasses and thin films, *Solid State Ionics*, **158**(1–2), 151–156 (2003).

Yang H., Downs R.T., Durt J.B., Costin G. Structure refinement of an untwinned single crystal of Ag-excess fizélyite, $Ag_{5.94}Pb_{13.74}Sb_{20.84}S_{48}$, *Canad. Mineralog.*, **47**(5), 1257–1264 (2009).

Yang S.-H., Li X.-H., Huang X., Liu W., Guo S.-P. $Ti_{0.85}Eu_3SiS_7$: the rare-earth/Ti based quaternary sulfide containing two variable valence elements, *J. Solid State, Chem.*, **311**, 123082 (2022a).

Yang Y., Guo Y., Zhang B., Wang T., ChenY.-G., Hao X., Yu X., Zhang X.-M. Lead tellurite crystals $BaPbTe_2O_6$ and PbV-TeO_5F with large nonlinear-/linear-optical responses due to active lone pairs and distorted octahedral, *Inorg.Chem.*, **61**(3), 1538–1545 (2022b).

Yang Y., Ibers J.A. Accidental silicon-containing compounds: crystal structures of $La_3Al_{0.44}Si_{0.93}S_7$, $BaSm_4(SiO_4)_3Se$, and monoclinic and orthorhombic $Ln_2(SiO_4)Te$ (Ln = Nd and Sm), *J. Solid State Chem.*, **155**(2), 433–440 (2000).

Yang Y., Song M., Zhang J., Gao L., Wu X., Wu K. Coordinated regulation on critical physiochemical performances activated from mixed tetrahedral anionic ligands in new series of $Sr_6A_4M_4S_{16}$ (A = Ag, Cu; M = Ge, Sn) nonlinear optical materials, *Dalton Trans.*, **49**(11), 3388–3392 (2020a).

Yang Y., Wu K., Wu X., Zhang B., Gao L. A new family of quaternary thiosilicates SrA_2SiS_4 (A = Li, Na, Cu) as promising infrared nonlinear optical crystals, *J. Mater. Chem. C*, **8**(5), 1762–1767 (2020b).

Yang Y., Wu K., Zhang B., Wu X., Lee M.-H. Infrared nonlinear optical polymorphs α- and β-$SrCu_2SnS_4$ exhibiting large second harmonic generation responses with requisite phase-matching behavior, *Chem. Mater.*, **32**(3), 1281–1287 (2020c).

Yan M., Sun Z.-D., Yao W.-D., Zhou W., Liu W., Guo S.-P. Highly distorted HgS_4 tetrahedra induced moderate second-harmonic generation response of $EuHgGeS_4$, *Inorg. Chem. Front.*, **7**(13), 2451–2458 (2020).

Yao G.-Q., Shen H.-S., Honig E.D., Kershaw R., Dwight K., Wold A. Preparation and characterization of the quaternary chalcogenides $Cu_2B(II)C(IV)X_4$ [B(II) = Zn, Cd; C(IV) = Si, Ge; X = S, Se], *Solid State Ionics*, **24**(3), 249–252 (1987).

Yao J., Deng B., Ellis D.E., Ibers J.A. New quaternary bismuth sulfides: syntheses, structures, and band structures of $AMBiS_4$ (A = Rb, Cs; M = Si, Ge), *Inorg. Chem.*, **41**(26), 7094–7099 (2002).

Yao J., Ibers J.A. $RbPbPS_4$, *Acta Crystallogr.*, **E60**(9), i108–i110 (2004).

Yashina L.V., Leute V., Shtanov V.I., Schmidtke H.M., Neudachina V.S. Comprehensive thermodynamic description of the quasiternary system $PbTe$–$GeTe$–$SnTe$, *J. Alloys Compd.*, **413**(1–2), 133–143 (2006).

Yelisseyev A.P., Isaenko L.I., Krinitsin P., Liang F., Goloshumova A.A., Naumov D.Yu., Lin Z. Crystal growth, structure, and optical properties of $LiGaGe_2Se_6$, *Inorg. Chem.*, **55**(17), 8672–8680 (2016).

Yeon J., Kim S.-H., Hayward M.A., Halasyamani P.S. "A" cation polarity control in $ACuTe_2O_7$ (A = Sr^{2+}, Ba^{2+}, or Pb^{2+}), *Inorg. Chem.*, **50**(17), 8663–8670 (2011).

Yeon J., Kim S.-H., Nguyen S.D., Lee H., Halasyamani P.S. New vanadium selenites: centrosymmetric $Ca_2(VO_2)_2(SeO_3)_3(H_2O)_2$, $Sr_2(VO_2)_2(SeO_3)_3$, and $Ba(V_2O_5)(SeO_3)$, and noncentrosymmetric and polar $A_4(VO_2)_2(SeO_3)_4(Se_2O_5)$ (A = Sr^{2+} or Pb^{2+}), *Inorg. Chem.*, **51**(1), 609–619 (2012).

Yin W., Feng K., Hao W., Yao J., Wu Y. Synthesis, structure, and properties of $Li_2In_2MQ_6$ (M = Si, Ge; Q = S, Se): a new series of IR nonlinear optical materials, *Inorg. Chem.*, **51**(10), 5839–5843 (2012a).

Yin W., Feng K., He R., Mei D., Lin Z., Yao J., Wu Y. $BaGa_2MQ_6$ (M = Si, Ge; Q = S, Se): a new series of promising IR nonlinear optical materials, *Dalton Trans.*, **41**(18), 5653–5661 (2012b).

Yin W., Iyer A.K., Li C., Lin X., Yao J., Mar A. Noncentrosymmetric selenide $Ba_4Ga_4GeSe_{12}$: synthesis, structure, and optical properties, *J. Solid State Chem.*, **241**, 131–136 (2016).

Yin W., Iyer A.K., Li C., Yao J., Mar A. Noncentrosymmetric chalcogenides $BaZnSiSe_4$ and $BaZnGeSe_4$ featuring one-dimensional structures, *J. Alloys Compd.*, **708**, 414–421 (2017).

Yin W., Lin Z., Kang L., Kang B., Deng J., Lin Zh., Yao J., Wu Y. Syntheses, structures, and optical properties of $Ba_4Ga_4SnSe_{12}$ and $Ba_6Ga_2SnSe_{11}$, *Dalton Trans.*, **44**(5), 2259–2266 (2015).

Yohannan J.P., Vidyasagar K. Syntheses and characterization of one-dimensional alkali metal antimony(III) thiostannates(IV), $A_2Sb_2Sn_3S_{10}$ (A = K, Rb, Cs), *J. Solid State Chem.*, **221**, 426–432 (2015).

Yohannan J.P., Vidyasagar K. Syntheses and structural characterization of non-centrosymmetric $Na_2M_2M'S_6$ (M, M' = Ga, In, Si, Ge, Sn, Zn, Cd) sulfides, *J. Solid State Chem.*, **238**, 147–155 (2016a).

Yohannan J.P., Vidyasagar K. Syntheses and structural variants and characterization of $AInM'S_4$ (A = alkali metals, Tl; M' = Ge, Sn) compounds; facile ion-exchange reactions of layered $NaInSnS_4$ and $KInSnS_4$ compounds, *J. Solid State Chem.*, **238**, 291–302/ (2016b).

Yousaf N., Khan W., Khan S.H., Yaseen M., Laref A., Murtaza G. Electronic, optical and thermoelectric properties of $SnGa_2GeX_6$ (X = S, Se) compounds, *J. Alloys Compd.*, **737**, 637–645 (2018).

Yuan F.-Y., Lin C.-S., Huang Y.-Z., Zhang H., Zhou A.-Y., Chai G.-L., Cheng W.-D. $BaCdGeSe_4$: synthesis, structure and nonlinear optical properties, *J. Solid State Chem.*, **302**, 122352 (2021).

Yu C.-Y., Wang M.-F., Chung M.-Y., Jang S.-M., Huang J.-C., Lee C.-S. Synthesis and characterization of a new quaternary selenide $Sr_3GeSb_2Se_8$, *Solid State Sci.*, **10**(9), 1145–1149 (2008).

Yurchenko O.M., Olekseyuk I.D., Parasyuk O.V., Pankevich V.Z. Single crystal growth and properties of $AgGaGeS_4$, *J. Cryst. Growth*, **275**(1–2), e1983–e1985 (2005).

Zargarova M.I., Babaeva P.K., Azhdarova D.S., Melikova Z.D., Mehtieva S.A. Investigation of the $CuInSe_2$–$InSe$ ($SnSe_2$, Bi_2Se_3) system [in Russian], *Neorgan. mater.*, **31**(2), 282–283 (1995).

Zargarova M.I., Kakhramanov K.Sh., Roshal' R.M. About chemical compatibility of nickel antimonide with lead chalcogenides [in Russian], *Izv. AN AzSSR. Ser. phys.-techn. i mat. nauk*, (4), 107–109 (1975a).

Zargarova M.I., Kakhramanov K.Sh., Roshal' R.M. Investigation of the PbS–$NiSb$ phase diagram [in Russian], *Izv. AN SSSR. Neorgan. mater.*, **11**(1), 165 (1975b).

Zargarova M.I., Kuliev A.Z., Kahramanov K.Sh., Magerova A.A. Investigation of the GeTe–NiSb system [in Russian], *Azerb. khim. zhurn.*, (*3*), 116–119 (1975c).

Zargarova M.I., Kuliev A.Z., Kakhramanov K.Sh., Roshal' R.M. Investigation of the PbSe–NiSb system [in Russian], *Azerb. khim. zhurn.*, (*6*), 138–140 (1975d).

Zeng H.-Y., Mao J.-G., Huang J.-S. Synthesis and crystal structure of $La_4S_3Si_2O_7$, *J. Alloys Compd.*, **291**(1–2), 89–93 (1999).

Zeng H.-Y., Zhao Z.-Y., Guo S.-P., Zheng F.-K., Guo G.-C., Huang J.-S. New quaternary sulfides in the AE–RE–Sn–S system (AE = alkaline-earth, RE = rare earth), *J. Alloys Compd.*, **514**, 135–140 (2012).

Zeng H.-Y., Zheng F.-K., Guo G.-C., Huang J.-S. Syntheses and single-crystal structures of La_3AgSnS_7, $Ln_3M_xMS_7$ (Ln = La, Ho, Er; M=Ge, Sn; $1/4 \leq x \leq 1/2$), *J. Alloys Compd.*, **458**(1–2), 123–129 (2008).

Zerouale A., Cros B., Deroide B., Ribes M. Electrical properties of Ag_7GeSe_5I, *Solid State Ionics*, **28–30**, Pt. 2, 1317–1319 (1988).

Zhai D., Bindi L., Voudouris P.C., Liu J., Tombros S.F., Li K. Discovery of Se-rich canfieldite, $Ag_8Sn(S,Se)_6$, from the Shuangjianzishan Ag–Pb–Zn deposit, NE China: a multi-methodic chemical and structural study, *Mineralog. Mag.*, **83**(3), 419–426 (2019).

Zhang C., Wang K.-N., Ji M., An Y.-L. Mild solvothermal syntheses of thioargentates A–Ag–S (A = K, Rb, Cs) and A–Ag–Ge–S (A = Na, Rb): crucial role of excess sulfur, *Inorg. Chem.*, **52**(21), 12367–12371 (2013).

Zhang C.-Y., Zhou L.-J., Chen L. Quaternary tellurides with different valent Ge centers: $Cs_2Ge_3M_6Te_{14}$ (M = Ga, In), *Inorg. Chem.*, **51**(13), 7007–7009 (2012a).

Zhang J.-H., Clark D.J., Brant J.A., Rosmus K.A., Grima P., Lekse J.W., Jang J.I., Aitken J.A. α-Li_2ZnGeS_4: a wide-bandgap diamond-like semiconductor with excellent balance between laser-induced damage threshold and second harmonic generation response, *Chem. Mater.*, **32**(20), 8947–8955 (2020a).

Zhang J.-H., Clark D.J., Brant J.A., Sinagra C.W., Kim Y.S., Jang J. I., Aitken J.A. Infrared nonlinear optical properties of lithium-containing diamond-like semiconductors $Li_2ZnGeSe_4$ and $Li_2ZnSnSe_4$, *Dalton Trans.*, **44**(24), 11212–11222 (2015).

Zhang J.-H., Clark D.J., Weiland A., Stoyko S.S., Kim Y.S., Jang J.I., Aitken J.A. $Li_2CdGeSe_4$ and $Li_2CdSnSe_4$: biaxial nonlinear optical materials with strong infrared second-order responses and laser-induced damage thresholds influenced by photoluminescence, *Inorg. Chem. Front.*, **4**(9), 1472–1484 (2017).

Zhang J.-H. Crystal structure and optical property of quaternary sulfide $Ba_{2.77}SnSb_{2.16}S_8$, *J. Struct. Chem.*, **59**(4), 955–960 (2018).

Zhang J.-H., Stoyko S.S., Craig A.J., Grima P., Kotchey J.W., Jang J.I., Aitken J.A. Phase matching, strong frequency doubling, and outstanding laser-induced damage threshold in the biaxial, quaternary diamond-like semiconductor $Li_4CdSn_2S_7$, *Chem. Mater.*, **32**(23), 10045–10054 (2020b).

Zhang R.-C., Yao H.-G., Ji S.-H., Liu M.-C., Ji M., An Y.-L. Copper-rich framework sulfides: $A_4Cu_8Ge_3S_{12}$ (A = K, Rb) with cubic perovskite structure, *Inorg. Chem.*, **49**(14), 6372–6374 (2010).

Zhang S., Mei D., Du X., Lin Z., Zhong J., Wu Y., Xu J. The structure and band gap design of high Si doping level $Ag_{1-x}Ga_{1-x}Si_xSe_2$ (x = 1/2), *J. Solid State Chem.*, **238**, 21–24 (2016).

Zhang S.-Y., Hu C.-L., Li P.-X., Jiang H.-L. Mao J.-G. Syntheses, crystal structures and properties of new lead(II) or bismuth(III) selenites and tellurite, *Dalton Trans.*, **41**(31), 9532–9542 (2012b).

Zhang Y., Mei D., Yang Y., Cao W., Wu Y., Lu J., Lin Z. Rational design of a new chalcogenide with good infrared nonlinear optical performance: $SrZnSnS_4$, *J. Mater. Chem. C*, **7**(28), 8556–8561 (2019).

Zhao D., Xia F., Chen G., Zhang X., Ma H., Adam J.L. Formation and properties of chalcogenide glasses in the $GeSe_2$–As_2Se_3–CdSe system, *J. Am. Ceram. Soc.*, **88**(11), 3143–3146 (2005).

Zhao D., Zhang X., Wang H., Zeng H., Ma H., Adam J.L., Chen G. Thermal properties of chalcogenide glasses in the $GeSe_2$–As_2Se_3–CdSe system, *J. Non-Cryst. Solids*, <u>354</u>(12–13), 1281–1284 (2008).

Zhao H.-J. Syntheses, crystals tructures, and NLO properties of the quaternary sulfides $RE_3Sb_{0.33}SiS_7$ (RE = La, Pr), *J. Solid State Chem.*, **227**, 5–9 (2015).

Zhao J., Mei D.J., Yang Y., Cao W.Z., Liu C., Wu Y.D., Lin Z.S. $Rb_{10}Zn_4Sn_4S_{17}$: a chalcogenide with large laser damage threshold improved from the Mn-based analogue, *Inorg. Chem.*, **58**(22), 15029–15033 (2019).

Zhen N., Nian L., Li G., Wu K., Pan S. A high laser damage threshold and a good second-harmonic generation response in a new infrared NLO material: $LiSm_3SiS_7$, *Crystals*, **6**(10), 121 (2016a).

Zhen N., Wu K., Wang Y., Li Q., Gao W., Hou D., Yang Z., Jiang H., Dong Y., Pan S. $BaCdSnS_4$ and $Ba_3CdSn_2S_8$: syntheses, structures, nonlinear optical and photoluminescence properties, *Dalton Trans.*, **45**(26), 10681–10688 (2016b).

Zhou J., Luo L., Chu Y., Wang P., Guo Z., Su X., Li J. Partial congener substitution induced centrosymmetric to noncentrosymmetric structural transformation and nonlinear optical properties of $PbSnSiS_4$, *J. Alloys Compd.*, **899**, 163366 (2022).

Zhou M., Dong C. Synthesis, crystal structure and physical properties of kiddcreekite Cu_6WSnS_8 and its congener Cu_6WSnSe_8, *J. Solid State Chem.*, **278**, 120918 (2019).

Zhou M., Gu Y., Ruan B., Yang Q., Chen G., Ren Z. Synthesis, crystal structure and physical properties of a novel quaternary selenide Cu_6GeWSe_8, *J. Solid State Chem.*, **301**, 122357 (2021).

Zhou M., Jiang X., Guo Y., Lin Z., Yao J., Wu Y. $Pb_{0.65}Mn_{2.85}Ga_3S_8$ and $Pb_{0.72}Mn_{2.84}Ga_{2.95}Se_8$: two quaternary metal chalcogenides with open-tunnel-framework structures displaying intense second harmonic generation responses and interesting magnetic properties, *Inorg. Chem.*, **56**(14), 8454–8461 (2017a).

Zhou M., Jiang X., Yang Y., Guo Y., Lin Z., Yao J., Wu Y. $K_2ZnSn_3Se_8$: a non-centrosymmetric zinc selenidostannate(IV) featuring interesting covalently bonded $[ZnSn_3Se_8]^{2-}$ layer and exhibiting intriguing second harmonic generation activity, *Chem. Asian J.*, **12**(12), 1282–1285 (2017b).

Zhou M., Li C., Li X., Yao J., Wu Y. $K_2Sn_2ZnSe_6$, $Na_2Ge_2ZnSe_6$, and $Na_2In_2GeSe_6$: a new series of quaternary selenides with intriguing structural diversity and nonlinear optical properties, *Dalton Trans.*, **45**(18), 7627–7633 (2016a).

Zhou M., Yin W., Liang F., Mar A., Lin Z., Yao J., Wu Y. $Na_2MnGe_2Se_6$: a new Mn-based antiferromagnetic chalcogenide with large Mn···Mn separation, *J. Mater. Chem. C*, **4**(46), 10812–10819 (2016b).

Zhou Y., Iyer A.K., Oliynyk A.O., Heyberger M., Lin Y., Qiu Y., Mar A. Quaternary rare-earth sulfides $RE_3M_{0.5}M'S_7$ (M = Zn, Cd; M' = Si, Ge), *J. Solid State Chem.*, **278**, 120914 (2019).

Zhukov E.G., Mkrtchian S.A., Dovletov K.O., Kalinnikov V.T., Ashirov A.A. The Cu_2SnSe_3–CdSe system [in Russian], *Zhurn. neorgan. khimii*, **27**(3), 761–762 (1982).

Zhukov E.G., Mkrtchian S.A., Dovletov K.O., Melikdzhanian A.G., Kalinnikov V.T., Ashirov A.A. The Cu_2GeSe_3–CdSe system [in Russian], *Zhurn. neorgan. khimii*, **29**(7), 1897–1898 (1984).

Zhu T., Huhn W.P., Wessler G.C., Shin D., Saparov B., Mitzi D.B., Blum V. I_2–II–IV–VI_4 (I = Cu, Ag; II = Sr, Ba; IV = Ge, Sn; VI = S, Se): chalcogenides for thin-film photovoltaics, *Chem. Mater.*, **29**(18), 7868–7879 (2017).

Zimmermann C., Dehnen S. $Cs_2[MnSnTe_4]$: Ungewöhnliche Synthese einer quaternären Phase mit eindimensionalen, ternären Anionensträngen, *Z. anorg. und allg. Chem.*, **629**(9), 1553–1556 (2003).

Zmiy O.F., Gulay L.D., Ostapyuk T.A., Klymovych O.S. Interaction of the components in the Ag_2Se–$SnSe_2$–As_2Se_3 system, *Chem. Met. Alloys*, **1**(2), 115–119 (2008).

Index